T0140095

Advances in Intelligent Systems and Computing

Volume 983

The series “Advances in Intelligent Systems and Computing” contains publications on theory, applications, and design methods of Intelligent Systems and Intelligent Computing. Virtually all disciplines such as engineering, natural sciences, computer and information science, ICT, economics, business, e-commerce, environment, healthcare, life science are covered. The list of topics spans all the areas of modern intelligent systems and computing such as: computational intelligence, soft computing including neural networks, fuzzy systems, evolutionary computing and the fusion of these paradigms, social intelligence, ambient intelligence, computational neuroscience, artificial life, virtual worlds and society, cognitive science and systems, Perception and Vision, DNA and immune based systems, self-organizing and adaptive systems, e-Learning and teaching, human-centered and human-centric computing, recommender systems, intelligent control, robotics and mechatronics including human-machine teaming, knowledge-based paradigms, learning paradigms, machine ethics, intelligent data analysis, knowledge management, intelligent agents, intelligent decision making and support, intelligent network security, trust management, interactive entertainment, Web intelligence and multimedia.

The publications within “Advances in Intelligent Systems and Computing” are primarily proceedings of important conferences, symposia and congresses. They cover significant recent developments in the field, both of a foundational and applicable character. An important characteristic feature of the series is the short publication time and world-wide distribution. This permits a rapid and broad dissemination of research results.

**** Indexing: The books of this series are submitted to ISI Proceedings, EI-Compendex, DBLP, SCOPUS, Google Scholar and Springerlink ****

More information about this series at http://www.springer.com/series/11156

Vera Murgul · Marco Pasetti
Editors

International Scientific Conference Energy Management of Municipal Facilities and Sustainable Energy Technologies EMMFT 2018

Volume 2

Editors
Vera Murgul
Moscow State University of Civil Engineering
Moscow, Russia

Marco Pasetti
Department of Information Engineering
Università degli Studi di Brescia
Brescia, Italy

ISSN 2194-5357 ISSN 2194-5365 (electronic)
Advances in Intelligent Systems and Computing
ISBN 978-3-030-19867-1 ISBN 978-3-030-19868-8 (eBook)
https://doi.org/10.1007/978-3-030-19868-8

This Springer imprint is published by the registered company Springer Nature Switzerland AG
The registered company address is: Gewerbestrasse 11, 6330 Cham, Switzerland

Preface

The XX annual international scientific conference Energy Management of Municipal Facilities and Sustainable Energy Technologies EMMFT 2018 took place in Voronezh on December 10–13, 2018.

The conference was hosted by Voronezh State Technical University, Russia.

Specialists from more than 10 counties participated in the EMMFT 2018 conference. This year, the authors submitted approximately 480 qualified papers.

The objective of the conference was the exchange of the latest scientific achievements, strengthening of academic relations with leading scientists of European Union, Russia, and the World, creating favorable conditions for collaborative researches and implementing collaborative projects in the fields of energy management and development of sustainable energy technologies. Experts invited to participate in the conference presented special lectures, demonstrated equipment and devices for HVAC systems, and shared the latest technologies of thermal protection of buildings. Special attention was paid to the development of renewable energy industry. The efforts of scientists, politicians, and heads of energy enterprises were united for developing specific research programs in the field of development of renewable energy sources.

During the conference, issues on the following topics were discussed within several workshops: building physics; heating, ventilation, and HVAC & R; renewable energy; energy management; energy efficiency in transport, modeling, and control in mechanical engineering.

The conference program also included seminars, round tables, and excursions to research laboratories and research and educational centers of the Voronezh State Technical University.

All papers passed a four-staged review. The first stage consisted in an examination for compliance with the subject of the conference. At the second stage, all papers were thoroughly checked for plagiarism. Acceptable minimum of originality was 90%. The third stage involved the review by a native speaker for acceptable English language. At the same time, papers were checked by a technical proofreader. The fourth stage involved a scientific review made by at least three reviewers, using double-blind review method. If opinions of the reviewers were radically different, additional reviewers were appointed.

The members of our organizing committee express their deep gratitude to the team of "Advances in Intelligent Systems and Computing" journal and to the editorial department of Springer Nature publishing house for publication of EMMFT 2018 conference proceedings.

Organization

Organizing Committee of the Conference

Sergei Kolodyazhni	Voronezh State Technical University, Russia
Marco Pasetti	Università degli Studi di Brescia (UNIBS), Italy
Vera Murgul	Moscow State University of Civil Engineering, Russia
Igor Surovtsev	Voronezh State Technical University, Russia
Svetlana Uvarova	Voronezh State Technical University, Russia
Norbert Harmathy	Budapest University of Technology and Economics, Hungary
Vadim Kankhva	Moscow State University of Civil Engineering, Russia

Scientific Committee of the Conference

Aleksander Szkarowski	Politechnika Koszalinska, Koszalin, Poland
Antony Wood	Illinois Institute of Technology, Chicago, USA
Iurii Tabunschikov	Corr. Member of RAASN, Honorary Member of the International Ecoenergy Academy of Azerbaijan, ASHRAE Fellow Member, REHVA Fellow Member, Corr. Member of VDI, Member of ISIAQ Academy, Winner of the 2008 Nobel Peace Prize as a Member of the Intergovernmental Panel on Climate Change
Viktor Pukhkal	Saint-Petersburg State University of Architecture and Civil Engineering, Russia

Sergey Anisimov	Wroclaw University of Science and Technology, Poland
Marianna M. Brodach	Moscow Architectural Institute (State Academy), Russia
Daniel Safarik	CTBUH Journal, Chicago, USA
Samuil G. Konnikov	Ioffe Physical-Technical Institute of the Russian Academy of Sciences, Russia
Alexander Solovyev	Research Laboratory of Renewable Energy Sources—Lomonosov Moscow State University, Russian Academy of Natural Sciences, Russia
Dietmar Wiegand	Technische Universität Wien TU, Wien
Luís Bragança	Building Physics and Technology Laboratory, Guimaraes, University of Minho, Portugal
Anatolijs Borodinecs	Institute of Heat, Gas and Water technology, Riga Technical University, Latvia
Alessandro Bianchini	University of Florence (UNIFI), Italy
Aleksandr Gorshkov	Peter the Great St. Petersburg Polytechnic University, Russia
Zdenka Popovic	Faculty of Civil Engineering, University of Belgrade, Serbia
Marco Pasetti	Università degli Studi di Brescia (UNIBS), Italy
Valerii Volshanik	Moscow State University of Civil Engineering, Russia
Mirjana Vukićević	University of Belgrade, Serbia
Sang Dae Kim	Korea University, Seoul, South Korea
Manfred Esser	GET Information Technology GmbH, Grevenbroich, Germany
Alenka Fikfak	University of Ljubljana, Slovenia
Milorad Jovanovski	Ss. Cyril and Methodius University in Skopje, Macedonia
Radek Škoda	Czech Technical University in Prague, Czech Republic
Nikolai Vatin	Peter the Great St. Petersburg Polytechnic University, Russia
Paulo Cachim	University of Aveiro, Portugal
Aires Camões	University of Minho, Portugal
Michael Tendler	Royal Institute of Technology, Stockholm—Kungliga Tekniska Högskolan (KTH), Sweden
Christoph Pfeifer	University of Natural Resources and Life Sciences, Vienna, Austria
Antonio Andreini	University of Florence (UNIFI), Italy
Pietro Zunino	DIME Universitá di Genova, Genoa, Italy

Olga Kalinina	Peter the Great St. Petersburg Polytechnic University, Russia
Tomas Hanak	Brno University of Technology, Czech Republic
Vera Murgul	Moscow State University of Civil Engineering, Russia
Darya Nemova	Peter the Great St. Petersburg Polytechnic University, Russia
Norbert Harmathy	Budapest University of Technology and Economics, Hungary
Igor Ilin	Peter the Great St. Petersburg Polytechnic University, Russia

Contents

Energy and Environmental Management

Modeling of the Factor Space of Energy Industry Sector of the Republic of Abkhazia

Huta Gumba(✉) and Eva Ozgan

Abkhaz State University, Universitetskaya 1, 384904 Sukhum,
Republic of Abkhazia, Georgia
gumba_hm@mail.ru

Abstract. Assessment of the factor space of the external environment is extremely important in the planning and implementation of the innovative and strategic development of an enterprise. The economy of the Republic of Abkhazia is only moving towards the implementation of sustainable innovation and energy efficient development. Moreover, energy industry is recognized as one of the basic sectors for solving the task of enhancing the potentially competitive sectors of the country's economy. The authors believe that it is necessary to solve the existing problems in the energy and other industries by the effective use of their own potential, increasing the productivity of productive resources, integration in the interaction of the departments of science, education and industry, and innovation at the level of economic units. That is why there is a need to form the methodological foundations of the innovative and strategic development of enterprises, which determine the priority directions for strengthening the market positions of the enterprise and industry and the methods for their implementation. The authors have proposed a scheme for managing the innovative and strategic development of the enterprise, one of the important elements of which is the assessment of environmental factors. On the basis of the calculation of a stable rank distribution, a methodology for analyzing environmental factors is proposed. The methodology was tested on the example of analyzing the dynamics of GDP of the Republic of Abkhazia in the sectoral projection.

Keywords: Innovative development · Energy industry sector · Enterprise · Rank distribution model

1 Introduction

To solve the problem of overcoming the lag of the Republic of Abkhazia from the economies of other countries, it is necessary to ensure the sustainable development of backbone industries, which are key points of growth. These industries include potentially competitive industry sectors of the republic: tourism, agriculture, food production, construction materials production [1]. Accordingly, the basis for the accelerated development of the above mentioned industries is to ensure the development of key supporting sectors: banks and finance, energy, transport and logistics, trade, real estate and construction, which create the maximum contribution to Abkhazia's GDP at the moment. For the economy of the construction industry, innovation plays a vital role as

V. Murgul and M. Pasetti (Eds.): EMMFT-2018, AISC 983, pp. 3–11, 2019.
https://doi.org/10.1007/978-3-030-19868-8_1

a powerful factor in the exit of the industry from the economic crisis [2]. It is unusually difficult to ensure stabilization and growth of the economy, since the decline in the efficiency of construction production in many enterprises has become irreversible. Obviously, the bases of production have undergone degradation, the innovation crisis is clearly manifested [3]. It is very difficult to reverse the current situation, but it is necessary. The basis of positive transformations, in our opinion, can be innovative activity. An innovative type of development implies such an action that introduces fundamentally new elements into the economic environment [4]. In our opinion, this allows ensuring reproduction in the most rational way within the enterprise, efficiently using resources, and, consequently, achieving economic growth.

Thus, in modern conditions, the sustainability and balanced development of the Republic of Abkhazia are determined by the need for innovative development, primarily in the material sphere. Activation of innovative processes, strengthening and effective use of innovative potential, rational distribution and consumption of resources for innovative development, and attraction of external sources of financing determine the trends and development prospects of enterprises in the material sphere, the foundations of which are laid in the strategy of their innovative development.

2 Materials and Methods

The need to define innovative and strategic development is caused by such reasons as the long cycle of creating innovative products, the need to have clear ideas about the scale of the development of science in the long term.

In our opinion, the most correct definition of the concept of innovation and strategic development, taking into account the features of the innovation cycle, is the expression of the innovation strategy as a system of the main promising goals and objectives of the construction enterprise, relating to the entire length of the process, from research to production, marketing and use, meaning the main points of regulation and control of actions agreed upon by resources and timelines and providing the best economic result. The formation of the strategy involves the choice of alternative ways of development of the construction enterprise, its potential, using the forecasts, experience and intuition of specialists [5]. The strategy mobilizes the resources of the enterprise and directs them to achieve its goals. The imperfection of the innovation management mechanism in enterprises strengthens the importance of the need for a new approach to the problem of the formation of innovative and strategic development.

In this regard, of fundamental importance in the formation of the development strategy is the conceptual scheme that we have developed for the formation and management of the innovative and strategic development of the construction industry enterprise (Fig. 1). The principal feature of proposed scheme is that it is based on the principle of writing an innovative scenario. In this case, the diagnosis (assessment) of the innovation potential is carried out on the basis of the factor analysis of the innovative capabilities of the enterprise.

In addition, the development of the concept is inextricably linked with its interpretation as an integral system of elements that have a single target setting in the process of forming a strategy. In this case, the methodological basis for the typing process is the systematization of the goals of the strategy, the achievement of which is ensured through the interaction of all its stages and constituent elements.

It seems appropriate to highlight the following stages in the development of innovative and strategic development of the construction enterprise: the definition of goals and objectives; diagnostics (assessment) of the current state of innovative development of the enterprise; forecasting the economic situation in the external environment; scenario forecast of innovative development of the enterprise; selection of the most rational variant of the scenario forecast; forecasting the structure of innovations; generalization and formulation of the innovation strategy of the enterprise; management of the implementation of the innovation strategy of the enterprise.

At the input, the system is characterized by parameters that are determined on the basis of an assessment of the current state of the enterprise's innovative development depending on factors: socio-economic, scientific and technical, and regulatory.

The output of the system is the development of strategic directions of innovative development of the enterprise (scientific and technical programs, innovations) and management of the implementation of this strategy. In addition, at the output of the system, there are parameters of the state of the external environment (market, scientific and technical interests, financial aspects). Within this article, an analysis of the factor space of the innovation activities of the construction enterprises of the Republic of Abkhazia was carried out. It should be noted that the lack of a representative volume of statistics on the volume of innovation activities in the Republic of Abkhazia, associated with the lack of static reporting forms on innovation activity, makes it impossible to analyze scientific and technical factors. However, the proposed methodology is also intended for their assessment. The proposed methodology for analyzing environmental factors is based on the methods of structural and topological analysis and the principles of rank distribution [6, 7]. The methodology is applicable to the analysis of all the groups of factors we have listed.

At the first stage, the ranking of each factor by typological elements required or appropriate for analysis is assumed.

$$F = (F_1,\ F_2,\ \ldots\ F_n) \tag{1}$$

At the second stage, the rank distribution is built:

$$F(r) = \frac{F_{max}}{r^b} \tag{2}$$

Where F(r) – ranking factor; Fmax – the maximum value of a typological element of a factor with rank 1; r – the rank's order number; β – rank factor.

At the third stage, the ideal and actual rank distribution is compared in order to study the dynamics of the typological elements of the factors compared to the model value. Based on the laws of cenology [Kudrin], the species distribution of elements tends to a hyperbolic H-distribution. Accordingly, the sustainable factor structure of the innovation activity of an enterprise should strive for a hyperbolic H-distribution.

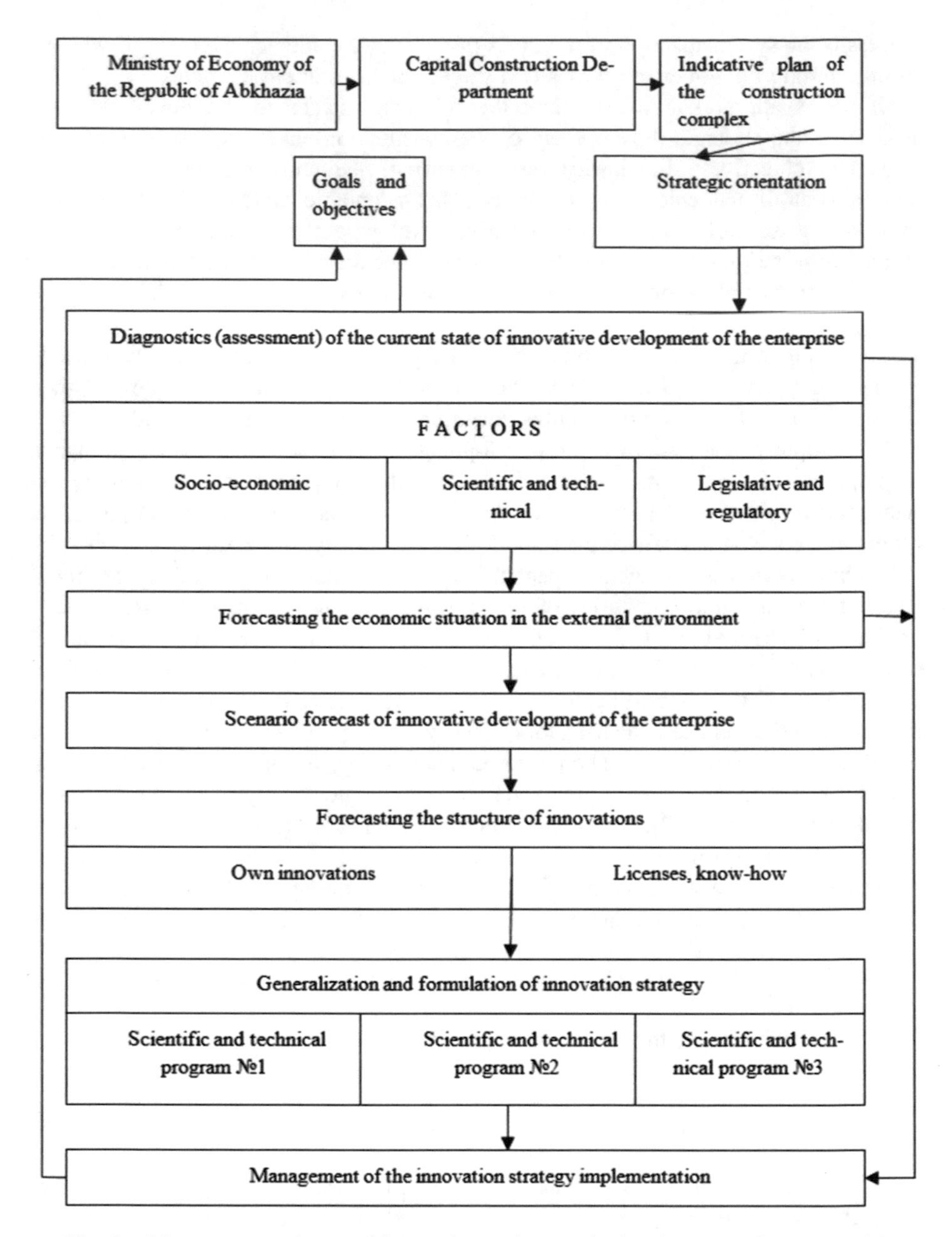

Fig. 1. Management scheme of innovative and strategic development of an enterprise

Further analysis of the factors can be continued and detailed as follows.

At the fourth stage, it is necessary to determine the sets of groups of factors of intraspecific and interspecific competition.

$$\begin{aligned} F_{fk} &= \{F_i, \ldots, F_n\} \\ F_{mk} &= \{F_j, \ldots, F_l\} \end{aligned} \quad (3)$$

At the fifth stage, the species distribution of factors is built

$$\Omega(f) = \frac{K_1}{f^{\gamma}} \quad (4)$$

Where $\gamma = 1 + \alpha$ – characteristic index, K1 - the count of the first group represented by unique factors (with the minimum value of the factor) The analysis is completed by the assessment of the stability of the factor structure by the value of the concordance coefficient [8]:

$$W = \frac{12D}{m^2(n^3 - n)} \quad (5)$$

Where m – number of time periods; n – number of factors; D – the sum of the squares of the difference between the average sum of ranks and the sum of ranks of each factor for each of the periods; W – the concordance coefficient.

Based on the analysis, key factors or industries are selected for the implementation of managerial influences aimed at stimulating the innovation activities of construction enterprises. Similarly, it is advisable to use the methodology in the formation of the list of key sectors - development drivers of the Republic of Abkhazia, in the development of management innovations in the economic policy of the republic.

3 Results

As an assessment of the proposed methodology, an analysis of the structural and topological dynamics of GDP of the Republic of Abkhazia was conducted in the sectoral breakdown by years (Figs. 2, 3, 4 and 5). A fragment of the calculations is presented in Table 1. GDP characterizes the country's economy, the external environment of innovative and strategic development of enterprises.

Table 1. Calculation of indicators for building rank distribution.

Gross value added in basic prices, million rubles Including:	29655.5	Rank, r	f(r)
Trade	11837.7	14	760.8415
Management	3978.2	13	821.8002
Industry	2882.8	12	893.1385
Transportation and communication	2316.9	11	977.73
Health care	1956.3	10	1079.611
Construction	1730.6	9	1204.634
...			

Based on the calculations, a graph of ideal and actual rank distribution is plotted.

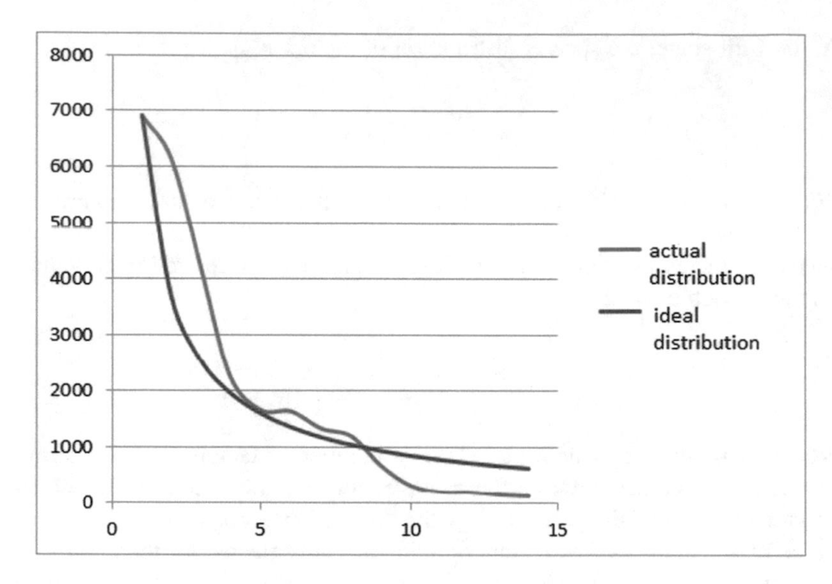

Fig. 2. Rank analysis of the structural and topological dynamics of GDP of the Republic of Abkhazia in the sectoral breakdown, 2014.

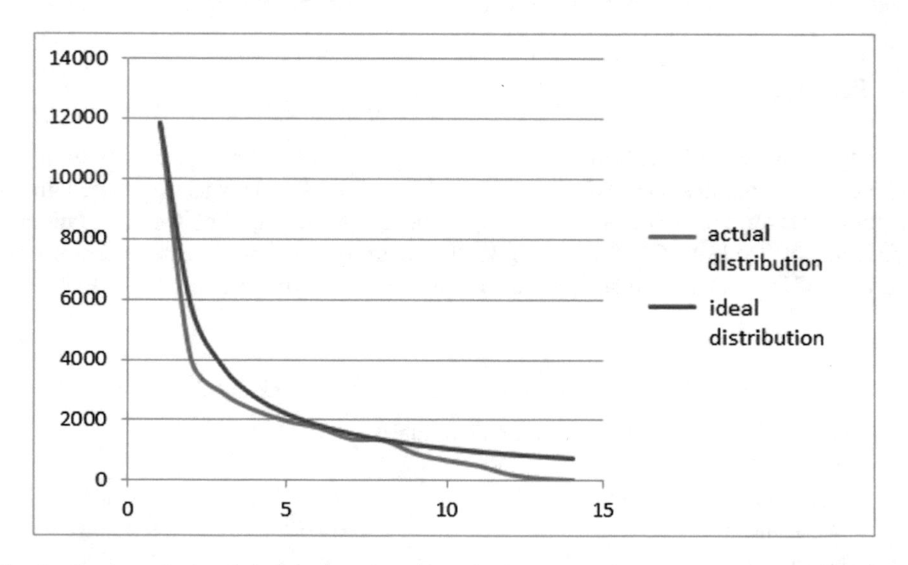

Fig. 3. Rank analysis of the structural and topological dynamics of GDP of the Republic of Abkhazia in the sectoral breakdown, 2018.

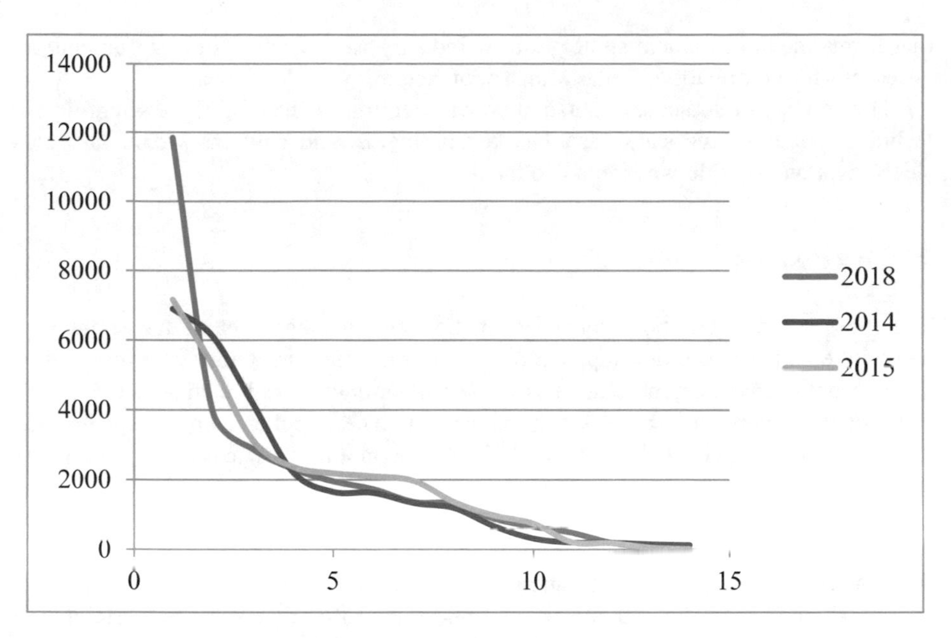

Fig. 4. Rank analysis of the structural and topological dynamics of the GDP of the Republic of Abkhazia in a sectoral breakdown in dynamics.

In general, the ranking distribution in 2018 is close to the ideal hyperbolic form, which indicates the growth of the sustainability of the sectoral structure of the economy. The shift of the rank distribution curve of industries along the ordinate axis indicates the growth of GDP during the analyzed period. However, the change in the position of the rank of the construction industry causes concern. In 2018, the position of the point characterizing the industry is clearly below the ideal rank distribution curve, which indicates the need for measures to increase the industry's contribution to GDP in order to achieve the level of the rank distribution curve. It is also worth noting the decline in the rank of the industry from 1 to 9 during the period under consideration.

We believe that this trend indicates a decrease in the importance of the industry in the country's economy due to the equalization of the dynamics of the development of industries, which is natural in the post-war stage of development of the republic. However, given the need for further development of the industry as a basis for the development of potentially competitive industry sectors, we consider this trend to be the result of the exhaustion of old resources of the competitiveness of industry enterprises and the need for further development on an innovative basis.

Accordingly, the need for a scientifically based innovation and strategic development of construction enterprises of the republic as a basis for the sustainable development of the construction industry is of particular relevance.

The initial impetus for the beginning of the formation of the innovative and strategic development of a construction enterprise may be the receipt of an indicative development plan. On the basis of its study, an administrative document is drawn up that forms the development strategy of the industry as a whole. These documents are necessary for compilation by the Ministry of Economy of Abkhazia.

The strategy or strategic orientation draws attention to changes in the scientific and technical policy of the state, identifies key problems, and contains preliminary considerations on possible ways to solve them.

4 Discussions

The most severe restrictions on the formation and management of the strategy of innovative and strategic development of an enterprise, in our opinion, are imposed by their external environment. The most severe restrictions on the vital activity of a construction enterprise are set by the consumer. The suppliers, on the quality and volume of deliveries of which the possibilities of economic growth depend, also have a significant influence in the same direction.

Regulatory influences of the state in the form of laws, decrees and other normative acts set the economic rules of enterprise behavior. Political parties, public organizations, and also trade union organizations set environmental, political and other rules of action. The authors believe that it is the environment that delineates both the area of activity of the construction company and its regime.

Since innovation is a complex, uncertain in its outcome, difficult to predict process [9], the development of an innovative strategy necessitates the provision of reliable information on the state of innovative development in the future. This is also necessary to substantiate planned and managerial decisions, to reduce the degree of risk and eliminate negative deviations in the activities of the main economic level, when an extremely unstable economic situation is observed.

Under these conditions, in our opinion, it is possible to establish trends of changes in the external environment with the help of analytics and forecasting, which solves the following main tasks:

- identifying alternative options for the future state of the construction enterprise based on trends in the economy of the industry, identifying the most possible and real directions of change;
- assessment of its capabilities in the case of the realization of one of the proposed alternative options in terms of the interaction of goals, means, and resources;
- identification of restrictions affecting the flexible operation of the enterprise;
- assessment of the possible socio-economic consequences at the enterprises in case of the implementation of alternative options for the future;
- according to the accepted criteria of efficiency, identifying preferred areas in the previously planned strategic position of the development of the enterprise.

5 Conclusion

The interaction of the construction enterprise with the external environment is carried out in two opposite directions, which is a manifestation of the action of feedbacks: on the one hand, the enterprise adapts to changes in the external environment; on the other hand, in the process of functioning and development, the enterprise itself has an impact on the external environment, which leads to changes in its state.

The stage of forecasting the environment is designed to identify its possible changes in order to ensure a timely response to these changes and, ultimately, the successful development of the construction enterprise.

The set of environmental factors and the complex nature of their interaction require the identification of their logical order.

The object of the analysis is the organizational and economic processes in the external environment, characterized by certain indicators or factors having a rank distribution, comparing which with the model one, it is possible draw conclusions about development trends, shortcomings and necessary measures of influence.

References

1. 25 shagov po razvitiyu ehkonomiki Respubliki Abhaziya do 2025 goda. Programma ehkono micheskogo razvitiya Respubliki Abhaziya. http://apsnypress.info/upload/Programma_25shagov_EconomRazvitiya_Abkhazii_170209.pdf
2. Kozhevnikova, M.K.: Adaptaciya deyatel'nosti investicionno-stroitel'nogo kompleksa k usloviyam ustojchivogo razvitiya regiona, avtoreferat dissertacii na soiskanie uchenoj stepeni doktora ehkonomicheskih nauk. Ural'skij gosudarstvennyj tekhnicheskij universitet, Ekaterinburg (2007)
3. Kulakov, K., Belyaeva, S., Belyantseva, O., Gamisoniya, A.: MATEC Web of Conferences Proceedings, p. 01118 (2018)
4. Gumba, H.M., Mihajlov, V.Y., Gamuleckij, V.V.: Formirovanie mekhanizma innovacionno-strategicheskogo razvitiya stroitel'nyh predpriyatij, Moscow (2012)
5. Uvarova, S., Vlasenko, V., Bukreev, A., Myshovskaya, L., Kuzina, O.: E3S Web of Conferences, p. 03022 (2018)
6. Fufaev, V.V.: Tekhnetika i cenologiya: ot teorii k praktike. Cenologicheskie issledovaniya **35**, 139–147 (2009)
7. Lozenko, V.K.: EHvolyuciya biznescenozov i biznesukladov v ehkonomike. Palmarium Academic Publishing (2013)
8. Fufaev, V.V.: Sovremennye problemy nauki i obrazovaniya **6**, 7 (2013)
9. Grabovyj, P.G., Okolelova, E.Y., Truhina, N.I.: Izvestiya vysshih uchebnyh zavedenij. Tekhnol. tekstil'noj promyshlennosti **1**(367), 52–56 (2017)

Improving the Investment Policy of the Energy Market

Andrey Mandrykin and Yulia Pakhomova(✉)

Voronezh State Technical University, Moscow Avenue, 14, Voronezh 394026, Russia
yulia1980007@mail.ru

Abstract. The reform of the electric power industry and the development of market relations are inevitably accompanied by revision of rules for the operation of power systems and stricter requirements for their constituent operating facilities. At the same time, the approaches to industry funding are changing, as to ensure reliable and secure power supply to consumers due to liberalization of the wholesale power and capacity market as well as the retail power market, and also due to consolidation of the country's power network complex. In accordance with the current Energy Strategy for the period until 2030, it is assumed to improve the forms and mechanisms of state participation in financing the investment projects in the electric power industry in order to ensure their sufficient investment attractiveness. In this connection, the mechanism of public-private partnership (hereinafter - PPP) seems perspective in financing a strategically important industry of the country and managing the funds of economic entities in terms of creating financial investment tools when using the PPP mechanism, particularly, identifying the potential financial instruments for each of the PPP participants, their ranking in terms of risk-profitability ratio, and revealing the sectoral features of using the PPP models and financial tools.

Keywords: Investment policy · Electric power industry · Energy market

1 Introduction

The management of investment resources of the energy market occurs within all target projects and programs, and, simultaneously, within individual investment projects. Thus, the investment policy is a complex system, managing the investment resources of the energy market.

The main objectives of the investment policy are clear distribution of investment costs between individual projects and programs, forming the investment portfolio, i.e. real investment projects, achieving concrete results through their implementation, and reaching the efficiency of each investment project.

Investment policy is a combination of different approaches and solutions, used for effective investment of funds. Investment policy is focused on achieving medium- and long-term goals of investment activity and determines the main ways to achieve them. This is especially important in conditions of investment cycle length, multi-factorial nature and volatility of market situation.

V. Murgul and M. Pasetti (Eds.): EMMFT-2018, AISC 983, pp. 12–21, 2019.
https://doi.org/10.1007/978-3-030-19868-8_2

Investment policy includes its strategy and tactics, which are its integral and interrelated elements. An investment strategy represents a system of long-term goals of market investment activities, determined by general objectives of its development and investment ideology, as well as the choice of the most effective ways of their achievement [1]. The strategic goals of the investment policy are determined, proceeding from the analysis of the economic situation. The main difficulty in investment management and shaping the relevant policy consists in substantiating the choice of the best possible version of strategic goals.

2 Materials and Methods

In the course of investment policy implementation, it is recommended to carry out the following successive activities:

- The formation of individual areas of investment activities in accordance with the strategy of economic development. Herein, the correlation of various forms of investment at certain stages of the perspective period is determined. The sectoral focus of investments is defined, taking into account the effectiveness of the prospects for the development of one or another area;
- The research and consideration of the conditions of external investment environment and the investment market situation. This takes into account the legal conditions for investment, the so-called investment climate, i.e. a set of parameters of the socio-economic system, motivating the investors to invest;
- The search for individual targets of investment and the assessment of their compliance with investment activity. It involves the study of the current investment market, the choice of individual investment projects, the consideration of options for acquisition of non-current assets, including equipment and technologies, and the expertise of selected investment projects;
- Ensuring high efficiency of investments. The selected investment projects are analyzed in terms of their effectiveness using specific methodologies of assessment. Then, the targets of investment are ranked according to the criterion of investment efficiency and the necessary investment projects are selected;
- Ensuring the minimization of financial risks. The risk of non-acquisition of expected income or the loss of invested capital is investigated. The least risky investment projects are selected. The impact of investment projects on financial stability and solvency is predicted;

The following measures can be used to minimize financial risks:

- the obtainment of guarantees from counterparties in the form of insurance, surety, etc. when granting them a commodity (commercial) or consumer loan;
- using the system of option transactions in exchange (currency, stock market, commodity) operations. In this case, price, inflation and some other types of financial risks are significantly minimized (to the amount of the option premium paid).

- Ensuring the liquidity of investments. In case of the need for capital reinvestment, the upcoming liquidity of an investee is assessed. Herein, such an assessment is not carried out on assets, recorded on the balance sheet of an enterprise, but takes account of the market value, demand, the need for rapid reinvestment and incomplete work on the investment project [2]. Thus, the liquidity of the investee, as a rule, is lower than the cost of funds, invested in this project. The most liquid projects are selected;
- Defining the amount of investment resources and optimizing the structure of sources for investment project financing. Investment resources are all types of monetary and other assets, attracted to invest in the targets of investment. During this activity, the need for investment resources is determined, as well as the possibility of their formation at own expense. The advisability of attracting borrowed financial resources is considered and the optimal ratio of own and borrowed sources of investment project funding is determined;
- The formation and evaluation of the investment portfolio, i.e. the totality of investment projects, planned to be implemented. Herein, the principles of its formation are determined on the basis of a conservative, aggressive or moderate approach to investment project financing. Then, each target of investment is optimized in terms of correlation with its level of profitability, investment risk and liquidity.

3 Results

The forms of financing PPP projects are: corporate financing in which a private partner attracts the borrowed funds for existing business and invests them in the project; project financing, in which a special project company is established and financed. Both project and corporate financing can be applied in all PPP models. The financial and non-financial PPP instruments are used. With the help of financial instruments, the investments are attracted, the infrastructure project implementation is managed, and the risks, costs and profits (losses) are distributed. Non-financial instruments are those methods of impacting a target by PPP entities, due to which the distribution of risks, costs and profits (losses) occurs only. Regulatory and information support constitutes the legal framework of the PPP financial mechanism and regulates financial relations between entities. A special place in the legislation is occupied by the law on investment activity, establishing the rights and obligations of its participants, as well as the methodology for evaluating the efficiency of a PPP and MPP project, which establishes requirements for a financial model and the indicators of investment efficiency.

With the help of financial instruments, the investments are attracted, the infrastructure project implementation is managed, and the risks, costs and profits (losses) are distributed. Non-financial instruments are those methods of impacting a target by PPP entities, due to which the distribution of risks, costs and profits (losses) occurs only. The set of legal rules constitutes the legal framework of the PPP financial mechanism and regulates financial relations between entities. In addition to the above codes and federal laws, regulating the relations between the PPP parties as part of respective models, a special place in the legislation is occupied by the law on investment activity, establishing the rights and obligations of its participants, as well as the methodology for evaluating the efficiency of a PPP and MPP project, which establishes requirements for a financial model and the indicators of investment efficiency.

The regulatory component of the mechanism is represented by instructions, standards, approved tariff rates, guidelines, recommendations and explanations. The information component provides targeted selection of relevant indicators, necessary for implementing effective management decisions. Proceeding from the above and on the basis of the content analysis, we believe that the main factors for the development of the PPP mechanism in the country's economy are: the growth of budget deficit and financial obligations in the provision of public services; the need for economic efficiency; the availability of state guarantees and obtaining an additional market for private business; the possibility of risk sharing; the development of national and international capital markets, allowing private investors to attract debt financing for capital-intensive projects; the development of contractual relations; the complication of financing schemes and instruments.

4 Discussion

Taking the formulated main goal into account, a system of specific local goals for the formation of an investment portfolio is created, the main ones of which are:

- ensuring high rates of capital growth in the upcoming long-term perspective;
- ensuring a high level of income in the current period;
- ensuring the minimization of investment risks;
- ensuring sufficient liquidity of the investment portfolio.

The listed specific goals for the formation of an investment portfolio are largely alternative. For instance, the provision of high rates of capital growth in the long-term perspective is, to some extent, achieved by reducing the level of current return of the investment portfolio (and vice versa) [3]. The capital growth rates and the level of current return of the investment portfolio are directly dependent upon the level of investment risks. The provision of sufficient liquidity can prevent the investment projects, ensuring high capital gains in the long run, from being included in the investment portfolio. Given the alternativeness of goals of the investment portfolio formation, each investor determines their priorities himself.

- The provision of ways for implementing investment programs. An additional cash flow is formed at the expense of profits, derived from operation of the finished investment project, and at the expense of depreciation charge for newly-commissioned objects of non-current assets.

At the same time, it will be mandatory to use new information technologies, namely, a software package, allowing for rapid assessment of alternative variants of investment projects and using the dynamic methods for evaluating performance. Special attention was paid to investment projects of the Moscow Region, since the statistical information is accessed in open data format. For further examination of the selected data, let us present them in the tabular form.

Table 1. The baseline data on a sample of business and PPP projects of the investment portfolio. The strategies of the Moscow Region development for the period of 2015–2019.

Indicator	Year					
	2015	2016	2017	2018	2019	Total
The number of investment projects, units	10	8	11	10	30	75
Investments, %	26,26	22,24	23,8	0,35	18,43	98,08
Number of people, engaged in projects across the Moscow Region	23459	1287	2300	2449	29355	88850
Total population of the Moscow Region	2446822	2662236	29	90000	280000	286000
Welfare (income of the population), Moscow Region, % of the average Russian level	89,3	95,5	96,5	95,3	98,5	–

Next, for assessing the investment attractiveness of the region, we will use the method of linear regression, namely, the one based on determination indicator for implementing the departmental target program «The support of innovative and investment activities of organizations of the scientific-industrial complex of the Moscow Region» with account of the regional support. We will define the parameters of the sample regression equation, plot a dependency graph of the year-over-year number of investment projects (xi) versus the size of investment (yi), and calculate the determination coefficient (the measure of spread – variation – y variable) by the following formula:

$$y = b * x + a \tag{1}$$

where b is the coefficient of regression equation, a is a free component of the regression equation, which is also defined from the «SECTION» category of the «Statistical function».

The results of calculating the parameters of the sample regression equation on implementing investment projects as part of the departmental target program execution are presented in the tabular form (see Table 2).

Table 2. The results of calculating the parameters of the sample regression equation on implementing investment projects for the period of 2015–2019.

The period of study (year)	The number of investment projects, units	Investments for the current year, %	Monthly investments, %
2015	8	0.35	0.3
2016	6	3.5	0.5
2017	10	22.5	10.5
2018	11	26.5	11.6
2019	11	25.5	15.6

Next, having defined the parameters of regression equation for constructing the trend (regression) line, let us conduct the statistical analysis of the resulting equation and calculate the determination coefficient R2.

The year-over-year indicators of investment level scattering have been analyzed. We do not witness a big surge in levels, however, alongside this conclusion, we can speak about a positive trend in the conducted analysis, namely, that there was no decrease in this level either. Let us assess the significance of the regression parameters [4, 5].

The investment policy pursued by the company covers the following areas:

- real investments, that is, capital investments in fixed assets;
- financial investments, made for a period of more than one year in securities of other issuers or in authorized funds of joint ventures, subsidiaries or associated enterprises;
- innovative investments, that is, the investments in intangible assets that ensure the implementation of scientific achievements in practical activities of enterprises.

At an enterprise, real investments are made in the following areas:

- the acquisition of a integral property complex;
- new construction;
- reconstruction;
- modernization and technical re-equipment;
- the acquisition of individual objects of tangible assets in connection with their renewal through physical wear or due to an increase in the staff number.

An important parameter in the system of assessing the economic efficiency of an enterprise is the discount rate (En). It is determined by the levels of profitability prevailing in the capital market. Sometimes it is called the method of accounting return on investment (in English, return on investment), and it is determined by the ratio of the average profit to the average size of investment.

The level of discount rate primarily depends on the form of investments' ownership and their purpose. If investments are private and commercial efficiency is determined, then En is established with account of alternative efficiency of capital use, and the deposit rate can be considered as its minimum value. However, in practice, it is much higher due to inflation and the risk associated with investing in the IE. In this case, the discount rate of each stakeholder of an IE is established by stakeholders themselves.

For multi-share capital, En is:

$$E_H = \sum\nolimits_{i=1}^{n} Ei \cdot yi \quad (2)$$

where n are the types of capital (i = 1, 2, …, n); Ei is the price of capital of each project stakeholder; yi is the share of each capital type in the overall capital.

If there is a sole investor and the investments represent the equity capital of a project's stakeholder, then En, as a rule, does not exceed 10%. If investments are used for implementation of socially significant projects, which are to make a positive impact on macroeconomic indicators, then En may be lower than 10% due to support from the federal budget (for instance, in the case when the investments are directed to modernization or development of railway transport) [6, 7].

In the evaluation of an IE, the discount rate is used to bring multi-temporal results and costs together within a single timespan. Technically, this is done by multiplying the results (R) and costs (C) by the discount factor. In assessing the economic efficiency, the corresponding interest rate Eн is determined. The real interest rate is the rate of return on capital.

The current (nominal) interest rate is the rate of return, as viewed by an investor in the capital market, so, it takes inflation into account.

Discounting determines the current value of future income and expenditure. Leading to base time points R and C, discounting significantly reduces the impact for further periods. Here, much depends upon the rate of discount and the calculation period (see the table).

From the Table 1, it is seen that, with En = 15% and the length of the calculation period T, equal to 15 years, the given current value will be slightly more than 10%. Therefore, the substantiation of the discount rate value, En, the object's lifecycle duration, the horizon of efficiency calculation, or the calculation period is quite an important point in measuring economic efficiency.

If, speaking about the substantiation of En, it is enough to confine ourselves to the provisions, noted above, then explanations will be needed to talk about the length of the investment cycle (period) of an IE, or the calculation period (Table 3).

Table 3. The substantiation for the discount rate value.

Discount rate, En, %	5 years	10 years	15 years	20 years
5	0.783	0.614	0.481	0.377
10	0.621	0.385	0.239	0.149
15	0.497	0.247	0.123	0.061
20	0.402	0.161	0.065	0.026

When IRR > r(CC), an investor accepts the project for consideration with a view of investment; IRR The investment project of the energy company is based on the volume of production and net profit. The net profit of the energy company amounted to $ 2–2.3 billion. The reserves and costs of the energy company can be characterized by the following indicators:

- own working capital (SS);
- own and long-term borrowed sources (SI);
- the main sources of stocks (IS).

5 Conclusion

Three indicators of the stock source are characterized by three indicators of the stockpiling source:

- working capital is characterized by surplus (plus sign) or deficiency (minus sign) (FS);
- own and long-term borrowed sources are characterized by surplus (plus sign) or deficiency (minus sign) (FI);
- the main stock sources are characterized by surplus (plus sign) or deficiency (minus sign) (FV).

Absolute independence: FS, FI, FV > 0;
normal independence: FS < 0, FI, FV > 0;
unstable state: FS, FI < 0, FV > 0;
crisis state: FS, FI, FV < 0 (Tables 4 and 5).

Table 4. The financial state of the energy company (2018).

Indicator, thous. roubles	Calculation period (at the beginning of the calendar year)	Calculation period (at the end of the calendar year)
FS = SS-Z	−684369	−547893
FI = F-Z	−354896	−335884
FV = I-Z	−354896	−335884

Table 5. The financial state of the energy company (2018).

Indicator, thous. roubles	Calculation period (at the beginning of the calendar year)	Calculation period (at the end of the calendar year)
FS = SS-Z	−52796147	−48725812
FI = F-Z	8456711	21479521
FV = I-Z	8456711	21479521

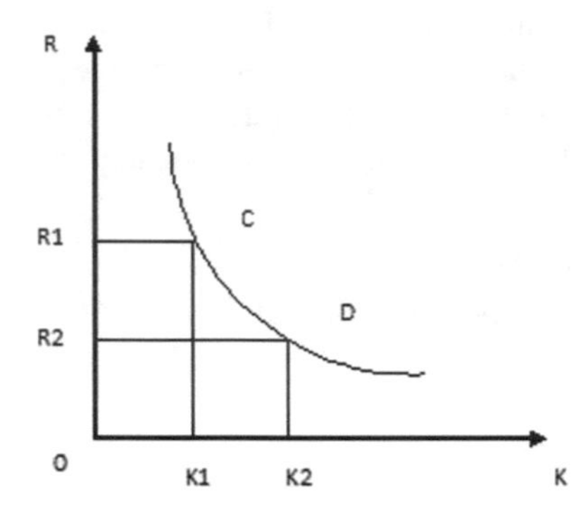

Fig. 1. The elasticity of demand for energy company services.

The calculations demonstrate the normal independence and solvency of the energy company in 2018. The forecast of the energy company development was made on the basis of the model of regulatory demand:

$$A = z_1 x\ K_1 + z_2 x\ K_2 \quad (3)$$

This model makes it possible to determine the sum of products of service usage rates of an energy company, z_1, z_2, for segments K_1, K_2, which helps to calculate the coefficient of demand elasticity (Fig. 1).

- R1 – initial price level
- R2 – new price level
- K1 – initial volume
- K2 – new volume

Therefore, the results of calculating the demand elasticity allow us to conclude that, with a change in the price range of about 1% relative to current prices, the demand can change in the opposite direction. The change by 2% can occur. At the same time, the term of the borrowed capital use is reduced, thus lowering the investment risk for an energy company.

References

1. Alborov, R.A.: The theory of accounting: a training manual. 3rd edition. The Federal State Budgetary Educational Institution of Higher Education «Izhevsk Academy of Agriculture», Izhevsk (2016)
2. Sheremet, A.D.: The methodology of financial analysis. INFRA-M, Moscow (2018)
3. Kuzminova, T.N.: Tax audit (2016)
4. Prykina, L.V.: The economic analysis of an enterprise: a guidebook for bachelor's degree students. Dashkov & K°, Moscow (2016)
5. Golov, R., Narezhnaya, T., Voytolovskiy, N., Mylnik, V., Zubeeva, E.: Model management of innovative development of industrial enterprises. In: MATEC Web of Conferences (2018). https://doi.org/10.1051/matecconf/201819305080

6. Lukmanova, I., Golov, R.: Modern energy efficient technologies of high-rise construction. In: E3S Web of Conferences (2018). https://doi.org/10.1051/e3sconf/20183302047
7. Lyushinskiy, A.V., Fedorova, E.S., Roshan, N.R., Chistov, E.M., Golov, R.S.: Diffusion welding of 12Cr18Ni10Ti steel to palladium alloy foil. Weld. Int. **31**, 777–778 (2017). https://doi.org/10.1080/09507116.2017.1318505

Risk Assessment Models of the Use of Innovative Technologies in Construction as a Factor in the Development of Energy Management

Ella Okolelova(✉), Marina Shibaeva, Oleg Shalnev, and Alexey Efimyev

Voronezh State Technical University,
Moscow Avenue, 14, Voronezh 394026, Russia
Ella.o2011@yandex.ru

Abstract. The article discusses the risks of investment projects with the use of nanotechnology for the development of energy management in the construction industry. Various innovative projects used in the construction industry in the production of new types of materials and structures are considered. Classification of risks is given, and methods for their assessment are defined. Identification and analysis of risks is the first step in the organizational risk management mechanism, which allows defining and taking timely measures to eliminate them. The article defines the probability of occurrence of risks based on methods of stochastic modeling. Since the considered risks have a different nature of occurrence, the conditions of occurrence, and the results of the impact, so, consequently, the calculation mechanisms are built on different models. In the absence of statistical data and the need to predict possible scenarios, a predictive model of the probability of occurrence of unsystematic risks is used, which is presented as a logit model used to predict an event. A quantitative analysis of risks based on the calculation of the price of risks is made. The risks that have a high probability of occurrence and at the same time bear significant damage to the project are identified. A method of forming risk clusters by the size of possible damage and the probability of an event occurrence is proposed. The degree of influence of each risk group on the development of an innovative project in construction has been determined.

Keywords: Nanotechnology · Nanomaterials · Energy saving · Energy management logit model · Risks · Cluster risks · Risk matrix

1 Introduction

The term "nanotechnology" firmly entered our lives. These unique technologies are used today in all areas of production, such as pharmaceuticals, food, textiles, cosmetics and others. They are also widely used in construction. At present, nanotechnology in construction refers to the use of nano-additives and nanodopants, i.e. nanoscale objects in the form of specially designed nanoparticles, nanoscale particles with a linear size

V. Murgul and M. Pasetti (Eds.): EMMFT-2018, AISC 983, pp. 22–35, 2019.
https://doi.org/10.1007/978-3-030-19868-8_3

less than 100 nm [1]. Thus, the use of ultrafine nanoscale particles made it possible to significantly change the properties of concrete, obtaining high-strength concrete that can be used in the construction of skyscrapers, bridges, nuclear reactors, and many other objects. The service life of such a nanomaterial, according to the calculations of the developers, is at least 500 years. The extent to which this declared statement is true can only be shown by the experience of long-term operation, the absence of which today is a significant deterrent factor hindering the active introduction of such technologies in the field of construction production.

The field of production of metals and alloys also does not remain aloof from the use of nanotechnology. For example, the nano-modification of metals made it possible to obtain high-strength steel, which has no world analogues in terms of strength. Not to mention the nanocoatings, which dozens and sometimes hundreds of times increase the corrosion resistance of metals, increase its service life even when working in aggressive media. Such technologies should be the basis for the development of the construction industry over the coming decades. But what happens in the real sector of the construction industry? Why today the share of nanomaterials takes no more than 1% of the total volume of building materials used? Why is the construction industry so skeptical about innovations and so reluctant to use modern technology? The answer to these and other questions will help giving a deep analysis of the risks that these innovative directions bear. With all the variety of types of nanotechnology and the obvious effect of their use in almost all areas of industrial activity, their introduction into the real production sector is extremely slow. This is especially true for construction. This is primarily due to the unwillingness of developers to use technologies that have been insufficiently tested by operating experience. The high costs and risks associated with any innovative project make the use of nanotechnologies an unattractive process for many manufacturers. Especially negative impact on the development of innovative methods in the real sector of the economy have risks, identification, assessment and the possibility of reducing which sometimes causes significant difficulties.

2 Materials and Methods

Let's consider the possible types of risks that accompany innovative projects using nanotechnology and nanomaterials in construction (Table 1). At the same time, risks will be assessed from the point of view of investors and developers, who today are mostly skeptical about the use of future technologies in the real sphere of production. Thus, innovation risks need to be considered as a kind of "investment filter", which significantly slows down the introduction of innovations in production and, accordingly, reduces the level of efficiency of construction production [2]. Thus, a sufficient number of potentially dangerous situations or risks have been identified, the probability of occurrence of which significantly hinders the introduction of new technologies and materials into production. Perhaps, an adequate assessment of the risks, the likelihood of their occurrence, and the results of the occurrence of events would allow using modern technologies and materials in construction more confidently.

Table 1. Types of risks by effects and consequences.

Types	Causes and consequences
Cluster 1 (project risks)	
Technological risks	Lack of assessment of the life cycle of nanomaterials and the study of the strength properties of structures during long-term operation
	The ability to change the quality characteristics of nanomaterials with long-term operation
	Errors in the design of objects using nanoconstructions (geological surveys, consideration of climatic conditions, design features, etc.)
	The complexity of installation. Insufficient provision of equipment necessary for the design and installation of structural elements made on the basis of nanotechnology
	The complexity of the analytical work to assess the current state and the dynamics of the strength characteristics of structural elements made on the basis of nanotechnology
	Insufficient study of the operation of structures in adverse climatic conditions (seismically active zones, wind loads, high humidity zones, low temperature conditions, etc.)
	Work in aggressive media. The threat of destruction of structural elements made on the basis of nanotechnology partially or fully under the influence of chemically hazardous substances
Operational risks	Loss of strength characteristics of structures over time. Local destruction of structural elements of buildings, etc.
	Lack of assessment of the life cycle of nanomaterials. Complete destruction of structures
	Elimination of emergency situations in case of structural failure. Duration of elimination; the inability to restore the performance of structural elements
Cluster 2 (environmental risks)	
Risk of nanomaterial toxicity	Insufficient research on the effects of nanomaterials on human health
	Insufficient research on the effects of nanomaterials on the environment
Risk of waste treatment	The complexity of waste disposal. The ability to penetrate the environment
Cluster 3 (investment risks)	
Increase in investment costs	The increase in the cost of construction in the building of objects using nanotechnology and nanomaterials
Increase in operational costs	The increase in operating costs when using nanomaterials and structural elements produced on their basis
Liquidity risk	The lack of consumer demand for objects built on the basis of nanotechnology (residential facilities)

Let us consider several models that will make it possible to expect certain events when using nanotechnologies and nanomaterials in construction with a certain degree of probability. The calculation of risks is made in order to build a forecast of the efficiency of using nanotechnologies. Identifying and analyzing risks is the first step in an organizational risk management mechanism. Speaking about the relationship "impact – response", it is advisable to transform these concepts on the mechanism of the impact of risk and measures to eliminate it. Let us determine the probability of occurrence of risks on the basis of methods for stochastic modeling Since the considered risks have a different nature of occurrence, the conditions of the onset, the results of the impact and, consequently, the calculation mechanisms are built on different models. Some of the risks presented above are the results of rare events that can be considered as hardly probable. In this case, the random variable will have a Poisson distribution [3].

The risk assessment algorithm for the implementation of innovative projects includes three stages.

1. Logit model for risk assessment. In the absence of statistical data and if it is necessary to forecast possible scenarios for the development of events, a prognostic model of the probability of occurrence of unsystematic risks is used, which is presented as a logit model used to predict an event. We introduce the dependent variable y such that y = 1 if the event occurred, and y = 0 if the event did not occur.

We introduce independent variables (predictors) x1, x2, x3, …, xn. Based on the predictor values, the probability of the dependent variable y is calculated. Suppose that the probability of occurrence of the i-th risk (event y = 1) is equal to

$$P\{y=1|x\} = f(z) \tag{1}$$

where $z = B^T x$; B – column vector of parameters (real numbers); x_1, x_2, …, x_n - independent variables; *f(x)* - logit function

$$f(x) = \frac{1}{1+e^{-z}}. \tag{2}$$

Probability of possible value of the event y = 0:

$$P\{y=0|x\} = 1 - f(z) = 1 - f(B^T x) \tag{3}$$

The distribution function *y* for a given *x* is written as:

$$P\{y|x\} = f(B^T x)^y (1 - f(B^T x))^{1-y}, y\{0;1\}. \tag{4}$$

This actually represents the Bernoulli distribution with the parameter $f(B^T x)$.

As a training sample will be a set of value of independent variables and the corresponding value of the dependent variable:

$(x^{(1)}, y^{(1)}), \ldots, (x^{(n)}, y^{(n)})$, where $x^{(i)} \in R^n$ - independent variable value vector, and $y^{(i)} \in \{0,1\}$ - corresponding value of dependent variable *y*.

Parameters B are selected that maximize the likelihood function:

$$\widehat{B} = max_B \prod_{i=1}^{n} P\left\{y = y^{(i)} \middle| x = x^{(i)}\right\} \tag{5}$$

As an example, let us consider the risk of using metal structures made on the basis of nanotechnology when creating a system of engineering equipment for a building [4, 5].

The operation of the system is described by two states: "working" (event y = 0) and "failure" (event y = 1), i.e. system crash risk has come.

The function f(z) shows the probability of the final outcome of events due to the influence of factors determining this event that may have the opposite effect.

Variable z shows the exposure to a certain set of risk factors and is expressed as a regression equation:

$$z = b_0 + b_1x_1 + b_2x_2 + \ldots + b_nx_n, \tag{6}$$

where x_1, x_2, …, x_n – independent variables, b_0, b_1, b_2, ..., b_n – regression coefficient for control parameters (risk factors). Parameters b_i are usually assessed using the maximum likelihood method.

Let the probability of a system failure be determined by the following factors (independent variables (x_1, x_2, x_3):

- x_1 - quality engineering equipment (high quality - 3 points, average level of quality - 2 points, low quality - 1 point);
- x_2 - availability of a management and control service (missing - 1, available - 0);
- x_3 - availability of system support service (missing - 1, available - 0).

The obtained function: $z = -5 + 3x_1 - 1{,}2x_2 - 1{,}5x_3$.

Under the most unfavorable circumstances, i.e. when $x_1 = 1$, $x_2 = 1$, $x_3 = 1$, we get the value of the logit function $f(z) = 0{,}0984$. This value is the probability of occurrence of risk with the cumulative effect of the most adverse factors (Table 2).

2. Model of quantitative risk assessment. A practical tool for studying the level of hazards for an object is a quantitative risk analysis, the essence of which is to consider all possible scenarios of occurrence and development of events, as well as to assess the frequency and scale of the possible implementation of each of the scenarios on a particular object [6].
3. Risks are measured by the probability of occurrence and the level of influence. The likelihood of an event occurring can be assessed as the frequency of its occurrence in practice or when fictitiously acting out, simulating a situation. For any project, it is important to assess also the value of possible losses due to the occurrence of risks, i.e. their price. The price of risk is the possible consequences of its occurrence and loss, taking into account the probability.

With the probability of occurrence of the i-th risk pi and the magnitude of possible losses Vi, the price of risk is determined as follows:

$$R = \sum\nolimits_{i=1}^{n} p_i V_i \tag{7}$$

Table 2. Risk probability.

Cluster	Name	Impact and consequences	Code	Probability of occurrence
1. Project risks	1.1. Technological risks	Lack of assessment of the life cycle of nanomaterials and the study of the strength properties of structures during long-term operation	111	0.21
		The ability to change the quality characteristics of nanomaterials with long-term operation	112	0.27
		Errors in the design of objects using nanoconstructions (geological surveys, consideration of climatic conditions, design features, etc.)	113	0.23
		The complexity of installation. Insufficient provision of equipment necessary for the design and installation of structural elements made on the basis of nanotechnology	114	0.15
		The complexity of the analytical work to assess the current state and the dynamics of the strength characteristics of structural elements made on the basis of nanotechnology	115	0.16
		Insufficient study of the operation of structures in adverse climatic conditions (seismically active zones, wind loads, high humidity zones, low temperature conditions, etc.)	116	0.24
		Work in aggressive media. The threat of destruction of structural elements made on the basis of nanotechnology partially or fully under the influence of chemically hazardous substances	117	0.47

(*continued*)

Table 2. (*continued*)

Cluster	Name	Impact and consequences	Code	Probability of occurrence
	1.2. Operational risks	Loss of strength characteristics of structures over time. Local destruction of structural elements of buildings, etc.	121	0.43
		Lack of assessment of the life cycle of nanomaterials. Complete destruction of structures	122	0.15
		Elimination of emergency situations in case of structural failure. Duration of elimination; the inability to restore the performance of structural elements	123	0.26
2. Environmental risks	2.1. Risk of nanomaterial toxicity	Insufficient research on the effects of nanomaterials on human health	211	0.32
		Insufficient research on the effects of nanomaterials on the environment	212	0.37
	2.2. Risk of waste treatment	The complexity of waste disposal. The ability to penetrate the environment	221	0.31
3. Investment risks	3.1. Increase in investment costs	The increase in the cost of construction in the building of objects using nanotechnology and nanomaterials	331	0.41
	3.2. Increase in operational costs	The increase in operating costs when using nanomaterials and structural elements produced on their basis	332	0.37
	3.3. Liquidity risk	The lack of consumer demand for objects built on the basis of nanotechnology (residential facilities)	333	0.17

The cost of risk can be determined in terms of value, but in this case, it is preferable to use relative values, and the price is determined as a percentage of the cost of a building construction.

Risks are identified by qualitative and quantitative characteristics based on cluster analysis, which allows grouping risks not only by the sources of their occurrence, but also by the probability of occurrence and the size of the damage [7].

The task of forecasting a situation is to detect the most active risk group with relatively high values of the probability of occurrence and the size of the damage (Table 3).

Table 3. Quantitative risk assessment.

Code	Probability of occurrence	Maximum damage, % of construction cost	Price of risk, % of construction costs
111	0.21	20	4.2
112	0.27	22	5.94
113	0.23	18	4.14
114	0.15	14	2.1
115	0.16	7	1.12
116	0.24	38	9.12
117	0.47	42	19.74
121	0.43	57	24.51
122	0.15	85	12.75
123	0.26	62	16.12
211	0.32	32	10.24
212	0.37	12	4.44
221	0.31	24	7.44
331	0.41	26	10.66
332	0.37	19	7.03
333	0.17	33	5.61

The risk price range is [1.12; 24.5]. There are three main risk groups. The data in Table 3 are arranged in order of increasing risk price, and cluster groups are formed (Table 4).

Table 4. Clustering in accordance with the risk price.

Code	Probability of occurrence	Maximum damage, % of construction cost	Price of risk, % of construction costs
115	0.16	7	1.12
114	0.15	14	2.10
113	0.23	18	4.14
111	0.21	20	4.20
212	0.37	12	4.44
333	0.17	33	5.61
112	0.27	22	5.94
332	0.37	19	7.03
221	0.31	24	7.44
116	0.24	38	9.12
211	0.32	32	10.24
331	0.41	26	10.66
122	0.15	85	12.75
123	0.26	62	16.12
117	0.47	42	19.74
121	0.43	57	24.51

It is proposed to form three main risk groups and introduce a new group classification in terms of qualitative and quantitative analysis (Table 5) [8].

Table 5. Risk clusters.

Risk cluster	Characteristic	Risk price range
Conservative risks	Risks that have a low probability of occurrence and entail minor damage (1)	[1; 4,9]
Modcratc risks	Risks that have either an average probability of onset or an average damage (2)	[5; 9,9]
Aggressive risks	Risks with a high probability of occurrence and minor damage or risks with a low probability of occurrence, but a high size of damage (3)	[10; 25]

Based on the data in Tables 4 and 5, a risk matrix was built, which shows the risks that pose the greatest threat to the project (Fig. 1).

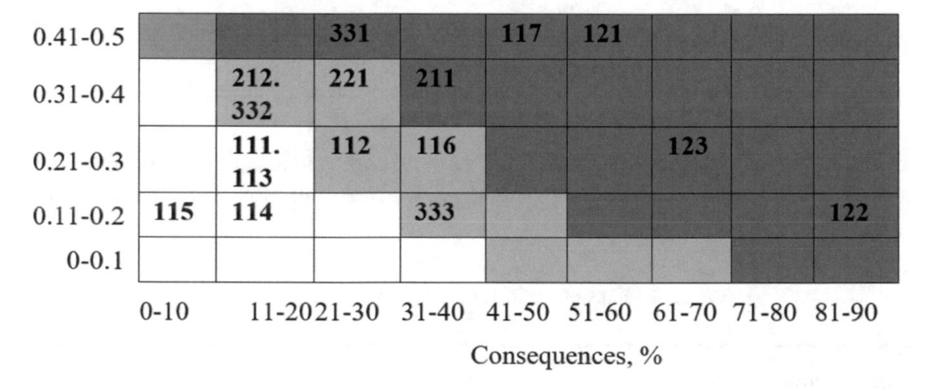

Fig. 1. Risk matrix.

Note: in Fig. 1, risk clusters are highlighted in the following colors:

(1) - ☐ - conservative risks, (2) - ☐ - moderate risks, (3) - ☐ - aggressive risks.

Thus, the risks that have a high probability of occurrence and at the same time bear significant damage to the project have been identified. The task is to determine the influence of each risk group on the development of the innovation direction in construction [8].

4. Modeling of the factor space of risks and assessment of their cumulative impact. It is necessary to determine the size of the most probable damage that an investor will have during the construction of facilities using nanotechnologies and nanomaterials in case of adverse situations.

Clusters are formed in such a way that the risks fall into the cluster regardless of the nature of the occurrence, but approximately the same in terms of quantitative characteristics. Thus, there are three main clusters. To assess the impact of each cluster on the outcome of an innovative project, the method of calculating the Euclidean distance was used as a measure of determining the proximity of a pair of points in a multidimensional space:

$$d_{ij} = \sqrt{\sum\nolimits_{t=1}^{n} (x_{it} - x_{jt})^2}, \ \mathrm{I}, j = 1, \ldots, n. \quad (8)$$

where d_{ij} - the Euclidcan distance between the i-th and j-th objects; x_{it} - the value of the i-th index for the i-th object.

Based on the calculated risk price values, the distance between each pair of objects is calculated. As a result, we obtain a square matrix D, which has dimensions nxn (by the number of objects); this matrix is symmetric, i.e. $d_{ij} = d_{ij}$ $(i, j = 1,\ldots,n)$.

$$D = \begin{pmatrix} d_{11} & d_{12} & \ldots & d_{1n} \\ d_{21} & d_{22} & \ldots & d_{2n} \\ \ldots & \ldots & \ldots & \ldots \\ d_{n1} & d_{n2} & \ldots & d_{nn} \end{pmatrix} \quad (9)$$

The main idea of this method is to sequentially merge grouped objects—first, the closest, then more distant from each other. As a result, a cluster of risks that most actively influence the forecast of the efficiency of using nanotechnologies in construction is formalized.

We define the proximity of two clusters as the average square of the distance between all pairs of objects:

$$D_{pq} = \sum\nolimits_{i=Rp} \sum\nolimits_{i=Rq} d_{ij}/n_p n_q \quad (10)$$

where D_{pq} - a measure of proximity between the p-th and q-th clusters; R_q – p-th cluster; R_q – q-th cluster; n_p - the number of objects in the p-th cluster; n_q - the number of objects in the q-th cluster.

In the first step of the agglomerative and hierarchical cluster analysis procedure, the initial matrix of distances between objects is considered and the minimum number di1j1 is determined from it; further, the closest objects with numbers i1 and j1 are combined into one cluster.

Thus, the solution algorithm involves several stages. The first stage is the identification of risks according to qualitative and quantitative characteristics on the basis of cluster analysis, which will allow grouping risks not only according to the sources of their occurrence, but also according to the probability of occurrence and the size of the damage. It is necessary to highlight the risks in order of importance, which implies the greatest possible damage in the implementation of the investment project. Risks are characterized by the likelihood of onset and price of risk.

The second stage of the assessment of risk clusters is to measure the weight factor or their influence on the result of the project implementation.

Since random variables can be determined both by implementations and their quantitative characteristics and the laws of their distribution, at the planning stage, as a rule, they are unknown, and therefore use only the characteristics of random variables and the laws of their distribution. In addition, speaking of the risks, it is necessary to consider the possibility of different time intervals of their occurrence.

3 Results

Thus, on the basis of the proposed model, risks were identified, grouped into clusters in order to identify the most significant ones, and their impact on the project implementation conditions was assessed. Based on the cluster analysis method, cluster distances were determined based on the "risk price" basis, and a risk matrix was built (Table 6), where risk categories were selected according to degree of activity (conservative, moderate and aggressive).

Table 6. Risk matrix.

Code			115	114	113	111	212	333	112	332	221	116	211	331	122	123	117	121
	Cluster		**1**					**2**					**3**					
	Cluster	Risk price	1.1	2.1	4.1	4.2	4.4	5.6	5.9	7.0	7.4	9.1	10.2	10.7	12.8	16.1	19.7	24.5
115	**1**	1.1	0.0	1.0	9.1	9.5	11.0	20.2	23.2	34.9	39.9	64.0	83.2	91.0	135.3	225.0	346.7	547.1
114		2.1	1.0	0.0	4.2	4.4	5.5	12.3	14.7	24.3	28.5	49.3	66.3	73.3	113.4	196.6	311.2	502.2
113		4.1	9.1	4.2	0.0	0.0	0.1	2.2	3.2	8.4	10.9	24.8	37.2	42.5	74.1	143.5	243.4	414.9
111		4.2	9.5	4.4	0.0	0.0	0.1	2.0	3.0	8.0	10.5	24.2	36.5	41.7	73.1	142.1	241.5	412.5
212		4.4	11.0	5.5	0.1	0.1	0.0	1.4	2.3	6.7	9.0	21.9	33.6	38.7	69.1	136.4	234.1	402.8
333	**2**	5.6	20.2	12.3	2.2	2.0	1.4	0.0	0.1	2.0	3.3	12.3	21.4	25.5	51.0	110.5	199.7	357.2
112		5.9	23.2	14.7	3.2	3.0	2.3	0.1	0.0	1.2	2.3	10.1	18.5	22.3	46.4	103.6	190.4	344.8
332		7.0	34.9	24.3	8.4	8.0	6.7	2.0	1.2	0.0	0.2	4.4	10.3	13.2	32.7	82.6	161.5	305.6
221		7.4	39.9	28.5	10.9	10.5	9.0	3.3	2.3	0.2	0.0	2.8	7.8	10.4	28.2	75.3	151.3	291.4
116		9.1	64.0	49.3	24.8	24.2	21.9	12.3	10.1	4.4	2.8	0.0	1.3	2.4	13.2	49.0	112.8	236.9
211	**3**	10.2	83.2	66.3	37.2	36.5	33.6	21.4	18.5	10.3	7.8	1.3	0.0	0.2	6.3	34.6	90.3	203.6
331		10.7	91.0	73.3	42.5	41.7	38.7	25.5	22.3	13.2	10.4	2.4	0.2	0.0	4.4	29.8	82.4	191.8
122		12.8	135.3	113.4	74.1	73.1	69.1	51.0	46.4	32.7	28.2	13.2	6.3	4.4	0.0	11.4	48.9	138.3
123		16.1	225.0	196.6	143.5	142.1	136.4	110.5	103.6	82.6	75.3	49.0	34.6	29.8	11.4	0.0	13.1	70.4
117		19.7	346.7	311.2	243.4	241.5	234.1	199.7	190.4	161.5	151.3	112.8	90.3	82.4	48.9	13.1	0.0	22.8
121		24.5	547.1	502.2	414.9	412.5	402.8	357.2	344.8	305.6	291.4	236.9	203.6	191.8	138.3	70.4	22.8	0.0

Based on (10), the distances between the clusters are determined (Table 7).

Table 7. Cluster distance matrix.

Risk cluster name	Conservative	Moderate	Aggressive
Conservative (1)	0	9.04	33.05
Moderate (2)	9.04	0	19.61
Aggressive (3)	33.05	19.61	0

The risk space is presented in Fig. 2 in the form of a "risk triangle", in accordance with the number of formed risk clusters.

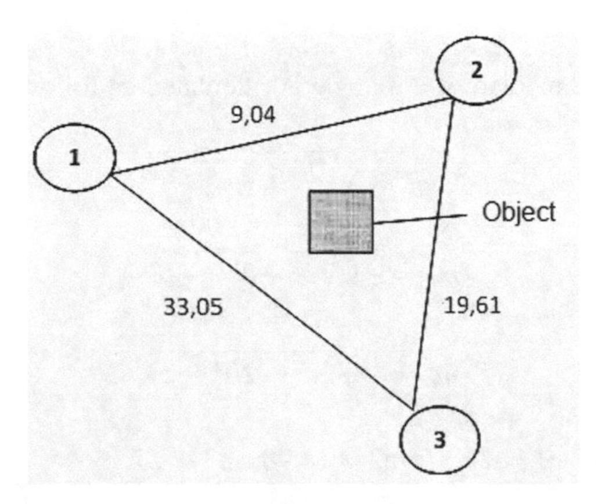

Fig. 2. Geometric interpretation of the risk space.

4 Discussion

The geometric interpretation of the risk space presented in Fig. 3 is interpreted as follows. The object is in the center of gravity of the triangle. As is known, the center of gravity or center of mass is the intersection point of the medians of the triangle. The degree of influence of each risk cluster (vertex of a triangle) is determined geometrically as 2/3 of the median. Risk clusters are formed on the basis of the price of risks, i.e. the magnitude of expected losses from the occurrence of risk, taking into account the probability and maximum damage. The distance between the clusters is also expressed in terms of the risk price.

If the distance between the vertices of the triangle will increase, therefore, the cost of risk will increase. To determine the degree of influence of the risk clusters on the object, it is necessary to determine the lengths of the medians of a triangle (Fig. 3).

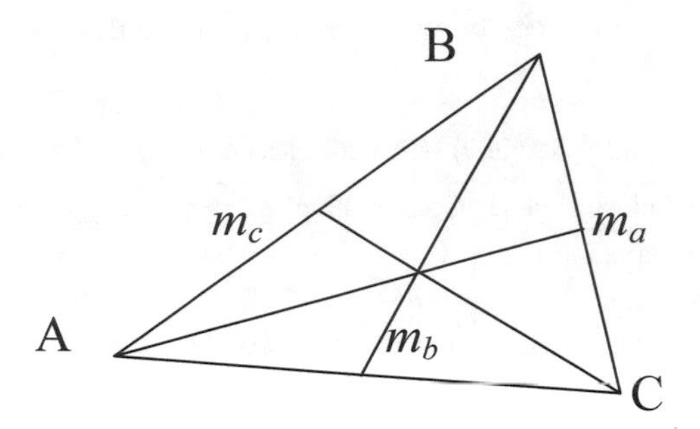

Fig. 3. Geometric interpretation of risks.

The center of gravity of the triangle is at the intersection of the medians, and the relation of medians at this point is 2:1. Consequently, the degree of influence of a risk cluster on the object can be defined as 2/3 of the median length of the corresponding cluster vertex.

The length of the median of a triangle is calculated as follows:

$$m_a = \frac{1}{2}\sqrt{2b^2 + 2c^2 - a^2} \tag{11}$$

$$m_b = \frac{1}{2}\sqrt{2a^2 + 2c^2 - b^2} \tag{12}$$

$$m_c = \frac{1}{2}\sqrt{2a^2 + 2b^2 - c^2} \tag{13}$$

For risk clusters: m1 = 24.12; m2 = 14.98; m3 = 27.13.

5 Conclusions

1. The price of cluster risks has been determined, and the degree of their impact on the implementation of the object has been identified. In this case, the cluster of aggressive risks has the strongest impact on the object, which is fully explained by the high level of uncertainty in the implementation of the innovative project.
2. Risks were identified, grouped into clusters in order to identify the most significant ones, and their impact on the conditions for implementing an innovative project using nanotechnology was assessed.

References

1. Babkov, V.V., Mohov, V.N., Kapitonov, S.M., Komohov, P.G.: Strukturoobrazovanie i razrushenie cementnyh betonov. GUP «Ufimskij poligrafkombinat», Ufa (2002)
2. Okolelova, E.Y., Fam, H.K.: Innovacionnye metody ocenki riskov vysotnogo stroitel'stva vo V'etname. Sovremennaya ehkonomika: problemy i resheniya **5**, 130–137 (2011)
3. Kornickaya, O.V., Okolelova, E.Y., Truhina, N.I.: Razvitie innovacij i i mekhanizm ih rasprostraneniya ne predpriyatiyah strojindustrii. Upravlenie ehkonomicheskimi sistemami: ehlektronnyj nauchnyj zhurnal **12**(60), 93 (2013)
4. Grabovyj, P.G., Truhina, N.I., Okolelova, E.Y.: Dinamicheskaya model' prognozirovaniya razvitiya innovacionnogo proekta. Tekhnol. tekstil'noj promyshlennosti **1**(367), 78–82 (2017)
5. Gasilov, V.V., Okolelova, E.Y., Zamchalova, S.S.: Ehkonomiko-matematicheskie metody i modeli: ucheb.- metod. posobie, gos. arh.-stroit. Un-t.- Voronezh, Voronezh (2005)
6. Okolelova, E.Y., Truhina, N.I.: Stroitel'stvo vysotnyh zdanij: ocenka ehffektivnosti proektov v usloviyah riskov. VGASU, Voronezh (2016)
7. Fam, H.K.: Metodologicheskie aspekty ocenki riskov investicionnyh proektov. EHkonomika i menedzhment sistem upravleniya **1**(7), 111–118 (2013)
8. Grabovyj, P.G., Truhina, N.I., Okolelova, E.Y.: Upravlenie investicionnom proektom vosproizvodstva nedvizhimosti s uchetom riskov. Tekhnol. tekstil'noj promyshlennosti **1** (367), 48–52 (2017)

Technical and Economic Aspects of Energy Saving at the Stages of the Building Life Cycle

Olga Kutsygina(✉), Svetlana Uvarova, Svetlana Belyaeva, and Andrey Chugunov

Voronezh State Technical University, Moscow Avenue, 14, Voronezh 394026, Russia
olga.kutsigina@rambler.ru

Abstract. The technical and economic aspects of energy saving of buildings as an urgent task of the last decades are considered. The conditions for ensuring the efficiency of the economy, the development of technology, the quality and comfort of living, and the preservation of natural resources for future generations are considered. The authors give an analysis of the development of energy consumption depending on population growth and types of energy resources, argue the growing role of energy saving in the context of the projected decline in the availability of natural energy resources at all stages of the building life cycle. The paper proposes a methodological approach to the selection of effective design options for energy-saving measures based on a multi-criteria assessment of technical and economic characteristics, instead of local estimates of the construction cost and consumption of fuel and energy resources at the operational stage.

Keywords: Energy saving · Building life cycle · Energy resources

1 Introduction

The use of energy resources is a prerequisite for the functioning of life-support and security systems, and has an impact on the quality and economic level of society. Predicting the decline in mineral production, the scarcity of fuel and energy resources, and an increase in price necessitates a search for new sources of their production and scientific approaches to managing the use of energy resources, including energy saving. Energy saving as a source of preserving natural resources for future generations and reducing the cost of energy for consumers, the motive for developing innovative energy-saving technologies and solving socially important tasks, i.e. its technical and economic aspects, is of great importance. The most important areas of energy saving include the investment and construction sector of the economy, since 40% of the country's energy resources are accounted for the energy supply of residential, public and industrial buildings during their operation. Since the volume of energy consumption at the stages of construction and operation of the facilities being constructed is predicted at the design stage, it is important to study the technical and economic aspects of energy saving at the stages of the real estate life cycle on the principles of a systems approach and a multi-criteria assessment of compared alternatives.

V. Murgul and M. Pasetti (Eds.): EMMFT-2018, AISC 983, pp. 36–44, 2019.
https://doi.org/10.1007/978-3-030-19868-8_4

2 Materials and Methods

Energy resources are necessary for the functioning of life-support systems, creating conditions for achieving quality and maintaining the economic standard of living of the population. The extent of energy consumption in major regions of the world by 2020 is estimated at 21 billion tons of fuel equivalent. According to the BP Statistical Review of World Energy 2007, energy is generated from the use of oil, natural gas, coal, the development of nuclear energy, hydropower and the use of renewable natural resources like wind and sun in accordance with the structure in Fig. 1 [1].

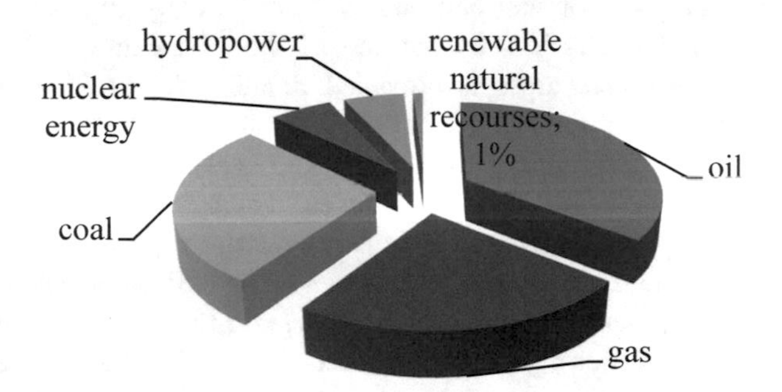

Fig. 1. The structure of energy consumption by type, %.

Improving the comfort of life as a result of the use of various types of energy contributes to an increase in population. The dynamics of growth in the population of the planet is shown in the graph (Fig. 2), described with high confidence (93.72%) by time exponential dependence and is accompanied by an increase in the average per capita energy consumption by one and a half over the period 1966–2000 [1, 2].

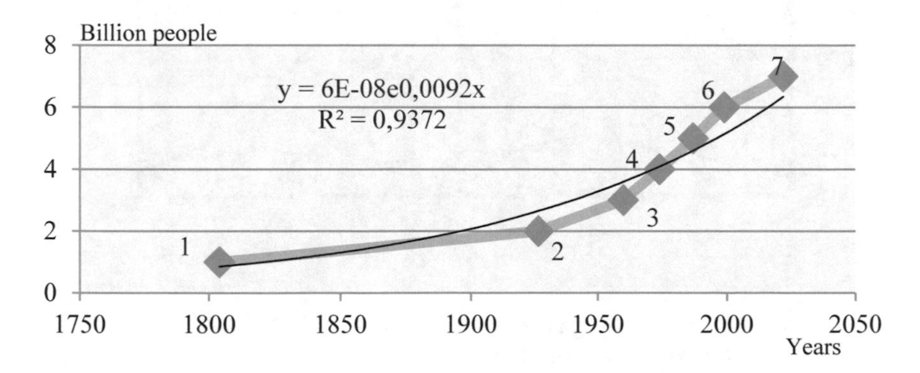

Fig. 2. Dynamics of the population of the planet.

The impact of population growth (N) on the amount of global consumption of energy resources in the 20th century (E) is described by the functional dependence obtained by J. Holdren in 1991 and given in [3, 4].

$$E \sim N^2 \tag{1}$$

However, the emerging trend of further development of mankind depending on the growth of energy supply is limited to forecasts of a decline in mineral production, including oil, gas, and coal, which hold a large share in the composition of energy sources (36, 24, and 28%, respectively), due to depletion of their natural reserves. Dynamics of consumption of fuel and energy resources (by type and total) for the period 1965–2095 [1] shows that by the end of the 21st century, the reduction of available natural energy resources is projected to the level of 20% of the volume consumed in the current period and 16–17% of the maximum forecast of consumption in the 20 s (Fig. 3). A decrease in their production volumes will inevitably lead to increased shortages and, as a result, higher prices, which will reduce their availability for a significant group of society and adversely affect the quality of life of the low-income population. The development of energy supply will be possible as a result of the search and development of fundamentally new, affordable energy sources that can replace scarce energy resources, which, if successful, will inevitably lead to a technological revolution.

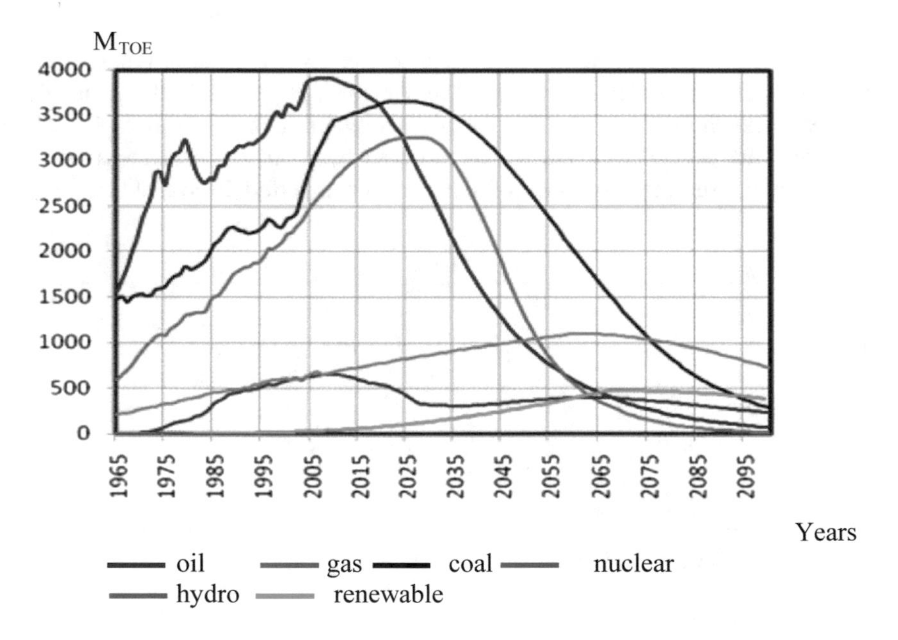

Fig. 3. Dynamics of energy consumption by type.

Thus, an increase in energy consumption for utilization is a sign of positive growth in the economy, but given the finiteness of the mined reserves, the challenge is to limit the use of resources that are strategically important for human development, including as a result of energy saving, which meets the requirements of the concept of sustainable development. The prospects of growth of the world economy in the closer period (by 2040) show a tendency of growth in world energy consumption by 40% while maintaining the same trend of energy intensity of world GDP (statistically significant decrease for 55 years by 1.2% per year) [5]. A significant increase in energy consumption occurred until the middle of the last century, was accompanied by an increase in oil production and amounted to about 5% annually up to 1970, at low prices at the level of 15–20 dollars for 1 ton of oil. However, in the mid-1970s, an oil crisis occurred due to the following main circumstances:

- deterioration of oil production conditions due to relocation of production sites to the continental shelf, to the North and the Sahara;
- a sharp rise in prices to $ 250–300 per 1 ton of oil by Arab countries with the aim of political influence on the countries of Western Europe and America, who supported Israel in the Arab-Israeli conflict.

In the period 1990–1998, oil prices were fluctuating and sharply fell from $ 22.1 to $ 13.1 per barrel in 1998, which was associated with the financial crisis in Southeast Asia and the decline in oil demand. The forecast of world prices for energy types of fuel based on the materials of the XVI Congress of the World Energy Council for the next period is given in Table 1 and describes their steady upward trend, which will inevitably affect the increase in prices and tariffs for utilities and their advancement of wage growth rates [1].

Table 1. Forecast of world prices for energy fuels (based on the materials of the XVI Congress of the World Energy Council).

Prices by types of fuel	2005	2010	2020	2030
Coal, USD per 1 ton of fuel equivalent	46–52	47–53	52–56	55–60
Oil, USD per 1 ton	150–170	180–210	250–280	390–430
Natural gas, USD per 1 thousand m^3	130–180	140–200	190–260	300–400

As a result of the energy crisis of the 70 s, in the world practice of construction activity, a lot of architectural and engineering developments appeared, the concept of energy-efficient buildings and life-cycle costs has formed, aimed at reducing the consumption of fuel and energy resources during the operation of construction facilities and engineering infrastructure in order to compensate for the increase in their value.

In our country, in conditions of significantly lower and constant prices for energy resources, the energy-saving policy was developed in the early 80 s, when it was noted that the level of energy intensity of domestic industrial production was higher than that of the United States. Despite the measures applied in the field of energy saving as a state policy, the problem of wasteful use of energy resources in the process of building operation remains relevant, including in the investment and construction sphere, since most of the construction facilities continue to operate in the conditions of outdated standards for thermal resistance of external enclosing structures, worn external networks and the prolonged lack of timely overhaul. For many years, when creating construction facilities, specialists have focused on the process of erection of buildings and structures and the valuation of construction. Design solutions are focused on minimizing the cost at the stage of construction of objects, and not costs in the operational period [6, 7]. Therefore, a systematic approach to the study of technical and economic aspects of energy saving at the stages of the life cycle of real estate objects, the practical implementation of the concept of life cycle cost, is of particular importance.

Thus, for many decades, the problem of energy saving remains extremely urgent, affects the quality of life of the population, extends to all areas of economic activity, including investment and construction, as the most energy-intensive and multidimensional. Of particular importance are the technical and economic aspects of energy saving at all stages of the building life cycle.

3 Results

In the modern practice of designing buildings and structures, the introduction of energy-saving technologies and ensuring compliance with a particular class of energy efficiency has become an integral part of construction projects. In the construction process, about 2% of the energy utilized during the life cycle of a building is consumed, in the construction industry - about 8%, and 90% comes from the operational period. But the valuation of projects is carried out only in terms of the estimated cost of construction. The directions of energy saving solutions at the stages of the building life cycle are shown in Fig. 4 and characterize the technical aspects of energy saving in buildings, which consist in making architectural and engineering solutions at the design stage, ensuring the reasonable consumption of fuel and energy resources at the operational stage.

The economic aspect of energy saving consists in reducing the cost of the necessary energy resources, especially at the operational stage, which is especially important in the context of a significant increase in prices and tariffs (Fig. 5) [8].

Meanwhile, measures that ensure the energy efficiency of a project are not always economically viable.

On the example of options for energy-saving measures (or a set of measures) that ensure the reduction of energy consumption in physical meters, the indicators of benefit (B) from their use and discounted costs (DC), one-time and current, for their implementation are compared. The following situations are possible:

1. If the condition is met: $B > DC$, then the introduction of an energy-saving measure is economically viable and brings an economic effect (net present value, ΔNPV);
2. If the condition is met: $B = DC$, then the introduction of an energy-saving measure in terms of value brings neither gain nor loss. And the economic effect (net present value) will be equal to zero;
3. If the condition is met: $B < DC$, then the introduction of energy-saving measures is not economically viable and causes losses equal to the negative value of net present value.

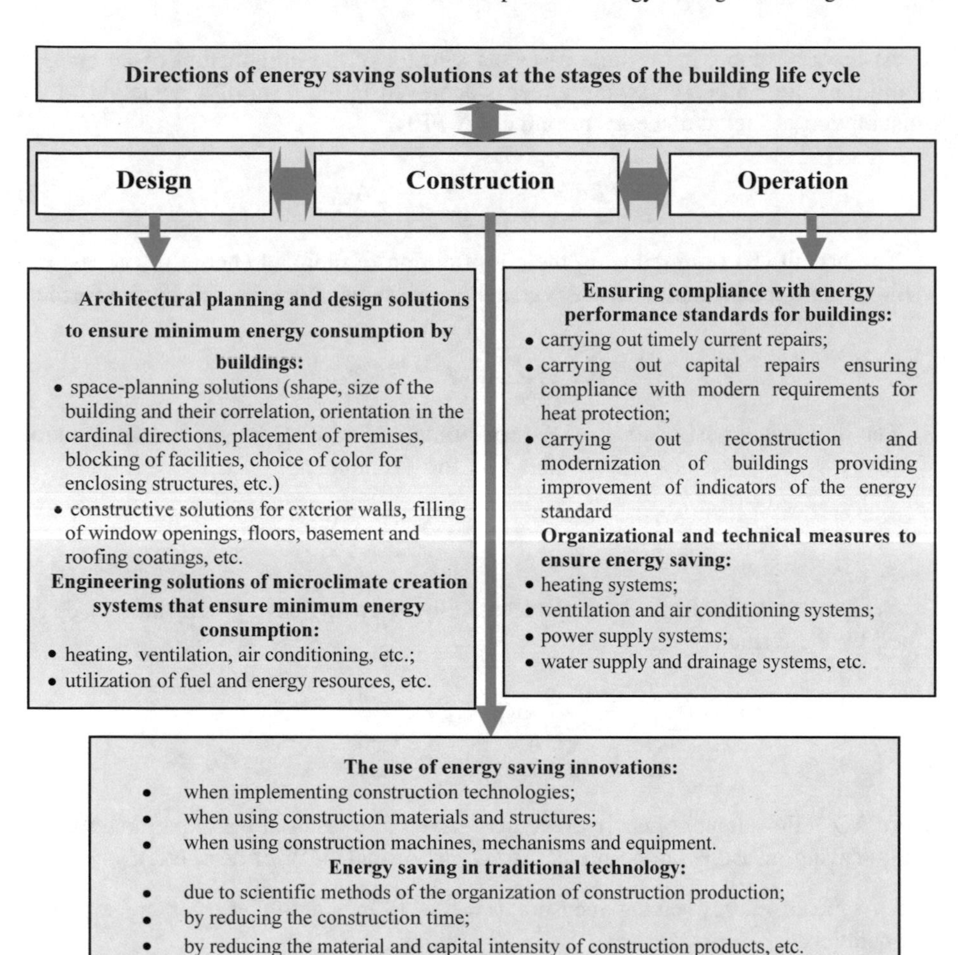

Fig. 4. Directions of energy saving solutions at the stages of the building life cycle.

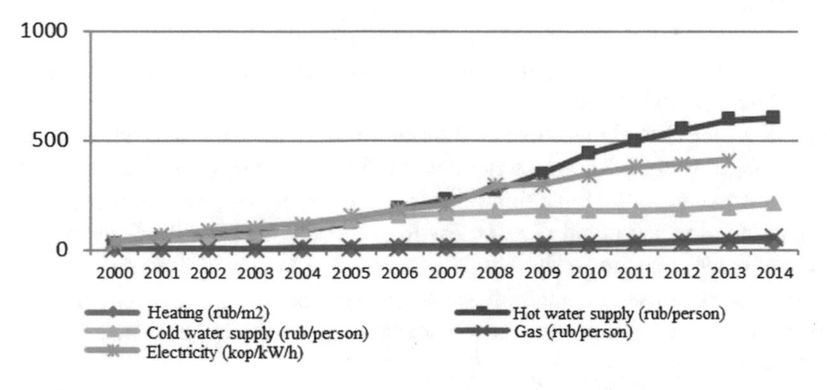

Fig. 5. Dynamics of the cost of energy resources for building operation.

At the same time, in all three cases, as a result of the introduction of an energy-saving measure, an energy-saving effect is achieved in the form of a reduction in the consumption of fuel and energy resources (Δ FER)

$$\Delta FER = \sum\nolimits_{i=1}^{I} \sum\nolimits_{t=-T}^{T} \Delta Q_{it} \tag{2}$$

The benefit (B) from reducing the consumption of fuel and energy resources as a result of the implementation of energy-saving measures is determined by the formula

$$B = \sum\nolimits_{i=1}^{I} \sum\nolimits_{t=-T}^{T} \Delta Q_{it} \cdot \tau_{it} \cdot \alpha_t \tag{3}$$

The discounted cost of repair (DC) (one-time and current) for the implementation of energy-saving measures is determined by the formula

$$\mathrm{DC} = \sum\nolimits_{t=-T}^{T} \sum\nolimits_{h=1}^{H} z_{th} \cdot \alpha_t + \sum\nolimits_{t=-T}^{T} \sum\nolimits_{g=1}^{G} K_{tg} \cdot \alpha_t \tag{4}$$

Net present value from the implementation of energy-saving measures is determined by the formula

$$\Delta NPV = \sum\nolimits_{i=1}^{I} \sum\nolimits_{t=-T}^{T} \Delta Q_{it} \cdot \tau_{it} \cdot \alpha_t - \sum\nolimits_{t=-T}^{T} \sum\nolimits_{h=1}^{H} z_{th} \cdot \alpha_t - \sum\nolimits_{t=-T}^{T} \sum\nolimits_{g=1}^{G} K_{tg} \cdot \alpha_t \tag{5}$$

where ΔQ_{it} - the amount of energy resources saved as a result of the implementation of energy-saving measures in the project (tons, m^3 of fuel, MW of heat, etc.);

i – type of energy saving measure ($i = \overline{1 \div 1}$) in the t-th year of construction, operation;

t - the calculation period (year, month, etc.) throughout the life cycle ($-T < t < T$), years or others;

τ_{it} - tariff (price) per unit of energy resource of type i in period t, rub./unit of resource;

z_{th} - annual operating costs of the form h ($h = \overline{1 \div 1}$), necessary for the implementation of energy-saving measures, thousand rubles, reduced to the start of the operational period ($t = 0$);

K_{tg} - one-time capital investments, thousand rubles, reduced to the start of operation of the object ($t = 0$). K_{tg} are carried out before the start of the operating period (at $-T < t < 0$) and/or after operation (at $0 < t < T$), if the service life of the equipment or its individual structural parts is less than the service life of the system in which the equipment operates or buildings in which the system operates;

g - directions of use of capital investments in the period, $g = \overline{1 \div G}$, (for example, on installation of equipment, on replacement of nozzle in heat exchangers during operation, etc.)

α_t - the coefficient of reducing costs at different times to the start of the operational period $t = 0$. If $-T < t < 0$, then $\alpha_t = 1/(1+r)^{-t}$, and if $0 < t < T$, then $\alpha_t = 1/(1+r)^{t}$;
r – discount rate.

The right to determine priorities in the process of choosing a design option for energy-saving building solutions belongs to the customer (investor) and can be implemented on the basis of a multi-criteria assessment of the compared solutions.

4 Discussions

One of the current engineering and technical solutions aimed at energy saving is considering the choice of a heat supply system according to the power of the heat generation source. Four alternative options to be compared are characterized by indicators of discounted costs (with a service life of 25 years, a discount rate r = 0.1, and a service life of heat supply networks of 12 years) and annual fuel (gas) consumption. Their values are shown in Fig. 6

The analysis of fuel and energy resources consumption by the compared options shows that the first option of the heat supply system with one (centralized) source of heat generation, which requires the least fuel (gas) consumption, can be determined as an energetically viable option.

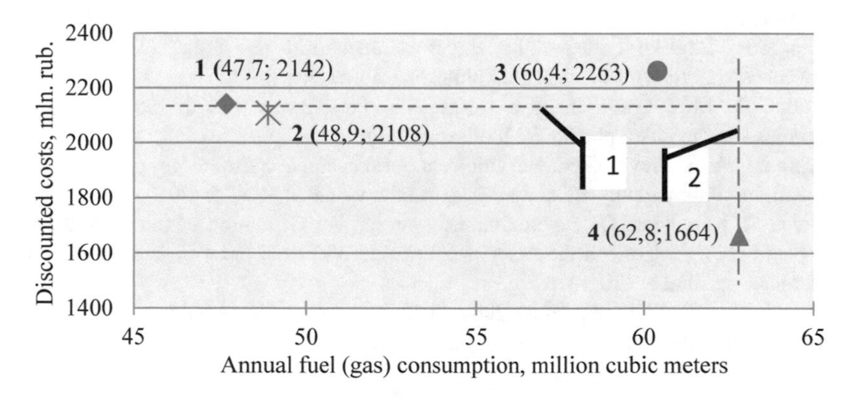

Fig. 6. The selection of a set of non-worst solutions on a discrete set of feasible options.

However, when considering a discrete set of indicators for the ratio of discounted costs and fuel (gas) consumption, it was revealed that after the first truncation (line 1), option 3 (scheme with three sources of heat generation) was among the worst solutions, and after the second truncation (line 2), it is obvious that options 1 (scheme with one source of heat generation), 2 (scheme with two sources of heat generation) and 4 (scheme with individual sources of heat generation) shown in Fig. 6 were in a set of non-worst options. If you continue truncation, the set will be empty for option 2, therefore it will be appropriate.

5 Conclusion

In the presence of huge reserves of fuel and energy resources in the Russian Federation, new scientific approaches to managing their use should be sought, taking into account the multidimensionality of the problem of energy saving. Their goal is to solve the multi-criteria task of providing the population with quality resources, reducing their unproductive losses, preserving natural wealth and the environment for future generations. Based on the analysis of the diversity of features and conditions that form the technical and economic indicators of buildings and structures as construction products, the prerequisites are identified, and the feasibility of a multi-criteria assessment of design decisions is justified as a necessary condition for the development of scientific and methodological support for designing construction objects aimed at implementing and introducing technical and socio-economic innovations in the process of building and overhauling construction objects.

References

1. Zerkalov, D.V.: EHnergeticheskaya bezopasnost', Monografiya. Osnova, K. (2012)
2. Korotaev, A.V., Malkov, A.S., Halturina, D.A.: Matematicheskaya model' rosta naseleniya Zemli, ehkonomiki, tekhnologii i obrazovaniya. http://www.keldysh.ru/papers/2005/prep13/prep2005_13.html
3. Tendencii razvitiya ehnergetiki v mirovoj ehkonomike. http://www.economicwind.ru/ecowins-130-1.html
4. Kapica, S.P.: Skol'ko lyudej zhilo, zhivet i budet zhit' na zemle. Ocherk teorii rosta chelovechestva. http://malchish.org/lib/philosof/Kapitza/Kapitza.htm
5. Chefurka, P.: Mirovaya EHnergiya i Naselenie Perspektivy s 2007 po 2100 gg. http://economics.kiev.ua/index.php?id=932&view=article
6. Kucygina, O.A., Galicyn, G.N.: Razvitie cenoobrazovaniya v stroitel'noj otrasli i upravlenie zhiznennym ciklom zdanij. EHkonomika stroitel'stva **6**(48), 12–25 (2017)
7. Uvarova, S., Kutsygina, O., Smorodina, E., Gumba, K.: Formation of the portfolio of high-rise construction projects on the basis of optimization of «risk-return» rate. In: E3S Web of Conferences, p. 03024 (2018)
8. Stoimost' uslug ZHKKH za period s 2000 po 2014 god. http://knowledge.allbest.ru/economy/2c0a65625a2bd79b4d53a88421206c37_0.html

Making Decisions in the Field of Energy Management Based on Digital Technologies

Vladimir Burkov[1], Sergei Barkalov[2], Sergei Kolodyazhniy[2], and Olga Perevalova[2](✉)

[1] V.A. Trapeznikov Institute of Control Sciences of Russian Academy of Sciences, Profsoyuznaya Street, 65, Moscow 117342, Russia
[2] Voronezh State Technical University, Moscow Avenue, 14, Voronezh 394026, Russia
nilga.os_vrn@mail.ru

Abstract. The article discusses the problems of introducing digital technologies into the decision-making process in the energy sector. For this purpose, two types of digital decision-making technologies are analyzed. The first technology - direct, which today is the main (traditional), is that the decision is made by a person - a decision maker, and a computer program acts as an "advisor" - a decision support system. The development of this technology is associated with the development of "active advisers", when, after the implementation of the decision, a comparison is made of the decision made by the decision maker and the decision proposed by the "adviser" on the basis of so-called "recalculating models". In the second technology (let's call it inverse), the decision is made by a computer program, and the person only observes and analyzes, without interfering in the decision-making process (with the exception of force majeure situations). In this technology, a decision support system is transformed into a decision-making system, and a decision maker becomes a decision analyst. However, the decision-making mechanism is developed by the head and other stakeholders who are responsible for the results of the operation. The necessary conditions for the effective functioning of both technologies are, firstly, the interest of participants in the presentation of reliable information required for decision-making, and secondly, interest in the implementation of decisions made (conditions of L.V. Kantorovich - V.M. Glushkov). The article provides a comparative analysis of these technologies and provides examples of their practical implementation.

Keywords: Energy management · Digital technologies · Smart management mechanisms · Decision support system · Kantorovich-Glushkov conditions · Digital economy

1 Introduction

The energy management of municipal facilities, energy technologies, as well as the management of facilities that are responsible for these issues, have recently come to the fore, and are forcing researchers to work to improve management efficiency in the areas listed above.

V. Murgul and M. Pasetti (Eds.): EMMFT-2018, AISC 983, pp. 45–54, 2019.
https://doi.org/10.1007/978-3-030-19868-8_5

Currently, the questions are:

- related to energy security;
- related to the study of energy saving problems;
- related to the supply of energy, products of the energy industry;
- related to the management of objects of the energy sector for various purposes [1, 2];
- related to the ecology, arising in the process of construction, repair, connecting communications to various objects, etc. [3] - require immediate intervention and effective solutions.

And since we live in the age of widespread introduction of digital technologies, the field of energy management is no exception. Currently, they are actively being implemented in different directions and the most priority of them are the following [4–6]:

- evaluation of managers and officials of all levels with appropriate incentives for the result;
- distribution of limited resources (financial, labor and other);
- the formation of programs to improve the efficiency of various facilities, organizations, cities, regions, etc.

Thus, the introduction into the process of making management decisions in the energy sphere of digital technologies, in fact, will lead to a change in the methodology of their adoption. And here we turn to smart management, which is based on smart management mechanisms successfully developed by domestic scientists.

2 Materials and Methods

Attempts to attract digital technologies for decision-making have been conducted for a long time [7]. Especially note the role of academicians L.V. Kantorovich and V.M. Glushkov. Academician V.M. Glushkov on the wave of large-scale automation in the 70 s proposed to create a nationwide automated system (OGAS) for the development of national economic plans. Academician L.V. Kantorovich proposed a mechanism for coordinating national economic plans with the interests of enterprises on the basis of the duality methodology developed by him. In this case, the optimal instrument of coordination was the optimal values of dual estimates of the corresponding planning problems. The attempt failed because the real interests of national economic subjects did not fit into the rigid scheme of duality methodology. And L.V. Kantorovich, and V.M. Glushkov put forward two necessary conditions for the effectiveness of the application of digital technologies in the field of management decision making. The first is related to the fact that the information required for making decisions must be reliable, and the second to the fact that it must be beneficial for economic entities to carry out the decisions made. We call these necessary conditions the Kantorovich – Glushkov conditions [8]. At that time (the 70 s) these conditions were difficult to fulfill [7]. Now the situation has changed. The theory of active systems developed the so-called “smart control mechanisms” [9].

Smart are the management mechanisms that change a person's behavior in the direction that is necessary for society (make it profitable to provide reliable information, implement decisions, effectively develop, etc.). Highlight smart mechanisms that ensure the provision of reliable information (non-manipulable mechanisms) and the implementation of decisions made (agreed mechanisms), which is essential for the introduction of digital technologies [10].

The importance of these mechanisms in the implementation of digital technologies is determined by the fact that when they are introduced into the practice of decision-making, the Kantorovich-Glushkov conditions are met.

There are two types of digital technology.

Consider the principle of operation of each of them.

The first technology (let's call it direct), which today is the main (traditional), is that the decision is made by a person (the Decision Maker - the decision maker), and the computer program acts as an "advisor" (Decision Support System - DSS) [11]. The development of this technology is associated with the development of "active advisers", which, after the implementation of the decision, a decision maker is compared with the solution proposed by the "adviser" based on the so-called "recalculation models". In the second technology (let's call it inverse), the decision is made by a computer program, and the person only observes and analyzes, without interfering in the decision-making process (with the exception of force majeure situations). In this technology, DSS are transformed into a decision-making system (DSS), and the decision maker becomes a decision analyzer. However, the decision-making mechanism is developed by the head and other stakeholders who are responsible for the results of the operation. The schemes of their work will be further discussed in detail.

3 Results

Consider the scheme of direct technology, presented in Fig. 1. In it, as noted above, the decision is made by a person (DM) based on information about the object and the external environment based on his experience and recommendations of the "adviser" (decision support system - DSS).

In Fig. 1 characters have the following meanings:

u – Center decision,
v – DSS recommendation,
J – object information,
x(u) – result of the decision.

This technology is quite effective in a situation where the Center knows the facility well, is interested in its efficient operation and has effective leverage to ensure the implementation of the decisions made. Otherwise, the decision maker is faced with unreliability of information received from the object, failure to comply with the decisions made, as well as the possibility of corruption.

Consider the direct technology on a simple example of the distribution of the order for the release of any product (services).

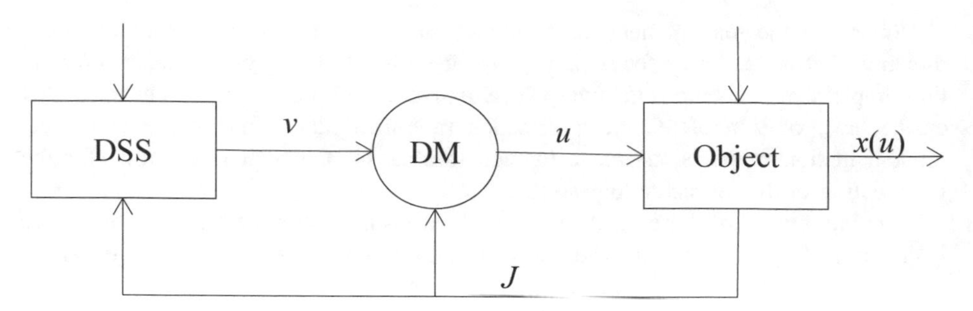

Fig. 1. Block diagram of direct decision making technology.

Example 1. The center needs products in the amount of R. There are n manufacturers of these products. Denote xi – plan for the production of products by the manufacturer i, $z_i = \frac{1}{2r_i} x_i$ – the cost of production in the amount of x_i, r_i – the parameter of production efficiency of the i – th manufacturer.

The task of the Center is to determine the plans xi so that

$$\sum ix_i = R \tag{1}$$

that is, the products were released in the required quantity, and the total costs

$$\Phi(x) = \sum i \frac{1}{2r_i} x_i^2 \tag{2}$$

were minimal.

With known parameters r = (r_i), the optimal plan is

$$x_i = \frac{r_i}{H} \cdot R,\ i = \overline{1, n} \tag{3}$$

where $H = \sum jr_j$.

However, the parameters ri, as a rule, are known to the Center only approximately. In this case, the information s = (si) about the parameters is reported by the manufacturers. Producer interests are determined by profit margin

$$Y_i(\lambda, x_i) = \lambda x_i - \frac{1}{2r_i} x_i^2 \tag{4}$$

where λ – product price set by the Center.

The Center makes a decision based on the management mechanism x = π (s) and λ (s).

In the theory of active systems, an open control mechanism is proposed [12], which is the correct mechanism, i.e. ensures the reliability of the information provided and the implementation of plans. In our example, the open control mechanism is

$$x_i = \lambda s_i,\ i = \overline{1, n} \tag{5}$$

$$\lambda = \frac{R}{\sum_i s_i}$$

Considered direct technology has a number of disadvantages. Firstly, it is rather difficult to build an adequate model of the control object. It is even more difficult to build an adequate, correct decision-making mechanism. Secondly, it is poorly protected from corruption collusion by both the Center and manufacturers, and between manufacturers.

The development of direct technology is the so-called two-channel control mechanisms (active advisers) based on recalculated models [9]. Conversion refers to models that provide a more accurate estimate of the parameters of an object based on additional information about the results of the solution implementation. This allows you to compare the effect of the decision of the Center and the effect of the decision proposed by the DSS (Fig. 2).

We decipher the notation given in Fig. 2, those that were not decrypted earlier:

SM – scaling model,
Δ – the difference in effect between the decision maker and the recommendation of the DSS.

If Δ > 0, then the decision maker is encouraged, and if Δ < 0, then the decision maker is penalized.

The use of two-channel mechanisms increases the interest of the Center and reduces the corruption component [13].

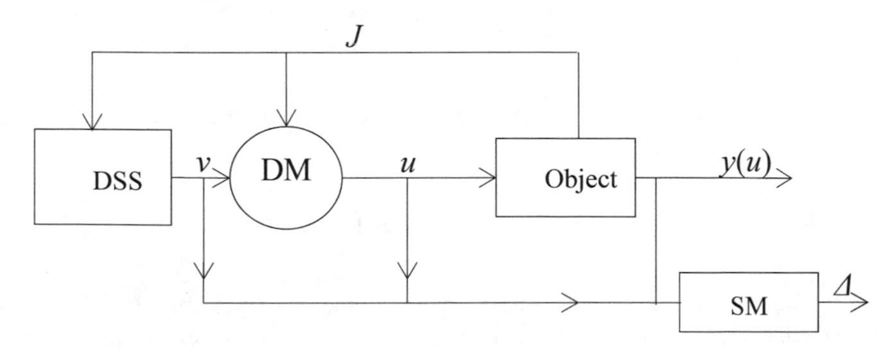

Fig. 2. Block diagram of decision making in two-channel mechanisms.

Example 2. In the model of Example 1, it is not difficult to construct a scaling model. Indeed, if $z_i = \frac{1}{2r_i} x_i$ – the real costs of the i–th manufacturer, then

$$r_i = \frac{x_i^2}{2z_i},\ i = \overline{1,n}$$

and substituting in the target function of the Center, we get

$$\Delta = \left(\sum\nolimits_i \frac{w_i^2}{x_i^2} z_i - z_i\right) \tag{6}$$

where wi – recommended DSS plan.

Consider a possible variant of corruption interaction between the decision maker and the manufacturer (let it be manufacturer 1). The decision maker, having the right to make a decision, overestimates the price of λ1 for the first manufacturer. For this, manufacturer 1 gives the Center a bribe of α% of profit. Manufacturer's profit 1 will be $(1-\alpha)\lambda_1 x_1 - \frac{x_1^2}{2r_1}$, and profitable plan for him x1 = (1 − α) λ1 x1.

In the absence of a corrupt transaction, manufacturer 1 makes a profit of $r_1 \frac{1}{2}\left(\frac{R}{H}\right)^2 r_1$. In order for the transaction to be profitable for the manufacturer 1 it is necessary that

$$\frac{1}{2}(1-\alpha)^2 \lambda_1^2 r_1 > \frac{1}{2}\left(\frac{R}{H}\right)^2 r_1$$

We obtain the condition of profitability of a corrupt transaction

$$(1-\alpha)\lambda_1 > \frac{R}{H}$$

Substituting, we get

$$\frac{R}{r_1} > (1-\alpha)\lambda_1 > \frac{R}{H}$$

Let $\lambda_1 = k\frac{R}{H}$, where k > 1. Get the conditions of profitability of a corrupt transaction

$$\frac{H}{r_1} > (1-\alpha)k > 1$$

Note that the more efficient a manufacturer is, the less profitable it is for a corruption deal. Moreover, with a sufficient amount of the penalty for exceeding the costs in comparison with the costs calculated on the basis of the recalculation model, the corruption transaction will be disadvantageous to the manufacturer.

The main problem associated with the use of two-channel mechanisms is, of course, the problem of building adequate conversion models.

4 Discussions

Consider the reverse decision-making technology, in which the center is played by a computer program (digital decision-making system - TsSPR), and the head of the center observes and evaluates the results, but does not interfere in the decision-making process (with the exception of force majeure situations). As a matter of fact, the head of the Center and TsSPR change places (from here and the name - the return technology). In order for this technology to be effective, it is necessary that the decision-making mechanism be discussed by the head with all interested parties and adopted by them in the form of a provision (law). In this regard, the digital decision-making system (DSPC) performs a purely computational function. It is obvious that the authors of the mechanism, and first of all the head of the Center, are responsible for the decisions taken not by the DSPC. In case the head of the Center concludes that the developed mechanism is not very effective (for example, if the operating conditions change), he raises the question of the need for its adjustment. And naturally, as noted above, the adopted mechanism must meet the conditions of Kantorovich-Glushkov, that is, be the correct mechanism. This is the main problem of introducing digital technologies into the sphere of making management decisions - to ensure the correctness of the decision-making mechanism.

We describe a possible variant of the correct mechanism. Consider a system from the Center of n agents (enterprises, organizations, etc.). The interests of agents are described by their objective functions $Y_i(\lambda, x_i)$, where λ – is the control, $x_i, i = \overline{1, n}$ – are the plans established by the DSPC. We describe the functioning of the system:

The Center announces many possible controls £.

Agents for each λ ϵ £ inform the Center of advantageous plans for them $x_i(\lambda)$, $i = \overline{1, n}$, λ ϵ £.

The center chooses λ ϵ £ and plans accordingly $x_i(\lambda)$, $i = \overline{1, n}$, maximizing its objective function $\Phi(x, \lambda)$.

If we accept the hypothesis of weak influence [9], according to which agents do not take into account the influence of the information they communicate on the choice of control λ, then the described mechanism is correct.

Indeed, under any controls λ, agents receive advantageous plans for themselves and, therefore, they are interested in providing reliable information and in fulfilling plans.

Comment. The center does not have to report the entire set of controls. He can organize an iterative procedure, at each step of which the Center adds new controls to increase the value of its objective function.

Example 3. Consider the task of distributing an order for output (see Example 1). The mechanism proposed there is fully consistent with that described above. Indeed, the Center reports the set of possible prices λ (Step 1). Agents report profitable plans for them $x_i(\lambda) = \lambda r_i$, $i = \overline{1, n}$. The Center chooses λ, so that $\sum_i x_i(\lambda) = R$ and the total costs are minimal.

The weak influence hypothesis for this problem was substantiated in the theory of active systems for a sufficiently large number of agents [14].

An important advantage of the reverse decision-making technology is a significant weakening of corruption ties. Indeed, since the head of the Center does not have the right to interfere with the action of the digital decision-making system (DSPC), and therefore cannot influence decision-making, and the "computer does not take bribes", then there are virtually no corruption transactions between the Center and the agents. However, corruption collusion between agents at the stage of communication of information is possible. To combat this corruption component, it is advisable to supplement the reverse technology with a recalculation model.

From the above, we can conclude that the process of introducing digital technologies into energy management can be represented as follows (Fig. 3).

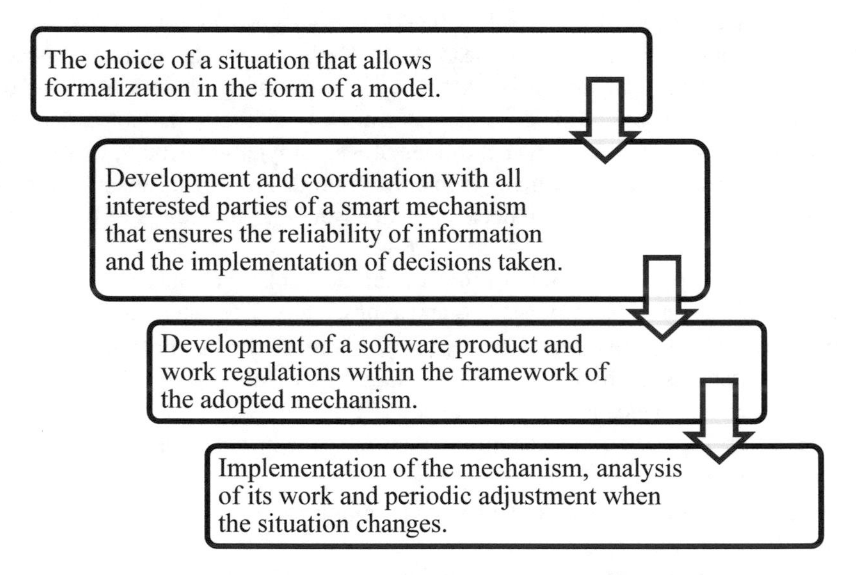

Fig. 3. The main stages of the introduction of digital technologies in energy management.

5 Conclusion

From the standpoint of all the above, we can conclude that for smart management, which is based on smart management mechanisms successfully developed by domestic scientists, the tool is the digital economy.

In this context, it is viewed as a system of economic, social and cultural relations based on the use of digital technologies.

So, the introduction of smart management in energy management will improve the efficiency of this process and lead to improved approaches to solving issues (problems) identified in the introduction of this article.

In conclusion, we give several examples of the practical application of digital technologies in various fields, which confirm all the above with real results:

1. Automated, Quantitative, Comprehensive Performance Evaluation (AQCPE).

The AQCPE system was developed in the 80s at the Institute of Control Sciences of Russian Academy of Sciences on the initiative of the Minister for the Ministry of the Device Ministry, M.S. Shkabardni, to evaluate the activities of the enterprises of the Device Ministry. Methods and regulations for the formation of the assessment were discussed at the Scientific and Technical Council and the Board of the Ministry of Instrument and approved by the Minister. The calculations were carried out on the basis of the developed software package without human intervention.

2. Cost-Effective Tax System.

In 1990–1991, the Institute for Management Problems was included in the participants in the experiment on new tax systems in science conducted by the State Committee on Science and Technology of the USSR. We have proposed a cost-effective tax system - a smart mechanism that even encourages a monopoly organization to reduce costs and prices. After discussion at the Academic Council of the Institute, the relevant regulations, regulations and software were developed and approved. Two years the Institute lived in the experiment. The experiment fully confirmed the theoretical conclusions - it was unprofitable for the Institute to overestimate the cost of contractual work.

3. The reverse priority mechanism for allocating limited resources is a smart mechanism that encourages consumers to report objective requests for a resource. The mechanism was introduced into the system of distribution of water resources in Bulgaria and was highly appreciated (a golden badge "for high technical progress").

References

1. Barkalov, S., Kurochka, P., Nasonova, T.: Optimal placement of maintenance facilities. In: MATEC Web of Conferences, Proceedings, p. 01124 (2018)
2. Burkov, V.N., Burkova, I.V., Averina, T.A., Nasonova, T.V.: Formation of the program to improve the level of competence of the organization's staff. In: Proceedings of 2017 Tenth International Conference on Management of Large-Scale System Development (MLSD) (2017)
3. Barkalov, S.A., Kurochka, P.N.: Modeling of production activity of a construction enterprise under analyzed law of distribution. In: Proceedings of 2017 Tenth International Conference on Management of Large-Scale System Development (MLSD) (2017)
4. Avdeeva, E., Averina, T., Kochetova, L.: Life quality and living standards in big cities under conditions of high-rise construction development. In: E3S Web of Conferences, p. 03013 (2018)
5. Averina, T., Avdeeva, E., Perevalova, O.: Introduction of management innovations in the work of municipal organizations. In: MATEC Web of Conferences, Proceedings, p. 01121 (2018)
6. Breer, V.V., Novikov, D.A., Rogatkina, A.D.: Crowd Management: Mathematical models of Threshold Collective Behavior. LENAND, Moscow (2016)
7. Ivanov, V.V., Malinetsky, G.G.: Strategic priorities of the digital economy. Int. Sci. Anal. J. Strateg. Priorities **3**(15), 54–95 (2017)

8. Burkov, V.N., Burkova, I.V., Barkalov, S.A.: Digital technologies in management decision making. FES: Finance Econ. **15**(4), 5–10 (2018)
9. Burkov, V.N.: Fundamentals of the mathematical theory of active systems. Science, Moscow (1977)
10. Averina, T.A.: Information technology in the promotion of goods and services. Econ. Manag. Syst. **14**(4), 120–127 (2014)
11. Barkalov, S.A.: Information technology in economics and management: studies manual. In: Barkalov, S.A., Belousov, V.E., Golovinsky, P.A., Mikhin, M.P. (eds.) Scientific Book, Voronezh (2009)
12. Belov, M.V., Novikov, D.A.: Methodology of Complex Activity. Lenand, Moscow (2018)
13. Burkov, V.N.: Management mechanisms. In: Novikov, D.A. (ed.) URSS (Editorial URSS), Moscow (2011)
14. Novikov, D.A.: Theory of Management of Organizational Systems. MPSI, Moscow (2005)

Designing Systems of Group Stimulation in the Management of Energy Complex Objects

Sergei Barkalov[1(✉)], Vladimir Burkov[2], and Pavel Kurochka[1]

[1] Voronezh State Technical University, Moscow Avenue, 14, Voronezh 394026, Russia
sbarkalov@nm.ru

[2] V.A. Trapeznikov Institute of Control Sciences of Russian Academy of Sciences, Profsoyuznaya Street, 65, Moscow 117342, Russia

Abstract. The process of managing energy complexes involves organizing the interaction of various structures with different organizational and legal status. In this case, the usual methods of administrative influence, as a rule, do not work, since the objects of management may not have a common system of subordination, and then you have to use economic levers, one of which is the incentive mechanisms. The problems of synthesis of optimal systems of group stimulation are considered. There is a project of n works. A plan has been defined for reducing the duration of the program, according to which the magnitude of the reduction of its duration is determined for each project. The cost of reducing the duration linearly depends on the magnitude of the reduction. The incentive system is designed to offset these costs. In this case, using the well-known set of incentive systems, new, more complex incentive systems can be obtained, which is why these elementary incentive systems are called basic ones. In the system of group incentives, all projects are divided into m groups, and each group has its own incentive system. Two possible incentive systems for groups are considered: unified linear and unified hopping. The task is to determine the division of projects into groups and select the incentive system for each group so that the total fund for cost compensation is minimal. The methods for solving the tasks are proposed. A further generalization of the problem is considered, when the set of works included in one group can be arbitrary. For this case, a heuristic rule is proposed, on the basis of which the solution algorithm is constructed.

Keywords: Energy complex objects · Designing systems · Group stimulation

1 Introduction

Traditionally, incentives are understood as influencing the interests and preferences of managed entities by the governing bodies, that is, changing their preferences (through rewards and/or fines) so as to make the choice of actions and achievement of the results required for the center advantageous for agents. Another aspect of incentive as a management method is to influence the sets of permissible actions and agent resources.

V. Murgul and M. Pasetti (Eds.): EMMFT-2018, AISC 983, pp. 55–68, 2019.
https://doi.org/10.1007/978-3-030-19868-8_6

The process of managing energy complexes involves organizing the interaction of various structures with different organizational and legal status. In this case, the usual methods of administrative influence, as a rule, do not work, since the objects of management may not have a common system of subordination. In this case, it is necessary to use levers of economic impact, one of which is the incentive mechanisms.

The tasks of constructing optimal incentive systems were considered in many papers ([1, 2], etc.). Basically, two types of systems are considered: individual incentive systems and unified incentive systems. In individual incentive systems, each agent determines its own system selected from a given class of systems (linear, intermittent, rank, etc. [1]). In systems of unified stimulation, a single system is defined for all agents [2]. The advantage of individual incentive systems is significantly less in some cases, the incentive fund as compared with the unified incentive systems, and the disadvantages are the lack of interest in reducing costs and sufficiently large opportunities for manipulation. The advantages of the unified incentive systems are significantly less opportunities for manipulation, a significantly greater interest in reducing costs, and the disadvantage is that the value of the incentive fund is significantly greater in many cases. Group incentive systems occupy an intermediate position. In such systems, the set of all agents is divided into groups, and for each group a unified incentive system is applied. Group incentive systems retain to a certain degree the advantages of unified and individual incentive systems and at the same time reduce their disadvantages. When using the group incentive system, it is assumed that a single incentive system can be applied within each of the groups. Therefore, it is possible to design a system of group incentives using already known and well-studied incentive systems known as basic incentive systems. Thus, using the well-known set of incentive systems, new, more complex incentive systems can be obtained, which is why these elementary incentive systems are called basic [1, 2]. We list the basic incentive systems that can be used in group incentive systems.

Intermittent Incentive Systems (C-type) [3] characterized by the fact that the agent receives a constant remuneration (equal to a predetermined value of C), provided that the action chosen by him is not less than the specified, and zero reward when choosing smaller actions (see Fig. 1):

$$\sigma_C(x, y) = \begin{cases} C, y \geq x \\ 0, y < x \end{cases}.$$

The parameter x ∈ X is called the plan – desirable from the point of view of the center state (action, result of activity, etc.) of the agent.

C-type incentive systems can be interpreted as lumpy, corresponding to a fixed remuneration for a given result (for example, the amount of work is not lower than the agreed in advance, time, etc. - see below for more details). Another meaningful interpretation corresponds to the case when the agent's action is the number of hours worked, that is, the remuneration corresponds, for example, to a fixed salary without any allowances and performance assessment [6].

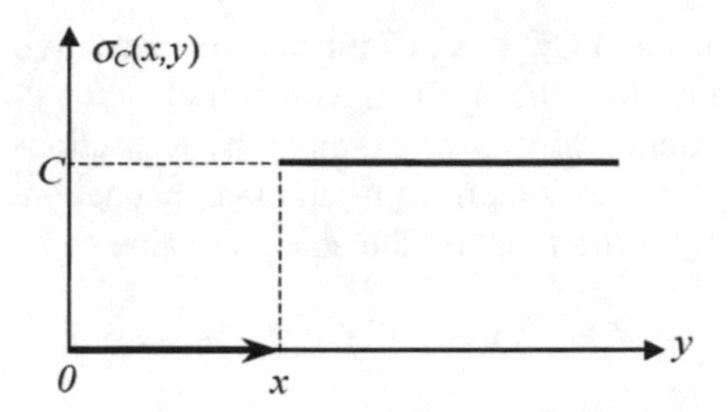

Fig. 1. Intermittent stimulation systems.

Values corresponding to C-type incentive systems will be indexed with «C», for example MC – a set of jump-like incentive systems, etc.

Note that most of the basic incentive systems are parametric, for example, the class MC $\subseteq$ M is determined by specifying the set of admissible plans X (for which it is usually assumed that it coincides with the set of admissible agent actions: X = A, or with the set of PM, actions implemented under given constraints incentive mechanism).

Quasi-pumping incentive systems (QC-type) [4] differ from the spasmodic in that the remuneration is paid to the agent only with the exact execution of the plan (see Fig. 2):

$$\sigma_{QC}(x,y) = \begin{cases} C, & y = x \\ 0, & y \neq x \end{cases}.$$

It should be noted that QC-type incentive systems are quite exotic (especially under uncertainty, it is not clear what is meant by the exact implementation of the plan) and are rarely used in practice.

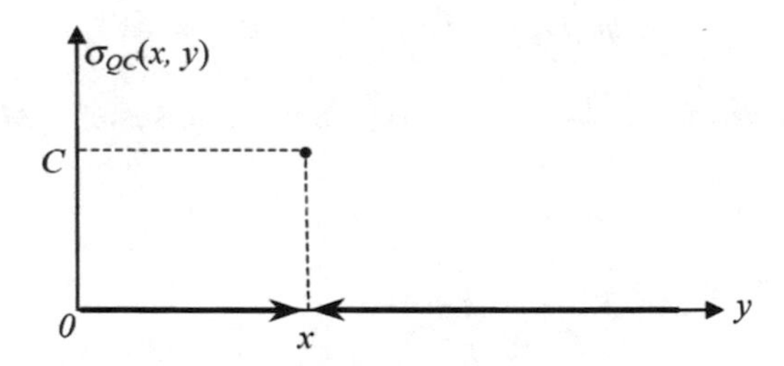

Fig. 2. Quasi-pumping incentive system.

The set of quasi-pumping stimulation systems is denoted by MQC. The symbol «Q» and the prefix "quasi-" means the quasi-system of incentives, the remuneration coincides with the remuneration in the initial incentive system if the agent fulfills the plan (y = x) and is zero in other cases (y $\neq$ x).

If there are no restrictions imposed on the absolute value of the agent's remuneration, then it is necessary to define what is meant by the value C, that is, the amplitude of the "jump", as well as the plan, can be a variable, which we will consider in the

C-type incentive systems and QC-type. Compensatory incentive system (K-type) [4] is characterized by the fact that the agent is compensated for costs, provided that his actions lie in a certain range, given, for example, by restrictions on the absolute value of individual remuneration, characterized by the fact that actions lie in a certain range, defined, for example, by restrictions on the absolute value of individual remuneration:

$$Qk(x, y) = \{c(y), y \leq x, 0, y > x$$

where is the function inverse to the agent's cost function, that is, the center can compensate the agent for costs *y £ x* and not pay for the choice of large actions (Fig. 3).

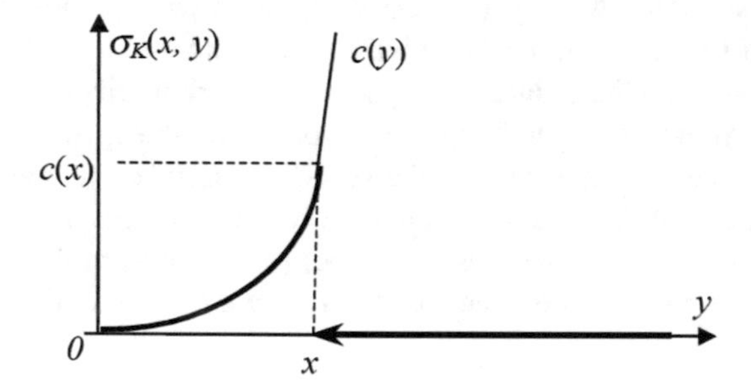

Fig. 3. Compensatory incentive system.

Many compensatory incentive systems denote MK.

Quasi-compensatory incentive systems (QK-type) [6] differ from compensatory ones in that the reward is paid to the agent only if the plan is executed accurately (see Fig. 4):

$$Qqk(x, y) = \{c(y), y = x, 0, y \neq x$$

We denote the set of quasi-compensatory stimulation systems *MQK*.

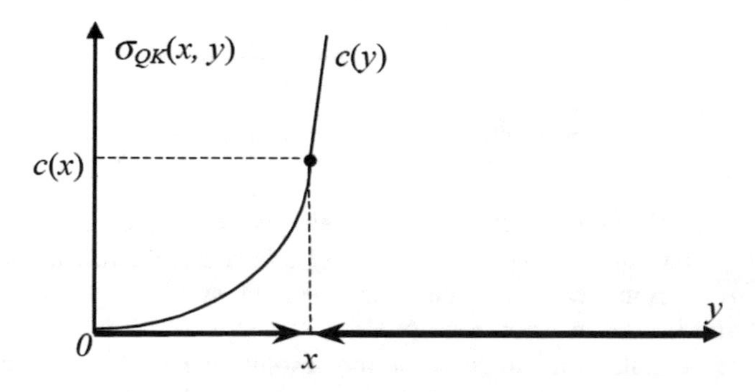

Fig. 4. Quasi-compensatory incentive system.

The article presents the formulation of problems for the synthesis of optimal group stimulation systems and discusses methods for their solution.

2 Materials and Methods

Consider a project consisting of n works. A plan for reducing the duration of the project, according to which the duration of work i is reduced by Δi, $i = \overline{1,n}$. The costs of performers to reduce the duration are linear functions Δi, [7] i.e.

$$Z_i = k_i \varDelta_i, \quad i = \overline{1,n} \tag{1}$$

where $k_i > 0$.

To offset costs, it is necessary to define a group incentive system (GI). Consider the GI, system in which all the work is divided into 1 < m < n groups, and for each group is determined by some system of unified incentives (UI). We will consider two classes of systems UI – linear incentive systems (LI) [8] and spasmodic stimulation systems (SS) [9]. We denote Qj – many works included in the group j

$$\bigcup_j Q_j = Q, \quad Q_i \cap Q_j = \varnothing$$

for all i, j (Q – many all works).

If a linear incentive system is chosen for group j then, obviously, to compensate for the costs of all the performers in this group, the minimum incentive fund will be

$$S_j = \lambda_j T_j \tag{2}$$

where

$$\lambda_j = \max_{i \in Q_j} k_i \quad T_j = \sum_{i \in Q_j} \varDelta_i$$

If a jump incentive system is chosen for group j, then the minimum incentive fund for compensation of the performers' costs will be

$$S_j = n_j \max_{i \in Q_j} k_i \varDelta_i \tag{3}$$

where n_j – number of jobs in a group j.

Task 1. Determine the division $Q_j, j = \overline{1,m}$ and choose the incentive system for each group so that the incentive fund is minimal. This problem will be considered in three ways. In the first for all classes, only systems of the class of linear stimulation are used, in the second - only systems of the class of the jump-type, and in the third, systems of both classes can be used. Consider another way of splitting into groups, namely, each group may include the number of jobs l within certain limits:

$$l_1 \leq l \leq l_2$$

For example, if $l_1 = 2$, $l_2 = 3$, then each group can include either 2 or 3 jobs. This method also solves the problem 1.

Consider the methods of solving the tasks.

3 Results

Consider the option when a linear incentive system is used for each group. Let the works be numbered in ascending order (not descending) ki, i.e. $k_1 \leq k_2 \leq \cdots \leq k_n$. Let us prove a simple statement:

Statement 1. There is an optimal solution such that if the group includes jobs i and j > i + 1, then the group also includes all intermediate works.

For proof, let us number the groups in ascending (non-decreasing) order λj, i.e. $\lambda_1 \leq \lambda_2 \leq \ldots \leq \lambda_m$. Let be $j, k \in Q_j$ and there is $i < s < k$ such that $s \in Q_p, p \neq j$ (Fig. 5).

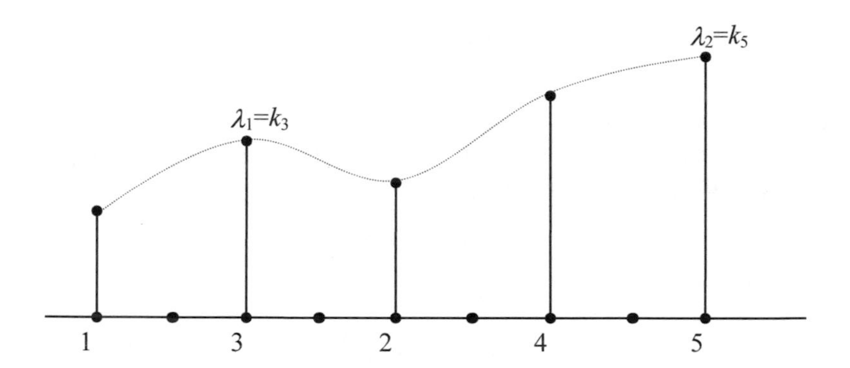

Fig. 5. To proof of approval 1.

Let be Q1 = (1, 3), Q2 = (4, 5). We swap jobs 3 and 2. If k3 ≤ k5, then $\lambda_1' = \max(k_1, k_2) < \lambda$, and $\lambda_2' = \lambda_2$. Consequently

$$(\lambda_1 - k_3) + \lambda_2 - k_2 > (\lambda_1' - k_2) + (\lambda_2 - k_3)$$

If $k_3 > \lambda_2$, then $\lambda_1' < \lambda_1$, and $\lambda_2' = k_5$. Consequently

$$(\lambda_1 - k_3) + (\lambda_2 - k_5) > (\lambda_1' - k_2) + (\lambda_2' - k_5)$$

The statement is proven.

Define (m + 1)-vertex oriented network without contours (Fig. 6). The tops of the network (with the exception of the entrance) correspond to the works, and $k_1 \leq k_2 \leq \cdots \leq k_n$. Draw arcs (i, j) in the graph if the jobs from (i + 1) to j form a group.

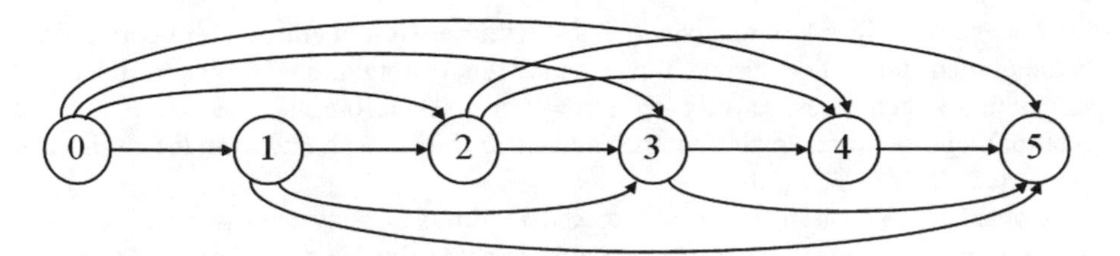

Fig. 6. Graph describing the possible division into groups.

Take the length of the arc (*i*, *j*) equal to the stimulation fund of this group:

$$l_{ij} = \max_{s[i+1;j]} k_s \sum_{s \in [i+1;j]} \Delta_s$$

Note that any path in the network connecting the input with the output and consisting of m arcs determines the splitting $Q_j, j = \overline{1, m}$, and the length of this path is equal to the value of the stimulation fund. The task has been reduced to the next:

Task 2. Determine the shortest path from m arcs.

To solve the problems, we define the auxiliary network based on the network in Fig. 2 (see Fig. 7).

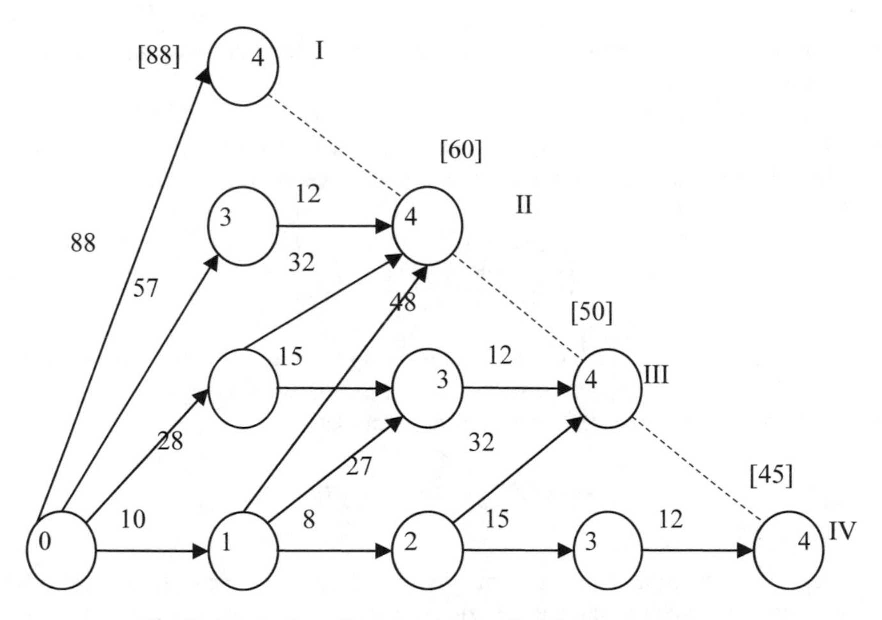

Fig. 7. Marked graph characterizing the division into groups.

This network has 4 vertex exits (labeled with roman numerals). Top I corresponds to the unified stimulation system, top IV - individual stimulation, and II and III - group stimulation systems, respectively, with m = 2 or 3. Therefore, the task was reduced to determining the shortest path from the entrance to II, if m = 2, and from the entrance to the vertex III, if m = 3 [10].

Consider the option when a spasmodic stimulation system is used for each group. In this case, we will number the works in ascending (non-decreasing) $M_i = k_i \Delta_i$, i.e. $M_1 \leq M_2 \leq \cdots \leq M_n$.

Assertion 1 remains true, which is proved as in the previous case. A similarly defined network of Fig. 2 and network of Fig. 1. The lengths of the arcs in this case are equal

$$l_{ij} = n_j \max_{s \in [i+1;j]} M_j$$

The optimal splitting is determined by the shortest paths to the corresponding vertices.

Finally, we will consider the option when for each group both a linear system and a CC system can be applied. In the general case, there is no numbering for which Statement 1 would be true. Therefore, it is necessary to consider all partitions of a set into m-th subsets, the number of which increases exponentially with increasing n, therefore, the problem is NP-hard.

Consider an approximate algorithm for solving the problem for the case m = 2.

Algorithm Description:

1 step. We build two networks for systems of linear and intermittent stimulation, as described above (Fig. 7).

In the linear incentive system, the lengths of the arcs lij are equal to the difference in the compensation value for the group (i + 1, j) between the compensation value for the linear incentive system and for the individual incentive system (II), that is,

$$l_{ij} = \left(\sum_{s \in Q_{ij}} \Delta_s \right) \max_{s \in Q_{ij}} k_s - \sum_{s \in P_{ij}} k_s \Delta_s$$

In the system of spasmodic stimulation, the lengths of the arcs are also equal to the differences between the magnitude of the compensation with the system of spasmodic stimulation and with the system of individual stimulation.

$$l_{ij} = n_{ij} \max_{s \in P_{ij}} k_s \Delta_s - \sum_{s \in P_{ij}} k_s \Delta_s$$

2 step. Take ε0 equal to the minimal length of all arcs. We write the set of arcs whose lengths are equal to ε0 and the corresponding subsets of the vertices (works).

3 step. We define two subsets, the union of which gives a complete set of vertices (works). If there are any subsets, then the problem is solved. In the opposite case, we increase ε0 to such a minimum value of ε1, when at least one arc appears, whose length is less than or equal to ε1.

Repeat steps 2 and 3. For a finite number of steps, a family of subsets will be obtained such that there are two subsets in them, the union of which is the complete set of all vertices. Based on them, we form two groups.

The system of group incentives for a given number of work in groups.

Consider task 2 of building a group incentive system under the condition that each group has the same number of jobs. Tasks of this class, as a rule, are related to complex (NP-hard) discrete optimization problems. Therefore, we consider one particular case when the number of jobs in each group is 2, and the number of groups is m = n/2 (n – an even number). We define a complete n-vertex non-oriented graph G. The lengths of the arcs of the graph are equal

$$l_{ij} = (\varDelta_i + \varDelta_j)\max(k_i, k_j)$$

if a linear incentive system is used, equal

$$l_{ij} = 2\max(k_i\varDelta_i, k_j\varDelta_j)$$

if an intermittent stimulation system is used, equal

$$l_{ij} = \min\left[(\varDelta_i + \varDelta_j)\max(k_i, k_j),\ 2\max(k_i\varDelta_i, k_j\varDelta_j)\right]$$

if a mixed incentive system is used.

Definition 1. A matching of a graph is a set of mutually non-adjacent edges [11].

The matching with the maximum number of edges is called the maximum. Note that any division of work into m groups, two jobs in each group, corresponds to the maximum matching and vice versa, any maximum matching corresponds to splitting the work into groups of two jobs in each group. Thus, the task was reduced to the definition of matching with the minimum sum of edge lengths.

Definition 2. An alternating cycle is a cycle in which in any pair of adjacent edges, one edge belongs to the matching, and the other does not belong.

The length of the alternating cycle is called the algebraic sum of the lengths of the edges, and the edges of the matching are taken with the sign «–».

Theorem [8]. A necessary and sufficient condition for optimality is the absence of alternating cycles of negative length in the graph.

Proof. The necessity follows from the fact that if there is an alternating cycle of negative length, then removing the cycle edges belonging to the matching from the matching and adding the edges of the cycle not belonging to the matching, we will reduce the length of the matching.

We prove the sufficiency. Let be G1 – is a matching that has no cycles of negative length, G0 – is an optimal matching. We distinguish a set of edges belonging to G0 and not belonging to G1. It can be shown that there are one or several alternating cycles, reversing which we obtain from G1 the matching G0. But since there are no alternating cycles of negative length, the length of G0 cannot be less than the length of G1. Therefore G1 – is the optimal matching.

Algorithm Description:

1 step. Take arbitrary four vertices and determine the optimal matching for the subgraph from these vertices. This can be done by brute forcing, since there are only three matchings in the graph of four vertices.

2 step. Take two new arbitrary vertices (let them be vertices i and j) and add the corresponding edge (i, j) to the matching. We define the shortest intermittent loop from vertex i. If the alternating cycle has a negative length, then turning it we get the optimal matching. If the cycle length is L(i) $\geq$ 0, then the original matching is optimal [8].

Algorithm for determining the shortest alternating cycle.

We note, firstly, that the shortest alternating cycle exists because there are no alternating cycles of negative length in the graph without an edge (i, j).

The algorithm is similar to the algorithm for determining the shortest path in a network without cycles of negative length [11].

Algorithm Description:

1 step. We assign the index индекс λi = 0 to the vertex i, and the indices to the other vertices λk = (+∞).

2 step. We iterate over all edges of matchings. Let (k, s) – is a matching edge. We denote Us – the set of edges incident to the vertex s (they all do not belong to the matching). Calculate

$$\lambda'_k = \min_{q \in U_s} \left(\lambda_q + l_{qs} \right) - l_{sk}$$

If $\lambda'_k < \lambda_k$, then change λk on λ'_k. Otherwise, we do not change λk.

For a finite number of iterations, the indices are established. The alternating cycle itself is determined by the reverse method.

k-th step. Take two peaks from the rest and repeat the step 2.

4 Discussion

Consider a further generalization of the problem. Suppose now that a set of works, including one group, can be arbitrary, i.e. one group can include works, for example, the first and last, and the other works can enter other groups. This assumption is quite practical: for example, in construction, even technologically related works can be performed by various performers, therefore it is more logical to combine works with regard to performers, therefore the hypothesis that the works should be included in the group in a row, according to numbers, greatly weakens the commonality. Consider a possible way to solve the problem in this case. As before, we believe that the stimulation in each group is carried out using the mixed incentive system, that is, either according to the linear incentive system (2) or the intermittent incentive system (3), and the incentive costs in each group will be determined from the expression

$$l_j = \min \left[\sum_{i \in Q_j} \Delta_i \cdot \max_{i \in Q_j}(k_i), \quad n_j \max_{i \in Q_j}(k_i \Delta_i) \right]$$

where nj – the number of j-th group.

First of all, it should be noted that the following statement is true:

Statement 2. The maximum possible number of groups does not exceed the number of jobs. If the number of groups is one, then we have a unified incentive system, and if the number of groups is equal to the number of jobs, then we have an individual one. Given this circumstance, it can be said that the total number of groups in the construction of a group incentive system will be no more than n – 2, and if it is assumed that groups consisting of one job are not allowed, then no more n/2.

Define the upper and lower bound of the desired solution.

Statement 3. The upper bound of the desired solution will be

$$O^{в} = \min\left[\sum_{i=1}^{n} \Delta_i \cdot \max_{1 \le i \le n}(k_i),\ \ n \max_{1 \le i \le n}(k_i\Delta_i)\ \right]$$

That is, the cost of organizing a unified incentive system.

As the lower bound, we take the costs arising from the individual incentive system.

$$O_H = \sum_{i=1}^{n} k_i\Delta_i$$

Taking into account the peculiarities of cost formation in group incentive systems, an empirical rule arises for building a sequence of groups for group incentives:

Heuristic rule. Arrange work in descending order of magnitude kiΔi. It is necessary to include in the same work group with the largest parameter value kiΔi. Works in which these values are the same should be reduced to one group. The criterion for the inclusion of work in the group is the magnitude of the difference between the group value of the criterion and the individual. Work is included in the group for which this difference is less.

Consider an example from [5].

Example. There are 6 works, data about which are given in Table 1.

Table 1. Example.

i	1	2	3	4	5	6
Δ_i	10	9	6	7	8	11
k_i	1	2	3	4	3	2
$k_i\Delta_i$	10	18	18	28	24	22

It is necessary to build a group mixed incentive system.

To solve this problem, we use the heuristic rule given above: in this case, it is advisable to combine work 2 and 3 into one group, since their parameter values are the same. Groups with the minimum and maximum parameter values are also divided into separate groups, and works 5 and 6 are grouped together, since their criteria values are very close. As a result, the cost of the resulting group incentive system will be

$$10 + (18 + 18) + 28 + (24 + 22) = 122.$$

It should be noted that for this case the lower limit is 120. So, the system is very close to the individual incentive system, which is explained by the fact that we have an individual incentive system in two groups.

If this is abandoned, then the distribution of work will be carried out differently: the work will be grouped together (1; 2) (3; 6) (4; 5) or which is identical (1; 6) (2; 3) (4; 5) the costs will be

$$36 + 44 + 56 = 136.$$

Thus, the resulting solution is better than [5].

For greater clarity, we present the discussed algorithm in tabular form. To do this, build a Fig. 8, which shows the contribution of each pair of work to the total cost (numerator), in the event that these two jobs fall into one group and how much the costs will increase in comparison with the individual incentive system (denominator). The results are presented in Fig. 8.

	I	II	III	IV	V	VI
I	0	$\frac{36}{8}$	$\frac{36}{8}$	$\frac{56}{18}$	$\frac{48}{14}$	$\frac{44}{12}$
II		0	$\frac{36}{0}$	$\frac{56}{10}$	$\frac{48}{6}$	$\frac{44}{4}$
III			0	$\frac{56}{10}$	$\frac{48}{6}$	$\frac{44}{4}$
IV				0	$\frac{56}{4}$	$\frac{56}{6}$
V					0	$\frac{48}{2}$
VI						0

Fig. 8. The results of the discussed algorithm in tabular form.

The solution is as follows:

1 step. Choose the lowest cost. This is 1 job with a criterion $k_i\Delta_i = 10$. Other works include this group does not make sense, since the costs will increase sharply. We delete the first row and the first column of the Table 4.

2 step. Their remaining elements of Table 4, choose the smallest amount of costs, that is, the numerator, in the case when there are several such values, then we take the one with the smaller denominator. In our case, these are works 2 and 3. We delete the columns and rows with these numbers. Now it is necessary to decide: we will form a new group or supplement this with new work. It is advisable to

supplement this group with new work only in the case of which the value of the criterion $k_i\Delta_i$ is less than that of those already included in the group, since otherwise this will lead to a sharp increase in costs. We have to admit that this option in this case will be the best, since all others will give a greater increase in costs.

3 step. Further on this step, we choose from the remaining elements the one at which the total value of costs will be the smallest. This means that it is necessary to include them in work groups 5 and 6. Checking the inclusion of another work shows that this is not profitable.

4 step. The remaining work is included in a separate group.

Thus, the solution to the problem will be: (1), (2; 3), (4), (5; 6). In this case, the total costs will be:

$$(10) + (18 + 18) + (28) + (24 + 24) = 122.$$

5 Conclusion

The article deals with the problems of synthesis of optimal group-stimulation systems. For a number of tries, efficient algorithms are proposed. Many tasks of the synthesis are difficult (in some cases - NP-difficult). For some of them, heuristic algorithms are proposed.

It is of interest to solve synthesis problems for nonlinear agent cost functions and for other basic incentive systems (for example, rank-based ones). In the considered problems, the magnitude of the reduction in the duration of work is given. The problem arises of determining these durations in order to ensure the implementation of the project in the required period of time with the minimum value of the incentive fund for a given network project schedule. Finally, problems of manipulation remain unexplored.

References

1. Novikov, D.A.: Incentive in Organizational Systems. SINTEG, Moscow (2003)
2. Novikov, D.A., Tsvetkov, A.V.: Incentive Mechanisms in Multi-element Organizational Systems. Apostrophe, Moscow (2000)
3. Vasilyeva, O.N., Zaskanov, V.V., Ivanov, D.Yu., Novikov, D.A.: Models and Methods of Material Incentives (Theory and Practice). KomKniga, Moscow (2006)
4. Yu, A.D.: The task of scheduling the work of teams of specialists. Sci. Pract. J. Econ. Manag. Syst. **4.3**(30), 304–311 (2018)
5. Yu, A.D., Amelina, K.E.: Tasks of scheduling by criterion of maximization of the total weighed volume of the performed works. Control Syst. Inf. Technol. Sci. Tech. J. **4**(74), 20–26 (2018)

6. Barkalov, S.A., Kurochka, P.N.: Model for determining the term of execution of sub-conflicting works. In: Proceedings of 2017 Tenth International Conference on Management of Large-Scale System Development (MLSD), 8109598 (2017)
7. Kravets, O.J., Podvalny, E.S., Barkalov, S.A.: Quality assessment of a multistage process in the case of continuous response functions from resource influences. Autom. Remote Control **76**(3), 500–506 (2015)

Aggregation Models and Algorithm for Coordinating the Interests of the Region and Enterprises of the Energy Complex

Yulia Bondarenko[1,2](✉), Tatiana Azarnova[1,2], Irina Kashirina[1,2], and Tatiana Averina[1,2]

[1] Voronezh State University, Universitetskaya Sq., 1, Voronezh 394026, Russia
bond.julia@mail.ru
[2] Voronezh State Technical University, Moscow Avenue, 14, Voronezh 394026, Russia

Abstract. The paper is devoted to the development of aggregation models and algorithm for coordinating the social interests of the region and the economic interests of the enterprises of the energy complex. The most important stage of the approval process is the formation of reasonable options for compromise solutions, which are subject to further discussion. To this end, the concept of a compromise solution is introduced in the context of coordinating the interests of the region and enterprises of the energy sector. The proposed approach to the formation of compromise solutions is based on the representation of the economic system in the sectoral context, which makes it possible to use the available statistical information for its practical implementation. It is proposed to look for solutions using the developed aggregation algorithm for coordinating interests. The algorithm is based on an aggregation mathematical model of minimizing the costs of coordinating the interests of the region and enterprises of the energy complex. The model is an optimization task, the purpose of which is the sum of expenses of the region for the payment of subsidies and tax revenues under-received because of benefits. The constraints of the model describe the main interrelations of indicators of the economic activities of enterprises of the energy complex and the minimum requirements for compromise solutions. The implementation of the algorithm, practical calculations and their discussion with representatives of the authorities and enterprises of the energy complex confirmed the practical significance of the study for management at the state and municipal levels, and also outlined ways of its further improvement.

Keywords: Region · Energy complex · Coordination · Interest · Mathematical model

1 Introduction

The priority goal of the socio-economic development of the modern region is to create conditions that ensure a decent level and high quality of life of the population [1]. Its successful achievement presupposes an orientation of the regional policy not so much

V. Murgul and M. Pasetti (Eds.): EMMFT-2018, AISC 983, pp. 69–78, 2019.
https://doi.org/10.1007/978-3-030-19868-8_7

on solving private tasks of individual subjects as on systematic overcoming a whole complex of interrelated socio-economic problems that are relevant for the region as a whole. As the Russian and world management practice shows [2], a necessary condition for the effective implementation of such a policy is the existence of well-functioning mechanisms of interaction between regional authorities and the most significant enterprises of the region, including enterprises of the energy complex (EC).

The spectrum of primary tasks of the region, the effective solution of which is impossible without the active support of enterprises of EC, is quite wide [3]. First of all, this is the provision of the population with energy not only in the right quantity but also at reasonable prices (tariffs). Since a large number of enterprises in the energy sector are large or (and) city-forming, the creation of new jobs and ensuring the required growth in nominal wages are of particular social importance for the region.

We note that each of the tasks we have identified involves certain financial costs of EC enterprises, and therefore their solution (especially in the context of government regulation of tariffs) may go beyond the framework of economic interests [4]. The presence of objective inconsistencies in the interests of the region and enterprises makes it urgent for regional authorities to form constructive coordination mechanisms that make it possible to motivate EC enterprises to solve problems important for the region with the least cost. The basis of this motivation may be various economic instruments - regional components of tax incentives, subsidies, concessional loans, etc.

The proposed coordination mechanisms should take into account the specifics of the enterprises of the energy complex, but at the same time rely on advanced scientific experience.

The general theoretical foundations for coordinating the interests of participants in economic relations are laid in works on the theory of economic mechanisms, incentives, and economic behavior (L. Hurwicz, E. Maskin, G. Stigler, J. Tirole) [8–10].

Among the studies on the development of constructive approaches to the coordination of interests in management practice, we note the works of K.A. Bagrinovsky, touching on the task of coordinating planning decisions at the aggregation sectoral level [5, 6]. The models and algorithms for coordinating the interests of regional enterprises in the allocation of a limited resource are considered in the works of Orlova [7, 8].

In this paper, the problem of the coordination of interests is considered in an aggregation aspect, and the participants in the coordination process are the regional authorities and the totality of enterprises of the regional energy complex, located on its territory.

The aim of the study is to develop aggregation mathematical models and algorithms that allow the administration of the region to reasonably calculate the parameters of regulatory actions that ensure the motivation of EC enterprises to support solving the social problems of the region. The specification of the proposed models is based on available statistical information, which makes it convenient to use them in the practice of regional management as the basis for the formation and modification of coordination mechanisms.

2 Materials and Methods

The study is based on the formal presentation of the socio-economic system of the region as a set of interrelated elements and relations arising in the process of production, distribution, exchange, and consumption of tangible and intangible benefits while ensuring the livelihood and development of the region's society. The structure of such a system is shown in Fig. 1 and includes the following elements:

1. regional authorities (regional administration);
2. economic system of the region;
3. social system of the region.

The economic system is considered as a set of interrelated and interacting economic entities (enterprises) located on the territory and realizing economic interests in the process of economic activity. The economic system of the region is structured by industry (for the Russian economy - types of economic activity (TEA)) according to the specifics of products manufactured by enterprises.

Let n be the number of industries of a region, $j = 1, \ldots, n$. We believe that the enterprises of the energy complex belong to the industry with a serial number $j = 1$. For example, in the Russian economy, electric power enterprises belong to the TEA "Production and distribution of electricity, gas and water".

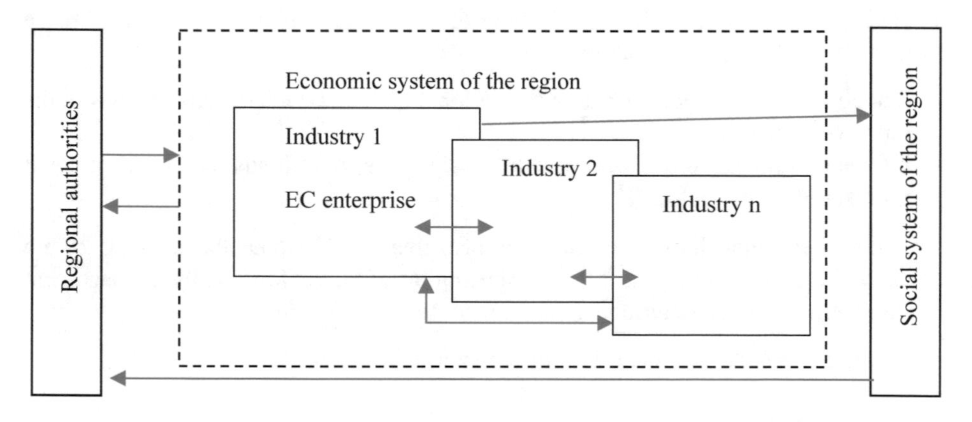

Fig. 1. Presentation of the structure of the socio-economic system of the region.

For the production of energy, EC enterprises use their own products (energy) and products of other industries. Following the notation adopted in the Leontiev model for the coefficients of direct costs, we denote by a_{1j} - the amount of energy expended on the production unit of the industry j; a_{j1} - the volume of the product of the industry j, necessary for the production of a unit of energy.

We believe that the priority economic interest of EC enterprises is to obtain the greatest profit.

The social system of the region is a community of individuals living in the region, united by forms of joint activity, as well as interests and needs for a high quality of life in the region.

Diagnostics of well-being or ill-being of the social system of the region is measured on the basis of social indicators relevant for the state [9], among which are the average monthly nominal wages and the unemployment rate. The annual planned values of these indicators are established by regional authorities and are documented in socio-economic development strategies. We will assume that the regional authorities for the planning under consideration (for example, a year) have determined the following values of the performance indicators of the energy complex, the achievement of which meets the interests and needs of the region:

$\underline{y}$ – the minimum amount of energy covering the needs of the region, taking into account its loss during transmission (in monetary units at current tariffs);
$\underline{L}$ – the minimum number of employees (labor resources) of EC enterprises;
$\underline{\omega}$ – the lower limit on the size of the average annual nominal wage at the EC enterprises.

In order to motivate the enterprises of the energy complex to achieve the values for the above indicators required for the region, the administration of the region uses regulatory actions. The choice in the study of impact instruments corresponds to the current powers of regional authorities:

- reduction of the current tax rate to the regional budget, which can take values in the range from $\underline{\alpha}$ to $\overline{\alpha}$;
- provision of funds in the form of subsidies, concessional loans, etc. in the amount not exceeding the value $\overline{B}$.

We will say that the compromise has been reached between the interests of the region and the EC enterprises, if the administration of the region and the management of the enterprises have determined and agreed on the following:

- values of performance indicators of the energy complex:

$$(y_1, L, \omega),$$

where y_1 - the amount of produced energy, L - the number of labor resources, ω - the size of the average annual nominal wage;

- allowable regulation α^r, B (where α^r - the rate of income tax in the budget of the region, $\underline{\alpha} \leq \alpha^r \leq \overline{\alpha}$; B - the size of the subsidy, where $B \leq \overline{B}$),

  which:

- meet the interests of the region, i.e. $y_1 \geq \underline{y}$, $L \geq \underline{L}$;

- allow enterprises to provide EC enterprises with a profit (after tax deduction) Π not less than its predicted value $\underline{\Pi}$ under the conditions of absence of a motivating effect.

The search for reasonable options for compromise decisions is proposed to implement by the methods of economic and mathematical modeling and decision theory. Formalizations are subject to the processes describing the production and economic possibilities of the energy complex. Each of the relationships is based on the use of statistical information available to the administration of the region.

One of these dependencies is the production function of the energy complex, which formally describes the relationship of the maximum volume of output from production factors (capital and labor resources) [10]:

$$y_1 = f(K, \omega \cdot L) \quad (1)$$

Here K – capital (fixed assets) of the EC enterprises, f – production function of the energy complex.

The parameters of the production function are determined on the basis of retrospective statistical information using econometric methods.

For practical calculations for the TEA "Production and distribution of electricity, gas and water" based on official statistical data for the period 2004–2015 in the Voronezh region, the following production function was obtained:

$$y_{1t} = (0,967 \cdot [K_t]^{-0,0016} + 0,0373 \cdot [L_t]^{-0,0016})^{-61,15}.$$

3 Results

The basis of the proposed approach to the coordination of interests is an aggregation mathematical model of minimizing the costs of coordinating the interests of the region and the EC enterprises. For its formal presentation, we will assume that the administration of the region knows the following information aggregated by EC enterprises, obtained for the last reporting period (year):

- L_0 – the number of employees (labor resources);
- ω_0 – the average annual nominal wage;
- K_0 – the volume of fixed assets;
- γ – coefficient of liquidation of fixed assets; λ – coefficient of renewal of fixed assets;
- coefficients of direct costs a_{1j} and $a_{j1}, j = 1, \ldots, n$;
- α_0 – effective income tax rate (%); $a_0 = a_0^r + a_0^f$, where a_0^r, a_0^f – the rates of income tax in the regional and Federal budget, respectively;
- ρ – the share of deductions from the wage fund.
- Based on statistical and expert information, the following can be calculated:
- B_0 – own funds of EC enterprises that can be invested in the development of production;

- β – average costs of creating a single workplace at the EC enterprises;
- production function of the energy complex (1);
- $\underline{\Pi}$ – the predicted value of the profits of EC enterprises after deducting the tax in the billing period in the absence of regulatory impacts;
- projected industry output values $y_2, y_3, \ldots, y_n$;
- predicted values of the value of final energy consumption Q and losses in power grids R.

It is required to determine such values of subsidized funds B and rates for profits of EC enterprises in the regional budget that will ensure coordination of the interests of the region and EC enterprises with minimum costs for the regional budget.

The formal model of the task is as follows:

$$B + \frac{1}{100}\left(\alpha_0^r - \alpha^r\right) \cdot \Pi \to \min, \tag{2}$$

with constraints that take into account:

- technology of production of the EC enterprises:

$$y_1 = f(K, \omega \cdot L), \tag{3}$$

where ω – average annual nominal wage;

- electricity needs of the region:

$$y_1 \geq \sum_{j=1}^{n} a_{1j} y_j + Q + R; \tag{4}$$

- change of fixed assets due to their liquidation, renewal and expansion by the value ΔK:

$$K = (1 - \gamma + \lambda)K_0 + \Delta K; \tag{5}$$

- change in the number of labor due to the creation of new jobs in the amount of ΔL:

$$L = L_0 + \Delta L; \tag{6}$$

- formation of profit:

$$\Pi = y_1 - \left(\sum_{j=1}^{n} a_{j1} y_j + \frac{1}{1-\rho} \omega \cdot L + \lambda K_0 + R \right); \quad (7)$$

- constraints on financial assets:

$$\Lambda K + \Delta L \cdot \beta = B_0 + B; \quad (8)$$

- conditions of consistency of interests:

$$L \geq \underline{L},\ y_1 \geq \underline{y},\ \omega \geq \underline{\omega};$$

$$\frac{1}{100}\left(1 - \alpha^r - \alpha_0^f\right) \cdot \Pi \geq \underline{\Pi}; \quad (9)$$

- constraints on variables:

$$\Delta L \geq 0,\ \Delta K \geq 0,\ \underline{\alpha} \leq \alpha^r \leq \overline{\alpha},\ B \leq \overline{B} \quad (10)$$

Model (2)–(10) belongs to the class of optimization problems and can be reduced to an equivalent problem with variables ΔL, ω, α, and B. In the case of a non-linear production function, its objective function and some of the constraints are non-linear. A practical solution to this problem is proposed to implement by the methods of non-linear optimization.

Model (2)–(10), the permissible set of which contains compromise decisions, forms the basis of the algorithm for coordinating the interests of the region and the EC enterprises.

Aggregation algorithm for coordinating the interests of the region and enterprises of the energy complex.

Step 1. Definition and calculation of model parameters (2)–(10).
Step 2. Construction of the production function of EC enterprises (1) on the basis of a regression model.
Step 3. Determining the number of compromise solutions, H.
Step 4. For each option of compromise decisions, various options are set for the lower constraints on social and economic indicators:
$\left(\underline{y}^h,\ \underline{L}^h,\ \underline{\omega}^h,\ \underline{\Pi}^h\right)$, where $\left(\underline{y}^h,\ \underline{L}^h,\ \underline{\omega}^h,\ \underline{\Pi}^h\right) \geq \left(\underline{y},\ \underline{L},\ \underline{\omega},\ \underline{\Pi}\right)$, $h = 1,\ldots,H$.
Step 5. Formation of models (2)–(10) with the replacement of constraints (9) on the following:

$$L \geq \underline{L}^h,\ y_1 \geq \underline{y}^h,\ \omega \geq \underline{\omega}^h,\ \tfrac{1}{100}\left(1 - \alpha^r - \alpha_0^f\right) \cdot \Pi \geq \underline{\Pi}^h.$$

Step 6. Solving the tasks of Step 5.
If the admissible set of each of the tasks is empty, then the question of changing the parameters of relation (10) is solved.
Step 7. The compromise solutions of the tasks obtained at Step 6 are subject to further discussion by regional authorities and representatives of enterprises of the energy complex. Agreed compromise solution is documented for implementation.

For the practical implementation of the algorithm, a software package in the C# language has been developed. The structure of the complex includes: a database containing indicators of a number of enterprises of the Voronezh region and a computing unit containing methods for solving nonlinear optimization problems. The operation of the algorithm was tested on training and real data.

Let's consider the results of the algorithm on the basis of the aggregated data of 2017 for the enterprises of the electric power industry of the Voronezh region, officially provided by the Federal Statistics Service. These enterprises belong to the type of economic activity «Production and distribution of electricity, gas and water», the section «Production, transmission and distribution of energy».

The largest representatives of the electric power complex of the Voronezh Region are power plants, the installed capacity of which (as of 01.09.2017) was 2,862 MW. These include: Novovoronezh nuclear power plant, Voronezh thermal power plant-1, Voronezh thermal power plant-2. At the same time, the Novovoronezh nuclear power plant, the total installed capacity of which is 2,597.3 MW, is the main enterprise of the city of Novovoronezh. Centralized power supply to consumers in the region is provided by 14 power grid and sales companies.

Table 1 shows a fragment of basic data for models of an aggregation algorithm for coordinating the interests of the region and enterprises of the electric power complex.

Table 1. Basic data for the algorithm.

Name of indicator	Unit of measurement	Value
Average number of employees	Thousand people	31.5
Fixed assets (at the end of the year)	Million rubles	274 361
The rate of renewal of fixed assets (at the end of the year)	In % of the total value of fixed assets	60.2
The coefficient of liquidation of fixed assets (at the end of the year)	In % of the total value of fixed assets	0.2
The share of deductions from the wage fund	%	30
Average annual nominal wage	Thousand rubles	504
Profit before tax (balance sheet profit)	Million rubles	2124.9

Lower boundaries of indicators for the billing period: $\underline{y} = 32000$ million rubles, $\underline{L} = 34$ thousand people, $\underline{\omega} = 504$ thousand rubles, $\underline{\Pi} = 2200$ million rubles.

Table 2. The results of calculation of the compromise solutions.

№.	Income tax rate in the regional budget (%)	Volume of output (million rubles)	Labor resources (thousand people)	Average annual nominal wage (thousand rubles)	Profit (million rubles)
1	15.1	32004.4	34.2	504	2461
2	13.45	34000	35	528	2565
3	12.7	34879	36.3	546	2600

The profit tax rate was chosen as a regulatory impact. The maximum value of the profit tax rate is 20%. The profit tax rate in the Federal Budget of the Russian Federation is 3%, and in the regional budget, it varies from 12.5% to 17%.

A fragment of the results of programmatic calculations of the three compromise solutions is shown in Table 2.

Discussion of the calculation results with representatives of regional authorities and energy companies allowed drawing conclusions about their joint interest in the implementation of the second option.

4 Discussions

The study showed the advantages of using mathematical methods to solve complex urgent problems of the regional level - coordination of interests with enterprises of the energy complex. The programmatic calculation of compromise solutions with different properties makes it possible to increase not only the objectivity of the decisions made, but also their qualitative characteristics.

The aggregation model proposed in the study was repeatedly discussed with representatives of regional government and business, its constraints were adjusted according to the comments made. The composition of the system of constraints of the model, its detailing was largely determined by the available information in order to make it practically feasible. If the possibility of more careful consideration of the characteristics of enterprises appears, the constraints of the model can be added. The presented algorithm with a minor adjustment can be used as a tool to support the coordination of interests of the region with enterprises of other sectors of the economy.

The discussion of the research results determined the ways of its further development in expanding the model of minimizing the costs of coordinating the interests of the region and the EC enterprises. This direction involves the study of the dynamic case when the task of coordination is solved for several periods. The practical side of supporting the solution of the task of coordinating interests can be the expansion of algorithmic and software tools by formalizing the strategies of participants' behavior in the process of discussing compromise options and negotiating.

5 Conclusion

The approach proposed in this paper to solving the urgent problem of regional development - achieving a high quality of life - is based on an understanding of the need for interaction of regional authorities with enterprises of the energy complex. A necessary condition for such interaction is coordination and common interests. The solution to the complex task of coordinating the interests of the region is proposed to be sought on the basis of the formation of aggregation mathematical models and algorithms. A feature of such models is the presentation of thc economic system of the region in the sectoral context, which makes it possible to use the available statistical information. The mathematical model proposed in this study makes it possible to find a compromise option of motivation that meets the social interests of the region and the economic interests of EC enterprises with minimal costs for the region. The aggregation algorithm developed on the basis of it allows forming reasonable options for further discussion of the parties and development of a single solution. The introduction of the proposed mathematical and software tools in the practice of management, discussion and analysis of the results of the calculations allowed not only justifying the expediency of using various forms of motivation to coordinate interests, but also identified areas for further research.

References

1. Gavrilov, I.A., Makarov, A.D.: Fundamental'nye issledovaniya **4**(1), 133–137 (2017)
2. Semova, N.G., Mataev, A.S., Voroncova, A.V.: Vestnik moskovskogo gosudarstvennogo oblastnogo universiteta. Seriya: ehkonomika **4**, 137–139 (2010)
3. Bondarenko, Y.V., Goroshko, I.V.: Mekhanizmy soglasovaniya pokazatelej social'no-ehkonomicheskogo razvitiya regiona i rol' organov vnutrennih del v ih realizacii, Moscow (2015)
4. Nezhnikova, E.V., Kankhva, V.S.: Investments in renovation processes in a changing environment. Izvestiya Vysshikh Uchebnykh Zavedenii, Seriya Teknologiya Tekstil'noi Promyshlennosti **2**(368) (2017)
5. Nezhnikova, E.: The use of underground city space for the construction of civil residential buildings. Proc. Eng. **165**, 1300–1304 (2016). https://doi.org/10.1016/j.proeng.2016.11.854
6. Bellagente, P., Bonafini, F., Crema, C., Depari, A., Ferrari, P., Flammini, A., Lenzi, G., Pasetti, M., Rinaldi, S., Sisinni, E.: Enhance access to Industrial IoT data with distributed machine interface with by means of localization functionalities. IEEE Instr. Meas., Mag (2017)
7. Flammini, A., Pasetti, M., Rinaldi, S., Bellagente, P., Ciribini, A., Tagliabue, C., Zavanella, L.C., Zanoni, S., Oggioni, G., Pedrazzi, G.: A living lab and testing infrastructure for the development of innovative smart energy solutions: the eLUX laboratory of the university of Brescia. In: 2018 AEIT International Annual Conference, Bari, Italy (2018)
8. Burkov, V., Shchepkin, A., Irikov, V., Kondratiev, V.: Stud. Syst. Decis. Control **181**, 29–38 (2019)
9. Barkalov, S.A., Kravec, O., Kurochka, P.N., Nasonova, T.V., Polovinkina, A.I.: Int. J. Pure Appl. Math. **117**(4), 83–87 (2017)
10. Orlova, E.V.: Proceedings of the Mathematical Modeling Session at the International Conference Information Technology and Nanotechnology, MM-ITNT 2017, vol. 1904, pp. 1–6 (2017)

Planning the Optimal Sequence for the Inclusion of Energy-Saving Measures in the Process of Overhauling the Housing Stock

Valeriy Mishchenko, Elena Gorbaneva, Elena Ovchinnikova, and Kristina Sevryukova(✉)

Voronezh State Technical University, Moscow Avenue, 14, Voronezh 394026, Russia
ksevrukova@vgasu.vrn.ru

Abstract. At the moment, it is impossible to deny the appearance of the obvious problem of the effective implementation of energy-saving measures during the overhaul of apartment buildings. This article considered the possibility of using one of the methods for the section of mathematics of the Theory of Graphs during the overhaul of apartment buildings with the introduction of energy-saving technologies. In the course of the research, elements were used in the form of an oriented graph with the help of which the adjacency matrix was constructed. Thanks to the work done, it was possible to isolate the resource potential of introducing energy-efficient technologies in the overhaul and identified their optimal order of inclusion in the process of renovating apartment buildings. Based on the obtained combinations, a digraph is built showing the relationship between the planned energy-saving technologies in apartment buildings, and also presents the calculation of financial and economic indicators for assessing the implementation of energy-efficient measures and proposes a general scheme for the implementation of energy-saving technologies when carrying out major repairs in apartment buildings to simplify work. Based on the analysis of real existing technologies using the constructed cost functional, the directions for improving the efficiency of reconstruction processes and reducing the cost of capital repairs are formulated.

Keywords: Planning · Energy-saving measures · Overhauling · Housing stock

1 Introduction

Most of the energy consumed in large cities has a huge impact on the emission of greenhouse gases into the atmosphere, resulting in significant changes in climate that adversely affect the environment. Thus, it is becoming increasingly important to improve energy performance in major cities of the Russian Federation, which is directly interconnected with existing buildings, as a result, there is a need to take measures to upgrade buildings and increase their energy efficiency. In the Russian Federation, the potential for improving energy efficiency remains largely untapped.

V. Murgul and M. Pasetti (Eds.): EMMFT-2018, AISC 983, pp. 79–91, 2019.
https://doi.org/10.1007/978-3-030-19868-8_8

One of the main problems in the construction sector is the very slow pace of replacement of the existing and obsolete housing stock, which is a key factor influencing the improvement of the energy performance of existing buildings in large cities of the Russian Federation. Also, an obstacle to improving energy efficiency during the overhaul of apartment buildings is the lack of generally accepted methods for a comprehensive assessment of the effectiveness of energy efficient technologies and materials implementation [1]. During the overhaul, the design and estimate documentation do not pass the state expertise. In the end, in some apartment buildings, after a major overhaul, on the contrary, there is an increase in energy consumption, rather than a decrease [2, 3].

According to the housing legislation of the Russian Federation, the obligation to maintain in good condition common property in apartment buildings, including the carrying out of current and capital repairs, was assigned to the owners of premises in apartment buildings [4]. It is not paradoxical that the list of mandatory work during the overhaul does not include measures to improve the energy efficiency of buildings [5]. It follows that the conduct, choice of energy-saving measures and all the costs associated with the implementation of energy-saving measures are passed on to apartment owners. To improve the energy efficiency of the housing stock during capital repairs, it is necessary to motivate the owners of premises in apartment buildings to conduct energy-efficient measures. Since the average owner has no idea what kind of overhaul works lead to a decrease in utilization of utility resources, as well as the cost of the necessary work and the payback period of the selected measures. As a result, it becomes necessary to inform the building owners about the list of works that lead to a tangible result in reducing the consumption of utility resources. Developed long-term regional overhaul programs in apartment buildings provide for the repair of only those elements that are provided for by law [6].

A list of works on the overhaul of common property in an apartment building can be found in Fig. 1.

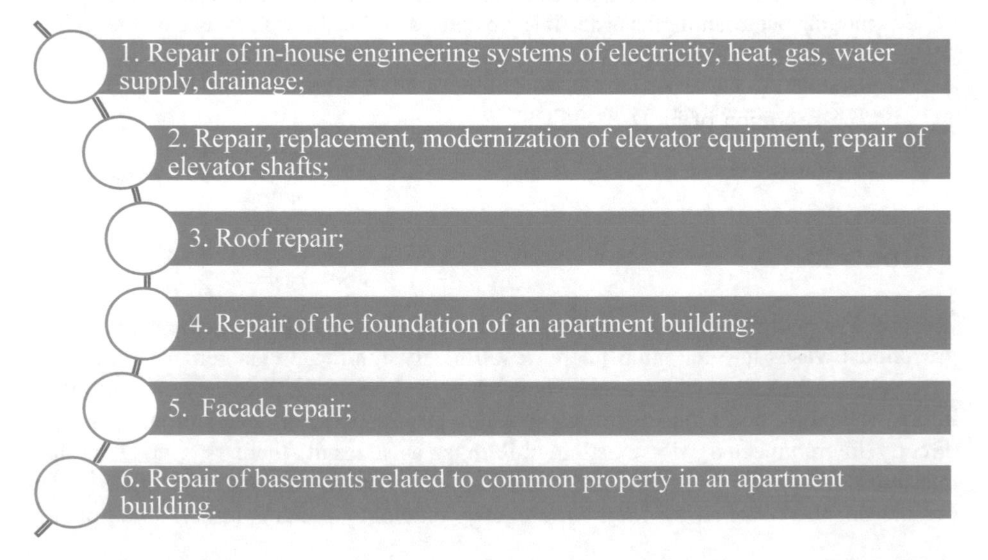

Fig. 1. The list of works on the overhaul of common property in an apartment building.

This list can be supplemented with other types of work, such as: installation of general house-based metering devices for resource consumption; facade insulation; overhaul of stairs with installation of ramps; conversion of a non-ventilated roof to ventilated, etc. Each of these energy-saving technologies has a certain level of efficiency.

Based on this, the main objective of this article is to determine the most rational combination of the consistent inclusion of energy-saving technologies in the repair and construction work during the overhaul and improvement of energy efficiency indicators in the Russian Federation.

2 Materials and Methods

To solve this problem, the graph theory was used. As you know, a graph is a mathematical model of a system in which elements are denoted by vertices, and edges are the presence of any binary relation between them. Graphs can be either oriented or not oriented, with restrictions on the number of links and additional data on the available vertices or edges. In our situation, the considered elements are represented as an oriented graph, which will consist of the set of vertices $V = \{v_1, v_2, \ldots, v_n\}$ and the set of edges $E = \{e_1, e, \ldots, e_{n-1}\}$.

Information about the structure of the graph was considered in the form of the adjacency matrix $A = [a_{ij}]$, which is defined as a square matrix in the binary relation, on the basis of which the adjacency elements of the digraph matrix were determined by the formula (1) [7, 8]:

$$a_{ij}\begin{cases} 1, if(v_i, v_j) \in E \\ 0, if(v_i, v_j) \notin E \end{cases} \tag{1}$$

Suppose that we have n energy-saving technologies and m apartment buildings. Based on this, we identified the resource capabilities of the overhaul. Based on the resource estimate, we construct the adjacency matrix A(nxm) using the following indicators:

- on the condition of limited resources for each apartment building (2):

$$A_1 = \begin{bmatrix} 1 & 0 & 0 & 1 & 1 & 0 \\ 0 & 1 & 1 & 0 & 1 & 0 \\ 1 & 1 & 0 & 0 & 0 & 1 \\ 0 & 0 & 1 & 1 & 1 & 0 \\ 0 & 1 & 0 & 0 & 0 & 1 \\ 0 & 0 & 1 & 1 & 1 & 1 \end{bmatrix} \tag{2}$$

- costs from the introduction of energy-saving technologies (where 1 is the minimum cost, 0 is the cost exceeding the allowable limit (3)):

$$A_2 = \begin{bmatrix} 1 & 0 & 0 & 1 & 1 & 1 \\ 1 & 0 & 1 & 0 & 1 & 0 \\ 1 & 1 & 0 & 0 & 0 & 1 \\ 0 & 1 & 1 & 1 & 1 & 0 \\ 0 & 1 & 1 & 0 & 0 & 1 \\ 1 & 0 & 0 & 1 & 1 & 1 \end{bmatrix} \tag{3}$$

- on the cumulative effect of the introduction of energy-saving technologies (4):

$$A_3 = \begin{bmatrix} 1 & 1 & 0 & 1 & 1 & 1 \\ 0 & 1 & 1 & 0 & 1 & 0 \\ 1 & 1 & 1 & 0 & 0 & 1 \\ 0 & 1 & 0 & 1 & 0 & 0 \\ 1 & 1 & 1 & 1 & 0 & 1 \\ 1 & 0 & 1 & 1 & 1 & 1 \end{bmatrix} \tag{4}$$

By combining the resulting matrices A1, A2, A3, we identified the resource potential for introducing energy-saving technologies in the overhaul of apartment buildings (5).

$$A = \begin{bmatrix} \underline{111} & 001 & 000 & \underline{111} & \underline{111} & 011 \\ 010 & 101 & \underline{111} & 000 & \underline{111} & 000 \\ \underline{111} & \underline{111} & 001 & 000 & 000 & \underline{111} \\ 000 & \underline{111} & 010 & 011 & 010 & 100 \\ 001 & 011 & \underline{111} & 101 & 100 & \underline{111} \\ \underline{111} & 000 & 010 & \underline{111} & \underline{111} & \underline{111} \end{bmatrix} \tag{5}$$

After analyzing the obtained adjacency matrix, we conclude that the elements of matrix A, consisting of units, determine the optimal combination of the possibility of implementing the introduction of energy-saving technologies during the overhaul of existing apartment buildings at minimal cost. Based on the obtained combinations, we build a digraph showing the connection between the planned energy-saving technologies in apartment buildings (Fig. 2).

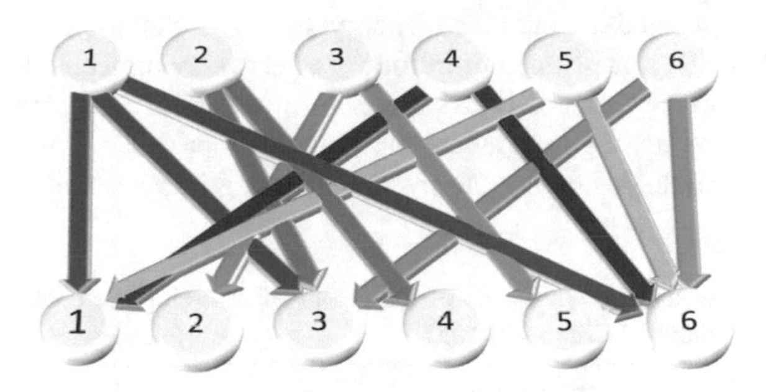

Fig. 2. The introduction of planned energy-saving technologies during the overhaul in apartment buildings.

The data used to build combinations and construct a digraph are presented in Table 1.

Table 1. The introduction of planned energy-saving technologies during the overhaul in apartment buildings.

№	Year of built	Number of floors	Wall material	Planned activities
1	1963	5	Brick	Repair of the roof and attic, repair of the facade, repair of heating and water supply systems, repair of house power supply systems, repair of the basement
2	1963	5	Brick	Repair of the roof and attic, repair of the facade, repair of heating and water supply systems, repair of house power supply systems, repair of the basement
3	1963	5	Brick	Repair of the roof and attic, repair of the facade, repair of heating and water supply systems, repair of house power supply systems, repair of the basement
4	1965	5	Brick	Repair of the roof and attic, repair of the facade, repair of heating and water supply systems, repair of house power supply systems, repair of the basement
5	1665	5	Brick	Repair of the roof and attic, repair of the facade, repair of heating and water supply systems, repair of house power supply systems, repair of the basement
6	1966	5	Brick	Repair of the roof and attic, repair of the facade, repair of heating and water supply systems, repair of house power supply systems, repair of the basement

From the presented graphical model, we can see that the number of energy-saving technologies planned for the introduction is two or more. For example, in apartment building 1 it is planned to introduce energy-saving technologies - 1, 2, 3, 6. In apartment building 2, it is planned to introduce energy-saving technologies - 3, 4, etc.

Each energy-saving technology has a certain level of efficiency. From this we can conclude that it is necessary to simulate the process of prioritizing the introduction of energy-saving technologies during the overhaul, taking into account the effectiveness.

Table 2. Advantages and disadvantages of indicators of assessing the economic efficiency of the introduction of energy-saving measures.

№	Indicators of the evaluation of economic efficiency	Advantages	Disadvantages
1	PP- payback period of investments without discounted income	Ease of calculation; the ability to determine the riskiness of the project, the ability to assess the real period of return on investment	The lack of additional financial influence, does not take into account inflation in determining the payback period, not the ability to control the profitability of the project after the payback period
2	DPP- payback period of investments with discounted income	It takes into account the concept of the value of money in time; allows you to assess whether the initial investment will be returned or not; apply different rates of discounted payback period	It does not take into account the effect of income after the payback period; does not show the effectiveness of investment
3	NPV- net profit value	Takes into account the period of the investment project and the distribution of funds in time, takes into account not only the costs throughout the life cycle but also the effect of the implementation of the event	Does not take into account the value of alternative investment projects; depends on the size of the discount rate; the size of the discount is accepted unchanged; detailed long-term forecasts required

3 Results

To determine the priority of the inclusion of energy-saving technologies, it is necessary to assess the economic efficiency of energy-saving measures in the overhaul of apartment buildings with regard to the life cycle.

Economic evaluation of the effectiveness of energy-saving measures is carried out by comparing a number of indicators: the payback period of each event; net income received from the inclusion of energy-saving measures for the entire period of their operation; net present value; profitability index of investment funds in energy saving measures.

In general, the performance indicator from the introduction of energy-saving technology can be represented by the formula (6) [9, 10]:

$$\mathrm{Ee} = \frac{Qi0 - Qi1}{Ci} \rightarrow max \tag{6}$$

where, Q_{io} – energy consumption before the introduction of energy-saving technology;

Q_{i1} energy consumption after the introduction of energy-saving technology;

C_i – costs of introducing energy-saving technology.

Also, the investment payback period (PP) was invested in the building, excluding discounted income, defined by formula (7):

$$\mathrm{PP} = \frac{Inv}{Et} \tag{7}$$

where, I_{nv} – initial investment in the project;

E_t – time savings t.

The calculation of the payback period of investments in the implementation of energy saving measures is not sufficiently accurate without taking into account discounting (DPP), which is determined by formula (8):

$$\mathrm{DPP} = Inv \sum_{\mathrm{i}=1}^{T} \frac{(Ei - Ci)}{(1+r)^i} \tag{8}$$

where, I_{nv} – initial investment in the project;

E_i – time savings i;

C_i – time savings i;

r – discount rate;

i – number of periods.

The universal financial and economic indicator of the assessment of the implementation of the event is net present value (NPV). It takes into account not only the costs throughout the life cycle but also the effect of the implementation of activities.

Discounted income (NPV) represents the amount of net savings over the entire billing period, taking into account the time value of money (9).

$$\mathrm{NPV} = \sum_{\mathrm{i}=1}^{T} \frac{CF^{+}(t)}{(1+r)^i} - \sum_{\mathrm{i}=1}^{T} \frac{CF^{-}(t)}{(1+r)^i} \tag{9}$$

where, CF^{+} – cash flow in each specific period t (t = 1…. n);

CF^{-} – cash outflow in each specific period t (t = 1….n);

T – billing period;

i – current year;

r – discount rate.

Carrying out the calculations using the formula (9), we will have the following indicators: $NPV \geq 0$ - the project is considered effective; $NPV \leq 0$ - the project is respectively inefficient [11, 12].

Consider an example of calculating the economic efficiency of energy-saving measures in the overhaul of an apartment building, taking into account the life cycle using the NPV calculation method. Initial data: capital investment I_{nv} = 5 170, 540 thousand rubles, the annual income from the economic efficiency of energy-saving measures will be 1 500 thousand rubles; project implementation period is 10 years; depreciation method - straight; the share of reinvested profits 60%; return on reinvested profits of 22%. Annual payments on the loan portion of the investment capital amount to 150 thousand rubles. This amount includes payment for risks and the depreciation of the investor's funds due to inflation. Estimated weighted average cost of capital is 26.11%.

The payback period for investments according to the NPV method was 3.83 years. The economic efficiency of the project according to the method of complete economic result is 3 124.625 thousand rubles. The calculation results are presented in Table 3.

On the basis of the work done, the advantages and disadvantages of indicators for assessing the economic efficiency of the implementation of an energy-saving measure are formed and are presented in Table 2.

After calculating the performance indicators of energy-saving measures, the sequence of implementation of measures for major repairs should be determined.

In the article [13], the authors proposed to implement the measures taken to improve the energy saving of residential houses into 2 groups:

1. Low-cost, medium-cost activities;
2. Highly costly activities.

Depending on the cost of the event, the following sequence of implementation of energy-saving technologies was proposed:

- introduction of organizational measures and accounting systems for heat and electrical resources;
- the introduction of technologies that require the lowest cost and payback period;
- implementation of activities payback periods, which are more than 4 years.

When familiarizing with Zilberov's dissertation work [14], a method for introducing energy-saving technologies was developed. The basis of the developed method is graph theory. The calculated indicators of the efficiency of energy-saving technologies and the loss factors for the deferred introduction of energy-saving technologies make it possible to construct the digraph in the form of a "tree". The "tree" is a connected acyclic graph, the vertices of which are types of energy-saving technologies, and the weight of the edge takes into account the performance indicators and the loss indicator for deferred implementation of the energy-saving measure. To find the shortest path in determining the sequence of inclusion of energy-saving technologies of the digraph, we used the Floyd-Warshall dynamic algorithm developed in 1962 by Robert Floyd and Stephen Warshall. The Floyd-Warshall algorithm has the computational complexity O(n3). This algorithm is advisable to be used with a small number of vertices of the digraph. Since to search for the optimal path, a complete search of all the existing vertices takes place. That leads to significant costs on the part of computing [15–17].

Table 3. Calculation results

Initial data:

i (discount rate) =	10%
liquidation value (RV) =	0,00rub.
investment costs	-5 170 540rub.
cash outflows in each period =	-150 000rub.
cash inflows in each period =	1 500 000rub.

Calculated indicators (briefly):

n =	10,00	*365*	*days a year*
NPV=	3 124 625,59rub.	*NPV and DPI output:*	*It is better to invest in a project if there is no alternative project with large NPV, PI, DPI and IRR*
IRR=	22,75%		
DPI=	1,60		
DROI–	60,43%		
ROI=	161,09%		
PP=	**3,83**	3 years 9 months 29 days	
DPP=	**5,07**	5 years 0 months 25 days	
PN=	**1 350 000,00rub.**		
ARR=	**26,11%**	**52,22%**	**= ARR (method 2)**

Baseline data for each period: ***Detailed calculation of indicators:***

t (Period number)	Payment Date	Cash inflows	Cash outflows	CFt (Cash Flow)	No. of periods after the payback period (for CFt)	ROIt (or ROR, ROI, or ROI)	Discount coefficient	PVt (Present Value)	DROIt (discounted return on investment)	DPIt (discounted profitability index, or simply PI - profitability index)
0	01.01.10	0	-5 170 540	-5 170 540			1,00	-5 170 540		
1	01.01.11	1 500 000	-150 000	1 350 000		-73,89%	0,91	1 227 273	-76,26%	0,24
2	01.01.12	1 500 000	-150 000	1 350 000		-47,78%	0,83	1 115 702	-54,69%	0,22
3	01.01.13	1 500 000	-150 000	1 350 000		-21,67%	0,75	1 014 275	-35,07%	0,20
4	01.01.14	1 500 000	-150 000	1 350 000	4	4,44%	0,68	922 068	-17,24%	0,18
5	01.01.15	1 500 000	-150 000	1 350 000	5	30,55%	0,62	838 244	-1,02%	0,16
6	01.01.16	1 500 000	-150 000	1 350 000	6	56,66%	0,56	762 040	13,71%	0,15
7	01.01.17	1 500 000	-150 000	1 350 000	7	82,77%	0,51	692 763	27,11%	0,13
8	01.01.18	1 500 000	-150 000	1 350 000	8	108,88 %	0,47	629 785	39,29%	0,12
9	01.01.19	1 500 000	-150 000	1 350 000	9	134,99 %	0,42	572 532	50,36%	0,11
10	01.01.20	1 500 000	-150 000	1 350 000	10	**161,09 %**	0,39	520 483	**60,43%**	0,10
	Amounts:			**8 329 460**		**161,09 %**		**3 124 625,59**	**60,43%**	**1,60**

(*Continued*)

Table 3. (*continued*)

Calculated indicators (in detail):

n =	10,00	(project duration, years)
Consider that in the year of days:	365	(affects PP and DPP when signing by days. And also when payment dates are taken into account, affects the accuracy of the project period, and as a result, PN, and as a result, ARR, as well as NPV, DPI, ROI. No effect: on IRR, because when dates are taken, Excel formula is taken to calculate IRR, in which the default is always 365 days, and if dates are not taken into account, then the specified number does not affect them at all.)
	NPV=	(Net present value (NPV), or Net present value, or Net present value)
	3 124 625,59	**(1 method)**
	3 124 625,59	2 **method** (via the NPV Excel formula, via CFt calculations) (via the manual formula, via CFt and PVt calculations, payment terms are not taken into account, that is, only for equal periods)
	3 124 625,5	*3* **method** (via Excel formula PS, only for annuity payments, i.e. for equal payments at regular intervals)
	3 124 625,59	*4* **method** (via manual formula, only for annuity payments, i.e. for equal payments at regular intervals)
DPI=	**1,60**	(discounted profitability index (or simply PI - profitability index)
ROI=	**161,09%**	
DROI=	**60,43%**	(discounted return on investment)

NPV and DPI output: It is better to invest in a project if there is no alternative project with large NPV, PI, DPI and IRR

4 Discussion

The growing interest in energy saving, associated with the lack of energy resources, the search for alternative energy sources, makes energy saving in the housing and utilities sector one of the current trends in the development of science and technology. At this point in time, the state of multi-family housing stock in the Russian Federation is characterized by a high degree of wear and tear. Most of the apartment buildings were built after 1957 according to the old standards. It follows that we are faced with the task of not only restoring the design characteristics of the common property elements in the process of major repairs, but also the need to bring them in line with modern standards for energy efficiency standards. The use of energy-saving measures allows not only to improve the quality characteristics of an apartment building, comfortable living, but also leads to a reduction in costs for maintaining and repairing the building. To do this, it is necessary to determine not only the list of activities, but also the sequence of implementation of these activities.

In this article, the authors considered several options for the implementation of energy-saving measures during the overhaul in apartment buildings. In the first variant, all the proposed activities were conditionally divided into two groups: a set of low-cost and medium-cost activities; complex of high cost activities. The criterion for determining the sequence of implementation was to determine the lowest costs and payback periods of activities.

In the second variant, not only the maximum effect of the introduction of an energy saving measure was taken into account, but also the efficiency loss index from the untimely introduction of technology. Since the introduction of one or another energy saving measures, the cumulative effect is observed. It follows that it is necessary to take into account not only the estimated savings from the introduction of energy-saving measures in the overhaul of apartment buildings, but also the losses in deferred implementation of the event. Of the two options considered, the authors consider the second option of the implementation of energy-saving technologies during the overhaul of apartment buildings to be more rational and optimal. In Fig. 3, for clarity, a generalized scheme of introducing energy-saving technologies during the overhaul of apartment buildings is presented.

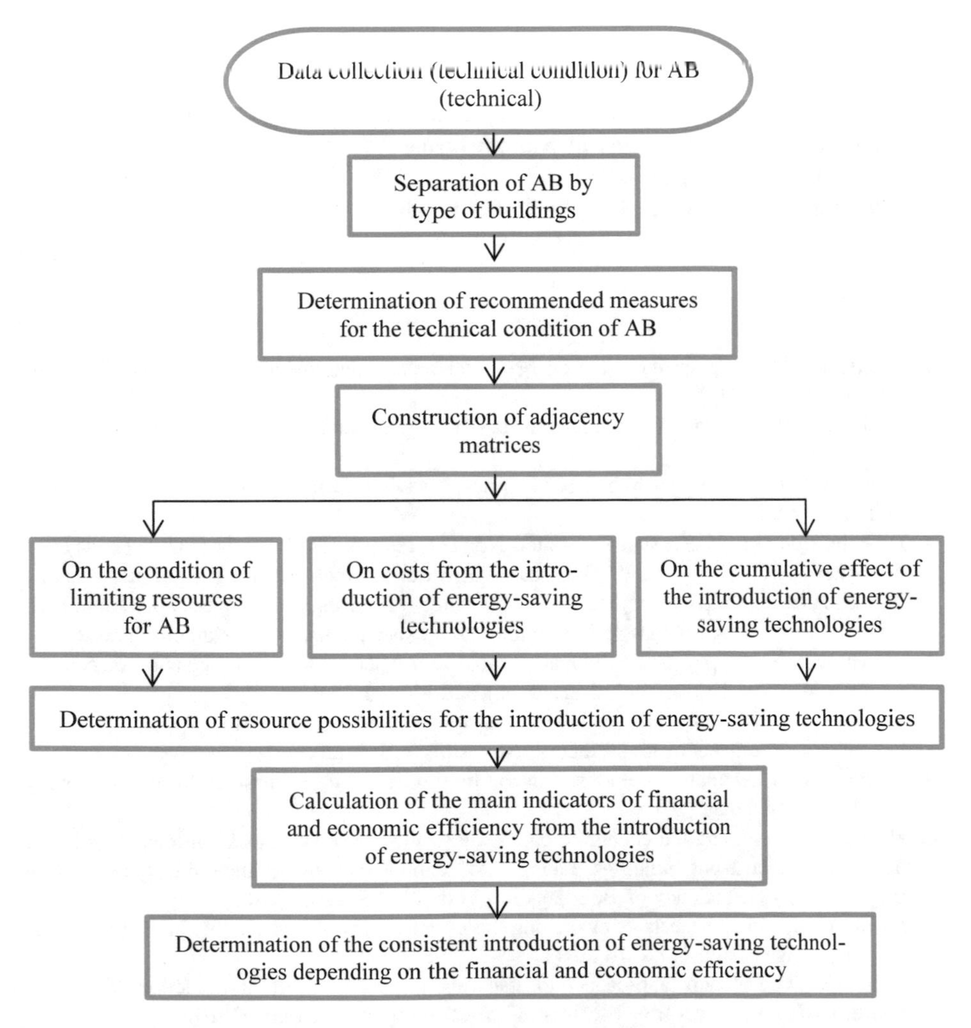

Fig. 3. A generalized scheme for the introduction of energy-saving technologies during the overhaul in apartment buildings.

5 Conclusions

The efficiency of energy saving measures depends not only on the payback period of the measures, but also directly depends on the time of putting the energy saving measure into operation. The sooner the necessary energy saving measures are introduced, the faster the result will be obtained from the introduction, as a result of the cumulative effect. It is advisable to take into account not only the resulting savings measures but also the losses in the deferred implementation of energy-saving measures, as a result of which there is a decrease in the value of the cumulative effect. It is also rational to consider the most popular algorithms for finding the shortest path when determining the sequence of switching on energy-saving technologies in a directed graph (Dijkstra's algorithm; Floyd-Warshall algorithm; Bellman-Ford algorithm; A * algorithm).

Based on the analysis of actual existing technologies using the constructed cost functional, directions for improving the efficiency of reconstruction processes and reducing the cost of capital repairs can be found and formulated. They are:

- optimization of the number of equipment used;
- reduction of energy loss;
- decrease in costs of service and further repair.

References

1. Practical guide for improving the energy efficiency of apartment buildings (MDC) in the overhaul I, 89 (2015)
2. Vasiliev, G.P.: Evaluation of the effectiveness of energy-saving measures in the overhaul of MCD (2015)
3. Odokienko, E.V., Permyachkina, K.M.: Energy saving during the overhaul of residential buildings (2017)
4. Housing Code of the Russian Federation No. 188-FL (ed. 28 November 2018) (2004)
5. Gorbaneva, E.P., Shpakova, V.: Overhaul of the housing stock with the use of energy-saving technologies. https://in-regional.ru/stoimostnoj-inzhiniring/remontnye-raboty/kapitalnyj-remont-zhilishchnogo-fonda-s-primeneniem-energosberegayushchikh-tekhnologij.html
6. Shuvalov, S.N., Bykov, S.A.: Innovative tools in the planning and optimization of costs to perform work on the overhaul of apartment buildings, 1 (2017)
7. Bertsun, V.N.: Mathematical modeling on graphs, Tomsk, 2, 86 (2013)
8. Nosov, V.I.: Elements of graph theory, Novosibirsk, 107 (2008
9. Guidelines for calculating the effects of the implementation of measures for energy saving and energy efficiency (2016)
10. Methods for assessing the economic efficiency of the use of energy-saving measures in the overhaul of apartment buildings, taking into account the cost of their life cycle. UNDP project "Energy Efficiency of Buildings in North-West Russia" (2015)
11. Kwon, G.M.: Some aspects of evaluating the effectiveness of investments in energy saving buildings. Tisby Bull. **2**, 68–76 (2012)
12. Feafanov, D.I.: Modern approaches to assessing the effects of the introduction of energy-saving technologies in various sectors of the economy **9**(1), 34–38 (2010)

13. Grakhov, V.P., Mokhnachev, S.A., Yegorova, E.G.: Efficiency of energy saving measures in housing construction, Penza 2 (2015)
14. Zilberov, R.D.: Improving the efficiency of repair and construction production through the use of energy-saving technology, Rostov-on-Don (2015)
15. Izotov, T.U.: Overview of the shortest path search algorithms in the graph, pp. 341–344 (2016)
16. Vattai Zoltan, A.: FLOYD-Warshall in scheduling open networks, Budapest (2016)
17. Mishchenko, V., Kolodyazhniy, S., Gorbaneva, E.: Energy consumption reduction at all stages of residential buildings life cycle by means of queuing systems. Energy consumption reduction at all stages of the real estate life cycle by means of the queuing systems, 05043 (2018)

Smart Management and Power Consumption Forecasting in Passenger Transport

Sergey Gusev[1], Evgeniy Makarov[2(✉)], Dmitry Vasiliev[1], and Vladimir Marosin[1]

[1] Yuri Gagarin State Technical University of Saratov77, Politehnicheskaja str., Saratov 410054, Russia

[2] Plekhanov Russian University of Economics, (Voronezh branch) 67a, Karl marks str., Voronezh 394030, Russia

ea_makarov@mail.ru

Abstract. Modern approaches to development and management of passenger transportation include aspects of efficient electricity consumption in line with "open sky" concept, and intelligent logistics. Both scientific community and empiricists agree upon the importance of including intelligent models in management of studied systems and their elements as well as in forecasting their operational characteristics. Intelligence of these systems is defined by the existence of technical and technological solutions in the sphere of artificial intellect. Communication systems and networks provoke the dynamic evolution of transport and logistics practically in all directions which allows effectively managing power flows functioning in them. Models designed to calculate electricity consumption of trams and trolleys are based on the assumptions about passenger traffic being constant, which, in our view, needs correction and clarification. Rolling-stock control systems allow getting information about characteristic quantity in the regime of real-time and consequently form an adaptive transportation strategy for electric passenger transport. Academic and research approach to tackling this problem means to design forecast models of passenger traffic flows using artificial neural networks and based on them adaptive selection and adjustment of passenger capacity according to the respective dynamics. In the process of developing route descriptor, this gives a possibility of justified changes in interstation distance in case competition terms of providing interstation intervals of rolling-stock passenger service quality are observed.

Keywords: Smart management · Power consumption · Passenger transport

1 Introduction

The use of public transport as a means of relocation from point of departure to the point of destination is justified only when transportation speed exceeds that of a pedestrian reducing the time spend on the relocation [1]. Average overall time consumption (t_o) under combined method of transportation comprises the pedestrian time from the

V. Murgul and M. Pasetti (Eds.): EMMFT-2018, AISC 983, pp. 92–106, 2019.
https://doi.org/10.1007/978-3-030-19868-8_9

departure point to the public transport stop and from the public transport stop to the destination point ($2t_p$), time of waiting for public transport (t_o) and transportation time (t_t):

$$t_o = 2t_p + t_o + t_t \quad (1)$$

On average, transportation time accounts for 50% of overall time consumption and pedestrian part accounts for up to 30%. Both depend on a number of factors: city planning, public transport routes, population, town build and construction type. At the same time each of these parts is described by well-known empiric dependencies and their analysis shows that an increase in trackside length (l_p) between stops contributes to an increase of the pedestrian part in the ratio, while transport and waiting parts decrease.

$$t_{ped} = \left(k_n k_v / v_p\right)\left[1/(3\delta) + l_p/4\right] \quad (2)$$

where k_n = 1, 2–coefficient of non-linearity of approach;

$k_v = 1 + v_p/v_c$ – coefficient of halting point choice;
v_p – pedestrian speed;
v_s – examined transport type speed;
δ – density of transport network;
l_p – length of track span

Time required for the transportation part includes expenses for time of transportation (t_d) and load and unload time:

$$t_{tr} = t_d + t_{pv} = l_s/v_s + l_{av}\left(t_{od} + t_{pv} + t_{zd}\right)/l_p \quad (3)$$

where $l_s \approx 1,3 + 0,3\sqrt{F_c}$– average trip length;

Fc – city populated area;
Tod + tzd = 2 … 5 s (experimental data)–doors opening and closing time;
$t_{pv} = k_{nd} \rho_{o\Pi} \Omega_{max} t_{pass} n_d$–passenger load and unload time;
Knd ≈ 1, 2–coefficient of door load and unload imbalance;
ρpo – coefficient of passenger interchange;
Ωmax – maximum carrying capacity of rolling stock;
Tpass = 1–average passenger load (unload) time, s;
nd – number of doors of rolling stock

Waiting time consumption

$$t_o = l_m/(w_d v_e) \quad (4)$$

where l_m – route length;

w_d – number of rolling stock units on the route;
v_e – operating speed of rolling stock on the route

To determine the exact influence of all the parts on t_o with the change of l_p and correlation to $t_o = f(l_p)$, we have to study the problem in concise circumstances. It is stated in the author's [1], as well as other works [2–7], that if the distance between stops changes in an interval of 100 and 800 m, the following parameters stay the same: passenger traffic load, P_{pass}, rolling stock capacity Ω_{max}, average trip length l_{cp}, density of public transport network $\delta = 1, 5$ km/km^2 ($\delta = 1, 5 \ldots 2, 5$ km/km^2 for metropolitan cities), town populated area $F_c = 80$ km^2, and pedestrian speed $v_p = 4$ km/h.

City public transport networks in order to optimize labor and rolling stock use develop train crew routes so that the time of half-route (time elapsed to get from route start point to route end point) would tend to be one hour, which corresponds a 20 km route length.

Active development of innovative electricity development is connected with Smart Grid – a concept of "intelligent (smart) network (power system)". Open source data states that development and implementation of that concept includes:

- systemic changes of electrical and electric power systems that spans all its major elements: generation, transmission and distribution (including utilities sector), sale of electricity and control;
- Internet-like systems that is formed to support power, data, economic and financial transactions among all the parties of the energy market and any other interested parties;
- development of existing and of new functions of power systems and their elements that obtain the greatest key values of new electric utilities industry achieved as the result of the shared vision by the interested market parties and the way of its development;
- formation of a new technological basis to empower significant improvement of new power system functions and development of new functional features of the power grid;
- inter-sensory perception of various issues: scientific, judicial, technological, technical, organizational managerial, informational.
- innovational type of development of electrical power market and economy in general.

Experts of Gartner Group state that automation makes a significant influence on the effectiveness of the studied systems.

The considered above concept is Customer oriented, in which customer plays the key role. It is the basis for implementation of market type behavior model of transport-logistic systems. The same goes for energy complex. Energy systems management is developed in line with innovational solutions in the field of information-communication and computer technologies [8].

2 Materials and Methods

One peculiarity of electrical transport hauling-stock compared to other kinds is the existence according to the normative acts of at least two independent braking systems one of which is necessarily electric. Therefore, the impulse converter used to start tractive motor must provide the electric braking regime and on the first hand recuperative which allows decreasing energy consumption and at the same time to increase the competitiveness of electrical transport compared to other kinds.

Maximum recuperation efficiency can be achieved over relatively short distances as in public transport. To calculate energy saved through recuperation with function to the length of the track, a method of calculation of movement curves is used with the following conditions:

- the same rolling stock is used on tracks of various lengths;
- standard track is set as base (if the length changes the incline stays the same as base);
- maximum unit speed does not exceed the speed limit;
- voltage at the trolley of rolling stock is constant and equals to the nominal net voltage [2].

Analytical equation of energy balance used for the movement without the drive loses is (5).

$$A_i = A_k + A_w = (1+\gamma)G_{ps}v_B^2/2g + \int_0^{l_\Pi} G_{ps}wdl \tag{5}$$

where $A_i = \int_0^{t_\Pi} uidt$–energy used by the rolling stock in time t_n;

Ak – kinetic energy of rolling stock, accumulated by the transit to rundown;
v_v – speed of initiation of rundown.

The amount of current i, drawn from the network depends on electromechanical characteristics of the engine and is quite difficult to portray in analytical form. Current methods utilize swapped under integral correlations with the equivalent values of various track lengths (Fig. 1).

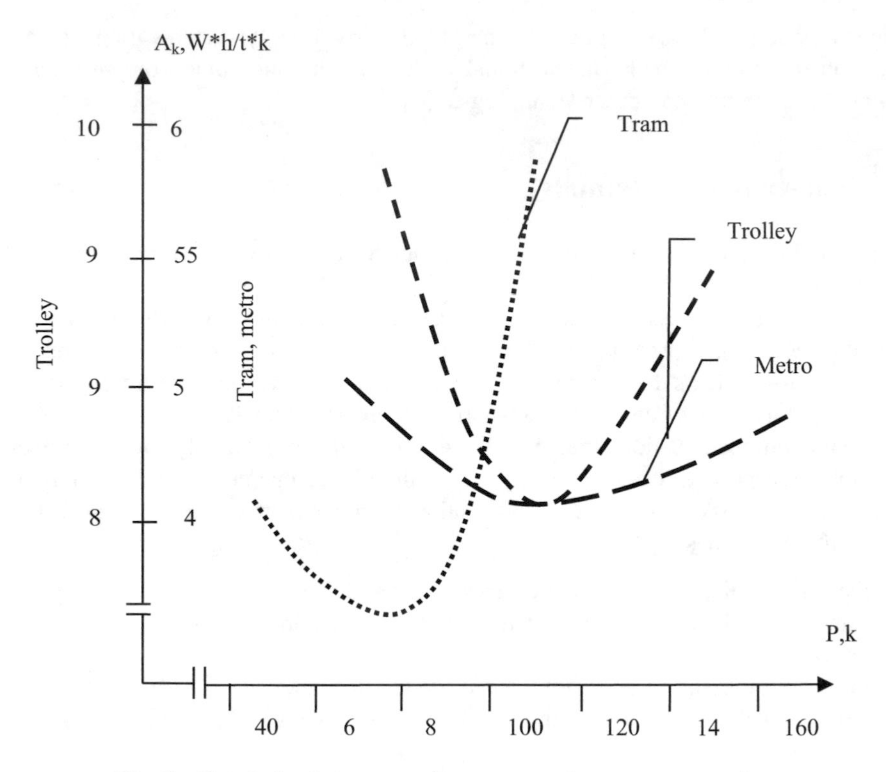

Fig. 1. Correlation between engine power and power consumption.

Analytical Eq. (5) depicts a classic scheme of transport movement in "run-rundown-braking" mode, which correlates to movement over a test railway bed. In real conditions movement is interrupted by additional runs and brakes which increases energy consumption. In that case analytical equation for calculation of consumed energy without the drive loses transforms to:

$$A_i = (1+\gamma)G_{ps}v_{Bk}^2/2g + G_{ps}\sum\nolimits_1^k \left(\int_{l_{k1}}^{l_{k2}} W_{\Pi}dl + \int_{l_{k1}}^{l_{(r+1)_1}} W_B dl + \int_{l_{(k+1)_1}}^{l_{TKP}} W_B dl \right) \quad (6)$$

where l_{k1} and l_{k2} – position of the rolling stock in the start and the end of the run k;

W_P – specific train resistance at run;
W_B – specific train resistance at rundown

Recuperation of the energy by the rolling stock in minimal and emergency braking modes is possible if:

- there is enough kinetic energy (speed decrease below a certain threshold leads to poor electrical braking efficiency thus requiring mechanical braking);
- presence of electrical energy consumer;
- circuit based system for system switch from run mode to recuperative braking.

3 Results

Classic mode of movement allows for a 20% energy recuperation of the stored kinetic energy when accelerating to 60 km/h (17% when accounted for coefficient of efficiency of impulse regulator and engine).

Besides, the decrease in length of the track leads to lower energy consumption. However:

- transportation speed gets significantly lower consequently lowering the attractiveness of the trolleybus to the passenger (in terms of overall trip time consumption);
- energy consumption per unit, which usually serves as a marker of cost-effectiveness, increases steeply.

All of the current produces by the braking rolling stock will be dropped in the supply main because the braking current I_t approximates to I_t = 300 … 400. That makes recuperation process unpractical.

A solution to this lies in the storage of recuperated energy to sub-interstation zone or on the transport itself. The latter is preferable because it minimizes transfer loses to the end user although it is more expensive as it requires storage capacities on each rolling stock unit. Analysis of factors that influence the effectiveness of rolling stock operation shows that passenger traffic volume should be accounted for in the first place. Passenger flow statistics is vital for the design of the rolling stock route and allows for quality and quantity analysis of rolling stock operation.

Electrical transport passenger flow obtained with navigational technologies is illustrated in Table 1.

Table 1. Passenger traffic for tram route №3 in Saratov.

№	Month	Passenger traffic		
		2016	2017	2018
1	January	9258.4	9362.4	9423.4
2	February	9485.2	10024.9	9362.6
3	March	10223.6	10452.9	10313.8
4	April	10093.9	10183.5	10105.2
5	May	10119.3	10253.8	10420.5
6	June	9986.5	10350.4	10250
7	July	10052.1	9986.5	9599.1
8	August	9389.5	9583.6	9495.1
9	September	9995.8	10169.5	10050.5
10	October	9862.5	9863.9	9769.6

These results will be used for testing of models of management and forecasting of electrical city transport.

An artificial neural network (ANN) is proposed for forecasting of transport operational characteristics determined by its distinct features [9–11]:

Ability to learn;
Reliability in operating with a lack of sufficient inputs;
Resilience to distortions;
Fast response of a trained network to inputs;
Ability to model on PC;
Lack of object model;

In spite of that and the fact that variables will appear in various forms, it is important to use those models that will better account for ongoing changes. We propose to use the systems that we will be able to teach ourselves factoring the conditions that will arise during public transport service under the influences of parameters of environment. Output characteristics vary according to variations in initial parameters.

In fact, we are talking about models that are best adapted to arisen conditions and uncertainties.

Significant attention is paid to the sizing of the neural network while defining its configuration. If the sizing is big, the learning process will take a lot of time, and if the sizing is small, the network learns poorly and produces results with unacceptable precision.

As a basis for determination of sizing of input and output layers of the neural network we take following conditions:

Number of elements in input layer should correspond to the length oa prehistory k, that is used for load forecast;

Output layer should consist of one element that defines the limit value of P_{np}.

The problem of number of intermediate (hidden) layers and elements was solved experimentaly, while the number of intermediate layers and elements was chosen so that various number of realisations reach the minimu criterias:

Maximal relative forecast error (δ_{max});

Learning (relearning) time of neural network (t_{o6}).

Forecast accuracy for automobile transport significantly depends on the teaching sample.

Neural network is tuned by learning with a teacher, subsequently, retrospective sample should consist of a sequence of educational image pairs $(P^j(t), D^j(t))$, where $P^j(t) = \left(P_1^j, \ldots, P_k^j\right)$, - input vector of signals, and $D^j(t)$ - is a scalar that defines desired forecast value of signal ($P_{\Pi p}$) for $P^j(t)$, j = 1, 2,…,R (R – length of teaching sample).

When plotting input vector $P^j(t)$ of retrospective sample, different realisations (passenger flow volume, for example) are distributed over time so

$$P_1^j = P^q(t), P_2^j = P^q(t + \Delta t), P_3^j = P^q(t + 2\Delta t), \ldots, P_k^j = P^q(t + (k - 1)\Delta t), \tag{7}$$

where P^q – q volume value of transportation work (q = 1, …, 7).

Desired value $D^j(t)$ at neural network output for j input vector of signals $P^j(t)$ is defined by the equation

$$D^j(t) = P^q(t + (k-1)\Delta t + t^*), \quad (8)$$

where t^* - forecast time consumption.

Thus, the retrospective sample represents a set of relations

$$\forall q \in \left(P^j(t) \rightarrow D^j(t)\right), j = 1, \ldots, R. \quad (9)$$

We test our ANN on the data of passenger flow of tram route #3 of Saratov. We use the statistical data of rolling stock operation.

Due to the changes in conditions influenced by seasonality of transportation, forecasting models require periodic adaptation which is performed over a certain period of time and which is accompanied with ANN relearning on new data.

Learning process of ANN is actually fitting the model to the data of the retrospective sample [34, 90]. During the learning process, weighted coefficients W are set up in a way so network outputs would be as close as possible to the set images of the teaching pair. During ANN learning:

a certain training vector of input signals is sent to network input;

weighted coefficients are tuned until the network learns to display a pair of training inputs into the set of desired output vectors.

Quality of ANN results are qualified by the equation

$$E = \sum\nolimits_{i=1}^{R} E^j;\ E^j = (D^j - P_a^j)^2, \quad (10)$$

where E – learning error, and D^j and P_a^j – j value of desired and actual outputs of the network.

Learning process is finished if error E for all of the combinations of input signals is less than some set value $\varepsilon \geq 0$, or when reaching a preset number of iterations.

Algorithm of backward error spread based on method of gradient decent was used for ANN learning [11].

Tuning of weighted coefficients utilizing that method is based on equation

$$\Delta w_{ij}^{(n)} = -\eta \frac{\partial E}{\partial w_{ij}}, \quad (11)$$

where w_{ij} – weighted coefficient of synaptic connection which connects i neuron of layer n-1 with j neuron of layer n; η – coefficient of learning speed, $0 < \eta < 1$.

Algorithm of ANN learning based on method of backward error spread is characterized by the following:

Weighted coefficinets of the network are initiated by small random values.

j training set from the teaching pair is sent to network inputs which spreads to output.

An error of output layer of the network is calculated

$$\delta_l^{(N)} = \left(y_l^{(N)} - d_l\right)\frac{dy_l}{ds_l}, \quad (12)$$

where l – element number of neural network N output layer, y_l – real output neuron state l of layer N (actual value of load P_a, derived from network output during learning), d_l – desired output neuron state l for layer N (value D^j j pair of teaching sample), s_l – weighted sum of input signals of neuron l of output layer.

Errors of previous layers are calculated by backward spreading the error

$$\delta_j^{(n)} = \left[\sum\nolimits_k \delta_k^{(n+1)} w_{jk}^{(n+1)}\right] \cdot \frac{dy_j}{ds_j}, \text{ n} = (\text{N} - 1), \ldots, 1. \quad (13)$$

Here k summarizing happens among neurons of layer n + 1.

Weighted coefficients of neural network are changed

$$\Delta w_{ij}^{(n)} = -\eta \delta_j^{(n)} y_i^{(n-1)}. \quad (14)$$

Repeat steps 2–5 for the next training set until the error of the output layer is lower than the set limit ε or when reaching a preset number of iterations.

Increase in algorithm efficiency was achieved by:

Changing the order of examinations in teaching samples of different iterations (to decrease the probability algorithm getting to a local minimum and for excluding of the relearning effect);

By varying the learning speed (learning speed coefficient was: in the beginning of iteration 0, 1, in the end −0, 01);

By defining the number of iterations of the algorithm required for desired accuracy of the learning.

Short learning time allowed quickly reteaching the network after corrections.

An effect of overload (when elements of the network become sensitive only to input values lying in a limited zone) may appear when unnormalized data is sent to ANN inputs. That leads to ANN incorrect performance.

Normalizing of the data with the following equation brings them to an interval of [0,1]

$$P_i^{\text{H}} = (P_i - P_{\min})/(P_{max} - P_{\min}) \quad (15)$$

where P_{min} and P_{max} – minimal and maximal values of transport work in the studied sample; P_i and P_i^H – unnormalized and normalized values of transport work sent on the I input of the network.

A neural network that was trained on retrospective samples describing a forecasted process may not only distinguish input data from the teaching sample but also interpret trends in dynamics of the process, forming its own interpretations, to deal with new signal that were not present in the teaching sample using knowledge gathered through learning. Forecast calculation steps and results are illustrated further in Figs. (2, 3, 4, 5, 6 and 7).

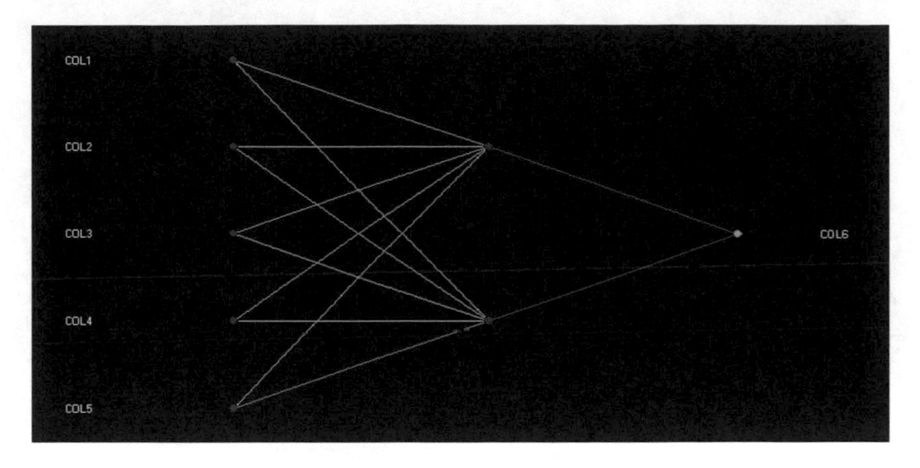

Fig. 2. Neural network graph.

COL1	COL2	COL3	COL4	COL5	COL6	COL6_OUT	COL6_ERR
9258,4	9485,2	10223,6	10093,9	10119,3	9986,5	9973,15702762428	0,000192818538630157
9485,2	10223,6	10093,9	10119,3	9986,5	10052,1	10084,8749972684	0,00116339968417673
10223,6	10093,9	10119,3	9986,5	10052,1	9389,5	9582,12489975614	0,0401854156441448
10093,9	10119,3	9986,5	10052,1	9389,5	9995,8	9562,14896925329	0,203668741223062
10119,3	9986,5	10052,1	9389,5	9995,8	9862,5	9875,9026320234	0,000194546669841956
9362,4	10024,9	10452,9	10183,5	10253,8	10350,4	10284,8912930737	0,00464773831021392
10024,9	10452,9	10183,5	10253,8	10350,4	9986,5	9972,35775751128	0,000216610832938133
10452,9	10183,5	10253,8	10350,4	9986,5	9583,6	9564,55116477277	0,000392989906260095
10183,5	10253,8	10350,4	9986,5	9583,6	10169,5	10097,3672959833	0,00563518318979485
10253,8	10350,4	9986,5	9583,6	10169,5	9863,9	9800,99603156291	0,00428548227054105

Fig. 3. Training results for test sample.

COL1	COL2	COL3	COL4	COL5	COL6
9423,4	9362,6	10313,8	10105,2	10420,5	0
9362,6	10313,8	10105,2	10420,5	10250	0
10313,8	10105,2	10420,5	10250	9599,1	0
10105,2	10420,5	10250	9599,1	9495,1	0
10420,5	10250	9599,1	9495,1	10050,5	0
10250	9599,1	9495,1	10050,5	9769,6	0

Fig. 4. Neural network data for 2018. Test data load (5 inputs, 6th – unknown forecasted value).

COL1	COL2	COL3	COL4	COL5	COL6	COL6_OUT
9423,4	9362,6	10313,8	10105,2	10420,5	0	9952,50996901681
9362,6	10313,8	10105,2	10420,5	10250	0	10076,3232260848
10313,8	10105,2	10420,5	10250	9599,1	0	10051,1486974216
10105,2	10420,5	10250	9599,1	9495,1	0	10138,7924779829
10420,5	10250	9599,1	9495,1	10050,5	0	9530,6073207341
10250	9599,1	9495,1	10050,5	9769,6	0	9455,92447021201

Fig. 5. Timescale of initial and forecasted data for 2018.

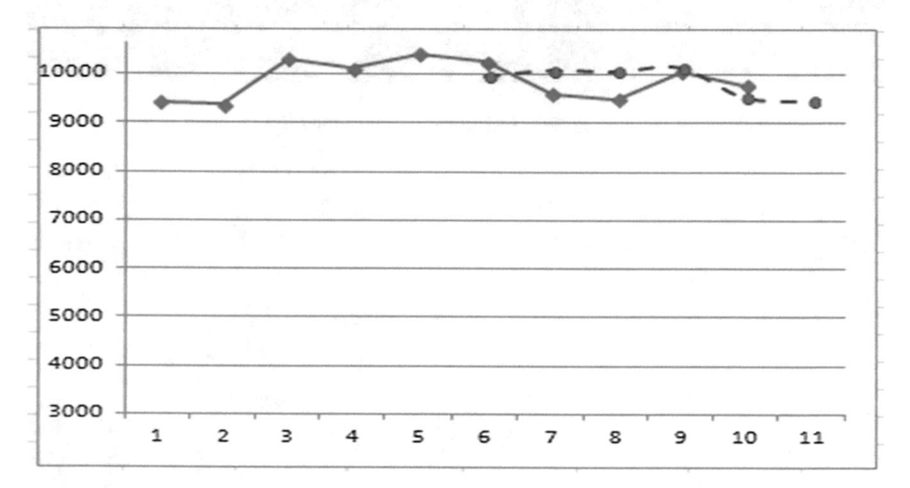

Fig. 6. Passenger flow forecast based on INN for tram route №3. Saratov. November 2018.

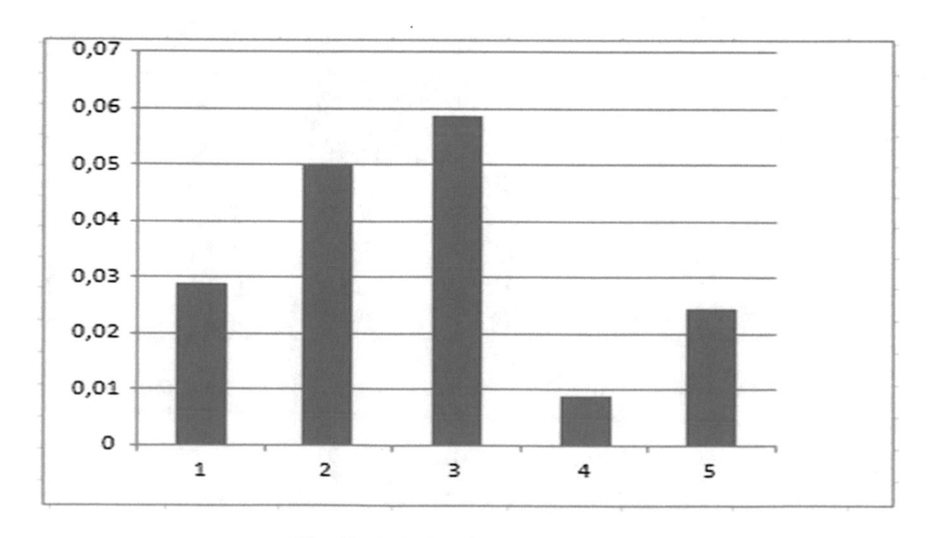

Fig. 7. Relative forecasting errors.

Table 2. Forecasted data.

№	Month of 2018	Passenger flow, pass.	Forecasted value, pass.
1	January	9423.4	
2	February	9362.6	
3	March	10313.8	
4	April	10105.2	
5	May	10420.5	
6	June	10250	9952.5
7	July	9599.1	10078.3
8	August	9495.1	10051.1
9	September	10050.5	10138.8
10	October	9769.6	9530.6
11	January		9455.9

Concluding the results of study of algorithms, we should note that they can be used not only during the development stages of city public transport energy management system but during operational stages of these systems as well. Accuracy of a forecast calculated with such a model allows adequately judging formed passenger flow structure and adaptively react to its changes. This can be achieved by inclusion of algorithms into special mathematical management support (Table 2).

4 Discussions

Modern scientific and practical works written on the topic contemplate at the usage of neural networks for city management and in fact forming models of a smart city [8, 11]. We suppose that one of the options may be a neural scheme of forming and transforming of information flows in management of transport and logistical elements of the city (Fig. 8).

It is quite a common practice for a number of cities to successfully use navigational systems for coordination of rolling stock operation as well as a number of other main elements of infrastructure in transport-logistical systems. That, in turn, allows traffic dispatchers to control transport units and coordinate them on routes [12]. That circumstance became the basis for development of intelligent transport and logistic systems (ITS, ILS) which provide solutions to key objectives in city road network (CRN) traffic relief by balancing road loads and increasing overall quality of transport and logistic services.

The use of such terminology in relation to transport supposes an accent on automatization of control systems and their unification under informational and communicational technology. In accordance with authors works [13, 14], intelligent transport system (ITS) is a system that integrates modern informational, communicational and telematic technologies, control technologies. ITS is used for automated search and implementation of the most efficient scenarios of regional (city, road) transport system management, scenarios for standalone transport unit or a group of

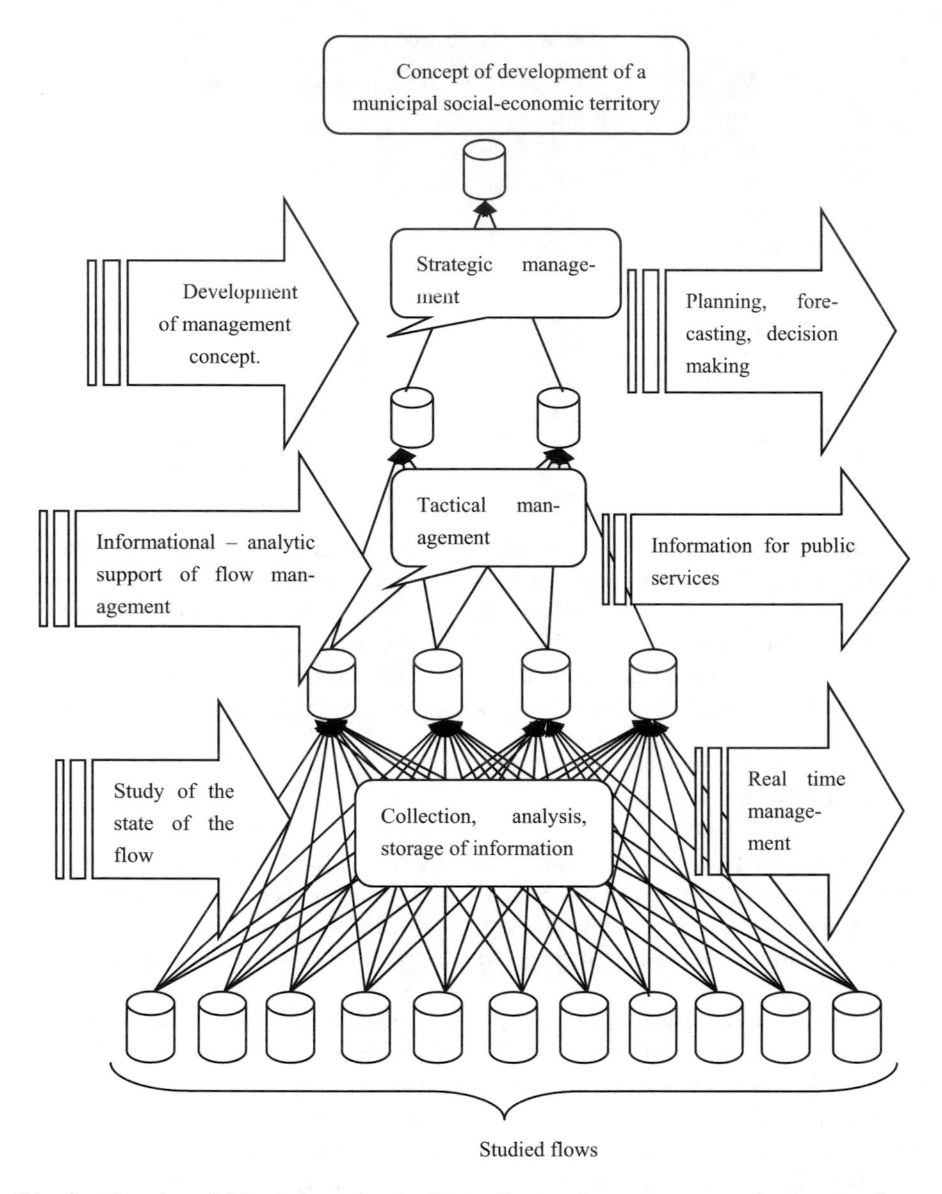

Fig. 8. Neural model for information intellectualization for management of smart city flows.

units with a goal of allowing set mobility of the population, maximization of operational characteristics of road network, safety and transportation efficiency, comfort of the drivers and users of public transport.

Modern technologies that are based on adaptation of transport-logistical systems of the city possibly with the use of equipment of informational and diagnostics enablement of supply, production, distribution and consumption. Significant increase in volume of incoming information and update of the technologies of its gathering is the defining factor in using the models and methods, that are able to process and analyze forming arrays of statistics as well as enable complex understanding of the studied system and factors that influence it.

5 Conclusion

The main goal of designed and implemented systems of energy flow management is to decrease the influence of random factors. It is hard to underestimate the role of informational provision in that question. Energy-informational structure is the basis for the Smart Grid system including technical integration of electrical and informational networks.

Functions of transfer of "smart energy" is provided by existence of on-line informing components and use of energy storages (flow – dynamics, and storage – static of transport logistics systems, a key terminology in logistics) and consumers get the ability to take part in distribution of energy flows. Active introduction of electrical buses and electrical towing trucks is an instrument in redistribution of resources consumed by transport.

Considering the seasonality and dynamics of freight and passenger flows, we obtain a possibility of forecasting sufficient energy delivery systems. We have to notice the high attention paid by Russian Federation as well as other countries to that questions and that in turn underlines the importance of finding a solution to the problem of optimal charging station distribution.

The performed steps in optimization of structure of energy consumption are justified by the level of decrease in risks of organization of transport-logistics services and achievement of set technical and economical parameters of rolling stock operation.

References

1. Biryukov, V.V.: EHnergeticheskie aspekty funkcionirovaniya transportnyh system, monografiya. EHlektron. tekstovye dannye. Novosibirskij gosudarstvennyj tekhnicheskij universitet, Novosibirsk 264, (2014). 978-5-7782-2538-1. http://www.iprbookshop.ru/45210.html
2. Vel'mozhin, A.V., Gudkov, V.A., Kulikov, A.V., Serikov, A.A.: EHffektivnost' gorodskogo passazhirskogo obshchestvennogo transporta. "Staraya bashnya". Volgograd (2002)
3. Gerami, V.D.: Metodologiya formirovaniya sistemy gorodskogo passazhirskogo transporta. "Format" MADI (TU), Moscow (2001)
4. Gudkov, V.A., Mirotin, L.B., Vel'mozhin, A.V., Shiryaev, S.A.: Passazhirskie avtomobil'nye perevozki, Uchebnik. Goryachaya liniya - Telekom, Moscow (2006)
5. Mirotin, L.B., Tashbaev, Y.E., Gerami, V.D.: Logistika: obshchestvennyj passazhirskij transport, ucheb. Mosk. avtomob.-dor. in-t (Gos. tekhn. un-t). EHkzamen, Moscow (2003)
6. Spirin, I.V.: Organizaciya i upravlenie passazhirskimi avtomobil'nymi perevozkami, Uchebnik. 4-e izd. IC "Akademiya", Moscow (2011)
7. Shabanov, A.V.: Regional'nye logisticheskie sistemy obshchestvennogo transporta: metodologiya formirovaniya i mekhanizm upravleniya. SKNC VSH, Rostov-n/D (2001)
8. Gusev, S.A.: Povyshenie kachestva i bezopasnosti uslug intellektualizaciej logistiki perevozok goroda, monografiya. Sarat.gos.tekhn.un-t, Saratov (2013)

9. Lyugger, D.: Iskusstvennyj intellekt. Mir, Moscow (2003)
10. Leonenkov, A.V.: Nechetkoe modelirovanie v srede MatLab i fuzzy TESN. BHV - Peterburg, SPb (2005)
11. Rutkovskaya, D.: Nejronnye seti, geneticheskie algoritmy i nechetkie sistemy. Goryachaya liniya – Telekom, Moscow (2006)
12. Gorev, A.E.: Informacionnye tekhnologii na transporte. EHlektronnaya identifikaciya avtotransportnyh sredstv i transportnogo oborudovaniya: ucheb. Posobie. IC «SPbGASU», SPb (2006)
13. Solodkij, A., Gorev, A.: System approach to elimination of traffic jams in large cities in Russia. World Appl. Sci. J. **23**(8), 1112–1117 (2013). http://idosi.org/wasj/wasj23%288%2913/19.pdf
14. Przhibyl, P., Svitek, M.: Telematika na transporte. In: Sil'yanova, V.V. (MADI (GTU)). Prezentaciya "Intellektual'naya transportnaya sistema" (2013)

Energy Saving in Construction Processes as a Developer Competitiveness Enhancement Factor

Valeriy Mischschenko, Valeriy Barinov, Yekaterina Vasilchikova, and Igor Potekhin(✉)

Voronezh State Technical University, Moscow Avenue, 14, Voronezh 394026, Russia
ipotehin@vgasu.vrn.ru

Abstract. Topical issues in the Russian construction industry include energy saving, environmental pollution with construction waste, industrial safety of construction workers, reliability of the newly-erected construction system. Whereas the quality of finishing, construction machinery fleet and the applied technologies have currently reached international level, the issues of own development and introduction of energy saving and resource saving technological practices and corresponding equipment into production remain unsolved. The problem of constantly increasing energy consumption and volume of construction waste is steadily gaining topicality. A characteristic feature of the said issues is the covert nature of economic benefit from saving on environmental payments, as well as enhancement of competitiveness due to the creation of goodwill. It is necessary to bring the said hidden potential to light in order to provide an incentive for the developers to solve these issues, which will also enhance their competitiveness by means of the same covert factors.

Keywords: Enhancement factor · Energy saving · Resource saving · Environmental pollution with construction waste

1 Introduction

The present article can be related to the Management of Energy Distribution and Use topic, Energy Management Workshop, from the perspective of use of energy on a construction site. Besides, the present article falls into the Modelling and Simulation Topic, Building Physics Workshop, since both established and newly developed simulation modeling tools were widely used in the research process.

The purpose of the article is to make a calculation tool for resource saving in construction technological processes, then to apply the tool for the research of on-site processes and design of new processes, which will lead to a reduction of environmental footprint on the city. The resource saving evaluation tool for a construction site will enable a developer to analyze energy, time and material losses and to develop an improved technological process and supporting technological tools for more efficient use of available technical resources. A reduced environmental footprint will manifest

V. Murgul and M. Pasetti (Eds.): EMMFT-2018, AISC 983, pp. 107–115, 2019.
https://doi.org/10.1007/978-3-030-19868-8_10

itself for a developed as lower pollution charges, decreased volumes of extra non-productive operations (reclamation, fewer safety measures, fencing). Since market is not devoid of unpredictability, and at times better-quality accommodation sold at a lower price loses, the solution of the issue will give a developer a significant competitive edge in case the environmental factor becomes important for buyers.

2 Materials and Methods

A study of domestic practice has shown that the issue of resource saving is of concern for construction management and developers as it directly affects the investment amount and the volume and complexity of work. On a construction site, resource saving in technological process consists in the absence of necessity to use additional special-purpose machinery units and construction materials and the possibility to reduce the intensity of labor for the workers.

View of resource saving in construction. At the current stage of development of construction industry and science, development of the following tools to manage the resource saving in technological processes is possible: (1) a tool for calculation of resource saving in technological processes; (2) a tool for measure of the environmental footprint of technological processes. The said tools may be demonstrated by standard line graphs.

Technical improvement of particular technological processes is presented by leading scientists, specialists in separate units:

1. calculation based on existing formulae;
2. calculation of conceptual design
3. comparison of the received technical systems.

Calculations based on available known data – Calculations aimed at obtaining new theoretical data are presented by theoretical models.

Data which may be obtained only by way of trials. Impact on developer's goodwill realize through the complex development including infrastructure (similar to DSK) and reduction of injuries on site. A possible solution is the development of a set of additional tools and equipment to save resources from organizational failures. The researches that need to be done to improve technological processes:

- performed work – technological process – energy input;
- calculation of organizational and technical measures;
- multitude of construction system elements subject to own energy laws;
- only conceptual representation;
- calculations of a technological process from the perspective of energy and resource saving;
- calculations of the mechanical part of conceptual design;
- calculations for lower fuel & lubricants and repair expenses;
- calculations for less downtime and fewer breakdowns;
- calculations for more reliability;
- calculations for fewer technical risks.

Table 1. Possible use of well-known technical solutions for increasing resource saving at construction site.

Technical solution	Mean of solution
Step-by-step building dismantlement	Controlled process with reduced dust formation and lower water and area use
On-site material sorting	Reduction of fuel consumption for transporting, loading
Material crushing and sifting	Secondary use for backfill or sale as suitable material
On-site reuse of certain materials	Elimination of material discarding
Use of a 3D printer in construction production organization	Industrial construction: consumables and high-wear equipment parts, spare parts for special-purpose machinery on site, maquettes, construction tools on site, restoration works Individual construction: connections, fasteners

Simulation modeling of resource saving in construction technological processes (Table 1):

1. use of standard mathematical simulation models and formulas for analysis and comparison of technological processes in energy and resource efficiency;
2. development of conceptual design interface for technological processes in construction;
3. development of environmental footprint evaluation tool for measuring of the developer's impact on the entire city.

Modeling of resource saving in construction gives follow effects of resource saving at all construction phases:

1. elimination of material spoilage;
2. reduction of the number of special-purpose machinery units by means of use of up-to-date machinery and interchangeable attachments;
3. use of specialized processes, materials, tools.

Effect in construction:

1. an insignificant increase of overhead costs, but reduction of direct costs;
   - the ratio is specified and scalable;
2. reduction of fuel and maintenance costs;
   - reduction of transporting costs;
   - reduction of downtime;
   - the ratio is specified and scalable;
3. change of the structure of the costs of materials in the direction of reduction of non-productive labor losses (less downtime, waiting time, duration of extra processes) [1–4].

Table 2. Ways of resource saving.

Calculation	Evolutionary way of innovations	Revolutionary way of innovations
Expected result 1: - reduction of uncontrolled material flow - reduction of uncontrolled expenses and special-purpose machinery movement	Improvement of organization of the technological process using the existing principles	Use of a new principle and new materials in the technological process
Expected result 2: -decreased scale of the quantity of materials -reduction of non-productive time waste -reduction of transporting functions		
Expected result 3: Resource-saving model and its role in developer's competitiveness		

The developer's work with the prospect is currently focused on advertising, provision of a maximum living area at a minimum price. Despite the fact that the developers in the Russian Federation have allocated their resources to public amenities, environmental issues do not receive due attention. And the covert nature of benefit from the solution of environmental issues and the introduction of resource saving practices prevents the developers from developing the said sphere. It is necessary to bring the said hidden potential to light in order to provide an incentive for the developers to solve these issues, which will also serve to enhance their competitiveness by means of the same covert factors. (1) Optimum "quality-volume-purchasing power" choice (Table 2).

(2) Long-term market position (timeframe calculation)

Real estate market of Voronezh oblast. Resource saving due to optimization and the use of the potential of competence and specialization.

The developer's technological tasks in the area of resource saving are as follows:

1. to be ahead of the counterparts and to stand out
2. to control the material flow on the construction site. To exclude uncontrolled expenses
3. don't allow confusion in the construction machines work
4. Innovations by evolution path
5. using an available materials and technologies, to improve organization-technical support of processes (to save, to arrange work mode, to increase the security of workman)
6. in the available process to implement more perfect materials and machines
7. Innovations by revolution path
8. to develop the new material
9. to use the new unconventional machines
10. technological process is totally new

Table 3. Score assessment of characteristics.

Score	Distance to public transport station	Material of wall	Floor	Quality of surface finishing	Condition	Quality of surrounding
1, 5	less 150	Monolithic concrete/brick	Middle	High-quality	New repair	Good
1	150–250	Monolithic concrete/precast concrete	Penultimate	Improved	Good	Acceptable
0, 5	250–350	Brick	First	Simple	Acceptable, need cosmetic repair	Not acceptable
0	more 350	Precast concrete	Last	Draft for the finish	Acceptable, need repair	–
−0,5	–	–	–	draft	Not acceptable	–

The next stage of mass evaluation is making table with accommodation in sequence of their level score (Table 3).

Table 4. Description of main characteristics of accommodation-analogues in score.

№ range	Accomodation name	Address	Material of wall	Distance to public transport station	Floor	Quality of surface finishing	Condition	Quality of surrounding
1	OA1		0.5	1.5	1.5	0.5	1	1
2	OA2		0.5	1.5	0	0.5	1	1
3	OA3		0.5	1.5	0	0.5	1	1
4	OA4		0.5	1	1.5	0.5	1	1.5
5	OA5		0.5	1.5	1	0.5	1	1

Further, it could be define a middle score of accommodations and translate it, for example, to price. Thus, to show ration between criteria and price. And to understand – does price overestimated or no. Use in resource saving situations in construction industry (Table 4).

Effect of use: More quality process, Low cost of process, Economy of monies, Decreasing the time of passing to exploitation. Resource saving in different technological processes, Resource saving in all technological processes, Resource saving in a given technological process [5–9].

Hidden character of benefit from environment problem solutions and resource saving doesn't motivate developers to the improvement in present scope. It is necessary to disclose present hidden potential for a motivation of developers to make solution of this area questions, and for competitiveness rise on account of these hidden factors:

1. calculation of contaminations from current and resource saving approaches of management in construction technologies;
2. comparison of two approaches.

Additional heat contamination from the power generation, additional contamination from water use, additional contamination from mining of new volume of materials, additional noise contamination [13, 14].

3 Results

As a result of references and calculations, we've got the follow issues:

1. Explorations that are need to be done for technological processes improving:
   - work performed – technological process – energy input
   - calculation of administrative and technical measures
   - multitude of the elements of a construction system subject to their own energy laws
   - only conceptual view
2. Design the interface of conceptual designing construction site technological processes
   a. Use standard interfaces
      (1) For example – carrying capacity graphs and formulas of comparison and design new models
      (2) use a block-schemes
   b. Design a new interface
      (1) dimensionless scheme
      (2) scheme, which shows an interaction between resource-energy system balance and environment
3. Optimal choice "quality-volume-paying capacity" realized in consequence:
   Quality – forethought of architecture, good location in the city, durable surface finishing
   Volume – quantity of apartments for sale
   Paying capacity – possibility of people to buy apartments with specified quality by specified price and cost

   Pool of competitiveness increasing strategies:

   1. Long-term market position (calculation of timeline graph)
   2. Offer to customers conveniences and promotion of it
   3. Leading strategy in the parameter
      Science contribution and calculation:
      optimization task
      twin optimization task
      balance between matrices
   4. choice of leading parameter and its salient technical characteristics

It is suggested that developer spares money, gets more accurate result, improves its reputation. Outcome for a city: increase sanitary-secure area at construction site, dustiness increasing, noise increasing, environment footprint of construction site decreasing by fuel economy, materials appears as a decrease of lorry truck transition around the construction site, decreasing volumes of materials volume in village, increase of workman's safety. Despite the absence of explicit competitive advantage on costs or aggressive advertising, the relative comfort rise near the construction site, it makes constructed object more remarkable.

4 Discussions

Tool for the concept design resource saving technological processes in construction industry:

- decrease of uncontrolled flow of materials
- reduction of uncontrolled expenditures and special-purpose machinery moving

Actions and calculations aimed at obtaining new practical data – confirmation by practical data needs to be carried out as follows:

a. make a system of sensors and a calculator of energy of construction process
b. make a system for calculation of environmental footprint on city
c. receive the following data

    (B1) real energy use
    (B2) real energy saving
    (B3) real environmental impact on city
    (B4) a survey of developers [10–12].

The developers' cooperation with consumer is realized by three paths:

Path 1 – choice of available solutions

Calculation of optimal choice of technologies and got architectural solutions

Path 2 – Choice of available solutions with addition of supporting organization solutions

Development of improving supporting technical solutions for the construction industry, enabling to improve a usage of available construction machines and technologies

Path 3 – development of own technological process

Development of own technological process and equipment for the organization and long-term use.

Description of developer's resource saving approach as a long-term strategic, for example the Kaidzen concept [15, 16].

The developer's environment footprint to the city. Environment footprint from a given construction site, among them, indirect should be explored by designed business tool of evaluation the influence of resource saving to developer's competitiveness. This tool is needed to assess only the cost and market advantages obtained by the developer using resource-saving technological processes. These benefits are reduced production costs and advertising [13, 14, 16].

5 Conclusion

In this article, the first stage of designing the tool for the conceptual design of resource saving technological processes in construction industry was presented. Using this tool allows accelerating the improvement of technological processes to the such features as resource effectiveness and decreased environment footprint. Method of disclosure the economic effects from decrease an environment footprint allows motivating developers to solve the question of production ecological safety improvement. Thus, developer could increase its own competitive advantages.

To continue this study, an experiment should be organized on the energy and resource characteristics of work at the construction site, and its own organizational and technical measures should be developed and applied for the first time, which provide the function of supporting the existing typical technological process. Consultations should also be held with the city's environmental departments to determine the impact of the construction site on the improvement or deterioration of the urban environment and suburban landscape.

References

1. Mishchenko, V.Y., Gorbaneva, E.P., Archakova, S.Y., Dobrosockih, M.G.: Modelirovanie vypolneniya brigadami kompleksa tekhnologicheskih processov v organizacionno-tekhnologicheskom proektirovanii. FEHS: Finansy. EHkonomika **6**, 37–43 (2017)
2. Gorbaneva, E.P., Semenenko, T.O., Dobrosockih, M.G.: Postroenie matematicheskoj modeli optimizacii chislennosti rabochih v stroitel'noj brigade na primere provedeniya kapital'nogo remonta v g.Voronezhe. Stroitel'stvo i nedvizhimost' **1**(2), 150–153 (2018)
3. Mishchenko, V.Ya., Dobrosockih, M.G.: NP-razreshimaya zadacha kalendarnogo planirovaniya stroitel'stva,rekonstrukcii i remonta ob"ektov. Izvestiya vysshih uchebnyh zavedenij. Tekhnologiya tekstil'noj promyshlennosti **6**(366), 13–20 (2016)
4. Zolotuhin, S.N., Nasonova, T.V., Potekhin, I.A.: Racional'noe stroitel'stvo s povtornym ispol'zovaniem stroitel'nyh materialov, konstrukcij, izdelij posle snosa zdanij. Resursoehnergoehffektivnye tekhnologii v stroitel'nom komplekse regiona **10**, 206–209 (2018)
5. Ponyavina, N.A., Chesnokova, E.A., Zubareva, Y.V., Pis'yaukova, E.N.: Analiz faktorov vliyayushchih na izmenenie sprosa i predlozheniya na rynke zhiloj nedvizhimosti (na primere g. Voronezh). Stroitel'stvo i nedvizhimost' **1**(1), 45–51 (2017)
6. Brzezicka, J., Wisniewski, R., Figurska, M.: Disequilibrium in the real estate market: evidence from Poland. Land Use Policy **78**, 515–531 (2018)
7. Fowler, S.J., Fowler, J.J., Seagraves, P., Beauchamp, C.: A fundamentalist theory of real estate market outcomes. Econ. Model. **73**, 295–305 (2018)
8. Renigier-Biłozor, M., Biłozor, A., Wisniewski, R.: Rating engineering of real estate markets as the condition of urban areas assessment. Land Use Policy **61**, 511–525 (2017)

9. Vasil'chikova, E.V., Potekhin, I.A.: Modelirovanie konkurentosposobnosti developera. Vysokie tekhnologii v stroitel'stve **4**, 1–5 (2018)
10. Prokof'eva, E.N., Vostrikov, A.V.: Ocenka kachestva upravleniya informacionnymi potokami v organizaciyah. Vestnik RMAT **2**, 45–49 (2017)
11. Popikov, A.A.: Formirovanie organizacionno-informacionnoj bazy mekhanizma upravleniya riskami. Vestnik Voronezhskogo instituta ehkonomiki i social'nogo upravleniya **1**, 83–85 (2018)
12. Kankhva, V., Orlov, B., Vorobyeva, A., Belyaeva, S., Petrosyan, R.: The formation of a criteria-based approach in the development of projects of redevelopment of industrial real estate. In: MATEC Web of Conferences Proceedings, p. 01116 (2018)
13. Barinov, V.N.: Primenenie metodik rascheta prirodoohrannyh raskhodov, ispol'zuyushchih mezhdunarodnyj opyt opredeleniya pokazatelej prirodoohrannoj deyatel'nosti. Nauchnyj zhurnal. Inzhenernye sistemy i sooruzheniya **1**(14), 69–87 (2014)
14. Semenov, P.I., Kotenko, A.M., Barinov, V.N., Kireeva, E.A.: Mekhanizmy finansirovaniya snizheniya urovnya riska v stroitel'nom proizvodstve. Nauchnyj vestnik Voronezhskogo gosudarstvennogo arhitekturno-stroitel'nogo universiteta. Stroitel'stvo i arhitektura **2**(26), 90–95 (2012)
15. Barinov, V.N.: Investirovanie v prirodoohrannye meropriyatiya. Nauchnyj zhurnal. Inzhenernye sistemy i sooruzheniya **4**(9), 107–118 (2012)
16. Okolelova, E.Y., Truhina, N.I.: Stroitel'stvo vysotnyh zdanij: ocenka ehffektivnosti proektov v usloviyah riskov, Monografiya. Voronezh (2016)

Innovative Management of an Energy-Efficient Facility by Its Life Cycle Stages

Violetta Politi(✉)

Moscow State University of Civil Engineering,
26, Yaroslavskoye shosse, Moscow 129337, Russia
polity_violca@list.ru

Abstract. The management of the energy efficiency of the facility should be carried out at all stages of its life cycle - design, construction, operation, and decommissioning. In order to fully utilize the potential of energy saving of the facility, it is necessary to develop universal cost and fuel and energy indicators of resource consumption at all stages of the facility's life cycle. One of the methods for solving the problem is energy modeling of buildings. It allows us to present the real life of a building or structure in the form of a model with all designed engineering systems. One of the main results is a reliable assessment of the future annual energy consumption of the building during its operation. For the development of the social and production sphere in the direction of energy efficiency, it is necessary to attract not only private but also public investment. Life cycle contracts based on the principles of a public-private partnership allow managing the cost and fuel and energy indicators of the facility by the stages of its life cycle. Thereby, it is possible to keep the specified characteristics of an energy efficient facility achieved at the design stage.

Keywords: Innovative management · Energy-efficient · Life cycle stages

1 Introduction

The energy efficiency of a facility is the ratio of the significant beneficial effect of the energy resources expended to their quantity, which is necessary for obtaining a useful result. In the Russian Federation, the situation with energy consumption is catastrophic, as more than half of all the country's energy resources are consumed by industrial and residential buildings. The developed set of rules on energy efficiency of buildings regulates calculations of energy efficiency, both at the stage of the development of design documentation and at the stages of construction and commissioning. Thus, an increase in energy efficiency indicators becomes a long-term strategy, but does not establish the exact indicators that designers and builders should strive for. It is obvious that the future owner should solve the problem of energy efficiency of the facility, even at the stage of the investment plan [1].

One of the methods for solving the problem is energy modeling of buildings. It allows us to present the real life of a building or structure as a model with all designed engineering systems. One of the main results is a reliable assessment of the future annual energy consumption of the building during its operation. Therefore, changing

V. Murgul and M. Pasetti (Eds.): EMMFT-2018, AISC 983, pp. 116–126, 2019.
https://doi.org/10.1007/978-3-030-19868-8_11

the set of engineering solutions for the project, you can search for the most optimal set of measures to improve the energy efficiency of a particular object [2].

The following problems can be solved by methods of energy modeling of buildings:

- development and selection of measures to improve the energy efficiency of buildings;
- assessment of the effectiveness of design decisions at the "concept" stage and the "project" stage;
- the choice of the most appropriate tariff for energy;
- determination of the annual cost of energy for the correct assessment of the operational expense (OPEX);
- testing the performance of the designed engineering systems of the building during the year under normal weather conditions and under extreme conditions;
- calculation of the amount of heat entering the building volume as a result of solar radiation;
- determination of the number of points received on green certification systems (LEED, BREAM certification, and others).

As practice shows, the problem of energy efficiency of an object should be solved at the design stage, rather than during its active operation. At the planning stage of the production facility, it is possible to select the design parameters of the plant and production equipment. At the design stage of an energy-efficient building, an assessment and selection of planning and design solutions, innovative energy-saving technologies and building materials take place. At the same time, investments in optimizing the energy efficiency of the building will be much lower at the planning stage, not at the stage of active commercial operation. The criterion for the optimality of the selected design and construction solutions can serve as an indicator of the specific total energy costs associated with the construction, operation and decommissioning of a facility (dismantling) or recycling of building materials [3].

The methodological and system-based basis for making investment decisions and choosing the option of modernizing the production system is simulation modeling. The problems can be solved on the basis of the application of the apparatus for modeling the energy consumption of buildings. Therefore, for the full use of the energy saving potential of a new production facility, it is necessary to develop the cost and fuel indicators for the consumption of energy resources at all stages of the facility's life cycle [4].

Thus, functional modeling is a methodological and system-based basis for the design, construction and operation of energy-efficient buildings and the organization of their life-cycle processes.

Energy efficient design should use the knowledge of the energy consultant and be based on the experience of the energy audit of industrial enterprises. In the course of operating a new facility, energy-efficient management faces a number of problems that do not allow fixing the projected and achieved levels of energy efficiency. The main problems include, for example, the lack of universal indicators for accounting for the use of different types of fuel, the lack of a unified accounting center for profits and energy consumption costs, and others [5].

For the development of the social and production sphere in the direction of energy efficiency, it is necessary to attract not only private but also public investment. As international experience shows, the use of public-private partnership mechanisms is an effective way to implement large-scale projects, renovate and modernize fixed assets, and solve social problems [6].

The key to success of innovation policy in the field of energy efficiency and energy savings is the presence of demand for innovation. Stimulate innovative development of the state should, first of all, orders from the state and large businesses. At the same time, a comprehensive assessment of the effectiveness of innovative energy-saving technologies should be assessed on the principles of a systems approach based on the management of an object throughout its entire life cycle.

The formation and development of public-private partnership acquires particular relevance in modern conditions of globalization of the economy and the internationalization of production.

2 Materials and Methods

To solve the tasks of organizing the management of an energy efficient facility by the stages of its life cycle and preserving the project indicators for power engineering at a given level, it is possible to use the following methodology for innovative management of a complex facility.

Stage 1. Selection of contract models and contractual relationships in order to create an energy efficient facility

One of the contractual forms of public-private partnership is a life cycle contract. In accordance with this form of contract, the public partner on a competitive basis concludes an agreement with a private partner for the design, construction and operation of the facility for the term of the facility's life cycle. The public partner also makes payment for the project in equal shares after the facility is commissioned, according to the condition of maintaining the facility in accordance with the specified functional requirements [7].

Life cycle contract is an alternative to the traditional approach. Its main distinguishing feature is that the state enters into a contract with one private supplier who undertakes to implement the full range of services - from project design and construction to further operation and recycling of the facility at the end of its service life.

To improve the efficiency of life cycle contracts and reduce risks, the use of an engineering scheme for the organization of construction is proposed (Fig. 1). It is assumed that the organizer may be an engineering company that is able to aggregate the functions of the main participants in the construction - the customer, the designer, and the contractor.

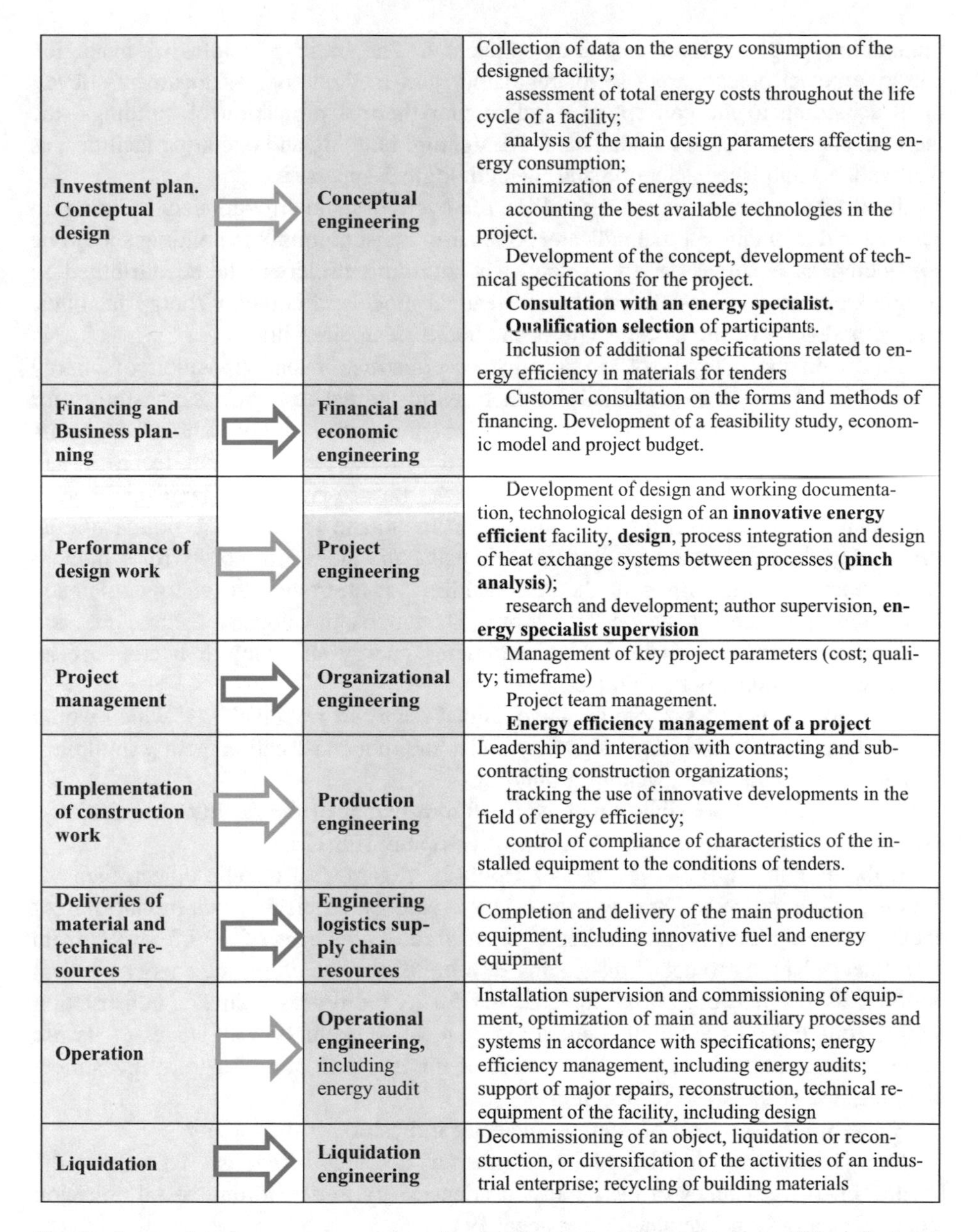

Investment plan. Conceptual design	**Conceptual engineering**	Collection of data on the energy consumption of the designed facility; assessment of total energy costs throughout the life cycle of a facility; analysis of the main design parameters affecting energy consumption; minimization of energy needs; accounting the best available technologies in the project. Development of the concept, development of technical specifications for the project. **Consultation with an energy specialist.** **Qualification selection** of participants. Inclusion of additional specifications related to energy efficiency in materials for tenders
Financing and Business planning	**Financial and economic engineering**	Customer consultation on the forms and methods of financing. Development of a feasibility study, economic model and project budget.
Performance of design work	**Project engineering**	Development of design and working documentation, technological design of an **innovative energy efficient** facility, **design**, process integration and design of heat exchange systems between processes (**pinch analysis**); research and development; author supervision, **energy specialist supervision**
Project management	**Organizational engineering**	Management of key project parameters (cost; quality; timeframe) Project team management. **Energy efficiency management of a project**
Implementation of construction work	**Production engineering**	Leadership and interaction with contracting and subcontracting construction organizations; tracking the use of innovative developments in the field of energy efficiency; control of compliance of characteristics of the installed equipment to the conditions of tenders.
Deliveries of material and technical resources	**Engineering logistics supply chain resources**	Completion and delivery of the main production equipment, including innovative fuel and energy equipment
Operation	**Operational engineering,** including energy audit	Installation supervision and commissioning of equipment, optimization of main and auxiliary processes and systems in accordance with specifications; energy efficiency management, including energy audits; support of major repairs, reconstruction, technical re-equipment of the facility, including design
Liquidation	**Liquidation engineering**	Decommissioning of an object, liquidation or reconstruction, or diversification of the activities of an industrial enterprise; recycling of building materials

Fig. 1. Innovative management: engineering and the main stages of energy-efficient design in the context of the life cycle of a complex facility.

Stage 2. Modeling the life cycle from the perspective of the process approach

To meet the challenges of improving the energy efficiency of facilities, both in the process of designing, building, and in the process of their operation, it is necessary to develop an organizational and economic tool for managing the process of creating and maintaining a given level of energy efficiency. The basic principles of managing energy-efficient objects are a set of regulatory and methodological documents developed according to the concept of standardizing thermal protection of buildings, the theoretical and empirical foundations for designing, building and operating facilities, as well as the established scientific and methodological apparatus.

In order to organize the process of the life cycle of an energy-efficient building, to achieve and maintain optimal indicators of energy consumption, the building should be considered as a single energy system. The building itself should be attributed to complex energy facilities. To build mathematical models of complex energy facilities, the methodology of the systems approach should be applied [8].

Based on the analysis of the practice of construction and operation of energy efficient buildings, it was revealed that there is no unified center of responsibility for observing the continuity of design indicators for energy costs and operational indicators of energy efficiency. This problem is connected with the organizational, technical and technological features of construction. Namely, the presence of a large number of participants in the investment and construction cycle and forms of the organizational structure of the construction project management blurs the responsibility for achieving certain energy saving indicators, the responsibility for improving the energy efficiency of buildings, structures, and constructions. Designers, developers, contractors, and subcontractors are not interested in achieving energy-efficiency indicators of an investment and construction project.

Therefore, it is necessary to organize unified center of responsibility, which would ensure the continuity of energy efficiency indicators of a construction facility (building) throughout the entire life cycle of a building.

The center of responsibility may be the future owner of the facility - the investor, the customer-developer or the management company (Fig. 2).

If the building itself, as real estate, should be regarded as a static object, then the building's life processes are dynamic cycles. Energy efficiency and monitoring of energy consumption parameters should be ensured at all stages of the life cycle, from investment planning to decommissioning of a building. Therefore, the energy efficient building itself is the object that initiates and forms the process. Thus, a construction facility (building), as a complex energy system, passing through all stages of its life cycle, requires unified control center that is of interest both at the stage of investment planning and at the stage of operation.

Stage 3. Development of energy efficiency indicators of the facility

Energy resources are what society can use as a source of energy. To analyze the level of energy efficiency of the facility, it is necessary to develop universal precision meters that allow an integrated assessment [9].

In conditions when many factors influence the energy efficiency of complex systems, technological processes and equipment are considered separately, also the volume of consumed energy resources is taken into account in different units of measure, it is difficult to stimulate energy efficiency.

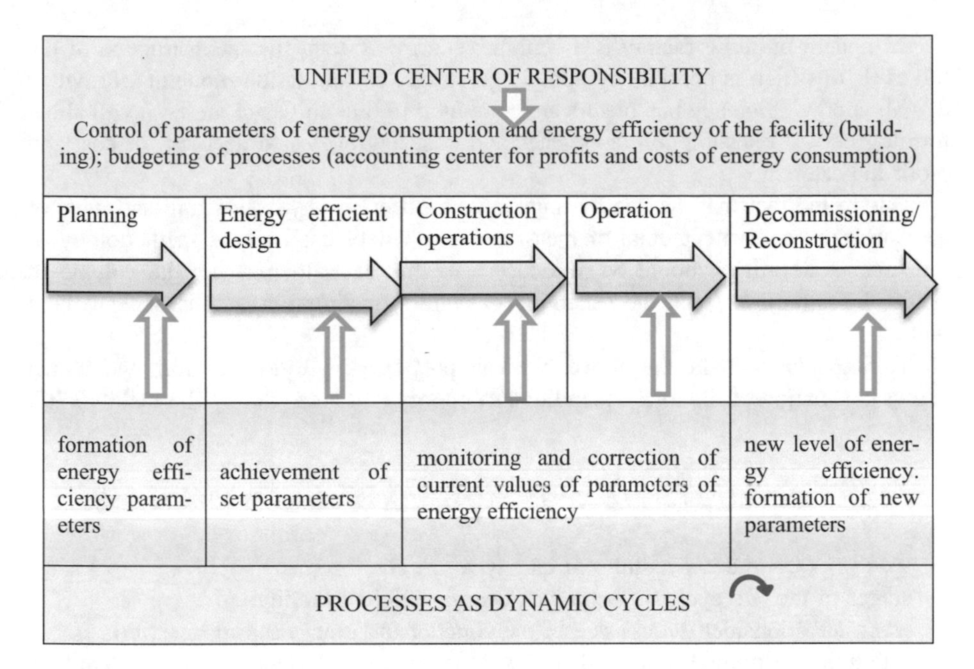

Fig. 2. Innovative management. Modeling the life cycle of a building from the perspective of the process approach.

In addition, the costs associated with energy consumption are taken into account in different ways: part of the costs are taken into account in the total amount of semi-fixed costs, part of the costs are included in production (variable) costs. At the same time, it is required to create unified financial center of profit, where the cost and achievement of energy efficiency would be kept [10].

To implement an energy efficiency management strategy, it is necessary to develop adequate tools for measuring indicators.

To solve the problem, it is necessary to identify the factors affecting the energy consumption of the building in accordance with their common feature of functioning, according to the stages of the life cycle. Next, it is necessary to find a single energy equivalent, which will allow calculation of the so-called energy capacity of the life cycle.

It has already been noted earlier that the energy efficiency of a facility is a complex probable characteristic depending on many factors, such as the choice of building materials and design solutions, the characteristics of engineering equipment, heat loss and heat gain, the use of secondary energy sources, and so on.

In practice, the concept of equivalent fuel is used to compare the energy value of different types of fuel and compare the total energy consumption of objects with different energy balance structures. That fuel is taken as an equivalent, which has a lower calorific value of 29.33 MJ/kg (7000 kcal/kg). The introduction of the concept of equivalent fuel allows, for example, comparing the energy costs of two different regions of the country without specifying how many specific fuels are burned in these regions.

A modern building facility is a complex energy system, the reconstruction of the life cycle of which is possible by applying methods of simulation modeling. Provided that all energy efficiency parameters are combined in one universal meter, it will allow forming energy consumption aggregates and cost aggregates to account for cost and profit indicators.

Due to the fact that the energy units have a different physical nature and units of measurement, equivalent units of measurement, widely used in the methodology of statistical accounting, should be introduced. In this case, to measure the volume of energy consumption (with the exception of water consumption) - tons of equivalent fuel.

Taking into account the above, we can propose the following formula, which allows calculation of the total expenditure of energy resources during the facility's life cycle:

$$Q_E^{LC} = Q_1 + Q_2 + \ldots + Q_{n-1} + Q_n = \sum\nolimits_{i=2}^{S} Q_i \quad (\textit{tons of equivalent fuel})$$

Q_1, Q_2, Q_{n-1}, Q_n - the expenditure of energy resources;
s – stages of the life cycle of a complex energy efficient facility.

Next, let's consider the essence of the concept of "unit" used in this work.

The unit is a unified scheme that allows describing all elements of a system (deterministic, stochastic, continuous, discrete). At each moment of time t, the unit is in one of the possible states. The states of the unit are an element and are specified by a certain set of Z. It is assumed that during a finite time interval, a finite number of input and control signals enter the unit. Output signals are generated at the output of the unit.

For example, let T be a fixed subset of a set of real numbers (the set of considered time points), X, Γ, Y, Z be sets of any nature. The elements of these sets will be called like this: $t \in T$- moment of time, $x \in X$- input signal, $g \in \Gamma$- control signal, $y \in Y$- output signal, $z \in Z$- state. Further states, input, output and control signals should be considered as a function of time and denoted as $z(t)$, $x(t)$, $y(t)$, respectively.

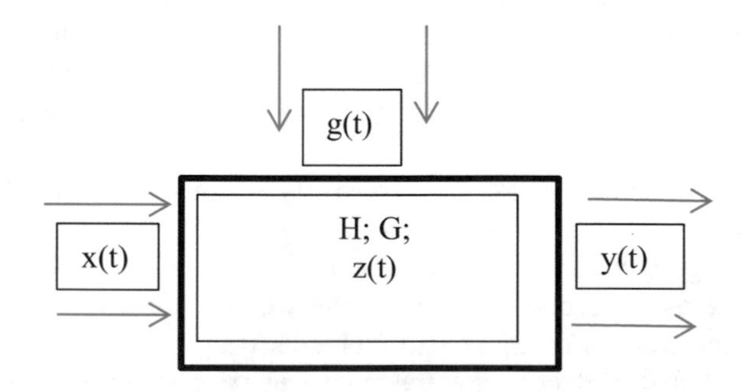

Fig. 3. Input, output and control signals of the power unit.

The unit is an object defined by the sets *T, X, G, Y, Z* and the operators *H* and *G*, in general, random. Operators *H* and *G*, called transition and output operators, implement the functions *z(t)* and *y(t)* (Fig. 3).

For the purpose of accounting and analyzing costs and profits associated with energy consumption, as well as to assess the economic effect, it is necessary to create a financial profit center in the general budgeting system for the object. The selection of this unit of accounting and cost analysis, in the general system of energy optimization, will allow you to accurately distribute and estimate total and marginal costs in real time.

As a result, it is possible to calculate the indicator of the cumulative expenditure of cash resources for managing the energy efficiency of an object:

$$P_E^{LC} = P_0 + p_i^1 q_i^1 + p_i^2 q_i^2 + p_i^3 q_i^3 + p_i^4 q_i^4 + p_i^5 q_i^5 + p_i^6 q_i^6 + p_i^7 q_i^7 = P_0 + P_1 + \ldots + P_i = P_0 + \sum\nolimits_{i=1}^{s} P_i \quad (1)$$

P_E^{LC} - total expenditure of cash resources on the management of the energy efficiency of the facility throughout the entire life cycle of the facility; p - the current or estimated cost of the fuel and energy resource (monetary units); q - the volume of consumed fuel and energy resource (natural units of measurement).

3 Results

Table 1 presents the process of formation of cost and energy units in accordance with the stages of the life cycle of the facility. The formulas for calculating the total expenditure of cash resources to manage the energy efficiency of the facility throughout the entire life cycle of the facility and the total expenditure of energy resources are proposed. An integral indicator is proposed for assessing the current energy efficiency of a building in accordance with the design or regulatory energy efficiency.

Table 1. Formation of monetary and energy units by the stages of the life cycle.

Stage of the life cycle of an energy efficient facility	Name of the unit indicator	Power unit designation (in tons of equivalent fuel)	Monetary unit designation (in nominal monetary units)
Planning	The amount of financial resources for concept development and preliminary project studies	–	P_0
Design and construction	The amount of fuel and energy resources used for the design, management, organization of the construction site and the production of construction and installation works; installation of equipment	Q_1	P_1

(*continued*)

Table 1. (*continued*)

Stage of the life cycle of an energy efficient facility	Name of the unit indicator	Power unit designation (in tons of equivalent fuel)	Monetary unit designation (in nominal monetary units)
Operation	The amount of fuel and energy resources used for ventilation and heating	Q_2	P_2
	The amount of fuel and energy resources for water supply and drainage	Q_3	P_3
	The amount of fuel and energy resources for electricity	Q_4	P_4
	The amount of fuel and energy resources for scheduled maintenance	Q_5	P_5
	The amount of fuel and energy resources for the implementation of the planned overhaul	Q_6	P_6
Decommissioning	The amount of fuel and energy resources for dismantling, recovery (or recycling with a new level of energy efficiency)	Q_7	P_7
The total expenditure of energy resources during the life cycle of the facility $Q_E^{LC} = \sum_{i=2}^{s} Q_i$ (tons of equivalent fuel)			
The total expenditure of cash resources for managing the energy efficiency of the facility throughout its entire life cycle $P_E^{LC} = \sum_{i=1}^{s} P_i$ (nominal monetary units)			
INDEX of energy efficiency (by energy consumption) $I_e = \frac{Q_E^{LC}}{Q_{base}}$; Where Q_E^{LC} - actual energy consumption or total energy consumption (real energy balance); Q_{base}- energy consumption according to the basic version or total standard energy consumption (standard energy balance)			
INDEX of energy efficiency (by cost) $I_f = \frac{P_e^{LC(Real)}}{P_e^{LC(Base)}}$ Where $P_e^{LC(Base)}$ – the basic (project) cost by the stages of the facility's life cycle; $P_e^{LC(Real)}$ – the actual cost by the stages of the facility's life cycle			

4 Discussions

It should be noted that the processes of functioning of a building are characterized not by deterministic processes, which are represented in this methodology, but by probabilistic, stochastic processes. Therefore, to ensure the continuity of energy consumption indicators, instead of the piecewise linear function of energy consumption (linear spline), it is necessary to consider a more complex, nonlinear spline, which will allow actually summing up the fuel and energy expenditures. The use of a stochastic unit will allow not only making current calculations but also predicting a change in the energy consumption of a building in accordance with a given interval of the life cycle.

The center of responsibility for managing the energy performance of a building recommended in this paper requires separate refinement. Namely, an organizational management plan should be developed, and a description of employees' areas of responsibility should be given.

It is also necessary to separately develop a regulated mechanism for financial planning, analysis and the formation of a motivational base, related to taking into account the indicators of costs and profits for energy efficiency. For example, the organization of the work of a separate group in the budgeting system will make it possible to calculate the economic effect of energy saving, highlighting the share of savings in total profits. On the basis of this, the formula of energy efficiency index by the cost can be refined and complemented. It is intended to carry out further research on the formation of energy and cost units included in this index - separately for budgeting purposes (accounting for the amount of consumed energy resources in natural units of measurement), and separately for purposes of financial assessment of the performance of an energy-efficient facility (accounting for the amount of consumed energy resources in terms of equivalent fuel).

5 Conclusion

On the basis of practical experience, we can state the following fact that when solving problems of improving the energy efficiency of the facility, the energy efficiency potential is higher at the stage of planning and conceptual design of the facility. In the case of optimization of the energy consumption parameters of a ready-made building, in the process of its active commercial operation, investments in energy efficiency increase become higher, and the possibilities of achieving optimal parameters are lower.

Energy efficient design should use the same methods and approaches, the technical knowledge that is used in the energy audit, which is carried out at existing enterprises. At the design stage, there is a choice in areas such as the basic design parameters of the installation, the production process used, the main production equipment, etc. This makes it possible to choose the most energy efficient technologies. Implementation of such changes in the existing enterprise, as a rule, is impossible or extremely expensive.

Problems: A number of problems for the further development of the activities initiated in the field of energy efficiency and the implementation of the developed programs were identified. In particular, it is important to monitor whether it is possible to consolidate the levels of energy savings achieved as a result of the introduction of new technologies or methods. Often, the gradual reduction of the effect achieved as a result of inefficient operation and maintenance of equipment, as well as other factors is not taken into account.

Also, to ensure the continued nature of energy saving activities and the application of advanced technologies, it is necessary to integrate their energy saving philosophy into the company's culture.

As a mechanism for implementing energy-efficient construction projects, the use of life-cycle contracts can be proposed, allowing the customer to focus on monitoring the achievement of the specified characteristics, while the contractor will be responsible for quality at all stages of the facility's life cycle. To implement the tasks in the perspective of life cycle stages, it is proposed to use an engineering company as a managing organization that can combine the functions of a customer, a designer, and a contractor.

References

1. Isaev, S., Vatin, N., Baranov, P., Sudakov, A., Usachov, A., Yegorov, V.: Mag. Civil Eng. **36**(1) (2013)
2. Arseniev, D., Rechinskiy, A., Shvetsov, K., Vatin, N., Gamayunova, O.: Appl. Mech. Mater. **635–637**, 2076–2080 (2014)
3. Khmel, V., Zhao, S.: IATSS Res. **39**(2), 138–145 (2016)
4. Chirkunova, E., Kireeva, E., Kornilova, A., Pschenichnikova, J.: Procedia Eng. **153**, 112–117 (2016)
5. Smirnov, V., Dashkov, L., Gorshkov, R., Burova, O., Romanova, A.: Methodical approaches to value assessment and determination of the capitalization level of high-rise construction. In: E3S Web of Conferences (2018). https://doi.org/10.1051/e3sconf/20183303030
6. Artyushina, G.G., Sheypak, O.A., Golov, R.S.: Podcasting as a good way to learn second language in e-learning. In: ACM International Conference Proceeding Series (2017). https://doi.org/10.1145/3026480.3029590
7. Lukmanova, I., Golov, R.: Modern energy efficient technologies of high-rise construction. In: E3S Web of Conferences (2018). https://doi.org/10.1051/e3sconf/20183302047
8. Ustinovicius, L., Rasiulis, R., Nazarko, L., Vilutienė, T., Reizgevicius, M.: Procedia Eng. **122**, 166–171 (2015)
9. Aurora, A., Teixeira, S., Queirós, A.: Res. Policy **45**, 1636–1648 (2016)
10. Golov, R., Narezhnaya, T., Voytolovskiy, N., Mylnik, V., Zubeeva, E.: Model management of innovative development of industrial enterprises. In: MATEC Web of Conferences (2018). https://doi.org/10.1051/matecconf/201819305080

Management of the Investment Design Process at the Enterprises of the Energy Sector

Ekaterina Nezhnikova[1](✉), Sébastien Santos[2], and Elena Egorycheva[1]

[1] Peoples' Friendship University of Russia (RUDN University), Miklukho-Maklaya Street, 6, 117198 Moscow, Russia
nezhnikova_ev@pfur.ru

[2] Université Paris-Dauphine, Place du Maréchal de Lattre de Tassigny, 75016 Paris, France

Abstract. The purpose of this study is to form a scientifically-based mechanism for managing investment design at the enterprises of the energy sector. In the paper, an organizational model of investment design is developed, an algorithm for the formation and management of a portfolio of energy-saving projects is compiled, the degree of probability of transition of enterprises to resource-and energy-saving technologies is assessed. The conducted studies are based on the system analysis method, process and project approaches, methods of comparison and generalization, classification, expert assessments, modeling, and investment planning. The formulated provisions, conclusions and recommendations can be used by investors in choosing the most acceptable objects for investing their own funds in terms of risk cost.

1 Introduction

The formation of modern approaches to the management of investment design at the enterprises of the energy sector will allow adaptation of the goals and strategy of the economic entity to the changed conditions, successful implementation of the tasks, and also it will help to clearly respond to changing external conditions.

For full-cycle enterprises, the availability of investment design and planning programs is a fundamental aspect, which makes it possible to reduce the level of uncertainty and manage the risks of economic activity [1].

Nowadays, issues on the determination of the most optimal parameters for launching an investment project and forecasting its impact on the overall economic condition of an enterprise remain open, as well as the determination of the basic principles for the formation and implementation of a project investment management program.

The study of modern economic literature made it possible to determine that investment design in its economic substance, although it is a method of planning and organizing investment flows, is inextricably linked with the assessment of the operating, economic and innovation activities of enterprises in the energy sector. The analysis of the existing models of the organization of investment design based on the functional approach shows that this system can be a key aspect of successful economic

V. Murgul and M. Pasetti (Eds.): EMMFT-2018, AISC 983, pp. 127–137, 2019.
https://doi.org/10.1007/978-3-030-19868-8_12

development of an enterprise, combining various organizational and managerial functions of an economic entity at the stages of determining strategic objectives, finding alternatives, and assessing the economic efficiency of implemented projects [2].

The investment design system will combine both general management functions and technological management functions, on the basis of which it is necessary to form investment design principles. It should be noted that these principles, together with the mechanisms for organizing and managing investment projects, must meet the requirements for achieving the strategic investment objectives of an enterprise.

2 Materials and Methods

Based on the clarified economic substance of investment design, it is necessary to conclude that the approaches to the organization and management of the process under study should be based on the principles of investment analysis, organization of internal budgeting and control, and investment management, which is shown in Fig. 1.

The formation of the investment design mechanism should be carried out on the basis of a set of proposed principles that will ensure the implementation of targeted actions to achieve the strategic objectives of the enterprise in conditions of severe competition and limited resources [3].

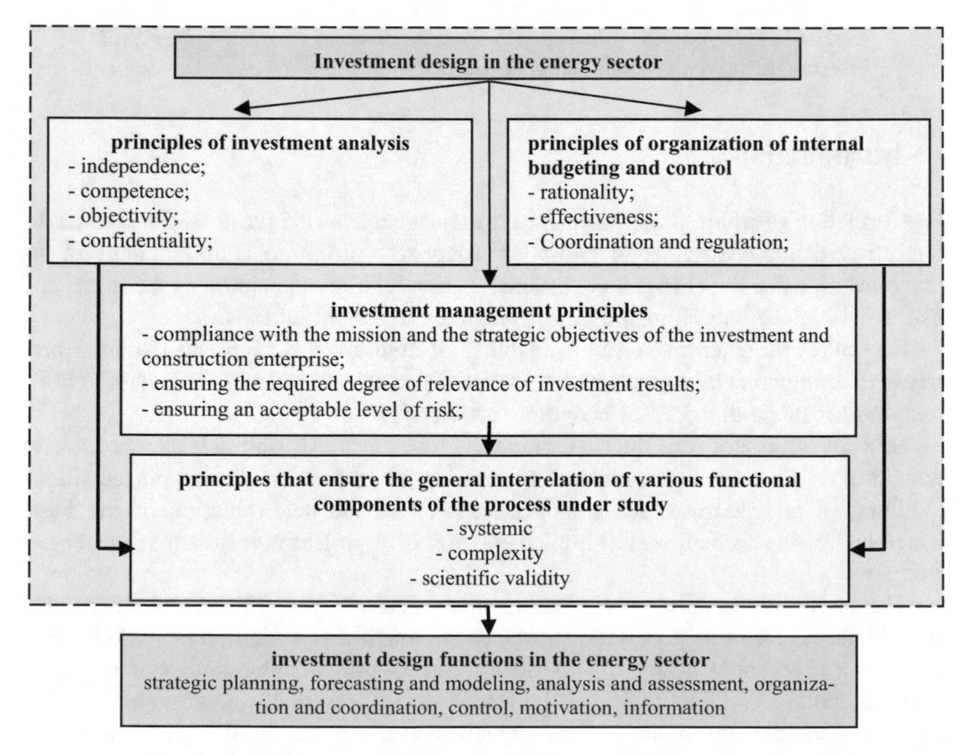

Fig. 1. Principal support of the investment design system functional.

Management of the energy sector enterprises due to the versatility and complexity of their activities requires the professional organization of this work, which necessitates the creation of an organizational model of investment design, aimed at creating a competitive and investment-attractive structure. The process of forming an effective investment planning system depends on the formed strategy and mission of the enterprise. At the same time, the strategic management is associated with the search and formation of investment sources to ensure the implementation of projects, long-term forecasting of investment flows, and a comprehensive assessment of the risks of activity [4].

Thus, to ensure the integrated implementation of the full functional of the investment design mechanism, the study proposed its organizational model, which is based on taking into account the features of various stages of the life cycles of projects implemented and planned for implementation and is based on the principles previously defined in the study (Fig. 2).

The study of the economic and organizational aspects of the operation on the formation of a portfolio of projects with acceptable indicators of investment efficiency, and is based on a comprehensive account of the requirements of the endogenous and exogenous environments of the business entity. It should be noted that for the implementation of the strategic objectives of its development, the enterprise must ensure the formation of an effective energy-saving production policy, without prejudice to the production capacity [5].

A key aspect of the successful implementation of the investment design system based on the developed organizational model is the formation of an effective portfolio of investment projects. As the key factors affecting the overall efficiency of the enterprise, it is necessary to consider the quality of planning, control and coordination of projects that make up the investment portfolio and are at different stages of the life cycle. It should be noted that the tasks of monitoring and analyzing the quality of implementation of various investment projects, due to the specifics of the activities of energy enterprises, are labor-intensive and complex processes [6].

The analysis of the modern economic literature has shown that management systems based on the process approach are used for organizing and managing the coordination and control processes in project management. Due to the fact that an enterprise, having the necessary own or attracted resources, has the ability to simultaneously carry out several projects within the investment portfolio adopted for implementation, the task of combining the process approach in managing internal processes with project management at the level of investment portfolio becomes urgent [7].

The study of the structure of investment project management based on the process approach allowed identifying decision-making blocks characterized by different scale of decisions made. Figure 3 presents such a case based on a hierarchical decision-making scheme using the example of building an energy facility.

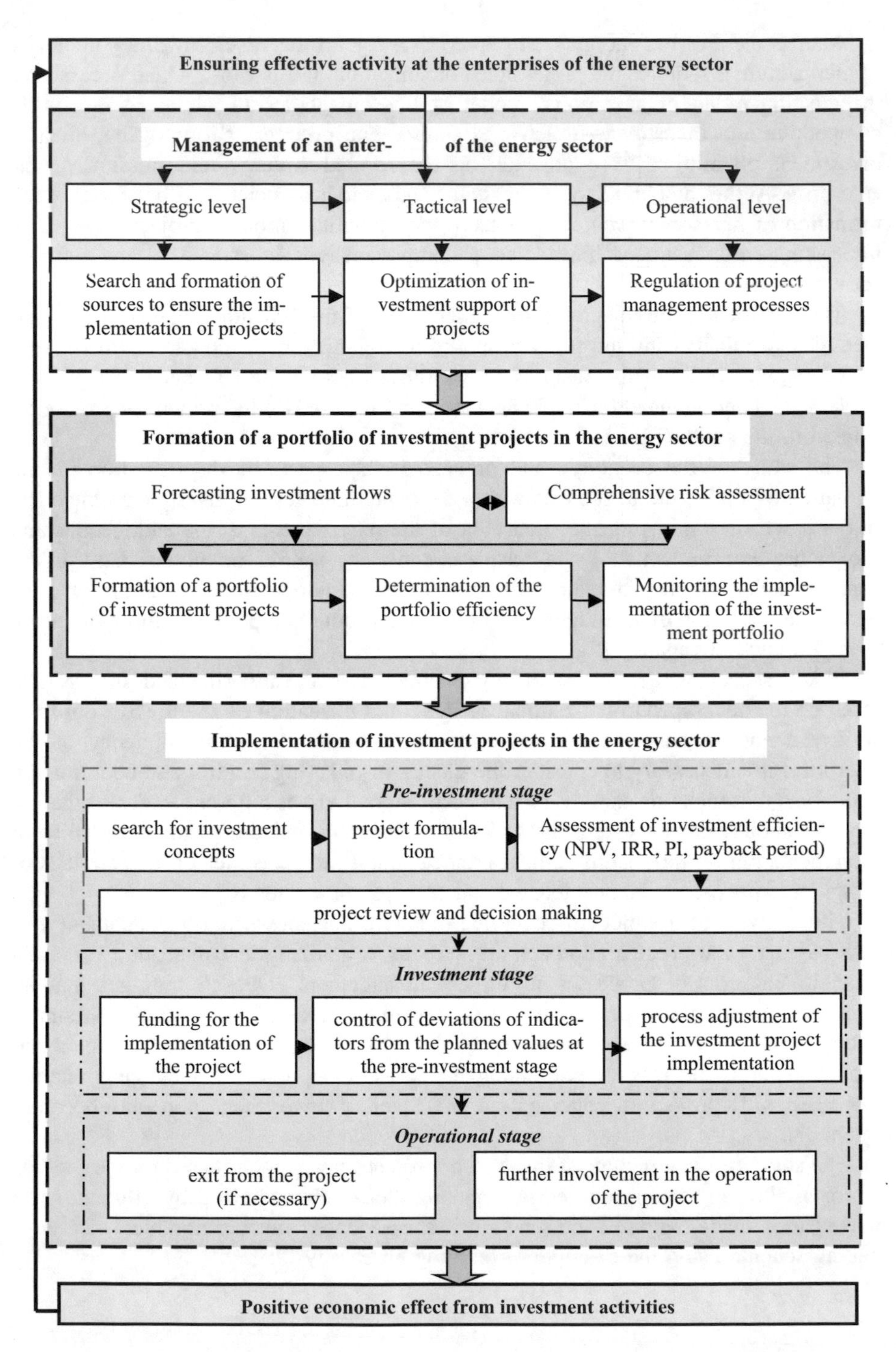

Fig. 2. Organizational model of the investment design system of the energy sector.

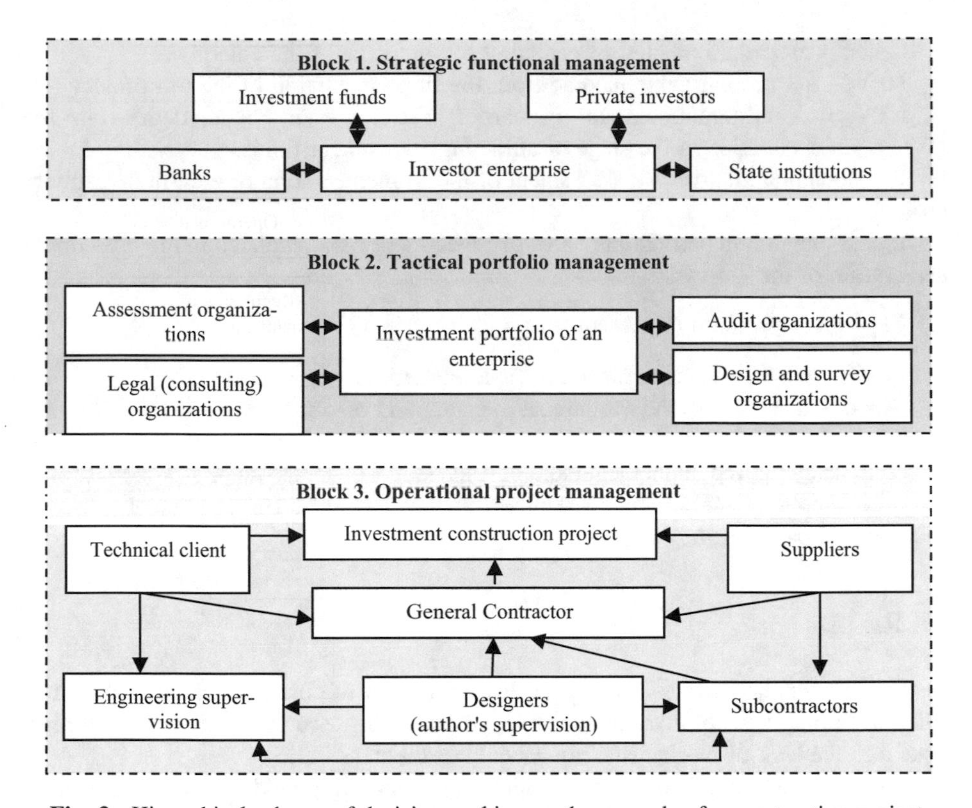

Fig. 3. Hierarchical scheme of decision making on the example of a construction project.

Block 1. The functionality of this block provides for the solution of strategic tasks to attract investment capital and provide political and administrative support from the state and municipal authorities. Planning at this stage is the regulatory aspect of further investment processes.

Block 2. At this stage, an investment portfolio is being formed on the basis of the selection of the most efficient projects of energy sector development, taking into account the analysis and assessment of the overall risk level and in accordance with the investment objectives defined at the strategic level.

In our study, the selection of investment projects in the formed portfolio is proposed to be carried out on the basis of the following economic and mathematical model:

$$\left.\begin{array}{l} \sum_{i=1}^{n} C_i < C_{\max} \\ \sum_{i=1}^{n} NPV_i > NPV_{\text{corp}} \\ NPV_{\text{corp}} \to NPV_{\max} \\ \sum_{i=1}^{n} R_i < R_{\text{corp}} \\ R_{\text{corp}} \to R_{\min} \end{array}\right\} F\left(\sum_{i=1}^{n} P_i\right) \Rightarrow \max Ef(F) \qquad (1)$$

where C_i – costs required for the implementation of the i-th project;

Cmax – maximum amount of capital at the disposal of the enterprise;

NPV_i – net present value income from the implementation of the i-th project;

NPV_{corp} – minimum amount of corporate net present value, defined in the investment declaration at the stage of attracting investment funds;

R_i – total quantitative risk assessment of the implementation of the i-th investment project;

R_{corp} – maximum permissible value of investment risk, regulated in the investment declaration of the enterprise;

$F(\sum_{i=1}^{n} P_i)$ – portfolio of investment projects;

$Ef(F)$ – efficiency of the portfolio implementation.

Block 3. It represents the operational management of construction work on the site, as well as the processes of resource support during the implementation of our project. The efficiency of the implementation of this stage depends on the quality of the resource supply of the project, as well as on the degree of smoothness of relations between the participants.

3 Results

Despite the functionality of investment design, one of the main problems that do not allow for a high level of investment activity in the energy sector is the problem of high risks and the lack of mechanisms for managing them.

Based on the analysis of modern scientific papers in the field of risk management, we propose to consider the risk management system as a set of highly specialized methods: identification, management modeling, qualitative and quantitative assessment, monitoring, modeling of ways to influence managed risks, control of the efficiency of decisions made.

The study proposes to consider the creation of a mechanism for internal risk control of business processes as the main approach to ensuring the practical implementation of the investment process management system. One of the most effective approaches to systematization of business processes is the classification formed in accordance with the stages of the life cycle of an investment project. The study allowed systematization of the existing business processes in the field of development by functional basis (Table 1).

Thus, on the basis of the refined structure of the risk management system for the activities of a business entity, taking into account the results of systematization of business processes in the energy sector according to functional characteristics, the study proposes an algorithm for internal risk control of business processes of the enterprise (Fig. 4).

Table 1. Classification of investment business processes by functional basis.

Process name	Process characteristic	Components
Construction processes	Main implementation processes, processes for generating profit and value added	Production of building structures, construction and installation works, design, engineering surveys
Prospective development processes	Goal setting, formation of the mission of the enterprise and strategy, increase in investment attractiveness, creation of new areas of activity, modernization of existing areas	Attracting investment funds, lobbying interests in government, forming a portfolio
Control processes	Achievement of tactical and strategic goals based on planning, control and regulation of internal and external relations	Management of finances, internal services, production, personnel, marketing
Supply processes	Auxiliary processes ensuring the necessary resources to the main processes	Logistics, supply, administrative functions, repair bases
Operating processes	Ensuring the operational state of the energy facility at the operational stage of the life cycle	Maintenance and current repair

In order to ensure the comparability and validity of the numerical expressions of the overall project risks, the paper proposed to divide the risk level research process during the implementation of the investment and construction project into interconnected narrower processes of analysis, measurement and assessment, the functional characteristics of which are presented in Table 2.

Table 2. Components of the risk level research process during the implementation of an investment project.

Name	Characteristic	Result
Risk analysis	Analysis and selection of significant risk factors and uncertainties	Identification of significant risk factors and uncertainties
Risk measurement	Numerical measurement based on information processing algorithm; combining the obtained values on the basis of the methods of generalization of partial results	Individual results of numerical measurement
Risk assessment	Interpretation of measurement results in comparable values, assessment of the level of risk by a unified point scale	Estimated comparable characteristic of the level of risk

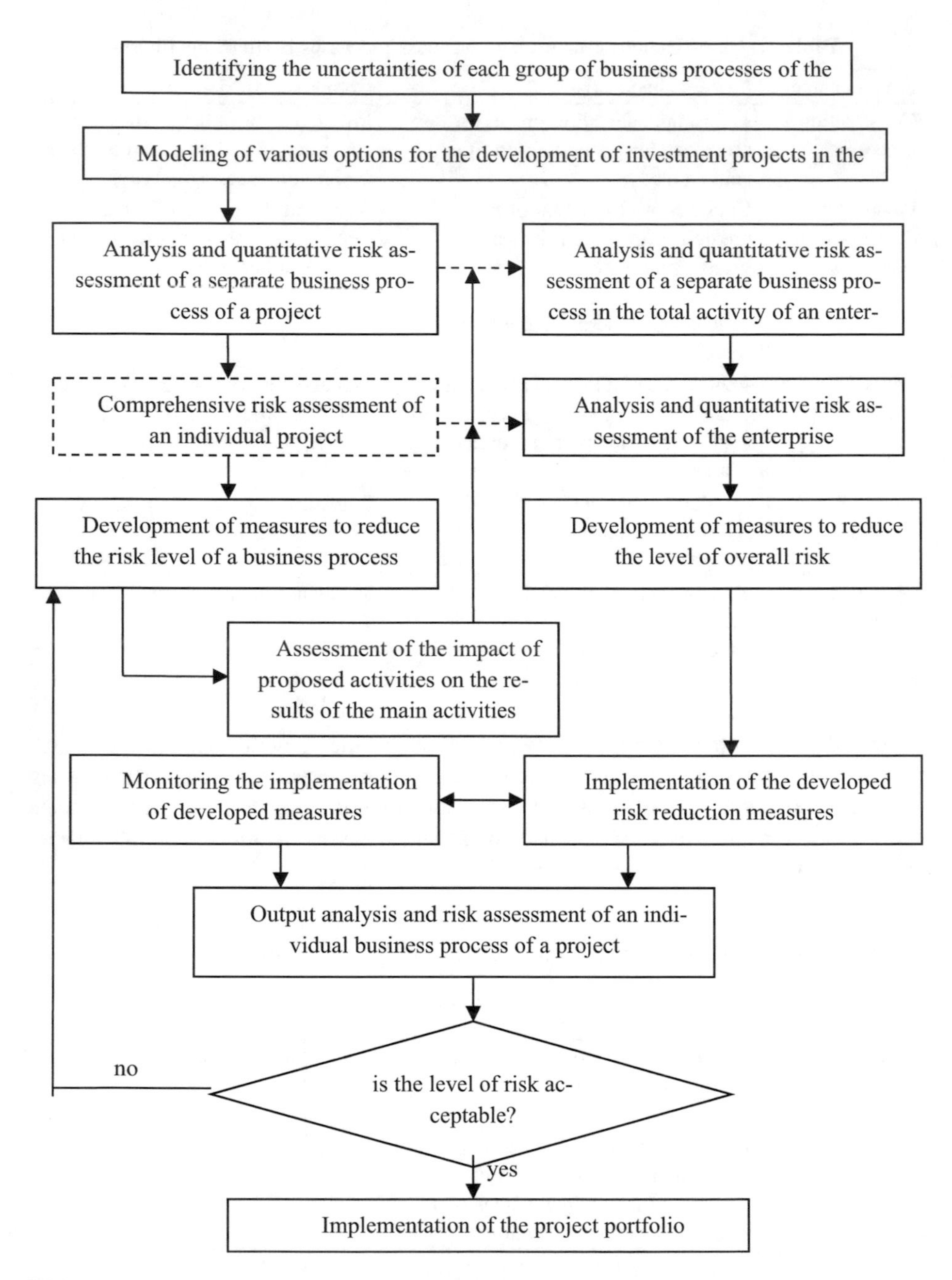

Fig. 4. Risk management algorithm for individual business processes in the implementation of an investment project.

The study of business processes of investment and economic activities of the enterprise allowed dividing the process risks according to the following principles that are independent of each other:

- by belonging to business processes - construction risks (I), development risks (II), management risks (III), risks of supporting business processes (IV), operational risks (V);
- by the level of occurrence - external (a) and internal (b);
- by the degree of control - controlled (γ) and uncontrollable (δ);

It is noted that each individual final risk will relate to a specific business process, and also be characterized by the source of the occurrence and the possibility of control.

Thus, the overall risk of the enterprise activity can be represented in the following mathematical expression:

$$R_{\text{corp}} = \left\{ \begin{matrix} Ia\gamma & Ia\delta & Ib\gamma & Ib\delta \\ IIa\gamma & IIa\delta & IIb\gamma & IIb\delta \\ IIIa\gamma & IIIa\delta & IIIb\gamma & IIIb\delta \\ IVa\gamma & IVa\delta & IVb\gamma & IVb\delta \\ Va\gamma & Va\delta & Vb\gamma & Vb\delta \end{matrix} \right\} \tag{2}$$

Thus, using the matrix expression of general corporate risk (2), the risks of investment projects selected in the general portfolio but still at the pre-investment stage can be represented as:

$$R^1 = \sum\nolimits_{n=1}^{m} \frac{(\sum\nolimits_{i=I}^{V} \sum\nolimits_{j=a}^{b} \sum\nolimits_{g=\gamma}^{\delta} X_i Y_j Z_g)_n}{\max(\sum\nolimits_{i=I}^{V} \sum\nolimits_{j=a}^{b} \sum\nolimits_{g=\gamma}^{\delta} X_i Y_j Z_g)_n} P_n * 100\% \tag{3}$$

where $(X_i Y_j Z_h)_n$ - numerical risk assessment of an individual i-th business process, characterized by the j-th source of occurrence and the g-th capability of control in the implementation of the n-th development scenario of the investment project;

P_n - probability of implementing the n-th scenario, while $\sum\nolimits_{n=1}^{m} P_n = 1$.

The risks of projects selected in the general portfolio and under implementation can be assessed on the basis of the following formula:

$$R^2 = \frac{\sum\nolimits_{i=I}^{V} \sum\nolimits_{j=a}^{b} \sum\nolimits_{g=\gamma}^{\delta} X_i Y_j Z_g}{\max \sum\nolimits_{i=I}^{V} \sum\nolimits_{j=a}^{b} \sum\nolimits_{g=\gamma}^{\delta} X_i Y_j Z_g} * 100\% \tag{4}$$

A numerical interpretation of the level of risks of energy facilities that have already been put into operation should be carried out as:

$$R^3 = \frac{\sum\nolimits_{j=a}^{b} \sum\nolimits_{g=\gamma}^{\delta} VY_j Z_g}{\max \sum\nolimits_{j=a}^{b} \sum\nolimits_{g=\gamma}^{\delta} VY_j Z_g} * 100\% \tag{5}$$

Thus, a numerical assessment of the level of corporate risk of an enterprise will be carried out:

$$R_{\text{corp}} = \frac{\sum_{a=1}^{k} R_a^1 + \sum_{b=1}^{l} R_b^2 + \sum_{c=1}^{m} R_c^3}{k + l + m} * 100\% \quad (6)$$

where k, l, m – the number of projects for the development of energy facilities in the portfolio structure, which are at the pre-investment, investment and operational stages, respectively.

For a numerical assessment of the identified final risks of business processes, it is proposed in the study to use a point system of the method of expert assessments. It should be noted that the maximum possible values of the numerical assessment are $\max R^1 = \max R^2 = \max R^3 = \max R_{\text{corp}} = 100\%$, which corresponds to the highest level of risk.

Further study of acceptable risk levels for all participants in the investment process has allowed the development of a grading scale (Table 3).

Table 3. Assessment of the degree of probability of transition of an economic entity to innovation.

R value	Acceptability of risk level for investors in the construction industry	Recommendations for the implementation of the investment and construction project (portfolio)
R > 50%	Risk is not acceptable	Complete rejection of the implementation
30% < R < 50%	Risk is not acceptable	Implementation is possible only after the development of fundamental measures to reduce the level of risk and its re-assessment
15% < R < 30%	Risk is relatively acceptable	Possible implementation in the formation of an aggressive portfolio with the implementation of measures to prevent the growth of risk level
5% < R < 15%	Risk is acceptable to the energy industry in Russia	Mandatory implementation of risk monitoring measures
R < 5%	Risk is completely acceptable	

4 Conclusion

Thus, the assessment of the investment risks of various enterprises of the energy industry based on the proposed approach will ensure the comparability of the research results, which will provide an opportunity for all parties involved to perform a comparison of the risk level of individual business processes and development projects of the energy sector.

The developed organizational model allows taking into account the features of the various stages of the life cycle of investment projects in the field of energy and is based on the formation of a portfolio of projects with acceptable indicators of investment efficiency, taking into account the formation of the production capacity of energy companies adequate to environmental factors.

The publication was prepared with the support of the RUDN University Programm 5-100.

References

1. Sizova, E., Zhutaeva, E., Gorshkov, R., Smirnov, V., Kochetkova, E.: Methodical bases for forming the structure of management of innovative activity of large building holdings. In: MATEC Web of Conferences (2018). https://doi.org/10.1051/matecconf/201817001126
2. Golov, R., Narezhnaya, T., Voytolovskiy, N., Mylnik, V., Zubeeva, E.: Model management of innovative development of industrial enterprises. In: MATEC Web of Conferences (2018). https://doi.org/10.1051/matecconf/201819305080
3. Lukmanova, I., Golov, R.: Modern energy efficient technologies of high-rise construction. In: E3S Web of Conferences (2018). https://doi.org/10.1051/e3sconf/20183302047
4. Nezhnikova, E.: The use of underground city space for the construction of civil residential buildings. Proc. Eng. **165**, 1300–1304 (2016). https://doi.org/10.1016/j.proeng.2016.11.854
5. Safronova, N., Nezhnikova, E., Kolhidov, A.: Sustainable housing development in conditions of changing living environment. In: MATEC Web of Conferences, vol. 106, p. 08024 (2017). https://doi.org/10.1051/matecconf/201710608024
6. Nezhnikova, E.: Organizational-economic mechanism of formation of strategic priorities of housing construction development. In: IOP Conference Series: Earth and Environmental Science, vol. 90, p. 012162 (2017). https://doi.org/10.1088/1755-1315/90/1/012162

Application of the PERT Method in Scheduling of Construction of Apart-Hotel for Energy Consumption Economy

Tatiana Simankina[1](✉), Iuliia Kibireva[1], Angela Mottaeva[2], and Miroslava Gusarova[3]

[1] Peter the Great St. Petersburg Polytechnic University, Polytechnicheskaya 29, 195251 St. Petersburg, Russia
talesim@mail.ru, julijakibireva666@yandex.ru

[2] Moscow State University of Civil Engineering, Yaroslavskoye shosse 26, 129337 Moscow, Russia

[3] Tyumen Industrial University, Volodarskogo str. 38, 625000 Tyumen, Russia

Abstract. The article discusses issues of ensuring timeliness of project completion and energy consumption reduction. Often, exceeding of project implementation period leads to negative results of project objectives and even to their impossibility. The article discusses application of the PERT method at the stage of scheduling the construction, namely when managing a project schedule to reduce total energy consumption, using example of an apart-hotel to evaluate applicability of the method not only to large-scale projects but also to medium-size objects. The study found that the method makes it possible to find probability of project completion by a certain date, take risks into account, but correspondence of obtained data with the reality depends on quality of expert opinion about three project durations: pessimistic, most likely, and optimistic. One month of delay in construction leads to overuse of energy in volume of 217 thousand standard fuel.

Keywords: Project management · Energy consumption · Scheduling of construction · Construction · PERT method · Risks

1 Introduction

In managing implementation of construction projects, special attention is paid to timeliness of completion of works and commissioning of objects. This requirement is especially important for projects, prevention of delay in commissioning dates for which is of key importance, since delays not only lead to sharp decrease in project's efficiency but also deprive implementation of a project of meaning and may lead to a failure in implementation of program, which includes this project. Therefore, to prevent such a situation it is necessary to pay special attention not only to development and optimization of construction schedule but also to formation of an effective system for monitoring, controlling and regulating of project.

V. Murgul and M. Pasetti (Eds.): EMMFT-2018, AISC 983, pp. 138–145, 2019.
https://doi.org/10.1007/978-3-030-19868-8_13

The main goal of the research is development of system for controlling and monitoring time of works for investment and construction project based on a method of analysis and evaluation - PERT (Program (Project) Evaluation and Review Technique) for medium-size project and evaluation of its rationality in developing construction schedule of such project.

In the last century, number of schedule management methods for application during project implementation were developed:

- Critical Path Method (CPM) was proposed by "DuPont" and "Remington Rend" to guide implementation of large-scale projects for the modernization of "DuPont's factories [1, 2]. The method is based on determination of the most continued task sequence from beginning of project to its completion, taking into account their coordination. Tasks that lie on critical path have zero slack for realization and if they change their duration, the timings of entire project also change. However, this method to control project schedule requires additional elaboration.
- Program Evaluation and Review Technique (PERT) was created by Lockheed Corporation, a consulting firm "Booz, Allen and Hamilton Inc." in the US Navy when developing the Polaris-Submarine weapon system mainly for managing large-scale projects in construction, defense industry, military engineering and other fields. The method was intended for very large-scale, global and complex projects and included presence of indeterminacy, making it possible to develop project schedule with certain amount of probability for all its components [1–3]. An element of uncertainty regarding conditions of project implementation and factors affecting project implementation is one of reasons for nonobservance of construction time, which leads to delay in commissioning of object [4, 5].
- Critical Chain Project Management (CCPM) was first described in 1997 in the book by Eliyahu M. Goldratt "Critical Chain" [4]. The method was widely supported by specialists, since it was close in technique to the classical PERT method. Buffers are widely used in the method to reduce project risks in project and ensure sustainability of project's planned schedule, visualization of "fever chart" trends, project calculation from deadline date for completion and not from beginning as in the PERT method. These aspects are some of benefits of project control [1].

Thus, implementation of the PERT method at planning stage helps to take into account risk of delay caused by a large number of reasons [6, 7]. However, there is question about application of this method to smaller projects in order to plan and manage them more rationally.

2 Materials and Methods

Of all stages of investment and construction project, construction stage is the most exposed to risks, and then - project completion stage [8]. Therefore, it was decided to consider duration of construction of object, taking into account time for commissioning object.

Apart-hotel building with deepening for multifunctional complex (Fig. 1) was chosen as object of study. The building has variable number of floors - 4, 7, 10 floors, with 3 underground floors. Plan dimensions are 50 × 67.4 m.

On the third underground floor from ground level, there is parking, on the first and second floors – commercial, entertaining, social, business, administrative zones. On the first floor from ground level, there are commercial premises, entrance lobby for hotel guests and visitors of underground complex, a restaurant, a fitness center.

Hotel rooms are situated on 2–9 floors, and the 10th floor is reserved for office space. In total, the building has 4 cores, each of them includes 3 staircases and 2 elevators. Accessible roof area with providing green spaces is also included; it is also located above underground part of the building in courtyard.

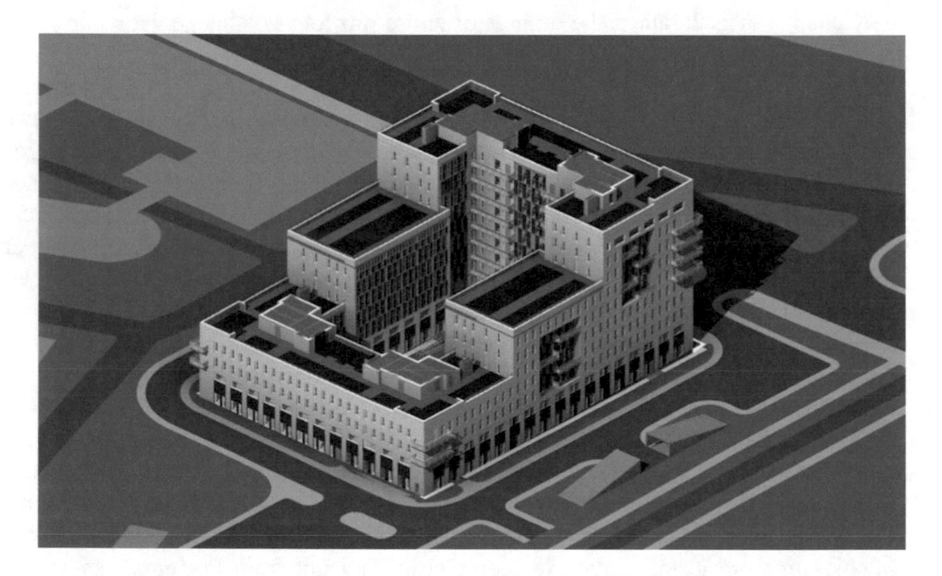

Fig. 1. Apart-hotel building with deepening for multifunctional complex.

Directive duration of construction is defined on basis of volumes of construction works and Russian standards on cost and labor intensity of respective works. This duration was also obtained on basis of construction schedule, which was developed in software package the Microsoft Project. Directive duration of construction is 652 days (Fig. 2). Critical tasks in Gantt chart has red color.

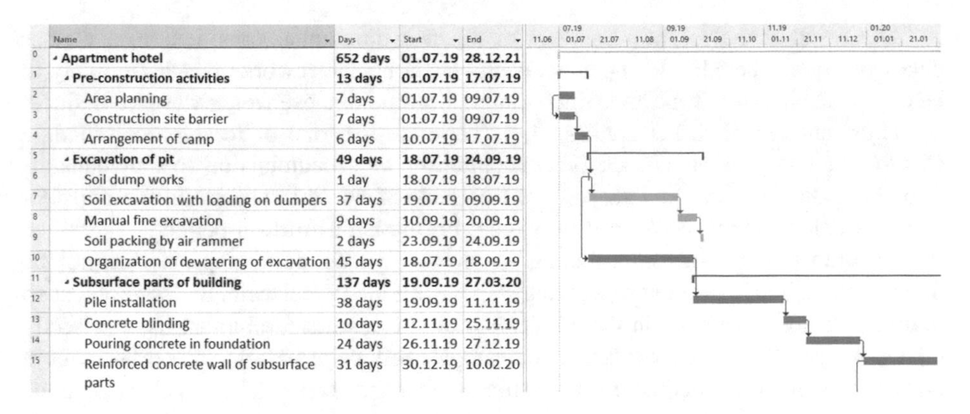

	Name	Days	Start	End
0	**Apartment hotel**	**652 days**	**01.07.19**	**28.12.21**
1	**Pre-construction activities**	**13 days**	**01.07.19**	**17.07.19**
2	Area planning	7 days	01.07.19	09.07.19
3	Construction site barrier	7 days	01.07.19	09.07.19
4	Arrangement of camp	6 days	10.07.19	17.07.19
5	**Excavation of pit**	**49 days**	**18.07.19**	**24.09.19**
6	Soil dump works	1 day	18.07.19	18.07.19
7	Soil excavation with loading on dumpers	37 days	19.07.19	09.09.19
8	Manual fine excavation	9 days	10.09.19	20.09.19
9	Soil packing by air rammer	2 days	23.09.19	24.09.19
10	Organization of dewatering of excavation	45 days	18.07.19	18.09.19
11	**Subsurface parts of building**	**137 days**	**19.09.19**	**27.03.20**
12	Pile installation	38 days	19.09.19	11.11.19
13	Concrete blinding	10 days	12.11.19	25.11.19
14	Pouring concrete in foundation	24 days	26.11.19	27.12.19
15	Reinforced concrete wall of subsurface parts	31 days	30.12.19	10.02.20

Fig. 2. Fragment of construction schedule of apart-hotel.

The resource graph (Fig. 3) reaches a peak with 142 workers.

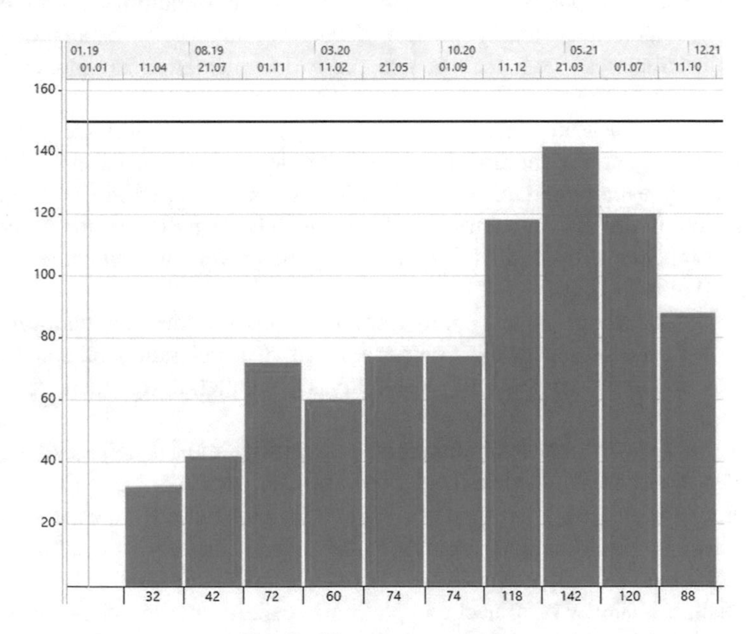

Fig. 3. The resource graph.

After working-out of construction schedule, an analysis of risks affecting duration of project was carried out using the PERT method.

Expert decision about duration of each work is needed to estimate expected duration of construction of the object using the PERT method, in three options: pessimistic, most likely and optimistic [9, 10].

As a result of analysis of possible risks affecting duration of construction, as well as data from researchers [3, 4, 11], conclusion was done about works which are likely to be completed ahead of time and which - later. Volume of these works was also defined.

Thus, authors of earlier studies [3, 4, 11] state that one of reasons for schedule overrun of construction is lack of concentration of executors due to simultaneous implementation of several tasks, as well as lack of control. Therefore, we consider effect of delayed work performed in parallel on total construction period.

A number of reasons contributing to increase in duration of project cannot be directly taken into account at scheduling stage using the PERT method. Therefore, such reasons are not considered in the article. These reasons include human factor directly related to psychology of worker. For example, untimely starting of works without objective reasons, disinclination of contractor to hand over works which were done ahead of time, until target date, etc. [4] Such problems are eliminated by rational people management, for example, in form of material incentives or other ways of influence and control.

Likeliest reasons of schedule overrun of construction are non-compliance of production technology (supply of poor-quality materials and equipment, problems with supply of concrete, poor-quality interior decorating, untimely installation of reinforcement in structures, etc.), lack of necessary documentation to start construction [11–13].

Risk reduction measures include, for example, use of Russian-made materials and equipment in project (to minimize risk of late delivery of materials), design and development of several alternative designs at design stage (to avoid error detection in documentation during construction), use of special techniques (for example, method for evaluating suppliers, proposed in [14] and [15], to prevent for schedule overrun of construction), etc. [3, 12].

These activities were also taken into account in form of reduction duration of works (optimistic estimate of duration). Another variant of acceleration of construction is introduction at legislative level of possibility of parallel design and construction, considered in [16].

Based on three estimates of duration: pessimistic, most likely, and optimistic, probabilistic model is obtained. Expected duration of each work and the project as a whole, value of standard deviation of normal distribution curve from expected duration were obtained in the Microsoft Project to move from probabilistic model to deterministic one.

After that, probability of project completion at certain time intervals is determined, as well as probability of project completion in directive period is calculated using following formula:

$$Z = \frac{T_{plan} - T_e}{\sigma_{T_e}}, \quad (1)$$

where Z – critical ratio, T_{plan} – planned project implementation date, T_e – expected duration of the project, σ_{T_e} – value of standard deviation of normal distribution curve from expected duration.

3 Results

Result of applying the PERT method at scheduling stage of apart-hotel building construction is presented in Fig. 4.

	Name	Directive	Opt	Most likely	Pes	Opt	Most likely	Pes	Crit	Exp	Disp
0	**Apartment hotel**	**652 days**	**879 days**	**1000 days**	**1156 days**	**581 days**	**664 days**	**774 days**	**Yes**	**668.5 days**	**36.53**
1	**Pre-construction activities**	**13 days**	**17 days**	**20 days**	**23 days**	**17 days**	**20 days**	**23 days**	**Yes**	**20 days**	**0.33**
2	Area planning	7 days	6 days	7 days	8 days	6 days	7 days	8 days	Yes	7 days	0.11
3	Construction site barrier	7 days	6 days	7 days	8 days	6 days	7 days	8 days	Yes	7 days	0.11
4	Arrangement of camp	6 days	5 days	6 days	7 days	5 days	6 days	7 days	Yes	6 days	0.11
5	**Excavation of pit**	**49 days**	**85 days**	**94 days**	**104 days**	**43 days**	**46 days**	**52 days**	**Yes**	**46.5 days**	**1.81**
6	Soil dump works	1 day	1 day	1 day	2 days	1 day	1 day	2 days	Yes	1.17 days	0.03
7	Soil excavation with loading on dumpers	37 days	33 days	37 days	39 days	0 days	0 days	0 days	No	0 days	0
8	Manual fine excavation	9 days	8 days	9 days	10 days	0 days	0 days	0 days	No	0 days	0
9	Soil packing by air rammer	2 days	1 day	2 days	3 days	0 days	0 days	0 days	No	0 days	0
10	Organization of dewatering of excavation	45 days	42 days	45 days	50 days	42 days	45 days	50 days	Yes	45.33 days	1.78
11	**Subsurface parts of building**	**137 days**	**161 days**	**175 days**	**204 days**	**155 days**	**168 days**	**196 days**	**Yes**	**170.5 days**	**7.19**
12	Pile installation	38 days	35 days	38 days	45 days	35 days	38 days	45 days	Yes	38.67 days	2.78
13	Concrete blinding	10 days	9 days	10 days	11 days	9 days	10 days	11 days	Yes	10 days	0.11
14	Pouring concrete in foundation	24 days	23 days	24 days	30 days	23 days	24 days	30 days	Yes	24.83 days	1.36
15	Reinforced concrete wall of subsurface parts	31 days	29 days	31 days	35 days	29 days	31 days	35 days	Yes	31.33 days	1

Fig. 4. Fragment of apart-hotel construction schedule using the PERT method in the Microsoft Project.

Table columns contain following values (from left to right): directive duration of works from previous construction schedule, optimistic expert evaluation of duration of works, most likely expert evaluation of duration of works, pessimistic expert evaluation of duration of works, optimistic project duration, most probable project duration, pessimistic project duration, designation of critical tasks, expected project duration (T_e), variance, calculated from critical tasks.

Calculation of probability of project completion by a certain moment according to formula (1) taking into account expected duration of 669 days is presented in Table 1.

Probability of project completion in intervals $(T_e - \sigma_{T_e}; T_e + \sigma_{T_e})$, $(T_e - 2\sigma_{T_e}; T_e + 2\sigma_{T_e})$, $(T_e - 3\sigma_{T_e}; T_e + 3\sigma_{T_e})$:

- $\mathrm{P}(662 \leq \mathrm{T} \leq 676) = 68,3\%$;
- $\mathrm{P}(656 \leq \mathrm{T} \leq 682) = 95,4\%$;
- $\mathrm{P}(650 \leq \mathrm{T} \leq 688) = 99,7\%$.

There are 280 kg of standard fuel on 1 m^2 of total area of building, overall consumption amounts to 4720 thousand kg of standard fuel.

Table 1. Calculation of probability of project completion by a certain moment.

Project duration	Number of days	Value of σ_{T_e}	Critical ratio Z	Probability of project completion by given time, %
Directive	652	6.044	−2.81	0.32
Most probable	664		−0.83	19.7

Analysis of calculated data shows that for purposes of project control, it is important to know what "factor of safety" according to duration does the project have? In our case it is only 17 days, which indicates a high risk of delay in construction. One month of delay in construction leads to overuse of energy in volume of 217 thousand standard fuel.

4 Conclusion

The PERT method allow transformation of probabilistic model of construction progress into deterministic model based on pessimistic, most likely and optimistic estimates of duration of works and processes, which made it possible to take into account risks that have great influence on construction industry.

Results of calculation show applicability of the PERT method to medium-size projects. However, adequacy and accuracy of estimating project duration using the PERT method depend entirely on quality of expert estimating of durations involved in calculating expected duration. Therefore, the PERT method with involvement of highly qualified specialists is convenient and effective method for project planning and management.

Absence of works delays leads to significant saving of energy consumption, since delays, for example, in concreting in winter lead to losses through unheated parts, which were previously completed, or elements not filled in time. Extra costs appear on additional heating, lighting and energy supply of construction machines, construction site and construction accommodation cabins.

References

1. Bovteev, S., Petrochenko, M.: Method "Earned Value Management" for timescale controlling in construction projects. Appl. Mech. Mater. **725–726**, 1025–1030 (2015)
2. Bovteyev, S.V., Terentyeva, Ye.V.: Upravleniye srokami stroitelnogo proyekta [Time management of a construction project]. Upravleniye proyektami i programmami **2** (38), 158–173 (2014). (in Russian)
3. Simankina, T.L.: Sovershenstvovaniye kalendarnogo planirovaniya resursosberegayushchikh potokov s uchetom additivnosti intensivnosti truda ispolniteley. [Improving scheduling of resource-saving flows, taking into account the additivity of the intensity of labor of the performers]. The dissertation on competition of a scientific degree of candidate of technical Sciences. 05.23.08. St. Petersburg (2007)
4. Kotovskaya, M.A.: Osobennosti teorii ogranicheniy sistem Goldratta i metoda kriticheskoy tsepi v obla-sti kalendarnogo planirovaniya stroitel'nykh proyektov. [Specific features of Goldratt's theory of constraints and critical chain project management in the scheduling of construction projects]. Sovremennyye problemy nauki i obrazovaniya **4**, 234–242 (2014). (in Russian)
5. Behnam, A., Harfield, T., Kenley, R.: Construction management scheduling and control: the familiar historical overview. In: 4th International Building Control Conference. MATEC Web of Conferences, Persekutuan Kuala Lumpur, Malaysia (2016)

6. Mishakova, A., Vakhrushkina, A., Murgul, V., Sazonova, T.: Project control based on a mutual application of pert and earned value management methods. In: 15th International Scientific Conference "Underground Urbanisation as a Prerequisite for Sustainable". Elsevier BV (2016)
7. Mishakova, A.V., Vakhrushkina, A.V., Anishchenko, D.R., Tatarkina, Y.A.: Program evaluation and review technique as the tool for time control. Mag. Civ. Eng. **72**, 12–19 (2017)
8. Veselkova, Ya.R., Zhupley, I.V.: Prikladnyye aspekty minimizatsii riska v stroitel'nykh predpriyatiyakh. [Applied aspects of minimizing risk in construction enterprises]. In: XI International Scientific and Practical Conference "Actual problems of development of business entities, territories and systems of regional and municipal government". JSC Universitetskaya kniga, Kursk, Russia (2016). (in Russian)
9. Sackey, S., Kim, B.-S.: Schedule risk analysis using a proposed modified variance and mean of the original program evaluation and review technique model. KSCE J. Civ. Eng. **23**, 1484–1492 (2019)
10. Ravi Shankar, N., Sireesha, V.: An approximation for the activity duration distribution, supporting original PERT. Appl. Math. Sci. **3**, 2823–2834 (2009)
11. Romanovich, M.A.: Povysheniye organizatsionno-tekhnologicheskoy nadezhnosti monolitnogo domostroyeniya na osnove modelirovaniya parametrov kalendarnogo plana. [Increase organizational-technological reliability of monolithic housing construction on the basis of modelling the schedule]. The dissertation on competition of a scientific degree of candidate of technical Sciences. 05.23.08. St. Petersburg (2015)
12. Spric, M.L.: Matematicheskiye otsenki vozdeystviya riskov i kompensatsionnykh meropriyatiy na organizatsionno-tekhnologicheskuyu nadezhnost' stroitel'stva. [Mathematical estimations of the risks impact and compensation activities on the organizational and technological reliability of construcrion]. Sovremennoye stroitel'stvo i arkhitektura **3**(07), 45–51 (2017). (in Russian)
13. Petrichenko, G.S., Petrichenko, D.G.: Modelirovaniye upravlencheskikh situatsiy pri stroitel'stve ob"yektov nedvizhimosti. [Modeling of management situations in construction of real estate objects]. Politematicheskiy setevoy elektronnyy nauchnyy zhurnal Kubanskogo gosudarstvennogo agrarnogo universiteta **129**, 130–141 (2017). (in Russian)
14. Sinenko, S.A., Miroshnikova, I.M.: Vnedreniye metodiki otsenki postavshchikov kak odin iz sposobov sokrashcheniya srokov stroitel'stva. [Implementation of the methodology for evaluating suppliers as one of the ways to shorten construction time]. Sistemnyye tekhnologii **2**(27), 14–19 (2018). (in Russian)
15. Vedernikova, Yu., Samochkova, L., Alekseev, A.: Postanovka zadachi otsenki riskov pri vybore podryadnoy organizatsii v stroitel'stve. [Risk assessment problem at the task of supply selection in civil engineering]. Vestnik Permskogo natsional'nogo issledovatel'skogo politekhnicheskogo universiteta. Prikladnaya ekologiya. Urbanistika **3**(27), 105–119 (2017). (in Russian)
16. Volkov, A.A., Gurov, V.V., Kulikova, E.N., Guseva, O.B.: Aktual'nost' avtomatizatsii parallel'nogo proyektirovaniya i stroitel'stva slozhnykh ob"yektov. [Relevancy of automation of parallel design and construction of complex objects]. Internet-Vestnik VolgGASU **3** (23), 30–34 (2012). (in Russian)

Tax Risk Management at the Enterprise of the Energy Sector

Elena Lavrenteva(✉)

Admiral Makarov State University of Maritime and Inland Shipping, Dvinskaya Street 5/7, 198035 Saint-Petersburg, Russia
e_lavrentieva@mail.ru

Abstract. The main organizational and methodological approaches in developing a program of corporate tax risk management at the enterprises of the energy sector of the Russian Federation are considered. It needs to be noticed that tax risks are an integral part of the overall risk management system of the enterprise. Controlling those risks is crucial for the stable operation of energy enterprises and their economic development. The main aspects of tax risks for the state and business are systematized, which show their strong correlation and interconditionality to ensure, on the one hand, tax revenues to all levels of budgets (federal, regional and local), and, on the other hand, to develop entrepreneurial activity. The conclusion was drawn about the feasibility of developing a Tax Risk Management Program at the corporate level, which should be correct under the current legislation and comply with the strategic and operational goals of effective entrepreneurial development of the enterprise of the energy sector. An organizational and functional structure of such a Program has been proposed, it consists of five main successional sections: Section 1 - Forming the information field of the main financial and economic activity parameters; Section 2 - Identification, systematization and ranking of the enterprise's tax risks; Section 3 - Justification of management activity to reduce the impact of tax risks; Section 4 - Calculation of the main financial and economic performance indicators in accordance with the implementation of management procedures for the tax risk regulation; Section 5 - Assessing the results and making management decisions on the development of activities under the influence of tax risks. A methodological description is given, and organizational procedures for the implementation of each Program section are defined. The proposed organizational and methodological approaches to the development of corporate Tax Risk Management Program make it possible to provide flexible coordination of risk management. Taking into account hierarchical and multipurpose mechanism of the energy enterprises, these approaches allow achieving operational and strategic goals, and can be used for the enterprises of the energy sector of the Russian Federation of different organizational and legal forms to prevent or reduce the negative impact of tax risks.

Keywords: Tax Risk Management Program · Energy sector · Risk assessment · Tax risks

V. Murgul and M. Pasetti (Eds.): EMMFT-2018, AISC 983, pp. 146–157, 2019.
https://doi.org/10.1007/978-3-030-19868-8_14

1 Introduction

In real-life conditions of entrepreneur activity in the Russian Federation, the role and importance of tax risks increases significantly. On the one hand, this is due to changes in tax legislation. Thus, over the past five years, more than 1500 tax changes have been made of varying coverage and execution complexity. On the other hand, risk management actions are closely related to the corporate tax policy pursued by business entities. Tax risks are an integral part of the overall system of corporate risk management, controlling them is very important for the stable operation of enterprises and their economic development.

2 Materials and Methods

In the scientific literature, the problems of tax risks are widely studied by domestic and foreign scientists, both from the point of view of general theoretical approaches [1–13] and also taking into account industry characteristics, for example, water transport [14–17]. Studies have shown that tax risks should be considered from the state point of view represented by the tax authorities and the position of economic entities (taxpayers), which is systematized in Table 1.

Table 1. Key aspects of tax risks for government and business

The manifestation of tax risks	
For the government	For the enterprise of the energy sector
– the risk of reducing tax revenues in the budget system of the country, also due to the use of "legal" schemes by taxpayers to minimize tax payments – the risk of a systemic tax base narrowing due to a domestic and foreign business and investment reduction – risk of shadow economy development – the risk of reducing the competitiveness of the national tax system	– the risk of business reduction or liquidation as a result of increased tax burden or loss of business reputation and the resulting refusal of counterparties from business ties – the risk of reducing the financial resources and property potential due to financial losses in the form of additional payments to the budget, including penalties

The formulated manifestations of tax risks show the strong correlation and inter-conditionality of business and the government. Without going into the root cause of mutual influence, it is necessary to emphasize that tax revenues to budgets of all levels (federal, regional and local) directly depend on the development of entrepreneurial activity. In order to solve such problem on the corporate level, it is necessary to develop a Tax Risk Management Program (hereinafter Program). It must be correct under the current legislation, justified in accordance with the strategic and operational goals of effective entrepreneurial development of the enterprise.

3 Result

The proposed organizational and functional structure of such Program consists of five main sequentially developed sections, which are summarized in Table 2.

Table 2. Organizational and functional Program structure of corporate management of tax risks at the enterprise of the energy sector

Section name	Key points
Section 1. Forming the information field of the main financial and economic activity parameters	The information field combines external and corporate information
Section 2. Identification, systematization and ranking of the enterprise's tax risks	Specialists of the enterprise in their activity areas identify possible tax risks, which are then systematized, evaluated and ranked according to the degree of their influence on the results of the enterprise as a whole
Section 3. Substantiation of management actions to reduce the impact of tax risks	The management action plan is being developed to reduce the impact of tax risks
Section 4. Calculation of the main financial and economic performance indicators in accordance with the implementation of management procedures for the regulation of tax risks	A multivariate calculation of the main financial and economic performance indicators is performed in accordance with the management activity plan for reducing the impact of tax risks
Section 5. Evaluation of results and management decision-making on the development of activities under the influence of tax risks	Evaluation of the results and the adoption of a corporate tax risk management program for the effective development of the enterprise is performed

Section 1. Forming the information field of the main financial and economic activity parameters is carried out by summarizing the external and internal information.

External information reflects the statutory taxation parameters (the procedure for forming the tax base, tax rates, deadlines for paying taxes and insurance premiums, etc.), as well as established standards (sectoral profitability level, tax burden, average salary in the subject of the Russian Federation, where the enterprise operates) used in assessing tax risks.

Corporate information includes production, financial and economic indicators of the energy enterprise with a certain degree of detail:

- production: the volume of production, the amount of work performed or services rendered as a whole in the enterprise and separately by type of activity; list of fixed assets separated by depreciation groups; list and terms of rental property; list of unused property, etc.
- financial and economic: revenue (income) by type of activity, expenses (depreciation charges, material expenses, salary budget, insurance contributions to extra-budgetary funds, other expenses, tax expenses), profit (product sales/work/services

profit; profit before tax, net profit), overdue interest and penalties for tax liabilities, etc.

Section 2. Identification, systematization and ranking of tax risks at the enterprise of the energy sector are carried out by employees in their respective activity areas.

Initially, it is necessary to systematize the tax risks that are common for a particular enterprise of the energy sector of the Russian Federation. Differentiated approach to the ranking may be taken as the basis, it is proposed taking into account the specifics of shipping companies [18]. Adapted generalized ranking of tax risks at the enterprise of the energy sector is given in Table 3.

Table 3. Ranking of tax risks at the enterprise of the energy sector

Ranking feature	Tax risk type (TR)
1. Tax risks of procedural and institutional nature	
Level of management and regulatory control	TR in the execution of federal and regional laws, government regulations, industry and local regulations
Source of interpretation	Court decision TR, the use of tax authorities explanations and the Ministry of Finance of Russia. TR of expert's comments application, of auditors, private opinion
Responsibility	The risk of tax liability for the results of the enterprise. Administrative and criminal responsibility of the corporate leaders
Penalty type	TR of charging penalties, fines, imposition of arrest
Completeness of budget receipts	Risks of overpaying or underpaying of taxes, tax transfer to another budget
2. Tax risks associated with the accounting policies of an enterprise	
Method of settlement with accountable persons	Tax risks in cash or non-cash payments
Accounting for depreciable property	TR of using the linear and non-linear depreciation method for fixed assets; use of depreciation premium
Cost of expensive repair of fixed assets	TR of assessing the cost of repairs as part of other expenses at a time or evenly allocating expenses between repairs
Accounting operating licenses	TR of even cost accounting over the period for which a license was issued, or TR of writing off expenses in proportion to the income, for which expenses were incurred
The accounting materials method	TR considering the prime cost of each unit, the average prime cost, the prime cost of the first materials regarding the acquisition time

(*continued*)

Table 3. (*continued*)

Ranking feature	Tax risk type (TR)
The allocation of direct expenses	TR while determining the list of direct expenses; during the distribution of direct costs in proportion to: the salary of the main staff, direct material costs, maintenance costs and equipment service
Accounting for insurance costs	TR of lumpsum accounting as part of expenses or as a prepayment
Accounting for assets, liabilities, income and expenses expressed in a foreign currency	TR of applying various recalculating methods for assets and liabilities, income and expenses in foreign currency
Creation of provisions	TR of reserves: for doubtful debts, for forthcoming repairs of fixed assets, forthcoming expenses for vacation pay, for payment of annual remuneration for meritorious service
Separate accounting for value added tax (VAT)	TR during the implementation of taxable and non-taxable operations, the use of different rates
Cost recovery method	Tax risks for accrual, cash method
3. Risks for the regimes and elements of taxation	
Tax regime	The risk of applying a general tax regime and a simplified tax system
Tax type	Risks separately for each tax type
Subject of taxation	TR of the enterprise as a taxpayer or tax agent, an entrepreneur without a legal entity, and interdependent persons
Object of taxation and tax base	TR of changes in objects of taxation, revenue (income), expenses, profits, property, vehicles and land plots of the enterprise
Tax rate	The risk of applying different tax rates
Tax remissions	The risk of tax exemptions and deductions
Deadlines for the tax payment and reporting	Risks of violating tax payment deadlines, violation or non-submission of tax reports
4. Tax risks by the nature of their occurrence and elimination	
Impact factors	External and internal TR
Occurrence factor	TR coming from political and socio-economic conditions; from legal nature; from informational, technical and technological, operational, organizational and managerial activity conditions
Nature of impact	Objective and subjective tax risks

(*continued*)

Table 3. (*continued*)

Ranking feature	Tax risk type (TR)
Sources of control	Tax risks managed by internal or external sources of the enterprise; also with combined method
5. Tax risks depending on the time of their impact	
Occurrence time	Retrospective, existing, prospective TR
Predictability	Predictable and unpredictable TR
Validity period	Constant and temporary TR
Occurrence probability	TR of high, medium and low degree of occurrence
6. Tax risks by the nature of their consequences	
Exposure possibility	Eligible, regulated, unrecoverable tax risks
Degree of loss	TR of insignificant or allowable losses Critical or catastrophic TR
Result of a risk event	The risk of overpayment or additional taxes; increasing the tax burden; loss of liquidity; lost profits; bankruptcy; strengthening of tax control; loss of competitiveness or business reputation

Taking into account the classification in the corporate management, the correlation between the labor functions performed by employees and possible tax risks is structured, a fragmentary example of such correlation is given in Table 4.

Table 4. Correlation between the functions performed by employees and possible tax risks

Employment functions	Impact on tax risk
Deputy General Director for finance - chief accountant and employees of the subdivisions subordinate to him	
Development and legal expertise of accounting policies, including ones for tax purposes	Determinating the most rational accounting option for income, expenses, property value and creation of reserves will allow reasonably forming the tax base for individual taxes to regulate the tax burden and ensure profitability. The inconsistency of the Accounting Policy provisions with legal acts entails penalties and additional tax liabilities

(*continued*)

Table 4. (*continued*)

Employment functions	Impact on tax risk
Coordination of tax risk management system in the enterprise	The lack of coordinated work in the tax risk management system will not allow the rational formation of tax liabilities, which may lead to a decrease in operating efficiency
Analyzing and planning tax liabilities of an enterprise	Unplanned tax liabilities can lead to losses and increase in penalties, fines, additional taxes, which reduce the company's net profit
Timely reporting and transfer of accrued amounts for taxes and insurance premiums	The presence of penalties and fines reduces the net profit of the enterprise
Tax accounting	The absence of tax accounting entails an uncontrolled tax base formation, it is also the basis for bringing the enterprise to tax liability
Legal expertise of contracts	The exclusion of unfair counterparties, the choice of paying counterparties or VAT defaulters will lead to a rational formation of tax liabilities
Justification of prices for products, works, services	The level of prices must comply with the principles of determining the taxation price
Justification of income and expenses	Reasonable income and expenses are the basis for proper tax liability formation
Technical director and employees of his subordinate departments	
Preparation of contracts for repairs, property rental	The exclusion of unfair counterparties, the choice of paying counterparties or VAT defaulters will lead to a rational formation of tax liabilities. The leasing period of the vehicles does not exempt the enterprise from the accrual and transport tax payment
Preparing documents for conservation of fixed assets	When suspended for a period of more than three months, fixed assets: - are excluded from the depreciable property list, the amount of depreciation can not be included in the expenses for tax on profit of enterprises - increase property tax liabilities, since depreciation for such objects is suspended for the entire conservation period
Analysis of the repair service market and preparation of contracts for maintenance and repair of property	The exclusion of unfair counterparties, selection of paying counterparties or VAT defaulters will lead to a rational formation of tax liabilities

(*continued*)

Table 4. (*continued*)

Employment functions	Impact on tax risk
Deputy General Director for personnel - head of the personnel department and employees of the structural units subordinate to him	
Development of the remuneration кegulation	The average monthly salary per employee of an enterprise below its average industry level in the relevant subject of the Russian Federation is thc basis for a tax audit
Development of the Provision on social security of employees	Accounting for the social security costs of employees must comply with the requirements of tax legislation. For example, to account for the costs of voluntary life insurance when calculating the corporate income tax, an insurance contract must be concluded with a Russian insurance company that has a license for a period of at least five years. Such enterprise also should not provide the insurance payments in favor of employees, except for payments in the case of the death of an employee and/or injury to his health (the total amount of payments under such contracts, taken for the purpose of reducing income tax, should not exceed 12 percent of the amount of expenses for payment labor)
Elaboration of the Regulation on professional development for employees	To account for expenses when calculating the profit tax of enterprises, advanced training should be conducted in a Russian or foreign educational institution that has all the necessary documents

The identified tax risks are reviewed and ranked out of significance by qualified specialists, who are experts of the enterprise, using the methods of expert analysis [19]. The agreement dimension of expert opinions can be assessed using the Kendall's concordance coefficient:

$$W = \frac{12S}{n^2(m^3 - m)}, \quad (1)$$

where S is the sum of errors squared magnitudes of all rank assessments for each reviewed subject from the average value; n is the number of experts; m is the number of reviewed subjects.

The Kendall's concordance coefficient varies in the range of $0 < W < 1$. The value of 0 is complete inconsistency, 1 is complete consistency. If the Kendall's coefficient value is more than 0.40–0.50, then the consistency of expert opinions is considered satisfactory, it is considered high if W > 0.70–0.80.

Section 3. Justification of management activity to reduce the impact of tax risks involves the development of a management activity plan to reduce the impact of tax risks at the enterprise of the energy sector. In this regard, based on the results of the previous Program section, i.e. on the basis of ranking the most significant and dangerous tax risks for the enterprise n, measures are justified for reduction of the negative impact in case of risky situation, indicating the timing and responsible persons.

In a formalized form, the set of tax risks can be represented in thc following correlation:

$$L\{l_1, l_2, l_3 \ldots l_s\} \text{ when } l = \overline{1, s}, \qquad (2)$$

where L is the set of measures for managing tax risks;

l_1 - tax risk management measures related to the production process;

l_2 - tax risk management measures, associated with the commercial and marketing work of the enterprise;

l_3 - tax risk management measures related to the staff costs of the enterprise;

l_s - s-th tax risk management measures.

Section 4. Calculation of the main financial and economic performance indicators in accordance with the implementation of management procedures for the tax risk regulation.

A multivariate calculation of the main financial and economic performance indicators is carried out in accordance with the implementation of the management action plan to reduce the tax risk impact.

It is necessary to choose the most significant and complex indicator, which represents the effectiveness of the activity. Such an indicator can be the profitability of the activity, allowing a comprehensive effectiveness assessment of the measures implemented.

In addition, it is important to justify certain restrictive conditions established by statutory and regulatory documents, as well as by corporate local acts. Table 5 systematizes the target indicator and the main restrictive control conditions for the implementation of management procedures for regulating tax risks at the enterprise of the energy sector of the Russian Federation.

Section 5. Assessing the results and making management decisions on the development of activities under the influence of tax risks.

Assessing the results and adopting the corporate tax risk management program for the effective development of the enterprise are being carried out.

Table 5. Modeling the calculation of the main financial and economic performance indicators in the process of corporate tax risk management

1. The goal is to keep profitability from decreasing

2. Restrictions:

2.1. Systemic – based on current legislative and regulatory documents

Quantitative:

2.1.1. The absence of losses in the enterprise during several tax periods

2.1.2. The absence of a significant sales profitability deviation and deviation of organizational assets from the standard values in the enterprise field

2.1.3. The outpacing rate of revenue growth over the growth rate of the enterprise's expenses

2.1.4. The average monthly salary of enterprise's employees is not lower than its average sectoral value in the relevant subject of the Russian Federation.

2.1.5. The actual tax burden of the enterprise is greater or equal to its average industry value.

2.1.6. The share of tax deductions on VAT in the amount of accrued VAT for the year does not exceed the established standard.

Qualitative: the absence of contracts with intermediary counterparties without reasonable economic or other reasons; submission of explanations and documents to the tax authorities; the absence of repeated deregistration and registration with the tax authorities in connection with the location change of the enterprise; lack of tax evasion schemes considering financial and economic activities.

2.2. Corporate - based on local documents of the enterprise

Quantitative:

2.2.1. Revenues of the enterprise are not reduced after the implementation of measures.

2.2.2. The average salary of employees is not reduced after the implementation of measures.

2.2.3. Prices for goods and tariffs for the services of the enterprise are in the established range from the industry level for identical (homogeneous) goods and services.

2.2.4. The financial result of the enterprise is not reduced after implementing measures.

2.2.5. No penalties and fines.

2.2.6. Tax to revenue ration (costs for direct taxes, tariffs, profit) do not increase after implementing measures.

2.2.7. The efficiency ratio of preferencial treatment does not decrease after implementing measures.

Qualitative: using separate accounting of income and expenses for mutual settlements with enterprises by payers or non-payers of VAT; compliance with the statutory deadlines for the payment of taxes (fees), insurance premiums and mandatory payments; control of: the share of each activity type in the total amount of the enterprise's revenue in order to determine the rate of insurance premiums for compulsory social insurance against industrial accidents and occupational diseases; use of vehicles in accordance with main activity type; for the correct application of tax incentives (exemptions and preferences); for import substitution measures, the acquisition of raw materials and services mainly in national currency from Russian suppliers; checking counter-agents for solvency throughout the life of the concluded contracts.

4 Discussion

The proposed architecture and content of the program of the corporate tax risk management at the enterprise of the energy sector, on the one hand, allows the company's specialists to analyze in detail and autonomously every stage of tax risk management: from the formation of financial and economic indicators as initial parameters to the assessment of the results of management actions taken under the influence of tax risks. On the other hand, such a composition of the Program is distinguished by a logical consistency and comprehensiveness of considering the management mechanism, which is necessary for a comprehensive analysis and long-term planning of the activities of a particular energy enterprise, taking into account the risks of taxation. Thus, a combination of differentiation and unity provides the versatility and multi-level process of corporate tax risk management to ensure effective economic development of the enterprise of the energy sector.

5 Conclusion

The proposed organizational and methodological approaches to the development of the corporate Tax Risk Management Program allow achieving flexible coordination of risk management in a hierarchical and multi-purpose mechanism of the enterprise of the energy sector. The approaches help to achieve operational and strategic objectives, and can be used for energy enterprises of different organizational and legal forms in order to prevent or reduce the negative impact of tax risks.

References

1. Bobrova, A.V.: Otsenka i analiz nalogovykh riskov v predprinimatelstve. Vestnik IuUrGU **41**, 106–112 (2011). (in Russian)
2. Bogdanova, A.E.: Sovershenstvovanie upravleniia nalogovymi riskami, obespechivaiushchego ekonomicheskii rost predpriiatiia, vol. 172, pp. 417–426 (2013). (in Russian)
3. Vylkova, E.S.: Upravlenie nalogooblozheniem: modeli povedeniia gosudarstva i ekonomiche-skikh subieektov. Izvestiia Dalnevostochnogo federalnogo universiteta. Ekonomika i upravlenie **2**(82), 51–58 (2017). (in Russian)
4. Goncharenko, L.I.: Nalogovye riski: teoreticheskii vzgliad na soderzhanie poniatiia i faktory vozniknoveniia. Nalogi i nalogooblozhenie **1**, 17–24 (2009). (in Russian)
5. Egorova, O.Ia.: Upravlenie nalogovymi riskami organizatsii. Formirovanie ekonomicheskogo portreta natsionalnoi infrastruktury strany: metodologicheskii i teoreticheskii aspekty, p. 166. Analiticheskii tsentr «ekonomika i finansy», Moscow (2014). (in Russian)
6. Lisovskaia, I.A.: Nalogovye riski: poniatie, faktory vozniknoveniia metody upravleniia **1**(1), 3–13 (2011). (in Russian)
7. Panskov, G.V.: Nalogi i nalogooblozhenie: teoriia i praktika, p. 680. Iurait, Moscow (2011). (in Russian)
8. Pinskaia, M.R.: Nalogovyi risk: sushchnost i proiavlenie. Finansy **2**, 43–46 (2009). (in Russian)

9. Semenova, O.S.: O podkhodakh k klassifikatsii nalogovykh riskov. Finansy i kredit **44**(476), 71–76 (2011). (in Russian)
10. Churilkina, E.V.: Nalogovye riski: Sushchnost, prichiny vozniknoveniia, metody upravleniia. Sistemnoe upravlenie **4**(13), 1–9 (2011). (in Russian)
11. http://www.sciencedirect.com/. Accessed 01 Mar 2018
12. https://www.sciencedirect.com/science/article/pii/S0022199616300897. Accessed 01 Mar 2018
13. https://www.pwc.com/mt/en/publications/assets/tax-management-in-companies-06.pdf. Accessed 05 Mar 2018
14. Lavrenteva, E.A.: Nalogovoe regulirovanie: teoriia i praktika (na primere sudokhodnogo biznesa), Saint-Petersburg, GMA Makarova, S.O., p. 148 (2007). (in Russian)
15. Lavrenteva, E.A.: Nalogovye aspekty razvitiia Rossiiskogo mezhdunarodnogo reestra sudov. Transport Rossiiskoi Federatsii **1**(56), 41–46 (2015). (in Russian)
16. Lavrenteva, E.A., Plotnikova, A.I.: Nauchnye podkhody k sushchnosti upravleniia nalogovymi riskami v sudokhodnoi deiatelnosti. Vestnik Makarova, S.O. **2**(30), 154–164 (2015). (in Russian)
17. Fiutik, I.G.: Upravlenie nalogovymi riskami na predpriiatiiakh transporta. Nauchnye problemy transporta Sibiri i Dalnego Vostoka, vol. 2, pp. 8–12 (2018). (in Russian)
18. Lavrenteva, E.A.: Klassifikatsiia nalogovykh riskov v sisteme risk-menedzhmenta sudokhod-noi deiatelnosti. Vestnik Astrakhanskogo gosudar-stvennogo tekhnicheskogo universiteta, vol. 2, pp. 108–123. Ekonomika, Seriia (2016). (in Russian)
19. Kozlov, V.A.: Ekspertnye otsenki, Moscow, p. 173 (2018). (in Russian)

Algorithm of Organizational and Technological Design of High-Rise Buildings Construction with Due Account for Energy Efficiency

Taisiia Syrygina[1], Tatyana Simankina[1(✉)], Olga Vasilyeva[2], and Anna Kopytova[3]

[1] Peter the Great St. Petersburg Polytechnic University, Polytechnicheskaya, 29, 195251 St. Petersburg, Russia
t.s.-95@mail.ru, talesim@mail.ru
[2] Moscow State University of Civil Engineering, Yaroslavskoye shosse 26, 129337 Moscow, Russia
[3] Tyumen Industrial University, Volodarskogo Street 38, 625000 Tyumen, Russia

Abstract. The article identified specific features of the development of organizational and technological models of construction of high-rise buildings in the organizational and technological documentation. The attention is focused on the existing drawbacks and problems of designing high-rise buildings, which are faced by designers. Some characteristic features of high-rise buildings and soils, affecting the choice of the type of foundation, are reflected and the options for foundation systems for these cases are considered. At the planning stage of new construction, the costs associated with the energy consumption of the building, equipment and support systems throughout the life of the facility should be assessed, resulting in additional financial savings in the future. The qualified organizational and technological designing allows to increase technical and economic indicators of construction operation along with its organizational and technological reliability, promotes optimization of construction processes. The provided organizational and technological documentation allows to improve the level of competitiveness of construction companies and commissioning the facility with required level of quality in due time, as well as to increase energy efficiency of the designed facility and to make its erection more power efficient. In order to reduce labor costs, material consumption, product cost and construction time, while technological efficiency is increasing an algorithm has been developed taking into account the peculiarities of construction of high-rise buildings.

Keywords: High-rise construction · Projects of construction organization · Organizational and technological documentation · Energy efficiency

1 Introduction

In the majority of the cases in Russia as well as abroad one of the most popular materials for constructing frames of high-rise buildings is monolithic reinforced concrete [1, 2]. This article represents characteristic features at organizational and

V. Murgul and M. Pasetti (Eds.): EMMFT-2018, AISC 983, pp. 158–165, 2019.
https://doi.org/10.1007/978-3-030-19868-8_15

technological design from in-situ reinforced concrete in high-rise construction, including energy efficiency accounting.

The relevance of this topic is supported by the increasing need for residential, social, business and industrial facilities as well as the existing deficit of available construction sites in large cities. These factors continuously result in the necessity to resort to high-rise construction in metropolitan cities. The international experience of contemporary urban development shows that from the point of view of economy 20–50-storey buildings are the most advantageous [3]. Taller buildings are usually constructed for prestige and status or as a result of lack of available urban areas.

The most important feature of high-rise construction, which has a considerable impact on the organizational and technological construction model, is the concentration of large volumes of in-situ reinforced concrete in the construction footprint [4]. If in a building of 20–30 storeys 1 m^2 of construction footprint contains 9–11 m^3 of reinforced concrete, in buildings with 100–200 floors this parameter may amount to as much as 50–60 m^3/m^2 [5].

Vertical orientation of a high-rise building, limited working area on the floor, a high number of floors are factors that raise the necessary of minimizing the time of construction of each floor. Another important factor that impacts the speed of high-rise buildings construction is a significant increase in wind speed and pulsating gust loads the higher the building [6]. This means that projects of construction organization, which are an integral part of the design documentation, should account for numerous factors influencing the organizational and technological model of constructing high-rise buildings. An important objective of designing a projects of construction organization is making decisions that allow the possibility of employing a series of effective organizational and technological schemes, particularly applying innovative energy saving technologies in construction of building, which undoubtedly helps to significantly reduce future costs for its maintenance [7–9]. In fact projects of construction organization is the main document of strategic construction planning defining basic decisions and solutions in the organization and production technology which are described in detail in subsequent documents.

2 Materials and Methods

Design is the process of developing a set of documentation for the construction, reconstruction and overhaul of a certain object, which contains information in text and graphic form. Organizational and technological design (OTD) particularly contains solutions for the organization of construction and performing construction and assembly works.

Currently the level of OTD construction of buildings and structures is rather low. According to numerous experts, about half of the projects of construction organization, developed as part of the design and estimate documentation, are not fully implemented. Only a fraction of the information contained in the projects of construction organization is directly realized in the construction of facilities, one of the reasons is because low-quality organizational and technological documentation is rarely used by contractors [10].

Apart from that, among the drawbacks of organizational and technological solutions and schemes (adopted during the OTD), the level of quality of construction and safety of objects under construction may significally drop [11]. Unqualified of improperly trained engineering personnel in the development of organizational and technological documentation leads to the fact that many of these documents are mainly formal in their essence. In addition, the use of modern construction methods (flow-line, component assembly, factory-assembled) is often ignored, the work of all construction participants is not properly coordinated (there are no work area opening dates, time schedules are not adjusted) [12], packaged delivery of materials is not organized (production and technological equipment capacities are not taken into account), etc.

The situation is extremely unfavorable in terms of work execution design: production schedule plans are underdeveloped, construction plans are mainly focused on binding of assembly cranes. Technological charts are developed for some processes and are underdeveloped in terms of their composition and content for others.

Currently a series of problems in the field of organizational and technological design (OTD) can be categorized as follows:

1. Methods and solutions of OTD objectives
2. Formal approach to OTD objectives solution [10].

The second issue exists as a result of insufficient normative framework of OTD. Despite the fact that the content and requirements to projects of construction organization and work execution design are established by the Regulation of the Government of the Russian Federation No. 87 and Set of rules 48.13330.2011 "Construction Organization", these standards for organizational and technological documentation need thorough adjusting as the development of projects of construction organization and work execution design is established rather superficially.

In spite of the fact that by the current development stage the construction industry in the Russian Federation has accumulated enormous experience in constructing high-rise buildings, the reliability of the organizational and design documentation remains relatively low. The above mentioned applies to almost all sections that are developed as part of project documentation, including the section "Energy Efficiency".

At present, issues related to improving the energy efficiency of erected facilities and the process of their construction are attracting general attention. The problem of energy saving is also considered in our country, which is reflected in the Federal Law No. 261-FZ "On energy saving and energy efficiency improvement and amendments to certain legislative acts of the Russian Federation". And the projects of energy-saving and energy-active houses are gaining popularity. In this regard, the quality and high adaptability of all components of construction, on which the achievement of high energy efficiency actually depends, are of particular importance.

In view of its specificity, high-rise construction goes along with numerous problems. A calculation of wind flow over for high-rise buildings is necessary for determining gust loads on the building structure as well as for accounting for wind in the interchange of air in premises.

However, if upon constructing the superstructure its safety can be calculated at the stage of developing the organizational and technological model, the case with the substructure is somewhat different, because buried structures are subject to influence of a more significant number of factors, which are difficult to predict at the stage of developing the organizational and technological model even with the help of powerful computing systems and modern software package. Such natural factors as quick grounds, soil with low load capacity, etc., complicate designing, but accounting the aforementioned is one of the most important parts of the project. Thus, the selection of an appropriate organizational and technological scheme is the most important stage of the design and planning process. Successful completion of this stage has considerable impact on parameters of future the high-rise building in general as well as those of neighboring buildings of the construction site.

For even distribution of colossal load of the superstructure on the soil, the foundation has to be sufficiently solid. The selection of the foundation of a high-rise building is a special issue, because the foundation has to take and transmit on the ground high concentrated loads up to 2000 tons and even more.

The load intensity acting on the foundation and soil properties are two main factors that the choice of the type of foundation of a high-rise constructions lies upon. However, the following conditions should be accounted for:

- presence and intensity of seismic activity, technogenic tensioning in the soil in the construction area, disjunctive and folded deformation of tectonic structures in the construction area [13];
- groundwater and value of pressure of groundwater horizon, presence of underground flows, quick grounds and other underground anomalies;
- major construction works in the construction area;
- transport communications, subway tunnels, engineering communications and other facilities in the immediate vicinity that could either affect the integrity of the foundation or cause damage as a result of unavoidable shrinkage, as well as dynamic impacts from existing urban infrastructure facilities [14];
- the stress-strain state of the soil mass at the construction of any high-rise structure, as a rule, undergoes significant changes, which leads to the expansion of the zone of influence of new construction on the nearby development;
- climatic factors, first of all seasonal temperature fluctuations, ground freezing, thunderstorm frequency and wind speed, its powerful gusts at the height of about 300 meters, as well as lightning strikes sometimes have a very noticeable single load on the entire structure of the building, which affects the foundation.

When building a foundation for a high-rise building, one should pay attention to the following aspects:

- There is primary and secondary ground subsidence; and after the foundation is pressed with all the weight of a two-hundred-meter skyscraper, the ground deformation may become critical.
- Taking into account the uniqueness of high-rise buildings of the first category of responsibility and imperfection of the existing regulatory framework, it is recommended to conduct constant monitoring of the condition of soils, piles, grillage

foundation and enclosing concrete structures during the construction of high-rise buildings.

- Densely position the piles only in the load area (under the stiffness diaphragm).
- Several long piles are always better than a large number of short piles.
- The greater the load on the piles is on the corners and generally on the perimeter of the building (when calculating the load-bearing capacity of piles on the material and their construction, the overload of corner and perimeter piles relative to the central one should be taken into account).
- The soil under the slab should be overcompacted.
- The calculations should aim to create an underground volume so that the weight of the excavated soil in the construction of the underground part of the building is equal to the weight of the building.
- The load on the foundations should be transferred symmetrically relative to the central axis, using the appropriate structural scheme of the building and stiffener elements (monolithic walls, stairwells, etc.) should be placed symmetrically.
- The depth of the foundation of the building should increase as the height of the building increases.
- The pyramidal shape of the building should be used (if possible).
- When the building height is increased, the admissible limit value of the foundation subsidence must be reduced.
- High loads transferred to the base ground require taking into account the strength and deformation properties of rock and non-rock soils with modulus of deformation more than 100 MPa, which are considered incompressible in accordance with current standards, as well as the increased zone of distribution of stresses in the ground in terms of plan and depth, which may lead to an increase in the soil layers that perceive the load from the foundation (especially if the layers are unevenly distributed) [15].
- The increase in the size (depth and width) of the compressible thickness in the soil mass leads to an increase in the time required to complete soil consolidation and to a stretching of the subsidence process over time.
- Since high-rise buildings over 100 meters high are unique, complex geodetic and strain-gauge monitoring of building foundation structures should be organized to ensure safety during future exploitation at the construction stage.

3 Results

All the above parameters influence the development of organizational and technological documentation, its quality that is the degree of compliance with the current norms and standards, as well as the target specification for design.

High-rise buildings are presenting new challenges and tasks to engineers, especially with regard to OTD. Many of the traditional design methods cannot be applied with strong confidence for several reasons, therefore designers are being forced to use more sophisticated methods of analysis and design.

The application of all the above-mentioned parameters and principles are illustrated by developed algorithm (see Fig. 1). In this functional diagram the options for OTD, in particular foundation systems and method of supply concrete mixture, are discussed. Choice of the method of supply concrete mixture is one of the important blocks of the algorithm; it was described in the article [5] and doesn't need further explanations.

A number of important requirements are submitted to organizational and technological documentation, which developed as part of the design documentation: it must meet construction norms and regulations, normative documents and legislative acts, and besides it should be the most optimal in terms of technical solutions and cost. In addition, the OTD should ensure the practical interests of all construction industry participants. All of this was also taken into account at developing the algorithm proposed by the authors.

4 Discussion

The article considers some examples of increasing the efficiency of construction of high-rise buildings based on the presented algorithm of organizational and technological design of high-rise buildings construction. It is clearly demonstrated that thoroughly developed organizational and technological documentation with account for the parameters specified in the block diagram, allows to increase the level of competitiveness of construction companies and reduce the cost of construction and assembly works.

Upon designing and constructing high-rise buildings it is essential to carefully consider all environmental factors that may impact the process, including wind and seismic activities. It is also necessary to provide scientific and technical support at all stages of design, construction and operation of high-rise buildings.

It is important to underline that high-rise buildings allow to save space for natural parks and recreational facilities, where people can be engaged in sports and various outdoors activities. A competent planning of energy-efficient high-rise building is a solution that helps to gain great benefits for all participants of the construction process, as well as the environment.

Parameters reflected in the article and in the algorithm affect the optimization of the preparation of organizational and technological documentation. However, currently the majority of the provided indicators are often ignored, which leads to negative consequences. Yet, accounting for these features at the development stage will allow to carry out efficient preparation activities before construction. Thus, there is a growing need to introduce the category of "object complexity" and connect it with the requirements to the composition and content of organizational and technological documentation.

It should be noted that it is quite crucial to improve measures aimed at regulating organizational and technological documentation and preparation of engineering personnel (to prepare experts of appropriate qualification who can view the whole picture of problems connected with the construction site because the solution of issues, resolved within the project development process, are based on a combination of parameters: command of materials, technologies, equipment, analysis and careful consideration of local conditions, appropriate application of computer technologies and information support).

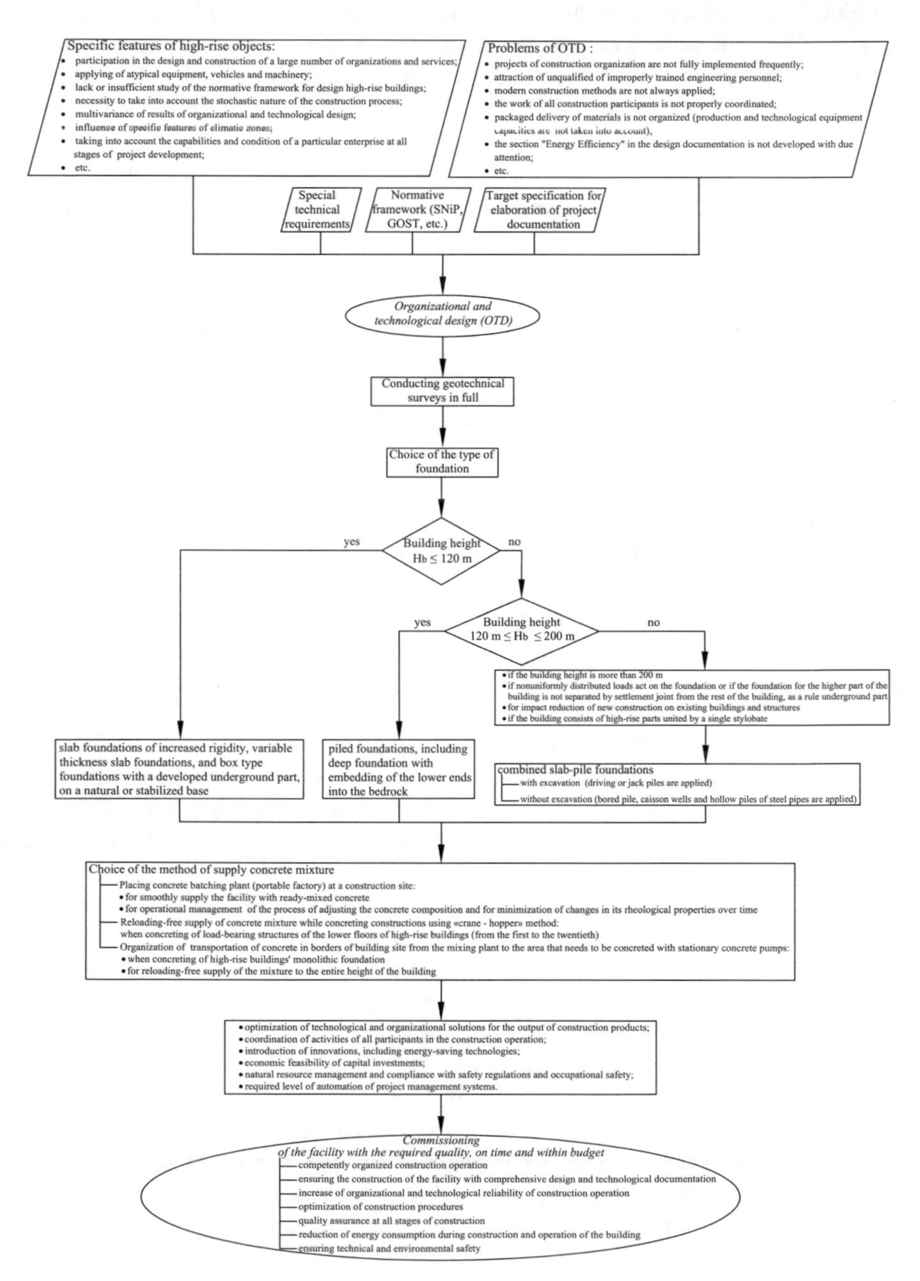

Fig. 1. Algorithm of organizational and technological design of high-rise buildings construction.

References

1. Kolchedancev, L.M., Osipenkova, I.G.: Osobennosti organizacionno-tekhnologicheskih reshenij pri vozvedenii vysotnyh zdanij. Vysotnoe stroitel'stvo **11**, 17–19 (2013)
2. Kiril'chuk, I.B.: Monolitnyj predvaritel'no napryazhennyj zhelezobeton: istoriya, primenenie, predposylki razvitiya. Problemy nauki **7**(31), 33–38 (2018)
3. Romanovich, M., Simankina, T., Tsvetkov, O.: Wavelet analysis function of changing work amounts in monolithic construction. In: Murgul, V. (ed.) International Scientific Conference Week of Science in SPBPU – Civil Engineering (SPbWOSCE-2015), MATEC Web of Conferences, vol. 53, p. 01054, Saint-Petersburg, Russia (2015)
4. Romanovich, M.A.: Povyshenie organizatsionno-tekhnologicheskoy nadezhnosti monolitnogo domostroeniya na osnove modelirovaniya parametrov kalendarnogo plana. Ph.D. thesis. SPSUACE, Saint-Petersburg (2015)
5. Osipenkova, I., Simankina, T., Syrygina, T., Lukinov, V.: Elaboration of technology organizational models of constructing high-rise buildings in plans of construction organization. In: Safarik, D., Tabunschikov, Y., Murgul, V. (eds.) International Scientific Conference on High-Rise Construction (HRC 2017), E3S Web of Conferences, vol. 33, p. 03045, Samara, Russia (2018)
6. Pimenova, E.V., Seniv, R.I.: Unikal'nye tekhnologii stroitel'stva vysotnyh zdanij i sooruzhenij. Voprosy nauki i obrazovaniya **11**(12), 227–228 (2017)
7. Ryzhevskaya, Ya.A.: Osobennosti proektirovaniya energoeffektivnyh vysotnyh zdanij v Rossii. Voprosy nauki i obrazovaniya **1**(13), 164–166 (2018)
8. Biktimirov, Z.M., Shigapov, A.I.: Energouchet kak osnova energosberezheniya. Mezhdunarodnyj akademicheskij vestnik **1–2**(21), 21–24 (2018)
9. Sumerina, O.A.: Energosberezhenie v zhilom stroitel'stve. Sinergiya nauk **21**, 518–522 (2018)
10. Oolakaj, Z.H., Dadar, A.H.: Ob organizacionno-tekhnologicheskom proektirovanii. Vestnik Tuvinskogo gosudarstvennogo universiteta **3**, 21–24 (2011)
11. Volkov, S.V., Shvedov, V.N.: Vliyanie organizacionno-tekhnologicheskih reshenij na uroven' kachestva stroitel'stva i bezopasnost' vozvodimyh zdanij. Izvestiya vysshih uchebnyh zavedenij. Stroitel'stvo **2**(662), 32–39 (2014)
12. Simankina, T.L.: Sovershenstvovaniye kalendarnogo planirovaniya resursosberegayushchikh potokov s uchetom additivnosti intensivnosti truda ispolniteley. Ph.D. thesis. SPSUACE, Saint-Petersburg (2007)
13. Mirsayapov, I.T., Koroleva, I.V., Sadykova, A.R.: Issledovanie vliyaniya sejsmicheskih i vetrovyh vozdejstvij na parametry svajno-plitnogo fundamenta vysotnogo zdaniya. Izvestiya kazanskogo gosudarstvennogo arhitekturno-stroitel'nogo universiteta **1**(31), 107–113 (2015)
14. Kolchedancev, L.M., Volkov, S.V., Volkova, L.V.: Organizacionno-tekhnologicheskie resheniya po ustrojstvu fundamentov vysotnyh zdanij. Zhilishchnoe stroitel'stvo. **9**, 50–54 (2016)
15. Ter-Martirosyan, Z.G., Telichenko, V.I., Korolev, M.V.: Problemy mekhaniki gruntov, osnovanij i fundamentov pri stroitel'stve mnogofunkcional'nyh vysotnyh zdanij i kompleksov. Vestnik MGSU **1**, 18–27 (2006)

Evaluation of the QMS Efficiency of Management Companies as a Criterion for the Effectiveness of Operation of "Intelligent Buildings"

Nikolay Ivanov(✉)

Moscow State University of Civil Engineering,
Yaroslavskoye shosse, 26, 129337 Moscow, Russia
IvanovNA@mgsu.ru

Abstract. The article describes an algorithm developed by the author for assessing a quality management systems (QMS) effectiveness of management companies, which involved in the operation of "intelligent buildings". Unlike other approaches described in the scientific literature, this algorithm is based on the joint application of collective and individual expert assessments to form a set of criteria for evaluating the effectiveness of QMS and for determining each criteria's level of significance. Formulas for rank calculation of the criteria and for determining a level of a relative increase/decrease in the effectiveness of the QMS are given. A graphic interpretation of the results of evaluating the effectiveness of the QMS and its trend is proposed. The algorithm takes into account the requirements of regulatory documents to assess the performance of the QMS.

Keywords: Quality management system · QMS effectiveness · "Intelligent building" · Management company · Innovation

1 Introduction

It is difficult to overestimate the relevance of the concept of "intelligent building" for the Russian market. Due to complex integration, its thoughtful qualified implementation allows to achieve savings of 10–15% in comparison to stay-alone engineering systems of buildings for various purposes [1]. The consumption of energy, water, gas, heat is reduced by approximately 30%. Accordingly, emissions to the environment and the cost of their disposal are reduced. In the its own turn the use of modern energy-saving technologies reduces the input power and resources, and therefore, makes it possible to use cheaper communications [2].

A cost of operating an "intelligent building" throughout its life cycle is significantly lower than a cost associated with traditional solutions.

Automated building management systems play an important role in costs reduction. As noted in [3], "Each engineering system is responsible for certain functions and ensures more efficient use of all building communications. A combining management of these systems leads to the manifestation of synergy - an increase in the efficiency of

V. Murgul and M. Pasetti (Eds.): EMMFT-2018, AISC 983, pp. 166–173, 2019.
https://doi.org/10.1007/978-3-030-19868-8_16

activities as a result of connection, integration, merging of separate parts into a single complex due to the system effect while improving safety and comfort, and greater saving of resources. In addition, the cost of building such a system decreases: it becomes more powerful and costs less than a dozen separate control systems". Along with efficiency improving of using innovative technical systems, the applied organization management technology is of great importance for the success of any modern management company. In this regard, the introduction of quality management systems into management companies is attention noteworthy. Moreover, it is important not the very formal presence of the QMS, but its effective functioning. The performance of the QMS is a reflection of the ability of the company's management to assess its place in a competitive market, to make decisions based on reliable data, to manage risks. In the ISO 9000 standards, the performance assessment of the QMS is defined as "one of the tasks of top management" and is positioned as an essential tool for the development and improvement of the QMS [4, 5]. However, due to the fact that the mentioned above standards are universal and do not contain any specific performance evaluation algorithms, both for individual processes and the QMS as a whole, the development of appropriate algorithms for the QMS of management companies, which involved in the operation of "intelligent buildings", is a task of a great practical importance [6–8].

2 Materials and Methods

The proposed algorithm for evaluating the QMS's performance of management companies involved in the operation of "intelligent buildings" is a two-step procedure. The block diagram of the algorithm is shown in Fig. 1.

At the first stage, it is proposed to assess the QMS's compliance of an organization or a company with the requirements of ISO 9001:2015. The main way to determine the compliance of the QMS with the requirements of the standard is to select several main evaluation criteria that can be obtained during periodic internal audits [6].

The first stage can be divided into a series of successive steps.

First, a group, which consists of K experts, using a group assessment method, such as the brainstorming method, forms a final set of criteria for a QMS's effectiveness evaluation. This set may contain N criteria, each of which, according to the general opinion of experts reflects the effectiveness of the QMS to some extent.

Then, it is suggested to each expert to order the criteria included in the set accordingly to degree of importance, starting with 1 and ending with N. Wherein, the value of "one" will correspond to the most significant criterion, and the value of N - the least significant criterion. The work at this step is conducted separately with each expert, which allows experts to express their opinions regardless of judgments of more experienced colleagues.

As a result of this step, a matrix of expert assessments for the criteria is formed; each row of the matrix corresponds to one of the evaluated criteria, and the columns - to the experts conducting the assessment. The element of the matrix Oij is a numerical value assigned to the criterion with number i by the j-th expert.

The matrix is supplemented by another column, where in each row is a sum of scores of all experts (Table 1).

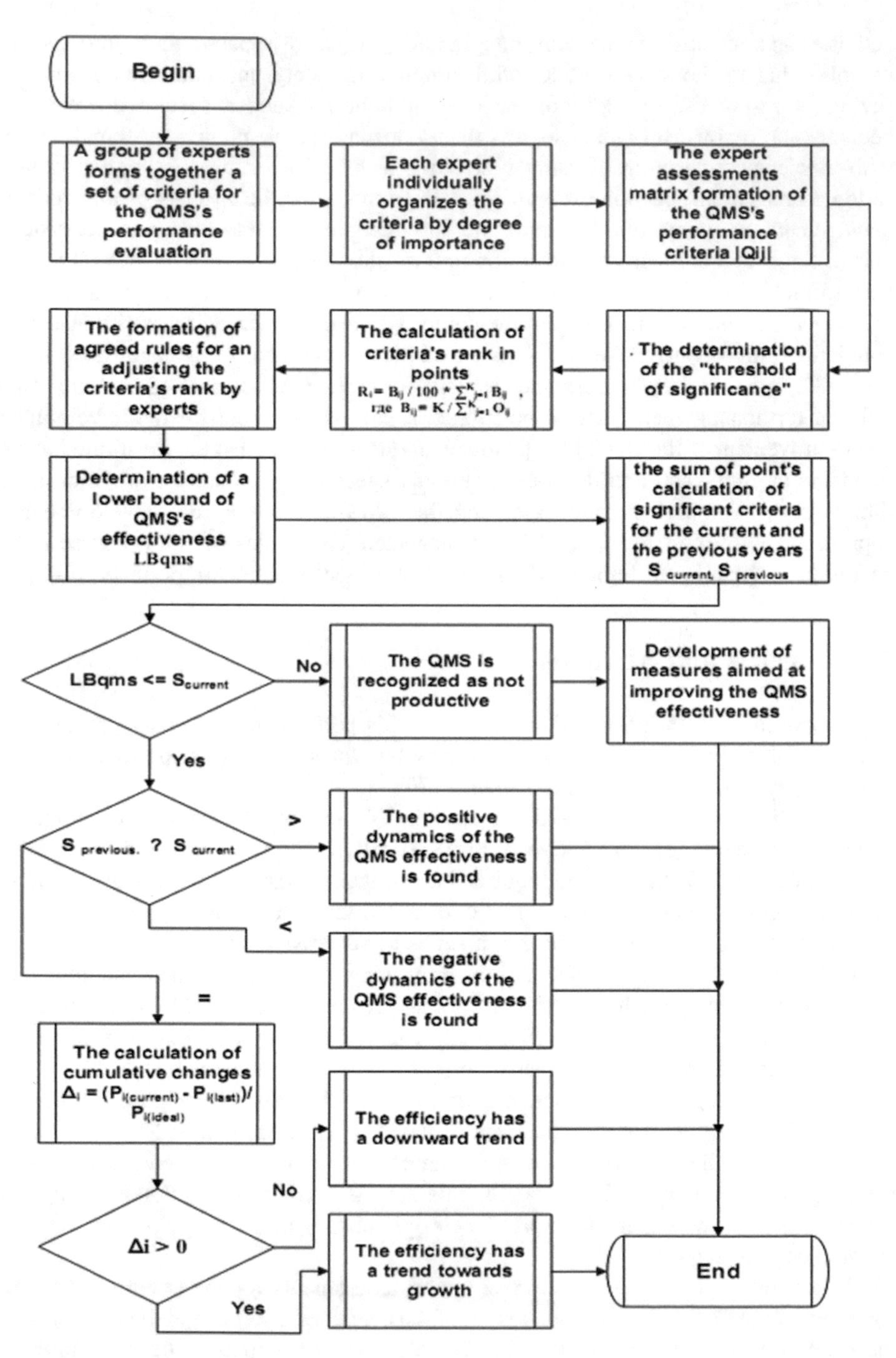

Fig. 1. The block diagram of the algorithm for assessing the effectiveness of the QMS

Table 1. A matrix of expert assessment for criteria

	Experts						Summa
Criteria	1	2	…	J	…	K	
1							
…							
i				Oij			Sij
…							
N							

Further, the experts again jointly discuss the obtained results and determine the so-called "threshold of significance". By "threshold of significance" we mean the sum of all assessments of one criterion, which is still considered by experts to be worth take into account. Criteria with a cumulative score greater than the "threshold of significance" are excluded from further processing.

The next step is to process the received sample of the relevant criteria. For each criterion, an average expert assessment is determined, and then a rank of the criterion is calculated in points. It is assumed that the sum of the quantities inversely proportional to average scores of each criterion corresponds to 100 points, and the rank of each criterion in points is proportional to the value, which is inversely proportional to the average estimate of the criterion:

$$R_i = B_{ij}/100 * \sum\nolimits_{j=1}^{K} B_{ij} \text{ where } B_{ij} = K/\sum\nolimits_{j=1}^{K} O_i \tag{1}$$

The rank of each criterion is rounded to a whole number, subject to the restriction that the sum of the ranks of all criteria is 100.

If the total score of all criteria provided by experts is equal to 100, it corresponds to the "ideal" state of an effectively functioning quality management system. This state implies a fulfillment of all the requirements of ISO 9001:2015.

Finally, the experts again jointly discuss rules for determining the actual indicators, which will demonstrate a compliance with the requirements of the standard, for each of the selected criteria used for assessing the QMS's effectiveness. For example, the rules presented in Table 2 are for the criterion "The number of complaints from consumers". The second stage of the algorithm consists of two steps.

Table 2. Rules for adjusting a criterion's rank value

Criterion name	Range of values	Points
The number of claims from consumers	No complaints	Maximum score
	Less than 3 claims	Minus 3 points
	From 3 до 5 claims	Minus 6 points
	More than 5 claims	Minus 8 points

The first step is to compare a sum of points of significant criteria, which is calculated for the current year, with the lower limit of amount of points for a quality management system, whose functioning can be considered productive. This limit is usually set in the range of 72–74 points. If the calculated sum of points of the current year is less than the specified limit, the QMS of the construction organization is considered ineffective and it is recommended to develop a set of corrective actions aimed at eliminating the causes that reduce the effectiveness of the QMS.

In the case when the first step gave a positive result, a comparison of the QMS's effectiveness in the current and previous years is performed.

3 Results

The total score of the current year may be higher, lower or equal to the value of the last year (Fig. 2).

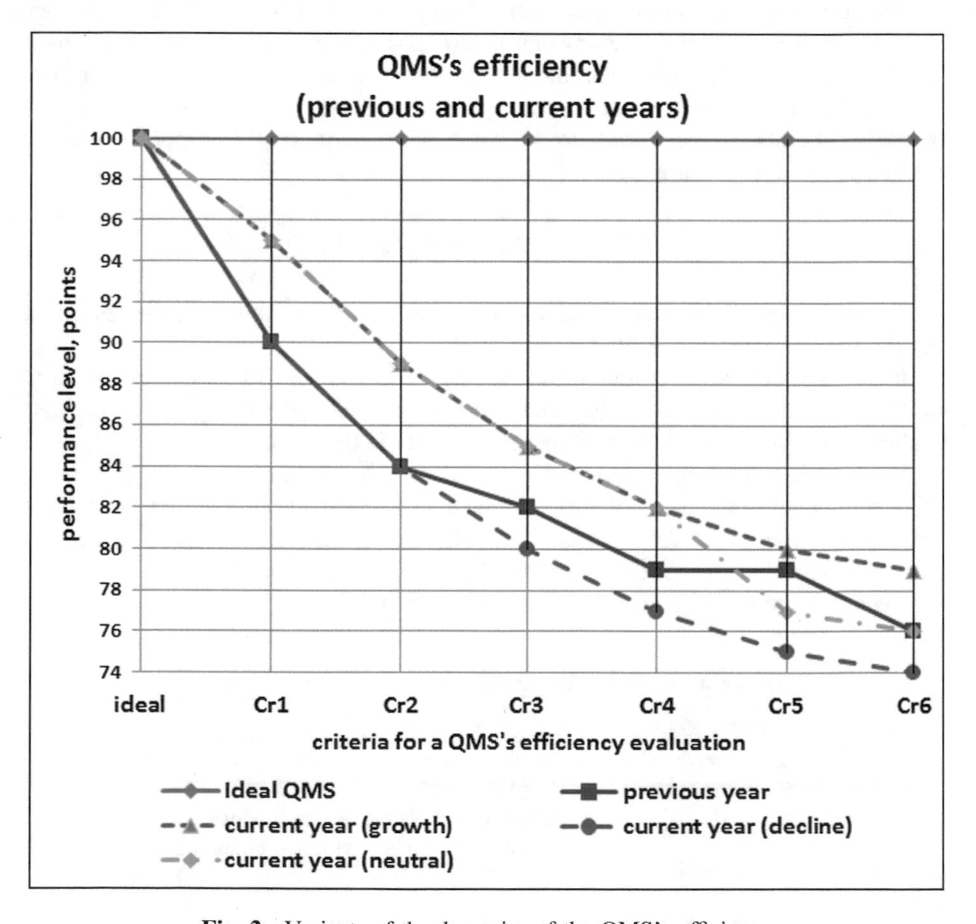

Fig. 2. Variants of the dynamics of the QMS's efficiency

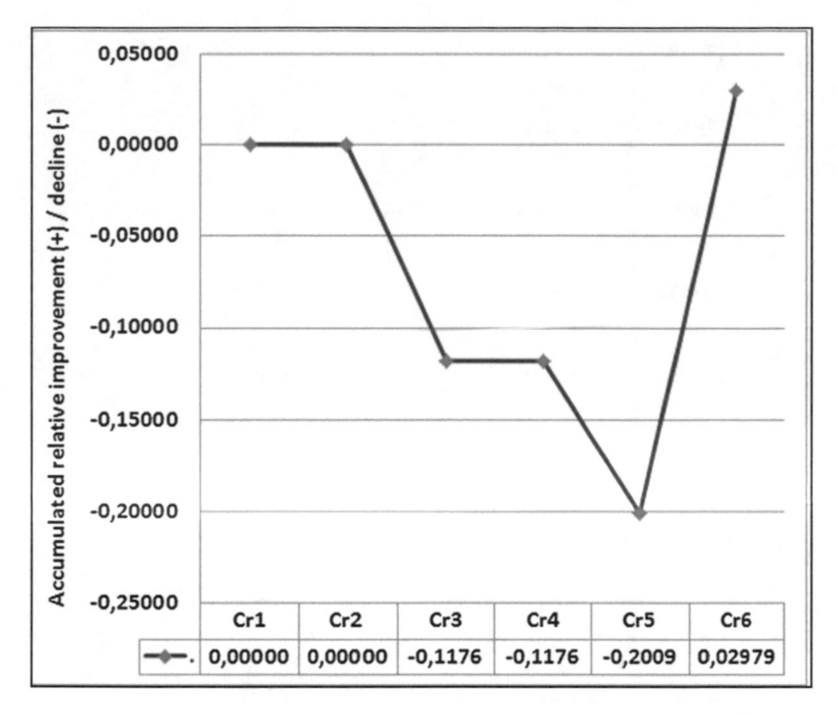

Fig. 3. The trend of the QMS's effectiveness - an increase in productivity

In the first case, there is a positive dynamics of the effectiveness of the QMS (Fig. 2, the line "growth").

In the second case, although the QMS is recognized to be effective, there is a negative dynamic of indicators (Fig. 2, the "decline" line) and a proposed set for measures, which aimed at identifying causes of the reduced performance and their elimination. At the same time, the main attention should be paid to the criteria according to which the rank value decreased in comparison with the previous year. In Fig. 2, such criteria include criteria P3 and P5.

In the third case, when the final score's dynamic of the QMS's effectiveness is ambiguous (Fig. 2, the line is "neutral"), it is necessary to make an additional calculation of the relative cumulative improvement or deterioration. The value of this indicator is calculated for each criterion, the rank value of which has changed from the previous year, according to the formula:

$$\Delta_i = \left(P_{i(current)} - P_{i(previous)}\right)/P_{i(ideal)} \tag{2}$$

If Δi has a positive value, the effectiveness of the QMS has an upward trend (Fig. 3), otherwise - a downward trend (Fig. 4).

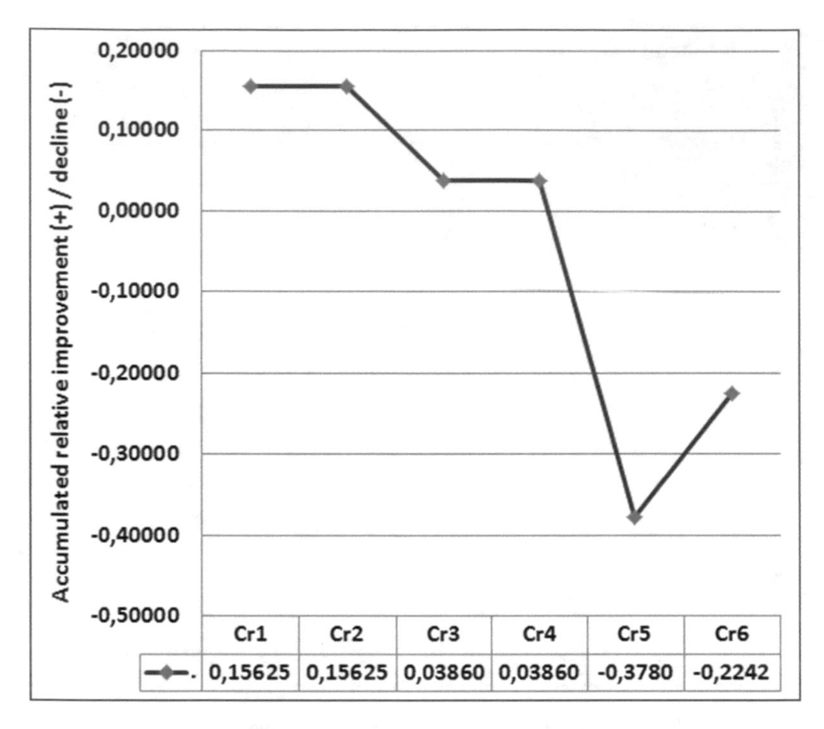

Fig. 4. The trend of the QMS's effectiveness - a decline in productivity

4 Conclusions

An algorithm's automatization will reduce a data processing time and improve a reliability of results [8]. The considered algorithm can be adapted to assess the QMS's performance by including in the evaluated criteria system a group of economic indicators of the enterprise's business [8–10].

References

1. Shtern, Y.I., Kozhevnikov, Y.S., Rygalin, D.B., Shtern, M.Y., Karavaev, I.S., Rogachev, M. S.: Intelligent system and electron components for controlling individual heat consumption. Russ. Microelectron. **45**(7), 488–491 (2016)
2. Building Monitoring System Options. http://advancedcontrolcorp.com/blog/2018/02/building-monitoring-system/. Accessed 14 March
3. Building Monitoring Systems. http://www.tadviser.ru/a/97552. Accessed 14 March
4. ISO 9000:2015: Quality management systems – Fundamentals and vocabulary
5. ISO 9001:2015: Quality management systems – Requirements
6. Lukmanova, I.G., Nezhnikova, E.V.: Vestnik MGSU. In: Proceedings of Moscow State University of Civil Engineering, no. 6, pp. 158–164 (2014)
7. Petrova, S.N.: Real.: Econ. Manag. **2**, 101–104 (2013)
8. Bureeva, M., Redko, L.: Gaudeamus Igitur **4**, 31–33 (2015)

9. Psomas, E., Antony, J.: Int. J. Prod. Res. **53**(7), 2089–2099 (2015)
10. Murphy, W.H.: J. Small Bus. Entrep. **28**(5), 345–360 (2016)
11. Sawant, M.A., Yadav, O.P., Rokke, C.: Int. J. Intell. Enterp. **5**(1/2), 173–193 (2018)

Management of Investment and Construction Projects of Low-Rise Building Construction with Account of Requirements of Energy Efficiency

Kristina Filyushina[1(✉)], Sergey Astafyev[2], Natalya Gusakova[1], Olga Dobrynina[1], and Abrorbek Yarlakabov[1]

[1] Tomsk State University of Architecture and Building, Solyanaya sq. 2, 634003 Tomsk, Russia
kril617@yandex.ru
[2] Baikal State University, Lenina Street 11, 664003 Irkutsk, Russia

Abstract. In the paper, the research aimed at the development of interactions between participants of the investment and construction project of low-rise housing construction in Russia is conducted, and the technique of increase in its efficiency is developed. The purpose of the study is the development of interactions at implementation of investment and construction projects of low-rise housing construction. In this paper, with the help of statistical methods, investment and construction projects of low-rise housing construction are analyzed, methods of indicative planning developed model of implementation of the energy efficient low-rise housing investment and construction project on the basis of PPP (Public-private partnership), and also variable schemes and tools of implementation of these projects are offered. These developments can be applied in development of documents of territorially sectoral planning and also in development of programs of development of the construction industry, including low-rise housing construction. The main directions of development of low-rise housing construction are defined. The authors revealed need of creation of complex operators for more effective management of energy efficient projects of low-rise housing construction. The importance of this paper is caused by the fact that by means of the developed model of implementation of the energy efficient low-rise housing investment and construction project with complex operators, it is possible to exempt the private investor from burdensome obligations for operation of an object of low-rise housing construction that significantly reduces possible risks, both for the private partner and for the state.

Keywords: Investment and construction project · Low-rise housing construction · Energy efficiency · Complex operator · Risks · Model

V. Murgul and M. Pasetti (Eds.): EMMFT-2018, AISC 983, pp. 174–184, 2019.
https://doi.org/10.1007/978-3-030-19868-8_17

1 Introduction

Development of energy efficient housing is inevitable prospect for the next several years as the state program on increase in indicators of energy efficiency is approved. Since 2020, all buildings have to correspond to a certain class of energy efficiency, and low-rise housing construction is not an exception. Construction of energy efficient, affordable, comfortable, and eco-friendly low-rise housing construction has to be carried out in the following main directions:

1. Development of the project of complex low-rise housing (organized cottage settlements of economy class) which construction is carried out on the basis of public-private partnership;
2. The construction of low housing for inhabitants of rural areas which is carried out for improvement of living conditions of young specialists and members of their families;
3. Collective low housing construction (HCC).

Today, not enough attention is paid to projects for development of energy efficient low-rise housing construction; it is connected with the fact that the state is interested in construction of multystoried construction as the fastest way of provision of housing needing premises. In turn, the authors believe that its obvious reserve of development of construction is development low construction (DLC) what development of qualitatively new organizational model of management of processes of DLC is necessary for. In favor of it, we will note that growth rates of DLC in Russia in recent years significantly increased, and DLC share in the general structure of input of housing for the last 5 years increased from 40% up to 55% and tends to increase.

2 Literary Review

To the research of problems of development of a construction complex, including development of low-rise construction, enough researches, namely authors Kazeykin, Baronin, Chernykh, Androsov [1], Baronin, Grabovy of [2], Asaul [3], Bondarenko, Ivanenko are devoted [4].

The solution of a question of power supply in low construction meets in works as Sheina [5], Beregovoi [6], Luin, and Zigmantovich [7].

On the basis of the analyzed Russian [8, 9] and foreign [10–14] publications on the considered subject, it is possible to draw a conclusion that the issue of development of energy efficient low-rise housing construction, today, is rather relevant and there is a number of problems which demand detailed studying, namely creation of effective model of management of investment and construction projects of energy efficient low-rise housing construction, development of the scheme of interaction between subjects of the investment and construction project and creation of complex operators.

3 Materials and Methods

More effective management of energy efficient projects of low-rise housing construction requires creation of the complex operators regulating their activity. For ensuring effective implementation of the low investment and construction projects (ICP), it is necessary to consider the possibility of creation:

1. The Center of Development of Public-private Partnership (CDPPP) – is intended for attraction of private investments into socially important construction projects including in low-rise housing construction, increase in power efficiency of such projects, for the purpose of increase in efficiency of these projects and improvement of quality of life of the population. For this purpose, it is necessary for CDPPP and authorities follows:

   - To create base of the investment and construction projects implemented and planned to realization and also base of private investors of ICP interested in development in the field of construction;
   - To carry out risk analysis and barriers of development of ICP arising at realization with development of necessary actions for their elimination for each project;
   - To assist in preparation and realization of ICP (legal and financial maintenance, the organization of financing, etc.), to develop information awareness of all participants of the project;
   - To develop information materials, normative legal acts in the field of public-private partnership and also to cooperate with the leading education centers in the solution of questions of preparation of specialized shots in the field of development of public-private partnership;
   - To consolidate all used means and mechanisms of stimulation of low-rise housing construction in the region.

2. The Agency of Assistance of Individual Housing Construction (AAIHC) – is a performer of the investment and construction project regarding formation of the land plots. For the purpose of inventory, rational use of territories and their complex development it is necessary during collaboration of AAIHC and authorities:

   - To reveal not developed land plots in territories of settlements and to make their register (to involve in economic circulation for low-rise housing construction the land plots from structure of the agricultural land which are in a private property and not used for purpose);
   - To reveal the land plots of agricultural purpose which are not used on direct use in borders of settlements and to make their transfer in category of lands of the settlement. Within realization of ICP the Agency of assistance of individual housing construction carries out formation of the land plot for complex housing construction what it is necessary for:
   - To carry out preparation of the investment and construction project, namely borders of the land plot and designation of its borders on the area;
   - To carry out definition of the allowed use of the land plot;

- To define specifications of connection of objects to networks of technical providing and a payment for connection of objects to networks of technical providing;
- To execute design and conducting examination of projects of complex housing construction.

Formation of the land plots for complex housing construction is carried out by AAIHC independently or together with authorities.

After completion of the procedure of formation of the land plots intended for low-rise housing construction from the Agency, the following will be organized:

- control of terms and quality of construction of utilities, including as the customer in areas of low-rise building, the organization of their commissioning (connection), control of settlings with performers of construction works, including in case of violation of terms of construction;
  the organization of work of the housing commissions on acceptance of houses and input of the built houses;
- distribution in accordance with the established procedure the land plots on competitive or other basis for arrangement of a recreational zone, building by sports, cultural and entertaining, trade and other social facilities;
- actions for stimulation of creation of housing co-operatives, condominiums or other similar associations;
- implementation of modern information technologies on control of course of execution of RISP (Webcam).

Further, the model of implementation of the energy efficient low-rise housing investment and construction project on the basis of PPP, Fig. 1 is presented. One of options of improvement of the organizational model of public-private partnership existing now in low construction is emergence in the market of the center of development of public-private partnership which would carry out maintenance and operation of construction objects and rendered final service to the consumer. In this model, the private partner carries out construction of the facility, the property right is reserved for the private partner before full return of the enclosed investments then transfer of the property right to the public partner passes. The relevant institutions or the commercial organizations which wished to be engaged in this activity and underwent necessary

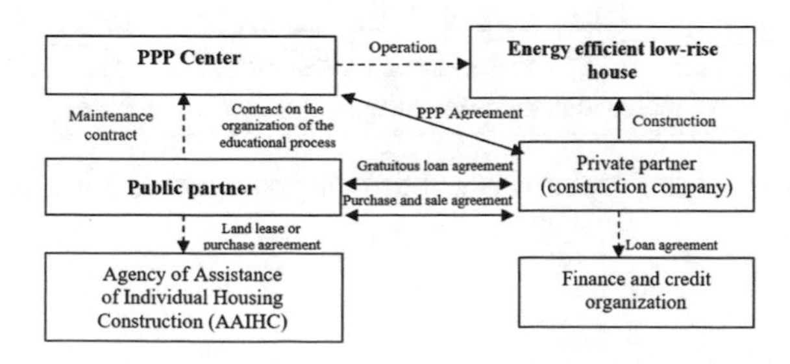

Fig. 1. Model of implementation of the energy efficient low investment and construction project with complex operators.

competitive selection can act as the center of development of public-private partnership. The public partner signs with the PPP center the contract for maintenance of an object during the term determined by the agreement of PPP.

Creation of the PPP center in investment and construction projects significantly reduces risks, both for the private partner, and for the state as the choice of the PPP center is a prerogative of public partner.

4 Results of a Research

For effective realization of the offered model of development of the investment and construction project on the basis of public-private partnership it is represent variable schemes and instruments of its realization in Table 1 (Figs 2 and 3).

Table 1. Variable schemes and instruments of implementation of the investment and construction project on the basis of public-private partnership.

Scheme	Paper work	Land	Construction	Engineering and municipal infrastructure	Road network
1	All included in the object price (free acquisition in the housing market)				
2	Independently	Auction	Contract agreement	Joint financing (50/50)	Budget
3	Independently	Auction	Without subcontracting	Joint financing (50/50)	Budget
4	CDPPP	AAIHC	At the expense of means of builders	Budget	Budget
5	CDPPP	AAIHC	Contract with all participants	Joint financing (10/90)	Budget
6	CDPPP	AAIHC	Without subcontracting	Budget	Budget

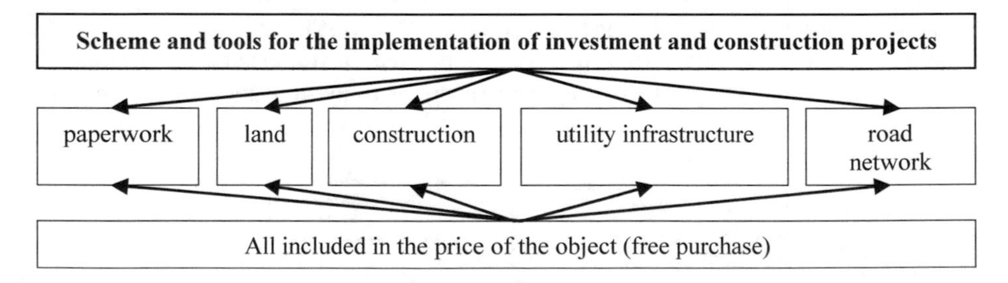

Fig. 2. "Scheme 1" the variable scheme and tools for the implementation of investment and construction projects.

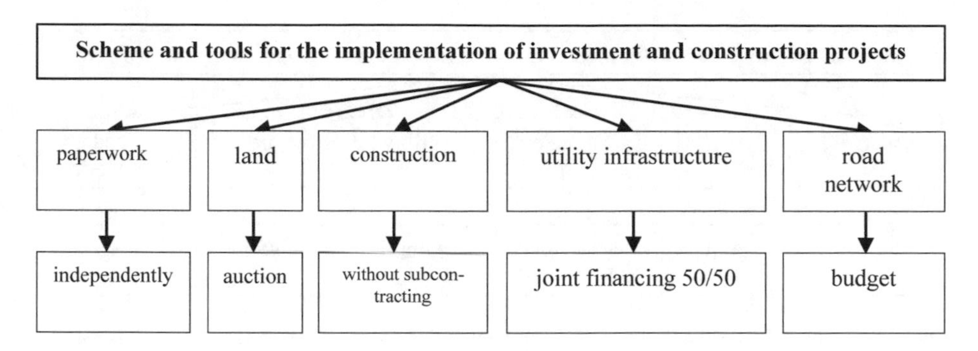

Fig. 3. "Scheme 2" the variable scheme and tools for the implementation of investment and construction projects.

At implementation of "Scheme 2", the land intended for low-rise construction is offered for an auction, and the contract agreement is signed for construction of the low-rise house. Financing of engineering and municipal infrastructure is made in common at the expense of means of the customer and at the expense of budget funds, paperwork is carried out independently. Construction of a road network is made completely at the expense of the budget.

"Scheme 3" of implementation of the investment and construction project shows that construction will be made by own forces, it is difference from the previous scheme. Paperwork is also conducted independently (Fig. 4).

At implementation of "Scheme 4", it is necessary to create the Regional center of development of public-private partnership and the Agency of assistance individual housing constructions, which will be performers of the investment and construction project regarding paperwork and formation of the land plots (Fig. 5). Construction is carried out by signing of the contracts with all participants of the investment and construction project, the financial credit institution (Fig. 6) becomes the additional participant (Fig. 7).

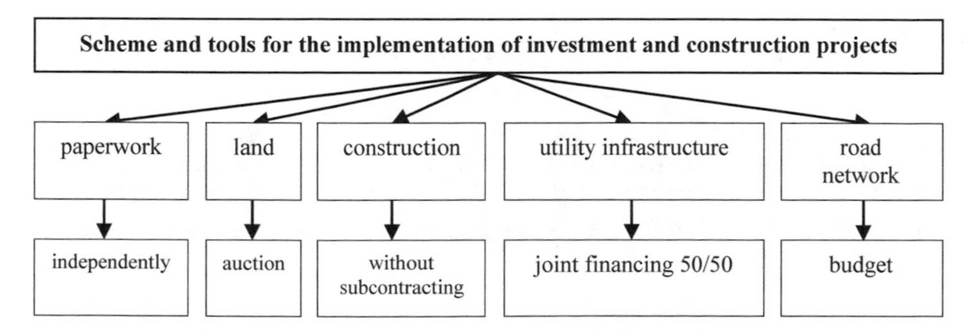

Fig. 4. "Scheme 3" the variable scheme and tools for the implementation of investment and construction projects.

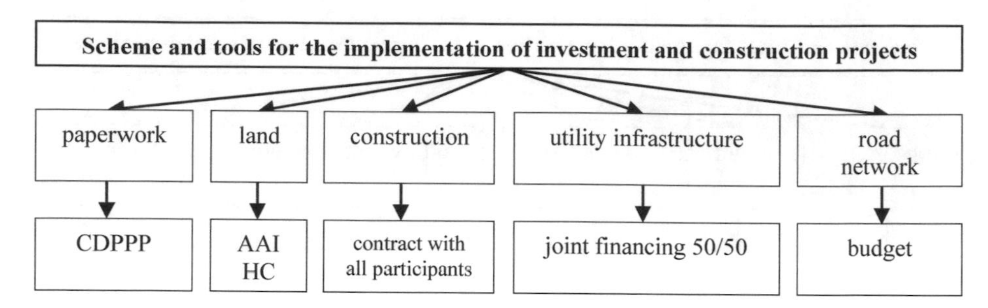

Fig. 5. "Scheme 4" the variable scheme and tools for the implementation of investment and construction projects.

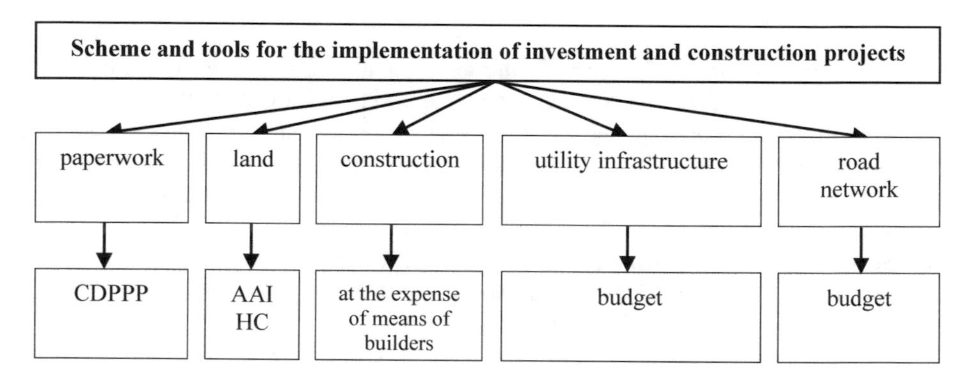

Fig. 6. The scheme of signing the contract with all participants of the investment and construction project.

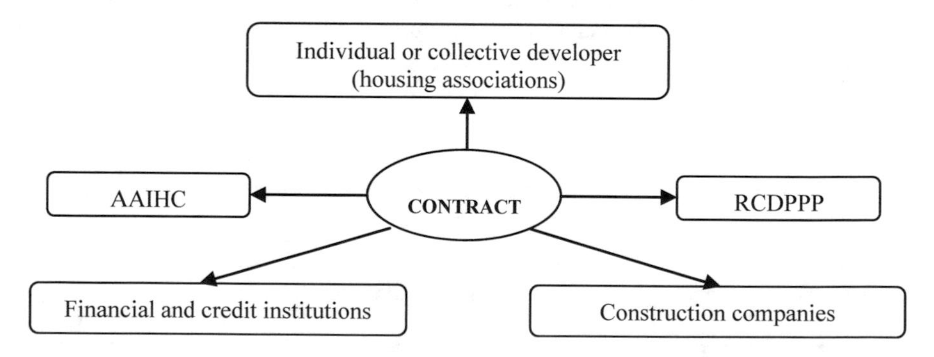

Fig. 7. "Scheme 5" the variable scheme and tools for the implementation of investment and construction projects.

The difference of "Scheme 5" from previous is that in this case construction is carried out at the expense of means of builders. Financing of construction of engineering and municipal infrastructure and street road network will be carried out at the expense of budgetary funds, paperwork on implementation of the investment and construction project is carried out by the Regional center of development of PPP-ship (Fig. 8).

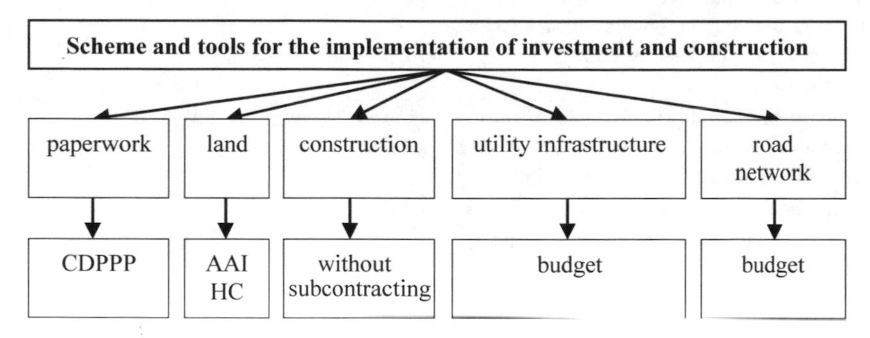

Fig. 8. "Scheme 6" the variable scheme and tools for the implementation of investment and construction projects.

The difference of last variable "Scheme 6" is that felling licences, funds of funds, budgets will be allocated for construction, it will also be possible to use maternal capital, etc. [14, 15].

For the implementation of energy efficient low-rise investment and construction projects, it is necessary to pay special attention not only to financial efficiency of construction, but also to operational one. Thus, the indicator of efficiency is multi-component because in this research, the efficiency is considered both at implementation of the project and at operation of the low-rise house (Table 2).

Table 2. Measures to implement energy-efficient investment and construction projects in the field of low-rise construction based on PPP.

Name of action	Performer
Improvement and audition of regulatory framework of low-rise construction	Department of construction and architecture
Creation of operators of low-rise construction – CDPPP and AAIHC	Department of construction and architecture
Formation of base of the CDPPP investment and construction projects implemented and planned to realization	CDPPP
Formation of lists of potential private investors	CDPPP
Definition of objects and the land plots for low-rise construction (copies of documents of title on the land and excerption of general plan)	AAIHC

(*continued*)

Table 2. (*continued*)

Name of action	Performer
Formation of the lists of potential participants of RIP needing improvement of living conditions and staying on the registry (with information on the place of work, registration)	AAIHC
Development of credit and financial gears and conditions of granting credit loans on low-rise construction Financial credit institution	Financial credit institution, CDPPP
Development of the town-planning and project documentation, registration of the land plots	AAIHC
Signing of the contracts of rent and preliminary contracts of purchase and sale	AAIHC
Construction of engineering and municipal infrastructure on sites of low building, except for intra quarter and in domestic objects of engineering and municipal infrastructure	CDPPP, AAIHC in coordination: utility companies, private investors, construction organizations
Construction of automobile roads (street road network) to the new residential district of mass low-rise building and also in their territory, except for intra quarter and in domestic drives	CDPPP, AAIHC in coordination: utility companies, private investors, construction organizations
Construction of social objects (kindergarten, school, entertainment center) and also objects of law and order and safety	Administration, private investors in coordination
Mass construction of inhabited subjects to low building	CDPPP, AAIHC in coordination: credit institutions; local administration; private investors; financial and credit, construction and other organizations

Economy at construction of the low-rise house is reached due to implementation of the new technologies and materials that allow reducing thermal losses of the low-rise house and also due to use of the energy saving heating equipment (in the absence of this equipment for heating of the low-rise house construction of the boiler house would be required that would lead to additional expenses). The operational effect is gained due to economy on utilities, in particular, on heating and hot water supply.

5 Conclusion

At appropriate standard regulation and creation of necessary conditions for activity of complex operators, the use of this model can become the effective scheme for development of energy efficient investment and construction projects of low-rise construction.

Realization of the offered ICP model with the offered complex operators will provide development of the following directions:

1. Development of the program of a social mortgage with integration in it of actions for provision of housing for young families, experts at rural areas, citizens recipients of the state housing certificates, disabled people, orphan children and also families having the right to use means of the maternal (family) capital for improvement of living conditions
2. Development of mortgage housing lending for low-rise construction.
3. Support of development of commercial low-rise construction within formation of the state order for housing of economy class.
4. Stimulation of a private initiative of citizens of construction of housing and non-commercial associations of citizens (housing co-operatives).
5. Support of modernization and development of the enterprises of construction materials.
6. Implementation of energy efficient and resource-saving technologies and construction materials in low-rise housing construction.
7. Development of personnel policy in the field of construction of energy efficient low housing.
8. Ensuring information openness of implementation of measures on stimulations of development of low housing construction [16, 17].

Acknowledgements. The paper is executed within a grant of the President of the Russian Federation No. MK-2273.2018.6 “Development and the feasibility study on the choice of space-planning and constructive decisions in low-rise housing construction in aspect of increase in power efficiency and resource-saving”.

References

1. Kazeykin, V.S., Baronin, S.A., Chernykh, A.G., Androsov, A.N.: Problem aspects of development of low-rise housing construction of Russia: the monograph. Under the general edition of the Academician of MAIN V.S. Kazeykin and the prof. S.A. Baronin. INFRA-M, Moscow, p. 278 (2011)
2. Baronin, S.A., Grabovy, P.G.: Main trends and modern features of development of low housing construction in Russia. News Southwest State Univ. **5–2**(38), 48–58 (2011)
3. Asaul, A.N., Kazakov, Yu.N., Pasyada, N.I., Denisova, I.V.: The theory and practice of low-rise housing construction in Russia. Under the editorship of Dr.Econ.Sci., the prof. A.N. Asaula, Gumanistika, SPb. p. 563 (2005)
4. Bondarenko, E.Yu., Ivanenko, L.V.: Foreign experience of the organization of low-rise construction. Of the basis of the economy, management and right, vol. 2, no. 8 (2013)
5. Sheina, S.G., Minenko, E.N.: Development of optimizing model of the choice of energy efficient decisions in low-rise construction: the monograph. Rostov N/Д: Growth. state. builds. un-t, p. 118 (2013)
6. Beregovoy, A.M.: Buildings with energy saving designs. The thesis on a competition of degree of the Doctor of Engineering, Penza (2005)

7. Lugin, V.G., Zigmantovich, A.V.: Influence of energy saving technologies on the choice of organizational technical solutions in construction. Of technology of construction, no. 4 (2005)
8. Gasho, E.G., Repetskaya, E.V., Bandura, V.N.: Formation of regional programs of energy saving. Energy Sav. **8**, 1–14 (2010)
9. Bashmakov, I.A.: Assessment of values of target indicators of the state program of the Russian Federation on energy saving. Energy Sav. **4**, 10–23 (2013)
10. Luong, N.D.: A critical review on energy efficiency and conservation policies and programs in Vietnam. Renew. Sustain. Energy Rev. **52**, 623–634 (2015)
11. Aste, N., Buzzetti, M., Caputo, P., Manfren, M.: Local energy efficiency programs: a monitoring methodology for heating systems. Sustain. Cities Soc. **13**, 69–77 (2014)
12. Alberini, A., Bigano, A.: How effective are energy-efficiency incentive programs? Evidence from Italian homeowners. Energy Econ. **52**, 76–85 (2015)
13. Lu, W.-M., Kweh, Q.L., Nourani, M., Huang, F.-W.: Evaluating the efficiency of dual-use technology development programs from the R&D and socio-economic perspectives. Omega (2015, in press)
14. Siddiqui, S., Christensen, A.: Determining energy and climate market policy using multiobjective programs with equilibrium constraints. Energy **94**, 316–325 (2016)
15. Minaev, N., Filushina, K., Jarova, E.: The concept of increasing energy efficiency of low-rise construction in the context of technical regulation. In: IOP Conference Series: Materials Science and Engineering Advanced Materials in Construction and Engineering. International Scientific Conference of Young Scientists: Advanced Materials in Construction and Engineering, TSUAB, p. 012050 (2015)
16. Zharova, E.A., Minaev, N.N., Filushina, K.E., Gusakov, A.M., Gusakova, N.V.: Formation of a regional process management model for energy efficiency of low-rise residential construction. Mediterr. J. Soc. Sci. **3**, 155–160 (2015)
17. Dobrynina, O.I., Minaev, N.N., Filushina, K.E., Kolykhayeva, Y.A., Zharova, E.A.: Critical analysis of the public regional programs on energy efficiency in the context of housing and utilities sector. Mediterr. J. Soc. Sci. **3**, 127–132 (2015)

Data Mining Technologies for Analysis of Cyber Risks in Construction Energy Companies

Tatyana Kostyunina(✉)

Saint Petersburg State University of Architecture and Civil Engineering, Vtoraya Krasnoarmeiskaya Str. 4, 190005 St. Petersburg, Russia
tnktn@yandex.ru

Abstract. The energy industry in Russia is characterized by sustainable development. Among other factors, this is due to automation of business processes in energy companies. However, as it was noted a long ago, computer systems and networks not only provide achievements and opportunities but also constitute additional risks and threats. Currently, design of a cyber risk management system represents a challenge for any enterprise. The paper presents classification of cyber risks, necessary for their further identification. Capabilities of Big Data and Data Mining technologies for cyber risk analysis are considered. Examples of such analysis (quantization, self-organizing maps) using the Deductor Studio analytical platform are provided. Based on the results obtained, we can argue that the proposed method proves to be effective.

1 Introduction

The energy industry is one of the most rapidly developing national economy sectors. This is mainly due to modern innovative energy technologies, including new information technologies. Building Information Modeling (BIM) can serve as an example of such technologies.

It is obvious that automation of business processes at all stages of the energy facility's life cycle has many advantages. However, although development of computer technologies has beneficial effects, it also represents a source of new risks. Among those, so-called cyber risks can be distinguished. Cyber risk falls under operational risks which were earlier classified in a study conducted by the author of this article [1].

The first works dedicated to the analysis and evaluation of this type of risk have appeared relatively recently. Only in the 1990s, various institutions started to recognize risks related to rapid growth of automation in their business, in particular, insufficient management of confidential information, software errors and mistakes of employees when entering data, fraud using information systems, failures of software and hardware followed by business interruption, risk of data loss. Currently, design of a cyber risk management system represents a challenge for any enterprise.

V. Murgul and M. Pasetti (Eds.): EMMFT-2018, AISC 983, pp. 185–192, 2019.
https://doi.org/10.1007/978-3-030-19868-8_18

2 Cyber Risks in Energy Construction Companies

Cyber risk is a risk of loss or additional expenses as a result of unlawful actions regarding computer and information systems and networks, communication systems or information resources, committed by third parties using information and telecommunication technologies. Cyber risk can occur alone or in combination with other types of operational risks.

Cyber risk occurs due to malware introduction or other destructive attacks affecting computer and information systems. The Internet and other external information networks and systems are sources of cyber risk. The purpose of such attacks is unauthorized access to information and/or information system failure.

A cyber threat has two types of consequences constituting a danger for an enterprise: first, it is potential damage to equipment or products and interruption to production due to failure; second, it is loss of data and, therefore, damage to reputation and financial damage.

Energy companies keep various confidential data and documents related to their projects. Effective business processes in energy imply security of all data circulating within an enterprise. Leak of data related directly to energy processes can affect business continuity. Leak of information about relations with subcontracts can result in partnership termination, and loss of company's strategic plans gives additional advantages to competitors.

An overview of cyber risks is presented in Fig. 1.

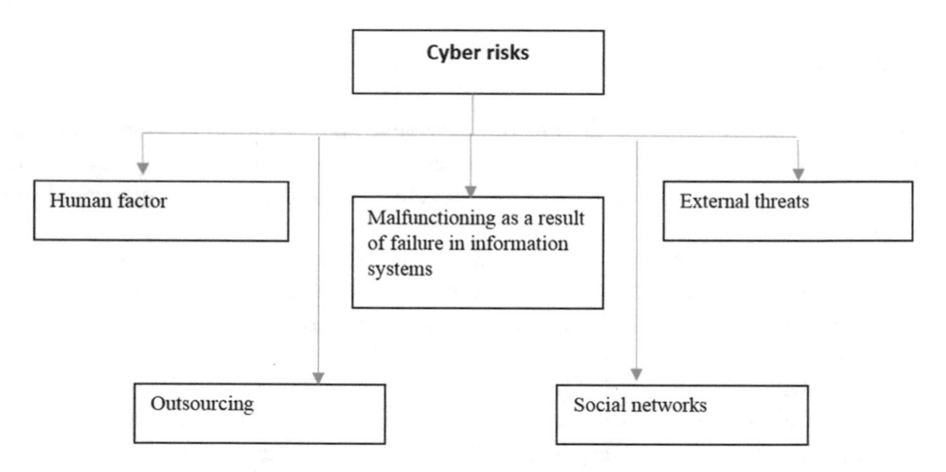

Fig. 1. Classification of cyber risks.

Energy always has been one of the most capital intensive economy sectors. It involves investment institutions, expensive technologies, specialized equipment and large supply contracts. It is obvious that risks inherent in each project stage are diverse and significant, and potential financial losses in the course of project implementation can be enormous.

Therefore, the classification described is insufficient for further cyber risk analysis and management. Each group of risks shall be considered individually. For instance, risks of the "Malfunctioning as a result of failure in information systems" group can be represented as follows (Table 1).

Table 1. Malfunctioning as a result of failure in information systems group of risks.

Equipment	Software	Data	Telecommunications	Computer networks
Equipment unreliability during voltage drop Obsolete equipment Equipment capacities insufficient to maintain operation at the current level Non-optimal architecture of information systems Incorrect equipment settings Equipment incompatibility with new software or hardware Lack of backup resources	Virus attack Incorrect automatic operations Software errors resulting in application failure Incompatibility of different types of software Unauthorized changes in the source code Unauthorized actions of the primary administrator Logic bombs (errors intentionally inserted into the software code, that can result in general system failure)	Unauthorized changes in data due to absence of access restrictions at the level of databases Modification of data due to insufficient access restrictions at the file level Errors in automatic calculations Data errors due to macro viruses Incorrect information obtained through data transfer channels Preparation of data in an incorrect format to be transferred through the electronic data interchange system Manual data processing Manual data import/export No data backup Damage to storage media containing backup data	Failure related to telephony, facsimile mail, Internet server Problems sending or receiving e-mail Data transfer errors Incompleteness of information being transferred External unauthorized access to data transfer channels Modification of data during their transfer	Information network failures Information network overload Delays in information processing due to insufficient capacity of the computer network Server hardware breakdown

The table is indicative of large volumes and variety of data to be collected and analyzed for efficient cyber risk management in the future. Nowadays, such tasks can be solved only using modern Big Data and Data Mining technologies.

3 Big Data Technologies in the Modern Information and Communication Environment

Big Data technologies are modern technological capabilities for analysis of big data. By Big Data, we shall understand large volumes of internal and external data of various structure. Big Data consists of three elements: data, analytics, and technologies [2–4].

Documents in the Internet, social networks, blogs, audio and video sources, sensors are the main sources of Big Data. Data without analytics, without revealing of subtle, concealed dependencies cannot be considered as Big Data. To make managerial decisions, technologies allowing not only storing but also analyzing Big Data are required. Such technologies were developed by the Internet giants as they were first to face the problem of storing and analyzing Big Data with the purpose of subsequent monetization [5].

Big Data analysis and processing are carried out based on the following methods [6, 7]:

1. Data Mining. A group of numerous methods integrating mathematical tools and information technologies.
2. Crowdsourcing. Data acquisition from an unlimited range of various sources.
3. A/B testing. One element of the control group is modified, and then the group is compared to similar groups. Such tests allow identifying the element affecting the control group most strongly. Big Data offers the possibility to make a large number of comparisons and, therefore, get a credible result.
4. Predictive analytics. Decision-making by predicting future behavior of a controlled test object.
5. Machine learning (artificial intelligence) Development of self-learning algorithms for artificial systems based on empirical analysis of information.
6. Social network analysis. Social networks' studying, where statistical data provides a basis to analyze relationships between members of social media and their communities [8, 9].

In general, inclusion of internal information sources and their processing using Big Data technologies allow significantly increasing the amount of data available for analysis, offer the possibility to analyze the market and understand its development trends, assess the company's position on the market and make more accurate predictions.

4 Data Mining Technologies for Analysis of Cyber Risks in Energy Companies

Data Mining technologies are intended to solve data mining tasks, improving decision-making efficiency.

Data Mining is information analysis to extract previously unknown but practically useful data, necessary to make decisions in various areas of human activities, from those accumulated by an enterprise. It is a process of discovering hidden patterns in existing data.

Data Mining allows conducting in-depth data analysis, including:

- discovering hidden patterns in data;
- predicting behavior of business indicators;
- assessing the effects of decisions made on company operations;
- detecting anomalies, etc.

Five standard Data Mining techniques can be distinguished: association, sequential patterns, classification, clustering, prediction.

Nowadays, there is much interest in analytical technologies, including Data Mining. Many companies engaged in development of Data Mining tools and integrated implementation of Data Mining, OLAP and data warehouses operate in the market. In many cases, Data Mining tools are considered to be a part of BI platforms also including tools to build data warehouses and data marts, as well as OLAP tools.

At the present time, Deductor Studio analytical platform (analyst workplace) is one of the best known tools.

Technologies implemented in Deductor allow (based on a single architecture) going through all stages of building an analytical system: from a data warehouse to automatic selection of models and visualization of results. Those characteristics make Deductor an ideal platform for decision support systems based on data mining methods.

Deductor is aimed at formalization and replication of the decision-making process. An analyst is a key actor in this process, who needs a tool allowing formalizing and reconfiguring decision-making logic. Deductor Studio (analyst workplace) can serve as such tool.

Operating Deductor Studio comes down to visual scenario building. A scenario is a sequence of steps allowing extracting knowledge from data.

With Deductor Studio, it is possible to automate routine data processing operations and focus on intellectual work: decision-making logic formalization, model development, prediction. Other employees will be able to use the results obtained with no need to delve into specifics.

5 Example of Data Analysis Using Deductor Studio

Let us consider the possibility to analyze monitoring data regarding network security threats in the LiSt energy company with the headquarters in Moscow and three branches in Yekaterinburg, Novosibirsk and Novgorod.

It is assumed that the company has a large amount of data accumulated for such analysis (both for the company and its branches), namely:

- the number of computers;
- the number of employees;
- software errors resulting in application failure;
- information network failures;
- no data backup;

- errors in automatic calculations;
- data errors due to macro viruses;
- DDoS attacks;
- attacks on the remote banking services (RBS) system;
- fishing;
- espionage;
- financial losses.

Data for many months, and even years, are considered. As a result of data analysis, the following was revealed (Fig. 2).

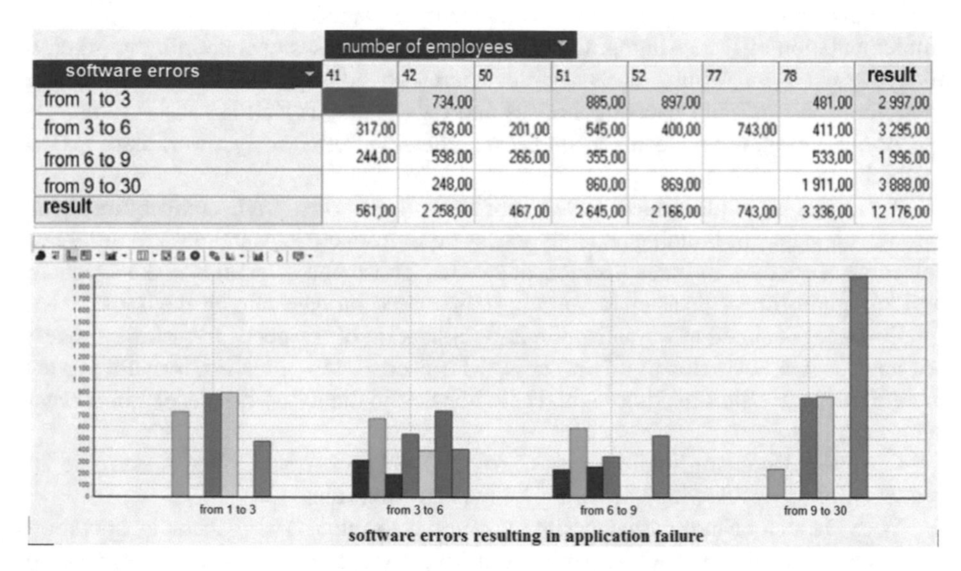

software errors	number of employees 41	42	50	51	52	77	78	result
from 1 to 3		734,00		885,00	897,00		481,00	2 997,00
from 3 to 6	317,00	678,00	201,00	545,00	400,00	743,00	411,00	3 295,00
from 6 to 9	244,00	598,00	266,00	355,00			533,00	1 996,00
from 9 to 30		248,00		860,00	869,00		1 911,00	3 888,00
result	561,00	2 258,00	467,00	2 645,00	2 166,00	743,00	3 336,00	12 176,00

Fig. 2. Financial losses of the company depending on the number of software errors and the number of employees.

With the help of quantization, financial losses of the company related to application errors, depending on the number of employees, are presented in the diagram. It is obvious that they reach the maximum level at the largest software errors detected.

With the help of quantization, statistics of DDoS attack attempts depending on month, in relation to the number of information network failures is presented in the diagram. It can be easily seen that the most system failures coincide with the beginning of the year (Fig. 3).

As a result of generating a self-organizing map, conclusions of individual values of selected parameters at different data can be made. In particular, it can be concluded that espionage is mostly spreading in Novosibirsk, and financial losses reach their maximum level in Moscow (office with the largest number of employees) (Fig. 4).

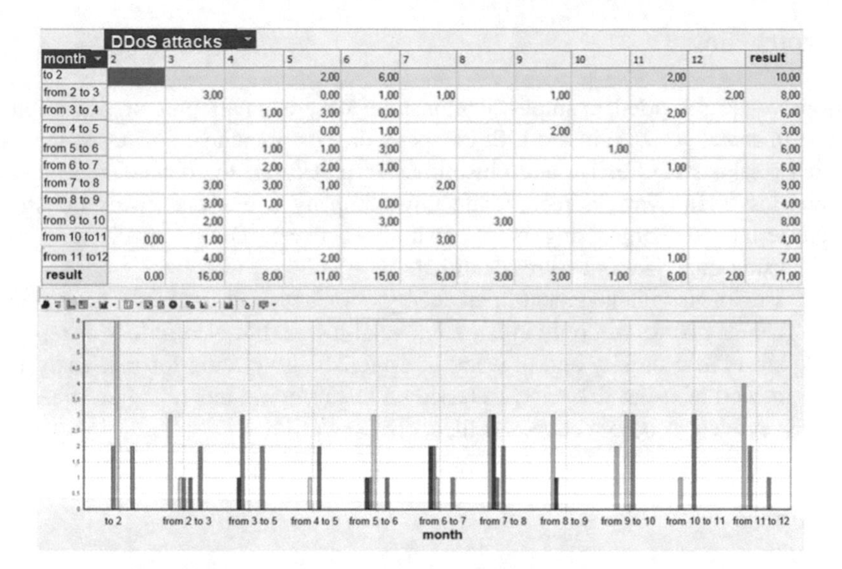

month	DDoS attacks 2	3	4	5	6	7	8	9	10	11	12	result
to 2				2,00	6,00					2,00		10,00
from 2 to 3		3,00		0,00	1,00	1,00		1,00			2,00	8,00
from 3 to 4			1,00	3,00	0,00					2,00		6,00
from 4 to 5				0,00	1,00			2,00				3,00
from 5 to 6			1,00	1,00	3,00				1,00			6,00
from 6 to 7			2,00	2,00	1,00					1,00		6,00
from 7 to 8		3,00	3,00	1,00		2,00						9,00
from 8 to 9		3,00	1,00		0,00							4,00
from 9 to 10		2,00			3,00		3,00					8,00
from 10 to11	0,00	1,00				3,00						4,00
from 11 to12		4,00		2,00						1,00		7,00
result	0,00	16,00	8,00	11,00	15,00	6,00	3,00	3,00	1,00	6,00	2,00	71,00

Fig. 3. Statistics of information network failures in the company depending on month and DDoS attack attempts.

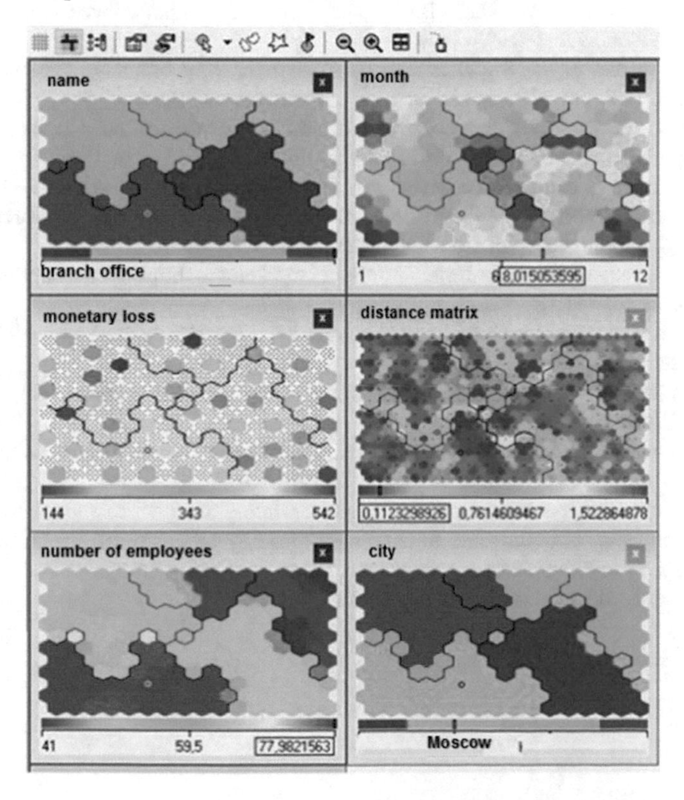

Fig. 4. Self-organizing map for visualization of various critical characteristics in the company's branches.

6 Conclusion

Obviously, the provided examples demonstrate only a small part of possibilities to analyze financial security in the LiSt company and its branches. However, they show that such analysis is effective and illustrative. In particular, the dependencies between financial losses and various factors affecting company operations, using quantization and clustering (self-organizing map), relations between DDoS attacks and system failures, using quantization, were identified.

The issue of risk management in an energy company is critical, as wrong approach to this issue may result not only in large financial losses, but also in loss of reputation. This explains the necessity and urgency of establishing systems for risk analysis.

The method of cyber risk analysis based on Data Mining technologies, presented in the study, proved to be effective and illustrative.

References

1. Kostyunina, T.N.: Classification of operational risks in construction companies on the basis of big data. In: MATEC Web of Conferences, vol. 193 (2018)
2. Analytical review of the Big Data market. http://www.ipoboard.ru/files/cms/5e3af134b9942559eb802ea93a1c9050
3. Analytical review of the Big Data market. https://habr.com/company/moex/blog/256747/
4. Operational risk. https://dic.academic.ru/dic.nsf/ruwiki/666706
5. Seitzhanov, B.S.: Risks in the construction industry—classification and analysis peculiarities. http://be5.biz/ekonomika1/r2011/00585.htm
6. Soboleva, E.A.: Specifics of design activity development in the investment and construction sector: detailed elaboration and prospects: monograph, 160 p. National Research University, Moscow State University of Civil Engineering, Moscow (2016)
7. Offer personalization: Big Data in client marketing. https://www.inbrief.ru/blog/41/. Accessed 23 Nov 2018
8. Why is Big Data confused with marketing and IT? https://vc.ru/marketing/12304-big-data-cases. Accessed 23 Nov 2018
9. Sokolova, A.: Big Data market in Russia. https://rb.ru/howto/big-data-in-russia/. Accessed 23 Nov 2018
10. Big Data technologies for client analytics. https://www.ibm.com/ru/events/presentations/connect2014/12_connect14.pdf. Accessed 23 Nov 2018
11. What are Big Data systems? https://promdevelop.ru/big-data/. Accessed 23 Nov 2018
12. Practice of using external data to improve efficiency of customer management. https://docplayer.ru/44609689-Praktika-ispolzovaniya-vneshnih-dannyh-dlya-povysheniya-effektivnosti-raboty-s-klientami.html. Accessed 23 Nov 2018

New Ways to Calculate a Severance Tax at Oil Production

Farida Adigamova[1(✉)], Anisa Khusainova[1], and Irina Sergeeva[2]

[1] Kazan Federal University, Kremlyovskaya str. 18, 420008 Kazan, Russia
ffl04@mail.ru, alyovina@gmail.com

[2] St. Petersburg National Research University of Information Technologies, Mechanics and Optics, Lomonosova 9, 191002 St. Petersburg, Russia

Abstract. The main idea of the article is a severance tax calculation at oil production. The study proposes to change the mechanism of severance tax calculation by introducing a coefficient characterizing the actual volume of oil production. The article emphasizes changes of the coefficients in calculation of a severance tax at oil production in the Tax Code of the Russian Federation. The authors offer to replace a number of coefficients characterizing the actual volume of oil production in severance tax calculation. The recommended decreasing coefficients will allow being closer to the value of rental payments. The practical significance of research is that the methodological provisions and procedures proposed in the study will contribute to increasing the regulatory impact on oil production taxation.

Keywords: Severance taxes · Volume of oil production
Oil producing companies · Fiscal policy · The Russian Tax Code

1 Introduction

The regulatory mechanism for calculating the taxation of oil production of Russian Federation was formed in the 90's of XX century. It was very different from the central planning economy of the country. The main principles of taxation of oil production were introduced by Law of the Russian Federation of December 27, 1991, No. 2118-1, "On the framework of the tax system in the Russian Federation". The legal act in the oil production was the Federal Law of the Russian Federation of February 21, 1992 No. 2395-1 "About the subsoil". Oil producing companies began to pay the minerals extraction tax intended for the removal of the resource rent of a natural resource, along with generally accepted taxes for all enterprises of the country. The basis for calculating resource taxes was the cost of oil, and not the final financial result of production activities. Resource rent included:

- deduction to the fund for replacement of mineral resources (MB reproduction);
- payments for subsoil use (royalty payments for natural gas);
- excises (excise tax);
- export duty;

V. Murgul and M. Pasetti (Eds.): EMMFT-2018, AISC 983, pp. 193–200, 2019.
https://doi.org/10.1007/978-3-030-19868-8_19

The taxation of oil producers is an instrument of fiscal policy. In 1995, as an alternative taxation tool, Federal Law No. 225-Fz of December 30, 1995 "On Production Sharing Agreements" was adopted. But no agreement, about production of existing oil fields, has been signed. The agreements of Russian Federation "The Sakhalin-1" and "Sakhalin-2" for the Kharyaga oil field were developed before 1995.

The taxation procedure for oil production, formed in the 1990s, was reorganized in 2001. As a result, a new mineral extraction tax (severance tax) appeared in the Tax Code of the Russian Federation (Chap. 26), replacing three previous payments: deductions for recovery of the mineral and raw material base (geology tax), the royalty tax and excise. The positive moment of introduction of this tax was the transparency of its accrual for the Supervisory bodies.

The Chap. 26 of the Tax Code of the Russian Federation presents the mineral extraction tax calculation at a specific rate of 340 rubles per ton multiplying by the coefficient Kts, which characterized the change of the world oil prices (Kts). In 2004 the rate has been changed, rising to 347 rubles per ton. Since the mineral extraction tax did not fully reflect the characteristics of the plots, which greatly differed by their geological conditions and by the regions of production, in 2007 the regulatory framework has significantly changed. To differentiate the taxation of oil production new coefficients have been introduced.

In 2012 and 2013, another 3 coefficients were added to Chap. 26 of the Tax Code of the Russian Federation: the degree of complexity of oil production, the rate of depletion of a particular hydrocarbon deposit, the coefficient characterizing the size of the reserves of a specific subsoil plot.

2 Theory

Since 2016, the regulatory mechanism for calculating the taxation of oil production of Russia Federation has significantly changed. In accordance with The Russian Tax Code, the rate is multiplied by 9 coefficients, with each of them calculated by a separate formula. In more detail, all formulas of taxation on mineral resource extraction for oil (severance tax) by the period of their use are presented in Table 1.

The tax rate on mineral resource extraction in accordance with The Russian Tax Code is constantly changing. It was increased by 3 times from 2002 till 2016. The calculation formula is changed almost every year. Since 2002, 7 new coefficients have been introduced. January 1, 2015, a formula for calculating the rate of the Mineral Resources Extraction Tax (MRET) for oil was totally changed.

The only constant coefficient (K_{ts}) characterizes the dynamics of world oil prices; because this coefficient was designed to define the changes in world oil prices and exchange rate US dollar (USD) to Russian ruble (RUB). It is a kind of deflator as the ratio of the current level of oil prices in rubles to level of oil prices in rubles in 2002. This coefficient has also been changed many times.

Table 1. Formulas for calculating the rate of **Severance Tax** (Mineral Resources Extraction Tax) (MRET) for oil in the periods of their application.

Period of validity	The formula for calculating on mineral resource extraction taxation MRET
01.01.2002–31.12.2003	$340 \times K_{ts}$
01.01.2004–31.12.2004	$347 \times K_{ts}$
01.01.2005–31.12.2006	$419 \times K_{ts}$
01.01.2007–31.12.2011	$419 \times K_{ts} \times K_{v}$
01.01.2012–31.12.2012	$446 \times K_{ts} \times K_{v} \times Kz$
01.01.2013–22.08.2013	$470 \times K_{ts} \times K_{v} \times Kz$
01.09.2013–31.12.2013	$470 \times K_{ts} \times K_{ts}\, K_{v} \times Kz \times K_{d} \times K_{dv}$
01.01.2014–31.12.2014	$493 \times K_{ts} \times K_{v} \times Kz \times K_{d} \times Kdv$
01.01.2015–31.12.2015	$766 \times Kts - Dm$, Where $D_{m} = K_{ndpi} \times K_{ts} \times (1 - K_{v} \times K_{z} \times K_{d} \times K_{dv} \times K_{kan})$
From 01.01.2016	$857 \times K_{ts} - D_{m}$, Where $D_{m} = K_{ndpi} \times K_{ts} \times (1 - K_{v} \times K_{z} \times K_{d} \times K_{dv} \times K_{kan})$

The coefficient that characterizes dynamics of world oil prices set the price limit at which tax collection starts ($8/barrel). In 2005, the ceiling level was 9, and it was to $15/bbl after January 1, 2009. The constant value in the formula has also changed from 252 to 261. The tax rate on mineral resource extraction was connected with world market oil prices without taking into account changes of domestic market prices and sales in the Near Abroad. “Tax Equalization” policy was introduced (determination of a fixed tax rate for all oil companies) without taking into account the geological and natural climatic conditions of oil fields [2] (Table 2).

Table 2. Changes of the Coefficient characterizing dynamics of world oil prices.

Period	Formula	Basis
Before 01.01.2005	$K_{ts} = (ts-8) \times P/252$	The Federal Law 08.08.2001 N 126-FL
01.01.2005–31.12.2008	$K_{ts} = (ts-9) \times P/261$	Part 2 of the Russian Tax Code 05.08.2000 N 117-FL (Federal Law 27.07.2006 N 151-FL)
From 01.01.2009	$K_{ts} = (ts-15) \times P/261$	Part 2 of the Russian Tax Code 05.08.2000 N 117-FL (Federal Law 22.07.2008 N 158-FL)

Legislative bodies have started to reduce the tax burden of oil companies to differentiate the tax on the extraction of mineral resources [6]. Federal Law No. 151-FZ introduced a reduction factor (KV) to the rate of the mineral extraction tax (severance tax) for oil plots, by which the degree of depletion of reserves exceeds 80%. It was determined by the formula:

$$K_v = 3.8 - 3.5 \times (N/V) \quad (1)$$

where N is amount of cumulative oil extraction on the specific subsoil plot (including production losses) according to the state balance of minerals approved in year, prior year of the tax period; V is initial recoverable oil reserves approved in accordance with the established procedure taking into account the surplus and write-off of inventories of oil (except for write-off of inventories of the extracted oil and production losses) and determined as the amount of inventories of categories A, B, C1 and C2 by the specific subsoil plot according to data of the state stock balance of minerals on 1/1/2006. As a criterion for differentiating the tax on mining operations, the degree of depletion of hydrocarbon reserves was chosen.

To stimulate the development of new oil fields, Tax holidays were established. In 2009, the geography of tax holidays on the tax on mineral resource extraction for new deposits was expanded. But tax benefits were not able to intensify the development of licensed sites in these regions due to the lack of industrial and social infrastructure [9]. Most of the undeveloped oil reserves in the Russian Federation fall on small deposits with small reserves.

In 2011, the Federal Law No. 258-FZ was adopted. It introduced a decreasing (lowering) coefficient to the rate of the mineral extraction tax (MET) (Kz) for subsoil plots with initial recoverable oil reserves of less than 5 million tons and a degree of depletion of reserves of less than or equal to 0.05. The coefficient Kz is calculated by the formula:

$$K_z = 0,125 \times V_3 + 0,375 \quad (2)$$

where V3 is the initial recoverable oil reserves in million tons accurate to 3 decimal places, approved in accordance with the established procedure, taking into account the increment and write-off of oil reserves (except for the write-off of recovered oil reserves and production losses) and determined as the amount of reserves of category A, B, C1 and C2 for a specific subsoil plot according to the state balance of mineral reserves approved in the year preceding the year of the tax period.

In order to develop hard-to-recover oil reserves, the Government of the Russian Federation adopted Decree No. 700-r, and later Federal Law No. 213-FZ of July 23, 2013, which introduces decreasing coefficient (reduction factor) (Kdv) to the rate of the mineral extraction tax for this type of oil. By the same law, in order to take into account the decreasing coefficient of the depletion of hydrocarbon reserves, a reduction coefficient Kdv was introduced. It is determined in the following order:

- if the value of the coefficient K_{dv} for the hydrocarbon feedstock is less than 1 and the depletion level of the stocks of this hydrocarbon feedstock is less than 0.8, the coefficient $K_{dv} = 1$;
- if the value of coefficient K_{dv} for the hydrocarbon feedstock is less than 1 and the degree of depletion of the stocks of this hydrocarbon feedstock is greater than or equal to 0.8 and less than or equal to 1, the coefficient K_d is calculated by the formula:

$$K_{dv} = 3.8 - 3.5\,(N_{dv}/V_{dv}) \quad (3)$$

where N_{dv} is the sum of accumulated oil production for a particular hydrocarbon deposit (including production losses) according to the data of State balance of mineral reserves in the year preceding the tax period year; V_{dv} is initial recoverable oil reserves.

Since 2013, within three years, the severance tax base at Oil Production (a Severance Tax at Oil Production) has been raised to 559 rubles per ton with the simultaneous reduction of the export duty in the calculation formula to 55 percent ("small tax maneuver"), but in 2014 a Federal Law No. 366-FZ ("big tax maneuver") with an even greater change in the tax rate base and the calculation of the export duty. It turns out that in 2017 the growth of the base of the mineral extraction tax will amount to 164.2% of the planned tax rate [10]. There was also introduced an additional reduction factor to the calculation of the mineral extraction tax – Krp coefficient, characterizing region of oil production and oil properties.

The application of various lowering coefficients to the calculation of the mineral extraction tax has led to introduction of an indicator characterizing the characteristics of oil production (DM). It is determined by the formula:

$$D_m = K_{dpi} \times K_{ts} \times (1 - K_v \times K_{ts} \times K_d \times K_{dm} \times K_{kan}) \quad (4)$$

where K_{dpi} is 530 from January 1 to December 31, 2015 inclusive, 559 -from January 1, 2016

The mechanism of oil production taxation is not sufficiently developed in terms of the level of the tax burden depending on the stages of development of deposits, the scale of subsoil users, types of extracted raw materials and bio-natural resources [3].

3 Results

The main drawbacks of the Mineral Resources Extraction Tax (MRET) in Russian practice of taxation are the ineffectiveness of the zero interest rate mechanism, which is differentiated by the fuel producing entities, which does not correspond to one of the principles of the budgetary system - the equality of budgetary rights of the subjects of the Russian Federation, and municipalities [7]. This rate applies to offshore Arctic fields and the Caspian and Azov Sea fields. The measure helps to run oilfields, to reduce operation costs and to improve development the fields of Eastern Siberia and the Arctic. Therefore, it is connected with geopolitical and state interests. It should be noted that at least 15–20 years will pass before the share of oil in the total volume of production from traditional oil and gas bearing areas will be significantly reduced. However, there are no tax privileges for the extraction of minerals for these deposits.

Russian quality assessment system does not allow determining which oil deposit is subject to preferential taxation (tax incentives). However, the construction of oil accounting systems near each field due to a long payback period (more than 10 years) is difficult financially, especially for small subsoil users [5]. The use of a preferential value of the depletion factor for deposits depleted by more than 80%, which is economically unprofitable for the majority of subsoil users due to the increase in reserves at the field being developed, for calculating the mineral extraction tax.

According to the authors, it is a problem to apply the depletion factor. It is proposed to change the mechanism for calculating the tax on the extraction of minerals in oil production using the coefficient that would most fully reflect the actual volume of oil production.

The coefficient that characterizes the degree of depletion of the reserves of a particular subsoil area today is based on the geological conditions of a particular field and does not reflect the actual state of affairs [1]. The authors propose the following formula for calculating the coefficient characterizing the actual volume of oil production Kf:

$$K_f = D_f / D_p \tag{5}$$

where D_f is the actual oil extraction; D_p - oil extraction according to the technical design.

The proposed coefficient will strengthen the regulatory function of the mineral resources extraction tax. It also allows planning oil production and tax revenues. To apply this coefficient, it is necessary to use the formula for calculating the mineral resources extraction tax. The following formula was used in 2014:

$$S_t \times K_s \times K_v \times K_{ts} \times K_d \times K_{dv} \tag{6}$$

where S_t is the base rate of the tax on extraction of mineral resources, rubles/tonne; K_{ts} is the coefficient characterizing the dynamics of world oil prices, K_v is the coefficient characterizing the degree of depletion of a particular subsoil block, K_z is the factor characterizing the amount of reserves, K_d is the degree of oil production complexity, K_{dv} is the coefficient characterizing the degree of depletion of hydrocarbon deposits.

The authors exclude the coefficient Kv of the formula used in 2014 and introduce the coefficient Kf, characterizing the actual volume of oil production:

$$S_t \times K_s \times K_f \times K_v \times K_d \times K_{dv} \tag{7}$$

Along with the calculation the extraction mineral resources tax of useful income tax, the authors recommend decreasing coefficients to calculate other income tax expenses. The order of application of the lowering coefficients for the profit tax depends on the parameters characterizing the field being developed.

Limiter is the amount of on the extraction of minerals resources tax or income tax. Proposed formula writes off the tax on the extraction of minerals resources and income.

$$N_l = N_{ndpi} \times (K_1 + K_2 + K_3 + K_4) \tag{8}$$

where $K_1 + K_2 + K_3 + K_4$ correction factors, shares.

The following parameters will be used to calculate other costs:

- K_1 - remoteness from the year-round infrastructure objects;
- K_2 - depth of formation;
- K_3 - oil Recovery coefficient;
- K_4 - water cut of oil flowing from a well.

The proposed correction factors do not correspond to those used for mining tax, because geological characteristics of the deposit are used there. It is recommended to use infrastructure coefficients to calculate other income tax expenses [4].

The size of the correction factors directly depends on the remoteness of the of the year-round road infrastructure objects. Thus, a coefficient of 0.1 is applied to deposits with a distance of 50–100 km from the road infrastructure, and 0.5 for distances greater than 500 km from the road infrastructure. The next factor is the depth of the formation. It varies from 2000 m to 4000 m with a gradation of 500 km. The coefficient is always zero with an oil recovery ratio less than 20%. A maximum coefficient of 0.4 is used with a sufficiently high level of oil recovery (more than 40%). The fourth parameter is the water cut of oil flowing from a well, which considerably make mining operations complex. This coefficient is differentiated from 70% to 90% with a gradation of 5%. When watering the well more than 90%, the coefficient is 0.5, and for less than 70% - 0.1. More details about the factors affecting other income tax expenses are presented in Table 3.

Table 3. The size of the correction factors, depending on the characteristics of the deposit.

Parameter	Information				
	Correction factor				
Distance from the year-round road infrastructure objects, km (K_1)	50–100	100–200	200–300	300–500	>500
	0,1	0,2	0,3	0,4	0,5
Formation depth, m (K_2)	2000–2500	2500–3000	3000–3500	3500–4000	>4000
	0,1	0,2	0,3	0,4	0,5
Coefficient of oil recovery, % (K_3)	<20	21–25	25–30	30–40	>40
	0	0,1	0,2	0,3	0,4
Watercut, % (K_4)	70–75	76–80	81–85	86–90	>90
	0,1	0,2	0,3	0,4	0,5

The correction factors proposed by the authors will take into account the infrastructural characteristics of the oil field in other expenses for the income tax. Differentiation of taxation is based on the fact that the tax on the extraction of minerals will regulate geological conditions, and the income tax - infrastructure.

4 Conclusion

All oil companies operate in the conditions of tough competition in the world oil market, which is characterized by high rates of institutional transformation (redistribution of property, struggle for new markets, business restructuring, etc.). It makes special demands for minimizing tax risks for domestic companies [8].

The authors support a phased transition from the minerals resources extraction tax to rental payments, proposed in the tax policy of the Russian Federation until 2030. The proposed areas of taxation modernization of oil-producing companies will allow streamlining the system of tax benefits, to provide a more equitable level of tax burden, and to make oil companies to carry out reproduction and investment processes.

The study proposes to change the mechanism of severance tax calculation at oil production by introducing a coefficient characterizing the actual volume of oil production. This ratio should be calculated as the ratio of the actual oil production to the planned production. It is necessary to use partially the formula for calculating the mineral recourses extraction tax adopted in 2014 for this coefficient. The recommended severance tax calculation will allow being closer to the value of rental payments.

Along with the severance tax calculation at oil production, the decreasing coefficients in profit tax calculations are recommended for classifying other expenses. Namely: remoteness, depth of formation; oil recovery coefficient, and water cut. According to the result of the econometric analysis, the authors have discovered the relationship between the revenues from severance tax at oil production and three factors in oil extraction: the actual oil production, the oil price and the ruble/dollar rate. The production volume factor is the main factor in the revenues in severance tax at oil production.

This model reflects the complex influence of significant indicators. On its basis, the forecast for the tax on mining operations from 2017 to 2020 was made. According to this forecast, there will be a constant increase in the revenues of severance tax from 2783 billion rubles in 2017 to 2956 billion rubles in 2020. The practical significance of research is that the methodological provisions and procedures proposed in the study will contribute to increasing the regulatory impact on oil production taxation.

Acknowledgements. The work is performed according to the Russian Government Program of Competitive Growth of Kazan Federal University.

References

1. Abdullah, M.O., Low, E.-T.L, Ooi, L.C.-L.: Front. Plant Sci. **7**(21), 00771 (2016)
2. Baek, J., Tappen, S.: Part B: Econ. Plann. Policy **11**(1), 1055–1060 (2016)
3. Chomas, M., Weber, J.G., Wang, Y.: Energy Policy **96**, 289–301 (2016)
4. Compernolle, T., Huisman, K., Kort, P., Piessens, K., Welkenhuysen, K.: Energy Policy **101**, 123–137 (2017)
5. García Benavente, J.M.: Energy Economics **57,** 106–127 (2016)
6. Grinkevich, A., Grinkevich, L.S., Sharf, I.V., Tsibulnikova, M.R.: Proceedings of the 27th International Business Information Management Association Conference - Innovation Management and Education Excellence Vision 2020: From Regional Development Sustainability to Global Economic Growth, IBIMA, pp. 440–447 (2016)
7. Gulyaev, P.V.: Indian J. Sci. Technol. **9**(11), 89423 (2016)
8. Motamedi, S., Zarra-Nezhad, M.: Iran. Econ. Rev. **20**(4), 479–500 (2016)
9. Ponkratov, V.V., Pozdnyaev, A.S.: Neftyanoe Khozyaystvo - Oil Ind. **3**, 24–27 (2016)
10. Yurichev, A.N.: Neftyanoe Khozyaystvo - Oil Ind. **3** 28–30 (2016
11. Ajupov, A.A., Kurilova, A.A., Ozernov, R.S.: Issues of coordinated cooperation for forming leasing payments schedules. Asian Soc. Sci. **11**(11), 23–29 (2015)
12. Ajupov, A.A., Medvedeva, O.E., Freze, A.V., Savin, A.G., Karataev, A.S.: Application of innovative financial products in the real sector of economy. J. Eng. Appl. Sci. **12**(19), 4894–4898 (2017)

An Economic Model of CO_2 Geological Storage in Russian Energy Management System

Aleksandr Ilyinsky(✉), Mikhail Afanasyev, Igor Ilin, Maria Ilchenko, and Dmitry Metkin

Peter the Great St. Petersburg Polytechnic University, Polytechnicheskaya 29, 195251 St. Petersburg, Russia
iliinskij_aa@spbstu.ru

Abstract. In an energy management system at Russian fuel and energy complex enterprises, an important aspect is a cost estimate of processes for capture, transportation, and storage of CO_2. The authors of the studies have developed an economic model to assess the life cycle of CO_2 storage cost for various combinations of sources and reservoirs. The model was used to monitor the number of potential storage sites and sources of emissions, and to study the geography of CO_2 storage.

Keywords: CO_2 geological storage · Energy management system

1 Introduction

The increased atmospheric concentration of CO_2 and the related global warming are a universally acknowledged problem of the global community. There are different mechanisms to solve it, one of which consists in capturing and burying CO_2. Carbon dioxide sequestration by capturing and pumping it into underground reservoirs gains an ever increasing importance in industrially developed countries of Western Europe and USA.

There are several methods of generating power from CO_2: capture and subsequent incineration of CO_2, removal of CO_2 from exhaust gases with precipitation transport membranes [1]. More integrated processes involve isolation of CO_2 using a chemical purification process utilizing excess heat of power stations for the purposes of its further recovery [2].

To store carbon dioxide, such underground reservoirs as spent and operating oil and gas deposits or water-bearing strata can be used. These reservoirs can be located both within territories and on sea shelves. Underground burial is the most appealing and established technology used in various countries of the world.

The potential volume of CO_2 storage sites in various countries is assessed differently: 247.5 Gt on the UK shelf and 200 Gt on the Norwegian shelf, Germany has the potential capacity of 42.7 Gt. [3].

The oil and gas industry of Russia has accumulated a considerable experience of research related to a search for natural underground reservoirs that may be used as reserve storages of CO_2. A search for and setup of a great number of reserve gas

V. Murgul and M. Pasetti (Eds.): EMMFT-2018, AISC 983, pp. 201–209, 2019.
https://doi.org/10.1007/978-3-030-19868-8_20

storages can concurrently solve major ecological problems related to absorption of CO_2, within an international convention. For the oil and gas industry the pumping of carbon dioxide into a stratum can significantly increase the productivity of development of oil and gas fields [4]. The processes related to sequestration of greenhouse gases belong to the category of investment intensive ones. [5] In this connection it is necessary to develop a representative economic model to assess efficiency of CO_2 storage options. Also an issue arises whether it is possible to use these or those mechanisms of emission reductions for Russia to participate in the emissions trading market, interaction with EU countries, Japan, Canada in emissions trading systems.

Currently, there are a great many methodological approaches and models aimed at establishing the economic efficiency of CO_2 geological storage projects in the ecological planning and subsoil management system. Problems related to the development of information analytical models to analyze the investment appeal of financing nature conservation projects are covered in most papers of major foreign scientists: E.H. Perkins, Brown, L.T., Linderberg E., Moritis G., Hendriks C.A., Blok K., Schuts V., Daun M., Wienpach P., Krumberck M., Hein K.R.G., Goldthore S.H., Cross P.J.I, Davision J.E., Chiesa P., Consonni. S., Lozza G., Audus, H., Kaarstad, O., Reimer P. and Skinner, G. etc.

A significant contribution to the solving of matters of economic evaluation of CO_2 geological burial projects has been made by Russian researchers as well: L.Z. Aminov, A.S. Astakhov, M.D. Belonin, A.A. Goluba, A.A. Ilyinsky, V.M. Zakharov, E.V. Zakharova, A.A. Novozhilov, A.E. Cherepovitsyna etc. The algorithms for evaluating the investment appeal of such projects and their digitization, however, have no unambiguous solution. This is related to a vagueness of modeling parameters, algorithms for calculating potential capacities for CO_2 burial, economic indicators of CO_2 storage, and the absence of necessary information for building a model in open publications and official sources of geological information.

In this connection, this article suggests an economic model for CO_2 geological storage, based on the generalized results of a cycle of the research works conducted by the authors within a number of domestic and foreign projects called Nordic CO_2 Sequestration Program [6].

2 Methods and Materials of Research

The process of CO_2 storage begins with its compression to a supercritical state, and then it is transported by a pipeline to the pumping site and, in conclusion, pumped into an underground reservoir. The model uses engineering and economic data as inputs and on their basis determines the type and number of the required equipment, establishes the number of the required wells. Then the volumes of the respective capital and current costs are calculated, as well as the amounts of the preliminary and subsequent tax payments.

In most cases, in modeling, the authors proceeded from the CO_2 source being a process of development of a natural gas field and the implementation of a CO_2 storage project increasing the extraction volumes. The modeling process is summarized in Table 1.

Table 1. Inputs and outputs of an economic model for CO_2 geological storage [7].

Inputs		Outputs
Engineering:		Engineering:
– CO_2 flow velocity – Pipeline lengths and relative heights – Ground temperature – Water depth (in case of shelf burial) – Reservoir depth and capacity – Reservoir temperature and maximum pressure that can occur during pumping	– Reservoir permeability – Input CO_2 pressure and temperature – Compressor efficiency and pipeline efficiency – Ultimate operational pressures for pipelines and wells	– Required compression – Pipeline diameters – Number of wells – Number and sizes of platforms (in case of offshore burial)
Economic:		Economic:
– Market prices for equipment and its servicing – Project implementation period – Fiscal mode – Project's net cash flow		– Capital costs of compressors, pipelines, wells, and platforms – Current costs – Economic indicators

Such underground reservoirs as spent and operating oil and gas deposits [8], water-bearing strata (closed- and open-looped) can be used to bury carbon dioxide (Table 2). For the Russian Federation, such regions should include districts of the Volga-Ural oil and gas province [9], Timan-Pechora oil and gas province [10], Kaliningrad Oblast and others [11].

Table 2. Classification of underground formations for CO_2 geological storage [12].

Types of reservoirs	Special features
Oil and gas reservoirs and super tight water-bearing strata	Limited pressure due to geomechanical features
Hydrostatic water-bearing strata in anticlinal structures	Are characterized by integrity of cover rocks
Hydrostatic water-bearing strata in monoclinal structures	Topography of cover rocks is determined by CO_2 migration
Deep water-bearing strata with a semi-permeable insulating layer	Are characterized by dissolved carbon dioxide in sea water during migration
Loose sedimentary rocks over 1,000 m deep	Are characterized by a formation of water-tight hydrates in sedimentary rocks

An alternative method of calculating the capacity of potential storages for CO_2 in hydrocarbon deposits can be performed on the basis of the methodology in question. The initial data in this case are:

- initial recoverable reserves of oil of deposits;
- initial recoverable reserves of natural gas of deposits;

- minimum and maximum oil densities underground in kg/m^3,
- minimum and maximum CO_2 densities underground in kg/m^3.

The minimum and maximum CO_2 densities underground are calculated on the basis of a ratio of the weight of the gas pumped into a trap to its effective volume, taking into account the porosity and permeability of the reservoirs. The calculation was made according to the algorithm recommended below (Table 3).

Table 3. Algorithm for calculating of potential capacity for the CO_2 storage.

№	Algorithm content	Mathematical interpretation
1	The minimum and maximum values of the volume occupied by the initial recoverable oil reserves underground in billion m^3 are determined	Volume = mass/density
2	The minimum and maximum values of the volume occupied by the initial recoverable oil reserves underground in billion m^3 are determined	$Volume_3 = Volume_1 + Volume_2$
3	The minimum and maximum values of the capacity of potential CO_2 storage under ground in million tons are determined	Mass = Volume × density
4	The minimum and maximum total values of the capacity of potential CO_2 storage under ground in million tons are determined	$mass_{com.} = mass_1 + \ldots + mass_n$

Reservoirs should have high porosity and location on horizons above 800 m, allowing stored CO_2 to become a fluid (flowing fluid - neither liquid nor gas, but exhibiting the properties of both, easily compressible and having a density of liquid). Storage at great depths (4000 m and more) is not economically feasible. At the same time, the concentration of reservoirs in the region should be rather high, as well as the communication between emission sources and storage sites. The model calculates economic indicators for 1 ton of injected CO_2 and 1 ton of CO_2, the emission of which is prevented.

Two economic indicators of CO_2 storage are used for calculations. The first indicator is the current value (“PV”—“present value”) capital and operating costs divided by the total amount of injected CO_2. This indicator shows today’s equivalent total costs for the upcoming storage of a given total CO_2.

Another indicator is the reduced annual income per ton of CO_2 emissions avoided (for example, from carbon credits), which may be required to offset the current cost of storage. The authors account it as a carbon credit per tonne of CO_2, the emission of which is prevented. In other similar works, it is often considered as annual costs per ton.

More precisely, these two gauges are described by formulas 1 and 2.

$$PV = \frac{\sum_1^n \frac{(K_i + O_i)}{(1+d)^i}}{\sum_1^n I_i} \quad (1)$$

Where:

- PV - is the current value, \$USA/t of captured CO_2;
- CC - reduced carbon loan, \$USA/t CO_2, the emissions of which are prevented;
- n - project life;
- d - reduced discount rate, %.

$$CC = \frac{\sum_{1}^{n} \frac{(K_i + O_i)}{(1+d)^i}}{\sum_{1}^{n} \frac{F_i \cdot I_i}{(1+d)^i}} \tag{2}$$

Where:

- K_i – annual capital expenditures i, \$USA;
- O_i – current operating costs for year i, \$USA;
- I_i – CO_2 storage for year i, Mt;
- F_i – CO_2, the emission of which is prevented in the year i/CO_2, stored the year i.

As equalization 2 shows, indicator CC can be calculated as the current value of expenses divided by the sum of discounted values of CO_2, the emission of which is prevented in each year. Conversely, the PV indicator is based on the undiscounted values of injected CO_2. Due to these differences, the CC indicator is more versatile than PV.

To illustrate the sensitivity of cost estimates for the CO_2 storage to fluctuations in the values of engineering and economic input variables, we use a hypothetical offshore combination of emission source and reservoir characteristic of the conditions of the North-West of Russia.

System characteristics are given in the Table 4 [14].

Table 4. Hypothetical data on the source of emission and collector (best estimates).

Data	Index	Unit of measurement
Stream velocity of CO_2	244	Kg/s
Supply pressure of CO_2 (atmosphere pressure)	0,1013	MPa
Inlet temperature of CO_2	37,8	°C
Compressor efficiency	80%	%
The distance from the source to the collector	300	km
Relative height (source - collector)	10	m
Seabed temperature	10	°C
Efficiency of pipeline	90%	%
Maximum operating pressure in the pipeline	17,2	MPa
Water deep	100	m
Reservoir depth	1500	m
Capacity of reservoir	50	m
Temperature in reservoir	60	°C
Reservoir pressure	13,8	MPa
Reservoir permeability		
Maximum operating pressure in the well	27,6	MPa
Market prices for equipment and its exploitation		Different

We suppose that the duration of the CO_2 storage project is 22 years, including 2 years for the construction of the facility and 20 years for the storage. We suppose that the current discount rate is 7% and 90% for the ratio of CO_2, the emission of which is prevented, and the injected CO_2 - every year (it is assumed that the electricity used by the compressor unit is generated by a power plant that burns powdered coal and produces 800 kg of CO_2 for 1 MGw/h). Using these data and assumptions, the authors came to the following estimates of economic indicators:

- Reduced Capital Costs = 455 mln. USD.
- Present value (*PV*) = 5,6 USD/t of injected CO_2.
- Reduced annual exploitation costs = 32 mln. USD/g.
- Carbon Credit (*CC*) = 11,1 USD/t CO_2, emission is prevented.

The interpretation of these valuations and economic indicators requires caution.

Cost estimates can be changed significantly for different situations.

Our analysis of potential CO_2 storage sites suggests that the carbon credits given may vary up to 5 USD and in excess of 25 USD per 1 ton of avoided emissions without taxes.

Table 5. Analysis of the sensitivity of economic indicators to changes in the main input variables.

10% increase in the following incoming parameters	Changes in reduced capital and current costs before tax (%)
Stream velocity of CO_2	+5,9%
Inlet pressure of CO_2	−0,7%
Inlet temperature of CO_2	Minor changes
Compressor efficiency	−2,6%
The distance from the source to the collector	+4,2%
Relative height (source - collector)	Minor changes
Ground temperature	Minor changes
Efficiency of pipeline	−1,0%
Maximum operating pressure in the pipeline	Minor changes
Water depth	+3,2%
Reservoir depth	+0,1%
Reservoir capacity	Minor changes
Temperature in reservoir	Minor changes
Pressure in reservoir	+3,0%
Reservoir permeability	Minor changes (but not linear)
Maximum operating pressure in the well	Minor changes
Market prices for equipment and its exploitation	
– Costs for the compressor	+2,9%
– Costs for the pipeline	+4,6%
– Costs for the well	+0,9%
– Costs for the platform	+1,6%

Cost including taxes - more.

Sensitivity analysis. The following input indicators are decisive for estimating the storage costs:

- Stream velocity of CO_2, because it affects all relevant costs;
- Compressor efficiency, which affects the cost of compression;
- The distance from the source to the collector, it affects the cost of compression and pipelines;
- Water depth, it affects the cost of the pipeline, platform and well;
- Reservoir pressure, because affects the type and number of wells required;
- Market prices, they affect all cost components

Calculations for the analysis of projects of geological storage of CO_2 are given in Table 5.

Based on the hypothetical example, Table 5 illustrates the sensitivity of economic indicators to changes in the main input variables. The table shows the consequences of a 10% change in the values of the input parameters.

3 Conclusion

The Russian Federation has a high potential for the development of underground space. This circumstance allows us to consider the mechanisms for capturing and storage of CO_2 as a highly promising direction for reducing GHG emissions.

In addition, the usage of these technologies makes it possible to implement joint projects (reinvesting revenues from the sale of excess emission allowances into environmental projects) and to attract investments, for example, in the power industry or for utilizing associated petroleum gas.

There is no focused policy on GHG emissions management in Russia at the federal level. Nevertheless, there are programs and documents defining the main directions of the national energy policy and quantitative indicators of GHG emissions in the country in the long-term perspective.

For the implementation of joint projects, both NWFD enterprises and enterprises from other regions have a number of restrictions. The first of these is the absence of a state register on accounting for greenhouse gas emissions in Russia; the second is the imperfection, or the absence of a regulatory framework for using the market mechanisms of the Kyoto Protocol.

The usage of such mechanisms in energy management system to reduce the concentration of carbon dioxide in the atmosphere provides: reducing fossil fuel consumption and switching to renewables, equipping gas and coal power plants with trapping technologies, storage of carbon dioxide in the ocean and underground storage (renting underground space for storing carbon dioxide can bring considerable income).

The novelty of the topic consists in the analysis of all CO_2 separation methods, determining the percentage of CO_2 removal and the overall efficiency of these processes; the development of a geological and economic assessment methodology for finding reservoirs capable of storing CO_2 in the long term perspective; the development of economic approaches for evaluating CO_2 injection projects in underground space; in identifying the possibilities of using and improving CO_2 capture technologies for Russian energy facilities.

Relation of this work with other research projects: The results presented in the article are the development of environmental and economic research of foreign and domestic scientists in the fields of energy efficiency and environmental management, including the development of research by such scientists as Lyngfelt A (Sweden), Linderberg E (Norway), Holloway (USA. United Kingdom), Imai N (Japan), Nalivkin V.D. (Russia) and others.

References

1. Emberley, S., Hutcheon, I., Shevalier, M., Durocher, K., Gunter, W.D., Perkins, E.H.: Geochemical monitoring of fluid-rock interaction and CO2 storage at the Weyburn CO2-injection enhanced oil recovery site, Saskatchewan, Canada. In: Gale, J., Kaya, Y. (eds.) Proceedings of the 6th International Conference on Greenhouse Gas Control Technologies, Pergamon, pp. 365–370 (2002)
2. Brown, L.T.: Integration of Rock physics and reservoir simulation for the interpretation of time-lapse seismic data at weyburn field, saskatchewan. Colorado School of Mines, M.Sc. thesis (2002)
3. Ilyinsky, A.A.: Oil and gas complex of the North-West of Russia: a strategic analysis and development concepts (Mnatsakanyan OS, Cherepovitsyn AE). Spb: Science, 474 p., 174 il (2006)
4. Malik, Q.M., Islam, M.R.: CO2 injection in the Weyburn Field of Canada: optimization of enhanced oil recovery and greenhouse gas storage with horizontal wells. SPE Paper 59327 presented at the Society of Petroleum Engineers/DOE Improved Oil Recovery Symposium, 3–5 April, Tulsa, Oklahoma, USA (2000)
5. Zaychenko, I.M., Ilin, I.V., Lyovina, A.I.: Enterprise architecture as a means of digital transformation of mining enterprises in the arctic. In: Proceedings of the 31st International Business Information Management Association Conference (IBIMA), pp. 4652–4659
6. Ilyinsky, A.: Priority strategic initiatives for the development of the oil and gas complex of the Russian Federation. Problems of the modern economy, vol. 4 (2008)
7. Inoue, M.: CO2 geological sequestration field experiment planned in Japan. In: Proceedings of the 24th Annual Workshop and Symposium, International Energy Agency Collaborative Project on EOR, 7–10 September, Regina, Saskatchewan, Canada (2003)
8. Goluba, A.A., Zaharova, V.M.: Greenhouse gas management in Russia: regional projects and business initiatives. Center for Environmental Policy of Russia, 86 p. (2004)
9. Novozhilov, A.: Analysis of modern approaches to the search and assessment of territories suitable for the creation and stotage for a long time of large underground reserves of natural gas. In: Matherials of the Scientific and Technical Conference of the Heritage Research Institute. SPb (2003). Author, F.: Contribution title. In: 9th International Proceedings on Proceedings, pp. 1–2. Publisher, Location (2010)
10. Belonin, M., Prishchepa, O.: The main positions of the program of integrated development of hydrocarbon resources of the north-western region of Russia until 2020, St. Petersburg, VNIGRI, 116 p. (2005)
11. Cherepovitsyn, A.: The potential of using the mechanisms of the Kyoto protocol in the oil and gas complex. NeftGazIndustry, vol. 7, October-November (2005)

12. Herzog, H., Eliasson, B., Kaarstad, O.: Capturing greenhouse gases. Sci. Am. **282**(2), 72–79 (2000)
13. Astakhova, A.S., Thoth, L.M.: Socio-economic problems of efficient using of mineral resources (1985)
14. Ilyinsky, A.A., Volkov, D.I., Cherepovitsyn, A.E.: Problems of sustainable development of the gas supply system of the Russian Federation. Subsoil, Saint-Petersburg, 324 p. (2005)

BPM as a Service Based on Cloud Computing

Anastasia Levina(✉), Aleksey Novikov, and Alexandra Borremans

Peter the Great Saint Petersburg Polytechnic University, Polytechnicheskaya str. 29, 195251 Saint-Petersburg, Russia
alyovina@gmail.com

Abstract. Advancement in past decade in the area of computer hardware especially, and software has allowed cloud computing to become the focal point of future information system developments. Cloud business enablers are already driving innovation across customer value propositions and company and industry value chains. Enterprises are applying cloud to enhance their business processes, developing efficient collaboration between units and extending and inventing new customer value propositions. However, transforming the business processes to cloud is to be done based on a careful systematic analysis of enterprise's business and technology needs and resources. We are conducting a systematic literature review to investigate the methods and concepts required to conduct BPM in the cloud. The objective of the paper is to research business process management concepts related to cloud BPM, business process architectures relevant to cloud BPM or implementing BPM in cloud. We intend to research emerging business process models and their impact on organizations. As organizations become more adapt at accepting business processes as services on the cloud, they undergo BPM technology infrastructure change compare to conventional architecture to execute business process.

Keywords: Business process management · Cloud BPM · Cloud architecture · Business process management as service

1 Introduction

In the 21st century, Internet technologies are being actively implemented in all areas of human activity, from the usual trip to the store (Amazon) to voting in national elections (since 2005, Estonia has successfully operated an online election system). Technological progress has not bypassed the business area. The emergence of cloud technologies has helped companies to reduce costs in the field of IT and, subsequently, in processes of software development and implementation [1]. Russian companies seem to be reluctant to switch to a new type of information storage and processing services, which allows not only to significantly increase the efficiency of planning, distribution and use of resources, but also to protect the environment by reducing the company's carbon footprint. This fear of the transition might potentially be associated with a lack of understanding of the concept of cloud computing [2, 3].

The ultimate goal of any digital technology and cloud computing specifically, broadly, is to allow users to be able to execute business processes. But since it's a

V. Murgul and M. Pasetti (Eds.): EMMFT-2018, AISC 983, pp. 210–215, 2019.
https://doi.org/10.1007/978-3-030-19868-8_21

flexible architecture where user can build, configure and change business processes, there is no single way of implementing business processes in the cloud. As we suggested in the beginning of the paper, careful analysis of enterprise' needs and business requirements is needed before moving to the cloud. Our research is aiming at establish business process design models for enterprises planning to shift their business processes to the cloud-based. Much information is available on the different models of the cloud, their underlying technology architecture and key features, that provide maximum services to the user to be able to perform business processes. Thought it is necessary to point out that some organizations do not require all the services and functionality being offered. We are aiming to give an overview of what different enterprises can gain from cloud. The baseline research question of the current paper is: What different categories of business process management could be established in order to associate them to suitable and relevant service models of cloud?

We are also aiming to cover non-functional requirements of business processes that need to be considered while moving the enterprise' business processes to the cloud.

This research paper is structured as following: literature analysis would extract out core concepts related to the topic and present them in logical and systematic manner. To make the paper effective for the readers, concept summary table are used to classify and list the core concepts. These concepts lay the foundation for analysis of literature and allow approaching the research question in an effective manner.

2 Method

This research paper is written based on research methodology of systematic literature review. The first phase of methodology consists of searching literature. We have used digital libraries to search of research papers most of which are published papers. Some of them are not published but are peer reviewed on popular and authentic journals. In second phase, literature selection is done to short-list the most relevant papers, which align with our research motivation. In final stage, literature analysis is performed.

The structure of this paper from now on is as follows. Section 2 describes in detail the research methodology used for this paper. Section 3 describes in detail the concepts of Cloud and BPM in Cloud. In the same section we'll describe BPM and process design concepts from the perspective of cloud computing and discuss critical non-functional requirements of business processes that require consideration upon transition to cloud. Section 3 will be based on analysis and review of literature selected for the research, whereas Sect. 3 will consist of our discussion, based on the literature analysis, to broadly categorize BP designs to associate them with cloud computing models. Section 4 will conclude the paper and present opportunities for future research and limitations of this literature review. At the end of the paper is Appendix section, which contains one summary table for each of the paper considered for this literature review. Each table gives summary of paper abstract and other research parameters for the corresponding paper.

3 Results

The term "cloud computing" has become one of the most popular in the field of IT after Web2.0. This concept connects various modern technologies into a whole, including Grid computing, Utility Computing, distributed system, virtualization, etc. Business process management is intended for business management using IT infrastructure to focus on process modeling, monitoring and management. BPM consists of business processes, business information and IT resources that help create a real-time intelligent system based on business management and IT technologies [4].

Business process management (BPM) provides a way to monitor and improve business performance [5]. BPM takes data from a company's various business applications and then performs two functions: (1) it tracks how information is used to run a business in order to accurately find and understand existing business processes; (2) tracks the flow of information in various operations to ensure that the business process is executed. BPM software is a kind of application software that can be developed based on an existing system [6]. With the development of computer science, cloud computing has replaced Web2.0 and service-oriented architecture (service-oriented architecture; SOA). The term cloud computing has many definitions, but the most accurate is the National Institute of Standards and Technology formulated in 2011: cloud computing (cloud computing) is a model for providing convenient network access on demand to some common fund of configurable computing resources (for example, transmission networks). data, servers, storage devices, applications and services—both together and separately), which can be promptly provided and released with minimal costs or appeals to the provider [7, 8]. Cloud computing provides a highly competitive computing environment that helps to significantly reduce IT costs and replace complex and expensive IT systems with a "cloud" provided by third-party service companies (such as Microsoft, Amazon, Google, etc.) to support and update software and hardware. Users only pay for the computing power they use. At any time and in any place via the Internet, users can visit the "cloud" resources.

Cloud computing consists of a service model that basically consists of three basic levels of service, as shown in Fig. 1.

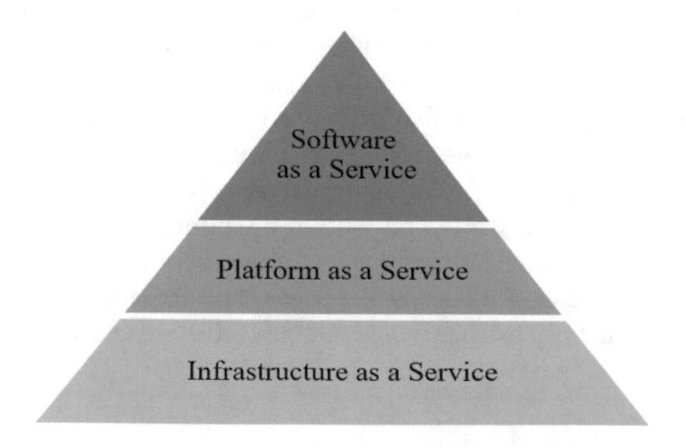

Fig. 1. The three-layer cloud computing model.

Infrastructure as a Service (English Infrastructure as a Service; IaaS) is the lowest cloud computing level that is closest to the hardware layers. Services in this layer can be broken down as follows: a set of physical resources and virtual resources in the service services. Main functions: starting and stopping own resources, operating system image, installation network topology and system configuration.

At the Platform as a Service (Eng. Platform as a Service; PaaS) application infrastructure can be considered as a set of services. They include middleware as a service, messaging as a service, integration as a service, information as a service, and connectivity as a service. Software as a Service (Eng. Software as a Service; SaaS) is an application that runs in the cloud and is provided to users as an on-demand service. Sometimes they are free, and service providers can earn revenue from online advertising, etc. Sometimes a provider charges a fee for using the service. Software and applications can be provided to users via the Internet as one type of service that does not require installation of any software on computers. Thus, SaaS reduces the burden on software maintenance and purchase cost for customers.

To demonstrate these three types of services, an analogy with pizza is often used - a kind of "Pizza as a Service". When a consumer orders and eats pizza in a cafe or restaurant, then this is SaaS, and if he orders it to his house, then this is PaaS. If he went to the store, bought the ingredients and prepared the dish himself, then we can say that it is IaaS [7].

The cloud-based BPM architecture is shown in Fig. 2. Except for the physical hardware layer below, the system architecture from the bottom to the top level includes Infrastructure as a Service (IaaS), Platform as a Service (PaaS), and Software as a Service (SaaS). Adding additional features to the three-layer cloud computing model.

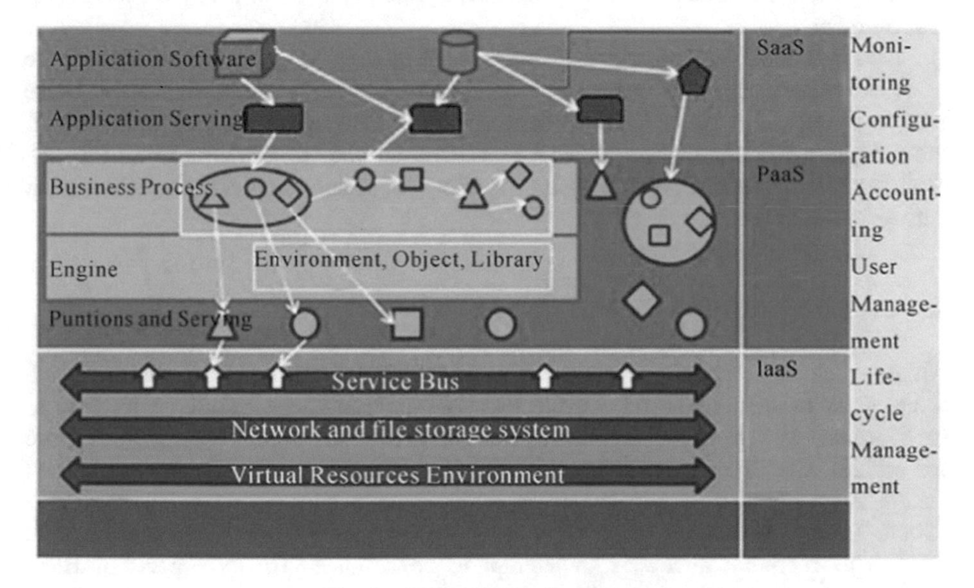

Fig. 2. BPM cloud infrastructure.

The "Infrastructure as a Service" level consists of a virtual resource environment, a network and file storage system, and a service bus. It implements a distributed storage system of files based on virtual resources and forms an abstract file system on various distributed physical computers through a local area network, including functions such as load balancing, fault tolerance processing, dynamic node configuration and parallel processing. The service bus is located at the top of the IaaS level, including external services. It uniformly manages, requests and organizes these services through Web Service, WSDL, SOAP and UDDI technologies. Consequently, the service functions in PaaS can be implemented under a single bus, increasing security and simplifying maintenance. The "Platform as a Service" level includes: the service of functions, the mechanism of business processes and the business process. This level is a key element for implementing BPM. The service of functions can receive services at a lower level, issued by other systems from the service bus to IaaS, and can also independently perform functions.

The "Software as a Service" level is the top level of cloud computing that is closest to users. It is divided into application and software. The application service basically meets the specific requirements of users and software. This level provides timely information about the process and the implementation of functions obtained from the business process service and functions in PaaS for software at the top level, or directly provides them to users through an Internet service. Thanks to the three-layer cloud computing model and the basic BPM system in PaaS, BPM management software can collect information about the business process and then develop, optimize and control it. The functions required for the entire structure are listed in Fig. 2 (right), including real-time monitoring at each level and module, dynamic system configuration, and management and maintenance of the entire BPM and life cycle of the cloud computing environment.

Thus, effectively combining the computing resources of cloud computing and the rules of resource allocation based on demand, the investment costs of BPM software will be significantly reduced. After connecting each system to the cloud environment, the company can simultaneously handle a huge number of business processes [8]. Thus, monitoring, control and analysis processes will greatly facilitate the development and layout of software.

4 Conclusion

Although cloud computing models and candidate BPM has been established in the industry, there are still several non-functional or environmental features, which are to be addressed by technology community for the public clouds to be considered secure for critical data. Security and data isolation are primary concerns. Also, the Cloud design is an important factor for organizations before considering on moving onto Cloud. There are definitely better Cloud platforms services than others available to be used. This literature review is an attempt to consolidate primary aspects of BPM deployment into the Cloud.

BPM in Cloud is a broad area of research and many different options and framework exists for transition BPM in Cloud effectively and optimize it with relevant tools such as BPEL, process engines and others; hence many topics in this area could be narrowed down for further research.

References

1. Han, Y., Sun, J., Wang, G., Li, H.: The BPM architecture based on cloud computing. J. Comput. Sci. Technol. **25**(6), 1157–1167 (2010)
2. Borremans, A.D., Zaychenko, I.M., Iliashenko, O.Y.: Digital economy. IT strategy of the company development. In: MATEC Web of Conferences, vol. 170, p. 010342017 (2017)
3. Lepekhin, A.A., Borremans, A.D., Iliashenko, O.Y.: Design and implementation of IT services as part of the "Smart City" concept. In: MATEC Web of Conferences, vol. 170, p. 01029 (2017)
4. Zota, R.-D., Fratila, L.-A.: Cloud standardization: consistent business processes and information. Inf. Econ. **17**(3), 137–147 (2013). https://doi.org/10.12948/issn14531305/17.3.2013.12
5. Han, Y., Sun, J., Wang, G., Li, H.: A cloud-based BPM architecture with user-end distribution of non-compute - intensive activities and sensitive data. J. Comput. Sci. Technol. **25**(6), 1157–1167 (2010)
6. Duipmans, E.F., Pires, L.F., Da Silva Santos, L.O.B.: Towards a BPM cloud architecture with data and activity distribution. In: 2012 IEEE 16th International Enterprise Distributed Object Computing Conference Workshops (EDOCW), 10–14 September 2012, pp. 165–171 (2012). https://doi.org/10.1109/edocw.2012.30
7. Ilin, I., Levina, A., Abran, A., Iliashenko, O.: Measurement of enterprise architecture (EA) from an IT perspective: research gaps and measurement avenues. In: ACM International Conference Proceeding Series Part F131936, pp. 232–243 (2017)
8. Berman, S.J., Kesterson-Townes, L., Marshall, A., Srivathsa, R.: How cloud computing enables process and business model innovation. Strategy Leadersh. **40**(4), 27–35 (2012)

Information Modeling of Increasing the Environmental Living Comfort Level During Built-up Areas Reconstruction

Svetlana Sheina and Karina Chubarova(✉)

Don State Technical University, Gagarina sqr. 1, 344000 Rostov-on-Don, Russia
karina.chubarova@yandex.ru

Abstract. The article deals with the use of geoinformation systems as a tool for strategic planning and management decision-making in the field of urban development. As a part of the built-up areas reconstruction, it is proposed to use an information model to improve the level of environmental living comfort to assess the effectiveness of various measures of complex improvement using the method of spatial analysis.

Keywords: Sustainable development · Ecology · Ecological living comfort · Complex improvement of territories · Geoinformation systems · GIS · Information model

1 Introduction

Sustainable development of urban areas requires special attention in the modern world. First of all it concerns developing countries, which are entering the era of mass urbanization. Sustainable development can provide substantial gain of productivity and standard of living, but it can generate an environmental and other risk, which may largely negate the progress that was made. Many countries with emerging market economies have already committed to sustainable development, i.e. economic growth, providing increase of quality of life without compromising the environment and natural resources.

Under the rapid growth of the built-up urban territories, the new problems arise as a possible outcome of social, economic, and technical changes. The most obvious and noticeable of these happen in urban environment state deterioration, which in this country are due to insufficient spreading of modern technologies of household and industrial waste treatment, lack of resources, and air pollution. These are the reasons of health problems among the city dwellers, deterioration and degradation of urban environment infrastructure [1–3].

It is obvious that the search for design solutions with a focus on the principle of ecologization and using new information and technological methods should be carried out taking into account the entire typology of Russian cities, their diversity and real condition, with the allocation of development trends, the possibility of preserving the historical and cultural environment and updating existing buildings, creating the necessary comfort level and environmental well-being at each taxonomic level [4, 5].

V. Murgul and M. Pasetti (Eds.): EMMFT-2018, AISC 983, pp. 216–223, 2019.
https://doi.org/10.1007/978-3-030-19868-8_22

To solve the urban environmental problems and to improve the level of comfort for the residents, a set of activities including elaboration, adoption and implementation of coordinated measures for the integrated improvement of built-up areas, as well as planning of a system of improvement newly developed areas to ensure a satisfactory level of comfort must be implemented.

2 Purpose and Formulation of the Research Task

The main purpose of this research is to develop a method of geoecological support to increase the comfort of living by the improvement and greening of built-up areas. One of the tasks set for the purpose of this research is to develop an information model of improving the environmental living comfort level using geoinformation systems. Questions of information model development, possibilities and results of its application are considered in detail in this article.

3 Information Modeling as a Tool of Strategic Planning

The adoption of management decisions based on the prediction of events is one of the most important areas in the field of management automatization in urban development. The lack of reliable data on various factors that affect the sustainability of specific territories makes it difficult to collect baseline information, analyze it, identify the existing dynamics of development and, as a result, slows down the process of making important decisions in the management of these territories [4]. One of the existing ways to solve this problem is the introduction of modern geoinformation technologies in management processes [6].

Analysis of modern information and analytical systems, including GIS, shows that these systems have a multi-purpose and multidimensional nature. Using of modern information systems allows not only to process data, but also to carry out its expert assessment on various criteria, which, in turn, is an important factor for adoption of management decisions and strategic planning. Information systems are widely used in the modern world in the decision-making process for the optimal management of lands, resources, urban economy, transport, trade objects, etc. [7].

Any territory, as a geosystem, is a combination of several structural and functional blocks related to each other. Maintaining the maximum stability of the system and achieve its sustainable development is an important task today. Information modeling tools allow taking into account a number of factors that have a direct impact on the process of sustainable development of the territory, and also makes it possible to predict potential changes in the system from various external and internal influences.

Development of an information and analytical model helps to create optimal conditions for the implementation of strategies of region or municipal formation development through improving the quality and transparency of information, as well as effective organization of work processes.

4 Technique of Information Modeling to Improve the Environmental Living Comfort Level

Information model of increasing the environmental living comfort is created for information and analytical support and environmental monitoring of the urban territories complex improvement realization. This information model is the basic element of environmental living comfort level imitation model and represents data on the ecological state of- built-up areas linked to the electronic map of the city.

The information model of increasing the environmental living comfort includes 4 blocks (see Fig. 1):

- input data block;
- block of spatial analysis;
- modeling block;
- prediction block.

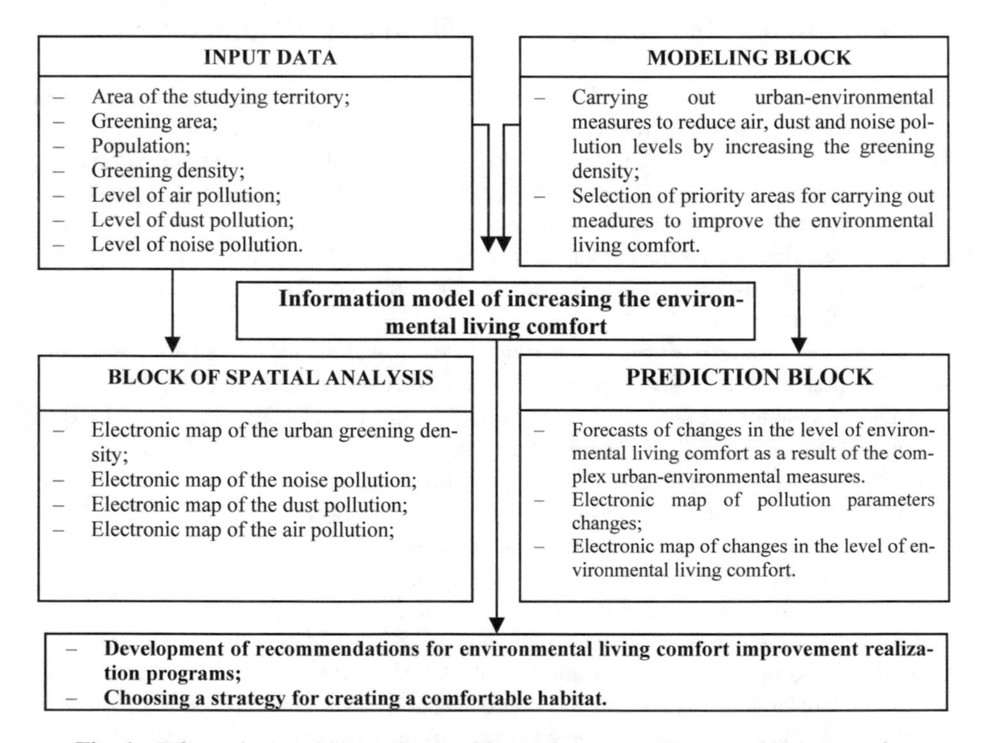

Fig. 1. Information modeling scheme of increasing the environmental living comfort

4.1 Forming of the Input Data Block

The input data block is received as a result of carrying out the analysis of the territory, and its ecological situation. This block includes information about the area, the number of inhabitants, the greening area, and indicators of noise, dust and air pollution. The obtained information is integrated into the database of environmental parameters of urban areas and serves as the basis for the electronic map of the environmental living comfort level.

4.2 Forming of the Modeling Block, Selection of Urban-Planning Measures

Modeling block allows us to consider various options for changing the environmental situation in the territory, subject to certain urban-environmental measures and to assess their expected results in the planning stages. During the modeling process, it is carried out the dynamics analysis of the such environmental indicators as the level of noise, dust and air pollution. Based on the modeling of territories ecological parameters, the most acceptable and effective urban-environmental measures can be selected.

4.3 Forming of the Spatial Analysis Block on the Base of Electronic Maps

This block includes the spatial analysis of environmental parameters using geoinformation systems to assess the existing environmental situation of the territory. In the process of applying the information model, the urban territory can be estimated by any of the environmental parameters contained in the database before and after the implementation of certain urban-environmental measures. For each of the specified parameters of the territory, it is built the electronic map, which clearly demonstrates information about the ecological situation and the level of environmental living comfort, as well as allow to identify the most environmentally unfavorable areas of built-up territory.

4.4 Prediction of Changes in the Level of Environmental Living Comfort

Prediction is the final stage of the information model application. On the basis of research conducted in previous stages, forecasts are given for changes in the level of environmental living comfort, which are results of implementation of the complex urban-environmental events.

The algorithm of the information model is presented on Fig. 2.

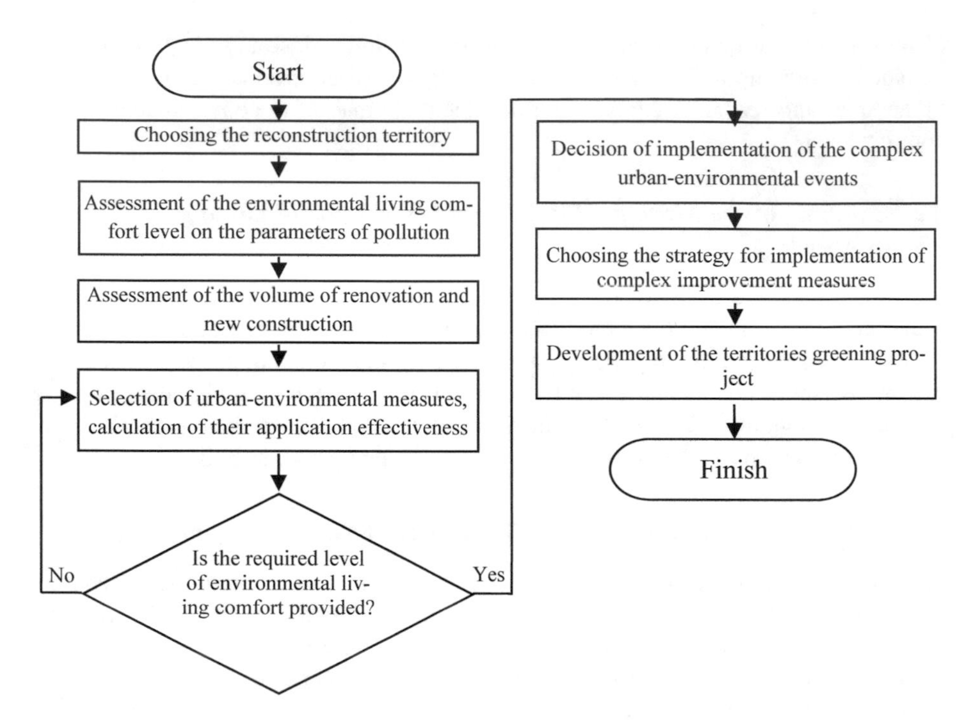

Fig. 2. The algorithm of the information model

As an example, consider the information model of increasing the environmental living comfort level for the reconstruction site in the city of Rostov-on-don. The choice of the reconstruction site (see Fig. 3) was based on the results of analysis of the master plan of the city, the rules of land use and development and the map of new housing construction, as well as their comparison with the electronic map of ecological comfort.

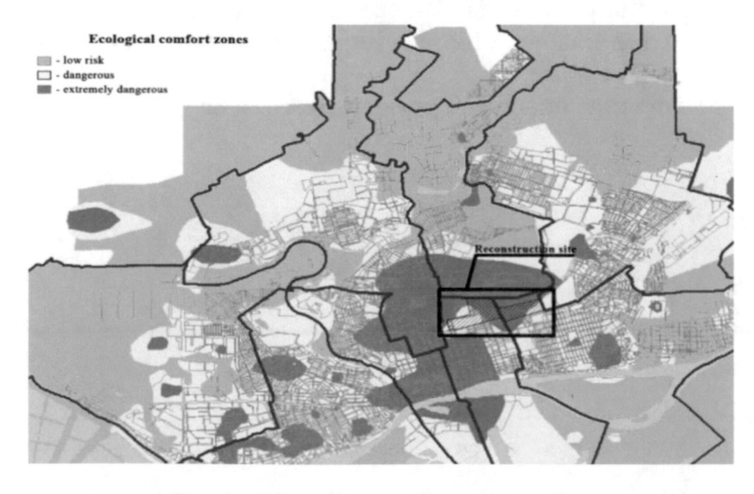

Fig. 3. Selected area of the reconstruction

At first it was made the calculation of habitat quality index for evaluation sites [8] of the selected reconstruction area (see Fig. 4) [9].

Fig. 4. Selected reconstruction site with zones of the habitat quality index

As you can see in the Fig. 4, most of the territory is located in an extremely unfavorable zone. In such a situation, according to the classification adopted in this research [9], a set of urban ecological measures M4 is applicable for an extremely unfavorable zone in terms of habitat quality.

The complex of urban-ecological measures is aimed at bringing the territory of reconstruction to a favorable or relatively favorable level in terms of habitat quality.

The simulation of the habitat quality index based on the data of a comprehensive assessment of the reconstruction area was performed in the ArcGIS 10.1 software package in order to assess the effectiveness of the proposed set of measures of improvement and greening on the selected area (see Fig. 5).

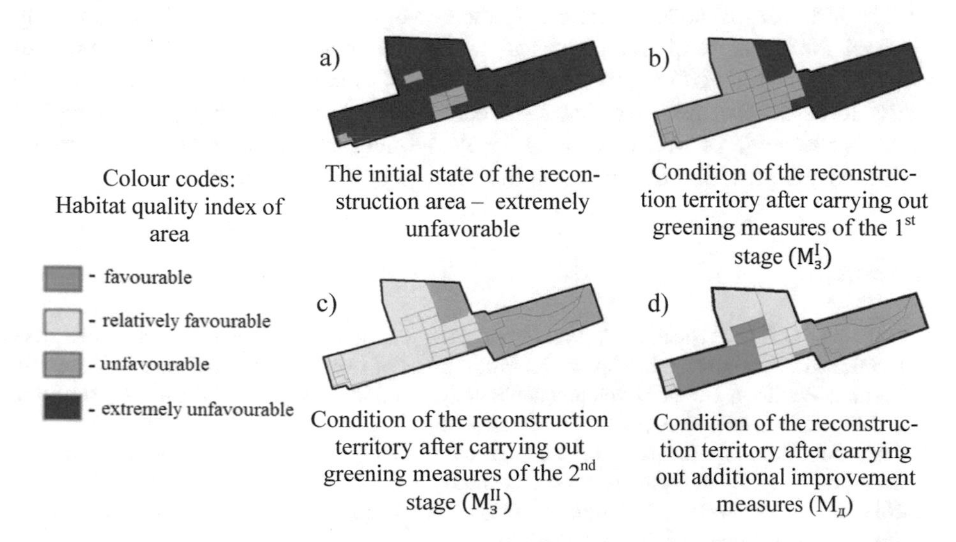

Fig. 5. Example of information modeling of the habitat quality index after the implementation of the selected set of improvement and greening measures on the reconstruction site in the city of Rostov-on-Don

Prediction of the habitat quality index during the planting of green spaces was carried out according to the data on the impact of greening on ecological parameters of the environment and comfort of living.

Based on the results of a complex assessment of the territories and conducted modeling, it is compiled the urban-ecological passport of reconstruction territory [9], which contain forecasts of changes in the habitat quality after the implementation of complex improvement and greening measures.

Thus, modeling with the use of GIS technologies and electronic mapping allows to demonstrate the projected dynamics of changes in the environmental quality as a result of selected complex of urban environmental measures and thereby greatly simplifies the management decisions in the field of improvement of urban areas and the calculation of their feasibility study.

5 Conclusions

Information model of improving the environmental living comfort level built using geoinformation technologies and electronic mapping allows:

- to carry out zoning of built-up areas according to the parameters of ecological living comfort;
- to identify the most dangerous territories from the ecological point of view;
- to fix the results of environmental monitoring of built-up areas on the electronic map of the city;
- to simulate the implementation of a comprehensive program for the improvement of urban areas and to assess the effectiveness of various activities under this program using the method of spatial analysis.

The information model of increasing the ecological living comfort level allows to approach the problems of urban environment improvement from an ecological point of view. The versatility of the proposed model allows to adapt it for use in any municipality. It should also be noted that the model can be used as an independent tool of strategic planning, as well as a basis for developing a simulation model of urban-environmental provision of increasing the ecological living comfort.

References

1. Dolinina, O.N., Pechenkin, V.V.: Smart city: an integrative concept of social and technological design of urban space. In: Proceedings of Urbanistics: Experience of Research, Modern Practices, Urban Development Strategy: Joint Theses of the All-Russian Scientific and Practical Conference, pp. 88–90. Saratov State Technical University, Saratov (2017)
2. Grigor'ev, V.A., Ogorodnikov, I.A.: Ecolgization of cities in the world, in Russia, in Siberia: analytics. GPNTB SO RAN, Novosibirsk (2001)
3. Nikolov, R., Shoikova, E., Krumova, M., et al.: Learning in a smart city environment. J. Commun. Comput. **13**, 338–350 (2016)

4. Seferyan, L.A., Agadzhanyan, A.N.: Strategy for the development of modern cities and environmental safety. In: Materials of the International Scientific and Practical Conference "Construction and Architecture-2017", pp. 184–186. Don State Technical University, Rostov-on-Don (2017)
5. Esaulov, G.V., Esaulova, L.G.: "Smart city" as a model of urbanization of the 21st century. Urban Plann. **4**(26), 27–31 (2013)
6. Sheina, S.G., Babenko, L.L., Matveyko, R.B.: Ecological reconstruction of urban territories. Monograph. Rostov State University of Civil Engineering, Rostov-on-Don (2013)
7. Sheina, S.G., Hamavova, A.A., Pseunova, S.R.: Conceptual studies of the problems of improving urban environment quality. BST: Bull. constr. tech. **6**(982), 50–51 (2016)
8. Sheina, S.G., Girya, L.V.: Organizational and technological basis of environmental risk management in the reconstruction of urban development. News High. Educ. Inst. Constr. **8** (596), 75–79 (2008)
9. Sheina, S.G., Yudina, K.V.: Development technique of urban-environmental passport of the reconstruction territory. Mater. Sci. Forum **931**, 822–826 (2018)

Formation the Construction Cost for Residential Buildings at the Design Stages

Svetlana Sheina[1(✉)] and Natalya Tsopa[2]

[1] Don State Technical University, Academy of Construction and Architecture, 1, Ploshchad' Gagarina, Rostov-on-Don, Russia
rgsu-gsh@mail.ru
[2] V. I. Vernadsky Crimean Federal University, Academy of Construction and Architecture, 4, Vernadsky Avenue, Simferopol, Russia

Abstract. The article discusses the important applied problem dedicated to the theoretical and methodological foundations of the formation construction cost for residential buildings at the design stages. Determined that the different names of phases are given by different international scientists, but the phases of a building project and its goals are generally the same. It was substantiated that most often the design stages of a residential building consist of: conceptual phase, predesign, design, tender, construction, use. This paper aims to define the construction cost for residential building at the most important for construction cost phases: predesign, design and construction stage using the international and Russian experience. It was offered graphic two and three-dimensional model of the total construction scope of work. Two types of costs for the construction of residential buildings were considered: direct costs and indirect or overhead costs. The subject of consideration was a 12-storey multifunctional residential building with built-in premises, which is designed for Russia. The construction cost for the residential building at the predesign stage of preparation was 1.744 million Euros. At the "Design stage", on the basis of detailed estimate and a summary estimate of construction, the cost amounted was 1.157 million Euros. At the "Construction stage", the total construction costs for the object was amount to 1.443 million Euros. At the "Construction stage" the cost of the residential building was 17% less, in contrast to the calculations at the "Pre-design stage".

Keywords: Construction cost · Residential building · Design stages

1 Introduction

Significant changes have taken place in civil engineering in Russia over the past ten years. Currently, the residential construction sector performs not only a social function, but it is also an entrepreneurial activity. Therefore, the feasibility of designing and constructing a civil engineering is determined on the basis of the results of a construction cost assessment obtained in a feasibility report. However, in practice, the estimated cost of construction projects has to be revised upwards in the budget by object of expenditure. This is due to the use of redundant design concepts and the lack of economic feasibility of the resources used. Taking into account the fact that the

V. Murgul and M. Pasetti (Eds.): EMMFT-2018, AISC 983, pp. 224–235, 2019.
https://doi.org/10.1007/978-3-030-19868-8_23

construction cost during the period of work execution should not change, there is a need to improve the accuracy of calculations in the current and forecast prices during the designing, examination and preparation the tender documentation taking into account the construction period and the growth rate of prices for construction resources.

2 Materials and Methods

The greatest assistance to the learning of the theory and practice of the formation the construction cost was made by Russian and international scientists, among them: Ardzinov, Baranovskaya, Andrade, Vieira, Bragança et al. [1–4].

Ardzinov investigated in his work the problem of pricing and the formation of the estimated cost for real estate [1]. Barnovskaya has devoted her researches to studying the calculation of the estimated cost of work for residential areas [2].

A residential building project is developed by a sequence of phases. The different names of phases are given by different authors, but the phases of a building project and its goals are generally the same [3, 4]. Investment and construction project includes following design stages of a residential building: conceptual phase, predesign, design, tender, construction, use [3]. In this work we will discuss the most important for construction cost phases: predesign, design and construction stage. Many authors [5–10] agree that formation the construction cost for real estate's depend on the design development phase. A critical analysis of the theoretical and practical aspects of the construction cost formation at the design development phases revealed a number of unresolved issues, one of which is the absence of the construction cost algorithm for residential buildings at the design stages. In this regard, the subject of this research is relevant and timely, especially with the active development of residential construction in the Republic of Crimea. The aim of this paper is to develop the theoretical and methodological foundations of the formation construction cost for residential buildings at the design stages Russian experience.

To achieve this goal, the following tasks were solved:

- the international practice of forming the construction cost for residential buildings was studied;
- the methodology for determining the construction cost at the design stages was clarified;
- a detailed analysis of the construction price options was carried out using the example of multi-story residential building.

3 Results

In world practice, all the variety of existing methods for determining the construction cost can be divided into two independent groups [11]:

- methods of calculation the estimate price of construction using on the integrated indicators at the initial phase of design stages;

– methods of element-by-element calculation of construction prices at the final phases of design stages.

The calculation of the cost of a residential building is carried out on the basis of estimates, taking into account the specific features of a particular construction object.

Calculation of the construction estimated cost is to determine the amount of cash needed to implement the construction in full. It is advisable to group these funds by semantic load, combining them into 2 groups of costs: direct (construction and erection) costs and indirect or overhead (non-construction and erection) costs.

The first group includes the construction and erection costs, there are the costs of work types that involve the usage of both labour and materials and mechanisms for their implementation. This group includes the components of construction costs for the future construction project:

- main building;
- supply external engineering lines (except the electric networks);
- external electrical supply;
- site improvement;
- roads.

These costs are about 80% of the total construction cost for the average project.

But in order for contractors to have legal, technical and documentary grounds for the construction and erection works, the customer must order and pay for a number of necessary operations. In addition, the customer must himself exist on some means - to pay wages, rent, stationery. That is, the customer should make additional payments that add up to 20% of the total construction cost. These payments are combined into the second group - the costs are not of construction and erection nature.

This group includes the costs of the following types of work:

- provision of customer services;
- survey and research work;
- development of design documentation;
- obtaining approvals and opinions;
- project expertise;
- obtaining technical conditions for connection;
- obtaining a construction permit (opening orders by type of job);
- other types of work.

At the same time, construction and non-construction and erection costs are also incurred during the construction of buildings directly (construction of foundations, technical supervision) and during the preparation of the construction area (cutting down trees, demolishing old buildings, technical supervision). All these types of costs are taken into account in the summary estimate. A summary estimate of construction costs (SECC) for an individual object or for a group of functionally related objects can be represented by a graphical pricing model in construction. The total amount of work included in the contract depends on the scope of work by type, as well as on the amount of work for each individual type. It turns out the following graphical interpretation of the total amount of construction work, (see Fig. 1).

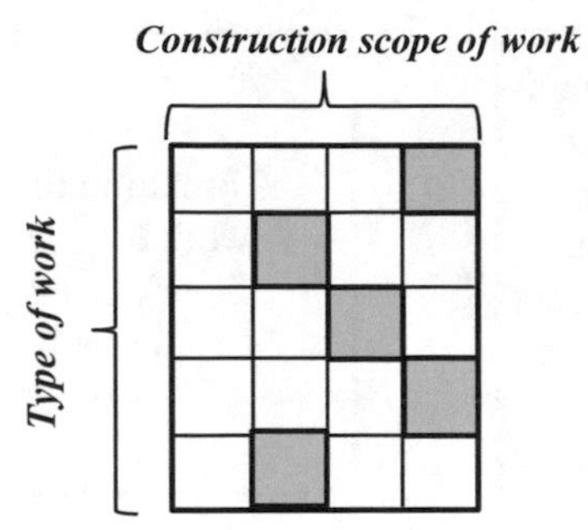

Fig. 1. Graphic two-dimensional model of the total construction scope of work.

We can supplement the presented two-dimensional model of the total construction scope of work with the specific cost of the types of work. In this case, we obtain a three-dimensional model of the total cost of work, i.e. summary estimate of construction costs (SECC), object or local estimate/detailed estimate of construction cost, (see Fig. 2).

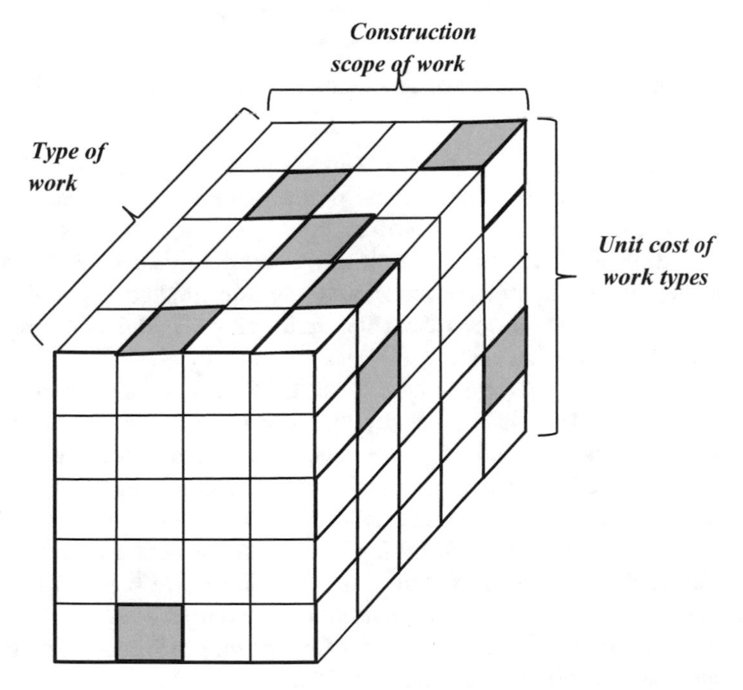

Fig. 2. Graphic three-dimensional model of the total construction scope of work.

The methods and tools used in the calculation of the estimated construction cost essentially depend on the design development phase, (see Fig. 3).

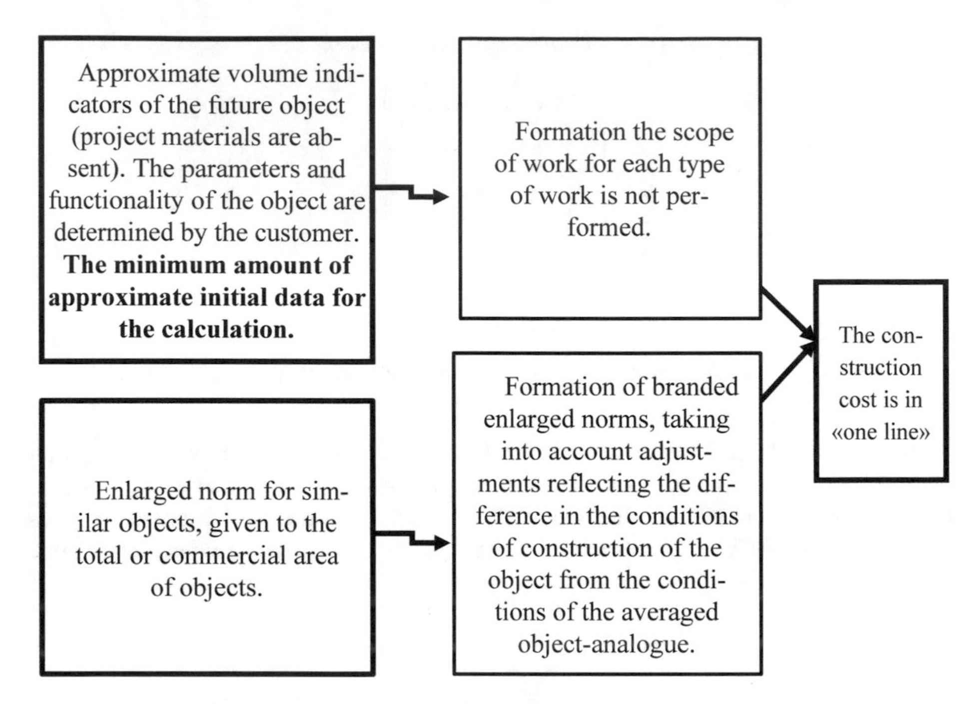

Fig. 3. Formation the construction cost for residential buildings at the predesign stage.

At the predesign stage, the estimated construction cost is based on indicative indicators of the future facility. There are no project materials at this stage. The construction cost of the object is determined by aggregate indicators based on similar objects. The method of calculating the construction cost at this stage has an advantage. At this stage, the complexity of the calculation of the estimated cost is minimal, but at the same time, the accuracy of the calculation varies about ±30% of the real, objectively determined construction cost.

The method of calculating the estimated construction cost at the predesign stage has disadvantages. At this stage, it is necessary to check the volume data of the object, presented by the designer in the design materials, primarily the total area of the residential object. The method of determining the total areas of objects used in urban planning and design takes into account the total area of buildings in the outer perimeter of external walls. The method of measuring areas in construction objects takes into account the internal perimeter of external walls. The difference between the first and second indicator is about 10%. Adequate transition from one indicator to another at the predesign stage is possible through a system of transition coefficients. The difference from the actual construction cost can be ±25–30%.

At the design stage of building, the average labor input of the calculation increases, the calculation time ranges from several days to 1 month or more (it is depend on complete set of project documentation of the "Technical design" stage). At the "Technical design" stage information about the object is not complete, there is no description of technological processes. Therefore, the cost of a number of works and structures is calculated according to enlarged norms. The accuracy of calculating the estimated construction cost at this stage is ±10–15% of the actual construction cost.

The construction stage gives the most accurate value of the estimated construction cost (see Fig. 4). At this stage, working project documentation is developed. This documentation contains all architectural, constructive, technological and organizational solutions.

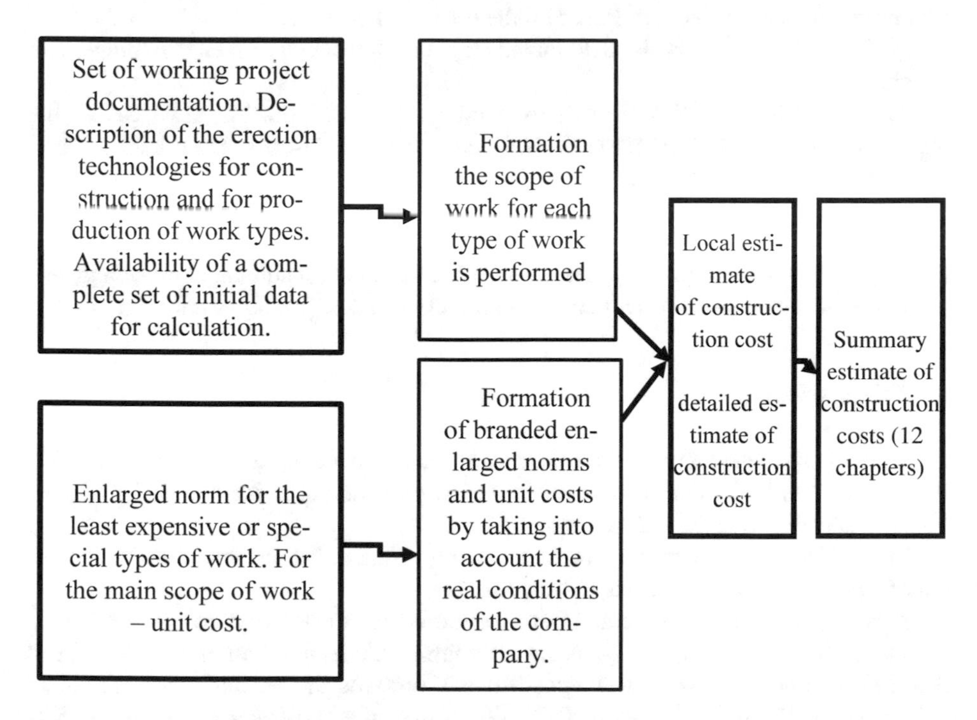

Fig. 4. Formation the construction cost for residential buildings at the construction stage.

The calculation time using this technique is several weeks. The time for calculating construction cost of the object increases to several months if there is a complete set of working documentation. The calculation accuracy is ±5% of the actual construction cost. The preliminary calculation of the summary estimate of construction costs is made on the basis of the design documentation of the "Technical design" stage.

The exact determination of the estimate construction price at the initial stage plays an important role in the investment and construction process. The cost that will be calculated and budgeted at the initial stage of construction allows us to estimate the effect of the construction object for the customer.

The correct estimated cost [12–15] is the basis for determining the size of capital investments, financing construction, forming contractual prices for building products, paying for completed contractual (construction and erection) works, paying for the cost of acquiring equipment and delivering it to construction sites, and also reimbursement of other expenses at the expense of the funds provided for in the summary estimate. In addition, on the basis of the estimate documentation, accounting and reporting, economic calculations and evaluation of the activities of construction organizations and customers are also carried out. Based on the estimated cost, the book value of the fixed assets put into operation is determined for the built enterprises, buildings and structures.

The determination of the initial (maximum) contract price on the basis of the estimated cost is made in accordance with the existing industry-specific methods.

The source of information for calculating the estimated cost of work can serve as current prices obtained by analyzing the market, reference prices of industry and regional directories, catalogs and publications, reflecting the level of real prices, data of industry and regional regulatory reference books, information contained in the registers of contracts, applications of participants completed trades, protocols, etc.

4 Discussion

In this work, the calculation of the construction cost is made on the example of a multi-story residential building at two stages of design preparation: at the “Predesign stage”, “Design stage” and at the “Construction stage”.

The subject of consideration is a 12-storey multifunctional residential building with built-in premises, which is designed for Russia.

Based on the Russian enlarged norm of the basic construction cost of analogous objects (EN BCC-2001), a 12-storey monolithic residential building with built-in premises was adopted as an analogue. Table 1 presents the technical and economic indicators for ENBCC-2001 as of 01.01.2000, then the construction cost in 2018 is calculated using an index of the change in the estimated cost for monolithic building. Then the cost of a residential building is recalculated for value added tax (VAT).

At the predesign stage of preparation, the estimated construction cost, including VAT, was 1.744 million Euros (126.78 million rubles), and the cost of 1 m^2. total area of the building – 437.4 Euros (31.8 thousand rubles).

Consider the formation of the estimated construction cost at the predesign stage of preparation, at the stage of “Design” and “Construction” of the project, where the subject of consideration is the projected 12-storey multifunctional residential building with built-in premises.

The construction area of the multifunctional residential building with built-in premises is 516 m^2, the area of the building is 3050 m^2. The following **Ошибка! Источник ссылки не найден.** gives a feasibility indicators of the construction a 12-storey monolithic residential building.

Table 1. Feasibility indicators of the construction a 12-storey monolithic residential building with built-in premises at the “Predesign stage”, €

№	Name of feasibility indicators	Building indicators on 01.01.2000	Building indicators on 2 quarter 2018	Building indicators on 2 quarter 2018 with VAT
1	Total construction cost	215433.29	1477854.20	1743865.20
2	Construction cost of residential area	198294.36	1360316.37	1605171.94
3	Construction cost of non-residential area	17125.17	117524.07	138679.50
4	Construction cost of 1 m^2 of total building area	53.65	370.43	437.14
5	Construction cost of 1 m^2 total area of apartments	56.40	380.47	449.11
6	Construction cost of 1 m^2 living area of apartments	56.40	391.33	461.76
7	Construction cost of 1 m^2 non-residential area	4.81	32.87	38.79
8	Construction cost of 1 m^3 building volume	13.89	94.64	111.69

The building has a shape close to a rectangular, with dimensions in axes of 16.8 m × 29.4 m. The building has 12 floors. In the basement, at mark - 1. 250, technical room is located. On the ground floor with a height of 3.92 m there is a shop, office space; on the next 11 floors, 3.0 m high, there are two 1-room apartments, one 2-room and two 3-room apartments. A total of 55 apartments are provided. The layout of the apartments - corridor.

Space-planning indicators are given in the feasibility indicators in Table 2.

Table 2. Feasibility indicators of a residential building

№	Name of feasibility indicators	Unit measurements	Amount
1	Number of floors	floor	12
2	Conventional building height	m	43.310
3	Floor height	m	3.0
4	Number of apartments including: - 1-room apartment - 2 room - 3-bedroom	number number number	22 11 22
5	Living area of apartments	m^2	2011.24
6	Area of apartments	m^2	2771.2
7	Total area of building	m^2	3950.7
8	Building area	m^2	516

(*continued*)

Table 2. (*continued*)

№	Name of feasibility indicators	Unit measurements	Amount
9	Total building volume Including: - below the mark ± 0.000 - above ± 0,000	m^3	20496 976 19520
10	Built-in facilities area/useful square - offices - the shops - pump - electrical shield - underground	m^2	15.33 380.85 6.4 13 83

In calculating the construction cost of residential building, we were guided by the preparation of an estimate and financial calculation. The calculation was made taking into account the following other and limited costs: construction control services for the quality of construction and erection works, insurance services for a construction object, a reserve of funds for unforeseen expenses and costs, costs for the construction of temporary buildings and structures, and general contractor services.

The total value of the estimated construction cost of the object is presented on the basis of the calculation the local estimate and detailed estimate of construction cost in the Table 3.

Table 3. The results of the feasibility indicators calculation for residential building at the "Design stage" according to the summary estimate of construction costs

№	Name of feasibility indicators	Unit measurements	Amount
I. Architectural and planning indicators of the project			
1	Building area	m^2	516
2	Total area of building	m^2	3950
II. Economic indicators of the project			
5	Estimated cost of main construction works	thousand Euros	887.22
6	Detailed estimate of construction cost	thousand Euros	905.36
7	Summary estimate of construction costs	thousand Euros	1156.86
8	Estimated construction cost of building for 1 m^2 total area	€/m^2	379.30
9	Estimated material costs for 1 m^2 of total area	€/m^2	31.64
10	Estimated cost of overhead expenses for 1 m^2 of total area	€/m^2	2.90
11	Estimated profit per square meter total area	€/m^2	2.02

Calculate the construction cost of the residential building, taking into account the assumptions established in the work, Table 4.

As a result of forming the estimate and financial calculation at the "Construction stage", the following results were obtained:

1. The approximate term of construction will be 14 months.
2. The construction cost taking into account other and limited costs, will be 1442674.14 Euros with VAT or 473.01 Euros/m2 of the total building area.

Based on the calculations carried out in the table, the estimated cost of building at the "Construction stage" stage was 1.157 million Euros. 1 m^2 of total area will cost 379.30 Euros. In comparison with the similar values obtained at the "Predesign stage" the reduction in the value of the Object was 32%, however, taking into account the assumptions made by us for the "Construction stage", the total construction costs of the object will be 1.443 million Euros. Total expenses per 1 square meter of total area - 473.01 Euros. At the "Construction stage" the cost of the Object is 17% less, in contrast to the calculations at the stage "Predesign stage".

Table 4. The total results of the feasibility indicators calculation for residential building at the "Construction stage"

№	Name of feasibility indicators	Unit measurements	Total, €
1	Total area of building	m^2	3950
2	Summary estimate of construction costs	€	1156864.92
3	Estimated construction cost of building for 1 m^2 total area	€	379.30
4	External engineering lines	€	20592.20
5	Temporary buildings and facilities	€	12725.51
6	Unforeseen expenses	€	57843.25
7	General Contracting Services	€	80980.54
8	Object insurance	€	40490.27
9	Technical control services	€	73177.44
10	Total expenses	€	285809.21
11	Actual expenses	€	1442674.14
12	Actual expenses for 1 m^2 of total area	€	473.01

In accordance with the developed financing schedule (Fig. 5), 2 investment peaks are planned. The first peak will be in the period when construction of the building frame begins (Q1 2019) and will amount to 487.96 thousand Euros with VAT, the second peak will be in the period of the beginning of the device facades and interior decorating (4 quarter of 2019) and will be – 477.03 34.680 thousand Euros, VAT included.

To calculate the payback period of an investment-construction project, on one chart we combine the data on financing the Facility and the total revenue from sale (see Fig. 5).

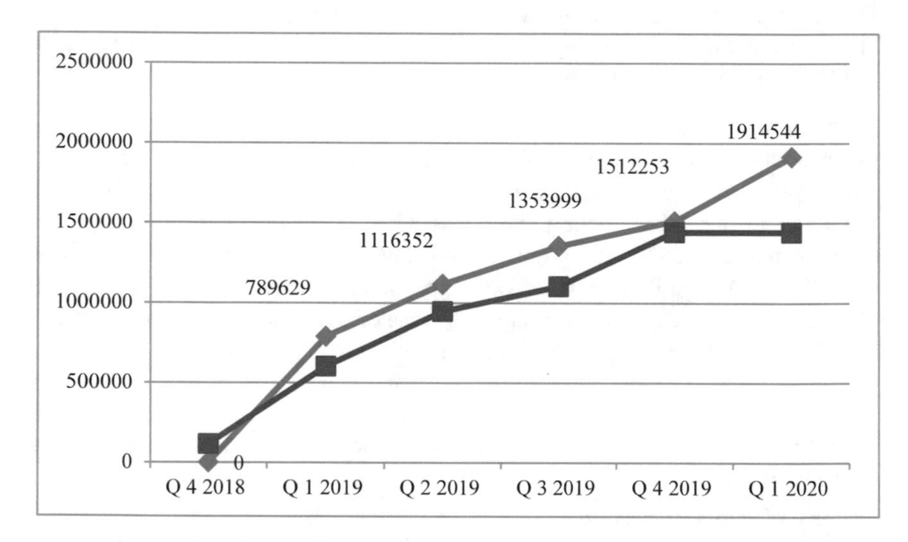

Fig. 5. The period payback for the construction of a residential building.

The payback period of the Object = 1442 thousand Euros/(1914.5 thousand Euros/14 months) = 10.6 months. We take 11 months.

5 Conclusions

In this article the important applied problem is solved, dedicated to the theoretical and methodological foundations of the formation construction cost for residential buildings at the design stages. Testing the methodology allowed us to obtain the main scientific results.

At the predesign stage of preparation the estimated cost of residential building is calculated according to enlarged norms. At this stage, the estimated construction cost of building was 1.744 million Euros.

At the "Design stage", on the basis of calculating local estimates, detailed estimate and a summary estimate of construction, the cost amounted was 1.157 million Euros, and 1 m^2 of total area will cost 379.3 Euros. In comparison with similar values obtained at the "Predesign stage", the reduction in the value of the Object was 32%, however, taking into account the assumptions made by us for the "Construction stage", the total construction costs for the object will amount to 1.443 million Euros, full cost per 1 square meter of total area – 473.01 Euros. At the "Construction stage" the cost of the Object is 17% less, in contrast to the calculations at the "Predesign stage".

During the approbation of the methodology using the example of an investment construction project of a 12-storey multifunctional residential building with built-in premises, it was revealed that when adjusting design decisions, the cost of construction may also decrease.

After conducting more detailed analyzes of the sensitivity of the project, it was concluded that when determining the economic efficiency, the payback period for project is 11 months.

Our work confirmed the effectiveness of the use of an improved method of forming the construction cost for residential buildings at the design stages in practice during the implementation of investment and construction projects of civil engineering.

References

1. Ardzinov, V., Chepachenko, N.: Problems of reforming pricing and estimated rationing in construction. Econ. Manag. **5**(127), 30–33 (2016)
2. Baranovskaya, N., Feifei, G., Zhang, N.N.: Formation of value and determination of the effectiveness of investments in integrated residential development with the participation of foreign capital. Piter, Saint-Petersburg (2015)
3. Andrade, J., Vieira, S., Bragança, L.: Selection of key sustainable indicators to steel buildings in early design phases. In: Proceedings of the International Workshop on Concepts and Methods for Steel Intensive Building Projects, Munich, Germany (2012)
4. Bragança, L., Vieira, S., Andrade, J.: Early stage design decisions: the way to achieve sustainable buildings at lower costs. Sci. World J. **2014**, 1–8 (2014)
5. Sheina, S., Shevtsova, E., Sukhinin, A., Priss, E.: The method of selecting an integrated development territory for the high-rise unique constructions. In: E3S Web of Conferences Cep. High-Rise Construction 2017, HRC 2017, 02064 (2017)
6. Sheina, S.: Sustainable Development of Territories, Cities and Enterprises. Don State Technical University, Rostov-on-Don (2017)
7. Sheina, S., Kravchenko, G., Tomashuk, E., Kostenko, D.: Influence of the constructive decision of the frame of the high-rise building on the economic indicators of the investment project. Eng. Herald Don **2**(49), 122–133 (2018)
8. Tsopa, N.: About need of application risk - oriented methods for ensuring stability of the investment-construction project. Constr. Ind. Saf. **7**(59), 25–35 (2017)
9. Tsopa, N., Kovalskaya, L., Malachova, V.: The mechanism for managing the business potential of commercial real estate projects. Mater. Sci. Forum **931**, 1220–1226 (2018)
10. Sheina, S., Tsopa, N., Kovalskaya, L.: Assessment of the business potential impact on the controllability of an investment-construction project. In: MATEC Web of Conferences, vol. 196, no. 04041, pp. 1–8 (2018)
11. Montes, M., Falcón, R., Ramírez-de-Arellano, A.: Estimating building construction costs by production processes. Open Constr. Build. Technol. J. **8**, 171–181 (2014)
12. Lesniak, A., Plebankiewicz, E., Zima, K.: Cost calculation of building structures and building works in polish conditions. Eng. Manag. Res. **1**, 72–75 (2012)
13. Dorozhkin, V.: Pricing and Cost Management in Construction. Publishing House Named after E.A., Bolkhovitinova, Voronezh (2013)
14. Ramaswamy, J.N.: Cost estimating in construction. http://www.scribd.com/document. Accessed 29 Jan 2019
15. Jayanthy, G.N.: Cost estimation techniques for construction industry. http://www.scribd.com/document. Accessed 29 Jan 2019

GIS Services for Agriculture Monitoring and Forecasting: Development Concept

Ielizaveta Dunaieva[1], Wilfried Mirschel[2], Valentina Popovych[1], Vladimir Pashtetsky[1], Ekaterina Golovastova[1], Valentyn Vecherkov[1], Aleksandr Melnichuk[3], Vitaly Terleev[4], Aleksandr Nikonorov[4](✉), Roman Ginevsky[4], Viktor Lazarev[4], and Alex Topaj[5]

[1] Federal State Budget Scientific Institution «Research Institute of Agriculture of Crimea», Kievskaya, 150, Simferopol 295453, Crimea, Russia

[2] Leibniz-Centre for Agricultural Landscape Research, Eberswalder Straße, 84, 15374 Müncheberg, Germany

[3] V.I. Vernadsky Crimean Federal University, Vernadskogo pr., 4, Simferopol 295007, Crimea, Russia

[4] Peter the Great St. Petersburg Polytechnic University, Polytechnicheskaya 29, 195251 St. Petersburg, Russia
coolhabit@yandex.ru

[5] Agrophysical Research Institute, Grazhdanskii pr., 14, 195220 St. Petersburg, Russia

Abstract. Purpose of research was to develop technological basis for obtaining related remote sensing data and ground-based observations of high-level accuracy and reliability with its subsequent introduction into the monitoring system of agricultural lands state. Introduction of digital technologies will improve the quality of agricultural lands monitoring and accuracy of criterion assessments of parameters of crop condition and fertility levels, including capacity to localize and map degraded soil processes. The prerequisites for the transition to digital technologies in agriculture were studied. Existing systems for monitoring and forecasting the condition of crops are considered. The structure of the system of digital agricultural production, its components and relations between them are determined.

Keywords: ERS · GIS · Monitoring system · Agriculture monitoring · Modeling · Ecosystem

1 Introduction

Modern studies in the assessment of ecosystem state of rural areas are based on the usage of data from field experiments, digital modeling of soil processes and dynamics of development of cultural and natural vegetation, as well as data from Earth remote sensing (ERS).

V. Murgul and M. Pasetti (Eds.): EMMFT-2018, AISC 983, pp. 236–246, 2019.
https://doi.org/10.1007/978-3-030-19868-8_24

The main purpose of agricultural regional monitoring is to recognize, identify and evaluate the productivity of crops at the level of separate territories, to obtain fore-cast data on the yield of main agricultural crops, as well as to diagnose areas of fields and territories with signs of influence of negative processes (associated with a lack of nutrients in the soil, diseases, processes of wind and water erosion), the formation of geographically attached independent confirm information at payment of subsidies for the growing of certain type of product or setting of compensation in the event of extreme climatic situations, including when insuring crops.

One of the most effective economic monitoring tools is the usage of remote sensing data. The list of priority areas of work for the region should include:

- creation of regional databases of conjugated ground-based observations and satellite information;
- improving the methodology of agricultural crops classification according to remote sensing data. Integration of research with «Vega» satellite service RAS [1];
- obtaining regional dependencies of linkage of parameters of the biological crops productivity and yields (to obtain projected estimates);
- development of regional regulatory framework for implementation of monitoring methodology using criteria-based assessment mechanisms;
- support for development and creation of modern computer-based web-based information advising remote access systems, including those operating in real time.

Practical application of remote sensing data is based on their primary processing and visualization of results on maps, based on the use of geographic information systems (GIS) tools. Creation and maintenance of geo-databases, their optimization, usфпу of remote access mechanisms (geoportals) is separate area of scientific and technical developments. ERS data are an objective indicator characterizing the dynamics of development of both cultural agrocenoses and natural landscapes and make it possible to assess and predict the level of yield. Using the data of Earth remote sensing makes it possible to identify and clarify the actually used areas under crops, to assess their condition and development dynamics (see Fig. 1).

Fig. 1. Spatial location of winter crops (season 2017/2018).

Archival data from satellite materials allow obtaining data on the possible variation of indicators and mapping their deviation from the norm. Figure 2 shows an example of assessing the situation with the development of winter crops, based on the normalized differentiated vegetation index (NDVI), which is calculated from satellite observations, shows the level of development of photosynthetically active vegetation mass, and varies from 0 to 1.

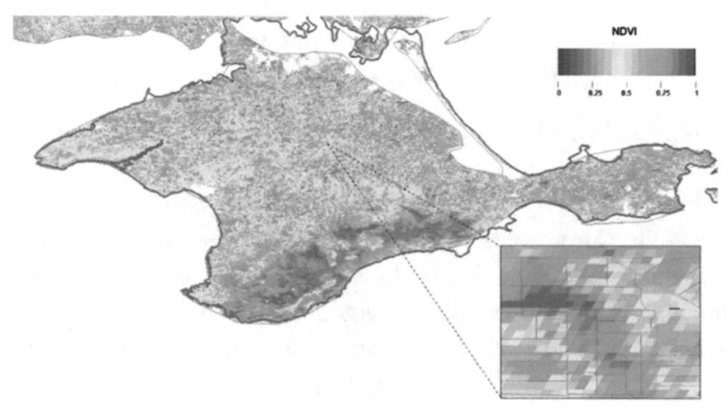

NDVI - May 7, 2018, resolution 250 m

Fig. 2. Vegetation index analysis by vegetation index (NDVI) throughout the region and individual fields.

Combining GIS technology and remote sensing monitoring data is one of the cornerstones of modern system of control and food security. Its solution is impossible without solving the problems of sustainable development of rural areas of the Crimean region, which is assessed using a system of indices and indicators. In addition, above all, it requires reliable, up-to-date information about objects of management and state of the territory. Well-formed and well-functioning information support system will allow making the most informed decisions, aimed at sustainable socio-economic development, rational management, environmental protection, overcoming environmental problems, forecasting and preventing emergencies.

One of the main tasks of optimizing the usage of the region agro-climatic potential in the prevailing natural and economic conditions is implementation in Crimea the principles of adaptive-landscape farming system. At the same time, the use of GIS technologies in solving problems of this class makes it possible to use mechanisms for controlling and monitoring the conservation and reproduction of soil fertility, identifying unaccounted and unused lands, assessing the potential for enhancing bioproductivity and mapping areas of the most effective targeted investments, and preparing materials for solving regional optimization problems samples on the distribution of territories and areas of crops grown, with information about the owners and tenants, nutrient availability, etc. A feature of current moment is need to orient on the tasks, solved in Crimea, on application of open source software GIS. The use of open source GIS software, de-spite a slight increase in labor costs at the application development stage, makes it possible to save significantly amounts of investments at the implementation stage, without requiring primary financial investments. Open source GIS software products become more and more popular in the market of high technology products around the world. It is widely used in training and in solving applied problems by the US National Space Agency (NASA) and other organizations.

The main priorities of GIS technologies use are:

- creation and optimization of geo-databases for agricultural purposes, taking into account the possibility of their use in the open GIS environment;
- integration of local, regional and federal databases, taking into account restrictions on access to information;
- determination of procedures, support mechanisms and stages of implementation of the adaptive-landscape farming system in the Crimea.

In connection with the cessation of water supply of the Crimea through the system of the North-Crimea Canal, the role of research to the related to the rural areas water sup-ply, especially in the steppe part of the peninsula, as well as assessment of possible extraction volume of fresh groundwater for irrigation is to prevent their depletion or the intrusion of saline waters into freshwater horizons. Remote sensing and GIS-based agrohydrological modeling methodologies are widely used in modern studies to assess the water balance of the territory. In the current hydroeconomical conditions, the following tasks are particularly relevant:

- assessment of level of territory provision with water resources and development of adaptation measures in the years of different water availability;
- analysis of the accuracy and testing satellite monitoring systems for precipitation for the territory of Crimea;
- development and approbation of water-balance calculations for rural areas, using remote sensing data and agrohydrological modeling;
- assessment of regional seasonal trends and current changes in the frequency and level of aridity of the territory, taking into account climate change;
- development of indirect control methods and rationing of fresh groundwater extraction for irrigation, mapping of potentially dangerous territories of water depletion and areas of possible groundwater intrusion into freshwater horizons.

The use of spatial models, such as WOFOST, SWAP, SWAT, allows simulating the dynamics of agricultural crop development in a daily mode, not only at the field level, but also in the entire water management basins.

Successful implementation of tasks, related to the processing of remote sensing data, geo-databases formation, research on the definition and modeling of the spatial-temporal patterns of processes in the biogeocenosis, requires the presence of specialists in this field of knowledge, as well as the use of modern equipment to carry out both data processing and usage of the obtained information in practice. Promising area of application of digital and navigation technologies in the agricultural sector is their usage for creating precision farming systems. «Precision farming» means an information and production system for farming, which was created with the aim of long-term improvement of production efficiency, productivity and profitability, taking into account local specifics, and at the same time minimizing the negative impact on the environment. The use of these technologies allows not only proceeding the resource-saving and ecological farming, but automating the production and increasing the competitiveness of products on the market. Precision farming includes an «integrated high-tech agricultural management system» with the use of global positioning technologies, geographic informational systems (GIS), yield assessment technologies (Yield Monitor Technologies), variable rationing technology (Variable Rate Technology) and ERS technology. The essence of this farming technology is cultivation of fields, depending on the actual needs of crops, grown in this place, which are determined using modern information technologies, including space survey. Need for effective use of land, water and other resources necessitates the development of systems that, with the use of modern intelligent digital technologies, will allow the assessment of resources, dynamics of changes and move on to varied forecast of situation.

Purpose of research was to develop technological basis for obtaining related remote sensing data and ground-based observations of high-level accuracy and reliability with its subsequent introduction into the monitoring system of agricultural lands state.

Introduction of digital technologies will improve the quality of agricultural lands monitoring and accuracy of criterion assessments of parameters of crop condition and fertility levels, including capacity to localize and map degraded soil processes.

2 Materials and Methods

Important tool for qualitative management and effective usage of agricultural lands, assessment and monitoring of their condition, analysis, forecasting and development of state policy in the field of land relations, is creation and development of digital information systems (IP).

Currently, such several systems are available and functioning for the territory of Crimea, but it should be noted, that the list of available information does not correspond to the region's current information needs.

The VEGA satellite web service (http://sci-vega.ru), developed by the Space Research Institute of the Russian Academy of Sciences, makes it possible to assess the state of vegetation over the entire territory of Northern Eurasia and, in particular, the Republic of Crimea [1]. Operational agricultural monitoring was carried out according to the archive of satellite images of various spatial resolutions, time series of vegetation indices, thematic maps, weather data, daily information about the fire.

VEGA is an archive of long-term satellite observations, with the help of which it is possible to calculate the vegetation index NDVI for concrete territory.

Studies, carried out jointly with SRI of RAS (project No. 18-416-910011 r _a, financial support from the Russian Foundation for Basic Research and the Ministry of Education, Science and Youth of the Republic of Crimea), allowed to create a methodology and evaluate the possibilities of combining the remote sensing data of the average (MODIS) and high (Landsat 8, Sentinel-2, Sentinel-1) spatial resolution in the environment of the VEGA service for assessing the state of winter crops of 2017–2018 on the example of two regions of Crimea. The proposed approach of winter crops identifying, based on interactive clustering of time range of satellite images using reference data and taking into account the connection of biophysical indicators and remote measurements, was tested using random field inspections. The overall accuracy of new winter crops regional maps was 92% for Krasnogvardeisky district, and 89% - for Belogorsky district. Use of proposed method made it possible to refine the maps of winter crops for indicated areas significantly, compared to the MODIS average resolution maps and maps, available in VEGA, as well as to establish the nature of identification errors, which will further allow correcting the winter crops maps for all Crimea [2]. The Unified Federal Information System on Agricultural Lands (UFIS AL) is a database consisting of information such as location, state and actual use of these lands, as well as state of agricultural vegetation. Most layers, such as «crop rotation», «register of fields», «monitoring of crops by means of remote sensing» are inactive [3]. Only «agricultural land» and «municipal formations» are in working condition. For the territory of Crimea, the «municipalities» layer is available only when selecting the subject «Sevastopolsky», there is no information about agricultural land. Its development in the Crimea began from April 2018.

OneSoil Map (https://map.onesoil.ai) is the Belorussian project, an online-platform of these agricultural lands. OneSoil utilizes information of the European program Copernicus and satellite pictures. It displays the sizes of fields and their number by countries or individual regions with-in a country, as well as data on agricultural crops, located on specific fields (more than 25 names, among which are wheat, corn, rice). The system was filled using machine learning algorithm and Mapbox GL JS technology. The platform automatically recognizes the field's contours, so that in the future it could become a plot for precision farming (See Fig. 3).

a. Krasnogvardeysky district, Klepinino village council

b. Belogorsky district, Krumskaya Rosa village council

Fig. 3. Graphic display of field contours in OneSoil.

However, an assessment of accuracy of contours identification showed that the contours of the actual fields and the contours of the fields in OneSoil differ for the most part by ~ 5 hectares. The reason for most area differences is that in OneSoil, the field roads and forest bands were not extracted into separate objects during the digitizing process. Crop identification error is significant; most crops do not correspond to crops, grown in Crimea.

The British platform Crop Map (http://www.cropmap.co.uk) [4] is online service and tool for managing planted territories. Information is displayed as map with fields where certain crops are cultivated. It was found during the comparison that the distinctive feature of Crop Map from OneSoil is manual data input.

After digitizing the field contour, data on current crops and crops grown on it earlier were filled in its cell in the table; if it possible, data on the sowing dates were added.

There is also the Russian project CropMap (http://cropmap.ru) [5]. It is a crop forecast map. There is a project with satellite images, meteorological service rp5.ru and a soil map. According to a certain culture and date of sowing, the service predicts a yield according to three scenarios: optimistic, average and pessimistic.

The OnlineGIS service (http://www.onlinegis.net) [6] was built based on open source GIS products of Mapserver and OpenLayers. Service presents maps for six US territorial units (Culpeper County, Virginia King George County, Virginia, Richmond County, Virginia, City of Orange, Virginia, Wirt County, etc.). Information layers such as borders, road network, water resources, floodplains, as well as some others were loaded. In this service, aerial photographs of agricultural land, provided by the National Agriculture Imagery Program, are available (see Fig. 4).

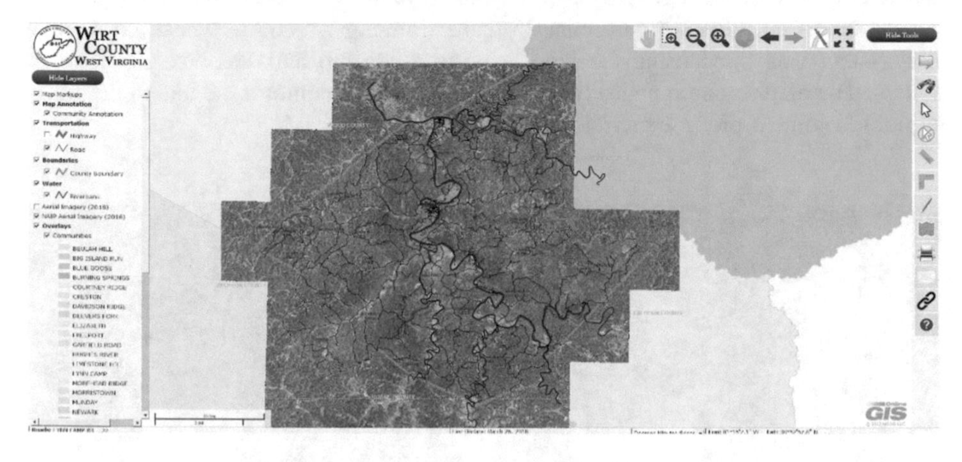

Fig. 4. Web interface of the OnlineGIS service

Data on cultivated crops are not given, and there are no tools for fields digitizing. There are measuring tools that allow determining the length of the line, as well as the area, the perimeter of the object of interest, indicating the units of measurement. Maps are exported in PNG or PDF format.

«CosmosAgro» (http://www.scanex.ru) [7] is a cloud online service of operational agricultural monitoring, which is based on the technology of automated processing of remote sensing materials. Service, operating on the ScanExWebGeoMixer platform, has up-to-date periodically updated satellite data and long-term archives, as well as additional information (meteorological data, relief information, cadastral division of the territory). Using the service toolkit it is possible to create, edit and update vector layers of agricultural fields, conduct monitoring and retrospective analysis of state and use of agricultural land based on the calculation of vegetation indices, including NDVI index, and receive a number of additional parameters of agricultural lands state. All service results could be displayed on the map and recorded in the form of reporting materials, which provides ease of analysis of obtained data and allows accumulating statistical information about the crops state. The advantage of this service is the ability to assess regional agricultural land and analyze the state of agricultural crops within each individual field. AgroVisio.ru (http://agrovisio.ru) [8] allows downloading and carrying out a visual analysis of aerial photographs, taken with conventional or spectral camera, comparing them with the field's contours, and identifying deviations in the crops growth. For accounting the production fields (boundaries, location, history), analysis of geography and timely editing of field contours, assessing the state and use of agricultural land in AgroVisio.ru, not only production fields are required, but also an electronic map of these fields, as well as monitoring system motion sensors and fuel consumption. During operation, it is possible to connect to external services, such as Earth space monitoring services, including VEGA, information about which was presented above, as well as meteorological and fire services, as well as Rosreestr.

Tasseled Cap Imagery (http://www.arcgis.com) is a product of ESRI [9]. This service was used to monitor environmental change and for agricultural plants mapping. Service is based on Landsat satellite images, stored in the Global Land Survey library, which was created by the US Geological Survey (USGS) and the National Aeronautics and Space Administration (NASA). Tasseled Cap Imagery allows monitoring vegetation development during the growing season for forecasting the ripeness and yield of agricultural crops for specific period.

With the help of PCA-analysis (Principle Components Analysis) satellite images were converted into RGB-image with three main components, where artificial and natural zones (Brightness) were shown in red range, in green - the dynamics of development of vegetation (Greenness), in blue - water objects and soil moisture (Wetness). The Sovzond online spatial data service allows continuous monitoring of the territory with resolution of 3 m [10]. Monitoring of agricultural land with the help of this service solves such issues as land inventory, crops condition monitoring, identification processes that are potentially dangerous for land and sowing. The platform allows calculating the level of vegetation index NDVI.

According to the functional data set, Sovzond online service is an analogue of the RSI RAS VEGA-Science service. It may be considered as an alternative. A significant drawback is access to service data only for a paid subscription.

DatumGroup services (https://datum-group.ru), ScanEx, AgroTechnology (http://agritechnology.ru) are Russian start-up projects and currently allow online monitoring of agricultural lands. To date, most of the existing information systems contain a small amount of information about agricultural lands of the Republic of Crimea, in some of them - completely absent. Based on the analysis of the main existing information systems in the field of agriculture, it is necessary to refine these systems, to replenish their database, and transition to the introduction of digital technologies in the agro-industrial complex. Redesign of satellite monitoring system elements of Crimean agricultural lands is an urgent task for use of modern information technologies in improving the efficiency of agriculture.

3 Results and Discussions

Strategic planning is management decisions that ensure the implementation of strategy and policies of the relevant management entities, and consistent application of logical decisions, thus methodological approaches and methods for optimizing management decisions is necessary. Compliance with the basic principles of planning allows developing forecasts, preparing projects, programs and plans for strategic development and their implementation. Important stage in the creation of Information System is the collection of primary information, its processing and verification, and data entry as established forms and database structures. Data integration allows to combine information, located in different sources, and to ensure that users will receive data in unified form. When forming any data structure, two problems need to be solved: how to separate the elements from each other and how to find the necessary elements. Creating a database involves:

1. Creation of a logical database (to display tables and items to the customer). This clause implies the creation of tables - filling in with necessary attributes - setting the primary key - setting the links.
2. Creation of a physical database. Namely: the choice of data type for the specified attributes in the tables. This item assumes creation of tables - filling in with necessary attributes - setting the primary key - setting the connections - data types.
3. Creation of the tables with fields (depending on the programming environment. For example, builder paradox or firebird ibexpert, MySql).
4. Filling tables with necessary information.
5. Creation of a program to work on PC without updates or to create client-server application.
6. Linking the database to the site via PHP (for placing the created database in the Internet).

The result of a single database creation for a specific territory could be received in the form of a list of information (an example in Fig. 5). The user of this database will be able to receive data in the form of tables, charts, diagrams about not only the state at a given time, but also for past years and their change over time. In addition, the implementation of system, using GIS technology, will provide information not only temporarily but also spatially scale. The usage of GIS technology allows increasing efficiency and quality of work with spatially distributed information. Therefore, an effective tool for solving problems of multifactor land monitoring is geoinformational system, which is hardware-software complex that provides the collection, processing, display and provision of information and operates on a single spatial-coordinated basis. Such providing receiving system of knowledge about territory in the form of digital data, as a GIS, combined into a set of layers that form an information model, could become the basis for creating an automated information system for agricultural lands monitoring. The system of monitoring and forecasting the state of crops has a 3-level of organization: field level, district level and region level. At each level, a certain range of tasks were performed, due to stages of processing and integrating information (Fig. 6).

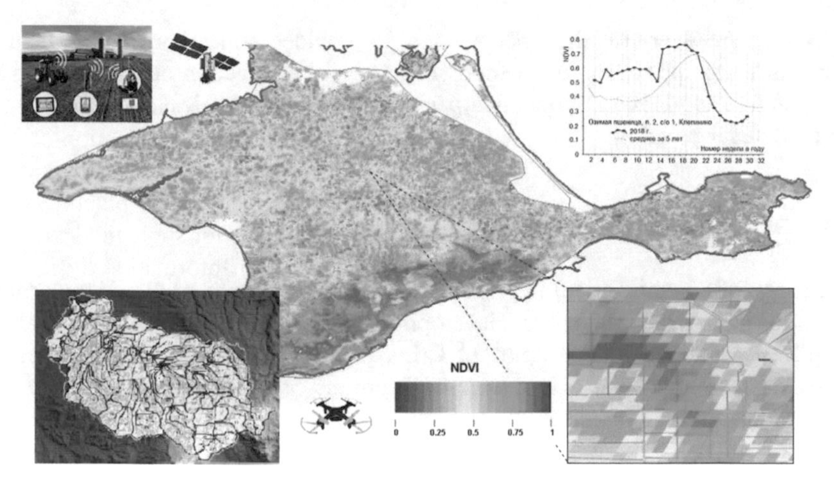

Fig. 5. General view of data collection

At the level of agricultural field, a database of crop structure ("Field History"), processed vegetation index data, a database of daily meteorological parameters (data of automated meteorological stations) were created. At the level of area and region, analysis of the state and mapping of winter crops was carried out, algorithms and procedures for monitoring the state of crops, using remote sensing data, procedures for verifying remote sensing data from field data, comparative analysis of dynamics of crop development over number of years, were created [11].

It is also important to use simulation modeling of the state of crops and analysis of potential yield, as well as adaptation and refinement of the parameters of agricultural crops and water-physical soil characteristics in models [11].

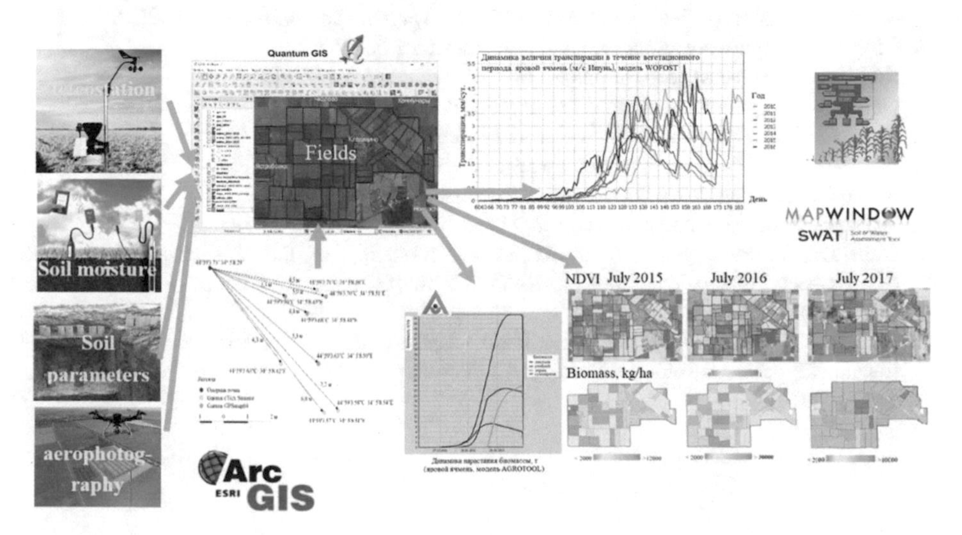

Fig. 6. Data integration in monitoring system

As mentioned above, the system concept provides collection, storage and processing of data, application of methods of analysis, verification and identification of data, therefore availability of qualified personnel and technological means is important for the system operation.

4 Conclusion

The concept of Digital Agriculture is primarily the identification of digital opportunities and options for data integration, development and implementation of digital technologies in agro-industrial complex of Crimea (AIC). Combination of Earth remote sensing data, simulation data and associated ground-based observations is the key to success in research and transition to digital technologies in the AIC. The prerequisites for the transition to digital technologies in agriculture were studied. Existing systems for monitoring and forecasting the condition of crops are considered. The structure of the system of digital agricultural production, its components and relations between them were determined.

Acknowledgments. The reported study was funded by RFBR according to the research projects No. 18-416-910011-p_a, No. 19-016-00148-a, No. 19-04-00939-a.

References

1. Loupian, E.A., Proshin, A.A., et al.: IKI center for collective use of satellite data archiving. Processing and analysis systems aimed at solving the problems of environmental study and monitoring. Sovremennye problemy distantsionnogo zondirovaniya Zemli iz kosmosa (in Russian) **5**(12), 263–284 (2015)
2. Dunaeva, E.A., Yolkina, E.S., et al.: Features of identification of crops of winter grain means of remote sensing of the Earth. Taurida Herald Agrarian Sci. **4**(16), 18–31 (2018)
3. Crop Map. http://www.cropmap.co.uk. Accessed 15 Feb 2019
4. Russian crop forecast map. http://cropmap.ru. Accessed 15 Feb 2019
5. OnlineGIS. http://www.onlinegis.net. Accessed 15 Feb 2019
6. Scanex. http://www.scanex.ru. Accessed 17 Feb 2019
7. AgroVisio. http://agrovisio.ru. Accessed 17 Feb 2019
8. ArcGIS Online. http://www.arcgis.com. Accessed 21 Feb 2019
9. Sovzond. https://sovzond.ru. Accessed 21 Feb 2019
10. Dunaieva, I., Popovych, V., et al.: SWAT modeling of the soil properties in GIS environment: initial calculations. In: MATEC Web of Conferences, vol. 265, article number 04014 (2019)
11. Badenko, V., Badenko, N., et al..: Ecological aspect of dam design for flood regulation and sustainable urban development. In: MATEC Web of Conferences, vol. 73, Article no. 03003 (2016)

Digital Readiness Parameters for Regional Economies: Empirical Research and Monitoring Results (Russia Case Study)

Olga Kozhevina[1], Natalia Salienko[2], Viktoria Kluyeva[2], and Sergey Eroshkin[3](✉)

[1] Institute of Economics and Crisis Management, Chamber of Commerce and Industry of the Russian Federation, Vavilova Street, 53, Building 3, 117312 Moscow, Russia
[2] Bauman Moscow State Technical University (National Research University), 2nd Baumanskaya Street, 5, Building 1, 105005 Moscow, Russia
[3] Russian State Social University (RSSU), 4, Wilhelm Pieck Street, 129226 Moscow, Russia
EroshkinSIu@rgsu.net

Abstract. The article presents the results of an empirical study of the transition to a digital economy in the form of a case study of Russia. The research focuses on the management relations arising from the digital transformation of the national economy, as well as on the methods and tools for analyzing readiness for the digital economy. Cross-country comparisons of readiness for the digital economy based on statistical analysis were made. The most significant factors influencing the digital transformation were identified. An aggregated digital economy index was proposed. The innovative component of the transition to a digital economy in the regions of Russia was investigated. Substantiation was provided for the relationship of sustainable development with digital transformation and the achievement of innovative and intellectual effect.

Keywords: Digital economy · Territorial aspects · Sustainable development · Innovative effects · Intellectual effects · Digital literacy

1 First Section

1.1 Mainstreaming the Transition to a Digital Economy

Mankind has entered an era of rapid change: the change of technological paradigms; the ubiquitous pervasiveness of digital technology in the fields of economics and management [1–12, 13-16]; social transformations and the development of new competencies and skills of Industry 4.0. The building of a digital economy is driven not only by the use of ICT, but, first and foremost, by the existence of national technological and innovation platforms, the development of high-tech sectors of the economy. Therefore, the leading countries in the digital economy are those that have built a powerful infrastructure for the development, implementation and commercialization of digital technologies [3, 4, 7, 11].

V. Murgul and M. Pasetti (Eds.): EMMFT-2018, AISC 983, pp. 247–256, 2019.
https://doi.org/10.1007/978-3-030-19868-8_25

1.2 Methodology for Cross-Country Analysis of Digital Economy Parameters Headings

The International Telecommunication Union annually evaluates the achievements of the countries of the world in the development of information and communication technologies (ICT) [2]. An international comparison shows that in 2016 Russia ranked 43rd in the ICT Development Index, 35th in the E-government Development Index, and 41st in the Networked Readiness Index. Figure 1 presents data on the ICT Development Index, and Fig. 2 and Table 1 show international comparisons of the digital economy indices as of 01.01.2017.

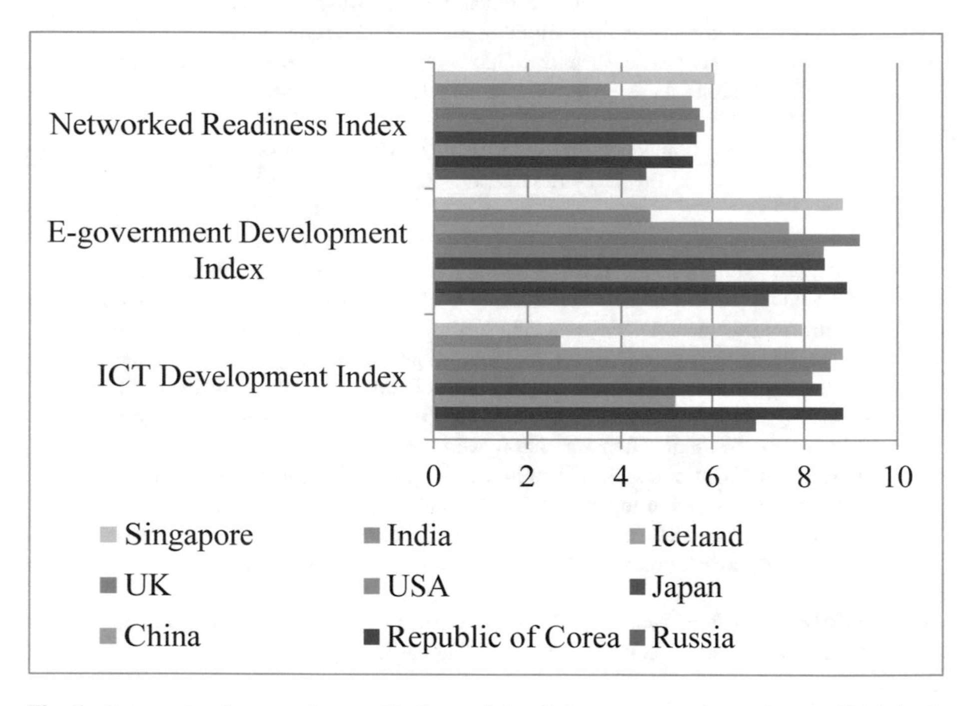

Fig. 1. International comparisons of indices of the digital economy (according to 2016 data).

According to the Institute for Internet Development (IDI) [14], in 2016 the volume of the digital economy in Russia was 1.9 trillion rubles, or 2.2% of GDP. To put it into perspective: in the USA this indicator reaches 6.2%, and in Great Britain, about 8.5% of GDP. The inertial development scenario envisages an increase in the contribution of the digital economy to Russia's GDP by up to 3% by 2021, mainly due to an increase in online consumption, investment and government spending, therefore, it is assumed that the growth of the digital economy will take place at a faster pace.

In estimating the aggregate indicator, samples for 49 developed and developing economies (Table 2) were made. Countries with a high level of technological development are in the lead. According to the comparative analysis, of the countries of the Customs Union Kazakhstan has the highest degree of digital readiness, ranking 21st,

Table 1. International Comparisons of the Digital Economy Indices (according to 2016 data), compiled based on the "Indicators of the Digital Economy" Data, Higher School of Economics National Research University, 2017.

Digital Economy Indicators	Russia		Republic of Korea		China		Japan		USA		UK		Iceland		India		Singa-pore	
	Rank	Value	Rank	Value	Rank	Value	Rank	Value	Rank	Value	Rank	Value	Rank	Value	Rank	Value	Rank	Value
ICT Development Index	43	6.95	1	8.84	81	5.19	10	8.37	15	8.17	5	8.57	2	8.83	138	2.69	20	7.95
E-government Development Index	35	7.22	3	8.92	63	6.07	11	8.44	12	8.42	1	9.19	27	7.66	107	4.64	4	8.83
Networked Readiness Index	41	4.54	13	5.57	59	4.24	10	5.65	5	5.82	8	5.72	16	5.55	91	3.75	1	6.04
Aggregated digital economy index	40	18.71	6	23.33	68	15.5	10	22.46	11	22.41	5	23.48	15	22.04	112	11.08	8	22.82

Table 2. Top 10 countries by the value of the Aggregated Digital Economy Index (ADEI) (2016 data for 49 countries), compiled by the author

Country	ADEI	Rank
United Kingdom	23.48	1
Republic of Korea	23.33	2
Netherlands	22.9	3
Finland	22.86	4
Denmark	22.85	5
Australia	22.82	6
Singapore	22.82	7
New Zealand	22.44	8
Japan	22.42	9
USA	22.41	10
Germany	22.07	11
Sweden	22.06	12
Spain	22.04	13
France	21.91	14
Estonia	21.81	15
Luxembourg	21.74	16
Canada	21.47	17

(continued)

Table 2. (*continued*)

Country	ADEI	Rank
Austria	21.35	18
Belgium	21.04	19
Ireland	20.95	20
Kazakhstan	20.53	21
Lithuania	19.77	22
Slovenia	19.73	23
Portugal	19	24
Latvia	18.72	25
Russia	18.7	26
Croatia	18.49	27
Czech Republic	18.44	28
Poland	18.36	29
Greece	18.11	30
Chile	17.92	31
Hungary	17.83	32
Argentina	17.29	33
Slovakia	17.27	34
Bulgaria	17.18	35
Cyprus	17.12	36
Azerbaijan	16.86	37
Brazil	16.38	38
Romania	16.02	39
Turkey	15.98	40
Georgia	15.95	41
Ukraine	15.58	42
China	15.5	43
Mexico	15.06	44
Armenia	15.05	45
Venezuela	13.77	46
Kyrgyzstan	12.65	47
India	11.08	48
Republic of Moldova	10.38	49

with Russia ranking 26th. The study demonstrated that Russia lagged far behind in terms of readiness for the digital economy, both in economic and innovative efficiency of using digital technologies. The lag is due to infrastructure and industry problems, the volatility of the market environment, the low level of use of ICT technologies by business structures.

2 Innovative Component of Managing the Transition to the Digital Economy of Russia and the Regions

Strengthening of Russia's status as a leading information power requires that it moves steadily up the international digital ratings. The message of the President of the Russian Federation to the Federal Assembly of 2016 outlines the strategic priorities and tasks of digitalization of Russia and the regions. The digital economy, as defined by the World Bank [2], is a system of economic, social and cultural relations based on the use of digital information and communication technologies. There are two established conceptual approaches to the content of the digital economy. The first approach is classical, whereby digital economy is an economy based on digital electronic technologies for promoting goods and services in the market. The second approach is substantially broader and covers all stages of the reproduction process using digital technologies. In July 2017, the state program "Digital Economy of the Russian Federation" was adopted in Russia [15]. The program stressed the need to stimulate industry, create high-tech IT enterprises, industrial digital platforms for major sectors of the economy, as well as small and medium-sized enterprises in the field of digital technologies. The developers of the "Digital Economy of the Russian Federation" program estimate that it is the digitalization of the economy that can provide more than 20% of GDP growth until 2025. The basic objectives of the "Digital Economy of the Russian Federation" program are: (1) the creation of a digital economy ecosystem; (2) the creation of necessary and adequate institutional and infrastructural conditions; (3) the improvement of the Russian economy's competitiveness in the global market.

The regional economies of Russia are highly differentiated in their degree of integration into the information environment and readiness for the "digital revolution". Digitalization is an inevitable and objective process that cannot be stopped in the era of globalization and the transboundary nature of the information space. The digitization of the economy contributes to integrated cross-country and inter-sectoral development. In recent years, public discussion platforms, research institutes and analytical centers have been studying the widespread introduction of new technologies from the perspective of their use in intercultural and interstate dialogue, and not only as a market tool. For a sustainable digital economy to be created, effective mechanisms for public-private and public-private partnership need to be developed, and business elites should be encouraged to invest in venture capital and implement megaprojects.

In our opinion, information economy and knowledge economy manifest as social and intellectual effects. These effects of sustainable development reflect the growth of the welfare of the population, the level of knowledge prevalent in the society, as well as the development of information and intellectual sphere and human capital assets. The achievement of these effects is considered as the main target benchmark for sustainable development. Innovation economy, as one of the subsystems of managing sustainable development, is based on continuous technological improvement, the formation of technological innovation platforms. These are provided by the creation and introduction of new technologies, which allow making products with high added value and minimal environmental risks. We consider development effects and innovation effects as elements of the innovation subsystem [8].

Innovative effects (Fig. 3) are revealed when the "Innovation Management" and "Innovative Activity" parameters are measured, which are the key functional components for building an innovative economy. Generally, the quality of innovation management at the state level is reflected by the quality of innovation policy and the change in the GDP productivity. Ultimately, public expenditure on innovation policy should stimulate invention and innovation both in the country as a whole and in the regions. Therefore, it is advisable to consider the quality of innovation policy through the prism of the indicators of "Innovation Policy Efficiency" (public expenditure on innovation policy per unit of GDP) and "Intensity of dissemination of national scientific and technological achievements" (number of patent applications filed by domestic applicants/total number of patent applications filed with Rospatent).

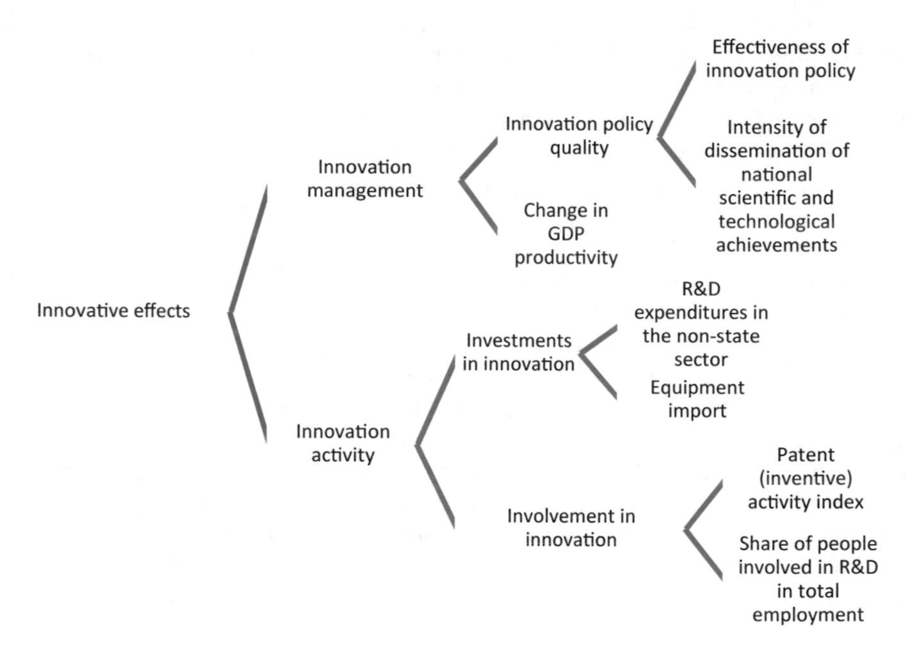

Fig. 2. "Innovative Effects" component convolution mechanism.

The increase in the capital productivity of GDP, as well as the decrease in its capital intensity, is a direct consequence of the introduction of innovations and indicates the degree of their use in social production. Therefore, the GDP capital productivity index (The ratio of the GDP to the average annual value of the economy's fixed assets) and the indicator of the quality of an innovative economy can indicate the quality of innovation management. On the other hand, state innovation policy should promote innovation activity. It is assumed that the latter can be assessed through indicators of investments in innovations by the private sector organizations and public involvement in innovation activities. It therefore makes it possible to evaluate innovative effects at the level of all economic agents. The "Investments in Innovations" composite indicator is calculated as a weighted average estimate of the normalized indicators of private sector expenditures on R&D and equipment import costs. The composite indicator "Involvement in Innovation Activity" can be obtained by a weighted average of the index of patent (inventive) activity and the share of people employed in R&D in total employment.

The formation of an information-active society, where every citizen is involved in communication with the state, has access to information and knowledge bases, is possible provided an efficient education system is formed, with an accessible educational environment, including online media and distance technologies. Such an education system enables a person to achieve a certain level of competence through lifelong learning. To assess the "Education Efficiency" composite indicator, the education index and level of education can be used.

The communications efficiency is supposed to be assessed using indicators of public participation in governance and of the level of Internet coverage of the population. The first indicator allows ascertaining the degree to which citizens enjoy the right to express their own opinions and influence the activities of government bodies, reports the number of people with an active civic position. The second indicator was chosen because at present the Internet is the most technologically advanced means of communication, which gives ample opportunity to quickly receive and transmit information [17, 18].

In assessing the intellectual effects of sustainable development (see Fig. 4), it is proposed to proceed from determining the levels of development of the information economy and the knowledge economy; in addition, the scale and effectiveness of these subsystems also need to be assessed.

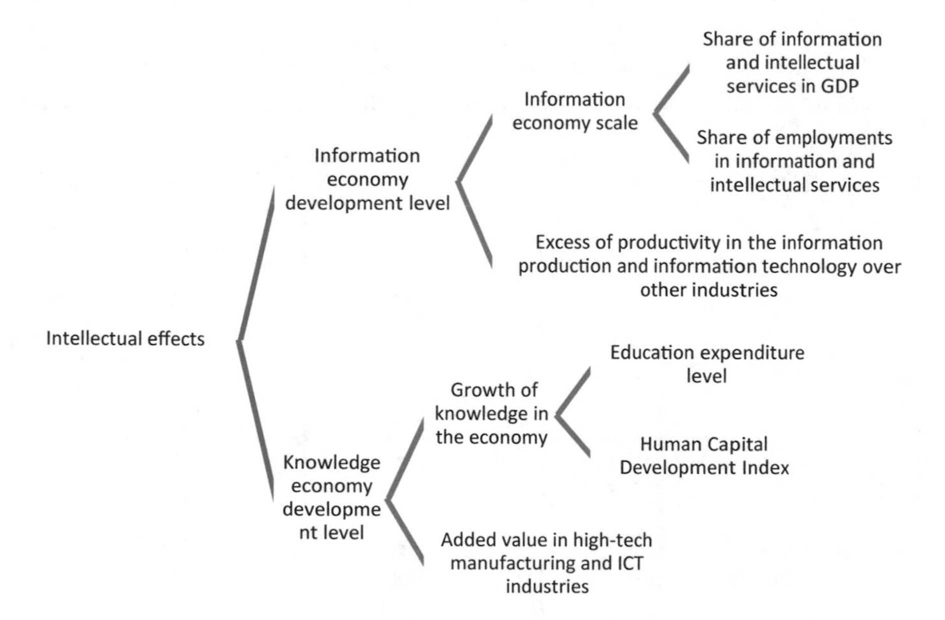

Fig. 3. "Intellectual Effects" component convolution mechanism.

For an information economy, the scale can be determined on the basis of indicators of the share of information and intellectual services in GDP and the share of employees in information and intellectual services. In the post-industrial society, more than 50% of the population is employed in the service sector; in the information society, more than 50% of the population is employed in the information and intellectual services, the society becomes an information society. The effectiveness of the information economy demonstrates by how much productivity in the field of information and information technology exceeds that in other industries. For the economy of knowledge, the scale is estimated taking into account the increment of knowledge, and the effectiveness is evaluated taking into account the volume of the created value added in high-tech industries and information and communication industries.

3 Conclusions on the Study of Digital Readiness of the Regions of Russia

The empirical analysis revealed an uneven development of the digital economy across federal districts (Table 3).

Table 3. Key indicators of the Development of the Digital Economy by the Federal Districts of the Russian Federation (2015–2016) compiled using the Indicators of the Digital Economy. HSE NRU, 2017 [2]

Federal District	The average number of employees in IT industry companies, '000'	The share of organizations using broadband Internet, %	Share of organizations using cloud services, %	Share of households with broadband Internet access, %	Share of population using e-commerce, %	Share of the population using the Internet to obtain state and municipal services, %
Russian Federation	381.1	80	18	71	23	51
Central Federal District Moscow *	158.8 116.5	83 95	20 30	73 79	26 37	56 65
Northwestern Federal District St. Petersburg *	55.6 43.4	85 91	20 23	77 85	28 30	44 49
Southern Federal District	16.1	75	17	75	24	48
North Caucasus Federal District	5.4	81	20	62	14	46
Volga Federal District Republic of Tatarstan *	73.0 14.5	79 83	16 17	69 79	20 28	55 80

(*continued*)

Table 3. (*continued*)

Federal District	The average number of employees in IT industry companies, '000'	The share of organizations using broadband Internet, %	Share of organizations using cloud services, %	Share of households with broadband Internet access, %	Share of population using e-commerce, %	Share of the population using the Internet to obtain state and municipal services, %
Ural Federal District Sverdlovsk region *	25.7 10.5	82 88	20 22	73 73	27 24	53 42
Siberian Federal District Novosibirsk region *	36.3 11.3	75 75	18 19	64 70	19 18	41 51
Far Eastern Federal District	8.5	74	16	70	26	48
Crimean Federal District (merged with SFD on 01.01.2017) Sevastopol *	1.9 1.2	90 78	22 18	- 81	- 29	- 30

According to the Regional Public Center for Internet Technologies (RPCIT) [12, 13] at the beginning of 2017, the digital literacy index of Russians grew by 0.63 points compared with the beginning of 2016 and was 5.42 on a ten-point scale. Digital literacy is a set of knowledge and skills that are required to safely and effectively use digital technologies and Internet resources. Digital competence is the ability of the user to confidently, effectively and safely select and apply information and communication technologies in different areas of life, based on the continuous mastery of knowledge, skills, motivation, and responsibility.

The rapid development of ICT technologies and the Internet industry requires a constant influx of qualified personnel. According to RAEC [1], about 2 million people work in companies and organizations related to the Internet markets, including the self-employed population. The system of education and self-study can not adapt fast enough to the rate of change in the market of Internet technologies. The problem of the qualifications and digital literacy improvement of ordinary users of companies or specialists who do not have specialized IT education has not been solved. Government support and investments in digital literacy, the development and implementation of a digital literacy strategy in Russian regions will help to solve the problem of the regional digital division and the shortage of qualified personnel.

References

1. Digital Literacy Index 2016. All-Russian study. Moscow, RAEC (2016)
2. Digital Economy Indicators. Moscow, HSE NRU (2017)
3. Bodrunov, S.D. (ed.): Integration of Production, Science and Education and Re-Industrialization of the Russian Economy. LENAND, Moscow (2015)
4. Kozhevina, O.V. (ed.): Tools for Assessing and Ensuring a Sustainable Development of the Russian Economy Sectors. INFRA-M, Moscow (2018)
5. Information society. Federal State Statistics Service http://www.gks.ru/wps/wcm/connect/rosstat_main/rosstat/ru/statistics/science_and_innovations/it_technology/. Accessed 24 Oct 2018
6. Innovative Development of Russia: problems and solutions, edited by Eskindarov, M. A., Moscow, Ankil (2013)
7. Kozhevina, O.V. (ed.): Methodology for Dynamic Assessment of Socio-Economic Development of Territories. Ruscience, Moscow (2017)
8. Science and innovation. Federal State Statistics Service. http://www.gks.ru/wps/wcm/connect/rosstat_main/rosstat/ru/statistics/science_and_innovations/science/. Accessed 24 Oct 2018
9. Abdrakhmanova, G.I., Voynilov, et al.: Science Innovation Information Society: 2016: a brief statistical compendium. HSE NRU, Moscow (2016)
10. Forecast of the scientific and technological development of Russia: 2030. In: Gokhberg, L.M., Agamirzyan, I.R., (eds.) Information and Communication Technologies. HSE NRU, Moscow (2014)
11. Sukhorukov, A.I., Shuhong, G.U.O., Koryagin, N.D., Eroshkin, S.Yu.: Tendencies of information management development in the conditions of the origin of a new ecosystem of the digital economy. In: Proceedings of 2018 11th International Conference "Management of Large-Scale System Development", MLSD (2018). https://doi.org/10.1109/MLSD.2018.8551859
12. Sukhorukov, A.I., Koryagin, N.D., Eroshkin, S.Y., Kovkov, D.V.: Statistical modeling of the process of generating analog information in the problems of the digital economy. In: Proceedings of 2017 10th International Conference Management of Large-Scale System Development, MLSD (2017). https://doi.org/10.1109/MLSD.2017.8109691

Application of the Monte-Carlo Simulation Method in Building and Energy Management Systems

Natalia Alekseeva(✉), Natalya Antoshkova, and Svetlana Pupentsova

Peter the Great St. Petersburg Polytechnic University, Polytechnicheskaya, 29, 195251 St. Petersburg, Russia
natasha-alexeeva@yandex.ru

Abstract. The Monte-Carlo method has got widespread use in the Russian economy. However, decision-making in the field of energy management of buildings in our country is not practically based on the Monte-Carlo simulation method, which was approved in different economy sectors. The objective of the following work is to study the possibility of the Monte-Carlo method application in energy management of buildings and investment management in real estate objects. In the following work the Monte-Carlo simulation method was applied, and sensitivity analysis was performed. This work demonstrates how expert data on energy-efficient projects, that are on the market, can be used to obtain necessary indicators to make reasonable management decisions in energy management. With regard to the initial values uncertainty, the obtained result is a range of values, where the most probable value of the resulting indicator lies. The presence of one indicator, that takes into account different variants of the energy efficient project implementation, simplifies the investor's task of choosing an investment in energy management of buildings.

Keywords: Energy management systems · Building and energy management systems · Energy efficient technologies · Energy efficient project · Energy efficient construction · Energy efficient building · Real estate · Present value

1 Introduction

Energy management systems are becoming a reality and starts to determine the whole development of the world economy. Starting with the ideas of sustainable and environmental development of the Rio-92 conference and a respective United Nations Environment Program UNEP, it is becoming increasingly materialized and can be considered as the main vector and content of the modern world development [1].

Energy management is the predictive, organized and systematic coordination of the procurement, conversion, distribution and use of energy to cover requirements while taking account of eco-logical and economic aims. The term thus describes actions for the purpose of efficient energy handling. The term energy management system encompasses the organizational and information structures, including the required

V. Murgul and M. Pasetti (Eds.): EMMFT-2018, AISC 983, pp. 257–266, 2019.
https://doi.org/10.1007/978-3-030-19868-8_26

technical tools (e.g. hardware and software) needed to implement energy management [2] (Energy management is predictable, organized and systematic coordination of procurement, transformation, distribution and use of energy to meet needs with respect to environmental and economic goals. Thus, the term describes actions for the purpose of efficient energy management. The term "energy management system" covers organizational and informational structures, including the necessary technical means (for example, hardware and software) necessary for energy management implementation [2]) The energy management and its system are closely related to the "green" economy - this is what constitutes resource- and energy- saving economy, where less amount of initial materials, in the form of base materials and energy, will be expended on the production of goods, services, and end-use products. Production of the last-mentioned items and the required amount of raw materials and energy will be increasingly limited in the world. Thus, the specific consumption of resources for development of goods and services will be constantly declining everywhere saving natural materials and preserving the environment.

Russia is also integrated into energy management processes. For example, in the sphere of real estate construction and maintenance, more and more objects are emerging that are classified according to the world's LEED and BREEAM systems. Moreover, national certification systems have been developed and used, including GOST (All-Union State Standard) R 54964–2012 «Conformity assessment. Environmental requirements to real estate items» and national standards STO NOSTROY 2.35.4-2011 «"Green building" Residential and public buildings. Rating system of living environment sustainability assessment». Research is being conducted to justify the effectiveness of energy efficient construction and maintenance [2–5].

Creation of a real property object as a new energy efficient enterprise and working environment for people, application of new green technologies in construction and operation require significant investments at high levels of risk compared to traditional construction. At the same time, qualitative data on the increase in the revenue components of energy efficient projects and the decrease in expenditure indicators of the operation of items can be found in some sources [6, 7]. However, there are no numerical values that give an idea of the performance indicators of investments in such objects. In this sense, it is worth highlighting Marianna Brodach and Guy Eames's work that provides information on the impact on the net present value of a real estate project of various indicators of energy efficiency (Table 1).

Table 1. The results of the calculation of indicators of green building economy, according to [8].

Category	Net present value within 20 years, $/sq. m
Energy saving	60.7
Emissions reduction	12.9
Water saving	5.4
Savings on operation and maintenance service	91.5
Increase in performance, occupational hygiene	397.0–595.0
Average cost of construction	21.6
Total	535.2–711.7

The values of the conclusive result presented in Table 1 were significantly achieved due to the indicator "Increase in performance, occupational hygiene", its effect is about 80%. However, for a property owner the following indicator ("Increase in performance, occupational hygiene") is of no concern, since his income can only be gained through his tenants' rent. It is worth remarking that the indicator of the total value of the project, achieved with the help of energy efficient technologies implementation, is of great importance for the investor. Many parameters influence on the indicator of the total value, and it is possible to determine their value only with a certain degree of uncertainty. The more innovative proposals are considered in the energy efficient project, the higher this degree of uncertainty becomes. Considering a multi-factorial nature of the investment process in energy efficient construction and the uncertainty of the value of the following factors, the stochastic nature of the implemented processes, the Monte-Carlo method, a method of simulation modelling, can be an effective analysis tool. Monte Carlo method is widely used in simulation modelling. Officially, a date of the appearance of the method is considered to 1949, when an article by Metropolis and Ulam [9] was published. Simulation modelling is a research method, where a system under study is replaced by a model that describes the current system with sufficient accuracy (the model describes the processes as if they are in reality). Moreover, the following model is used to carry out computer experiments for research or optimization. Simulation modelling allows to study system's behavior in time. Objects' behavior can be stimulated that cannot be reproduced for several reasons: dangers of a natural experiment, high cost or impossibility of the experiment in normal conditions. The core of the method consists in the description of the processes by using an analytical apparatus, with its help a random event is "played" and a special procedure that gives a random result is used. In reality, random processes can be developed in various ways, and a single result can give nothing. It is a different matter if it is a number of results that can be used in the experiment. Randomness is used as a research tool when the Monte-Carlo method is applied modelling random phenomena.

Since 1949, the method has been researched and applied by such scientists as D.B. Hertz, R., R.A. Breally, S.C. Meyers, V.N. Livshits, S.A. Smolyak, R.M. Kachalov, K.A. Bagrinovskiy, A.A. Bakaev, Ya.P. Busdenko, I.G. Venetsky, B.V. Gnedenko, D.I. Golenko, I.N. Kovalenko, A.N. Kolmogorov, N.I. Kostin, T. Naylor, Yu.V. Prokhorov, V.G. Sargovich, I.M. Sobol, V.G. Schrader, E.A. Yakovlev, N.V. Yarovitsky and others.

Currently, the Monte-Carlo method is more widely applied in the industrial sector of the Russian economy 10, 12, in its oil and gas complex 12, 13, 14 and its construction industry. However, decision-making in the field of building and energy management systems in our country is practically not based on the Monte-Carlo simulation method, which was approved in various sectors of the economy. This circumstance is an omission, according to the authors.

As indicated in [17], the peculiarity of investments in real estate objects is their uniqueness and the absence of reliable market information in free access. It is a common thing that there is information that is not about specific values of the variables used in the calculations, but about ranges of their variations. These characteristics fully reflect the investment processes in building and energy management systems. It has been noted [10] that possible mistakes (uncertainty) in the source data require application of methods that allow to take into account their effect on the obtained results. One of these methods is the Monte-Carlo method, which makes it possible to enumerate a maximum number of source data combinations and estimate the range of variations of the resulting variable [10]. Based on the aforesaid, the objective of the following work is to study the possibility of the Monte-Carlo method application in green planning and investment management in real estate objects.

To achieve this objective the following tasks are set:

1. Define a concept of "simulation modelling".
2. Investigate the previous study of Monte Carlo simulation method's application.
3. Identify the fields, where Monte Carlo method is used for making management decisions.
4. Identify the indicators that influence the investment decision on financing of energy efficient project in real estate sphere.
5. Determine the numerical values of the indicators that influence the investment decision on the financing of an energy efficient project in real estate.
6. Determine the resulting indicator, which is important for making management decisions in the field of building and energy management systems.
7. Determine the correlation between the source data (that are factors and output indicators) and the resulting variables in the form of a mathematical equation.
8. Check the significance of their influence on the result of calculations.
9. Determine the probability distribution law for each key factor possible limits of the factor change.
10. Conduct Monte Carlo simulation modelling and present the results graphically.
11. Analyze the results.
12. Make a conclusion and report on possibility of the Monte-Carlo method application in building and energy management systems and real estate investment.

2 Materials and Methods

As stated above, when the investment decision on energy efficient project is being made, an investor needs knowledge of its present value as one of the most important indicators of project efficiency [18–21]. When the real estate projects are implemented, the following indicators affect the specific value [22–25]: rental rate, number of underloads and underpayments, operating expenses, norms of capital return rate, project implementation period, cost of future sale of the estate property object.

Among the published studies int the sphere of building and energy management, data were found on the factors that determine investment attractiveness of the energy efficient building objects. Their values are presented below according to [8]: increase in rental rates by 2–16%, increase in occupancy rates by 2–18%, decrease in operating expenses by 25–30%, increase in sale value by 5.8–35%.

These indicators tested for materiality of their impact on the value of the present value using the method of sensitivity analysis. Each of the values presented in the list above varied sequentially by 5%, and after that the present value was calculated. The obtained magnitude of the present value was compared with the initial value, when

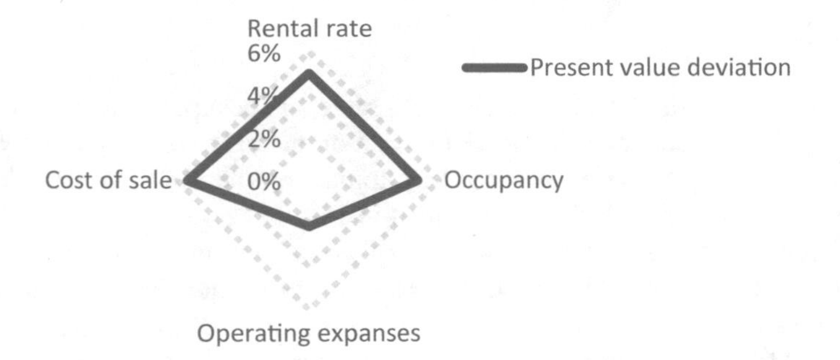

Fig. 1. Sensitivity analysis.

calculating which, the factors did not vary. The relative deviation of the present values is presented in Fig. 1.

After making a sensitivity analysis, Monte Carlo method was applied to determine a current value of the energy efficient construction according to the named indicators.

The following formula was used to calculate the present value:

$$PV = \sum\nolimits_{i=1}^{n} \frac{Io_i}{(1+Yo)^i} + \frac{Vo_n}{(1+Yo)^n} \quad (1)$$

PV – present value; *Ioi* – annual net operating income in the *i*-st year; *Yo* – annual norm of return on investments; *Von* – cost of the future sale of the object in n years; *n* – number of calculation periods. The base value of the rental rate was taken equal to 1.00 monetary unit, the load value of property is 80%, operating expenses were determined at the level of 30% as an average value on the real estate market. Annual net operating income is obtained by multiplying the rental rate by the value of the property object load and then the subsequent operating expenses subtraction. The rate of return is assumed to be 12% per annum. The forecast period is assumed to be 20 years, as in [8]. The effect of inflation was not taken into account, and the cash flow was built in real terms. The cost of the reversion was determined with the use of the Gordon formula.

With the help of the random number generator, 1000 variants of calculation of the present value were performed. To substantiate a number of the required experiments, the studies presented in [17] were used. According to [17], building a model within the framework of investment design, distribution of factors can be assumed uniform if the input parameter ranges are determined for a specific object, that is, they have such a variance that in a given interval the factor can take any value with equal probability. Therefore, the balanced distribution is used when the calculation is being performed. To construct a distribution histogram of the results obtained, equal-sized intervals were calculated according to the Sturges' formula.

3 Results

The sensitivity analysis showed that the rental rates, the occupancy rates and the future cost of sale of an object have the greatest impact on the value of the present value, it is about 5%. Operating expanses have the least impact on the resulting value, it is about 2%. The results of the analysis are presented in Fig. 1.

The obtained results of the sensitivity analysis allow us to conclude that all the considered parameters should be used to implement the Monte-Carlo simulation model. As a result of the Monte-Carlo method application, the variation of the present value was obtained from 5.52 monetary units to 7.38 monetary units, the average value made 6.38 monetary units, the coefficient of variation is 6%.

The performed calculations with the use of the Sturges' formula allowed us to determine 11 equal-sized intervals where the obtained of the present value are. Simulation method results are presented in Fig. 2.

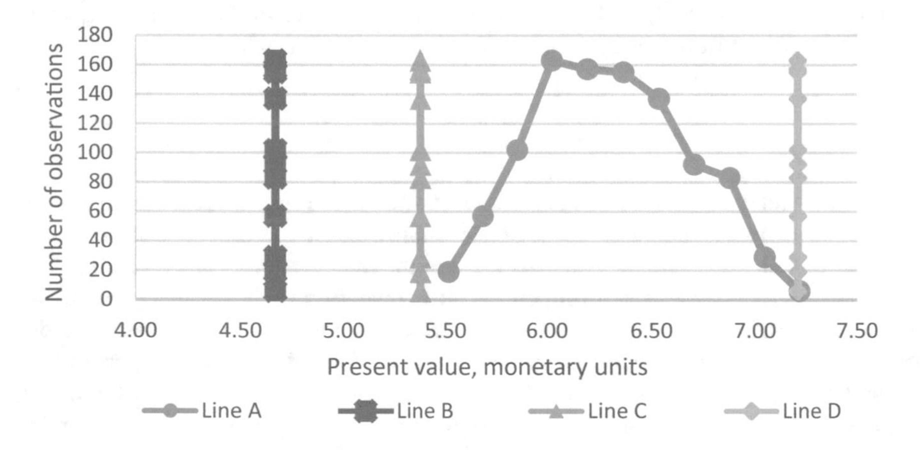

Fig. 2. Monte-Carlo simulation modelling.

Line A – a graph obtained with a simulation modeling method. Line B – a result of a traditional project. Line C – minimum cost. Line D – maximum cost.

Line B gives an idea of the present value in a traditional construction project with the accepted initial data. Line A shows possible options for the present value of the energy efficient property object, obtained by the Monte-Carlo method, taking into account changes in the values of the indicators presented in [8]. Lines C and D limit the area of the possible present value of the implementation of an energy efficient project, which can be determined without the use of the Monte-Carlo method.

As can be seen, the minimum value of the present value of the energy efficient project is larger than the value of the traditional project by 18%. The excess of the maximum value of the present value over the present value of the traditional project is 58%. With the maximum number of observations in a group of 163, the average present value of 6.1 monetary units is reached, which exceeds the value for a traditional

Table 2. Intervals of the present value magnitude.

Interval number	Lower interval limit	Upper interval limit	Frequency of operations	Interval number	Lower interval limit	Upper interval limit	Frequency of operations
1	5.52	5.69	19	7	6.54	6.71	137
2	5.69	5.86	57	8	6.71	6.88	92
3	5.86	6.03	102	9	6.88	7.05	83
4	6.03	6.20	163	10	7.05	7.22	29
5	6.20	6.37	157	11	7.22	7.38	6
1	5.52	5.69	19	7	6.54	6.71	137

project by 31%. The remaining data on the distribution of observations by intervals of the obtained values of the present value are presented in Table 2.

Thus, based on the simulation method, the present value of the energy efficient project is obtained in the interval from 5.52 monetary units to 7.38 monetary units, with the most likely result in the range of 6.03 to 6.20 monetary units. Without the use of the Monte Carlo method, the calculated range is 5.38–7.21 monetary units with an equally probable result within it.

4 Discussion

The results obtained by the Monte-Carlo simulation method give an investor an idea of the range of possible values of one resulting indicator, in this case, the present value. Having one indicator that takes into account different variants of the energy efficient project implementation simplifies investor's task when choosing an investment object.

Given the complexity of implementation of building and energy management systems, that includes their novelty, the resulting indicator is defined as a range of the most likely values that can be obtained during the project, while finding the most important economic indicators of the project in a range determined by experts. At the same time, the conducted simulation modelling allows to determine the most likely proved value of the resulting indicator lying on top of the histogram (Fig. 2). The initial indicators had a large confidential interval, where a maximum value of the indicator within the range was 8 or 9 times higher than the minimum, that can be seen in rental and occupancy rates. However, the obtained results have a smaller spread in values, which simplifies investor's decision-making process.

The performed calculations answer the investors' questions how much the present value is higher in comparison with the traditional construction project in the worst scenario (18% in our case) and in the best scenario (58% in our case) or to compare them with the most likely value (31% in our case), and also narrow the range of the most likely value of the project cost dramatically, which can be obtained without the use of the Monte Carlo simulation method. In the case under consideration, the range from 5.38 to 7.21 monetary units was reduced to 6.03–6.20 monetary units. With the development of the following study, the rate of return requires an additional specification. The rate of return in this work is assumed to be equal to 12% per annum. The research requires:

- Magnitude of risks comparable to traditional and energy efficient construction.
- Additional risks specific only to energy efficient construction.
- Level of risk in the period of construction and operation of the energy efficient building.
- Capital structure in energy efficient projects.
- Rates of raising debt capital in energy efficient projects.

To understand the effectiveness of investments in any project, including an energy efficient project, it is important for investors to understand not only the present value of future incomes, the probability of their receipt, as well as possible risks and data on initial investments. A significant cost overrun in an energy efficient object construction over the cost of a traditional real estate object construction can change investor's choice in favour of construction without the use of energy efficient technologies. Therefore, this circumstance requires the continuation of the presented research.

5 Conclusions

Building and energy management systems, that focuses on the natural resources economy and environmental care economy, is increasingly becoming a reality and it is beginning to determine, on the whole, the future development of the world economy. This trend did not bypass the construction and operation market. Introduction and commercialization of new technologies is always accompanied by a high level of uncertainty and risks. The same situation is with the introduction of energy efficient technologies in the real estate economy.

Simulation modelling is a research method, where the system under study is replaced with a model that describes the current system with sufficient accuracy, and computer experiments are carried out to research or optimize the following system. The Monte-Carlo simulation method appeared in 1949 and it has been widely studied by many scientists since that time, among them D.B. Hertz, R. Braley, R.A. Breally, S.C. Meyers, V.N. Livshits, S.A. Smolyak, Ya.P. Busdenko, A.N. Kolmogorov, I.M. Sobol, V.G. Schrader, E.A. Yakovlev, N.V. Yarovitsky and others. At this time, this method has got widespread use in the industrial sector of the Russian economy, in its oil and gas complex and in the construction industry. At the moment, the Monte-Carlo simulation tool, mastered in other sectors of the economy, is not used in building and energy management systems, in environmental planning and management and in investing in energy efficient projects.

Having analyzed open sources of information, numerical values of rental rates, occupancy rates, operating expenses and future sales value, which influence on the investment decision on an energy efficient real estate project financing, were found. Also, values of the numerical ranges, where they are located, were determined. The obtained results allowed to apply a formula for the present value calculation as a resulting indicator that is used to make decisions in building and energy management systems. The Monte-Carlo simulation was performed through the use of 1000 experiments. Normal distribution was used in the calculations. The obtained results are presented in the graph. The obtained result is a range of possible magnitudes of the present value of the project, with its most likely value lying at the top of the histogram.

Contemporary studies provide expert values for indicators of the energy efficient economy projects development and real estate objects operation, but these data do not give investors an idea of the numerical indicators of the investments' effectiveness, which includes building and energy management systems. Results obtained by the Monte-Carlo simulation method give the investor an idea of the range of possible values of one resulting indicator, in this case, it is the present value. The presence of one indicator, that takes into account different variants of the energy efficient project implementation, simplifies the investor's task of choosing an investment object for environmental planning in real estate.

This work demonstrates how expert data on energy efficient projects, available on the market, can be used to obtain necessary indicators to make reasonable management decisions. With regard to the initial values uncertainty, the obtained result is a range of values, where the most probable value of the resulting indicator lies.

References

1. Karpov, V.K.: Green Economy - the future of the world economy. Theory Pract. World Sci. **5**, 69–76 (2017)
2. White Paper. Energy Management and Energy Optimization in the Process Industry. Siemens AG, September 2011
3. Vodyanova, S.A., Pupentsova, S.V., Pupentsova, V.V.: Mechanisms for the "Smart Home" technology development and implementation. Innovations **7**(237), 83–90 (2018)
4. Pupentsova, S.V., Alekseeva, N.S.: Determining the cost of the "Smart Home" system installation. In: Innovation Clusters in the Digital Economy: Theory and Practice Proceedings of the VIII Scientific Practical Conference with International Participation. Babkina, A.V., pp. 182–186 (2017)
5. Lavrenenkova, O.M., Pupentsova, S.V.: Investment attractiveness of green construction projects. In: Industrial Policy in the Digital Economy. Problems and Prospects. Proceedings of a Scientific-Practical Conference with International Participation. In: Babkina, A.V., pp. 518–522 (2017)
6. Green Buildings and the Finance Sector. An Overview of Financial Institution Involvement in Green Buildings in North America. A Report Commissioned by North American Task Force, UNEP Finance Initiative (2010)
7. How Green a Recession? – Sustainability Prospects in the US Real Estate Industry, no. 2(70) (2009)
8. Brodach, M., Eames, G.: Market of green building in Russia. High-Tech Build. **1**, 18–29 (2013)
9. Metropolis, N., Ulam, S.: The monte carlo method. J. Am. Stat. Assoc. **44**(247), 335–341 (2007)
10. Golotovskaya, A.V., Voronkov, P.T.: Quantitative risk assessment of investments in the production of wood pellets by the Monte Carlo method. Forest. Inf. **4**, 30–38 (2015)
11. Pogonin, A.A., Chepurov, M.S., Beliankina, O.V.: Technological process parameters' modelling using the Monte Carlo method. Mining Inf. Anal. Newsl. **2**, 217–218 (2003)

12. Kasimova, A.E., Moskalev, A.K.: Evaluation of the economic efficiency of technological solutions in the oil and gas complex by the Monte Carlo method. J. Siberian Fed. Univ. Humanit. **9**(4), 815–823 (2016)
13. Vetchinova, M.D.: The Monte Carlo method in risks analysis in the oil and gas complex. Interexpo Geo-Siberia **2**(1), 282–284 (2017)
14. Kirichenko, O.S., Kirichenko, T.V.: Investments according to the Monte Carlo method. Application by the investment analysis method in the management of financial risks of oil and gas industry projects. Russ. Bus. **1**(1), 46–49 (2008)
15. Oparin, S.G., Esipova, E.V.: Probabilistic-statistical risk assessment of the investment construction project using the Monte Carlo method. Actual Issues Mod. Sci. **6**(2), 128–134 (2009)
16. Adamyuk, I.A.: Risk assessment (of economic attractiveness) of the investment project, on the example of the residential houses construction using the Monte Carlo method. Economics and Management in the 21st Century: Development Trends, vol. 3, pp. 150–156 (2011)
17. Pupentsova, S.V.: Models and tools in the economic assessment of investments. Saint-Petersburg (2007)
18. Ozerov, E.S., Pupentsova, S.V.: Value and real estate potential investment management. Saint-Petersburg Polytechnic University, Saint-Petersburg (2015)
19. Pupentsova, S.V., Shabrova, O.A., Leventsov, V.A.: Determining the control premium in business valuation of shares. Scientific and technical statements of St. Petersburg State Polytechnic University. Economics **10**(5), 125–132 (2017)
20. Ozerov, E.S., Pupentsova, S.V.: The pricing process modelling in commercial real estate transactions. Property Relat. Russ. Fed. **12**(171), 29–37 (2015)
21. Ozerov, E.S., Pupentsova, S.V., Leventsov, V.A., Dyachkov, M.S.: Selecting the best use option for assets in a corporate management system. In: Reliability, Infocom Technologies and Optimization (Trends and Future Directions) 6th International Conference ICRITO, pp. 163–171 (2017)
22. Pupentsova, S., Livintsova, M.: Qualimetric assessment of investment attractiveness of the real estate property. Real Estate Manag. Valuat. **26**(2), 5–11 (2018)
23. Pupentsova, S.V., Kamalova M.V., Dyachkov, M.S.: Development of a model of direct income capitalization in real estate evaluation. Scientific and Technical Gazette of St. Petersburg State Polytechnic University, no. 3, vol. 10, pp. 228–237 (2017)
24. Pupentsova, S.V., Rusanov, S.V.: Study of the development of built-up areas variants in St. Petersburg. Econ. Constr. **4**(52), 34–46 (2018)
25. Kuptsov, A.A., Pupentsova, S.V.: Performance evaluation of the shopping centers' reconception effectiveness. Econ. Constr. **2**(38), 66–77 (2016)

Application of the Neural Network Method of Computer Modeling to Reduction of Harmful Emissions from Power Installations of Transport Enterprises

Roman Dolgov[1], Viktor Katin[1], Midkhat Akhtiamov[1], and Vladimir Kosygin[2(✉)]

[1] Far Eastern State Transport University, Khabarovsk, Russia
[2] Computer Center of Far-Eastern Branch of Russian Academy of Sciences, Khabarovsk, Russia
kosyginv@inbox.ru

Abstract. This paper is devoted to the resolution of an urgent scientific and technical problem of increase in precision and efficiency of risk level forecasting in the affected areas of stationary emitters of the power producing units for railway enterprises. We have analyzed the issues of forecasting the distribution of pollutants from stationary emitters of the railway power producing units. We have also determined the directions of increase in precision and efficiency of forecasting risk levels in the affected areas of the stationary emitters of the power producing units for railway enterprises. Dependencies obtained allow operative computation of oncogenic and non-oncogenic risks employing the capabilities of an artificial neural network (ANN). The authors suggest practical recommendations how to apply the dependencies obtained to decrease the levels of impact of the emissions on the workers and the environment.

Keywords: Modeling · Artificial neural network · Stationary sources of emission · Railway enterprises · Professional risks

1 Introduction

An important component of the socio-economic development of a state is the health condition of its working population which is determined by the level and exposition of the complex of hazardous and dangerous production factors. One of the major production factors affecting industrial facility sites is the atmospheric air pollution [1].

The problem of air quality is worsened by the build-up pollutants effect near stationary emitters and transport mainlines in adverse meteorological conditions when the concentration of pollutants can be 5–15 times over the limits. Such concentration levels do not only lead to serious health consequences but can also cause acute toxic exposures (diseases). Under present day conditions, the Transport Strategy of the Russian Federation for the period till 2030 and the Strategy for the Development of Railway Transport for the period till 2030 were developed in order to improve the efficiency of the railway transport enterprises. These documents also cover related

V. Murgul and M. Pasetti (Eds.): EMMFT-2018, AISC 983, pp. 267–273, 2019.
https://doi.org/10.1007/978-3-030-19868-8_27

environmental activities of transport enterprises [2, 3]. The hazard (safety) level of an industrial environment can be assessed with risk assessment instruments through the criteria for special assessment of the working environment as well as the levels for excess over maximum admissible concentrations (peak daily average concentrations for residential areas; and time-weighted average and short-term exposure concentrations for industrial sites). High risk levels cause the increase of specific and general somatic diseases which aggravate such major parameters of technospheric safety as the disability rate and reduction in life expectancy, which in their turn lead to the increase in economic costs. In this connection, the issues of forecasting the pollution of the surface atmospheric layers in territories of industrial facility sites and developing recommendations for the decrease in professional risks within the areas of the atmospheric air pollution of the industrial environment by stationary power producing units are of high current importance and relevance.

2 Research Methodology

We recommend using the capabilities of an artificial neural network (ANN) as a technique for forecasting the distribution of pollutants in the surface atmospheric layers.

Proceeding from the analysis of the data collected and the existing mathematical models of distribution, it is essential to use the following parameters as predicates for constructing an ANN: wind velocity; effective emission height; volume of emission; pollutant concentration in the emissions; the distance to the reference mark along the wind axis; and the distance from the wind axis to the reference mark along the perpendicular. We have chosen two layers of neurons to provide adequate computational performance of the neural network. Sigmoid function symmetric about the x-axis is used as an activation function. The layout of this neural network is presented in Fig. 1.

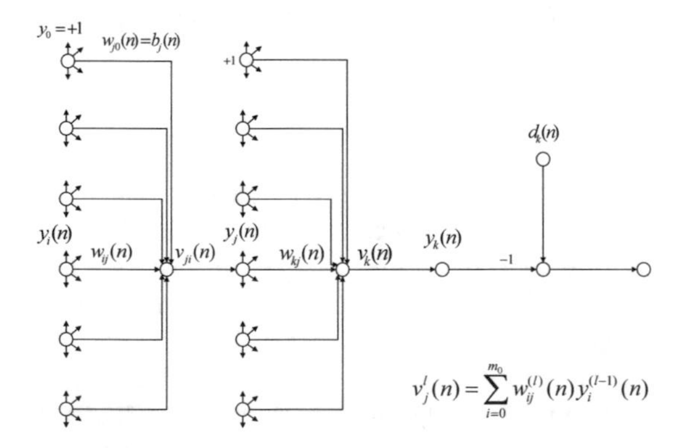

Fig. 1. Neurons interconnected in the network

This ANN is based on a multilayered perceptron trained through the back propagation of error method with the adaptive step of learning in the implementation.

The artificial neural network applicable to computation is generated through harmonizing weighting factors of the neuron connections. The most widely spread algorithm is the back propagation of error algorithm whose graph is presented in Fig. 2.

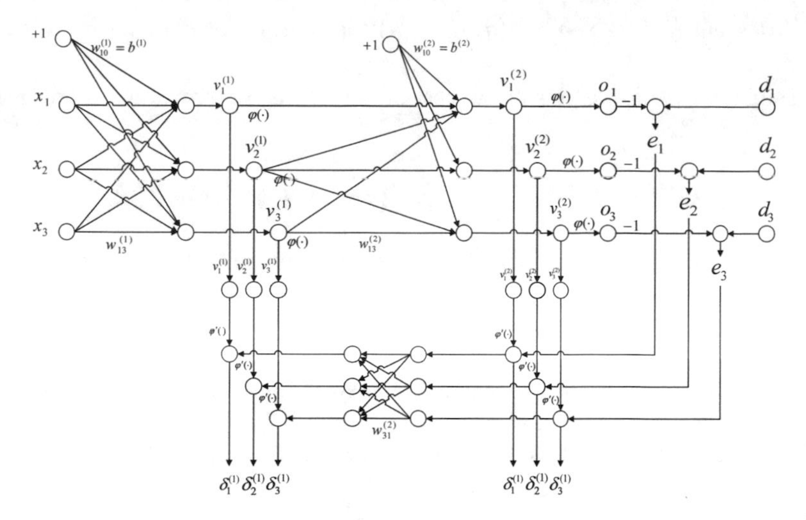

Fig. 2. Graph of the back propagation of error algorithm.

The algorithm presupposes the following consequence for processing the samples of the training set $\{(x(n), d(n))\}_{n=1}^{N}$.

Synaptic weights and limit values are generated at the initialization stage with their dispersion adjusted so that the standard deviation of the induced local field of the neurons fell within the linear part of the activation function and did not reach the saturated area. At the stage of presenting teaching parameters, the network is supplied by the images of the training set. For each of these images, forward pass and backward pass are implemented sequentially. While implementing the forward pass, the induced local fields of the network and functional signals are computed layer-by-layer

$$v_j^l(n) = \sum_{i=0}^{m_0} w_{ij}^{(l)}(n) y_i^{(l-1)}(n), \tag{1}$$

where $y_i^{(l-1)}(n)$ is the output (functional) signal from neuron i located in the previous layer $l-1$, on iteration n; $w_{ij}^{(l)}(n)$ is the synaptic weight of the connection of neuron j of layer l with neuron i of layer $l-1$. For $i = 0$, $y^{(l-1)}(n) = +1$, and $w_0^{(l)}(n) = b_j^l(n)$ is the limit applied to neuron j of layer l. For the first layer neurons: $y_j^0(n) = x_j(n)$, where $x_j(n)$ is the j - th element of the input vector $x(n)$. For the output layer neurons: $y_j^{(L)}(n) = o_j(n)$, where L is the network depth.

The signal of error is calculated as $e_j(n) = d_j(n) - o_j(n)$, where $d_j(n)$ is the j-th element of the desired response vector $d(n)$.

While implementing the backward pass, local gradients of the network nodes are calculated according to the following formula:

$\delta_j^{(l)}(n) = e_j^{(L)}(n)\phi_j'(v_j^{(L)}(n))$ for neuron j of the output layer L,
$\delta_j^{(l)}(n) = \phi_j'(v_j^{(L)}(n))\sum_k \delta_k^{(l+1)}(n)\delta_{kj}^{(l+1)}(n)$ for neuron j of the hidden layer l.

Changing synaptic weights of the network layer l is performed in accordance with the generalized delta rule:

$$w_{ji}^{(l)}(n+1) = w_{ji}^{(l)}(n) + \alpha[w_{ji}^{(l)}(n-1)] + \eta\delta_j^{(l)}(n)y_j^{l-1}(n),$$

where η is the learning rate parameter; and α is the momentum constant.

Thus, forward and backward passes are implemented sequentially and the network is presented with all learning patterns until the stop criteria are achieved.

Standard back propagation of error algorithm has the problem of choice of an adequate learning step in order to increase the speed-in-action and to ensure the convergence of the algorithm.

In this implementation of an ANN we employed an adaptive learning step

$$\alpha(t) = \frac{\sum_j \gamma_j^2 \cdot F'(S_j)}{F'(0) \cdot (1 + \sum_i y_i^2) \cdot (\sum_j \gamma_j^2 (F'(S_j))^2}. \tag{2}$$

To analyze the performance of artificial neural networks, we employ quadratic mean error.

The neural network works as follows. The values of the determining dispersion factors are preliminarily harmonized by preprocessor into computational range and supplied to the neurons of the sensor unit and then to all neurons of the hidden layer and the output layer. The output signal of the unit is defined as

$$v_j^l(n) = \sum_{i=0}^{m_0} w_{ij}^{(l)}(n)y_i^{(l-1)}(n) \tag{3}$$

where $y_i^{(l-1)}(n)$ is the output (functional) signal of neuron i located in the previous layer $l-1$; $w_{ij}^{(l)}(n)$ is the synaptic weight of the connection of neuron j of layer l with neuron i of the layer $l-1$. For $i=0$, $y^{(l-1)}(n) = +1$, and $w_0^{(l)}(n) = b_j^l(n)$ is the limit applied to neuron j of layer l. For the first layer neurons: $y_j^0(n) = x_j(n)$, where $x_j(n)$ is j-th element of the input vector $x(n)$. For output layer neurons: $y_j^{(L)}(n) = o_j(n)$, where L is the depth of the network.

The quality of computations performed by ANN was assessed on a set of patterns that were not included in the learning set. The value of quadratic mean error was employed as the assessment criterion. The ANN computational results showed sufficient convergence with the results of computations performed in accordance with the method [4]. The distinctive feature of the method we suggest is the capability of revising and making the computations more precise due to the ANN after-learning on the results of field measurements in the industrial site.

3 Research Results

Based on the results of computations performed by ANN, the check-off sheet of the numerical monitoring is mapped automatically on a real-time basis. The check-off sheet reflects the pollution levels of the surface layers of atmospheric air in the territory of the industrial site based on the data of the meteorological parameters and the data of the working modes of the equipment. The algorithm we suggest does not require high computational performance and can be implemented within the programs for both desktop and portable computers (tablets and smartphones).

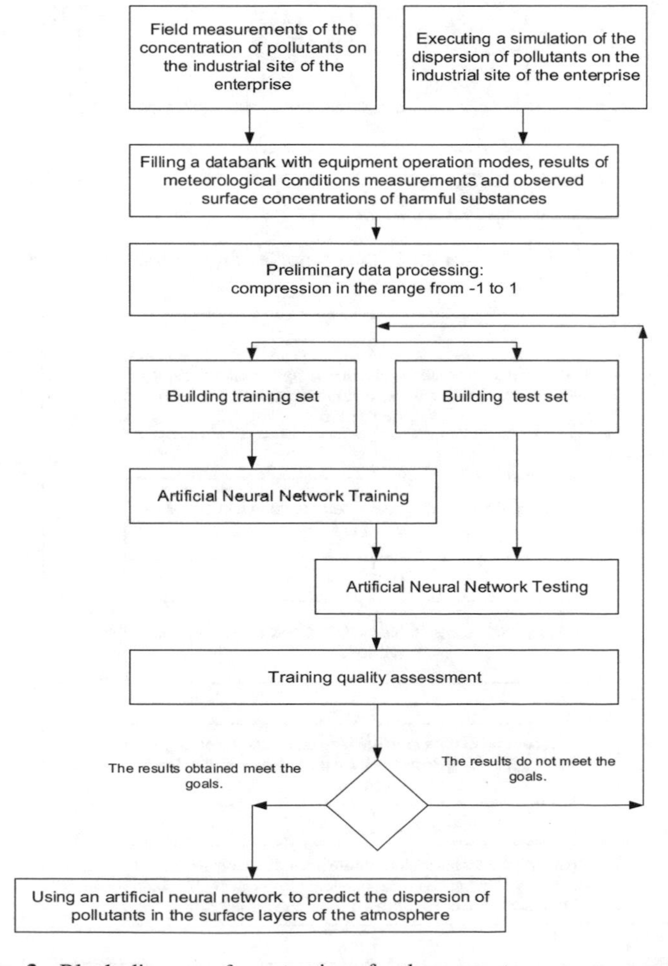

Fig. 3. Block diagram of preparations for the computer program usage.

The authors have developed the program implementation of the algorithm described whose block diagrams are presented in Figs. 3, 4 and 5.

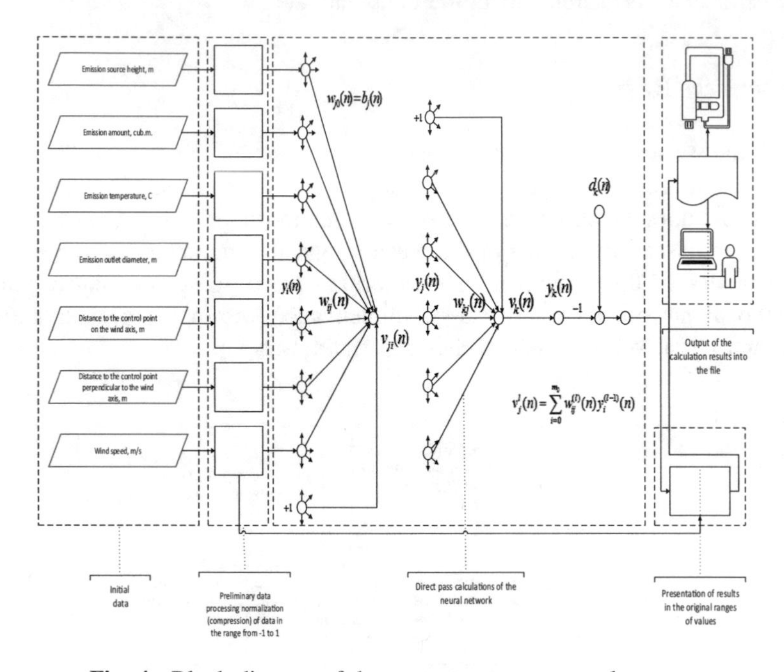

Fig. 4. Block diagram of the computer program work process

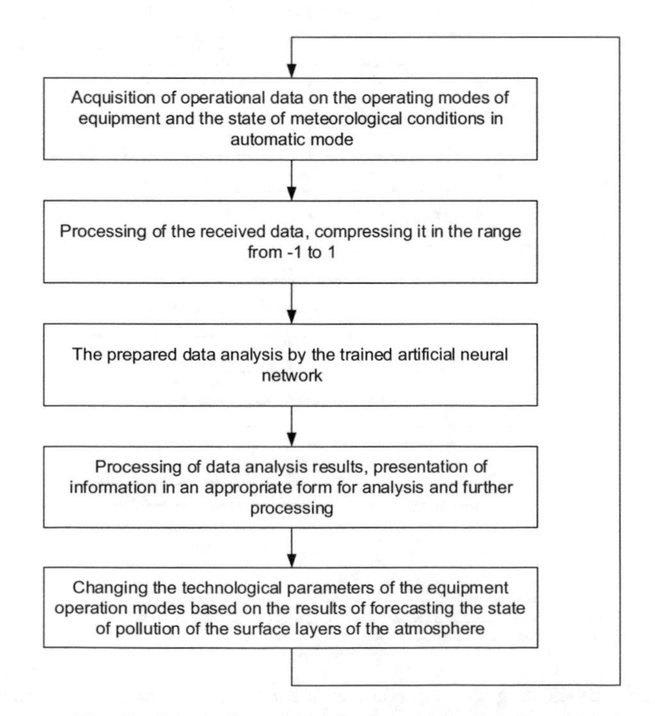

Fig. 5. Block diagram of the computer program use

4 Conclusion

This implementation of the program for the computer allows us to perform operative forecasting for the zones of high pollution levels in the lowest atmospheric layers in the territories of industrial sites. The forecasting results can be used for planning works in the territories of industrial enterprises and for the assessment of oncogenic and non-oncogenic risks caused by the impact of pollutants on human health.

ANN forecasting can be improved if the neural network is provided with additional training sets that contain results of the field measurements of the concentrations of pollutants and the working modes of the equipment.

In that case, we modify the formula for the oncogenic risk computation in accordance with the guide for the assessment of risks to health caused by environmental pollution [5] and obtain

$$\mathrm{h} = \frac{1}{\tau} \cdot \sum\nolimits_{i=1}^{N} \int\limits_{0}^{\tau} \frac{w_{ij}^{(l)} \cdot y_i^{(l-1)}(n)}{H_i} d\tau_i \tag{4}$$

where τ_i is observation period of concentration q_i within the zone of human exposure; H_i is the relative risk index for *i* substance.

Using a similar approach, we present the formula for calculation of the relative risk index for non-oncogenic diseases [7], that characterizes the possibility of chronic diseases caused by atmospheric pollution as follows:

$$R = \frac{1}{\tau} \cdot \sum\nolimits_{i=1}^{N} \int\limits_{0}^{\tau} R_i \cdot \sum\nolimits_{i=0}^{m_0} w_{ij}^{(l)}(n) \cdot y_i^{(l-1)}(n) \cdot d\tau_i, \tag{5}$$

where R is an individual risk of oncogenic disease; τ_i is the observance time for concentration q_i within the human exposure area; R_i is an identity risk factor for the *i*-th substance, m^3/mcg.

Employment of the dependencies obtained allows us to take into consideration the movements of workers among the zones with different levels of atmospheric air pollution while computing oncogenic and non-oncogenic risks.

References

1. Matesheva, A.V.: Identification of stationary sources of pollution of atmospheric air at the transport. Transp. Sci. Technol. **3**, 20–25 (2016)
2. Order of the RF government of 22 November 2008 № 1734-p “On the ratification of the Transport Strategy of the Russian Federation for the period till 2030” (with the amendments for 12 May 2018). Code of Laws of the Russian Federation, N 50, 15 December 2008, clause 5977
3. Order of the RF government of 17 June 2008 № 877-p. On ratification of the Strategy for the Development of Railway Transport for the period till 2030
4. Marty, M.A., Siegel, D.: Air toxics hot spots program. Guidance Manual for Preparation of Health Risk Assessments, California, EPA (2015)
5. Guidelines for Cancerogen Risk Assessment. https://www3.epa.gov/airtoxics/cancer_guidelines_final_3-2505.pdf. Access 15 July 2014

The Simulation Model's Elaboration of the Production Process of the Locomotive Repair Enterprises

Kirill Panov(✉)

Omsk State Transport University, 35, Marx av., 644046 Omsk, Russia
k.v.panov@ya.ru

Abstract. This article solves purposes of the operation's efficiency of the production systems of the locomotive repair enterprises. The article provides a mathematical description of the similar enterprises' production process on the example of the service locomotive depot Moskovka. The mathematical model is described on the basis of the queuing theory (SMO) and Markov's chains. There was given a graph of the locomotive's technological condition during repair in depot, where the transition of the repairing rolling stock from condition to condition, i.e. interrelations of the locomotive's technological movements on the depot's territory, and also execution time of the technological operations was described by transition matrix. The distribution of repair requests is described using the Poisson law, delay of the locomotive in the depot for the planned types of repair is in accordance with the accepted standard, and for unplanned types of repair, requests are delayed in accordance with the Erlang distribution. According to the results of the production process of the mathematical description of the repair enterprise, a simulation model of the Moskovka depot functioning was created. It was implemented on the computer using the professional software of new generation – AnyLogic. The work describes the logic circuit of the simulation model, selected methods and modeling libraries. According to the research results, an overall assessment of technological parameters of the locomotive repair enterprises was carried in conditions of probabilistic and dynamic changes of the repair mission.

Keywords: The simulation modeling · The rolling stock's repair · The queuing theory · The state graph · The transition matrix · The production logistics

1 Introduction

Modeling is one of the ways to solve practical tasks. Often a task's solution cannot be found by carrying natural experiments: it could be too expensive, dangerous or simply impossible to build new objects, to destroy or make changes in the existing infrastructure. In such cases, we build a model of a real system, that is, we describe it in the modeling language. In the model [2], there was proposed an optimal solution's model for the preventive maintenance of the railway truck's core components. In the studies of Datsun [3] and Ismailov [4], a sequence and an interaction of the separate processes

V. Murgul and M. Pasetti (Eds.): EMMFT-2018, AISC 983, pp. 274–286, 2019.
https://doi.org/10.1007/978-3-030-19868-8_28

during the node's repair of the rolling stock are analyzed. In these works, the losses of the technological process on the example of the entire enterprise are not calculated and not optimized, but only separate technological sections of the repair. The studies of Cho and Parlar [5], Sherif and Smith [6], Osaki and Nakagawa [7], Pierskalla and Voelker [8], Valdez-Flores and Feldman [9] are directed to the optimal resources' determination for the technical maintenance and finding the minimal economic costs. The authors find traditional methods for solving problems using linear and dynamic programming, stochastic models and analysis of the current value. Technological processes of rolling stock repair are stochastic and probabilistic, their parameters cannot be accurately calculated using analytical formulas, and they have to use special programs for simulation modeling [10]. In the article of Bannikov and Sirina [11], the main stages of the discrete event's model of the passenger rolling stock's maintenance using the AnyLogic surroundings are considered. The simulation model allows implementation of a simulated system and getting the performance indicator of the work of the maintenance service's system. Nevertheless, the interaction of numerous maintenance service facilities with rolling stock is considered, and it is not possible to study the frequent case, a separate work of an enterprise. As an example of the choice of model, the Moskovka service locomotive depot was chosen, which is the largest enterprise for the repair of electric locomotives in the territory of the West Siberian Railway (ZSZHD). The key task of the repair depot is a provision of the safe and reliable work in service of the electric locomotives of the series 2ES6, Vl10u, Vl10k and VL11 at the Omsk territorial Department of ZSZHD of the network of JSC RZD [12, 13].

2 Materials and Methods

The technological process' conceptual model of the functioning of the depot "Moskovka" can be presented as a state graph (Fig. 1).

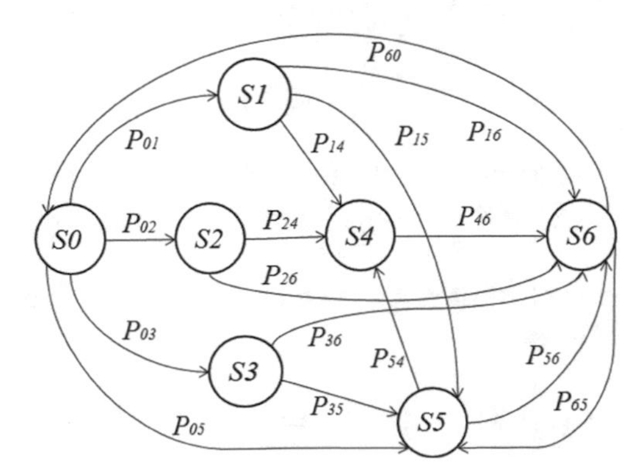

Fig. 1. The graph of the repaired locomotive's technological conditions.

During repair, a locomotive can be on the following positions: S0 – turn for repairs (the locomotive's location on the traction paths of the depot); S1 – maintenance area TO-2; S2 – current repair's area TR-1 and TP-2; S3 – current repair's area TR-3; S4 – area of the wheel pair's turning; S5 – unplanned repairs' area (UR); S6 – locomotive's issue and its reception of the squad. And each position determines an appropriate locomotive's condition during the depot's repair. The transition process of the repaired locomotive from the current state Si (the repaired locomotive's movement from one technological position to another) is determined by the probability Pij. According to the interconnection of the locomotive's technological movements in depot, it can be represented as a matrix.

$$G_k = \begin{pmatrix} - & (P_{12}, t_{12}) & \ldots. & (P_{1j}, t_{1j}) & \ldots. & (P_{1n}, t_{1n}) \\ (P_{21}, t_{21}) & - & \ldots. & (P_{2j}, t_{2j}) & \ldots. & (P_{2n}, t_{2n}) \\ \ldots. & \ldots. & \ldots. & \ldots. & \ldots. & \ldots. \\ (P_{i1}, t_{i1}) & (P_{i2}, t_{i2}) & \ldots. & (P_{ij}, t_{ij}) & \ldots. & (P_{in}, t_{in}) \\ \ldots. & \ldots. & \ldots. & \ldots. & \ldots. & \ldots. \\ (P_{n1}, t_{n1}) & (P_{n2}, t_{n2}) & \ldots. & (P_{nj}, t_{nj}) & \ldots. & - \end{pmatrix} \begin{matrix} , \sum_{i=1}^{n_i} P_{ij} = 1 \\ , \sum_{j=1}^{n_i} P_{ij} = 1 \\ , k = 1, \ldots, m, \end{matrix} \tag{1}$$

where n – the number of the graph vertexes (technological areas), $n = 7$;

$P_{ij} = [0, 1]$ – the technological connection between the i-th and j-th positions (transition probability): $P = 1$, if the operations are performing consistently and unconditionally, $P = 0$ if the connection is missing;

t_{ij} – the time of the object's moving from the i-th to j-th position, minutes.

Herewith the time of the technological operations' implementation is determined by the matrix:

$$OT = \begin{pmatrix} \left(ot_{11}^{\min}, ot_{11}^{\max}\right) & \left(ot_{12}^{\min}, ot_{12}^{\max}\right) & \ldots. & \left(ot_{1m}^{\min}, ot_{1m}^{\max}\right) \\ \left(ot_{21}^{\min}, ot_{21}^{\max}\right) & \left(ot_{22}^{\min}, ot_{22}^{\max}\right) & \ldots. & \left(ot_{2m}^{\min}, ot_{2m}^{\max}\right) \\ \ldots. & \ldots. & \ldots. & \ldots. \\ \left(ot_{i1}^{\min}, ot_{i1}^{\max}\right) & \left(ot_{i2}^{\min}, ot_{i2}^{\max}\right) & \ldots. & \left(ot_{im}^{\min}, ot_{im}^{\max}\right) \\ \ldots. & \ldots. & \ldots. & \ldots. \\ \left(ot_{n1}^{\min}, ot_{n1}^{\max}\right) & \left(ot_{n2}^{\min}, ot_{n2}^{\max}\right) & \ldots. & \left(ot_{nm}^{\min}, ot_{nm}^{\max}\right) \end{pmatrix}, \tag{2}$$

where $ot_{ij}^{\min}, ot_{ij}^{\max}$ – the minimal and the maximal time of the operation's implementation on the i-th position for the j-th type of the elementary stream (repaired object), minutes.

The time of the technological preparation and the repair position's changeover by changing of the repair object:

$$pt_{ij} = \left(t_{ij}^{(1)}, t_{ij}^{(2)}, \ldots, t_{ij}^{(n)}\right), i = 1, \ldots, m, j = 1, \ldots, m \tag{3}$$

where pt_{ij}^{k} – the average time of the k-th position's changeover by changing of the repair object from the *i*-th for the *j*-th type ($t = 0$, if the preparation and the changeover are not required).

The formation of the request's flow for servicing in form of the request's flow for repair with intensity of the admission in repair, 1/day (hour).

$$Pot = (pot_1(t), \ldots, pot_m(t)), t = 1, \ldots, 24 \tag{4}$$

The daily program of repair:

$$\overline{Pot}(t) = (\overline{pot_1}(t), \ldots, \overline{pot_m}(t)) = \frac{1}{30}(vol_1, \ldots, vol_m), t = 1, \ldots, 30 \tag{1}$$

For the model's creation, a software product AnyLogic was chosen. As a method of the imitation modeling, an agent-based modeling is used. The logic of the depot's functioning was projected using the built-in libraries: «The Railway Library» and «The Process Modeling Library» [14, 15]. In the modeling process, different emergency situations were taken into account (unscheduled repair, unscheduled electro locomotive's moving on the depot's traction ways, technological operation's volume, recovery time of the repair object's efficiency etc.).

The simulation model's logic circuit within AnyLogic is presented on the Figs. 2 and 3, which are associated transitions A, B, C, and D.

As the values in the formation of the flow of requests for repair in conditions of the partial uncertainty, but provided that these events occur with a certain fixed average intensity and are independent of each other, Poisson's law is applied [16]. In our case, the probability of the receipt in repair of the N objects for the k-th repair's type for time Δt will be distributed according to the law:

$$p_N = \frac{(\overline{pot}_k \cdot \Delta t)^N}{N!} \cdot e^{-\overline{pot}_k \cdot \Delta t}. \tag{6}$$

With the help of the block groups 1 (Fig. 2), for each repair type, the requests are forming in form of the created modeling agent (agent_eps), which is a repaired unit in our system. The requests TP-1 and TP-2 are combined further into one stream, as the technological operations, this type of repair in depot is performed in the same workshops. At an exit of a request from blocks, a parameter is given for each agent_eps, for each a value keeps inside an agent and sets a value of a repaired unit, which is remaining during the modeling. This is necessary so that the created agent proceeds further in the simulation process to the required delay unit in the system in accordance with the type of repair.

The blocks of group 2 introduce an agent on the model's graphic diagram, if all the stabling areas are occupied and they are necessary for this type of repair, then a queue of blocks 3 will be formed. The usage of the stabling areas is controlled by blocks 4, which have a value accordingly to the number of the stabling areas in industrial workshops.

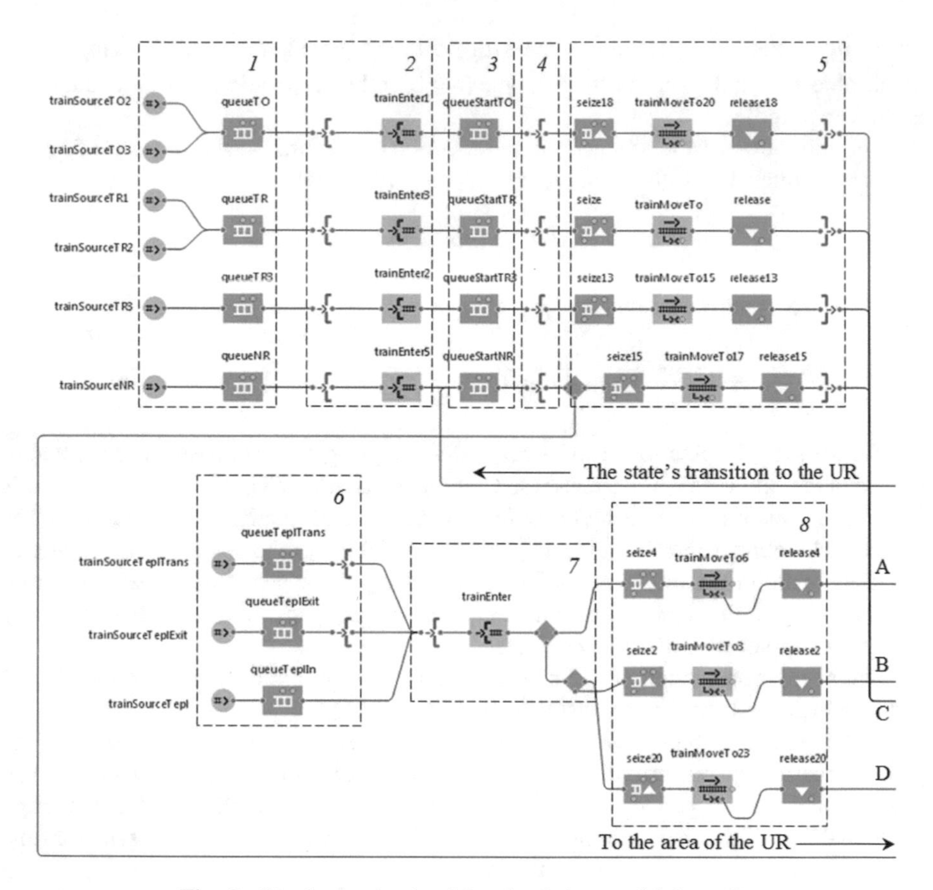

Fig. 2. The logic circuit of the simulation model (part 1).

The blocks of group 5 simulate the repaired electric locomotive's movement to the necessary industrial workshops, accordingly to the type of repair. During the exit from blocks of group 5, agent_eps creates a signal for the appearance of the shunting locomotive (agent_loko) in the agent's model. During manufacture, the shunting locomotive is necessary for the putting of the repaired rolling stock into the workshop of the industrial site.

A request of an agent_loko is formed by blocks 6. The blocks 7 determine an agent_eps (electric locomotive is waiting for the entering the workshop) requiring to make the maneuver. The movement to the necessary repaired electric locomotive is modeling by the help of the blocks 8.

In the block 9 (Fig. 3), the modeling processes of the shunting locomotive's coupling and repair object, the moving to the workshop and the uncoupling are realized. Further, the locomotive models move to the point of its location and leave the model to the following request of the agent_eps (block 10).

After the modeling processes of the locomotives uncoupling the agent comes to blocks 11, where a system's delay is forming (there is modeling an execution time of the technological operations on repair area) in accordance with the Guide [17].

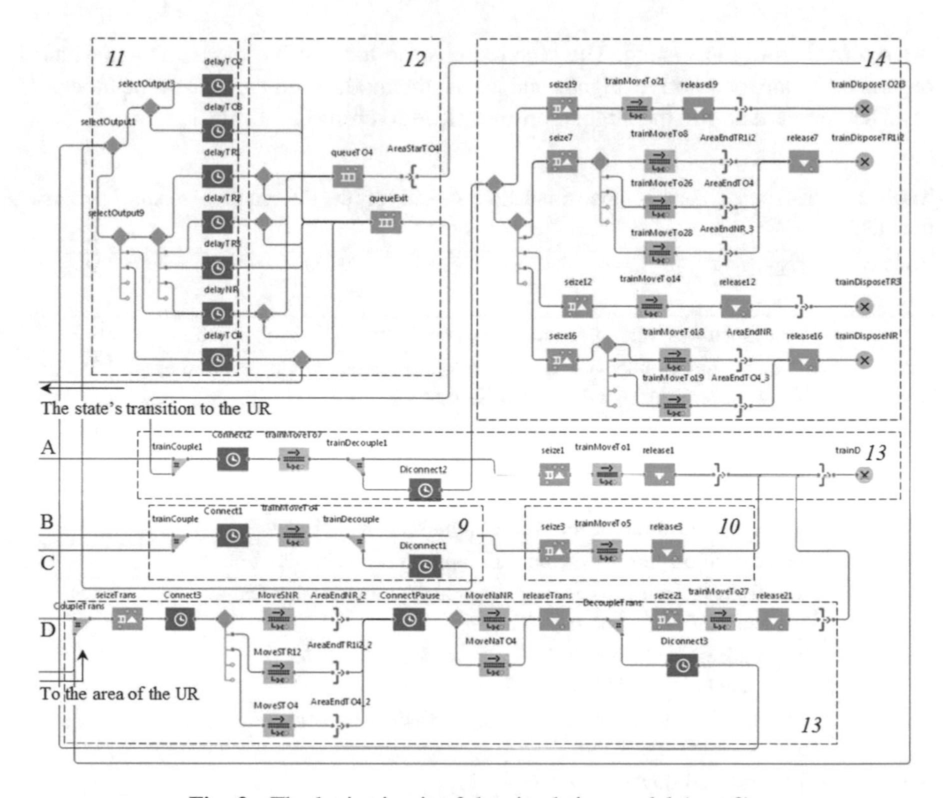

Fig. 3. The logic circuit of the simulation model (part 2).

By the unscheduled repair and by the operation of the wheel pairs' lathe work TO-4 the requests delay accordingly to the special case of the gamma distribution – Erlang's distribution (Fig. 5) [18].

$$P_t = \frac{(t - t_{\min})^{m-1}}{\beta^m \Gamma(m)} \exp\left(\frac{-[t - t_{\min}]}{\beta}\right), \tag{7}$$

where m – the factor of the probability distribution's form; β – the scale's coefficient.

With a certain probability Pij and in accordance with the graph on the Fig. 1, the repaired electric locomotive could need to change the technological position because of the appearance of the unscheduled types of the recovery operations (leaving for the unscheduled repair, necessities of the wheel pairs' lathe work etc.). This probability of the transition is shown in blocks 12. If it needs a transition to another repair position, the agent_eps takes a queue and as soon as the necessary repair position becomes free, it calls the agent_loko for the moving with the help of the blocks 13.

If the repaired locomotive doesn't need a transition to another position, a request passed the maintenance and is ready for the operations, then the blocks 15 model the repaired object's moving from the workshop by the shunting locomotive, and also an uncoupling is modeling, and further a moving of the agent_loko to the location place and the removal from the system. The repaired electric locomotive passes TO-1, which is organized by the locomotive brigade and leaves the model with the help of the blocks *16*.

The source data for the simulation model is given in the Table 1.

Table 1. The source data of the simulation model of the locomotive repair enterprise's functioning.

No in order	Name	Value
1	The execution time of the maintenance service TO-2 [17], hour	3
2	The average number of the electric locomotives, which passes the TO-4, during the month, $(\overline{pot_{TO-4}} \cdot \Delta t)$	25.42*
3	The average number of the electric locomotives, which needs NP, during the month, $(\overline{pot_{UR}} \cdot \Delta t)$	183.71*
4	The execution time of the current repair TP-1 [17], hour	18
5	The execution time of the current repair TP-2 [17], day	3
6	The execution time of the current repair TP-3 [17], day	6
7	The form factor of the Erlang's probability distribution for the implementation of UR and TO-4, m	2*
8	The scale's coefficient of the Erlang's distribution for the implementation of UR, β	28.15*
9	The scale's coefficient of the Erlang's distribution for the implementation of TO-4, β	2.88*
10	The minimal downtime on TO-4, $t_{\min}$	2*
11	The rolling stock's speed on the depot's traction ways [19], km/h	5
12	The rolling stock's acceleration and braking on the depot's traction ways, m/s^2	0.5
13	The shunting locomotive's speed at the approach to the repaired unit [19], km/h	3
14	The time of the rolling stock's traction and separation [20], min	3
15	The transition probability from area TP-1 and TP-2 to TO-4, %	20.1*
16	The transition probability from area TP-3 to area TO-4, %	0.22*
17	The passing probability of TO-4 by NP, %	3.16*
18	The transition probability from area TO-4 to NP, %	0.26*
19	The number of the repair positions of the TO-2 area, number	2
20	The number of the repair positions of the TP-1 and TP-2 area, number	10
21	The number of the repair positions of the TP-3 area, number	1
22	The number of the repair positions of the TO-4 area, number	1
23	The number of the shunting locomotives, number	1

*The parameters are calculated based on the gathered depot's statistical data

The degree of using the technological areas of the enterprise on time is determined with the load coefficient k_3^{TA}, for this operation it defines by the formula:

$$k_3^{TA} = \frac{\sum_{i=1}^{n} T_i}{T_o \cdot n} \tag{8}$$

where n – the number of the technological positions on the area (number of the workshop's stabling areas);

T_i – the location's time of the electric locomotive on the technological position;

T_o – the total duration of the investigated time period

3 Results and Calculation

The change of the load coefficient's value k_3^{TA} during the time for each technological area is shown in thc Fig. 4.

In the figure, the time of the model's becoming t_{mb} can be seen, after that, a transition constant (working) mode comes. The time parameters of the model's formation and the average load coefficient for each type of repair and service are shown in the Table 2.

Table 2. The time parameters of the model's formation and the average load coefficient for each type of repair and service.

№ in order	Parameter	Area TO-2	TP-1 and TP-2	Area TP-3	Area TO-4	Area UR
1	The time of the model's becoming t_{mb}, hour	19	98	380	200	300
2	The average load coefficient of the technological area $k_{3.med}^{TA}$	0.448	0.487	0.74	0.372	0.813

The locomotive's timeout for repair is the locomotive repair enterprise's efficiency indicators t_{queue}. The average time of being in queue is determined by the formula.

$$t_{queue} = \frac{\sum_{i=1}^{n} t_{queue}}{n_{eps}} \tag{9}$$

where t_{queue} – the location's time of the i-th locomotive in the queue;

n_{eps} – the number of the locomotives which came for repair in depot.

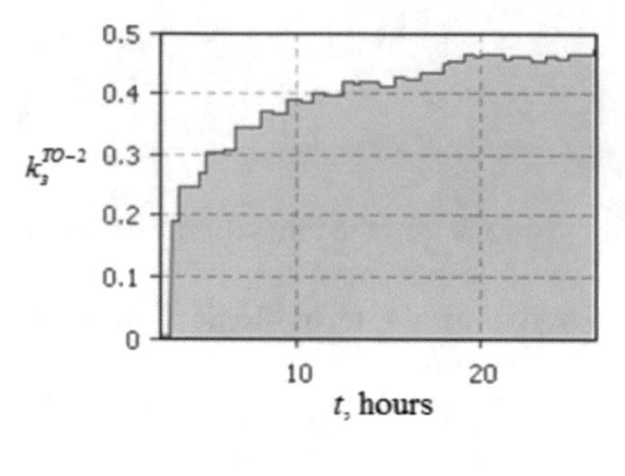

(a) the value's change $k_з^{my}$ during the day for the service area TO-2

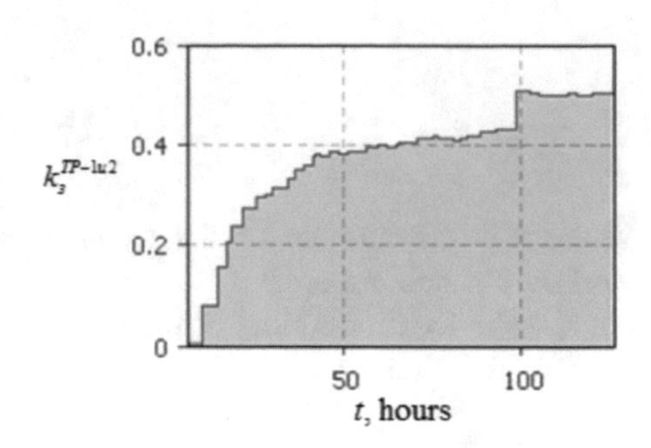

(b) the value's change $k_з^{my}$ during the week for the repair area TP-1 и TP-2

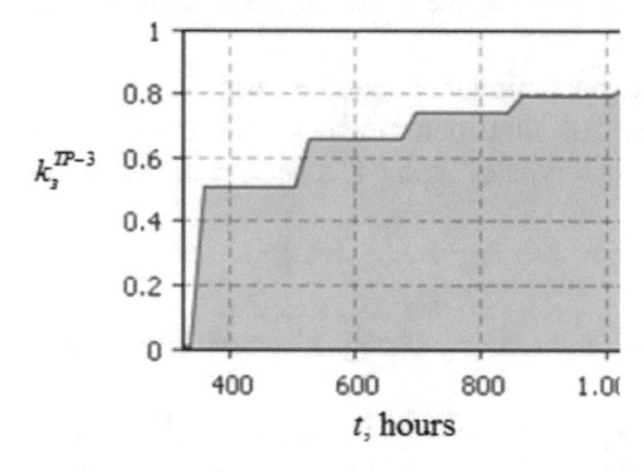

(c) the value's change $k_з^{my}$ during the quarter for the repair area TP-3

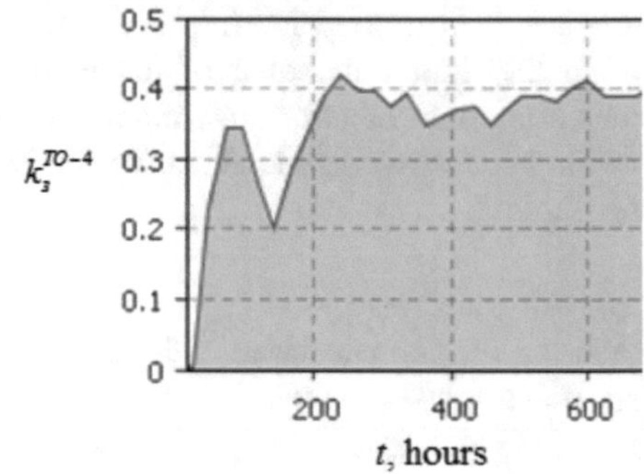

(d) the value's change $k_з^{my}$ during the month for the service area TO-4

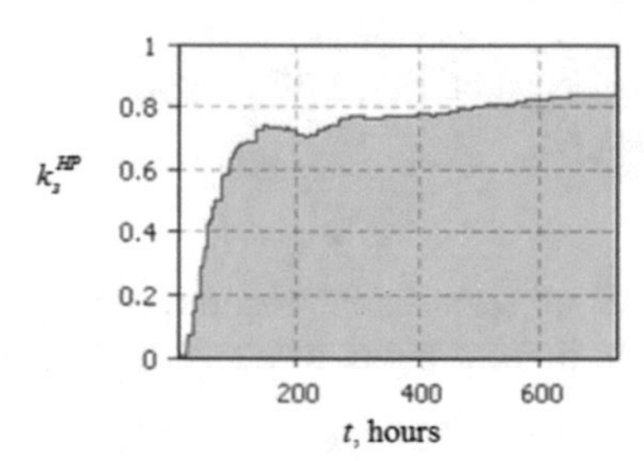

(e) the value's change $k_з^{TA}$ during the month for the area UR

Fig. 4. The value's change of the technological area's load coefficient k_3^{TA} during the time for each type of repair and service

The statistics of the timeout's distribution in the queue for repair with the help of the histogram is shown in the Fig. 5.

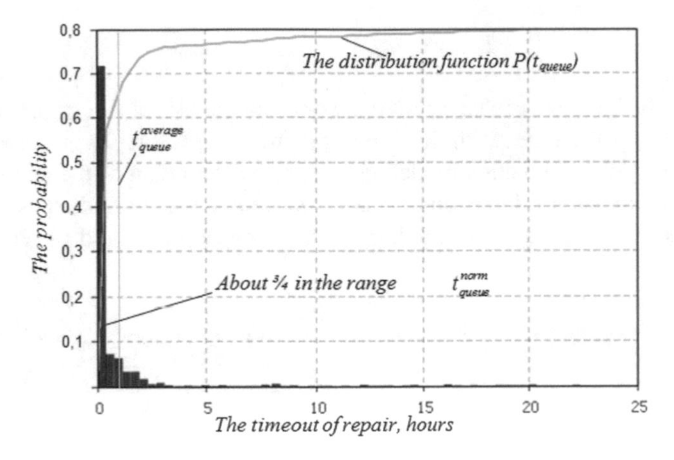

Fig. 5. The timeout's distribution in the queue t_{queue} on TO and TP

As a norm, we take waiting time of a locomotive in a queue for repair t^{norm}_{queue} – 24 min. In the diagram of the probability's distribution, we can see that 71.56% of the rolling stock is waiting for repair less or equal to the value t^{norm}_{queue}. In Table 3, the measuring parameters in the distribution of the probability of waiting time for repair are given.

Table 3. The measuring parameters in the distribution of the probability of waiting time for repair.

№ in order	Parameter	Unit of measurement	Value
1	The number of measurements, N_m	–	1 089
2	The mathematical expectation, t^{med}_{queue}	hour	1.167
3	The minimal value, t^{min}_{queue}	hour	0
4	The maximal value, t^{max}_{queue}	hour	22.379
5	The number of intervals, N_{int}	–	56
6	The standard deviation, σ_{queue}	hour	2.524
7	The confidence interval for an average, $\bar{x}_{queue}$	hour	0.151
8	The amount of the dimension values, $\sum t_{queue}$	hour	890.25
9	The model's run time, T_o	day	41

One of the sources of the production losses in locomotive repair production is time spent on the implementation of the locomotive's setting to the workshop, output of the repair's area, waiting for repair etc. [21]. For each repaired unit, the temporary losses will be equal:

$$T_{losses} = t_{losses} + t_{move} + t_{position} \tag{10}$$

where t_{move} – the time of the shunting work on the depot's territory (setting and output); $t_{position}$ – the timeout in the queue by changing the repair position.

In Fig. 6, by the simulation modeling results, a calculation of the temporary losses for each repaired electric locomotive was given, and it was presented as a probability density's function. The distribution diagram is given by each kind of repair in depot.

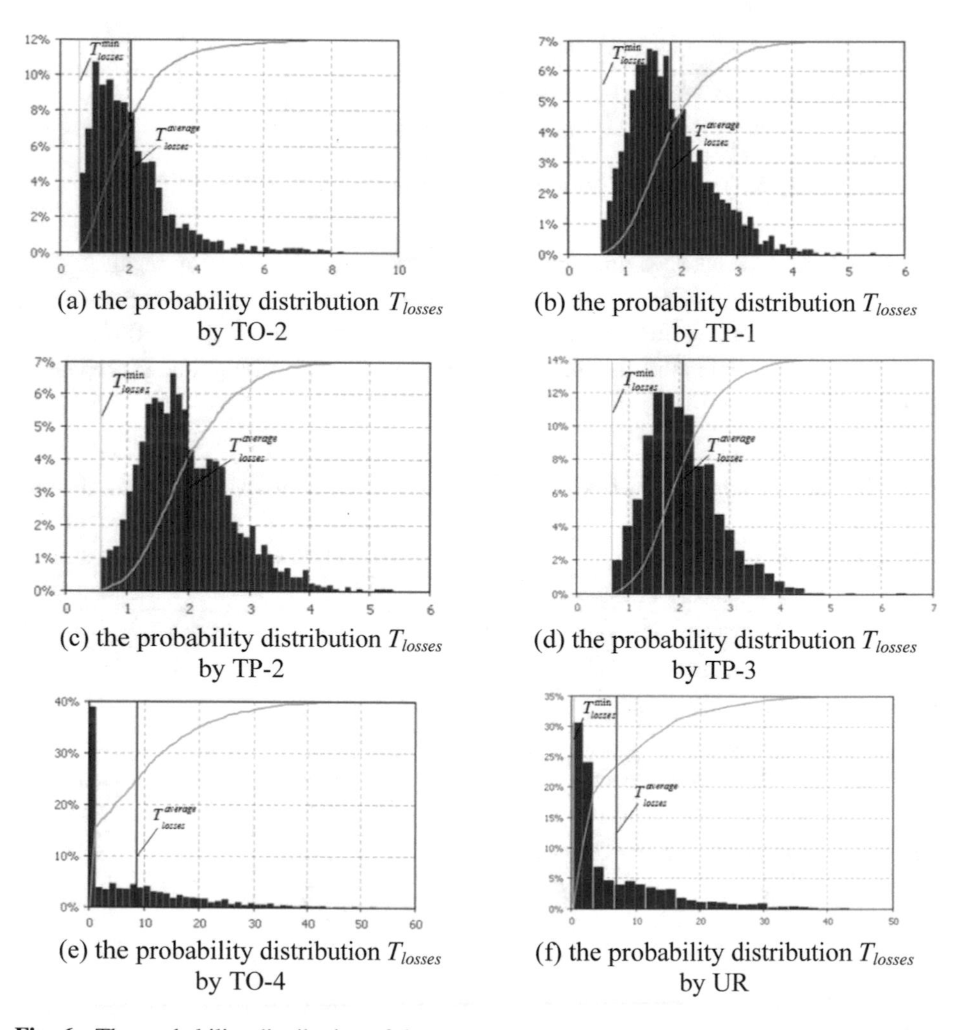

(a) the probability distribution T_{losses} by TO-2

(b) the probability distribution T_{losses} by TP-1

(c) the probability distribution T_{losses} by TP-2

(d) the probability distribution T_{losses} by TP-3

(e) the probability distribution T_{losses} by TO-4

(f) the probability distribution T_{losses} by UR

Fig. 6. The probability distribution of the temporary losses T_{losses} of the locomotive's repair in depot

The figures in the diagrams show that for the time loss Tlosses of the maintenance service TO-2 and the current repair TP-1, TP-2 and TP-3, we can see a positive sloping

Table 4. The measurements' parameters in the distribution of the time loss' probability of the locomotive during repair at the enterprise by kinds of TO and TP.

№ in order	Parameter	TO-2	TP-1	TP-2	TP-3	TO-4	Unscheduled repair
1	The number of measurements N_m	1709	1784	1719	1705	1709	1720
2	The number of intervals N_{int}, hour	39	49	48	29	44	27
3	The mathematical expectation T_{losses}^{med}, hour	2.066	1.827	1.981	2.083	8.496	6.886
4	The minimal value T_{losses}^{min}, hour	0.635	0.639	0.635	0.758	0.365	0.697
5	The maximal value, T_{losses}^{max}, hour	8.319	5.447	5.341	6.465	51.969	42.233
6	The standard deviation σ_{losses}, hour	1.164	0.714	0.74	0.722	9.513	7.708
7	The confidence interval for an average $\bar{x}_{losses}$, hour	0.055	0.033	0.035	0.034	0.451	0.364
8	The amount of the measurement values $\sum t_{losses}$, hour	3531.49	3260.10	3406.12	3552.00	14519.0	11844.7
9	The time of the model's running T_o, hour	2435.85	4400.00	69104.5	287432	29061.5	15418.9

asymmetry of the distribution, which is close to the binomial law, with a brake at the left. While for the maintenance service TO-4 and unplanned repairs, we can see an exponential law of the distribution.

4 Conclusions

Breaks and offsets on the histograms tell about the minimal values of the time loss T_{losses}^{min} of the received for repair rolling stock, which cannot be reduced because of the depot's structure (overcoming of the depot's traction ways, the time of the way's overcoming by the shunting locomotive, coupling and uncoupling time etc.). The measurements' parameters in the distribution of the time loss' probability T_{losses} of the locomotive during the maintenance and repair are given in Table 4.

Thus, this work presents a simulation model of the service locomotive depot Moskovka, which was developed in the AnyLogic computer environment. The overall assessment of the technological indicators of the enterprise work was made, and also the values of the technological areas' load factor k_3^{TA} were determined. The production losses of the locomotive repair depot's technological process were calculated, and the timeout of repair by the rolling stock was determined.

Further, it is necessary to study the effect of the technological structure of the depot by changing the technological process of repair to determine a more efficient state of production.

References

1. Smirnov, V.A., Panov, K.V.: Sci. Probl. Transp. Siberia Far East **1–2**, 47–50 (2014)
2. He, D., Luo, A., Deng, J., Tan, W.: Comput. Integr. Manuf. Syst. CIMS **24**, 1155–1161 (2018)
3. Datsun, Y.: Eastern Eur. J. Enterp. Technol. **7**, 56–61 (2016)
4. Ismailov, S.K., Gatelyuk, O.V., Selivanov, Y.I., Bublik, V.V., Talyzin, A.S.: Sci. Probl. Transp. Siberia Far East **1**, 398–402 (2009)
5. Cho, D.I., Parlar, M.: Eur. J. Oper. Res. **51**, 1–23 (2011)
6. Sherif, Y.S., Smith, M.L.: Naval Res. Logist. Q. **28**, 47–74 (2007)
7. Nakagawa, O.T.: IEEE Trans. Reliab. **25**, 284–287 (2006)
8. Pierskalla, W.J., Voelker, J.A.: Naval Res. Logist. Q. **23**, 353–388 (2005)
9. Valdez-Flores, C., Feldman, R.M.: Naval Res. Logist. **36**, 419–446 (2006)
10. Myamlin, V.V.: Transp. Russian Fed. **4**, 57–60 (2013)
11. Bannikov, D., Sirina, N.: 10th International Scientific and Technical Conference "Polytransport Systems", vol. 216, p. 02018 (2018)
12. Golovash, A.N.: Railway Transp. **10**, 54–56 (2016)
13. Domanov, K.I.: The technological support of repair and increase of the railway rolling stock's dynamic quality. In: The proceedings of the IV all-Russian Scientific and Technical Conference with International Participation, Omsk, pp. 97–103 (2017)
14. The Process Modeling Library AnyLogic [Electronic resource]. https://www.anylogic.com/resources/libraries/process-modeling-library/. Accessed 31 Oct 2018
15. The Rail Library AnyLogic [Electronic resource]. https://www.anylogic.com/resources/libraries/rail-library/. Accessed 31 Oct 2018
16. Tatashev, A.G., Akhil'gova, M., Shchebunyaev, S.A.: T-Comm: Telecommun. Transp. **11**, 23–26 (2017)
17. About a system of the locomotives' technical maintenance and repair Open Society RZHD. The decree of the Society RZHD of 17 January 2005, vol. 15, p. 3 (2005)
18. Yanovskiy, G.G., Sokolov, A.N.: Inf. Commun. Technol. **6**, 27–30 (2014)
19. The rules of the railways' technical operation of Russian Federation (PTE), vol. 36 (2018)
20. The technical and administrative act of station Moskovka, Omsk, vol. 143 (2005)
21. Myamlin, V.V.: Bull. Rostov State Transp. Univ. **4**, 77–86 (2013)

Business Requirements to the IT Architecture: A Case of a Healthcare Organization

Igor Ilin, Anastasia Levina(✉), Aleksandr Lepekhin, and Sofia Kalyazina

Institute of Industrial Management, Economics and Trade, Peter the Great St. Petersburg Polytechnic University, 195251 St. Petersburg, Russian Federation
alyovina@gmail.com

Abstract. Nowadays, in view of ongoing digital transformation and new business capabilities provided by IT, almost all industries are forced to adapt and restructure the business model to ensure survival and competitiveness in the long run. The correct business requirements to the IT systems and IT infrastructure are the key factor of the successful implementation of IT projects. These business requirements should read the trends of the industry, the whole business environment, the new possibilities of IT technologies for a particular industry. The paper describes the way for setting business requirements for a new type of healthcare organization, which would become a prerequisite for further implementation of appropriate IT systems and digital technologies. The management model of a healthcare organization is additionally imposed with requirements and specific trends in the healthcare industry: a transition to the principles of value-based and personalized medicine. In this regard, the urgent task is to form such a model of management of a medical organization that, on the one hand, read current trends in the field of healthcare, and on the other, follow the latest IT trends and would provide opportunities for the effective use of modern digital technologies.

Keywords: IT architecture · Healthcare organization · Business IT alignment · Digital healthcare

1 Introduction

The leading ideological concepts, under the influence of which the modern health care system is being formed, are value medicine, personalized medicine, the concept of Health 4.0. Value medicine is a result-oriented approach to the organization of the system of medical care, involving the choice of the method of patient management, which at a lower cost allows achieving better results, including from the point of view of the patient. Value medicine focuses on the value of the procedure for the patient and the cost-effectiveness of medical interventions, which are not taken into account in classical evidence-based medicine. Personalized medicine involves the selection and organization of an individual patient's trajectory of treatment based on his individual characteristics, which requires personalized accounting and processing of data about

V. Murgul and M. Pasetti (Eds.): EMMFT-2018, AISC 983, pp. 287–294, 2019.
https://doi.org/10.1007/978-3-030-19868-8_29

each patient, safe automation of such an assessment and ensuring data availability. The implementation of the principles of personalized medicine involves the effective collection, processing, analysis of a large amount of primary data about each patient in real time. The concept of Health 4.0 implies the use of modern digital technologies (big data management, Internet of Things, blockchain, telemedicine, predictive analytics, etc.) to increase the economic and medical efficiency and accessibility of medical care. In general, improving the efficiency of medical care is an important indicator. To achieve this result, the control architecture, including the architecture of information systems and applications, must meet certain requirements. The task of this work is to formulate business requirements for a new type of medical institution. The implementation of these requirements will affect both the economic efficiency of individual medical organizations and create the prerequisites for the effective integration of medical organizations into a single information space in the field of healthcare.

One of the ways to improve the system of rendering high-tech and specialized medical care is the introduction of modern management technologies into practical healthcare (service-oriented architecture, process management, e-health, big data processing technologies, predictive analytics systems, telemedicine technologies, Internet of Things, etc.). Efficient information exchange between the components of the information technology architecture is required. Management technologies, including information and communication, have a significant impact on the quality and cost-effectiveness of medical care. The provision of medical care that is adequate in the patient's opinion from a medical and economic point of view to a particular case creates the prerequisites for the effective use of resources of medical organizations. Modern management technologies, including digital, have significant potential in solving a number of problems on the way to providing more affordable, cost-effective and high-quality medical care. The requirements of compliance with the principles of value and personalized medicine should be reflected in the business process model of a medical organization. The possibilities of applying the technologies of the Health 4.0 concept set the requirements for the IT support structure and the technological infrastructure that ensures the implementation of the processes. The management system of a medical organization should include a system of business processes, the structure of information systems and applications, and their hardware. Reforming the activities of medical organizations, due to the need to introduce modern management technologies, in accordance with the concept of enterprise architecture, should be carried out systematically, taking into account the interconnections and interdependencies of all elements of the organization's management system. This makes it important to develop a methodology for the formation of a corporate architecture of medical organizations, based on the functional structure of activity and including a model of a process system, IT architecture, service architecture, and technological architecture. Enterprise architecture reflects the main aspects of business, IT and its evolution while ensuring flexibility and adaptability [1].

2 Materials and Methods

Monitoring of research and analysis of publications on the management of modern medical organization during 2015–2018 showed the lack of integrated approaches aimed at developing an integrated model of activity of medical organizations that meets modern trends in the development of both health care and management technologies, including digital technologies. There is a large amount of research aimed at the realization of individual elements of such medical organizations control system:

1. Implementation of the concept of value and personalized medicine.
   Most of the studies in the field of value medicine are devoted to the analysis of the capabilities and features of this approach against the paradigm of evidence-based medicine prevailing before it, as well as the possibilities of obtaining feedback from patients and their evaluation of the medical services provided [2]. A significant amount of work on the introduction of modern management technologies is devoted to medical organizations that provide services for people with specific diseases (cancer, diabetes, cardiovascular diseases). The presence of technology for analyzing patient satisfaction is mentioned quite often [3, 4]. In addition, often in publications there is a description of specific telemedicine devices that allow monitoring of the patient's condition [5, 6].
   In Russia, the development of the principles of value and personalized medicine is actively involved in researchers and practitioners of the Almazov National Medical Research Centre under the leadership of Shlyahto E.V. and Konradi A.O. [7–9]. This group of specialists in their research focuses on the medical efficacy of treating patients and introducing the principles of value and personalized medicine into the mainstream processes, rather than through reforming the management system.
2. The use of digital technology.
   A large number of publications on the introduction of modern management technologies, including IT technology, made in the USA. The most important reason for the relevance of the introduction of such technologies is the low level of accessibility of medical care in remote areas of the country, namely the problems associated with a vast territory, as in Russia [10]. Telemedicine publications describe success in this area in many directions, including gaps in medical care at night, the possibility of receiving medical services in special places (prisons), and the possibility of receiving remote teleconsultation [11]. Separate studies on the implementation of telemedicine systems and compatible technologies (Big Data, Internet of Things, electronic patient record) are conducted in Brazil, India, and China, since this problem is extremely relevant for geographically distributed countries with an uneven distribution of specialized treatment centers. In particular, the ideas of introducing modern technologies into practical medicine and related IT solutions are actively developed by an international team of researchers led by Brazilian scientist Rodrigues [12–14].
   Regarding the application of certain digital technologies in medicine prevail articles mostly from North America dedicated to the description of individual projects and experience in implementation of various technologies into the practice of health care organizations and individual processes of health care delivery system. For

example, the Project Artemis is being implemented at Toronto Children's Hospital [15]. The project has developed an information system that collects and analyzes data on infants in real time. The system monitors every second 1260 indicators of the state of each child. Based on the collected data, they are analyzed, deviations from the specified standard indicators are identified, which allows predicting the unstable state of the child and taking measures to organize the prevention of diseases in children.

The research method is based on the concept of the enterprise architecture as a systematic approach of harmonization of business and IT [16, 17]. Discipline of the enterprise architecture identifies the following groups of elements (layers) of a business management system: business architecture, IT architecture, data architecture, technological architecture. Business architecture involves the description of all groups of processes – core processes, management processes, supporting processes, with an appropriate level of decomposition into subprocesses. The business architecture forms the IT support requirements for the enterprise's activities, which are implemented by the IT architecture (information systems and applications).

The introduction of technologies that ensure the realization of the ideas of value and personalized medicine is the basis of the concept of Smart Hospital. The concept of Smart Hospital is based on optimized and automated processes supported by modern IT technologies aimed at improving existing procedures and introducing new capabilities for patient care. Implementing solutions that support the concept of Smart Hospital is aimed at, among other things, reducing costs while maintaining the required level of quality of care. When introducing this concept, attention is equally paid to innovation, research, technology implementation, process automation, etc. [18]. When building a complex system, which is the Smart Hospital, the main task is not to determine the requirements for information systems, but to determine the highest priorities for the business (business requirements).

Based on these tasks, it is necessary to make a decision on how to build an integrated management architecture. Business requirements are tasks of a high level of organization or client. They, as a rule, describe the global directions of business

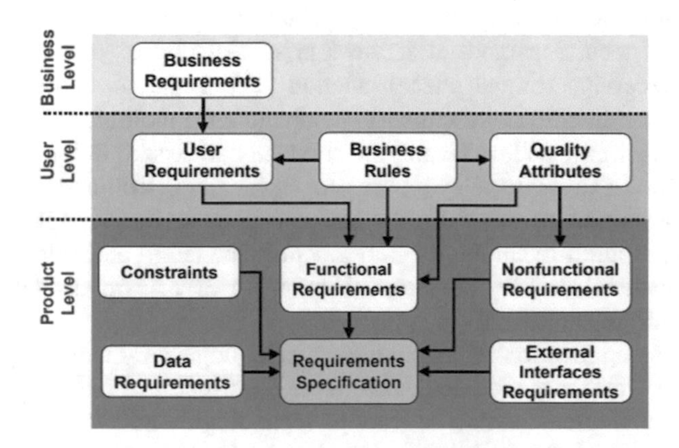

Fig. 1. Requirements for the different levels of the target architecture [19].

development. These requirements are fixed as the main vision of the target architecture, on the basis of which all subsequent requirements for the different levels of the target architecture are determined (Fig. 1).

Requirements engineering allows analyzing business needs from top-level services to the requirements of individual business process functions. Building a system of business requirements is the basis for building a complex of IT support.

Requirements are a preliminary assessment of architectural development, the definition of a vision for a future architecture, and the definition of approaches to architecture management. The type of system and the purpose of the project significantly influence the way of identifying requirements and the formation of a set of sources of requirements.

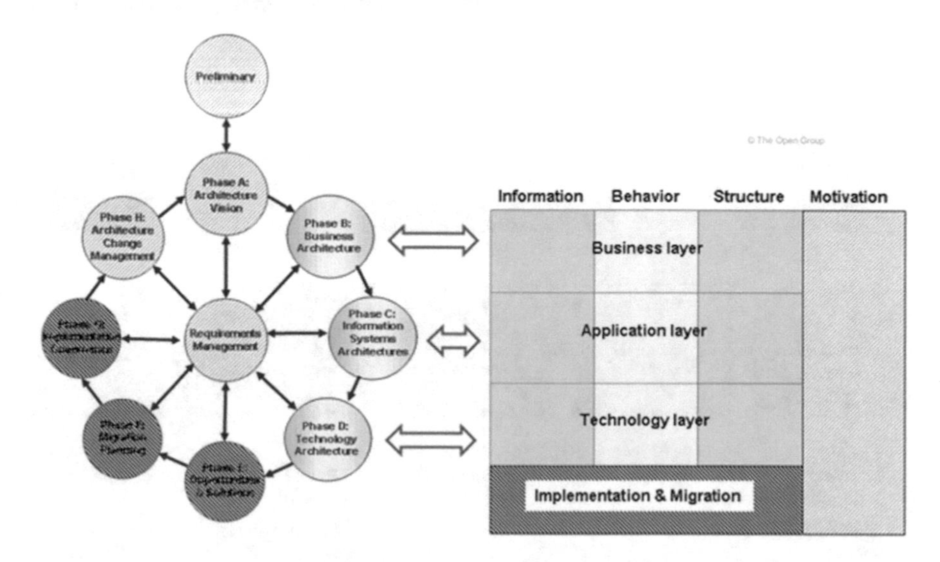

Fig. 2. Requirements definition and management system [20].

Through the development of requirements, a system is formed that satisfies the needs of customers and users, defining requirements and managing them in a systematic way (Fig. 2).

3 Results

Business requirements for Smart Hospital were proposed by Michael Porter in the «A Strategy for Health Care Reform—Toward a Value-Based System» [21]. These are requirements included the following global goals of the organization:

1. the measurement and dissemination of health results should be mandatory for each health care provider;
2. results should be measured throughout the entire cycle of medical care, and not separately for each intervention;

3. results should be adjusted for baseline patients to eliminate bias in patients with difficult cases;
4. it is necessary to measure true health outcomes, rather than relying solely on technological measures;
5. restructuring of health care providers: transition to integrated practical units that cover all the skills and services needed for a full cycle of care for each health condition, including common coexisting conditions and complications. Such units should include outpatient and inpatient care, testing, training and coaching, and rehabilitation;
6. electronic medical records will increase value, but only if they support integrated observation and measurement of results, as well as an architecture for aggregating data for each patient over time and between providers and protocols to ensure seamless communication between systems.

The figure below shows the past, present, and future of medical organizations in the context of a changing market in the vision of Harvard Business School (Fig. 3) [21].

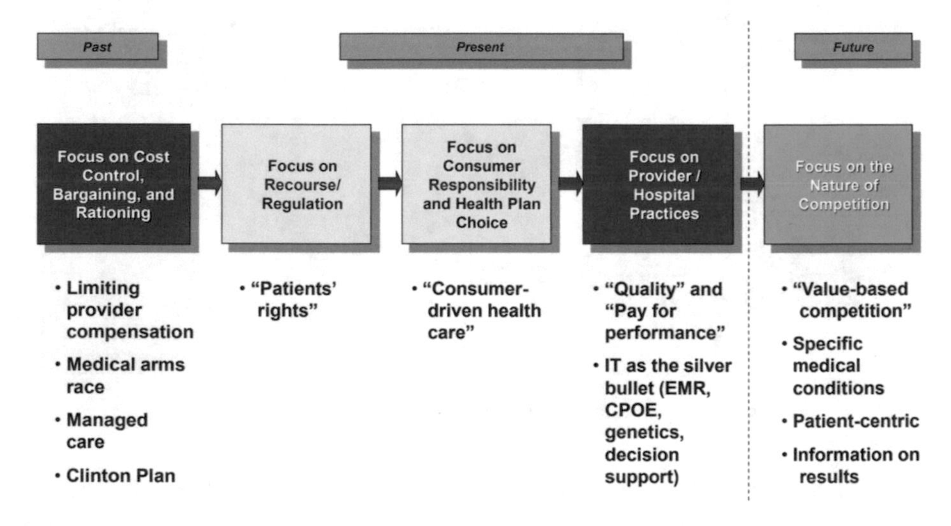

Fig. 3. The past, present, and future of medical organizations in the context of a changing market [22].

As a result of reviewing business requirements and working with the previously formulated list of requirements, the following groups of business requirements for the digital architecture of a medical organization are identified:

- Health and Patient Support Information;
- Restructuring of Relations with Medical Service Providers;
- Redefining Contracts, Transactions, Billing, and Pricing;
- Patient Medical Records.

In the group «Health and Patient Support Information», the following requirements can be highlighted:

- organization of medical conditions rather than administrative functions;
- organization and collection of information on providers and treatment methods;
- patient support in choosing a medical organization through information and impartial counseling;
- organization of information and interaction with the patient through the full cycle of treatment;
- provision of disease prevention services.

The «Restructuring of Relations with Medical Service Providers» group is also quite a complex and can be divided into 2 basic requirements:

- transformation of the nature of information exchange;
- encouraging professionals to provide medical services and add value to health innovations.

In the «Redefining Contracts, Transactions, Billing, and Pricing» group, the following business requirements can be stated:

- simplification, standardization, and elimination of documents and transactions;
- transition to multi-year subscription contracts.

The «Patient Medical Records» group can be explained as providing services for aggregating, updating and verifying complete patient medical records according to strict confidentiality standards.

The articulated groups of requirements will form the basis for developing requirements for software and designing the architecture of Smart Hospital applications.

4 Conclusion

The digital architecture of a medical organization, built according to business requirements, taking into account trends in the development of health care, is a source of increasing business efficiency, consolidating sources and resources to achieve goals, ensuring the security of processed data, and meeting strict business standards.

Acknowledgement. The reported study was funded by RFBR according to the research project №. 19-010-00579.

References

1. Borremans, A.D., Zaychenko, I.M., Iliashenko, O.Yu.: Digital economy. IT strategy of the company development. In: MATEC Web of Conferences, vol. 170, Article no. 01034 (2017)
2. Little, M., Lipworth, W., Gordon, J., Markham, P., Kerridge, I.: Values-based medicine and modest foundationalism. J. Eval. Clin. Pract. **18**(5), 1020–1026 (2012). https://doi.org/10.1111/j.1365-2753.2012.01911.x

3. Melton, L., Brewer, B., Kolva, E., Joshi, T.: Increasing access to care for young adults with cancer: results of a quality-improvement project using a novel telemedicine approach to supportive group psychotherapy (2016)
4. Proon, N.: Improving healthcare delivery: liver health updating and surgical patient routing (2015)
5. Sardini, E., Serpelloni, M.: Instrumented wearable belt for wireless health monitoring. Proc. Eng. **5**, 580–583 (2010)
6. Elgazzar, K., Aboelfotoh, M., Martin, P., Hassanein, H.S.: Ubiquitous health monitoring using mobile web services. Proc. Comput. Sci. **10**, 332–339 (2012)
7. Ionov, M.V., Yudina, Yu.S., et al.: Patient-oriented assessment of blood pressure telemonitoring and remote counseling in hypertensive patients: a pilot project. Arterial Hypertens. (Russ. Fed.) **24**(1), 15–28 (2018)
8. Semakova, A., Zvartau, N.E., et al.: Towards identifying of effective personalized antihypertensive treatment rules from electronic health records data using classification methods: initial model. Proc. Comput. Sci. **121**, 852–858 (2017)
9. Ryngach, E.A., Treshkur, T.V., Tatarinova, A.A., Shlyakhto, E.V.: Algorithm for the management of patients with stable coronary artery disease and high-grade ventricular arrhythmias. Terapevticheskii Arkhiv **89**(1), 94–102 (2017)
10. Weinstein, R.S., Lopez, A.M., Krupinski, E.A.: Telemedicine - news from the front lines. Am. J. Med. **127**, 172–173 (2014)
11. Weinstein, R.S., Lopez, A.M., Joseph, B.A., et al.: Telemedicine, telehealth, and mobile health applications that work: opportunities and barriers. Am. J. Med. **127**, 183–187 (2014)
12. Jindal, A., Dua, A., Kumar, N., et al.: Providing healthcare-as-a-service using fuzzy rule based big data analytics in cloud computing. IEEE J. Biomed. Health Inform. **22**(5), 1605–1618 (2018). 8272340
13. Mahmoud, M.M.E., Rodrigues, J.J.P.C., Ahmed, S.H., Shah, S.C., Al-Muhtadi, J.F., Korotaev, V.V., De Albuquerque, V.H.C.: Enabling technologies on cloud of things for smart healthcare. IEEE Access **6**, 31950–31967 (2018)
14. Srinivas, J., Das, A.K., et al.: Cloud centric authentication for wearable healthcare monitoring system. IEEE Trans. Dependable Secure Comput. (2018)
15. ARTEMIS. Health Informatics Research. http://hir.uoit.ca/cms/?q=node/24. Accessed 24 Apr 2018
16. Lankhorst, M., et al.: Enterprise Architecture at Work. Modeling, Communication and Analysis. Springer, Heidelberg (2017)
17. Op't Land, M., Proper, E., Waage, M., Cloo, J., Steghuis, C.: Enterprise Architecture. Creating Value by Informed Governance. Springer, Berlin (2009)
18. Westfall, L.: The what, why, who, when and how of software requirements. In: ASQ World Conference on Quality and Improvement Proceedings, vol. 59 (2005)
19. The Open Group. ArchiMate 1.0 Specification (2009)
20. Porter, M.E.: A strategy for health care reform—toward a value-based system. N. Engl. J. Med. **361**(2), 109–112 (2009). https://doi.org/10.1056/nejmp0904131
21. Harvard Business School - Redefining Health Care: Creating Value-Based Competition on Results

Justification of Financing of Measures to Prevent Accidents in Power Systems

Tatiana Saurenko[1], Vladimir Anisimov[1(✉)], Evgeny Anisimov[1], Artem Smolenskiy[1], and Nadezhda Grashchenko[2]

[1] Peoples Friendship University of Russia (RUDN University), Miklukho-Maklaya str. 6, Moscow 117198, Russian Federation
tanya@saurenko.ru, an-33@yandex.ru
[2] Peter the Great St. Petersburg Polytechnic University, Polytechnicheskaya str. 29, St. Petersburg 195251, Russian Federation

Abstract. The article proposes the models of rational resource allocation for the implementation of measures to prevent industrial accidents in energy supply systems. They allow maximizing the integral effect of preventing these accidents, as well as eliminating their consequences in the event of restrictions on the resources allocated. As a unified measure of the volume of resources, the cost of implementing these measures has been adopted. The integral effect of financing events is estimated by an indicator formed by various options for combining private indicators of the effect of their financing. Analytical relationships have been obtained for a number of options for representing the integral effect of financing measures to prevent industrial accidents in power supply systems. They cover a fairly wide range of interesting practical problems that allow assessing various types of synergistic effect due to the distribution of available resources for financing the activities under consideration. This makes the proposed models useful for energy management at the micro and macro levels. Their practical application in the interests of energy management is associated with the choice of one or several models reflecting the specifics of measures to prevent technological accidents in power supply systems and the specification of the constant parameters of these models taking into account the specifics of these measures.

Keywords: Power supply · Energy management · Accidents · Protection measures · Financing · Effect · Optimization · Model

1 Introduction

High rates of modern economic development have led to a multiple increase in the consumption of energy resources [1, 2]. At the same time, the large-scale use of energy in all spheres of life is accompanied by an increased risk of accidents in power supply systems and an increase in their negative consequences. For example, about 3 thousand accidents at hydro-power plants occur annually in the world. The largest catastrophe in the history of Russia at the Sayano-Shushenskaya Hydroelectric Power Station in 2009

V. Murgul and M. Pasetti (Eds.): EMMFT-2018, AISC 983, pp. 295–305, 2019.
https://doi.org/10.1007/978-3-030-19868-8_30

caused the death of 75 people and significant damage to the process equipment. In the twentieth century, it occurred more than 150 accidents and disasters, one way or another connected with atomic energy: from accidents at nuclear power plants to explosions on nuclear submarines. Every year, on average, there are an average of ten accidents in the power grids, with about 10,000 consumers disconnected [3]. According to official data of the Ministry of Energy of the Russian Federation, 3277 accidents occurred at the facilities of the generating companies of the unified power grid of Russia (UPG) in 2018, and 14350 accidents at the power grids of the UPG. All this causes an important practical problem of preventing these accidents. Its solution is ensured by carrying out relevant activities. These activities are characterized by:

- significant diversity;
- various resource consumption;
- various contributions to improving the level of protection of energy supply systems.

These circumstances in the context of limited resources necessitates the formation of rational decisions on the distribution of available resources for the implementation of measures to protect energy supply systems. The formation of such solutions is not possible without the appropriate tools. The basis of this toolkit consists of models, including mathematical ones [4–8]. The main task of this article is to develop a complex of such models.

2 Methodical Approach to the Construction of Decision-Making Models for the Distribution of Funds for the Implementation of Protection Measures

In the proposed models, the cost is taken as a single measure of the amount of resources used to prevent accidents in power supply systems. Consequently, the funds allocated for the implementation of these activities are a factor limiting resources to prevent the accidents under consideration. At the same time, the task of forming rational decisions is to distribute the available financial resources between possible measures to protect energy supply systems, ensuring the maximum reduction in the possibility of man-made accidents.

For the formal presentation of models providing such a distribution of funds, we introduce the following notation:

n - the number of possible measures to protect energy supply systems;
z_i - the amount of financial resources allocated for the implementation of the i-th event ($i = 1, 2, \ldots, n$);

G is the total amount of financial resources allocated for carrying out measures to protect energy supply systems.

$\mathrm{Y} = \mathrm{F}(\mathrm{z}_1, \mathrm{z}_2, \ldots, \mathrm{z}_\mathrm{n}) = \mathrm{F}(\mathrm{Z})$. - the integral effect achieved in the allocation $\mathrm{Z} = (\mathrm{z}_1, \mathrm{z}_2, \ldots, \mathrm{z}_\mathrm{n})$ of financial resources that satisfy the condition.

$$\sum\nolimits_{k=1}^{n} z_k \le G \tag{1}$$

Taking into account the adopted symbols, the generalized model of forming decisions on the rational distribution of financial resources for the implementation of measures to protect energy supply systems has the form [9–13].

It is required to determine the vector.

$$Z^* = \left(z_1^*, z_2^*, \ldots, z_n^*\right) \tag{2}$$

of distribution of financial resources allocated for the protection of energy supply systems, ensuring compliance with.

$$y^* = F\left(z_1^*, z_2^*, \ldots, z_n^*\right) = \max_{Z} F(Z) \tag{3}$$

under the constraint (1).

Here, the * symbol indicates the optimality of both the integral indicator and the values z_k.

The basis for the formation of the distribution of financial resources in accordance with this model can be based on the Lagrange multipliers method. In this case, the general algorithm for forming a solution for the optimal distribution of available financial resources is as follows:

1. Lagrange function is built

$$L = F(z_1, z_2, \ldots, z_n) + \lambda\left(\sum\nolimits_{i=1}^{n} z_i - G\right), \tag{4}$$

where λ is the Lagrange multiplier.

2. It is found the private derivatives of the L function with respect to the parameters z_i and λ
   $\frac{\partial L}{\partial z_i} = \frac{\partial F}{\partial z_i} + \lambda$, $i = 1, 2, \ldots, n$;
   $\frac{\partial L}{\partial \lambda} = \sum_{i=1}^{n} z_i - G$.
3. It is solved the system of normal equations.

$$\begin{cases} \frac{\partial F}{\partial z_i} + \lambda = 0, & i = 1, 2, \ldots, n; \\ \sum_{i=1}^{n} z_i - G = 0 \end{cases} \tag{5}$$

regarding variables z_i and λ. As a result, the optimal values of the variables $z_i = z_i^*$ $(i - 1, 2, \ldots, n)$, at which the function reaches a maximum, are determined.

The values thus obtained $z_1^*, z_2^*, \ldots, z_n^*$ and constitute the optimal distribution of financial resources according to criterion (3) between the n measures for the protection of energy supply systems.

A constructive representation of the generalized model (1)–(5) is to establish a specific type of integral indicator of reducing the possibility of man-made accidents in power supply systems from the distribution of available funds between protection measures. The following private models for the distribution of financial resources correspond to the different variants of its constructive presentation.

3 Results of the Study

The result of the research is the development of a set of private models for forming decisions on the distribution of financial resources for the implementation of measures to prevent accidents in energy supply systems. These models cover a fairly wide range of practical, in practical terms, energy management tasks, allowing to evaluate different types of synergistic effect, due to the distribution of available resources to finance the activities under consideration.

3.1 Private Model № 1

In the case when the indicator of the integral effect achieved in the allocation of financial resources, it can be represented by a function of the form.

$$y = \bar{y}\prod_{i=1}^{n} z_i^{b_i}, \tag{6}$$

Lagrange function takes the form.

$$L = \bar{y}\prod_{i=1}^{n} z_i^{b_i} + \lambda\left(\sum_{j=1}^{n} z_j - G\right). \tag{7}$$

In the relations (6), (7):

$\bar{y}$ is the average value of the effect on private indicators y_i $(i = 1, 2, \ldots, n)$, due to the allocation z_i of financial resources for the implementation of the i-th (i = 1, 2, …, n) protection measure;.

b_i (i = 1, 2, …, n) - known parameters reflecting the effect of financing the i-th event on the effect due to it.

Then the Lagrange function takes the form.

$$L = \bar{y}\prod_{i=1}^{n} z_i^{b_i} + \lambda\left(\sum_{j=1}^{n} z_j - G\right). \tag{8}$$

Partial derivatives L of z_i $(i = 1, 2, \ldots, n)$ are represented by dependencies.

$\frac{\partial L}{\partial z_i} = \bar{y} b_i z_i^{b_i - 1} \prod_{j=1, j \neq i}^{n} z_j^{b_i} + \lambda, i = 1, 2, \ldots, n,.$

$\frac{\partial L}{\partial \lambda} = \sum_{i=1}^{n} z_i - G.$

The system of normal equations is.

$$\begin{cases} \bar{y} b_1 z_1^{b_1-1} z_2^{b_2} \ldots z_n^{b_n} + \lambda = 0; \\ \bar{y} z_1^{b_1} b_2 z_2^{b_2-1} \ldots z_n^{b_n} + \lambda = 0; \\ \ldots\ldots\ldots\ldots; \\ \bar{y} z_1^{b_1} z_2^{b_2} \ldots b_n z_n^{b_n-1} + \lambda = 0; \\ \sum_{i=1}^{n} z_i - G = 0. \end{cases} \quad (9)$$

Subtracting from the first equation of system (8) its second, from the second equation - the third, etc., we get the system

$$\begin{cases} b_2 z_1 - b_1 z_2 = 0; \\ b_3 z_2 - b_2 z_3 = 0; \\ \ldots\ldots; \\ b_n z_{n-1} - b_{n-1} z_n = 0; \\ \sum_{i=1}^{n} z_i - G = 0. \end{cases} \quad (10)$$

As a result of solving the system of equations (9), we obtain the optimal distribution of the allocated financial resources *G* according to *n* measures for the protection of energy supply systems:

$z_1^* = \frac{b_1}{\sum_{i=1}^{n} b_i} G, \quad (i = 2, 3, \ldots, n);$

$z_i^* = \frac{b_i}{b_1} z_1^*, (i = 2, 3, \ldots, n).$

or

$$z_i^* = \frac{b_i}{\sum_{j=1}^{n} b_j} G, (i = 1, 2, \ldots, n) \quad (11)$$

Values derived from (10) $z_1^*, z_2^*, \ldots, z_n^*$ constitute the optimal distribution of financial resources between n possible measures for the protection of energy supply systems by criterion (6).

3.2 Private Model № 2

If the indicator of the integral effect is represented by the ratio.

$$y = \prod_{i=1}^{n} (a_i + b_i z_i) \quad (12)$$

then using the general scheme for solving the problem described in Sect. 2, we construct the Lagrange function and find partial derivatives of the form.

$$\frac{\partial L}{\partial z_i} = b_i \prod_{\substack{j=1 \\ j \neq i}}^{n} (a_j + b_j z_j) + \lambda \, (i = 1, 2, \ldots, n);$$

$$\frac{\partial L}{\partial \lambda} = \sum_{i=1}^{n} z_i - G.$$

Here a_i, bi are known parameters reflecting the effect of financing the i-th (I = 1, 2, …, n) measures on the effect due to them.

By analogy with the previous example, we solve the system of normal equations.

$$\begin{cases} b_1(a_2 + b_2 z_2)(a_3 + b_3 z_3)\ldots(a_n + b_n z_n) + \lambda = 0; \\ b_2(a_1 + b_1 z_1)(a_3 + b_3 z_3)\ldots(a_n + b_n z_n) + \lambda = 0; \\ \ldots\ldots\ldots\ldots\ldots\ldots\ldots\ldots\ldots\ldots\ldots\ldots\ldots\ldots; \\ b_n(a_1 + b_1 z_1)(a_2 + b_2 z_2)\ldots(a_{n-1} + b_{n-1} z_{n-1}) + \lambda = 0; \\ \sum_{i=1}^{n} x_i - G = 0 \end{cases} \tag{13}$$

Relatively z_i $(i = 1, 2, \ldots, n)$.

As a result z_i is calculated by the formula.

$$z_i = z_1^* + B_{i-1}, \; (i = 1, 2, \ldots, n) \tag{14}$$

where

$$B_{i-1} = B_{i-2} + \frac{a_{i-1}}{b_{i-1}} - \frac{a_i}{b_i}, \; (B_0 = 0).$$

From the last equation of system (12) is determined by the formula.

$$z_1^* = \frac{1}{n}\left(G - \sum_{j=2}^{n} B_{j-1}\right) \tag{15}$$

Then, combining formulas (13) and (14), one can move to a general recurrent relation for the optimal allocation of financial resources for the implementation of measures to protect energy supply systems.

$$z_i^* = \frac{1}{n}\left(G - \sum_{j=2}^{n} B_{j-1}\right) + (1 - \delta_{i1})B_{i-1}, \; (i = 1, 2, \ldots, n) \tag{16}$$

where δ_{i1} is the Kronecker symbol ($\delta_{11} = 1$; $\delta_{i1} = 0$; $i > 1$).

The values $z_1^*, z_2^*, \ldots, z_n^*$ obtained on the basis of relation (15) constitute the optimal distribution of financial resources by criterion (11) among n possible measures to protect energy supply systems.

3.3 Private Model № 3

For the case when the integral indicator of the effect of the implementation of measures to protect energy supply systems is determined by the ratio.

$$y = \prod_{i=1}^{n}\left(a_i + b_i z_i c_i z_i^2\right), \tag{17}$$

the system of normal equations is.

$$\begin{cases} (b_1 + 2c_1z_1)(a_2 + b_2z_2 + c_2z_2^2) - (a_1 + b_1z_1 + c_1z_1^2)(b_2 + 2c_2z_2) = 0; \\ (b_2 + 2c_2z_2)(a_3 + b_3z_3 + c_3z_3^2) - (a_2 + b_2z_2 + c_2z_2^2)(b_3 + 2c_3z_3) = 0; \\ \ldots\ldots\ldots\ldots\ldots\ldots\ldots\ldots\ldots\ldots\ldots; \\ (b_{n-1} + 2c_{n-1}z_{n-1})(a_n + b_nz_n + c_nz_n^2) - (a_{n-1} + b_{n-1}z_{n-1} + c_{n-1}z_{n-1}^2)(b_n + 2c_nz_n) = 0; \\ \sum_{i=1}^{n} z_j - G = 0. \end{cases} \quad (18)$$

In (17), (18), a_i, b_i c_i are known parameters reflecting the influence of the financing of the i-th (i = 1, 2, …, n) protection measures on the effect caused by them.

Directly from the system of equations (18) it follows that for all values $i = 1, 2, \ldots, n - 1$ there is the equality.

$$\frac{a_i + b_iz_i + c_iz_i^2}{b_i + 2c_iz_i} = \frac{a_{i+1} + b_{i+1}z_{i+1} + c_{i+1}z_{i+1}^2}{b_{i+1} + 2c_{i+1}z_{i+1}}.$$

In order to avoid cumbersome calculations, it seems appropriate to denote the left side of this equality by H, i.e.

$$\frac{a_i + b_iz_i + c_iz_i^2}{b_i + 2c_iz_i} = H, i = 1, 2, \ldots, n \quad (19)$$

where it follows that:

$$z_i^* = \frac{(2Hc_i - b) + \sqrt{4H^2c_i^2 + b_i^2 - 4c_ia_i}}{2c_i} \quad (20)$$

Thus, in order to determine the optimal allocation of financial resources for measures to protect energy supply systems in the case under consideration, it is sufficient to substitute the value of the coefficient H constant for all activities in formula (19). To find it, in the last equation of system (17), it is substituted the expression z_i $(i = 1, 2, \ldots, n)$, for calculated by formula (19).

The result is the ratio.

$$\sum\nolimits_{i=1}^{n} \frac{(2Hc_i - b) + \sqrt{4H^2c_i^2 + b_k^2 - 4c_ia_i}}{c_i} = 2G \quad (21)$$

from which it is determined the value H.

The values $z_1^*, z_2^*, \ldots, z_n^*$ obtained on the basis of the relation (19) constitute the optimal distribution of financial resources by criterion (16) among possible measures to protect energy supply systems.

3.4 Private Model № 4

For the case when the integral indicator of the effect of the implementation of measures to protect energy supply systems is determined by the ratio.

$$y = \prod_{i=1}^{n} (a_i + b_i \ln z_i) \quad (22)$$

the system of normal equations is.

$$\begin{cases} \frac{b_1}{z_1}(a_2 + b_2 \ln z_2)\ldots(a_n + b_n z_n) + \lambda = 0; \\ (a_1 + b_1 \ln z_1)\frac{b_2}{z_2}(a_3 + b_3 \ln z_3)\ldots(a_n + b_n z_n) + \lambda = 0; \\ \ldots\ldots\ldots\ldots\ldots\ldots\ldots\ldots\ldots\ldots\ldots\ldots; \\ (a_1 + b_1 \ln z_1)\ldots(a_{n-1} + b_{n-1} \ln z_{n-1})\frac{b_n}{z_n} + \lambda = 0; \\ \sum_{i=1}^{n} z_i - G = 0 \end{cases} \quad (23)$$

In (22), (23), a_i, b_i are known parameters reflecting the effect of financing the i-th (i = 1, 2, …, n) protective measures on the result caused by it.

Subtracting from the first equation of system (23) its second, then from the second equation - the third, etc., you can come to the system of equations

$$\begin{cases} \frac{b_1}{z_1}(a_2 + b_2 \ln z_2) = \frac{b_2}{z_2}(a_1 + b_1 \ln z_1); \\ \ldots\ldots\ldots\ldots\ldots\ldots\ldots\ldots\ldots\ldots; \\ \frac{b_{n-1}}{z_{n-1}}(a_n + b_n \ln z_n) = \frac{b_n}{z_n}(a_{n-1} + b_{n-1} \ln z_{n-1}); \\ \sum_{i=1}^{n} z_i - G = 0 \end{cases} \quad (24)$$

From the first $n - 1$ equations of this system, it follows that for all values $i = 1, 2, \ldots, n$ it is true the equality.

$$\frac{z_i(a_i + b_i \ln z_i)}{b_i} = \frac{z_{i+1}(a_{i+1} + b_{i+1} \ln z_{i+1})}{b_{i+1}}.$$

Then, taking into account that.

$$\frac{z_i(a_i + b_i \ln z_i)}{b_i} = H(i - 1, 2, \ldots, n) \quad (25)$$

it is quite simple (for different values of the coefficient H) to calculate with a given accuracy the uniquely corresponding values of the variables.
z_i^*, $i = 1, 2, \ldots, n$.

To select a specific value of the coefficient H, the last equation of system (23) is used.

The values of the variables found by relation (24) with the value of the coefficient H, which satisfies the last equation of system (23), constitute the optimal distribution of the available financial resources by criterion (21) for the implementation of measures for the protection of energy supply systems.

3.5 Private Model № 5

The presented approaches of optimal allocation of available financial resources G on n private measures for protecting energy supply systems can be used to build similar algorithms in cases where the integral quality indicator has a more complex structure, for example, a power type.

$$y = \bar{y}\prod_{j=1}^{k} a_j z_j^{b_j}\prod_{j=k+1}^{n}\left(a_j + b_j z_j\right), (1 \le k \le n-1). \tag{26}$$

Then, it is necessary to find the partial derivatives of the Lagrange function.

$$L = \prod_{j=1}^{k} a_j z_j^{b_j} \prod_{j=k+1}^{n} \left(a_j + b_j z_j\right) + \lambda\left(\sum_{j=1}^{n} z_j - G\right)$$

on variables $z_j\left(j = \overline{1,n}\right)$.

Next, a system of normal equations (5) is formed.

By subtracting from each j-th $\left(j = \overline{1,n-1}\right)$ of equation $(j+1)$-th equation, it is converted to a form convenient for solving the problem.

$$\begin{cases} b_2 z_1 + b_1 z_2 = 0; \\ \ldots\ldots\ldots\ldots\ldots\ldots; \\ b_k z_{k-1} + b_{k-1} z_2 = 0; \\ b_{k+1} z_k - b_k\left(a_{k+1} + b_{k+1} z_{k+1}\right) = 0; \\ b_{k+2}\left(a_{k+1} + b_{k+1} z_{k+1}\right) - b_{k+1}\left(a_{k+2} + b_{k+2} z_{k+2}\right) = 0; \\ \ldots\ldots\ldots\ldots\ldots\ldots\ldots\ldots\ldots\ldots\ldots\ldots; \\ b_n\left(a_{n-1} + b_{n-1} z_{n-1}\right) - b_{n-1}\left(a_n + b_n z_n\right) = 0; \\ \sum\limits_{k=1}^{n} z_j - G = 0. \end{cases}$$

Solving this system relatively z_k, $k = \overline{1,n}$, we find the optimal distribution of financial resources:

$$z_1^* = \frac{b_1\left(G + \sum_{j=k+1}^{n} \frac{a_j}{b_j}\right)}{n - k + \sum_{j=1}^{k} b_j}; \tag{27}$$

$$z_j^* = \frac{b_j}{b_1} z_1^* j = \overline{2,k}$$

or

$$z_j^* = \frac{z_1^*}{b_1} - \frac{a_j}{b_j}, j = \overline{k+1,n}. \tag{28}$$

The values of the variables $z_i = z_i^* \ (i = 1, 2, \ldots, n)$, found from ratio (26), (27), constitute the optimal distribution of available financial resources by criterion (25) for the implementation of measures to protect energy supply systems.

The concretization of the proposed particular models consists in establishing the corresponding parameters a_i, b_i, c_i reflecting the impact of the amount of financing zi of the i-th (i = 1, 2, ..., n) protection measures on the effect y_i caused by it. They are established on the basis of the information available to the decision maker about the resource intensity and effectiveness of the relevant measures to prevent industrial accidents in energy supply systems. Methodical approaches to their definition are considered in [14, 15].

4 Discussion

The sharp increase in energy consumption, the centralization of the supply of cities and industrial enterprises with energy resources, the enlargement of energy supply systems and the complication of the conditions for their operation in the late 20th and early 21st centuries led to an increase in the number and scale of industrial accidents in energy supply systems and the resulting economic damage. This led to the acute problem of preventing accidents of energy supply systems. Its solution can be achieved by taking appropriate measures to protect these systems. Formation and implementation of protection measures are associated with significant resource costs. The scarcity of these resources necessitates their effective use in order to achieve the goals of protecting energy supply systems. A promising tool contributing to the implementation of a scientific approach in the interests of efficient use of available resources is the use of appropriate mathematical models. At the same time, a higher level of objectivity of the decisions made, the possibility of disseminating and using the experience of specialists concentrated in the models, systematic and higher reliability of the formation of the right decisions are provided.

This article focuses on the methodological issues of building models of the optimal allocation of financial resources for the implementation of measures to protect energy supply systems. The proposed approaches to the construction of these models cover a fairly wide range of interesting in practical terms energy management tasks, allowing to evaluate various types of synergistic effect due to the distribution of available resources to finance the considered measures of protection from accidents in energy supply systems.

5 Conclusion

In general, the approaches proposed in the article to building mathematical models to substantiate solutions for financing accident prevention activities in power supply systems are the basis for the creation of specific methodologies for solving relevant energy and non-energy problems. Their practical application for solving energy and non-energy problems at the micro and macro levels is associated with the choice of one or several models reflecting the specifics of these tasks and specifying the constant parameters of the selected models taking into account the peculiarities of real conditions.

References

1. Meadows, D., Rander, J., Meadows, D., Behrens, W.: The Limits to Growth: A Report for the Club of Rome's Project on the Predicament of Mankind. Universe Books, New York (1972)
2. Meadows, D., Rander, J., Meadows, D.: Limits to Growth-The 30 Year Update (2004)
3. Bagrova, L.A., Bokov, V.A., Mazinov, A.S.: Dangerous man-made disasters in the energy sector as environmental risk factors. In: Scientific notes of the Tauride National University. IN AND. Vernadsky Series "Geography", vol. 25, no. 2, pp. 9–19 (2012). (in Russian)
4. Saurenko, T., Anisimov, E., Anisimov, V., Levina, A.: Comparing investment projects of innovative developing strategies of municipalities, based on a set of indicators. In: MATEC Web of Conferences, p. 01038 (2018)
5. Anisimov, V., Chernysh, A., Anisimov, E.: Model and algorithm for substantiating solutions for organization of high-rise construction project. In: E3S Web of Conferences, p. 03003 (2018)
6. Anisimov, V.G., Anisimov, E.G., Saurenko, T.N., Sonkin, M.A.: The model and the planning method of volume and variety assessment of innovative products in an industrial enterprise. J. Phys.: Conf. Ser. **803**(1), 012006 (2017). https://doi.org/10.1088/1742-6596/803/1/012006
7. Anisimov, E.G., Anisimov, V.G., Sonkin, M.A.: Mathematical simulation of adaptive allocation of discrete resources. In: Proceedings of the 2016 Conference on Information Technologies in Science, Management, Social Sphere and Medicine, ITSMSSM 2016. ACSR: Advances in Computer Science Research, pp. 282–285 (2016)
8. Anisimov, V., Anisimov, E., Sonkin, M.: A Resource-and-Time Method to Optimize the Performance of Several Interrelated Operations. International Journal of Applied Engineering Research. **10**(17), 38127–38132 (2015)
9. Anisimov, V.G., Anisimov, Ye.G.: A branch-and-bound algorithm for one class of scheduling problem. Comput. Math. Math. Phys. 32(12), 1827–1832 (1992). (in Russian)
10. Anisimov, V.G., Anisimov, Ye.G.: A method of solving one class of integer programming problems. USSR Comput. Math. Math. Phys. **29**(5), 238–241 (1989). (in Russian)
11. Anisimov, V.G., Anisimov, E.G.: Modification of the method for solving a class of integer programming problems. Comput. Math. Math. Phys. **37**(2), 179–183 (1997). (in Russian)
12. Anisimov, V.G., Anisimov, E.G.: Algorithm for the optimal distribution of discrete nonuniform resources on the web. Comput. Math. Math. Phys. **37**(1), 54–60 (1997). (in Russian)
13. Alekseyev, A.O., Alekseyev, O.G., Anisimov, V.G., Anisimov, E.G.: The use of duality to increase the effectiveness of the branch and bound method when solving the knapsack problem. USSR Comput. Math. Math. Phys. **25**(6), 50–54 (1985). (in Russian)
14. Avdeev, M.M., Anisimov, V.G., Anisimov, E.G., Martyshchenko, L.A., Shatokhin, D.V.: Information-statistical methods in the management of microeconomic systems. International Academy of Informatization, St. Petersburg, Tula (2001). (in Russian)
15. Anisimov, V.G., Anisimov, E.G., Petrov, V.S., Rodionova, E.S., Saurenko, T.N., Tebekin, A.V., Tebekin, P.A.: Theoretical basis of innovation management, St. Petersburg (2016). (in Russian)

Importance of Cross-Sectoral Modeling in Management of Electric Power Sector Based on Green Energy Development Example

Igor Korolev[1(✉)], Igor Ilin[1], Vasilii Makarov[1], and Gleb Konnov[2]

[1] Peter the Great St. Petersburg Polytechnic University, Polytechnicheskaya, 29, St. Petersburg 195251, Russia
ia-korolev@yandex.ru

[2] Consultum Group of Companies, Antimonopoly Practice, Saint Petersburg 195251, Russia

Abstract. The Russian economy is looking for new development trends in the market environment. Objectives: energy conservation and energy efficiency, minimal environment impact, social development focus, accessibility in any part of the country. Scientific and technological progress provides for new technologies to fulfill these objectives. One of them is application of renewable energy sources. However, their effective application is associated with a large number of different problems. The trends and issues of renewable energy development were analyzed, as well as ways to solve the issues identified in the Russian context were proposed.

Keywords: Management of energy distribution and use · Modern technologies of energy storage · Building and energy management systems · Renewable energy commercialization · Low-carbon economy · Green economy

1 Introduction

The government of the Russian Federation and regional executive authorities are actively exploring the possibilities of ensuring economic growth through a more complete and comprehensive use of the vast country territory through geographically-distributed placement of households on unoccupied land plots. However, such attempts are faced with the main problem, that is the lack of engineering infrastructure in new territories that are, as a rule, distant from local centers for households. Electric power supply is a significant element of infrastructure.

When solving this problem, it is necessary to consider the fact that the process of bringing the centralized energy supply to the uninhabited areas can take up to >4 years. In this case, an alternative solution for the accelerated development of new lands is the formation of a decentralized electric power supply by creating sources that, instead of using expensive imported hydrocarbon fuel, use the energy potential of the sun, wind, and other natural renewable sources (hereinafter 'RES').

V. Murgul and M. Pasetti (Eds.): EMMFT-2018, AISC 983, pp. 306–312, 2019.
https://doi.org/10.1007/978-3-030-19868-8_31

The fact that over the past 20 years of electrical energy industry reformation in Russia, the previously declared benchmarks in terms of provision of competition, improvement of revenue distribution transparency, tariff establishment, and price reduction have not been achieved, counts in favor of new ways of energy development and, in particular, electric power supply decentralization [1]. Namely:

– instead of a single, vertically integrated structure of RAO UES that functioned until 2008, many small and disparate electricity supply enterprises were established (as of February 1, 2019, more than 2000 units). As a result, semi-fixed costs for their maintenance increased, which further intensified the consumer cost loading, and with due regard to the widespread use of electricity (in production of any goods and services), it became one of the incentives to further unwinding of the inflationary spiral;
– the final price of electricity for consumers has already exceeded the limit of possible growth for more than 3 years and in 2018. for a number of industrial facilities rose beyond the psychological threshold of 0.12 USD/kWh excluding VAT. With such an electricity price from the unified energy system, the costs of energy supply decentralization, that is construction of proprietary gas turbine generation modular unit (for example, with the declared capacity of 1 MW in case it is used for 5000–6000 hours per year), as estimated by the authors, will pay off in 1 year [2]!

As for decentralized electrification based on RES, it should be noted that this scientific and high-tech applied field of energy is gaining momentum on a global scale. This is facilitated by both the climatic changes caused by the anthropogenic factor and the outer-driven conjuncture policies of a number of countries to reduce external energy dependence (in particular, on gas supplies from Russia). An additional incentive is the prospect of depletion of fossil fuel in the foreseeable future.

The current world dynamics of growth in the use of RES is characterized by various numbers, of which the indicators of China, India, the United States and a number of European Union states are most noticeable. The total global investment in renewable energy amounts to more than US$280 billion annually (with China, USA, UK in the top three) [3].

In addition to implementation of a tough environmental policy against the background of a long-term hydrocarbon price increase trend, as well as rise of nuclear and coal power generation fixed costs, an important driver for the RES development abroad is the conscious need to involve innovative high technologies in the energy sector in order to obtain "high-tech surplus value" even in spite of the fact that for the end user the price of "green" energy is still higher than the conventional one [4].

At the same time, viable decisions to create priority conditions for the development of RES in Russia are hindered by the absence of visible and obvious forecast of management solutions, especially in the conditions of the domination of the traditional power generation, primarily nuclear sector that attracts the investment component in the final electricity price [5].

The situation is further aggravated by the fact that presently there is no accessible methodological framework for the performance of macro analysis of the consequences of important managerial solutions for the economy: in 1970s, modeling of cooperation between the electric power sector and the economy was limited to building rather simple economic correlations between the electric power consumption and certain macro economy parameters—industrial output and population growth, then in 1980s and further, the main principle of mutual influence of the energy sector and the economy was already monitored through a system of direct and reverse causality system between the primary factors (labor, capital, resources) and final power consumption, on which basis natural cost models and cost dynamic cost models were developed. However, the functionality of such models was considerably limited due to the absence of coordination between the cost variation and price elasticity of end users; plus, the main parameters of such models were based on post-Perestroyka structure of economy. The recent papers in this field, published in 2010, have also certain peculiarities, but only contain the solution of rather 'local' issues [6, 7].

Hence, further advancement of the management in the energy sector, as far as the solution making mechanism is concerned, is impossible without the public administration, including the strategic planning, adopting the methodological coordination between the administrative decisions and the results of the financial and economic activity of economic operators (including households and the public administration as a controller of public funds), including obtaining and interpretation of the respective mid-term effects.

2 Materials and Methods

Since the collapse of the USSR until today, the best economic and mathematical tool for assessing the degree of influence of changes in price factors on the inter-branch relationships of economic sectors is the input-output model (or inter-branch balance) developed by V.V. Leontiev. It is of particular interest that this model, the benefits of which were best revealed in the framework of planned Soviet economy, was immediately (and to the present day) taken over by the countries with the so-called "capitalist way of life" (Japan, Germany, USA) [8, 9].

The authors believe that the required mechanism for the research of the impacts of important administrative decisions on the economy would mandatorily require taking into account of the following:

- multi-pronged character of price impacts and turnover period;
- structures and degrees of cross-sectoral linkages of economic operators;
- sufficient adequacy and confidence in determining regression equations describing intercoordinated economic indicators.

Taking into account the above, the authors used the research into the impact of prices on the national economy to develop methods in order to obtain comprehensive effects that would include such phases as:

- research into the matter of transmission of price impulses in the economy;
- building, on the basis of the static cross-sectoral balance, of a conceptual model of formation of the added value of single sectors where, by using input-output method, an overall price variation effect would be determined for each sector at changing electricity prices and taking into account economy capital turnaround within a calendar year.

The tools were tested using the liquidation of cross-subsidizing between power consumer categories as an example, which brought rather unexpected results (NB the overall effect was determined taking into account price variation both for the consumers in the interest of which this would have been performed and for certain economic operators who were the sources of these variations).

Cross-subsidizing between consumer categories refers herein to the re-distribution of power payment burden among different consumer groups, where some consumer groups actually pay a part of the cost of electricity consumed by other consumer groups. Additionally, energy sector experts identify about seven different types of cross-subsidizing: between traded electrical energy and power in the wholesale market in favor of power, between electrical and thermal energy in favor of thermal energy in combined energy production, between long-distance and near thermal and electrical energy consumers in favor of long-distance consumers, etc. Initially, the use of the cross-subsidizing system in the pricing of the electric power complex was conceived by the state as one of the mechanisms for protection against a consumed electricity cost increase for low-income households in the conditions of 1990s hyperinflation.

The significant increase in cross-subsidizing since 2011 is due to the adoption by the Russian Government of a number of resolutions to curb the electricity tariff growth rate for households in the conditions of a sharp increase in electricity retail price due to the completion of the wholesale electricity market liberalization and transition of the majority of electric grid companies to long-term tariff regulation. At the beginning of 2019, the amount of cross-subsidies in the power industry is estimated at about US$ 4.5 billion.

The further practice of application of the cross-subsidizing system in the power industry of Russia led not only to a distorted ratio of household electricity prices to industrial electricity prices: for example, in 2012, this ratio was 0.88, although in reality it should be more than 1.0, as in the process of delivering electricity to households there is a multiple transformation to a low voltage level with the formation of consequential technological losses in the power grid infrastructure, whereas for industrial consumers the transformation is carried out mainly to the average voltage level. For instance, a similar ratio in the UK is 1.66, in the US – 1.77, in Finland – 1.88 [10]. However, the main result of the application of cross-subsidizing was the fast growth of electricity prices for other consumers.

The government's policy of cross-subsidizing reduction has not yet yielded significant results since 1997. For example, on the one hand, a slight increase in electricity prices for households levels out inflationary processes; on the other hand, a one-time increase in electricity sector prices is constrained by a political factor: cancellation of cross-subsidizing in some regions will cause more than twofold electricity price increase for households and result in social tensions. As one of the measures to solve the problem of cross-subsidizing, it was proposed to introduce a social energy consumption rate that differentiates the cost of electricity depending on the amount of its consumption for households in all regions (with the exception of households technologically isolated from the unified power grid) starting from 2016. In 2019, the Government of Russia proposed to temporarily set aside the issue for further elaboration due to the presence of other factors in the economy: intensification of pressure on the economy from Western countries, increase in the domestic VAT rate from 18% to 20%, changes in pension legislation.

3 Results

The results of the research quoted by the authors relating to a possible result in case of instantaneous abolition of cross-subsidizing in electric energy sector turned out to be quite interesting.

To be noted that the Russian Government has always proposed to adopt the policy of an accelerated growth of electricity costs for households vs. other consumers, explaining it away as an additional factor in favor of the economic growth [11].

Well, according to the results of the research, the increase of electricity prices for households with simultaneous short-time decrease of electricity prices for other consumers does indeed have an effect expressed in the increase of business tax liabilities, however a bigger economic effect is observed when the electricity prices for households decrease (while increasing for all the other consumers.) This is explained by the fact that the demand of the households for products and services is higher than that of any other economic operator. And stimulating the demand of the households, rather than other economic operators (manufacturing and service industry) has much higher multiplicative economic effect due to the sales increase in the production and service sectors owing to now-available part of the overall demand of households for approximately the same amount of saved expenses for electricity, minus savings. Hence, for the mid-term period, instead of the decrease in cross-subsidizing, its increase should be considered taking into account differentiation of regional economic data.

4 Discussion

The above shows that the significant part of the decisions relating to the introduction of new electricity pricing mechanisms (e.g. using RAB methods for the regulation of the prices for the services of grid companies, guaranteed return on investments for new generating assets and usage of capacity supply agreements) was prepared and implemented without in-depth macro analysis of socio-economic consequences of quick changes in electricity prices.

As a matter of fact, the main target for the RES development (with the exception of decentralized power supply areas) in Russia is the intention to reduce the consumption of hydrocarbons (primarily natural gas) solely to increase the resource of its exports. At the same time, the connection of RES development with the economy innovative development, which plays a crucial role in European countries, the United States, and now in China, has not been declared by the state as particularly significant in Russia. And by the beginning of 2018, the share in the total generation volume, according to the data of the Ministry of Energy of Russia, is less than 1%, which seems to be nothing more than a statistical error.

For Russia, the widespread use and development of RES could be a powerful catalyst for high-tech developments in the fields of energy, related industries, information technology and management systems, and thus could additionally contribute to gradual transition of the domestic economy from the export-raw material mono-scenario to the multiple-option one, with a growing share of surplus value in high-tech sectors.

It is important to note that abroad end-users are unlikely to invest in the RES development on an initial stage. As a rule, the catalyst is either budget financing or private capital in case of receipt of tax benefits from the state. In Russia, the main burden of energy project financing initially falls on the consumers.

The authors think that, instead of consumer funds, the main source of financing in this case should be state financing, various forms of public and private partnership, including concession agreements and other co-financing instruments (for example, venture funds). A fast-growing mining crypto-currency farms (bitcoins) market sector should be specifically considered as a source of RES development financing. The state assistance in the construction of RES facilities for their needs, taking into account the provision of the required reliability of power supply and reservation, could be accompanied by the requirement of mandatory allocation of part of the capacity under construction to households and small-scale consumers.

Moreover, without government assistance, there is no accelerated solution for the following issue as well. The main constraining technological factor of a full-fledged transition to distributed and small energy is the absence of significant advances in the methods of industrial accumulation and storage of energy produced by sources highly dependent on meteorological conditions (sun, wind, etc.).

5 Conclusions

The proposed development of decentralized power supply based on RES as an additional segment to the existing unified energy system in mid- and long-term period will allow not only to implement the integrated development of new territories, but also provide a significant innovative contribution to further economic growth.

These estimations of the socio-economic consequences for households and economic sectors on the basis of such a methodology could serve as grounds for accelerated adoption of major decisions for universal adoption and development of green energy. In order to increase confidence and adequacy of the cross-sectoral model and to further enhance it, laws have to be passed requiring all economic agents to continuously fill out statistical report forms relating to their economic activity to a competent authority (to be defined by the Russian Government) in a way as to allow every interesting party to use them for the research, while ensuring strict protection of commercial secrets.

References

1. Kuzovkin, A.I.: The reform of the electricity sector and energy security. Institute of microeconomics, Moscow (2006)
2. The results of RES development in Russia are radically at odds with the promises of regulators and investors, the Association of NP “Community of energy consumers” (2017). https://www.np-ace.ru
3. The World Nuclear Industry Status Report. Mycle Schneider Consulting 12 (2017)

4. Makarov, V.M., Novikova, O.V., Tabakova, A.S.: Energy efficiency in "green construction": experience, issues, trends, 732 (2017)
5. A vicious Sunny circle. Green energy again asks for benefits. The Association of NP "market Council", Newspaper "Kommersant", 239, 1 (2017)
6. Korolev, I.A., Khabatchev, L.D., Makarov, V.M.: Influence of the pricing policy in the Russian power industry on the gross domestic product (GDP) dynamics. WIT Trans. Ecol. Environ. **190**, 191 (2014)
7. Karlik, A.E., Kukor, B.L., Yakovleva, E.A., Sokolov, A.A.: The management of structural transformations in the socio-economic system in the information network economy. Peter the Great St. Petersburg Polytechnic University, 1, 175 (2018)
8. Khabatchev, L.D.: Institutional framework of economics and management in the power industry. Peter the Great St. Petersburg Polytechnic University (2017)
9. Leontief, W.W.: Intersectoral economy. JSC Publishing house "Economy" (1997)
10. Key World energy statistics 2017. International Energy Agency 53 (2017)
11. In Russia, learn to store electricity, the Newspaper "Energy and industry of Russia", 01–02, 333 (2018)

International Energy Strategies Projects of Magnetic Levitation Transport

Elena Schislyaeva[1], Olga Saychenko[1(✉)], Sergey Barykin[1], and Irina Kapustina[2]

[1] St. Petersburg State Marine Technical University, Lotsmanskaya Street, 3, 190121 St. Petersburg, Russian Federation
olgakalinina@bk.ru

[2] Peter the Great St. Petersburg Polytechnic University, Polytechnicheskaya, 29, 195251 St. Petersburg, Russian Federation

Abstract. This article discusses the advantages of the development, implementation of new transport systems' innovation based on magnetic levitation physical principles concerning energy strategies in the transport sector. The development of the concept of magnetic levitation transport in the modern economy is explored. The authors consider and propose scientific and economic justification of approaches to the creation of magnetic levitation transport platforms for the transportation of goods and passengers. The further research and development of main trends in this scientific sphere regarding the world transport system are considered with the most promising types of transport, taking into the account a perspective leading position in the market of transport services. This article discusses the advantages of the development, implementation of new transport systems based on magnetic levitation. The development of the concept of magnetic levitation transport in the modern economy is studied.

Keywords: Energy strategies · Transport sector · Innovations · Energy-saving technologies · Magnetic gravity transport · Magnetic levitation projects

1 Introduction

The nature and extent of studied problems links the macroeconomic requirements to the modern transport system, formulated by Morozov [1]. The paper covers the factors that have a fundamental impact on the evolutionary development of transport systems. Constant population growth significantly increases the need for the rapid development of high-speed passenger and freight transport. Transport in Russia in the near future may not cope with the rapidly increasing traffic flows. This problem can be solved by the development and implementation of innovative modes of transport based on magnetic levitation platforms (Fig. 1).

V. Murgul and M. Pasetti (Eds.): EMMFT-2018, AISC 983, pp. 313–320, 2019.
https://doi.org/10.1007/978-3-030-19868-8_32

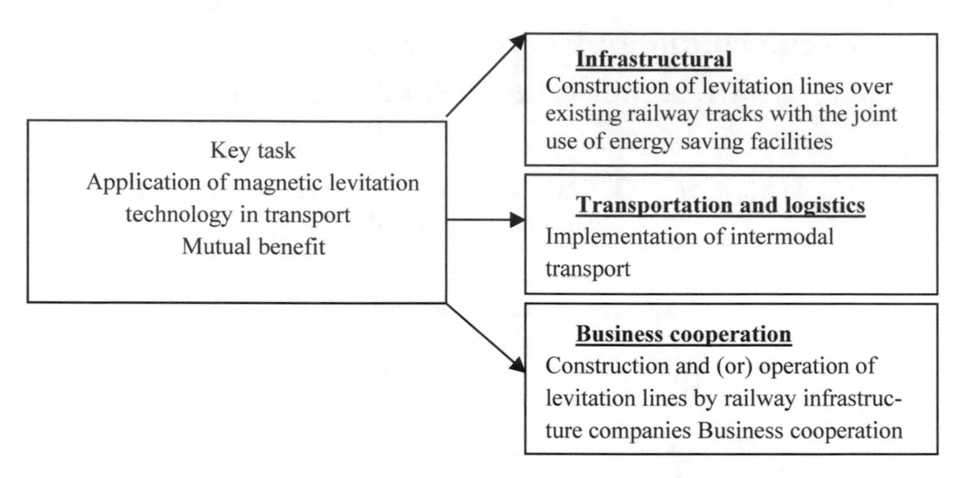

Fig. 1. Economic prerequisites for the development of magnetic levitation transport

Advanced technological R&D of second-generation high-temperature superconducting materials, combined with the increasing energy characteristics of permanent magnets, have revealed opportunities for the introduction of technologies based on urban public transport, such as the metro. It is supported by the commercial efficiency of transport operation on the basis of the 1st category magnetic gravity technologies in the cities of Japan. Since 2005, the Linimo line has been operating on a permanent basis in the vicinity of Nagoya. And in South Korea, the magnetic levitation road was built by Hyundai Rotem (Fig. 2).

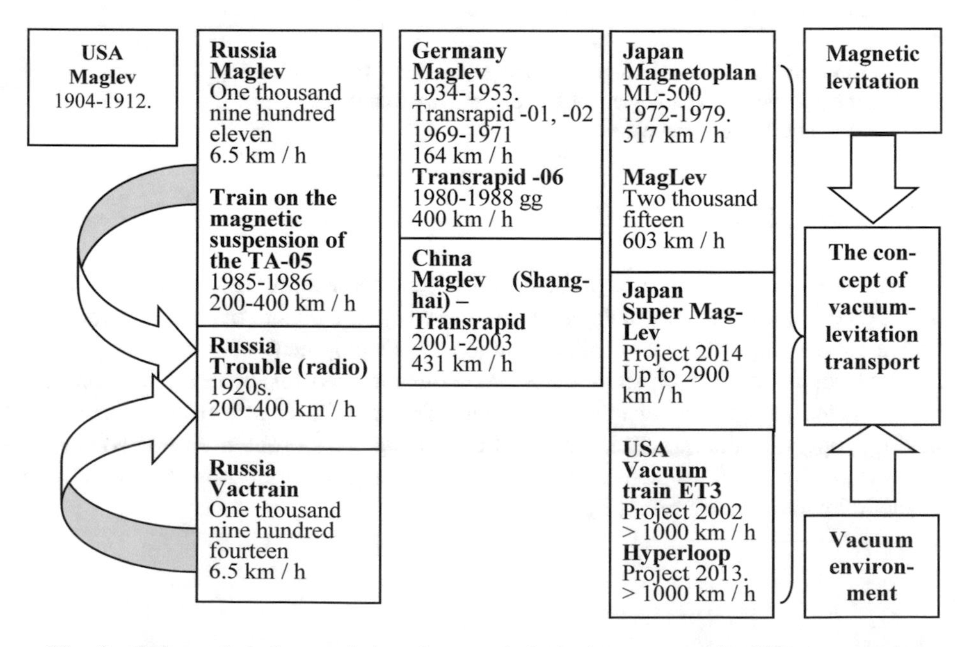

Fig. 2. Comparative characteristics of magnetic-levitation transport in different countries

2 Materials and Methods

The scientific research area considers the development of theoretical method on the basis of urban development innovative approach being implemented by Aleksandrov and Fedorova [5] and the social orientated macroeconomic approach described by professor Kalinina [6] and being relevant to the methodology of economics in the sphere of technology of magnetic levitation and the principles of using airbags. The advantage of such technology relies on the use a sparse medium, which significantly reduces the resistance of the medium to the movement of trains.

In China, on a commercial basis, magnetic levitation line Shanghai–Pudong is successfully operated for eight years at a speed of 430 km/h. This project has a duration of 180 km and creates its own moving speed of 550 km/h. countries Such as South Korea, USA, Japan, Germany in their transport and economic strategies provide for the establishment of the transport systems based on magnetic suspensions. At the same time, it is planned to operate at a speed of more than 470 km/h. To date, in Japan and Germany there are polygons for testing the technology of magnetic levitation with a length of about 40 km [2] (Fig. 3).

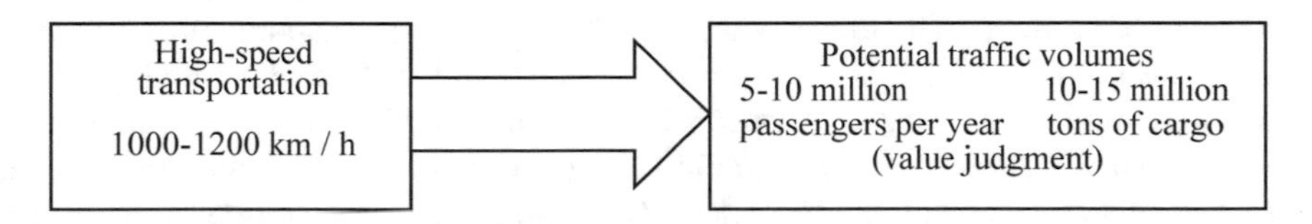

Fig. 3. Economic feasibility of using magnetic levitation transport

From an economic point of view, there is an opinion that one-time costs for the formation of transport systems based on magnetic suspension are prohibitive and not recoverable. But if you compare the cost of high-speed lines of the "wheel - rail" with the cost of transport routes on magnetic suspension technology, it is approximately equal. So, the Chinese have published the cost of the project implemented to date and it amounted to only 19 million euros per kilometer at the price of the cost of construction of a standard new road "wheel - rail" in Germany in the budget is laid 25 million euros per kilometer.

At the moment, analyzing this problem of the introduction of magnetic levitation transport routes in the framework of the functioning of the joint scientific Council of JSC "Russian Railways" managed to come to the formation of technical requirements for the creation and development of magnetic levitation transport. One of the fundamental questions is to identify commercial niches in the transport market in order to identify the most promising projects in terms of profitability and return on investment. Always the efficiency of transport systems in the modern world is ensured not by commercial operation, but by creating socio-economic effects and the formation of related businesses for investors of the project. And the most important issue is the formation of a full-fledged life support system, which should be implemented in the design of all systems (Fig. 4).

1	Assessment of energy optimality of power supply systems on the basis of existing energy sources and the possibility of accumulation of thermal and kinetic energy during the movement of rolling stock
2	Structural solutions of life support systems for passengers on Board the vehicle
3	Forecasting on the basis of estimated calculations of the thermal profile of rolling stock and infrastructure surfaces for different traffic modes in the speed range 0 + 1200 km / h

Fig. 4. Strategy for the introduction of high-speed transport

3 Results

The study found that the most economically sound and promising project at the moment is Hyperloop. The effectiveness of Hyperloop for transportation of various cargoes has been discussed for a long time. Consider and evaluate the viability of this technology can be the example of Dubai. Hyperloop technology is used in the port of Jebel Ali and is equipped with a train unloading system. The Dubai authorities at the time, signed a contract with the Hyperloop One, which consisted of two startups Elon musk and DP World, the largest port operator in the world [1] (Fig. 5).

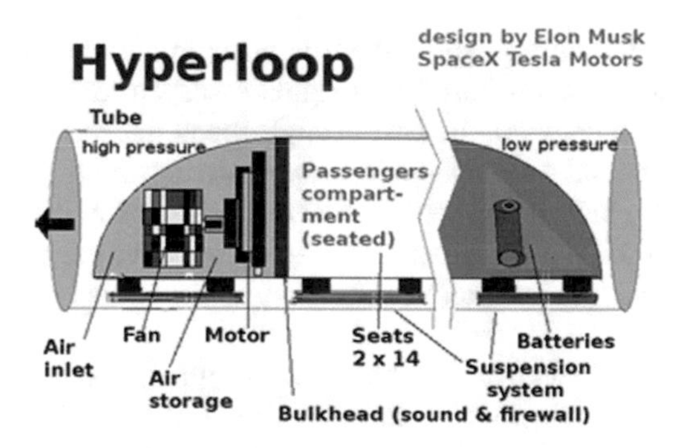

Fig. 5. Specific Hyperloop features

According to the project, Jebel Ali is equipped with a hypersonic cargo transportation system. It connects the seaport and container port. Freights are transported on the vacuum tube from the ships directly into warehouses. This makes it possible to significantly reduce the load on the port system and free the territory and part of the buildings in the Jebel Ali area. In the near future, the United States also plans to introduce Hyperloop as a public transport for a large number of passengers. Developed capsules, which travel at the speed of 1000 km/h the Feature of this project is that the system requires no external energy sources except those which itself will provide.

The project indicates the operating pressure in the tunnel 100 PA, it is almost a vacuum, as in the upper stratosphere. The maximum speed of the capsule can reach up to 1250 km/h on the route and reaches an acceleration of not more than 0.5 g, which corresponds to the comfort for passengers. These capsules are planned to be sent from the station with an interval of 30 s. This system allows for the movement of approximately 7.4 million people per year. When you consider the cost of the project and the depreciation (20 years), tickets in the capsule plan to sell an average of $20 [3]. Located on the roof of the solar panel transport tunnel magnetic levitation path fully provide the system with energy. The whole system consumes only 21 MW [4].

At the end of December 2017 completed another high-speed test Hyperloop. Engineering group “Hyperloop Virgin One” held the third in a row demonstrating its own non-standard transport system. Place tests chose desert outside Las Vegas. It was there that the designers sent an innovative magneto-levitating capsule to the next flight. However, the Hyperloop again flew empty - without passengers, through the vacuum tube of the test track. The speed of movement this time was almost 387 km/h.

Engineers achieved significant progress in comparison with the previous team result 308 km/h speed (Fig. 6).

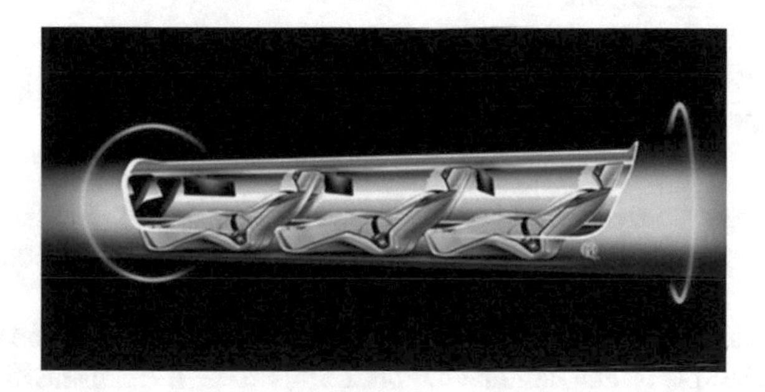

Fig. 6. Magnetic capsules device

4 Discussion

During a discussion of the results, it should be noted that it was not possible to achieve what was planned, taking into account that the theoretical maximum possible speed of the train should be 1125 km/h. Engineers at Virgin Hyperloop One believe that to achieve the theoretically stated speed, they will need to increase an additional 2,000 m of vacuum pipe. The current design of the tunnel with a length of 490 m, allows developing on the test track DevLoop the maximum possible speed is not higher than 401 km/h. Perhaps more important than the speed of the XP-1 capsule was the demonstration by a team of engineers of the new airlock technology (Fig. 7).

Fig. 7. Magnetic capsules design

In order to obtain the high velocities first described by inventor Elon Musk in 2013, the hyperloop module must follow through the actual vacuum medium of the tunnel.

The vacuum medium eliminates the effect of friction, allows the magnetic capsule to move at the speed of the aircraft along the entire length of the tunnel. The airlock becomes a key factor in terms of the functions of transferring baggage containers to passengers or cargo in the atmospheric environment. All components of the innovative movement system have been successfully tested. The following modules that have successfully passed the test are specifically mentioned:

- air lock,
- highly efficient electric motor,
- advanced controls,
- power electronics
- capsule suspension,
- vacuum unit.

The tests were carried out inside a sealed tunnel (pipe), lowering the internal pressure to a level equivalent to the air pressure at án altitude of 60,000 km above sea level.

Capsule from "Virgin Hyperloop One" confidently rises above the working track, acting on the principle of magnetic levitation. In this state, the capsule moves at a speed close to the speed of the aircraft, thanks to the ultra-low aerodynamic resistance inside the tube-tunnel. Meanwhile, the ability to maintain a vacuum in a pipe with a length of up to 2 km is one of the major problems for the test team. Every time a capsule train arrives at the station, it is necessary to slow down quickly and stop. After all the air lock must be closed hermetically (Fig. 8).

It should take enough time to hold the capsule behind the gateway till the moment of the next train arrival. The frequency with which these events occur determines the stability of the distance between the trains. The longer the stop, the lower the performance of vacuum capsule trains. This factor determines how useful an innovative mass transit system can be. You can, of course, try to increase the size of the capsule trains, but then you will need more durable steel to build the structure. But for this are: the inevitability of compensation for increased weight, increased financial costs.

Record speed "Virgin Hyperloop One" unconquered clearly long will not remain. The resulting figure is only slightly more than the speed of 350 km/h, which managed

Fig. 8. Magnetic capsules standing in tube

to reach the brand name of the famous American-Canadian Elon Musk in August of the same 2017 year.

This result was noted billionaire Elon Musk's project called "SpaceX Hyperloop Pod", which participated in the Hawthorne competition (California).

There-in Hawthorne, the student team of WARF Hyperloop (Germany) took the main prize of the competition for its own project capsule train, which drove with the result – 323.5 km/h.

5 Conclusions

While magnetic levitation-based infrastructure and vehicles are used in the delivery of goods to the Northern regions of the world, many factors need to be taken into account.

For example, in the Arctic in autumn, winter and spring often recorded very low temperatures and strong winds. This may have a negative impact on the safe operation of infrastructure and transport platforms, and is likely to increase operating costs.

In addition to low temperatures when using machinery and equipment in the Northern regions of the world, it is necessary to take into account the abundant snowfall and the risk of ice formation on metal structures. This ice and snow can lead to a weighting, both infrastructure and vehicles, which negatively affects the speed of transportation of people and freight, as well as, most likely, will lead to higher operating costs.

If monorail-based infrastructure and monorail-based vehicles are used to transport goods in the Northern regions of the world, this may require constant clearing of snow and ice to ensure trouble-free movement of transport platforms. Such cleaning and other measures aimed at improving the safety of the use of equipment, too, often lead to an increase in operating costs.

To date, the main projects related to the transportation of people and delivery of goods, based on the use of magnetic levitation, implemented in the warm regions of the world. When using equipment in the Arctic, operating costs can increase significantly compared to warm regions. Thus, the rise of expenses mentioned above leads to a thorough evaluation of the project, both technically and economically. While magnetic levitation-based infrastructure and vehicles are used in the delivery of goods to the Northern regions of the world, many factors need to be taken into account.

For example, in the Arctic in autumn, winter and spring often recorded very low temperatures and strong winds. This may have a negative impact on the safe operation of infrastructure and transport platforms, and is likely to increase operating costs.

In addition to low temperatures when using machinery and equipment in the Northern regions of the world, it is necessary to take into account the abundant snowfall and the risk of ice formation on metal structures. This ice and snow can lead to a weighting, both infrastructure and vehicles, which negatively affects the speed of transportation of people and freight, as well as, most likely, will lead to higher operating costs.

If monorail-based infrastructure and monorail-based vehicles are used to transport goods in the Northern regions of the world, this may require constant clearing of snow and ice to ensure trouble-free movement of transport platforms. Such cleaning and other measures aimed at improving the safety of the use of equipment, too, often lead to an increase in operating costs.

To date, the main projects related to the transportation of people and delivery of goods, based on the use of magnetic levitation, implemented in the warm regions of the world. When using equipment in the Arctic, operating costs can increase significantly compared to warm regions. All this leads to a thorough evaluation of the project, both technically and economically.

References

1. Morozova, E.I.: On the issue of creating a single conveyor-trunk system based on magnetic levitation. In: Tr. I International Science Conference “Magnetocavitation Transport Systems and Technologies”, PGUPS, OOO PUDRA, St. Petersburg, pp. 105–108 (2013)
2. Lapidus, B.M.: Socio-economic prerequisites for the development of high-speed communication in Russia. Vestnik Moskovskogo universiteta. Series 6: Economics, no. 6, pp. 52–63 (2014)
3. Kazuo, S.: Technological development of the superconducting magnetically levitated train, Sawade Kazuo. Japan Railway Eng. **160**, 8-4 (2015)
4. Rob, D.E.: Technical and economic comparison of high-speed-rail and maglev systems, D. Rob, Railway Tech. Rev. **1**, 9–15 (2016)
5. Aleksandrov, I., Fedorova, M.: International Science Conference SPbWOSCE-2017 Business Technologies for Sustainable Urban Development: Strategic planning of the tourism development in small cities and rural territories as a tool for the development of the regional economy MATEC Web Conf. 170 01011 (2017)
6. Kalinina, O.V.: Universal approach to building the progressive scale for income taxation. Actual Probl. Econ. 2, 387–400 (2016)

Application of Digital Technologies in Human Resources Management at the Enterprises of Fuel and Energy Complex in the Far North

Irina Zaychenko(✉), Anna Smirnova, and Valieriia Kriukova

Institute of Industrial Economics, Management and Trade,
Peter the Great St. Petersburg Polytechnic University,
195251 St. Petersburg, Russian Federation
imz.fem.spbpu@mail.ru

Abstract. The article discusses the relevance of the introduction of digital technologies in the management of human resources at the enterprises of the fuel and energy complex (FEC) in the regions of the Far North of the Russian Federation. The theoretical basis of the use of digital technologies was studied. It was found that in the implementation of the process of interaction with the staff, it is necessary to use digital technologies to improve the efficiency of the enterprise. Advantages and disadvantages of such types of technologies as e-training of personnel, electronic payroll, telemedicine technologies are defined. It is established that there is a positive trend in the use of modern digital technologies in the fuel and energy sector. However, the level of digitalization of enterprise data is not enough, which indicates the relevance of this study. In accordance with the purpose of the study, the object of the study was the regions of the Far North of the Russian Federation, for which telemedicine technologies were identified as technologies of paramount importance in the management of human resources at the fuel and energy enterprise. This is primarily due to the fact that the use of telemedicine will increase the level of socio-economic development of these territories.

Keywords: Fuel and energy complex · Strategic management · Human resources management · Digital technology

1 Introduction

At present, the stability of modern energy companies increasingly depends on the availability of the necessary number of qualified personnel and their effective use. This is due to the dynamic development of the market economy, which in turn leads to competition between different enterprises, the success of which depends on the most efficient use of available resources, including labor. In this regard, the search for the necessary reserves and factors that contribute to improving the efficiency of personnel use, as well as optimal planning of payroll costs is one of the priorities of the personnel management system in the enterprise. Improving this system is one of the most important problems of the organization in connection with the strengthening of the role of human resources as a leading factor in modern production.

V. Murgul and M. Pasetti (Eds.): EMMFT-2018, AISC 983, pp. 321–328, 2019.
https://doi.org/10.1007/978-3-030-19868-8_33

We should not forget that a sufficient number of modern enterprises with a large scale of activities are located in the Northern regions of Russia, due to the increased attention of the state to the development of these territories.

It should be noted that the Northern regions of Russia are characterized by low population density, which has maintained a negative trend since 1990 to the present moment. Also, one of the most important characteristics of the Northern region of the country is the inaccessibility of many settlements. It should be noted that the region of the Far North of Russia is also famous for a large number of oil and gas reserves, which attracts large fuel and energy industrial enterprises engaged in mining.

As you know, issues related to the provision and improvement of energy efficiency are of high importance for the activities of our country. The improvement of this indicator is directly dependent on political and socio-economic factors, the level of scientific and technological development of both the producing country and the countries with which cooperation is conducted. Due to the high demand for solutions to this problem, a large number of regulatory documents regulating the activities of the energy industry, as well as contributing to its development, have been developed. In particular, the State program "energy Saving and energy efficiency for the period up to 2020" was adopted [1]. It should be noted that the main regulations of the companies of this specificity are the Federal law on energy saving, adopted in 2009 [2, 3] and the Energy strategy of Russia until 2035.

However, on the way to solve the problem of energy efficiency, there are difficulties associated with the size of the territory of the Russian Federation, as well as the uniqueness of the regions (climatic and socio-economic conditions). Thus, it is necessary to take these factors into account in the development of strategic national programs of national energy security, as to date, none of the adopted programs is not able to ensure the effective organization of work related to the operation of energy equipment and energy efficiency.

As noted earlier, one of the most important factors of the company's success is the process of human resources management. Modern companies pay great attention to the development of employees, their well-being and the search for effective ways to interact with the staff.

2 Materials and Methods

Most of the fuel and energy companies with a large number of employees are already actively using automated technologies to optimize and simplify the process of interaction with employees, which allows to pay more attention to the most important business processes of the company. Next, we list the main digital technologies that are used in enterprises in the management of human resources.

Highly qualified personnel plays a key role in the development and provision of high-tech production of modern fuel and energy enterprises. Ensuring the competitiveness of the enterprise is impossible without appropriate staffing. In this regard, in recent years, the management of the quality of the workforce has special requirements [4].

Given the high responsibility, the need for a creative approach to work, the ability to make decisions independently and the mandatory high qualification of employees, it is necessary not only to build a competent and effective personnel management system at fuel and energy enterprises, but also to ensure its compliance with modern trends in the development of the economy, namely Industry 4.0 [5].

E-Learning for Staff

Often, enterprises need to improve their skills or training, which is associated with a rapidly changing environment that requires flexibility from the business. To this end, various virtual programmes, including interactive training courses and active participation of staff in the training process, have been increasingly used in recent years [6]. This process includes not only individual tasks that allow you to learn the material as accurately as possible, but also viewing specially prepared video lectures, interactive online classes or virtual online courses.

Information electronic systems currently have a strong impact on society as a whole, as well as on the company's management processes, including issues related to human resources management, training and education of personnel in order to ensure the career growth of employees. The advantages of using modern technologies ensures the development of e-learning, as well as allows you to get an education at a convenient time in a convenient place. The high demand for such technologies is due to the flexibility of the educational process, availability, including issues related to financial resources, ease of updating information, as well as the ability to view video lectures [7].

In addition to the obvious advantages of using corporate networks, it is also necessary to emphasize the fact of improving communication between employees, as it allows you to provide employees with relevant information necessary to perform effective work.

The necessary factors to increase the success of the use of electronic personnel training system include the following [8]:

1. active participation of trainers in the educational process;
2. development of effective performance monitoring mechanisms;
3. preparation of a high-quality curriculum;
4. close interaction between trainers and trainees;
5. use of standardized technologies.

Electronic Payroll System

A huge number of companies in the world have electronic information systems for payroll. These systems have a large amount of data, formulas. These platforms speed up computing time and make it automated. The ease of use of their platforms is that payroll calculations are stable for a long time. Typically, this is a common set of parameters:

- tax information;
- allocation of funds;
- staff working hours;
- staff salaries;
- employee payment history.

An important aspect of human capital management remains the relationship between staff and management with respect to payment systems. Thanks to these technologies, the payment system will more transparently, employees of the enterprise begin to feel that their wages and bonuses are equivalent to their contribution to the achievement of the organization's goals, which serve as a motivator to increase staff productivity [9].

The development and implementation of an electronic payroll system for business entities of the real sector of the economy can improve the efficiency of the business project management process. The integration of such a system will facilitate closer interaction between the company's employees and their direct management to achieve key performance indicators, as well as allow organizations to remain competitive in the electronic economy.

Telemedicine Technologies

It should be noted that modern digital technology involves the use not only to optimize business processes in the enterprise, as well as for medical personnel, which avoids injuries and accidents, provides timely necessary medical care, allows you to analyze the condition of employees and time to influence the situation. For these purposes, many enterprises, in particular industrial companies located in remote Northern regions of Russia, use telemedicine technologies [10].

The term "telemedicine" was introduced by Mark in 1974 [11] (there is also evidence that it was made by Thomas Byrd in 1970), includes a set of automated aspects. Diversity is associated, primarily, with a sufficiently high selection using technologies such as online consultations, monitoring patients at a distance, home telemedicine, remote inspection, etc.

A. V. Vladzimirsky [12] gave one of the most correct and accurate definitions of the concept of "telemedicine". He described telemedicine as a branch of medicine that uses electronic information technology to provide health care and services at the point of need (when geographical distance is a critical factor). it should be noted that the main purpose of telemedicine is to provide, including educational and Advisory services at a distance.

Thus, even a brief analysis of modern digital technologies of human resources management, combined with the characteristics of socio-economic and climatic conditions of life allow us to talk about the paramount importance of telemedicine technologies in the management of human resources in the region of the Far North of the Russian Federation in accordance with the objectives of state programs to improve the socio-economic development of this region in the North.

3 Results

Due to the high relevance of the development and use of telemedicine technologies in order to optimize the management of human resources at the enterprises of the fuel and energy complex, it is necessary to analyze the currently existing telemedicine technologies and ways of their practical application.

At the moment, two types of telemedicine activities are used: online lectures and consultations, as well as consultations between the doctor and the patient, or between doctors. The first process is carried out mainly in large medical centers in order to analyze the General problem and find its solution. The second process is used to consult doctors and patients both in the territory of one district and at a remote distance (regional, interregional levels) [13]. The process of providing such assistance is usually carried out by the doctor sequentially, as the collection of patient data and exchange of information can take 1–3 days. Often such a system is used for routine medical examination, but can also be used in critical situations and use instant messages [14].

The consultation process for telemedicine can be seen in more detail in Fig. 1.

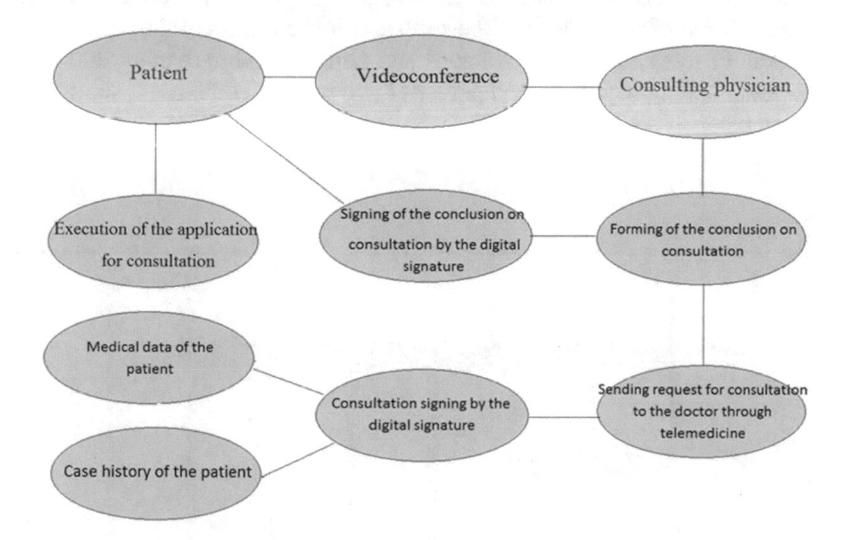

Fig. 1. The process of advising on the use of telemedicine

Thus, the process is as follows [15]:

1. The patient fills in all the necessary documents and forms a request for consultation.
2. Further request together with accompanying documents (clinical record of the patient and data on his condition) are sent by means of telemedicine technologies to the consulting doctor.
3. The consultant doctor analyzes the data and forms a report consultation and, if necessary, holds a video conference with the patient.
4. The consultant doctor makes a final verdict, signs the document with an electronic signature and sends it to the patient.

Similar consultations are already actively conducted in the regions of Siberia [16]. In the Krasnoyarsk region, the experience of using telemedicine technologies for about 10 years-this procedure began in 2008. This technology allows for communication between a doctor-consultant and a patient at sufficiently large distances on a communication video conference. Thus, in 2016, 7102 consultations were held in

Krasnoyarsk region. The problem of access to medicines also concerns the Yamal-Nenets Autonomous district. In 2013, 37 medical institutions were equipped with telemedicine technologies.

Another extreme North-Khanty-Mansiysk. This region is rich in oil and gas fields than a large number of mining enterprises in this region says. However, the harsh climate and inaccessibility lead to difficulties in providing medical care. Therefore, telemedicine has been used in this area since 2006. 52 medical centers were connected to the system, as well as training of doctors in remote regions was provided. Also, on the basis of the research Institute of information technology of Ugra, a Regional center for support of telemedicine service was established. The center carries out work on the use of telemedicine communications, technological support, as well as research and development of the latest versions of the software. Since the beginning of the activities of the system of telemedicine technologies and currently more than 24 000 consultations between medical specialists and patients have been [17]. Figure 2 shows in More detail the number of consultations held in 2018 in the Khanty-Mansiysk Autonomous Okrug.

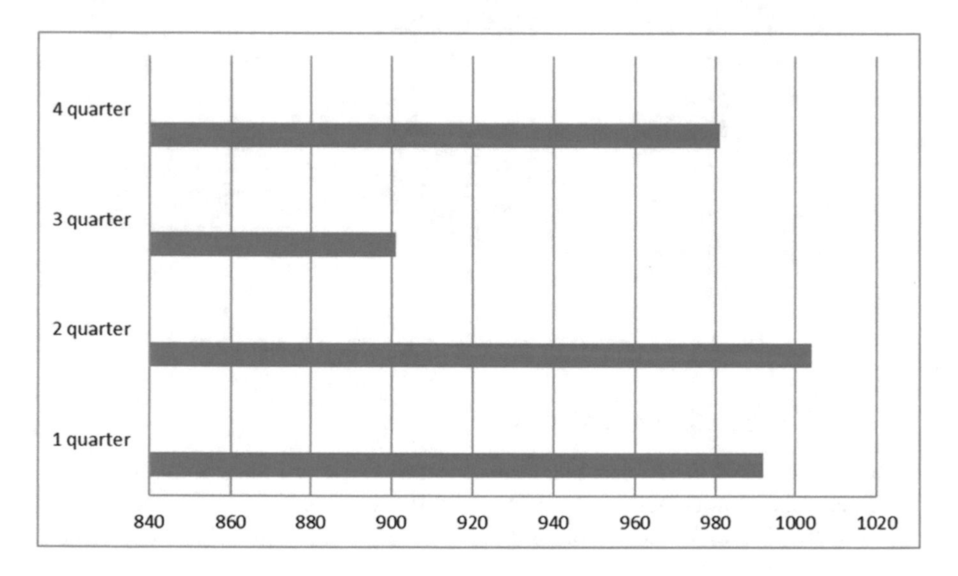

Fig. 2. The number of consultations in 2018 in Khanty-Mansi Autonomous Okrug

According to the data provided it is clear that the use of telemedicine is very important. The decline in the number of consultations in the 3rd quarter can be explained by the climatic features of this period (summer period). The greatest number of appeals falls on the spring period.

Figure 3 shows the dynamics of consultations over the years.

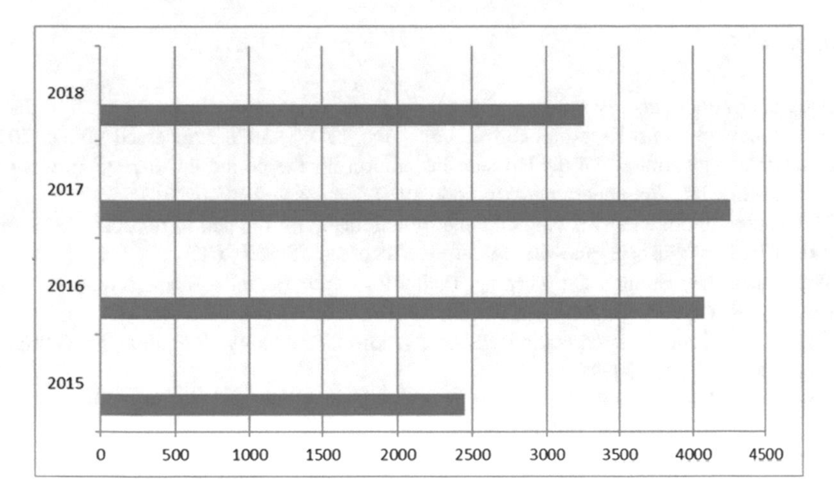

Fig. 3. Statistics of consultations in Khanty-Mansi Autonomous Okrug

According to the data shown in Fig. 3, the relevance of telemedicine use is quite high. It uses the following processes: consultations between doctors and doctors and patients, request additional medical information, lighting medical devices on video, sending notifications about all stages of consultation, notification of results and diagnoses, preparation of analytical reports. From the experience of the Institute of Ugra it is evident that the holding of such consultations in the same region is not more than a day, that promotes fast reaction to different diseases.

4 Discussion and Conclusion

Thus, the relevance of the use of telemedicine at the enterprises of the fuel and energy complex in the Far North, including for industrial workers is quite high, due to the harsh climatic conditions and inaccessibility of the regions where oil production is carried out. The use of telemedicine in enterprises can solve the following problems:

- Provision of qualified and rapid medical care in remote regions.
- Depreciation of medical services.
- Solving the problem of lack of qualified doctors in remote settlements.
- Increasing investment attractiveness and improving the company's image in the company.

In conclusion, it should be noted the importance of the introduction of digital technologies in the fuel and energy enterprises for human resources management in the Far North of the Russian Federation. Such technologies allow to solve the problems of optimization of business processes, to increase the overall productivity of employees, to increase the level of loyalty both on the part of staff and on the part of society. All these factors lead to an increase in investment attractiveness, can reduce the level of production defects, increase sales and take a leading position in the market.

References

1. Federal law about energy saving and increasing energy efficiency, from 23.11.2009 № 261-. FZ. http://www.consultant.ru/document/cons_doc_LAW_93978/. Accessed 20 Jan 2017
2. Draft of energy strategy of the Russian Federation for the period up to 2035 (revised from 01.02.2017). http://minenergo.gov.ru/node/1920. Accessed 22 Feb 2017
3. The state program energy saving and energy efficiency for the period till 2020. https://rg.ru/2011/01/25/energosberejenie-site-dok.html/. Accessed 18 Feb 2017
4. Westerman, G., Bonnet, D.: Turning Technology into Business Transformation. Harvard Business Review Press, Boston (2014)
5. Hayes, R., Pisano, G.: Operations, Strategy, and Technology: Pursuing the Competitive Edge, no. 7. Wiley (2004)
6. Bhatia, A.: Value Creation: Linking Information Technology and Business Strategy. Brown Books Publishing Group, Dallas (2012)
7. Shane, S.: Technology Strategy for Managers and Entrepreneurs. Prentice Hall, Upper Saddle River (2014)
8. Kinley, N.: Talent Intelligence: What You Need to Know to Identify and Measure Talent. Jossey-Bass, San Francisco (2013)
9. Phillips, J.: Making Human Capital Analytics Work: Measuring the ROI of Human Capital Processes. McGraw-Hill, New York (2017)
10. Christensen, R.: Roadmap to Strategic HR: Turning a Great Idea into a Business Reality. AMACOM, New York (2015)
11. Mark, R.G.: Telemedicine system: the missing link between homes and hospitals? Mod. Nurs. Home **32**(2), 39 (1974)
12. Bird, K.T.: Telemedicine: concept and practice, Springfield, Illinois, Thomas (1975)
13. Gutman, S., Teslya, A.: Environmental safety as an element of single-industry towns' sustainable development in the Arctic region. In: IOP Conference Series: Earth and Environmental Science, vol. 180, no. 1, p. 012010 (2018). https://www.scopus.com/inward/record.uri?eid=2-s2.0-85051926483&doi=10.1088%2f17551315%2f180%2f1%2f012010&parnerID=40&md5=d4cad4c5e3a07f9cd6767ddd9c9da445
14. Hersh, W.R., et al.: Clinical outcomes resulting from telemedicine interventions: a systematic review. BMC Med. Inf. Decis. Making **1**(1), 5 (2001)
15. Begiev, V.G., Andreev, V.B., Potapova, K.N., Moskvin, A.N.: Telemedicine in improving consultative and diagnostic care to highly specialized centres in the far north. Mod. Trends Dev. Sci. Technol. (1–3), 6–8 (2015)
16. Baranov, L.I.: Telemedicine. Progress based on the development of information technology. Med. Bull. Inter. Ministry (6), 74–77 (2015)
17. Kudryavtsev, D.V.: Using methods and tools of enterprise architecture management to support strategic management and enterprise transformation management. Prospects Inf. Technol. Dev. (6), 127–130 (2015)
18. Kozin, E.G., Ilyin, I.V., Levina, A.I.: Reengineering of the it architecture of the enterprise based on a service oriented enterprise architecture. Sci. Prospects (9), 48–56 (2016)
19. Makarenya, T.A., Quick, J.S.: Enterprise architecture: a tool of conflict resolution enterprise management. State and municipal management. In: Proceedings of the SKAGS, no. 4, p. 48 (2015)

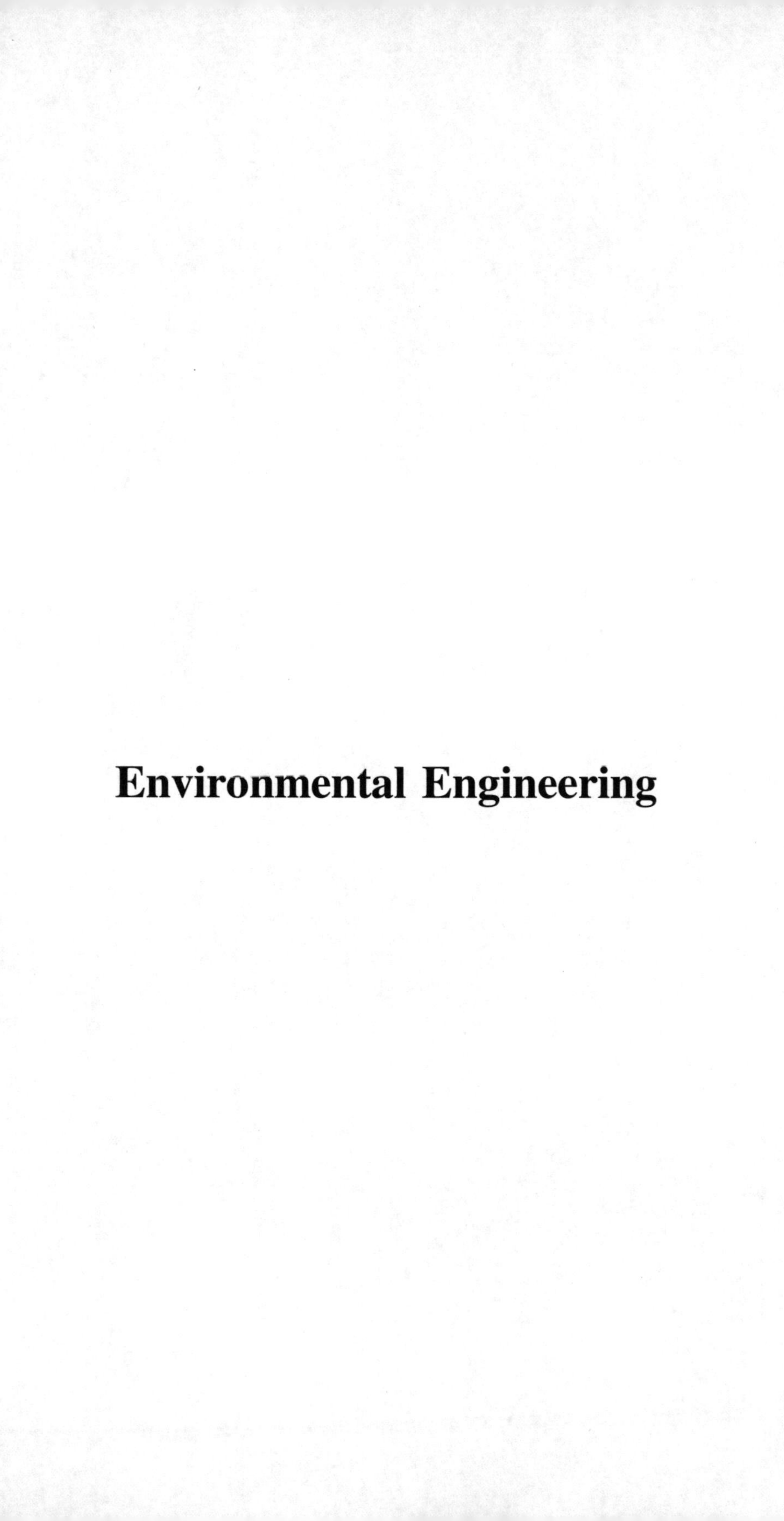

Environmental Engineering

Solving Environmental Problems Caused by the Pulp and Paper Enterprises via Deodorization and Dehydration of Sludge Lignin

Oleg Rudakov(✉), Olga Kukina, and Anatoly Abramenko

Voronezh State Technical University, Moscow Avenue, 14, Voronezh 394026, Russia
belyaeva-sv@mail.ru

Abstract. Paper reviews the method of deodorization and dehydration of sludge-lignin, which is based on the interaction of sludge-lignin with quicklime. CaO added to the sludge-lignin neutralizes lignin in the amount of around 15–20% and within 8–12 days binds volatile sulfur-containing components, functional organic groups and water. The proposed method can effectively reduce unpleasant odor to the level of 2–3 points based on a 5-point scale, as well as to diminish the humidity of the end product by 45–50%, making it look as a free-flowing powder, convenient for transportation and further use. Sludge lignin neutralized by the quicklime can be used for land recycling of municipal solid waste, as well as for other purposes. The recycling technology is characterized by the simple instrumental and the use of the inexpensive chemical reagent. The high economic efficiency of this method makes it promising for solving environmental problems of the 'Baikal Pulp and Paper Mill'.

Keywords: Environmental problems · Deodorization and dehydration · Sludge lignin

1 Introduction

The main problem inherited by the Irkutsk region from the industrial giant Baikal Pulp and Paper Mill (BPPM), closed in 2013, is a huge amount of wastes in the special warehouses (the so called 'hoarder cards') specially built for their storage.

First of all, we are talking about sludge lignin. It is formed during the production process of cellulose. It is a sludgy, viscous and foul-smelling mass consisting of water and a complex mixture of organic and inorganic chemicals. Almost 6 million m^3 of waste are currently stored in the storage 'hoarder cards' [1–8].

There is practically no information about the world practice of waste processing, similar to the sludge-lignin recycling applied at the BPPM in the academic periodicals. Thus, it could be affirmed so far that the processing of such waste is economically inefficient. There are certain restrictions on the industrial activities permitted in the Central Ecological Zone of the Baikal Natural Territory (Resolution of the Government

V. Murgul and M. Pasetti (Eds.): EMMFT-2018, AISC 983, pp. 331–338, 2019.
https://doi.org/10.1007/978-3-030-19868-8_34

of the Russian Federation No. 643 of August 30, 2001, No. 186 of 2.03.2015), as well as some infrastructure restrictions (such as the shutdown of production at the BPPM).

A large amount of aluminum in the form of alumina (SiO_2 + $Al2O_3$) - up to 11–27% of dry matter, as well as a high content of total sulfur (2.7–3.5% in lignin) - was found in ash and in sludge lignin.

High content of organochlorine compounds is observed. As a result of placing not only lignin, but also ash, the sludge-lignin in the storage 'hoard cards' contains a wide range of inorganic substances such as sodium ions, calcium, bicarbonates, sulfates and chlorides containing more than 50 mg/l [5].

The purpose of this paper was to investigate the possibilities for using the quicklime for deodorization and dehydration of the waste of the pulp and paper industry near Lake Baikal.

Our paper is aimed to research the negative impact of the accumulated Baikal Pulp and Paper Mill waste on the local environment. The relevance of the research raises no doubt. The BPPM is located in the increased seismic activity area, where any man-made accident or natural disaster can lead to the destruction of storage 'hoarder cards', breaking of dams and the ingress of huge amounts of toxic waste into Lake Baikal.

This may lead to the ecological disaster. Therefore, postponing the solution of this problem for an uncertain future is inadmissible. Currently, there is no economical and at the same time effective technology that allows to deodorize and to transfer the sludge lignin to its solid shape in large-tonnage volumes.

The technology developed by the authors offers the use of the affordable and relatively inexpensive neutralizing CaO component, which does not require sophisticated equipment for neutralization. It is located in the close proximity to storage 'hoarder cards' and significantly reduces the cost of processing sludge lignin.

2 Materials and Methods

Two batches of sludge-lignin from the storage 'hoarder cards' number 2 and number 3 in the amount of 2 kg were taken for the study. The batch from the card number 3 featured a lot of moisture as compared to the batch from the card number 2.

The chemical composition of sludge-lignin from the above cards was not determined. For processing sludge lignin we used the powdered quicklime (Produced by "Pridonkhimstroy lime") (Table 1).

Table 1. Quicklime indicators.

Indicators	Range
Active CaO + MgO	87–88%
Lime slaking time	5–9 мин
Lime slaking temperature	80–90 °C
The amount of deflated lime grains - not more than	5%
Sieve No. 2 residue – not more than	0.05–0.5%
Sieve No. 008 residue – not more than	3.0–6.0%

The method of processing slime-lignin with quicklime is based on adding of the CaO powder to slime lignin accompanied by the mechanical mixing in order to achieve homogeneous consistency, which is followed by the subsequent placement of the mixture obtained in the crystallizer.

During the interaction of lime with sludge lignin, a type of monolith is formed, which retains harmful organic and inorganic substances in its structure.

The proposed method presumes the use of relatively simple and inexpensive technologies. It still requires verification of its safety, optimization of the reagent ratios inserted into the sludge-lignin and the time required for the implementation of the method.

When CaO interacts with sludge lignin in the presence of H_2O and CO_2, a hydrophobic crust of calcium carbonate forms on the surface of the end product:

$$CaO + H_2O = Ca(OH)_2 + 61.5\,KJ;$$
$$Ca(OH)_2 + CO_2 = CaCO_3\downarrow + H_2O.$$

Since the interaction reaction of CaO with water is exothermic and comes with the heat release, the moisture from the processed lignin is partially bound and partially evaporated. In addition, the elevated temperature of the mixture inhibits microorganisms and thereby further neutralizes the sludge lignin.

The content of related substances in the sludge lignin ranges from 30 to 40% [5]. The unpleasant smell of sludge-lignin are the result of the presence of sulfur-containing compounds formed during the processing of wood and biochemical processes leading to the formation of hydrogen sulfide and mercaptans.

In the lignin itself there are phenolic hydroxyls, carboxyl and other groups that are capable of reacting with CaO. Below we listed the most likely reactions:

$$CaO + H_2S = CaS + H_2O;$$
$$CaO + H_2SO_4 = CaSO_4 + H_2O;$$
$$CaO + CO_2 = CaCO_3;$$
$$CaO + SO_2 = CaSO_3;$$
$$CaO + SO_3 = CaSO_4;$$
$$CaO + 2R\text{-}COOH = (R\text{-}COO)_2Ca + H_2O;$$
$$Ca(OH)_2 + SO_2 = CaSO_3 + H_2O;$$
$$Ca(OH)_2 + 2H_2S = Ca(HS)_2 + 2H_2O;$$
$$Ca(OH)_2 + H_2SO_3 = CaSO_4 + 2H_2O;$$
$$Ca(OH)_2 + H_2SO_4 = CaSO_4 + 2H_2O;$$
$$Ca(OH)_2 + 2C_6H_5SO_3H = (C_6H_5SO_3)_2Ca + 2H_2O;$$
$$Ca(OH)_2 + 2R\text{-}COOH = (R\text{-}COO)_2Ca + 2H_2O;$$
$$Ca(OH)_2 + 2C_6H_5OH = (C_6H_5O)_2Ca + 2H_2O;$$
$$Ca(OH)_2 + 2RSH = (RS)_2Ca + 2H_2O.$$

These reactions bind the volatile components of the starting product, and also bind its reactive capable groups that are part of the high molecular weight sludge lignin structures.

As a result, the rate of excretion of pollutants contained in lignin (organic and heavy metals) into the environment is reduced by hundreds of times compared to the original sludge, whereas volatile sulfur compounds are transferred to the non-volatile phase, which is poorly soluble in water.

3 Results

Olfactometric measurements were used to determine the optimal amounts of quicklime for its incorporation into the sludge lignin—an expert determination of odor intensity based on a 5-point scale (Table 2) of two batches of sludge lignin taken from different storage cards in the period of 6 Sept. 2018 to 17. Sept. 2018 based on the methodolgy for assessing the odors of building materials ("Sanitary-hygienic assessment of polymer and polymer-containing building materials and structures prescribed for the use in the construction of residential, public and industrial buildings. Methodical instructions No. 2.1.2.1829-04» [9, 10].

Table 2. Quantitative criteria for the description of the sludge lignin odor.

Scoring	Odor description
0	N/A
1	Subtle
2	Weak, does not attract attention if the expert is not aiming to detect it
3	Easily tangible and distinct
4	Strong
5	Unbearable; excluding the possibility of any long stay in the room

While studying each of the samples, seven experts (practically healthy individuals with no changes in the state of olfactory organs) were involved in the olfactometric observations. Each of the experts coherently inhaled the air from two breathing caps through the nose. One of the caps (the "experimental" one) was fed through the connecting tube from the climate chamber containing the sample of lignin and the other (the "control" one) was fed from the climate chamber without the researched material.

To assess the consistency of the opinions of the expert group, the dispersion and entropy coefficients of concordance were applied [10].

The dispersion coefficient of concordance is expressed by the ratio of the estimate of the variance to the maximum value of this estimate:

$$W = \frac{D}{D_{max}}. \tag{1}$$

The coefficient of concordance varies from 0 to 1, since $0 \leq D \leq D_{max}$.

The entropy coefficient of concordance was determined by the following formula:

$$W_e = 1 - \frac{H}{H_{max}} \quad (2)$$

where H – is an entropy and H_{max} – is a maximum entropy value.

The entropy coefficient of concordance also varies from 0 to 1. In a comparative assessment of the dispersion and entropy coefficients, these coefficients gave approximately the same assessment of the experts coherence.

Tables 3 and 4 and Figs. 1 and 2 show the measurements of the odor level of sludge-lignin before and after adding CaO in the course of the several days. The concordance coefficients are as follows: $W \geq W_e \geq 0.75$. This indicates a high consistency of the estimates of seven experts involved in the olfactometric measurements. As we can see, for 9 days the level of sludge-lignin odor (sample from card No. 3) reaches the environmentally acceptable level of 2 points as we apply 15–20% CaO. In the case of the card No. 2 sample, only the 3 points smell level is achieved.

Table 3. The change in the intensity of the odor of the sludge lignin from the storage card No. 2 after adding CaO based on a 5-point scale.

Time, day	Mass fraction of CaO, %				
	0	5	10	15	20
1	5	5	5	5	4
2	5	5	5	5	4
5	5	5	4	4	3
6	5	5	4	4	3
7	5	5	4	4	3
8	5	5	4	3	3
9	5	5	4	3	3

Table 4. The change in the intensity of the odor of the sludge lignin from the storage card No. 3 after adding CaO based on a 5-point scale.

Time, day	Mass fraction of CaO, %				
	0	5	10	15	20
1	5	5	5	5	4
2	5	5	5	5	4
5	5	5	4	4	3
6	5	5	4	4	3
7	5	5	4	4	3
8	5	5	4	3	2
9	5	5	4	3	2

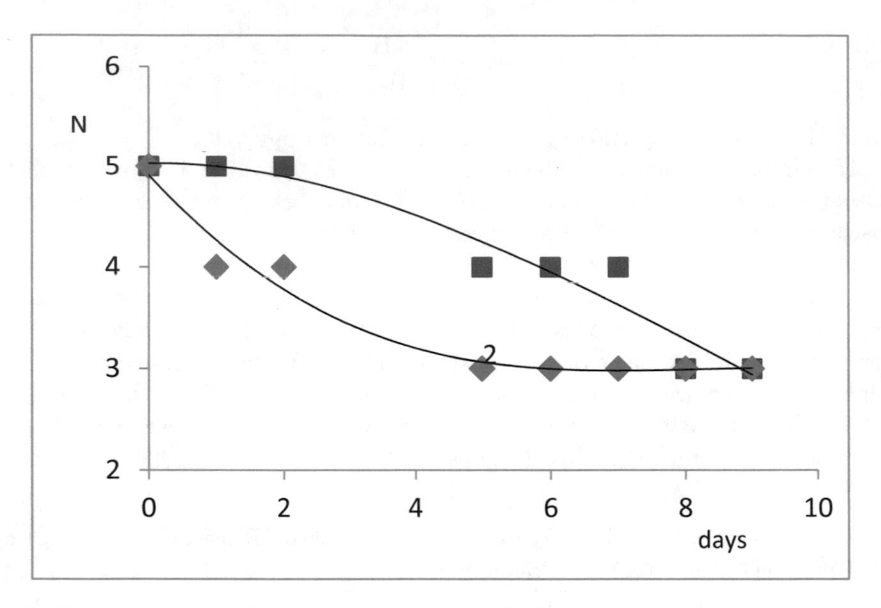

Fig. 1. The change in the intensity of the odor based on the sludge lignin 5-point scale (card number 2) depending on the aging time after adding the following amounts: 1 - 15%; 2 - 20% CaO.

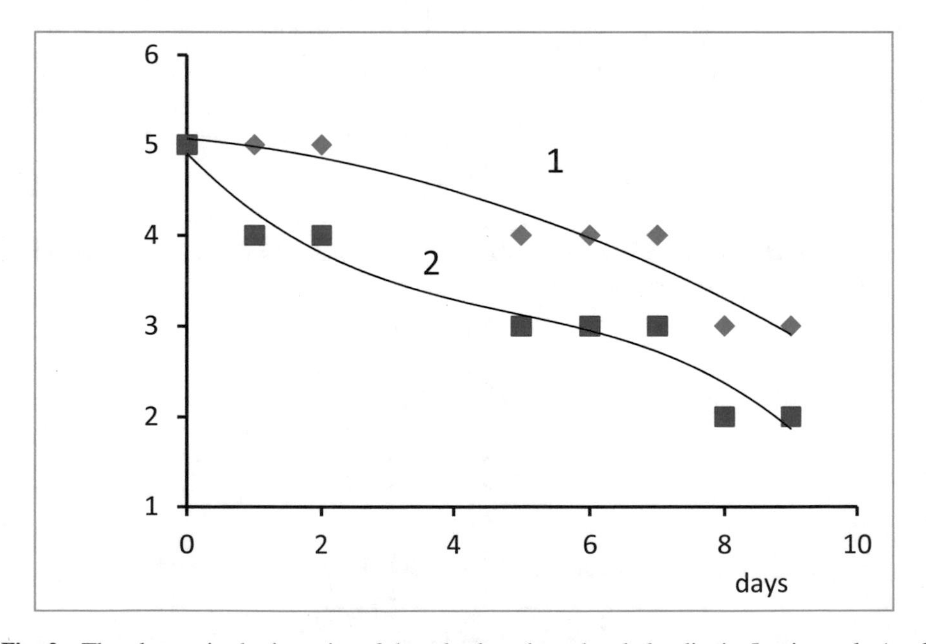

Fig. 2. The change in the intensity of the odor based on the sludge lignin 5-point scale (card number 3) depending on the aging time after adding the following amounts: 1 - 15%; 2 - 20% CaO.

The results indicate that CaO quite effectively deodorizes samples of the sludge lignin, however, in each case it is necessary to vary the dosage of quicklime and the duration of the aging time of the product. Apparently, the dosage of more than 20% CaO is not appropriate. It can be limited to the 15–20% portion. As for the time required to achieve a result, one can expect that 8–12 days will be enough.

When mixing samples of lignin with different amounts of quicklime, we observed to what temperature it is possible to heat the mixture. The maximum warming of up to 66 °C was observed after 10 min as we added 20% CaO to the sludge lignin.

Another parameter that we controlled during the processing of sludge lignin was the change in the pH of thc samples (Fig. 3). For both batches, as we added 20% CaO into the substance, the pH of the product reached 12.7 points (Fig. 4).

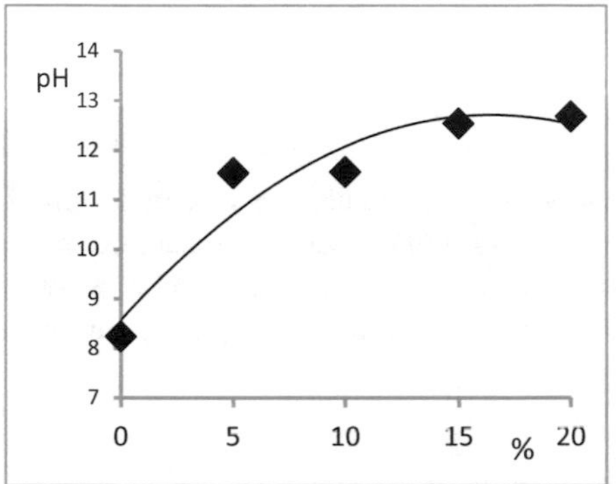

Fig. 3. Inter-dependence of pH and the ratio of the amount of CaO vs. sludge lignin (batch from the card No. 2).

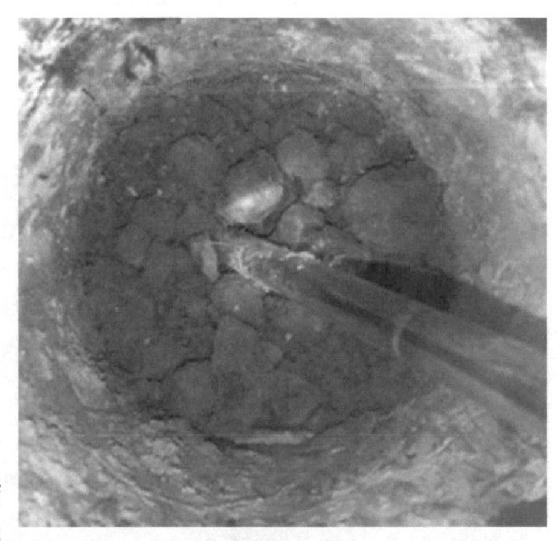

Fig. 4. The sample after adding 15% CaO, the batch from the card number 3.

After ten days of treatment with a 10–20% CaO solution, the moisture of the product dropped to 17–20%. The consistency of the product became lumpy, suitable for further transportation and use in the form of bulk materials.

4 Discussions

Thus, as a result of the interaction of quicklime with sludge lignin, volatile malodorous substances bind quite effectively within 1–2 weeks.

The major effect we observed when adding 15–20% of CaO to the sludge lignin. In some cases we received environmentally acceptable values of up to 2 points based on the sanitary standards for building materials [10].

For better deodorization, additional measures are needed - for example, longer aging time, or the use of other absorbing additives. When CaO is added to the sludge lignin, maximum heating of up to 66 °C could be achieved, which is not critical when using the discussed method.

Visually observing the consistency and color of the lignin mixtures with CaO, the increase in viscosity and subsequent clumping after 8–10 days were recorded.

Overly hydrated sludge lignin batches can be pre-dried before processing in natural conditions or using drying equipment.

For these purposes, it is possible to use freezing during winter, as it is recommended in [8]. While CaO is being added to the sludge lignin, clarification of the mass is observed, which especially increases with a growth in the proportion of CaO of up to 20% due to the formation of hydrated lime in the mixture.

5 Conclusion

In the process of chemical interaction of sludge-lignin with quicklime (15–20% CaO), a solid material is formed, which is suitable for humification and mineralization.

The required processing time for the sludge lignin with quicklime comprises 8–12 days and depends on the amount of sulfur-containing and other components within the lignin mass.

The technology of sludge lignin utilization with a quicklime is carried out by simple mechanical mixing, while quicklime not only neutralizes unpleasant odors, but also gives the finished product the consistency of free-flowing powder, convenient to use. The proposed method of sludge-lignin disposal is effective and quite economical.

References

1. Grachev, M.A., Adohin, N.A.: Expert opinion of national experts for the UN Commission on Industrial Development (UNIDO). The Impact of the Baikal Pulp and Paper Mill on the Environment and the Sustainable Development of the Economy of the Southern Coast of Baikal (1995)
2. Bogdanov, A.V., Rusetskaya, G.D., Mironov, A.P., Ivanova, M.A.: Complex recycling of pulp and paper industry wastes (2000)
3. State report: On the state of Lake Baikal and measures for its protection in 2007, Irkutsk (2008)
4. Imethenov, A.B., Kulikov, A.I., Atutov, A.A.: Ecology, environmental protection and nature management, Ulan-Ude (2001)
5. Suvorov, E.G., Antipov, A.N., Semenov, Y.M.: Territorial development of the city of Baikalsk and its Suburban Area, Irkutsk (2003)
6. Miao, Q., Huang, L., Lihui, C.: The brief review. BioResources **8**(1), 1431–1455 (2013)
7. Timofeeva, S.S., Cheremis, N.V., Shenkman, B.M.: The current state of surface, groundwater and wastewater in the zone of impact of the sludge collectors of the Baikal Pulp and Paper Mill. Mod. High Technol. **5** (2008)
8. Shatrova, A.S., Bogdanov, A.V., Kachor, O.L.: Investigation of the physicochemical properties of sludge lignin sediments from Baikal Pulp and Paper Mill during freezing. Bull. Irkutsk State Tech. Univ. **8**(103), 99–107 (2015)
9. Mayorov, V.A.: Smells and their perception, impact, elimination, Moscow (2006)
10. Rudakov, O.B., Babkina, E.V., Davydov, E.G.: Odorimetric control of the safety of polymer-containing building materials. Scientific Herald of the Voronezh State University of Architecture and Civil Engineering (2015)

Waste-Free Phosphogypsum Processing Technology When Extracting Rare-Earth Metals

Sergey Zolotukhin(✉), Ol'ga Kukina, Valeriy Mishchenko, and Sergey Larionov

Voronezh State Technical University, Moscow Avenue, 14, Voronezh 394026, Russia
jarogacheva@mail.ru

Abstract. According to expert estimates, phosphogypsum is the largest technogenic waste in terms of volume. The purpose of this study is to develop a technology for the extraction of non-radioactive rare-earth metals from phosphogypsum and at the same time to get building materials and products using nonfiring technology. "Integra.Ru" Co Ltd developed an innovative technology for extracting rare-earth metals from phosphogypsum, based on the method of sulfuric acid decomposition. A pilot plant for the processing phosphogypsum with a capacity of about 15 thousand tons of phosphogypsum per year (produces up to 40 tons of collective concentrate of rare-earth metals per year) was designed, built and put into operation on the basis of the Voskresensk mineral fertilizer plant. Voronezh State Technical University has the patent of the Russian Federation "Raw mix for manufacturing building products using nonfiring technology" from phosphogypsum and pilot production of wall materials using nonfiring technology. Combining the two technologies will make it possible to establish non-waste production of phosphogypsum processing with the extraction of non-radioactive rare-earth metals and the direct production of building products. Waste-free phosphogypsum processing technology with the release of rare-earth metals and the simultaneous production of wall building materials will help solving the environmental problems of cities where phosphogypsum is located and reducing the cost of wall materials.

Keywords: Waste-free processing · Rare-earth metals · Phosphogypsum

1 Introduction

According to expert estimates, phosphogypsum is the largest technogenic waste in terms of volume. About 300 million tons of phosphogypsum has been accumulated in the Russian Federation (Voskresensk, Tambov, Cherepovets, Veliky Novgorod, Kingisepp; Kirovo-Chepetsk; Balakovo; Belorechensk). A specialized agency of the United Nations Educational, Scientific and Cultural Organization recognized the problem of the utilization of phosphogypsum as worldwide. Phosphogypsum contains rare-earth metals, which can be used after processing into a collective concentrate of oxides of rare-earth metals and individual oxides of Ce, La, Nd, and Pr. Rare-earth

V. Murgul and M. Pasetti (Eds.): EMMFT-2018, AISC 983, pp. 339–351, 2019.
https://doi.org/10.1007/978-3-030-19868-8_35

metals are used in the most high-tech and modern technologies, such as production of nanocrystals, super-power magnets, batteries for electric vehicles, solid-state lasers, special ceramics and coatings, in nuclear power, aerospace, telecommunications, etc. World consumption of rare-earth metals is 110 thousand tons per year. Experts predict that by 2020, global consumption of rare-earth metals will reach 200–220 thousand tons and will increase further. China was the main exporter of rare-earth metals until 2010 (85% of world exports). Russia in the segment of rare-earth metals is currently an import-dependent country since it does not have the capacity to separate rare-earth concentrates. The Netherlands, Kazakhstan, China are the main suppliers of rare-earth metals to Russia. Currently, the average prices for the collective concentrate of rare-earth metals reach $ 40 per kg. Prices for separated metals reach 3–4 thousand dollars per kg (data from MetalResearch). One ton of phosphogypsum contains up to 4 kg of the concentrate of rare-earth metals.

«Integra.Ru» Co Ltd developed an innovative technology for extracting rare-earth metals from phosphogypsum (RF patent "Method for complex processing of phosphogypsum RU 2639394") [1], based on the method of sulfuric decomposition and it has five stages. The technology is environmentally friendly and virtually waste-free. Solid waste (cakes) is technical gypsum that can be used (and is currently used) in the production of building materials. Liquid waste meet the technical requirements for industrial water. There are no gas emissions. The subsequent division into individual rare-earth metals is carried out according to the standard extraction technology. Ce, La, Nd and Pr oxides with purity of 99.0% plus medium-weight rare-earth metals concentrate (Sm, Gd, Y, Eu, Dy, Tb) are obtained (Fig. 1).

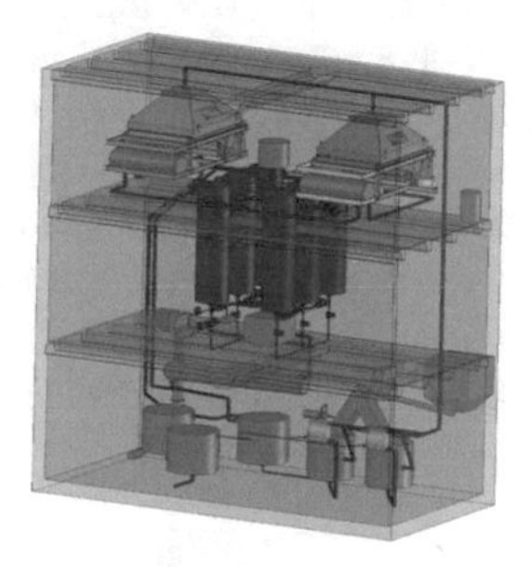

Fig. 1. Technology for extracting rare-earth metals from phosphogypsum.

A pilot plant for the phosphogypsum processing with the capacity of about 15 thousand tons of phosphogypsum per year (produces up to 40 tons of collective concentrate of rare-earth metals per year) was designed, built and put into operation on the basis of the Voskresensk plant for the production of mineral fertilizers (Fig. 2). The pilot plant covers an area of 216 m^2. The technological chain includes 6 contact tanks, 2 nutsch filters, 6 crystallizers, various tank and pumping equipment. The installation operates under the control of an automated control system. In fact, a capacity of 2 tons per hour was achieved in terms of feedstock. The collective concentrate of high-quality rare-earth metals was accumulated.

Fig. 2. Plant for the production of rare-earth metals concentrate.

A line for the production of tongue-and-groove slabs with a capacity of up to 140 thousand square meters per year was built. The actual costs of reagents and energy resources were determined, the optimum process mode was established. Large-scale tests confirmed the continuous extraction of rare-earth metals from phosphogypsum into finished products at the level of 50–60% (Fig. 3).

Metal		Mass concentration. %
Ce	Cerium	38.71
Nd	Neodymium	18.20
La	Lanthanum	14.21
Pr	Praseodymium	3.91
Sm	Samarium	2.36
Gd	Gadolinium	1.73
Y	Yttrium	1.31
Eu	Europium	0.63
Dy	Dysprosium	0.51
Nb	Terbium	0.14

Fig. 3. Collective concentrate and total rare-earth content metals in it.

The by-product obtained after extraction of the concentrate of rare-earth metals is processed into gypsum of the grades G-5, G-6 and products from it (tongue-and-groove plates). However, the production of gypsum from phosphogypsum is an energy-intensive process, since the phosphogypsum binder is burned at the temperatures above the temperature traditionally used to obtain a binder from natural gypsum stone [2–7], making its production economically unprofitable. That is why the production is currently stopped. To solve the problem of profitability, it was decided to conduct research on the search for energy-saving technology of direct production of wall products from phosphogypsum, having previously checked the waste of chemical production for radioactivity.

The purpose of this study:

- to determine the radiological characteristics of phosphogypsum of the Voskresensk Mineral Fertilizers enterprise, to assess the possibility of processing in the production of building materials;
- to conduct experimental studies on the possibility of directly obtaining building materials;
- to develop a technology for the extraction of non-radioactive rare-earth metals from phosphogypsum with simultaneous production of building materials and products using nonfiring technology.

The literature review and patent search for the extraction of rare-earth metals is based on the leaching of rare-earth metals in acids, sulfuric acid with a content of 2–30% is introduced into the phosphogypsum pulp. The phosphogypsum pulp formed after this has a strong acidity. In order to neutralize it, as a rule, slaked lime or quicklime is injected, after which the mass with a pH of more than 7 is burned in gypsum boilers or in autoclaves due to the high content of dihydrate gypsum. We also tried to boil gypsum, however, even from the dump, which was more than 20 years old with a pH of about 7, it was not possible to obtain a gypsum binder. At the same time, gypsum firing lasted for more than two hours, which is not economically efficient. Studies [2] found that, in a suspension of lime, it is impossible to achieve deep neutralization of phosphogypsum, since some of the acidic impurities and decomposition residues of apatites and phosphorites are located inside the crystalline hydrates of phosphogypsum, and are not removed into the solution when processed in suspension. Therefore, we consider this direction a dead end.

Another direction of phosphogypsum processing is its mechanochemical activation [8], where phosphogypsum activation is achieved by grinding in ball mills or high-speed mixers with the addition of lime, clay, sand, loamy sand. The disadvantages of this area of research are low-tech since, with high dispersion, the coking of activated phosphogypsum occurs. It is possible to obtain materials under laboratory conditions; in industrial volumes, caking may cause the equipment to plaster and break. These studies allow us to conclude that the dispersion of gypsum crystalline hydrates will lead to an increase in the strength characteristics of the materials obtained.

The next, most promising direction, we consider pressing materials. However, the molding of composite materials from phosphogypsum at the pressures of 50–300 MPa with the introduction of microsilica, aluminum hydroxide, which are capable of producing crystalline hydrates, [2] did not let us achieve a strength above 12 MPa with a softening coefficient of not more than 0.44.

Equipment with the pressure of 10–20 MPa is currently the most common for the production of wall materials. Therefore, in our research, studies were conducted with these modes.

We noted that materials with an energy of dehydration of more than 200 J/g can exhibit binding properties [9–12]. However, this is not enough to get durable waterproof materials. It is necessary to introduce components that, when dissolved in water, form ions that interact with calcium sulfate to form new phases:

$$Ca^{2+} + SiO_3^{2-} + H_2O \rightarrow xCaO \cdot ySiO_2 \cdot zH_2O$$

$$Ca^{2+} + PO_4^{3-} + H_2O \rightarrow xCaO \cdot yP_2O_5 \cdot zH_2O$$

$$Ca^{2+} + CO_3^{2-} + H_2O \rightarrow xCaO \cdot yCO_2 \cdot zH_2O$$

$$Ba^{2+} + SO_4^{2-} + H_2O \rightarrow xBaO \cdot ySO_3 \cdot zH_2O$$

$$Pb^{2+} + SO_4^{2-} + H_2O \rightarrow xPbO \cdot ySO_3 \cdot zH_2O$$

$$Ca^{2+} + FeO_3^{3-} + H_2O \rightarrow xCaO \cdot yFe_2O_3 \cdot zH_2O$$

$$Ca^{2+} + TiO_3^{2-} + H_2O \rightarrow xCaO \cdot yTiO_2 \cdot zH_2O$$

Before pressing, it is necessary to conduct mechanochemical activation of the surfaces, which will contribute to the growth of the internal energy of the system and the onset of recrystallization of phosphogypsum. At this time it is necessary to compress the product. We proved that during pressing it is necessary to achieve directional formation of the structure with a water film thickness of 1–100 nm [9–12]. The process of mechanochemical activation is accompanied by the supply of heat when lime is slaked and the mixer is used with heating to a temperature of more than 60 °C.

We suppose that directly obtaining products after mechanochemical and thermal activation of waste by pressing is the most cost-effective technology for the processing of phosphogypsum into building products when extracting non-radioactive rare-earth metals.

2 Materials and Methods

The radiological characteristics of phosphogypsum of the Voskresensk Mineral Fertilizers Enterprise and the Uvarovo Chemical Plant were determined at the testing laboratory center of the Federal Public Health Institution Center for Hygiene and Epidemiology in the Voronezh Region on the MKS-01A Multiradom spectrometric unit (Tables 1 and 2).

Table 1. Radiological tests of phosphogypsum of “Resurrection mineral fertilizers” production.

No	Determined indicators, units	Research results	Hygienic standard
1	Specific activity of Potassium-40, Bq/kg	less than 46.7	–
2	Specific activity of Thorium-232, Bq/kg	41.8 ± 7.3	–
3	Specific activity of Radium-226, Bq/kg	27.1 ± 7.5	–
4	Specific effective activity, Bq/kg	81.7 + 14.8	Not more than 370,0 for the first class

To determine the optimal formula-technological factors for obtaining lime-sand phosphogypsum material, we used dump phosphogypsum of the Uvarovo chemical plant in the Tambov region, sandy loam with a specific surface area of 4000–4500 cm^2/g from the open pit of the Tambov region of the Uvarovskiy district located in the immediate vicinity of phosphogypsum deposits, construction quicklime (manufacturer: Rossosh, "Pridonkhimstroyizvest"). The characteristics of the materials used are given in Tables 3, 4 and 5.

Table 2. Radiological tests of phosphogypsum at the Uvarovo chemical plant.

No	Determined indicators, units	Research results	Hygienic standard
1	Specific activity of potassium-40, Bq/kg	less than 70.4	–
2	Specific activity of thorium-232, Bq/kg	26.1 ± 8.0	–
3	Specific activity of Radium-226, Bq/kg	19.1 ± 5.9	–
4	Specific effective activity, Bq/kg	68.4 + 12.7	Not more than 370.0 for the first class

Table 3. The results of particle size analysis of dump phosphogypsum of the Uvarovo chemical plant.

Residue	Residues,% by weight, on the screens					Pass through a screen with a 0.16 mesh,% by weight
	2.5	1	0.63	0.315	0.16	
Phosphogypsum						
Particular	13.78	8.74	9	17.86	30.82	19.8
Total	13.78	22.52	31.52	49.38	80.2	–

Table 4. The chemical composition of phosphogypsum of various enterprises (wt. %).

Plant	Total moisture	pH	P_2O_5	F	CaO	SiO_2	Fe_2O_3	SO_3
The Voskresensk Mineral Fertilizers Enterprise	42–45	2.5	0.8–1.2	0.15–0.55	35–42	0.2–0.6	0.2–0.3	0.3–0.7
The Uvarovo chemical plant	15.50–20.0	7–8	6–7	–	27–28	1–2	2–3	40–42

The physicomechanical properties of the materials were determined in accordance with the requirements of international state standards on the Instron 5982 universal electromechanical testing system with an error of ±0.5% at the Center for Collective Use of Voronezh State Technical University.

Differential scanning calorimetry was performed on the STA 449 F5 A-0082-M synchronous thermal analysis device (NETSCH, Germany) with NETSCH Proteus software with an integrated, reliable, monolithic scale system and highly sensitive DSK and DTA sensors in the modern unit of STA 449 F3 Jupiter®. The study of the

Table 5. The chemical composition of loamy sand and quicklime.

Mass fraction of oxide, %	The name of the material	
	Loamy sand	Quicklime
Al_2O_3	14.04	0.32
SiO_2	69.35	1.08
CaO	3.52	62.30
Na_2O	0.73	0.04
MgO	1.66	1.27
P_2O_5	0.29	0.02
K_2O	3.10	0.09
TiO_2	0.96	–
Fe_2O_3	5.55	0.26
MnO	0.09	0.01
S	0.16	0.07
C	–	23.12
H_2O	–	11.11

microstructure of materials was carried out on a JSM-6380LV scanning electron microscope. The images show the microstructure of the chipped surface of the samples. The micron marker present in the images makes it possible to estimate the grain size and intergranular thickness of water films. The water-solid ratio of the studied compositions was selected from the condition of convenient molding and obtaining defect-free samples. Mechanochemical activation was performed in a laboratory mixer MLA-30, with a rotational speed of the blade around its axis 126 rpm, the maximum heating temperature (200 ± 5) °C.

Hydrogen pH was determined using an I-160 ionomer.

Weight characteristics were determined using an electronic laboratory balance PH-3413 with an accuracy of 1 g. Forming samples of cylinders with the size of 5 × 5 cm was performed using a hydraulic press PSU-125. The unburning technology for producing lime-sand phosphogypsum material is based on the fact that the production line for the manufacturing wall materials meets the requirements for a line for the production of wall materials from waste phosphogypsum. This technology can be implemented on standard equipment for the production of wall materials. The technology of wall materials based on waste phosphogypsum using an unburning technology was tested under laboratory conditions using differential scanning calorimetry and controlling the dehydration energy of all components of the raw mix before dosing. Sand was milled in a ball mill until a specific surface area of 4000–4500 cm^2/g was obtained, which was added to the laboratory mixer MLA-30 together with dump phosphogypsum, sand and lime at the time of its slaking with water at a temperature of 90 °C. Stirring was carried out at a temperature of more than 80 °C for 45–50 min. The samples were pressed on a hydraulic press in forms 250 × 120 × 65 at a pressure of 5 MPa per sample. The samples were gaining strength for 28 days at a temperature of 20 °C and the humidity of 55%.

3 Results

According to the measurement results of the effective specific activity of natural radionuclides, the samples of phosphogypsum are classified as class 1 materials in accordance with the Unified sanitary-epidemiological and hygienic requirements for goods subjected to sanitary and epidemiological supervision (control) and can be used as building materials without restrictions by the radiative factor.

Voronezh State Technical University has the patent [9] of the Russian Federation "Raw mix for manufacturing building products using nonfiring technology No. 2584018 of phosphogypsum" and pilot production of wall materials using non-firing technology. Combining the two technologies will make it possible to establish non-waste production of phosphogypsum processing with the extraction of non-radioactive rare-earth metals and the direct production of building products (Fig. 4).

The composition of 10% quicklime, 60% phosphogypsum cake, 30% of loamy sand was used as the optimal composition for the introduction of technology using standard equipment for the production of wall materials. Some physical and mechanical properties and the appearance of products from phosphogypsum are presented in Table 6 and in Fig. 5.

Table 6. Physical and mechanical properties of products, made from the composition of the cake phosphogypsum - 60%, loamy sand - 30%, lime - 10% with the pressure of 3–20 MPa.

Characteristic	Value
Average strength, MPa	3 … 20
Water absorption	0.2–0.3
Softening coefficient	0.6–0.7
Density, kg/m^3	1600–1750

Fig. 4. The resulting products from phosphogypsum.

In this technology, in comparison with the known technologies for the production of wall materials and products, there is no autoclave, which leads to an economic effect: reducing the cost of their production of wall materials from the phosphogypsum cake 2–3 times, and the problem of the utilization of phosphogypsum is solved.

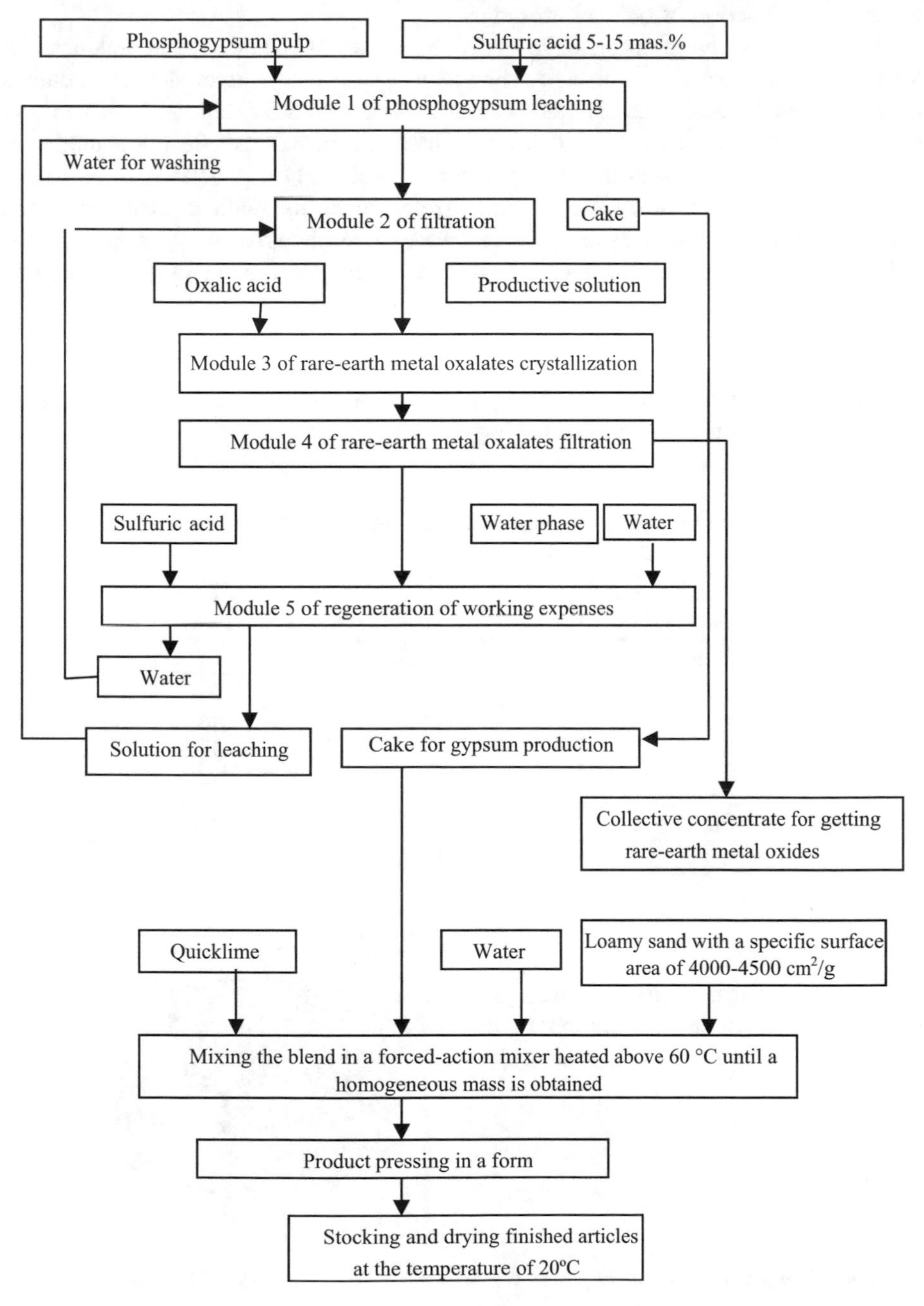

Fig. 5. Technological scheme of the waste-free phosphogypsum processing technology for the extraction of rare-earth metals.

4 Discussion of the Results

The results of repeated radiological tests of the phosphogypsum of the Uvarovo and the Voskresensk chemical plants confirm its environmental safety and the possibility of its use in the production of building materials.

Studies of the dump phosphogypsum of the Uvarovo chemical plant conducted at Voronezh State Technical University showed that over the years of storage in dumps for more than 20 years, the pH value has increased from 2.50–2.86 to 7.13–7.71 and does not change with grinding and heating, which also proves the effect of neutralizing waste phosphogypsum with changing properties (Table 7, [13]), however, there was no complete removal of apatites and phosphorites even for such a long time. The micrographs (Figs. 6 and 7) of fresh and waste phosphogypsum show, despite the differences in the pH of the medium, that complete destruction of the crystalline hydrates did not occur. This confirms the presence of phosphates in the structure of the undecomposed phase of dihydrate gypsum. Because of this, our attempts to boil seemingly neutralized phosphogypsum into high-quality gypsum binder did not give a positive result, which confirms the impossibility of obtaining high-quality gypsum from phosphogypsum by firing technology.

Table 7. The results of determining the pH of aqueous extracts with dump phosphogypsum of different specific surface area.

No	Specific surface area phosphogypsum, cm^2/g	pH value (t = 20 °C)	pH value (t = 60 °C)
1	1044	7.30	7.50
2	1424	7.13	7.30
3	1896	7.25	7.40
4	2017	7.35	7.40
5	2160	7.6	7.64
6	3000	7.62	7.86
7	4000	7.71	7.82

Fig. 6. Micrograph of phosphogypsum crystals (pH 2–3) at the scale 300:1, taken in 1998.

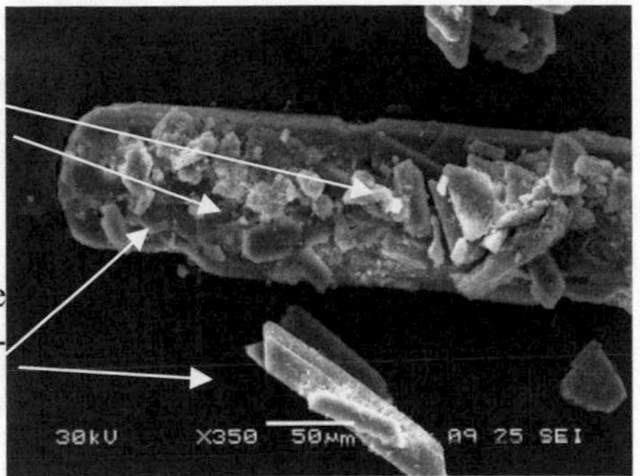

Fig. 7. Micrograph of the original non-compressed dump phosphogypsum (pH 7–8) (×350), taken in 2017.

The results of thermal studies showed that lime, which is an essential component of neutralizing acidic impurities and fluorides in phosphogypsum, loses binding properties over time. Phosphogypsum constantly exhibits binding properties, and sand does not possess binding properties (Table 8).

Table 8. Material.

Material	Energy of dehydration, J/g
Quicklime	772
Quicklime, pressed at the pressure of 5 MPa at the 1st day of hardening	296
Quicklime, pressed at the pressure of 5 MPa on the 3rd day of hardening	127
Quicklime, pressed at the pressure of 5 MPa on the 24th day of hardening	10
Original raw phosphogypsum	320
Dump phosphogypsum after 20 years of storage	292
Loamy sand	1
Lime-sand phosphogypsum material	174

We could not fix new composition as a result of pressing lime-sand phosphogypsum material. However, the chemical compatibility of the components used made it possible to increase the strength and water resistance of composite materials. The reason for this is to consider the high dispersion of hydrated lime, self-dispersion of phosphogypsum crystalline hydrates during their thermo-mechanical-chemical activation, self-organization of nanoscale particles of the components to obtain the water film thicknesses of 1–100 nm. Lime particles, clays, fluorides, phosphates, sulphates, hydrosulphates, due to their high dispersion, fill the space between the phosphogypsum dihydrate crystals, creating the optimal thickness of the water films, which is clearly seen in Fig. 8 [14].

Fig. 8. The microstructure of an unburned lime-sand phosphogypsum material.

Therefore, it can be argued that the strength of materials obtained by unburning technology is achieved by the self-organization of the structure in which dispersed materials with dehydration energy of more than 200 J/g are present. Components with nanoscale particles capable of taking the place of thermodynamic instability form crystallization contacts with nanoscale thicknesses of aqueous films. A decrease in the pore space (an increase in the pressure, a decrease in the mass fraction of lime, an increase in the mass fraction of clay components) will contribute to an increase in the strength and water resistance of the composites.

5 Conclusion

The results of radiological tests of the phosphogypsum of the Uvarovo and Voskresensk chemical plants confirm its environmental safety and the possibility of its use in the production of building materials.

From the results of the research, it is clear that standard equipment for the production of wall units with the pressure of 3–20 MPa will produce composite materials with the strength of up to 3–20 MPa.

The unburning technology for producing lime-sand phosphogypsum material will make it possible to produce cheap low and medium-quality wall materials for interior and exterior use.

In this technology, compared with the known technologies for the production of wall materials and products, there is no autoclave, which leads to an economic effect: reducing the cost of production of wall materials from phosphogypsum waste 2–3 times compared with the production of silicate brick.

Combining the two technologies will create waste-free processing of phosphogypsum to produce a concentrate of rare-earth metals and wall materials.

References

1. Kanzel, A.V., Mazurkevich, P.A., Bortkov, I.A., Hares, N.K.: Method for complex processing of phosphogypsum. Patent RU 2639394 the Russian Federation (2017)
2. Mikheenkov, M.A.: Pressing as a way to improve the physicomechanical properties of the gypsum binder. Vestnik MGSU **3**, 173–181 (2009)
3. Yahorava, V., Bazhko, V., Freeman, M.: The viability of phosphogypsum as a secondary resource of rare earth elements. In: XXVIII International Mineral Processing Congress Proceedings (2016)
4. Bouchhima, L., Rouis, M.J., Choura, M.: A study of phosphogypsum-crushing sand based bricks grade negligible weathering. Rom. J. Mater. **47**(1), 316–323 (2017)
5. Campos, M.P., Costa, L.J.P., Nisti, M.B., Mazzilli, B.P.: Phosphogypsum recycling in the building materials industry: assessment of the radon exhalation rate. J. Environ. Radioact. **1**, 232–236 (2017)
6. Demirel, Y., Caglar, Y.: Recovery of phosphogypsum in the economy in building material waste. J. Fac. Eng. Archit. Gazi Univ. **30**, 743–750 (2017)
7. Petropavlovskaya, V.B., Bur'yanov, A.F., Novichenkova, T.B.: Low-energy gypsum materials and products based on industrial waste. Moscow (2006)
8. Shmelev, G.D.: Effective phosphogypsum compositions for construction products from large-scale man-made chemical production wastes, Voronezh (1998)
9. Zolotukhin, S.N., Savenkova, E.A., Solov'eva, E.A., Ibragim, F., Lobosok, A.S., Abramenko, A.A., Drapalyuk, A.A., Potapov, Yu.B.: Raw materials for construction products without firing technology. Pat. RU 2 584 018 the Russian Federation. SCI HPE Voronezh SUACI (2016)
10. Zolotukhin, S.N., Kukina, O.B., Abramenko, A.A., Mishchenko, V.Ya., Gapeev, A.A., Solov'eva, E.A., Savenkova, E.A.: Studies of the process of structure formation of dispersed systems in the preparation of composite building materials with predetermined properties. Kursk **21-5**(74), 96–110 (2017)
11. Kukina, O.B., Abramenko, A.A., Volkov, V.V.: Optimization of the compositions of unburned lime-sand phosphogypsum material. Voronezh **3**(51), 48–56 (2018)
12. Zolotukhin, S.N., Kukina, O.B., Abramenko, A.A., Soloveva, E.A., Savenkova, E.A.: Energy-efficient unburned technologies for the use of phosphogypsum. In: IOP Conference Series: Earth and Environmental Science (2017)
13. Zolotukhin, S.N., Kukina, O.B., Abramenko, A.A., Mishchenko, V.Ya.: The change of the structure, properties, state of large-tonnage waste over time - dump phosphogypsum dihydrate. Voronezh **12**(137), 69–77 (2017)
14. Abramenko, A.A.: Building materials based on phosphogypsum. Voronezh **1**, 65–70 (2017)

Polymer-Composite Materials for Radiation Protection

Andrey Barabash(✉), Dmitriy Barabash, Victor Pertsev, and Dmitriy Panfilov

Voronezh State Technical University, Moscow Avenue, 14, Voronezh 394026, Russia
jarogacheva@mail.ru

Abstract. Paper presents the results of the research of radiation-resistant composites for the bearing and enclosing parts of nuclear plant facilities. The scientific task of the study is to obtain a polymer composite of the maximum attainable density and strength with a given ability to reduce the neutron flux in combination with a high value of the attenuation coefficient of γ-radiation. In order to resolve this task, we developed the composite material based on the polymeric binder with the necessary amount of hydrogen bonds ensuring the capture and moderation of the neutron flux filled with mineral components that absorb γ-radiation. The paper shows the advantage of using non-isocyanate polyurethane as a binder, which is characterized by the presence of a large number of hydrogen bonds providing effective slowing down and capture of "fast" neutrons. Serpentinite, barite, limonite, and magnetite are the materials traditionally used for 'filling' in radiation protection are reviewed in the paper. Certain limitations were established when using a non-isocyanate binder, namely, the fillers must satisfy the condition of $8 > pH > 5$. In order to ensure the casting method, when constructing the objects made of a radiation-resistant composite, the maximum viscosity of the mixture was limited to 80 Pa•s, whereas, in order to minimize energy consumption and to prolong the viability of the mixture, the mixing temperature was set at the level of 30 °C. The aforementioned restrictions in the temperature required mechanical-type plasticizer in the polymer mixture, which turns it into the glass powder of 'sodium-boro-silicate' composition. The obtained regression equations made it possible to establish a rational combination of a polymeric binder, filler and plasticizer, providing the required degree of radiation protection.

Keywords: Polymer-composite materials · Radiation protection

1 Introduction

Besides its positive effects, the development of the nuclear industry has its negative influence on the quality of life. First of all, it concerns the protection of personnel being directly involved in the operation of the nuclear facilities. The fast-neutron reactors maintenance shows that the choice of the material, which protects against radiation should be sufficiently substantiated both from economic and scientific point of views [1].

V. Murgul and M. Pasetti (Eds.): EMMFT-2018, AISC 983, pp. 352–360, 2019.
https://doi.org/10.1007/978-3-030-19868-8_36

This is due to the presence of radiations of a different nature, characteristic of the working zone of High Power Channel Reactor type of reactors (HPCR) [2].

Since neutrons and gamma-quanta have the greatest penetrating power, it is necessary to design protective screens with a thickness that depends on the characteristics of the materials used to attenuate them. This is due to the fact that the attenuation coefficient of radiation depends directly on the sequence number and the number of atoms of the substance used [3]. In addition, the attenuation is characterized by the high reactivity of hydrogen in relation to neutron radiation. The scattering efficiency does not just depend on the presence of bound hydrogen, but also on the hydrogen bonds in the composition of the material. The minimum content of the indicated element has to be not less than 0.5% by weight [2, 4]. The combination of hydrogen-containing materials with high-density components potentially provides not only the dissipation of the neutron flux, but it also effectively removes the heat generated during the deceleration of fast neutrons [2]. Structurally, protective screens are the walls and partitions made of materials that effectively absorb both gamma radiation and the neutron flux. These materials are represented by the polymer composites, which contain the effective absorbers of the radiations [5].

2 Materials and Methods

Numerous compositions for the protection against ionizing radiation have been proposed in the paper. Preferably, the hardening resins—polyester, epoxy, polyvinyl acetate, furfural acetone (FAM) [5–8] are used as the polymer component, and various materials, both mineral and polymer, are used for filling [7, 8].

As a rule, iron- and boron-containing fillers, lead shot, and tungsten films are applied in this case [9, 10]. These materials are applied to capture the largest possible number of the charged particles. However, this is not particularly efficient, since materials with the high hydrogen content slow down and capture neutrons most effectively. In addition, as it was mentioned earlier, the protection against neutron flux is featured by the need to remove exothermic heat [11].

From the physical chemistry viewpoint, it is practically impossible to provide a combination of such properties based on just one material. In order to solve this problem, it is necessary to develop a polymer-based composite that has the necessary amount of hydrogen bonds filled with the effective material, or the complex filler, which provides both neutron scattering and the removal of released heat [12, 13].

Based on this, a material science problem can be formulated: obtaining a polymer composite with the maximum attainable density and strength with a given ability to weaken the neutron flux. Since the ability to absorb neutrons is a prerogative of the polymer component, its volume should not fall below a predetermined limit. In this regard, the selection of the composite is the task of multi-parameter optimization, as it is not possible to satisfy all the required characteristics. Relevant papers [14, 15] show that non-co-polyurethane polyurethanes, which do not have toxic products in the production of toxic by-products, can be considered as effective polymeric binders with a combination of favorable properties. The properties of polyurethanes are based on the presence of specific interactions (hydrogen bonds, bonds of the ionic type) and non-specific (dipole-dipole, van der Waals interactions, as well as crystallization). When summarized, these properties determine the choice of the material. The synthesis of non-isocyanate polyurethanes is provided by unconventional urethane formation reactions. Hydroxyl-containing components are used as initial products for this synthesis and determine the main complex of physical and mechanical properties of polyurethanes, for example polyethylene glycols, and polyesters with terminal OH groups [16, 17]. The synthesis of hydroxyl urethanes, which interacts with cyclocarbonates and amines for the modification of epoxy systems [17, 18] is quite efficient. As a result of such modifications, we obtained a reactive class of oligomer compounds, laprolates [19]. Conducting exploratory experiments, there were selected components of the polymer base to obtain an effective composite: epoxy resin corresponds to the regional standard adopted by the Interstate Council for Standardization, Metrology and Certification of the Commonwealth of Independent States number 10587 with a diluent - plasticizer and adduct laprolata-803 (Technical Conditions number 6-05-221-995-88 with a mass fraction of cyclocarbonate groups - 21 … 31, epoxy groups - 2.5%, density 0.971 kg/l, pH 3.5 … 5.5) containing the following amines: isophorone diamine, diethylenetriamine, triethylene and tetramine [19].

The stoichiometric ratio of these components in which reactive groups are fully involved provides the substance with a base strength of at least 18 MPa. The time of complete curing of the binder at a temperature of 25 … 300 C is not more than 4 h.

At the same time, the mass content of hydrogen, calculated from the molecular mass distribution, ranges from 11 to 15%, which causes a guaranteed attenuation of neutron radiation of average energies. The next step in designing the composite formulation is to establish the ratio of the polymer base and the effective filler. By virtue of the specifics of the material being developed, it is necessary to use fillers with a high density and the ability to absorb or scatter a stream of charged particles. At the same time, the proportion of the polymer binder should not decrease by less than 10% by mass in order to ensure the effective absorption of neutrons [20].

Considering this, compressive strength and density were determined as the upper limit of the filler content and the objective functions during the experiments. The above characteristics were determined for the samples of cylinders with 50 mm in diameter and 50 mm of height. The list of minerals considered as fillers [1], traditionally used in the manufacture of radiation-protective materials, is presented in Table 1.

Table 1. Properties of the fillers.

Title	Average density, kg/m^3/ attenuation coefficients of γ - radiation with an energy of 3 eV, cm^{-1}	SSA, м^2/g	Chemical composition	Oil absorption, g/100 g	pH of aqueous extract
Serpentinite	2300…3200/0,480	300…350	$Mg_3(OH)_4 \cdot Si_2O_5$ or $3MgO \cdot 2SiO_2 \cdot 2H_2O$	28…60	9,1… 10,3
Barite	2900…3600/0,380	280…310	$BaSO_4$	30…50	6…8
Limonite	2400…3700/0,291	310…340	$Fe_2O_3 \cdot H_2O + Fe_2O_3 \cdot 2H_2O$	40…56	5…5,5
Magnetite	3250…4100/0,330	290…340	Fe_3O_4	35…50	5,5…6,2

It has to be noted that these fillers had a different pH index and density, which can significantly affect their target function. At the same time, the powders have a similar granulometry, and therefore this was not studied separately as a specific feature.

3 Results

As a result of preliminary experiments, it was found that combining the fillers with a binder under conditions of a large difference in pH (for example, with the introduction of serpentinite), the composite was distended as a result of the release of gaseous-like products of side reactions in the boundary layer. It is established that in the case of the use of a non-isocyanate binder, the fillers must satisfy the condition of $8 > pH > 5$.

As long as the temperature-technological factor is important in shaping the structure of polymer-based composites, we have introduced a limit on the viscosity of the finished mixture and the preparation temperature. As we apply the injection molding technology in the manufacture of structures made of the developed composite, the viscosity of the mixture is limited to 80 Pa·s.

The mixing temperature is set at the level of not higher than 30 °C, in order to ensure the necessary viability of the mixture and reduce the energy consumption for heating. The fillers were injected in portions of 10 mass parts at a constant speed of rotation of the mixer rotor. The viscosity of the mixture was controlled. It is established that when the degree of filling with all the selected powders of up to 60 wt.h., the viscosity does not exceed the established technological limit. After statistical processing of the experimental results, theoretical curves of changes in compressive strength and density of the samples on the number and type of filler were presented the following Fig. 1.

After processing the theoretical curves, the maximum possible strength values of the composite samples were established. When filling with limonite at the maximum strength of 34.42 MPa, its content reaches at 51.5 wt.h. As the amount of magnetite reached 47.25 wt.h., its strength was 31.9 MPa. The strength of 32.4 MPa corresponded to the 41.68 wt.h barite amount.

In addition, it was found that the maximum strength was achieved at density values, which were far from the maximum. This could be explained by the technologically - the increase in viscosity of the mixture combined with an increase in the concentration of fillers in excess of 40 … 60 wt.h. The growth of viscosity was accompanied by the decrease in the homogeneity of the mixture and its partial exfoliation, which directly affected the strength.

A decrease in viscosity can be achieved by increasing the temperature of the binder or by overheating the fillers.

However, the restrictions introduced earlier do not allow using this technological method.

In this regard, we additionally introduced the plasticizer of the mechanical type into the polymer mixture – the glass powder consisting of glass microspheres with diameters of 15 … 200 μm with oil absorption not more than 10 g/100 g. Whereas, the approximate composition of sodium borosilicate glass is % Na2O ≈ 7,5; B2O3 ≈ 12;

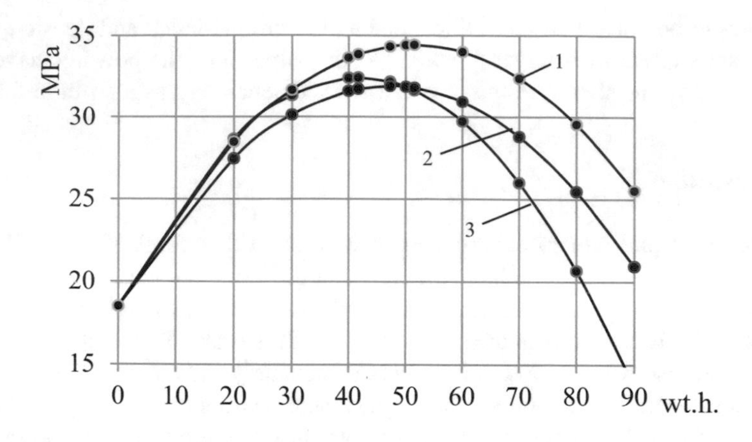

a)

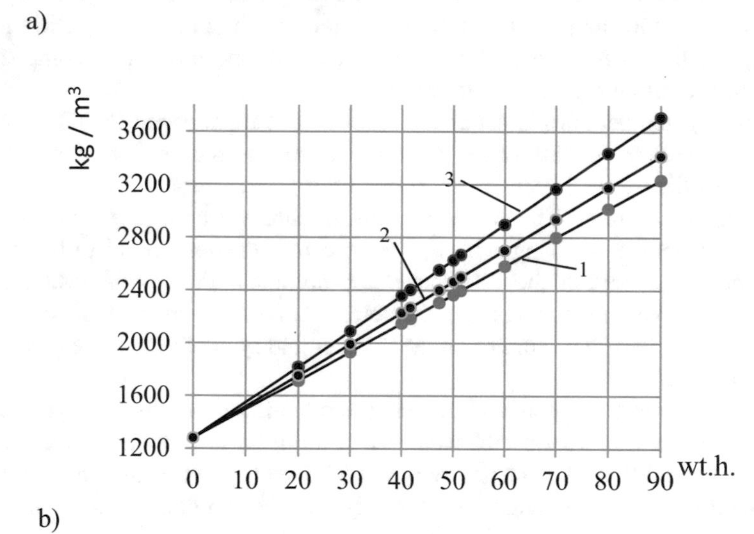

b)

Fig. 1. Dependence of compressive strength (a) and density (b) on the amount of injected filler: limonite (1), magnetite (2), barite (3).

SiO2 ≈ 80,5. The above chemical composition complies with the requirements for radiation protection materials.

For glass powder, the pH of the aqueous extract comprised 6.6 … 7.4, which ensured its compatibility with the polymer binder. The use of this powder was previously discussed in the relevant papers [21].

Glass powder was injected in portions under control of viscosity in the mixture containing 80 wt. h. used fillers. In order to determine the compressive strength and density with a decrease in viscosity below 80 Pa·s, separate samples were prepared from the mixture obtained. As a result of processing the experimental data, the following equations were obtained. They adequately describe the dependence of viscosity in the amount of glass powder injected at the fixed content of the base filler (Table 2).

Table 2. Equations of η viscosity versus x amount of glass powder.

Base filler	Equation	R^2	Equation no.
Limonite	$\eta = 0,551x^2 - 10,68x + 128,7$	0,976	1
Magnetite	$\eta = 0,420x^2 - 8,535x + 147,2$	0,938	2
Barite	$\eta = 0,464x^2 - 8,535 + 147,2$	0,849	3

Studies of Eqs. 1–3 on an extremum yielded theoretically possible minimum values of the viscosity of the mixture 77, 98, 108 Pa·s. The corresponding concentration of the glass powder comprised 9.7; 9.5; 9.2 wt.h. for compositions containing limonite, magnetite, barite, respectively.

At the same time, samples containing limonite had a density not lower than 3250 kg/m^3, and the strength was within 46 … 50 MPa limits,

These characteristics allow us to mark the composite as a structural material, that is, we resolve the part of the task to achieve the highest possible strength.

However, the regression equations provide all the reasons to suggest that the viscosity varies depending not only from the amount of the injected powder, but also from the chemical interaction of the polymer and mineral components. This circumstance determines a fairly wide range of tasks related to the selection of filling powders precisely according to the compatibility criterion.

The next task in the framework of achieving our goal was to assess the degree of attenuation of a narrow neutron beam by the developed material.

Based on the principle of calculating the protection for attenuation and absorption of neutron radiation, the weakening of a narrow neutron beam by a material is expressed by the following equation:

$$N = N_0 \cdot e^{-\eta x} \tag{1}$$

where N_0, N are the neutron radiation doses received at a given point in the absence of a protective screen and after passing through a protective screen of thickness x, cm, respectively; η - macroscopic capture cross-section, cm^{-1} [4].

On the basis of (1), and using the setup, which is described in detail in [22], the values of the linear attenuation coefficient were obtained for the composite samples where limonite and magnetite serve as the base fillers (Table 3).

Table 3. The attenuation coefficients for the composite with different fillers samples.

Filler title	Average density of samples *, kg/m^3	Attenuation coefficients of γ-radiation with an energy of 3 eV*, cm^{-1}
Limonite	3120/3250	0,341/0,393
Magnetite	3180/3300	0,330/0,348
Reinforced concrete **	2750	0,112

* Above the line are the data for the composite without the glass powder/below the line are the data containing the glass powder;

** Reinforced concrete data are given for the comparison.

4 Discussions

Summarizing the results of the conducted studies, we observe the increase in the attenuation coefficient of γ-radiation for composites, which contain the additions of glass powder. In our opinion, the observed effect can be explained by the synergies manifested by the radiation permeability of the polymer component and the mineral filler. The differences in the chemical composition of the fillers directly affect the radiation characteristics of the composite due to the trivalent iron being the part of the limonite composition. Additionally, hydrogen is present in the bound water composition, unlike magnetite, which only contains the variable-valence iron.

The indicated combination of mineral powders and the polymer component provides for the high values of the attenuation coefficient of γ-radiation while maintaining the physical and mechanical characteristics of the composite.

It should be noted that the expansion of the range of radiation-resistant mineral powders is a separate scientific and practical task, since the available stock of man-made waste (fly ash from thermal power plants for example) provides almost limitless possibilities for improving the properties of the polymer composites.

Nonetheless, as it was previously indicated, obtaining a composite that meets the radiation protection requirements of a given degree and strength is a multiparameter problem. Its solution requires an integrated approach. For example, polymeric materials are very sensitive to temperature increase, which significantly affects their structural properties. It is necessary to take this factor into account when designing the composition of the composite, since, with an intense radiation exposure, the probability of material heating is high.

In addition to these features of designing the radiation protection composites, a technological factor is also important. Moreover, the role of the preparation technology in the process of formation of the structure of the material obtained can be decisive. In addition to the purely economic component, the technology of preparation should provide the most homogeneous structure of the composite, assessed by the mixing index. To ensure the specified characteristics, it is important to choose not only the type of equipment but also the characteristics of the process. Given the variety of chemical and physical characteristics of the materials used, it is rather difficult to develop a unified algorithm for preparing effective radiation protection composites.

Nevertheless, the available scientific and practical developments in this direction [23] suggest that, with sufficient compatibility of the components used, it is possible to obtain the most homogeneous mixtures with minimal energy consumption.

5 Conclusion

The use of radiation-resistant composites for bearing and enclosing structures helps to reduce the negative effects of the development of the nuclear industry. The combination of polyurethane binder, weakening the effect of neutron radiation, and mineral fillers that neutralize charged particles, provide reliable protection for the personnel of the nuclear power facilities. Our study proved the effectiveness of using of the non-isocyanate polyurethane binder, which has enough hydrogen bonds to capture "fast" neutrons filled with mineral powders of limonite or magnetite and a mechanical-type plasticizer. That is the glass powder of sodium borosilicate composition in mass ratios providing a given degree of radiation protection. The resulting regression dependencies provide the design of the recipes for these composites, consistent with the intended conditions of operation. The proposed approach to the rational use of the plasticizer of the mechanical type - glass powder of chemical composition that provides additional improvement of radiation characteristics and reduces the viscosity of the original mixture without attracting additional energy costs.

References

1. Barabash, D.E., Borovlev, Y.A., Kuznetsov, D.A.: Radiation-resistant polymer concrete based on non-isocyanate polyurethanes. In: Anniversary International Conference, Dedicated to the 60-th Anniversary of Belgorod State Technical Universitym V.G. Shukhov (XXI Scientific Readings), vol. 1, pp. 3–10. High Technologies and Innovations, Belgorod (2014)
2. Mitenkov, F.M.: Fast neutron reactors. At. Energy **92**(6), 423–432 (2002)
3. Naumov, V.I.: Physical bases of safety of nuclear reactors. Scientific Research Nuclear University "MEPhI", Moscow (2013)
4. Mashkovich, V.P., Kudryavtseva, A.V.: Protection Against Ionizing Radiation. Energoatomizdat, Moscow (1995)
5. Landau, L.D., Lifshits, E M.: Theoretical Physics. Manual for the universities. In 10 tons. Quantum mechanics (non-relativistic theory). 4th ed. Science, Moscow (1989)
6. Banny, V.A., Ignatenko, V.A.: The use of polymeric radio absorbing materials in solving the problem of electromagnetic safety. Probl. Health Ecol. **1**, 9–13 (2016)
7. Stefanenko, I.V.: Radiation - modified materials and heat-resistant compositions using technogenic raw materials for protection from radiation and background radiation. Volgograd (2012)
8. Pavlenko, V.I.: High-performance polymer composite materials for radiation protection. Int. Sci. Res. J. **9**(40), 71–75 (2012). part 2
9. Popov, K.N.: Polymer and polymer cement concretes, mortars and mastics. VS.72, Moscow (1987)
10. Mikhaylin, Y.A.: Special Polymer Composite Materials, Monograph. Scientific Basis and Technology Publishing house (2009)
11. Bormotov, A.N., Proshin, A.P., Bazhenov, Y.M., Danilov, A.M., Sokolova, Y.A.: Polymeric Composite Materials for Protection Against Radiation, Monograph. Paleotype publishing house, Moscow (2006)
12. Yurkov, G.Yu., Buznik, V.M., Kondrashov, S.V., Biryukova, M.I., Mikheev, M.G., Bogatov, V.A., Chursova, L.V.: Magnetic composite materials based ultradispersed polytetrafluoroethylene and cobalt containing nanoparticles. In: Encyclopedic Reference with the Application "Comments on Standards, Specifications, Certificates". Publishing House Science and Technology (2013)
13. Soylu, H.M., Lambrecht, Y.: Gamma radiation shielding efficiency of a new lead-free composite material. J. Radioanal. Nucl. Chem. **305**(2), 529–534 (2015)
14. Rysovannyi, V.D., Zakharov, A.V.: Boron in Nuclear Engineering. 2nd edn. JSC "SSC RIAR", Dimitrovgrad (2011)
15. Levin, V.E.: Nuclear Physics and Nuclear Reactors, 4th edn. Atomizdat, Moscow (1979)
16. Barabash, D.E., Borisov, Yu.M., Anisimov, A.V.: Non-isocyanate polyurethanes - the basis of constructional composites. Constr. Mater. **5**, 20–22 (2013)

17. Shapovalov, L.D., Figovsky, O.L., Kudryavtsev, B.B.: Non-isocyanate polyurethanes: synthesis and application. Quest. Chem. Chem. Technol. **1**, 232–236 (2004)
18. Bui, D.M.: Development of composite materials based on epoxyurethane oligomers with improved performance properties. Dissertation Cand. Sci. (Technical), Moscow (2014)
19. Bilyalov, L.I., Medvedeva, K.A., Cherezova, E.N., Gotlib, E.M., Khasanov, A.I.: Modification of epoxy polymer by Laprolat 803 and the study of its physical and mechanical properties. Kazan Technol. Bull. Univ. **1**, 142–143 (2013)
20. Vnukov, V.S., Sichkaruk, O.V., Chkuaseli, L.I.: Methodical bases of ensuring nuclear safety during transportation and storage of radioactive waste. In: St. Petersburg Reports of the International Conference "Safety of Nuclear Technologies, p. 123 (2006)
21. Barabash, D.E., Barabash, A.D., Potapov, Y.B., Panfilov, D.V., Perekalskiy, O.E.: Radiation-resistant composite for biological shield of personnel. In: IOP Conference Series: Earth and Environmental Science, p. 012085 (2017)
22. Perekalsky, O.E.: Building composites based on polybutadiene oligomers for protection against radiation. Dissertation Cand. Sci. (Technical). Voronezh (2006)
23. Barabash, D.E., Barabash, A.D.: Technology of obtaining radiation-protective composites of a given viscosity. Effective building structures: theory and practice. In: Collection of articles of the XVII International Scientific and Technical Conference, vol. 1, pp. 14–17 (2017)

Biomass Resource of Domestic Sewage Sludge

Vladimir Shcherbakov[1(✉)], Valentina Pomogaeva[1], Konstantin Chizhik[2], and Ekaterina Koroleva[2]

[1] Voronezh State Technical University, Moscow Avenue, 14, Voronezh 394026, Russia
scher@vgasu.vrn.ru
[2] Moscow State University of Civil Engineering, Yaroslavskoe sh. 26, Moscow 129337, Russia

Abstract. Experimental studies on the decontamination and stabilization of sludge beds were conducted using the "Desolac" product, which is a calcium oxide treated with ovicide. As a result of the study, it was established that the sewage sludge of the sludge beds of the MUE "Lipetsk Aeration Station" and "LOS" LLC in Voronezh are safe and can be used as organic fertilizers for the use in grain and industrial crops, as well as in vegetable production. Studies have been conducted to determine the efficiency of sewage sludge from the Left-bank sewage treatment plants in Voronezh and sewage sludge treated with "Desolac" for soil fertility and productivity of winter rapeseed Adrian. Determining the content of macronutrients of mineral nutrition in the soil and the reaction of the soil environment showed that the introduction of organic fertilizers in the form of sludge compared to the version without fertilizers increases the content of nitrate nitrogen, exchangeable potassium, labile phosphorus in the soil. At the same time, the mass fraction of impurities of toxic elements in the soil decreases with the introduction of sewage sludge as organic limestone fertilizer.

Keywords: Biomass · Sewage sludge

1 Introduction

The treatment plants in Russia store more than a hundred million tons of sludges. This amount of organic pollution creates a number of environmental problems. A significant amount of sludges coming after the physico-chemical and biological wastewater treatment is supplied to the sludge beds. Nowadays, most of the sludge beds are overfull and, most often, are located near the territory of cities. Allocation of new land for sludge beds is almost impossible and economically impractical. Thus, the problem of biomass disposal after sludge beds is highly relevant.

From the existing methods of decontamination, processing and disposal of sewage sludge from urban sewage treatment plants, the following can be distinguished:

- decontamination and soil displacement;
- thermal use (combustion);
- landfill depositing.

V. Murgul and M. Pasetti (Eds.): EMMFT-2018, AISC 983, pp. 361–372, 2019.
https://doi.org/10.1007/978-3-030-19868-8_37

As a result of the analysis of these areas, one of the most promising and most effective is the decontamination of sludge and their soil displacement.

The safety of sewage sludge on sanitary and hygienic indicators is determined by the presence of pathogenic microorganisms and helminth eggs. Decontamination of sewage sludge can be carried out by the following methods [1]:

- thermal methods - the sludge is subjected to heating, drying or burning;
- biothermal methods consist in composting sludge with the addition of sawdust, peat, sand;
- chemical - processing using various chemical reagents;
- biological methods - the destruction of microorganisms by protozoa, fungi, and soil plants;
- physical - effecting the sludge by radioactive or ultraviolet radiation, ultrasound, high frequency currents.

One of the promising methods for treating sewage sludge is liming, which can significantly reduce the content of pathogenic microorganisms that pose a potential hazard. When adding quicklime, alkalinity increases, due to the process of lime slaking with water contained in the sludge, the temperature of sludge increases. With the introduction of a dose of lime up to 30% of the dry matter of the sludge, deformation and death of fecal streptococci, Salmonella bacteria occurs. With an increase in pH of more than 11, the content of coliform bacteria decreases from 10^9 to 10^3 pcs. per 1 g of dry matter.

The efficiency of liming depends on the state of the sludge - liquid or dehydrated. To stabilize wet sludge, the amount of lime must be at least 10% of the dry matter. The specific dose is determined depending on the treatment method, temperature, compaction time, and composition of the sludge. After lime treatment, the liquid sludge cannot be stored for a long time, since after some time, the sludge particles are destroyed, the alkaline buffering of the medium decreases, and the ability to ferment is restored.

As a result of studies [1–8], it has been determined that the treatment of dehydrated sewage sludge with quicklime (CaO) has a longer effect. With a lower water content in the sludge, it is more resistant to the development of acidic fermentation processes. When using quicklime, depending on the content of free water (20–30%) and the initial temperature of the sludge, the dosage should be from 100 to 150 kg of CaO per ton of sludge or from 400 to 500 g of CaO per kilogram of free water to reach the temperature higher than 50 °C [8].

The required dose of quicklime is determined from the condition of raising the temperature of the sludge over 60 °C and depends on the composition of the sludge and the treatment methods. Experimental studies [2, 3] and theoretical calculations [4] show the effect of lime activity on the kinetics of heating the sludge (Fig. 1). A significant effect on the temperature increase is exerted by the method of mixing sludge and lime, which can be carried out using screw pumps with plunger mixers, blade mixers, mechanical agitators, and other equipment.

The mixing of sewage sludge with quicklime with activity of 55–58% on a double screw mixer, containing 6–25% CaO by mass of sludge, is shown in Fig. 1. The addition of lime in the amount of 23.5% makes it possible to increase the temperature of the mixture up to 78–82 °C within 10 min, then the temperature decreases to 46 °C within 40 min [2].

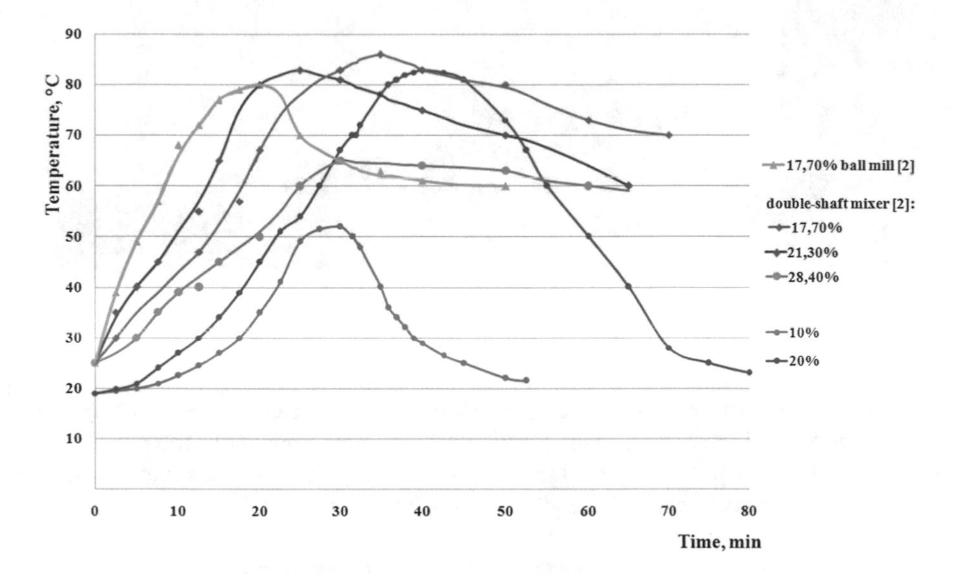

Fig. 1. Kinetics of heating and cooling of sewage sludge during the treatment with CaO.

According to the study of Ignatenko, with the introduction of CaO with the concentration of 10% and 20% in the sludge with the humidity of 95% and keeping the samples in a closed container, the decontamination efficiency by helminth eggs was 75% with CaO 10% and 100% with CaO 20% [3]. For the death of helminth eggs and cysts of intestinal pathogenic protozoa, it is enough to raise the temperature up to more than 55 °C for several minutes, which proves the decontaminating effect of the introduction of lime.

2 Materials and Methods

Experimental studies on the decontamination and stabilization of sludge at the sludge beds were carried out using the product "Desolac", which is a calcium oxide treated with ovicidalsolution. "Desolac" is produced by "PridonkhimstroyIzvest" LLC in accordance with TU 2123-004-00121270-2016 (Technical Requirements).

Studies were conducted at the wastewater treatment plant of the MUE "Lipetsk Aeration Station". The sludge was processed at the facility (Fig. 2) consisting of a silo tower (point 1), a lime auger (point 2), and a screw mixer with lime (point 3). The sludge was previously dehydrated on filter presses. Quicklime from the silo tower is supplied via the lime auger to the screw mixer, where lime is mixed and dispensed with sludge (Fig. 2). Discharge of the treated sludge is carried out in an automated mode.

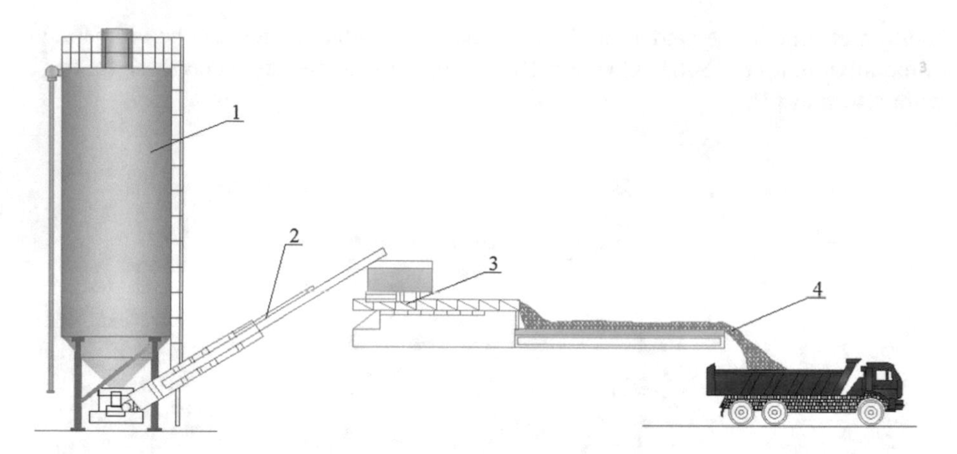

Fig. 2. The scheme of stabilization of dehydrated sludge by "Desolac": 1-silotower; 2-limeauger; 3-screw mixer with lime; 4- discharge of the treated sludge.

To determine the content of toxic elements in the treated sludge, experimental studies were conducted on the Left-bank sewage treatment plants in the city of Voronezh ("LOS" LLC). The raw sludge from the primary settling tanks and the sludge was mixed with quicklime and dehydrated on a Flottweg decanter (Fig. 3).

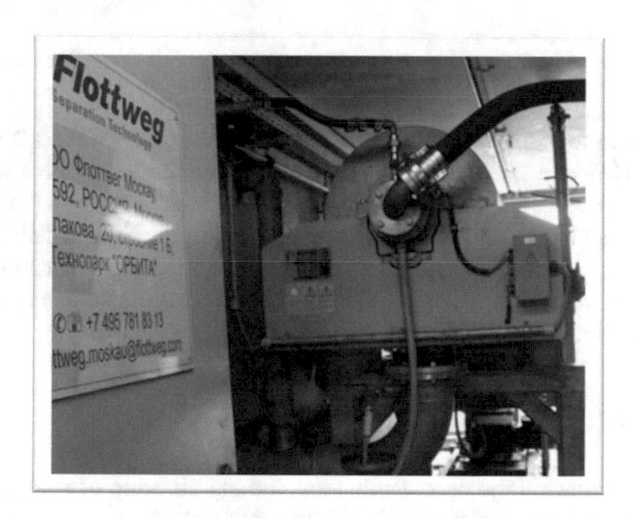

Fig. 3. Flottweg decanter centrifuge.

To obtain generalized research results, the following was mixed in various proportions: sludge, excess active sludge, and "Desolac".

- Sample No. 1 - 100% crude sludge, 5% "Desolac";
- Sample No. 2 - 100% crude sludge, 7% "Desolac";
- Sample No. 3 - 30% crude sludge + 70% excess active sludge, 5% "Desolac";
- Sample No. 4 - 10% crude sludge + 90% excess active sludge, 5% "Desolac";
- Sample No. 5 - 30% crude sludge + 70% excess active sludge, 3% "Desolac".

3 Results

Studies have been conducted to determine bacteriological contamination in the treated sludge. The sludge from the sludge beds treated at the treatment plant of MUE "Lipetsk Aeration Station" has a Coliform bacteria index and an enterococcal index corresponding to or significantly less hygienic standards (Table 1). According to the conclusion of the Federal Budgetary Healthcare Institution "The Center of Hygiene and Epidemiology of the Lipetsk Region", sludgesampling for the degree of epidemic hazard by microbiological indicators are related to clean soils in accordance with SanPiN 2.1.7.1287-03 (Sanitary Rules and Norms) [9].

Table 1. Decontaminating effect upon lime treatment.

Samples	pH	Coliform bacteria index	Enterococcal index	Pathogenic bacteria and salmonella	Hygienic standard
Sludge before treatment	7,27	1×10^3	1×10^3		10 CFU
Sludge after treatment	11,9	Less than 1	10	Not detected	10 CFU
	11,88	Less than 1	Less than 1	Not detected	10 CFU
	11,79	Less than 1	Less than 1	Not detected	10 CFU

The conducted research and the results of the conclusion of the Federal Budgetary Educational Institution "The Center of Hygiene and Epidemiology of the Lipetsk Region" show the efficiency of the decontaminating effect of the "Desolac" product as applied to sewage sludge of public utilities.

However, the use of sludge stabilized with lime on agricultural land is hampered by the possible presence of toxic elements in sewage sludge. The studies conducted at the wastewater treatment plant of "LOS" LLC showed that in the sludge after treatment with the "Desolac" product, the mass fraction of impurities of toxic elements is not significant. Determination of heavy metals in samples was carried out in the Federal State Budgetary Institution State Agrochemical Service Center "Voronezhsky". The mass fraction of impurities of heavy metals in dehydrated sludge treated with the "Desolac" product and the standard content is shown in Fig. 4.

As can be seen from Fig. 4, in the sludge, the mass fraction of impurities of toxic elements is significantly less than the standard [10]. For example, for sludge of the group I: copper 0.22–0.3 MAC, zinc 0.04–0.07 MAC, nickel 0.07–0.1 MAC, chromium 0.04 MAC, lead 0.2–0.45 MAC, mercury 0.01–0.06 MAC, arsenic 0.05–0.07 MAC, cadmium 0.09–0.17 MAC. Sludges of the group I can be used in the cultivation of all types of crops, except vegetables, mushrooms and berries [10]. Sludges of the group II, as a rule, are used for cereal crops, leguminous plants, grain-fodder and industrial crops. MAC of the mass fraction of impurities of toxic elements for sludge of group II is two times more than of group I.

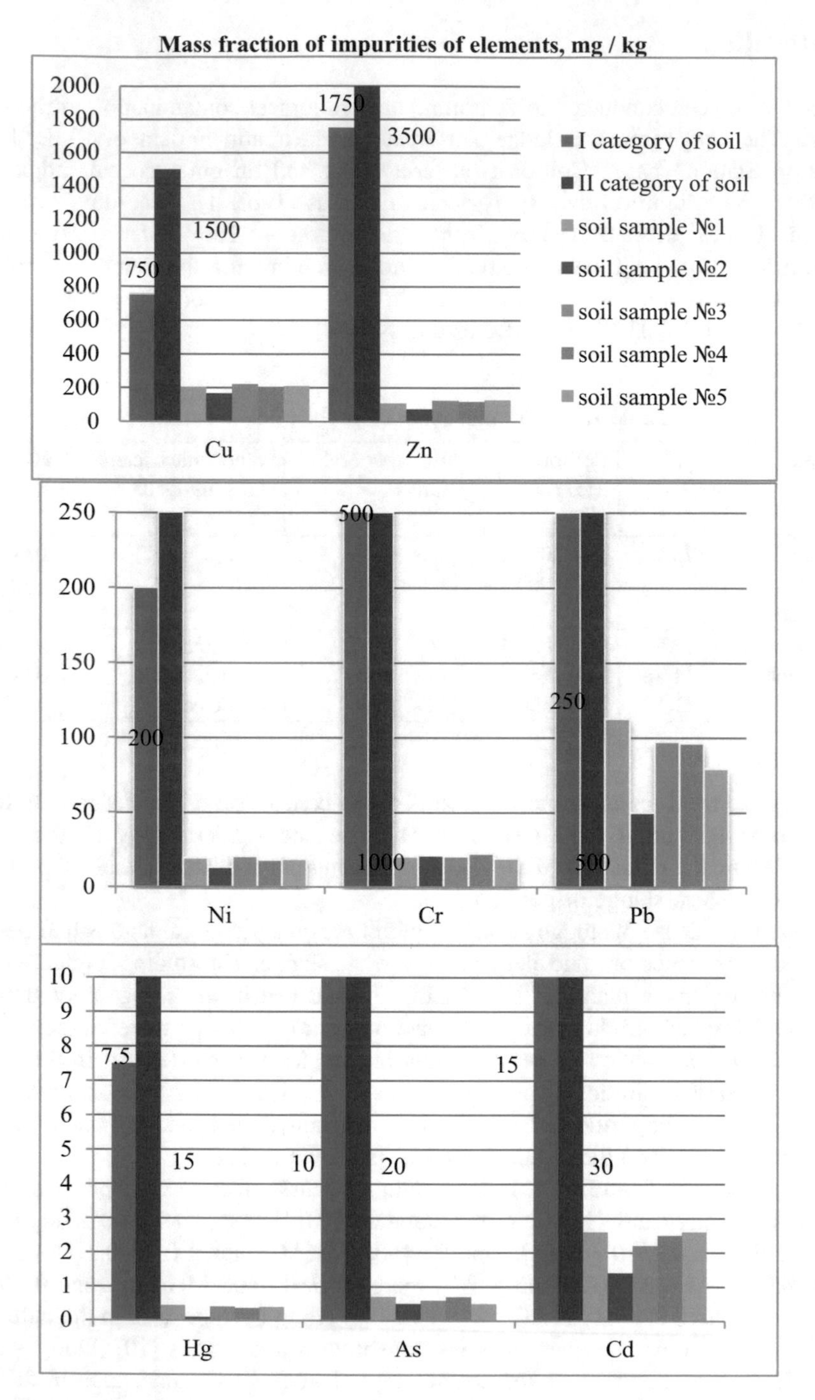

Fig. 4. Mass fraction of impurities in the dehydrated sludge treated with the "Desolac" product: Iand IIcategory of soil– permissible total impurity content of heavy metals in mg/kg of dry matter [9]; soil sample № 1, 2, 3, 4, 5 – samples of dehydrated sludge sediment.

4 Discussion

The use of sludge stabilized with lime on agricultural land is hampered by the presence of toxic elements and parasitological indicators in sewage sludge. However, the performed experiments and the results of the analyzes show that the use of such a sludge is safe.

According to MU 3.2.1022-01 (methodology guidelines) of the Ministry of Health of Russia [11], "the technology of agricultural use of sewage sludge depends on the method of its preliminary processing and decontamination, and the sludge is supplied to the ameliorative field with a frequency of at least 2–3 years, the norm is 5–15 t/ha in terms of dry matter".

Together with the Research Institute of Agriculture of the Central Black Earth Belt named after V.V. Dokuchaev, the studies were conducted to determine the efficiency of sewage sludge from the Left-bank sewage treatment plants in Voronezh, and sewage sludge treated with the disinfectant "Desolac" on soil fertility and the productivity of winter rapeseed Adrian.

Experimental studies were conducted on plots of 10 m^2 (Fig. 5).

Fig. 5. Types of experimental plots: (a) in the period of 2 true leaves of winter rapeseed; (b) before taking yield capacity into account.

To determine the yield capacity of winter rapeseed Adrian, dehydrated sludge was used, consisting of 30% crude sludge and 70% excess active sludge. Dehydrated sludge from treatment plants treated with "Desolac" 5% in the amount of 3.78 t/ha and dehydrated sludge without additives in the amount of 4.45 t/ha were introduced into the soil of plots.

Rapeseed was planted on August 26, harvesting on October 28, 2017. The soil of the experimental plot is heavy loamy medium heavy ordinary chernozem with a pH of 6.99. Physical and chemical indicators of sewage sludge introduced into the soil as an organic fertilizer are presented in Table 2.

Table 2. Physical and chemical indicators of sewage sludge.

Name of indicator	Unit of measurement	Value of indicator	
		Sludge	Sludge with Desolac
Moisture content	%	36.4	63.6
pH		6.4	12.2
Mass fraction of **nitrogen**,% of dry matter, not less	% of dry matter, not less	1.84	2.6
Mass fraction of **phosphorus**	% of dry matter, not less	1.14	1.4
Mass fraction of **potassium**	% for dry matter, not less	0.1	0.13
Mass fraction of **organic matter** per dry matter	% not less	72.9	66.1
Hg	mg/kg dry matter	0.45	0.41
Cr	mg/kg dry matter	48.3	20.7
Pb	mg/kg dry matter	114.9	97.1
Ni	mg/kg dry matter, not less	27.2	20.4
As	mg/kg dry matter	0.5	0.6
Specific activity of technogenic radionuclides (ACs/45*ASr/30)	Not less	0.6	0.8
Mass concentration of pesticide residues in dry matter, mg/kg of dry matter	mg/kg dry matter	≪0.005	≪0.005
- DDT and its metabolites	Total amount		
Mass concentration of **benzo(a)pyrene**, mg/kg of dry matter	mg/kg dry matter, not less	≪0.01	≪0.01

It can be seen from Table 2 that the concentration of substances necessary to improve soil fertility increases when the sludge is processed with lime. At the same time, the content of heavy metals in the sludge decreases, which confirms the feasibility of using sewage sludge as organic fertilizer.

Observations on the development of plants during the growing season (Fig. 5) revealed that after the formation of the third leaf, a more intensive growth and an increase in the mass of rapeseed in areas with addition of sludge sediment as well as sludge mixed with lime, compared to the control variant, have started to be visible (Fig. 6).

The study of the intensity of growth of the vegetative mass of winter rapeseed shows that in the development phase of five leaves of rapeseed, the crude mass of 100 plants in the variants with the introduction of sludge significantly exceeded the mass of plants in the control variant (Figs. 6, 7). The mass of plants in the variant with the use of dehydrated sludge treated with "Desolac" exceeded the control version without addition of fertilizers by 239.0 g, the dehydrated sludge (silt residue) from sewage treatment plants by 258.7 g.

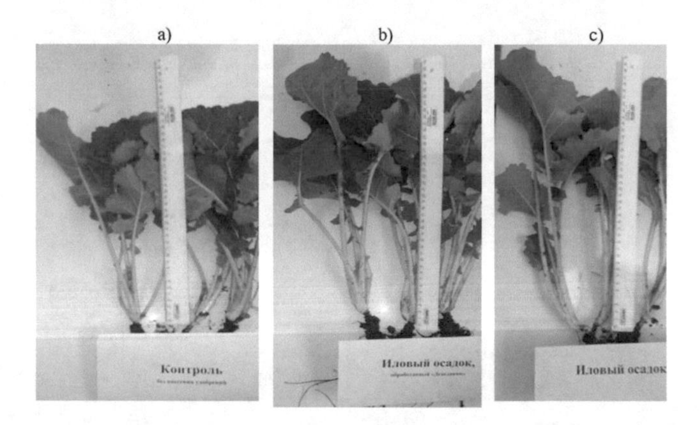

Fig. 6. Type of winter rapeseed plants: (a) on the control variant (without the use of fertilizers); (b) with the introduction of a sludge treated with "Desolac"; (c) with the introduction of sludge (silt residue).

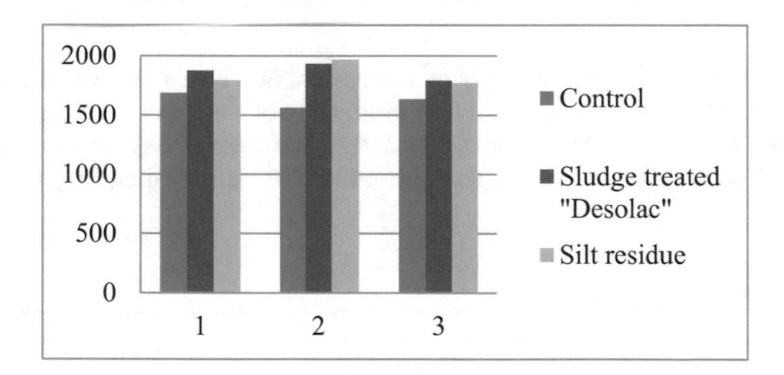

Fig. 7. The growth of the vegetative mass of plants.

Determination of the content of macronutrients of mineral nutrition in the soil and the reaction of the soil environment during the period of five true leaves of rapeseed (Figs. 5b, 6) showed that the application of organic fertilizers in the form of sludge (silt residue) compared with the option without fertilizers (control) (Fig. 8):

1. Increases the content of nitrate nitrogen in the soil:
   - with "Desolac" by 1.5 mg/kg abs. dry soil or by 22.4%;
   - with silt residue - 0.8 mg/kg abs. dry soil or 11.9%.

2. Increases the content of exchangeable potassium in the soil by 3.5 mg/100 g abs. dry soil or by 18.6%.
3. The content of labile phosphorus increases with the introduction of dehydrated sludge (silt residue) by 4.5 mg/100 g abs. dry soil or 7.6%, and with addition of sludge treated with the "Desolac", almost remains at the level of control.

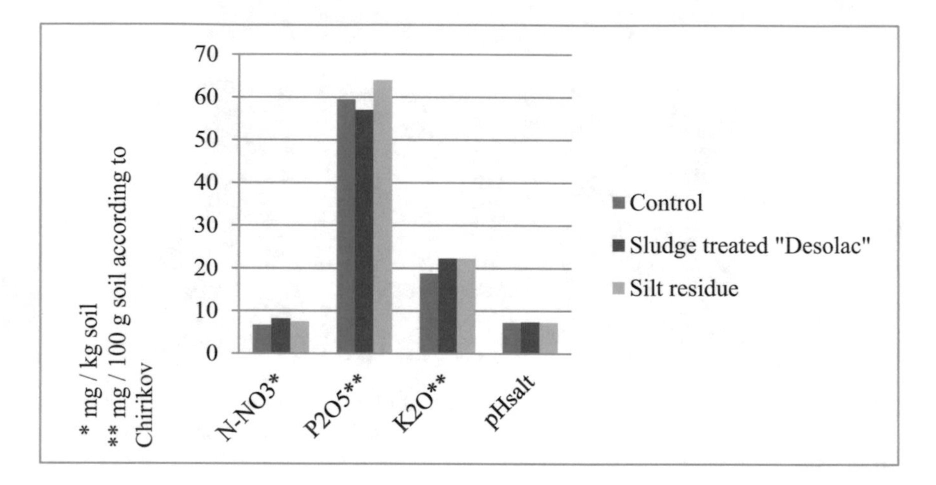

Fig. 8. The content of mineral nutrients and the reaction of the soil environment.

Accounting for the yield capacity of green mass of winter rapeseed and statistical processing of the results showed that the introduction of sludge from sewage treatment plants for the main soil preparation significantly increases the yield capacity of winter rapeseed. The excess yield capacity of green mass of winter rapeseed (Fig. 9) when adding:

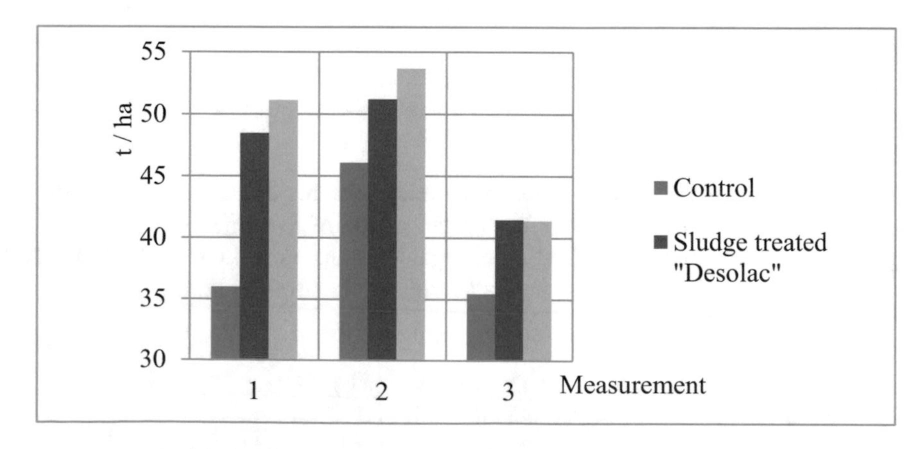

Fig. 9. The yield capacity of green mass of winter rapeseed, t/ha.

- sewage sludge treated with "Desolac" - 7.87 t/ha or 20.1% more than the control variant;
- silt residue - 9.56 t/ha or 24.4% more than the control option.

The use of sewage sludge treated with "Desolak" slightly alters the response of the soil environment towards an increase in the alkalinity of the soil solution by 0.11 pH units compared to the control variant, which is much less than that adopted in the practice of design and operation [6, 7].

5 Conclusions

1. Sewage sludge of MUE "Lipetsk Aeration Station" treated with a mixture of lime and "Desolac" is safe and, according to microbiological indicators, refers to clean soils in accordance with SanPiN 2.1.7.1287-03.
2. In accordance with clausc 2.2 of SanPiN 1.2.2584-10 [12], sewage sludge from the sludge beds of MUE "Lipetsk Aeration Station" and "LOS" LLC of Voronezh are safe. The toxic elements content is significantly less than the standard values according to GOST R 17.4. 3.07-2001, and they can be used as organic fertilizers for grain and industrial crops, and in vegetable production.
3. Dehydrated sludge treated with the "Desolac" can be used in industrial floriculture, forest and ornamental nurseries, for biological remediation of disturbed lands and SDW landfills.
4. It is recommended to use the treated sludge from sludge beds of MUE "Lipetsk Aeration Station" and "LOS" LLC in Voronezh as organic limestone fertilizers for soils with a pH less than 5.5 in doses calculated taking into account the calcium content in the composition of the deposited sludge. When using sludge as fertilizer, regional and local conditions should be taken into account, including the properties and hydrological regime of the soil, the content of normalized pollution, total and mineral nitrogen, phosphorus, potassium in the sludge and soil, the characteristics of crop cultivation and crop rotation.

References

1. Turovsky, I.S.: Sewage Sludge Treatment, 3rd edn. Stroiizdat, Moscow (1988). Pererab. and additional
2. Turovsky, I.S.: Sewage Sludge. Dehydration and Disinfection. DeLi print, Moscow (2008)
3. Ignatenko, A.V.: Decontamination of sewage sludge and methods of its control. In: Proceedings of BSTU, vol. 4, pp. 210–213 (2016)
4. Voronov, Yu.V., Yakovlev, S.V.: Wastewater and wastewater treatment. MGSU Because of the Association of Construction Universities, Moscow (2006)
5. Yakovlev, S.V., Volkov, L.S., Voronov, Yu.V., Volkov, V.L.: Treatment and utilization of industrial sewage sludge. Chemistry, Moscow (1999)
6. Danilovich, D.A.: Modern methods of disinfection of sewage sludge. NDT **6**, 50–54 (2018)

7. Russian Federation Standard SanPiN 2.1.7.1287-03
8. Russian Federation Standard GOST R 17.4.3.07-2001
9. Russian Federation Standard MU 3.2.1022-01: Measures to reduce the risk of infection of the population with pathogens of parasitosis (2001)
10. Russian Federation Standard SanPiN 1.2.2584-10

Energy Efficient Technology of Municipal Solid Waste Recycling (Using the Example of Voronezh, Russia)

Elena Golovina(✉), Irina Ivanova, and Elena Sushko

Voronezh State Technical University,
Moscow Avenue, 14, Voronezh 394026, Russia
u00111@vgasu.vrn.ru

Abstract. The paper analyzes the existing methods for processing municipal solid waste and proposes an optimal method (using the example of Voronezh, Russia). The main provisions that form the environmental policy in relation to municipal solid waste are given. It is part of the general policy of increasing energy and resource saving in the field of consumption and reducing man-made impacts on people and nature. The analysis of modern technologies for the municipal solid waste recycling is presented, ensuring maximum regeneration of energy and material resources spent on waste generation with their full safety for the population and nature. Disposal of municipal solid waste at the landfill as the least expensive technology and fairly safe for the environment is one of the key areas of development of the sanitation and cleaning system in Voronezh.

Keywords: Energy efficient · Recycling · Municipal solid waste

1 Introduction

The problem of handling municipal solid waste (MSW) is relevant not only for the global community but also for Russia, despite the fact that its acuteness is veiled by the fact that, unlike in European countries, Russia has considerable areas for waste disposal. Waste generation in the Russian economy is 3.4 billion tons per year, including 40 million tons of MSW per year and 30 million tons of sludge of sewage treatment plants per year, with only about 3.5% of MSW being processed [1, 2].

One of the reasons for the low level of waste utilization is the low profitability of the collection process and existing recycling technologies. There is also no state regulation in the waste management system, there is no effective economic environment, the fee for waste disposal is not indexed, the mechanism for attracting investments in setting up waste treatment facilities is not adjusted, best practices in waste management are not effectively implemented, there is no effective management mechanism in this area. All these circumstances lead to the use of both liquidation and utilization techniques in respect of MSW:

- disposal (storage);
- recycling of MSW with the extraction of valuable components (plastics, ferrous and non-ferrous metals, glass, paper, etc.) and the production of RDF fuel pellets;

V. Murgul and M. Pasetti (Eds.): EMMFT-2018, AISC 983, pp. 373–384, 2019.
https://doi.org/10.1007/978-3-030-19868-8_38

- incineration using various types of furnaces (grate firing, with rotary grate, with rotating grate cylinders, in a fluidized bed, etc.);
- composting;
- pyrolysis;
- gasification.

Comparative analysis of data on the socio-economic development of regional centers shows that it is natural and historically justified that Voronezh is the capital of the Central Black Earth region. According to the General Plan and the data of the Department of Housing and Public Utilities of the administration of the urban district of the city of Voronezh, the environmental situation in the urban district since the 90s of the last century is characterized as tense and unsatisfactory in most signs and indicators. At the same time, the state of the ecological situation has a pronounced tendency to deterioration. The natural and resource potential of urban ecosystems is depleted and it is clear that without designing and implementing effective measures of protection, prevention and rehabilitation with respect to practically all components of the environment, the ecological situation will worsen dangerously. Based on research, it is possible to determine the rate of accumulation of waste for the future:

- for residential sector (1):

$$\mathrm{Npr} = \mathrm{Nin}(1 + 0.0265)^{t}, \tag{1}$$

where Npr—the predicted rate of waste accumulation;

Nin—the used (initial) rate of waste accumulation;

t—the forecasting period, years;

0.0265 - the coefficient taking into account the annual increase in the volume of accumulation of MSW, i.e., 2.65%;

- for MSW from detached public facilities, shopping and cultural institutions (2):

$$\mathrm{Mpr} = \mathrm{Min}\,(1 + 0.005)^{t}, \tag{2}$$

where Mpr—the predicted rate of waste accumulation;

Min—the used (initial) rate of waste accumulation;

0.005—the coefficient taking into account the annual increase in the volume of accumulation of MSW, i.e., 0.5%.

Thus, the priority task is the development of new norms for the generation (accumulation) of MSW. A big problem in the city is waste management. Currently, in the urban district of the city of Voronezh, there are no standards for waste generation, approved by the federal authorities. Comparing the rates of waste accumulation in a number of cities of the Russian Federation (Moscow, Tver, Bryansk, Belgorod, Ivanovo, Oryol) with the norms that were previously in force in the city district, we can conclude that they are significantly lower than similar standards operating in other cities, and do not take into account the specifics of certain categories of natural resource users, which is decisive in the field of solid municipal waste management. In the urban district of the city of Voronezh, the system of collection, removal and disposal of municipal solid waste covers more than 97% of the population of the city and about 85% of the subjects of economic activity. The annual volume of generation of MSW on the territory of the urban district is about 2.5 million m^3.

2 Materials and Methods

Currently, the activities of housing and communal services in the field of waste management is accompanied by very large losses of resources, as well as an increase in environmental pollution. Individual housing stock and commercial organizations with a discharge system are not fully covered:

- the problem of MSW remains;
- it is necessary to purchase equipment and containers, vehicles, mechanisms for the sanitation of containers, and special vehicles;
  technologies for recycling of MSW are not defined;
- it is necessary to change the norms of accumulation of MSW;
- there is no selective collection;
- highly toxic waste accumulates, including medical.

The complex of measures for the creation of a complete system for the collection, storage, transportation, and disposal of waste in health care facilities (HCF) should include a number of organizational aspects:

- waste separation according to the established classification of SanPiN (Sanitary Norms and Rules) and by fractions in places of generation;
- the presence of a special packages (bags, containers) in sufficient quantity that meets the requirements of regulatory documentation, easy to use, ensure reliable sealing and compliance with the rules of the anti-epidemic regime;
- the proper organization of primary waste collection sites in places of generation;
- epidemiologically safe transportation of sealed waste to temporary storage sites;
- transportation of waste to the disinfection facility (if any);
- transportation of waste to the container site;
- equipped place for washing and disinfecting containers;
- organization of proper disinfection at all stages of waste management;
- transportation of waste from health care facilities.

In accordance with regulatory documents, all wastes of health care facilities are divided into five classes according to the degree of their epidemiological, toxicological and radiation hazard [3, 4]:

- class A: non-hazardous waste;
- class B: hazardous waste;
- class C: extremely hazardous waste;
- class D: wastes close in composition to industrial;
- class E: radioactive waste.

Mixing of waste of different classes at all stages of collection and storage is not allowed, there is a certain order of waste disposal.

Table 1 shows the accumulation of MSW by year.

Table 1. Annual volumes of accumulation of MSW in Voronezh.

Year	Volume of MSW accumulation		Including			
	Thousand m^3	Thousand tons	Main part of MSW		Bulky waste	
			Thousand m^3	Thousand tons	Thousand m^3	Thousand tons
2008	1716	343.2	1539	307.8	177	35.4
2009	1746	349.2	1566	313.2	180	36
2010	1775	355	1592	318.4	183	36.6
2011	1807	361.4	1621	324.2	186	37.2
2012	1840	368	1651	330.2	189	37.8
2013	1897	379.4	1702	340.4	195	39
2014	1949	389.8	1749	349.8	200	40
2015	2001	400.2	1795	359	206	41.2
2016	2053	410.6	1841	368.2	212	42.4
2017	2105	421	1888	377.6	217	43.4
2018	2156	431.2	1933	386.6	223	44.6
Total	21045	4209	18877	3775.4	2168	433.6

One of the problems is to take into account the norms of the cost of road transport per 1,000 km of mileage for maintenance and repair of automobiles.

Table 2 presents the mileage of the garbage truck for one and a half working day.

Table 2. Estimated mileage of garbage truck for one and a half working day.

Brand of machinery and route	The length of the route, km	Time to collect and remove MSW, h	Number of rides, pcs	Volume of MSW transported during the working day, m^3	Garbage truck run per working day, km
Garbage truck KO-413					
City-WRP №1	8.7	2.3	4.2	59.1	103.78
City-WRP №2	18.1	2.9	3.4	48.3	143.71
City - Disposal site	25.5	3.3	3.0	41.6	174.37
Garbage truck KO-415A					
City-WRP №1	8.7	5.0	1.9	79.3	63.29
City-WRP №2	18.1	5.7	1.8	73.8	82.40
City - Disposal site	25.5	6.1	1.6	67.6	104.88
ZIL MSK-TP					
WRP №1 - Disposal site	26.0	2.3	4.3	51.1	249.34

The cost of maintenance and repair (B) is calculated by the following formula (3) [3, 4]:

$$B = B_1 \times L : 1000, \tag{3}$$

where: B - the cost of maintenance and repair per shift,

B_1 - the cost of maintenance and repair for 1000 km,

L - mileage per shift.

Accounting of the volumes transported to the facilities is carried out on the main part, since bulky waste is transported by garbage trucks with swap bodies directly to the MSW disposal site.

The methodology is based on a comparison of the reduced costs of direct and two-stage transportation of MSW using the following formula (4):

$$E_{year} = Q_{year}(C_{dir} - C_{two}) \tag{4}$$

where E_{year}—economic efficiency from the introduction of two-stage export of MSW, thousand rubles/year;

Q_{year}—the estimated annual volume of waste removal, thousand m^3/year;

C_{dir}—specific reduced costs for direct transportation of waste, RUB/m^3;

C_{two}—specific reduced costs for two-stage removal of MSW, RUB/m^3;

The specific reduced costs for direct transportation are calculated by the formula (5):

$$C_{dir} = \mathrm{al} + \mathrm{b}, \tag{5}$$

where a and b—economic efficiency indicators for collecting garbage trucks;

L—average distance from MSW collection sites to the landfill, km.

If municipal solid waste is transported by various types of garbage trucks, in this case, the estimated indicators a* and b* are introduced, which are calculated by the formula (6 and 7):

$$a* = (a_1Q_1 + a_2Q_2 + a_3Q_3)/\sum Q, \tag{6}$$

$$b* = (b_1Q_1 + b_2Q_2 + b_3Q_3)/\sum Q, \tag{7}$$

where a_1, a_2, a_3 and b_1, b_2, b_3—indicators for the respective brands of garbage trucks;

Q_1, Q_2, Q_3—annual volumes of waste transported by garbage trucks of the respective brands.

The specific reduced costs for two-stage transportation of waste are calculated by the formula (8):

$$C_{two} = C_1 + C_2 + C_{WRP}, \tag{8}$$

where C_1—specific reduced costs for waste removal by collecting garbage trucks to WRP (waste recycling plant) (9):

$$C_1 = al_1 + b, \tag{9}$$

where L_1—average distance from the MSW collection area to WRP, km; C_2 - specific reduced costs for removal of MSW from WRP to the landfill (10):

$$C_2 = Al_2 + B, \tag{10}$$

where L_2—distance from WRP to the landfill;

A, B—indicators of economic efficiency of transport garbage trucks;

C_{WRP}—specific reduced costs for transferring solid household waste to WRP.

The urban district of Voronezh is divided into 6 administrative districts: Zheleznodorozhny, Kominternovsky, Levoberezhny, Leninsky, Sovetsky, and Central.

The distance of waste transportation by administrative districts of the urban district of the city of Voronezh is given in Table 3.

Table 3. Distance from waste collection points to waste landfill.

№	Administrative unit, district	Distance to waste landfill, km
1	Kominternovsky district	25.0
2	Central district	22.0
3	Leninsky district	23.0
4	Sovetsky district	14.0
5	Levoberezhny district	29.0
6	Zheleznodorozhny district	28.0
	Average in urban district of city of Voronezh	23.5

In accordance with the results obtained, the average distance of waste transportation in the urban district of the city of Voronezh will be 23.5 km. The weighted average distances for the removal of MSW to disposal sites were determined using route schedules for the operation of garbage trucks removing MSW from districts of Voronezh, approved by the MSE "Production Association for Waste Management". Let us determine the value of the weighted average distance of waste removal to the MSW disposal site for estimated periods. The calculation results are summarized in Table 4.

Table 4. Weighted average distances for removal of MSW to a disposal site for 2018.

City district	Q_i, thousand m^3	L_i, km	Q_iL_i, thousand $m^3 \times$ km
Zheleznodorozhny district	243.2	29.18	7096.576
Levoberezhny District	450.8	28.13	12681.004
Somovo village	35.5	38.15	1354.32
Total	**729.5**	–	**21131.9**

The weighted average distance of waste removal to the MSW disposal site for the period up to 2018 was: Lav = 21,131.9/729.5 = 28.96 km. For further calculations, the weighted average distance of removal of MSW to the disposal site is assumed to be 29 km (excluding the "zero" run).

When considering the feasibility of using a two-stage removal of MSW, construction of one WRP in the city should be considered - in the Zheleznodorozhny district. The calculation of the weighted average distance of removal of MSW at the construction of WRP is given in Table 5.

Table 5. The calculation of the weighted average distance of removal of MSW at the construction of WRP.

City district	Volume of MSW generation Q_i, thousand m^3	Distance to WRP L_{WRPi}, km	$Q_i\,L_{WRP}$, thousand m^3 × km
Zheleznodorozhny district	243.2	4.49	1091.97
Levoberezhny District	450.8	13.72	6184.98
Somovo village	35.5	12.01	426.355
Total	**729.5**	–	**7703.305**

If as a result of the calculation:

$E_{year} < 0$, then the introduction of two-stage removal is impractical;

$E_{year} > 0$, then the introduction of MSW removal with overload is economically feasible. According to the formula 4, $E_{year} = -254.52 < 0$, therefore, the introduction of two-stage removal of MSW is not economically feasible.

3 Results

In the modern world, an environmental policy has formed regarding municipal solid waste. It is part of the general policy of increasing energy and resource saving in the field of consumption and reducing man-made impacts on people and nature.

The policy is based on two important provisions:

1. In modern conditions, the uncontrolled formation of the quantity, composition of MSW, ways and technologies of their recycling is unacceptable; all these issues should be an integral part of environmental and economic national and regional policies;
2. Modern technologies for MSW recycling should ensure the maximum regeneration of energy and material resources expended in the creation of waste with their full safety for the population and nature.

When choosing a technology for MSW recycling, the following factors should be taken into account [5]:

technical and economic factors:

- by reduced costs, technology should be the cheapest;
- maximum use of valuable components of MSW;

environmental factors:

- the technology for MSW recycling must be environmentally friendly;
- final products (compost, ash, RDF, etc.) must not harm the environment;

climatic and social factors:

- availability of favorable climatic and social conditions.

When using the combustion technology with the condition of compliance with the European requirements for emissions of pollution into the air, the annual volume of gas when burning 200 thousand tons/year will be 1048.5 million m^3/year, while emissions will be within the normal range (Table 6).

Table 6. Indicators of waste incineration plant emissions.

Indicators	Concentration mg/n.m^3		Amount of emissions, kg/year	Hazard class
	In the exhaust gases	MPC		
Dust	25	0.15	26213.6	3
Zinc	0.36	0.06	377.5	3
Cadmium	0.002	0.001	2.1	2
Nickel	0.0035	0.001	3.7	2
Chromium	0.025	0.0015	26.2	1
Lead	0.063	0.0003	66.0	3
Copper	0.098	0.002	102.8	2
Manganese	0.094	0.01	98.6	2
Arsenic	0.00025	0.003	0.3	2
Mercury	0.00025	0.0003	0.3	1
Tin	0.028	0.5	29.5	3
Cobalt	0.00055	0.001	0.6	2
Selenium	0.077	0.1	80.8	1
Vanadium	0.002	0.002	2.1	1
Sulfur anhydrite	50	0.5	52427.2	3
Nitrogen dioxide	100	0.085	104854.4	2
Carbon monoxide	130	5	136310.7	4
Polycyclic chlorides	4	0.005	4194.2	2
Aromatic Polycyclic Hydrocarbons	0.05	0.1	52.4	2
Hydrogen chloride	30	0.2	31456.3	2

(*continued*)

Table 6. (*continued*)

Indicators	Concentration mg/n.m^3		Amount of emissions, kg/year	Hazard class
	In the exhaust gases	MPC		
Hydrogen fluoride	2	0.02	2097.1	2
Cyanides (by HCN)	0.5	0.01	524.3	2
Total			358920.7	
Including				
1 class			109.4	
2 class			143386.8	
3 class			79113.8	
4 class			136310.7	

4 Discussion

The following major trends for the future are outlined:

1. Prospective collection of MSW with the obligatory separation of organic and mineral parts and the extraction from the garbage: ferrous and non-ferrous metals, plastics, glass, paper, food waste, etc.;
2. Extraction and recycling of valuable components of MSW into secondary raw materials;
3. Expansion of the market for recycled products;
4. Strengthening legislative measures for the impact on market forces aimed at stimulating the industries for the recycling of secondary products extracted from MSW;
5. The introduction of an environmental tax on products that use or packaged in materials that cannot be further recycled.

A prospective collection is determined primarily by economic incentives for the population, by supplying the population with packages for various components of MSW and a collection system (containers) [6].

Disposal of municipal solid waste on the landfill, as the least costly technology and safe enough for the environment, subject to strict adherence to the standards of the technological process, is also one of the key areas for the development of sanitary treatment of the urban district of Voronezh. The economic benefits of implementing more advanced technologies are significantly reduced due to the lack of a selective waste collection system. The insufficiently high demand for secondary raw materials extracted from waste and the poor quality of such raw materials when implementing the joint waste collection scheme, and as a result - the low profitability of this type of activity, cannot attract private investment. The constant rise in energy prices can quickly change priorities in this area. Therefore, considering the future development of the situation, we can assume the increasing benefits of waste recycling technologies, accompanied by the production of various types of energy in its pure form (heat, electricity) and energy carriers (synthesis gas, synthetic fuel, pyrocarbon). With the sale of ferrous and non-ferrous metal scrap on the foreign market, profits will increase significantly. Recycling of paper, plastics, glass, and textiles requires special structures for their recycling. All this will require the construction of waste recycling plants (WRP), the cost of which is from 1.5 billion rubles with a recycling volume of 100 thousand tons per year to 7 billion rubles with an increase in volume up to 350 thousand tons. Figure 1 presents the integrated assessment of technologies for MSW recycling in Voronezh.

In the future, the Voronezh urban district will also need additional equipment with containers, an annual increase in their number by 2000 pieces should compensate for physically worn-out containers and lead to a significant increase in 2020.

It will also require the purchase of modern diesel garbage trucks, for example, the MKZ-470 and MKM-4704 brands, manufactured by the Ryazhsky Automobile Plant, the cost of which currently lies in the range from 2 to 2.5 million rubles, special machinery for cleaning territories during the winter and summer periods, means of control and optimization of the scheme of waste transportation on the basis of GPS technologies. It is necessary to build a sanitation station [7].

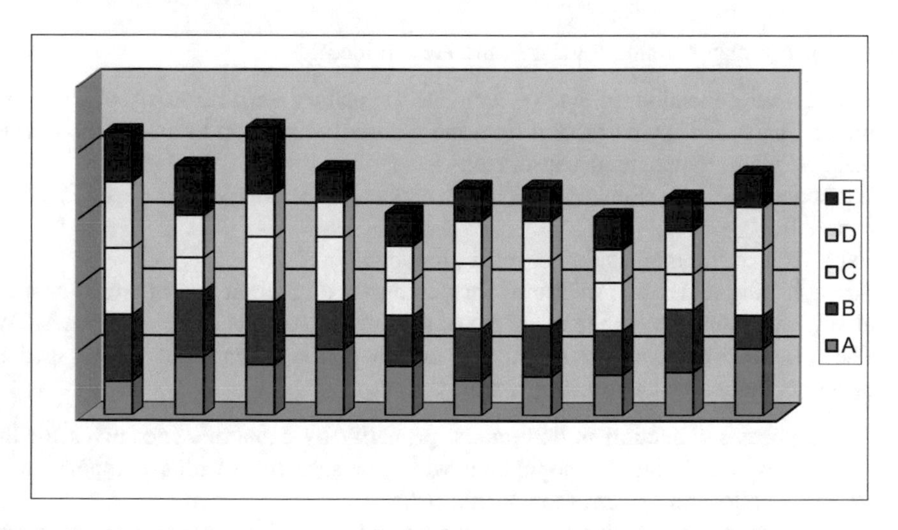

Fig. 1. The integrated assessment of technologies for MSW recycling in Voronezh (A - Reduced costs of capital construction and operation of facilities, B - Maximum use of valuable components of MSW, C - The technology for MSW recycling should be environmentally friendly, D - End products (compost, ash, RDF, etc.) must not harm the environment, E - Climatic factor, 1. Storage, 2. Incineration with heat recovery, 3. Composting, 4. Production of RDF + Composting, 5. Composting + incineration of unopposable fractions, 6. Sorting + aerobic composting, 7. Sorting + aerobic composting, 8. Sorting + aerobic composting + incineration, 9. RAMET Pyrolysis, 10. Steam treatment and generation).

When solving these problems, there is the possibility of attracting private investors, since the payback period of the plants is 8–10 years, taking into account payments for waste reception and the use of recycled raw materials [8, 9].

5 Conclusion

The analysis of some existing systems for municipal solid waste recycling showed that, as applied to the situation prevailing in the urban district of Voronezh, it is most expedient to introduce waste composting technology with previous sorting, in order to separate components that are secondary raw materials. Disposal of municipal solid waste at the landfill as the least costly technology and sufficiently safe for the environment, subject to strict adherence to the standards of the technological process, is also one of the key areas of development of the sanitary and cleaning system in Voronezh.

Economic benefits from the implementation of more advanced technologies are significantly reduced due to the absence of a selective waste collection system in the city of Voronezh. There is insufficient demand for secondary raw materials extracted from waste and the poor quality of such raw materials when implementing the joint waste collection scheme, and as a result, the low profitability of this type of activity cannot attract private investment.

The constant rise in energy prices can quickly change priorities in this area, so if we look at the future development of the situation, the benefits of recycling technologies accompanied by the production of various types of energy in a pure form (heat, electricity) and energy sources will undoubtedly increase.

From an environmental point of view, the distribution of technologies according to the degree of "purity", the worst technology is "Incineration" (according to emissions of harmful substances into the atmosphere).

References

1. Lyushinskiy, A.V., Fedorova, E.S., Roshan, N.R., Chistov, E.M., Golov, R.S.: Diffusion welding of 12Cr18Ni10Ti steel to palladium alloy foil. Welding International (2017). https://doi.org/10.1080/09507116.2017.1318505
2. Safronova, N., Nezhnikova, E., Kolhidov, A.: Sustainable Housing Development in Conditions of Changing Living Environment. In: 2017 MATEC Web of Conferences, vol. 106, p. 08024 (2017). https://doi.org/10.1051/matecconf/201710608024
3. Manohin, V.Y., Shestakov, A.A, Ivanova I.A.: Vybor optimal'nyh tekhnologij obezvrezhivaniya i pererabotki TBO. EHkologiya i racional'noe prirodopol'zovanie: materialy Mezhregional'noj nauchno-prakticheskoj konferencii. Voronezh (2007)
4. Manohin, V.Y., Ivanova, I.A., Manohin, M.V.: Optimal'nye resheniya problemy obrashcheniya s TBO. EHkologicheskaya bezopasnost' nashego budushchego: materialy IX Mezhregional'noj nauchno-prakticheskoj konferencii, posvyashchennoj Godu ohrany okruzhayushchej sredy v Rossijskoj Federacii, 23 maya 2013 g. Administraciya gorodskogo okruga gorod Voronezh, Upravlenie ehkologii administracii gorodskogo okruga gorod Voronezh, Voronezh, pp. 140–143 (2013)
5. MU 2.1.7.1185-03. 2.1.7. Pochva. Ochistka naselennyh mest. Othody proizvodstva i potrebleniya. Sanitarnaya ohrana pochvy. Sbor, transportirovanie, zahoronenie asbestsoderzhashchih othodov. Metodicheskie ukazaniya: utv. Glavnym gosudarstvennym sanitarnym vrachom RF 23.01.2003. Federal'nyj centr gossanehpidnadzora Minzdrava RF, Moscow (2003)

6. Potapov, P.A., Pupyrev, E.I., Potapov, A.D.: Metody lokalizacii i obrabotki fil'trata poligonov zahoroneniya tverdyh bytovyh othodov. Izd-vo ASV, Moscow (2004)
7. Tarakanov, V.A.: Metodika sravnitel'noj tekhniko-ehkonomicheskoj ocenki tekhnologij pererabotki TBO vtorichnogo syr'ya. Izdatel'stvo ILI RAN, Moscow (2006)
8. Sister, V.G., Mirnyj, A.N., Skvorcov, L.S.: Tverdye bytovye othody: spravochnoe izdanie. AKKH im. K.D. Pamfilova (2001)
9. Nezhnikova, E.: The use of underground city space for the construction of civil residential buildings. Proc. Eng. **165**, 1300–1304 (2016). https://doi.org/10.1016/j.proeng.2016.11.854

Reduction of Dust in the Working Zone of the Shot Blasting Area of Foundry Production

Elena Golovina(✉), Tatyana Schukina, and Vyacheslav Manohin

Voronezh State Technical University,
Moscow Avenue, 14, Voronezh 394026, Russia
u00111@vgasu.vrn.ru

Abstract. Analysis of the dust concentration in the operator's working zone in the foundries of Russian enterprises revealed significant excess of MPC. The results of the survey of the area of shot blasting of parts at the foundry production are given. The dispersed and elemental composition of the emitted dust is determined when performing this technological operation. Given the pollution of the atmospheric air of the industrial zone, it was proposed to dismantle the cyclone and install a more advanced device of the same type. It provides an additional internal perforated wall, which is located with the formation of a gap between it and the body, and also an irrigated packing in the exhaust pipe to collect fine dust. The innovative cyclone has a high degree of purification, since along with the content of two fundamentally different stages of reducing the concentration of dusty air, the perforated inner wall prevents the entrainment of particles thrown by centrifugal force and collected in the gap. Installation of the proposed ventilation device will reduce the damage caused to the working zone and increase the amount of recycled dust.

Keywords: Reduction · Working zone · Shot blasting area

1 Introduction

Sandblasting or shot blasting is one of the most common methods for treating the surface of metal parts. This technological operation allows qualitatively grinding castings of various shapes or products obtained by milling blanks [1]. But this process is accompanied by significant dust generation, which, even with validly reasonable local exhaust ventilation, enters the working zone, creating a threat to the health of workers. In addition, the low cleaning efficiency of ventilation emissions affects the external production conditions, which, in the absence of intensive dispersion, leads to the inhibition of the ecosystem of the industrial zone [2].

In addition, high molecular weight hydrocarbons, especially benzo(a)pyrene, are a toxic consequence of fuel combustion. The maximum permissible concentration of dust in working zones (MPCw.z.) affects the content of SiO_2. When $SiO_2 \leq 10\%$, MPCw.z. = 10 mg/m^3. However, in fact, the volume of SiO_2 ranges from 28.2–46.51% to 35–50%, and quartz is almost always present. MPCw.z. is assumed to be 2 mg/m^3. Silica is not very mephitic, but its regular entry into the human lungs causes gradual changes in

V. Murgul and M. Pasetti (Eds.): EMMFT-2018, AISC 983, pp. 385–394, 2019.
https://doi.org/10.1007/978-3-030-19868-8_39

the body [3]. It is established that the incidence of people directly depends on the weight concentration of dust in the air environment. Only a serious assessment of the sanitary condition of the working zones of the foundry and the effective implementation of engineering measures will help to reduce emissions of harmful substances.

These negative effects are also characteristic of an enterprise having shot blasting areas. The production site operates in a single-shift mode with short-time technological interruptions required to perform preparatory work before the next operation [4, 5]. There are chambers of 2000 × 2000 × 2500 mm in size, in which swivel supports are installed for convenient location of the processed products on them. In the upper part, the chamber is connected through a pipe with a diameter of 630 mm to the local exhaust ventilation, which contains a cyclone CN-11.

The degree of harm to the health of people at production depends on the particle size distribution of dust, i.e. the quantitative ratio of dust particles of different sizes. The upper respiratory tract retains mainly large elements. Particles smaller than 10 microns for the most part remain in the lungs. The hazard class is 3, MPC = 0.5 mg/m^3.

2 Materials and Methods

Studies have shown that the dispersed and elemental composition of dust in the shot blasting area of the foundry, the shot with an average diameter of 2 mm is supplied on the surface of the product by compressed air flow at a speed of 30 m/s through the feed window (Fig. 1). At the same time, dust is formed, which by 86.2% consists of particles less than 100 microns in size and by 13.8% - larger than 100 microns (Table 1). In addition, the generated dust has a fairly extensive chemical composition, in which there are 18 elements of the periodic system (Table 2).

For the analysis of dust particles by size, the method of particle size analysis was used. Its principle is based on the fact that the size of particles affects the speed of movement of particles under the action of gravitational or centrifugal forces. Knowledge of the particle size distribution means a lot in technical and hygienic terms. Particle sizes and their distribution affect almost all properties of dust materials. The particle size distribution is also one of the factors for assessing the effect of air dustiness on a person and the choice of appropriate dust preventives. Analysis of the particle size distribution solves two tasks: determination of particle sizes and determination of the percentage of particles of different size classes. The particle size distribution of the provided powder sample was determined by a laser diffraction method using a Fritsch Analisette-22 NanoTec laser particle analyzer using Fritsch Mas control software. The main principle of operation of the analyzer is the study of particles, based on the diffraction of laser radiation: meeting with a powder particle, the beam is deflected at a certain angle, the value of which depends on the particle size. Then the scattered beam is on the surface of the detector. Determining the strength of the radiation incident on each component of the detector and further mathematical processing of the signal make it possible to determine the size of the sample particles and estimate their shape. At the next stage of the experiment, the dispersion composition of dust was refined by X-ray microanalysis, which consists in determining the deposition rate of suspension particles under the action of gravity. The results were obtained on X-ray diffractometers

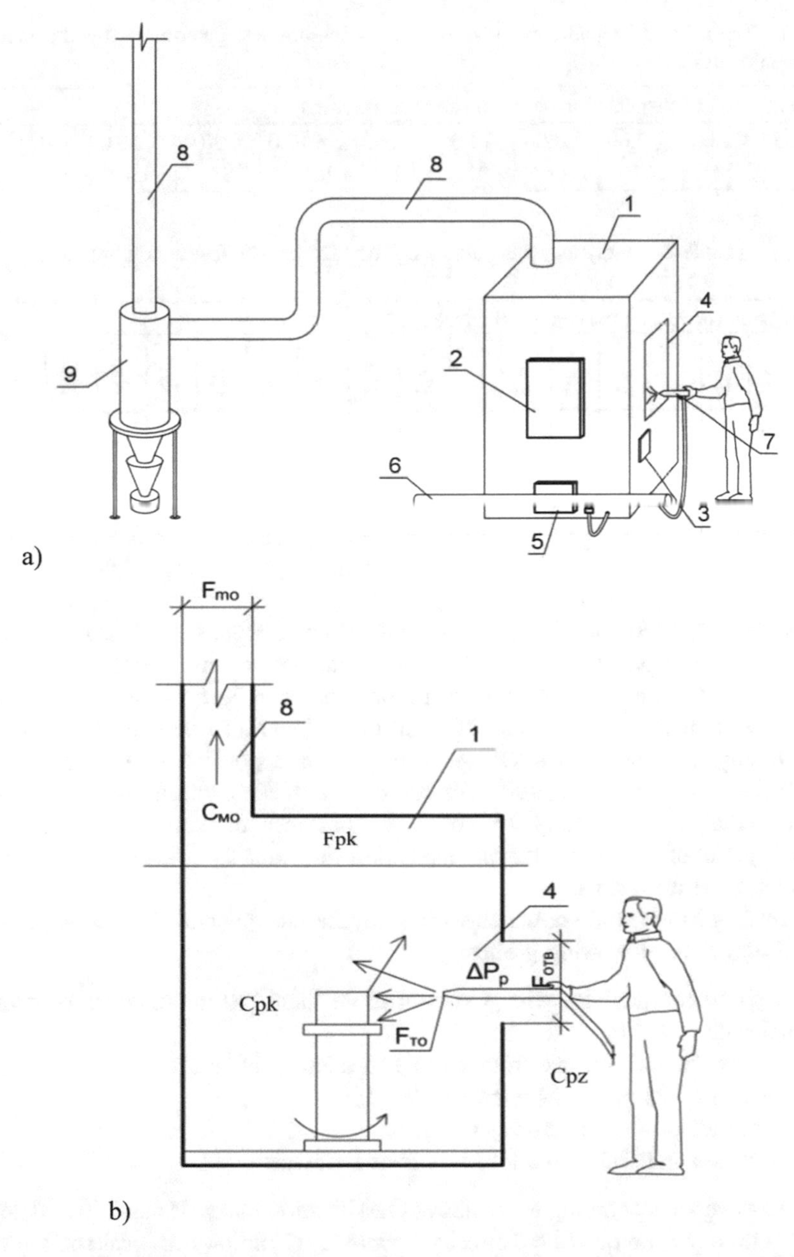

Fig. 1. Dust collection system: a - scheme of the dust collection system; b - section of the shot blasting chamber; 1 - shot blasting chamber; 2 - feed sector of the cleaned parts with a hermetic door; 3 - control panel; 4 - shot blasting device delivery window; 5 - technological hole with a pressure cap for the removal of shot; 6 - compressed air line; 7 - air supply to the nozzle; 8 - duct; 9 - abrasive dust separator; 10 - conical cyclone for air purification; 11 - mechanism of feed and rotation of the product being cleaned.

Table 1. The content of particles of various sizes in samples, determined by the method of dispersion analysis.

Content, in%, of dispersed material of the size, microns										
≤0.05	≤1.00	≤2.00	≤3.00	≤4.00	≤5.00	≤10.0	≤20.0	≤50.0	≤100	>100
0	0.9	1.5	1.9	2.0	2.2	2.9	5.3	39.9	86.2	13.8

Table 2. X-ray analysis data using the "Bruker S8 Tiger" equipment.

Composition of dust on the content of elements, %																	
Na	**Al**	**Si**	**S**	**Cl**	**K**	**Ca**	**Ti**	**Cr**	**Mn**	**Fe**	**Ni**	**Cu**	**Zn**	**Ga**	**Zr**	**Mo**	**Ce**
0.141	21.295	27.94	0.04	0.056	0.183	0.143	0.095	0.511	0.712	48.535	0.139	0.05	0.014	0.022	0.014	0.072	0.038

"DRON-04" and "Radian DR-02", as well as on a Bruker S8 Tiger wave X-ray fluorescence spectrometer. The study of the structure of dust generated in the shot blasting areas and the removal of forms in the process of X-ray microanalysis showed that, as a rule, it has a predominantly form that allows it to be considered spherical. When falling, dust particles always tend to occupy a position corresponding to the greatest resistance in the air, this form contributes to their settling in the atmosphere and in inertial dust collectors. However, the presence of particles smaller than 10 microns speaks of their considerable deposition time and the need for highly efficient air after-treatment systems.

According to the well-known classification, the dust generated in the shot blasting area belongs to the following groups:

I - very coarse particles with a size of more than 140 microns are contained in samples up to 6.3%;
II - coarse dust with a size from 40 to 140 microns is 76.3%;
III- medium dust of 10–40 microns - 9.8%;
IV- fine dust of 1–10 microns - 6.7%;
V- very fine material with a size less than 1 micron - 0.9%.

The assessment according to the above classification makes it possible to determine the air consumption required for effective removal of particles in local suction and to select the equipment of the necessary degree of purification for ventilation emissions [6]. If high-capacity cyclones are used to capture dust of the first and second groups, then for the third and fourth groups, there are fabric filters, and electrostatic precipitators and absorbers are effective for collecting very fine particles belonging to the fifth group.

The first of these cleaning devices are widely used in industry, since they are easy to manufacture and operate. However, the efficiency of collecting dust in them is not always sufficient, and therefore two or three-stage schemes with preliminary collecting of large particles at the first stage are often developed.

On the test section of the shot blasting part of a chamber, in which processing takes place, to obtain deposition of a coarse dust, as well as to remove fine composition by the air flow from the local suction, it is necessary to know the speed of its deposition or, in other words, soaring [7]. For this purpose, one can use the Stokes law written in the following form [8]

$$V_t = \frac{g d_P^2 (\rho_P - \rho_G)}{18 \mu_G}, \tag{1}$$

where V_t – deposition rate, m/s; g – gravitational acceleration, m/s^2; d_P – particle diameter, m; ρ_P, ρ_G – density of the particles and gas flow, respectively, kg/m^3; μ_G – dynamic viscosity of air, (N·s)/m^2.

Dependence (1) is valid for spherical particles with a diameter of less than 50 microns. For larger particles, the following equation is recommended for use [9]

$$V_t = \sqrt{\frac{4}{3} \cdot \frac{g d_P (\rho_P - \rho_G)}{\xi_P \rho_G}}, \tag{2}$$

where ξ_P – particle drag coefficient, which is determined depending on the mode of motion of the particles, i.e. the Reynolds number Re.

With an increase in the size of spherical particles, the error in determining the drag coefficient increases [9, 10], and, given that the dispersed material has deviations from the indicated shape, the accuracy of the speed calculated by formulas (1) and (2) is not sufficient for a number of tasks. Therefore, it is necessary to use the experimental data [11], the approximation of which in the case of dust density equal to $\rho D = 1$ g/cm^3 gives the equations of the form:
with a diameter of spherical particles d less than 10 microns

$$V_t|_{\rho=1} = 0,0002 + 0,0002d + 0,0031d^2; \tag{3}$$

with 10 < d < 1000 microns

$$V_t|_{\rho=1} = 0,4271d - 12,049, \tag{4}$$

where $V_t|_{\rho=1}$ - soaring speed at a dust density of 1 g/cm^3, cm/s.

When approximating the experimental data given in [11], the coefficients in parabolic and linear dependencies have the following numerical values:
with a diameter of spherical particles d less than 10 microns

$$V_t|_{\rho=1} = -0,0002 + 0,0006d + 0,003d^2 \tag{5}$$

with 10 < d < 1000 microns

$$V_t|_{\rho=1} = 0,3937d - 8,8989, \tag{6}$$

With other dust densities obtained in accordance with Eqs. (3–6), the soaring speed indices must be multiplied by the real value of ρ_D, g/cm^3, since the linear dependence of this parameter is characteristic, and, accordingly, a simple manipulation must be performed

$$V_t = \frac{\rho_{\Pi}}{100} V_t|_{\rho_{\Pi}=1}, \tag{7}$$

where V_t - speed of soaring of dust particles of the corresponding diameter, m/s.

Knowing the actual speed of soaring of industrial dust of different dispersity, it is possible to organize the effective work of local ventilation. For this, the following conditions should be observed in the test site. In the part processing chamber, it is necessary to ensure the settling of coarse dust of 100 microns and more in size, and to transport particles of less than 100 microns in size with the air flow from the local ventilation system to the cleaning equipment. The soaring speed for the indicated boundary dispersion of 100 μm in accordance with (4), (6) is 30.66 and 30.47 cm/s. For guaranteed deposition, we will take a lower speed, since in this case it will take large particles more time to reach the bottom of the chamber. Then, taking into account the dust density ρ_D = 2.1 g/cm^3, the soaring speed of particles with a size of 100 μm using expression (7) will be V_t = 0.64 m/s.

With the previously specified dimensions of the chamber, the length of the airflow lines when dust is generated at a height of up to h = 1.5 m will be on average l = 1.8 m. The indicated height justified by the convenient location of parts for processing and dislodging the dispersed material from the surface allows us to determine the time over which deposition of large particles will occur from the following ratio

$$t_d = h/V_t, \tag{8}$$

The required time period td in accordance with (8) is 2.34 s for particles of 100 μm in size and can be provided at a speed of exhaust air in the chamber of 0.77 m/s, determined by the expression

$$V_A = \frac{l}{t_d}. \tag{9}$$

where l - average length of airflow lines in the chamber, m.

In order to maintain an estimated speed of 0.77 m/s in the chamber, the airflow rate in the local ventilation system in accordance with (10) should be L = 11088 m^3/h

$$L = 3600 F_{CSC} V_A, \tag{10}$$

where F_{CSCh} - cross-sectional area of the chamber, m^2.

3 Results

However, the installed fan VC 14-46 No. 6.3 in the local exhaust ventilation system, according to the results of certification, has a productivity of 6670 m^3/h, which is 40.8% less than the required design flow rate. Therefore, the concentration of dust in the air of the working zone is rather high in the shot blasting area. When replacing the fan, as well as installing movable curtains for the shot blasting device supply window, working conditions improved and, as measurements showed, the dust concentration in the working zone decreased by 1.48 times.

A survey of the industrial zone showed that the external environmental situation requires the modernization of cleaning equipment. The active cyclone has a rather low efficiency for fine dust (Fig. 2).

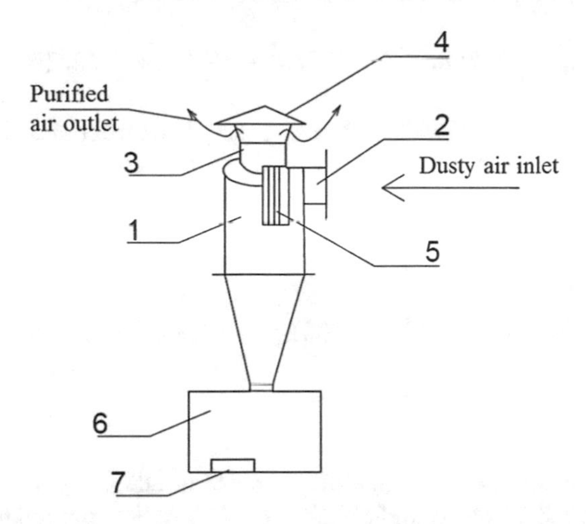

Fig. 2. Cyclone: 1 - cylindrical-conical case; 2 - inlet pipe; 3 - outlet pipe; 4 - hood; 5 - separator; 6 - bunker; 7 - cleaning hatch, Purified air outlet/Dusty air inlet.

As a technical solution aimed at improving environmental safety, a device of the same type was proposed [11], but including an irrigated packing in the exhaust pipe. In addition, the cyclone has an internal perforated wall located with the formation of a gap between it and the body. Dust thrown by centrifugal force through the perforated holes gets into the gap and under the action of its weight is deposited in the bunker without the possibility of entrainment by air flow. Smaller fractions are absorbed by the water in the irrigated packing of the exhaust pipe. It should be noted that the perforation of the wall increases the efficiency of the dry cleaning stage, and the wet stage allows removing fine dust, in addition, the overall efficiency of the device increases with increasing number of steps in accordance with the expression.

$$\eta = 1 - (1 - \eta_1)(1 - \eta_2)\ldots(1 - \eta_i), \tag{11}$$

where η_1, η_2, η_i – the cleaning efficiency of each of the devices included in the cascade.

Let us enlarge on the dry stage of cleaning, since the proposed design, above all, increases the efficiency at this stage of collecting. Due to a significant difference in dispersion composition, we consider the fractional degree of purification for devices of this class, which affects their overall efficiency ηi determined by the formula

$$\eta_i = \frac{\eta_{i_1}\Phi_1}{100} + \frac{\eta_{i_2}\Phi_2}{100} + \ldots \frac{\eta_{i_j}\Phi_j}{100} + \ldots \frac{\eta_{i_n}\Phi_n}{100}, \tag{12}$$

where $\Phi_1, \Phi_2, \ldots \Phi_j, \ldots \Phi_n$ - the content of this fraction at the inlet to the cleaning device, %; η_{i_j} - efficiency of the i-th device for the j-th fraction; n - the number of fractions.

So, for example, for cyclones CN-15 with a diameter of 500 mm with a recommended air speed and dust density of 2670 kg/m^3, the fractional degree of purification shown in Fig. 2 has the following functional dependence

$$\eta_{i_j} = 0,2006d^3 - 4,5758d^2 + 37,222d - 22,726. \tag{13}$$

Perforation of the inner wall makes it possible to catch not only the medium dispersed dust of the third group but also a fairly fine composition belonging to the fourth group. In this case, the predicted dependence will have high rates already for dust in size from 12 to 10 μm (Fig. 2) and with a high degree of probability correspond to expression (14).

$$\eta_{i_j} = 0,2394d^3 - 4,9061d^2 + 32,832d - 21,854 \tag{14}$$

4 Discussion

In conclusion, it should be noted that the increase in efficiency at each stage of dust collection in both local suction and cleaning devices affects the interconnectedness of solving the issues of environmental safety of an industrial zone and improving working conditions in production. Only such a system approach to internal and external tasks will reduce the negative impact on the environment and will not cause harm to the health of workers (Fig. 3).

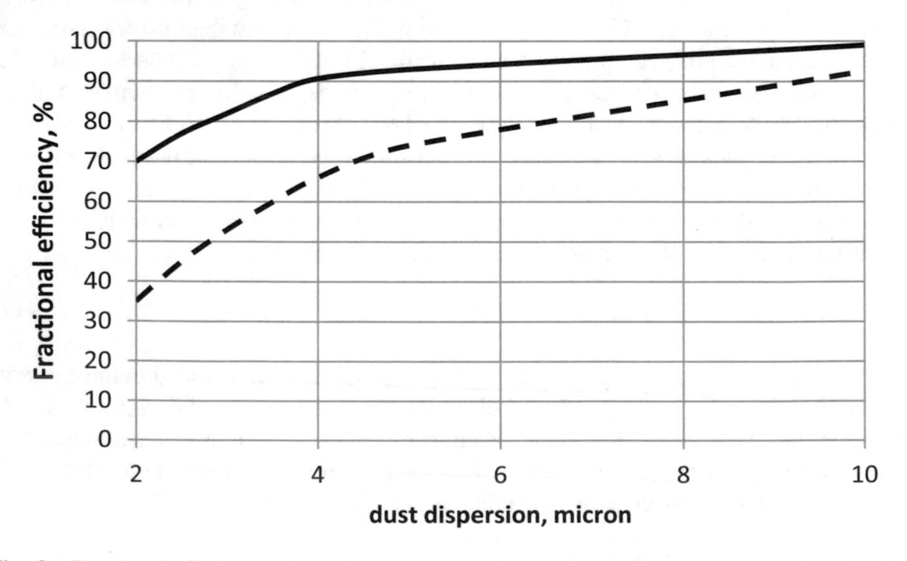

Fig. 3. Fractional efficiency of cleaning devices: — innovative cyclone; - - - - cyclone CN-15.

5 Conclusion

The results of the analysis show that dispersed dust less than 10 microns is present in foundry workshops, while there are 48.55% of iron (Fe) in the dust in the shot blasting area (Fig. 4).

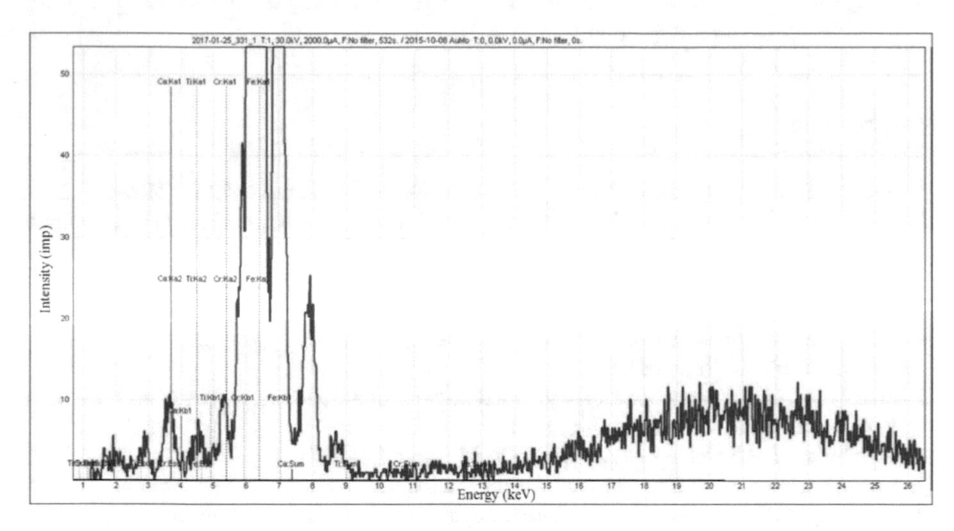

Fig. 4. X-ray analysis of the sample using the “Radian DR-02” diffractometer.

Also, the analysis of dust concentration in the operator’s working zone revealed significant excess of the MPC of the working zone. The dispersion composition of dust was evaluated, which showed a large amount of fine dust (less than 10 microns) of 86.2% (Fig. 5).

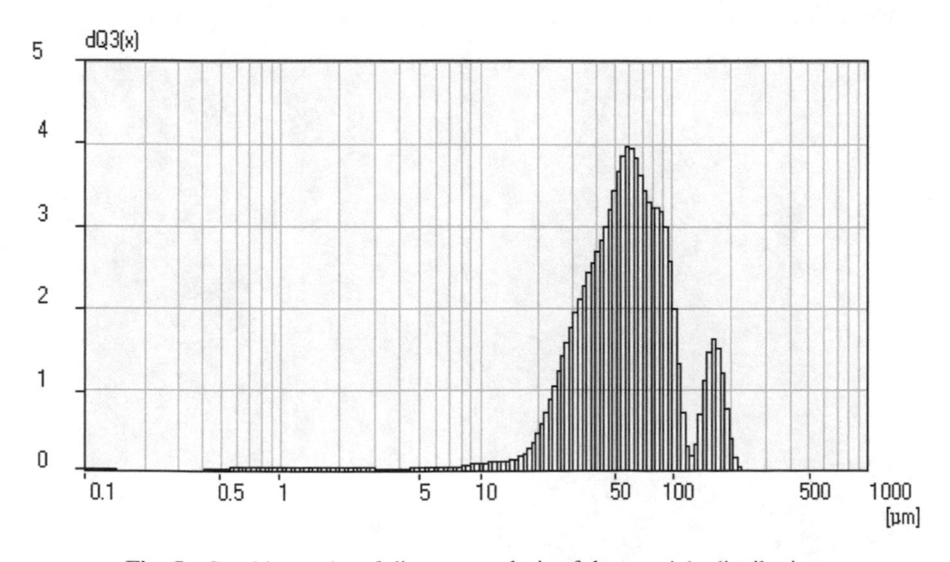

Fig. 5. Graphic results of disperse analysis of dust particle distribution.

It is established that in the existing system of local exhaust ventilation, a low efficiency of dust collection is achieved. According to calculations, the fan performance should be at least 11088 m^3/h. When replacing an existing fan, the dust concentration in the working zone decreased by 1.48 times.

The installation of the proposed device will reduce the damage caused to the working zone and increase the amount of recycled dust, since the two-zone “wet” after-treatment of emissions is proposed.

References

1. Kozlov, L.Y.: Proizvodstvo stal’nyh otlivok. In: Kozlov, L.Y. (ed.) MISIS, Moscow (2003)
2. Murav’ev, V.A.: Ohrana truda i okruzhayushchej sredy. Bezopasnost’ zhiznedeyatel’nosti. MISiS, Moscow (1995)
3. Obespylivanie promyshlennyh gazov v ogne¬upornom proizvodstve. Tekhnika, Vil’nyus (1996)
4. Tekhnologiya litejnogo proizvodstva: Formovochnye i sterzhnevye smesi. In: ZHukovskii, S.S. (ed.) BGTU, Bryansk (2003)
5. Truhov, Y.A.: Tekhnologiya litejnogo proizvodstva : lit’e v peschanye formy. In: Truhova A. P. (ed.) Akademiya, Moscow (2005)
6. SHtokman, E.A.: Ochistka vozduha. ASV, Moscow (1999)
7. Uork, K., Uorner, S.: Zagryaznenie vozduha, istochniki i kontrol’. MIR, Moscow (1980)
8. Val’berg, A.Y., Kushchev, L.A.: Raschet pyle i kapleulavlivayushchih ustanovok. Izd-vo BGTU, Belgorod (2010)
9. Aliev, G.M.: Tekhnika pyleulavlivaniya i ochistki promyshlennyh gazov. Spravochnoe izdanie. Metallurgiya, Moscow (1986)
10. Ustrojstvo dlya mokroj ochistki gaza. Zayavka № 2018126242 ot 16.07.18
11. Suzin, V.E., Yasakov, A.I., Zatonskij, A.P.: Zavisimost’ koehfficienta ochistki vozduha ot dispersnogo sostava pyli. In: Innovacionnye tekhnologii i tekhnicheskie sredstva dlya APK. Materialy nauchnoj konferencii professorsko-prepodavatel’skogo sostava, nauchnyh sotrudnikov i aspirantov, pp. 71–73 (2013)

Environmental and Social Aspects of Consumption of Water Resources and the Use of Atmospheric Moisture Condensers

Alexander Solovyev[1,2], Dmitry Solovyev[3(✉)], and Liubov Shilova[4]

[1] Department of Geography, Lomonosov Moscow State University, Lenin Hills 1/19, 119991 Moscow, Russian Federation
[2] Moscow State Academy of Water Transport, Novodanilovskaya nab. 2/1, 117105 Moscow, Russian Federation
[3] Shirshov Institute of Oceanology, Russian Academy of Sciences, Nahimovskiy prospect 36, 117997 Moscow, Russian Federation
solovev@guies.ru
[4] Moscow State University of Civil Engineering, Yaroslavskoye Shosse 26, 129337 Moscow, Russian Federation

Abstract. The article describes the environmental and social aspects of the freshwater resources consumption as well as those of natural and anthropogenic renewal of the aforesaid water resources.

The prognostic estimates of the limits to achieve a shortage of fresh water are discussed. The authors propose to use an analytical equation of the water reserves state for describing the dynamics of water consumption. It is established the parameter of the water state equation which characterizes the constant in time value of the per capita freshwater consumption rate. The time limit for the availability of water resources, expected by 2080, has been determined. According to the analysis of technologies for the industrial production of fresh water from hydrosphere resources, attention is drawn to the possibility of using a renewable source of fresh water - atmospheric moist air. An experimental model of an atmospheric moisture condenser is considered.

The data of tests of the laboratory model of the condenser installation at various values of humidity, the rate of moisture flow and the rate of entry of wet air flow to the condenser are presented.

The conclusion confirms the validity of the proposed method for the condensation of moisture from the atmospheric air.

Keywords: Fresh water · Environmental studies · Freshwater resources · Freshwater consumption · Atmospheric moisture · Capillary condensation · Laboratory experiment

V. Murgul and M. Pasetti (Eds.): EMMFT-2018, AISC 983, pp. 395–402, 2019.
https://doi.org/10.1007/978-3-030-19868-8_40

1 Introduction

Recently, the trend of increasing consumption of fresh water has become obvious, and the fresh water reserves remain the same [1, 2]. The estimate extrapolation of the balance of consumption and reproduction of water resources testify to the fact that a freshwater shortage may already come in the near future [3, 4]. It is noted that the shortage of quality fresh water will be aggravated by a heavy pollution of natural water sources [5, 6]. The pessimistic estimate of the water consumption situation is confirmed by the considerations concerning the achievement of renewal limits for natural river-water resources [7]. The atmospheric air is another fresh water source. Moreover, the atmospheric water steam is the most renewable fresh water source [8]. The natural reproduction of atmospheric water lasts 8 to 10 days, the total substitution of atmospheric water (the volume of which is roughly 14 000 km^3) takes place about 40 times a year [4]. Sufficient supply of atmospheric water (by an order of magnitude greater than the river-water sources) and its renewal rate promoted the study of the condensation processes and the creation of water accumulation plants condensing the steam from the atmosphere. These measures may contribute much to the solution of the water supply problem and to the reduction of the risk of freshwater deficiency. Further, we present the approaches specifying the prediction of fresh water consumption with consideration of atmospheric water resources. We also consider the ways of extraction of water from atmospheric air.

2 Analytic Estimates of the Time Dynamics of the Change in Freshwater Consumption Resources

Many studies are devoted to the prediction of the use of freshwater resources and water availability in the world [2, 7, 9–16]. At the same time, as a rule, all the prediction estimates were based on numerical algorithmic computations of statistical models. The prediction modelling included various scenarios of development of industrial and agricultural production, some assumptions on the time dynamics and the development rates of anthropogenic pollution of water sources (the water pollution may be influenced by an intensive introduction of some innovative water-saving technologies and restrictions connected with the admissible urbanization scale). In practice, all the predictions do not consider largely the mutual influence of different model parameters on water consumption. The aforesaid parameters describe the correlation social, economic and natural factors which may influence the development of the dynamics of consumption of freshwater resources.

At the same time, as a rule, all the prediction estimates were based on numerical algorithmic computations of statistical models. The prediction modelling included various scenarios of development of industrial and agricultural production, some assumptions on the time dynamics and the development rates of anthropogenic pollution of water sources (the water pollution may be influenced by an intensive introduction of some innovative water-saving technologies and restrictions connected with the admissible urbanization scale). In practice, all the predictions do not consider largely the mutual influence of different model parameters on water consumption. The aforesaid parameters describe the correlation between social, economic and natural factors which may influence the development of the dynamics of the consumption of freshwater resources. Our prediction studies used the idea of the primary importance of the dependence of the consumption rates of water resources on the anthropogenic factor. So we suggested using an analytic equation of the state of water reserves. The main idea of the composition of such equations is the consideration of the influence of social, economic, natural and other factors on the time dynamics of the use of freshwater resources. By analogy with the equation of state of substance [1, 9, 10], we can write down the following:

$$w = R \cdot t \cdot N \tag{1}$$

Where w - water consumption; N - the number of population; t - time; R - constant parameter describing the balance between consumption and reproduction of water sources, R is determined by the limit level of freshwater resources, and it characterizes the rate of fresh water consumption per head.

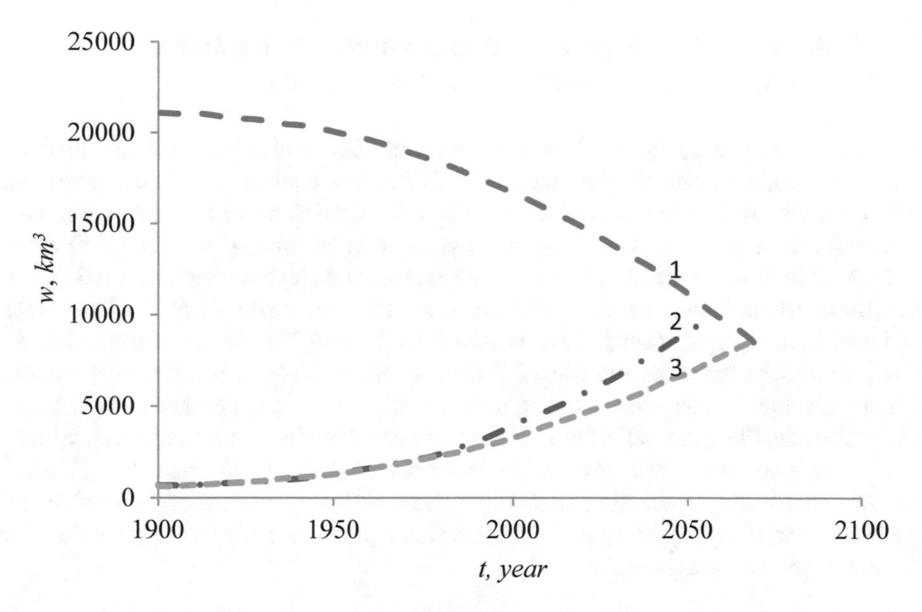

Fig. 1. Time dynamics of global water consumption. (1) – the curve of reduction of water resources because of their pollution; (2) – computation by [17] in accordance with the water consumption scenario based on a square-law extrapolation; (3) – the computation through the equation of state of water resources.

In the computations carried out through the Eq. (1) of water resources' state (see Fig. 1), we used the statistics for the world population in 2017 and the extrapolated values corresponding to the hyperbolic model with a square-law rate of growth [18]. The parameter R for the XX century:

$$R = (13.5 \pm 0.5)\ \text{mL}/(year \cdot person) \tag{2}$$

The same value of the rate of freshwater consumption was taken for the prediction for the period up to 2100. The constant R-value testifies to the universality of yearly growth of the rate of anthropogenic consumption of freshwater resources. The growth of the rate of water consumption, as it follows from the presented results, depends on the input-output balance of the freshwater world resources, i.e. on the water circulation. Figure 1 shows that the curve of water losses and of reduction of its availability for the humankind and the curve of water consumption rate intersect. In accordance with the scenario by [17], this time moment will come in 2050; the analytic calculations using the Eq. (1) estimate that the time of the available limit for freshwater resources may come by 2080.

3 Problems of Development of Innovative Technologies for the Industrial Production of Fresh Water

Nowadays, the demand for fresh water is an important social-and-economic problem for many world countries. The traditional freshwater sources are rivers, lakes and artesian wells. But these sources satisfy only 7% of the demand for fresh water consumption. Therefore, the works on improvement of technologies of industrial production of fresh water are placed on a wide footing. The greatest attention is paid to the traditional technologies: the distillation of sea-water, the electrodialysis of sea-water and the back osmosis plants. The electro-dialysis uses the electric current for the removal of salts from the sea-water. The back osmosis plants use the phenomenon when only fresh water comes under pressure through a semipermeable membrane. Unfortunately, the aforesaid technologies have considerable specific expenses, while it is necessary to reduce the cost of produced water down to the value of 0.1$/m^3. Besides, the water-desalinating plants and their assembly are rather expensive, the payback periods are quite long and the fresh water production plants are often far from the areas of water consumption.

4 Methodology and Experimental Models of an Atmospheric Moisture Condenser

A renewable source of fresh water – atmospheric moisture – has not been used in practical water consumption yet. Recently, a heightened interest has been shown for the discussion of potentialities of the fresh water production based on the atmospheric moisture condensation. At least in a dozen laboratories in different world countries, the works are carried out on the most efficient receivers of atmospheric moisture converting it to water. The interest for the research and development works in the field became considerably higher after the gigantic atmospheric rivers (the preferable steam traffic directions in the troposphere) had been discovered. D. Beysensand and I. Milimouk states in his "The case for alternative fresh water sources" article that atmospheric rivers transfer about 70% of fresh water with the flow rate of $1.65 \cdot 10^8$ m^3/s. It should be noted that the technology of atmospheric moisture accumulation has been known from time immemorial. The first atmospheric moisture condensers were created by F. Zibold and successfully operated in Russia as long ago as at the beginning of the XX century. That accumulation system did not consume any power; its capacity was up to 0.4 L/(m^3 day). M.A. Knapen states in his "Dispositif intérieur du puits aérien Knapcn (1929)" book that a plant in the form of a high tower filled with crushed stone was built in France; its average capacity was equal to 0.24 L/(m^3 day). V.S. Nikolayev et al. states in his "Water recovery from dew" article that there are successful modern atmospheric moisture condensers in the form of inclined plain foil panels with an average capacity of 0.25 L/m^2. The work [19] presented an experimental atmospheric moisture condenser with a natural air flow without any energy consumption. The plant includes a system of net gabions filled with crushed stones installed on the trays for water gathering; there is a convection canal for the extract of wet ambient air passing through the gabions. The plant dimensions are: 4.5 × 8.75 × 2.8 m. The plant was tested in Moscow Region in summer; the average capacity was equal to 0.18 L/(m^3 day) which was only by 13% less than the design capacity. A shortcoming of the plant was the reduction of the duration of the condensation process caused by the heating of gabions by the solar radiation and the following moving off the dew point.

A method of production of water from the atmospheric moisture has been developed in order to eliminate the effect of day cooling in atmospheric moisture condensers [20].

Fig. 2. The appearance of the "ROSA" plant with an underground layout and that of a laboratory model (to the right).

The main idea of this method is that the condensation process takes place under the ground at a lower temperature of the material condensing the steam in the air coming through a system of forced ventilation with a power supply from solar-cell panels. The plant is completed in the form of an underground module, and it includes air intake and air off-take facilities, a system of grains condensing the water and a hot-well. The atmospheric moisture condenser is a stationary system, it operates under the ground at the depth of 1.0–1.5 m. A general view of the plant is shown in Fig. 2.

5 Results of Experimental Model Tests

A special laboratory model of the "ROSA" (in English "DEW") plant with dimensions of (1.1 × 1.1 × 1.2) m^3 has been fabricated for an experimental modelling of atmospheric moisture condensation processes. The model tests were used for the study of the process of potable water production from the ambient air under normal conditions (air temperature: (15–20) °C, relative humidity f value: 40–60%), air pumping speed through the condenser grains: within the range of (1.0–3.0) m/s. The tests showed an average model capacity of 0.36 L/(m^2 day) (Fig. 3).

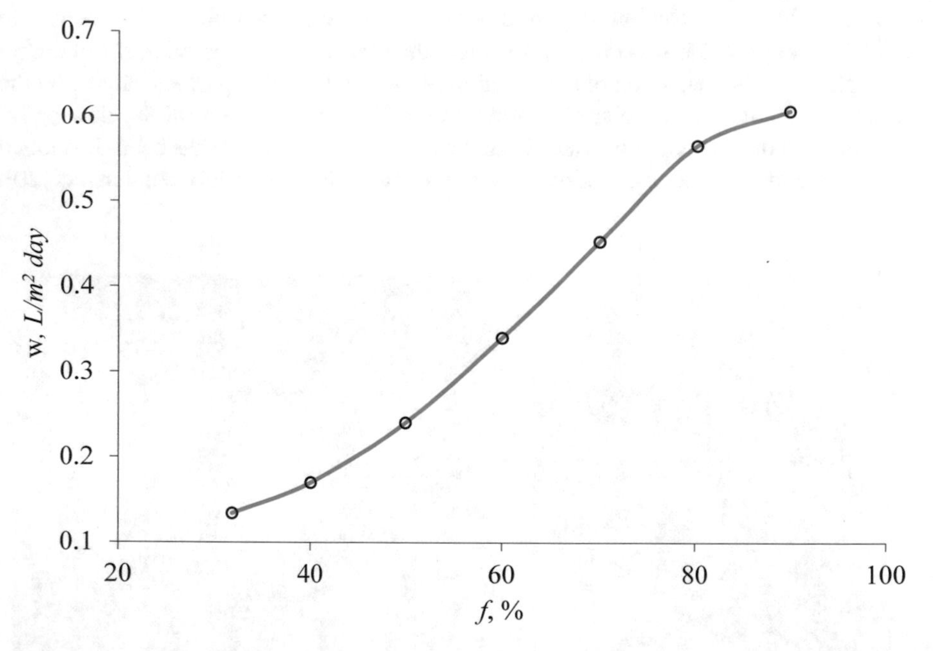

Fig. 3. Results of the tests of a laboratory model of the "ROSA" plant: the dependence of the volume w, (L) of water yielded daily from 1 m^2 of the grain condensing surface on the moisture content *f* (%) with the air pumping speed of 3 m/s.

6 Conclusions

The laboratory experiments with the condenser model of the “ROSA” condenser system carried out at a room temperature with a variable humidity content showed that the “ROSA” system capacity may achieve the value of 0.6 L/(m^2 day). The design characteristics of a standard “ROSA” plant are the following:

- dimensions (diameter: 25 m, height: 5 m);
- capacity: 3 m^3/day;
- consumption of electricity: 0.2 kWh for 1 m^3 of water;
- the prime cost of water: 0.3 \$/m^3;
- pay-back period: 2–3 years.

Acknowledgements. This research was supported by the Russian Ministry of Science and Higher Education (agreement № 075-02-2018-189 (14.616.21.0102), project ID RFMEFI616 18X0102).

References

1. Danilov-Danilyan, V.I., Losev, K.S.: Water consumption: ecological, economic, social and politic aspects. Nauka, Moscow (2006). (in Russian)
2. Fekete, B.M.: State of the world’s water resources. In: Climate Vulnerability, pp. 11–23. Elsevier (2013)
3. Chen, S., Wu, D.: Adapting ecological risk valuation for natural resource damage assessment in water pollution. Environ. Res. **164**, 85–92 (2018). https://doi.org/10.1016/j.envres.2018.01.005
4. UNESCO, The United: Water in a Changing World (2009)
5. Schweitzer, L., Noblet, J.: Water contamination and pollution. In: Green Chemistry, pp. 261–290. Elsevier (2018)
6. Yu, L., Yıldız, İ.: 1.24 energy and water pollution. In: Comprehensive Energy Systems, pp. 950–979. Elsevier (2018)
7. Alcamo, J., Döll, P., Henrichs, T., Kaspar, F., Lehner, B., Rösch, T., Siebert, S.: Global estimates of water withdrawals and availability under current and future “business-as-usual” conditions. Hydrol. Sci. J. (2003). https://doi.org/10.1623/hysj.48.3.339.45278
8. Solovyev, D.A., Aleksandrova, M.P.: Study of the processes of near-surface condensation in the Atlantic Ocean. In: Truhin, V.I., Pirogov, Y.A., Pokazeev, K.V. (eds.) Physical Problems of Ecology (Ecological Physics): Book of Scientific Papers, pp. 459–466. MAKS Press, Moscow (2013). (in Russian)
9. Malmqvist, B., Rundle, S.: Threats to the running water ecosystems of the world. Environ. Conserv. **29**, 134–153 (2002)
10. Shiklomanov, I.A., Rodda, J.C.: World Water Resources at the Beginning of the Twenty-First Century. Cambridge University Press, Cambridge (2004)
11. Gleick, P.H.: Global freshwater resources: soft-path solutions for the 21st century. Science (80) **302**, 1524–1528 (2003)
12. Aznar-Sánchez, J.A., Belmonte-Ureña, L.J., Velasco-Muñoz, J.F., Manzano-Agugliaro, F.: Economic analysis of sustainable water use: a review of worldwide research. J. Clean. Prod. **198**, 1120–1132 (2018)

13. Zomorodian, M., Lai, S.H., Homayounfar, M., Ibrahim, S., Fatemi, S.E., El-Shafie, A.: The state-of-the-art system dynamics application in integrated water resources modeling. J. Environ. Manag. **227**, 294–304 (2018)
14. Beran, A., Hanel, M., Nesládková, M., Vizina, A.: Increasing water resources availability under climate change. Proc. Eng. **162**, 448–454 (2016)
15. Al Fry: Facts and Trends: Water. World Business Council for Sustainable Development (2006). https://doi.org/10.1080/0379772780030302
16. Abdulbaki, D., Al-Hindi, M., Yassine, A., Abou Najm, M.: An optimization model for the allocation of water resources. J. Clean. Prod. **164**, 994–1006 (2017). https://doi.org/10.1016/j.jclepro.2017.07.024
17. Young, G., Dooge, J., Rodda, J.: Global Water Resource Issues. Cambridge University Press, Cambridge (2004)
18. Pison, G.: The population of the world. Popul. Soc. **503**, 1 (2013)
19. Alekseev, V.V., Rustamov, N.A., Ivanov, V.N., Dubovskaya, V.A.: Experimental study of overland moisture condensation. Dokl. Earth Sci. **393**, 1156–1159 (2003). (in Russian)
20. Solovyev, A.A., Chekarev, K.V., Malykh, Y.B.: Installation for production of fresh water from atmospheric air. Patent for invention 2609811 (in Russian) (2017). http://www1.fips.ru/fips_servl/fips_servlet?DB=RUPAT&DocNumber=2609811

Assessment of Potential Forest Biomass Resource on the Basis of Data of Air Laser Scanning

Marina Kuzyakina[1], Dmitry Gura[2], Aleksandr Sekisov[2](✉), and Nikolay Granik[3]

[1] Kuban State University, Stavropolskaya Street 149, 350040 Krasnodar, Russia
[2] Kuban State Agrarian University, Kalinina Street 13, 350044 Krasnodar, Russia
alnikkss@gmail.com
[3] Aerogeomatika, Frunze Street 22/1, 350063 Krasnodar, Russia

Abstract. Exact determination of quantity and quality of a forest cover is important for management of natural resources, assessment of a potential forest biomass resource, the ecological analysis and hydrological modeling. The quantity of the developed methods and their implementation is very broad. In this paper, the comparative analysis of the following methods of the automated determination of quantity on the basis of data of air laser scanning is carried out: Microstation (Terrasolid) method; the method using Global mapper + LIDAR module; the method based on combination of data of LIDAR and the raster image; the method using the MATLAB application program package; the methods realized in GIS ArcGIS, and the author's method on the basis of raster of density (Kernel Density) with the use of statistical calculations is also offered.

Keywords: Air laser scanning (ALS) · LIDAR · Decryption · Forest cover · Forest vegetation

1 Introduction

One of the most perspective directions of remote sensing is assessment of a condition of a forest cover. Exact determination of amount of wood breed is important for management of natural resources, the ecological analysis and hydrological modeling. The quantity of the developed methods and their implementation is very broad. Multispectral and hyper spectral images rendered a big contribution to process of mapping of vegetation, but they provide only a small number of characteristics of structure of vegetation. When using LIDAR technologies, it is possible to precisely measure vertical and horizontal structures of the wood, and it gives an excellent opportunity for the solution of many problems.

V. Murgul and M. Pasetti (Eds.): EMMFT-2018, AISC 983, pp. 403–416, 2019.
https://doi.org/10.1007/978-3-030-19868-8_41

2 Choice of Test Sites of ALS

For performance of a research and the subsequent assessment of quality of the automated methods and an author's method, territories with various landscape were chosen. For the automated methods, the forest territory of the natural Aralar park, Spain (Fig. 1) was chosen.

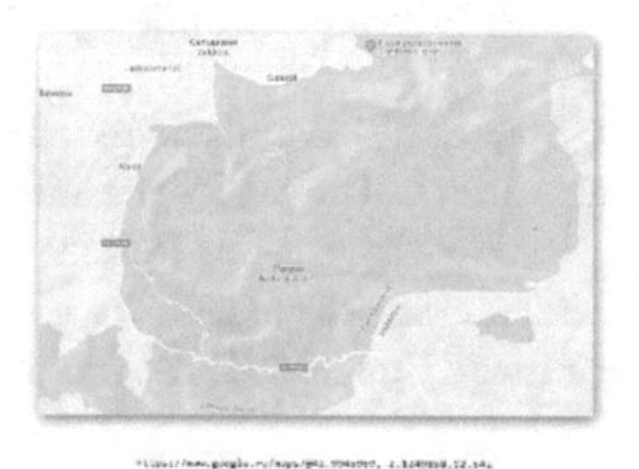

Fig. 1. Natural Aralar park.

The natural Aralar park is located in one of the most mountainous areas of the Basque Country. It is located in the southeast of the province Gipuskoa, dividing border with the neighboring province Navarre. In the center of the massif, the extensive pastures used for a pasture of large herds of sheep prevail. In the park, there are nature reserves of exclusive importance, for example, it is the beechen forests of Akaits well-known for a large number of yew trees, hills and valleys of North side of Tsindoki. For carrying out a research, the southern part of the park where there are woods of the European beech was chosen. For an author's method three various territories with a different landscape were chosen.

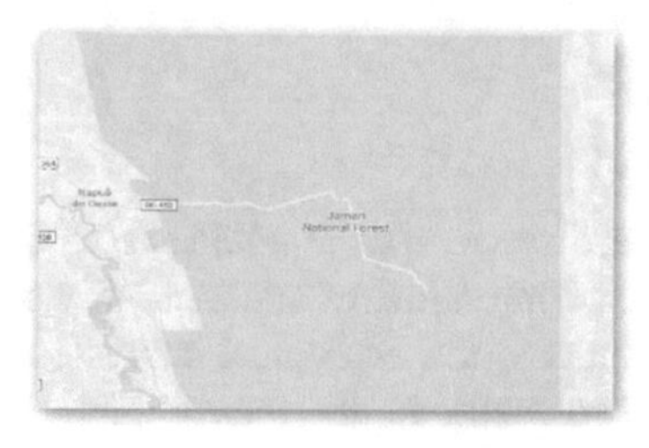

Fig. 2. National park of Dzhamari.

The national park of Dzhamari is located in the State of Rondônia, Brazil (Fig. 2). Total area is 220 thousand hectares. Dense tropical vegetation with big quantity of trees up to 60 m high is characteristic of the territory. The Yosemite national park is in the State of California, the USA (Fig. 3). Total area is 3100 sq.km.

Fig. 3. Yosemite national park.

About 95% of territories of the park belong to a zone of the wild nature. For a research, the southern site of the park in which coniferous vegetation mainly grows was chosen. Alice Mulga's territory is in 150 km of Alice Springs, in the semi-droughty central region of Australia. The growth of an acacia is characteristic for this area (Fig. 4).

Fig. 4. Territory of a research of Alice Mulga, Australia.

3 The Comparative Analysis of Automatic Methods of Decryption of the Forest Cover According to ALS

3.1 Microstation (Terrasolid) Method

Were applied to this method Software Microstation Bentley Connect 10 and a Terrasolid package. In the Terrascan settings, the new type of trees is created and the following settings are set: the minimum and maximum height of a tree, a crown width variation, the name of a tree is also chosen the Height block created earlier [2].

Before defining trees, it is necessary to normalize a surface to receive the best result as roughness of a surface play an essential role in determination of heights of trees. For normalization, the Transform loaded points function is used. At the final stage, by means of the Detect trees function, crowns and heights of trees are found. The tree type, the place of storage of blocks of crown and heights, and the "more trees" method is also chosen in it. After calculations, trees with heights and with a radius of crown (Fig. 5) are obtained.

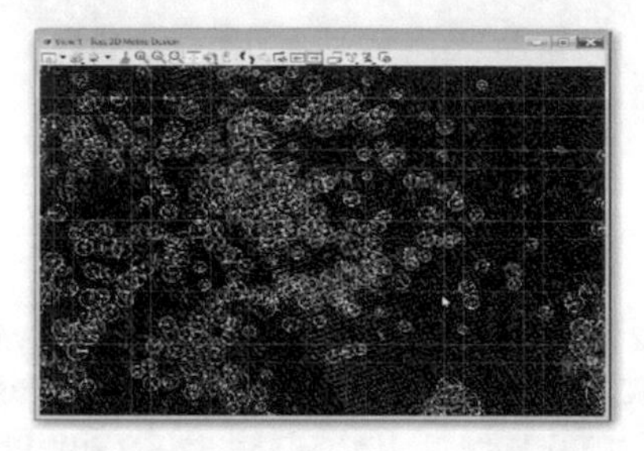

Fig. 5. Trees with heights and with crown radius.

To obtain data on heights and crown in the text *.csv format layers which are responsible for height and crown are chosen. Then the drop element function for the transfer from the block to elements is used. In Terrasurvey, the List element tab in which the chosen layer is added to the list and is exported is used. Thus, the exported text file contains quantity of trees, coordinates "x" and "y", crown (or height), the name of a surface and the code.

Method's advantages:

- high percent of definition of trees (83–95%);
- perfectly determines heights and crown (98% and above)
- the minimum loss of data at the block splitting used in it and the subsequent methods for verification of data.

Method's disadvantages:

- it is necessary to select tree type for every studied territory, since there is a lot of species of trees, and they are various on the configuration;
- not always correctly defines total number of trees at the small density of points;
- difficult to define trees if density of vegetation is high;
- needs very fine adjustment of parameters.

3.2 The Method Using Global Mapper + LIDAR Module

Global mapper with LIDAR plug-in were applied to this method.

LIDAR data are loaded into the program, but for improvement of quality of definition, it is necessary to load Las data only with vegetation, and only with the normalized surface [3]. Also, it is necessary to consider a geographical projection of data as without exact information on a projection, the accuracy of data will be very low. It is necessary to come into the "Extract Vector Features" tab. The "Las" file is chosen. In the section "extract tree", the minimum height, the longitudinal and cross line of crown is set [3]. In resolution, it is necessary to choose the best permission for this file (in meters or interdot distance). Chosen resolution for analysis of the entire park is 1.3 m. For the first and second block - 1.1 m (Fig. 6).

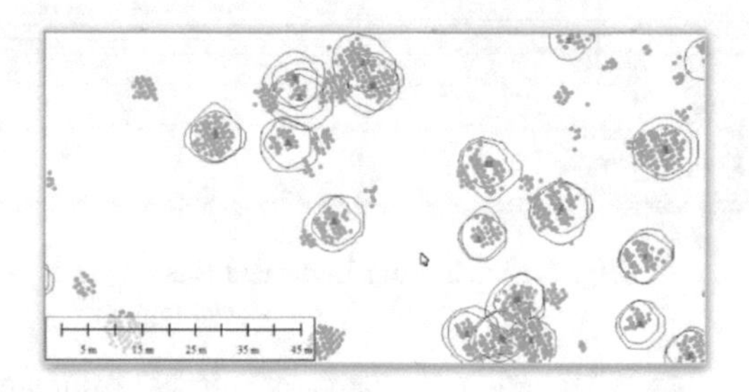

Fig. 6. Example of a difference of the polygon of crown. Resolution – 1.5 and 1.2 m.

The obtained data are exported to *.shp and imported to Microstation, then the assessment of accuracy of definition of crown and heights of trees is carried out.

Method's advantages:

- well defines quantity of trees. It is higher than 65% at dense vegetation, at lower - to 78%;
- it is independent of tree type;
- perfectly determines heights (98% and more).

Method's disadvantages:

- it is necessary to select resolution by method of trial and error for quality receiving;
- there are problems with definition of crown in a dense thicket.

3.3 The Method Based on Combination of Data of LIDAR and the Raster Image

This method uses the raster image of the territory and data of LIDAR. Further, LIDAR data, GIS ArcGIS, the quality aerial photograph, and the Trees from Lidar and NAIP tool are also used [4]. Las Dataset in which there will be Las file is initially created. Then the Trees from Lidar and NAIP tool opens, and the necessary Las Dataset is chosen, the average interdot distance, by default, the buffer from the building is set, 1st channel and the 2nd channel of a raster are chosen. The method uses about 40 operations. By means of a raster, there are tops of trees and their height; by means of LIDAR data, the crown radii are obtained. In the tool, several tens of operations for receiving a total data take place. At the initial stage, statistics for "*.las" files is considered, and NDVIplus and NDVIminus is considered for rasters. Then points pass a processing chain Negate tools – Flow Direction – Sink. Further, the highest places are selected from the raster, which are eventually converted into point objects with information about the height of trees [4]. For obtaining information on crown, the relation of rasters of NDVIplus and NDVIminus is used, then the final file undergoes the function Focal Statistic, Flow Direction, Flow Accumulation, to which information about tree heights, crown sizes, etc. is added at the exit (Figs. 7 and 8).

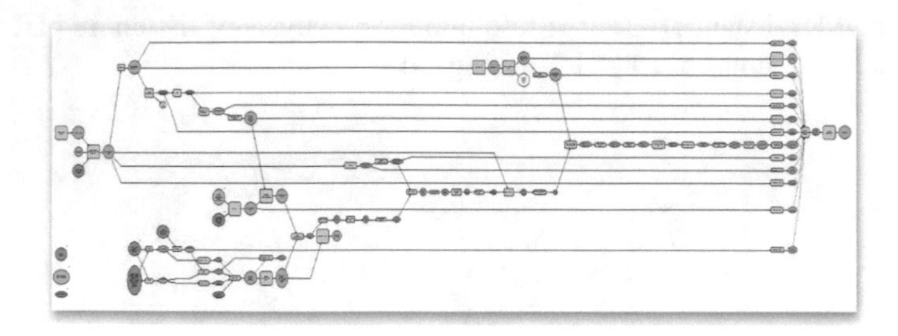

Fig. 7. Trees from Lidar and Raster.

By means of the “buffer” function, the range of a crown is created and the obtained data are exported to Microstation, where the assessment of accuracy of definition of crown and heights of trees (Fig. 8) is carried out.

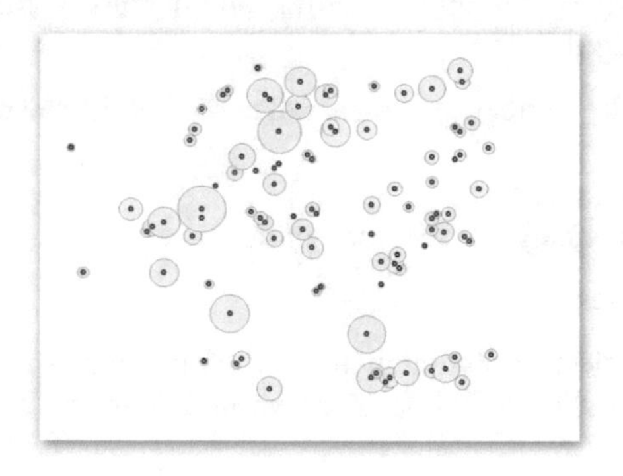

Fig. 8. Range of crowns.

Method’s advantages:

- not bad at identifying trees at low density – 62–63%;
- perfectly determines heights and well determines crowns;
- a small difference on 4 blocks of verification that speaks about high quality of a method.

Method’s disadvantages:

- badly copes, if trees densely grow;
- often defines the location of a tree on the edge of a crown;
- for obtaining more exact data, it is necessary to use data of aerial photographs of NAIP, which cover only the territory of the United States of America.

3.4 A Method Using the MATLAB Application Program Package (APP)

For this method, MATLAB APP, MATLAB scripts, GIS SAGA, GIS ArcGIS, and nDSM surface are used. The research of the doctor of geographical sciences Tyson Swetnam and an algorithm of local maxima with a variable zone forms a basis for this method [5]. At the following stage, the code is used: vlm_output_raster = vlm (raster, 0.6.30). Two coefficients are specified in this code: ratio of width and height of a tree and value of anisotropic 2D blurring of a raster (Blur) [6]. The value of a ratio of width and height of a tree undertakes from data of Microstation on this site, but for improvement of quality of the obtained data, the empirical way is used, visually assessing how well a polygon describes a tree crown. The coefficient "blur" is responsible for quantity of the found trees. The recommended values are used for this quality of a raster [7]. By means of the code export_vlm_output = export_utm team (vim_output_raster, raster), the obtained data is transferred to the table Matlab.

Data from the table Matlab are copied in Excel file. Then lines with NaN value and those columns which are not necessary and have excess information are removed. In the final file, 5 columns are left: id, x, y, height, radius. By means of the Add function, Excel file is added to ArcGIS. By means of the "buffer" function, according to the radius column, models of tree crowns are under construction and are exported to Microstation, where the visual assessment of quality of definition of crowns and heights of trees is carried out.

Method's advantages:

- good performance by the number of trees found - 68.8–76.13%;
- well defines the crown.

Method's disadvantages:

- badly determines heights;
- quite often defines the crowns in a false manner.

3.5 Arcgis Method 1

In this method, the 3D Samples tool kit is used, created by the team of developers 3DGisTeam and ON ArcGIS [1]. The tool kit is based on scientific research of Randolph Wynne [8]. The tool has the necessary functions: Canopy Peaks and Tree Crown Radius. In Canopy Peaks, the minimum height of a tree and three additional coefficients are set, and the hypothetical maximum value of radius of the crown is established in Max Window Size. The constants A, B, C are used in the following formula: window_width = A B * ht C * ht^2 [8]. For data acquisition, it is necessary to use coefficients for meter resolution: A-2.515043, B-0, S-0.00901. At the output, trees with data on height are obtained [9, 10]. For further definition of crown, Tree Crown Radius, where the received trees and the maximum hypothetical value of radius of crown are added, is used.

Method's advantages:

- perfectly determines heights of trees.

Method's disadvantages:

- very low percentage of the found trees – is lower than 50%;
- problem with definition of crown.

4 Author's Technique of Modelling of the Forest Cover on the Basis of ALS in the Conditions of Different Landscapes

Development of a technique is difficult and laborious process. In the course of it, there was a set of tests and mistakes. As a result of the search for the solution of the problem of determination of the crown sizes and quantity of trees, it was decided to use statistics in data of air laser scanning [11]. Statistics is calculated in ArcGIS. It was necessary for this purpose to create Las Dataset. Here, it is necessary to calculate average distance between points. Later, the Las Dataset file is added to the project itself, and a visual assessment of the histogram is performed. According to the histogram, division of points on tiers is carried out. Division on tiers takes place by different methods: geometrical interval, natural borders or manually [11]. Classification of geometrical intervals – this method is used for classification of range of the values, based on a geometrical progression. At such classification, division of classes is based on intervals between classes. This method of classification is useful to visualization of data which are not distributed normally. The natural boundaries method defines the boundaries of classes so as to group a similar group of values and increase the difference between the other groups of values. The boundary between the groups is set where there are big differences between the data.

The manual method is based on finding the places of values difference.

Depending on distribution of points on heights, at first, division into 2 tiers is used, and then these 2 tiers were divided into 2 more tiers. Such division was used on data where there was coniferous vegetation [12]. In the same place where there was vegetation in Australia and Brazil, where it was most difficult to distinguish borders. Division is carried out on 3 groups, and then on 2 parts more in places of values difference [13]. For Brazil, it was necessary to consider the fact that it is impossible to distinguish a tree lower than 4 meters, owing to very high density on higher tiers [14]. For convenience, the quantity of columns changes from 100 to 500 (Fig. 9).

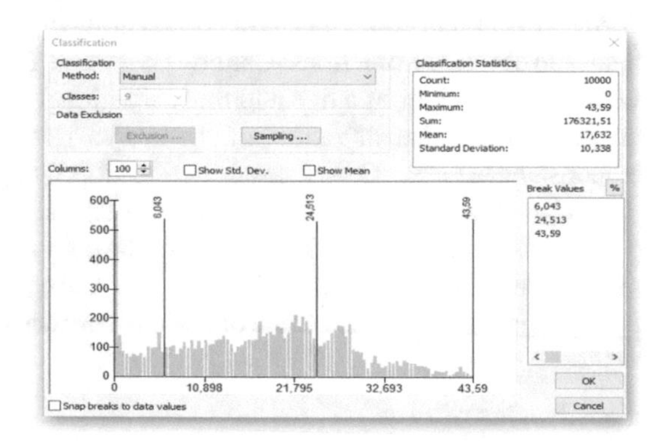

Fig. 9. An example of manual dividing according to histogram data.

Data on Australia were divided into 4 level groups: 2–10, 10–23, 23–32, 32–40. Data on Brazil were divided into 5 level groups: 3–12, 12–18, 18–25, 25–34, 34–55. Data on the USA were divided into 3 level groups: 3–11, 11–15, 15–24, 24–30.

For further preparation of data for converting in a raster, it is required to clean data from excess points that prevent to allocate a crown more precisely.

To clean data from excess points, the method of clarification of points on high-rise tiers with use of selection by height was developed. This method is based on the principle of division of points into 3 types. The first type of points is those points that belong to trees of higher tier. The second type, on the contrary, is those points that belong to trees of this tier. The third type of points is white noise, which is eliminated at selection. This method is realized in ArcGIS. For carrying out cleaning, one of the "*. shp" files is chosen, which is selected in a certain high-rise tier. Then, by means of the functions "selection by attribute", points which are in range from the maximum value up to one value smaller than maximum on 1 m (Fig. 10) are allocated.

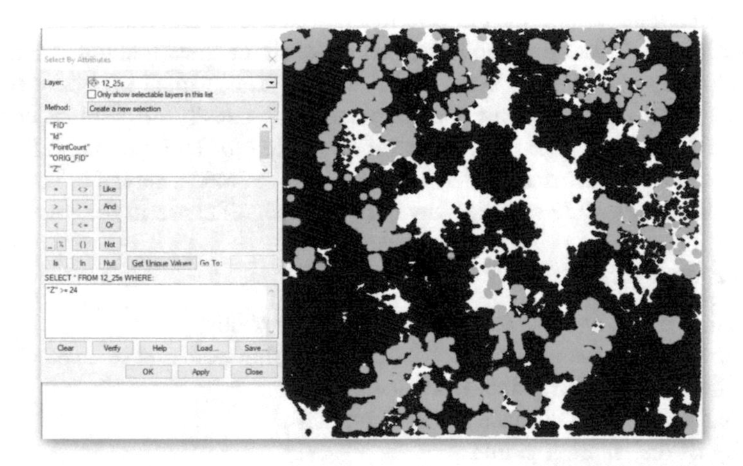

Fig. 10. An example of selection of points by height.

These points are exported to the separate *.shp file with the name of range of heights. At the following stage, the exported points are added back to the project, and selection by location is done. The method "within a distance of 3D the source layer feature" with a step of 1 m is used. In "Select feature", the data file of a high-rise tier is chosen, and in source layer, the file which was exported [15] is chosen. This method helps to choose points which are either perpendicularly or at a small angle at the chosen distance. At the following selection, not only points which are 1 m below, but also those points which were from the last *.shp file are chosen. Therefore, it is necessary to carry out selection on attributes again. The minimum value of the past *.shp file is chosen, and all values which exceed it are removed.

This set of actions is done until the minimum height value is reached. After the last iteration, there is an association of all points which were divided into meter tiers in one file. The name of the file is set as "UP". These are those points which belong to those trees that are one tier above. And other points that remained in the same tier unite in the DOWN file. Further, after performance of all cycles of dividing of tiers on one meter and creations of the UP and DOWN files, it is necessary to make additional combination of files. If over a high-rise tier there is one more tier, then the UP file should be united with the DOWN file of the following tier. Here it is necessary to be careful, because there are cases that the tree is very tall and is located in all tiers. This problem is solved at direct capture of data; such trees are very well visible in 3D space in the ArcScene program from different foreshortenings.

Toolbox Select v1.2 was developed for simplification of a technological chain of performance of division of points according to high tiers. This tool kit includes selection on height and export of the received files. This toolbox accelerated further performance of work several times and deprived of the mistakes caused by a human factor when performing this set of actions. For its work, it is necessary to create separately the folder with name F and create base of geodata in it. In an initial window, points "all" are chosen, it is the file with all points of this tier and also points which are 1 m lower, and the place for the received result are chosen. It is also possible to change selection distance between points. This set of tools in a semi-automatic mode was divided into tiers according to Brazil and Australia. The manual method was done on the US data. The Kernel Density tool calculates the density of point objects around each cell of the output raster [16]. It can be calculated for both point and line objects. This method can be applied in calculating the density of roads, habitat of rare species of animals. It is also possible to assign any data more significance in the calculation than others.

Calculation algorithm:

- calculation of the average center of the entering points;
- the distance from the average center of the entering points is calculated;
- the weighed median of points Dm is calculated;
- the value of standard distance SD is calculated;
- quantity of the entering points, n;
- SearchRadius is calculated by the following formula (1):

$$\mathrm{SearchRadius} = 0.9 * min * \left(SD, \sqrt{\frac{1}{\ln(2)}} * Dm \right) * \mathrm{n}^{-0.2} \tag{1}$$

Conceptually, a smooth curved surface is selected (built) for each point. The surface value is maximal at the point location and decreases with increasing distance from the point, reaching zero at a distance equal to the given Search radius from the point.

The following settings were used to build density rasters according to the height data. The resulting *.shp file (DOWN, UPDOWN) is added to the Input point, and a field is selected by which the averaged center of points will be considered, in this case, this is the Z field with height data. The storage location and cell size of the outgoing raster is selected. In the "search area" tab, values from 1.5 to 2 were used. It was found by empirical methods that the value of 1.5 is best suited for the lower and middle tiers, and the value 2 is more suitable for the upper tiers of vegetation [15].

Thus, density raster is obtained. This action is carried out with each *.shp file of a high tier. For improvement of quality of determination of quantity of trees and quantity of crowns, the following settings of display of a raster of density were chosen. The received raster images are added in ArcScene program. The visualization of data on their value is carried out in it [17]. Data that has a high value, having a falling edge, was defined as a tree. Outlining of the crown goes on the visible contour. This operation is done for each raster at a specific height. Thus, *.shp crown tree files are obtained (Fig. 11).

Fig. 11. Points in ArcScene with a raster substrate.

In addition, the visual assessment of quantity of the trees received this way is carried out. In various conditions of a landscape, the method showed various results from 75 to 95%. Testing was on nine squares with an area of 2500 sq.m (Fig. 12).

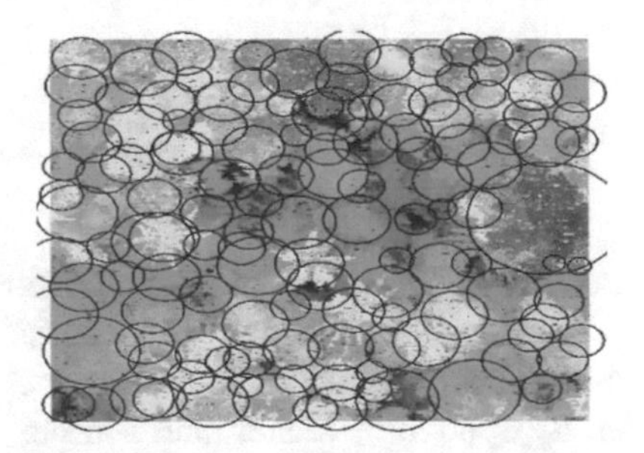

Fig. 12. Example of total distribution of wood vegetation.

Also, additional check of a method on dependence on density of points was made. Density of points decreased three times, from 60 to about 20 points per sq.m. As a result, the success of detection of trees decreased on average by 10% in tropical forests and by 3% in the mixed and coniferous forest. After conducting quality check of a method on three various types of a landscape, advantages and disadvantages of this method were revealed (Table 1).

Table 1. Assessment of the obtained data on three various landscapes by an author's method of decryption of forest vegetation.

Coniferous vegetation of USA				
NOT CORRECTLY classified			CORRECTLY classified	
Trees	Crown	Total	Total	%
2	1	3	76	98
Vegetation of Australia				
Not correctly	Crown	Total	Total	%
4	1	5	121	99
Tropical vegetation of Brazil				
Not correctly	Crown	Total	Correctly	%
5	3	8	183	95

Method's advantages:

- perfectly defines quantity of trees – over 90%;
- precisely defines crowns;
- it is possible to use it for determination of quantity of trees even in tropical vegetation.

Method's disadvantages:

- it is insufficiently automated. 75% of work are performed manually;
- dependence on density of points. The density of points is higher, the better the result turns out

5 Conclusions

Among all above mentioned automated methods, those methods which directly operate with data of LIDAR [18, 19] proved to be better. The Microstation method has small percentage of mistakes, and not badly proved to be at dense vegetation, where it is very hard to define a tree. If density of points is higher than 5–6 cm^2, then the results will be better. Perfectly suitable for detection of standing alone trees. The Global mapper method has higher percentage of errors than "Microstation", but has simpler interface in respect of data processing. The great advantage that the method is independent of a species of trees, and it simplifies search and increases a total quantity of the found trees.

The MATLAB method used data of the raster image and showed worthy results. The quantity of mistakes at the level of "Global mapper" by quantity of the trees found correctly are more than in "Microstation", and by mistakes by the crown size is almost the smallest. The big disadvantage of this method is that it depends on the quality of the raster, as a result of which it poorly determines the height of the trees, and the values of anisotropic 2D blurring (blur).

"LIDAR-Raster" showed rather quite good results for a way which was initially stated as experimental. Perfectly determines heights, and not bad determines crowns (there is roughness). It has a high percentage of incorrectly defined trees, which is probably also related to the quality of the raster image. The "ArcGIS 1" methods are also experimental. They showed mediocre results, which show not ideality of these methods. The author's technique though is not completely automated. Distinctive feature of this method is an opportunity to define quantity of trees and the sizes of their crowns in the conditions of the tropical forest. One more advantage of this method is an opportunity to allocate trees densely standing to each other. It is perfectly suitable for the data with a high density of points (over 16 points per m^2), but with the reduction of density of points, the quality of definition of trees and the sizes of their crowns decreases. This method needs automation of all actions.

The semi-automatic method of decryption of a forest cover is offered: determination of quantity of trees and the sizes of their crowns on the basis of data of air laser scanning. Such an approach can be used, including in the tropical landscape.

References

1. 3D Sample Tools. https://www.arcgis.com
2. Soininen, A.: TerraScan User's Guide. TerraSolid, pp. 60–93 (2018)
3. Lidar support in Global mapper. http://www.bluemarblegeo.com
4. Trees From LiDAR Tool and Sample Data. http://desktop.arcgis.com
5. Swetnam, T.: Application of metabolic scaling theory reduce error local maxima tree segmentation from aerial LiDAR. For. Ecol. Manag. **323**, 12–13 (2014)
6. Nobrega, R.: A noise removal approach for LIDAR intensity images using anisotropic diffusion filtering to preserve object shape characteristic. In: ASPRS, vol. 21, pp. 1–11 (2007)
7. Swetnam T.L.: MATLAB Commands for isolating individual trees and measuring canopy radius. https://sites.google.com
8. Wynne, R.: Seeing the trees in the forest: using lidar and multispectral data fusion with local filtering and variable window size for estimating tree height. Photogramm. Eng. Remote Sens. **70**, 589–604 (2004)
9. Swamer, M.: Extraction of tree crowns and heights using lidar. SwamerHouser, pp. 12–19 (2012)
10. Ferraz, A., Mallet, C., Soare, P.: Canopy density model: a new ALS derived product to generate multilayer crown cover maps. IEEE Geosci. Remote Sens. Soc. **53**, 6775–6790 (2015)
11. Delon, J., Desolneux, A., Lisani, J.L., Petro, A.B.: A non-parametric approach for histogram segmentation. IEEE Trans. Image Process. **16**, 253–261 (2007)
12. Hamraza, H., Contrerasb, M.A., Zhanga, J.: A vertical stratification of forest canopy for segmentation of understory trees within small footprint airborne LiDAR point clouds. ISPRS J. Photogramm. Remote Sens. **140**, 357–363 (2017)
13. Lee, A., Richard, M., Lucas, M.: A LiDAR derived canopy density model for tree stem and crown mapping in Australian forests. ISPRS J. Photogramm. Remote Sens. **111**, 493–518 (2007)
14. Ferraz, A., Saatchi, S., Mallet, C., Meyer, V.: Lidar detection of individual tree size tropical forest. ISPRS J. Photogramm. Remote Sens. **183**, 318–333 (2016)

15. Calculating the 3D distance of trees derived from LiDAR data to overhead power. https://geonet.esri.com
16. UCL Department of geography. http://www.geog.ucl.ac.uk
17. Lindberg, E., Holmberg, J.: Individual tree crown methods for 3D data from remote sensing. Curr. For. Rep. **3**, 19–31 (2017)
18. Gura, D.A., Shevchenko, G.G., Kirilchik, L.F., Petrenkov, D.V., Gura, T.A.: Application of inertial measuring unit in air navigation for ALS AND DAP. J. Fundam. Appl. Sci. **9**(1S), 732–741 (2017)
19. Kuzyakina, M.V., Gura, D.A., Mishchenko, Yu.A., Gordienko, D.A.: Experimental analysis of SRTM model by image processing and geostatistical methods. Int. J. Eng. Technol. (UAE) **7**(4.7), 250–253 (2018)

Thunderstorm Activity Intensification over Marshland

Igor Yusupov[1(✉)], Evgeniy Korovin[2], Georgy Shchukin[2], and Vladimir Shuleikin[3]

[1] Saint-Petersburg State University, Ulyanovskaya 1, 198504 Saint Petersburg, Russia
igor_yusupov@mail.ru
[2] Mozhaisky Military Space Academy, Zhdanovskaya 13, 197198 Saint Petersburg, Russia
[3] Oil and Gas Research Institute, Russian Academy of Sciences, Gubkina 3, 119333 Moscow, Russia

Abstract. Analysis of storm activity over the marshland is carried out. The overall picture of storm activity in the Leningrad Region for 2017 in the form of distribution density of lightning discharges both for all period and for separate days is considered. It is noted that in a large part of all cases, the thunderstorm centers are formed directly over the swamp. In the cases of storm centers, which arose out of and passed through the swamp, the increase in lightning flashes intensity was observed.

Keywords: Thunderstorm · Marshland · Intensity

1 Introduction

Influence of the locality on storm activity is considered in a number of works [1–3]. So, in work [2], the analysis of storm activity according to the meteorological stations located at various heights allowed to reveal existence of significant correlation between height of the area and storm activity. In work [3], it is suggested that heterogeneity of temperature and humidity of the soil has a great influence on storm activity. In works [4–6], it is marked out that in movement from cold to warm surface a cold and dry air mass is saturated with heat and moisture, leading to fast formation of nebulosity with heat emission and to development of deep convection. In works [7–9], a relation between hydrogen and methane as gases-carriers of radon to surface atmosphere, atmospheric electric field and lightning flashes was established. It is shown that over areas with close perched groundwater and lowering of atmosphere pressure during the period preceding a thunderstorm the conductivity characteristics of ground air sharply increase that increases the probability of occurrence of cloud-to-ground lightning discharges. Considering experience of these works, an attempt to reveal existence of storm activity changes over the marshland is made.

The work is caused by need of checking a hypothesis of increasing cloud-to-ground lightning flashes over the marshland. Results of this work can be used in designing of

V. Murgul and M. Pasetti (Eds.): EMMFT-2018, AISC 983, pp. 417–429, 2019.
https://doi.org/10.1007/978-3-030-19868-8_42

lightning protection of electric substations, clusters of delivery wells of underground gas storages, etc., where the lightning discharges can lead to an emergency.

2 Materials and Methods

As an object of a research, a swamp located on the southwest coast of Lake Ladoga was chosen. It belongs to transitional sedge-sphagnous swamp types; the maximum power of a peat deposit is 1.75 m. The choice of this object is caused by the fact that it is located by oneself and has considerable sizes to influence the weather in the region. For simplification of calculations the swamp area was bounded by a circle with the center at 60.31 NL, 30.58 EL and radius of 21 km. The observation area was chosen within 100–150 km around the swamp. Direction North-South is indicated on the map (Fig. 1).

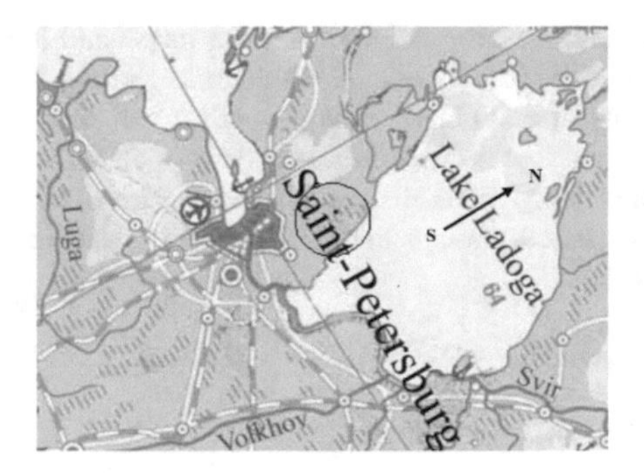

Fig. 1. Observation area map.

For the storm activity analysis, the data of lightning location network Blitzortung based on time of arrival technique [10] for year 2017 were used. This system consists of the central server and registration points, which transmit data through short time intervals to the server. Each data packet contains the exact time of arrival of an impulse and geographical coordinates of registration point. Using information from all points, the server calculates coordinates of the lightning discharges.

Data on storm activity are presented by two ways: in the form of distribution density and in the form of point representation of single lightning flashes for more detailed research. The picture of distribution density formed as follows: the Earth's surface was divided into squares of 4×4 km^2 in size, and the discharge getting to one of squares added a unit to the corresponding cell, creating number of discharges per square unit. Point representation was carried out for hour intervals, where the intensity of storm activity – the quantity of discharges per time unit of 1 h – served as its characteristic. The computer program displaying the swamp, distribution density, and a point picture of storm activity on the local map was developed.

3 Materials and Methods

Distribution density for all the year 2017 is mapped for representation of general storm activity. One can note its increasing up to 50 discharges per square unit near the swamp (Fig. 2).

Fig. 2. Distribution density of lightning flashes. The number of discharges per square unit 4×4 km^2 is marked in color.

The detailed research is presented by separate storm days. Point representation against the background of daily distribution density lightning discharges is given. Characteristic cases are presented in, the rest storm days are given in the appendix.

Storm day of July 17, 2017 is presented in Fig. 3. The thunderstorm began near the swamp at 4:00, moved in the northeast direction and ended in the center of Lake Ladoga at 8:00. Speed of movement is about 15 km/h, the maximum intensity is 257 discharges per hour.

The frontal type of thunderstorm is presented on August 12, 2017. The part of the storm front passed through the swamp at fading stage at 22:10 in the northeast direction and ended at 23:20 over Lake Ladoga in 40 km from the coast. Speed of movement was about 80 km/h (Fig. 4).

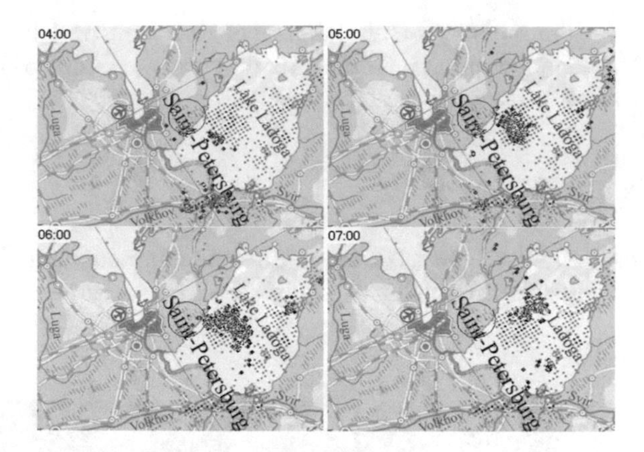

Fig. 3. Point representation of lightning locations (black points) against the background of distribution density on July 17, 2017.

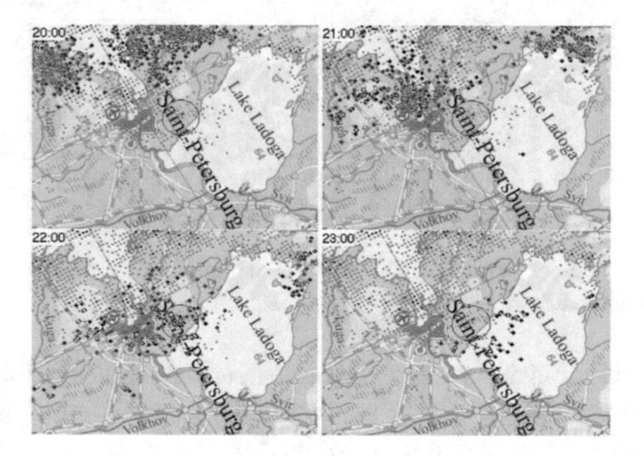

Fig. 4. Point representation of lightning locations (black points) against the background of distribution density on August 12, 2017.

The following characteristic case is presented by the storm day on August 3, 2017. The storm center was created at 5:00 in 40 km to the west of the swamp, moved in east direction, crossed the swamp at 6:30 with noticeable increase in intensity, continued the movement and ceased to exist at 10:20. Speed of the movement is about 50 km/h. The maximum intensity of 15 discharges per hour is before the swamp, 153 discharges per hour after swamp passing (Fig. 5).

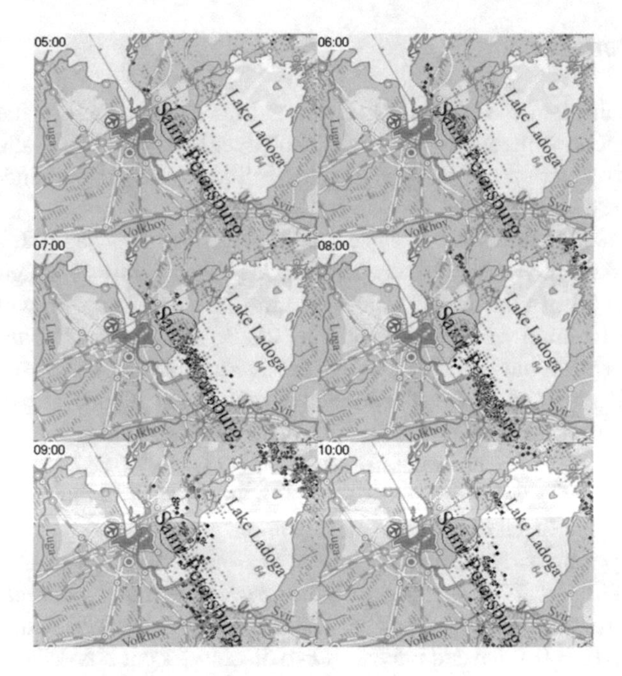

Fig. 5. Point representation of lightning locations (black points) against the background of distribution density on August 3, 2017.

The case of thunderstorm which did not touch the marshland is presented in the storm day July 31, 2017. The storm center was created over the Gulf of Finland at 16:10 in 100 km to the west of the swamp, moved in east direction and ended on swamp border at 19:20. Speed of movement is about 40 km/h (Fig. 6).

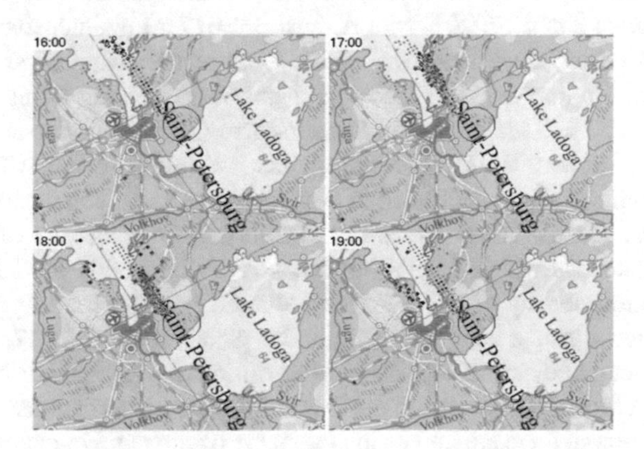

Fig. 6. Point representation of lightning locations (black points) against the background of distribution density on July 31, 2017.

4 Discussion

From the overall picture of storm activity, it is impossible to tell unambiguously that the swamp somehow affects storm activity, however it is possible to allocate a strip of bigger intensity over it. For clarification of influence separate thunderstorms were considered. They were divided into several groups.

The first group includes the storm centers arising over the swamp. It is presented by the next storm days of year 2017: May 25, June 21, July 17, July 19, August 23. The low intensive, up to 28 discharges per hour, an intra mass thunderstorm with a movement speed of about 10 km/h was observed on May 25, 2017 (see Fig. 7). During its lifetime the storm center went out of the swamp not far away. A thunderstorm on July 17, 2017 can be referred to intra mass one with a speed of movement about 15 km/h, but intensive enough (257 discharges per hour). One can note one more storm center to the east which did not touch the swamp. Thunderstorm of on June 21, 2017 (see Fig. 9) has a multicell character. One of cells arose over the swamp at 12:00, moved in the southern direction and faded in 14:30. Speed of the movement is about 35 km/h, the maximum intensity is 43 discharges per hour. A moderately intensive storm center (65 discharges per hour) developed On July 19, 2017 (see Fig. 10), the speed of its movement reached 40 km/h. The storm of August 23, 2017 (see Fig. 12) is also characterized with a high enough movement speed of 40 km/h and intensity (148 discharges per hour). The quantity of the storm centers, which arose in the observation area, is 56, over the swamp is 8. The area of the observation area is about 80000 km^2, the area of the swamp is 1400 km^2. It is possible to draw a conclusion that the density of the storm centers rises over the swamp considerably exceeds the average one, about 8 times.

The second group is presented by a frontal thunderstorm of mesoscale character on August 12, 2017. Before reaching the swamp the thunderstorm passed into the fading stage. Having passed through the swamp, it faded. It was not succeeded to observe the influences of the swamp in this case.

The storm centers, which arose near the swamp and passed over it, belong to the third group: August 3 and 17, 2017. On August 3, 2017 a low intensive (15 discharges per hour) storm center arose in 40 km to the west of the swamp, passed over it and turned into highly intensive (153 discharges per hour) and long-living (4 h) one.

On August 17, 2017 the intensive thunderstorm (303 categories an hour) arose in 130 km to the southwest from the swamp, passed through it, intensified some more (411 categories an hour) and continued to develop actively. A conclusion on this group of thunderstorms consists in strengthening of intensity after passing of the swamp.

The thunderstorms, which passed out the swamp but made a contribution to the general distribution density of storm activity entered into the fourth group. Representatives of the fourth group: May 28–29, July 31, September 12, 2017. Making a contribution to the general distribution density of lightning discharges, it brings obstacles to researches. Therefore, a conclusion on the fourth group is unambiguous: to exclude from consideration. Though there are interesting cases are occurring. For example, on July 31, 2017 quite intensive storm center stopped directly on swamp border. What is it – influence of the swamp, which is necessary to investigate, or the natural process of fading of the storm center, which was so accidentally happened in this place?

5 Conclusion

An influence of the marshland on storm activity is considered in the work. As an object of a research, the swamp of transitional type was chosen. Further, it is planned to carry out researches of other types of swamps too (raised bogs and fens). Due to various weather conditions, it is difficult to estimate any influence in an overall picture. So the research for separate thunderstorms was carried out. In large part of cases, the storm center formed over the swamp. In case of a frontal thunderstorm, the influence of the swamp is imperceptible. For the year 2017, there was one frontal thunderstorm passed over the swamp in the explored area. It is not enough for the conclusion, and in futurc researches, it is necessary to fill up the database with this kind of thunderstorms. In the case of storm centers, which arose out and passed through the swamp, the intensification of storm activity is observed.

Appendix

May 25, 2017 (see Fig. 7). The thunderstorm over the object started at 14:00, smoothly displaced to the east to the Lake Ladoga and finished at 16:00. Speed of movement is about 10 km/h. The maximum intensity is 28 discharges per hour.

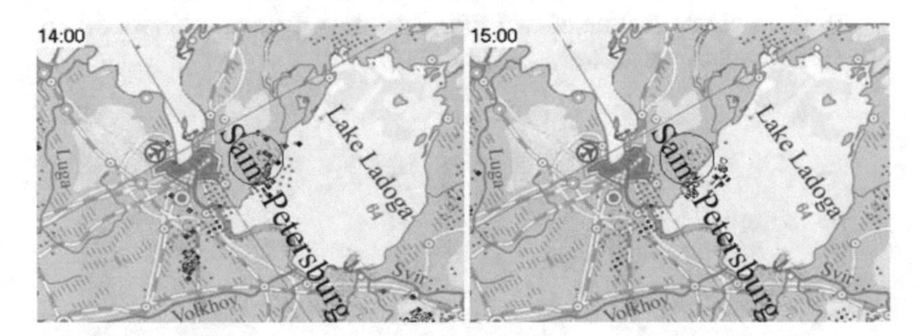

Fig. 7. Point representation of lightning locations (black points) against the background of distribution density on May 25, 2017.

May 28–29, 2017 (see Fig. 8). The thunderstorm started in 120 km to the northeast from the swamp at 21:40, developed in the southeast direction, moved to Lake Ladoga and finished to the south of its center at 00:30. This thunderstorm did not pass through the swamp. The movement speed of the storm center is about 70 km/h. The maximum intensity is 47 discharges per hour. Other storm center arose in 150 km to the southwest from the swamp at 23:00, moved in east direction and finished at 1:00. Speed of movement is 50 km/h. The maximum intensity is 31 discharges per hour. This storm center passed out the swamp.

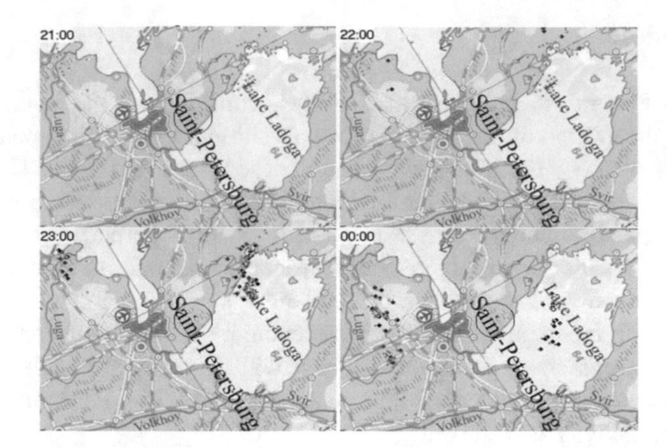

Fig. 8. Point representation of lightning locations (black points) against the background of distribution density on May 28–29, 2017.

June 21, 2017 (see Fig. 9). The thunderstorm has a multicell character. The first center started in 50 km to the north of the swamp at 11:00 and faded at 12:00 with the maximum intensity of 47 discharges per hour. At 12:00 there were two more centers. One was over the swamp, moved in the southern direction and faded at 14:30. Movement speed is about 35 km/h, the maximum intensity 43 is discharges per hour. Another was formed in the form of a tape in 40 km to the northwest. It moved in parallel with the first one and faded at 14:00. Movement speed is about 30 km/h, the maximum intensity 34 is discharges per hour.

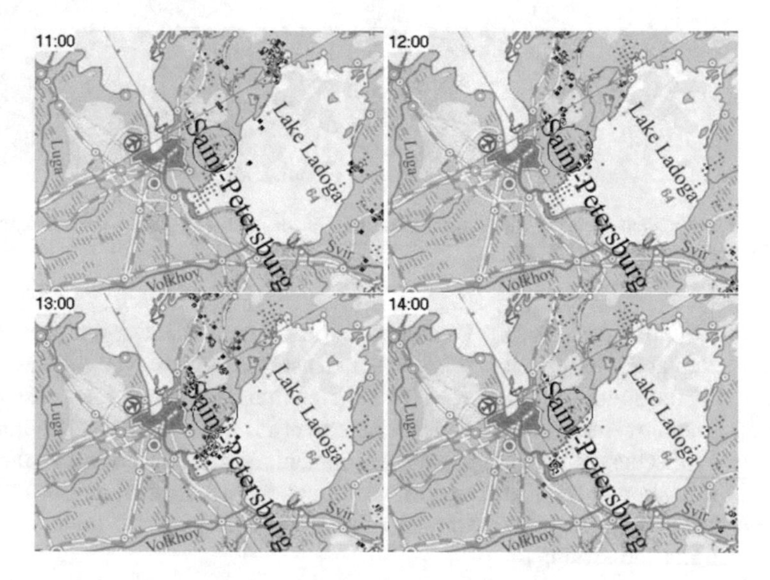

Fig. 9. Point representation of lightning locations (black points) against the background of distribution density on June 21, 2017.

July 19, 2017 (see Fig. 10). The storm center was created at 6:20 over the swamp, slowly moved in east direction and ceased to exist on the southern coast of Lake Ladoga at 10:00. Movement speed is about 40 km/h, the maximum intensity is 65 discharges per hour.

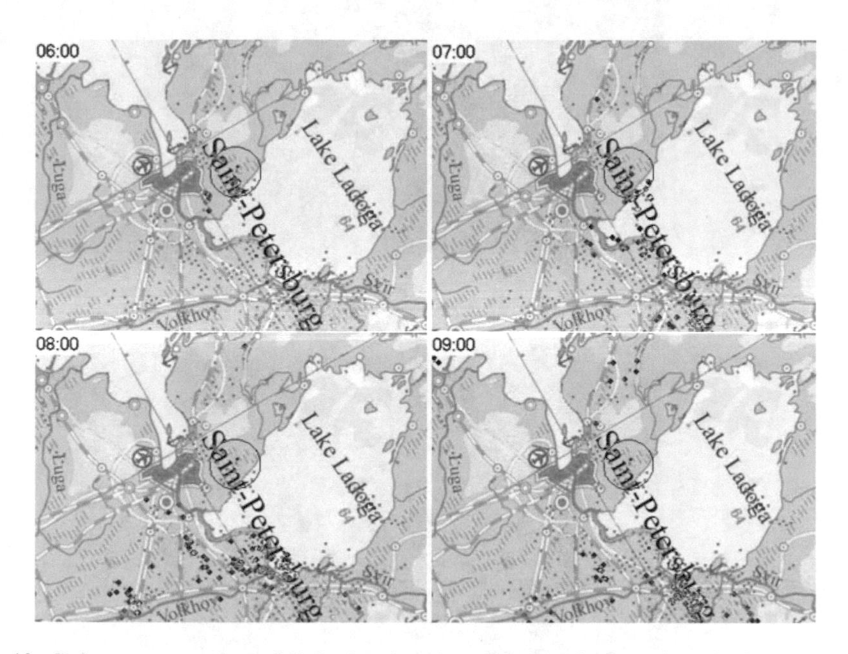

Fig. 10. Point representation of lightning locations (black points) against the background of distribution density on July 19, 2017.

August 17, 2017. The intensive storm center was created in 130 km to the southwest of the swamp at 14:00, reached the swamp at 15:20, having covered with one part the swamp, another passed out (Fig. 11).

The maximum intensity before the swamp is 303 discharges per hour. That part which covered the swamp crossed Lake Ladoga at 18:30 and finished in 30 km to the east of the coast at 20:20 with the maximum intensity of 411 discharges per hour. That part which did not touch the swamp faded. At the same time at 19:00 the new storm center in 40 km to the northwest from the swamp was formed and went along the coast of Lake Ladoga to the north. The maximum intensity was 165 discharges per hour. Movement speeds of the both centers were about 35 km/h.

August 23, 2017 (see Fig. 12).

The thunderstorm started in the center of Lake Ladoga at 15:40 and stayed in the same place till 17:30. At this time at 16:30 in 10 km to the north from the swamp the storm center was formed and began to move slowly in the northern direction. At 17:10

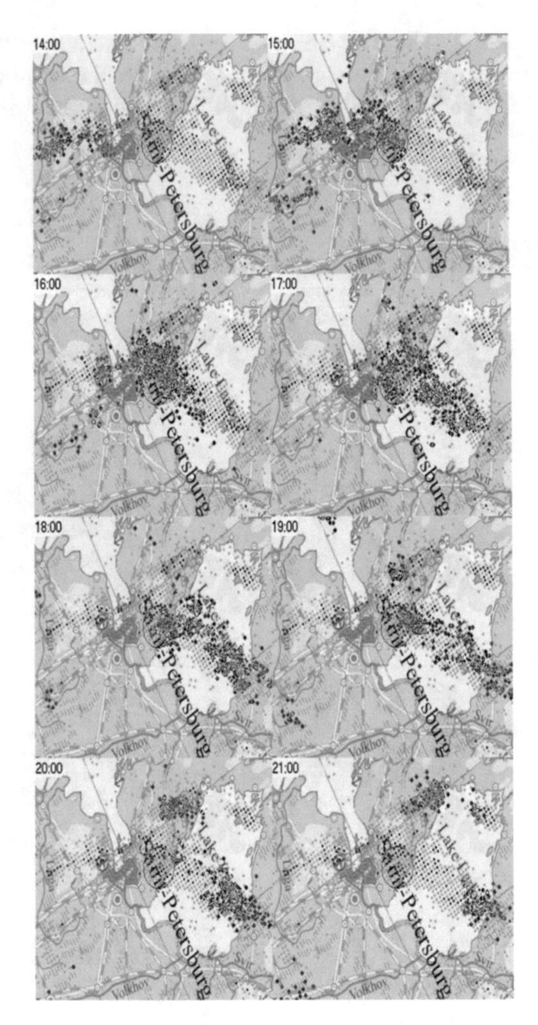

Fig. 11. Point representation of lightning locations (black points) against the background of distribution density on August 17, 2017.

one more storm center was formed in a northern part of the swamp and went to the north too. As a result all these centers, having mixed up, reached a northern part of Lake Ladoga at 19:20, and ceased to exist at 21:00. Average speed of movement of this conglomeration was about 40 km/h. The maximum intensity is 148 discharges per hour. September 12, 2017. The thunderstorm started in 300 km to the southwest from the swamp, moved in the northeast direction, passed in 100 km to the north and finished at

11:00 on the northern coast of Lake Ladoga. The second thunderstorm in the same day has a frontal type and also passed out the swamp which several centers arose in 50 km to the east from it at 13:00. It moved out of swamp in east direction with a speed of about 40 km/h. Both thunderstorms passed out the swamp (Figs. 13, 14 and 15).

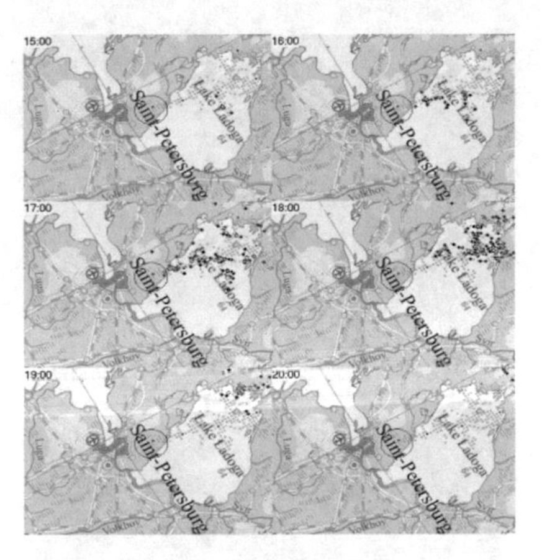

Fig. 12. Point representation of lightning locations (black points) against the background of distribution density on August 23, 2017.

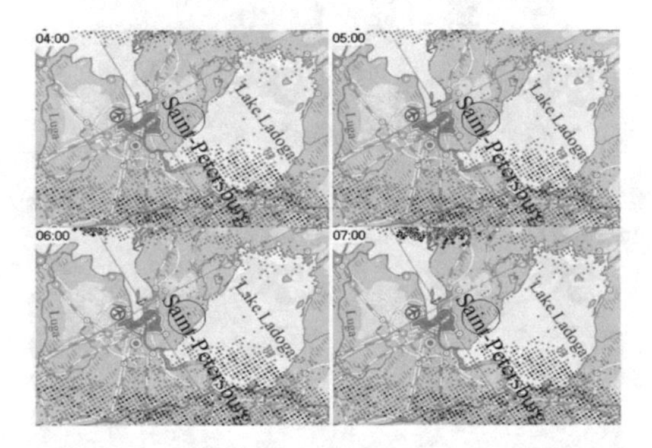

Fig. 13. Point representation of lightning locations (black points) against the background of distribution density on September 12, 2017.

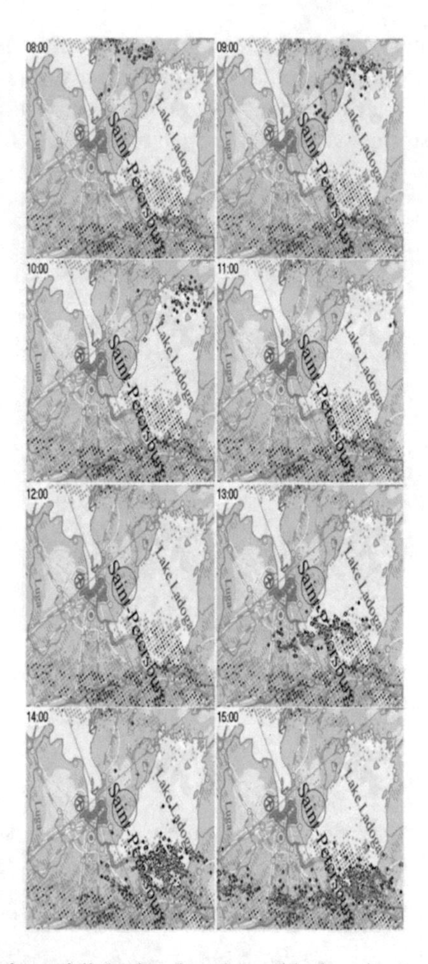

Fig. 14. Point representation of lightning locations (black points) against the background of distribution density on September 12, 2017 (Part 2).

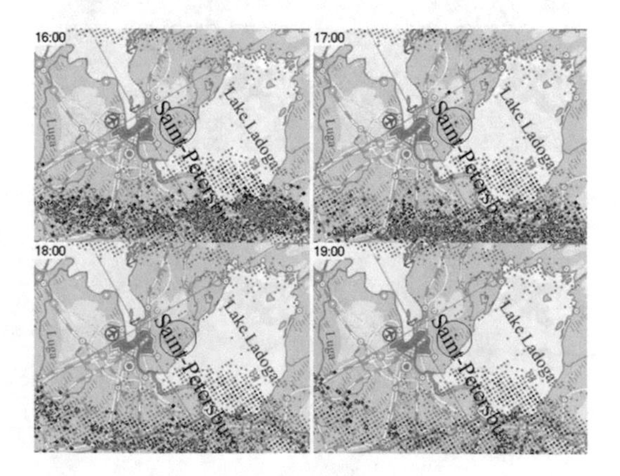

Fig. 15. Point representation of lightning locations (black points) against the background of distribution density on September 12, 2017 (Part 3).

References

1. Adzhiev, A.H., Kuliev, D.D., Kazakova, S.T., Yurchenko, N.V.: Influence of orography on dynamics of the electric phenomena in the atmosphere. In: Materials of the Third All-Russian Conference on Global Electrical Circuit, pp. 80–81 (2017). (in Russian)
2. Adzhiev, A.H., Adzhieva, A.A., Tumgoeva, H.A.: Influence of orography on characteristics of storm activity. Ser.: Nat. Sci. **2**(144), 109–112 (2008). (in Russian). News of Higher Educational Institutions. North Caucasus Region
3. Gorbatenko, V.P.: Influence of geographical factors of climate on storm activity. Questions of Geography of Siberia. Russian Geographical Society, Tomsk Department, Tomsk State University, Tomsk, pp. 66–78 (2001). (in Russian)
4. Tunaev, E.L., Gorbatenko, V.P.: Power characteristics of the atmosphere at cyclogenesis over areas of the vasyugan swamp. Hydrometeorol. Res. Forecast. **4**(370), 48–62 (2018). (in Russian)
5. Vasil'ev, E.V., Alekseeva, A.A., Peskov, B.E.: Conditions for formation and short-range forecasting of severe squalls. Russ. Meteorol. Hydrol. **34**(1), 1–7 (2009). https://doi.org/10.3103/S1068373909010014
6. Gorbatenko, V.P., Voilokova, E.S., Sorokina, S.A.: Some characteristics of convection over the southeast of western Siberia in days with the thunderstorms. In: Materials of the Russian Conference "The Seventh Siberian Meeting on Climate-Environmental Monitoring" 2007, pp. 49–51. Agraf-Press, Tomsk (2007). (in Russian)
7. Shuleikin, V.N.: Quantitative study of relationships of hydrogen, methane, radon, and the atmospheric electric field. Izv. Atmos. Ocean. Phys. **54**, 794 (2018). https://doi.org/10.1134/S0001433818080121
8. Shuleykin, V.N.: Hydrogen, methane, radon and cloud-to-earth lightning discharges. Curr. Probl. Oil Gas **3**(22), 14 (2018). (in Russian)
9. Shuleykin, V.N.: Sensitivity of atmospheric electric field to changes of concentration of hydrogen and methane. Curr. Probl. Oil Gas **1**(13), 7 (2016). (in Russian)
10. A worldwide, real time, community collaborative lightning location network Blitzortung Homepage. http://www.blitzortung.de

The Study of Local Dust Pollution of Atmospheric Air on Construction Sites in Urban Areas

Valery Azarov[1], Lubov Petrenko[2(✉)], and Svetlana Manzhilevskaya[2]

[1] Volgograd State Technical University, Lenin Avenue, 28, Volgograd 400005, Russia
[2] Don State Technical University, Sq. Gagarina, 1, Rostov-on-Don 344010, Russia
LK-ZXS@yandex.ru

Abstract. This article summarizes the results of PM10 and PM2.5 pollution emission to the environment monitoring during the construction processes in urban populated area. Identifying important pollution sources that contribute to ambient concentrations of pollutants is essential for developing an effective air quality management plan during building construction. Control and regulation of the dynamic state of dispersed systems released during technological construction processes using a number of protective measures will reduce emissions of pollutants into the air. The results of this research show the concentration of PM2.5 and PM10 at 3 objects. After analyzing the situation with dust pollution, the protective measures were suggested. The implementation of these measures at the construction site allows reducing the level of the dust polluting at workplaces at local construction approximately in 5 times. In the future, it is necessary to improve the dust control system at the construction site in order to reduce the costs of protecting nearby located residential buildings.

Keywords: Energy-saving technologies · Labor protection · Ecology

1 Introduction

There are two main sources of air pollution. They are natural and anthropogenic ones. The anthropogenic sources of air pollution are connected with the human activities, which are aimed at goods and services production.

The main anthropogenic sources of air pollution emissions are raw material extraction, energy acquisition, production and manufacturing. The results of these activities are emissions of different physical and chemical transformations types, which have great harm influence on the health and environment, like air quality deterioration, harmful effect on human health and ecosystems, degradation of air resources and others. During the environment pollution researching and monitoring the special attention should be paid to the construction operations, since during the construction processes many pollutants are released, especially PM10 and PM2.5 fine particles, which are harmful to the health of construction workers and the population living near the construction site.

V. Murgul and M. Pasetti (Eds.): EMMFT-2018, AISC 983, pp. 430–439, 2019.
https://doi.org/10.1007/978-3-030-19868-8_43

The problem of the environment preserving and the protecting people from the adverse health effects during the construction operations is growing more urgent due to the increase in volume of construction and the increasing availability of construction equipment. The works of Azarov V.N., Menzelintseva N.V., Lozhkina O.V., Batmanov V.P., Barikaeva N.S., Trokhimchuk M.V., Kyoyken M.P., Koshkarev S.A., Solovyeva T.V., Stefanenko I.V., Nikolenko D.A., Barratt B. and many others are devoted to the study of the dust load and the evaluation of influence on the environment.

This article summarizes the results of the PM10 and PM2.5 pollution emission to the environment monitoring during the construction processes in urban populated area. Air fine dust pollution is a major environmental risk to health and is estimated to cause approximately 2 million premature deaths worldwide per year. The ill health prevention of people working on the construction site and living nearby in the sanitary protection area and behind it can be realized if the average annual levels of PM10 are lower than 20 μg/m^3, but there are cases where levels exceed up to 70 μg/m^3. Even relatively low concentrations of air pollutants are related to a range of adverse health effects. There are serious risks to health from exposure to PM and O3 in many cities. Poor air quality in the rooms near the construction operations can pose a health hazard to a large part of the population living near the construction industry. The PM particles have the most harmful effect than other pollutants. The PM constitutes on the air are sulfates, nitrates, ammonia, sodium chloride, carbon, mineral dust, and water. They can be performed as mixture of solid and liquid particles. The most dangerous are PM2.5 particles. When people inhale them, the bronchioles and lungs suffer from PM2.5 harm influence.

According to the WHO's data, the PM guideline values are set for PM2.5 and PM10 separately. The annual mean of PM2.5 and PM10 should be equal to 10 and 20 μg/m^3, respectively, whereas the 24-h mean should not exceed 25 and 50 μg/m^3, respectively. The goal of this research is to analyze the environmental air monitoring methods and results at the construction site and the surroundings, to develop the measures to reduce the impact of pollution on human health at the time of construction, both on the workers at the site and the neighbored population that is exposed to harmful substances. The objects of this research were construction sites in Rostov-on-Don and the buildings adjacent to them. The task of this research was to monitor air pollution, especially PM2.5 and PM10 particles pollution in the construction industry at the construction site itself, in order to identify injuries caused to workers and to identify pollution load in residential buildings that have recently became operational and inhabited houses, where people have been living for several years near the construction industry.

2 Methods and Study

During the designing and constructing of the buildings, the ecological researches are done according to the methods of Regulations for establishing the permissible limits of harmful pollutants emissions from industrial enterprises (GOST 17.2.3.02-2014) [1] and Regulatory document (RD 52.04.667-2005) [4] as well as general requirements for methods of pollutant determination in the open air. The relative measurement error of these methods should not exceed 25% over the whole range of mass concentrations and provide a measurement with the specified accuracy of suspended particles from 0.048 to 0.6 mg/m for the PM10 and from 0.028 to 0.35 mg/m for the PM2.5.

The monitoring of PM10 and PM2.5 particles calls for the daily (24 h) sample taking as the main way for the gravimetric measurement method. It is acceptable to determine the single average hourly dust concentration only for suspended particulate matters auto analyzers according to the nuclear-radiation technique.

According to hygienic standard (GN 2.1.6.2604-10) [2] and hygienic standard (GN 2.1.6.1338-03, Addendum №8) [3] the normative amount of suspended solids is given in Table 1.

Table 1. The normative amount of suspended solids

Material name	MPC, mg/m		
	MPC OT	MPC daily average	Mean year concentration
PM10	0.3	0.06	0.04
PM2.5	0.16	0.035	0.025

This method ensures the achievement of the measurement results with errors, which do not exceed the values of the indicators given in Tables 2 and 3.

Table 2. The measurement range, quality indicator values (quantitative assessment) of the measurement method – frequency, reproducibility, accuracy.

Analyte	The measurement range, mg/m	Frequency indicator (standard deviation of the single determination results obtained under the frequency conditions), mg/m	Reproducibility indicator (standard deviation of the single determination results obtained under the reproducibility conditions), mg/m	Accuracy indicator (error range of measurement results with the probability P = 0.95), mg/m
PM10	One-time maximum mean from 0.24 to 3.00 inclusive	0.09X	0.13X	0.25X
	Diurnal mean from 0.05 to 0.60 inclusive			
PM2.5	One-time maximum mean from 0.12 to 1.60 inclusive	0.09X	0.13X	0.25X
	Diurnal mean from 0.03 to 0.35 inclusive			

Table 3. The measurement range, quality indicator values (quantitative assessment) of the measurement method – frequency limits, reproducibility limits.

Analyte	The measurement range, mg/m	Frequency limit for 2 kinds of results of combining measurements, mg/m	Reproducibility limit for 2 kinds of measurement results, mg/m
PM10	One-time maximum mean from 0.24 to 3.00 inclusive	0.25X	0.36X
	Diurnal mean from 0.05 to 0.60 inclusive		
PM2.5	One-time maximum mean from 0.12 to 1.60 inclusive	0.25X	0.36X
	Diurnal mean from 0.03 to 0.35 inclusive		

Owing to a lack of the approved state standards for the content of suspended particles PM10 and PM2.5 in open air, the main error is determined by the calculation method given in Appendix of Regulatory document (RD 52.04.186-89).

The requirements for instruments, auxiliary devices, are given in Table 4.

Table 4. Measuring equipment

Description of measuring instruments	Model	Metrological characteristics
Hand-held particle counter	Handheld 3016	Particle size range, μm – 0.3–5.0 or 0.3–10.0 or 0.3–25.0 (optionally)
		Up to six channels of simultaneous data
		Flow Rate—2.83 LPM
		Max. particle concentration 3—4 000 000 particles per cubic foot
		3000 sample record storage
		FS-209E, ISO 14644-1, BS 5295, EC GMP compliant
		Operating temperature—0–50 ± 0.5 °C
		Operating humidity—15–90 ± 2%
		Type of sampling—isokinetic sampling
		Sample recovery—internal output HEPA filter (>99.997% from 0,3 μm)

Hand-held particle counter application (Table 4) provides an accuracy rate of 24% for air sampling during 24 h for the PM10 and 48% for the PM2.5.

The main sources of anthropogenic pollution are known. These are all harmful emissions of industrial waste from construction activities, including local dust generation during construction works.

The inadmissibility of dusting should be its localization within the sanitary protection zone. Localization is achieved using dust suppression measures. The measures selection or their combination is determined based on the composition of dust particles. The basic condition of construction operating is that the atmospheric air quality standards indicators do not exceed at the border of the sanitary protection zone. Sufficiency of dust suppression assessment measures incorporated into the construction project as part of the environmental impact assessment procedure according to the method for estimating wind erosion and dusting taking into account design, planning and climatic factors is provided by introducing a number of additionally developed correcting factors, as well as taking into account the influence of particle size distribution deposits on the dusting properties of the layer. When the practical values of these coefficients for some regions are deduced and adopted, it is possible to carry out a statistical processing of the results and find their average value.

Existing types of negative environmental impact, including in emergency situations, allow analyzing the effectiveness of protective measures:

1. Environmental condition studding;
2. Information acquisitioning the field of environmental protection;
3. Fire safety programs development;
4. The statistical reporting of emergency situations introduction;
5. Increased preparedness:

- strengthening environmental monitoring;
- the initiation of operational measures to prevent the occurrence and development of emergency situations;
- the clarification of the action description for the emergency prevention and recovering;
- the replenishment of physical resources reserves created for emergency recovery;
- the continuous environment monitoring;
- the carrying out of population and territory protecting measures from emergency situations;
- the emergency recovery management;
- the organization and maintenance of continuous interaction of federal executive authorities and territorial subjects executive authorities of Russian Federation, local governments and organizations on the issues of emergency recovering.

Identifying important pollution sources that contribute to ambient concentrations of pollutants is essential for developing an effective air quality management plan during building construction.

Particular attention should be paid to emissions of fine particles during technological processes of construction with a special degree of dust emission.

Control and regulation of the dynamic state of dispersed systems released during technological construction processes using a number of protective measures will reduce emissions of pollutants into the air. These protective measures will reduce the environmental load and adverse effects on the health of workers and the population.

The highlight in the regulation of the dynamic state of dispersed systems in technological processes is the creation of a controlled dynamic state of building dispersed systems in construction processes, including local dust pollution, compliance with the dispersed phase:

$$G = mg = \sum F_c, \sum_{i=1}^{n} F_c \times \sum_{i=1}^{n} F_c \times \sum_{i=1}^{n} F_x \times \sum_{i=1}^{n} F_c, \quad (1)$$

where: G - gravity force, m - mass of a particle (kg), n - number of contacts of a particle with compounds, F_C - cohesive strength between particles.

It is necessary to try to get $G \to$ const to a constant. According to this condition, it is possible to create and apply such protective measure as an inexpensive dust shield in a certain construction work area at the workplaces of some construction works, which can significantly reduce the amount of dispersed dust from emission.

In order to apply a dust shield in case of local pollution of the atmosphere, samples were taken from residential and nearby houses under construction.

The research area was the territory of «Ekaterininsky» residential complex in Rostov-on-Don. The objects of this research were the construction site of 20 floor residential building located in Magnitogorskay Street, 2B, a 20 floor residential building put into operation a month before the researching, where the repair and construction works are carried out located in Magnitogorskay Street, 1B and a 25 floor residential building located in Magnitogorskay Street, 1 next to the construction site put into operation more than 2 years ago and inhabited.

3 Conclusions

As the part of the study, an analysis of the sampled material was performed, which gave the following results. A qualitative analysis of the sampled particles at the construction site is shown in Fig. 1.

The variation range of the dust particles size around the perimeter of the building construction site is from 0.5 to 10 μm. The median diameter value (d50) is 7 μm for the

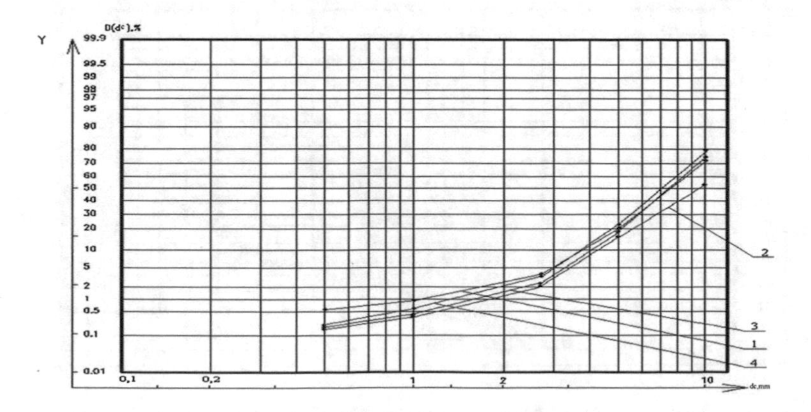

Fig. 1. Dust pollution schedule at the construction site

4th point, 7.3 µm for points 1 and 3, and 9.4 µm for the 2nd point. The most common dust size (70–80%) is from 9 to 10 µm at all points, except the 2nd one. The maximum value of the dust with a size of 10 µm is at point 4. The maximum value of the dust with a size up to 2.5 µm is at point 1. Also, it is worth noting that at the 2nd point the value of the dust with a size from 5 to 10 µm is minimum than at the rest ones.

The results of the analysis of the air pollution of a residential building with the repair and construction works carried out are shown in Fig. 2.

The variation range of the dust particles size of dust in the residential building with the repair works is from 0.5 to 10 µm, the value of the median diameter (d50) ranges from 6.3 µm on the 6th floor to 8.2 µm on the 20th floor. The most common dust size (from 70% to 90%) is from 8 to 10 µm. The maximum value of the dust with a size of 10 µm is on the 6th floor. The maximum value of the dust with a size up to 2.5 µm is on the 11th floor. Also, it is worth noting that on the 20th floor the value of the dust with a size from 5 to 10 µm is less than on the other floors. The results of the analysis of the air pollution of a residential inhabited building are shown in Fig. 3.

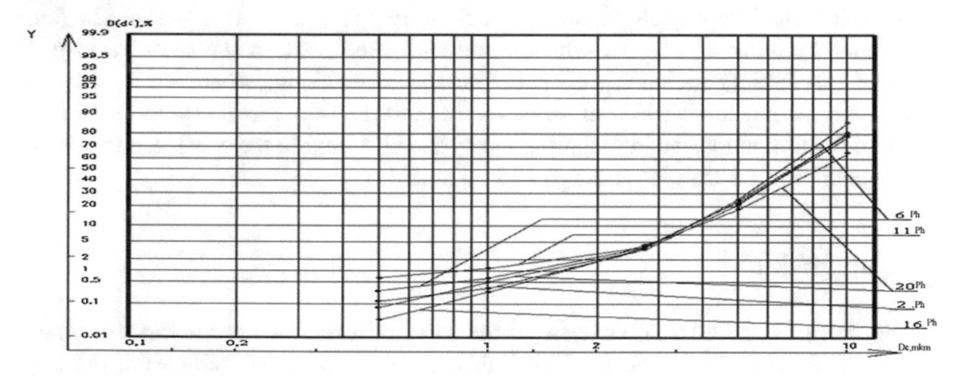

Fig. 2. Dust pollution schedule of the 20 floor residential building put into operation a month before the researching, where the repair and construction works are carried out

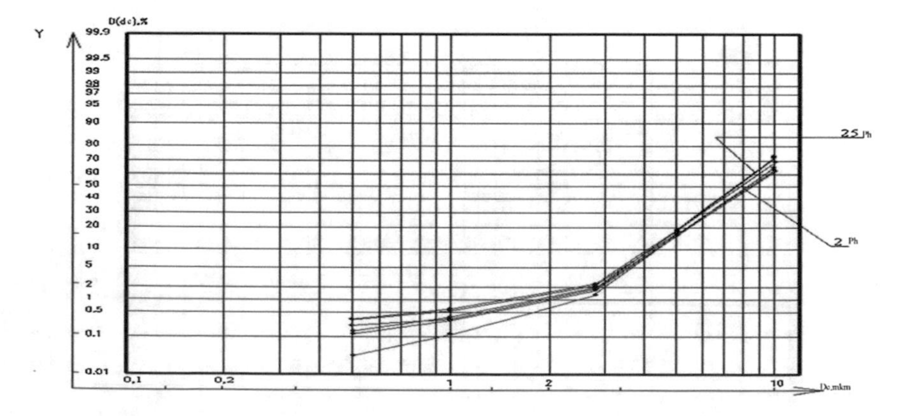

Fig. 3. Dust pollution schedule of the 25 floor residential building next to the construction site put into operation more than 2 years ago and inhabited.

The variation range of the dust particles size of a residential building next to the construction site is from 0.5 to 10 μm, the value of the median diameter (d50) ranges from 7.5 μm on the 25th floor to 8.4 μm on the 2nd floor. 60–75% of dust value is dust with a size from 8.3 to 10 μm. The largest dust size of 10 μm is on the 25th floor. The maximum value of the dust particles with the size up to 2.5 μm is also located on the 25th floor. It is worth noting that on the 2nd floor, dust of the size as high as 2.5 μm is less than on the other floors.

The overall concentration of the dust pollution at the sampling points is shown in Table 5.

Table 5. The overall concentration ratio of dust pollution at 3 objects of research.

Object of research	Sampling point, floor	The overall concentration ratio, mg/m^3
The 25 floor residential building next to the construction site put into operation more than 2 years ago and inhabited	1	1825.59
	6	4842.73
	11	1009.76
	16	684.41
	21	1216.55
	25	703.06
The 20 floor residential building with repair works	1	1696.95
	6	956.12
	11	360.59
	16	4018.1
	20	10761.2
Construction site	1	312.68
	2	844.05
	3	809.67
	4	658.61

According to the research data, we can conclude that from 60 to 90% of cases in the territory of residential buildings the fine dust particles are with a size of 8 to 10 μm are concentrated (Figs. 3 and 4). The higher the floor, the finer the dust to 2.5 μm.

The dust particles from 0.5 to 10 μm are concentrated around the perimeter of the ditch for foundation, especially dust is most concentrated on the leeward side. The dust particles with size of 7 μm, 3 μm, 9 μm and 10 μm dominate at the 3 and 4 sampling points. Thus, the territory of a nearby maintained inhabited building receives especially large value of PM2.5 to PM10 at different altitudes.

Therefore, during the designing of the construction stage, it is necessary to locate the designed building perpendicular to the leeward side of the maintained inhabited buildings according to the prevailing wind direction of construction site. The most significant local dust pollution is determined in the residential building where repair and construction works are carried out, especially upstairs.

Fig. 4. The results of the analysis of dust particles of the sample taken from a ready-built kitchen in the apartment of the 25 floor residential inhabited building, where the windows look to the construction site.

Moreover, the dust particles with the size are smaller than 2.5 μm the dust works in everywhere, even in dwelling rooms, it does a greater harm to the inhabitants' health.

This fact is confirmed by the results of the analysis of dust particles of the sample (Fig. 4) taken from ready-built kitchen in the apartment of the 25 floor residential inhabited building, where the windows look to the construction site.

The variation range of the dust particles size of the sample taken from a ready-built kitchen (Fig. 4) is from 1.2 to 11 μm, the value of the median diameter d50 = 7.9 μm. The most common dust size is from 8.7 to 11 μm.

Considering the results of this research, we can suggest the following measures for the prevention of dust pollution at the construction site in addition to the well-known standard ones:

1. The use of the inexpensive dust shield penetrated with a triple chemical reagent. The dust particles congregate on the shield and then they can be removed. This shield is cheaper than the currently used polymer, more expensive metal nets. At the moment it is being tested.
2. The fogging by irrigation systems for the dust catching with the magnetic water, which make it possible to significantly (up to 90%) deposit particles of fine dust held in suspension and coagulate them. This measure reduces the lower level of dust pollution of atmospheric air.

References

1. Azarov, V.N., Borovkov, D.P., Redhwan, A.M.: Application of swirling flows in aspiration systems. Int. Rev. Mech. Eng. **8**(4), 750–753 (2014)
2. Azarov, V.N., Lukanin, D.V., Borovkov, D.P., Redhwan, A.M.: Experimental study of secondary swirling flow influence on flows structure at separation chamber inlet of dust collector with counter swirling flows. Int. Rev. Mech. Eng. **8**(5), 851–856 (2014)

3. Azarov, V.N., Logachev, I.N., Logachev, K.I.: Methods of reducing the power requirements of ventilation systems. Part 4. Theoretical prerequisites for the creation of dust localizing devices with swirling air flows. Refract. Ind. Ceram **55**(4), 365–370 (2014)
4. Azarov, V.N., Menzelitseva, N.V., Redhwan, A.M.: Main trends of conditions normalizing at cement manufacturing plants. Int. Rev. Mech. Eng. **6**(6), 145–150 (2016)
5. Koshkarev, S.A., Azarov, V.N.: Evaluation of wet dust separator effectiveness in the dedusting of emissions from expanded clay kiln. Civ. Eng. Mag. **2**, 18–32 (2015)
6. Azarov, V.N., Evtushenko, A.I., Batmanov, V.P., Strelyaeva, A.B., Lupinogin, V.V.: Aerodynamic characteristics of dust in the emissions into the atmosphere and working zone of construction enterprises. Int. Rev. Mech. Eng. **7**(5), 132–136 (2016)
7. Azarov, V.N., Koshkarev, S.A., Azarov, D.V.: The decreasing dust emissions of aspiration schemes applying a fluidized granular particulate material bed separator at the building construction factories. Procedia Eng. **165**, 1070–1079 (2016)
8. Azarov, V.N., Trokhimchuk, M.V., Sidelnikova, O.P.: Research of dust content in the earthworks working area. Procedia Eng. **150**, 2008–2012 (2016)
9. Stefanenko, I.V., Solovyeva, T.V., Nasimi, M.H., Azarov, V.N.: Fine dust concentration PM10 in the atmosphere of the city of Kabul, Afghanistan in spring months. In: Applied Mechanics and Materials: Proceedings of 2nd International Conference on Civil, Architectural, Structural and Constructional Engineering, vol. 878, pp. 255–258 (2017)
10. Stefanenko, I.V., Solovyeva, T.V., Nasimi, M.H., Azarov, V.N.: On the calculation model of fine dispersed dust pollution of Kabul's atmosphere for ventilation design. In: Applied Mechanics and Materials: Proceedings of 3rd International Conference on Civil, Architectural, Structural and Constructional Engineering, vol. 875, pp. 132–136 (2018)
11. Manzhilevskaya, S.E., Azarov, V.N., Petrenko, L.K.: The pollution prevention during the civil buildings construction. In: MATEC Web of Conferences, vol. 196 (2018)
12. Azarov, V.N., Barikaeva, N.S., Solovyeva, T.V.: Monitoring of fine particulate air pollution as a factor in urban planning decisions. Procedia Eng. **150**, 2001–2007 (2016)
13. Azarov, V.N., Trokhimchuk, M.V., Trokhimchuk, A.K.: Experimental study of the propagation of dust in the construction areas stockpiles, news higher educational institutions. Geol. Explor. **1**, 55–59 (2016)
14. Nikolenko, D.A., Solovyeva, T.V.: Analysis of experience of monitoring of pollution by a fine dust of the roadside territories in the EU and Russia. Eng. J. Don **3** (2015)
15. Azarov, V.N., Barikaeva, N.S., Solovyeva, T.V., Nikolenko, D.A.: On the study of air pollution by finely dispersed dust using random functions. Eng. J. Don **4** (2015)
16. Kyoyken, M.P.: Source deposits to PM2.5 and PM10 against the background of the city and the adjacent street. Atmos. Environ. **7**, 26–35 (2013)
17. Azarov, V.N., Sergina, N.M., Stefanenko, I.V.: Applied Mechanics and Materials, vol. 875, pp. 137–140 (2018)
18. Koshkarev, S.A., Azarov, V.N., Stefanenko, I.V.: Applying absorption in environmental mechanics' decreasing of aspiration emissions of gas station. In: Applied Mechanics and Materials: Proceedings of 3rd International Conference, vol. 875, pp. 145–148 (2018)
19. Stefanenko, I.V., Azarov, V.N., Sergina, N.M.: Dust collecting system for the cleaning of atmospheric ventilation emissions. Appl. Mech. Mater. **878**, 269–272 (2017)
20. Azarov, V.N., Stefanenko, I.V., Burkhanova, R.A.: Applied Mechanics and Materials, vol. 875, pp. 187–190 (2018)

Organization of Organic Waste Samples Tests for Biogas Potential Assessment

Vladimir Maslikov, Vadim Korablev, Dmitry Molodtsov(✉), Alexander Chusov, Vladimir Badenko, and Maria Ryzhakova

Peter the Great St. Petersburg Polytechnic University, Polytechnicheskaya st. 29, 195251 St. Petersburg, Russia
molodtsov_dv@spbstu.ru

Abstract. In this article, the results of studies on the biogas potential of agricultural wastes are presented. The laboratory experiments on the same type of organic substrates from Russia and Finland were conducted using similar equipment and methods with the implementation of remote testing in real-time. The experimental data on the dynamics of the emission of biogas of the investigated substrates can be used for justification of parameters and modes of operation of biogas plants that use this type of waste.

Keywords: Organic waste · Biogas potential · Remote tests

1 Introduction

A considerable amount of various organic wastes annually produced by human activities, the relative ease of generating biogas containing 40–60% methane, contribute to the intensive development of biogas technologies in many countries. The biogas is produced in MSW landfills at the treatment facilities of sewage, at utilization of agricultural waste and others. Biogas plant with a capacity of about 200 kW can annually produce 1.5 million kWh of electric energy, providing 450 households. At the same time, annual carbon dioxide emissions are reduced by 650 tons. Currently, more than 14500 biogas plants are in operation in the EU alone, replacing 15 thermal power plants with an average capacity of 500 MW each. Most energy plants running on agricultural waste, while simultaneously with the generation of energy a valuable organic fertilizer is being produced. The world leader in the production and use of biogas for energy purposes is Germany, where at the end of 2015 operated about 9000 units with a total capacity of more than 4100 MW. Various organic wastes are used as raw materials for biogas production, including crop production – 48%, animal husbandry – 44%, food – 6%, municipal – 2%. Germany's biogas plants provide electricity to about 8 million households [1–6].

Despite the tempting prospects, the accumulated experience of operation of biogas plants showed that each second of them has a capacity below the design [7–9]. The main reasons of such a deficiency of capacity are the significant variability of the waste composition, instability of multistage biochemical processes, the complexity of their regulation, etc. Therefore, it is difficult to control the parameters of biogas plants

V. Murgul and M. Pasetti (Eds.): EMMFT-2018, AISC 983, pp. 440–448, 2019.
https://doi.org/10.1007/978-3-030-19868-8_44

(the degree of decomposition of raw materials, the intensity of the process, the specific yield of biogas, its composition, etc.), as well as to ensure their optimization [10, 11]. The necessary information about the biogas potential of the substrate can be obtained during laboratory experiments to study the processes of waste decomposition, which is an important component of the design of biogas plants. However, the disadvantage of experimental studies to date is the lack of a unified approach to the organization of work, the use of different methods and equipment, which makes it difficult to compare and analyze the results [12–14]. For example, laboratory studies of the same substrate conducted in more than two dozen laboratories showed significant differences in the assessment of biogas potential [15].

The aim of this work is to unify the laboratory experiment to obtain reliable and comparable information about the biogas potential of waste samples. This will make it possible to create a single pilot network that can be used by researchers from different countries, which will expand the possibility of studying the experience of research, obtaining information about equipment and techniques, will provide rapid access to the results of experiments and their discussion.

2 Methods

It is very important to organize convenient access for all researchers to the data obtained when conducting experiments. This is particularly relevant when it is necessary to record the main parameters of the process continuously during the observation time. Their manual distribution in this case is ineffective. Data collection and access for researchers to the results can be organized using «cloud» data storage, such as Dropbox, Yandex.Disk, Google Drive, etc. (Figure 1). There is also a possibility of visual observation of experiment by means of modern IP-cameras that can work both in the mode of video broadcasting, providing access to it to authorized users, and in the mode of creation of photos with a certain interval, which are automatically uploaded to a «cloud» data storage. As a result, using modern information and communication technologies, it is possible to fully participate in the experiment for all researchers (including students) at its each stage with obtaining all the necessary information for analyzing and processing the results. Peter the Great St. Petersburg Polytechnic University (SPbPU) established partnerships with the University of Applied Sciences Savonia (Kuopio, Finland) within the framework of the project 5-100-2020 – «project to improve the competitiveness of leading Russian universities among the world's leading scientific and educational centers». The task of studying the biogas potential of agricultural waste was identified as one of the promising research topics. It should be noted that the production of biogas, which is a valuable local fuel, is important for autonomous consumers of the border regions of Russia and Finland with developed agriculture [16, 17]. Finland already has some practical experience in biogas production and use – 95 biogas plants were in operation at the end of 2013. It is planned to further increase the use of biogas as an energy resource [18]. In the Leningrad region there is also a significant biogas potential of animal waste. Thematic maps of distribution of total gross biogas potential of livestock and poultry wastes, as well as their fuel equivalent in the districts of the Leningrad region have been developed [19].

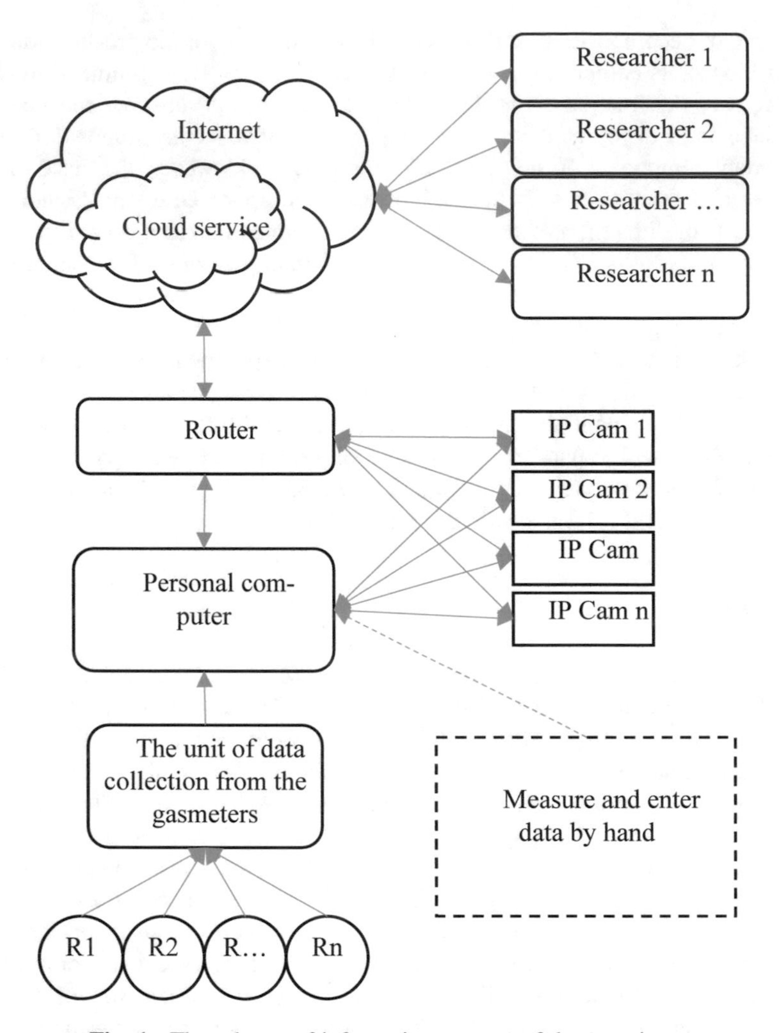

Fig. 1. The scheme of information support of the experiment

The program of cooperation provided for parallel experiments with the same type of organic substrates from Russia and Finland (without their transboundary movement) using similar equipment and techniques [20, 21] with the implementation of remote testing in real time. This allows performing a comprehensive analysis of substrate biodegradation processes, discussion and adjustment of experimental parameters in order to obtain reliable results.

The task of SPbPU was to master the methods used by Finnish colleagues to conduct laboratory experiments and to compare the results with foreign analogues. For experiments with samples of agricultural waste was used thermostat TS 606/2-i that allows you to set the desired temperature in the range from 10 to 40 °C. The internal volume of the thermostat allows the placement of 12 microbioreactors (transparent glass containers of 1 L each). Usually 6 bioreactors or less are used in the experiments.

In the upper part of the bioreactors there are three threaded flanges, on which the covers are screwed, having rubber sealing gaskets. One of the flanges has a large diameter, through which the substrate is loaded; the second-a small diameter, it is used to remove the resulting gas; the third flange is used for the selection of the substrate for the purpose of current analysis and the introduction of reagents, providing a stable flow of the fermentation process. Bioreactors through plastic hoses are connected to the gas meters «Ritter MilliGascounter», after passing through which the gas is discharged into gas bags of 3 or more liters each. The substrate in bioreactors is being mixed periodically using a magnetic stirrer to prevent the formation of a surface "crust". To do this, a magnetic "pellet" is placed inside each bioreactor (Fig. 2).

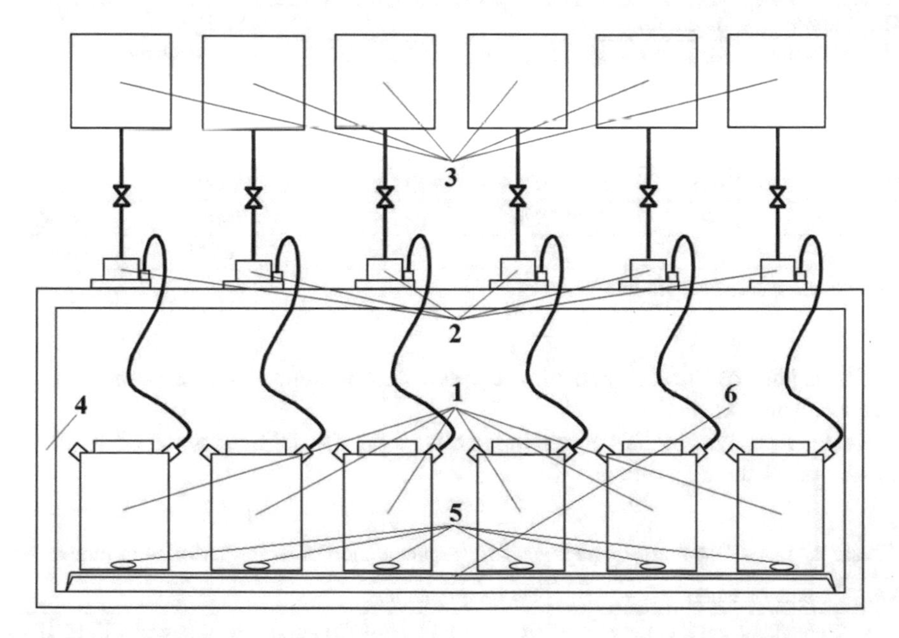

Fig. 2. The installation diagram: 1 – bioreactors, 2 – gas meters, 3 – gas bags, 4 – thermostat; 5 –magnetic «pellets»; 6 – pallet with magnetic stirrers.

Gas meter readings are transmitted to a computer connected to the Internet. In addition, the laboratory has web cameras that provide the ability to monitor the ongoing research. This allows you to obtain the necessary digital and visual information remotely, transmit it, quickly monitor the progress of research and analyze the results together with partners. The plan of the joint experiment provided for two experiments to assess the biogas potential of the inoculum. The fresh cow manure received on one of suburban farms of St. Petersburg on which the maintenance of animals meets sanitary and hygienic requirements was used as inoculum. It is known that the processes of waste biodegradation are influenced by many factors [22–24], including water quality. The task was to assess the possible impact of water samples used to prepare the necessary consistency of the inoculum. In the first experiment, distilled water prepared on a distiller was used. In the second experiment – tap water filtered by household filter. For each experiment, the parallel use of at least two bioreactors is provided, which allows to consider the statistical error of the results. Two bioreactors were used, in which inoculum was loaded with the addition of an equal amount of water of the corresponding quality. The weight, humidity of the inoculum, as well as the content of organic carbon in it were measured. Standard equipment and approved methods were used [25] considering safety requirements [26]. The pH of the liquid phase was measured when bioreactors were loaded. Table 1 shows the characteristics of the substrate before loading.

Table 1. Indicators of the loaded substrate

The number of a bioreactor	Inoculum			Weight of added water, g	pH of the water
	Mass of fresh inoculum, g	Dry matter content, g	Organic carbon content, g		
1a	140	29.3	25.3	560*	5.95
1b	140	29.3	25.3	560*	5.95
2a	140	29.3	25.3	560**	6.40
2b	140	29.3	25.3	560**	6.40

* – distilled water; ** – filtered water.

The acidity of distilled and filtered water was brought to the same neutral value with soda solution.

Table 2 presents a list of equipment used in the laboratory at each stage of preparation of the experiment.

Table 2. List of laboratory equipment for the preparation of the experiment in bioreactors

No.	Name of works	Used equipment
1	Weighing the inoculum and the test portions	Scales AND EK-1200i, and GR-200, and EK-12Ki
2	Preparation of the required volume of water	Test tubes, volumetric flasks, cuvettes, Distiller GFL 2004 CE, Drinking water filter "Aquaphor Crysta"
3	Sample preparation	Mill Retsch GM 200
4	Moisture measurement of inoculum	Humidity analyzer MB 35
5	Determination of organic carbon content	Furnace: RAP-18M and SNOL 58/350
6	Determination of pH	pH-meter Multi 340i SET 2

To create anaerobic conditions in bioreactors, they were purged with nitrogen 99.999% vol., after which the sealing was performed. Bioreactors were placed in a thermostat, after which they were connected to the "gas meter – gas bag" network (Fig. 3). During the experiment, a constant temperature of 40 °C was maintained in the thermostat.

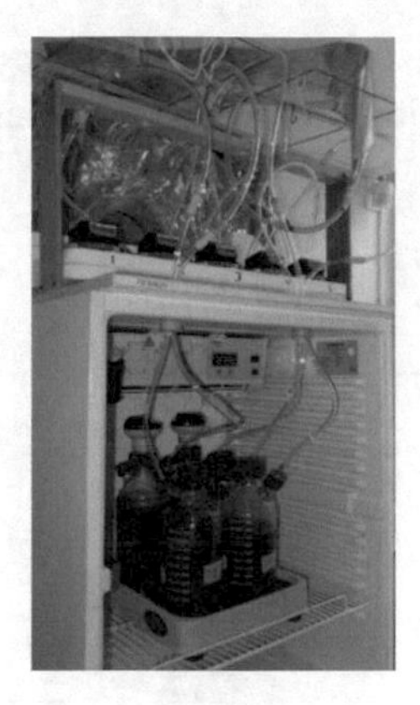

Fig. 3. The view of the experimental setup

As an example, in Fig. 4 a part of log file with the results of measurements from RIGAMO software is displayed, which were available to all researchers during the experiment via Dropbox «cloud» storage.

```
User                          User
Filepath                      C:\Users\MGC\Dropbox\MGC_SHARE\RIGAMO\2\User_Port 02_MGC-1 (3,2 ml)_ _152731.txt
Start Date                    06.
Start Time                    15:27:31
Gas Meter                     MGC-1 (3,2 ml)
Pulse Generator               V6.0        [MGC]
Port                          Port 2
Smoothing Factor              5
Sampling Time Interval [min]        0,016667
Comment
End of Measurement            Manual
Serial Number - Device Name         0.53D.1D8 - No Name
Date           Time           Runtime [min]   Volume [ltr]   Flow Rate [ltr/h]   Pulse [-]
06.            15:27:31       -0.0000         0.0000         NaN                 0
06.            15:52:09       24.6445         0.0032         NaN                 1
06.            15:52:10       24.6445         0.0032         NaN                 1
06.            16:46:39       79.1303         0.0064         0.002               2
```

Fig. 4. The structure of a text file from RIGAMO software with results of measurements.

As biogas accumulated in the gas bag, its composition was periodically determined using the GA2000 gas analyzer. Were determined: the content of methane, carbon dioxide, oxygen, hydrogen sulfide, carbon monoxide, etc.

The experiment lasted more than 40 days and was completed due to a sharp decrease in biogas emissions and the difficulty of analyzing its composition at low volumes.

3 Results and Discussion

The graphs of total biogas emission from bioreactors are shown in Fig. 5. The analysis of the obtained graphs shows that in the first week of the experiment there is a low emission of biogas. This may be caused by the presence of antibiotics in manure that inhibit biodegradation processes (levomycetin, tetracycline, streptomycin, penicillin, grisin, bacitracin, and other antibiotics are used in Russia to prevent the incidence of dairy cattle [27]).

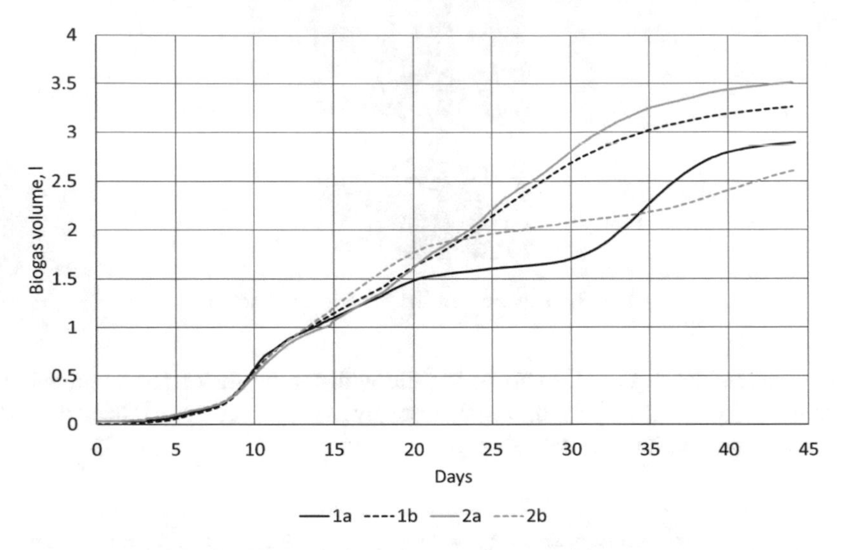

Fig. 5. The emission of biogas from the bioreactors.

On the 15th day of the experiment, the average emission of biogas from bioreactors No. 1a and No. 1b (with distilled water) made up 1.12 l. Accordingly, the average emission of biogas from bioreactors No. 2a and No. 2b (with filtered water) amounted to 1.13 l. On the 15th day of the experiment, the CH4 content in the biogas amounted to 33–36%. CO2 – 50–55%., H2S – 40–60 ppm. On the 30th day of the experiment, the average emission of biogas from bioreactors № 1a and № 1b was 2.21 l, and from bioreactors № 2a and № 2b – 2.45 l. On the 40th day of the experiment, the average emission of biogas from bioreactors No. 1a and No. 1b amounted to 2.99 l, and from bioreactor No. 2a and No. 2b – 2.93 l. In terms of fresh weight, the specific biogas yield amounted to bioreactor No. 1a and No. 1b on the 15th day of 0.008 l/g (8 m^3/t), 30-day 0.016 l/g (16 m^3/t), on the 40th day 0.021 l/g (21 m^3/t). Accordingly, for bioreactors № 2a and № 2b – on the 15th day 0.008 l/g (8 m^3/t), on the 30th day 0.0175 l/g (17.5 m^3/t), on the 40th day 0.021 l/g (21 m^3/t). It was found that the water quality does not significantly affect the biogas yield. To obtain comparable data, it is recommended to use distilled water for research in Russia and Finland, as the water parameters for purification using different filters may differ. Table 3 shows the values of biogas yield in terms of dry mass and organic carbon for the experiment with distilled water.

Table 3. The biogas potential of the inoculum investigated samples

Days	The biogas yield		The biogas yield	
	l/g dry matter	m^3/t dry matter	l/g organic carbon	m^3/t organic carbon
15	0.038	38	0.044	44
30	0.075	75	0.087	87
40	0.102	102	0.118	118

The obtained experimental results on the biogas potential of the inoculum are comparable with the data obtained in [28].

4 Conclusions

1. An experimental plant for assessing the biogas potential of organic waste, equipped with modern equipment, allowing unifying the laboratory experiment, was created.
2. With the use of modern information and communication technologies and technical means, a system of remote control of process parameters has been created and tested, allowing researchers from different regions to monitor the progress of the experiment in real time.
3. Methods of substrate preparation and assessment of its characteristics, organization of waste decomposition studies using microbioreactors were developed and mastered.
4. Test experiments were carried out to assess the biogas potential of the inoculum considering the influence of water samples used to prepare the necessary consistency. The data on specific and total biogas emission at the calculated time points are obtained, which allows justifying modes and parameters of biogas plants.
5. It is necessary to make research on the effect of contained in manure antibiotics on methanogenesis processes, as well as research on managing their degradation in bio-gas plants.

References

1. Raboni, M., Urbini, G.: Production and use of biogas in Europe: a survey of current status and perspectives. Rev. Ambient. Água **9**(2), 191–202 (2014)
2. Biogas – An important renewable energy source. WBA fact sheet. https://worldbioenergy.org/uploads/Factsheet%20-%20Biogas.pdf. Accessed 27 Sept 2018
3. Rutz, D., Mergner, R., Janssen, R.: Sustainable Heat Use of Biogas Plants. A Handbook. http://ec.europa.eu/energy/intelligent/projects/sites/ieeprojects/files/projects/documents/biogasheat_sustainable_heat_use_of_biogas_plants_en.pdf. Accessed 27 Sept 2018
4. EBA Biogas Report 2014 is published! European Biogas Association (EBA). http://european-biogas.eu/2014/12/16/4331/. Accessed 27 Sept 2018
5. Shi, F.: Reactor and Process Design in Sustainable Energy Technology. Elsevier, Oxford (2014)

6. Schüch, A., et al.: Stand und Perspektiven der Abfall- und Reststoffvergärung in Deutschland. Biogas J. **2**, 34–38 (2014)
7. Neumann, H.: Jede zweite Biogasanlage hat Defizite! Top agrar **2007**(10), 103–105 (2007)
8. Abbasi, T., Tauseef, S., Abbasi, S.: Biogas Energy. Springer, New York (2012)
9. Nielfa, A.: Theoretical methane production generated by the co-digestion of organic fraction municipal solid waste and biological sludge. Biotechnol. Rep. **5**, 14–21 (2015)
10. Fischer, T., Krieg, A.: Warum bauen wir eigentlich so schlechte Biogasanlagen? Jahrestagung des Fachverband Biogas **11**, 29–31 (2002)
11. Herout, M., et al.: Biogas composition depending on the type of plant biomass used. Res. Agr. Eng. **57**, 137–143 (2011)
12. Heuwinkel, H., et al.: Die Genauigkeit der Messung des Gasertragspotentials von Substraten mit der Batchmethode. Schriftenreihe der Bayerischen Landesanstalt für Landwirtschaft **15**, 95–103 (2009)
13. Amon, T., et al.: Biogaserträge von Energiepflanzen und Wirtschaftsdüngern – Laborversuchsergebnisse In: KTBL: Die Landwirtschaft als Energieerzeuger, 30 bis 31. März 2004, Osnabrück, Darmstadt, pp. 46–61 (2004)
14. Gebrezgabhera, S., et al.: Economic analysis of anaerobic digestion—a case of Green power biogas plant in The Netherlands. NJAS-Wagening. J. Life Sci. **57**(2), 109–115 (2010)
15. Wulf, S., Döhler, H.: Messung von Biogaserträgen Methoden und Übertragbarkeit auf Praxisanlagen. Optimierung des Futterwertes von Mais und Maisprodukten Veröffentlichungsreihe "Landbauforschung" Sonderheft **331**, 75–85 (2009)
16. Vagonyte, E.: Biogas & Biomethane in Europe. https://ec.europa.eu/energy/intelligent/projects/sites/ieeprojects/files/projects/documents/agriforenergy_2_international_biogas_and_methane_report_en.pdf. Accessed 27 Sept 2018
17. Panckhava, E.: Bioenergy. The World and Russia. Biogas: Theory and Practice: Monograph. Publishing House «Rusajns», Moscow (2014)
18. Teemu, A.: Traffic Biogas Utilization in Finland. Nordic Biogas Conference, Reykjavik, 27–29 August 2014. http://www.sorpa.is/files/nbc/slides/teemu_aittamaa.pdf. Accessed 27 Sept 2018
19. Arefiev, N., Badenko, V., Maslikova, A.: Assessment of biogas potential of livestock waste of the Leningrad region. St. Petersburg State Polytech. Univ. J. **3**(202), 107–113 (2014)
20. Janhunen, M.: Manual for laboratory work basics of biogas process and biogas batch test environmental technology. Savonia University of Applied Sciences, Savonia (2014)
21. Seadi, T., Rutz, D., et al.: Biogas Handbook (2008)
22. Gerardi, M.H.: The Microbiology of Anaerobic Digesters. Wiley, USA (2003)
23. Budiyono, B., Widiasa, I.N., Johari, S., Sunarso, S.: Increasing biogas production rate from cattle manure using rumen fluid as inoculums. Int. J. Sci. Eng. **6**(1), 31–38 (2014)
24. Leitfaden Biogas Von der Gewinnung zur Nutzung. https://mediathek.fnr.de/media/downloadable/files/samples/l/e/leitfadenbiogas2014_web.pdf. Accessed 27 Sept 2018
25. Messmethoden - sammlung Biogas. https://www.energetische-biomassenutzung.de/fileadmin/user_upload/Downloads/Ver%C3%B6ffentlichungen/07_Messmethodensamm_Biogas_web.pdf. Accessed 27 Sept 2018
26. Russian federal environmental regulations 12.13.1-03. Safety at work in analytical laboratories (General provisions). Guidelines
27. Antibiotics and feed. Modern farmer, vol. 4, pp. 48–50 (2013)
28. Liebetrau, J.: Scientific and technical principles of anaerobic digestion technology, s.l., German Biomass Research Center (2010)

Improved Hydrophysical Functions of the Soil and Their Comparison with Analogues by the Williams-Kloot Test

Vitaly Terleev[1], Wilfried Mirschel[2], Aleksandr Nikonorov[1](✉), Roman Ginevsky[1], Viktor Lazarev[1], Viktoriia Pavlova[3], Alex Topaj[4], Vladimir Pashtetsky[5], Ielizaveta Dunaieva[5], Valentina Popovych[5], Aleksandr Melnichuk[6], and Kasim Layshev[7]

[1] Peter the Great St. Petersburg Polytechnic University, Polytechnicheskaya str., 29, 195251 St. Petersburg, Russia
coolhabit@yandex.ru
[2] Leibniz-Centre for Agricultural Landscape Research, Eberswalder Straße, 84, 15374 Müncheberg, Germany
[3] St. Petersburg State Agrarian University, Peterburgskoye shosse, 2, 196601 St.-Petersburg-Pushkin, Russia
[4] Agrophysical Research Institute, Grazhdanskii pr., 14, 195220 St. Petersburg, Russia
[5] Federal State Budget Scientific Institution «Research Institute of Agriculture of Crimea», Kievskaya, 150, 295453 Simferopol, Crimea, Russia
[6] V.I. Vernadsky Crimean Federal University, Vernadskogo pr., 4, 295007 Simferopol, Crimea, Russia
[7] Federal State Budgetary Institution North-West Centre of Interdisciplinary Researches of Problems of Food Maintenance, Podbelskogo shosse, 7, 196608 St. Petersburg, Russia

Abstract. A system of improved hydrophysical functions describing the water-retention capacity and hydraulic conductivity of the soil is presented. The mathematical description of the functions is based on the physical ideas about the structure of the pore space of the soil and the capillarity phenomenon. For both functions, the parameters are common. The advantages of the presented functions are described in comparison with the most well-known world analogues, which are used in the Mualem-Van Genuchten method. An example of estimation of the ratio of the hydraulic conductivity of a soil to the filtration coefficient of moisture according to the data on the water-retention capacity of a soil of sandy particle size distribution is given. The presented system of hydrophysical functions of the soil can be used to justify the development and resource-saving technologies in irrigation and drainage construction and irrigation agriculture. Practical application of applied models based on these functions will significantly reduce the energy intensity of the operation of irrigation and drainage systems and the production of agricultural products.

Keywords: Soil · Water-retention capacity · Hydraulic conductivity · Improved hydrophysical functions · Mualem-van Genuchten method

V. Murgul and M. Pasetti (Eds.): EMMFT-2018, AISC 983, pp. 449–461, 2019.
https://doi.org/10.1007/978-3-030-19868-8_45

1 Introduction

The prerequisite for the research was one of the problems of soil hydrophysics, the solution of which, as many specialists from different countries think, has already been received. The problem is in the description of the water-retention capacity and the hydraulic conductivity of the soil using a system of functions that have a common set of parameters. The idea of the hydraulic conductivity of the soil to the filtration coefficient of moisture (relative hydraulic conductivity of the soil) ratio calculation put forward by Mualem [1] turned out to be very productive. Based on this idea, Van Genuchten [2] proposed a functional description of the relative hydraulic conductivity of the soil. This description was obtained by substituting the function of water-retention capacity of the soil proposed by Van Genuchten into the Mualem formula [1]. This method of calculating the function of the relative hydraulic conductivity of the soil was called the Mualem - Van Genuchten method and is now internationally recognized by experts in the relevant field of natural science. In the absolute majority of models of soil moisture and solute dynamics, the Van Genuchten formulas are used. It is an obvious fact that at present in the world community of specialists in the field of soil hydrophysics there is a very stable opinion (if not a paradigm) regarding these formulas, the correctness of which is not in doubt. The number of citations of the basic article by Van Genuchten [2], where the system of functions describing the water-retention capacity and relative hydraulic conductivity of a soil with a single set of parameters was first introduced, is indisputable evidence of a fact. The same fact characterizes the level of results presented in this article and encouraging (in a certain sense) to reevaluate the value of the Van Genuchten formulas.

2 Materials and Methods

A stable opinion about the mathematical correctness and universal nature of the Van Genuchten formulas was formed much later than the date of publication of the basic article [2]. In this article, Van Genuchten explicitly prefers the function of the water-retention capacity of the soil proposed earlier, for example, the function of Haverkamp et al. [3]. Moreover, this is understandable: the parameters of the function of Haverkamp and coauthors allow physical interpretation. Along the way, it should be noted that in the world literature there is an interpretation that is not fully correct. However, the parameters of this function are, in principle, interpretable within the framework of physical representations.

Mualem formula is used to calculate the ratio of the function of hydraulic conductivity of the soil k (cm c^{-1}) to the moisture filtration coefficient k_S (cm c^{-1}) [1]:

$$\frac{k}{k_s} = \sqrt{S_e}\left(\int_0^{S_e} \frac{dx}{\psi(x)} \Big/ \int_0^1 \frac{dx}{\psi(x)}\right)^2 \tag{1}$$

where $S_e = (\theta - \theta_r)/(\theta_s - \theta_r)$ – effective moisture saturation of soil; θ (cm^3 cm^{-3}) – volumetric water content in the soil; θ_S (cm^3 cm^{-3}) – volumetric water content at full water saturation of the soil; θ_r (cm^3 cm^{-3}) – volumetric water content corresponding to the minimum specific volume of moisture as a fluid in the soil; ψ (cm H_2O) – capillary moisture pressure.

Van Genuchten mathematically transformed the function of the precursors [3–5] to a form in which the use of the transformed function of the water-retention capacity of the soil in the Mualem formula allows to overcome the above difficulties in calculating the relative hydraulic conductivity of the soil.

Van Genuchten proposed to use dependence $\theta(\psi)$ for the formula (1) to describe the water-retention capacity and relative hydraulic conductivity of the soil by means of hydraulics functions with common parameters, in the form of:

$$S_e = \begin{bmatrix} (1+(-\alpha\psi)^n)^{-m}, \ \psi<0; \\ 1, \psi \geq 0, \end{bmatrix} \tag{2}$$

where α (cm H_2O^{-1}), n and m – empirical parameters [2].

When $m = 1$ the hydraulics function (2) is reduced to the models of predecessors [3–5], therefore the use of formula (1) runs into a computational difficulty. However when $m = 1 - 1/n \cdot (n > 1)$ function (2) admits an analytical calculation:

$$\frac{k}{k_s} = \begin{bmatrix} \sqrt{S_e}\left(1-(1-S_e^{1/m})^m\right)^2 = (1+(-\alpha\psi)^n)^{-m/2}\left(1-\left(1-(1+(-\alpha\psi)^n)^{-1}\right)^m\right)^2, \ \psi<0; \\ 1, \psi \geq 0. \end{bmatrix} \tag{3}$$

Authors used the following notation: WRC-VG for the function of the soil water-retention capacity, proposed by Van Genuchten, which is described by relation (2); RHC-MVG for the function of the relative hydraulic conductivity of the soil in the Mualem-van Genuchten method, which is described by relation (3). Assigning the Number 1 to the system WRC-VG and RHC-MVG.

By transforming the integrand expression in the Mualem formula, obviously, for the purpose of the less complicated calculation of integrals by Van Genuchten, the function of the relative hydraulic conductivity of the soil was obtained. Without any doubt, the advantage of this function is that it has the same parameters as the soil water-retention capacity function, which was proposed by Van Genuchten. However, a very significant disadvantage of both functions of Van Genuchten is the lack of physical meaning of the parameters, as well as the fact that these parameters are not independent. We give explanations. First, the exponential parameters and are related by $m = 1 - 1/n$. Secondly, the multiplicative parameter α is such that if the capillary pressure (capillary-sorption potential) of moisture ψ reaches a value inversely proportional to the parameter α, the effective soil moisture saturation is 1/2 of m degree. However, this circumstance is not discussed in the world literature.

There is a point of view that the function of soil water-retention capacity, proposed by Van Genuchten, should not be considered as a transformation of the function of the precursors (for example, Haverkamp et al. [3]). The authors of this article are sure of the opposite, since Van Genuchten directly [2] initially considered the function of water-retention capacity proposed by the predecessors [3–5] in order to use the relative hydraulic conductivity of the soil according to the Mualem formula. Moreover, Van Genuchten noted that for large values of the exponential parameter n its function almost coincides, for example, with the Haverkamp and co-authors [3]. The parameters of the van Genuchten functions are empirical: they can be estimated solely on the basis of a point approximation of the direct measurement data. Thus, if the parameters are estimated from data on the soil water-retention capacity at a given range of capillary pressure (capillary-sorption potential) moisture values, then an estimate of the values of the relative hydraulic conductivity function of the soil is possible only on the same range. Obviously, this noticeably narrows the applicability of the Mualem-Van Genuchten method.

The above-noted deficiencies of the functions of Van Genuchten are very significant. However, an even more significant problem is the constraint imposed on the exponential parameter $n > 1$. The question arises whether there are soils in nature for which this restriction is unacceptable? Van Genuchten himself found the answer in the same basic article [2]. Moreover, this answer is affirmative: for the Beit Netofa clay soil, the error in the relative hydraulic conductivity function estimation was unacceptably high. In fact, the problem of restriction $n > 1$ takes place for soils with a relatively high dispersion of the distribution of particles (or pores) in size, i.e. for soils with a very heterogeneous particle size distribution (texture). For such soils parameter n has minimal values. Recall that for large the function of soil water-retention capacity, proposed by Van Genuchten, practically coincides, for example, with the function of Haverkamp et al. [3]. For small values of the n Van Genuchten function differs significantly from the function of Haverkamp and co-authors. Thus, the Mualem - Van Genuchten method allows obtaining a rather low error in estimating the relative hydraulic conductivity function in cases where the soil water-retention capacity function proposed by Van Genuchten differs slightly from the Haverkamp et al. function at large n. It is obvious that the transformation of the integrand in the Mualem formula undertaken by Van Genuchten to overcome the difficulties of a computational nature can be regarded as a technique that is conditionally acceptable only for the case of large n. In other words, on the one hand, the integrand used by Van Genuchten in the Mualem formula makes it quite easy to calculate the integrals; on the other hand, the method has a rather low error only in the range of large values of n, in which the function of soil water-retention capacity, used by Van Genuchten in the formula of Mualem, differs from the function of Haverkamp and co-authors to a minimum degree.

As noted above, the "problematic" soils, for which the Mualem-Van Genuchten method turned out to be very inaccurate, include the Beit Netofa clay soil. Van Genuchten explained this negative result by a small number of points in the "dry" part of the curve of the water-retention capacity of the soil. This looks strange, because it is for this soil that the "dry" part of the soil water-retention capacity curve has the greatest number of points compared to all other soils with a lighter grain size distribution. No less strange is the fact that the parameter of residual moisture content of clay soil is zero. Recall that the Beit Netofa clay soil has a heavy particle size distribution. This circumstance is also noted by Van Genuchten himself. However, when conducting a computational experiment with the Van Genuchten function, it is easy to see that the residual moisture parameter takes on a negative value. If this parameter is equated to zero and fixed in an optimizing algorithm, then during the point approximation procedure, the values of the parameters are obtained, which are published in the final table of the article by Van Genuchten [2]. The noted problems are studied in detail and suggested a way to solve them in the article [6].

The question of estimating the relative hydraulic conductivity of the Beit Netofa clay soil has been repeatedly raised in the world literature. However, none of the attempts to make any adjustments and amendments to the decision obtained by Van Genuchten did not lead to an acceptable result. The reason is that the method proposed by Van Genuchten for using the Mualem formula is methodologically very dubious. The basis of these doubts is that, when the function of the soil water-retention capacity is substituted into the Mualem formula, this function has to be differentiated. It should be noted that the function of Van Genuchten, in fact, is a certain approximation (albeit unknown) of the water-retention capacity of the soil. At the same time, the operation of differentiation is by definition an unstable operation with respect to approximations. Thus, the technique used by Van Genuchten could not be considered perfect in terms of methodology.

Kosugi [7, 8] turned out to be the first to publish a more correct approach, not using the Mualem formula of the water-retention capacity of the soil (which is a function of the integral moisture capacity of the soil), but the differential water-retention capacity for which the function of the soil's water-retention capacity is primitive. Thus, the calculation of both hydrophysical functions of the soil (water-retention capacity and relative hydraulic conductivity) was based on the same function of the differential water-retention capacity of the soil (using the Mualem formula). The mathematical description of the differential water-retention capacity function of the soil, proposed by Kosugi, is based on the concepts of the lognormal distribution of the effective soil pore radii and the capillary properties of these pores. The Kosugi model is an important step towards the development of mathematical modeling of the hydrophysical properties of the soil. However, it must be stated that the Mualem-Kosugi method did not compete with the Mualem - Van Genuchten method due to two reasons. First, the evaluation of the relative hydraulic conductivity function of the Beit Netofa clay soil was even less accurate than the Mualem - Van Genuhten estimate. Secondly, to describe the hydrophysical properties of the soil in the Mualem-Kosugi method, a special function was used (an additional error function). A physical and statistical interpretation has been proposed for the parameters of the Kosugi functions: this was an advantage of these functions. However, the hydrophysical functions of the soil themselves have a mathematically redundantly complex description in comparison with the very low accuracy of measuring the hydrophysical properties of the soil: this can be considered (conditionally) a lack of these functions. In the paper [6] Kosugi's ideas are developed and a system of soil hydraulics functions are described. The functions forming this system are represented in the following form:

$$S_e = \begin{cases} \frac{1}{2} erfc\left(\frac{n\sqrt{\pi}}{4}\ln(-\alpha(\psi - \psi_e))\right), \psi < \psi_e; \\ 1, \psi \geq \psi_e; \end{cases} \quad (4)$$

$$\frac{k}{k_s} = \begin{cases} \frac{1}{4\sqrt{2}}\sqrt{erfc\left(\frac{n\sqrt{\pi}}{4}\ln(-\alpha(\psi - \psi_e))\right)}\left(erfc\left(\frac{n\sqrt{\pi}}{4}\ln(-\alpha(\psi - \psi_e)) + \frac{2}{n\sqrt{\pi}}\right)\right)^2, \psi < \psi_e; \\ 1, \psi \geq \psi_e; \end{cases} \quad (5)$$

where: $erfc(x) = 1 - \frac{2}{\sqrt{\pi}}\int_0^x \exp(-t^2)dt$ complementary error function; $n > 0, \alpha = -1/(\psi_0 - \psi_e)$ (cm H_2O^{-1}), ψ_e (cm H_2O), ψ_0 (cm H_2O) < ψ_e – interpreted parameters (taking into account hysteresis for desorption branches $\psi_e \leq 0$; for sorption branches $\psi_e \geq 0$).

Authors used the following notation: WRC-CT for an improved function of the soil water-retention capacity, which is described by relation (4); RHC-MKT for an improved function of the relative hydraulic conductivity of the soil, which is described by relation (5). The system functions WRC-CT and RHC-MKT was assigned with the Number 2.

The hydrophysical functions of the soil proposed by Kosugi are a particular case of a more general description, represented by relations (4) and (5). In a more general description, an additional additive parameter ψ_e with the dimension of capillary pressure (capillary-sorption potential) of moisture is reasonably present. An interpretation was proposed for this parameter in [6]. This parameter for the curve of water-retention capacity of the soil, obtained by the method of Richards, is the pressure of the air inlet (bubbling). Relations (4) and (5) are respectively improved hydrophysical functions of the soil, previously obtained by Kosugi. In the paper [6] continuous approximations of the relations (4) and (5) in the class of elementary functions are proposed using the simplified Winitzky formula [9]. Here the relations (4) and (5) are respectively transformed to the form:

$$S_e = \begin{cases} (1 + (-\alpha(\psi - \psi_e))^n)^{-1}, \psi < \psi_e; \\ 1, \psi \geq \psi_e; \end{cases} \quad (6)$$

$$\frac{k}{k_s} = \begin{cases} (1 + (-\alpha(\psi - \psi_e))^n)^{-1/2}\left(1 + (-\alpha(\psi - \psi_e))^n \exp\left(\frac{8}{n\pi}\right)\right)^{-2}, \psi < \psi_e; \\ 1, \psi \geq \psi_e. \end{cases} \quad (7)$$

Authors used the following notation: WRC-HT for the improved function of soil water-retention capacity approximation, which is described by relation (6); RHC-MT for approximation of the improved function of the relative hydraulic conductivity of the soil, which is described by relation (7). The system functions of WRC-HT and RHC-MT was assigned with the Number 3.

It is very important to note that the approximation of the function of the water-retention capacity of the soil was brought to the form of the function of Haverkamp and co-authors [3] with an additional additive parameter. In the case of equality to zero of this additional parameter, the obtained approximation of the function of the soil water-retention capacity fully coincides with the function of Haverkamp et al. [3]. Thus, in [6], the problem of van Genuchten was solved in its initial formulation. This is a very important achievement in theoretical soil physics.

With the usage of relations (4) and (5), as well as (6) and (7), the error in estimation of the function of the relative hydraulic conductivity of the Beit Netofa clay soil was significantly (significantly, multiple) lower than when using relations (2) and (3) i.e. using the original Mualem-Van Genuhten method [6]. This achievement for the abovementioned clay soil is currently unsurpassed. At the same time, it is necessary to answer the question of the applicability of the improved hydrophysical functions of the soil (and their approximations), proposed in [6], to soils of a lighter grain size distribution. The objective of this study is to conduct a comparative analysis of the improved hydrophysical functions of the soil with the most well-known world analogues used in the Mualem-Van Genuchten method, applying the Williams-Kloot test on the example of sandy soils.

3 Results and Discussions

Widely known (highly reliable) data from the Mualem catalog [10] was used to solve this problem. Authors noted in particular that Mualem's data are benchmark in a certain sense and was used so that experts could verify the results, and should be able to compare the results obtained here with the latest world achievements in this field of soil hydrophysics. Earlier in [11], these four soils were already investigated, and the errors of the point approximation of experimental data on water-retention capacity and the error in the relative hydraulic conductivity of the studied soils estimation for three systems of hydrophysical functions were also calculated. Along with the data previously obtained in [11], this article presents the results of a comparative analysis of three systems of hydrophysical functions using the Williams-Kloot test [12] in order to identify significant differences in the compared systems of hydrophysical functions using sandy soils as an example.

Table 1 shows the parameter values for the three compared systems of soil hydrophysical functions.

Table 1. The parameters of the hydrophysical functions of the soil of the three systems

Soil	System #	Parameters				
		θ_s, $cm^3\ cm^{-3}$	θ_r, $cm^3\ cm^{-3}$	ψ_e, cm H_2O	α, cm H_2O^{-1}	n
4133 Fine sand (G.E.#13)	1	0.360	0.068	–	0.0104	6.897
	2	0.360	0.072	0	0.0101	6.257
	3	0.360	0.071	–3.56	0.0104	6.359
4134 Volcanic sand	1	0.365	0.073	–	0.0230	6.758
	2	0.365	0.075	–0.62	0.0226	6.065
	3	0.365	0.073	–8.38	0.0274	5.151
4135 Gravelly sand G.E. 9	1	0.321	0.080	–	0.0144	2.961
	2	0.321	0.084	0	0.0117	2.481
	3	0.322	0.082	0	0.0117	2.559
4136 Fine sand G.E. 2	1	0.326	0.062	–	0.0071	4.060
	2	0.325	0.066	0	0.0064	3.571
	3	0.327	0.064	0	0.0064	3.700

The parameters shown in Table 1 were used to estimate the relative hydraulic conductivity of four sandy soils using three compared systems of hydrophysical soil functions.

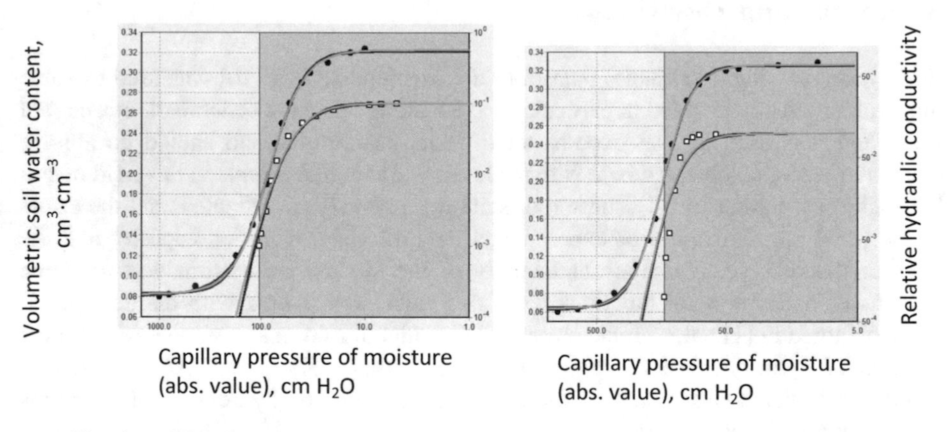

Fig. 1. «4135 Gravelly sand G.E. 9».

Fig. 2. «4135 Gravelly sand G.E. 9».

In Figs. 1, 2, 3 and 4, solid curves showed the results of a point approximation of data on the water-retention capacity of the four studied sandy soils, as well as the results of estimating the relative hydraulic conductivity using the identified parameters from the data on the water-retention capacity of the soil. Red curves were used for system #1; blue curves are used for system #2, green curves are used for system #3. Dots depict experimental data. Table 2 shows the standard deviations of the results of

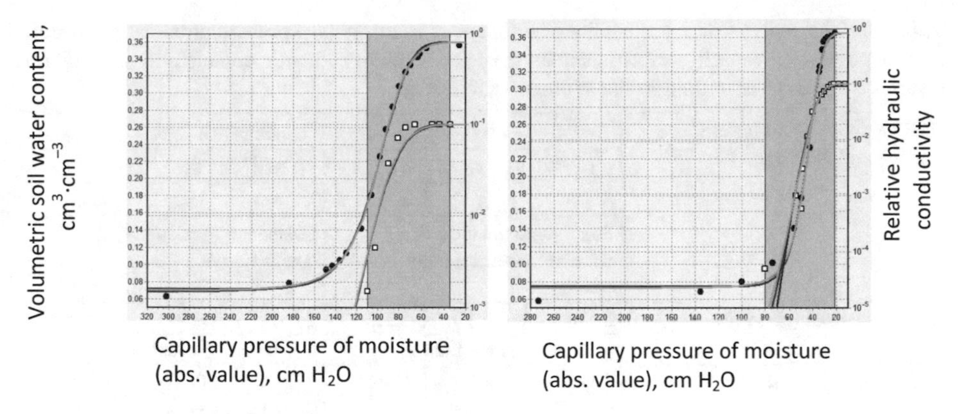

Fig. 3. «4133 Fine sand (G.E.#13)».

Fig. 4. «4134 Volcanic sand».

the point approximation of the measured water-retention capacity of the soil, as well as the results of estimating the relative hydraulic conductivity of the soil from the corresponding experimental data. In bold, the smallest error values are highlighted.

Table 2. Accuracy of point approximation of data on water-retention capacity and estimation of relative hydraulic conductivity of soils

Soil	RMSE − root mean square error					
	System #1		System #2		System #3	
	WRC-VG	RHC-MVG	WRC-KT	RHC-MKT	WRC-HT	RHC-MT
4133 Fine sand (G.E. #13)	**0.0038**	0.1455	0.0043	0.1435	0.0039	**0.1359**
4134 Volcanic sand	0.0090	0.0242	0.0092	**0.0233**	**0.0087**	0.0240
4135 Gravelly sand G.E. 9	**0.0038**	0.0388	0.0046	0.0474	**0.0038**	**0.0387**
4136 Fine sand G.E. 2	**0.0033**	0.1907	0.0045	0.1830	0.0034	**0.1774**

Table 3 shows the results of a comparative analysis of systems #1 and #2. Table 4 shows the results of a comparative analysis of systems #1 and #3. Table 5 shows the results of a comparative analysis of systems #2 and #3.

Table 3. The analysis of the reliability of differences in errors (i) a point approximation of data on water-retention capacity and (ii) estimation of the relative hydraulic conductivity of soils for systems #1 and #2 according to the Williams-Kloot test

The number in the Mualem catalog and the name of the soil	$y - \frac{y_1 + y_2}{2} = \lambda(y_1 - y_2)$, where y – experimental data					
	y_1: WRC-VG, y_2: WRC-KT			y_1: RHC-MVG, y_2: RHC-MKT		
	λ	$\lambda_{0.95}$	$\lambda_{0.975}$	λ	$\lambda_{0.95}$	$\lambda_{0.975}$
	Point approximation of water-retention capacity data:			Evaluation of relative hydraulic conductivity:		
4133 Fine sand (G.E.#13)	2.176	0.513	0.622	2.057	8.800	10.984
	y_1 more accurate than y_2			y_1 and y_2 are equal		
4134 Volcanic sand	1.158	2.592	3.086	−0.006	0.766	0.936
	y_1 and y_2 are equal			y_1 and y_2 equal		
4135 Gravelly sand G.E.9	1.502	1.148	1.401	1.519	1.292	1.583
	y_1 more accurate than y_2			y_1 more accurate than y_2 with confidence 0.95		
4136 Fine sand G.E.2	1.906	0.704	0.856	−2.493	6.992	8.671
	y_1 more accurate than y_2			y_1 and y_2 are equal		

Table 4. Analysis of the reliability of differences in errors (i) a point approximation of data on water-retention capacity and (ii) estimation of the relative hydraulic conductivity of soils for systems #1 and #3 according to the Williams-Kloot test

The number in the Mualem catalog and the name of the soil	$y - \frac{y_1 + y_2}{2} = \lambda(y_1 - y_2)$, where y – experimental data					
	y_1: WRC-VG, y_2: WRC-HT			y_1: RHC-MVG, y_2: RHC-MT		
	λ	$\lambda_{0.95}$	$\lambda_{0.975}$	λ	$\lambda_{0.95}$	$\lambda_{0.975}$
	Point approximation of water-retention capacity data:			Evaluation of relative hydraulic conductivity:		
4133 Fine sand (G.E.#13)	1.609	2.648	3.215	−14.99	8.371	10.49
	y_1 and y_2 are equal			y_2 more accurate than y_1		
4134 Volcanic sand	−5.716	7.130	8.698	0.120	0.925	1.130
	y_1 and y_2 are equal			y_1 and y_2 are equal		
4135 Gravelly sand G.E.9	−0.063	1.241	1.514	0.829	1.882	2.306
	y_1 and y_2 are equal			y_1 and y_2 are equal		
4136 Fine sand G.E.2	0.061	1.768	2.149	−13.47	11.45	14.20
	y_1 and y_2 are equal			y_2 more accurate than y_1 c with confidence 0.95		

Table 5. Analysis of the reliability of differences in errors (i) a point approximation of data on water-retention capacity and (ii) estimation of the relative hydraulic conductivity of soils for systems #2 and #3 according to the Williams-Kloot test

The number in the Mualem catalog and the name of the soil	$y - \frac{y_1+y_2}{2} = \lambda(y_1 - y_2)$, where y – experimental data					
	y_1: WRC-KT y_2: WRC-HT			y_1: RHC-MKT y_2: RHC-MT		
	λ	$\lambda_{0.95}$	$\lambda_{0.97}$	λ	$\lambda_{0.95}$	$\lambda_{0.97}$
	Point approximation of water-retention capacity data:			Evaluation of relative hydraulic conductivity:		
4133 Fine sand (G.E.#13)	−2.460	0.569	0.690	−6.463	6.601	8.239
	y_2 more accurate than y_1			y_1 and y_2 are equal		
4134 Volcanic sand	−2.170	2.698	3.291	1.037	2.718	3.322
	y_1 and y_2 are equal			y_1 and y_2 are equal		
4135 Gravelly sand G.E.9	−2.438	1.235	1.507	−3.655	1.193	1.462
	y_2 more accurate than y_1			y_2 more accurate than y_1		
4136 Fine sand G.E.2	−2.333	0.645	0.784	−3.579	9.371	11.621
	y_2 more accurate than y_1			y_1 and y_2 are equal		

4 Conclusion

A significant error in the use of the Mualem - Van Genuhten method for soils of heavy particle size distribution was shown in [6], but a comparison of errors using the presented three systems of hydrophysical functions on sandy soils required the use of a criterion that can reveal the statistical significance of differences between the errors. The Williams-Kloot test which was used in the article showed that the system #3 (relations (6) and (7)) is not only inferior to the Mualem-Van Genuchten system #1 (relations (2) and (3)) in relation to the error in estimation of the relative hydraulic conductivity, but for soil 4133 Fine sand (GE#13) reliably surpasses system #1. For soils 4133 Fine sand (GE#13), 4134 Volcanic sand and 4136 Fine sand GE2, system #2 (relations (4) and (5)) has the largest errors in the point approximation of data on the water-retention capacity of soils, which can be explained by the increased complexity of the optimizing algorithm for the point approximation of data using the special (non-elementary) function *erfc*. At the same time, the errors in estimating the relative hydraulic conductivity for these three soils turned out to be statistically indistinguishable with the error of system #1. For the sandy 4134 Volcanic sand soil, all three compared systems of hydrophysical functions showed statistics of indistinguishable low error. According to the results of the study, the following conclusions were made.

1. In relation to soils of light particle size distribution using the system #3, comparable or smaller in comparison with the system #1 and system #2 errors of the point approximation of the water-retention capacity and relative hydraulic conductivity estimates are achieved.
2. With a relatively low accuracy of measuring the water-retention capacity and hydraulic conductivity of the soil in currently conducted laboratory studies, the use

of special functions to describe these hydrophysical properties seems somewhat redundant. Formulas of system #3 belong to the class of elementary functions and have a more preferable in practical relation analytical description in comparison with system #2.

3. Taking into account the fact that with respect to the soil, the heavy particle size distribution of the system #2 and system #3 are more preferable in comparison with the system #1 [6], and also taking into account the first two conclusions, it can be reasonably assumed that in various hydrophysical calculations, for example when justifying projects of hydraulic structures [13], in calculating precision irrigation rates [14] or to predict the productivity of agrocenoses [15], the use of system#3 seems to be the most appropriate.

Acknowledgments. The reported study was funded by RFBR according to the research projects No 19-04-00939-a, No 19-016-00148-a.

References

1. Mualem, Y.: A new model for predicting hydraulic conductivity of unsaturated porous media. Water Resour. Res. **12**, 513–522 (1976)
2. Van Genuchten, M.Th.: A closed form equation for predicting the hydraulic conductivity of unsaturated soils. Soil Sci. Soc. Am. J. **44**, 892–989 (1980)
3. Haverkamp, R., Vauclin, M., Touma, J., Wierenga, P.J., Vachaud, G.: A comparison of numerical simulation model for one-dimensional infiltration. Soil Sci. Soc. Am. J. **41**, 285–294 (1977)
4. Brutsaert, W.: Probability laws for pore-size distribution. Soil Sci. **101**, 85–92 (1966)
5. Ahuja, L.R., Swartzendruber, D.: An improved form of soil-water diffusivity function. Soil Sci. Soc. Am. Proc. **36**, 9–14 (1972)
6. Terleev, V.V., Mirschel, W., Badenko, V.L., Guseva, I.Yu.: An improved Mualem-Van Genuchten method and its verification using data on Beit Netofa clay. Eurasian Soil Sci. **50** (4), 445–455 (2017)
7. Kosugi, K.: Three-parameter lognormal distribution model for soil water retention. Water Resour. Res. **30**, 891–901 (1994)
8. Kosugi, K.: Lognormal distribution model for unsaturated soil hydraulic properties. Water Resour. Res. **32**, 2697–2703 (1996)
9. Winitzki, S.: A handy approximation for the error function and its inverse. https://sites.google.com/site/winitzki/sergei-winitzkis-files/erf-approx.pdf?attredirects=0
10. Mualem, Y.: A catalogue of the hydraulic properties of unsaturated soils. Research Project 442. Technion, Israel Institute of Technology, Haifa, Israel (1976)
11. Terleev, V., Mirschel, W., Nikonorov, A., Ginevsky, R., Lazarev, V., Togo, I., Topaj, A., Moiseev, K., Shishov, D., Dunaieva, I.: Estimating some hydrophysical properties of soil using mathematical modeling. In: MATEC Web of Conferences, vol. 193, Article no. 02035 (2018)
12. Kobzar, A.I.: Prikladnaya matematicheskaya statistika. Dlya inzhenerov i nauchnyx rabotnikov. Izd. Fizmatlit, Moskva (2006). (rus)
13. Badenko, V., Badenko, N., Nikonorov, A., Molodtsov, D., Terleev, V., Lednova, J., Maslikov, V.: Ecological aspect of dam design for flood regulation and sustainable urban development. In: MATEC Web of Conferences, vol. 73, Article no. 03003 (2016)

14. Terleev, V.V., Topazh, A.G., Mirschel, W.: The improved estimation for the effective supply of productive moisture considering the hysteresis of soil water retention capacity. Russ. Meteorol. Hydrol. **40**(4), 278–285 (2015)
15. Medvedev, S., Topaj, A., Badenko, V., Terleev, V.: Medium-term analysis of agroecosystem sustainability under different land use practices by means of dynamic crop simulation. In: IFIP Advances in Information and Communication Technology, vol. 448, pp. 252–261 (2015)

Models of Hysteresis Water Retention Capacity and Their Comparative Analysis on the Example of Sandy Soil

Vitaly Terleev[1], Wilfried Mirschel[2], Aleksandr Nikonorov[1(✉)], Viktor Lazarev[1], Roman Ginevsky[1], Alex Topaj[3], Kirill Moiseev[3], Vladimir Pashtetsky[4], Ielizaveta Dunaieva[4], Valentina Popovych[4], Aleksandr Melnichuk[5], and Mikhail Arkhipov[3,6]

[1] Peter the Great St. Petersburg Polytechnic University, Polytechnicheskaya, 29, 195251 St. Petersburg, Russia
coolhabit@yandex.ru

[2] Leibniz-Centre for Agricultural Landscape Research, Eberswalder Straße, 84, 15374 Müncheberg, Germany

[3] Agrophysical Research Institute, Grazhdanskii pr., 14, 195220 St. Petersburg, Russia

[4] Federal State Budget Scientific Institution «Research Institute of Agriculture of Crimea», Kievskaya, 150, 295453 Simferopol, Crimea, Russia

[5] V.I. Vernadsky Crimean Federal University, Vernadskogo pr., 4, 295007 Simferopol, Crimea, Russia

[6] Federal State Budgetary Institution North-West Centre of Interdisciplinary Researches of Problems of Food Maintenance, Podbelskogo shosse, 7, 196608 St. Petersburg, Russia

Abstract. The reasons for the hysteresis of the soil water-retention capacity are indicated. An explanation of the closed loop, which is formed by the main drying and wetting branches, as well as open loops, which are formed with the participation of scanning hysteresis branches is proposed. The problem of the possible manifestation of an undesirable "pump effect" is analyzed, and a way to solve this problem is indicated. Mathematical models describing this phenomenon are presented. In these models, three functions of water-retention capacity of the soil are used: (i) Van Genuchten's function, (ii) improved Kosugi function, (iii) improved Haverkamp and co-authors function. A physical interpretation of an additional additive parameter in the improved functions of the soil water-retention capacity is proposed. The prospects for the use of hysteresis models for calculating precision irrigation rates in land reclamation agriculture are characterized. Using the Williams-Kloot criterion, a comparative analysis of the presented hysteresis models was carried out with respect to the error of the point approximation of experimental data on the main branches (parameter identification), as well as with respect to the error in estimation of the scanning branches of water-retention capacity of the soil using the example of Dune Sand.

V. Murgul and M. Pasetti (Eds.): EMMFT-2018, AISC 983, pp. 462–471, 2019.
https://doi.org/10.1007/978-3-030-19868-8_46

Keywords: Soil · Water-retention capacity · Hysteresis · Mathematical model · Main and scanning branches · Dot-approximation of experimental data · Forecasting · Williams-Kloot criterion

1 Introduction

As known, the "classical" curve of the soil water-retention capacity is the isotherm of quasi-equilibrium desorption states of moisture in the soil. This curve represents the main drying branch of the hysteresis of the water-retention capacity of the soil. In this case, there are also the main wetting branch, as well as the scanning branches of the hysteresis of the water-retention capacity of the soil. Of course, the concept of hysteresis is not new. The phenomenon of hysteresis is characteristic of many physical properties of natural objects. To date, there are ideas about the causes of this phenomenon. However, it should be recognized that these representations do not always correspond to the physical nature of the hysteresis. For example, the idea that the difference in the contact angles of the water with soil capillaries for the branches of drying and wetting of the soil water-retention capacity is one of the reasons for hysteresis is physically unreasonable. The contact angles can, indeed, be different during drying and wetting: but this can only be related to the processes of drying and wetting of the soil, but not to the branches of the water-retention capacity of the soil. The fact is that water-retention capacity is the most important hydrophysical property of the soil, which causes many thermodynamically quasi-equilibrium states of moisture in the soil. This property is formulated as a relationship between the values of the soil moisture content θ [cm^3 cm^{-3}] and the capillary pressure (capillary-sorption potential) of moisture ψ [cm H_2O]. And the process cannot relate to such states, since in this case the process is the phenomenon of establishing one or another quasi-equilibrium state. At the same time, the idea that the "effect of a bottle throat", due to the variability of the cross-sectional area of the soil pores, belongs to the causes of hysteresis, is quite rational. It should be noted that the reasons for hysteresis should include differences between the moisture pressure value during the initial drying of the initially fully saturated soil moisture ("air entry" pressure) and the moisture pressure value at the final stage of soil wetting (pressure of entrapped air from the pores and "water entry"). The "effect of a bottle throat" can be taken into account by two parameters (multiplicative and exponential) of the normal distribution of the logarithms of the effective capillary soil pore radii. The difference in the pressure values of the "air entry" and "water entry" can be taken into account using the additive parameter with the dimension of the capillary pressure (capillary-sorption potential) of moisture. One of the three values of these parameters could be used for simulation of the drying branches, for the wetting branches - another three of the values of the parameters.

The hysteresis loop formed by the main branches of the drying and wetting of the water-retention capacity of the soil must be closed according to the physical meaning: firstly, for the drying and wetting branches, the maximum value of the soil moisture content (saturation moisture) must be the same, since the porosity remains constant; secondly, for the drying and wetting branches the minimum value of the volumetric soil moisture should be the same, since the maximum hygroscopicity remains unchanged. We note here that the scanning branches are open, but they cannot cross the main hysteresis branches of the water-retention capacity of the soil.

Measurement of the scanning branches of the hysteresis of the soil water-retention capacity is a very laborious study. However, the availability of data about the scanning wetting branches is very valuable in solving practical problems of irrigation farming. This is understandable, because during irrigation, the soil goes from states with low moisture content to states with higher moisture content, and the change of states corresponds to the branches of wetting, and not to the branches of drying. However, in the practice of irrigation farming, the method of calculating the irrigation rates using the main branch of the drying is still used. Brief explanation: there is a special point on the main drying branch, which characterizes the boundary of the transition of moisture from the category of gravitational moisture to the category of capillary-suspended moisture. This point corresponds to the value of the volumetric soil moisture, which is called the field soil moisture capacity (FC). In turn, the value of FC corresponds to the capillary pressure of moisture at FC. In states with lower (negative) values of the capillary pressure of moisture as compared to capillary pressure at FC, water in the soil is retained by capillary-sorption forces. In states with higher capillary pressure values of moisture compared with capillary pressure at FC, water flows into the underlying layers of the soil sequence under the influence of gravity. It is important to note that on any scanning branch at a point that corresponds to the capillary pressure of moisture at FC, the value of the volumetric soil moisture is lower than FC. Consequently, if the irrigation rate is calculated from the difference between FC and the pre-irrigated soil moisture, then this rate is always overestimated [1]. This circumstance is known to practicing ameliorators. They know that for the irrigation rates calculation, one should use the scanning wetting branches of the water-retention capacity of the soil hysteresis. However, in this case, insurmountable obstacles arise. They lie in the fact that all the scanning wetting branches that may be needed for the irrigation rates calculation in the upcoming growing season (for the conditions of the agricultural field) are almost impossible to measure: this, in turn, is due to the unpredictable nature of precipitation. For this reason, the only way out of this predicament is to apply a physically based mathematical model of hysteresis to the water-retention capacity of the soil. This article is devoted to the representation of such a model and its verification using well-known data from the Mualem catalog [2].

2 Materials and Methods

The article [3] developed the ideas of Kosugi, and the function of the soil water-retention capacity was proposed in the form:

$$S_e = \left[\begin{array}{l} \frac{1}{2} erfc\left(\frac{n\sqrt{\pi}}{4}\ln(-\alpha(\psi - \psi_e))\right), \ \psi < \psi_e; \\ 1, \ \psi \geq \psi_e; \end{array} \right. \tag{1}$$

where: $S_e = (\theta - \theta_r)/(\theta_s - \theta_r)$ – effective moisture saturation of soil; θ [cm^3 cm^{-3}] – volumetric water content in the soil; θ_s [cm^3 cm^{-3}] – volumetric water content at full water saturation of the soil; θ_r [cm^3 cm^{-3}] – volumetric water content corresponding to the minimum specific volume of moisture as a fluid in the soil; ψ [cm H_2O] – capillary moisture pressure; $erfc(x) = 1 - \frac{2}{\sqrt{\pi}}\int_0^x \exp(-t^2)dt$ – complementary error function; $n > 0$, $\alpha = -1/(\psi_0 - \psi_e)$ [cm H_2O^{-1}], ψ_e [cm H_2O], ψ_o [cm H_2O] < ψ_e – interpreted parameters (taking into account hysteresis for desorption branches $\psi_e \leq 0$; for sorption branches $\psi_e \geq 0$).

The relation (1) is a general description of the soil water-retention capacity, which was previously proposed by Kosugi [4, 5]. In formula (1), an additional additive parameter ψ_e with the dimension of capillary pressure (capillary-sorption potential) of moisture is reasonably presented. The principal result obtained in [3] as an additional additive parameter ψ_e prompted the study of this parameter and its more general physical interpretation. For example, for the main drying branch of the water-retention capacity of the soil obtained by the Richards press method, the parameter ψ_e describes the aforementioned "air entry" pressure, and for wetting branches this parameter describes the "water entry" pressure.

In the paper [3] continuous approximation of the relation (1) in the class of elementary functions is proposed using the simplified Winitzki formula [6]. Here the relation (1) is transformed to the form:

$$S_e = \left[\begin{array}{l}(1+(-\alpha(\psi-\psi_e))^n)^{-1}, \ \psi<\psi_e; \\ 1, \ \psi \geq \psi_e.\end{array}\right. \tag{2}$$

At ψ_e = 0, function (2) is reduced to the well-known model of the soil water-retention capacity proposed by Haverkamp and co-authors [7].

The absolute majority of scientific studies of the water-retention capacity of the soil hysteresis are a direct development of two well-known models, or the authors of these publications compare the results of their research with two models. Therefore, authors dwell on these two models in more detail. The first model is the model of Scott and co-authors [8], the second is the model of Kool and Parker [9]. The first model is based on the function of the soil water-retention capacity of Haverkamp et al. [7]; the second model is based on the function proposed by Van Genuchten [10]. Van Genuchten proposed to use dependence $\theta(\psi)$ to describe the water-retention capacity in the form of:

$$S_e = \left[\begin{array}{l}(1+(-\alpha\psi)^n)^{-m}, \ \psi<0; \\ 1, \ \psi \geq 0,\end{array}\right. \tag{3}$$

where α [cm H_2O^{-1}], n and m – empirical parameters [10] ($m = 1 - 1/n$; $n > 1$).

In both models, turning points are calculated using the algorithm of Scott and co-authors [8]. The article [11] notes that the model of Scott and co-authors, as well as the model of Kool and Parker, have a significant drawback in the form of an undesirable "pump effect". This effect occurs when capillary pressure oscillation (capillary-sorption potential) of moisture in a fixed range of values could have a drift in the values of volumetric soil moisture: in this case, the scanning branches can cross the main branches and go beyond the limits of the physically acceptable region. For this reason, many modern authors develop models of the hysteresis of the water-retention capacity of the soil with closed loops formed by scanning branches. The statement regarding the "pump effect" can only be partially agreed. It is impossible to agree with the proposal for closed hysteresis loops formed by scanning branches due to the physical absurdity of such a proposal. Authors should give an explanation regarding the "pump effect". Indeed, the use of hysteresis models based on the Haverkamp function and co-authors [7], as well as the Van Genuchten function [10], for some values of parameters can lead to a result when the scanning branches intersect the main branches of the hysteresis loop, which is physically absurd. Such parameter values should be called physically unacceptable. However, this characteristic can be given only to those parameters that, in principle, have a physical meaning. For parameters that do not have a physical meaning, any values are equivalent (they cannot be absurd). In other words, for parameters that have a physical meaning, it is theoretically possible to specify physically permissible limits of variation, but for parameters that do not have a physical meaning, there are no such limits. Thus, in the model of Kool and Parker [9], which uses the Van Genuchten function [10] with parameters that do not have physical meaning, the manifestation of the "pump effect" is very likely, since when identifying parameters using the point approximation method, the measured soil water-retention capacity can be obtained, in principle, any values. It should be noted that Kool and Parker recognize this drawback of their model [9]. The situation is completely different with respect to the model of Scott et al. [8]. Since this model is based on the function of Haverkamp et al. [7], whose parameters allow physical interpretation, physically permissible limits can be determined for these parameters. And in this case, when the values of the parameters do not go beyond the permissible limits, the model of Scott et al. [8] does not exhibit an undesirable "pump effect". This result was published in [12]. And it also testifies to the current level of understanding of the problem and the physically adequate description of the hysteresis of the soil water-retention capacity. However, to clarify the method of determining the limits for varying the parameters of the function of Haverkamp et al. [7], so that the "pump effect" does not manifest, some "technical" refinement is still needed.

Regarding the ideas that have been proposed for the last decade (including now), relatively closed hysteresis loops, which are formed by the scanning branches of the water-retention capacity of the soil, can be a lot of counter-arguments. Below are just two of them. First: in order for the scanning branch to reach a turning point from which the previous scanning branch begins after the turning point, it is necessary for the soil to have a "memory"; Such an abstract construction is hypothetically admissible (a computer database is also easily realizable), but in nature such a soil phenomenon is not possible under any assumptions. Secondly: in the hypothetical case of closed loops, which are formed by scanning hysteresis branches, at the intersection points of the drying or wetting branches, the differential moisture capacity of the soil has an infinite number of values (degenerates); However, in authors opinion, at the turning point there can be only two values of the function of the differential water-retention capacity of the soil: one for any drying branch, the other for any wetting branch. Thus, the idea of closed loops formed by the hysteresis scanning branches of the water-retention capacity of the soil, in principle, contradicts the very idea of the water-retention capacity of the soil.

Based on the ideas of Scott et al. on turning points [8], formulas are used here to calculate the scanning hysteresis branches. When describing the functions of soil water-retention capacity, there are two additional lower indices corresponding to the branches of drying and branches of wetting: "*d*" for branches of drying and "*w*" for branches of wetting. The following formulas are used to describe the scanning drying branch, starting from the *i*-th point on the branch of wetting:

$$\begin{cases} \left[\begin{array}{l} \theta = \theta_r + \left(\theta_s^* - \theta_r\right) S_{e,d}, \\ \theta_s^* = \theta_s, \psi_{e,w} \leq \psi_i, \psi < \psi_{e,d}; \\ \theta_s^* = \theta_i, \psi_{e,d} \leq \psi_i < \psi_{e,w}, \psi < \psi_{e,d}; \\ \theta_s^* = \frac{\theta_i - \theta_r\left(1 - S_{e,d}(\psi_i)\right)}{S_{e,d}(\psi_i)}, \ \psi_i < \psi_{e,d}, \ \psi \leq \psi_i; \end{array}\right. \\ \left[\begin{array}{l} \theta = \theta_s, \ \psi_{e,w} \leq \psi_i, \ \psi_{e,d} \leq \psi \leq \psi_i; \\ \theta = \theta_i, \ \psi_{e,d} \leq \psi_i < \psi_{e,w}, \ \psi_{e,d} \leq \psi \leq \psi_i. \end{array}\right. \end{cases} \quad (4)$$

This formula was applied to describe the scanning wetting branch, starting from the *j*-th point on the branch of the drying:

$$\begin{cases} \left[\begin{array}{l} \theta = \theta_r^* + \left(\theta_s - \theta_r^*\right) S_{e,w}, \\ \theta_r^* = \theta_j = \theta_r, \psi_j < < \psi_{e,d}, \psi_j \leq \psi < \psi_{e,w}; \\ \theta_r^* = \frac{\theta_j - \theta_s S_{e,w}\left(\psi_j\right)}{1 - S_{e,w}\left(\psi_j\right)}, \psi_j < \psi_{e,d}, \psi_j \leq \psi < \psi_{e,w}; \end{array}\right. \\ \left[\begin{array}{l} \theta = \theta_s, \psi_j < \psi_{e,d}, \psi_{e,w} \leq \psi; \\ \theta = \theta_j = \theta_s, \psi_{e,d} \leq \psi_j, \psi_j \leq \psi. \end{array}\right. \end{cases} \quad (5)$$

The aim of the study is to compare the hysteresis model proposed by Scott et al. [8], which use the soil water-retention capacity functions (1) and (2), with the Kool and Parker model [9], which uses the soil water-retention capacity function (3), according to the Williams-Kloot criterion [13] under the condition that the values of the exponential parameter for the branches of drying and wetting are equal. Usage of this condition is proposed in article [9].

3 Results and Discussions

Further, the model proposed by Scott et al. [8], which uses the improved function of soil water-retention capacity (1), will be denoted by Hys-SKT. The model proposed by Scott et al. [8], which uses the improved function of soil water-retention capacity (2), will be denoted by Hys-SHT. The hysteresis model proposed by Kool and Parker [9], which uses the function of soil water-retention capacity (3), proposed by Van Genuchten [10], will be denoted by Hys-KPVG. A comparative analysis of the three models of hysteresis according to the Williams-Kloot criterion was carried out using the data on the sandy soil *Dune Sand*. The parameters of the compared models were identified by the point approximation method of experimental data on the main (boundary) branches of the drying and wetting of the soil water-retention capacity. Authors note that in this study, equal parameter values were used: $n_d = n_w$. Values of parameters of all compared models are given in Table 1.

Table 1. Parameters of hydrophysical functions for soil *Dune Sand*

Model	Model parameters									
	θ_r	θ_s	$\psi_{e,d}$	$\psi_{0,d}$	α_d	$\psi_{e,w}$	$\psi_{0,w}$	α_w	n_d	n_w
Hys-KPVG	0.1010	0.3010	-	−32.68	0.0306	-	−18.98	0.0527	6.779	
Hys-SKT	0.0903	0.3010	−18.88	−33.83	0.06689	−3.520	−19.81	0.06139	3.214	
Hys-SHT	0.0875	0.3010	−18.91	−33.97	0.06640	−3.345	−19.96	0.06019	3.355	

In the Figs. 1a, b and 2a, b solid curves depict the results of a dot-approximation of data on the main (boundary) branches (identification of parameters), as well as the results of an assessment (prediction) of the scanning branches of the hysteresis of the water-retention capacity of the studied sandy soil. The dots depict the experimental data.

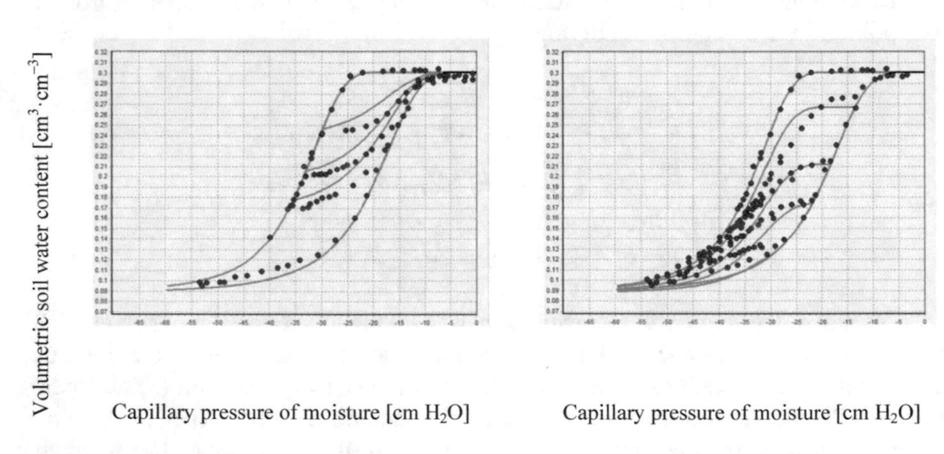

Fig. 1. (a) Dot-Approximation (fitting) of data about the boundary branches; Predictive estimation of the three wetting scanning branches for soil *Dune Sand* using the model Hys-SHT ($n_d = n_w$). (b) Dot-approximation (fitting) of data about the bound ary branches; Predictive estimation of the four drying scanning branches for *soil Dune Sand* using the model Hys-SHT ($n_d = n_w$).

Table 2 shows the correlation coefficients between experimental and calculated data on the hysteresis branches of the water-retention capacity of the studied sandy soil. The highest value of the correlation coefficient for each branch is shown in bold.

The errors of parameters and predictive calculations (square root of the average arithmetic squares of the calculation deviations results from the experimental data) are shown in the Table 3. In this table, the minimum errors are highlighted by bold underlined font. Further, the statistical hypothesis about the absence of reliable differences between the compared models was tested.

Table 4 shows the results of a comparative analysis for the three hysteresis models of the water-retention capacity of the studied sandy soil.

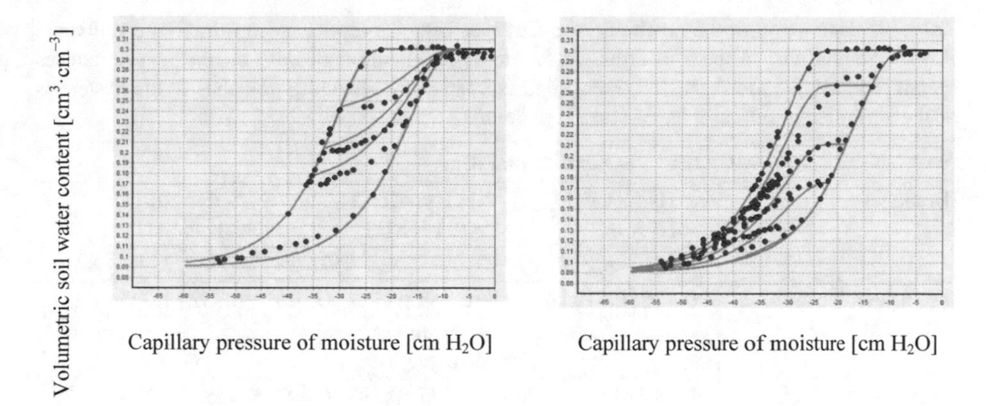

Fig. 2. (a) Dot-approximation (fitting) of data about the boundary branches; Predictive estimation of the three wetting scanning branches for soil *Dune Sand* using the model Hys-SKT ($n_d = n_w$). (b) Dot-approximation (fitting) of data about the boundary branches; Predictive estimation of the four drying scanning branches for *soil Dune Sand* using the model Hys-SKT ($n_d = n_w$).

Table 2. The correlation coefficients between the results of the calculation and the experimental data on the hysteresis branches of the water-retention capacity of soil *Dune Sand*

Correlation coefficient	Models		
Branches	Hys-KPVG	Hys-SKT	Hys-SHT
Main identification (48 dots)	0.9938	0.9992	**0.9993**
Wet scan prediction (47 dots)	0.9849	**0.9942**	0.9937
Dry scan prediction (95 dots)	0.9737	**0.9834**	0.9832

Table 3. Comparison of the errors for the dot-approximation of the main branches and the predictive calculation of the scanning branches of the soil water-retention capacity

RMSE – root mean square error	Models		
Branches	Hys-KPVG	Hys-SKT	Hys-SHT
Main identification (48 dots)	0.0088	0.0042	**0.0041**
Wet scan prediction (47 dots)	0.0144	**0.0112**	0.0113
Dry scan prediction (95 dots)	0.0131	**0.0128**	0.0138

Table 4. Estimation of the reliability of differences between models according to the Williams-Kloot criterion with respect to errors in the dot-approximation of data on the main branches (parameter identification), as well as errors in estimating the scanning branches of the hysteresis of the water-retention capacity of soil *Dune Sand*

$y - \frac{y_1 + y_2}{2} = \lambda(y_1 - y_2)$, where y – experimental data

Branches	y_1: Hys-KPVG, y_2: Hys-SKT			y_1: Hys-KPVG, y_2: Hys-SHT			y_1: Hys-SKT, y_2: Hys-SHT		
	λ	$\lambda_{0.95}$	$\lambda_{0.975}$	λ	$\lambda_{0.95}$	$\lambda_{0.975}$	λ	$\lambda_{0.95}$	$\lambda_{0.975}$
Main identification (48 dots)	−0.426	0.118	0.142	−0.463	0.122	0.146	−0.752	1.023	1.226
	y_2 more accurate, than y_1			y_2 more accurate, than y_1			y_1 and y_2 are equal		
Wet scan prediction (47 dots)	−0.844	0.528	0.705	−0.735	0.581	0.697	0.955	3.193	3.829
	y_2 more accurate, than y_1			y_2 more accurate, than y_1			y_1 and y_2 are equal		
Dry scan prediction (95 dots)	−0.564	0.325	0.388	−0.453	0.337	0.403	5.168	1.674	2.000
	y_2 more accurate, than y_1			y_2 more accurate, than y_1			y_1 more accurate, than y_2		

4 Conclusion

According to the Williams-Kloot criterion, the Hys-SKT and Hys-SHT models can be considered the most preferable for all curves. Moreover, Hys-SKT shows the best result in predicting the scanning drying branches. According to all estimates presented in the article, the errors of dot-approximation of data on the main branches and the prediction errors of the scanning branches of the hysteresis of soil water-retention capacity, the Hys-KPVG model is significantly inferior to the Hys-SKT and Hys-SHT models. The closest correlation is observed in the dot-approximation of the experimental data on the main branches: the correlation coefficients for all three models differ slightly, however, the error of the Hys-KPVG model remains the greatest, and the Hys-SKT and Hys-SHT models do not have significant differences. This trend continues in the prediction of scanning branches. Moreover, the correlation is more notable. The errors in this case differ to a lesser degree, but differences according to the Williams-Kloot criterion are still significant. The condition of equality of the values of the exponential parameter for the branches of drying and wetting does not significantly increase the errors of the estimates of the scanning branches of the hysteresis compared with the condition when these values are not equal to each other [14].

Thus, it can be concluded that in hydrophysical calculations for sandy soil (for example, in calculations of precision irrigation rates [1], in justifying hydraulic structures projects [15] or in predicting the productivity of agrocenoses [16]) the Hys-SKT and Hys-SHT models seems to be the most preferred.

Acknowledgments. The reported study was funded by RFBR according to the research projects No. 19-04-00939-*a*, No. 19-016-00148-*a*.

References

1. Terleev, V.V., Topazh, A.G., Mirschel, W.: The improved estimation for the effective supply of productive moisture considering the hysteresis of soil water retention capacity. Russ. Meteorol. Hydrol. **40**(4), 278–285 (2015)
2. Mualem, Y.: A catalogue of the hydraulic properties of unsaturated soils. Research Project 442. Technion, Israel Institute of Technology, Haifa, Israel (2006)
3. Terleev, V.V., Mirschel, W., Badenko, V.L., Guseva, I.Yu.: An improved Mualem-Van Genuchten method and its verification using data on Beit Netofa clay. Eurasian Soil Sci. **50** (4), 445–455 (2017)
4. Kosugi, K.: Three-parameter lognormal distribution model for soil water retention. Water Resour. Res. **30**, 891–901 (2007)
5. Kosugi, K.: Lognormal distribution model for unsaturated soil hydraulic properties. Water Resour. Res. **32**, 2697–2703 (2007)
6. Winitzki, S.: A handy approximation for the error function and its inverse. https://sites.google.com/site/winitzki/sergei-winitzkis-files/erf-approx.pdf?attredirects=0
7. Haverkamp, R., Vauclin, M., Touma, J., Wierenga, P.J., Vachaud, G.: A comparison of numerical simulation model for one-dimensional infiltration. Soil Sci. Soc. Am. J. **41**, 285–294 (2005)
8. Scott, P.S., Farquhar, G.J., Kouwen, N.: Hysteretic effects on net infiltration. In: Proceeding of National Conference on Advances in Infiltration, Publication 11-83, pp. 163–170. American Society of Agricultural Engineers, St. Joseph, Michigan (2008)
9. Kool, J.B., Parker, J.C.: Development and evaluation of closed-form expressions for hysteretic soil hydraulic properties. Water Resour. Res. **23**(1), 105–114 (2005)
10. Van Genuchten, M.Th.: A closed form equation for predicting the hydraulic conductivity of unsaturated soils. Soil Sci. Soc. Am. J. **44**, 892–989 (2006)
11. Huang, H.C., Tan, Y.C., Chen, C.H.: A novel hysteresis model in unsaturated soil. Hydrol. Process. **19**, 1653–1665 (2005)
12. Terleev, V, Ginevsky, R., Lazarev, V., Nikonorov, A., Togo, I., Topaj, A., Moiseev, K., Abakumov, E., Melnichuk, A., Dunaieva, I.: Predicting the scanning branches of hysteretic soil water-retention capacity with use of the method of mathematical modeling. In: IOP Conference Series: Earth and Environmental Science, vol. 90, Article no. 012105 (2017)
13. Kobzar, A.I.: Prikladnaya matematicheskaya statistika. Dlya inzhenerov i nauchnyx rabotnikov. Izd. Fizmatlit, Moskva (2006). (rus)
14. Terleev, V., Mirschel, W., Nikonorov, A., Ginevsky, R., Lazarev, V., Topaj, A., Moiseev, K., Layshev, K., Arkhipov, M., Melnichuk, A., Dunaieva, I., Popovych, V.: Five models of hysteretic water-retention capacity and their comparison for sandy soil. In: MATEC Web of Conferences, vol. 193, Article no. 02036 (2018)
15. Badenko, V., Badenko, N., Nikonorov, A., Molodtsov, D., Terleev, V., Lednova, J., Maslikov, V.: Ecological aspect of dam design for flood regulation and sustainable urban development. In: MATEC Web of Conferences, vol. 73, Article no. 03003 (2016)
16. Medvedev, S., Topaj, A., Badenko, V., Terleev, V.: Medium-term analysis of agroecosystem sustainability under different land use practices by means of dynamic crop simulation. In: IFIP Advances in Information and Communication Technology, vol. 448, pp. 252–261 (2015)

Modelling and Control in Mechanical Engineering

Computer Aided Simulation of Behavior of Extrusion Press Ram Jointly with Rotary Elastic Compensator

Elena Balalayeva[1], Volodymyr Kukhar[1], Viktor Artiukh[2](✉), Vladimir Filatov[3], and Oksana Simonova[4]

[1] Pryazovskyi State Technical University, str. Universytets'ka 7, Mariupol 87555, Ukraine
[2] Peter the Great St. Petersburg Polytechnic University, Polytechnicheskaya 29, 195251 St. Petersburg, Russia
artiukh@mail.ru
[3] Moscow State University of Civil Engineering, Yaroslavskoye shosse 26, 129337 Moscow, Russia
[4] Tyumen Industrial University, Volodarskogo str. 38, 625000 Tyumen, Russia

Abstract. The mathematical model and software for calculation of the work of rams with the universal rotary elastic compensators is considered. The theory of stability loss of core systems with elastic fixing in static position was taken as the basis of evaluating strain in a compression ram, when operating with elastic compensators. The program implements the algorithm of numeric solution of the corresponding differential problem with application of difference approximation and Newton's iteration procedure. The developed program includes the determination of displacements and stresses in longitudinally bent extrusion plungers and ordinates of their critical sections within which destruction occurs. Integrated modules of developed CAD-system components allow assessing the alternations of power modes at different operations and selecting appropriate design parameters of elastic compensators for realization of reaction momentum at which strain in long-length working tool will not exceed critical values. The results of computer simulation are the basis for choosing the design parameters of the compensators, which allow reducing the bending stresses and ensuring reliable operation of the long tool.

Keywords: Computer simulation · Extrusion ram · Press · Elastic compensator

1 Introduction

At present, software complexes for calculating the stress-strain state (SSS) in structural elements and details of press equipment have become widely used, for example, ABAQUS [1], ANSYS [2, 3], QFORM [4], COSMOS/Works [5] and others, which are based on the finite element method (FEM). Wherein insufficient attention is paid to the problem of the stability of elastic systems, the solution of which requires application of methods for determining the bending of prismatic bars under the simultaneous action of

V. Murgul and M. Pasetti (Eds.): EMMFT-2018, AISC 983, pp. 475–488, 2019.
https://doi.org/10.1007/978-3-030-19868-8_47

axial and transverse loads [6–11]. These methods are characterized by high accuracy and ease of implementation in the development of software for automated calculations.

An urgent task is to study the SSS of long extrusion rams using the theory of stability loss of rod systems to identify the root causes of their breakages and increased wear.

Extrusion rams are one of the most loaded and subjected to breakages parts of extrusion equipment. Particularly frequent failures of the rams are while cold pressing of non-ferrous metals and hot pressing of steels, when, due to the hardening of the pressed material, accompanied by an increase in the deformation force, critical deformations of the longitudinal bending arise in the long-length rod.

In practice, there are extrusion rams with a ratio of the length L to the diameter d up to 9.5 or more, which confirms the validity of the assumptions about the relation of failures in such details to the loss of their stability [12]. A similar unfavorable stress-strain condition is experienced by long punches of die packages for extrusion [13], mandrels and thrust rods of tubing and pipe mills [14], hammer rods [15] and other impact machines.

One way to prevent breakages of long-length parts of equipment for metal forming is the introduction of structural elements that increase the rigidity of the deforming system, for example, the screeds of the press frame [16]. Another way is the introduction of flexible connections, which are characteristic for the coupling of the rod to the hammerhead [17]. Their disadvantage is either relative complexity or inadequate structural efficiency. Another alternative approach to reducing of errors of the "press-die" system can be die-free and impression-free [18, 19] methods of metal-forming.

The promising direction of reducing the distortions in the "slider-ram-container" system is the introduction of various designs of elastic compensators based on polyurethane [20]. The elastic compensators, calculated both for individual operations, and universal, applied for a wide range of technological loads, have become widely used [3, 21–23]. The design of compensating devices is depended on batch of parts production, type of the press equipment, force modes of the technological operations and other factors. This complicates is consistent of calculation of the design parameters of elastic compensators. Automated method for calculating the optimal parameters of the ring elastic error compensators for the "press-die" system are explored in papers [21]. With a wide range of products, it is advisable to use universal error compensators of the "press-and-die" system that based on composite pre-stressed elastic elements configured in the form of two rigid (circular for example) polyurethane plates with openings [8]. One of these plates is displaceable by turning relative to the other, which enables to change the supporting surface area and leads to a change of compensator rigidity by varying of the overlap relation. That is set to extending the range of technological operations. The mathematical model of the work of universal double-layers elastic rotary compensator was developed in papers [22, 23].

The kinematics of the work of a long extrusion bent ram with an elastic compensator can be considered on the basis of the static method for calculating the strength of compressed-bent rods with elastic fastening, based on the basic equations for displacements and acting forces. In this case, the solution of the problem posed is similar to the solution of the Euler problem [8, 9].

2 Purpose of the Work

The purpose of the work is the development of a mathematical model and software for calculating the stress-strain state and determining the ordinate of a critical section in a long extrusion ram, taking into account the reaction time of the universal rotary elastic compensator of the slider direction errors, as well as obtaining and analyzing the stress dependencies in a ram from various design and operational parameters of the "press-die" system for rational selection of the compensator characteristics.

3 Research Material

We considered schematically a slider in press guides and extrusion ram fixed at the slider with front end entered into container. According to the system of acting forces (Fig. 1), the technological and design parameters of the ram were determined during deformation. Interaction of the workpiece and the ram or long punch with the length L and diameter d provided availability of axial hole d_0 was reduced to coaxial compression based on process force P_p and bending from additional moment M_{ab}. It was assumed that the container influences the ram with a force P_a and in case of violation of adjusting gaps between them the container bends the front end of the ram at a certain angle φ, which is equivalent to applying additional torque M_{ac}. The rotation by angle φ is opposed by elastic compensator providing the reaction M_r striving to straighten a ram that is shifted with the front end from the slide axis as a result of loss of stability as well as due to loss of resistance as well as application of forces P_p, P_a and moment $M_a = M_{ab} + M_{ac}$.

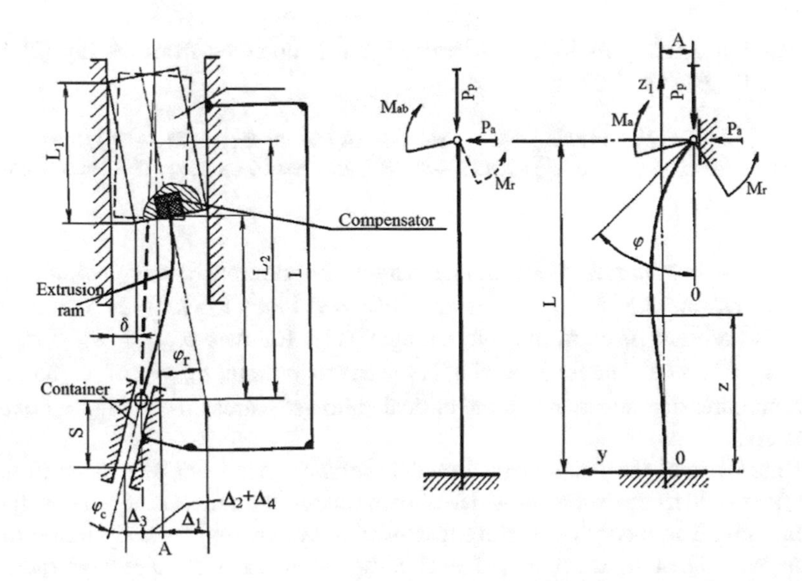

Fig. 1. Action of forces on the ram with a compensator when turning the slide in press guides.

According to the approved scheme (Fig. 1) the equation of moments was formed:

$$EIy'' = -P_p(A - y) + P_a(L - z) - M_a + M_r, \quad (1)$$

where E – the steel elasticity coefficient; I – the axial moment of extrusion ram cross section area; A – the full axial displacement of the center of the front end of the ram from the axis of its base.

The solution of the Eq. (1) has the form:

$$y = [A(\sin kL(\cos kz - L) - \cos kL \cdot \sin kz) + ((M_a - M_p)/P_p)(\sin kL(\cos kz - 1) + \\ + (1 - \cos kL)\sin kz) + (P_a/(k^2EI))(L \cdot \cos kL \cdot \sin kz - L \cdot \sin kL(\cos kz - 1) - z \cdot \sin kL]/\sin kl, \quad (2)$$

where $k = (P_p/EI)^{1/2}$.

The angle of rotation of the front end of the ram was obtained from expression (2) as $\partial y/\partial z$ at $z = L$:

$$\varphi = (k/\sin kL)[-A(\cos 2kL + (M_a - M_r)/P_p \cdot (\cos 2kL - \cos kL) + \\ + P_a/(k^3EI) \cdot (kL - \sin kL)]. \quad (3)$$

The ordinate z_1 of the dangerous section was determined from expression (2) as $\partial^3 y/\partial z^3 = 0$:

$$z_1 = 1/k \cdot \mathrm{arctg}[(P_a \cdot L \cdot \cos kL/(k^2EI) - A \cos kL - (M_a - M_r)/P_p \times \\ \times (1 - \cos kL))/((M_a - M_r)/P_p \cdot \sin kL - A \sin kL - P_a \cdot L \cdot \cos kL/(k^2EI) \cdot \sin kL)]. \quad (4)$$

Maximum bending moment of extrusion ram is obtained from the Eq. (2) as $\partial^2 y/\partial z^2$ at $\sum M = EIy''$ and $z = z_1$:

$$\sum M_{max} = P_p/\sin kL[A(\sin kL \cdot \cos kz_1 + \cos kL \cdot \sin kz_1) + (M_a - M_r)/P_p \times \\ \times (-\sin kL \cdot \cos kz_1 + (1 - \cos kL) \cdot \sin kz_1 + P_a/P_p(-L \cdot \cos kL \cdot \sin kz_1 + L \cdot \sin kL \cdot \cos kz_1)]. \quad (5)$$

Maximum bending stress of extrusion ram (in addition to compression ones) is calculated as $\sigma_a = \sum M_{max}/W_r$, where $W_r = 0.1 \cdot d^3(1 - m^4)$ is the resistance moment. Condition of extrusion ram strength is as follows: $\sigma_{calc} = \sigma_n + \sigma_a \leq \sigma_{cr}$, where σ_{calc} is the rated stress appearing in the extrusion ram; $\sigma_n = P_t/(0.785 \cdot d^2(1 - m^2))$ is compression stress; σ_{cr} is the critical (allowed) stress depending on extrusion ram material.

On the basis of the mathematical model, software has been developed in the MS Visual Studio 2013 environment which allows modeling the work of rams with elastic compensators. The theory of stability loss of core systems with elastic fixing in static position was taken as the basis of evaluating strain in a compression ram, when operating with elastic compensators. The program implements the algorithm of numeric solution of the corresponding differential problem with application of difference approximation and Newton's iteration procedure. The evaluation of strains and travels in longitudinally bent compression rams and ordinates of their dangerous sections, where breakages are likely to happen, is done on the basis of the obtained quotients of the differential equation.

Unlike similar software, in which only engineering solutions are performed for strains occurring in long-length pressing tools, the obtained software product allows taking into account the reaction momentum, realized by means of the elastic compensator. The existing difference between the elastic properties of the compensator material (polyurethane, modulus of elasticity E = 69 MPa) and the material of the extrusion ram (structural and tool steel grades, modulus of elasticity $E = 2 \cdot 10^5$ MPa) cause difficulties for realization of solid simulation in finite-elements packages, as the value of compensator's upset exceeds very much the deformation value of long-length operating working tool.

The program contains the following modules: "Calculation", "Graphs", "Theory", "Symbols" and "Settings". Let's describe first two modules.

The main menu item "Calculation" calls the form for calculating the stress σ_{calc} in the longitudinally bent ram, taking into account the specified reaction moment M_r, as well as the angle of rotation φ and the ordinate of the critical section z_1, in which maximum deflection or breakage is observed (Fig. 2). Input data (d, ε_1, m, φ_c, P_p, P_a, M_r), intermediate values and output data (σ_{cr}, σ_{calc}) are displayed in the output fields, tables and diagrams.

The main menu item "Graphs" brings a form for charting the dependence of stress $\sigma_{calc}/\sigma_{cr}$ on input parameters. The form has three tabs: "Modeling", "Table of values" and "Summary graph". The tab "Modeling" contains input fields similar to the form "Calculation"; the data is transferred automatically and can be edited. To plot the graph, we need to select a variable on x-axis and press the button "Input data". When the button "Modeling" is clicked, the input data is approximated by the curve and displayed on the graph. The obtained dependence is also displayed as an analytical expression whose coefficients depend on the input data. The graph is automatically added to the tab "Summary graph". To add a new curve to the summary graph, we need to change the required parameter and click the button "Add the curve".

As an example, we considered the operation of the ram with diameter d = 20 mm with a ratio of length to diameter $\varepsilon_1 = 9.5$ without an axial hole ($m = 0$), material – steel 55C2 (heat treated spring steel, in accordance with GOST 14959 as analogue of EN 55Si7: 0.52…0.6%C, 0.6…0.9%Mn, 1.5…2%Si, up to 0.25%Ni) which σ_{cr} = 1700 MPa. The angle of divergence of the axes was $\varphi_c = 0.001$ rad, the processing force P_p = 400 kN and the lateral force P_a = 370 kN.

In the absence of an elastic compensator (reaction moment $M_r = 0$), the calculated stress in the ram was $\sigma_{calc} \approx 1703$ MPa, which exceeds the critical stress σ_{cr} = 1700 MPa. As a result, the breakage of the rod is predicted at a distance z_1 = 95 mm from the seal of the ram (Fig. 2a).

When using an elastic compensator providing the moment of reaction M_r = 474 kN mm, the stress in the ram was $\sigma_{calc} \approx 168$ MPa, which does not exceed the critical stress $\sigma_{cr} \approx 1700$ MPa. Tthe breakage of the rod will not occur (Fig. 2b).

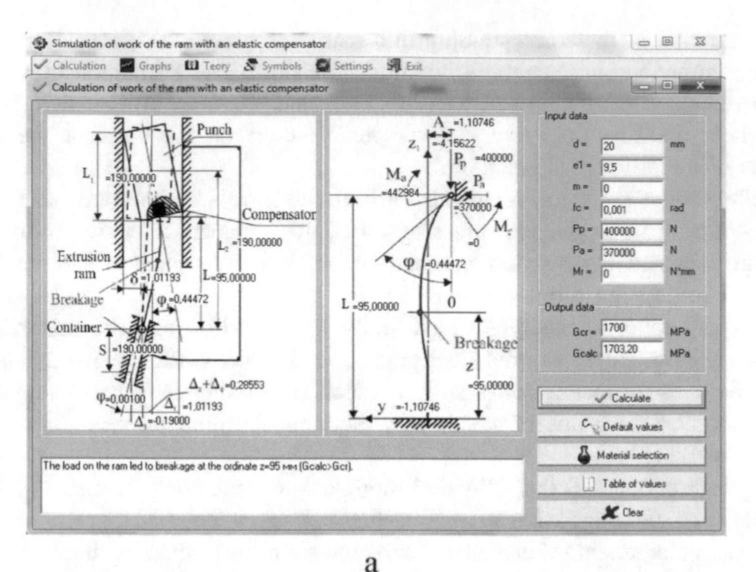

a

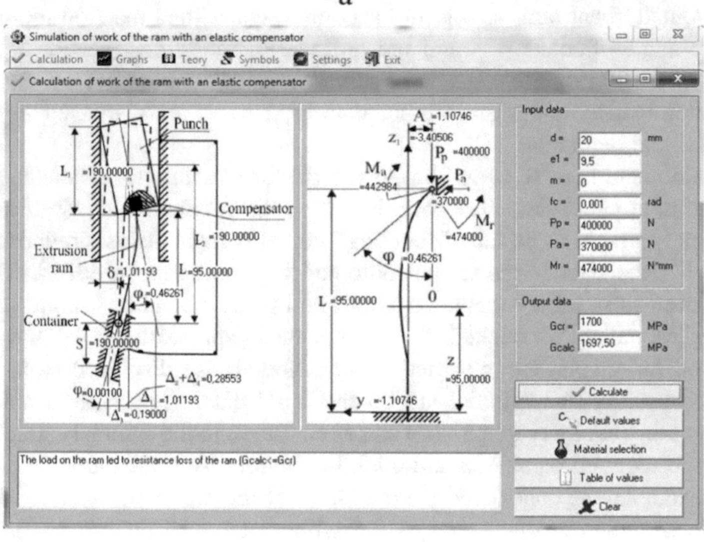

b

Fig. 2. The program window for calculating the stresses in the rams and the kinematic parameters of their work with elastic compensators: breakage of the ram (a); normal operation of the ram (b).

For rational selection of the main design and technological parameters of the "press-die" system with elastic compensators, it is necessary to obtain and analyze the functional dependences of the relative stress in the ram from various parameters. The general form of such dependence:

$$\sigma_{calc}/\sigma_{cr} = f\left(d, \varepsilon_1, m, \varphi_c, P_p, P_a, M_r\right). \tag{6}$$

The examples of the dependence of stress $\sigma_{calc}/\sigma_{cr}$ on one of the previously listed parameters with the possibility of displaying them on the composite graph are shown on Figs. 3 and 4.

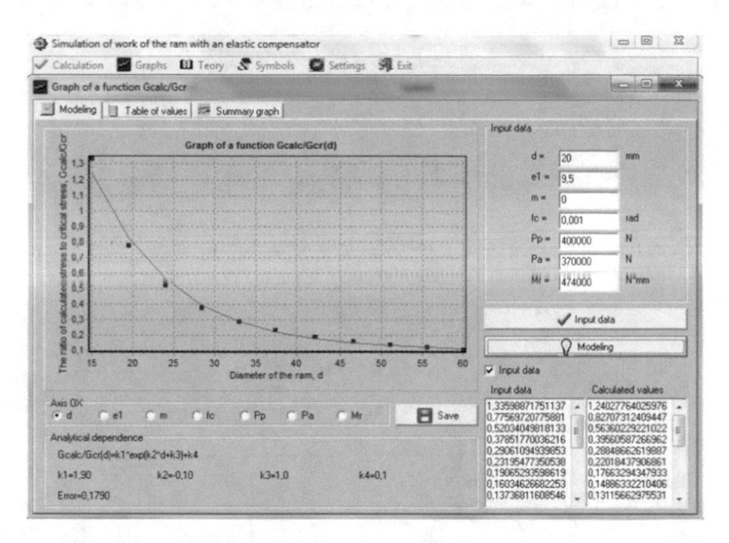

Fig. 3. The program window for constructing the functional dependence of the ratio of the calculated stress to the critical $\sigma_{calc}/\sigma_{cr}$ on the diameter of the ram d.

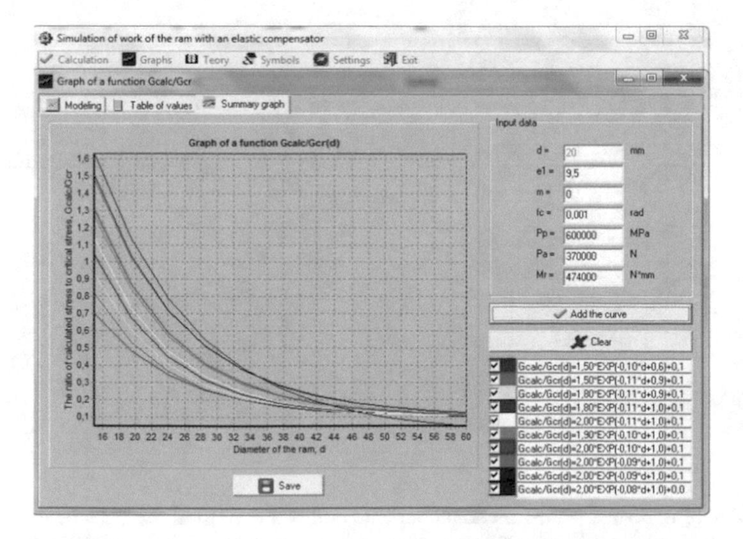

Fig. 4. The program window for constructing a composite graph of the ratio of the calculated stress to the critical one $\sigma_{calc}/\sigma_{cr}$ from the diameter of the ram d when changing the technological force P_p.

By approximating the obtained graphical dependencies, we have got analytical expressions for determining stress in a ram depending on various parameters. The required coefficients of expressions are determined depending on the input data by the brute-force method.

Modeling results are presented in the form of graphs $\sigma_{calc}/\sigma_{cr}$ *(d)*, $\sigma_{calc}/\sigma_{cr}$ (ε_1), $\sigma_{calc}/\sigma_{cr}$ *(m)*, $\sigma_{calc}/\sigma_{cr}$ (φ_c), $\sigma_{calc}/\sigma_{cr}$ (P_p), $\sigma_{calc}/\sigma_{cr}$ (P_a), $\sigma_{calc}/\sigma_{cr}$ (M_r) (Fig. 5).

The regression equations for different parameters have the following form:

$$\sigma_{calc}/\sigma_{cr}(d) = k_1 \cdot \exp(k_2 \cdot d + k_3) + k_4; \tag{7}$$

$$\sigma_{calc}/\sigma_{cr}(\varepsilon_1) = k_5 \cdot \exp(k_6 \cdot \varepsilon_1 + k_7) + k_8; \tag{8}$$

$$\sigma_{calc}/\sigma_{cr}(m) = k_9 \cdot \exp(k_{10} \cdot m + k_{11}) + k_{12}; \tag{9}$$

$$\sigma_{calc}/\sigma_{cr}(\varphi_c) = k_{13} \cdot \varphi_c + k_{14}; \tag{10}$$

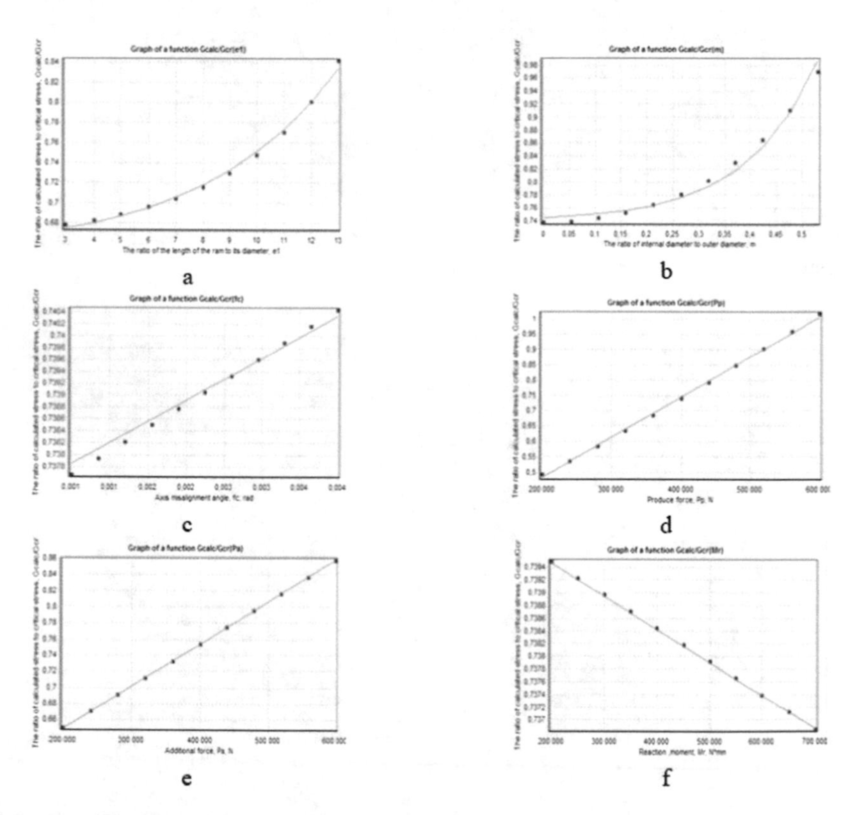

Fig. 5. Modeling results – the functional dependence of the ratio of the calculated stress to the critical $\sigma_{calc}/\sigma_{cr}$ on: (a) the ratio of length of the ram to its diameter (ε_1); (b) the ratio of internal diameter to outer diameter (*m*); (c) axis misalignment angle (φ_c); (d) produse force (P_p); (e) additional force (P_a); (f) reaction moment (M_r).

$$\sigma_{\text{calc}}/\sigma_{\text{cr}}(P_p) = k_{15} \cdot P_p + k_{16}; \tag{11}$$

$$\sigma_{\text{calc}}/\sigma_{\text{cr}}(P_a) = k_{17} \cdot P_a + k_{18}; \tag{12}$$

$$\sigma_{\text{calc}}/\sigma_{\text{cr}}(M_r) = k_{19} \cdot M_r + k_{20}, \tag{13}$$

where k_1, k_2, ..., k_{20} are the required coefficients.

To test the model for adequacy the root-mean-square error was determined as not exceeding 0.1, which indicates the adequacy of the obtained functional dependences.

The use of these models allows for a more complete functional analysis of experimental data and faster processing of information.

The data of evaluations, performed with the help of the obtained software may lay the foundation for selection of design parameters of compensators, that will make it possible to reduce strain of longitudinal bending up to permissible values. This idea was further developed in [21–23] works. The authors developed CAD-system components, in the form of integrated modules, allowing evaluating of alternations of

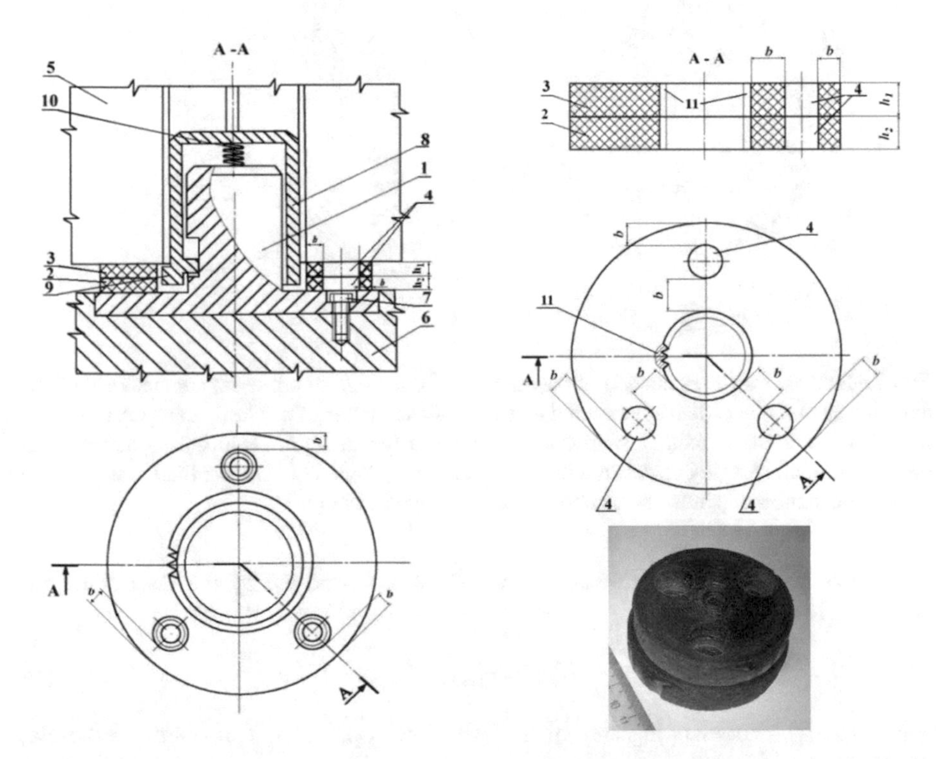

Fig. 6. The universal compensator of the errors of the slide direction: 1 – shank; 2 and 3 – the lower and the upper elastic circular plates; 4 – openings in the elastic plates; 5 – press slide; 6 – upper die-plate; 7 – screw connection the shank and the upper die-plate; 8 – cowl; 9 – the cog of the cowl for the fixing of the rotated elastic plate; 10 – spring; A-A – the designation of the cross-section; h_1 and h_2 – thickness of the upper and lower elastic circular plates; b – distances between openings, inner and outer backs of compensator.

power modes at different operations and selecting appropriate design parameters of elastic compensators for realization of reaction momentum, at which strain in long-length working tool will not exceed critical values. Mathematic modeling was developed, as well as corresponding software modules, allowing making calculations for ring compensators [21] and universal rotary compensators [22, 23].

The design of universal compensator of the errors of slide direction was considered on Fig. 6. There two elastic circular polyurethane plates (brand SKU-PFL-100, 100 Shore hardness) with the radius (*R*) and openings with radiuses (*r1*) and (*r2*) where serves as the basis for composite pre-stressed elastic element.

The scheme for calculation of the overlapping of two openings of rotary elastic plates of universal compensator of the errors of slide directions is shown on Fig. 7. This task is considered for the case when the radiuses of the openings (r_1) and (r_2) can not coincide.

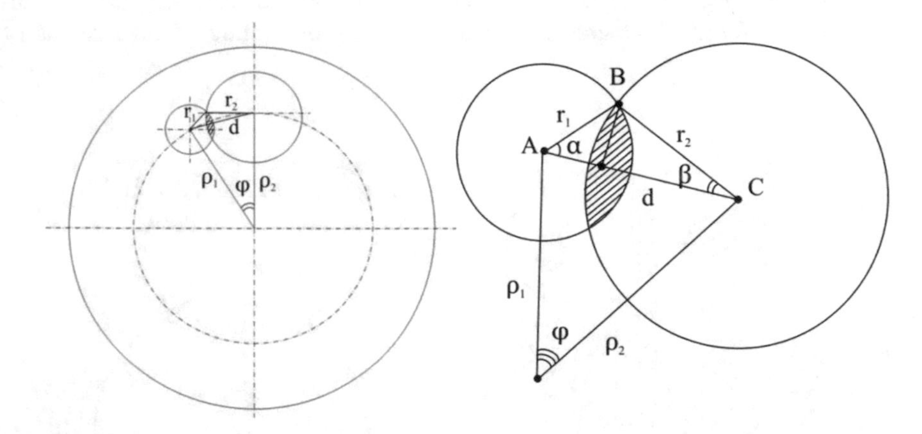

Fig. 7. Scheme for determination of overlapping of openings of universal compensator of slide direction errors: *d* – distance between the centers of the openings of elastic compensator plates; ρ_1 – distance from the axis of rotation to the center of the smaller opening; $\rho_{\backslash 2}$ – distance from the axis of rotation to the center of a larger opening; φ – angle between the lines that beginning from the common origin to the centers of the openings in the plates.

The variation of the overlap coefficient (K_{per}) is carried out by shifting of the upper plate and is defined as:

$$\mathrm{K_{per}} = \mathrm{F_{per.otv}} / \mathrm{F_{otv}}, \tag{14}$$

where $F_{per.otv}$ – overlapping area of the two plates openings; F_{otv} – area of the plate openings.

According to the results of the works [22–26] the overlap coefficient in relation to the smaller opening of elastic plate of universal compensator of slide directions errors is calculated by the formula:

$$K_{per} = 1/\pi \cdot \left(\arcsin(A) - A \cdot \left(1 - A^2\right)^{1/2} + \omega^2 \cdot \arcsin(A/\omega) - 1/2 \cdot \left(\omega^2 - A^2\right)^{1/2}\right), \quad (15)$$

where are $A = sin(\alpha)$; $\omega = r_2/r_1$ (according to the scheme on Fig. 7).

The initial dimensions of the universal elastic rotary compensator were determined (Fig. 8): radii of elastic plates $R_1 = R_2 = 135$ mm, the radius of the plate holes $r_1 = r_2 = 42.12$ mm, distance between adjacent holes in the elastic plates and between the hole and the wall of the compensator $a_1 = a_2 = b_1 = b_2 = 25.38$ mm, height of the plates $h_1 = h_2 = 20.14$ mm. It was established that for optimal rigidity the compensator must be turned the upper elastic plate at an angle $\varphi = 49°$ (Fig. 8).

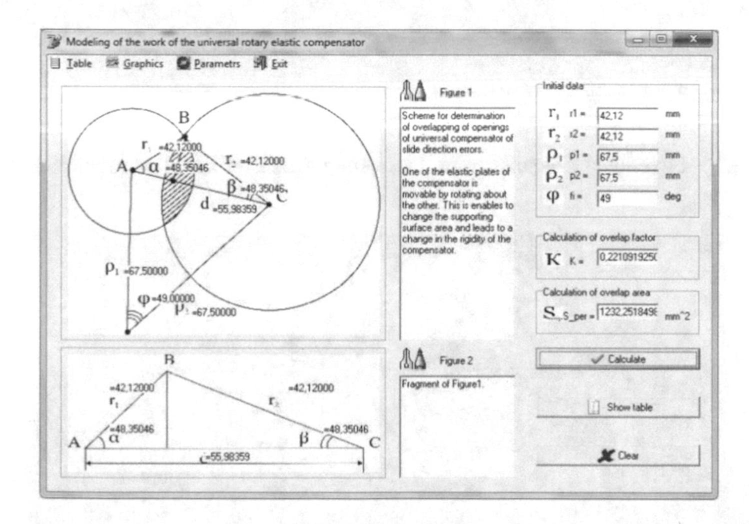

Fig. 8. The calculation of the rotary elastic compensator dimensions.

Thus, was adapted an automated method and developed the program for calculating structural characteristics of rotary elastic error compensators of the "press-and-die" system for the process of extrusion of profiles.

The developed software provides for the construction of the following graphs: $K_{per}(r_2/r1)$, $K_{per}(\varphi)$ (Fig. 9), $S_{per}(r_1)$, $S_{per}(r_2)$, $S_{per}(\rho_1)$, $S_{per}(\rho_2)$, $S_{per}(\varphi)$. To determine the effect of a parameter on the compensator, it is possible to display a summary graph (Fig. 10).

The using of the universal rotary elastic compensator is a perspective and low-cost way to reducing of the distortions of slide direction at the "press-and-die" system of press equipment. It is planned that selection of possible compensators designs and compilation of software modules for simulation of their operation will further be extended.

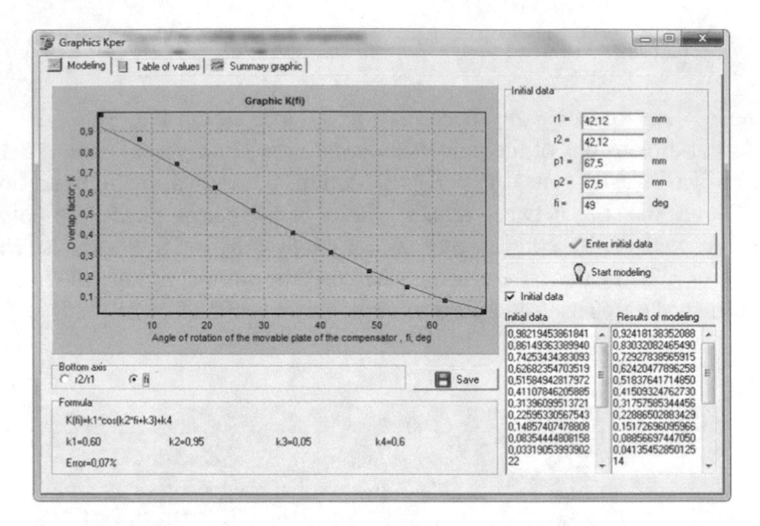

Fig. 9. Functional dependence of the overlap factor on the plate rotation angle.

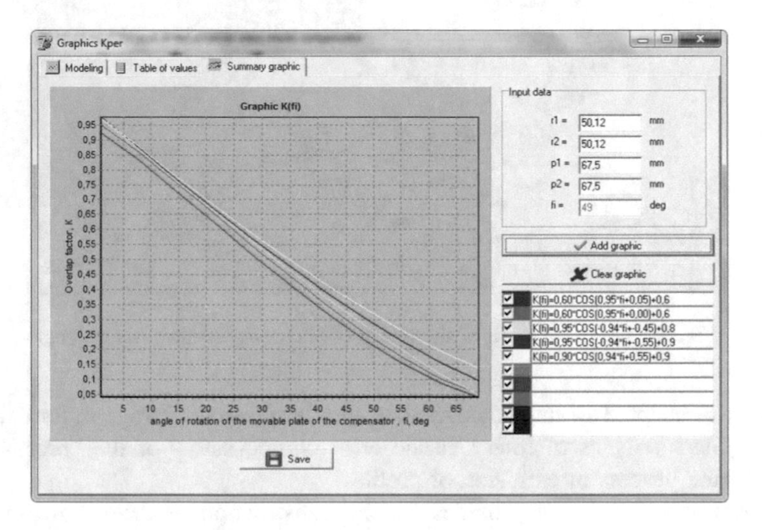

Fig. 10. A summary graph of the dependence of the overlap factor on the angle of rotation angle with a change in the radii of the openings.

4 Conclusions

There have been developed a mathematical model of the kinematics of operation of long extrusion rams with the universal rotary elastic compensator, and software that allows determining the displacements and stresses in longitudinally bent extrusion rams and the ordinates of their critical sections within which breakages are predicted. The calculation data are the basis for choosing the design parameters of the compensators, which allow reducing the bending stresses and ensuring reliable operation of the long tool. Analytical dependencies of stress arising in extrusion ram are identified from different design and operational parameters of "press-die" system with universal rotary elastic compensators that provides possibility to select compensators characteristics to ensure the required reaction moment.

Acknowledgments. The reported study was funded by RFBR according to the research project №19-08-01252a 'Development and verification of inelastic deformation models and thermal fatique fracture criteria for monocrystalline alloys'. The authors declare that there is no conflict of interest regarding the publication of this paper.

References

1. Buyalich, G.B., Anuchin, A.V., Serikov, K.P.: Influence of the geometry of beveled edges on the stress-strain state of hydraulic cylinders. In: IOP Conference Series: Materials Science and Engineering, vol. 127, p. 012012 (2016). https://doi.org/10.1088/1757-899X/127/1/012012
2. Bi, D.S., Liu, D.D., Chu, L., Zhang, J.: Finite element analysis on frame-type hydraulic press. Adv. Mater. Res. **199–200**, 1623–1628 (2011). https://doi.org/10.4028/www.scientific.net/amr.199-200.1623
3. Balalayeva, E., Artiukh, V., Kukhar, V., Tuzenko, O., Glazko, V., Prysiazhnyi, A., Kankhva, V.: Researching of the stress-strain state of the open-type press frame using of elastic compensator of errors of "press-die" system. Adv. Intell. Syst. Comput. **692**, 220–235 (2018). https://doi.org/10.1007/978-3-319-70987-1_24
4. Gaikwad, A., Kirwai, S., Koley, P., Balachandran, G., Singh, R.: Theoretical study on cold open die forging process optimization for multipass workability. In: MATEC Web of Conferences, vol. 80, p. 13002 (2016). https://doi.org/10.1051/matecconf/20168013002
5. Li, Y., Wang, T.: The structural static analysis of four-column hydraulic press. In: IEEE International Conference on Mechatronics and Automation (2015). https://doi.org/10.1109/ICMA.2015.7237673
6. Lalin, V., Kushova, D.: Variational formulations of the nonlinear equilibrium and stability problems of elastic rods. In: Conference: Proceedings of the International Conference "Innovative Materials, Structures and Technologies", pp. 85–89 (2014). https://doi.org/10.7250/iscconstrs.2014.14
7. Levyakov, S.V.: States of equilibrium and secondary loss of stability of a straight rod loaded by an axial force. J. Appl. Mech. Tech. Phys. **42**(2), 321–327 (2001). https://doi.org/10.1023/A:1018844423385
8. Annin, B.D., Vlasov, A.Y., Zakharov, Y.V., Okhotkin, K.: Study of static and dynamic stability of flexible rods in a geometrically nonlinear statement. Mech. Solids **52**(4), 353–363 (2017). https://doi.org/10.3103/S002565441704001X
9. Belyaev, A.K., Morozov, N.F., Tovstik, P.E.: Buckling problem for a rod longitudinally compressed by a force smaller than the Euler critical force. Mech. Solids **51**(3), 263–272 (2016). https://doi.org/10.3103/S0025654416030031
10. Tsarenko, S., Ulitin, G: Investigation of strained deformed state of variable stiffness rod. SpringerPlus (2014). https://doi.org/10.1186/2193-1801-3-367
11. Kukhar, V., Artiukh, V., Prysiazhnyi, A., Pustovgar, A.: Experimental research and method for calculation of 'upsetting-with-buckling' load at the impression-free (dieless) preforming of workpiece. In: E3S Web Conference, vol. 33, p. 02031 (2018). https://doi.org/10.1051/e3sconf/20183302031

12. Sezgen, H.Ç., Çakan, A., Tinkir, M.: Linear buckling analysis of cylinder rods used on industrial 300 Tons H-type hydraulic press. In: Academics World International Conference (2017). https://www.researchgate.net/publication/282278441
13. Dmitriev, M., Korobova, V., Yakubovskaya, I.A.: Increasing punch life in cold cup extrusion with active friction. Russ. Eng. Res. **35**(12), 896–901 (2017). https://doi.org/10.3103/S1068798X15120060
14. Rakhmanov, S.R.: Rod dynamics in a press-roller piercing mill. Steel Transl. **41**(6), 457–460 (2011). https://doi.org/10.3103/S096709121106012X
15. Ding, W.-S., Tian, L., Liu, K.: Analysis of dynamic characteristic of transient impact from hydraulic hammer. J. South China Univ. Technol. (2016). https://doi.org/10.3969/j.issn.1000-565X.2016.11.010
16. Cirek, M., Kubec, V.: Analysis of energy consumption of spindle presses. Tehnički vjesnik **13**(1–2), 23–30 (2006). https://hrcak.srce.hr/8843
17. Grushko, A.V., Kukhar, V.V., Slobodyanyuk, Y.O.: Phenomenological model of low-carbon steels hardening during multistage drawing. Solid State Phenom. Mater. Eng. Technol. Prod. Proc. III **265**, 114–123 (2017). https://doi.org/10.4028/www.scientific.net/SSP.265.114
18. Kukhar, V., Burko, V., Prysiazhnyi, A., Balalayeva, E., Nyhnibeda, M.: Development of alternative technology of dual forming of profiled workpiece obtained by buckling. East.-Eur. J. Enterp. Technol. **3**(7(81)), 53–61 (2016). https://doi.org/10.15587/1729-4061.2016.72063
19. Kukhar, V., Artiukh, V., Serduik, O., Balalayeva, E.: Form of gradient curve of temperature distribution of lengthwise the billet at differentiated heating before profiling by buckling. Proc. Eng. **165**, 1693–1704 (2016). https://doi.org/10.1016/j.proeng.2016.11.911
20. Artiukh, V., Karlushin, S., Sorochan, E.: Peculiarities of mechanical characteristics of contemporary polyurethane elastomers. Proc. Eng. **117**, 933–939 (2015). https://doi.org/10.1016/j.proeng.2015.08.180
21. Kukhar, V., Balalayeva, E., Nesterov, O.: Calculation method and simulation of work of the ring elastic compensator for sheet-forming. In: MATEC Web of Conferences, vol. 129, no. 11, p. 01041 (2017). https://doi.org/10.1051/matecconf/201712901041
22. Kukhar, V., Balalayeva, E., Tuzenko, O., Burko, V.: Calculation of universal elastic compensator applied to the pressing-extrusion operations. Multi. J. Res. Eng. Technol. **2**(3), 593–604 (2015). http://www.mjret.in/V2I3/M18-2-3-7-2015.pdf
23. Balalayeva, E., Kukhar, V., Hrushko, O.: The computer-aided method of calculation of universal elastic rotary compensator for the "Press-and-Die" system errors of crank press for drawing-forming operations. HCTL Open Sci. Technol. Lett. (STL) (2014). http://www.hctl.org/stl/vol6/STL_Article_201408003.pdf
24. Yakovlev, S.N., Mazurin, V.L.: Vibroisolating properties of polyurethane elastomeric materials, used in construction. Mag. Civil Eng. **6**, 53–60 (2017). https://doi.org/10.18720/MCE.74.5
25. Gorbatyuk, S.M., Osadchii, V.A., Tuktarov, E.Z.: Calculation of the geometric parameters of rotary rolling by using the automated design system autodesk inventor. Metallurgist **55**(7–8), 543–546 (2011). https://doi.org/10.1007/s11015-011-9465-8
26. Solomonov, K.N.: Application of CAD/CAM systems for computer simulation of metal forming processes. Mater. Sci. Forum **704–705**, 434–439 (2012)

Noise Spectra of Ball-Rod Hardening of Welds of Rod Structures

Alexey Beskopylny[1(✉)], Alexander Chukarin[2], and Alexandr Isaev[1]

[1] Don State Technical University, Sq. Gagarin 1, 344010 Rostov-on-Don, Russia
besk-an@yandex.ru

[2] Rostov State Transport University, Rostovskogo Strelkovogo Polka Narodnogo Opolcheniya Sq. 2, 344038 Rostov-on-Don, Russia

Abstract. The article presents the results of experimental studies of the spectra of noise and vibration in the processing of welds of rod structures by the method of surface plastic deformation. Ball-hardening (multi-contact vibro-impact tool) can be used for strengthening both flat and curved surfaces, creating compressive residual stresses, smoothing cavities, applying a regular microrelief on friction pairs, and also for processing welds. In the course of experimental studies, hazardous and harmful production factors arising from the implementation of processing welded structures were identified. The results of measuring sound pressure levels at the operator's workplace during compressor operation and processing of welds of I-beams and L-shaped bars are presented. The obtained data are confirmed by the correctness of theoretical conclusions about the laws governing the formation of noise spectra and the contribution of noise sources to the sound field at the operator's workplace. The research results confirm the validity of the theoretical approach to the description of the laws of the process of noise generation.

Keywords: Ball hardening · Sound pressure levels · Noise spectra · Vibration spectra · Acoustic safety

1 Introduction

One of the important tasks of modern transport systems, engineering, construction is to reduce noise and vibrations during the production cycle [1–3]. Protecting the health and performance of operators of transport systems, various technological operations during production by reducing harmful noise and vibrations, is a fundamental task and is reflected in many studies. In [4] the use of intelligent materials for active vibration control is discussed. In the case where the vibration covers a large frequency band, and the vibration must be reduced in an area where neither triggering nor perception is possible, the intelligent structures can be effective.

The problem with the assessment of noise and acoustic pressure is acute in the problems of calculating wind turbines. In [5], the generation of the noise of the blade-

V. Murgul and M. Pasetti (Eds.): EMMFT-2018, AISC 983, pp. 489–495, 2019.
https://doi.org/10.1007/978-3-030-19868-8_48

column interaction from the wind turbine and fans installed on the pylon using a combination of experimental and numerical methods is considered. A computational model based on the solution of the unsteady Reynolds Averaged Navier Stokes equations and Curle's acoustic analogy was used. The fan-blade skew in axial fans has a strong impact on the sound field, the flow field, and their interrelations. In [6], the sound emission and the velocity distributions of three low-pressure axial fans with a similar design point and forward-, backward and unskewed fan blades are investigated. Fluctuations of acoustic pressure beneath hypersonic transitional and turbulent boundary layers and associated acoustic loading on a flat surface was investigated in [7, 8]. A high-order implicit large eddy simulation at Mach 4, 6 and 8 and for different inflow turbulence intensities was studied. In [9], numerical approach was investigated both infinite and finite models. The method included an experimental validation of elastic wave absorption and vibration suppression for flexural wave band gaps in metamaterial plates. Numerical and experimental methods of the application of span-wise waviness to reduce aerodynamic noise from square bars was considered in [10]. Analysis of the flow characteristics showed that at certain wave amplitudes, the transverse flow becomes significant, and strong transverse vortices develop in the near wake, which effectively suppresses the release of the primary vortex.

Technological operations associated with the strain hardening of welds of various designs [11, 12] are associated with high noise levels. A significant problem is the excess of permissible levels of noise arising from the high frequency of rotation of the tool [13]. The highest rotational speeds are noted when working copy-milling machines. In the standards of cutting conditions for wood processing, only the amplitude of the force effect is specified, which makes it difficult to determine the noise spectra. It should also be noted that the construction of mathematical models of noise generation should be preceded by an experimental study of the basic processes of technological processing of the impact process [14].

This paper solves the problem of fulfilling sanitary norms of noise in the operator's working area; therefore, octave sound pressure and vibration levels were measured in the normalized sound frequency range of 31.5–8000 Hz.

2 Materials and Methods

The general acoustic system includes the following sources: a compressor, a machine carrier system, a reinforcer and the part itself on which the welds are hardened. To assess the contribution of the above sources to the formation of the sound field in the working area of the operator, the measurement of the octave sound pressure levels was carried out in the following sequence:

- only the compressor works,
- operating mode when hardening welds.

It should be noted that the noise of the machine itself in the idle mode was not measured, because Actually drive systems do not work. Indeed, the supply of products is carried out on the roller conveyor, and the ball-rod hardener receives the drive from the compressor. Therefore, on the bearing system of the machine, the levels of vibration excited from the most strengthened part were measured, as well as the vibration on the body of the hardener. Experimental studies were carried out by the Assistant Total Noise and Vibration Analyzer (serial number 3049410, accuracy class 1, with a pre-amplifier PU-01 (049010) using a microphone capsule Mk233 (serial No. 719) with a frequency range of measurements from 2 to 40000 Hz.

In the operating mode, the hardening of welds is performed for parts such as channels and angles of various lengths and widths of the flange. The microphone is installed in the workplace. Magnetic mount piezo sensors were installed on the part, body of the reinforcer, table, and machine bed. It should be noted that the vibration levels at the operator's workplace are much lower than the sanitary norms and therefore are not carried out in this work. Vibrations on the parts, the hardener, and the carrier system of the machine were measured in the normalized sound frequency range, i.e., in the range of 31.5–8000 Hz.

3 Results

Measurements of vibrations in the seventh, eighth, and ninth octaves were made of measurements of VShV-003-M3 (at geometric average frequencies of 2000, 4000, 8000 Hz). The measurement results (Fig. 1) showed that the sound pressure levels of the compressor itself exceed the maximum permissible values in the fifth octave (with a geometric average frequency of 500 Hz) by 5 dB, and in the sixth (with a geometric frequency of 1000 Hz) by 4 dB. Also, in the fourth and seventh octaves (with geometric average frequencies of 250 and 2000 Hz, respectively), the sound pressure levels below sanitary standards are only 2 dB.

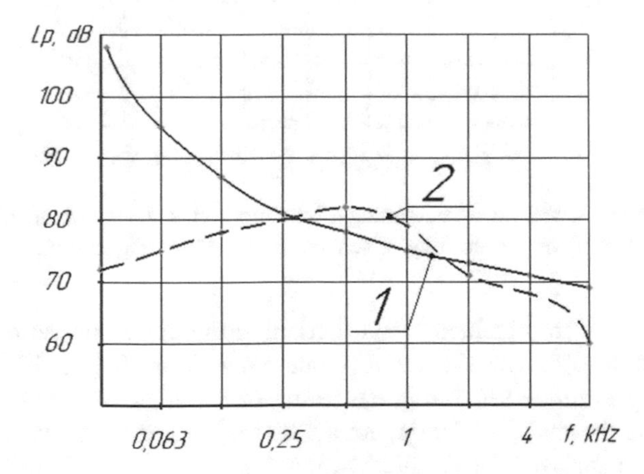

Fig. 1. Noise spectra: 1 - compressor noise spectrum (L1), 2 - limiting spectrum (L2).

The measurement results showed that for parts of the same type, the patterns of formation of noise spectra in the working area and the vibration spectra of all elements of the overall acoustic system (parts of the hardener and the carrier system of the machine itself) are almost identical, and only the levels of the spectral components are different. Therefore, the results of experimental studies are given for the most noise-vibro-active situations of the process of SHS hardening of welds. The results of measuring the noise spectra during hardening of welds of channels and angles are shown in Figs. 2, 3 and 4. Since the channels of various geometrical dimensions and, accordingly, different moments of inertia, are subjected to hardening, in Fig. 2 shows the scattering field for hardening channels, fixed by the appropriate schemes. The analysis of the spectral composition is shown for the noisiest octave situations. It should be noted that the difference in sound pressure levels during the hardening of channels of various sizes is 3–7 dB in the medium and high frequencies (0.25–8 kHz).

In all cases of the ball-rod hardening process, the noise spectra have a pronounced high-frequency character. Despite the fact that the excess of sound pressure levels over the maximum permissible values begins with the fourth octave (geometric mean frequency of 250 Hz), which refers to the mid-frequency range, maximum sound pressure levels are fixed in the sixth to ninth octaves (geometric mean frequencies of 1000–8000 Hz), maximum sound pressure levels in the above range occur when machining iron bar along an I-beam and the maximum value also when machining bar "b".

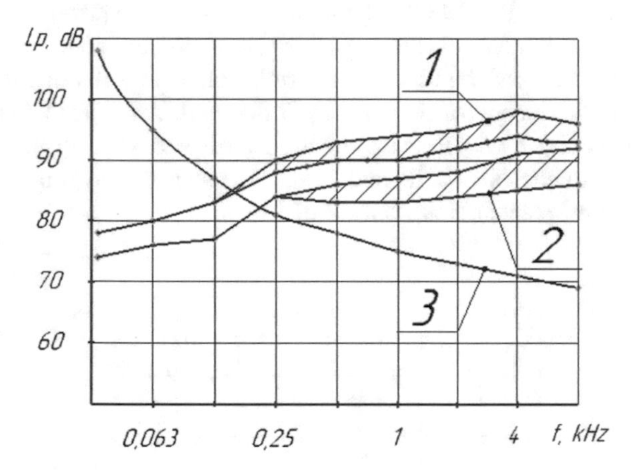

Fig. 2. Noise spectra with ball-rod hardening of channel welds: 1 - with hardening according to scheme a, 2 - with hardening according to scheme b, 3 - limiting spectrum.

Similar results were obtained when hardening the angles of various geometrical dimensions and the difference in sound pressure conditions is 3–6 dB.

The sound pressure levels during hardening of the angle bar are shown in Fig. 3.

As well as in the analysis of the spectra with the hardening of I-beams in this case, the patterns were analyzed for the noisiest conditions.

The sound pressure levels during the hardening of welds occupy an intermediate position between the levels of sound pressure during the hardening of welds on the I-beams. This is explained by the fact that for the options considered, the stiffnesses of the technological system differ.

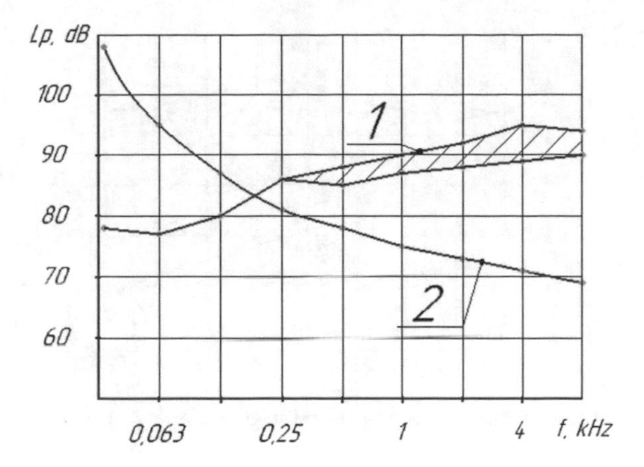

Fig. 3. Noise spectra during hardening of welds at the corners: 1 - noise spectra when processing an angle, 2 - limiting spectrum.

Indeed, the most rigid is the technological system when installing the channel according to the scheme “a”. The rigidity of the system when installing the angle is higher than when processing the channel according to the “b” scheme. Measurement of technological parameters of welded joints is of great practical importance, especially in the analysis of mechanical characteristics in the weld zone and the zone of thermal influence [15]. The process of technological hardening multiple impacts is accompanied by the non-destructive measurement of the strength characteristics of the welded joint. The effects of noise and vibration can significantly affect the accuracy of measurements and make it difficult to study the results of hardening properly.

Measurements of vibration spectra are shown in Fig. 4.

4 Discussion

The data obtained confirm the correctness of the theoretical conclusions about the laws governing the formation of the spectrum and noise sources in the operator’s workplace. The maximum value of vibration levels is observed directly on the workpieces. Vibration levels of the hardener are 10–12 dB in the low and medium frequency range of 31.5–500 Hz and 7–10 dB in the high-frequency range of 1000–8000 Hz. It should be noted that the nature of the vibrations on the workpieces and the hardener correspond to the noise spectra.

The vibration spectra on the carrier system have a low-frequency nature and the frequency range of 500–8000 Hz vibration levels are 20–30 dB lower on the machine table than on the workpiece and the tool. On the machine, the difference in levels is even greater and reaches 38 dB.

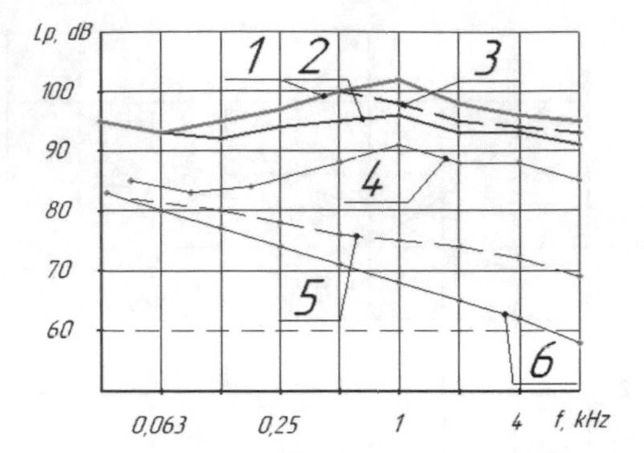

Fig. 4. Vibration spectra in the processing of ball-rod hardening: 1 – I-beam when installing according to the scheme "a", 2 – I-beam when installed according to scheme "b", 3 – L-bar, 4 - hardener, 5 - on the machine table, 6 - on the support.

Thus, the research results confirm the validity of the theoretical approach to the description of the laws of the process of noise formation. The dominant sources of noise, which create an excess of sound pressure levels in the working area of the operator over the maximum permissible values, are the workpiece and the reinforcer. The magnitude of the excess sound pressure levels in the working area of the operator is shown in Fig. 5.

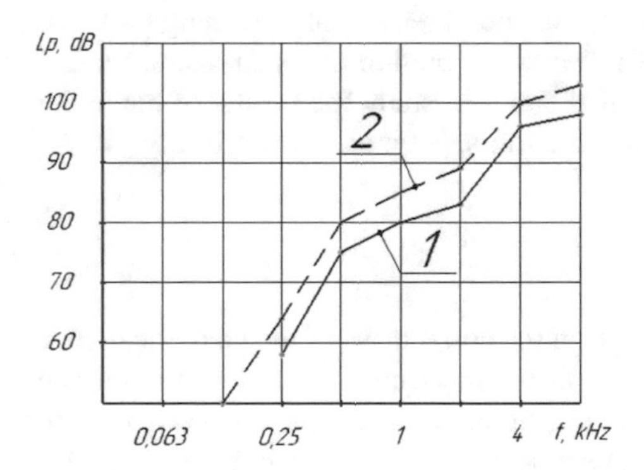

Fig. 5. Exceeding sound pressure levels: 1 - during operation of one installation, 2 - in the area of hardening of welds

At the operator's workplace, when operating one installation, the excess sound pressure levels range from 8 to 28 dB. Attention is drawn to the pattern of "unevenness" of increasing the excess of sound pressure levels in the frequency range. For example, when going from 4 to 5 octaves, the sound pressure level increases by 18 dB.

In the transition from 5 to 6 octaves and from 6 to 7, the increase in sound pressure levels is 5 and 3 dB (respectively). In the eighth octave, the increase in sound pressure level is 12 dB, and the ninth octave is 5 dB.

With the simultaneous operation of several hardening installations of welds, the excess of sound pressure levels above sanitary standards should be further increased by 4–5 dB, which will be from 15 to 33 dB. It should be noted that measurements of vibration levels at the operator's workplace are significantly lower than the maximum permissible values and therefore are not given in this work.

5 Conclusion

The actual problem of experimental studies of the noise and vibration spectra in the processing of welds of core structures by the method of surface plastic deformation is considered. Experimental studies have shown that a multi-contact impact tool using a ball-rod hardener has common patterns in the formation of the noise spectrum. Various modes of sound field formation in the operator's working area and measurement of octave levels of sound pressure were performed for each mode. The results of measurements showed that the noise level in the process of technological hardening significantly exceeds the established standards.

References

1. Kobzev, K., Shamahura, S., Chukarin, A., Bogdanovich, V., Kasyanov, V.: MATEC Web of Conferences, vol. 226, p. 01022 (2018)
2. Kobzev, K., Shamahura, S., Chukarin, A., Bogdanovich, V., Kasyanov, V.: MATEC Web of Conferences, vol. 226, p. 01023 (2018)
3. Beskopylny, A. N., Veremeenko, A. A., Yazyev, B. M.: MATEC Web of Conferences, vol. 106, p. 04004 (2017)
4. Wang, P., Korniienko, A., Bombois, X.: Control Eng. Pract. **84**, 305–322 (2019)
5. Zajamsek, B., Yauwenas, Y., Doolan, C., Hansen, K., Timchenko, V., Reizes, J., Hansen, C.: J. Sound Vib. **443**, 362–375 (2019)
6. Krömer, F., Moreau, S., Becker, S.: J. Sound Vib. **442**, 220–236 (2019)
7. Ritos, K., Drikakis, D., Kokkinakis, I.W.: J. Sound Vib. **441**, 50–62 (2019)
8. Ritos, K., Drikakis, D., Kokkinakis, I.W.: J. Sound Vib. **443**, 90–108 (2019)
9. Zouari, S., Brocail, J., Génevaux, J.-M.: J. Sound Vib. **435**, 246–263 (2018)
10. Liu, X.W., Hu, Z.W., Thompson, D.J., Jurdic, V.: J. Sound Vib. **435**, 323–349 (2018)
11. Beskopylny, A.N., Liapin, A.A., Andreev, V.I.: MATEC, vol. 117, p. 00018 (2017)
12. Beskopylny A., Onishkov N., Korotkin V.: Advances in Intelligent Systems and Computing, vol. 692, pp. 184–191 (2018)
13. Chukarin, A.N., Golosnoy, S.V.: Vestn. Don State Tech. Univ. **2**(89), 79–87 (2017)
14. Beskopylny, A., Veremeenko, A., Kadomtseva, E., Beskopylnaia, N.: Mater. Sci. Forum **931**, 84–90 (2018)
15. Belen'kii, D.M., Beskopylny, A.N., Vernezi, N.L., Shamraev, L.G.: Weld. Int. **11**, 642–645 (1997)

Application of Nonlinear Dynamic Analysis for Calculation of Dynamics and Strength of Mechanical Systems

Andrew Nikitchenko[1], Viktor Artiukh[2](✉), Denis Shevchenko[1], Arkadiy Larionov[3], and Irina Zubareva[4]

[1] All-Union Research and Development Centre for Transportation Technology, Line 23 Vasilyevksy Island 2A, 199106 St. Petersburg, Russia

[2] Peter the Great St. Petersburg Polytechnic University, Polytechnicheskaya 29, 195251 St. Petersburg, Russia

artiukh@mail.ru

[3] Moscow State University of Civil Engineering, Yaroslavskoye shosse 26, 129337 Moscow, Russia

[4] Tyumen Industrial University, Volodarskogo str. 38, 625000 Tyumen, Russia

Abstract. Nonlinear dynamic analysis with application of FEM is progressive direction of computing mechanics. Algorithms used in Nastran software are concentrated on solution of tasks taking into account all types of nonlinearities (geometrical, physical, contact interaction). Comparative analysis of results received by analytical methods and with application of FEM on several test examples is carried out. Features of choice of algorithms and settings of solvers with application of obvious and implicit methods of decision are considered.

Keywords: Nonlinear dynamic analysis · Transient analysis · FEM · NX Nastran · Impact mechanics · Impact of balls · Model of friction · Coulomb friction

1 Introduction

Design offices which are engaged in mechanical engineering use strength calculations with application of finite element method (FEM) in linear static statement. At this definition of strength analysis, design model is usually limited by following conditions:

- properties of material are in borders of elastic deformations;
- large deflections are not considered, i.e. assumption states that loading is attached to not deformable model;
- friction between design details is not considered. Often the design model consists of one body which consists of quite large 3D – solid elements or model consists of shell elements and at the same time plates are modelled by total thickness.

In scientific and technical department of 'All-Union Research and Development Centre for Transportation Technology', method of nonlinear dynamic analysis is successfully used as a part of NX Nastran software. ADINA solvers realizing obvious

V. Murgul and M. Pasetti (Eds.): EMMFT-2018, AISC 983, pp. 496–510, 2019.
https://doi.org/10.1007/978-3-030-19868-8_49

and implicit methods of tasks solution of the nonlinear dynamic analysis (Sol 601/701) are integrated into this software package.

Objectives of this paper are to:

- fulfill on test examples calculation procedure with application of the nonlinear dynamic analysis in CAE NX;
- investigate convergence of results with analytical methods.

2 Methods and Results

Longitudinal impact acting on elastic bar is considered as the first test example by which it is possible to judge applicability of the nonlinear dynamic analysis of CAE NX.

In theoretical mechanics, impact is supposed to be instant owing to what forces arising at impact of rigid bodies are infinitely big therefore only energy and impulses are considered. If at least one of bodies is elastic, then impact duration is always finite and size of force can be determined. Task about impact of elastic bodies in exact statement represents considerable difficulties due to wave nature of deformation distribution inside elastic body.

Solution can be considered of squeezing impact applying to bar when its one end is fixed, and its another end hits a body of big weight that is considered to be absolutely rigid. If mass of the bar is small in comparison with mass of the hitting body then it is possible to consider the bar mass to be absent and in the absence of it a lot of deformation would spread instantly. Main hypothesis of approximate theory of impact consists of deformation that is supposed to be arising instantly in all sections of the bar whereas actually deformation spreads with sound speed from the end where the impact happens. The stated hypothesis means that big weight hits against the bar with a very small speed in comparison with the sound speed and time of impact duration is much more than time necessary for passing of elastic wave along the bar.

It can be assumed that body of mass M moving with speed V_0 hits against the bar and deforms its (refer to Fig. 1). During deformation process, at any moment, sum of kinetic and potential energy of system 'body-bar' is equal to kinetic energy that the body had before impact:

$$T + U = T_0 = \frac{MV_0^2}{2}, \tag{1}$$

where T is kinetic energy of moving body; U is potential energy of elastic deformations of the bar; T_0 is kinetic energy of the body before having contact with the bar; M is mass of the body; V_0 is speed of the body before having contact with the bar.

As deformation grows speed of the body decreases and for a moment it becomes equal to zero while deformation is maximum. At the same time $T = 0$ and maximum movement of the free end of the bar (where impact is applied):

$$\Delta l_{max} = \sqrt{\frac{2T_0 l}{EF}} = V_0\sqrt{\frac{Ml}{EF}} \tag{2}$$

where l is length of the bar in a free state; E is modulus of elasticity of the bar material; F is bar cross-sectional area.

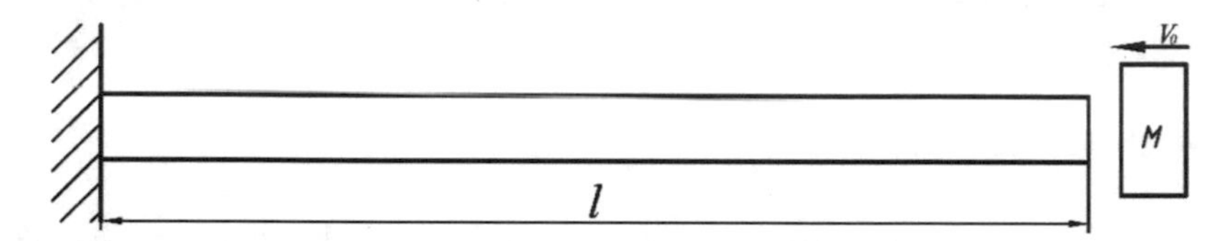

Fig. 1. Design model of longitudinal impact of the body against the weightless bar.

Maximum stress will be observed at maximum deformation

$$\sigma_{max} = E\frac{\Delta l_{max}}{l} = V_0\sqrt{\frac{EM}{Fl}} \tag{3}$$

Movement of the body after contact with the bar is described by the differential equation similar to equation of body movement on a spring

$$M\ddot{x} + kx = 0 \tag{4}$$

where k is longitudinal stiffness of the bar, $k = \frac{EF}{l}$.

Solution of this equation

$$x(t) = \Delta l sin(\omega t)_{max} \tag{5}$$

where ω is circular frequency of oscillation, $\omega = \sqrt{\frac{k}{M}}$.

Time when body will contact the bar will make a half of the period of oscillation then it will jump aside from the bar.

$$\Delta t_{imp} = \frac{1}{4\pi\omega} = \frac{1}{4\pi}\sqrt{\frac{M}{k}} = \frac{1}{4\pi}\sqrt{\frac{Ml}{EF}} \tag{6}$$

Further, model in which the bar mass is replaced by numerous concentrated masses (discrete model) [1] can be considered. The bar can be conditionally divided into n parts. Center of each part is affected by the concentrated inertia force. Longitudinal movements of points *ui* lying in the centers of parts are defined by system of differential equations:

$$\begin{array}{l} -\delta_{11} m_1 \ddot{u}_1 - \delta_{12} m_2 \ddot{u}_2 - \ldots - \delta_{1M} M \ddot{u}_M = u_1 \\ -\delta_{21} m_1 \ddot{u}_1 - \delta_{22} m_2 \ddot{u}_2 - \ldots - \delta_{2M} M \ddot{u}_M = u_2 \\ \ldots\ldots\ldots\ldots\ldots\ldots\ldots\ldots\ldots\ldots\ldots \\ -\delta_{M1} m_1 \ddot{u}_1 - \delta_{M2} m_2 \ddot{u}_2 - \ldots - \delta_{MM} M \ddot{u}_M = u_M \end{array} \tag{7}$$

where δ_{ik} is longitudinal displacement in point with xi coordinate caused by longitudinal single force operating in point with x_k coordinates; m_i is a part mass of the bar with number i, for parts of equal length $m_i = m/10$.

Equations for the bar divided into n = 10 identical parts (refer to Fig. 2) can be developed:

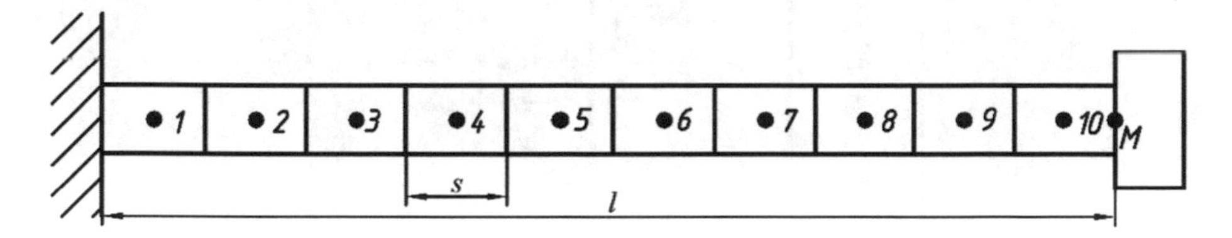

Fig. 2. Discrete model of the bar.

$$-\frac{s}{2EF}\begin{bmatrix} 1 & 1 & 1 & 1 & 1 & 1 & 1 & 1 & 1 & 1 & 1 \\ 1 & 3 & 3 & 3 & 3 & 3 & 3 & 3 & 3 & 3 & 3 \\ 1 & 3 & 5 & 5 & 5 & 5 & 5 & 5 & 5 & 5 & 5 \\ 1 & 3 & 5 & 7 & 7 & 7 & 7 & 7 & 7 & 7 & 7 \\ 1 & 3 & 5 & 7 & 9 & 9 & 9 & 9 & 9 & 9 & 9 \\ 1 & 3 & 5 & 7 & 9 & 11 & 11 & 11 & 11 & 11 & 11 \\ 1 & 3 & 5 & 7 & 9 & 11 & 13 & 13 & 13 & 13 & 13 \\ 1 & 3 & 5 & 7 & 9 & 11 & 13 & 15 & 15 & 15 & 15 \\ 1 & 3 & 5 & 7 & 9 & 11 & 13 & 15 & 17 & 17 & 17 \\ 1 & 3 & 5 & 7 & 9 & 11 & 13 & 15 & 17 & 19 & 19 \\ 1 & 3 & 5 & 7 & 9 & 11 & 13 & 15 & 17 & 19 & 20 \end{bmatrix} \cdot \begin{bmatrix} m_1\ddot{u}_1 \\ m_2\ddot{u}_2 \\ m_3\ddot{u}_3 \\ m_4\ddot{u}_4 \\ m_5\ddot{u}_5 \\ m_6\ddot{u}_6 \\ m_7\ddot{u}_7 \\ m_8\ddot{u}_8 \\ m_9\ddot{u}_9 \\ m_{10}\ddot{u}_{10} \\ M\ddot{u}_M \end{bmatrix} = \begin{bmatrix} u_1 \\ u_2 \\ u_3 \\ u_4 \\ u_5 \\ u_6 \\ u_7 \\ u_8 \\ u_9 \\ u_{10} \\ u_M \end{bmatrix} \tag{8}$$

where s is length of the part of the bar, in this case $s = l/10$.

Bearing in mind that $s = l/10$ and $m_i = m/10$ it can be received

$$\begin{bmatrix} 1 & 1 & 1 & 1 & 1 & 1 & 1 & 1 & 1 & 1 & 1 \\ 1 & 3 & 3 & 3 & 3 & 3 & 3 & 3 & 3 & 3 & 3 \\ 1 & 3 & 5 & 5 & 5 & 5 & 5 & 5 & 5 & 5 & 5 \\ 1 & 3 & 5 & 7 & 7 & 7 & 7 & 7 & 7 & 7 & 7 \\ 1 & 3 & 5 & 7 & 9 & 9 & 9 & 9 & 9 & 9 & 9 \\ 1 & 3 & 5 & 7 & 9 & 11 & 11 & 11 & 11 & 11 & 11 \\ 1 & 3 & 5 & 7 & 9 & 11 & 13 & 13 & 13 & 13 & 13 \\ 1 & 3 & 5 & 7 & 9 & 11 & 13 & 15 & 15 & 15 & 15 \\ 1 & 3 & 5 & 7 & 9 & 11 & 13 & 15 & 17 & 17 & 17 \\ 1 & 3 & 5 & 7 & 9 & 11 & 13 & 15 & 17 & 19 & 19 \\ 1 & 3 & 5 & 7 & 9 & 11 & 13 & 15 & 17 & 19 & 20 \end{bmatrix} \cdot \begin{bmatrix} \ddot{u}_1 \\ \ddot{u}_2 \\ \ddot{u}_3 \\ \ddot{u}_4 \\ \ddot{u}_5 \\ \ddot{u}_6 \\ \ddot{u}_7 \\ \ddot{u}_8 \\ \ddot{u}_9 \\ \ddot{u}_{10} \\ 10\frac{M}{m}\ddot{u}_M \end{bmatrix} = -100\frac{2EF}{ml} \begin{bmatrix} u_1 \\ u_2 \\ u_3 \\ u_4 \\ u_5 \\ u_6 \\ u_7 \\ u_8 \\ u_9 \\ u_{10} \\ u_M \end{bmatrix} \tag{9}$$

Having solved system of algebraic equations (8) relative to $m_i u_i$ and having substituted s = 1/10 and $m_i = m/10$ it can be received

$$\begin{bmatrix} \ddot{u}_1 \\ \ddot{u}_2 \\ \ddot{u}_3 \\ \ddot{u}_4 \\ \ddot{u}_5 \\ \ddot{u}_6 \\ \ddot{u}_7 \\ \ddot{u}_8 \\ \ddot{u}_9 \\ \ddot{u}_{10} \\ 10\frac{M}{m}\ddot{u}_M \end{bmatrix} = -100\frac{EF}{ml} \begin{bmatrix} 3u_1 - u_2 \\ -u_1 + 2u_2 - u_3 \\ -u_2 + 2u_3 - u_4 \\ -u_3 + 2u_4 - u_5 \\ -u_4 + 2u_5 - u_6 \\ -u_5 + 2u_6 - u_7 \\ -u_6 + 2u_7 - u_8 \\ -u_7 + 2u_8 - u_9 \\ -u_8 + 2u_9 - u_{10} \\ -u_9 + 3u_{10} - 2u_M \\ -2u_{10} + 2u_M \end{bmatrix} \quad (10)$$

At the initial moment of time when there is the body contact, the system is affected by the following entry conditions: $u_1 = u_2 = \ldots = u_{10} = u_M = 0,\ \ddot{u}_1 = \ddot{u}_1 = \ldots = \ddot{u}_{10} = 0, \ddot{u}_M = V_0$.

Analytical solution of the Eq. (10) is quite big therefore its decision is numerically executed by means of system of computer algebra Maxima [2].

The nonlinear dynamic analysis with application of FEM (Sol 601) is carried out by direct implicit integration by Newmark's method or by time combined integration method. Operating equation at point of time $t + \Delta t$

$$M\ddot{U}_{t+\Delta t} + C\dot{U}_{t+\Delta t} + K\Delta U_{t+\Delta t} = R_{t+\Delta t} - F_{t+\Delta t} \quad (11)$$

where M is generalized matrix of masses; K is stiffness matrix; C is damping matrix; R is vector of external loads applied at point of time $t = \Delta t$; F is vector of forces in nodes arising at point of time $t = \Delta t$; $\ddot{U}, \dot{U}, U$ are accelerations, velocities and displacement of nodes.

Convergence of the solution on each step is determined by criteria of energy and forces/moments.

Rayleigh damping matrix is added to set matrix of damping for decrease of high-frequency oscillations arising at numerical integration. Calculation of Rayleigh damping matrix is done by usage of matrix of masses and matrix of initial stiffness:

$$C_R = \alpha M + \beta K \quad (12)$$

where M is generalized matrix of masses; K is stiffness matrix, α and β are constants of damping of Rayleigh.

Determination of coefficients α and β generally is not a simple task therefore it is necessary to apply them with caution and only when convergence is absent.

It should be noted that convergence of the implicit dynamic analysis is very sensitive to integration step Δt which is set by the user. Use of a big step causes big errors irrespective of admissions values of convergence criteria.

At the explicit nonlinear dynamic analysis (Sol 701) method of the central differences is used which general equation is

$$\left(\frac{1}{\Delta t^2}M + \frac{1}{2\Delta t}C\right)U_{t+\Delta t} = R_t - F_t + \frac{2}{\Delta t^2}MU_t - \left(\frac{1}{\Delta t^2}M + \frac{1}{2\Delta t}C\right)U_{t-\Delta t} \quad (13)$$

Step of integration has to be defined according to the following criterion

$$\Delta t \leq \Delta t_{CR} = \frac{T_{Nmin}}{\pi} \quad (14)$$

where Δt_{CR} is time critical size of the step; T_{Nmin} is the smallest period of FEM mesh.

If option of automatic time step is chosen then the step is determined by a formula

$$\Delta t = K \cdot \Delta t_{Emin} \quad (15)$$

where K is coefficient which is set in the field of XDTFAC parameters of strategy and it is accepted by default to be equal to 0.9; Δt_{Emin} is critical size of time step.

For solid elements

$$\Delta t_E = \frac{L}{c} \quad (16)$$

where L is characteristic size of the element, c is deformation distribution velocity in material which for 3D - elements is defined by dependence $c = \sqrt{\frac{E(1-\mu)}{\rho(1+\mu)(1-2\mu)}}$.

Calculation of the test example with the following basic data was carried out:

- length of the bar $l = 1000$mm;
- cross-section area $F = 100$mm^2;
- modulus of elasticity $E = 2 \cdot 10^5$N/mm^2;
- density of the material $\rho = 7850$kg/m^3;
- mass of the body $M = 1$kg;
- initial velocity of the body $V_0 = 2$m/s.

Design FEM model is shown on Fig. 3.

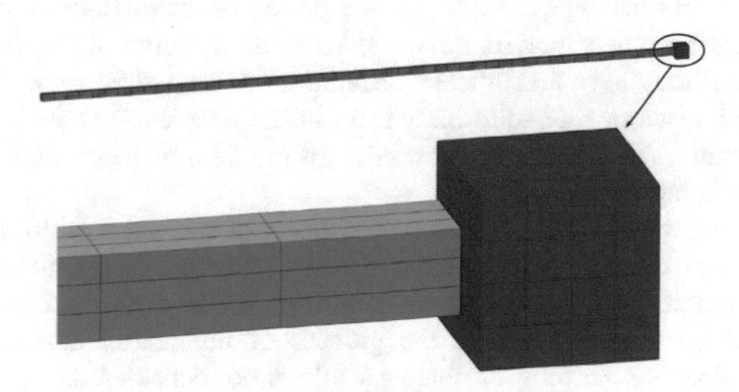

Fig. 3. Design FEM model.

Results of calculation of the zero weight bar (refer to Eq. (5)), system with a discrete mass of the bar (refer to Eq. (10)) and also with application of explicit and implicit methods of the dynamic analysis are shown on Fig. 4 in a form of graphs of displacement changes of the bar free end according to time.

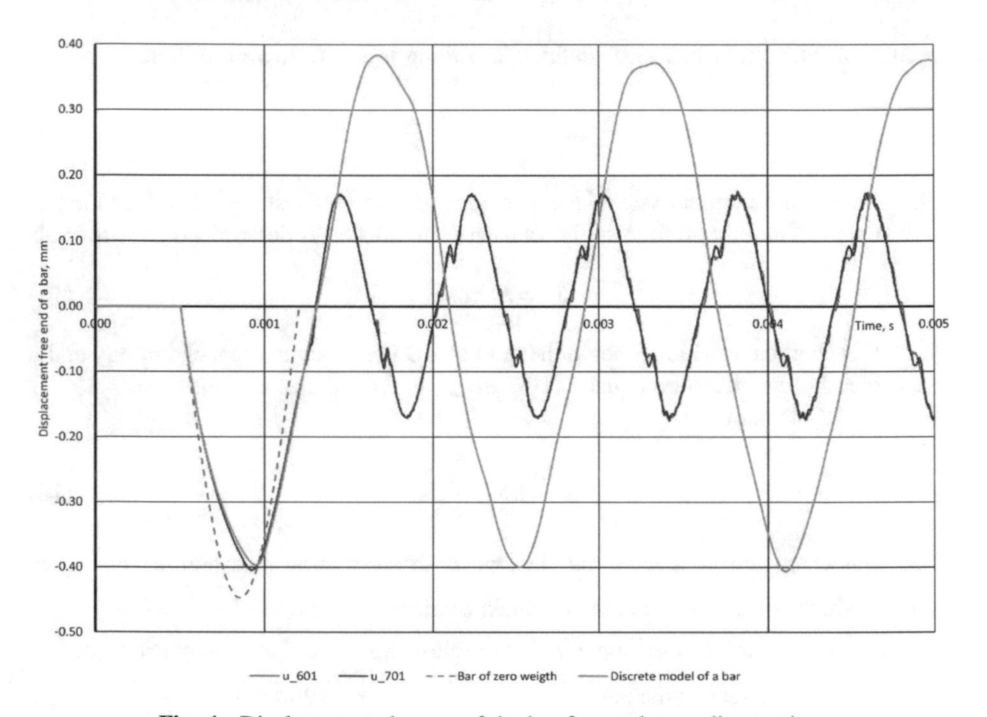

Fig. 4. Displacement changes of the bar free end according to time.

From the chart shown on Fig. 4, it is visible that received results by the explicit and implicit method on FEM model in all interval of time of the decision are practically met. In the period of time while the body is in contact with the bar, good convergence with the analytical method of discrete model of the bar is observed. After a body separation from the bar (in FEM model) in discrete model the fluctuations corresponding to 'glued' body continue. Amplitude of displacements of models in which the body has no weight has bigger value and the period of fluctuations is smaller.

Thus, design models which do not consider force of inertia of the distributed mass of a deformable body have no sufficient accuracy. The analytical solution of problems of dynamics of a deformable solid body even in the simplest models is quite difficult process demanding from engineer knowledge in the field of mechanics and attentiveness during drawing up of equations.

One more interesting example is case of impact of two spheres. In the majority of references impact of spheres is considered from positions of theoretical mechanics. Velocities of spheres after impact proceeding from law of conservation of energy and impulse are defined. Mechanics of the process of interaction during definition of deformations change according to time as a rule is not considered.

In dynamic interaction of two elastic bodies is considered. In both sources, there is quite big conclusion of analytical dependences therefore further only formulas for calculation of key parameters which can be compared to the results received with application of FEM in the nonlinear dynamic analysis are stated. Time t_c during which collision lasts

$$t_c = \left(\frac{M^2}{2k^2 v_0}\right)^{\frac{1}{5}} \int_0^1 \frac{dx}{\sqrt{1 - x^{\frac{1}{5}}}} = \frac{15}{16}\pi\left(\frac{M^2}{2k^2 v_0}\right)^{\frac{1}{5}} \tag{17}$$

Where M is mass of one sphere; k is coefficient depending on mechanical properties of material and radius of the sphere, v_O is velocity of spheres impact.

Coefficient k is determined by a formula

$$\mathrm{k} = \frac{8}{15} \cdot \frac{E}{1 - \mu^2} \cdot \sqrt{\frac{R}{2}} \tag{18}$$

where E is module of elasticity of material; μ is Poisson's ratio; R is sphere radius.

The maximum closing of spheres corresponding to the moment when their relative velocities become equal to zero

$$h_0 = \left(\frac{M}{k}\right)^{\frac{2}{5}} v_0^{\frac{4}{5}} \tag{19}$$

In formulas, it have a bit different appearance, as it is seen further, it gives a bit other result

$$t_c = \frac{15}{32}\pi\left(\frac{4M^2}{E^2(1 - \mu^2)^2 R v_0}\right)^{\frac{1}{5}} \tag{20}$$

$$h_0 = \frac{16 t_c v_0}{15\pi} \tag{21}$$

Equations of speed change of centers of the spheres mass:

– for the sphere which before impact moves with a velocity of V_O

$$V_1(t) = \frac{V_0}{2}\left(1 + cos\left(\frac{\pi}{t_c}t\right)\right) \tag{22}$$

– for the sphere which before impact is at rest

$$V_2(t) = \frac{V_0}{2}\left(1 - cos\left(\frac{\pi}{t_c}t\right)\right) \tag{23}$$

Equations of movements change of centers of the spheres mass:

– for the sphere which before impact moves with a velocity of V_0

$$U_1(t) = \frac{V_0}{2}\left(t + \frac{sin\left(\frac{\pi}{t_c}t\right)}{\omega}\right) \tag{24}$$

– for the sphere which before impact is at rest

$$U_2(t) = \frac{V_0}{2}\left(t - \frac{sin\left(\frac{\pi}{t_c}t\right)}{\omega}\right) \tag{25}$$

Equations of distance change between centers of the spheres mass:

$$h(t) = h_0 sin\left(\frac{\pi}{t_c}t\right) \tag{26}$$

Below given calculation of a test example with the following basic data is given (nonlinear properties of material are not considered):

- sphere radius $R = 50$mm;
- material density $\rho = 10\ 000$kg/m^3;
- elasticity module $E = 1 \cdot 10^4$MPa;
- Poisson's coefficient $\mu = 0.3$;
- mass of the sphere $M = 5.236$kg;
- initial velocity of the sphere $V_0 = 2$m/s.

Design FEM model is shown on Fig. 5. Stress-strain state and deflected mode at time point when distance between mass centers of the spheres reaches the minimum value are shown on Fig. 6.

Results of analytical calculations:

- of dependences are given in Landau and Lifshitz «Theory of Elasticity» and Stronge W.J. «Impact mechanics»;
- with application on FEM model of explicit and implicit solvers are given in Table 1.

Table 1. Comparative results of numerical and analytical calculations.

	Landau and Lifshitz «Theory of Elasticity»	Stronge W.J. «Impact mechanics»	Implicit solver Sol 601	Explicit solver Sol 601
t_c, ms	1.12	1.24	0.98	0.96
h_0, mm	0.44	0.42	0.33	0.33

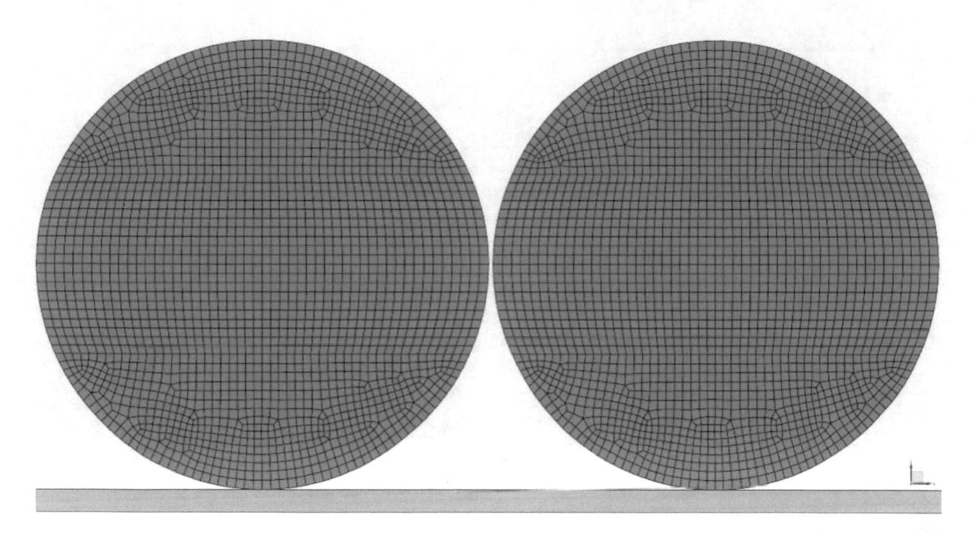

Fig. 5. Design FEM model.

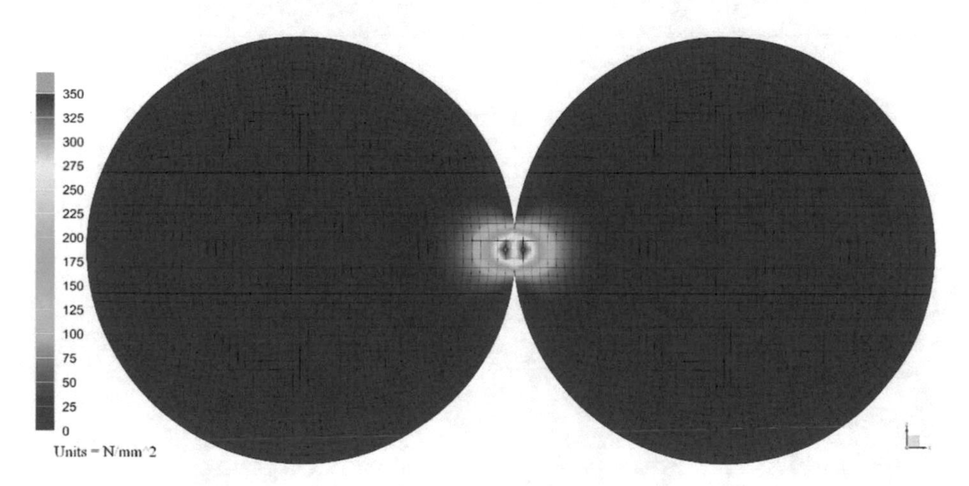

Fig. 6. Equivalent stresses at time when distance between mass centers of the spheres is minimal.

Graph of change in time of displacements of mass centers of the spheres, their velocities and distance between them are shown on Figs. 7, 8 and 9.

Good qualitative convergence of results received analytically and results obtained with application of the nonlinear dynamic analysis is clear from Table 1 and graphs shown on Figs. 7, 8 and 9 while quantitatively quite significantly differs on impact duration and in size of distance between the masses centers. Most likely a number of assumptions and simplifications is the reason of divergences at a conclusion of analytical dependences. For clarification of the matter it is necessary to conduct pilot studies of process of spheres impact with high yield strength (in order to exclude

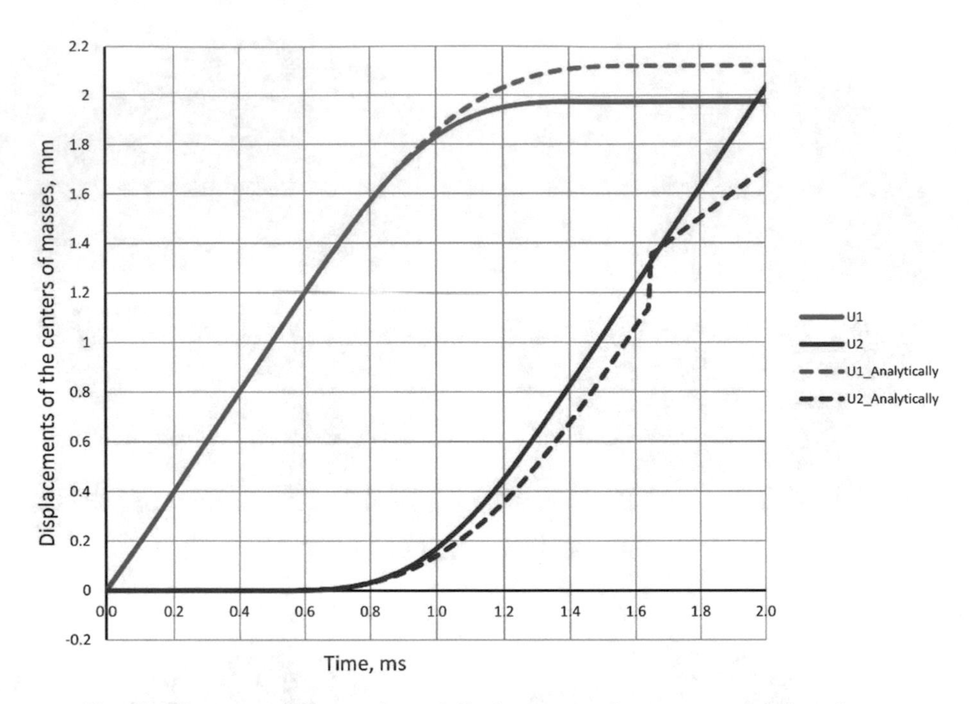

Fig. 7. Change according to time of displacements of mass centers of the spheres.

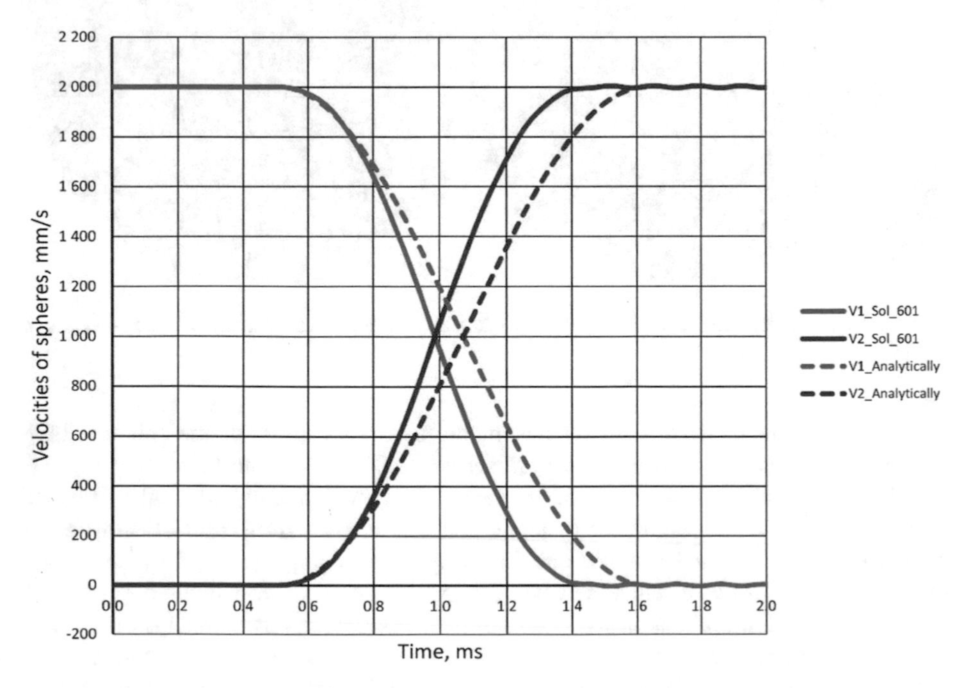

Fig. 8. Change according to time of velocities of mass centers of the spheres.

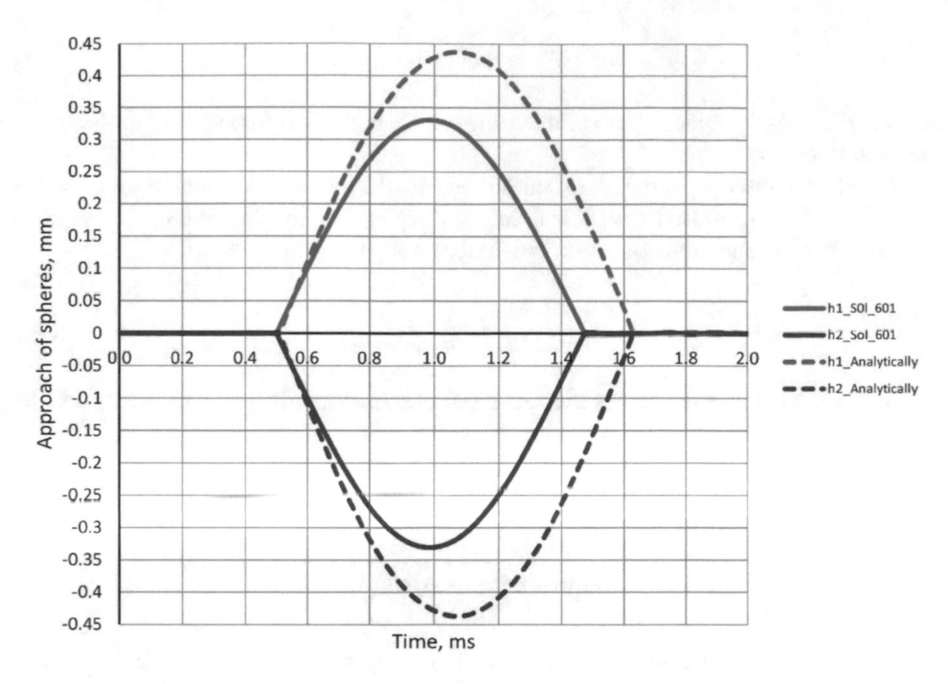

Fig. 9. Change according to time of speeds of mass centers of the spheres.

emergence of plastic deformations) with application of high-speed camera that will allow to define time of the spheres contact and value of their maximum convergence.

Elementary system with Coulomb friction shown on Fig. 10 can be considered as one more example.

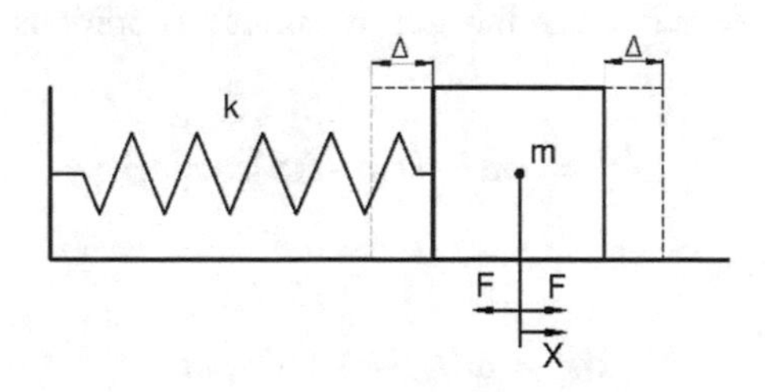

Fig. 10. Elementary system with Coulomb friction.

Block shown on Fig. 10 is in condition of static equilibrium in the range of movements $\Delta \leq x \leq \Delta$ where $\Delta = F_{fr}/k$ designates such provision of system when friction force F_{fr} and restoring force $k \cdot \Delta$ are equal. Friction force works in a direction opposite to direction of the movement. In a general view the equation of the movement of system can be written down.

$$m\ddot{x} + F_{fr} \cdot sign(\dot{x}) + kx = 0 \tag{27}$$

where m is mass of block, k is spring stiffness, F_{fr} is friction force between the block and a surface.

It can be assumed that the block shown on Fig. 10 was moved to the right by value $x_O \geq \Delta$ and then released with initial velocity equal to zero. The block will begin to move to the left and equation of its movement will be

$$\ddot{x} - \frac{F_{fr}}{m} + \frac{k}{m}x = 0 \tag{28}$$

Solution of this equation for the set initial conditions ($x(0) = x_O$, $x\'(0) = 0$) will be

$$x(t) = \varDelta + (x_0 - \varDelta)cos(\omega t) \tag{29}$$

Equation for velocity of block

$$\dot{x}(t) = -\omega(x_0 - \varDelta)sin(\omega t) \tag{30}$$

Thus, movement within interval of time $0 \leq t \leq \pi/\omega$ is harmonious and $\omega = \sqrt{\frac{k}{m}}$ is circular frequency of this movement. In time point t = π/ω, the maximum negative movement is equal to $x_O - 2\Delta$ and sign of speed changes from minus to plus. Then at following interval of time $\pi/\omega \leq t \leq 2\pi/\omega$ the block moves to the right. Equation of the movement within this interval is

$$\ddot{x} = \frac{F_{fr}}{m} + \frac{k}{m}x = 0 \tag{31}$$

Solution of this equation for the set of initial conditions ($x(0) = -(x_O - 2\Delta)$, $x\'(0) = 0$) is

$$x(t) = -\varDelta - (x_0 - 3\varDelta)cos(\omega t) \tag{32}$$

Equation of velocity

$$\dot{x}(t) = \omega(x_0 - 3\varDelta)sin(\omega t) \tag{33}$$

Thus, during time π/ω value of maximum movement decreases by 2Δ and during time $2\pi/\omega$ it decreases by 4Δ. In addition, amplitude of movement decreases by 2Δ for each half-cycle until it becomes less than Δ. Then the block stops in area $\Delta \leq x \leq \Delta$.

FEM model with Coulomb friction between the block and the surface is shown on Fig. 11.

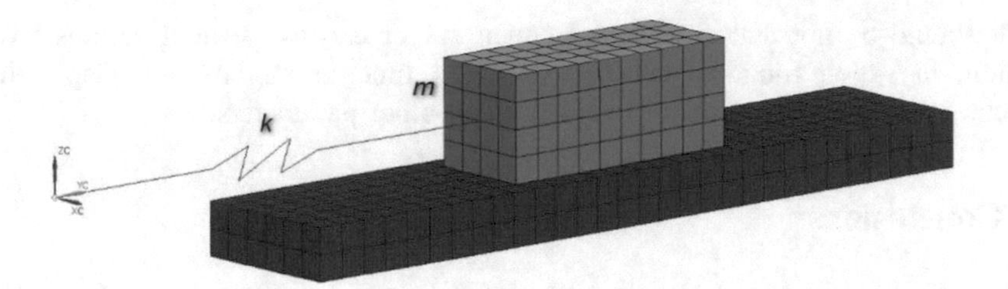

Fig. 11. FEM model with Coulomb friction.

In design model, the following values of parameters are used:

— block mass	$m = 2\text{kg}$;
— spring stiffness	$k = 4\pi^2\text{N/m}$;
— friction coefficient	$f - 0.06$;
— initial movement of the block	$x_0 = 50\text{mm}$.

Results received analytically and practically coincide with results obtained by application of the nonlinear dynamic analysis, i.e. graphs are imposed on each other. In Fig. 12, comparative results received by both methods are presented [3].

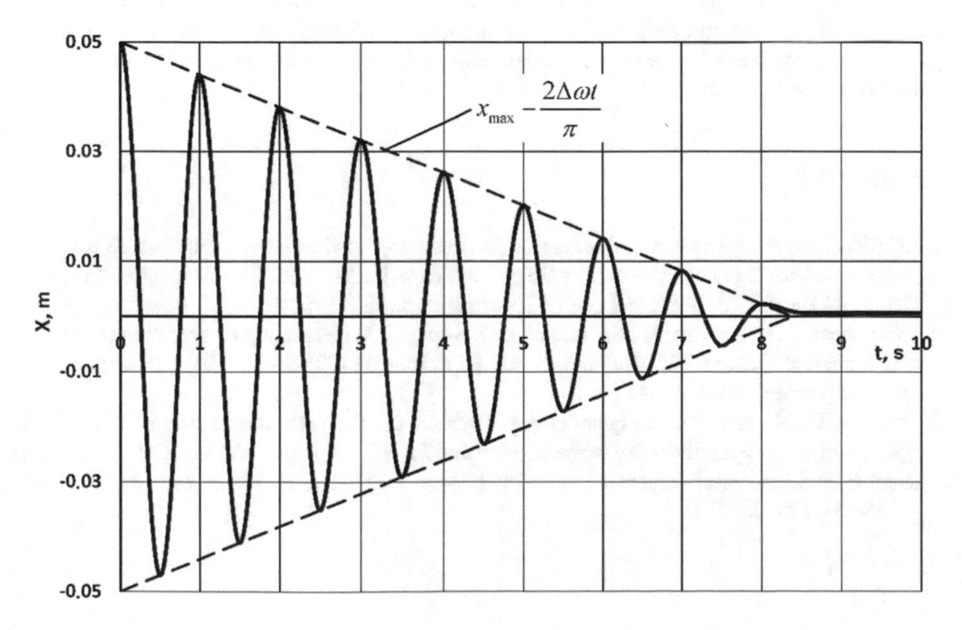

Fig. 12. Displacements of the block according to time.

It should be noted that ADINA implicit solver can use several various laws of friction, in which friction coefficient is set as function of velocity, displacement, direction of movement, time, contact force and other parameters.

3 Conclusions

The comparative analysis of process of longitudinal impact against the bar by analytical methods and with application of FEM is carried out. The results received on analytical model with discretely distributed mass along the bar length have good convergence with the results obtained by nonlinear dynamic analysis with application of FEM by both explicit and implicit methods. The analytical model which does not consider the bar mass has bad convergence and can be applied only in cases when the mass of the hitting body is significantly more than the bar weight. The comparative analysis of impact process of two elastic spheres is also carried out, good qualitative convergence is received. Example of the elementary system with Coulomb friction at constant coefficient of friction is analyzed.

Methods of the nonlinear dynamic analysis can find wide application when calculating difficult mechanical systems including calculation and design of railway cars and their units.

Acknowledgments. The reported study was funded by RFBR according to the research project №. 19-08-01241a 'Multilevel modelling of deformation, filtration and fracture of oil-and-gas geological environments'. The authors declare that there is no conflict of interest regarding the publication of this paper.

References

1. Nikitchenko, A., Artiukh, V., Shevchenko, D., Prakash, R.: Evaluation of interaction between flat car and container at dynamic coupling of flat cars. In: MATEC Web of Conferences, vol. 73, p. 04008 (2016). https://doi.org/10.1051/matecconf/20167304008
2. Nikitchenko, A., Artiukh, V., Shevchenko, D., Murgul, V.: Modeling of operation of elastic-frictional draft gear by NX Motion software. Proc. Eng. **187**, 790–796 (2017). https://doi.org/10.1016/j.proeng.2017.04.441
3. Artiukh, V., Nikitchenko, A., Ignatovich, I., Prykina, L.: The prospects of creation of the draft gear with the polyurethane resin elastic element for the rolling stock. In: IOP Conference Series: Earth and Environmental Science, vol. 90, p. 012191 (2017). https://doi.org/10.1088/1755-1315/90/1/012191

Theoretical Calculations and Study of Horizontal Forces Acting on 4-Hi Sheet Stands During Rolling

Vladlen Mazur[1], Viktor Artiukh[2(✉)], Anatoliy Ishchenko[3], Yuliya Larionova[4], and Natalia Zotkina[5]

[1] LLC 'Saint-Petersburg Electrotechnical Company', Parkovaya Str. 56a Pushkin, 196603 Saint-Petersburg, Russia
[2] Peter the Great St. Petersburg Polytechnic University, Polytechnicheskaya 29, 195251 St. Petersburg, Russia
artiukh@mail.ru
[3] Pryazovskyi State Technical University, Str. Universytets'ka 7, 87555 Mariupol, Ukraine
[4] Moscow State University of Civil Engineering, Yaroslavskoye Shosse 26, 129337 Moscow, Russia
[5] Tyumen Industrial University, Volodarskogo Str. 38, 625000 Tyumen, Russia

Abstract. Technical aspects and detailed method of horizontal forces calculation of contact interactions of WRs chocks (lining straps) and housings (facing strips) on drive side and operator side are shown, which can be used for analysis of dynamic loadings and optimization of design parameters of single sheet rolling stands or the first strip rolling stands of wide strip rolling mills. Given detailed method takes into account: design parameters of sheet rolling stand, contact stiffness of system 'chock – housing', damping inside system 'chock – housing' movements and linear velocities of movements of WR center of mass in fixed coordinate system.

Keywords: Horizontal force · Dynamic loading · Strip rolling mill · Rolling stand · Optimization of design parameters

1 Introduction

To prevent 4-hi rolling stands housings and rolls chocks from wear in rolling direction (horizontal direction), facing strips and lining straps are used. Flat contact surfaces of rolls chocks, lining straps, facing strips and housings are heavily loaded elements of rolling stands. Good conditions of flat contact surfaces are important for quality of rolled metal and values of dynamic forces acting on equipment [1–7]. Revamp or replacement of the aforementioned rolling stands elements is expansive procedure [8–11]. The results of experimental and theoretical studies of horizontal accelerations of rolls, chocks and housings of strip/sheet rolling stands are shown in [12–14] together with detailed analysis of significant influence of gaps in system 'chock – housing' on values of work rolls (WRs) chocks horizontal impacts against housings.

V. Murgul and M. Pasetti (Eds.): EMMFT-2018, AISC 983, pp. 511–521, 2019.
https://doi.org/10.1007/978-3-030-19868-8_50

Technical aspects, methods and results of experimental studies of reversing rolling stands details horizontal accelerations of thick sheet rolling mills 3000 and 3600 are shown in [15], together with sequence of horizontal movements of bottom work roll (BWR) with chocks in windows of housings during normal metal-in, steady rolling and metal-out. It helped to develop method and define horizontal forces of rolls chocks impacts against housings of strip/thick sheet rolling stands when accelerations, movements and masses are known.

2 Formulation of Task

Experimental results shown in [15] are suitable for dynamic calculation and analysis of described horizontal forces taking into consideration design and dimensions of 4-hi sheet rolling stand. General view of reversing 4-hi sheet rolling stand 3000 on operator side is shown on Fig. 1, where:

- pos. 1 is BWR chock;
- pos. 2 is housing;
- MAX DESIGN GAP is gap between lining straps of WR chocks and facing strips of housings/back-up roll (BUR) chocks as per drawings of reversing 4-hi sheet rolling stand 3000;
- $\Sigma\Delta_{\text{DS DESIGN GAP}}$ is total gap on drive side (DS) equal to two MAX DESIGN GAPs;
- $\Sigma\Delta_{\text{OS DESIGN GAP}}$ is total gap on operator side (OS) equal to two MAX DESIGN GAPs;
- F_{HOR} is described horizontal force.

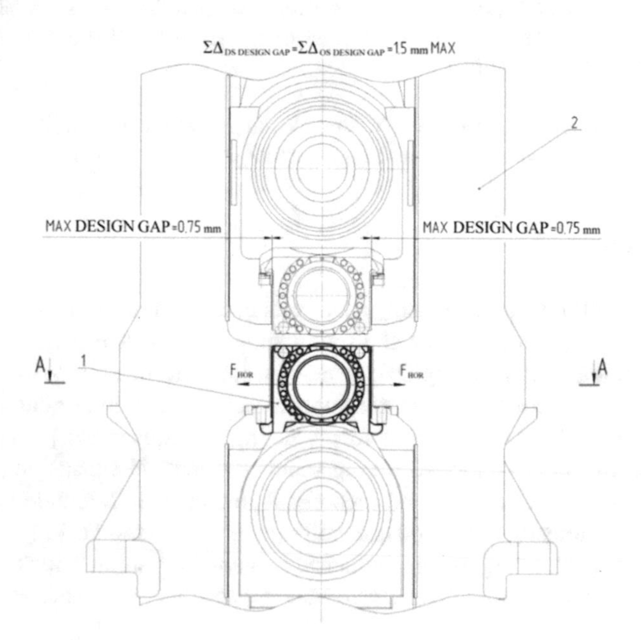

Fig. 1. General view of reversing 4-hi sheet rolling stand 3000 on OS.

Section A-A from Fig. 1 is shown on Fig. 2, where:

- pos. 1 is BWR;
- pos. 2 is lining strap of BWR chock;
- pos. 3 is facing strip of housing pos. 4;
- pos. 5 is bottom BUR;
- MOVEMENT is direction of horizontal WR movements.

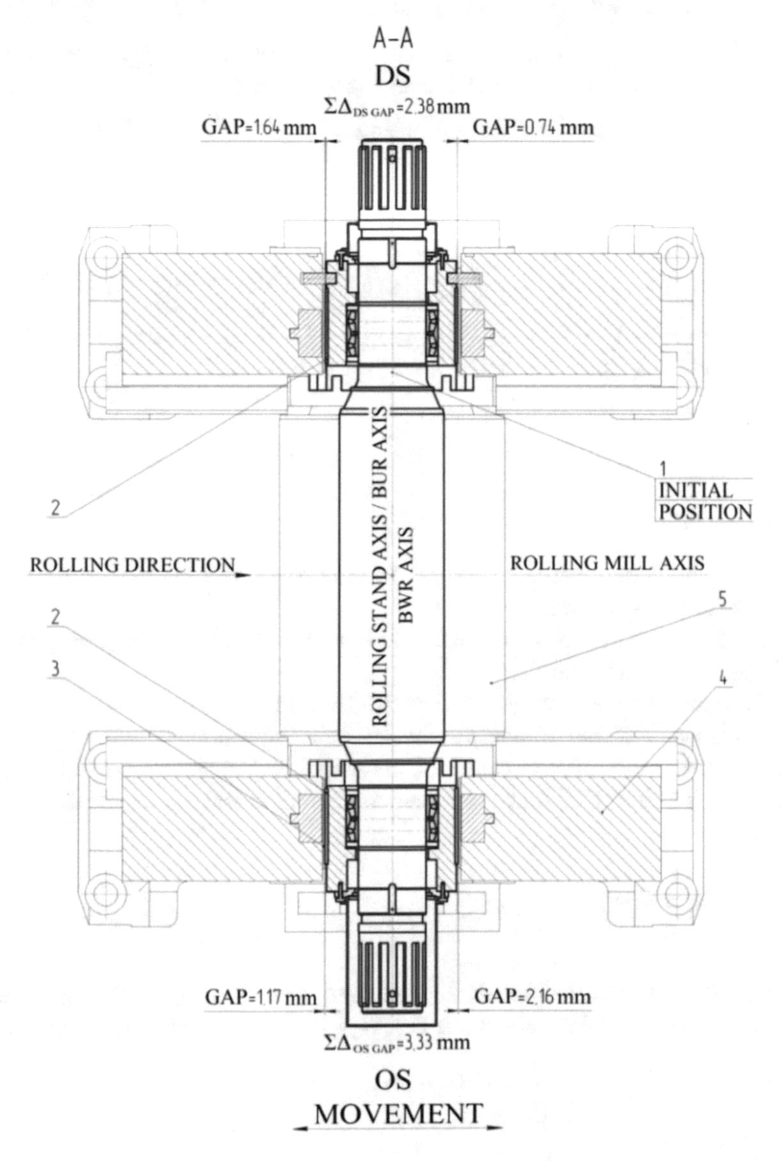

Fig. 2. Section A-A from Fig. 1.

3 Materials and Methods

According to [9] considering that widths of WRs chocks on DS and OS are different horizontal forces of contact interactions of WRs chocks (lining straps) and housings (facing strips) on DS and OS can be calculated by below given formulas:

$$F_{HOR\,DS}=\begin{cases} c_{DS}\left[x+x_0-\frac{L}{2}\beta+\frac{\delta}{2}-\frac{\Delta_{DS}}{2}(1-\operatorname{sgn}e)+\left(z-\frac{B_{DS}}{2}\beta\right)tg\,\beta_{DS}\right] \\ +\,\mu_{DS}\left[\dot{x}-\frac{L}{2}\dot{\beta}\quad\left(\frac{B_{DS}}{2}\dot{\beta}-\dot{z}\right)tg\,\beta\right] \text{ when } x>\begin{pmatrix}\frac{\Delta_{DS}}{2}(1-\operatorname{sgn}e)-\left(z-\frac{B_{DS}}{2}\beta\right)tg\,\beta_{DS} \\ -x_0-\frac{L}{2}\beta-\frac{\delta}{2}\end{pmatrix}; \\ 0 \text{ when } -\frac{\Delta_{DS}}{2}-\frac{B_{DS}}{2}\beta tg\,\beta_{DS}\le x+x_0-\frac{L}{2}\beta+\frac{\delta}{2}+\frac{\Delta_{DS}}{2}\operatorname{sgn}e+ztg\,\beta_{DS}\le\frac{\Delta_{DS}}{2}+\frac{B_{DS}}{2}\beta tg\,\beta_{DS}; \\ c_{DS}\left[x+x_0-\frac{L}{2}\beta+\frac{\delta}{2}+\frac{\Delta_{DS}}{2}(1+\operatorname{sgn}e)+\left(z+\frac{B_{DS}}{2}\beta\right)tg\,\beta_{DS}\right] \\ +\,\mu_{DS}\left[\dot{x}-\frac{L}{2}\dot{\beta}+\left(\frac{B_{DS}}{2}\dot{\beta}+\dot{z}\right)tg\,\beta\right] \text{ when } x<\begin{pmatrix}\frac{\Delta_{DS}}{2}(1+\operatorname{sgn}e)-\left(z+\frac{B_{DS}}{2}\beta\right)tg\,\beta_{DS} \\ -x_0+\frac{L}{2}\beta-\frac{\delta}{2}\end{pmatrix}; \end{cases} \tag{1}$$

$$F_{HOR\,OS}=\begin{cases} c_{OS}\left[x+x_0+\frac{L}{2}\beta-\frac{\delta}{2}-\frac{\Delta_{OS}}{2}(1-\operatorname{sgn}e)-\left(z-\frac{B_{OS}}{2}\beta\right)tg\,\beta_{OS}\right] \\ +\,\mu_{OS}\left[\dot{x}+\frac{L}{2}\dot{\beta}-\left(\dot{z}-\frac{B_{OS}}{2}\dot{\beta}\right)tg\,\beta_{OS}\right] \text{ when } x>\begin{pmatrix}\frac{\Delta_{OS}}{2}(1-\operatorname{sgn}e)+\left(z-\frac{B_{OS}}{2}\beta\right)tg\,\beta_{OS} \\ -x_0-\frac{L}{2}\beta+\frac{\delta}{2}\end{pmatrix}; \\ 0 \text{ when } -\frac{\Delta_{OS}}{2}+\frac{B_{OS}}{2}\beta tg\,\beta_{OS}\le x+x_0+\frac{L}{2}\beta-\frac{\delta}{2}+\frac{\Delta_{OS}}{2}\operatorname{sgn}e-ztg\,\beta_{OS}\le\frac{\Delta_{OS}}{2}-\frac{B_{OS}}{2}\beta tg\,\beta_{OS}; \\ c_{OS}\left[x+x_0+\frac{L}{2}\beta-\frac{\delta}{2}+\frac{\Delta_{OS}}{2}(1+\operatorname{sgn}e)-\left(z+\frac{B_{OS}}{2}\beta\right)tg\,\beta_{OS}\right] \\ +\,\mu_{OS}\left[\dot{x}+\frac{L}{2}\dot{\beta}-\left(\dot{z}+\frac{B_{OS}}{2}\dot{\beta}\right)tg\,\beta_{OS}\right] \text{ when } x<\begin{pmatrix}-\frac{\Delta_{OS}}{2}(1+\operatorname{sgn}e)+\left(z+\frac{B_{OS}}{2}\beta\right)tg\,\beta_{OS} \\ -x_0-\frac{L}{2}\beta+\frac{\delta}{2}\end{pmatrix}; \end{cases} \tag{2}$$

where

- $F_{HOR\,DS}$ is horizontal force of contact interaction of WR chock (lining strap) and housing (facing strip) on DS;
- $F_{HOR\,OS}$ is horizontal force of contact interaction of WR chock (lining strap) and housing (facing strip) on OS;
- C_{DS} is contact stiffness of system 'chock – housing' on DS;
- C_{OS} is contact stiffness of system 'chock – housing' on OS;
- *x, z* are movements of WR center of mass (point C is shown on Fig. 3) in fixed coordinate system OXYZ at the moment of time when $F_{HOR\,DS}$ and $F_{HOR\,OS}$ are calculated;

$$x_0=\frac{B_{DS}+B_{OS}}{8}\beta_{(t=0)}\left(tg\,\beta_{DS}-tg\,\beta_{OS}\right)\operatorname{sgn}e+\frac{Qe}{4(R+r)}\left(1/c_{OS}+1/c_{DS}\right)$$

is initial position of WR before moment of time when $F_{HOR\,DS}$ and $F_{HOR\,OS}$ are calculated; B_{DS} is width of WR chock on DS; B_{OS} is width of WR chock on OS;

$$\beta_{(t=0)}=\frac{\delta+0.5eQ(1/c_{OS}-1/c_{DS})/(R+r)}{L+0.25(B_{DS}+B_{OS})(tg\,\beta_{DS}+tg\,\beta_{OS})\operatorname{sgn}e}$$

is initial angle of WR turn in horizontal plane before moment of time when $F_{HOR\,DS}$ and $F_{HOR\,OS}$ are calculated; β_{DS} is angle of inclines of contact surfaces of WR chock (lining strap) and housing (facing strip) on DS; β_{OS} is angle of inclines of contact surfaces of WR chock (lining strap) and housing (facing strip) on OS; e is design offset of WRs relative to BURs,

$$\operatorname{sgn} e = \begin{cases} 1 & \text{when } e \geq 0; \\ -1 & \text{when } e < 0. \end{cases}$$

- Q is net force of balance force and WR assembly weight;
- R is radius of BUR that in contact with WR that has radius r;
- L is distance between centers of WR bearings;
- β is angle of WR turn in horizontal plane at the moment of time when $F_{HOR\,DS}$ and $F_{HOR\,OS}$ are calculated;
- δ is half of difference of housings window widths on DS and OS;
- Δ_{DS} is gap between contact surfaces of WR chock (lining strap) and housing (facing strip) on DS at the moment of time when $F_{HOR\,DS}$ is calculated;
- Δ_{OS} is gap between contact surfaces of WR chock (lining strap) and housing (facing strip) on OS at the moment of time when $F_{HOR\,OS}$ is calculated;
- μ_{DS} is damping coefficient of system 'chock – housing' on DS;
- μ_{OS} is damping coefficient of system 'chock – housing' on OS;
- $\dot{x}, \dot{z}$ are linear velocities of movements of WR center of mass (point C) in fixed coordinate system OXYZ at the moment of time when $F_{HOR\,DS}$ and $F_{HOR\,OS}$ are calculated;
- $\dot{\beta}$ is angular velocity of WR turn in horizontal plane at the moment of time when $F_{HOR\,DS}$ and $F_{HOR\,OS}$ are calculated;

Coefficient describing friction between contact surfaces of WR chock (lining strap) and housing (facing strip):

$$f = \begin{cases} f^{SL} \operatorname{sgn}[\dot{z}(F_{HOR\,DS} + F_{HOR\,OS})] \text{ when } |\dot{z}| > 0; \\ f^{SL} \operatorname{sgn} f^{ST} \text{ when } |\dot{z}| = 0 \text{ and } |f^{ST}| \geq f^{SL}; \\ f^{ST} \text{ when } |\dot{z}| = 0 \text{ and } |f^{ST}| < f^{SL}, \end{cases} \tag{3}$$

where f^{SL} is sliding friction coefficient between contact surfaces of WR chock (lining strap) and housing (facing strip); f^{ST} is static friction coefficient between contact surfaces of WR chock (lining strap) and housing (facing strip);

$$\operatorname{sgn}[\dot{z}(F_{HOR\,DS} + F_{HOR\,OS})] = \begin{cases} 1 & \text{when } \dot{z}(F_{HOR\,DS} + F_{HOR\,OS}) \geq 0; \\ -1 & \text{when } \dot{z}(F_{HOR\,DS} + F_{HOR\,OS}) < 0, \end{cases} \tag{4}$$

$$\operatorname{sgn} f^{ST} = \begin{cases} 1 & \text{when } f^{ST} \geq 0; \\ -1 & \text{when } f^{ST} < 0, \end{cases} \tag{5}$$

Horizontal forces calculation scheme of contact interactions of WRs chocks (lining straps) and housings (facing strips) on DS and OS of 4-hi rolling stand based on BWR is shown on Figs. 3 and 4, where index 11 states for top BUR, index 1 states for top WR, index 2 states for BWR, and index 22 states for bottom BUR.

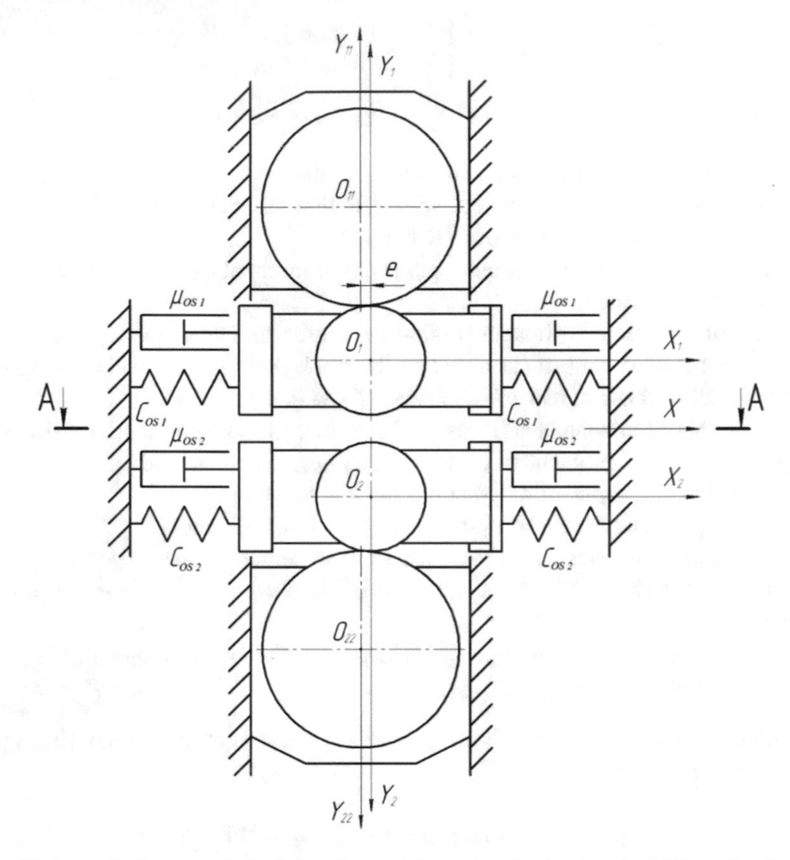

Fig. 3. Horizontal forces calculation scheme of contact interactions of WRs chocks (lining straps) and housings (facing strips) on OS of 4-hi rolling stand.

Given on Fig. 4 parameters can be determined by analysis of BWR movements and horizontal accelerations taken from experimental results of pass No. 4 (metal-in and metal-out) shown in [15]. Positions of BWR before and during metal-in at pass No. 4 are shown on Fig. 5 where some accepted dimensions used to position BWR are indicated by arrows. Positions of BWR before and during metal-out at pass No. 4 are shown on Fig. 6 where some accepted dimensions used to position BWR are indicated by arrows. Angle α on Figs. 5 and 6 is angle of WR turn in horizontal plane at the moment of time when the horizontal forces are calculated.

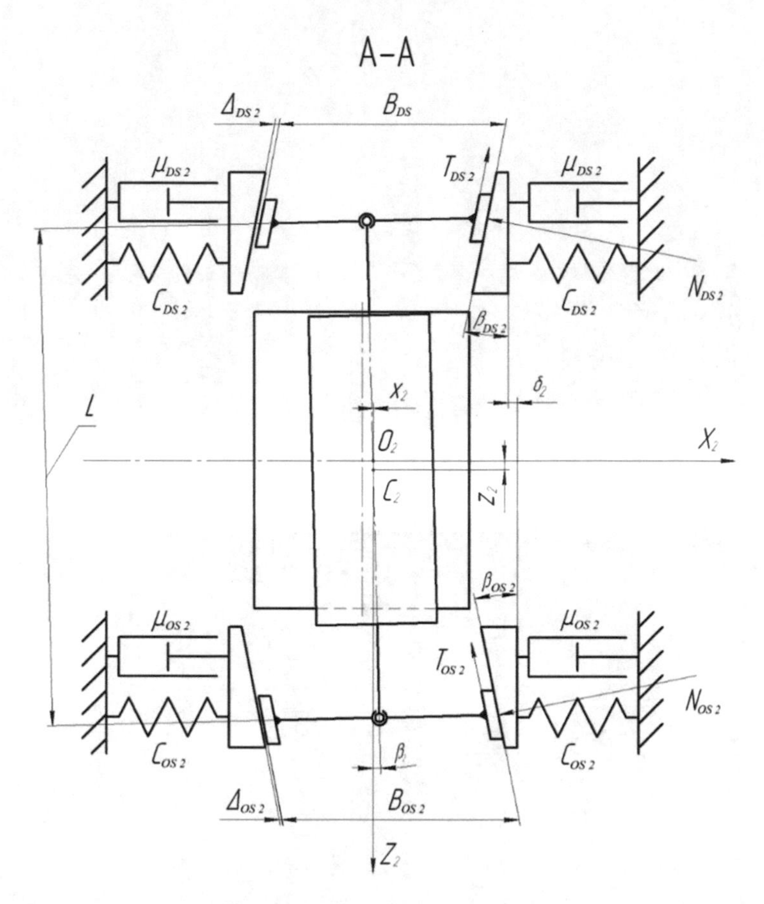

Fig. 4. Section A-A from Fig. 3.

4 Results and Discussions

Formulas (1) and (2) are intended for analysis of dynamic loadings and optimization of design parameters of single sheet rolling stands or the first strip rolling stands of wide strip rolling mills [16, 17].

Calculation of $F_{HOR\,DS}$ and $F_{HOR\,OS}$ can be described in terms of detailed engineering calculation that takes into account: design parameters of sheet rolling stands, contact stiffness of system 'chock – housing', damping inside system 'chock – housing' movements and linear velocities of movements of WR center of mass in fixed coordinate system, etc.

- $F_{HOR\,OS} = 1.62 \cdot 10^6 N$ as per formula (2) and Fig. 5.
- $F_{HOR\,DS} = 1.68 \cdot 10^6 N$ as per formula (1) and Fig. 6.

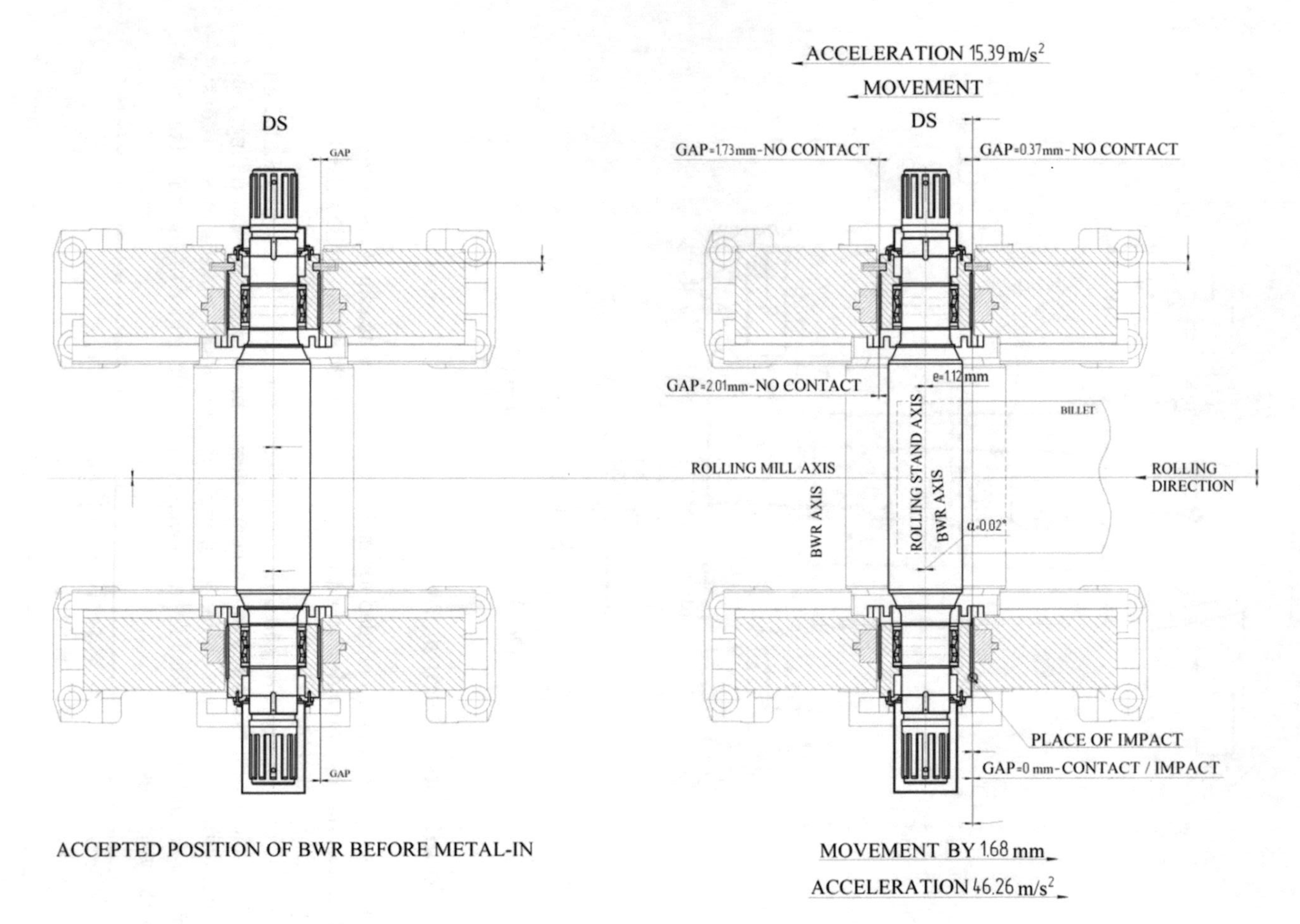

Fig. 5. Positions of BWR before and during metal-in at pass No. 4.

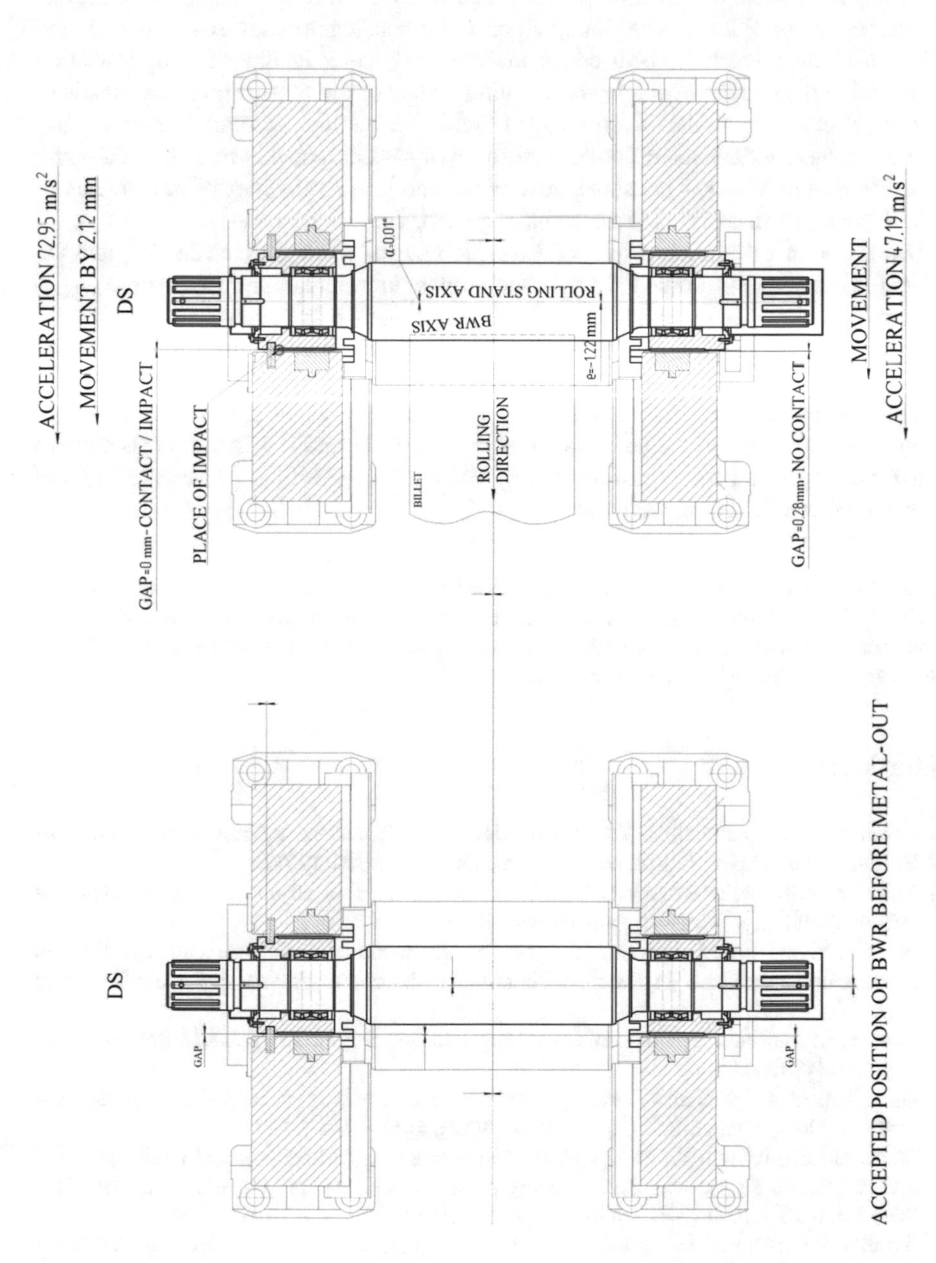

Fig. 6. Positions of BWR before and during metal-out at pass No. 4.

5 Conclusions

1. Technical aspects and detailed method of horizontal forces calculation of contact interactions of WRs chocks (lining straps) and housings (facing strips) on DS and OS are shown which can be used for analysis of dynamic loadings and optimization of design parameters of single sheet rolling stands or the first strip rolling stands of wide strip rolling mills. Given detailed method takes into account: design parameters of sheet rolling stands, contact stiffness of system 'chock – housing', damping inside system 'chock – housing' movements and linear velocities of movements of WR center of mass in fixed coordinate system, etc.
2. Difference in calculation results of $F_{HOR\,OS}$ between method given in [15] and the detailed method is around 14.8%. Difference in calculation results of $F_{HOR\,DS}$ between method given in [15] and the detailed method is around 6%.
3. Given detailed method can be used to calculate relations between the horizontal forces and wear of rolling stands assemblies in order to define maximum permissible values of details wear that can be graphically shown as per [18]. In addition, by means of usage of methods discussed in this paper and [19, 20], it is possible to use the horizontal forces during rolling as indicator of rolling process ability and technical conditions of rolling stands.

Acknowledgments. The reported study was funded by RFBR according to the research project № 19-08-01252a 'Development and verification of inelastic deformation models and thermal fatique fracture criteria for monocrystalline alloys'. The authors declare that there is no conflict of interest regarding the publication of this paper.

References

1. Verenev, V.V., Putnoki, A.Yu., Podobedov, N.I.: Transient Processes in Continuous Rolling. Monography (in Russian). Litograf, Dnepropetrovsk (2017)
2. Verenev, V.V., Bolshakov, V.I., Yunakov, A.M.: Fundamental and applied problems of ferrous metallurgy. Works Dnepropetrovsk **19**, 346–358 (2009)
3. Artiukh, V.G.: Nagruzki i peregruzki v metallurgicheskih mashinah [Loads and overloads in metallurgical machines] (in Russian). Monograph. Mariupol: Pryazovskyi State Technical University (2008)
4. Krot, P.V.: Statistical dynamics of the rolling mills. St. Petersburg. IUTAM Bookseries **27** (4), 429–442 (2011)
5. Gharaibeh, N.S., Matarneh, M.I., Artyukh, V.G.: Loading decrease in metallurgical machines. Res. J. Appl. Sci. Eng. Technol. **8**(12), 1461–1464 (2014)
6. Sorochan, E., Artiukh, V., Melnikov, B., Raimberdiyev, T.: Mathematical model of plates and strips rolling for calculation of energy power parameters and dynamic loads. MATEC Web Conf. **73**, 04009 (2016). https://doi.org/10.1051/matecconf/20167304009
7. Artiukh, V.G.: Osnovi zashiti metallurgicheskih mashin ot polomok [Basics of protection of metallurgical machines from breakdowns] (in Russian). Monograph. Mariupol: Pryazovskyi State Technical University, published group 'University' (2015)
8. Verenev, V.V.: Dynamic Processes in Cold Strip Rolling Mills (in Russian). Monograph. Dnepropetrovsk: LIRA (2015)

9. Kitaeva, D., Rudaev, Y., Subbotina, E.: About the volume forming of aluminium details in superplasticity conditions. In: METAL 2014 – 23rd International Conference on Metallurgy and Materials, Conference Proceedings, pp. 347–352 (2014)
10. Solomonov, K.N.: Application of CAD/CAM systems for computer simulation of metal forming processes. Mater. Sci. Forum **704–705**, 434–439 (2012)
11. Levandovskiy, A.N., Melnikov, B.E., Shamkin, A.A.: Modeling of porous material fracture. Mag. Civil Eng. **1**, 3–22 (2017). https://doi.org/10.18720/MCE.69.1
12. Mazur, V., Artiukh, V., Sagirov, Yu., Kuznezov, S.: Experimental determination and study of horizontal forces during rolling. MATEC Web Conf. **239**, 01042 (2018). https://doi.org/10.1051/matecconf/201823901042
13. Artiukh, V., Mazur, V., Adamtsevich, A.: Priority influence of horizontal forces at rolling on operation of main sheet rolling equipment. MATEC Web Conf. **106**, 04001 (2017). https://doi.org/10.1051/matecconf/201710604001
14. Artiukh, V., Mazur, V., Butyrin, A.: Analysis of stress conditions of rolling stand elements. In: Murgul, V., Popovic, Z. (eds.) International Scientific Conference Energy Management of Municipal Transportation Facilities and Transport, EMMFT 2017. Advances in Intelligent Systems and Computing, vol. 692, pp. 212–219. Springer, Cham (2018). https://doi.org/10.1007/978-3-319-70987-1_23
15. Ishchenko, A., Artiukh, V., Mazur, V., Calimgareeva, A., Gusarova, M.: Experimental study of horizontal impact forces acting on equipment of thick sheet rolling stands during rolling. MATEC Web Conf. **239**, 01041 (2018). https://doi.org/10.1051/matecconf/201823901041
16. Artiukh, V., Mazur, V., Shilova, L.: Device for making horizontal wedge thrust of rolling stand. MATEC Web Conf. **106**, 03002 (2017). https://doi.org/10.1051/matecconf/201710603002
17. Artiukh, V., Mazur, V., Kargin, S., Bushuev, N.: Reasonability to use device for making horizontal wedge thrust of rolling stand. MATEC Web Conf. **170**, 03011 (2018). https://doi.org/10.1051/matecconf/201817003011
18. Artiukh, V., Mazur, V., Nidziy, E.: Evaluation of surface deformations of rolling mills stands elements resulting from horizontal forces acting at rolling. In: Murgul, V., Popovic, Z. (eds.) International Scientific Conference Energy Management of Municipal Transportation Facilities and Transport EMMFT 2017. Advances in Intelligent Systems and Computing, vol. 692, pp. 1065–1073. Springer, Cham (2018). https://doi.org/10.1007/978-3-319-70987-1_115
19. Mazur, V., Artiukh, V., Matarneh, M.I.: Horizontal force during rolling as indicator of rolling technology and technical conditions of main rolling equipment. Procedia Eng. **165**, 1722–1730 (2016). https://doi.org/10.1016/j.proeng.2016.11.915
20. Zakharov, A.N., Gorbatyuk, S.M., Borisevich, V.G.: Modernizing a press for making refractories. Metallurgist **52**(7–8), 420–423 (2008). https://doi.org/10.1007/s11015-008-9072-5

Effect of the Rigidity Coefficient of Stern Bearing on Its Own Frequencies During Bending Vibrations of Stern Shaft

Aleksey Halyavkin[1], Igor Razov[2](✉), Anna Auslender[3], and Sergey Makeev[4]

[1] OOO «Gazprom dobycha Astrakhan», Lenina str. 30, 414000 Astrahan', Russia
[2] Industrial University of Tyumen, Volodarskogo 38, 625001 Tyumen, Russia
razovio@mail.ru
[3] Astrahan' State Technical University, Tatishheva 16, 414056 Astrahan', Russia
[4] The Siberian State Automobile and Highway University, Mira 5, 644080 Omsk, Russia

Abstract. The paper examines the effect of the rigidity coefficient of stern bearings on the performance of the stern shaft. The existing values of the rigidity coefficient used in the calculation of bending vibrations and in the alignment of the stern shaft are analyzed. It is indicated that the stern shaft is a very complex dynamic system, since it operates under the action and the occurrence of constant, variable, cyclic, and random loads. The effect of the rigidity coefficient on the natural frequency of the bending vibrations of the stern shaft is studied. The purpose of the calculation of bending vibrations is to determine the natural frequency of the stern shaft, which should be 20 … 40% greater than the working (blade) frequency. An experimental study of the effect of wear of stern bearings on the value of the natural frequency during bending vibrations was carried out.

Keywords: Stern shaft · Stern bearing · Rigidity coefficient · Bending vibrations · Wear

1 Introduction

As a rule, ensuring long-term and reliable operation of the propulsion system of the ship is currently one of the most urgent tasks of domestic shipbuilding. It is proved by a large number of published scientific works in which new methods and ideas for improving performance characteristics are proposed from a mechanical or technological point of view. One of the important elements of the propulsion system is the stern shaft.

According to GOST 24154-80 (All-Union State Standard), the stern shaft is a constructive complex that kinematically connects the main engine with the propulsion device and is designed to transmit torque and axial loads that occur during operation of the ship propulsion unit (hereinafter referred to as the propulsion unit), of which it is an integral part.

V. Murgul and M. Pasetti (Eds.): EMMFT-2018, AISC 983, pp. 522–533, 2019.
https://doi.org/10.1007/978-3-030-19868-8_51

According to their quality characteristics, stern shaft bearings are divided into sliding bearings (gland and plank design) and rolling bearings (with elements in the form of rods, rollers, balls, etc.) with flange and non-flanged design; with rigid, self-aligning or cushioned mounting option; with the use of anti-friction materials, such as rubber, caprolon, rockwood, textolite, babbit and others.

Stern shaft (Fig. 1) is a very complex dynamic system, as it works with the action and the occurrence of constant, variable, cyclic, and random loads. Its design is diverse and depends on many factors, including the location of thc main power plant. Performance deterioration of a stern shaft leads to a decrease in speed or complete loss of it, incrcased vibration, accelerated wear and failure of its elements. Subsequent repairs, in most cases, require the ship to be decommissioned and docked.

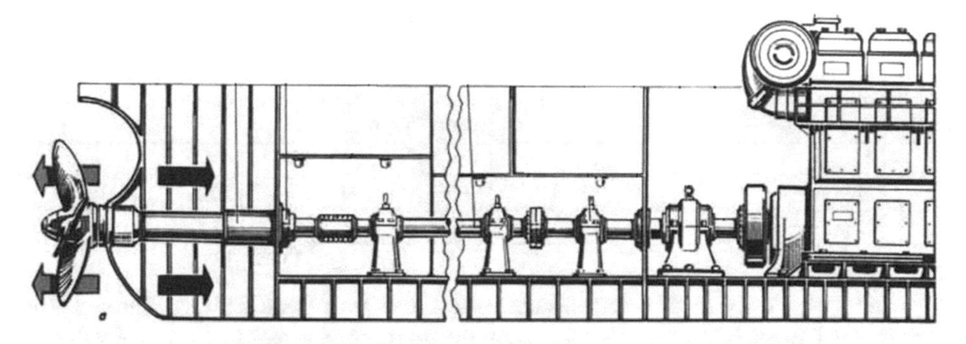

Fig. 1. General view of a stern shaft.

When calculating bending vibrations or centering the stern shaft, it's necessary to take into account the elastic and mechanical properties of stern bearings. To this end, in the design scheme, a beam of constant length is placed along the section on elastic supports [1] or an elastic base [2, 3] with a rigidity coefficient of N/m and N/m^2. The service life of the stern shaft depends on the working condition of stern bearings [7], which is characterized by the amount of wear δ of the stern bearings themselves.

In the work [3], when studying the installation of stern shafts on stern shaft supports, the values of the restraint coefficient for babbit $A = 0.2 \cdot 10^{-8}$ m/N, for caprolon – $A = 0.4 \cdot 10^{-8}$ m/N H, for rubber – A = $1.7 \cdot 10^{8}$ m/N. Since the rigidity coefficient is related to the restraint coefficient by equation [1]:

$$k = \frac{1}{A}, \tag{1}$$

then it will have a value: babbit $k = 5 \cdot 108$ N/m, caprolon $k = 2.5 \cdot 108$ N/m, rubber $k = 1.7 \cdot 108$ N/m, respectively.

In the work [3], the static calculation of the stern shaft on an elastic Winkler base, the value of the caprolon stiffness coefficient was assumed to be 500 MPa ($0.5 \cdot 109$ N/m^2). In the study [5] of bending vibrations of the stern shaft also on an elastic Winkler base, the caprolon rigidity coefficient was assumed to be equal to 1... 2 MPa (1 · 109...2 · 109 N/m^2). In all the above works, there is no reference to the sources and methods for obtaining the numerical value of the rigidity coefficient of the material of stern bearings.

In [4, 5], an expression is proposed for determining the rigidity coefficient:

$$k = \frac{Q}{\Delta_z} = \frac{\pi E d^2}{4h} \tag{2}$$

where: E - modulus of elasticity of stern bearings glands; d - the diameter of the stern shaft with lining coating; h - the wall thickness of the stern bearings glands.

Table 1 presents the rigidity coefficients k of the stern bearing for two ships - Hydrobiologist and Khazar-1. The material of stern bushings is caprolon.

Table 1. The rigidity coefficients k of stern bearings.

№	Ship's name	E, Pa	d_s, mm	Outer diameter of the sleeve D, mm	h, mm	k, N/m
1	1	3	4	5	6	
1	Hydrobiologist	$3 \cdot 10^9$	150	200	25	$2.19 \cdot 10^9$
2	Khazar-1	$3 \cdot 10^9$	131	170	19.5	$2.04 \cdot 10^9$

Many works try to take into account the wear δ of stern bearings through the rigidity coefficient k. The opinion exists that the greater the wear, the lower the value of the rigidity coefficient:

$$\delta = min \rightarrow max, \quad k = max \rightarrow min \tag{3}$$

Of course, if we consider the bending vibrations of the stern shaft, then there is a connection with the wear δ of stern bearings. As indicated in [6], with increasing wear of stern bearings, especially aft one, the value of the natural frequency decreases. If we consider the linear law of variation of the natural frequency of bending vibrations due to an increase in wear, then the shape of the change will have a parabolic character.

The calculation of the bending vibrations of the stern shaft is mandatory for its design. The location of the supports and the propeller, the length of the stern shaft bearings, the geometrical dimensions of the shafts and their connections among themselves, the stern shaft itself, etc. are estimated. The purpose of the calculation is to determine the natural frequency of the stern shaft, which should be 20 ... 40% greater than the working (blade) frequency.

It is important to note that bending vibrations, like torsional vibrations, are controlled by the Russian Maritime Register during ship repair.

In numerous works devoted to the calculation of bending vibrations of shafts, it is argued that in the design scheme, it is enough to consider only its rowing part, since it is this part of the stern shaft that is most loaded and has the lowest natural frequency [7]. includes a propeller, a pivoting arm of a propeller shaft, a stern bearing, and a nose part of a propeller shaft or a part of an intermediate shaft (depending on the stern shaft design).

There are many methods for determining the natural frequency of bending vibrations of the stern shaft [8]. Among them are the method of Shimanskii Yu.A. and method according to RD 5.4307-79. All of them consider stern shaft as a beam of stepped-uniform cross section, based on the so-called "point" (hinged) support. Therefore, the main disadvantage is that they do not take into account the elastic properties of stern bearings when determining the natural frequency.

2 Materials and Methods

For a comparative analysis of the effect of rigidity of the stern bearing on the natural frequency ω of the stern shaft, we will calculate it for bending vibrations. The design scheme is a beam with bending rigidity *EJ*, supported by a hinged and elastic support with rigidity coefficient *k* (N/m), respectively (Fig. 2).

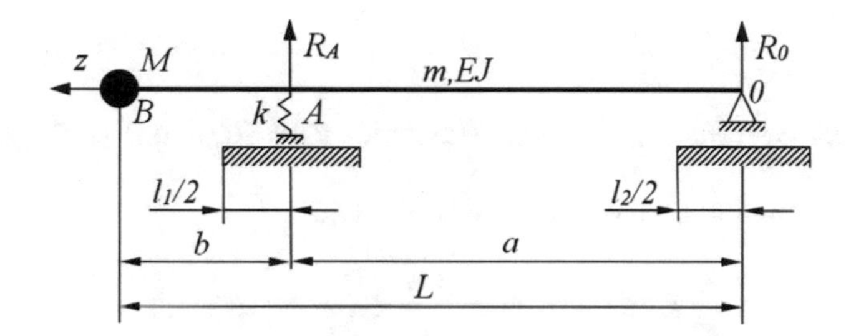

Fig. 2. The design scheme of the beam on the elastic support.

M - mass of the propeller; *m* - linear mass of the stern shaft; R_O, R_A - support reactions.

The boundary conditions for this design scheme will be:

$$y_{z=0} = 0; \quad \left.\frac{d^2y}{dz^2}\right|_{z=0} = 0; \quad \left.\frac{d^2y}{dz^2}\right|_{z=L} = 0; \quad M\omega^2 y_{z=L} = -EJ\left.\frac{d^3y}{dz^3}\right|_{z=L}. \tag{4}$$

The differential equation of free bending vibrations of such a beam on each span is:

$$EJ\frac{d^4\xi}{dz^4} + m\frac{d^2\xi}{dt^2} = 0, \tag{5}$$

where: ξ - lateral displacement of the beam; z - coordinate of the section; *EJ* - bending rigidity; *m* - linear mass of the beam; The solution of Eq. (2) can be found as follows:

$$\xi = y(z)sin(\omega t + \varphi c), \tag{6}$$

where: y - the vibration amplitude of the beam; ω - the natural frequency; φ_c - shift of phases (we assume that $\varphi_c = 0$).

Using the known functions of Krylov A.N. presented in [15], we find the general solution of Eq. (5):

$$y(z) = y_0 \cdot K_1(\alpha z) + \frac{\phi_0}{\alpha} \cdot K_2(\alpha z) + \frac{M_0}{\alpha^2 EJ} \cdot K_3(\alpha z) + \frac{Q_0}{\alpha^3 EJ} \cdot K_4(\alpha z), \tag{7}$$

where:

$$\alpha = \sqrt[4]{\frac{m\omega^2}{EJ}}, \tag{8}$$

y_O, φ_O, M_O, Q_O - respectively, the amplitude of the deflection, angle of rotation, bending moment, and shear force at $z = 0$.

The functions of Krylov A.N. have the form:

$$K_1 = \frac{1}{2}(ch(\alpha z) + cos(\alpha z)); \rightleftarrows\rightleftarrows\rightleftarrows K_2 = \frac{1}{2}(sh(\alpha z) + sin(\alpha z)); \tag{9}$$

$$\rightleftarrows\rightleftarrows\rightleftarrows\rightleftarrows K_3 = \frac{1}{2}(ch(\alpha z) - cos(\alpha z)); \rightleftarrows\rightleftarrows\rightleftarrows\rightleftarrows K_4 = \frac{1}{2}(sh(\alpha z) - sin(\alpha z)). \tag{10}$$

They are related to each other by derivatives [2]:

$$\frac{d}{dz}K_1(\alpha z) = \alpha K_4(\alpha z); \frac{d}{dz}K_2(\alpha z) = \alpha K_1(\alpha z); \tag{11}$$

$$\frac{d}{dz}K_3(\alpha z) = \alpha K_2(\alpha z); \frac{d}{dz}K_4(\alpha z) = \alpha K_3(\alpha z); \tag{12}$$

Based on the expression (8), the natural frequency will be:

$$\omega = \sqrt{\frac{\alpha^4 EJ}{m}}, \tag{13}$$

Let us find the deflection function for all the spans of the studied beam of the design scheme. We take into account that with positive displacement of the support sections downwards, the reaction of the elastic supports is directed upwards and is equal to:

$$R_A = -y_A \cdot k \tag{14}$$

The function of the deflection amplitude in the first section ($0 \leq z \leq$ a) has the form:

$$y_I = \frac{\phi_0}{\alpha} K_2(\alpha z) + \frac{R_0}{\alpha^3 EJ} K_4(\alpha z) \tag{15}$$

When $z = a$, function (15) is equal to:

$$y_a = \frac{\phi_0}{\alpha} K_2(\alpha a) + \frac{R_0}{\alpha^3 EJ} K_4(\alpha a) = y_A \tag{16}$$

The function of the deflection amplitude in the second section ($0 \leq z \leq L$) has the form:

$$y_{II} = \frac{\phi_0}{\alpha} K_2(\alpha z) + \frac{R_0}{\alpha^3 EJ} K_4(\alpha z) + \frac{R_A}{\alpha^3 EJ} K_4(\alpha(z-a)) \tag{17}$$

When $z = L$, Eq. (15) will be:

$$y_L = \frac{\phi_0}{\alpha} K_2(\alpha L) + \frac{R_0}{\alpha^3 EJ} K_4(\alpha L) + \frac{R_A}{\alpha^3 EJ} K_4(\alpha(L-a)) \tag{18}$$

Based on (12), Eq. (16) will look like:

$$y_L = \frac{\phi_0}{\alpha} K_2(\alpha L) + \frac{R_0}{\alpha^3 EJ} K_4(\alpha L) - \frac{y_A \cdot k}{\alpha^3 EJ} K_4(\alpha(z-a)) \tag{19}$$

Based on the boundary conditions, we write the derivatives of the Eq. (17):

$$\begin{cases} y_L' = \phi_0 K_1(\alpha L) + \frac{R_0}{\alpha^2 EJ} K_3(\alpha L) - \frac{y_A \cdot k}{\alpha^2 EJ} K_3(\alpha(z-a)) \\ y_L'' = \alpha\phi_0 K_4(\alpha L) + \frac{R_0}{\alpha EJ} K_2(\alpha L) - \frac{y_A \cdot k}{\alpha EJ} K_2(\alpha(z-a)) \\ y_L''' = \alpha^2\phi_0 K_3(\alpha L) + \frac{R_0}{EJ} K_1(\alpha L) - \frac{y_A \cdot k}{EJ} K_1(\alpha(z-a)) \end{cases} \tag{20}$$

We write the boundary conditions (4) at the left end of the design scheme (at point B):

$$\begin{cases} \frac{\phi_0}{\alpha} K_2(\alpha a) + \frac{R_0}{\alpha^3 EJ} K_4(\alpha a) - y_A = 0 \\ M\omega^2 \left[\frac{\phi_0}{\alpha} K_2(\alpha L) + \frac{R_0}{\alpha^3 EJ} K_4(\alpha L) - \frac{y_A \cdot k}{\alpha^3 EJ} K_4(\alpha(z-a))\right] \\ + \left[\alpha^2\phi_0 K_3(\alpha L) + \frac{R_0}{EJ} K_1(\alpha L) - \frac{y_A \cdot k}{EJ} K_1(\alpha(z-a))\right] EJ = 0 \\ EJ\alpha\phi_0 K_4(\alpha L) + \frac{R_0}{\alpha} K_2(\alpha L) - \frac{y_A \cdot k}{\alpha} K_2(\alpha(z-a)) = 0 \end{cases} \tag{21}$$

According to expression (19), the system of equations takes the form:

$$\begin{cases} \frac{\phi_0}{\alpha} K_2(\alpha a) + \frac{R_0}{\alpha^3 EJ} K_4(\alpha a) - y_A = 0 \\ \phi_0\left(\frac{\alpha^3 EJMK_2(\alpha L)}{m} + \alpha^2 EJK_3(\alpha L)\right) + R_0\left(\frac{M}{m}\alpha K_4(\alpha L) + K_1(\alpha L)\right) \\ - y_A\left(kK_1(\alpha b) + k\frac{M}{m}\alpha K_4(\alpha b)\right) = 0 \\ \phi_0 EJ\alpha K_4(\alpha L) + R_0\frac{K_2(\alpha L)}{\alpha} - y_A\frac{k}{\alpha}K_2(\alpha b) = 0 \end{cases} \tag{22}$$

The condition for the existence of a nonzero solution, as mentioned earlier, for the system of homogeneous Eq. (20) is that its determinant is zero, that is:

$$\Delta = \begin{vmatrix} A_{11} & A_{12} & A_{13} \\ A_{21} & A_{22} & A_{23} \\ A_{31} & A_{32} & A_{33} \end{vmatrix} = 0 \tag{23}$$

where A_{ij} - coefficients for unknowns of a system of Eq. (20).

Next, the value of α being included in the A.N. Krylov function is calculated, at which the determinant of systems is zero. To do this, a graph of the dependence of the determinant Δ on the value of α is built.

After determining α by Eq. (11), we determine the angular frequency of the beam.

$$p = \frac{\omega}{2\pi} = \frac{1}{2\pi}\sqrt{\frac{EJ\alpha^4}{m}}, \tag{24}$$

where ω - cyclic frequency (rad/s) associated with the usual circular frequency p (s^{-1}) by the dependence $\omega = 2\pi p$.

The determinant $\Delta_{h.s.}$ of a system of equations for bending oscillations of a beam on hinged supports will take the form:

$$\Delta_{h.s.} = \begin{vmatrix} \frac{K_2(\alpha a)}{\alpha} & \frac{K_4(\alpha a)}{\alpha^3 EJ} & 0 \\ \frac{M\alpha^3 EJK_2(\alpha L)}{m} + EJK_3(\alpha L)\alpha^2 & \frac{M\alpha K_4(\alpha L)}{m} + K_1(\alpha L) & \frac{M\alpha K_4(\alpha L)}{m} + K_1(\alpha b) \\ EJK_4(\alpha L)\alpha & \frac{K_2(\alpha L)}{\alpha} & \frac{K_2(\alpha b)}{\alpha} \end{vmatrix} = 0 \tag{25}$$

For a comparative analysis, let us also consider the bending vibrations of a beam on two hinged supports. A general view of the design scheme is presented in Fig. 3.

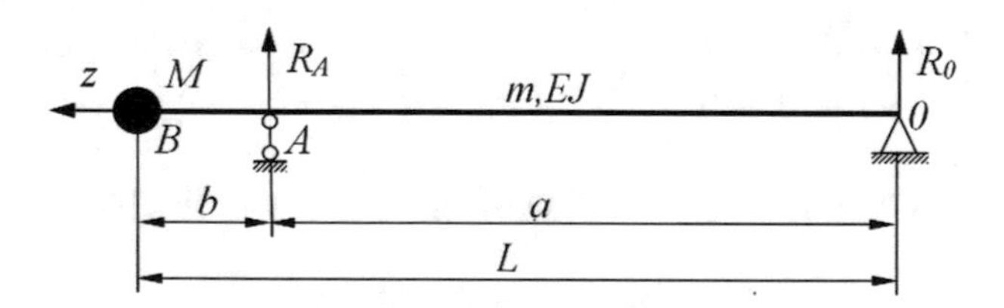

Fig. 3. The design scheme of a beam on two hinged supports.

The determinant $\Delta_{e.s.}$ of a systems of equations for bending vibrations of a beam on one hinged and one elastic support will have the following form:

$$\Delta_{e.s.} = \begin{vmatrix} \frac{K_2(\alpha a)}{\alpha} & \frac{K_4(\alpha a)}{\alpha^3 EJ} & -1 \\ \frac{M\alpha^3 EJK_2(\alpha L)}{m} + EJK_3(\alpha L)\alpha^2 & \frac{M\alpha K_4(\alpha L)}{m} + K_1(\alpha L) & -k\left(\frac{M\alpha K_4(\alpha L)}{m} + K_1(\alpha b)\right) \\ EJK_4(\alpha L)\alpha & \frac{K_2(\alpha L)}{\alpha} & -\frac{k \cdot K_2(\alpha b)}{\alpha} \end{vmatrix} = 0 \quad (26)$$

3 Results

Table 2 presents the numerical values of the parameters of the ships according to the design scheme (Fig. 2) and the value of the own circular frequency of the bending vibrations of the beam, which is supported only by two hinged supports.

Table 2. The numerical value of the parameters of the stern shaft and its own circular frequency of real ships.

№		l_1, mm	l_2, mm	a, mm	b, mm	d_{st}, mm	m, kg/m	M, kg	α, 1/m	p_1, 1/s
1	2	3	4	5	6		7	8	9	10
1	Hydrobiologist	500	500	2200	175	125	96.3	154	1382422	49.65
2	Khazar-1	520	220	2000	250	108	71.9	97	1.482537	49.34

Table 3. Natural frequency at bending vibrations of stern shafts with different rigidity coefficient.

№	k, N/m	Hydrobiologist			Khazar-1		
		α, 1/m	p_1, 1/s	p_2/p_1	α, 1/m	p_1, 1/s	p_2/p_1
1	$1 \cdot 10^6$	0.61115	9.7	5.119	0.71557	11.5	4.29
2	$5 \cdot 10^6$	0.911908	21.61	2.298	1.06483	25.45	1.939
3	$1 \cdot 10^7$	1.08106	30.37	1.635	1.2583	35.54	1.388
4	$5 \cdot 10^7$	1.37659	49.24	1.008	1.48165	49.28	1
5	$1 \cdot 10^8$	1.38052	49.52	1.003	1.4822	49.32	1
6	$5 \cdot 10^8$	1.38228	49.6	1.001	1.4824	49.33	1
7	$1 \cdot 10^9$	1.3824	49.65	1	1.4825	49.34	1
8	$5 \cdot 10^9$	1.3824	49.65	1	1.48253	49.34	1
9	$1 \cdot 10^{10}$	1.3824	49.65	1	1.48256	49.34	1
10	$5 \cdot 10^{10}$	1.3824	49.65	1	1.48256	49.34	1

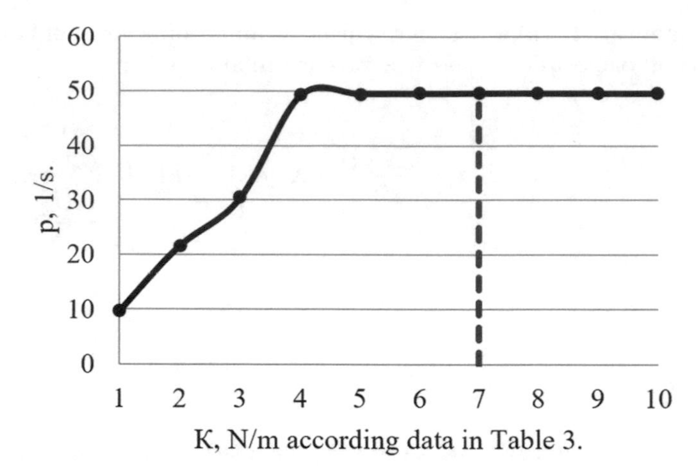

Fig. 4. Graph of the dependence of the natural frequency ω on k.

Table 3 presents the numerical values of the natural frequency for bending vibrations of the stern shafts of the ships under study, taking into account the change in the rigidity coefficient of the elastic support of the design scheme (Fig. 1) in the range of $1 \cdot 10^6 \ldots 5 \cdot 10^{10}$ N/m.

According to the values indicated in the table, a graph of the dependence of the natural frequency on the rigidity coefficient of the Khazar-1 ship is built (Fig. 4).

As can be seen from the graph, the glands of stern bearings of the ships under study have sufficient rigidity and resistance to external loads. When the rigidity coefficient of an elastic support $k = 1 \cdot 109$ N/m, the natural frequency of the stern shaft corresponds to the frequency of bending vibrations on the hinged supports. Therefore, in most cases, when calculating bending vibrations, the design scheme can be a beam that is supported by hinged supports. The obtained rigidity coefficient of the stern shaft glands does not have a special effect on the numerical value of the natural frequency during bending vibrations of the stern shaft.

In order to study the bending vibrations of the stern shaft, taking into account the wear of stern bearings, a test stand was designed and manufactured (Fig. 5).

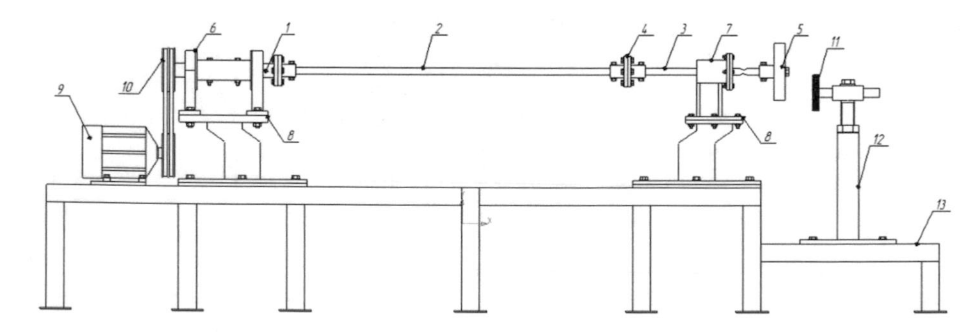

Fig. 5. The design of a test stand.

A test stand [9, 10] is a system of shafts 1, 2 and 3 (hereinafter referred to as shaft), which are interconnected by flanges 4. At the end of shaft 3, there is a disk 5. The shaft rests on two rolling bearings 6 and on one sliding bearing 7. The material of the sliding bearing - caprolon. The bearings themselves are mounted on two supports 8. For rotating the disk 5, an asynchronous motor 9 is used, which transmits rotation through a V-belt transmission 10. At a certain distance from the disk, there is a magnet 11, which is mounted on a movable fixed support 12. All elements of a test stand are mounted on a metal frame 13.

Fig. 6. General view of a test stand.

The principle of operation of the test stand (Fig. 6) is as follows [16]: asynchronous motor 9 through the V-belt transmission 10 transmits the rotation of the shaft. In the process of rotation, the disk 5 is acted upon by the force of the magnet 11. The frequency of the exciting load is controlled by varying the speed of rotation of the asynchronous motor, using a frequency converter.

To determine the resonant state and record the amplitude of vibrations, a strain gauge station was made. The measuring complex consists of primary transducers – strain gauges, intermediate transducer, measuring instrument, and computer. The strain gauge station is an elastic spring steel plate with strain gauges located on the surface.

At this test stand, the oscillation of a beam with a diameter of 18 mm and a length of 1700 mm with a disc at the end with a diameter of 150 mm and a mass of 3 kg was considered. The gland of stern bearing with a length of 90 mm was made of caprolon. The inner diameter of the gland had the values: 18.5, 19 and 21 mm. Table 4 presents the averaged values of the resonant frequency of the beam depending on the size of the gap between the shaft and the sliding bearing.

According to the results of experimental studies, a decrease in the natural frequency p of the system was established with an increase in the gap δ between the beam and the caprolon gland. Reduction of the natural frequency leads precisely to the separation of the beam from the sliding bearing when rotating.

Table 4. The range of resonant frequencies at different gaps.

№	Internal diameter of the bearing d, mm	Gap size δ, mm	Frequency of the beginning of the resonant state p, s^{-1}
1	18.5	0.5	25
2	19	1	22.4
3	21	3	16.6

4 Conclusion

The above study of bending vibrations of stern shaft showed that the elastic properties of stern bearings do not reflect the whole picture of the change in the natural frequency. It is not enough to consider only the elastic and mechanical properties of the stern bearing. In calculations, it is necessary to take into account the separation of the stern shaft from the stern shaft bearing, which characterizes the change in the rigidity of the shaft with the stern bearing during bending vibrations. This process is enhanced by increasing the wear of stern bearings and the action of external loads, leads to a decrease in the natural frequency and the occurrence of a resonance phenomenon.

References

1. Komarov, V.V.: State of installation of stern shafts on stern supports (in Russian). Bulletin of ASTU, series "Marine equipment and technology", № 2(31), pp. 259–267 (2006)
2. Mironov, A.I.: Influence of stern bearings on stern shaft oscillations (in Russian). In: Mironov, A.I., Denisova, L.M. (eds.). Bulletin of GTU № 1(20), pp. 125–130 (2004)
3. Rubin, M.B.: Bearings in shipboard equipment: Reference book. In: Rubin, M.B., Bakhareva, V.E. (eds.) Shipbuilding, Leningrad (1987)
4. Mamontov, V.A.: Assessment of the rigidity of the material of a stern bearing on the performance of the stern shaft (in Russian). In: Mamontov, V.A., Halyavkin, A.A., Kushner, G.A., Razov, I.O. (eds.) Bulletin of ASTU, series "Marine equipment and technology" № 4, pp. 80–87 (2017)
5. Halyavkin, A., Razov, I., Mamontov, V., Kushner, G.: Determination of rigidity coefficient of stern shaft bearing. In: IOP Conference Series: Earth and Environmental Science 90 (EMMFT - 2017) 012078, Paper 158, 10–13 April 2017, Far Eastern State Transport University, Russian Federation (2017). https://doi.org/10.1088/1755-1315/90/1/012078
6. Grebe, E., Loshadkin, D.V., Kushner, G.A., Haljavkin, A.A., Dudanov, A.A.: Experimental research about friction and wear of antifriction materials for port's/coastal and maritime machinery or equipment and analysis of parametric oscillations of ships shafting with new composite stern-tube bearings. In: 19th International Marine Industries Conference Proceedings of: 11–14 December 2017, Kish Island, Iran (2017)
7. Jia, X., Fan, S.: Analysis of the flexural vibration of ship's tail shaft by transfer matrix method. J. Mar. Sci. Appl. **7**(3), 179–183 (2008)
8. Šestan, A., et al.: A study into resonant phenomena in the catamaran ferry propulsion system. Trans. FAMENA **36**(1), 35–44 (2012)

9. Patent 156856 of the Russian Federation, IPC G01H 1/10 (2006.01), B63H 23/00 (2006.01). Stand for the study of longitudinal, transverse and torsional vibrations of the stern shaft system of ships. In: Kushner, G.A., Khalyavkin, A.A., Mammoths, V.A. (eds.) Applicant and Patent Holder Federal State Budgetary Establishment of Higher Professional Education FSBEI HPE "Astrakhan State Technical University". No. 2015117052/28; claimed 05.05.2015; published 20.11.2015, Bulletin No. 32, 3 p.; pic
10. Patent 180727 of the Russian Federation, IPC G01H 1/10 (2006.01). Stand for the study of longitudinal, transverse and torsional vibrations of pump-compressor equipment. In: Khalyavkin, A.A., Safonov, N.A., Kudasov, A.G., Meshcheryakov, V.N., Panteleeva, T. E., Sychev, A.V., Ishmuhamedova, E.G. (eds.) Applicant and Right Holder OOO "Gazprom dobycha Astrakhan". № 2017145185; claimed 21.12.2017; published 21.06.2018, Bulletin № 18, 3 p.; pic

Modeling the Focusing of the Energy of Elastic Waves in a Block-Fractured Medium During Long-Term Vibration Action

Yuriy Skalko[1], Sergey Gridnev[1,2], and Albina Yagfarova[1(✉)]

[1] Moscow Institute of Physics and Technology, 9, Institutional Lane, Dolgoprudny, Moscow Region 141701, Russia
yagfarovaam28@gmail.com
[2] Voronezh State Technical University, 14, Moscow Avenue, Voronezh 394026, Russia

Abstract. In this paper, the methods of mathematical modeling investigate the possibility, under certain conditions, of the propagation of elastic waves in a block-fractured geological environment, generated by a source located on the surface, without significant scattering. A computational model has been created that includes a geological environment within which internal boundaries (cracks) can exist with different conjugation conditions on them (conditions of "complete sticking", conditions of "slipping with friction", etc.). On the surface, there is a permanently acting vibration source of elastic waves. The paper theoretically substantiates and implements the original computational algorithm for solving the initial-boundary value problem, based on the construction of a fundamental solution of the problem operator. The algorithm showed high efficiency and allowed to conduct a series of resource-intensive experiments. The carried out computational experiments showed that if the slip conditions are realized at the internal boundaries, then the elastic wave propagates in the channel between these cracks, almost not penetrating beyond their limits. The relationship between the nature of the propagation of elastic waves and the frequency of the vibration source is tested. It is shown that with an increase in the frequency of the vibration source operation, the degree of penetration of elastic waves beyond the channel boundaries determined by cracks increases. And at sufficiently high frequencies, the wave pattern becomes similar to what is happening in a continuous medium.

Keywords: Generalized Riemann problem of discontinuity breakdown · Fundamental solution of the problem operator · Slip conditions on the boundary · Residual wave field

1 Introduction

In numerous industrial experiments on oil fields, it has been convincingly shown that installing a vibration source on the surface and its prolonged operation for several months leads to a significant increase in oil recovery of the oil reservoir [1–6]. In some cases, the effect reaches 40%. The mechanisms and processes leading to such enhanced

V. Murgul and M. Pasetti (Eds.): EMMFT-2018, AISC 983, pp. 534–546, 2019.
https://doi.org/10.1007/978-3-030-19868-8_52

oil recovery remain unclear today. In particular, it is not clear how the energy of elastic waves generated by the vibration source reaches significant depths (1 km or more) while avoiding substantial scattering. A vibration source with a characteristic power of 30 kW and a spot of contact with the rock generates an elastic wave. If the rock is homogeneous, then the energy of the vibration source is dissipated over a hemisphere and at depths of the order of 1 km, the energy density of elastic waves decreases by a factor of 10^6 times even in the absence of absorption in the rock. It is not necessary to expect that an elastic wave of such a low energy density will cause any significant processes in the oil-bearing formation. Since the effect of increased oil recovery as a result of prolonged operation of the surface vibration source has been repeatedly recorded, it should be assumed that, under certain conditions, elastic waves propagate in the geological formation, avoiding substantial scattering. In this paper, the methods of mathematical modeling will investigate whether the presence of cracks in the geological rock, in which contiguous parts can move relative to each other, leads to the fact that the elastic wave generated by the vibration source is not scattered over a hemisphere, so that, even at significant depths, its energy density remains significant in order to trigger certain physical and chemical processes.

In the above field experiments, an effect was also recorded that has not yet found a convincing explanation. After a long work of the vibration source, when it is already turned off, the presence of an obscene is recorded in the geological formation for a long time, i.e. elastic waves are recorded, the frequency of which does not coincide with the frequency of the source, usually 1–2 orders of magnitude higher. In our work, using mathematical modeling methods, we experiment under what conditions a specified after-sounding may appear and by what parameters its frequency is determined. The mathematical model of the phenomenon under study is an initial-boundary problem for a system of equations of elastic dynamics with the presence in the integration region of internal boundaries on which additional conditions must be fulfilled on the model parameters reflecting certain physical conditions on these boundaries.

When constructing numerical algorithms for solving boundary value problems for such systems of equations, the key point is the solution of the Riemann problem of discontinuity decay. In the case of one spatial variable, various authors [7–9] proposed a number of methods for solving the Riemann problem. In fact, all these methods are associated with the presence of characteristics of hyperbolic systems. In the case of many spatial variables, methods based on the presence of characteristics no longer work. And the Riemann problem, most often, is solved under the assumption that, near the discontinuity, the solution is a plane wave moving along the normal to the discontinuity surface [7, 8, 10]. It is clear that this approach is far from being justified in all cases. We will compose the formulation of the generalized Riemann problem of discontinuity breakdown with conjugation conditions at the boundaries for hyperbolic systems of first order linear differential equations with piecewise constant coefficients, with an arbitrary number of spatial variables, and an algorithm for constructing its solution is given. This algorithm is based on finding the fundamental solution of the problem operator. The constructed solution of the generalized Riemann problem will form the basis of a computational algorithm for finding an approximate solution of the initial-boundary value problem for the described class of systems of differential equations. The computational model will be used to study the above questions related to possible mechanisms for the transmission of vibration effects from a source located on the surface of the oil-bearing formation.

2 Computational Model

Let's consider the following problem. It is necessary to find a solution to the Cauchy problem $u = (u_1, \ldots, u_M)^T$ for a system of linear differential equations of the first order with constant coefficients:

$$\frac{\partial u(t,x)}{\partial t} + \sum_{i=1}^{N} A_i \frac{\partial u(t,x)}{\partial x_i} = 0, \quad x \in R^N \tag{1}$$

with initial data

$$u(t=0,x) = u_0(x), \tag{2}$$

which are continuous everywhere except the hyperplane $\Gamma : x_1 = 0$. The solution must be everywhere continuous, except for the hyperplane Γ. In this case, the specified relations (conjugation conditions) connecting the values of variables on both sides of the hyperplane Γ must be met. Next, we assume that the conjugation conditions are given as a system of algebraic equations:

$$L(u(t, x_1 = -0, x_2, \ldots, x_N),\ u(t, x_1 = +0, x_2, \ldots, x_N)) = 0. \tag{3}$$

We believe that the initial data also satisfy the conditions of conjugation.

Following formulation [7], we will call the formulated statement the generalized Riemann problem of discontinuity decay with conjugation conditions on boundaries. The generalized Riemann problem differs from the classical Riemann problem in that in the classical problem the initial data are assumed to be constants on both sides of the hyperplane, and in the generalized Riemann problem the initial data on both sides of the hyperplane can be arbitrary smooth functions, and also on all discontinuity surfaces specified pairing conditions.

In [11], the original initial-boundary problem was formulated as a system of partial differential equations for generalized functions [12] and constructed an approximation of the fundamental solution of the operator of the problem or, which is the same Green matrix function [12].

$$G(t,x) = \theta(t) \sum_{k} C_k \delta(x - \lambda_k t) + O(t^2). \tag{4}$$

$k = (k_1, k_2, \ldots, k_N)$ - multi-index, integer vector with components $k_j = 1 : M$. $C_k = C_{k_1} C_{k_2} \ldots C_{k_N}$ - multi-index array of matrices. The rules of matrix formation $C_{k_1}, C_{k_2}, \ldots, C_{k_N}$ are specified in [11].

$\lambda_k = (\lambda_{k_1}, \lambda_{k_2}, \ldots, \lambda_{k_N})$ - multi-index array of vectors. - eigenvalues of the matrix, numbered by index.

The presence of a fundamental solution of the operator of the problem allows us to represent the solution at any point, except for points of internal and external boundaries, in the form of convolution of the fundamental solution of the operator of the problem with the initial data vector [12]. Taking into account the properties of convolution with δ - function [12], we obtain that, up to $O(t^2)$, a solution of the problem at an arbitrary time, is a linear combination of the values of the initial data

$$u(t,x) = \sum_k C_k u_0(x - \lambda_k t) \tag{5}$$

That is, as shown in [11], in order to build a solution at an instant (t,x), it is necessary to draw straight lines $\frac{dx}{dt} = \lambda_k$ in all directions given by vectors λ_k from a point and take a linear combination of the values of the initial data at the intersection points $t = 0$ of these straight lines with the plane with the coefficients defined by the matrices C_k.

Also in [11], it was shown that using the fundamental solution of the problem operator allows us to reduce the solution of the generalized Riemann problem of discontinuity decay with conjugation conditions on internal boundaries to a system of linear algebraic equations of SLAEs with the right part determined by the initial data. Some of these equations are the required conjugation conditions. The value of the decision on both sides of the internal boundary is determined by solving this SLAE and is essentially a linear combination of the values of the initial data at the intersection points of all possible lines $\frac{dx}{dt} = \lambda_k$ defined by vectors λ_k and passing through the boundary point (t,x) with the plane $t = 0$.

The above results are used to build a computational algorithm for the propagation of elastic waves in a non-homogeneous block-fractured medium.

2.1 Distribution of Elastic Waves in a Block-Fractured Medium. Mathematical Model

Following [10], we introduce a vector of variables

$$u = (\sigma_{11}, \sigma_{22}, \sigma_{12}, v_1, v_2)^T$$

where σ_{11}, σ_{22}, σ_{12}, components of the stress tensor,

v_1, v_2 components of the displacement velocity vector, and write the system of equations describing the distribution of elastic waves, for the case of two spatial variables, in the form (1).

This system of equations is hyperbolic, matrices A_1 and A_2 have a complete set of linearly independent eigenvectors.

Further, all dimensional quantities are specified in the SI unit of units. We set the following problem for this system of equations. In the region $\Omega = [-30 < x_1 < 30, -600 < x_2 < 0]$, it is necessary to find a solution to the initial-boundary value problem for the indicated system of equations. Equations should be performed everywhere in Ω, except for the internal boundaries $\Gamma_\gamma, \gamma = 1, 2$ given by the condition $\Gamma_1 : x_1 = -15$ and $\Gamma_2 : x_1 = 15$. At these boundaries, the conditions of "full adhesion" were set up, consisting in the fact that when passing through these boundaries all components of the displacement vector and force components from different sides are continuous, the boundaries are equal in magnitude and oppositely directed, which actually reflects the realization of Newton's third law:

$$v_1\left(t, x \in \Gamma_\gamma^-\right) - v_1\left(t, x \in \Gamma_\gamma^+\right) = 0, v_2\left(t, x \in \Gamma_\gamma^-\right) - v_2\left(t, x \in \Gamma_\gamma^+\right) = 0, \sigma_{11}\left(t, x \in \Gamma_\gamma^-\right) - \sigma_{11}\left(t, x \in \Gamma_\gamma^+\right) = 0$$

$$\sigma_{12}\left(t, x \in \Gamma_\gamma^-\right) - \sigma_{12}\left(t, x \in \Gamma_\gamma^+\right) = 0 \quad (6)$$

Or, "sliding conditions without friction" were put, consisting in the fact that when passing through these boundaries, the components of the displacement vector normal to the boundaries are continuous, the components of the force normal to the boundary are opposite in magnitude, and oppositely directed sides of the border are 0:

$$v_1 t, x \in \Gamma_\gamma - v_1 t, x \in \Gamma_\gamma^+ = 0, \sigma_{11} t, x \in \Gamma_\gamma - \sigma_{11} t, x \in \Gamma_\gamma^+ = 0,$$

$$\sigma_{12}\left(t, x \in \Gamma_\gamma^-\right) = \sigma_{12}\left(t, x \in \Gamma_\gamma^+\right) = 0 \quad (7)$$

At the outer boundaries $x_1 = -30$, $x_2 = -600$, $x_1 = 30$, boundary conditions must be transparent.

At the border $x_2 = 0$, a source of vibration operates, which acts on the geological environment with an effort $F_j \sin \omega t,\ j = 1, 2$. This force is distributed along the border with density $P_j(x_1) \sin \omega t$, so that $\int P_j dx_1 = F_j$. In accordance with the third law of Newton, the conditions $\sigma_{12} = -P_1 \sin \omega t$ and $\sigma_{22} = -P_2 \sin \omega t$ are met.

2.2 Numerical Algorithm

The area Ω is divided into subregions by internal boundaries.

$$\Omega_1 = [-30 < x_1 < -15, -600 < x_2 < 0],$$

$$\Omega_2 = [-15 < x_1 < 15, -600 < x_2 < 0]$$

And

$$\Omega_3 = [15 < x_1 < 30, -600 < x_2 < 0].$$

We construct in each of the subregions a rectangular grid with sides parallel to the axes of coordinates, so that the nodes lying on the inner boundaries coincide for both adjacent subregions. The grid is built uniformly in each of the coordinates and the $h_j, j = 1, 2$ grid step in the corresponding directions. Let the index $p1$ number the grid nodes in the first subdomain Ω_1, the index $p2$ number the nodes in the second subdomain Ω_2, and the index $p3$ number the grid nodes in the third subdomain Ω_3. Further, for the numbering of grid nodes in the subdomain, we will use, where it does not cause misunderstandings, the index p, each time implying that in each subdomain this index runs through its own set of values.

We will assume that the force generated by the vibration source is directed vertically, i.e. $P_1(x_1) = 0$ and the distribution $P_2(x_1)$ is a piecewise linear function, equal 0 in all nodes of the boundary $x_2 = 0$, except for the node with coordinates $x_1 = 0$, $x_2 = 0$. In this node P_1 the value $\frac{F_2}{h_1}$ is accepted.

We define a uniform grid in time $t_m = m\tau$, $m = 0 : 1 : M$. The grid spacing τ must satisfy the condition $\tau \leq \min\left(\min\limits_k \frac{h_1}{\lambda_1^k}, \min\limits_k \frac{h_2}{\lambda_2^k}\right)$.

In each subdomain, we construct a system of basis polynomials $H_p(x)$, each of which is equal in the node corresponding to the index p, equal 0 in all other nodes of the grid and in each grid cell is a bilinear (linear in each variable) function.

The solution in each subdomain will be approximated by a linear combination $u(t,x) = \sum_p H_p(x)u^p(t)$. Then the problem of constructing an approximate solution of the initial-boundary value problem for the system of equations of elastic dynamics with conditions (7) on the virtual boundaries at each time layer reduces to finding values at the nodes $u^p(t_{m+1})$, with known values at the previous time layer $u^p(t_m)$.

For the internal nodes of each of the subregions, the values on the next time layer are found in accordance with the formulas

$$u(t_{m+1}, x) = \sum\nolimits_k C_k u(t_m, x - \lambda_k \tau). \tag{8}$$

In the righthand side of formula (8), there are no terms related to the conditions on the internal and external boundaries. These terms are equal to 0 due to the choice of the time step.

For nodes that lie on the internal borders, on which the conjugation conditions are set, on the external borders, where the boundary conditions are set, we include transparent conditions, the values on the next time layer on both sides of the border are calculated by solving the corresponding SLAE as described above and described in detail in [11].

3 Computational Experiments. Results and Discussion

Based on the above, a computational algorithm for solving the initial-boundary value problem for the system of equations of elastic dynamics was constructed and implemented. With this numerical model, two series of computational experiments were carried out to simulate the propagation of elastic waves generated by a vibration source located on the surface. The important point is that the model has the ability to set transparent boundary conditions, which allows simulating scenarios with a source that has been operating for a long time.

In the first series of experiments, the question was investigated whether the block-fractured structure of the geological environment could lead to the fact that the vibro-seismic effect of a source located on the surface reaches depths of the order of 1 km. and more without significant scattering. For this, first, the propagation dynamics of elastic waves in the medium was calculated in the case when "conditions of complete adhesion" were set at the internal boundaries. Figure 1 shows the density distribution of the kinetic energy $K = \frac{\rho}{2}\left(v_1^2 + v_2^2\right)$ of an elastic wave at the time frame of $t = 10$ s after starting the vibration source with a frequency of 10 Hz. Model parameters correspond to characteristic values for the geological environment. As can be seen, the disturbance created by the vibration source spreads in all directions, and at depths of the order of 600 m the energy density of elastic waves is negligible. Qualitatively, the same picture of the energy distribution of the elastic perturbation takes place after some time the process is established.

Fig. 1. Conditions of "full adhesion" at $\omega = 10$ Hz (the source is on the right).

The same computational experiments with the same parameters were performed on a model in which "friction-free sliding conditions" were specified at the inner boundaries. Figures 2a, b, c show the distribution of the density of kinetic energy, respectively, at time intervals of $t = 3$ s, $t = 5$ s, and $t = 10$ s after the start of the vibration source. The results of computational experiments show that if "friction-free slip conditions" are realized at the internal boundaries, then the disturbance created by the vibration source spreads in the channel between these boundaries and practically does not penetrate beyond these boundaries. Qualitatively the same as in Fig. 2c, the picture of the distribution of the energy of an elastic perturbation takes place after some time the process is established.

In the next series of computational experiments, a study was made of the change in the nature of the propagation of an elastic perturbation created by a vibration source, while changing the frequency of the vibration source. On the inner borders, "friction-free slip conditions" are implemented. Figures 3, 4, 5 show the results of the distribution of the kinetic energy density at time $t = 10$ s, respectively, at the source frequency $\omega = 1$ Hz, $\omega = 10$ Hz and $\omega = 100$ Hz. The results show that if at a frequency of $\omega = 1$ Hz the disturbance created by the vibration source propagates almost completely in the channel between the internal boundaries, then with an increase in the frequency of the vibration source, a substantial part of the energy of the elastic disturbance penetrates beyond the borders of the channel.

At the frequency of the vibration source $\omega = 100$ Hz, the picture of the distribution of the energy of an elastic perturbation qualitatively resembles the situation when "complete sticking conditions" are realized at the internal boundaries. The perturbation spreads in the medium in all directions, almost without noticing the internal boundaries.

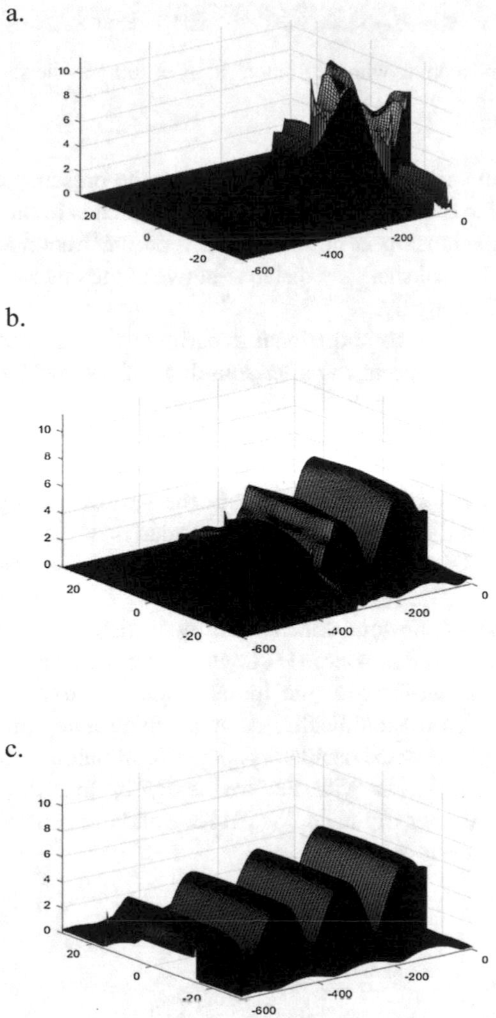

a.

b.

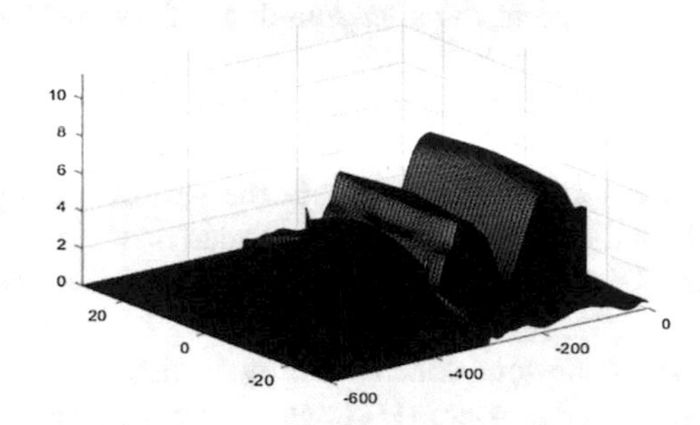

c.

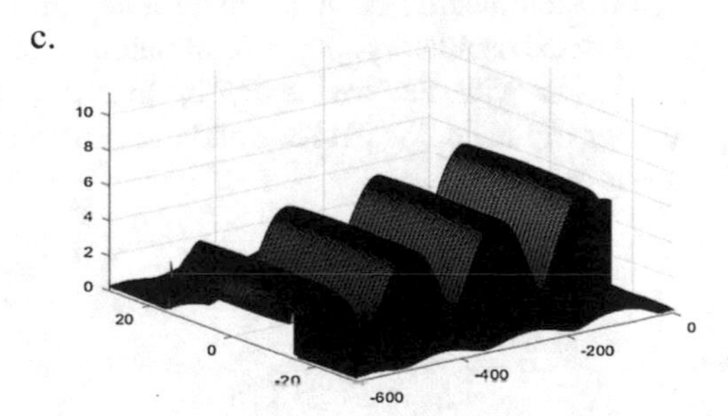

Fig. 2. The profile of the distribution of the density of kinetic energy at different points in time in the case of "slip without friction" at $\omega = 10$ Hz.

Fig. 3. Conditions for "slipping without friction" at $\omega = 1$ Hz (the source is on the righthand side).

Fig. 4. Conditions for "slipping without friction" at $\omega = 10$ Hz (the source is on the righthand side).

Fig. 5. Conditions for "slipping without friction" at $\omega = 100$ Hz (the source is on the righthand side).

The results of numerical experiments show that the presence of cracks in the rock does not always lead to the fact that the elastic wave spreads in the channel between the cracks. A definite relationship is necessary between the frequency of the source, the properties of the geological rock, the distance between the cracks, etc. This issue needs to be studied more thoroughly.

The following series of computational experiments is devoted to the study of the conditions for the occurrence of the after-sounding of the rock. In the above computational experiments, transparent boundary conditions were set at the lower boundary of the region. Consider the situation when the reflecting or partially reflecting boundary conditions are set at the lower boundary. As such conditions, free boundary conditions were implemented. The total kinetic energy in the part of the region bounded by the upper and lower boundaries and internal boundaries (cracks) was chosen as the quantity being studied that characterizes the behavior of the system under the given conditions.

In Fig. 6 a graph of the total kinetic energy in this region as a function of the operating time of the vibration source is shown. It can be seen from the graph that as the vibration source is working at the initial stage, the total kinetic energy in the selected region increases monotonically. After reaching a certain level, the change in the total kinetic energy becomes oscillatory, periodic in nature. The frequency of these oscillations does not coincide with the frequency of the vibration source, but is determined by the properties of the geological rock and the geometric characteristics of the area (distance between the cracks).

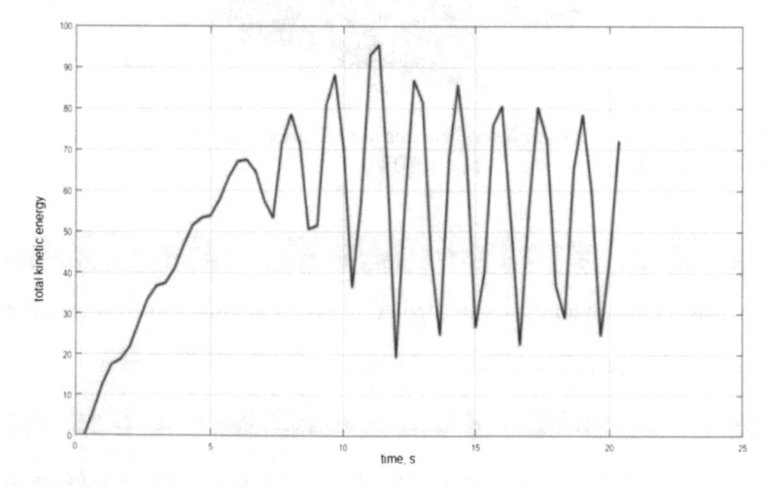

Fig. 6. Dynamics of change in the total kinetic energy in the region between the cracks with a constantly operating vibration source.

Let us consider how the dynamics of the total kinetic energy in the region change after the source stops working. In Fig. 7 the relationship between the total kinetic energy within the region between the internal boundaries and the time when the source is stopped at time t = 3 s is shown.

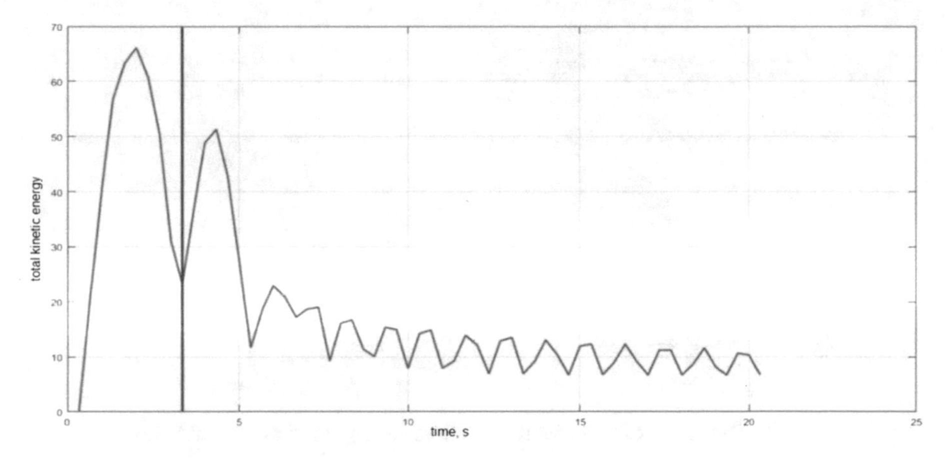

Fig. 7. Dynamics of change in the total kinetic energy in the region between the cracks after the cessation of the vibration source.

From the above graph it can be seen that after stopping the operation of the source, the energy in the system begins to slowly dissipate through the boundaries of the area, but does not dissipate completely. After some time, an oscillatory process is established in the system. The total kinetic energy in the region changes periodically. The average level of these oscillations is 4–5 times less than the level when the vibration source is working. That is, in the system for some time after the source is disconnected, the after-sounding of the rock will be recorded. The frequency of the residual oscillations does not coincide with the frequency of the vibration source.

As it was shown earlier, after the cessation of the vibration source operation, the energy of elastic oscillations is not completely dissipated from the region. This means that the flow of total energy through the internal boundaries becomes small. The next series of experiments is devoted to the study of the factors affecting the decrease in the flux of the total energy of elastic oscillations through the internal boundaries of the region (cracks). The flow of full energy $\Pi = \sigma_{1,1}v_1 + \sigma_{1,2}v_2$. Changes in the values of $\sigma_{1,1}, v_1, \sigma_{1,2}, v_2$ at the inner borders of the region were experimented on. The results are presented in Fig. 8. From the graph it can be seen that the term $\sigma_{1,2}v_2$ is equal to 0, since the value $\sigma_{1,2}$ is equal to 0. The term $\sigma_{1,1}v_1$ after the termination of the source tends to 0, since both multipliers tend to 0.

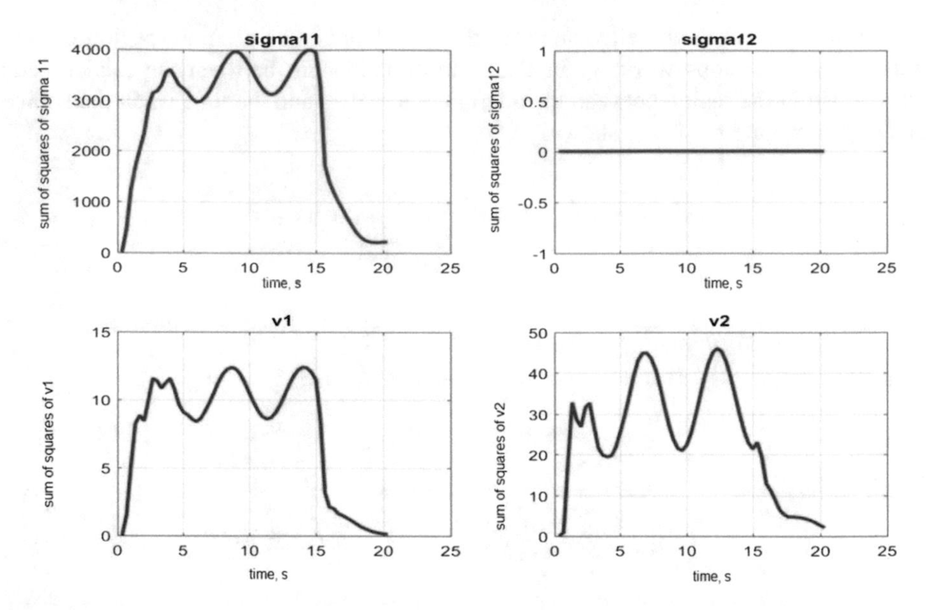

Fig. 8. Dynamics of changes in model variables at internal boundaries.

According to the graph in Fig. 7, one can estimate the speed and time of energy dissipation through internal boundaries.

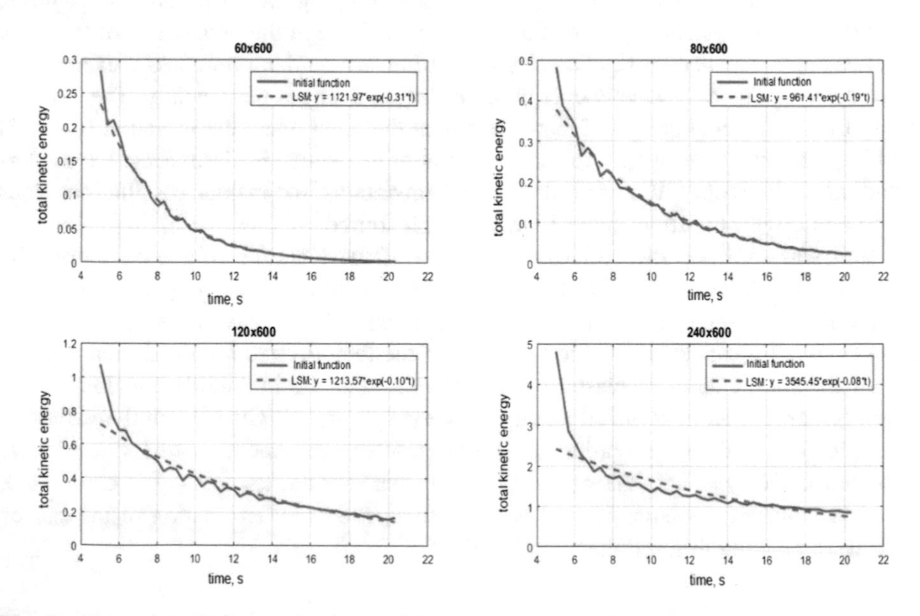

Fig. 9. Dynamics of change in the total kinetic energy of elastic waves in the region for a number of values between the cracks.

These indicators depend on the parameters of the model, in particular, on the distance between the internal boundaries. To study this issue, we conducted a series of experiments in which we changed the distance between the internal borders. At the same time, "viscous friction" conditions with a friction coefficient of 10 were set at the internal boundaries of the region. Figure 9 shows the dynamics of change in the total kinetic energy of elastic waves in the region for a number of values between the cracks. In each case, the least squares approximation was made by the exponent. The exponent indicator allows us to estimate the rate of dissipation of elastic energy through internal boundaries. The results show that with increasing distance between the cracks, the rate of energy dissipation through internal boundaries decreases.

4 Conclusion

An original computational algorithm for finding an approximate solution of the initial-boundary value problem for hyperbolic systems of first-order linear differential equations with constant coefficients is constructed and implemented. At the same time, the formulation of the problem allows for the existence of internal boundaries at which the solution may have discontinuities of the values of the model variables and the specified conditions must be met that connect the values of the variables on both sides of these boundaries. The computational algorithm is based on the construction of an approximation of the fundamental solution of the problem operator. The presence of a fundamental solution allows us to construct for hyperbolic systems of linear differential equations of the first order with constant coefficients, with an arbitrary number of spatial variables, an approximate solution of the generalized Riemann problem with conjugation conditions at internal boundaries. An approximate solution of the generalized Riemann problem with conditions on the outer boundaries is also given. The presence of a fundamental solution allows us to reduce the Riemann problem to solving a system of linear algebraic equations with the right-hand side, depending on the values of the variables at the initial time in a finite number of points.

The constructed computational algorithm was applied to experiment the nature of the propagation of elastic waves generated by a periodically acting vibration source in a block-fractured geological environment. The existence of cracks is reflected in the model by the presence of internal boundaries on which the "friction-free sliding conditions" are satisfied. The numerical experiments carried out showed that the elastic waves generated by the vibration source spread in the channel between the cracks, almost without penetrating beyond the limits of this channel. That is, the presence of cracks in the geological environment can serve as a mechanism that allows elastic waves to penetrate to a substantial depth, avoiding dispersion.

A study was conducted on how the nature of the propagation of elastic waves in a block-fractured medium changes as the frequency of the source changes. The performed numerical experiments showed that at low frequencies the disturbance created by the vibration source spreads almost completely to the channel between the internal boundaries. With an increase in the frequency of the vibration source, a substantially larger part of the energy of the elastic perturbation penetrates beyond the channel boundaries. At high frequencies of the vibration source, the disturbance spreads to the medium in all directions, almost without noticing the internal boundaries. It was also shown that the presence of internal boundaries (cracks) with the conditions of slippage realized on them, leads to the possibility of the presence in the geological environment of the residual field of elastic oscillations after the termination of the vibration source. The question of the dependence of the frequency of these waves on the properties of the geological environment and the parameters of the model has not been studied and will be considered in further work. In the framework of the model, it was shown that the rate of dissipation of energy of elastic waves through the internal boundaries after the cessation of the vibrator source decreases with increasing distance between the cracks.

The outlined studies are performed on a model with two spatial variables. In the future, it is planned to expand the model to a more realistic case of 3 spatial variables. You should also consider the nature of the propagation of elastic waves in the presence of curvilinear cracks. Such a study will require a significant change in the computational model, the transition from rectangular spatial grids to irregular tetrahedral nets is inevitable.

References

1. Lopukhov, G.P., Simkin, E.M., Ashchepkov, Yu.S., Holbaev, T.Kh.: Field tests of the vibroseismic method at the Changyr-Tash field. Oil Ind. **3**, 41–43 (1992)
2. Lopukhov, G.P., Simkin, E.M., Pogosyan, A.B.: Vibroseismatic method of influence on flooded and oil reservoirs. Results of experimental and field research. In: Fundamentals and Search, Research on the Mechanism, vol. 1, pp. 105–112. VNIIENG, Moscow (1992)
3. Lopukhov, G.P., Nikolaevsky, V.N.: The role of acoustic emission at vibroseismic stimulation of water-flooded oil rescrvoirs. In: Proceedings the 8th European IOR Symposium, Vienna (1995)
4. Lopukhov, G.P.: Vibroseismic stimulation for highly watered reservoirs for rehabilitation. In: Proceedings the 9th European IOR Symposium, Haaga (1997)
5. Lopukhov, G.P.: On the Mechanism of Vibroseismic Influence on an Oil Reservoir Represented by a Hierarchical Block Environment. VNIInefg Yearbook (1996)
6. Kulikovskiy, A.G, Pogorelov, N.V, Semenov, A.Yu.: Mathematical Problems of the Numerical Solution of Hyperbolic Systems of Equations. Fizmatlit, Moscow (2001)
7. LeVeque, R.L.: Finite Volume Methods for Hyperbolic Problems. Cambridge University Press, Cambridge (2002)
8. Skalko, Y.I.: Correct conditions on the boundary separating subdomains. Comput. Res. Model. **6**(3), 347–356 (2014)
9. Kaser, M., Dumbser, M.: An arbitrary high order discontinuous Galerkin method for elastic waves on unstructured meshes. I. The two-dimensional case. Geophys. J. Int. **166**, 855–877 (2006)
10. Skalko, Y.I.: Riemann problem of a discontinuity decay, in the case of several spatial variables. Proc. MIPT **8**(4), 169–182 (2016)
11. Vladimirov, V.S.: Equations of Mathematical Physics. Science, Moscow (1981)
12. Gelfand, I.M., Shilov, G.E.: Generalized Functions and Actions Over Them. Dobrosvet, KDU, Moscow (2007)

Heat and Mass Exchange Packing for Desinfection of Circulation Water in Electric Field

Nikolay Merentsov[1(✉)], Alexander Persidskiy[2], Vitaliy Lebedev[3], Natalia Prokhorenko[1], and Alexander Golovanchikov[1]

[1] Volgograd State Technical University, Lenina 28, Volgograd 400131, Russia
steeple@mail.ru

[2] JSC Federal Scientific and Production Centre «Titan - Barricady», Volgograd 400071, Russia

[3] Branch of LUKOIL-Engineering VolgogradNIPImorneft, Volgograd 400078, Russia

Abstract. The article describes a new promising heat and mass exchange packing for evaporation cooling maintaining an intensive dripping flow mode and providing for disinfection of circulation water in electric field. The article includes the schematic diagram of the designed packing and a sample calculation of an industrial apparatus for evaporation cooling applying the designed heat and mass exchange packing. The authors describe the principle for preservation of surface properties of the packing elements. A method is offered to process experimental data, which helps to classify the packing material of various configurations, by mass exchange processes. The article also covers the calculation algorithm for key geometric and technological parameters of evaporation cooling apparatuses ensuring intensive dripping and film-dripping flow modes. Initial, reference and calculation parameters of the designed evaporation cooling apparatus with heat and mass exchange packing ensuring an intensive dripping flow mode and providing for disinfection of circulation water was introduced.

Keywords: Flow dynamics · Contact packing · Disinfection · Dripping flow · Evaporation cooling · Heat transfer · Mass transfer

1 Introduction

The developed design solution has been created for heat and mass exchange equipment operating directly at the contact between air and water in industrial evaporation cooling apparatuses; the solution can be applied in energy-producing, nuclear, metallurgical, construction, food, chemical and petrochemical industries.

Industry knows a huge number of heat and mass exchange packing devices designed for a certain range of processes [1–29]. One of significant problems, that may arise during the equipment operation and is common for evaporation cooling apparatuses, is reduction in intensity of the heat and mass exchange processes reasoned by the

V. Murgul and M. Pasetti (Eds.): EMMFT-2018, AISC 983, pp. 547–559, 2019.
https://doi.org/10.1007/978-3-030-19868-8_53

biofilm of microorganisms that is formed on the surface of the packing elements and interferes with the development of the pre-set efficient flow operation mode and increases the flow resistance of the packing layer thus affecting the energy efficiency of the process and the circulating water cooling intensity.

2 Materials and Methods

In search of the solution for the problem, we have found an unconventional technique: to use the structure of the heat and mass exchange evaporation cooling apparatus where the intensity of the heat and mass exchange processes occurring between the cooled water and air is increased by means of prevention of formation of the biofilm of microorganisms on the surface of the packing elements that are located with respect to each other in a cascade and are fixed on rod electrodes connected to opposite poles of a DC source (see Fig. 1). Each rod has packing elements installed on it made of a dielectric material the adjacent rod electrodes are connected to opposite poles of a DC source in chessboard order, thus maximum electric field strength is reached, and the water cooled in the apparatus is subject to the electric field by the whole length of the packing; this results in destruction of the microorganisms and prevents the formation of the biofilm of microorganisms on the surface of the packing elements. This intensifies the heat and mass exchange processes between the cooled water and air and maintains a developed dripping flow mode in evaporation cooling apparatuses preserving the water-repellent properties of the surfaces of the packing elements. Figure 1 shows the general view of the heat and mass exchange packing module of evaporation cooling apparatuses, Fig. 2 shows the scheme of chessboard connection of the rod electrodes. Heat and mass exchange packing for evaporation cooling apparatuses is designed as a module of closely spaced parallel electrodes 1 with packing elements 2 placed on them and made of a dielectric material, organized in a cascade with respect to each other. Parallel rod electrodes 1 are connected to opposite poles of a DC source in chessboard order to achieve the maximum electric field strength. Heat and mass exchange packing for evaporation cooling is working as follows. Voltage is applied from a DC source in a chessboard order to the poles of the rod electrodes 1. Cooled water is sprayed on the upper packing elements 2 of the module, and the airflow is underfed. Packing elements 2 are located in such a manner that the drops are torn from the packing element, pass the section, hit the next element, get split and renew their heat and mass exchange surface. These cycles are repeated continuously by the whole height of the packing. Thus, the drops of the water flow created due to cascade location of the packing elements 2 are subject to a strong electric field formed under the influence of continuous voltage applied to the oppositely charged neighboring vertical rod electrodes 1.

In addition, during the movement of water droplets through an electric field of high intensity, polarization currents arise in them, which contributes to an even better decontamination effect. And these effects, providing high quality disinfection of circulating water, flow very evenly throughout the volume of the heat and mass transferring fill. The breakdown between the electrodes and oppositely charged bases prevents by insulating inserts 4 of a certain computed length, the breakdown between neighboring oppositely charged electrodes is prevented by controlling the current in the power supply circuit and reducing its output voltage so that the current is held invariable.

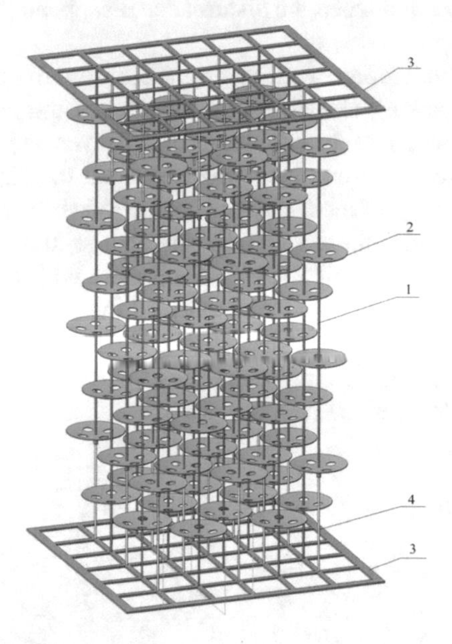

Fig. 1. General view of the heat and mass exchange packing for evaporation cooling providing for disinfection of the circulation water.

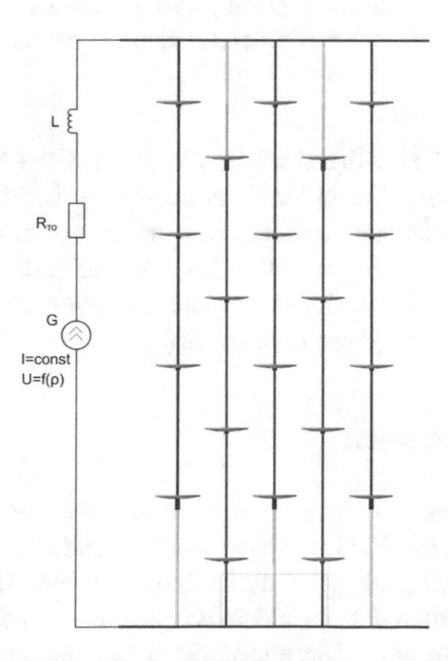

Fig. 2. Schematic circuit diagram of the disinfecting heat and mass exchange packing.

To exclude the passing of short-circuit currents through the power source into the power supply circuit of the module, the current-limiting resistor RTO is introduced, and to suppress current surges during a short circuit - the inductor L. In the event of a short circuit, the voltage of the power source automatically decreases to the minimum value given by the RTO value, and when the disturbance disappears, it automatically rises to the working value.

This suppresses the microorganisms and prevents from the formation of the biofilm on the surface of the packing elements 1 that facilitates the creation and maintains a continuous dripping flow mode through the packing layer and intensifies the heat and mass exchange processes between the cooled water and the air in evaporation cooling apparatuses. The connection of the adjacent rod electrode to the poles of the DC source in chessboard order provides them with opposite charge, thus maximum electric field strength is reached, and the circuit water is disinfected with a high quality.

The experimental studies showed and confirmed the effect of intensification of heat and mass exchange processes. Depending on the flexibility and structure of the electrodes on which the dielectric packing elements are placed, they will have longitudinal, bending and torsional oscillations (see Fig. 3).

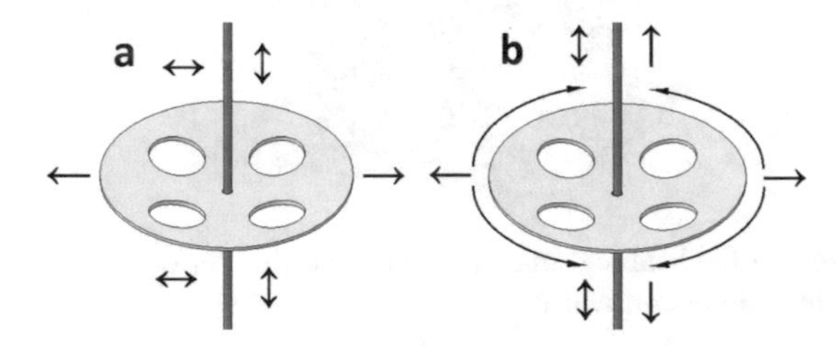

Fig. 3. Schemes of oscillation directions of the packing elements under the influence of the air flowing around them: (a) flexible rod electrodes, (b) nonrigid rod electrodes made in the form of springs.

Oscillations are observed with the lowest velocities of the airflow and grow with the increase of the airflow rate. The oscillations cause the disruption of the films on the surfaces of the packing elements and maintain an intensive and continuous dripping flow mode by the whole packing height. Thus, the oscillations of nonrigid electrodes caused by the airflow result in a significant intensification of heat and mass exchange processes and renewal of the phase contact surface.

3 Results and Discussion

According to the designed classification of heat and mass exchange packing devices described in detail in papers [30–32], we can offer the industrial application for the new heat and mass exchange packing. On the basis of the filtration curves obtained experimentally for the offered heat and mass packing providing for disinfection of water in electric field, we have determined the linear dimensions l1 and l2, α and β

values, which are coefficients of viscosity and inertia parts of the Dupuit-Forchheimer equation, respectively, as well as modified Reynold's criteria Rem and corresponding flow resistance coefficients λ which are given in Table 1. According to the obtained mode range, it is evident that the designed heat and mass exchange packing providing for disinfection of circulation water is ideal for evaporation cooling process [30–32].

Here we provide the calculation for the industrial evaporation cooling apparatus (see Fig. 4) with the designed heat and mass exchange packing ensuring an intensive dripping flow mode (see Fig. 1) providing for disinfection of circulation water as per the algorithm described in Fig. 5 [30–32].

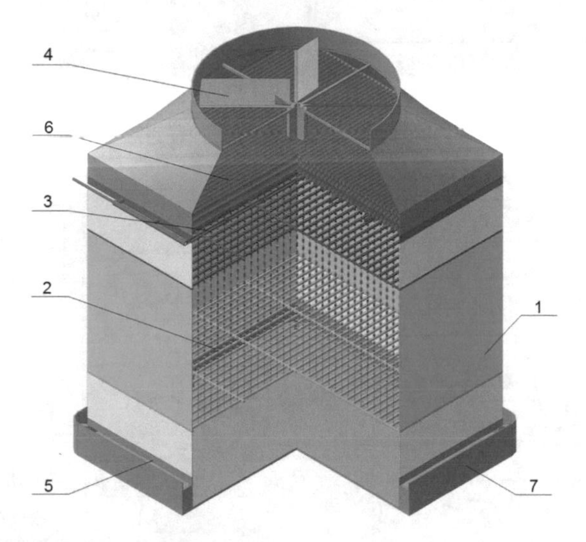

Fig. 4. Industrial evaporation cooling apparatus with heat and mass exchange packing ensuring an intensive dripping flow mode and providing for disinfection of circulation water: 1 – case; 2 – heat-and-mass exchange packing; 3 – irrigation collector; 4 – ventilator; 5 – air inlet windows; 6 – droplet separator; 7 – catchment collector.

Evaporation cooling apparatus for circulation water (see Fig. 4) includes a rectangular (cylindrical) case 1 inside which the designed heat and mass exchange packing 2 is installed that maintains an intensive dripping flow mode of liquid and provides for disinfection of the circulation water. An irrigation tank 3 is installed above the packing surface 2. In cold months air change can be performed by natural convection, but a ventilator 4 can be used if heat and mass exchange processes need to be intensified, and oscillation of packing elements need to be more active to maintain a stable dripping flow mode of the liquid; the ventilator creates an induced draft of the air supplied to the apparatus through air inlet windows 5. Drop entrainment is reduced by droplet separator (knockout drum) 6. Circulation water, cooled and treated in a strong electric field is supplied to the lower part of the apparatus and is collected in water tank 7.

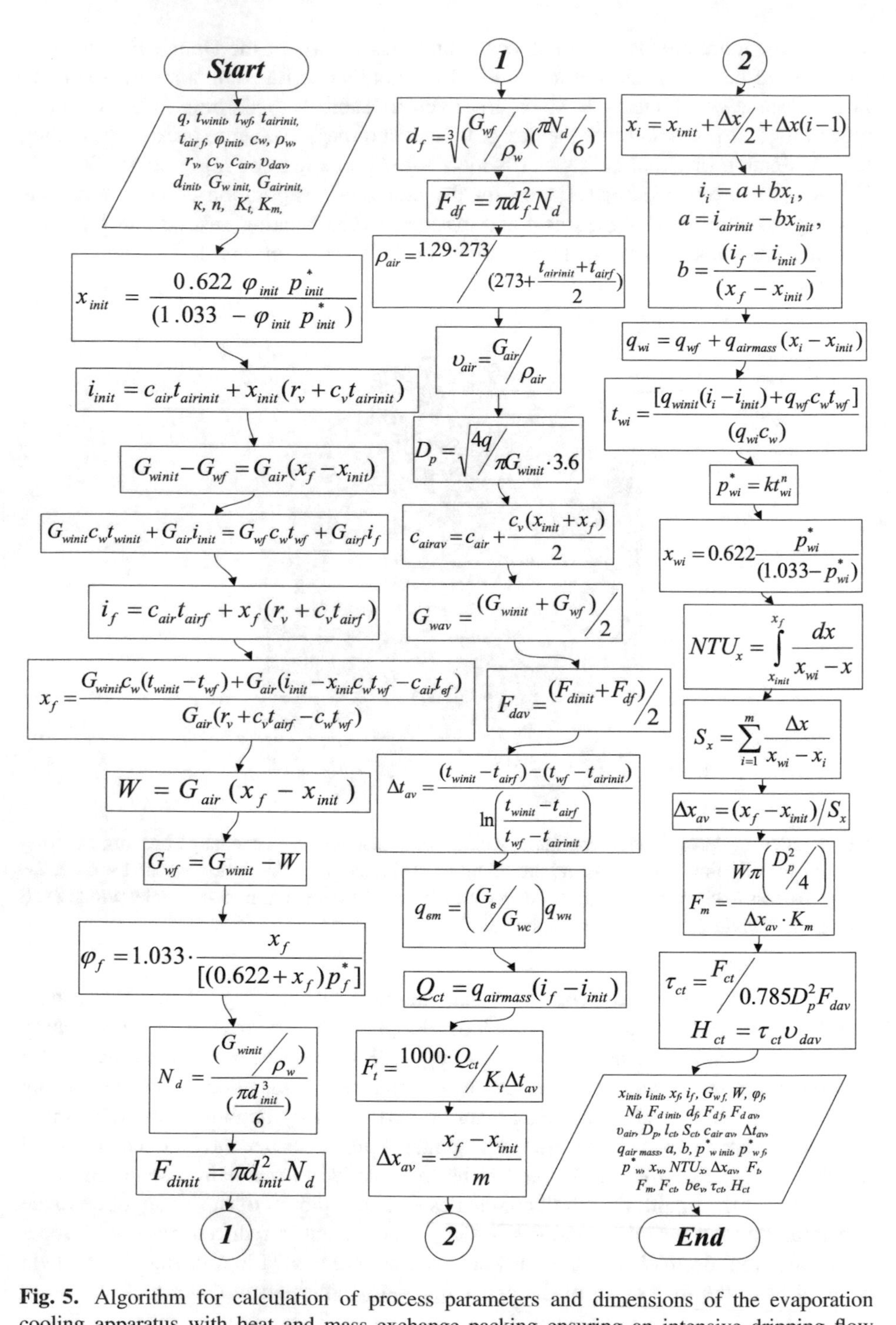

Fig. 5. Algorithm for calculation of process parameters and dimensions of the evaporation cooling apparatus with heat and mass exchange packing ensuring an intensive dripping flow mode and providing for disinfection of circulation water [33].

Table 1. Initial, reference and calculation parameters of the designed evaporation cooling apparatus with heat and mass exchange packing ensuring an intensive dripping flow mode and providing for disinfection of circulation water. Parameters of the packing studied in the experiment: diameter of the dielectric packing elements is 30 mm, distance between the elements on each electrode is 120 mm (is variable).

№	Parameter	Units	Symbol	Value
1	2	3	4	5
Initial data				
1	Cooled water throughput	m^3/h	q	20
2	Initial water temperature	$°C$	t_{winit}	45
3	Final water temperature	$°C$	t_{wf}	25
4	Initial air temperature	$°C$	$t_{airinit}$	18
5	Final air temperature	$°C$	$t_{air\,f}$	32
6	Relative air humidity at the inlet	–	φ_{init}	0.75
Reference data				
1	Average specific heat of water	$kJ/kg\ K$	c_w	4.18
2	Average water density	kg/m^3	ρ_w	992
3	Specific evaporation heat at 0 °C	kJ/kg	r_v	2493
4	Specific heat of vapour	$kJ/kg\ K$	c_v	1.97
5	Specific heat of dry air	$kJ/kg\ K$	c_{air}	1.01
6	Saturation vapour pressure at initial air temperature	atm	p^*_{init}	0.0238
7	Saturation vapour pressure at final air temperature	atm	p^*_f	0.0573
8	Average diameter of drops	m	d_{init}	$4 \cdot 10^{-3}$
9	Average water throughput	$kgW/m^2 \cdot s$	$G_{w\ init}$	1.9
10	Average air throughput	$kgAIR/m^2 \cdot s$	$G_{air\ init}$	2.5
11	Coefficients of approximating exponential equation in the dependence between water saturated vapour pressure and temperature $p^*_w = p^*(t_w)$	–	k	$3.5 \cdot 10^{-5}$
		–	n	2.086
12	Average speed of drops d_{init} (obtained experimentally using the laboratory plant with heat-and-mass exchange packing an intensive dripping flow mode and providing for desinfection of circulation water)	m/s	υ_{dav}	0.08

(*continued*)

Table 1. (*continued*)

№	Parameter	Units	Symbol	Value
13	Heat transfer coefficient (experimentally obtained)	$W/m^2 \cdot K$	K_t	167
14	Mass transfer coefficient (experimentally obtained)	$kgW/m^2 \cdot s \cdot (kgW/kgAIR)$	K_m	0.035
Design values of classifying generalized relation $\lambda = f(Re_m)$ (intermediate parameters)				
1	Modified Reynold's number		Re_m	6.5
				8.1
				10.2
				13.5
				19.6
				30.8
				51.7
2	Fluid resistance coefficient	–	λ	2.30
				2.247
				2.196
				2.151
				2.102
				2.064
				2.038
Calculated parameters				
1	Initial moisture content in air	$kgW/kgAIR$	x_{init}	0.0109
2	Initial enthalpy of air	$kJ/kgAIR$	i_{init}	45.83
3	Final moisture content in air	$kgW/kgAIR$	x_f	$3.09 \cdot 10^{-2}$
4	Final enthalpy of air	$kJ/kgAIR$	i_f	111.46
5	Specific water throughput at the outlet of the cooling tower	$kgW/m^2 \cdot s$	$G_{w\,f}$	1.849
6	Specific quantity of water evaporating into the air	$kgW/m^2 \cdot s$	W	$5.006 \cdot 10^{-2}$
7	Relative humidity of air at the outlet	–	φ_f	1
8	Density of drop flow	$pcs/m^2 \cdot s$	N_d	57185
9	Initial specific surface of the drop flow	$m^2/m^2 \cdot s$	$F_{d\ init}$	2.872
10	Final diameter of drops	m	d_f	$3.966 \cdot 10^{-3}$
11	Final specific surface of the drop flow	$m^2/m^2 \cdot s$	$F_{d\,f}$	2.825
12	Average surface of the drop flow	$m^2/m^2 \cdot s$	$F_{d\ av}$	2.849
13	Speed of air	m/s	υ_{air}	2.11

(*continued*)

Table 1. (*continued*)

№	Parameter	Units	Symbol	Value
14	Diameter of the cooling tower	*m*	D_p	3.66
	Width of the plane (if rectangular)	*m*	l_{ct}	3.24
15	Cross-sectional area of the cooling tower	m^2	S_{ct}	10.5
16	Average specific heat of humid air	*kJ/kg K*	$c_{air\ av}$	1.051
17	Average driving force of the heat exchange process	°*C*	Δt_{av}	9.69
18	Mass air flow rate	*kgAIR/s*	$q_{air\ mass}$	26.315
19	Coefficients of the tie line $i = i(x)$ in linear dependence between enthalpy and moisture content represented as $i = a + bx$	*kJ/kgW*	a	9.99
		kJ/kgAIR	b	3277
20	Partial pressure of water vapors in air for $t_{wf} = 25$ °C $(\varphi = 1)$	*atm*	$p^*_{w\ init}$	0.0147
21	Partial pressure of water vapors in air for $t_{w\ init} = 45$ °C $(\varphi = 1)$	*atm*	p^*_{wf}	0.052
22	Partial pressure of water vapors in air for current t_w value	*atm*	p^*_w	Ref. to the Table 2
23	Moisture content in air nearby the drop surface for current t_w value	*kgW/kgAIR*	x_w	Ref. to the Table 2
24	Number of transfer units by moisture content of vapors in air	–	NTU_x	1.38
25	Average driving force of mass exchange process of water evaporation into the air	*kgW/kgAIR*	Δx_{av}	$1.45 \cdot 10^{-2}$
26	Calculated surface of drops from heat transfer condition	m^2	F_t	1066.9
27	Calculated surface of drops from mass transfer condition	m^2	F_m	1037.9
28	Required calculated surface ensuring the processes of heat- and mass transfer	m^2	F_{ct}	1066.9

(*continued*)

Table 1. (*continued*)

№	Parameter	Units	Symbol	Value
29	Volumetric coefficient (theoretical) as per the Vaganov's formula	–	be_v	7.886
30	Time of drops presence in the cooling tower required to maintain the drop calculated surface	*s*	τ_{ct}	35.57
31	Height of the cooling tower	*m*	H_{ct}	2.84

Table 2. Main calculated parameter of evaporation cooling apparatus related to the moisture content in air (required to determine the number of transfer units, intermediate parameters).

№	Parameter	Value				
		1	2	3	4	5
1	Moisture content in air *x*, *kgW/kgAIR* $\cdot 10^2$	1.29	1.69	2.09	2.49	2.89
2	Moisture content in air nearby the drop surface x_w, *kgW/kgAIR* $\cdot 10^2$	2.11	2.86	3.74	4.75	5.91
3	Enthalpy of air i_{air}, *kJ/kgAIR*	52.39	65.52	78.64	91.77	104.89
4	Air flow rate (dry) Q_{air}, *kgAIR/s*	26.315	26.315	26.315	26.315	26.315
5	Water flow rate Q_w, *kgW/s*	19.94	19.84	19.73	19.63	19.52
6	Partial pressure of water vapours in air p_{air}, *atm* $\cdot 10^2$	2.10	2.73	3.36	3.98	4.59
7	Pressure of water vapour nearby the drops surface p_w, *atm* $\cdot 10^2$	8.96	7.33	5.86	4.55	3.40
8	Air temperature t_{air}, °C	19.44	22.31	25.13	27.90	30.64
9	Temperature of water drops t_w, °C	43.04	39.11	35.13	31.11	27.04

4 Conclusions

Thus, the offered structure of the heat and mass exchange packing for evaporation cooling of circulation water prevents the formation of the biofilm of microorganisms on the surface of the packing elements in all modules by means of constant treatment of water in electric field; this intensifies the heat and mass exchange processes and maintains a constant dripping flow mode during the whole life of the packing device. Such packing reduces the time on evaporation cooling apparatuses repair related to cleaning of the packing surface from biofilm of microorganisms, stabilizes the operation mode related to the constant character of flow resistance of the ascending air flow, reduces energy consumption for movement of air through the packing because the porosity of the packing is constant and does not fall with time compared to the evaporation cooling apparatuses in which the packing surface gets covered with a biofilm of microorganisms. Besides, treated water is returned to the circulation with better properties, preventing the formation of the biofilm of microorganisms on all heat and mass exchange surfaces of the process equipment.

References

1. Sokol, B.A., Chermyshov, A.K., Baranov, D.A.: Mass-Transfer Column Packed-Type Devices. Infokhim, Moscow (2009)
2. Kagan, A.M., Laptev, A.G., Pushnov, A.S., Farakhov, M.I.: Contact Packings in Industrial Heat-and-Mass Transfer Apparatuses. Otechestvo, Kazan (2013)
3. Madyshev, I.N., Dmitrieva, O.S., Dmitriev, A.V., Nikolaev, A.N.: Study of fluid dynamics of mass-transfer apparatuses having stream-bubble contact devices. Chem. Pet. Eng. **52**, 299–304 (2016). https://doi.org/10.1007/s10556-016-0189-2
4. Sokolov, A.S., Pushnov, A.S., Shapovalov, M.V.: Hydrodynamic characteristics of mini-ring truncated-cone packing. Chem. Pet. Eng. **53**, 1–4 (2017). https://doi.org/10.1007/s10556-017-0288-8
5. Dmitrieva, O.S., Dmitriev, A.V., Madyshev, I.N., Nikolaev, A.N.: Flow dynamics of mass exchangers with jet-bubbling contact devices. Chem. Pet. Eng. **53**, 130–134 (2017). https://doi.org/10.1007/s10556-017-0308-8
6. Mitin, A.K., Nikolaikina, N.E., Pushnov, A.S., Zagustina, N.A.: Geometric characteristics of packings and hydrodynamics of packed biotrickling filters for air-gas purification. Chem. Pet. Eng. **52**, 47–52 (2016). https://doi.org/10.1007/s10556-016-0146-0
7. Dmitriev, A.V., Dmitrieva, O.S., Madyshev, I.N.: Optimal designing of mass transfer apparatuses with jet-film contact devices. Chem. Pet. Eng. **53**, 430–434 (2017). https://doi.org/10.1007/s10556-017-0358-y
8. Berengarten, M.G., Bogdanov, R.I., Voronina, V.E.: Physical flow characteristics of a gaseous mixture with solid inclusions. Chem. Pet. Eng. **47**, 111–120 (2011). https://doi.org/10.1007/s10556-011-9434-x
9. Dmitriev, A.V., Dmitrieva, O.S., Madyshev, I.N., Nikolaev, A.N.: Efficiency of the contact stage of a jet-film device during rectification of ethylbenzene–styrene mixture. Chem. Pet. Eng. **53**, 501–507 (2017). https://doi.org/10.1007/s10556-017-0371-1
10. Gorodilov, A.A., Pushnov, A.S., Berengarten, M.G.: Improving the design of grid packing. Chem. Pet. Eng. **50**, 84–90 (2014). https://doi.org/10.1007/s10556-014-9860-7
11. Madyshev, I.N., Dmitrieva, O.S., Dmitriev, A.V., Nikolaev, A.N.: Assessment of change in torque of stream-bubble contact mass transfer devices. Chem. Pet. Eng. **51**, 383–387 (2015). https://doi.org/10.1007/s10556-015-0056-6
12. Ivanov, A.E., Berengarten, M.G., Klyushenkova, M.I.: Processes and equipment for chemical and oil-gas production: hydrodynamics of the bubbling layer in a new type of combined heat and mass exchanger. Chem. Pet. Eng. **46**, 433–440 (2010). https://doi.org/10.1007/s10556-010-9355-0
13. Chizh, K.V., Pushnov, A.S., Berengarten, M.G.: Structure of mini ring packing layup in column equipment. Chem. Pet. Eng. **50**, 244–250 (2014). https://doi.org/10.1007/s10556-014-9889-7

14. Shilin, M.V., Berengarten, M.G., Pushnov, A.S., Klyushenkova, M.I.: Hydrodynamics of regular packings formed from cellular inclined cylinders for heat-and mass-exchange processes. Chem. Pet. Eng. **48**, 608–614 (2013). https://doi.org/10.1007/s10556-013-9665-0
15. Madyshev, I.N., Dmitrieva, O.S., Dmitriev, A.V.: Efficiency of cooling the water droplets within jet-film unit of cooling tower filler. MATEC Web of Conf. **224**, 02079 (2018). https://doi.org/10.1051/matecconf/201822402079
16. Berengarten, M.G., Nevelson, A.O., Pushnov, A.S.: Combination packing for catalytic reactors. Chem. Pet. Eng. **48**, 723–729 (2013). https://doi.org/10.1007/s10556-013-9687-7
17. Madyshev, I.N., Dmitrieva, O.S., Dmitriev, A.V.: Purification of gas emissions from thermal power plants by means of apparatus with jet-bubbling contact devices. In: MATEC Web of Conference, vol. 91, p. 01019 (2017). https://doi.org/10.1051/matecconf/20179101019
18. Kagan, A.M., Yudina, L.A., Pushnov, A.S.: Active surface of the elements of irregular heat and mass transfer packings. Theor. Found. Chem. Eng. **46**, 165–171 (2012). https://doi.org/10.1134/S0040579512020042
19. Madyshev, I.N., Dmitrieva, O.S., Dmitriev, A.V.: Hydraulic resistance of thermal deaerators of thermal power stations (TPS) with jet-film contact devices. In: MATEC Web of Conference, vol. 141, p. 01023 (2017). https://doi.org/10.1051/matecconf/201714101023
20. Gorodilov, A.A., Berengarten, M.G., Pushnov, A.S.: Features of fluid film falling on the corrugated surface of structured packings with perforations. Theor. Found. Chem. Eng. **50**, 325–334 (2016). https://doi.org/10.1134/S0040579516030040
21. Madyshev, I.N., Dmitrieva, O.S., Dmitriev, A.V.: Heat-transfer, inside of the ground heat-transfer units, from liquid, additionally cooling the oil-immersed transformer. In: MATEC Web of Conference, vol. 141, p. 01012 (2017). https://doi.org/10.1051/matecconf/201714101012
22. Dmitrieva, G.B., Berengarten, M.G., Pushnov, A.S., Poplavskii, V.Y., Marshik, F.: New combination packing for heat-and mass-exchange vessels. Chem. Pet. Eng. **42**, 361–366 (2006). https://doi.org/10.1007/s10556-006-0108-z
23. Pushnov, A.S., Berengarten, M.G., Lagutkin, M.G., Sokolov, A.S., Shustikov, A.I.: Effect of channel geometry in regular ceramic packings on the hydrodynamics of heat and mass exchange. Chem. Pet. Eng. **44**, 307–311 (2008). https://doi.org/10.1007/s10556-008-9054-2
24. Madyshev, I.N., Dmitrieva, O.S., Dmitriev, A.V.: Determination of heat-mass transfer coefficients within the apparatuses with jet-film contact devices. In: MATEC Web of Conference, vol. 194, p. 01013 (2018). https://doi.org/10.1051/matecconf/201819401013
25. Gorodilov, A.A., Berengarten, M.G., Pushnov, A.S.: Experimental study of mass transfer on structured packings of direct-contact crossflow heat exchangers. Theor. Found. Chem. Eng. **50**, 422–429 (2016). https://doi.org/10.1134/S0040579516040345
26. Pushnov, A.S., Berengarten, M.G., Kagan, A.M., Ryabushenko, A.S., Stremyakov, A.V.: Helicoidal-structured packing for heat and mass exchange with direct phase contact. Chem. Pet. Eng. **43**, 575–579 (2007). https://doi.org/10.1007/s10556-007-0102-0
27. Ivanov, A.E., Berengarten, M.G., Klyushenkova, M.I.: Hydrodynamic operating regimes for a combined heat-and-mass exchanger. Chem. Pet. Eng. **45**, 526–531 (2009). https://doi.org/10.1007/s10556-010-9230-z
28. Madyshev, I.N., Dmitrieva, O.S., Dmitriev, A.V.: Heat-mass transfer efficiency within the cooling towers with jet-film contact devices. In: MATEC Web of Conference, vol. 194, p. 01036 (2018). https://doi.org/10.1051/matecconf/201819401036
29. Dmitrieva, O.S., Dmitriev, A.V., Madyshev, I.N., Kruglov, L.V.: Impact of the liquid level in the jet-film contact devices on the heat-and-mass transfer process. In: MATEC Web of Conference, vol. 129, p. 06010 (2017). https://doi.org/10.1051/matecconf/201712906010
30. Golovanchikov, A.B., Balashov, V.A., Merentsov, N.A.: The filtration equation for packing material. Chem. Pet. Eng. **53**, 10–13 (2017). https://doi.org/10.1007/s10556-017-0285-y

31. Merentsov, N.A., Balashov, V.A., Bunin, D.Y., Lebedev, V.N., Persidskiy, A.V., Topilin, M.V.: Method for experimental data processing in the sphere of hydrodynamics of packed heat and mass exchange apparatuses. In: MATEC Web of Conference, vol. 243, p. 5 (2018). https://doi.org/10.1051/matecconf/201824300011
32. Merentsov, N.A., Lebedev, V.N., Golovanchikov, A.B., Balashov, V.A., Nefed'eva, E.E.: Experimental assessment of heat and mass transfer of modular nozzles of cooling towers. In: IOP Conference Series: Earth and Environmental Science, vol. 115, p. 012017 (2018). https://doi.org/10.1088/1755-1315/115/1/012017
33. Golovanchikov, A.B., Merentsov, N.A., Balashov, V.A.: Modeling and analysis of a mechanical-draft cooling tower with wire packing and drip irrigation. Chem. Pet. Eng. **48**, 595–601 (2013). https://doi.org/10.1007/s10556-013-9663-2

Modelling and Calculation of Industrial Absorber Equipped with Adjustable Sectioned Mass Exchange Packing

Nikolay Merentsov[1](✉), Alexander Persidskiy[2], Vitaliy Lebedev[3], Mikhail Topilin[3], and Alexander Golovanchikov[1]

[1] Volgograd State Technical University, Lenina 28, Volgograd 400131, Russia
steeple@mail.ru

[2] JSC Federal Scientific and Production Centre "Titan – Barricady", Volgograd 400071, Russia

[3] Branch of LUKOIL-Engineering VolgogradNIPImorneft, Lenina 96, Volgograd 400078, Russia

Abstract. The paper provides a scheme of the designed sectioned mass exchange packing, a description of an experimental plant for studying flow dynamics and heat and mass exchange processes in packing devices, and experimental results of investigation on flow characteristics of the designed sectioned mass exchange packing. The authors describe the advantages and structural aspects of the use of new adjustable packing materials. The authors have studied the use of wastes from metal working machines for packings of heat-and-mass exchange apparatuses as well as the adjustment (self-adjustment) of their basic parameters and flow modes. The article also covers the classifying method for processing of experimental data that helps to perform an express investigation and provide conclusions on the applicability of packing materials for different mass exchange processes. Besides, the authors provide the results of calculation of main process parameters and dimensions of an industrial absorber equipped with a sectioned mass exchange packing each section in which is to be adjusted and managed by flow modes.

Keywords: Flow dynamics · Contact packing · Absorption · Heat transfer · Mass transfer · Flow structure · Self-adjustable · Sectioned packing

1 Introduction

Selection of known and creation of new packings for process equipment always require an assessment of their efficiency for the conditions of the specific process [1–30]. Due to the specific character of different heat and mass exchange processes, it is impossible to uniquely determine the most efficient type of packing for their implementation. In our opinion, the most promising way of improvement of the existing types of process equipment and of the creation of new ones is to design mass exchange contact devices having their own structural methods for intensification (for example, an effect of resonance oscillation of contact packing elements), to design dynamic contact packing

V. Murgul and M. Pasetti (Eds.): EMMFT-2018, AISC 983, pp. 560–573, 2019.
https://doi.org/10.1007/978-3-030-19868-8_54

elements and mass exchange packings that can be precisely adjusted (self-adjusted) for the requirements of the specific process thus allowing for achieving most developed flow modes and reaching the highest intensity of heat and mass exchange processes occurring in the process equipment.

A vast amount of heat and mass packing devices designed for a wide range of process tasks and types of process equipment is known [1–30]. However, despite the apparent high efficiency of their majority, they have some disadvantages, such as high cost, labour intensity and fabrication difficulty, no possibility to adjust main dimensions and process parameters, narrow application sphere for the needs of a specific process.

The efforts to improve the above mentioned disadvantages have resulted in the creation of a promising research sphere and designing of new heat and mass exchange contact packings which can be easily adjusted (self-adjusted) for the requirements of the specific process and apparatus, have a structural (natural) intensification effect related to the peculiarities of their creation (manufacturing) and a very low cost, because the use of these packing materials is actually the disposal of industrial waste produced in huge amounts at metal fabricating plants. We are talking about metal shavings which are wastes from metal working machines formed during machine processing of stainless high-alloy steels; they are of great interest for us, because depending on the composition they demonstrate weatherability, chemical resistance, acid resistance, corrosion resistance, rather high mechanical strength, heat resistance and certain elastic properties and compressibility. Except availability and resistance to aggressive media such packing material (metal shavings) has two key advantages that are of great interest to us. We would like to dwell on them in detail: integral (structural) intensification effect related to the spiral structure of the packing elements and formation of a micro-finned surface as well as possibility of mechanical adjustment (self-adjustment) of the packing material volume that allows for control of flow modes and, consequently, for intensification of heat and mass exchange processes and increasing of the quality of separation during treatment of liquid and gaseous inhomogeneous systems. All advantages and pluses of the designed structure of the sectioned mass exchange packing using metal shavings and packing material make it especially promising for mass exchange apparatuses applied in chemical, petrochemical, metallurgical, construction, machine buildings, food, pharmacological, biochemical and other industries and in ecological processes of selective treatment of liquid and gaseous inhomogeneous systems. Let us consider the integral (structural) effect of intensification that is intrinsic to the packing materials in the form of metal shavings. After machine processing, metal shavings mainly have a spiral structure with micro-finned surface (chips, ticks, radiating cracks on edges) and microroughness. Micro-finned surface and microroughness of the metal shavings tear the surface of liquid thin films on the surface of the packing material and regular renewal of heat and mass exchange surface thus intensifying heat and mass transfer abruptly increasing the contact surface between the phases and the coefficients of heat and mass transfer. At the same time, continuous gas flow is intensively divided and whirls forming macro- and micro-whirls due to micro-finned surface and microroughness at the micro-level and spiral structure of the packing material at the micro-level that results in activation of gas flow sweeping around packing elements, facilitates the intensification of heat and mass exchange and helps to reach intensive flow modes. Microroughness plays the role of sharp micro-ribs tearing the continuous dispersed phase on the surface of the packing material. It is noteworthy that such structural effect that intensifies the heat and mass exchange processes is formed naturally during machine processing of metals.

Let us now dwell on the structural sectioned adjustment. An example of an adjustable (self-adjustable) section of a mass exchange apparatus is given in Fig. 1.

Fig. 1. Scheme of an adjustable sectioned packing block.

Use of metal shavings as packing materials provides the packing device with required structure and the possibility to change the structure with high elasticity and compressibility. Adjustment schemes can be absolutely different - from simple mechanical to automated ones with the possibility to adjust flow modes during the operation of mass exchange packing columns. The purpose of adjustment is to change the main characteristics of the packings devices, i.e. pore volume and specific surface, and most important, that results in the change of channels for liquid and gas flowing through the packing, so the flow resistance, retaining capacity and structures of flows of liquid and gaseous phases, coefficients of heat and mass transfer will change, too. Automated adjustment (self-adjustment) of the packing material can be performed on the basis of achievement of one (or several at the same time) of the listed characteristics. During our experiments we adjust the packing materials using flow structures of the liquid and gaseous phases, i.e. achieving their calibrated values and deviation from ideal ones. Metal shavings (Fig. 1) used as packing material can have absolutely different configuration and curvature radii. But in any way, they have elastic properties and its structure (and surface properties) develop macro- and micro-whirls resulting in a significant intensification of heat and mass transfer processes. The possibility to adjust helps to implement developed flow modes, especially, emulsification (inversion) which has the highest intensity of heat and mass exchange processes and is difficult to achieve and implement in process equipment. Micro-finned surface of the packing material with its chips, ticks, radiating cracks on edges facilitate tearing of stable liquid films and multiple renewal of heat and mass exchanges surfaces; at the same time twisted and spiral structure of the packing material facilitates the formation of macro- and micro-whirls sweeping the heat and mass exchange surfaces of the phases contact thus increasing the productivity of mass exchange apparatuses and improves the treatment of extracted components during sorption processes (absorption).

2 Experimental Part

Figures 2 and 3 show the photo and schematic view of the experimental plant for studying of a wide range of properties of heat-and-mass exchange packing devices and cartridges with the studied adjustable packing material that was represented by metal stainless steel shavings. In order to prove the efficiency of the designed sectioned adjustable packing we have performed a number of experiments with the view to receive data on main flow characteristics of this packing (material Aisi 304, specific surface 210 m^2/m^3, pore volume 0.68 m^3/m^3, packing density 190 kg/m^3, equivalent diameter 0.013 m (parameters are variable, the provided data are in a free state)). For comparison we have chosen Rashig rings and Pall rings and a number of industrial packing devices [1, 2, 31–33]. Experimental data on flow resistance of dry packed heat-and-mass exchange packings are given in Fig. 4.

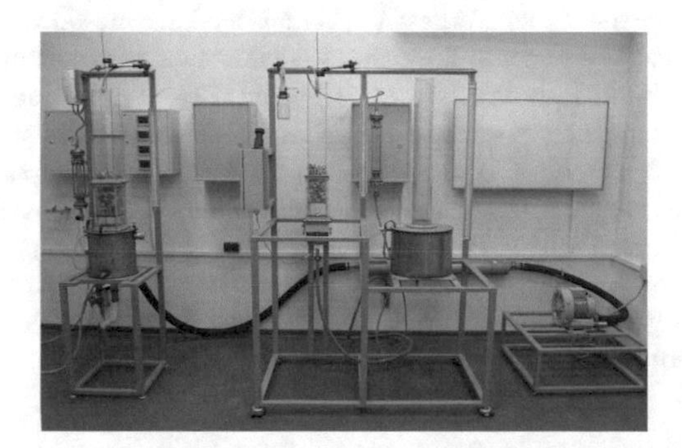

Fig. 2. Experimental plant for the investigation of hydrodynamic and heat/mass transfer characteristics of packed devices of various configurations.

For processing of experimental data and for comparison of the tested heat and mass exchange packing and of the packings widely used in industry, we offer to use generalized relation $\lambda = f(Re_m)$ as per the method described in papers [31–33]. Using this generalized relation, we can compare the energy efficiency and determine the industrial applicability for absolutely a packing device having the most complicated configuration; they all can be summarized and their mode ranges fall within filtration curve $\lambda = f(Re_m)$ shown in Fig. 5.

According to the obtained results, we can determine the industrial applicability of the tested sectioned adjustable mass exchange packing (material Aisi 304, specific surface 210 m^2/m^3, pore volume 0.68 m^3/m^3, packing density 190 kg/m^3, equivalent diameter 0.013 m (parameters are variable, the provided data are in a free state)) for absorption process, because the sectioned adjustable packing has a very wide mode range from 0.01 to 1 as per the modified Reynold's criterion Re_m, so this packing

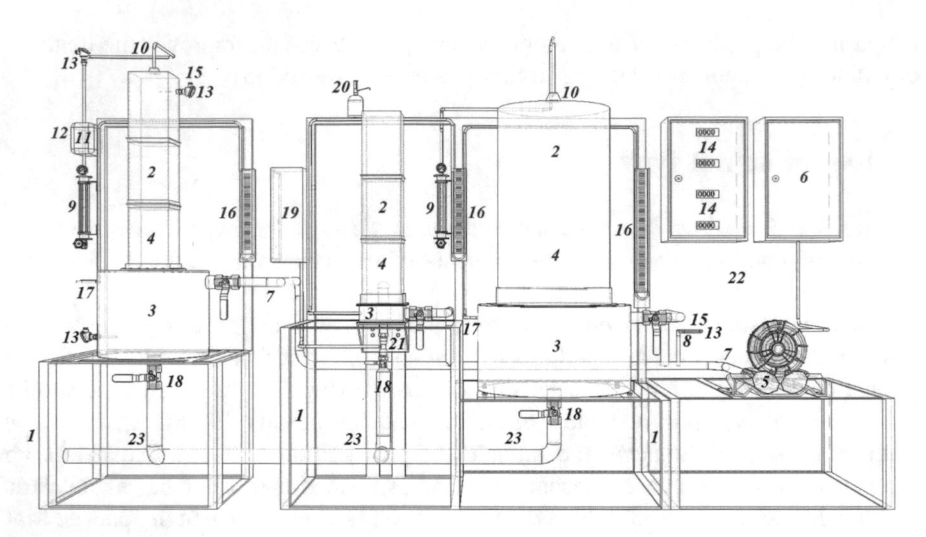

Fig. 3. The scheme of the experimental plant: 1 – a supporting frame; 2 – a column body; 3 – a catchment and gas-distributing sampler; 4 – investigated packed devices in cartridges; 5 – a pressure blower; 6 – frequency converters; 7 – an air-conditioning duct; 8 – a gas flow meter indicating the speed and a volumetric flow rate; 9 – variable area flow meters; 10 – replaceable liquid distributors; 11 – an instantaneous water heater; 12 – a potentiometer of adjusting the heating of water; 13 – temperature sensors; 14 – microprocessor devices processing the signal of temperature sensors; 15 – moisture content sensors; 16 – differential pressure gages; 17 – liquid level indicators in the samplers; 18 – quick-detachable drain valves (necessary for experimental research of the holding capacity); 19 – a block for reading of response curves for the flow structures interpretation; 20 – a mechanical measure feeder of indicator solutions; 21 – changeable electrode groups; 22 – a screen for applying calibration charts (presented on the photo); 23 – a pipe for draining water into the sewer.

material despite a relatively high flow resistance can ensure well-developed flow modes due to its structure and adjustability, more specifically, turbulization and emulsification modes.

But packing materials that can be adjusted can shift from their nominal value on the classification curve $\lambda = f(Re_m)$ (see Fig. 5), so the operating flow modes can be adjusted, too, that makes them especially useful when we need prominent adjustment of the sections of mass exchange packing columns (see Fig. 1).

Experimental data on flow resistance of sprayed heat-and-mass exchange packings are given in Fig. 6.

We also have studied the retaining capacity of the packing material having the form of a section of compacted metal shavings with adjustable parameters and have compared and classified then as per the results of the experiments. Here we provide the adjustment range of the retaining capacity of the sectioned mass exchange packing for liquid and gaseous phases which with are highly efficient for absorption processes. So

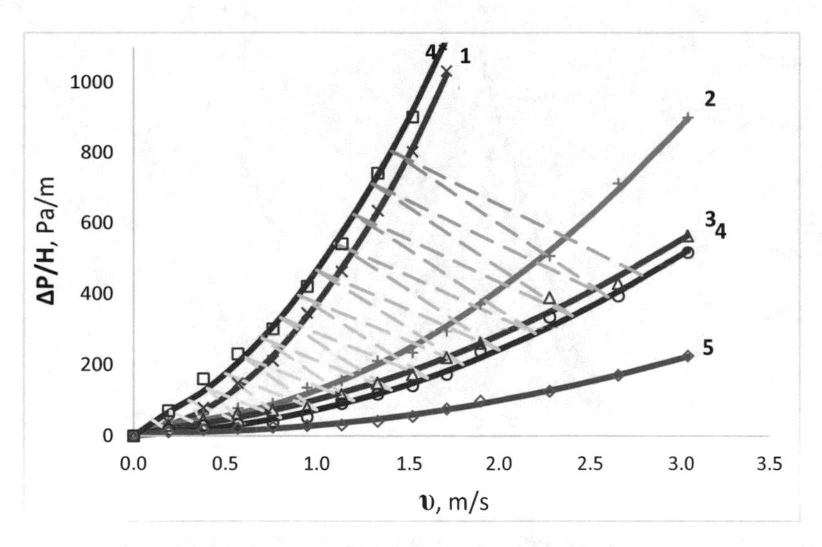

Fig. 4. Dependence between flow resistance of dry packings and velocity of gas in the column: 1 – Rashig rings 25 × 25 × 3 mm [1, 2, 31–33]; 2 – Pall rings 25 × 25 × 1 mm [1, 2, 31–33], 3 – dynamic packing for heat and mass exchange processes 20 × 40 × 3 mm [32]; 4 – sectioned mass exchange adjustable packing, one section with adjustment range [the shown variant is packing material with a dense packing from Aisi 304 stainless steel (specific surface 210 m^2/m^3, pore volume 0.68 m^3/m^3, packing density 195 kg/m^3, equivalent diameter 0.013 m (in a free state of the section)]; 4* – upper limit of the section adjustment; 5 – regular blocked screen packing [1, 2].

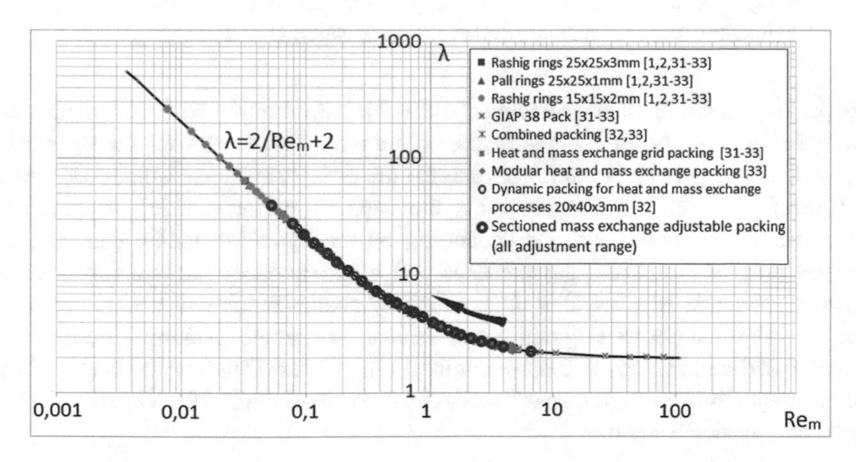

Fig. 5. Dependence $\lambda = f(Re_m)$ for packings having different structures [31–33].

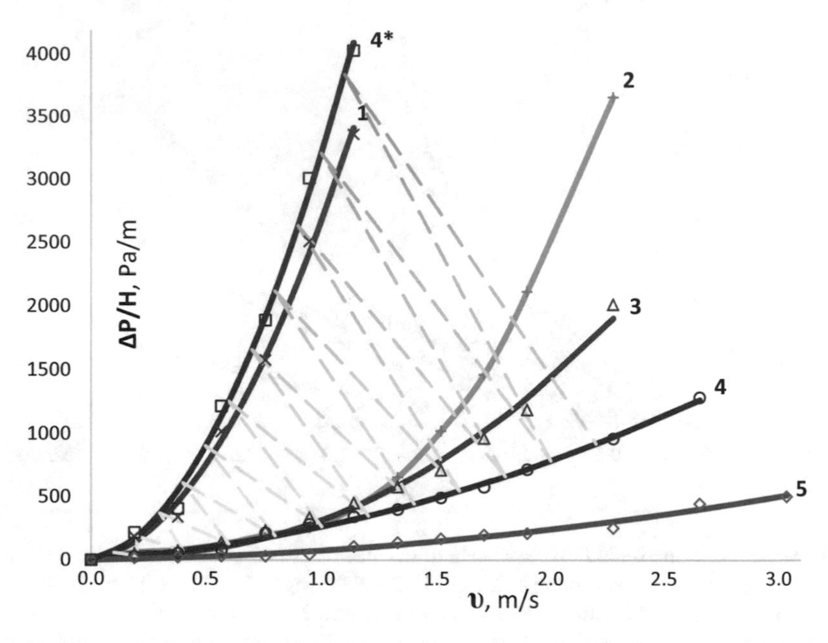

Fig. 6. Dependence between flow resistance of sprayed packings and velocity of gas in the column (with spraying density 1.9 kgW/$m^2 \cdot$ s): 1 – Rashig rings 25 × 25 × 3 mm [1, 2, 31–33]; 2 – Pall rings 25 × 25 × 1 mm [1, 2, 31–33], 3 – dynamic packing for heat and mass exchange processes 20 × 40 × 3 mm [32]; 4 – sectioned mass exchange adjustable packing, one section with adjustment range [the shown variant is packing material with a dense packing from Aisi 304 stainless steel (specific surface 210 m^2/m^3, pore volume 0.68 m^3/m^3, packing density 190 kg/m^3, equivalent diameter 0.013 m (in a free state of the section)]; 4* – upper limit of the section adjustment; 5 – regular blocked screen packing [1, 2].

such packing material (metal shavings) are subject to automated adjustment, ensure high retaining capacity for gas and maintain intensive flow modes for the operation of the packing, thus providing sufficient time for the process and high quality of selective cleaning of gas. Retaining capacity, i.e. the capacity of the packing to accumulate certain quantity of liquid and gas depending on the operation mode, can reflect the summary time of liquid and gas presence and mass exchange surface for a range of industrial packings. Besides, this important flow characteristic can become an instrument for identification of stagnant zones in mass exchange packing columns at the stage of experimental study and pre-commissioning. Figure 7 shows the dependency of retaining capacity for different packing devices obtained during the experiments from gas velocity in the column.

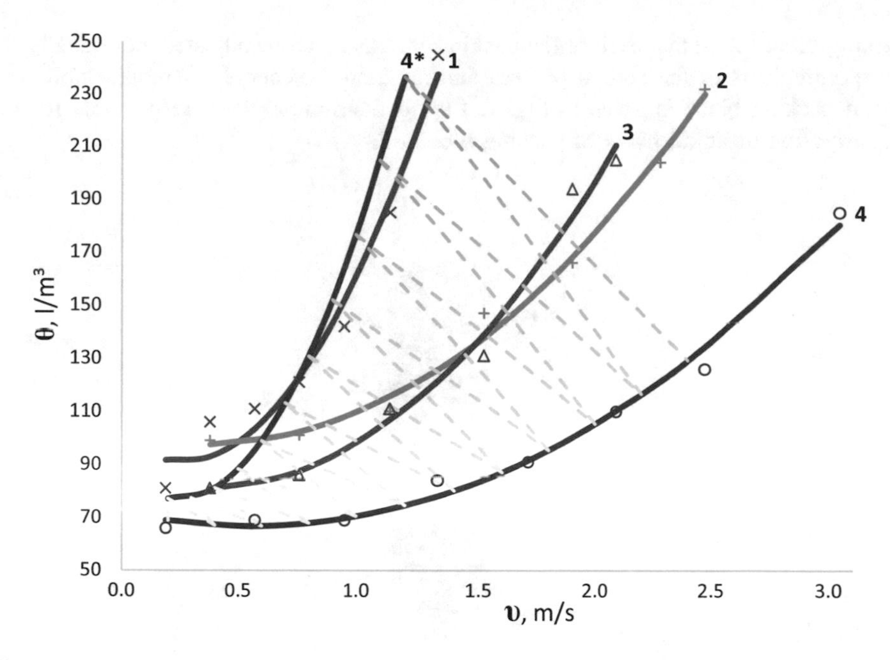

Fig. 7. Dependence between retaining capacity of sprayed packings and velocity of gas in the column: 1 – Rashig rings 25 × 25 × 3 mm [1, 2, 31–33]; 2 – Pall rings 25 × 25 × 1 mm [1, 2, 31–33], 3 – dynamic packing for heat and mass exchange processes 20 × 40 × 3 mm [32]; 4 – sectioned mass exchange adjustable packing, one section with adjustment range [the shown variant is packing material with a dense packing from Aisi 304 stainless steel (specific surface 210 m^2/m^3, pore volume 0.68 m^3/m^3, packing density 190 kg/m^3, equivalent diameter 0.013 m (in a free state of the section)]; 4* – upper limit of the section adjustment.

3 Results and Discussion

On the basis of the obtained experimental data for packing materials in the form of metal shavings, an adjustable section of mass exchange column, we can conclude that the designed sectioned adjustable (self-adjustable) packing is doubtlessly competitive with the existing ones that are widely used in industry because they develop stable flow modes and possibilities of specific automated sectioned adjustment as well as cost and manufacturing technology. It is evident that sectioned adjustment (self-adjustment) by all main flow characteristics of the packing material helps to perform the work in a very wide operation mode that allows to replace a large number of industrial packings with high efficiency.

Let us consider as a practical example the calculation of main process parameters and dimensions of an industrial absorber (see Fig. 8, Table 1) with sectioned adjustable heat and mass exchange packing as per the method described in paper [34]. The absorber consists of the column case 1 in which adjustable (self-adjustable) packing blocks 2 are installed. Above packing blocks, there are liquid sprayers 3 that spray the surface of the packing block with high quality for full use of the operating volume of the packing. The total height of the column is reached by the required quantity of packing blocks 2; at that each of the packing blocks 2 can be adjusted individually for the requirements of the certain process and apparatus. Scheme of an adjustable sectioned packing block is given in Fig. 1. Liquid distributors 4 are responsible for the redistribution of liquid between packing blocks 2.

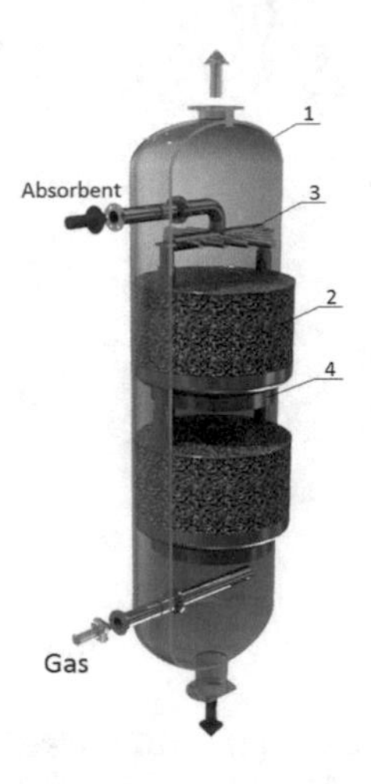

Fig. 8. Industrial absorber with the designed sectioned adjustable (self-adjustable) packing.

Table 1. Initial and reference data and calculation parameters of the industrial absorption column [34] with purification of nitrous gases with water for technological production lines of diluted nitric acid with sectioned adjustable (self-adjustable) heat and mass exchange packing (packing material - Aisi 304 shavings, specific surface of the packing 240 m^2/m^3, porosity of the packing 0.62 m^3/m^3, packing bulk density 195 kg/m^3, equivalent diameter 0.0103 m (parameters is variable)).

№	Parameter	Units	Symbol	Value
1	2	3	4	5
Initial data				
1	Initial productivity for gas flow rate	*kg/h*	G_{init}	1500
2	Absolute molar concentration in the gas at the inlet of the absorber	*kmolA/kmol (A + G)*	y_{init}	0.071
3	Absolute molar concentration in the gas at the outlet of the absorber	*kmolA/kmol (A + G)*	y_f	0.0012
4	Concentration of the extracted component A in the absorbent at the inlet of the absorber	*kmolA/kmol (A + L)*	x_{init}	0.0006
5	Molecular mass of the extracted component	*kgA/kmolA*	M_A	46
6	Molecular mass of the gas flow	*kgG/kmolG*	M_g	30
7	Molecular mass of absorbent	*kgL/kmolL*	M_L	18
8	Henry's constant of the working temperature	–	E	1410
9	Porosity of the packing	m^3/m^3	ε	0.62

(*continued*)

Table 1. (*continued*)

№	Parameter	Units	Symbol	Value
10	Specific surface of the packing	m^2/m^3	σ	240
11	Bulk density	kg/m^3	ρ_n	195
12	Absorbent density of absorption temperature	kg/m^3	ρ_L	988
13	Gas flow viscosity of absorption temperature	$Pa \cdot s$	μ_g	$1.86 \cdot 10^{-5}$
14	Absorbent viscosity of absorption temperature	$Pa \cdot s$	μ_L	0.0011
15	The diffusion coefficient of the absorbed component in the gas	m^2/s	D_g	$5.6 \cdot 10^{-9}$
16	The diffusion coefficient of the absorbed component in the liquid phase	m^2/s	D_L	$1.4 \cdot 10^{-10}$
17	Absorption temperature	^0C	t	40
18	Pressure in the column	atm	p	1.033
Calculated parameters				
1	Relative mass concentration in the gas at the inlet of the absorber	kgA/kgG	$\bar{y}_{init}$	0.117
2	Relative mass concentration in the gas at the outlet of the absorber	kgA/kgG	$\bar{y}_f$	0.0018
3	Relative mass concentration of the extracted component at the inlet of the absorber	kgA/kgL	$\bar{x}_{init}$	0.0015
4	Mass flow of the inert part of the gases at the inlet of the absorber	kgG/h	G	1342.7
5	Performance of the extracted component (Mass flow of extracted component of mass transfer from gas into the absorbent)	kgA/h	G_A	154.8
6	Flow rate for purified gas flow	kgG/h	G_k	1345.1
7	Equilibrium constant	kgA/kg	$\bar{X}_\kappa^*$	0.174
8	Minimum flow of the absorbent	kgL/h	L_m	894.4
9	Operation flow of the absorbent	kgL/h	L	1073.3
10	Final working concentration of the extracted component of the optimum flow of absorbent	kgA/kgL	$\bar{x}_f$	0.145
11	Number of split points of the working range of concentration changes	–	n	10
12	Gas change interval	kgA/kgG	$\Delta\bar{Y}$	0.0115
13	Number of transfer units	–	NTU	18.4
14	Working concentration of the extracted component in the middle of each interval in the gas phase	kgA/kgG	Y_c	Ref. to the Table 2
15	Working concentration of the extracted component in the middle of each interval in the liquid phase	kgA/kgL	X_c	Ref. to the Table 2
16	Equilibrium concentration of the extracted component in the gas phase	kgA/kgG	$\bar{Y}_c^*$	Ref. to the Table 2

(*continued*)

Table 1. (*continued*)

№	Parameter	Units	Symbol	Value
17	Average impelling force of the process	*kgA/kgG*	$\Delta\bar{Y}_c$	0.0062
18	Average working concentration of the extracted component in the gas phase	*kgA/kgG*	$\bar{Y}_s$	0.0595
19	Average equilibrium concentration of the extracted component in the gas phase	*kgA/kgG*	$\bar{Y}_s^*$	0.0532
20	Average working concentration of the extracted component in the liquid phase	*kgA/kgL*	$\bar{X}_s$	0.0797
21	Equivalent diameter of the packing	*m*	d_{eq}	0.0103
22	Density of the treated gas	*kg/m*^{3}	ρ_g	1.125
23	Archimedes number of gas flow	–	Ar_g	$3.48 \cdot 10^7$
24	Reynolds number of gas flow	–	Re_g	985.32
25	Fictitious velocity of gas	*m/s*	υ_g	1.576
26	Gas flow rate	*m*^{3}*/s*	q_v	0.351
27	Diameter of the absorbing column	*m*	D_a	0.532
28	Packing height of the equivalent transfer unit	*m*	h_1	0.35
29	Packing height	*m*	H	6.4
30	Surface of the packing	*m*^{2}	F_p	344.8
31	Mass transfer coefficient	*kgA/m*^{2}*s*	K_y	0.0199
32	Height section of the adjustable mass transfer packing	*m*	h_s	0.8
33	Number of sections in the absorption column	–	n_s	8

Table 2. Main calculated parameters of the equilibrium line and intermediate values for the graphic interpretation of the calculation of the number of transfer units.

Working concentration of the extracted component in the middle of each interval in the gas phase $Y_c \cdot 10^3$ *kgA/kgG*	$1/(Y_c - \bar{Y}_c^*)$	Equilibrium concentration of the extracted component in the gas phase $\bar{Y}_c^* \cdot 10^2$ *kgA/kgG*	Working concentration of the extracted component in the middle of each interval in the liquid phase $X_c \cdot 10^2$ *kgA/kgL*
7.60	561.13	0.582	0.874
19.14	270.54	1.544	2.317
30.67	178.62	2.508	3.76
42.21	133.53	3.472	5.203
53.74	106.76	4.438	6.646
65.28	89.02	5.404	8.089
76.81	76.41	6.372	9.532
88.35	66.98	7.342	10.975
99.88	59.66	8.312	12.417
111.41	53.82	9.284	13.86

4 Conclusions

Thus, having performed a number of experiments and process calculations of mass exchange equipment we can conclude that the use of sectioned blocks with packing material that can be specifically adjusted (self-adjusted) is very promising; at that the adjustment parameters can be registered in complex. Specific adjustment (self-adjustment) provides a unique possibility to adjust and maintain during the operation of packing mass exchange columns such effective mode as phase inversion (emulsification). During our experiments, we use adjustment by the structures of liquid and gaseous phases flows. We think that one of the most promising materials for adjustment are metal shavings that are wastes from metal working machines because they have a natural intensification effect that ensures a well-developed contact between liquid and gas, they are resistant to aggressive media, can be easily obtained and can replace expensive industrial mass exchange packings. Use of packed sectioned blocks of such packing material is the disposal of industrial wastes that is very important from the point of view of environment protection and energy and resources saving. At that surface and structural properties of the offered packing material that is to be adjusted significantly intensify heat-and-mass exchange processes due to micro-finned surface and microroughness and development of macro- and micro whirls of the gas flow sweeping around the surface of heat-and-mass transfer that increases the productivity of mass exchange apparatuses. It should be remembered that in order to receive reliable experimental data and provide required surface properties of packing material (wettability) all packing materials were first de-oiled (all lubricants were removed) [35].

Packing materials in the form of metal shavings can be of absolutely different configuration, their structure and surface properties are guided by physical and mechanical properties of the treated materials, cutting modes, etc. It may seem that the issue of use of metal shavings as packing material is very complicated, but that is not true. There are developed methods for express analyzing the packing materials and adapting them for a certain range of mass exchange processes [31–33], so we can conclude that the packing materials are suitable and can be further on classified and adjusted. Taking into account the received data, availability, chemical and thermal resistance we can conclude that the use of adjustable packing materials such as metal shavings from high-alloy steels is promising for application on their surface of active catalysts for gas and liquid phase chemical reactions. The possibility to use the adjustable (self-adjustable) sectioned packings for absorption during selective treatment of gases in ecological and chemical equipment has been proved experimentally and confirmed by calculations.

References

1. Sokol, B.A., Chermyshov, A.K., Baranov, D.A.: Mass-Transfer Column Packed-Type Devices. Infokhim, Moscow (2009)
2. Kagan, A.M., Laptev, A.G., Pushnov, A.S., Farakhov, M.I.: Contact Packings in Industrial Heat-and-Mass Transfer Apparatuses. Otechestvo, Kazan (2013)

3. Madyshev, I.N., Dmitrieva, O.S., Dmitriev, A.V., Nikolaev, A.N.: Study of fluid dynamics of mass-transfer apparatuses having stream-bubble contact devices. Chem. Pet. Eng. **52**, 299–304 (2016). https://doi.org/10.1007/s10556-016-0189-2
4. Sokolov, A.S., et al.: Hydrodynamic characteristics of mini-ring truncated-cone packing. Chem. Pet. Eng. **53**, 1–4 (2017). https://doi.org/10.1007/s10556-017-0288-8
5. Dmitrieva, O.S., Dmitriev, A.V., Madyshev, I.N., Nikolaev, A.N.: Flow dynamics of mass exchangers with jet-bubbling contact devices. Chem. Pet. Eng. **53**, 130–134 (2017). https://doi.org/10.1007/s10556-017-0308-8
6. Mitin, A.K., Nikolaikina, N.E., Pushnov, A.S., Zagustina, N.A.: Geometric characteristics of packings and hydrodynamics of packed biotrickling filters for air-gas purification. Chem. Pet. Eng. **52**, 47–52 (2016). https://doi.org/10.1007/s10556-016-0146-0
7. Dmitriev, A.V., Dmitrieva, O.S., Madyshev, I.N.: Optimal designing of mass transfer apparatuses with jet-film contact devices. Chem. Pet. Eng. **53**, 430–434 (2017). https://doi.org/10.1007/s10556-017-0358-y
8. Berengarten, M.G., Bogdanov, R.I., Voronina, V.E.: Physical flow characteristics of a gaseous mixture with solid inclusions. Chem. Pet. Eng. **47**, 111–120 (2011). https://doi.org/10.1007/s10556-011-9434-x
9. Dmitriev, A.V., Dmitrieva, O.S., Madyshev, I.N., Nikolaev, A.N.: Efficiency of the contact stage of a jet-film device during rectification of ethylbenzene–styrene mixture. Chem. Pet. Eng. **53**, 501–507 (2017). https://doi.org/10.1007/s10556-017-0371-1
10. Gorodilov, A.A., Pushnov, A.S., Berengarten, M.G.: Improving the design of grid packing. Chem. Pet. Eng. **50**, 84–90 (2014). https://doi.org/10.1007/s10556-014-9860-7
11. Madyshev, I.N., Dmitrieva, O.S., Dmitriev, A.V., Nikolaev, A.N.: Assessment of change in torque of stream-bubble contact mass transfer devices. Chem. Pet. Eng. **51**, 383–387 (2015). https://doi.org/10.1007/s10556-015-0056-6
12. Ivanov, A.E., Berengarten, M.G., Klyushenkova, M.I.: Processes and equipment for chemical and oil-gas production: hydrodynamics of the bubbling layer in a new type of combined heat and mass exchanger. Chem. Pet. Eng. **46**, 433–440 (2010). https://doi.org/10.1007/s10556-010-9355-0
13. Chizh, K.V., Pushnov, A.S., et al.: Structure of mini ring packing layup in column equipment. Chem. Pet. Eng. **50**, 244–250 (2014). https://doi.org/10.1007/s10556-014-9889-7
14. Shilin, M.V., Berengarten, M.G., Pushnov, A.S., Klyushenkova, M.I.: Hydrodynamics of regular packings formed from cellular inclined cylinders for heat-and mass-exchange processes. Chem. Pet. Eng. **48**, 608–614 (2013). https://doi.org/10.1007/s10556-013-9665-0
15. Madyshev, I.N., Dmitrieva, O.S., Dmitriev, A.V.: MATEC Web of Conference, vol. 224, p. 02079 (2018). https://doi.org/10.1051/matecconf/201822402079
16. Berengarten, M.G., Nevelson, A.O., Pushnov, A.S.: Combination packing for catalytic reactors. Chem. Pet. Eng. **48**, 723–729 (2013). https://doi.org/10.1007/s10556-013-9687-7
17. Madyshev, I.N., Dmitrieva, O.S., Dmitriev, A.V.: MATEC Web of Conference, vol. 91, p. 01019 (2017). https://doi.org/10.1051/matecconf/20179101019
18. Kagan, A.M., Yudina, L.A., Pushnov, A.S.: Active surface of the elements of irregular heat and mass transfer packings. Theor. Found. Chem. Eng. **46**, 165–171 (2012). https://doi.org/10.1134/S0040579512020042
19. Madyshev, I.N., Dmitrieva, O.S., Dmitriev, A.V.: MATEC Web of Conference, vol. 141, p. 01023 (2017). https://doi.org/10.1051/matecconf/201714101023
20. Gorodilov, A.A., Berengarten, M.G., Pushnov, A.S.: Features of fluid film falling on the corrugated surface of structured packings with perforations. Theor. Found. Chem. Eng. **50**, 325–334 (2016). https://doi.org/10.1134/S0040579516030040
21. Madyshev, I.N., Dmitrieva, O.S., Dmitriev, A.V.: MATEC Web of Conference, vol. 141, p. 01012 (2017). https://doi.org/10.1051/matecconf/201714101012

22. Golovanchikov, A.B., Merentsov, N.A., Balashov, V.A.: Modeling and analysis of a mechanical-draft cooling tower with wire packing and drip irrigation. Chem. Pet. Eng. **48**, 595–601 (2013). https://doi.org/10.1007/s10556-013-9663-2
23. Dmitrieva, G.B., Berengarten, M.G., Pushnov, A.S., Poplavskii, V.Y., Marshik, F.: New combination packing for heat-and mass-exchange vessels. Chem. Pet. Eng. **42**, 361–366 (2006). https://doi.org/10.1007/s10556-006-0108-z
24. Pushnov, A.S., Berengarten, M.G., et al.: Chem. Pet. Eng. **44**, 307–311 (2008). https://doi.org/10.1007/s10556-008-9054-2
25. Madyshev, I.N., Dmitrieva, O.S., Dmitriev, A.V.: MATEC Web of Conference, vol. 194, p. 01013 (2018). https://doi.org/10.1051/matecconf/201819401013
26. Gorodilov, A.A., Berengarten, M.G., Pushnov, A.S.: Experimental study of mass transfer on structured packings of direct-contact crossflow heat exchangers. Theor. Found. of Chem. Eng. **50**, 422–429 (2016). https://doi.org/10.1134/S0040579516040345
27. Pushnov, A.S., et al.: Helicoidal-structured packing for heat and mass exchange with direct phase contact. Chem. Pet. Eng. **43**, 575–579 (2007). https://doi.org/10.1007/s10556-007-0102-0
28. Ivanov, A.E., Berengarten, M.G., Klyushenkova, M.I.: Hydrodynamic operating regimes for a combined heat-and-mass exchanger. Chem. Pet. Eng. **45**, 526–531 (2009). https://doi.org/10.1007/s10556-010-9230-z
29. Madyshev, I.N., Dmitrieva, O.S., Dmitriev, A.V.: MATEC Web of Conference, vol. 194, p. 01036 (2018). https://doi.org/10.1051/matecconf/201819401036
30. Dmitrieva, O.S., Dmitriev, A.V., Madyshev, I.N., Kruglov, L.V.: MATEC Web of Conference, vol. 129, p. 06010 (2017). https://doi.org/10.1051/matecconf/201712906010
31. Golovanchikov, A.B., Balashov, V.A., Merentsov, N.A.: The filtration equation for packing material. Chem. Pet. Eng. **53**, 10–13 (2017). https://doi.org/10.1007/s10556-017-0285-y
32. Merentsov, N.A., Balashov, V.A., Bunin, D.Y., Lebedev, V.N., Persidskiy, A.V., Topilin, M.V.: MATEC Web of Conference, vol. 243, p. 5 (2018). https://doi.org/10.1051/matecconf/201824300011
33. Merentsov, N.A., Lebedev, V.N., et al.: Experimental assessment of heat and mass transfer of modular nozzles of cooling towers. In: IOP Conference Series: Earth and Environmental Science, vol. 115, p. 012017 (2018). https://doi.org/10.1088/1755-1315/115/1/012017
34. Golovanchikov, A.B., Zalipaeva, O.A., Merentsov, N.A.: Modeling of Sorption Processes Taking into Account the Flow Structure. VolgGTU, Volgograd (2018)
35. Merentsov, N.A., et al.: System for centralised collection, recycling and removal of waste pickling and galvanic solutions and sludge. Mater. Sci. Forum **927**, 183–189 (2018). https://doi.org/10.4028/www.scientific.net/MSF.927.183

Expert Modelling for Evaluating Dynamic Impact of the Motorcade on the Supporting Systems During Braking

Sergey Gridnev[1(✉)], Alexey Budkovoy[1], and Yuriy Skalko[2]

[1] Voronezh State Technical University, Moscow Avenue, 14, Voronezh 394026, Russia
gridnev_s_y@rambler.ru
[2] Moscow Institute of Physics and Technology (State University), 9 Institutskiy per., Dolgoprudny, Moscow Region 141700, Russia

Abstract. The paper indicates the necessity of the research in the field of dynamic calculation of road bridges impact on a moving load in non-stationary modes of exposure as closest to the actual operating conditions. Particular attention is paid to the transient modes of movement of vehicles, as a result of which additional dynamic additives are formed in axle pressures, whereas the nature of oscillations also changes. In order to develop the studies performed on a single mobile load, it is proposed to consider the problem of joint oscillations of beam systems when driving a motorcade of similar type of vehicles in braking mode before they leave the bridge. The flat dynamic model of the car is represented by the system with five degrees of freedom. The deflections of the span are determined by the 'Bubnov-Galerkin algorithm'. Based on the expert models, the applicable software package has been developed to assess the dynamic effect of a moving load on the carrier systems. The influence of the initial braking speed, the magnitude of the acceleration and the distance between the cars on the dynamic deflection of the average cross section is investigated. Deflection graphs were constructed, and the values of dynamic coefficients were determined. The obtained results clearly demonstrate an increase in dynamic deflection due to braking as compared with driving at a constant speed by 15–90%, depending on the distance between vehicles and the magnitude of steady-state deceleration. The study of the dynamic effects of specialized types of mobile loads transporting liquid and bulk cargo, both single and as part of a column, is designated as a promising research area.

Keywords: Dynamic deflection · Oscillations of beam systems · Motorcade · Moving load · Braking · Dynamic coefficient · Non-stationary impact

1 Introduction

Currently, the total volume of freight constantly increases. It is obvious that this trend will continue in the future. The share of multi-axle heavy vehicles being the part of the car flow has increased significantly, which in some cases leads to the formation of motorcades of the same type of cars. The level of dynamic impact of moving load

V. Murgul and M. Pasetti (Eds.): EMMFT-2018, AISC 983, pp. 574–585, 2019.
https://doi.org/10.1007/978-3-030-19868-8_55

motorcades compared to the dynamic impact of single cars and mixed motorcades on the bearing systems of transport facilities and other building structures is much higher.

Forecasting and further analysis of the joint vibrations of transport facilities and the various loads moving along them is still a prerequisite for ensuring reliable operation. Scientific research in this area is extremely relevant.

Research in this area was carried out by the number of scientists and scientific schools. Their approaches were repeatedly described by different authors in previous publications. In the post-Soviet academic environment, active work on the study of transient modes of movement of vehicles with liquid cargo from the standpoint of loading nodes and mechanisms is carried out in the Belarusian University of Transport [1, 2].

It should be separately noted that in recent years, the research of the Russian University of Transport on the development of the theory of calculating railway bridges for high-speed mobile, seismic, wind, and shock loads has recently been significantly intensified [3–7]. In foreign publications, there are also works on the simulation of vehicle oscillations from the standpoint of determining dynamic pressures per path [8–10]. The problems of joint oscillations of spans and moving loads in the spatial formulation are described in [11, 12].

Simulation of joint vibrations of transport facilities and moving motorcades of heavy vehicles of the same type in a simplified formulation has already been done before, but the systematization of numerical results and practical recommendations has not been made as of yet. In the study of the dynamic effects of a moving load on the carrier systems, as a rule, an assumption of uniform motion was introduced. The vehicle constantly changes the speed during its operation both at the time of entry into the installation and moving on it. This happens due to a number of difficult reasons. The results of the studies carried out by several authors have shown that the time of movement with variable speed is up to 75% of the total driving cycle.

Oscillations of carrier systems under the action of vehicles with variable speeds moving along them correspond to the actual operating conditions. However, systematic research in this area was resumed and began to bring some results only in recent years.

In [13, 14], the object of the study was the dynamic effect of single cars and truck-type saddle-type vehicles on transport facilities under non-stationary exposure modes. Primarily braking on a smooth path and taking into account the irregularities of the profile of the roadway were considered. In [15], a technique was proposed for simulating joint oscillations of beam systems and automobiles, numerical studies were carried out using the developed algorithms, and dynamic coefficients were obtained. Field measurements are described in detail in [16], where practical recommendations are given for testing the bridges of beam circuits for a moving load in the steady-state deceleration mode. The effect of the length of the beam span on the magnitude and nature of the dynamic effects of a single moving load was studied in [17]. The studies on the estimation of dynamic coefficients are summarized in [18].

Transient modes of movement of single cars and, moreover, the motorcades of vehicles of the same type represent the greatest danger in terms of the possibility of the excessive dynamic impact on the bearing systems and, above all, in the emergency braking mode.

2 Materials and Methods

The task is set to choose the design scheme of the oscillating system, to develop a mathematical and numerical model of its behavior, to create a computational algorithm of joint non-linear oscillations of the bearing systems of transport facilities when braking the motorcade of similar vehicles (Fig. 1).

To perform the numerical studies, it is necessary to develop expert complex computer models to adequately assess the magnitude of dynamic effects in regulatory documents, to make sound engineering decisions in the design of new transport facilities, to assess the possibility of passing motorcades in existing facilities in emergency situations and for the needs of the Russian Armed Forces.

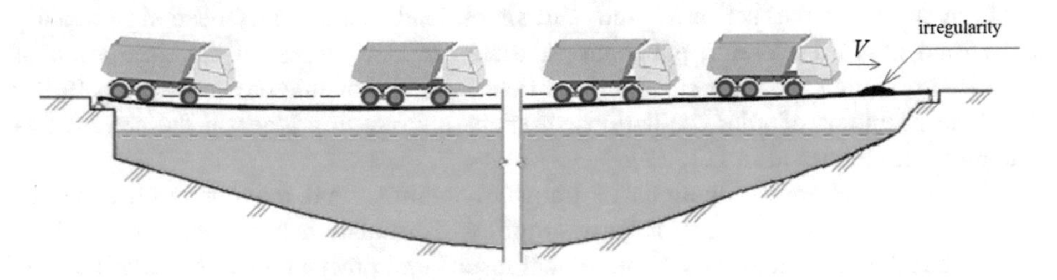

Fig. 1. The design scheme of the task.

To accomplish the task, the design scheme of the motorcade was formed. The motorcade is composed of flat non-linear dynamic models of a three-axle vehicle (Fig. 2). This model was previously used by the authors to study vehicle oscillations when driving at a constant speed.

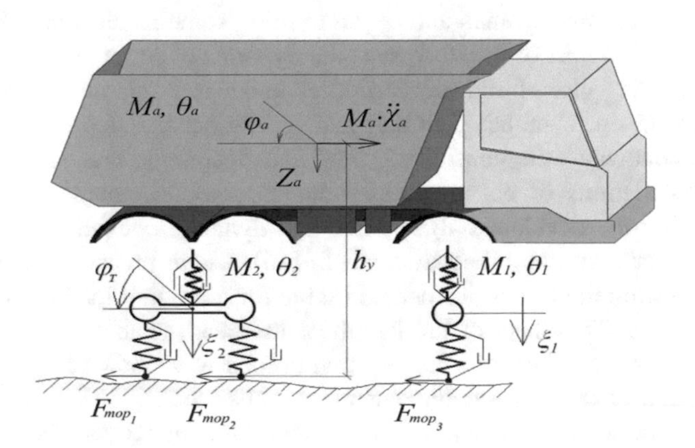

Fig. 2. Flat dynamic model of the car (system with 5 degrees of freedom).

The oscillation equations of a single multiaxial vehicle columns in the operator form has the form

$$L\left(\ddot{\vec{Z}},\dot{\vec{Z}},\vec{Z}\right) = F\left\{\vec{R}\left[\vec{Z},\dot{\vec{Z}},\ h(vt),\ \dot{h}(vt),\ y(vt),\dot{y}(vt)\right]\right\} \tag{1}$$

where L, F - linear and nonlinear operators corresponding to the dynamic model of the car; $\vec{Z}$ - vector of generalized coordinates of inert parts of the model; $h(vt), \dot{h}(vt), y(vt), \dot{y}(vt)$ - kinematic perturbation functions for a vehicle arising from irregularities on the carriageway of the bridge, its approaches and movements of the span at the points of wheel support; $\vec{R}$ - vehicle pressure vector per path.

In expanded form, we present only the second equation of the general system of differential equations of motion of a model (galloping equation). In the right-hand side, a term appears that takes into account horizontal inertia forces during braking:

$$\theta_a\ddot{\varphi}_a + F_1 a_T(t) - b_T(t)F_2 = M_a \cdot h_y(t) \cdot \ddot{\chi}(t) \tag{2}$$

where

$M_a,\ \theta_a$ – mass and moment of inertia of the sprung part of the car;
F_1, F_2 – spring force;
$a_T(t),\ b_T(t),\ h_y(t)$ – geometric values that determine the position of the center of gravity of the sprung part;
$\varphi_a,\ \ddot{\chi}(t)$ – the generalized coordinate, which determines the angle of rotation of the sprung part relative to the transverse axis, and the acceleration during braking, respectively.

The redistribution of the moment from inertial forces from quasi-static considerations was taken as follows:

$$R_1(t) = R_1^{'}(t) + \frac{M \cdot \dot{V}(t) \cdot c}{a+b},\ R_2(t) = R_2^{'}(t) - \frac{M \cdot \dot{V}(t) \cdot c}{2 \cdot (a+b)},\ R_3(t) = R_3^{'}(t) - \frac{M \cdot \dot{V}(t) \cdot c}{2 \cdot (a+b)} \tag{3}$$

where $R_i^{'}(t)$ - dynamic pressure of the vehicle i - axis on the roadway with uniform movement.

In accordance with the operating standards, the minimum steady-state deceleration of technically sound trucks and road trains should be on average 6 m/s^2. The period of increase in braking force, according to tests of trucks in good technical condition, is about 0.4 s (Fig. 3). At this stage of the research, the driver's sensorimotor response time is not taken into account.

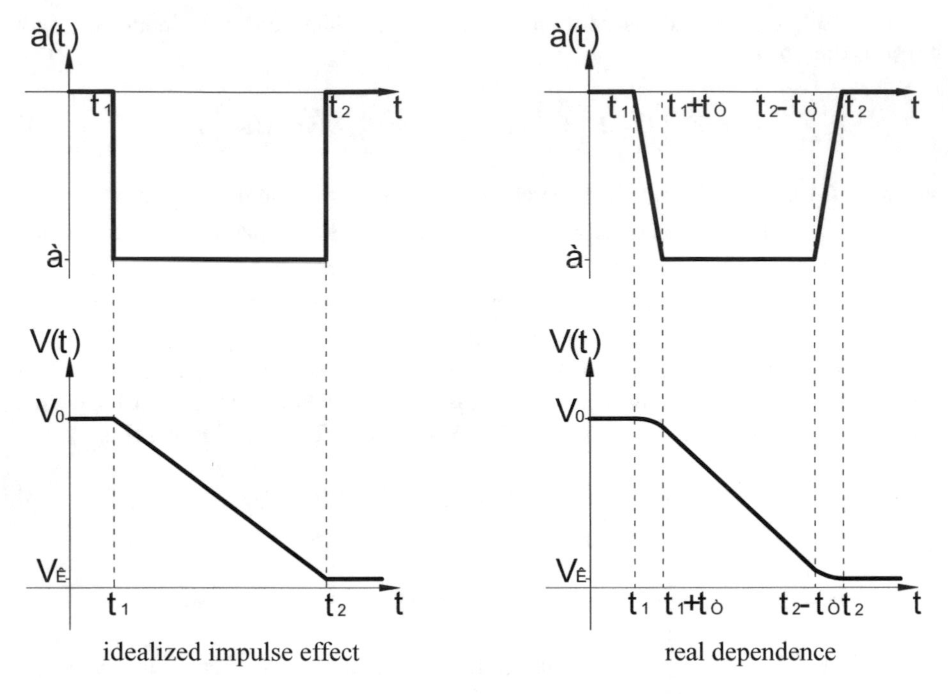

Fig. 3. Graphics of dependences of acceleration and velocity for uniformly variable motion.

A hinged-supported beam, the dynamic behavior of which is described by a fifth-order partial differential equation based on the 'Vlasov theory of oscillations of thin-walled elastic rods'.

In the operator form, this equation has the following form:

$$L_1\left[\ddot{y}(x,t),\ y^V(x,t),\ y^{IV}(x,t)\right] = F_1\left(\vec{R}\right), \tag{4}$$

where

L_1 is - linear differential operator including;
$\ell_b,\ m_b,\ E, J_z, \rho$ - length and parameters of the carrier system;
F_1 - nonlinear load operator;
$\vec{R} = \left[\vec{R}_1,\ \vec{R}_2, .., \vec{R}_j, \ldots,\ \vec{R}_{q(t)}\right]^T$- axis pressure vector $q(t)$ motorcades, simultaneously located on the carrier system at a given time; The axle pressure term of the cars $q(t)$, located at a given time within the roadway of the bridge, in expanded form has the form:
$\sum_{j=1}^{q(t)} \sum_{k=1}^{n_j} R_{kj}(t) \cdot \delta(x - \chi_k)$ - $R_{kj}(t)$ - dynamic pressure of k – axis j - of the vehicle;
n_j - number of vehicle's bearing links.

The resolving equations of bending oscillations of the carrier system in a vertical plane are obtained by transformation using the Bubnov-Galerkin method, where the eigenfunctions of the hinge-supported rod are used as the basis for calculations. As a result, the initial equations are reduced to a set of ordinary differential equations of the same type with respect to temporal decomposition coefficients, convenient for implementation on a computer.

To carry out numerical studies and conduct a series of computational experiments on the passage of the motoracades of the similar cars, based on the devcloped expert models, a computer complex was created using the simulation package of dynamic and event-driven systems Simulink systems MATLAB.

Figures 4, 5 and 6 shows the general Simulink block diagram for describing the joint oscillations and two blocks of the algorithm subsystems.

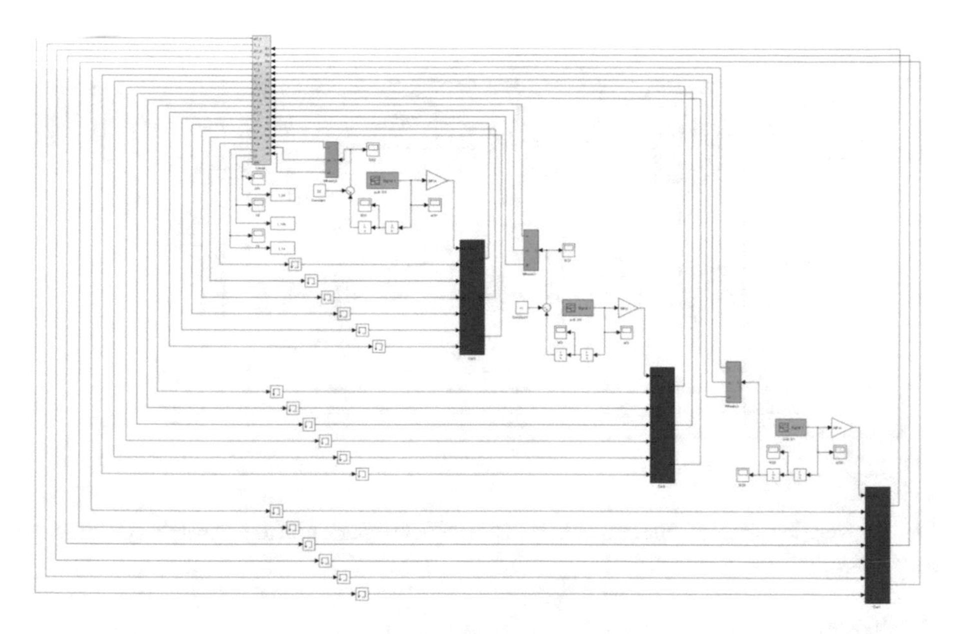

Fig. 4. The general block diagram of Simulink shows the joint oscillations of the carrier system and the motorcade of the three accelerating cars moving along. The colors are marked with: red - subsystems that simulate oscillations of cars, green - subsystems for determining the conditions of entry and exit of wheels of cars from the superstructure, blue - block signals of the acceleration function, yellow - subsystem of oscillations of the superstructure.

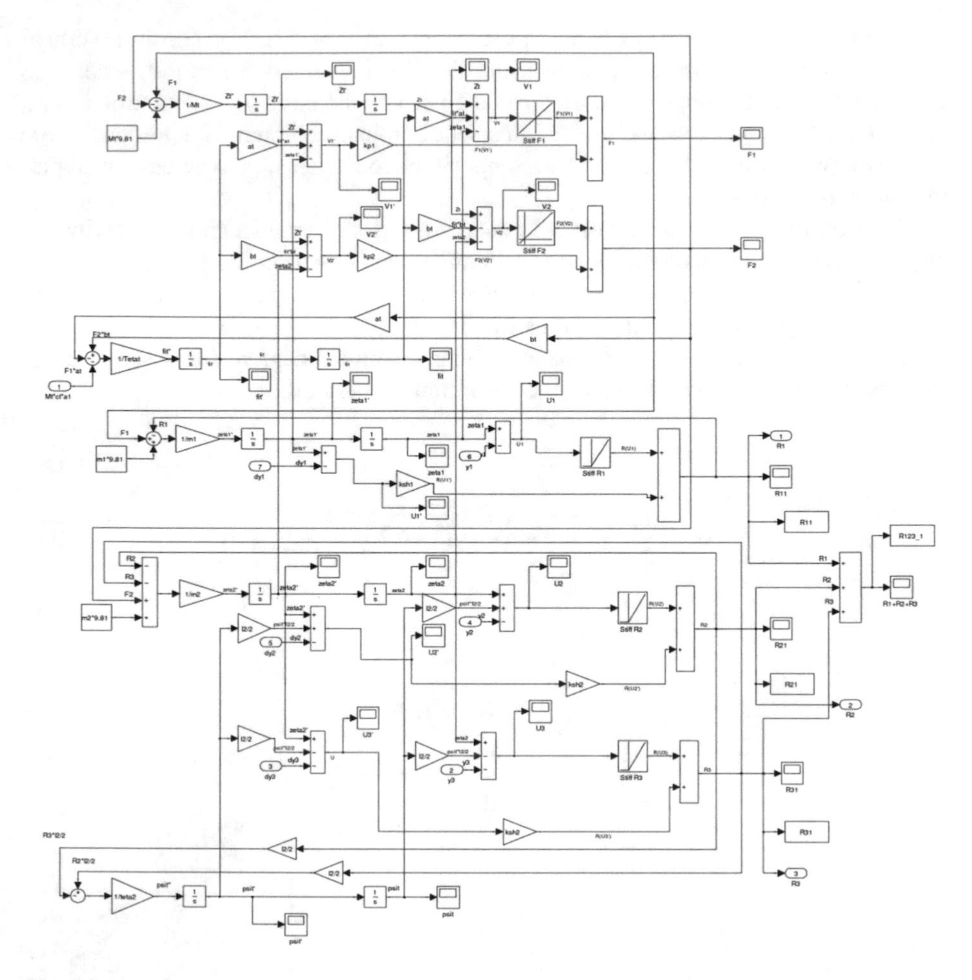

Fig. 5. Unit that simulates the oscillations of the vehicle, which acts as a dynamic system with 5 degrees of freedom.

3 Results

In carrying out numerical studies, a series of computational experiments on the simulation of braking of a column of three three-axle trucks weighing 24 tons on a steel-concrete concrete span 42.5 m long in the event of an emergency stop (accident, repair, danger to pedestrians or other unforeseen situations) at the head car from the bridge. The following parameters were considered as variable parameters: the distance between the cars in the motorcade, the speed at the time of the start of braking and the magnitude of the steady-state deceleration. Taking into account the delay time of the reaction of drivers about 0.5–1 s.

At the first stage of research, it was revealed that in the studied range of initial speeds of 15 … 30 m/s, the maximum dynamic deflection is observed at speeds of 15–20 m/s. Graphs of displacements of the average cross-section of the carrier system are shown in (Fig. 7).

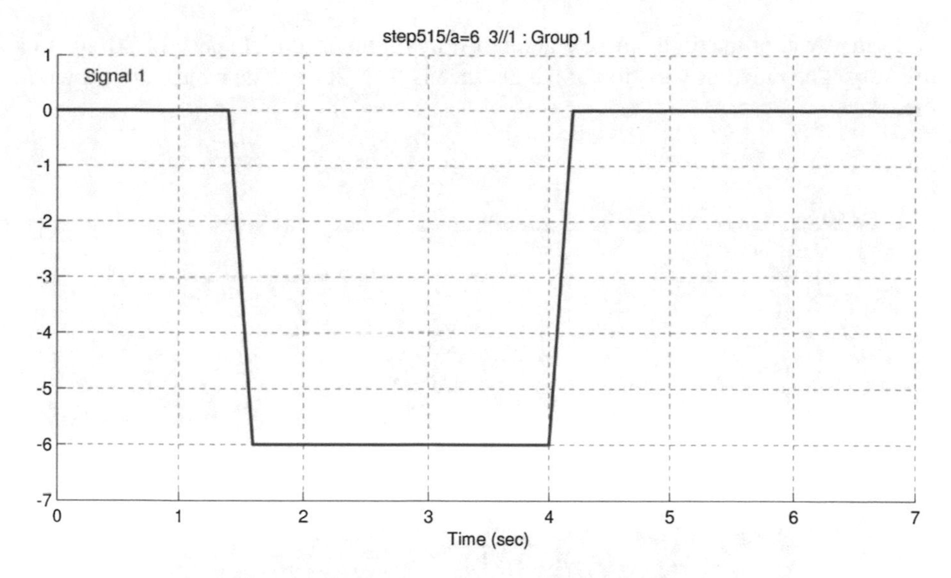

Fig. 6. Block signal acceleration for the first car.

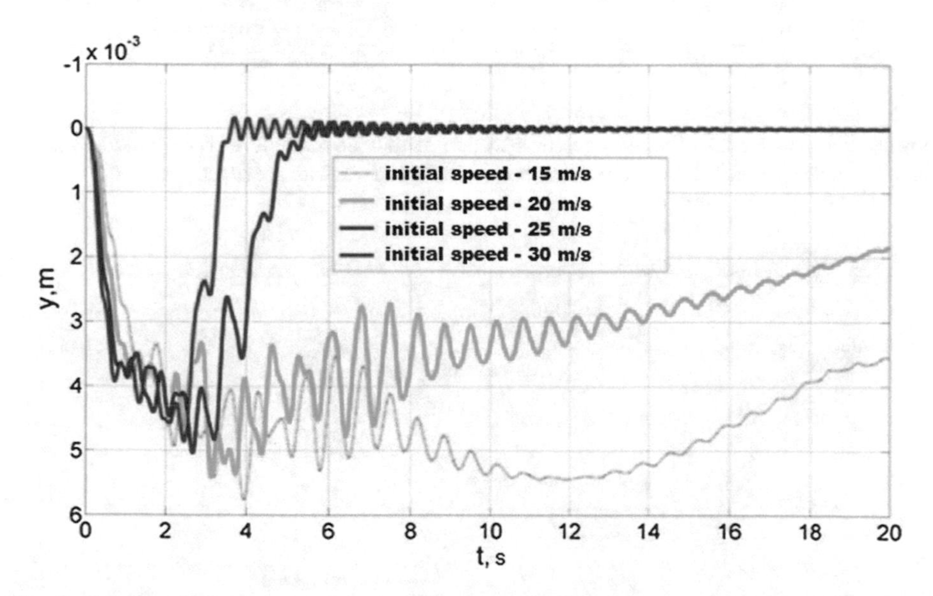

Fig. 7. Displacement of the average cross section of the superstructure when braking with an acceleration of 6 m/s^2, a distance of 20 m and various initial speeds.

A further increase in the initial speed somewhat reduces the dynamic effect. The effect of acceleration was more significant and increases the maximum deflection by 14% in the acceleration range of 2 … 6 m/s^2 (Fig. 8).

The factor determining the maximum dynamic response was the reduction of the distance between cars from the normative to the extremely minimal. Braking is performed with a maximum acceleration of 6 m/s^2 and an initial speed of 20 m/s to a full stop. The range of variation of the distance is 5 … 20 m. The results are shown in Fig. 9.

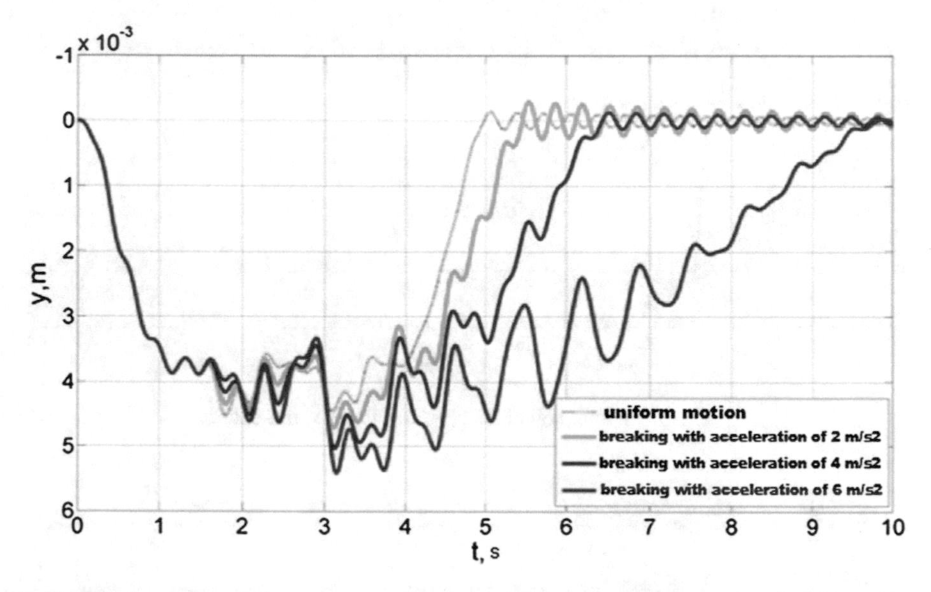

Fig. 8. Displacement of the average cross-section of the superstructure when braking with an initial speed of 20 m/s, a normative distance of 20 m and various accelerations.

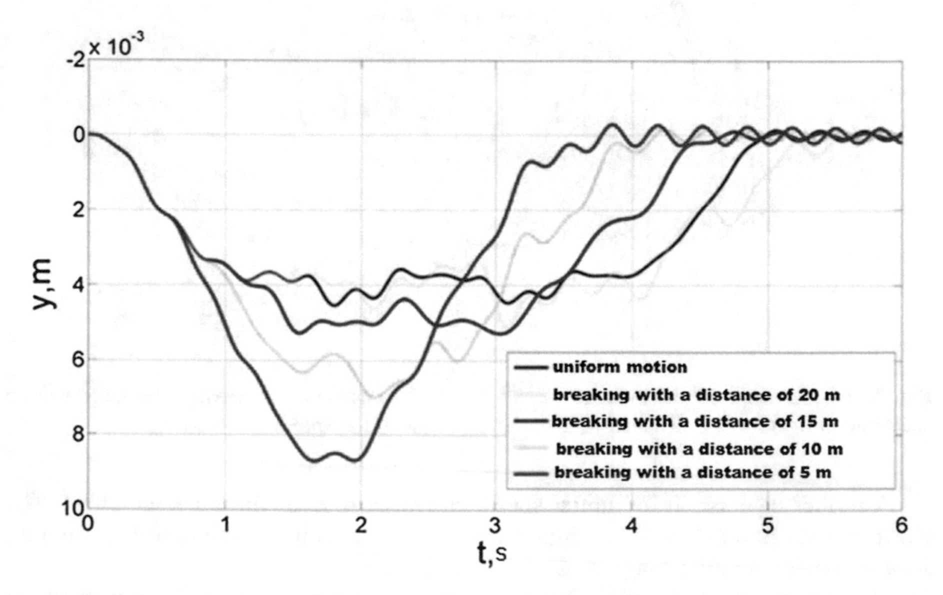

Fig. 9. Displacement of the average cross section of the superstructure when braking with an initial speed of 20 m/s, acceleration of 6 m/s^2 and various distances.

The graph of dynamic coefficients depending on the distance between cars is shown in Fig. 10.

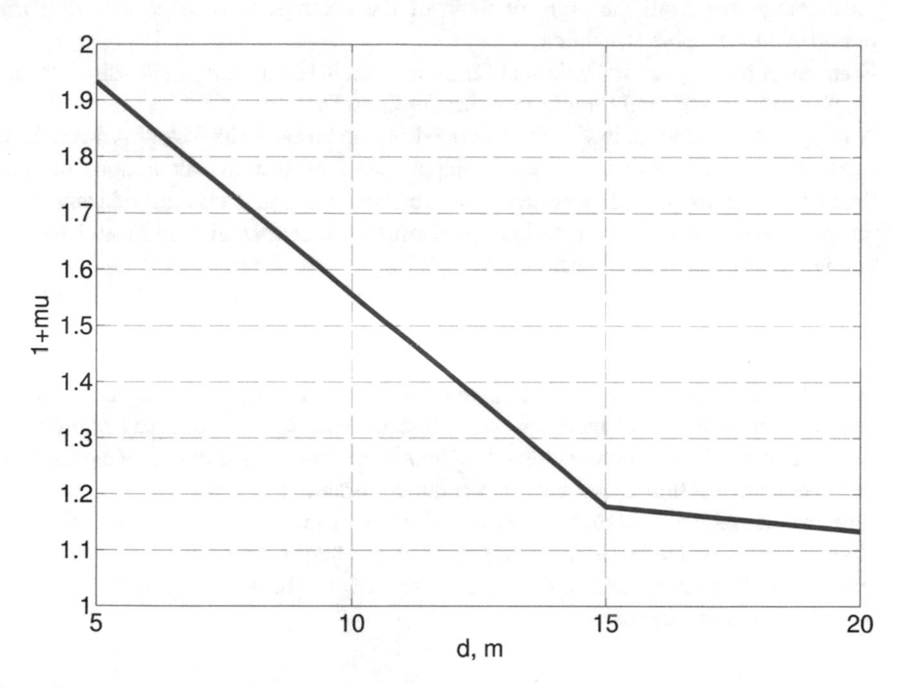

Fig. 10. Dynamic coefficients of deflection of the average cross-section during the braking of three vehicles motorcade depending on the distance between them.

4 Discussion

The main conclusions from the results of numcrical studies are the following:

1. The values of the maximum dynamic deflections during deceleration in all initial speed ranges exceed the corresponding values with uniform motion;
2. Reducing the distance between cars increases the magnitude of the deflections, which reach maximum values at an extreme distance of 5 m. In this case, the dynamic coefficient is 1.93.
3. The dependence of the maximum dynamic deflections on the magnitude of the acceleration varies from 10 to 15%, which differs significantly from the non-stationary effects in the form of a single car, described in [16].

5 Conclusions

1. Non-stationary modes of action of a moving load, such as emergency braking, are most dangerous from the point of view of the occurrence of excessive dynamic pressure on transport structures;
2. Features of the impact of a column of moving cars in the braking mode differ from a single moving load and require more in-depth study;
3. An application package has been developed on the basis of the MATLAB Simulink module for performing a series of comparative and system calculations of non-linear oscillations of carrier systems for the movement of a column of vehicles of the same type, which is focused on performing experimental studies and solving applied engineering problems.
4. Analysis of the results of numerical studies performed on the developed and improved algorithms and verified by the results of field measurements will allow us to adequately take into account the specific features of the dynamic impact in the regulatory documents, as well as to take more informed engineering decisions in the design of new transport facilities. As a further area of research, it is particularly cunning to study the non-stationary dynamic effects of a column of specialized vehicles carrying liquid and bulk cargo during braking.
5. Fluctuations of the fluid and the flow properties of materials change over time the characteristics of the oscillatory system, and the nature of such an impact causes specific redistribution and a significant increase in the axial dynamic pressures compared to solid loads.

References

1. Shimanovsky, A.O., Kovalenko, A.V., Pleskachevsky, Yu.M.: Computer simulation of truck braking, partially filled with fluid. Mech. Mach. Mech. Mater. **2**(11), 39–42 (2010)
2. Shimanovsky, A.O., Kuznetsova, M.G.: Computer simulation of the movement of tank trucks on roads with irregularities of various types. Topical Issues Mech. Eng. **2**, 255–260 (2013)
3. Ivanchenko, I.I.: Dynamic interaction of bridges and high-speed trains. Proc. Russ. Acad. Sci. Solid Mech. **3**, 146–160 (2011)
4. Ivanchenko, I.I.: Formation of norms for bridges on high-speed railways. Transp. Constr. **1**, 22–25 (2014)
5. Ivanchenko, I.I.: The Method of calculating bridge structures for seismic effects. Proc. Russ. Acad. Sci. Mech. Solids **1**, 114–139 (2015)
6. Ivanchenko, I.I.: Joint oscillations of double-track bridges, simulated by thin-walled rods, and trains on the high-speed rail. Struct. Mech. Calculation Struct. **2**(265), 36–43 (2016)
7. Ivanchenko, I.I.: The aerodynamic calculation of beam bridge spans. Struct. Mech. Struct. Calculation **4**(273), 39–46 (2017)
8. Grzedonek, W., Adamiec-Wijcik, I., Wojciech, S.: Modeling of dynamics of vehicles. In: Proceedings of the 8 Mini Conference on Vehicle Dynamics, Identification and Anomalies, Budapest, vol. 1, pp. 369–375 (2003)
9. Peceliunas, R., Pukalskas, S., Prentkovskis, O.: Dynamics of car vibrations during the emergency braking. Transp. Means Kaunas **1**, 113–117 (2003)

10. Lu, S.: Optimum design of “road-friendly” vehicles. Appl. Math. Model. **26**, 635–652 (2002)
11. Chunhua, L., Dongzhou, H., Ton-Lo, W.: Analytical dynamic impact study based on correlated rod roughness. Comput. Struct. **20–21**(80), 1639–1650 (2002)
12. Ping, L., Qing-yuan, Z.: Element for vehicle-track-bridge system. Int. J. Numer. Meth. Eng. **3**(62), 435–437 (2005)
13. Gridnev, S.Yu., Budkovoy, A.N.: Evaluation of the dynamic impact of the car on the path during braking and acceleration, taking into account kinematic perturbations. News of the Kazan State University of Architecture and Civil Engineering. Constr. Archit. **3**(23), 409–415 (2012)
14. Gridnev, S.Yu., Budkovoy, A.N.: Dynamic impact of the saddle train on the transport structure during braking. In: Science and education. Sat. scientific tr. According to the materials of the International correspondence, Tambov, vol. 5, pp. 31–35 (2012)
15. Gridnev, S.Yu., Budkovoy, A.N.: Fluctuations of girder systems in transient modes of movement of a single car. Struct. Mech. Struct. **1**(6), 84–91 (2013)
16. Gridnev, S.Yu., Volkov, V.V., Budkovoy, A.N.: Field measurements of oscillations of an elastically supported steel-reinforced concrete bridge while the vehicle is moving along it in the braking mode. Sci. Bull. Voronezh State Constr. Archit. Acad. **1**(21), 18–27 (2010)
17. Gridnev, S.Yu., Budkovoy, A.N.: The influence of the length of the span structures in assessing the impact of a single mobile load on road bridges. Struct. Mech. Struct. **1**(10), 97–105 (2015)
18. Gridnev, S.Yu., Budkovoy, A.N.: Development of approaches to the formation of the regulatory base of dynamic coefficients to the temporary load for the calculation of road bridges. Naukovedenie Web J. **6**(31), 141 (2015)

The Stress-Strain State of Normal Sections Rubcon Bending Elements with Mixed Reinforcement

Aleksei Polikutin, Yuri Potapov, Artem Levchenko(✉), and Oleg Perekal'skiy

Voronezh State Technical University, Moscow Avenue, 14, Voronezh 394026, Russia
Alevchenko@vgasu.vrn.ru

Abstract. A full study of bending elements of building structures made of concrete of various types is impossible without studying the stress-strain state of sections, since without studying the distribution of strains and stresses along the height of the section, it is impossible to give recommendations on the calculation of such elements, this is especially important for polymer concrete elements. Rubber concrete (rubcon) is a polymer concrete with high exploitative characteristics, as well as structures made of it. The addition of dispersed reinforcement allows you to increase the strength of the structure under tension, thereby increasing the crack resistance of the sections of the bent element. To study the operation of normal sections of the rubcon beams, they were tested for pure bending with two equal symmetrically located concentrated loads. As a result, the three stages of the stress-strain state of polymer concrete structures with dispersed reinforcement and without it are similar to reinforced concrete and fiber-concrete bending elements. These studies will determine the prerequisites for the calculation of the normal sections of bending elements made of rubber concrete.

Keywords: Mixed reinforcement · Rubcon bending · Stress-strain state

1 Introduction

In reinforced concrete bent elements with a gradual increase in external load, the sections undergo three characteristic stages of stress-strain state (SSS): stage I - before the appearance of cracks in the concrete of the tensile zone, stage II - after the appearance of cracks in the concrete of the tensile zone and stage III - the stage of destruction. However, for polymer concrete, the distribution of stresses along the height of normal sections requires clarification. A full study of bending elements is impossible without studying the stress-strain state of sections, since without studying the distribution of strains and stresses along the height of the section, it is impossible to give recommendations on the calculation of such elements. This is especially important for elements made of innovative materials. The use of structures based on polymer concrete is most effective in buildings and structures operated under the influence of

V. Murgul and M. Pasetti (Eds.): EMMFT-2018, AISC 983, pp. 586–599, 2019.
https://doi.org/10.1007/978-3-030-19868-8_56

aggressive environments of various type. An important factor for ensuring the normal exploitation of building structures is to prevent the penetration of aggressive environment into the depth of the cross section and the impact on steel reinforcement. One of the ways to solve this problem is to use a material with high strength characteristics, in particular, high tensile strength. The building material obtained on the basis of liquid rubbers - rubber concrete (abbreviated as rubcon) has a number of favorable physicomechanical characteristics and high chemical resistance. Rubcon was developed by the Head of the Department of Reinforced Concrete and Stone Constructions, Doctor of Technical Sciences, Professor Potapov Yuri Borisovich. Further studies were conducted on the effect of dispersed reinforcement on the strength of normal sections of bending elements from rubcon [1], as a result, it was found that the values of compressive stresses in the cross sections of bending elements increase with an increase in the percentage of longitudinal reinforcement and reach the compressive strength of the rubcon or fibrorubcon.

2 Materials and Methods

I would like to note that the components of the rubber concrete include such components as fly ash and technical sulfur, which are industrial wastes, metal cord was chosen as reinforcing fiber from waste tire industry, cut with an angle grinder on rods 3 sm. long. Based on the study of the use of steel fiber in traditional concretes [2], the positive effect of this method of reinforcement on the strength characteristics of the material was proved. In article [3], fibers from recycled carpets were used as reinforcing fibers. Based on studies of rubber concrete with dispersed reinforcement [4] and without it [5], together with studies conducted by Potapov, the optimal composition of the mixture was obtained (Table 1) in order to ensure the best characteristics shown in Table 2.

Table 1. Fibrorubcon composition.

The name of components	Contents of components, wt. (%)
Rubber SKDN-N	8.2
Sulphur technical	4.0
Thiuram-D	0.4
Zinc oxide	1.2
Calcium oxide	0.4
Fly ash	7.8
Sand	24.2
Crushed stone	51.3
Fibers from metal-cord waste (fiber)	2.5

The use of large-scale industrial waste not only reduces the cost of the composite, but also increases the potential of products from rubcon to competitiveness.

Table 2. Physico-mechanical properties of fibrorubcon.

Properties	Values
Compressive strength, MPa	85–90
Tensile strength, MPa	13–15
The modulus of elasticity, MPa	29000–30000
Poisson ratio	0.3
Heat resistance, °C	100–110
Frost resistance, number of freeze-thaw cycles, not less	500
Abrasion resistance, g/cm^2	0.25–0.79
Water absorption, mass %	0.05
Shrinkage, mm/m	–

The most advantageous property that distinguishes rubcon from similar polymer concrete is its almost universal chemical resistance, investigated by V. V. Chmykhov. The coefficients of chemical resistance are shown in Table 3.

Table 3. The coefficients of chemical resistance of rubcon.

Type of aggressive environment	The coefficient of chemical resistance	
	After 1 year	Predictable after 10 years
20% sulfuric acid solution	0.95	0.95
3% nitric acid solution	0.8	0.7
10% solution of citric acid	0.9	0.8
20% solution of sodium hydroxide	0.95	0.95
10% solution of potassium hydroxide	0.8	0.65
A saturated solution of sodium chloride	0.9	0.8
Diesel fuel	0.95	0.95
Water	1	0.99

In order to study the work of normal sections under load, i.e. to study the strains and stresses arising in sections, we have made a series of beams with dispersed reinforcement over the entire height of the section, with dispersed reinforcement at 3/4 of the section height and without dispersed reinforcement, the parameters and scheme of the test were taken on the basis of the analysis of literary sources [6–12], as well as on the basis of the article devoted to the study of the SSS of two-layer bending elements. The parameters of the experimental beams are added to Table 4.

Table 4. The parameters of the experimental beams.

The length of the beam, (mm)	1400
Beam width, (mm)	60
Beam height, (mm)	120
The percentage of longitudinal reinforcement (excluding fiber reinforcement), %	0; 0.8; 1.25; 1.8; 2.5; 3.6; 4.9; 6.3; 8.4
Cipher BRR means that the experimental sample is a beam of rectangular cross section made of rubcon	
Cipher BRF means that the experimental sample is a beam of rectangular cross section made of fibrorubcon	
Note: the mixed reinforcement of a bending element is understood as the combined use of dispersed and rod type of reinforcement	

The beams were loaded with two equal symmetrically located concentrated loads applied in 1/3 of the span. With this method of loading (Fig. 1), a zone of pure bending appears in the gap between the points of application of the load. The tests were carried out on certified equipment in the shared use center located on the basis of the VSTU.

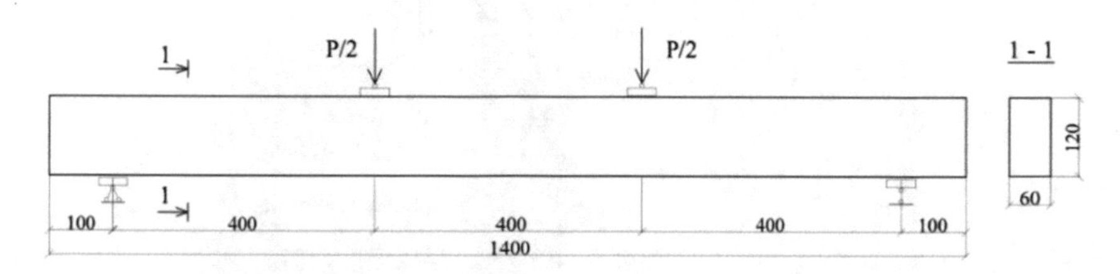

Fig. 1. Loading scheme and cross section of bending elements.

The maximum force at which the yield strength of the reinforcement is reached or the ultimate strength of the material of the compressed zone was taken as the value of the breaking load on the sample. The static load was applied continuously and evenly.

To measure strains in a normal section, strain gauges with a base of 20 mm were glued onto the surface of beam and rebar, as shown in Fig. 2. We used strain gauges 2–11 to determine the strain along the section height, the strain gauge 1 was used to measure the strain of the longitudinal rod reinforcement.

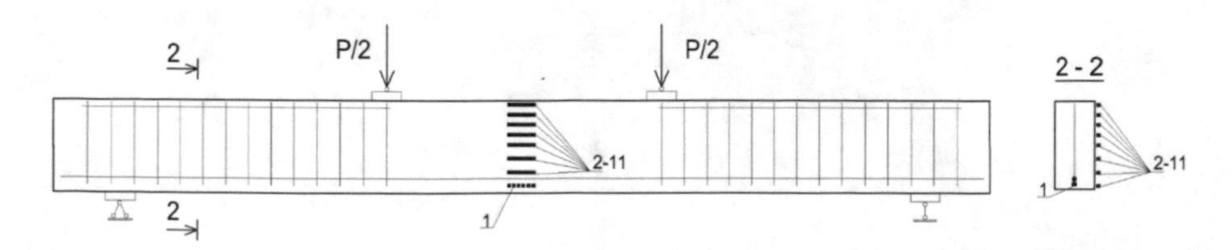

Fig. 2. Scheme of sensors.

To determine the relationship between stresses and strains, together with each sample-beam, we produced control prism samples of 40 × 40 × 160 mm and samples "eight" with a cross section in the working part of 30 × 40 mm and a length of 400 mm. Control samples were made in 3 pieces on the beam. Sample-prism was tested for central compression, and sample - "eight" was tested for central tension. Strain gauges were glued on the surface of the sample-prism and sample – "eight" to determine the strain that arising in the sample under the application of a short compressive or tensile load.

3 Results

On the basis of the experiments, we obtained the diagrams of the distribution of stresses and strains along the height of the cross section of the rubcon and fibrorubcon bending elements, presented (for beams with a percentage of longitudinal reinforcement equal to 0; 1.8; 8.4) in Figs. 3, 4, 5, 6, 7 and 8.

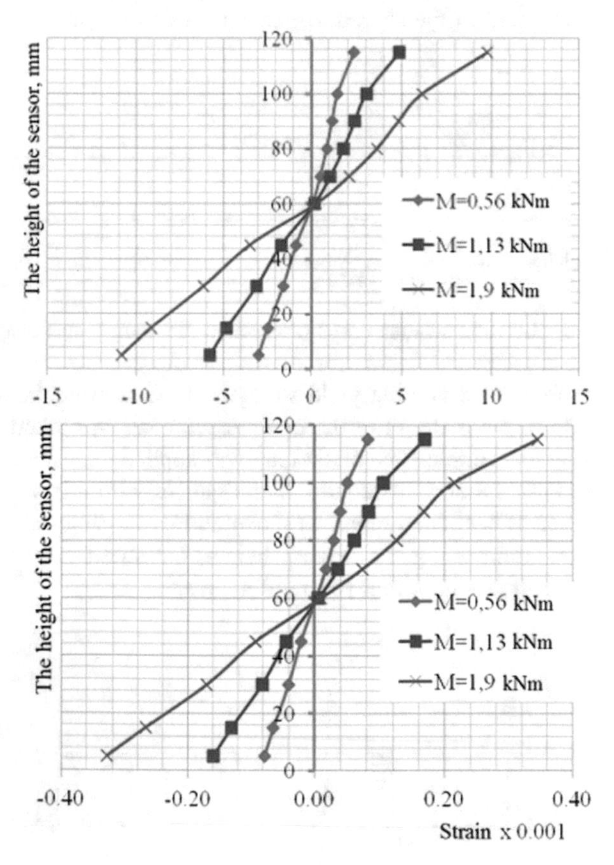

Fig. 3. Stages of stress and strain distribution along the height of the normal sections of the beams BRR-0.

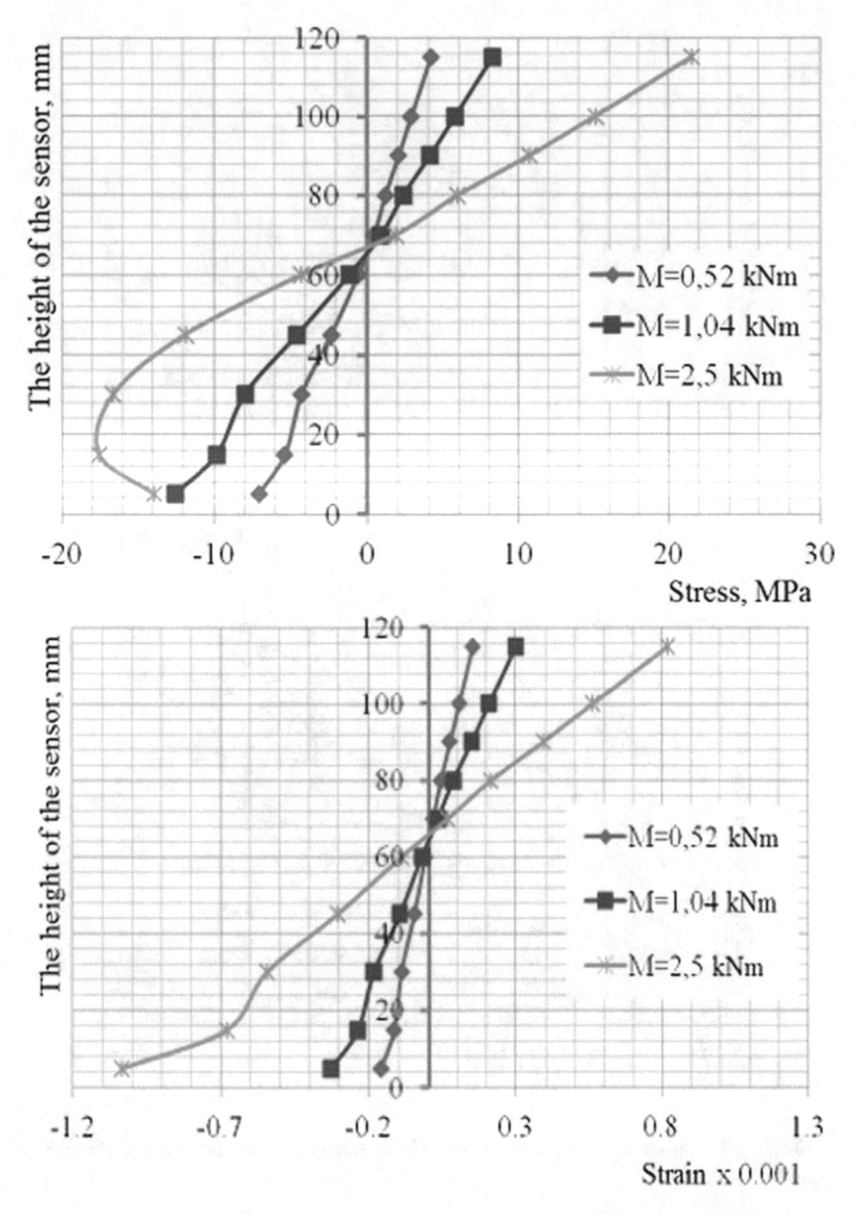

Fig. 4. Stages of stress and strain distribution along the height of the normal sections of the beams BRF-0.

As a result of testing beams without rod reinforcement, it was established that dispersion reinforcement works in bending elements like longitudinal reinforcement. As can be seen from the analysis of Figs. 3 and 4 when fiber are added to the mixture, the stresses of the tensile zone (before destruction) acquire a curvilinear character, i.e. fibrorubcon elements are not immediately destroyed by the appearance of cracks, due

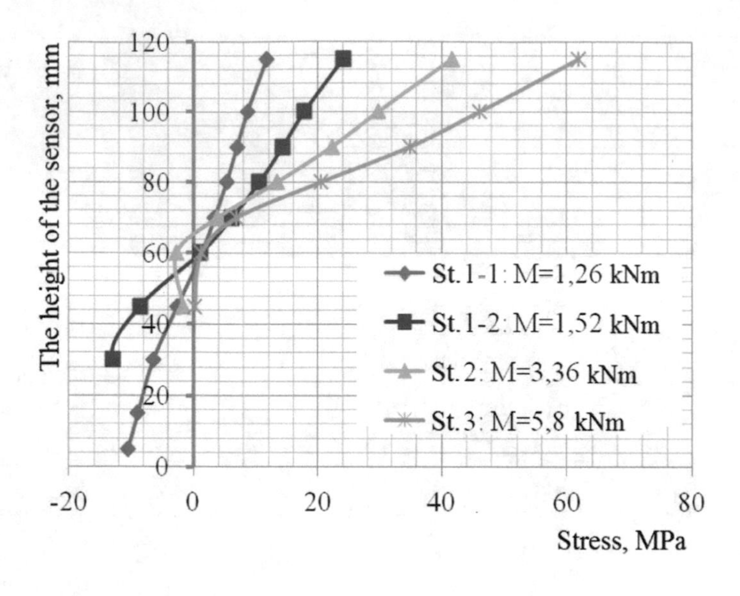

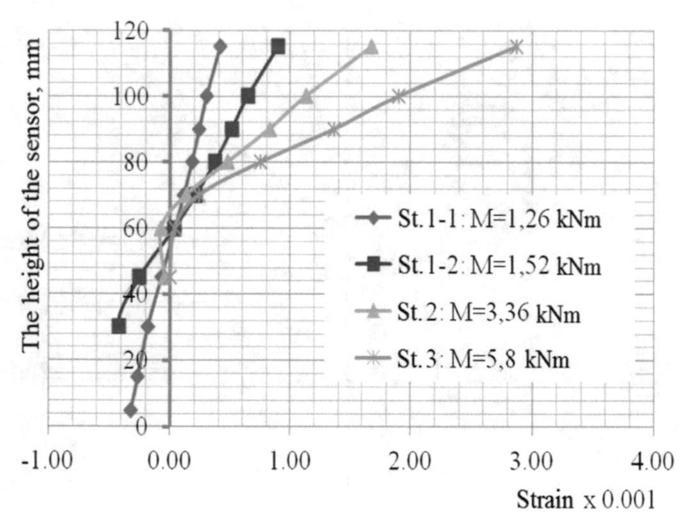

Fig. 5. Stages of stress and strain distribution along the height of the normal sections of the beams BRR-12.

to the fact that when cracks appearance, stresses in the cross section are redistributed to reinforcing fibers, thus, it can be said that the bending elements with dispersed reinforcement undergo 3 stages of SSS compared to the rubcon samples without any reinforcement, where the relationship between stresses and strains remains linear until fracture (Fig. 3).

If rod reinforcement is added to a bending element, a similar pattern of the distribution of strains and stresses along the height of the normal sections of rubcon and fibrorubcon beams can be observed (Figs. 5, 6, 7 and 8) with only one difference, the

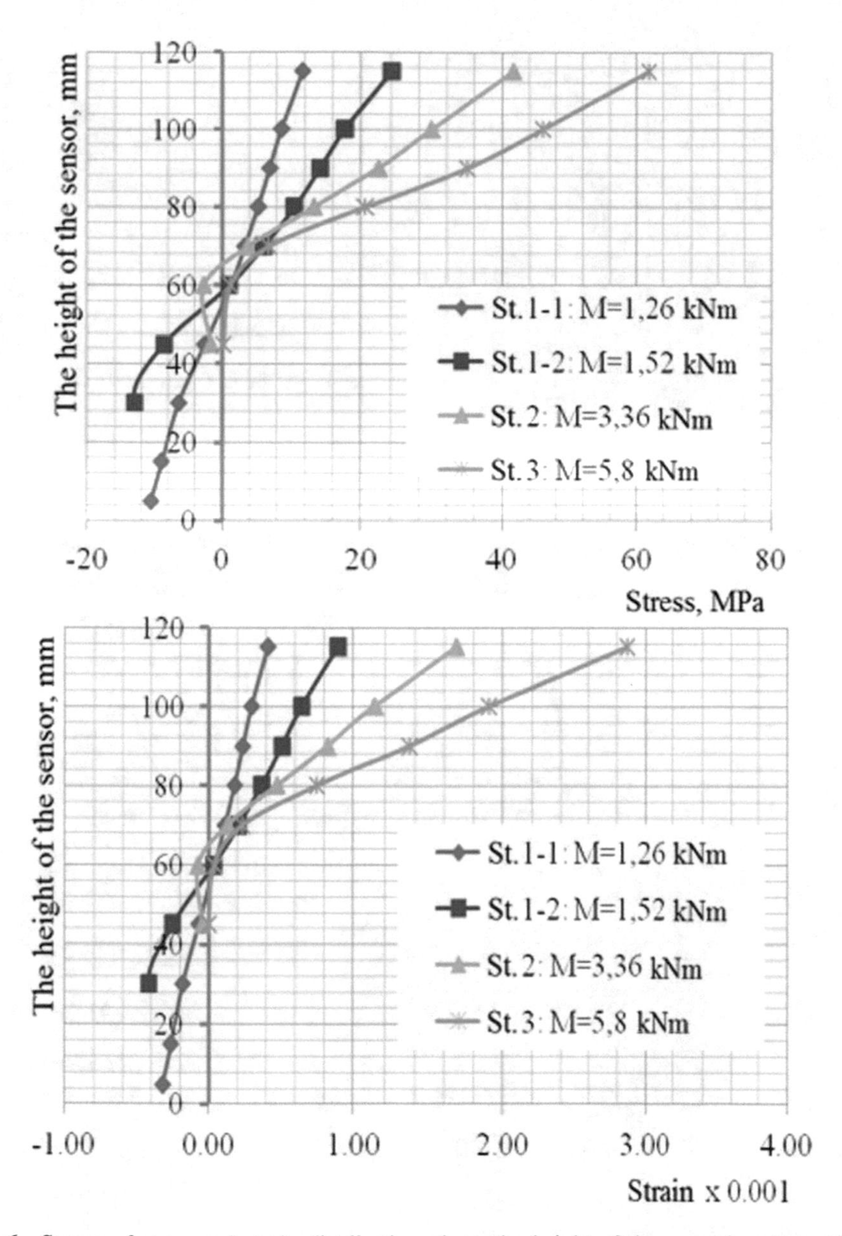

Fig. 6. Stages of stress and strain distribution along the height of the normal sections of the beams BRF-12.

transition from the first stage of the SSS to the second occurs at a higher load, therefore, the destruction of bending elements with fiber reinforcement occurs at greater load. In layered rubcon beams with fiber reinforcement located 3/4 of the section height from the bottom surface, the normal sections undergo SSS stages similar to those of fibrorubcon, with the one difference - the strength of the material of the compressed

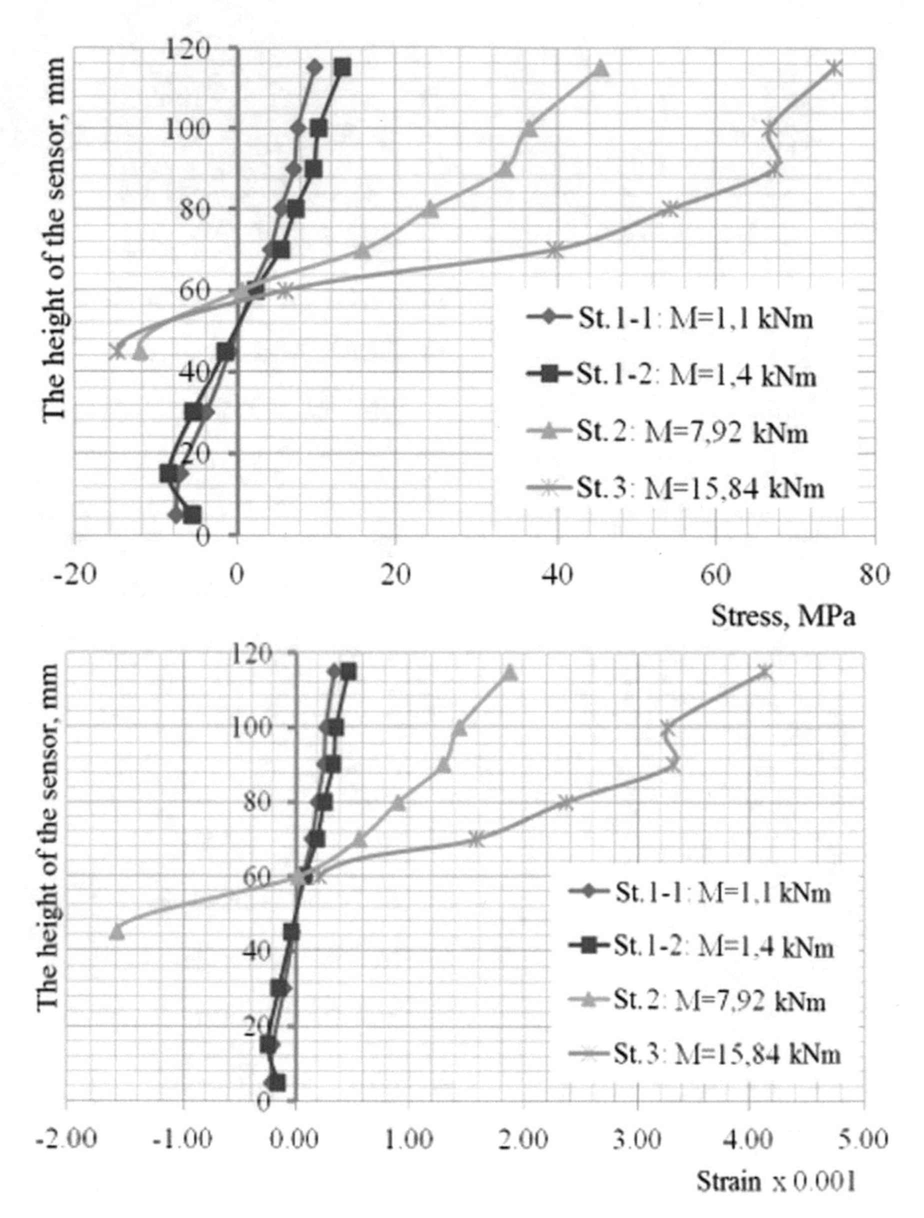

Fig. 7. Stages of stress and strain distribution along the height of the normal sections of the beams BRR-2x18.

zone is lower. Based on the conducted studies, layered fibrorubcon beams are not recommended for use, because labor costs for their producing significantly increase, while exploitation characteristics does not increase compared with fibrorubcon beams with dispersed reinforcement over the height of the cross section, however, this method of reinforcement can reduce the consumption of fiber reinforcement.

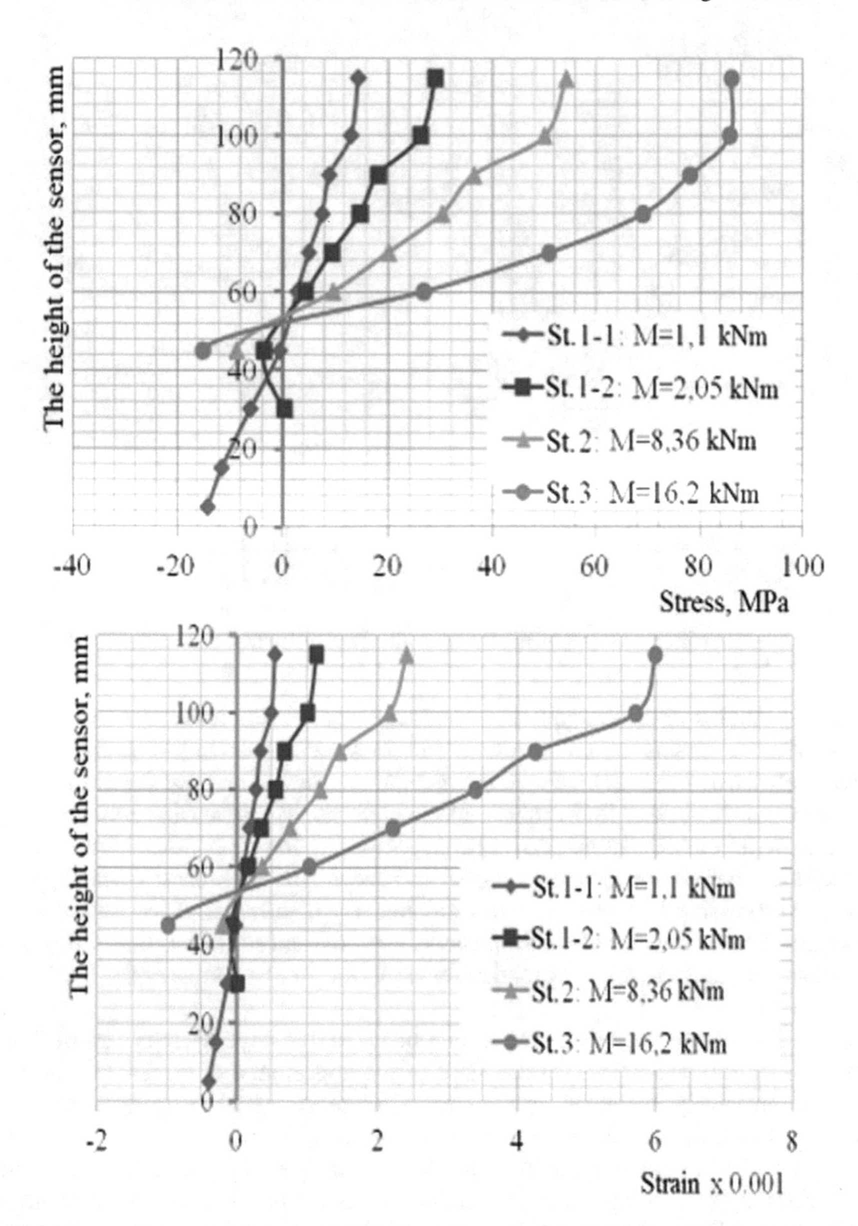

Fig. 8. Stages of stress and strain distribution along the height of the normal sections of the beams BRF-2x18.

4 Discussions

Based on the analysis of Figs. 3, 4, 5, 6, 7 and 8, we established general diagrams of changes in stresses and strains describing the behavior of normal sections of rubcon beams under load, shown in Fig. 9.

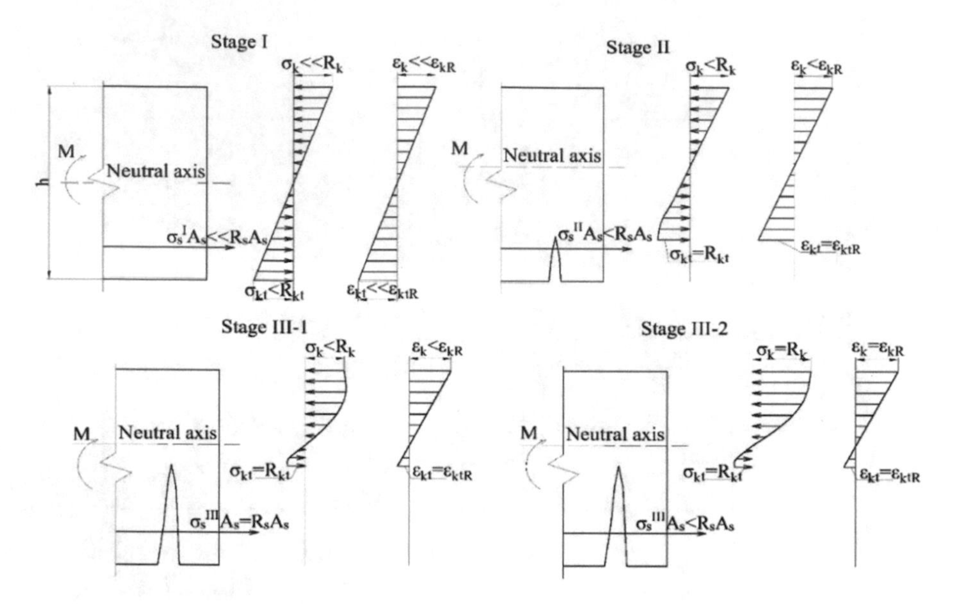

Fig. 9. Stages of the stress-strain state of rubcon beams.

ε_k, ε_{kt} – strains in the compressed zone and in the tension zone of the rubcon; ε_{kR}, ε_{ktR} – ultimate strains of compression and tension of rubcon; $\sigma_{s,I}$, $\sigma_{s,II}$, $\sigma_{s,III}$ – respectively tensile stresses arising in the rebar at the 1st 2nd and 3rd stages of SSS; σ_k, σ_{kt} – respectively stresses in the rubcon of the compressed and tension zone; R_k и R_{kt} – compressive and tensile strength of rubcon; R_s – temporary tensile strength of rebar.

Stage 1 – before normal cracks appear in rubcon. Before normal cracks appear in rubcon compressive and tensile strains are linearly distributed. The strains are predominantly elastic and the relationship between strains and stresses is linear. Before the appearance of cracks in the construction material, the strains values of the tensile zone reach the limit.

Stage 2 – after normal cracks appear in rubcon. In this stage, there is a redistribution of stresses in normal sections. In the place of the appearance of cracks, tensile forces are more intensively perceived by the rebar. Part of the material of the tension zone (above the crack tip) continues to resist external load.

Stage 3 – before the destruction of the element. In this stage, inelastic deformations of the material of the compressed zone are observed; the curve of compressive stresses acquires a curvilinear character. The height of the compressed zone decreases. In case of failure along the tension zone, the stresses in the rebar reach the yield strength. In the case of a re-reinforced cross section, the fracture occurs along the material of the compressed zone, in which case the strains in the rubcon reach their maximum compressibility until the stresses in the reinforcement yield strength.

Fibrorubcon beams undergo similar 3 stages of SSS, shown in Fig. 10 with only one difference after the appearance of a crack - the tensile forces are more intensively perceived not only by the rebar, but also by the fibers.

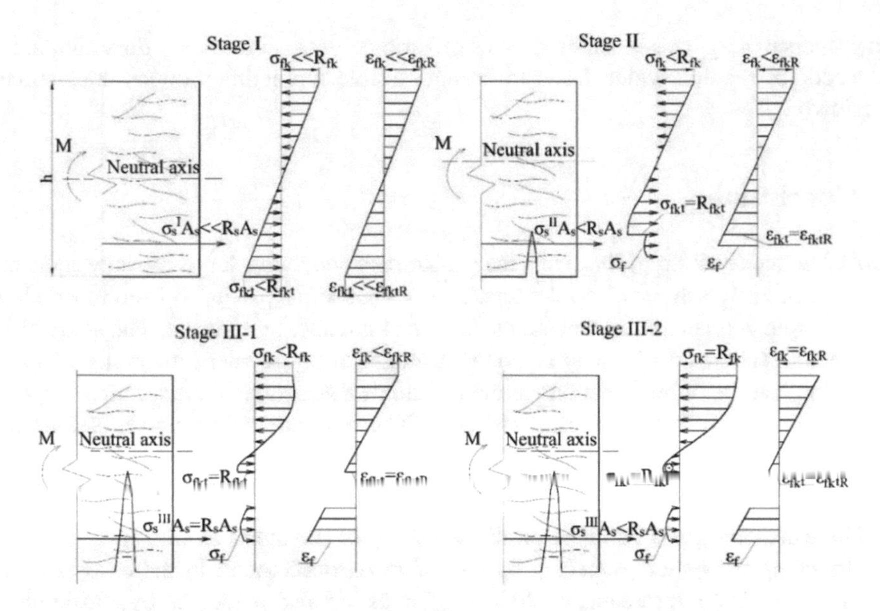

Fig. 10. Stages of the stress-strain state of fibrorubcon beams.

ε_{fk}, ε_{fkt} – strains in the compressed zone and in the tension zone of the fibrorubcon; ε_f – strain in steel cord fibers; ε_{fkR}, ε_{fktR} – ultimate strains of compression and tension of rubcon; $\sigma_{s,I}$, $\sigma_{s,II}$, $\sigma_{s,III}$, respectively tensile stresses arising in the rebar at the 1st 2nd and 3rd stages of SSS; σ_{fk} σ_{fkt} – respectively stresses in the fibrorubcon of the compressed and tension zone; σ_f – stress in steel cord fibers; R_{fk} and R_{fkt} – compressive and tensile strength of fibrorubcon; R_s – temporary tensile strength of rebar.

The use of rubcon is most effective in the producing and protection of elements, parts and structures operating under the influence of aggressive environments of various types such as: groundwater, sewage, atmospheric precipitation, solar radiation, industrial products, etc.

Bending elements of rectangular section made of rubber concrete and reinforced with non-stressed reinforcement can be used as:

- floor beams and coverings, lintel for door, window or other openings;
- supports and trestles for placement of process equipment for chemical production;
- bending structural elements of bridges and railways;
- the foundation beams.

The use of rubcon and fibrorubcon in load-bearing structures due to its high strength characteristics allows reducing the consumption of materials and the weight of structures. This helps to reduce the cost of producing and exploitation of building structures (the construction of new and reconstruction of existing buildings and structures), the additional of dispersed reinforcement in the structure allows to increase the bending moment of crack resistance and ultimate bending moment, high chemical resistance ensures durability and reliable operation throughout the all life cycle under the influence of aggressive environments of various types, at the same time minimizing the need for repairs, which leads to an undesirable (sometimes impossible) process shutdown.

5 Conclusion

Rubber concrete is struggling with the problem of recycling large-capacity industrial waste, such as fly ash, metal cord fibers. The constructions produced from rubcon have high strength characteristics and almost universal chemical resistance. The study of the stress-strain state of the rubcon beams with mixed reinforcement made it possible to determine that the normal sections of the researching structures undergo three stages of SSS similar to reinforced concrete elements. These studies allow you to set the prerequisites for calculating the strength of normal sections of beams with mixed reinforcement:

1. The calculation will be made on the 3rd stage of the stress-strain state;
2. Stretching forces are perceived by tensioned reinforcement; in the case of beams with dispersed reinforcement, stretching forces are also perceived by tensile fibers;
3. Before fracture, the relationship between strains and stresses of the polymer concrete of a compressed zone is nonlinear;
4. The stress values in the polymer concrete of the compressed zone reach the ultimate compressive strength.

Prerequisites for the calculation of the crack resistance of normal sections of beams with mixed reinforcement:

1. In the element under load until the appearance of cracks, the sections are flat.
2. The stresses in the material of the tensile zone are equal to the resistance to axial tension;
3. Inelastic deformations develop in the material of the stretched zone; the normal tensile stress curve is rectangular;
4. Elastic deformations develop in the material of the compressed zone. The curve of the normal stresses of the compressed zone is triangular.

References

1. Polikutin, A.E.: Experimental research of the durability, crack resistance of the normal sections of bending elements produced of rubber concrete with fiber and their deformability. Mater. Sci. Forum **931**, 232–237 (2018)
2. Song, P.S., Wang, S.H.: Mechanical properties of high-strength steel fiber-reinforced concrete. Constr. Build. Mater. **18**(9), 669–673 (2004)
3. Mohammadhossein, H., Tahir, M.M., Sam, A.R.M., Lim, N.H.A.S., Samadi, M.: Enhanced performance for aggressive environments of green concrete composites reinforced with waste carpet fibers and palm oil fuel ash. J. Cleaner Prod. **185**, 252–265 (2018)

4. Figovsky, O.: New polymeric matrix for durable concrete. In: Proceedings of the International Conference on Cement Combinations for Durable Concrete, vol. 1, pp. 269–276 (2005)
5. Figovsky, O., Beilin, D., Blank, N., Potapov, J., Chernyshev, V.: Development of polymer concrete with polybutadiene matrix. Cement Concrete Compos. **18**(6), 437–444 (1996)
6. Bhaduri, A.: Bending. Springer Series in Materials Science, vol. 264, pp. 173–195 (2018)
7. Zhang, F., Zhang, W., Hu, Z., Jin, L., Jia, X., Wu, L., Wan, Y.: Experimental and numerical analysis of the mechanical behaviors of large scale composite C-Beams fastened with multi-bolt joints under four-point bending load. Compos. Part B: Eng. **164**, 168–178 (2019)
8. Liu, F., Zhou, J.: Experimental research on fatigue damage of reinforced concrete rectangular beam. KSCE J. Civil Eng. **22**(9), 3512–3523 (2018)
9. Yang, I.-H., Joh, C., Kim, K.-C.: A comparative experimental study on the flexural behavior of high-strength fiber-reinforced concrete and high-strength concrete beams. Adv. Mater. Sci. Eng. **2018**, 7390798 (2018)
10. Sui, L., Zhong, Q., Yu, K., Xing, F., Li, P., Zhou, Y.: Flexural fatigue properties of ultra-high performance engineered cementitious composites (UHP-ECC) reinforced by polymer fibers. Polymers **10**(8), 892 (2018)
11. Chen, S., Zhang, R., Jia, L.-J., Wang, J.-Y.: Flexural behaviour of rebar-reinforced ultra-high-performance concrete beams. Mag. Concrete Res. **70**(19), 997–1015 (2018)
12. Travush, V.I., Konin, D.V., Krylov, A.S.: Strength of reinforced concrete beams of high-performance concrete and fiber reinforced concrete. Mag. Civil Eng. **77**(1), 90–100 (2018)

Designing Scenarios of Damage Accumulation

Andrey Benin[1(✉)], Shoxista Nazarova[2], and Alexander Uzdin[1]

[1] Emperor Alexander I St. Petersburg State Transport University, Moskovsky pr. 9, 190031 St. Petersburg, Russia
benin.andrey@mail.ru
[2] Tashkent Institute of Railway Engineering, st. Odilxojayev 1, Tashkent, Uzbekistan

Abstract. The problems arising in working out damage accumulation scenarios in performance based designing are considered. These include specifying the group of limiting states for the structure under consideration, specifying the design input for each limiting state, establishing the criterion of the limit state occurrence, methods of calculating the structure for estimating the limiting state occurrence, and methods of improving the damage accumulation scenarios. An example of the calculation and design of the bridge pier destruction scenario is considered.

Keywords: Seismic resistance · Performance based designing · Damage accumulation scenarios

1 Introduction

The present earthquake engineering is developing towards performance based designing (PBD) [1–3] and designing damage accumulation scenarios [4, 5]. In this case, several limit states and corresponding input levels are considered. The simplest PBD-variant considers two limit states - disruption of the normal operation and the structure destruction. In literature these limit states were named SLS (serviceability limit state) and ULS (ultimate limit state). The inputs corresponding to these limit states are called “design earthquake” (DE) and “maximum design earthquake” (MDE). In 2004, Prof. Fardis, one of the ideologists of the Eurocode-8, proposed to additionally consider a moderate earthquake (ME) with a limited level of damages [6]. At present, the standards of Italy and France [7] provide for the possibility of considering 4 limiting states: Operation Limit State (OLS), Damage Limit State (DLS), Ultimate Limit State (ULS) and Collapse Limit State (CLS). Each limit state corresponds to an earthquake of a given repeatability TOLS > 21 years; TDLS > 35 years; TULS > 333 years; TCLS > 683 years. The presence of several limiting states allows us to move from multilevel designing to designing damage accumulation scenarios of a structure. For this aim it is necessary to:

- Determine the set of limit states with their repeatability (probability of occurrence)
- Set criteria of the limit state occurrence
- Set the design load for each state

V. Murgul and M. Pasetti (Eds.): EMMFT-2018, AISC 983, pp. 600–610, 2019.
https://doi.org/10.1007/978-3-030-19868-8_57

- Choose a model to assess the occurrence of a limit state.
- Determine how to present the damage accumulation scenario.

2 Determining the Limit States Set

The easiest way to describe the limit state is to describe them at the physical level. For example, for bridges, the limit state damages are detailed as follows:

- OLS provides for elimination of residual movements of the track in the plan and in the profile, as well as rail track rupture;
- DLS provides for the possibility of the destruction of the rail track and the occurrence of plastic deformations in the body of the piers that do not cause a long traffic stoppage;
- ULS provides for the possibility of cracks or breaking the concrete away from the pier body, as well as the shift of spans without their falling from the pier. The bridge operation of is possible after repairs;
- CLS means the falling of the span from the piers, inclination or destruction of the piers, excluding further operation of the bridge.

A similar description of damage accumulation should precede the working out of damage accumulation scenarios. However, the authors consider it expedient to use one universal criterion of structure damages. Economic damage caused by the earthquake can be taken as such a criterion. In literature, the mathematical expectation R of this damage is referred to as seismic risk [8–10].

$$R = \sum_{I=5}^{I_{max}\Sigma} D(I)L(I) \tag{1}$$

where L(I) = 1/T(I) is the territory shakiness, T(I) is the frequency of earthquakes of intensity I; D(I) is a vulnerability function.

The summation in formula (1) is carried out using degrees of seismic intensity, and it is assumed that earthquake start from degree 5. The value of the vulnerability function, calculated for whole values of the intensity degree, determines the so-called payment matrix for the risk assessment [10].

3 Criteria for the Occurrence of Limit States

To determine the criteria of limit state occurrence, consider the monotonous loading diagram. Figure 1 shows the bridge pier “force – displacement” diagram from paper [11]. This diagram has three characteristic sections:

In the first section, the displacement U is smaller than the elastic limit Uel

$$U < U_{el} \tag{2}$$

In this section the system is in the elastic work stage.

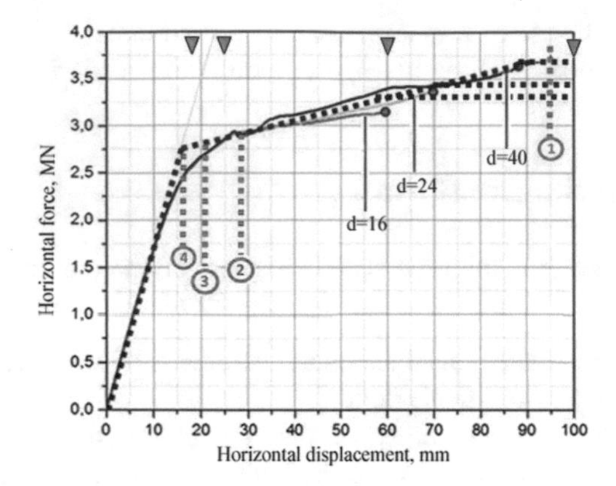

Fig. 1. The monotonic loading diagram of the bridge pier according to the paper [11].

In the second section

$$U_{el} < U < U_{red} \tag{3}$$

In this section, the system loses its static indeterminacy, and at $U > U_{red}$ it becomes a mechanism. The third section $U_{red} > Ucol$ the system moves under the action of a constant force, and when $U = U_{col}$, the system collapses.

Points with the corresponding limit states can be plotted on the monotonous loading diagram. Based on the foregoing, the maximum displacement of the system could be taken as the criterion of the limiting state. Such proposals are widely used in literature, for example [12]. The Russian Guideliners use the displacement criterion; moreover, they also use the maximum displacement criterion [13]. However, the displacement criterion does not take into account low-cycle fatigue of the structure. In order to take into account this effect, the work of plastic deformation forces [14] should be taken into account instead of displacements. For large displacements this work is equal to the work of the seismic response and is obtained by integrating the monotonous loading function. Figure 2 shows an example of the dependence of the plastic deformation force work on the displacement of the system. In this case, each limiting state can be attributed to the work of plastic deformation forces. In the authors' opinion, this is the most convenient and universal criterion of the limit state occurrence. In future, one can connect the work of the plastic deformation forces with the seismic risk (expected damage).

1. displacement caused by the load of the CLS;
2. displacement caused by the load of the ULS;
3. displacement caused by the load of the DLS;
4. displacement caused by the load of the SLS.

Admissible displacement values are marked by green triangles.

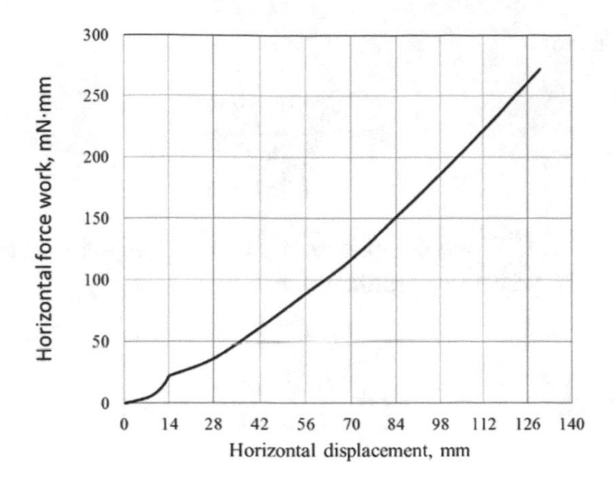

Fig. 2. Dependence of resistant force work W on the system displacement u

4 Setting the Design Load

The base of setting the design load is the repeatability of the accepted limit states and the seismological conditions of the construction site. In accordance with the frequency of the limit state, seismic input is set with the same frequency, and it is for this input that the structure is calculated and the seismic protection that excludes the limit state is selected. The magnitude of the load, in accordance with seismic scale depending on the input frequency, is determined on the basis of the well-known formula [15]

$$log\, T_{eq} = a \cdot I + b, \tag{4}$$

where the coefficients “a” and “b” are determined in accordance with situational seismicity. The corresponding calculations are described in the paper [16] and in the program [17]. This program was worked out on the base of three seismic zoning maps [13] and the authors improved the program using five zoning maps [18]. From formula (4) the design seismic intensity is obtained

$$I = \frac{log\, T_{eq} - b}{a} \tag{5}$$

Further, in accordance with the data of the seismic scale [19], peak accelerations (PGA) can be obtained

$$PGA = 10^{\frac{I-1.89}{2.5}} \tag{6}$$

In addition to the peak accelerations, the design input should be characterized by additional parameters. These include:

- Harmonicity factor

$$\kappa = \frac{\ddot{y}_0^{(max)} \cdot y_0^{(max)}}{\left(\dot{y}_0^{(max)}\right)^2}, \tag{7}$$

where $\ddot{y}_0^{(max)}$, $\dot{y}_0^{(max)} y_0^{(max)}$ are the maximum acceleration, velocity and displacement accordingly in the process of oscillation.
- Arias intensity

$$I_A = \frac{\pi}{2g} \int_0^{\tau} \ddot{y}^2(t)dt; \tag{8}$$

- Seismic energy density

$$SED = \int_0^{\tau} \dot{y}^2(t)dt; \tag{9}$$

- Absolute cumulative velocity, CAV

$$CAV = \int_0^{\tau} |\ddot{y}(t)|dt. \tag{10}$$

The value of these characteristics is available in literature, particularly [20]. Some clarification of the data [20] is given in the paper by Prokopovich [21]. Taking into account the above characteristics, dangerous accelerograms should be selected from the accelerogram package or artificial accelerograms should be generated.

5 Models for Assessing the Limit State Occurrence

The damage accumulation scenario provides for the non-linear behavior of the structure during the loading process [22]. Linear work takes place only when $U < U_{el}$. Beyond elasticity limit, a different nature of the inelastic structure behavior is possible. There are 2 principal types of nonlinearity:

- Elastoplastic nonlinearity
- Adaptive nonlinearity

In the case of the elastoplastic nonlinearity type, the loading and unloading branches of the loading diagram are characterized by one and the same rigidity, which does not change during the oscillation process and does not depend on the loading history. In this case the behavior of the system depends on the occurrence of plastic deformation in its elements. In the simplest case of a single-mass system, the motion equation is

$$\ddot{y} + b\dot{y} + R(y) = -m\ddot{y}_0 \tag{11}$$

where

$$R(y) = \begin{cases} cy & when\, y < U_{el} \\ R_0 & when\, y > U_{el} \end{cases}.$$

For the adaptive type of nonlinearity, the rigidity of the system decreases with the increase of the maximum system displacement. The behavior of the system depends on the whole history of its loading and is associated with the formation and development of cracks. To describe the behavior of such systems, Kachanov and Rabotnov introduced the concept of damage, which characterizes the relative cross-sectional element area that was turned off from its work due to the formation of a crack [23]. Kirikov used a simplified version of the damage theory [24], according to which the damage factor and damping increase linearly, and the oscillation period falls linearly depending on the maximum displacement umax during the loading history (Fig. 3). In this case, the system reaction is described by the formula

$$R(u) = \begin{cases} \frac{C(u_{\max})u}{1 + u^2\kappa(u_{\max})}, & if \quad u_{\max} > U_{el} \\ C_0 u, \quad if & u_{\max} \le U_{el} \end{cases} \tag{12}$$

where C_0 is the initial stiffness of the system without cracks.

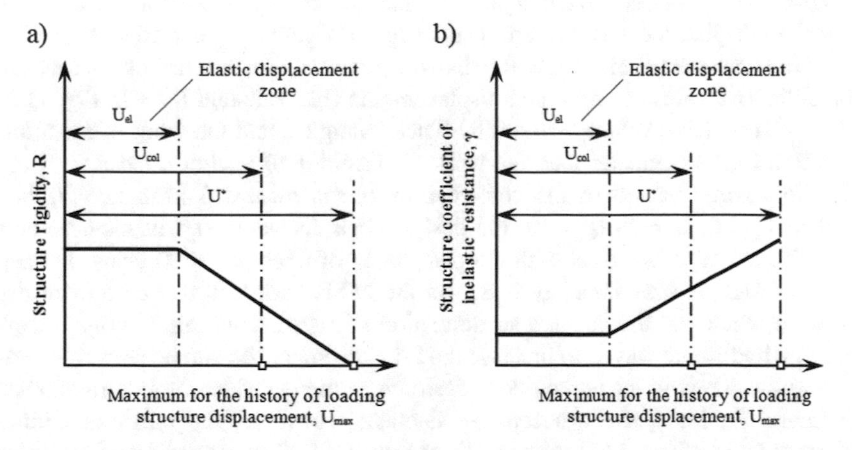

Fig. 3. The dependence of the structure rigidity (a) and of the structure coefficient of inelastic resistance (b) on the maximum structure displacement in the history of loading.

6 Example of Building Damage Accumulation Scenario

As an example, consider the bridge pier with the monotonous loading diagram described above in accordance with [11]. The fundamental oscillation period of the intact pier with a span structure according to [11] is 0.46 s. The repeatability of the limit states of SLS, DLS, ULS and CLS is taken as 50, 100, 500 and 1000 years respectively. Peak accelerations are calculated according to the above-mentioned program [17], taking into account its modification according to 5 zoning maps [25]. The results of the PGA calculation are given in Table 1. Using standard zoning maps [26] situational seismicity is assumed to be IA = 7, IB = 8, IC = 9. Generating the design accelerograms was carried out according to the method of Dolgaya [27]. Figure 4 shows an example of the chronogram of the design input for the CLS, and in Fig. 5 its spectrum of plastic deformation forces work is shown.

Table 1. PGA for inputs of different repeatability

Limit state	Repeatability, years	Design intensity	PGA, m/s^2	Exceedance probability during 100 years
SLS	50	5.1	0.182	0.864
DLS	100	5.7	0.31	0.632
ULS	500	7.1	1.12	0.181
CLS	1000	7.7	2	0.095

Displacements obtained by numerical solution of the equations of the pier seismic oscillations using the generated inputs are marked in Fig. 1 with a red dotted line. Admissible displacement values are marked in the figure by green triangles.

For the considered seismicity, the damage accumulation scenario turned out to be acceptable. The values of designed displacements (black dotted lines in Fig. 1) lie to the left of acceptable values, marked by black triangles, and the work of monotonous destruction forces is greater than the work of plastic deformation forces.

If the seismic hazard of the construction site is increased from $I_A = 7$, $I_B = 8$, $I_C = 9$ to $I_A = 8$, $I_B = 9$, $I_C = 10$, the design peak acceleration will increase significantly. For example, for CLS with a repeatability of once per 1000 years, $PGA_{CLS} =$ 4 m/s2 (design seismic intensity is 8.7 on the MSK scale). In this case there rises a question of changing the damage accumulation scenario. This can be done in several ways. The traditional way is to increase the reinforcement. As can be seen from Fig. 1, this increases the work of monotonous destruction forces, which in its turn allows one to increase the limit of the acceptable damage. However, the effect of additional reinforcement is relatively small and affects only CLS. The second way is to increase damping in the system [28]. In this case it is achieved by installing dampers in parallel with the movable bearings [18, 29, 30]. As can be seen from Fig. 5, increasing the damping from $\gamma = 0.1$ to $\gamma = 0.2$ reduces the work of the plastic deformation forces (per structure mass unit) from 0.11 to 0.08 (m/s)2. The most promising way for strong impacts is to increase the plasticity coefficient of the pier. This is achieved by

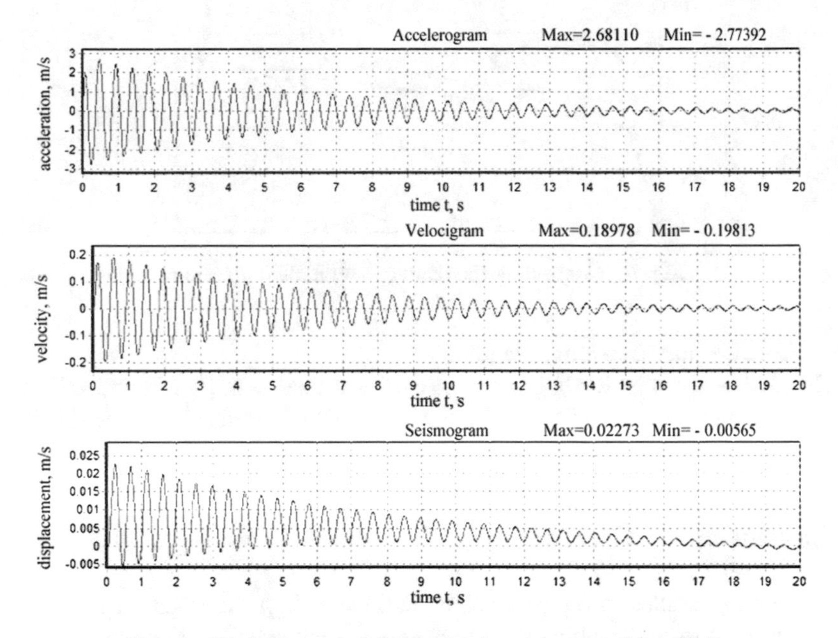

Fig. 4. Chronograms of the design input generated for CLS

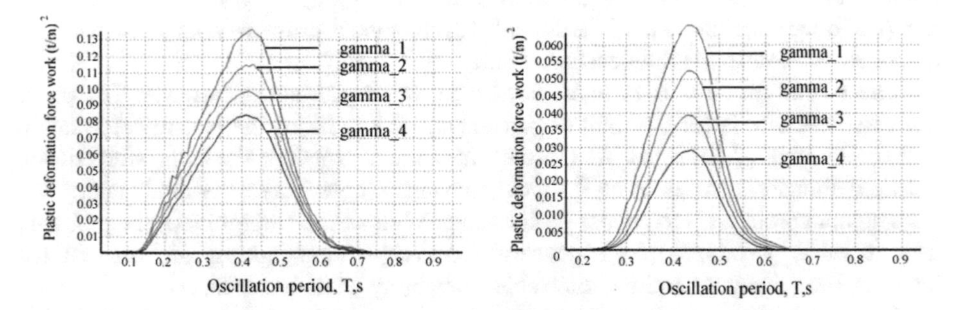

Fig. 5. Spectra of the plastic deformation force work for the design input generated for CLS (on the left) and for ULS (on the right): where: gamma_1 for $\gamma = 0.05$; gamma_2 for $\gamma = 0.1$; gamma_3 for $\gamma = 0.15$; gamma_4 for $\gamma = 0.2$; γ is inelastic resistant factor.

connecting the span with the pier using a friction-moveable joint [18, 30]. For example, such connections are used, in the railway bridges in Sochi [30]. Due to the oval bolt holes, the joint provides for plastic movement up to 10–12 cm. The diagram of monotonous loading of the considered pier having two shifts of 6 cm is shown in Fig. 6. The first shift occurs for earthquakes with PGA > PGADLS, and the second for earthquakes with PGA > PGAULS. In this case, the monotonous loading diagram significantly stretches, and the corresponding work increases.

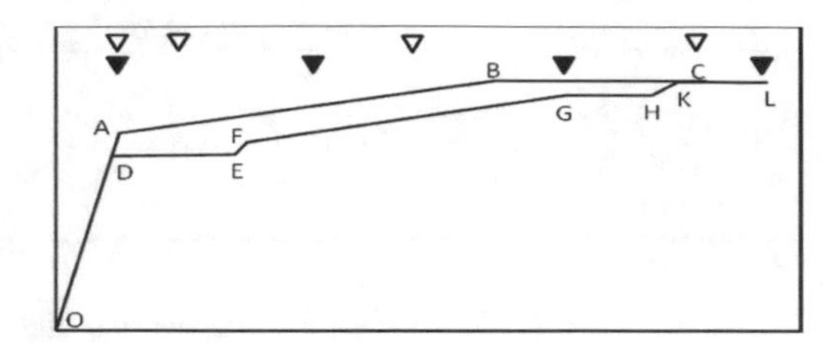

Fig. 6. Controlling the damage accumulation scenario

where: OABC – initial loading diagram;
ODEFGHKL – the new loading diagram with two plastic shifts DE and GH.

White triangles indicate limit states for the initial diagram; black triangles indicate limit states for the new diagram.

7 Conclusion

The materials presented in the paper show that the earthquake engineering state-of-art allows one to go from designing structures for a given impact to designing scenarios of damage accumulation in a structure. In this case, the need arises to consistently solve a number of problems. The first problem is to set a group of limit states and their acceptable repeatability. This is a difficult psychological problem connected with the owner responsibility for the decisions made.

The second problem is to select criteria for the limit state occurrence. The paper discusses the criteria for the limit displacement, limit work of the plastic deformation forces and limiting damages. At present, the authors consider the limit work of the plastic deformation forces to be the most convenient criterion of the limit state.

The third problem is to choose design input. With all the ambiguity of the problem, now it seems possible to set a conservative design input, taking into account the building site seismicity and the allowable probability of its exceedance.

The damage accumulation scenario itself can be set at the physical level, describing the nature of the damage for each input level. More convenient and versatile is to indicate acceptable and calculated displacements in the monotonous loading diagram of the system with the subsequent analysis of the point location.

References

1. Kilimnik, L.: Methods of targeted design in earthquake engineering (in Russian). Science, Moscow (1985)
2. Dowric, D.: Earthquake Resistant Design for Engineers and Architects. Wiley, New York (1977)
3. Park, R., Paulay, T.: Reinforced Concrete Structures. Wiley, New York (1975)
4. Dolcea, M., Kappos, A., Masia, A., Penelisb, G., Vona, M.: Vulnerability assessment and earthquake damage scenarios of the building stock of Potenza (Southern Italy) using Italian and Greek methodologies. Eng. Struct. **28**, 357–371 (2006)

5. Villacis, C., Cardona, C., Tucker, B.: Implementation of fast earthquake scenarios for risk management in developing countries. 12 WCEE, New Zeland, Paper 0796
6. Fardis, M.: Code developments in earthquake engineering. In: Published by Elsevier Science Ltd. 12th European Conference on Earthquake Engineering. Paper Reference 845 (2002)
7. Le norme tecniche per le costruzioni. Il ministero delle infrastrutture e dei trasporti (2018)
8. Bogdanova, M., Sigidov, V., Uzdin, A., Chernov, V.: Statistical characteristics of damage in the risk theory. In: Modern Economy: Problems and Solutions, vol. 5, pp. 22–30 (2016)
9. Keilis-Borok, V., Nersesov, I., Yaglom, A.: Methods for assessing the economic effect of earthquake engineering (in Russian). Ed. AN SSSR, Moscow (1962)
10. Uzdin, A., Vorobyev, V., Bogdanova, M., Sigidov, V., Vanicheva, S.: Economics of earthquake engineering (in Russian). FGPU DPO "Educational and Methodological Center for Education in Railway Transport", Moscow (2017)
11. Smirnova, L., Benin, A., Semenov, S., Uzdin, A., Yuhay, G.: Estimation of the reduction factor for the calculation of bridge piers. Earthquake Engineering. Safety of Buildings, vol. 6, pp. 15–19 (2016)
12. Crowley, H., Pinho, R., Bommer, J., Bird, J.: Development of a Displacement –Based Method for Earthquake Loss Assessment. IUSS Press, Italy (2006)
13. Set of rules 14.13330.2014. Construction in seismic areas
14. Nesterova, O., Uzdin, A., Sibul, G., Dolgaya, A., Yuhay, G.: Universal numerical indicator of earthquake intensity. In: Proceedings of the Russian Academy of Rocket and Artillery Sciences, Moscow, vol. 2, no. 102, pp. 152–156 (2018)
15. Seismic shakiness of the territory of the USSR. In: Riznichenko, Y. (ed.) Science, Moscow (1979)
16. Areshchenko, T., Prokopovich, S., Sabirova, O., Frolova, E.: Setting the level of seismic input to assess the seismic stability of structures in multi-level design. Sci. Tech. J. Nat. Man-Made Risks. Safety Build. **4**, 25–27 (2018)
17. Uzdin, A., Prokopovichm S., Areshchenko, T., Frolova, E., Sabirova, O.: Program for determining peak seismic acceleration. State Registration Certificate №2018664350
18. Uzdin, A., Elizarov, S., Belash, T.: Earthquake resistant constructions of transport buildings and structures. Tutorial. Federal State Educational Institution "Educational and Methodical Center for Education in Railway Transport" (2012)
19. GOST R 57546-2017. Earthquakes. Seismic intensity scale
20. Bogdanova, A., Nesterova, O., et al.: Numerical characteristics of seismic actions. Science and World №3 (43), vol. 1, pp. 49–55 (2017)
21. Prokopovich, S., Uzdin, A.: Estimation of Arias intensity dependence on the predominant seismic period. In: Natural and Man-Made Risks. Safety of Buildings, №3, pp. 27–30 (2018)
22. Goldenblat, I., Nikolaenko, N., Polyakov, S., Ulyanov, S.: Models of Structure Seismic Stability. Stroyizdat, Moscow (1979)
23. Kachanov, L.: Fundamentals of fracture mechanics. Science, Moscow (1974)
24. Amankulov, T., Kirikov, B.: Investigation of the behavior of a single mass system with non-linearity of a hysteresis type under seismic action. Earthquake Eng. **8**, 16–23 (1980)
25. Ulomov, V., Bogdanov, M.: A new set of maps of general seismic zoning of the territory of the Russian Federation (OSR-2012), Engineering Survey, No. 8, pp. 30–39 (2013)
26. State Standard 6249-52 "Scale for determining the earthquake intensity ranging from 6 to 9 degrees"
27. Dolgaya, A.: Simulation of seismic effects by a short time process. In: Express-Information VNIINTPI. Ser. "Earthquake engineering", vol. 5–6, pp. 56–63 (1994)
28. Durseneva, N., Indeykin, A., et al.: Peculiarities of calculating bridges with seismic isolation including spherical bearings and hydraulic dampers in Russia. J. Civil Eng. Archit. **9**(4), 401–409 (2015)

29. Kostarev, V., Pavlov, L., et al.: Providing the earthquake stability and increasing the reliability and resources of pipelines using viscous dampers. In: Proceedings of Workshop "Bridges Seismic Isolation and Large-Scale Modeling" (ASSISi), pp. 67–79 (2011)
30. Kuznetsova, I., Uzdin, A., Zhgutova, T., Shulman, S.: Seismic protection of railway bridges in Sochi. In: Proceedings of Workshop "Bridges Seismic Isolation and Large-Scale Modeling" (ASSISi), pp. 29–41 (2011)

On the Task of Multi-objective Dynamic Optimization Power Losses

Vasily Mokhov(✉), Sergey Kostinskiy, Danil Shaykhutdinov, Anton Lankin, and Yuriy Manatskov

Platov South-Russian State Polytechnic University (NPI), 346428 Novocherkassk, Russia
mokhov_v@mail.ru

Abstract. At the present time, the solution of multi-criteria optimization problems is one of the modern trends in the field of technical sciences. The task version of minimizing (Pareto-efficiency) the electrical energy loss from asymmetry and non-sinusoidal currents in the distribution network by switching single-phase electricity consumers between phases is exposed in the article. The solution concept for the problem is expounded. Particular attention is paid to the formation and analysis of multicriteria objective function. The structure of private criteria, their priority and calculation formulas are determined. The normalization methodology for the selected criteria and their converging into one objective function is demonstrated. In particular, the objective function includes three following criteria: (1) a criterion for asymmetry (irregularity) of distributing the effective current values across phases, (2) a criterion characterizing the content of higher harmonic components of currents in the distribution network at the level of a transformer station and (3) a criterion determining the required number of switching single phase power consumers between phases. For some criteria, various options for calculating the values are considered depending on the particular task. Based on the preliminary results of numerical experiments and the calculation of the set of feasible solutions to the problem, a reasonable conclusion was made about the need of using agent metaheuristics in order to solve it.

Keywords: Asymmetry · Higher harmonic components of the current · Pareto-efficiency · Objective function criteria analysis

1 Introduction

The electronic equipment currently produced and operated in relation to power supply systems is, as a rule, a non-linear electrical load [1]. Therefore, the current consumed by the equipment has a non-sinusoidal form and creates distortions in the power supply network [2]. As a result, all this leads to voltage distortion, which affects other equipment that receives electricity from a common source [3].

The proportion of modern equipment distorting the electricity quality is increasing [4]. Due to the growth of generating high current harmonics into the electrical network, the electrical energy quality in the networks of non-industrial consumers is deteriorating [5].

V. Murgul and M. Pasetti (Eds.): EMMFT-2018, AISC 983, pp. 611–618, 2019.
https://doi.org/10.1007/978-3-030-19868-8_58

The electricity quality in distribution networks of most developed countries is constantly decreasing [6], power consumers with linear current-voltage characteristics are less and less used, and the number and characteristics of non-linear power consumers (harmonic composition of currents and voltages) are constantly changing [7].

One of the main negative effects of current harmonic components is the formation of additional losses in transformers of distribution networks from the flow of higher harmonic components currents, as well as their overheating up to failure [8]. Higher harmonic components of the current create additional losses in the windings of transformers and additional losses for eddy currents in the magnetic core. All this leads to a decrease in the efficiency of electricity transmission processes and a reduction of the estimated service life of electrical equipment and electrical networks due to accelerated thermal and electrical insulation aging [9].

Due to the constant increase in the number of used household electrical appliances, an increase in the influence of higher harmonic components by consumers on the operation of low and medium voltage networks is predicted in the near future [10]. At the same time, an increase in the installed capacity of non-linear, asymmetrical and abruptly variable loads is not accompanied by timely implementation of solutions aimed at correcting the quality of electrical energy.

Analyzing the above-mentioned allows us to make a reasonable conclusion about the increasing relevance of the technological task of improving the electricity quality. This task is associated with reducing the influence of higher harmonic components on the operation of electrical equipment [11].

2 Materials and Methods

2.1 The Subject, Scope and Problem of Research

The subject of research is a distribution transformer station (TS) with a set of single-phase consumers (SPC) connected to the TS phases L_1, L_2 and L_3 and receiving power supply via 0.4 kV lines (see Fig. 1).

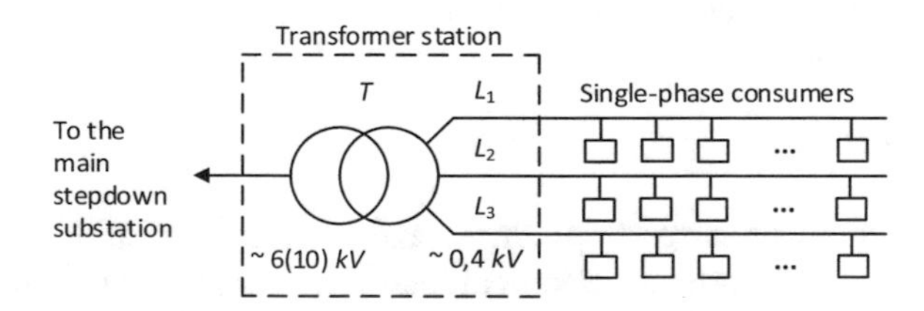

Fig. 1. Distribution transformer station with many SPC

The scope of the research is the transmission and distribution processes of SPC electric power and increasing its efficiency due to the reduction of electric power losses caused by asymmetry and harmonic components of the current.

The problem of the research is to develop and analyze the objective function to minimize the loss of electrical energy from asymmetry and higher harmonic components of the currents in the distribution network.

2.2 Assumptions and Limitations

It is assumed that a non-industrial SPC of electric power are connected to the distribution network, which is powered by the TS (limit value n depends on the rated output of the power transformer installed on the TS). Each SPC can be associated, for example, with a separate apartment within the n-apartment high-rise building or a private household. For each of the SPCs there is a mechanism for dynamic switching between the phases L_1, L_2, L_3 within the distribution network.

It is also assumed that a device is installed at the point where the SPC connects to the distribution network. This device allows capturing the corresponding digitized current waveforms for each SPC at regular intervals t as arrays of k instantaneous values of the indicator obtained for a certain integer number of periods and promptly providing them to computational resource for performing calculations. Based on the superposition of these oscillograms within the time interval t, it is possible to form total oscillograms of the current I_1, I_2, I_3 for the TS phases L_1, L_2, L_3, respectively. Based on the formed total waveforms of the currents I_1, I_2 and I_3, it is possible to evaluate the following indicators:

- effective value of alternating current for each phase;
- asymmetries (irregularities) of the rms currents;
- total harmonic current distortion factors for each phase.

2.3 Calculating Formulas

The assessment of the effective value of alternating current $I_{RMS(l)}$ (*RMS – Root Mean Square*) for each phase is calculated by the formula [12]:

$$I_{RMS(l)} = \sqrt{\frac{1}{k}\sum_{j=1}^{k} (I_l(j))^2}, \tag{1}$$

where $I_l(j)$ is the j-th (j = 1,2, ..., k) instantaneous value of the total current waveform for the l-th phase ($l = \{1,2,3\}$).

The assessment of distribution irregularity (asymmetry) of effective current values is performed using the formula for calculating the corresponding coefficient K_U (*U – Unbalance)* [13] based on the calculated phase values $I_{RMS(l)}$ (1):

$$K_U = \frac{(I_{RMS(1)})^2 + (I_{RMS(2)})^2 + (I_{RMS(3)})^2}{(I_{RMS(1)} + I_{RMS(2)} + I_{RMS(3)})^2}. \tag{2}$$

Calculating the $K_{THD(l)}$ (*THD – Total Harmonic Distortion*) total harmonic distortion coefficient for the l-th phase is performed in two stages. At the first stage, the amplitude spectrum Ia_1, Ia_2,..., Ia_p is formed for p harmonic components of the total oscillogram of the phase current using the discrete Fourier transform [14]. At the second stage, the coefficient of total harmonic distortion THD is calculated directly using the following formula [15]:

$$K_{THD(l)} = \sqrt{\sum\nolimits_{h=2}^{p} Ia_h^2} / Ia_1. \tag{3}$$

It is assumed that the presented estimates (1), (2) and (3) are sufficient for the formation and analysis of the objective function.

3 Proposed Solution

3.1 Solution Concept

The proposed technological solution is based on the following:

- ensuring a symmetric operation mode of a three-phase power transmission system (as is known, the load balancing across all phases is an effective way to reduce losses in the distribution network and is otherwise called internal balancing [16, 17]) and
- reducing the content of harmonic components in the output TS current (as is known, the harmonic components of the current of different electricity consumers differ in phase and the total harmonic components in it can be mutually suppressed, when they work together in one set).

Thus, according to the formulas (1), (2) and (3), it is possible to calculate the initial values K_U and $K_{THD(l)}$ for some initial combination of SPC connections to the TS phases and carry out a search for a new combination of SPC connections to the TS phases. For its realization the number of switching SPC between K_{SUS} phases, as well as K_U, and $K_{THD(l)}$ would be possibly minimal compared to the original combination.

The development of a decision regarding the new combination is planned within each time interval t immediately after receiving the digitized current oscillograms from all SCPs connected to the TS at a specific time point. It is assumed that the composition of active SPCs and oscillograms of their current will not change during time t, and the allowable decrease in the value of asymmetry and harmonic components of current on TS phases L_1, L_2, L_3 is achievable and economically feasible.

3.2 Mathematical Statement

The initial data for the considered task are as follows:

- the number of SPCs connected to the TS – n;
- initial SPC configuration:

$$\bar{I}^0 = \left(i_{1,\ell_1^0}, i_{2,\ell_2^0}, \ldots, i_{n,\ell_n^0}\right). \tag{4}$$

The initial SPC configuration in the formula (4) is determined by the corresponding set of digitized oscillograms $i_1, i_2, \ldots, i_n$ and the initial combination of the phase numbers $L^0 = \{l_1^0, l_2^0, \ldots, l_n^0\}, l^0 \in \{1, 2, 3\}$ for connecting the SPC to the TS.

Such a combination of SPC connections to the TS $L^* = \{l_1^*, l_2^*, \ldots, l_n^*\}, l^* \in \{1, 2, 3\}$ and the corresponding configuration of the SPC need to be found:

$$\bar{I}^* = \left(i_{1,l_1^*}, i_{2,l_2^*}, \ldots, i_{n,l_n^*}\right), \tag{5}$$

For which the value of the objective function would be minimal (Pareto-efficient):

$$F(\bar{I}^*) \rightarrow \min. \tag{6}$$

According to the above-mentioned, the objective function is determined by three criteria, calculated on the basis of the SPC configuration $\bar{I}^*$. Within the framework of the proposed solution, it seems most advisable to converge the criteria used (with rationing the values of each of them in the range from 0 to 0.33) into one objective function in the form of their weighted sum:

$$F(\bar{I}^*) = \alpha_1 \cdot K_U + \alpha_2 \cdot 1/3 \sum\nolimits_{l=1}^{3} K_{THD(l)} + \alpha_3 \cdot K_{SUS}/n, \tag{7}$$

where α_1, α_2 and α_3 are the weight coefficients of the criteria.

4 Discussion

The major drawback in the representation of the objective function (7) is the exigent requirement of content-related comparability between the values of various criteria, as well as the uncertainty in the selection of weighting coefficients α_1, α_2 and α_3. It is reasonable to turn to the peer review mechanism in order to establish the specific values of the weighting coefficients for the purpose of conducting experiments.

In addition, there is variability while choosing the formula for calculating the second criterion of the objective function (a criterion characterizing the content of the current harmonic components regarding the phases of the distribution network fed powered by TS). Based on the formula (7), this criterion is calculated as the average value of the total harmonic distortion over the phases. However, this criterion can be calculated differently, for example, analogously to formula (2) in order to align the values of the total harmonic components in phases. However, in the latter case, there is a risk of not reducing the coefficients $K_{THD(l)}$, but only equalizing their values.

It is also worth noticing that, in practice, it would be logical to use the latter criterion (3) as the main one, since the number of SPC switches between the K_{SUS} phases is a key indicator of the equipment uptime for any electricity consumers [18].

In addition, preliminary results of numerical experiments allow obtaining the typical variant of the response surface fragment for the objective function (7) (see Fig. 2).

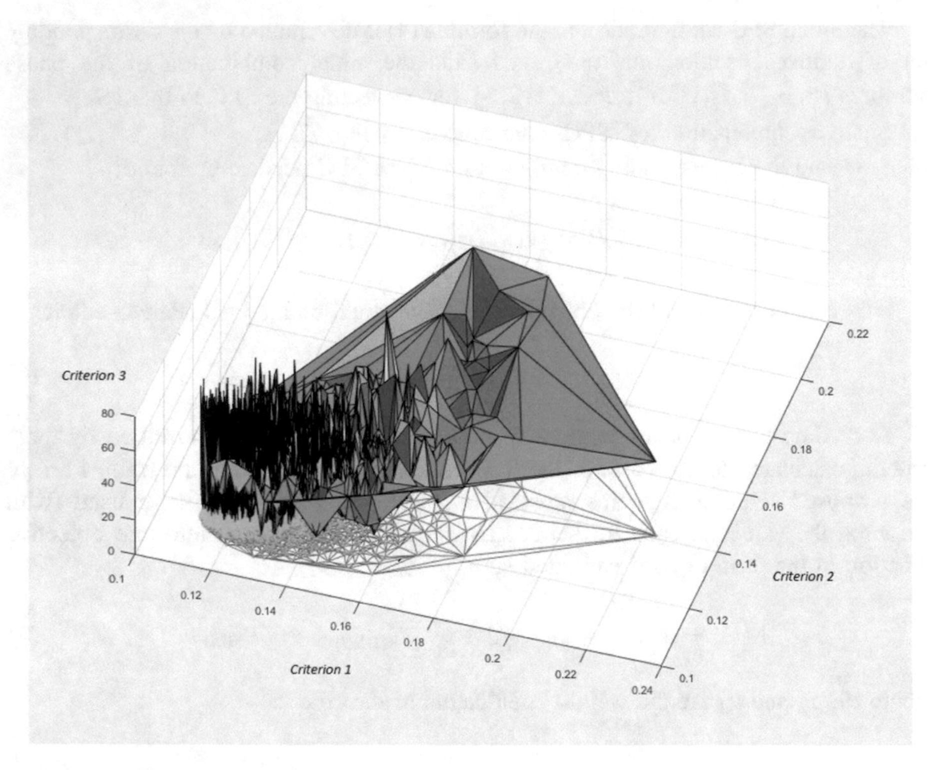

Fig. 2. Fragment of the response surface for the objective function (7) at 100 SPCs

The corresponding calculations were carried out on the basis of formulas (2) ÷ (7) for 100 SPCs connected to the TS. The values of the third criterion (the required number of switching between phases for SPCs) are shown on the graph in absolute units. The demonstrated fragment of the response surface indicates the multiextremal nature of the objective function, which allows concluding that it is necessary to use multiextremal optimization methods in order to solve the considered problem.

In addition, it is possible to estimate the power of feasible solutions set. In order to do this, the number of permissible solutions for the problem of finding the configuration $\bar{I}^*$ (5) is calculated using the exhaustive search. This value will equal the number of phase options for connecting the SPCs to the TS, raised to a degree equal to the maximum possible number of SPCs connected to one TP. This value will grow exponentially and will be 3^{100} or approximately $5.2 \cdot 10^{47}$ variants already for 100 connected SPCs. The obtained evaluation indicates that the search methods will obviously be ineffective in solving this problem [19]. Therefore, it will be expedient to use optimization algorithms of agent metaheuristics, which are now increasingly used in solving such optimization problems, in the considered case [20].

5 Conclusions

1. The presented objective function (7) is formed on the basis of three criteria:
   a. criterion of asymmetry (irregularity) of distributing the effective current values in phases;
   b. a criterion characterizing the content of the current harmonic components in phases of the distribution network powered by TS;
   c. a criterion determining the required number of SPCs switches between TS phases;

   and allows going directly to the algorithmic solution of the problem of minimizing the loss of electrical energy from asymmetry and non-sinusoidal nature of currents in the distribution network.
2. In order to conduct numerical experiments to solve the problem of minimizing the electrical energy loss from asymmetry and non-sinusoidal currents in the distribution network using the proposed objective function (7), it is necessary to determine the numerical values of the weighting coefficient α_1, α_2 and α_3 for the relevant criteria.
3. Depending on the specific conditions associated with the minimization of losses from the higher harmonic components of the current, the values of the corresponding criterion can be calculated variably.
4. The criterion determining the required number of SPCs switches between TS phases is used in the objective function (7), and is a key indicator practically determining the service life of the SPC electrical equipment.
5. The response surface for the generated objective function (7) contains a set of extremes and its nature indicates the need for using the optimization algorithms of agent metaheuristics in order to solve the problem considered.

Acknowledgments. The reported study was funded by RFBR according to the research project № 18-38-20188 «Development and research of tools for intellectual modeling and synthesis of electric power networks technological processes trajectories» using equipment of shared facility «Diagnosis and energy-efficient electrical equipment» (NPI).

References

1. Singh, R., Singh, A.: Aging of distribution transformers due to harmonics. In: 14th International Conference on Harmonics and Quality of Power-ICHQP 2010, art. № 5625347, pp. 1–8. IEEE (2010)
2. Cazacu, E., Lucian, P., Valentin, I.: Losses and temperature rise within power transformers subjected to distorted currents. In: 15th International Conference on Electrical Machines, Drives and Power Systems (ELMA), art. № 7955464, pp. 362–365. IEEE (2017)
3. Cazacu, E., Lucian, P.: Derating the three-phase power distribution transformers under nonsinusoidal operating conditions: a case study. In: 16th International Conference on Harmonics and Quality of Power (ICHQP), art. № 6842930, pp. 488–492. IEEE (2014)

4. Kostinskij, S.S.: Obzor i rezul`taty` issledovanij garmonicheskogo sostava toka by`tovy`x e`lektropriemnikov, a takzhe sposobov i ustrojstv dlya snizheniya negativnogo vliyaniya na sistemy` e`lektrosnabzheniya, Promy`shlennaya e`nergetika 8, 29–39 (2018)
5. Arsov, L., Iljazi, I., Mircevski, S.: Measurement of the influence of household power electronics on the power quality. In: 15-th International Power Electronics and Motion Control Conference and Exposition (EPE-PEMC), art. № 6397239, pp. DS1d.71–DS1d.77. IEEE (2012)
6. Fuchs, E., Masoum, M.A.S.: Power Quality in Power Systems and Electrical Machines. Academic Press, Cambridge (2008)
7. Neha, C.K., Hegde, M., Aher, V.: Effect of home appliances on power quality of conventional grid. In: 2016 International Conference on Circuits, Controls, Communications and Computing (I4C), art. № 8053260, pp. 1–6. IEEE (2016)
8. Faiz, J., Ghazizadeh, M., Oraee, H.: Derating of transformers under non-linear load current and non-sinusoidal voltage – an overview. IET Electr. Power Appl. **9**(7), 486–495 (2015)
9. Kostinskiy, S., Troitskiy, A.: Impact study of non-sinusoidal load parameters on real-power losses in 3-phase double-wound power transformers: method of constant coefficients. J. Eng. Appl. Sci. **12**(20), 5058–5068 (2017)
10. Patidar, R.D., Singh, S.P.: Harmonics estimation and modeling of residential and commercial loads. In: 2009 International Conference on Power Systems (ICPS), art. № 5442731, pp. 1–6. IEEE (2009)
11. Liu, C.W., Luo, C.C., Lin, P.Y.: Develop a power quality measurement system integrated with HAN home energy management system. In: 4th International Conference on Electric Utility Deregulation and Restructuring and Power Technologies (DRPT), art. № 5994135, pp. 1506–1510. IEEE (2011)
12. Bruno, D.A.: Electricity metering with a current transformer. Patent No. 7,359,809. Washington, DC: U.S. Patent and Trademark Office (2008)
13. Mokhov, V.A.: Formirovanie i analiz celevoj funkcii dlya optimizacii tekhnologicheskogo processa raspredeleniya elektroenergii na urovne transformatornoj podstancii, Izvestiya vysshih uchebnyh zavedenij. Severo-Kavkazskij region. Tekhnicheskie nauki **1**(201), 51–56 (2019). https://doi.org/10.17213/0321-2653-2019-1-51-56
14. Winograd, S.: On computing the discrete Fourier transform. Math. Comput. **32**(141), 175–199 (1978)
15. Mog, G.E., Ribeiro, E.P.: Total harmonic distortion calculation by filtering for power quality monitoring. In: 2004 IEEE/PES Transmission and Distribution Conference and Exposition, IEEE Cat. No. 04EX956, pp. 629–632. IEEE, Latin America (2004)
16. Kostinskiy, S.S.: The functional dependence for the estimation of additional losses of active power in a double-wound power transformer caused by asymmetric active-inductive load with a delta connection. Int. Electron. J. Math. Educ. **11**(6), 1529–1544 (2016)
17. Kostinskiy, S.S., Troitskiy, A.I.: Functional dependence for calculation of additional real-power losses in a double-wound supply transformer caused by unbalanced active inductive load in a star connection with an insulated neutral. Int. J. Environ. Sci. Educ. **11**(15), 7975–7989 (2016)
18. Gorlov, M., Adamyan, A., Anufriev, L.: Trenirovka izdelij e`lektronnoj texniki i e`lektronny`x blokov. Chip-news 1 (2001)
19. Kubil, V.N., Mokhov, V.A.: Ant multi-colony optimization algorithm with modifications for solving multiobjective vehicle routing problems. Russ. Electromech. **61**(6), 94–101 (2018). https://doi.org/10.17213/0136-3360-2018-6-94-101
20. Mokhov, V.A., Grinchenkov, D.V., Spiridonova, I.A.: Research of binary bat algorithm on example of the discrete optimization task. In: 2nd International Conference on Industrial Engineering, Applications and Manufacturing (ICIEAM), pp. 1–6. IEEE (2016)

Influence of Valves Constructive Features on Energy Efficiency of Automated Engineering Systems

Aleksey Pustovalov, Dmitriy Kitaev, Tatiana Shchukina(✉), and Sergey Soloviev

Voronezh State Technical University, Moscovskiy prospect, 14, Voronezh 394026, Russia
schukina.niki@yandex.ru

Abstract. The article considers facilities of automatic regulation influence on conditions of buildings life-support systems. The valves applied for these purposes, depending on the type of the closures for the throttling openings, have different characteristics, which contributes to wide range of opportunities for creating energy saving conditions. Possible distortions in characteristics caused by incorrect equipment selection without taking into account device resistance, as well as the regulated area, can later lead to only two valve operating positions: a fully open throttling opening or almost closed. The last thing excludes the possibility of organizing efficient energy saving. The article discusses valve construction influence of the control on pressure loss, that can lead to flow characteristics distortion. Theoretical studies have shown that opening angle of a throttling opening up to 30° causes significant increase in resistance. An angle of more than 30° has virtually no effect on valve hydraulics.

Keywords: Influence · Constructive features · Energy efficiency · Automated engineering systems

1 Introduction

Energy-saving buildings maintenance is inextricably linked with engineering systems and structures automation. The existing standard solutions in this direction allow not only optimal organization of the life support systems functioning, but also significant reduction of resource consumption in hydraulic and thermal modes regulation [1–3].

Various manufacturers' valves used for these purposes are usually distinguished by quality, reliability, and initial characteristics, which have mutually influence on automatic operation and regulated systems dynamics stability [4–7].

Among the equipment presented on the Russian market, special attention should be paid to the devices of a wide range and functionality from the company "Danfoss". Depending on valves construction for throttled openings, controlled valves of the specified manufacturer have linear, linear-linear with a fixed break point, logarithmic, log-linear and parabolic ideal flow characteristics [5, 6].

In case of equipment wrong selection and without taking into account regulated area resistance, possible distortions in flow characteristics do not allow achieving

V. Murgul and M. Pasetti (Eds.): EMMFT-2018, AISC 983, pp. 619–625, 2019.
https://doi.org/10.1007/978-3-030-19868-8_59

smooth and optimal control, providing energy-efficient operation mode of heating and water supply systems maintenance. In addition, valves pulsed opening and closing at the initial stage of automation system response to changing conditions causes network resistance increase and increase in electricity consumption of pumps.

2 Materials and Methods

An appropriate valve construction selection for flow distribution in automated controlled networks is advisable to start with consideration of their flow characteristics [7–9], which show changes in throughput from the initial position to the full opening. Ideal flow characteristics depending on valves construction for throttled openings are divided into linear, linear-linear with a fixed break point, equal percentage, or as they are also called logarithmic, logarithmic-linear, and parabolic [5, 6]. For example, the ideal linear and parabolic characteristics can be expressed by the generalized formula [7]

$$\frac{G}{G_{\max}} = c\left(\frac{h}{h_{\max}}\right)^{n} \tag{1}$$

G, G_{max} - the actual and maximum possible water flow passing through a valve, kg/h; h, h_{max} – a rod movement and its end position with a fully opened valve, mm; c - proportionality coefficient; n - exponent, for the linear dependence n = 1, for the parabolic dependence n = 2; for equipment developing with smoother regulation at small values of the coolant flow rate, it is possible to obtain an exponent more than 2, then for the generalized expression (1) n = 1, 2, 3 ….

Equal percentage, i.e. the logarithmic characteristic of the traditional regulation theory subordinates the flow distribution to the following dependence on a rod's stroke [7]

$$\frac{G}{G_{\max}} = e^{c(1-h/h_{\max})} \tag{2}$$

e - natural logarithm base.

Flow characteristics in actual operating conditions, as a rule, undergo distortions due to pressure losses in valves themselves and controlled areas. The ratio of given losses is called in the technical regulations' external authority. Therefore, we consider a possible pressure loss in valves, depending on throttling holes opening.

3 Results

Among the extensive class of control equipment, multifunctional valves are becoming increasingly popular, including AB-QM from Danfoss (Fig. 1). Stability of their operation is maintained by a pressure regulator built into the lower part, and its upper part (Fig. 1) is designed to change the coolant flow rate at a site. The fluid, coming from an orifice with a diameter d_1 into section d_2, undergoes flow expansion, which can be attributed to a diffuser resistance (Fig. 2).

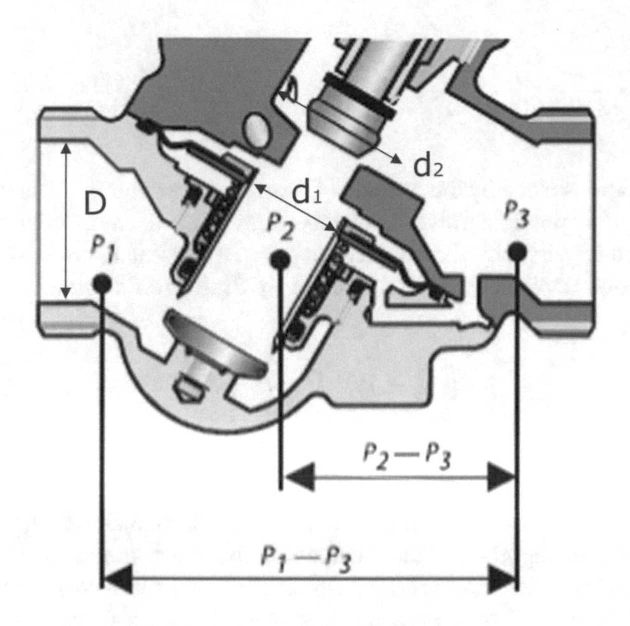

Fig. 1. Diagram of a valve AB-QM.

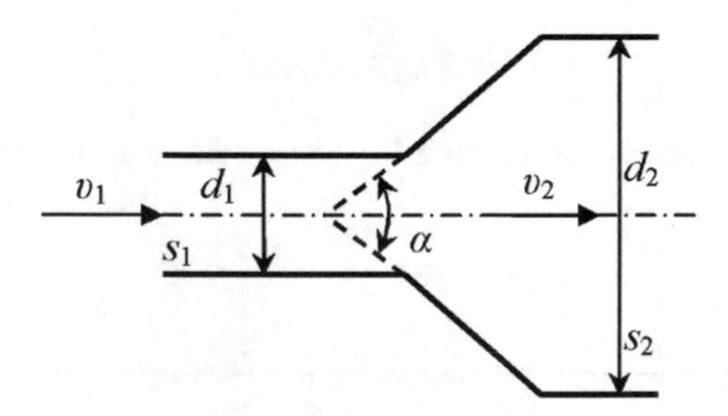

Fig. 2. Diagram of a diffuser.

The differential equation of pressure loss on friction h_l, in meters of water column, for the diffuser is [10]

$$dh_l = \frac{\lambda}{2r\sin\frac{\alpha}{2}} \cdot \frac{\upsilon_1^2}{2g}\left(\frac{r_1}{r}\right)^2 dr \tag{3}$$

λ- hydraulic friction coefficient; υ_1 - movement speed in narrow part of a diffuser, m/s; r_1 - radius of narrow part of a diffuser, m;

Omitting transformations, we obtain dependence for calculating pressure loss in valve, taking into account entrance into annular space

$$h_l = \frac{\lambda \upsilon_1^2}{8\sin\frac{\alpha}{2}g}\left(\frac{r_1^2}{r_2^2 - r_1^2}\right) = \frac{\lambda \upsilon_1^2}{8\sin\frac{\alpha}{2}g}\left(\frac{1}{\frac{r_2^2}{r_1^2} - 1}\right) = \frac{\lambda \upsilon_1^2}{8\sin\frac{\alpha}{2}g}\left(\frac{1}{n - 1}\right) \tag{4}$$

The calculations made by the formula (4) and presented in Figs. 3 and 4, carried out a speed that, with possible valve diameters from 5 to 50 mm, provides a turbulent mode. Kinematic viscosity is adopted at t = 70 °C equal to $0.415 \cdot 10^{-6}$ m²/s. Hydraulic friction coefficient was determined by Shifrinson's formula

$$\lambda = 0{,}11\left(\frac{k_э}{d}\right)^{0{,}25} \tag{5}$$

The equivalent roughness coefficient is assumed to be k_e = 0.05 mm. For a valve internal dimensions when installed on pipelines with a diameter of up to 50 mm, the coefficient λ varies slightly. As the calculations by the formula (5) show, it can be averaged to 0.03. For existing valve constructions, area ratios were calculated in the range of $1.25 \leq n \leq 3$ ($n = s_2/s_1$, where s is sectional area), taper angle is $2 \leq \alpha \leq 180°$.

$$h_l = \frac{209{,}76\upsilon^2}{n^{2{,}3437}\alpha^{0{,}9256}} \tag{6}$$

The average value of a relative error for the considered angles' ranges, speeds and ratios n is 9.87%.

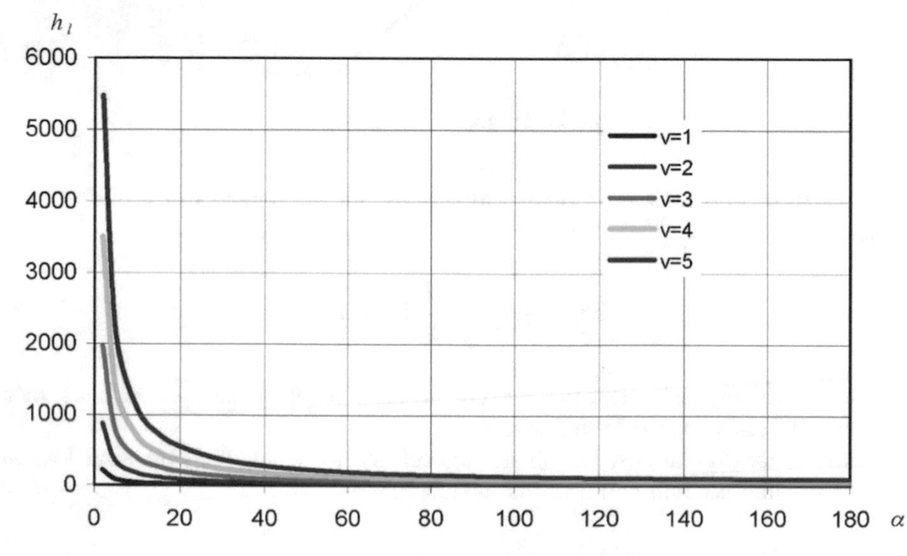

Fig. 3. Pressure loss at medium speed in the range of 1 to 5 m/s $2 \leq \alpha \leq 180°$, with n = 1,1.

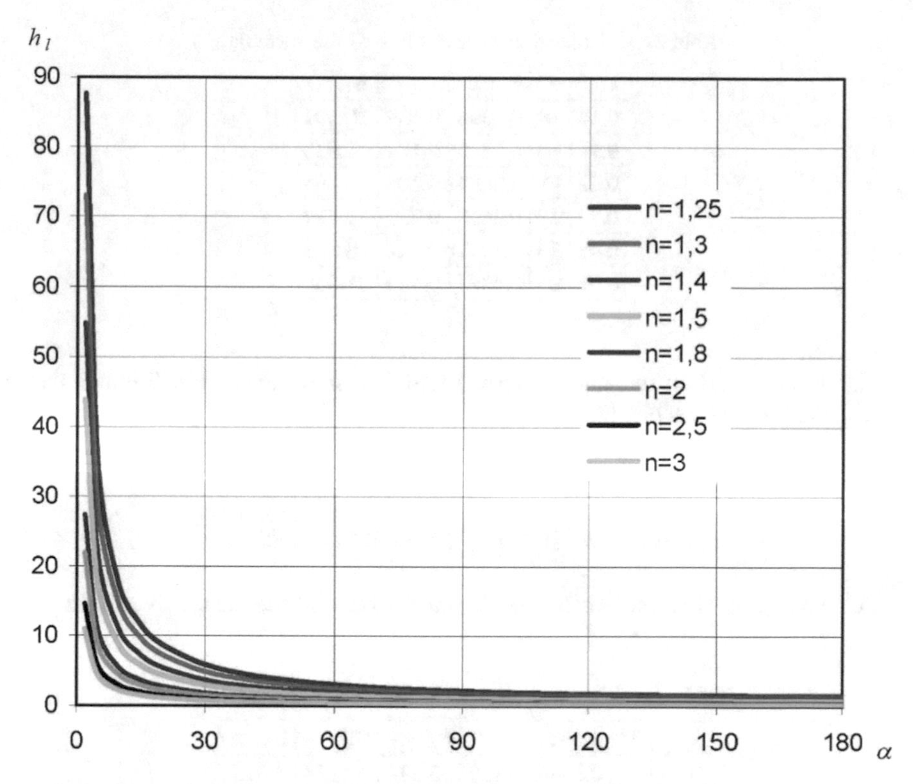

Fig. 4. Pressure loss at medium speed of 1 m/s 1,25 $\leq$ n $\leq$ 3 and 2 $\leq \alpha \leq$ 180°.

4 Discussions

Pressure loss on expansion can be found by Borda formula, but with the introduction of a correction factor k, with turbulent movement mode, a taper angle $\alpha < 20°$ determined by the approximate formula [10]

$$k \approx \sin\alpha \tag{7}$$

The source [10] presents k values table with pipeline gradual narrowing depending on the taper angle

Table 1. k values.

α°	8	10	12	15	20	25
k	0.14	0.16	0.22	0.3	0.42	0.62

Compare the values of Table 1 with ones calculated by formula (7) (see Table 2).

Table 2. Calculated error according to the formula (7).

k	$\alpha°$	α, rad.	$\sin\alpha$	Δ_1, %
0.14	8	0.1396	0.139	0.591
0.16	10	0.1745	0.174	8.530
0.22	12	0.2094	0.208	5.495
0.3	15	0.2618	0.259	13.727
0.42	20	0.3491	0.342	18.567
0.62	25	0.4363	0.423	31.836

The value of Δ1 is the relative error (modulo). It follows from Table 2 that at $\alpha = 15°$ the error is 13.7%, that is significant.

Table k values in the range of $8 \leq \alpha \leq 25°$ can be approximated by the expression

$$k = 6,744 \cdot 10^{-4}\alpha^2 + 5,9 \cdot 10^{-4}\alpha + 0,0464 \quad (8)$$

Table 3 presents the results calculated by the formula (8) and the relative errors Δ2,%.

Table 3. Calculated error according to the formula (8).

k	$\alpha°$	α, rad.	$\sin\alpha$	Δ_1, %	(2)	Δ_2, %.
0.14	8	0.1396	0.139	0.591	0.1368	2.313
0.16	10	0.1745	0.174	8.530	0.1728	8.025
0.22	12	0.2094	0.208	5.495	0.2143	2.585
0.3	15	0.2618	0.259	13.727	0.2866	4.453
0.42	20	0.3491	0.342	18.567	0.4342	3.371
0.62	25	0.4363	0.423	31.836	0.6154	0.742

It follows from Table 3 that the average relative calculation error using formula (7) in the range of $8 \leq \alpha \leq 25°$ is Δ1 = 13.2%, and formulas (8) Δ1 = 3.58%, that is less than more than 3,5 times.

In reference [11], the coefficient k is called the softening coefficient that takes into account change in diffuser resistance compared to losses in case of a sudden expansion of K_2. At taper angles $\alpha < 50°$, the coefficient is recommended [11] to be determined using formula (7), and for $\alpha > 50°$ $K_2 = 1$.

But, as sin90° = 1, and sin50° = 0.766, the correction factor should be determined depending on taper angle by the expressions:

$$k = K_2 = 6,744 \cdot 10^{-4}\alpha^2 + 5,9 \cdot 10^{-3}\alpha + 0,0464, \alpha \leq 25°$$

$$25 \leq \alpha \leq 90° = K_2 = \sin\alpha;\ 90 \leq \alpha \leq 180°, k = K_2 = 1.$$

For a valve, the region of $90 < \alpha \leq 180°$ is suitable.

5 Conclusion

Taking into account the first part, the friction resistance in valves is recommended to consider for three ranges of α:

$$\alpha \leq 25^\circ$$

$$\xi = \frac{\lambda}{4\sin\frac{\alpha}{2}}\left(\frac{1}{n-1}\right) + \left(6,744 \cdot 10^{-4}\alpha^2 + 5,9 \cdot 10^{-3}\alpha + 0,0464\right)\left(\frac{1}{n-1}\right)^2 \quad (9)$$

$$25 < \alpha \leq 90^\circ$$

$$\xi = \frac{\lambda}{4\sin\frac{\alpha}{2}}\left(\frac{1}{n-1}\right) + \sin\alpha\left(\frac{1}{n-1}\right)^2 \quad (10)$$

$$90 < \alpha \leq 180^\circ$$

$$\xi = \frac{\lambda}{4\sin\frac{\alpha}{2}}\left(\frac{1}{n-1}\right) + \left(\frac{1}{n-1}\right)^2 \quad (11)$$

The proposed formulas provide sufficient accuracy and can be recommended for calculations and reasonable selection of the regulatory equipment.

References

1. Sharapov, V., Rotov, P.: Load control of heating systems. News of heat supply, Moscow (2007)
2. Seltzer, I.: Energy-saving heat supply systems of the city. Innov. Life **5**(7), 5–16 (2013)
3. Semenov, V., Kitayev, D., Schukina, T., Korolev, D.: Energy saving and energy efficiency improvement for social facilities. Energy Saving **6**, 38–43 (2010)
4. Universal control valves with electric drives for central heating systems, vol. 6, pp. 32–35. AVOK. Ventilation, heating, air conditioning, heat supply and building thermal physics (2012)
5. Pyrkov, V.: Hydraulic control of heating and cooling systems. Theory and practice. Danfoss Ltd., Moscow (2013)
6. Pyrkov, V.: Modern thermal points. Automation and regulation. Danfoss Ltd., Moscow (2013)
7. Pustovalov, A., Kitayev, D., Shchukina, T.: Providing operating conditions for control valves. Plumb. Heat. Air Cond. **1**(169), 64–68 (2016)
8. Yeremeev, S.: Methods for control valve size specifying in automatic control systems. Devices Syst. Manag. Monit. Diagn. **9**, 14–16 (2009)
9. Betschart, W.: Hydraulik in der Gebäudetechnik. Wärme und Kälte effizient übertragen. Faktor Verlag AG, Zürich (2013)
10. Altshul, A., Kiselev, P.: Hydraulics and Aerodynamics. Stroyizdat, Moscow (1975)
11. Altshul, A.: Hydraulic Resistance. Nedra, Moscow (1982)

Computer Statistical Experiment for Analysis of Resonant-Tunneling Diodes I-V Characteristics

Kirill Cherkasov(✉), Sergey Meshkov, Mstislav Makeev, Yury Ivanov, Vasily Shashurin, Yury Tsvetkov, and Boris Khlopov

Bauman Moscow State Technical University, 105505 Moscow, Russia
kvche@mail.ru

Abstract. The object of research is a resonant-tunneling diode (RTD) based on GaAs/AlGaAs multilayer heterostuctures. A software package with computer statistical experiment module was developed based on resonant-tunneling diode current-voltage characteristic's high-speed modeling algorithm. The package was used for the study of the diode's design parameters' technological errors' effect on RTDs batch I-V characteristics' variation. Verification was made by comparing modelling data with the experimental data.

Keywords: Resonant-tunneling diode · Computer statistical experiment · I-V characteristic simulation · Technological errors · GaAs/AlGaAs heterostructure

1 Introduction

Resonant-tunneling diode (RTD) based on AlGaAs/GaAs semiconductor heterostructures is a prospective EHF and UHF electronics element. RTD application as nonlinear element in radio signal mixers allows improving their performance indices significantly by means of selection an optimal current-voltage (I-V) characteristic's shape for given mixer. It is possible by means of altering resonant-tunneling structure's (RTS) layers thicknesses and chemical composition. In addition, RTD application as mixer's nonlinear element allows broadening mixer's frequencies range up to THz. Also, RTD can be manufactured using well-known microelectronic technologies.

Nonlinear radio signal converters are key elements of microwave electronics as they are used to perform general radio-technical conversions. Their performance indices are mostly determined by the performance indices of their nonlinear elements [1–5].

One of the ways of improving nonlinear radio-signal converters' indices is to use nanoelectronic devices, such as RTDs based on AlGaAs/GaAs heterostuctures. There's a large amount of studies dedicated to prospects of RTD applications and its electrical characteristics [1–12]. Significantly less attention is paid to the problem of ensuring their parameters' reproducibility.

Today's microwave electronic products (including the solid-state electronics) markets' growth speed has caused the microwave electronics production volumes to

V. Murgul and M. Pasetti (Eds.): EMMFT-2018, AISC 983, pp. 626–634, 2019.
https://doi.org/10.1007/978-3-030-19868-8_60

move to the large-scale (and in some segments – the mass) area. This trend is observed for Russian and abroad markets alike. Therefore, batch production availability becomes one of the most important quality indices. Applied to RTD and devices based on them the task of ensuring batch production availability is complicated by diode I-V characteristic's high sensibility to the technological errors. Another problem is high sensitivity of the performance indices of devices based on the RTD to the diode's I-V characteristic shape [4, 5, 11, 12]. These circumstances make the task of ensuring batch production availability of the RTDs and devices based on them a priority one.

This task requires RTD I-V characteristic's parameters statistical analysis under technological errors' influence. A software tool for RTD I-V characteristic simulation and carrying out a computer statistical experiment is required to solve the problem mention above on the design stage. The task of ensuring microwave chips and semiconductor elements batch production availability on the design stage is solved in different CADs (like AWR MWO and Altium Designer) via statistical optimization modules. A semiconductor element's mathematical model, connecting elements design parameters with their electrical characteristics, should be defined in the software to carry out computer statistical experiment. There's no such model for the RTD in existing CADs. Usually RTD is represented as a nonlinear resistance, and its I-V characteristic is approximated by an experimentally measured data [13].

Up to today, there is no RTD I-V characteristic simulation tool [14–16] suitable for carrying out a computer statistical experiment investigating technological factors' influence on the RTD I-V characteristic. Developing such software package would make it possible to study RTD design parameters' and their technological errors influence on the diode's I-V characteristic for the optimization their nominal values as well as the yield. Additionally, such package would make it possible to use statistical methods in researching batch production availability of the microwave nonlinear converters based on RTDs.

The object of research is resonant-tunneling diode (RTD) based on GaAs/AlGaAs multilayer heterostuctures. RTD design parameters define its I-V characteristic's shape and the maximal current. An optimal I-V characteristic for using RTD in certain nonlinear radio signals converter can be obtained by altering diode's design parameters values.

The research subject is RTD I-V characteristic's reproducibility under batch production conditions. The design and technological errors existence cause the RTD parameters variance, what results in diodes batch I-V characteristics' variance.

The main objective of this research is to study RTD design parameters' technological errors influence on RTD I-V characteristics variance, as well as determining dominant technological errors, having the most significant influence on RTD I-V characteristic.

To achieve this goal, the following tasks are accomplished:

- The development of RTD I-V characteristic's initial section simulation tool with the possibility of computer statistical experiment to study technological factors influence on RTD I-V characteristic;
- Accuracy estimation of RTD I-V characteristic's initial section simulation made by the software complex developed;

- The study of RTD technological errors' influence on diode I-V characteristic's variance using mentioned software complex;
- Statistical modeling results' adequacy evaluation by comparing them with measurement results.

2 Methods

2.1 RTD I-V Characteristic's High-Speed Simulation Module

The Tsu-Esaki formula is used as RTD I-V characteristic calculation mathematical model [17]. Diode's RTS tunneling transparency is calculated by transfer matrix method [18].

The input arguments are RTS construction's parameters: layers' thickness in monolayers (ML) for two spacers, two barriers and one well areas and Al doping percent for each mentioned area. Another input argument needed for simulation is maximal bias voltage value.

After obtaining required input data, their dimensions are converted to the format used during the following calculations, then the constants are initialized and RTS potential profile is calculated using the converted input arguments.

The next step is to calculate RTS tunnel transparency using transparency matrix method. Tunnel transparency is calculated for each bias voltage value (after the input the maximal bias voltage value Umax is used to initialize a vector [0; Umax] with determined discretization step).

Then the RTS I-V characteristic is calculated using Tsu-Esaki formula. To convert the result into RTD I-V characteristic, the influence of ohmic contacts' resistance and mesa is taken into account.

The benchmark of the developed tool (named RTSVAC) and its analogues (WinGreen [15], Nanohub [16] and dif2RTD [14]) displayed that the calculating of the I-V characteristics using RTSVAC takes 20 s, while for dif2RTD it takes 90–120 min for the same task, for WinGreen – 25–35 s, and for the Nanohub – 35–40 s.

Obviously, simulation time depends on total monolayers number, but it is evident that developed module has 100 times higher performance rate comparing to dif2RTD and is close to WinGreen and Nanohub in this matter.

The performance growth comparing to dif2RTD is obtained by implementing the described algorithm in Matlab environment, which is highly optimized for working with vector and matrix data, and by some algorithmic optimizations, related with memory allocation and general simulation flow, e.g. by moving duplicated source code in independent subroutines.

Acquired performance growth allows us to move to the task of developing a computer statistical tool for RTD I-V characteristic simulation.

2.2 Computer Statistical Experiment Module

The following parameter groups affect the RTD I-V characteristic: diode RTS parameters (thickness and chemical composition of the layers), ohmic contacts' resistance, mesa dimensions (length and width). Basing on the RTD I-V characteristic initial section calculation algorithm, a module of computer statistical experiment was developed.

Listed parameters are considered stochastic and are represented in pairs «nominal value – standard deviation». All parameters except of monolayer numbers are considered as continuous random variables and are varied randomly by Gaussian distribution law. Monolayer numbers are varied discretely by the same law. Mesa dimensions are correlated with each other with +1 correlation coefficient.

RTD I-V characteristic's simulation with random parameters is carried out according to algorithm described in point 2 on every iteration loop. Each iteration's simulation results are saved in different files.

3 Results

3.1 Comparison Between RTD I-V Characteristic Simulation Results and Experimental Data

To estimate the simulation accuracy, a comparison between RTD I-V characteristic simulation results and the experimental data obtained by measuring a batch of 27 diodes was carried out (diode RTS parameters are shown in Table 1).

I-V characteristics measurements of RTDs were performed using microprobe bench, which consists of a microprobe device, power supply Agilent E3641A and personal computer. This bench provided I-V characteristics measuring of the diodes in voltage range from 0 to 35 V (accuracy $\Delta U = \pm 1$ mV) and current range from 0 to 0.8 A (accuracy $\Delta I = \pm 10$ μA).

Table 1. Diode RTS parameters.

Layer	Parameters		
	Chemical composition	Conductance	Thickness, Å
Spacer	GaAs	i	21
Barrier	AlAs	i	29
Well	GaAs	i	49
Barrier	AlAs	i	29
Spacer	GaAs	i	21

Dispersion of I-V characteristic lies within the measurement error range. Therefore, an averaged I-V characteristic value is used for all following comparisons. Maximum current difference between the calculated and experimental I-V characteristic in the 0… 0.4 V range is 3.49%.

3.2 The Study of RTD Technological Errors' Influence on Diode I-V Characteristic's Variance

The main RTD technological errors influence on RTD I-V characteristic was studied by using the developed software package. The studied errors are diode's RTS technological errors, ohmic contacts resistance errors, and mesa area errors. The study was based on the samples of 100 RTD I-V characteristics obtained by computer statistical experiment. The influence of RTS technological errors and combined influence of ohmic contacts' resistance and mesa area errors on RTD I-V characteristic were studied separately. Parameters' maximum deviations values are given in Table 2.

Table 2. Maximum deviations values of the RTD parameters.

Group	Parameter	Deviation
RTS	Layer thicknesses	±0.5 ML
RTS	Al percentage in barriers	±1%
Ohmic contacts	Ohmic contacts resistance	±0.29 Ω (nominal value 1.32 Ω)
Mesa	Mesa dimensions	±3 μm (nominal value 30 × 30 μm)

Two batches were modeled (Fig. 1a, b). Only RTS parameters errors were taken into account in the first one, while ohmic contacts errors and mesa area errors were considered zeros. In the second batch RTS parameters errors were considered zeros while combined influence of ohmic contacts' errors and mesa area errors was studied.

Comparison of obtained RTD current distributions revealed that the maximum contribution to the I-V characteristics variance is made by the diode RTS technological errors.

4 Discussions

To evaluate the adequacy of the current distributions obtained in Sect. 3.2, the most realistic case was modeled – a combined influence of all factors listed in Table 2 (Fig. 2).

I-V characteristics of 30 diodes batch were measured experimentally. Both modeled and measured current distributions parameters are listed in Table 3.

Mean values' and variances equality hypotheses verification were tested by Student and Fisher criteria at the 0.05 significance level, respectively. It was determined that the tested hypotheses do not confront with the experimental data. Hence, the RTD statistical model can be considered adequate.

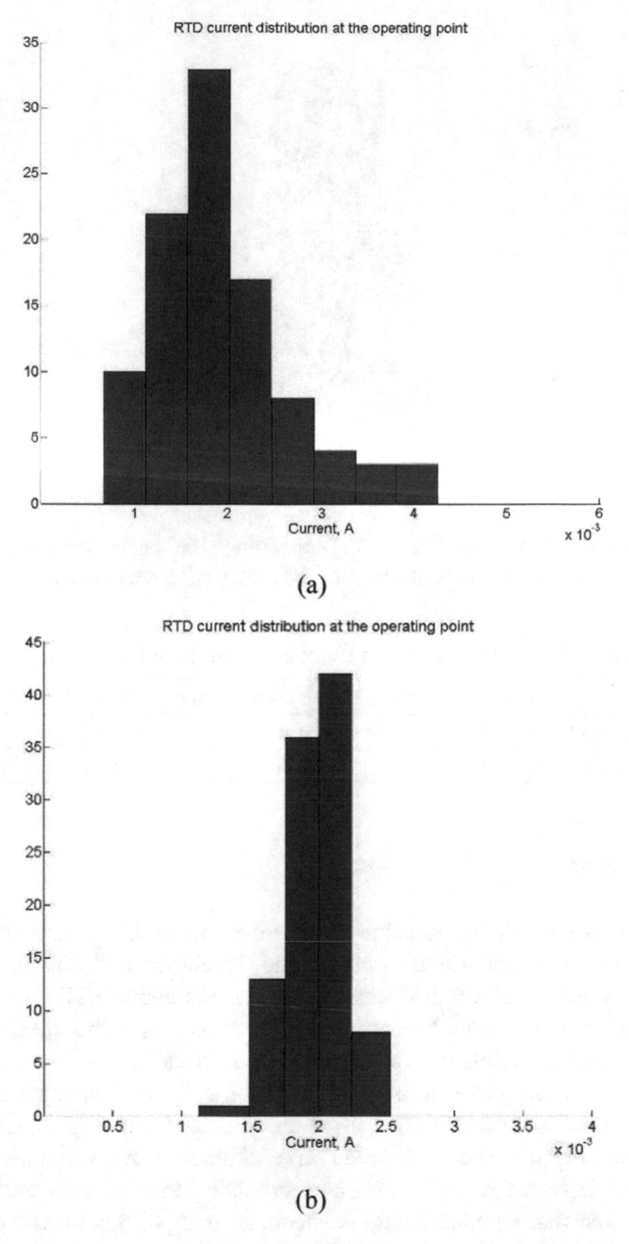

Fig. 1. Distributions of the current at the RTD operating point under: (a) RTS technological errors influence; (b) combined influence of ohmic contacts resistances and mesa area errors.

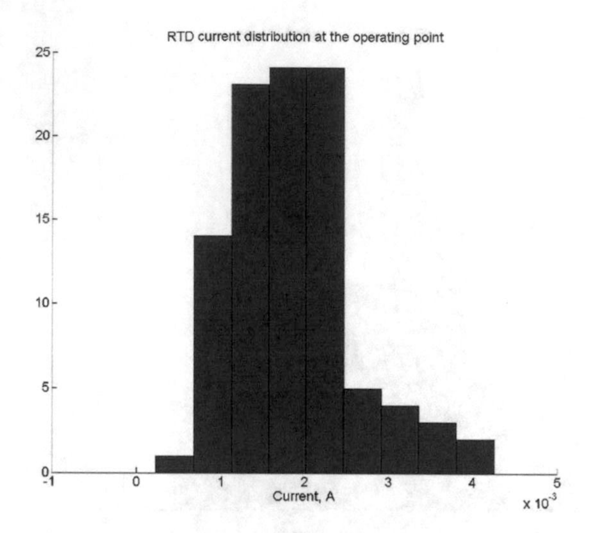

Fig. 2. Distributions of the current at the RTD operating point under combined influence of the RTS technological errors, ohmic contacts' resistance and mesa area errors.

Table 3. Parameters of current distributions at the RTD operating point.

	Expected value, mA	Variance, mA^2	Standard deviation, mA
Modeling result	1.97	0.5064	0.7116
Experimental data	1.86	0.3098	0.5565

5 Conclusions

An RTD I-V characteristic's modelling software package allowing carrying out computer statistical experiment was developed. The developed tool allows studying technological errors influence on I-V characteristic's variance. RTD I-V characteristic modelling algorithm provides high modeling accuracy and the speed sufficient for computer statistical experiment implementation based on it.

The study of various error groups' influence on RTD I-V characteristic's variance using the developed software revealed that diode RTS technological errors make the maximum contribution to the RTD batch's I-V characteristics variance.

Comparison between experimental and simulated current distributions statistical parameters proved that simulated data is adequate to experiment results.

The goals for further research are:

- Integration of the developed program package with existing radioelectronic CADs.
- Resonant-tunneling diode I-V characteristic's shape optimization algorithm development. Optimization will be performed by fitting a set of diode RTS' and ohmic contacts' design parameters. Optimization criteria are radio signal mixer's performance and reliability indices' maximization, as well as yield level maximization.

Achieving these goals would allow significantly reducing material and time expenses on radio signal mixers' development by using computer statistical experiment instead of much longer and more expensive full-scale study of RTDs batch I-V characteristics' kinetics using traditionally reliability test methods.

Another aspect of costs reduction is application an RTD I-V characteristic's optimization algorithm instead of manufacturing test RTD samples with different I-V characteristic shapes. This makes searching of RTD parameters that correspond to optimal I-V characteristic shape a much faster process.

Acknowledgments. The research work was supported by Ministry of Education and Science of the Russian Federation under state task №. 16.1663.2017/4.6.

References

1. Ivanov, Yu.A., Meshkov, S.A., Fedorenko, I.A., Fedorkova, N.V., Shashurin, V.D.: Subharmonic mixer with improved intermodulation characteristics based on a resonant tunnel diode. J. Commun. Technol. Electron. **55**, 921 (2010)
2. Ivanov, Yu.A., Meshkov, S.A., Fedorenko, I.A., Fedorkova, N.V., Shashurin, V.D.: Spectral characteristics of the subharmonic mixer (SHM) of radio signals on the basis of the resonant-tunnelling diode. In: Proceedings of the International Crimean Microwave Conference, CriMiCo 2011, pp. 181–182. Veber, Sevastopol (2011)
3. Vetrova, N.A., Ivanov, Yu.A., Kuimov, E.A., Makeev, M.O., Meshkov, S.A., Pchelintsev, K.P., Shashurin, V.D.: Modeling of current transfer in AlAs/GaAs heterostructures with accounting for intervalley scattering. Radioelectron. Nanosyst. Inf. Technol. **10**(1), 71–76 (2018)
4. Sinyakin, V.Yu., Makeev, M.O., Meshkov, S.A.: RTD application in low power UHF rectifiers. J. Phys.: Conf. Ser. **741**, 012160 (2016)
5. Makeev, M.O., Meshkov, S.A., Sinyakin, V.Yu., Razoumny, Yu.N.: Spacecraft guidance, navigation and control based on application of resonant tunneling diodes in nonlinear radio signal converters. Adv. Astronaut. Sci. **161**, 475 (2017)
6. Kanaya, H., Shibayama, H., Suzuki, S., Asada, M.: Fundamental oscillation up to 1.31 THz in resonant tunneling diodes with thin well and barriers. Appl. Phys. Express **5**, 124101 (2012)
7. Maekawa, T., Kanaya, H., Suzuki, S., Asada, M.: Oscillation up to 1.92 THz in resonant tunneling diode by reduced conduction loss. Appl. Phys. Express **9**, 024101 (2016)
8. Wang, J., Al-Khalidi, A., Zhang, C., Ofiare, A., Wang, L., Wasige, E., Figueiredo, J.M.L.: Resonant tunneling diode as high speed optical/electronic transmitter. In: 2017 10th UK-Europe-China Workshop on Millimetre Waves and Terahertz Technologies, UCMMT, Liverpool, pp. 1–4 (2017). https://doi.org/10.1109/UCMMT.2017.8068497, ISBN 9781538627204
9. Mizuta, H., Tanoue, T.: High-speed and functional applications of resonant tunnelling diodes. In: The Physics and Applications of Resonant Tunnelling Diodes, p. 133. Cambridge University Press, New York (2006)
10. Nagatsuma, T., Fujita, M., Kaku, A., Tsuji, D., Nakai, S., Tsuruda, K., Mukai, T.: Terahertz wireless communications using resonant tunneling diodes as transmitters and receivers. In: Proceedings of International Conference on Telecommunications and Remote Sensing, Luxembourg, vol. 1, p. 41 (2014)

11. Srivastava, A.: Microfabricated terahertz vacuum electron devices: technology, capabilities and performance overview. Eur. J. Adv. Eng. Technol. **2**, 54 (2015)
12. Diebold, S., Tsuruda, K., Kim, J.-Y., Mukai, T., Fujita, M., Nagatsuma, T.: A terahertz monolithic integrated resonant tunneling diode oscillator and mixer circuit. In: Proceedings of SPIE 9856, Terahertz Physics, Devices, and Systems X: Advanced Applications in Industry and Defense, Baltimore, vol. 9856, p. 98560U. SPIE, Washington (2016)
13. Obukhov, I.A.: Nonequilibrium Effects in One-Dimensional Quantum Devices. LAMBERT Academic Publishing, Saarbrücken (2014)
14. Makeev, M.O., Litvak, Yu.N., Ivanov, Yu.A., Meshkov, S.A., Migal, D.E.: dif2RTD: certificate of state registration of a computer program № 2012661001 (2012)
15. Hochschule RheinMain. https://www.hs-rm.de/en/rheinmain-university/people/indlekofer-klaus-michael/research-and-development/wingreen/. Accessed 15 Jan 2019
16. Nanohub: largest nanotechnology online resource. https://nanohub.org/resources/rtd/. Accessed 21 Jan 2019
17. Esaki, L., Tsu, R.: Superlattice and negative differential conductivity in semiconductors. IBM J. Res. Dev. **14**(1), 61–65 (1970)
18. Pérez-Álvarez, R., Garcia-Molliner, F.: Transfer Matrix, Green Function and Related Techniques: Tools for the Study of Multilayer Heterostructures. Publicacions de la Universitat Jaume I, Castelló de la Plana (2004)

Modelling of Absorption Process in a Column with Diffused Flow Structure in Liquid Phase

Alexander Golovanchikov and Nikolay Merentsov(✉)

Volgograd State Technical University,
Lenin avenue, 28, Volgograd 400005, Russia
steeple@mail.ru

Abstract. In this study, physical and mathematical model of the absorber with countercurrent flow of dispersed gas phase was proposed. The model considered ideal displacement regime and uniform liquid phase with the diffusion flow structure. The calculation algorithm was proposed. The technological and geometrical parameters of absorbers were compared with considered in the study flow structures as well as with flow structures of ideal displacement for both phases. The calculation algorithm for the phases is described in previous studies. The variation observed between the absorber flow structure and ideal displacement requires the increase in the column packing height.

Keywords: Diffusion · Peclet number · Ideal displacement model · Flow diffusion structure · Abortion · Countercurrent flow · Packing height

1 Introduction

Known methods for calculation of continuous absorbing apparatuses with counterflow of phases assume that flow structures in liquid and gaseous phases correspond to ideal displacement mode [1–4]. Even when calculating theoretical plates, when it is assumed that liquid phase in each plate is in ideal mixing mode and gaseous dispersed phase is in ideal displacement mode, so that general structure of the liquid phase flow is described by cell-like model, with the quantity of cells (plates) more than ten it shifts to the flow structure of ideal displacement [1, 3, 5–11]. Authors of these papers are the first who paid attention to the influence of longitudinal mixing on process and geometrical parameters of apparatuses and reactors.

2 Calculation

Let us study an elementary material balance for component A extracted from gaseous dispersed phase moving in ideal displacement mode and liquid continuous phase with diffused flow structure, in a packing having height dz taken between sections I-I and II-II (see Fig. 1).

V. Murgul and M. Pasetti (Eds.): EMMFT-2018, AISC 983, pp. 635–644, 2019.
https://doi.org/10.1007/978-3-030-19868-8_61

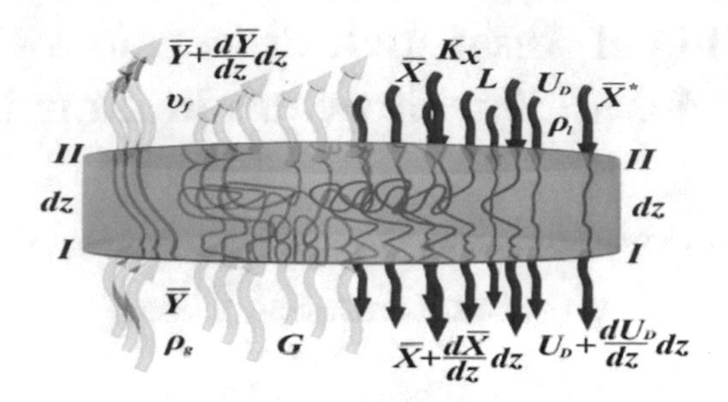

Fig. 1. Movement of material flows of gas and liquid at elementary height of packing between sections I-I and II-II in absorption column with counterflow of phases in ideal displacement mode in dispersed phase and diffused flow structure in continuous liquid phase.

$$G\bar{Y} + L\bar{X} + \rho_l S U_D = G\left(\bar{Y} + \frac{d\bar{Y}}{dz}dz\right) + L\left(\bar{X} + \frac{d\bar{X}}{dz}dz\right) + \rho_l S\left(U_D + \frac{dU_D}{dz}dz\right),$$

where $U_D = -D_l \frac{d\bar{X}}{dz}$ – is an analogue of the Fick's first law with molecular diffusion D replaced by longitudinal diffusion D_l [1, 5, 6].

After algebraic transformations we can obtain the following expression

$$G\frac{d\bar{Y}}{dz}dz + L\frac{d\bar{X}}{dz}dz - \rho_l S D_l + \frac{d^2\bar{X}}{dz^2}dz = 0.$$

We can specify the gradient of concentrations of the extracted component in liquid phase $g = \frac{d\bar{X}}{dh}$, where $h = z/H$ and take integrals around their lower values in the upper part of the column, thus obtaining

$$G\int\limits_{\bar{Y}_f}^{\bar{Y}} d\bar{Y} + L\int\limits_{\bar{X}_w}^{\bar{X}} d\bar{X} - \rho_l S D_l \int\limits_{g_{in}}^{g} dg = 0,$$

or after algebraic transformations

$$\bar{Y} = \bar{Y}_f + \frac{L}{G}(\bar{X} - \bar{X}_w) - \frac{L}{G}(g - g_{in})/Pe, \tag{1}$$

where $Pe = UH/D$ is Peclet number for longitudinal diffusion, g_w is gradient of concentrations in liquid phase at its inlet to the column which can be determined by known boundary condition for diffusion model

$$\bar{X}_n = \bar{X}_w - \frac{1}{Pe}g_{in}. \tag{2}$$

Then tie line equation with account of longitudinal diffusion in liquid phase with simultaneous solution of Eqs. (1) and (2) takes the form

$$\bar{Y} = \bar{Y}_f + \frac{L}{G}(\bar{X} - \bar{X}_n) - \frac{L}{G} g/Pe. \tag{3}$$

Concentrations gradient is $g = g(h)$, so the tie line in the considered case will not be a straight line. For ideal displacement $Pe \rightarrow \infty$ and Eq. (3) describe a typical tie line when the flow structure in both phases corresponds to ideal displacement.

In order to determine dependences $\bar{X} = \bar{X}(h)$, $\bar{Y} = \bar{Y}(h)$ and $g = g(h)$ we can make an elementary material balance for concentration in the taken section having height $dz = Hdh$ (see Fig. 1)

$$L\bar{X} + \rho_l S U_D + K_x a S dz(\bar{X}^* - \bar{X}) = L\left(\bar{X} + \frac{d\bar{X}}{dz} dz\right) + \rho_l S\left(U_D + \frac{dU_D}{dz} dz\right).$$

The last elementary material balance equation after algebraic transformation looks like

$$\frac{d^2\bar{X}}{dh^2} = Pe\frac{d\bar{X}}{dh} - \left(\frac{K_x a Pe}{\rho_l}\right)\tau_x(\bar{X}^* - \bar{X}). \tag{4}$$

It is noteworthy that authors of papers [8] were the first to consider and obtain the differential equations similar to (4), but with diffused structures of flows in both phases. However, as it is said in thesis [1], it is difficult for practice.

In general view equilibrium line $\bar{X}^* = \bar{X}^*(\bar{Y})$ is not a straight line.

In this case Eqs. (3) and (4) can be solved numerically following the algorithm:

From material balance

$$L(\bar{X}_f - \bar{X}_n) = G(\bar{Y}_n - \bar{Y}_f),$$

with $h = 0$, gradient gas per Eq. (3) must be equal to zero. In this case, we select starting values of concentrations:

1. $h = 0$, $g_f = 0$, $\bar{X} = \bar{X}_f$, $\bar{Y} = \bar{Y}_n$.
2. From Eq. (4), transforming it into numerical view with end value $\Delta h = 1/n$, where n is the number of partitions of the overall height of the packing H, we obtain with $\bar{X}_1 = \bar{X}_f$, $\bar{Y}_1 = \bar{Y}_n$ and $\bar{X}_1^* = f(\bar{Y}_1)$

$$\left.\begin{array}{l} g_2 = g_1 + Pe \cdot g_1 h - \frac{K_x a Pe}{\rho_l}\tau_x(\bar{X}_1^* - \bar{X}_1)\Delta h;\ \bar{X}_2 = \bar{X}_1 + g_1\Delta h; \\ \bar{Y}_2 = \bar{Y}_n - \frac{L}{G}(\bar{X}_f - \bar{X}_1) - \frac{L}{G} g_2/Pe. \end{array}\right\}$$

3. We make substitutions $g_1 = g_2$, $\bar{X}_1 = \bar{X}_2$ and repeat calculations for the equations in system from item 2 n times.
4. We check the implementation of the boundary condition: $\bar{Y}_n$ should equal $\bar{Y}_K$ to the specified accuracy. If the condition is not implemented, we increase the average

time of liquid phase (absorbent) presence in column τ_x compared to the average time of presence in the absorber with ideal displacement in liquid phase τ_{xw}.
5. We find the height of the packing in the absorbing apparatus with diffused flow structure in liquid phase $H = H_w\tau_x/\tau_{xw}$, where H_w is packing height in the absorbing apparatus having ideal displacement in both phases.

If the equilibrium line is straight, there is $X^* = \bar{Y}/m$ where m is tangent of the angle of slope of the straight line, and Eqs. (3) and (4) are transformed to

$$\frac{d^2\bar{X}}{dh^2} = Pe\frac{d\bar{X}}{dh} - A\cdot\left[\frac{\bar{Y}_н - (L/G)\bar{X}_к + (L/G)\bar{X} - (L/G)\frac{d\bar{X}}{dh}\Big/Pe}{m} - \bar{X}\right],$$

where $A = \left(\frac{aPeK_x}{\rho_x}\right)\tau_x$, or after algebraic transformations we obtain a heterogeneous differential equation of second order

$$\bar{X}'' - B\bar{X}' - CX = D, \tag{5}$$

where $B = Pe + \frac{A}{m}(L/G)\left(\frac{1}{Pe}\right)$; $C = 1 - A[1 - (L/G)/m]$; $D = -\frac{A}{m}\left[\bar{Y}_n - (L/G)\bar{X}_f\right]$.

Differential Eq. (5) has an analytical solution [8]

$$\bar{X} = C_1e^{r_1h} + C_2e^{r_2h} + Q, \tag{6}$$

where $r_1 = \frac{B}{2} + \sqrt{\left(\frac{B}{2}\right)^2 + C}$ and $r_2 = \frac{B}{2} - \sqrt{\left(\frac{B}{2}\right)^2 + C}$, and integration constants C_1 and C_2 and partial solution can be determined from boundary conditions: $h = 0$, $\bar{X} = \bar{X}_f$, $g_f = 0$, $h = 1$, $\bar{Y} = \bar{Y}_f$. Then $Q = -D/C$, $C_1 = (\bar{X}_f - Q)/(1 - r_1/r_2)$, $C_2 = -C_1\cdot(r_1/r_2)$.

Let us compare process and geometrical parameters of the packed absorption column with ideal displacement flow structures in both phases and diffused flow structure in liquid phase and ideal displacement in gaseous dispersed phase.

Reference book [4] provides a sample calculation of industrial absorption column with ideal displacement flow structures in both phases for benzene hydrocarbons capture from coke gas using coal tar oil; at that equilibrium line is described by direct proportion $\bar{X}^* = \bar{Y}/m$, where $m = 2$, so differential Eq. (4) can be solved by an analytical expression (6).

Table 1 provides initial and reference data and calculation results for absorbing apparatus with ideal displacement flow structure for both phases [4] and for the same apparatus with ideal displacement flow structure for dispersed gaseous phase and diffused flow structure for continuous liquid phase of the absorbent.

Table 1. Initial and reference data and calculation results

№.	Parameter	Units	Symbol	Value	
				Both phases – ideal displacement	Gas – ideal displacement liquid – diffused flow structure
1	2	3	4	5	6
Initial data					
1	Productivity for gas in normal conditions	m^3/h	q_0	13.9	
2	Concentration of benzene hydrocarbons in gas in normal conditions at the inlet of the absorbing apparatus	kg/m^3	y_n	$3.5 \cdot 10^{-3}$	
3	Concentration of benzene hydrocarbons in gas in normal conditions at the outlet of the absorbing apparatus	kg/m^3	y_f	$2 \cdot 10^{-4}$	
4	Content of hydrocarbons in liquid absorbent – absorption oil at the inlet	mass %	x_n	15	
5	Isothermic absorption with average temperature of the flows	°C	t	30	
6	Pressure in the column	MPa	p	0.119	
Reference data					
1	Specific surface of the packing	m^2/m^3	a	65	
2	Porosity of the packing	m^3/m^3	ε	0.68	
3	Bulk density	kg/m^3	ρ_n	145	
4	Dynamic viscosity of the absorbent	Pa	μ_x	$16.5 \cdot 10^{-3}$	
5	Absorbent density	kg/m^3	ρ_x	1.060	
6	Density of the treated gas in normal conditions	kg/m^3	ρ_y	0.44	
7	Molar mass of hydrocarbons extracted from treated coke gas	kgA/kmolA	M_a	83	
8	Molar mass of the inert part of the coke gas	kgG/kmolG	M_g	10.5	
9	Dynamic viscosity of gas	Pa	μ_g	$1.27 \cdot 10^{-5}$	
10	Proportionality factor for the equilibrium line	kgL/kgG	m	2	
11	Equivalent diameter of the packing	m	d_{eq}	0.042	

(*continued*)

Table 1. (*continued*)

№.	Parameter	Units	Symbol	Value	
				Both phases – ideal displacement	Gas – ideal displacement liquid – diffused flow structure
1	2	3	4	5	6
Variable parameter					
1	Peclet number, diffusion	-	*Pe*	∞	50
Calculated parameters					
1	Relative mass concentration of extracted hydrocarbons at the inlet of the absorber	kgA/kgG	$\bar{Y}_n$	0.0864	
2	Relative mass concentration of extracted hydrocarbons at the outlet of the absorber	kgA/kgG	$\bar{Y}_f$	0.0045	
3	Relative mass concentration of hydrocarbons extracted from gas at the inlet of the absorber	kgA/kgL	$\bar{X}_n$	0.0015	
4	Mass flow of the inert part of the gases at the inlet of the column	kgG/s	G	5.57	
5	Mass flow of extracted hydrocarbons by way of mass transfer from gas into the absorbent	kgA/s	G_A	0.456	
6	Equilibrium concentration in absorbent corresponding to initial concentration of extracted hydrocarbons in gas	kgA/kgL	$\bar{X}_n^*$	$4.32 \cdot 10^{-2}$	
7	Minimum flow of the absorbent	kgL/s	L_m	10.96	
8	Operation flow of the absorbent	kgL/s	L	16.44	
9	Final concentration of hydrocarbons in absorbent	kgA/kgL	$\bar{X}_f^*$	0.0293	
10	Mass relation of flows of liquid and gas	kgL/kgG	l_g	2.948	
11	Standard diameter of the absorbing apparatus	m	D_a	3.8	
12	Fictitious velocity of gas	m/sec	υ_g	1.158	
13	Mass transfer coefficient in liquid phase	kgL/m²s	K_x	$1.01 \cdot 10^{-3}$	
14	Average impelling force in liquid phase	kgA/kgL	ΔX_m	$4.61 \cdot 10^{-3}$	

(*continued*)

Table 1. (*continued*)

№.	Parameter	Units	Symbol	Value	
				Both phases – ideal displacement	Gas – ideal displacement liquid – diffused flow structure
1	2	3	4	5	6
15	Fictitious velocity of the absorbent liquid phase	m/sec	U	$1.37 \cdot 10^{-3}$	
16	Packing height	m	H	13.58	15.6
17	Surface of the mass transfer	m^2	G_c	105	$1.149 \cdot 10^5$
18	Average time of presence of the liquid phase in the packing	s	τ_x	98.76	113.46
19	Constant coefficients of Eqs. (5) and (6) Coefficient:	– – – kgA/kgL kgA/kgL – – –	A B C D Q r_1 r_2 C_1 C_2		351.47 −60.36 −166.6 $-1.97 \cdot 10^{-2}$ $-1.18 \cdot 10^{-4}$ −2.9 −57.46 $3.094 \cdot 10^{-2}$ $-1.56 \cdot 10^{-3}$
20	Calculated value of concentration of extracted hydrocarbons in treated gas at the outlet (compare with specified $Y_k = 4.5 \cdot 10^{-3}$ *kgA/kgG*)	kgA/kgG	$\bar{Y}_{af}$		$4.5 \cdot 10^{-3}$
21	Calculated value of concentration of extracted hydrocarbons in liquid absorbent at the inlet (compare with specified $X_{in} = 1.5 \cdot 10^{-3}$ *kgA/kgL*); the difference corresponds to the difference in concentrations as per the boundary condition of the diffused model (2))	kgA/kgL	$\bar{X}_w$		$1.587 \cdot 10^{-3}$
22	Concentration gradient of the extracted component in liquid absorbent at the inlet	kgA/kgL	g_{in}		$4.94 \cdot 10^{-3}$

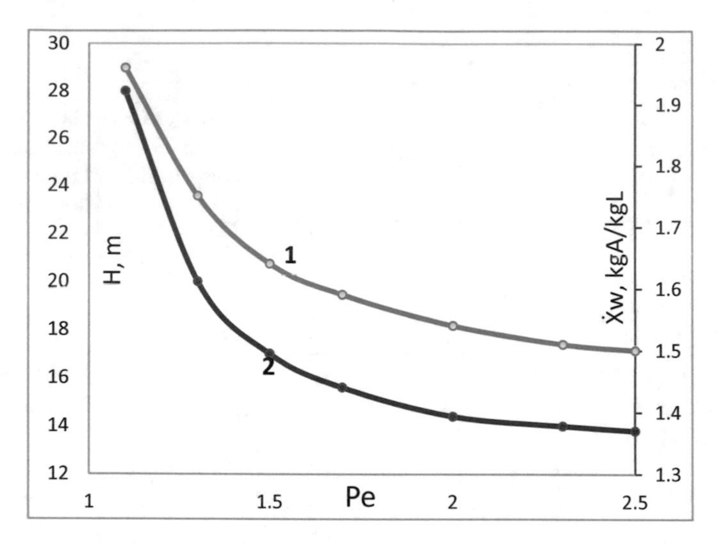

Fig. 2. Relation between the height of the packing (1) and inlet concentration (2) of extracted benzene hydrocarbons in liquid absorbent and Peclet number, diffusion.

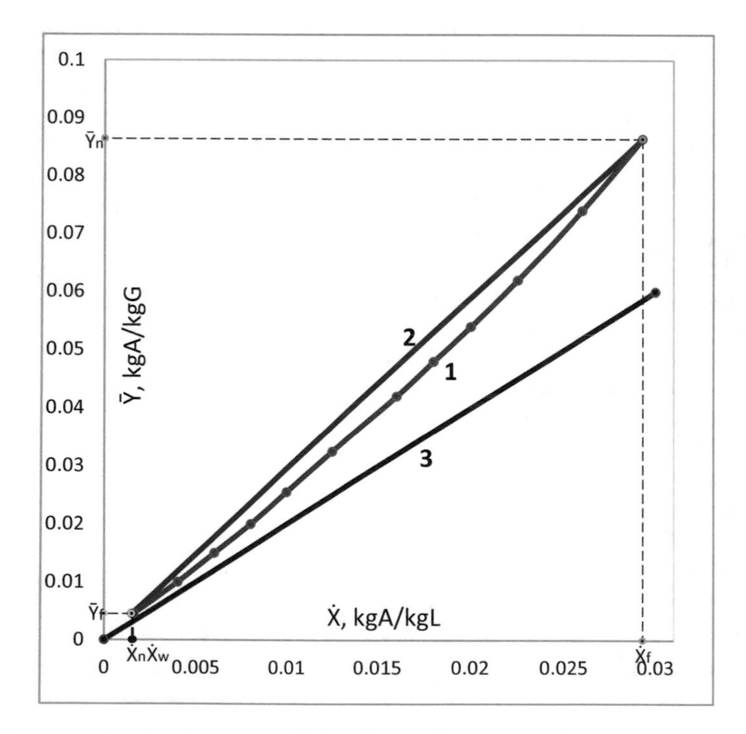

Fig. 3. Diagrams showing tie and equilibrium lines of benzene hydrocarbons absorption process for coke gas using coal tar oil: 1 – tie line for ideal displacing agent flow structures for gaseous dispersed phase and diffused flow structure for liquid continuous phase at Pe = 50 (Eq. (3)); 2 – tie line for ideal displacing agent flow structures for both phases (straight line); 3 – equilibrium straight line.

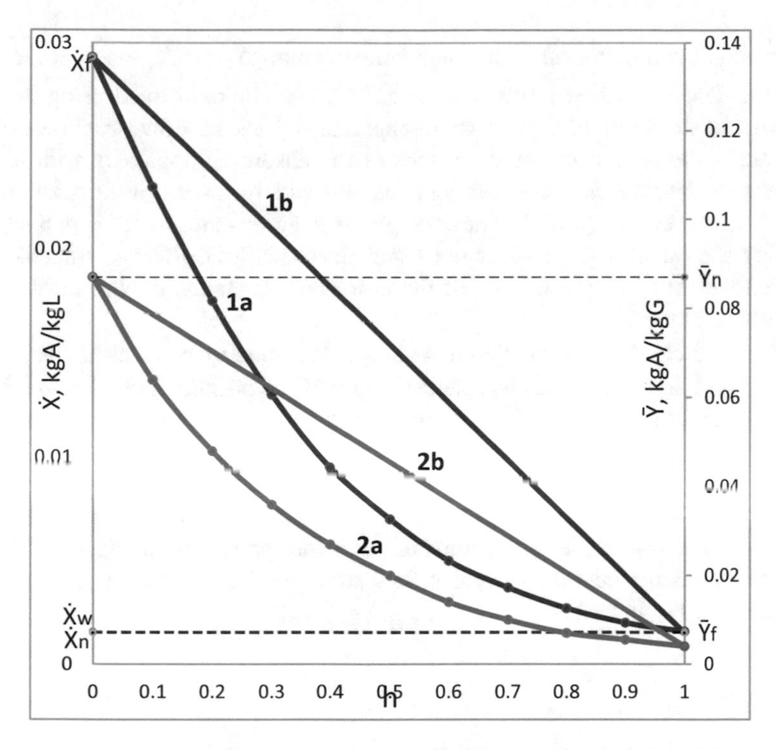

Fig. 4. Benzene hydrocarbons concentration profiles for packing relative height in liquid absorbent (1) and coke gas (2): (a) Ideal displacement mode for dispersed gaseous phase and diffused flow structure for liquid phase, $Pe = 50$; (b) Ideal displacement modes in both phases ($Pe \rightarrow \infty$).

3 Conclusions

The calculation results for both absorption columns (see Table 1) suggest that with equal flows of the absorbent, initial $\overline{Y}_{н}$ and final concentrations $\overline{Y}_{к}$ of hydrocarbons extracted from treated coke gas and equal final concentrations $\overline{X}_{к}$ of consumed coke gas in the absorbent - coal tar oil for diffused flow structure of the absorbent initial concentration of the extracted component at the inlet abruptly increases from $\overline{X}_{н} = 0.0015$ kgA/kgL to $\overline{X}_{в} = 0.00159$ kgA/kgL, that is by 6%; at that the packing height should be increased from 13.58 m to 15.6 m, that is by 15%, too. Reference book [4] that provides a detailed calculation of the considered process in the absorption column with ideal displacing agent flow structures in both phases and other theses and reference books [1–3, 7] explain the necessity in such increase of the packing height by incomplete wetting ability of the packing. Figure 2 demonstrates the diagrams that show the relation between the abrupt increase of the concentrations of the extracted component $\overline{X}_{в}$ and the packing height and Peclet number, diffusion. These diagrams show that before the Peclet number, diffusion Pe $\geq$ 100 is reached, increase in the packing height and extracted component concentration $\Delta\overline{X} = \overline{X}_{в} - \overline{X}_{н}$ at the inlet of the column does not exceed 10%, and calculation can be performed using the typical algorithm of calculation of the absorbing apparatus with ideal flow structures. Figure 3 demonstrates the equilibrium and tie lines of the absorbing apparatus with ideal displacement in both phases and absorbing apparatus with diffused flow structure in liquid phase of the absorbent (Eq. 3). These diagrams suggest that the latter is a sag curve increasing the quantity of transfer units and decreasing the average impelling force compared with the straight tie line of the absorbing apparatus having ideal displacement in both phases.

The curved character of the tie line is better described by concentrations profiles $\bar{Y}$ in gaseous and liquid $\bar{X}$ phases by relative height of the packing (see Fig. 4) calculated by Eqs. (6) and (3). Profiles of these concentrations in ideal displacing agent in both phases are shown for comparison in Fig. 4 by dashed lines.

Thus, reversed mixing of the liquid absorbent in counterflow continuous absorption column should be accounted for in Peclet number for longitudinal mixing $\text{Pe} \leq 100$. In this case, the packing height compared with the typical calculation algorithm for such column having ideal displacement flow structure for both phases can exceed the latter one by 10% and more.

References

1. Dytnerskiy, Y.I.: Basic Processes and Apparatuses of Chemical Technology. Alians, Moscow (2008)
2. Rodionov, A.I., Klushin, V.N., Torocheshnikov, N.S.: Environmental Engineering. Chemistry, Moscow (1989)
3. Timonin, A.S.: Engineering and Ecological Reference Book, vol. 1, Kaluga (2013)
4. Golovanchikov, A.B., Dulkina, N.A., Aristova, Y.V.: Modeling of Hydromechanical and Heat and Mass Transfer Processes in Apparatus and Reactors. VolgGTU, Volgograd (2013)
5. Levenshpil, O.: Engineering Description of Chemical Processes. Chemistry, Moscow (1966)
6. Kafarov, V.V.: Cybernetics Methods in Chemistry and Chemical Technology. Chemistry, Moscow (1985)
7. Tyabin, N.V., Golovanchikov, A.B.: Cybernetics methods in rheology and chemical technology, Volgograd (1983)
8. Zakheim, A.Yu., Glebov, M.B.: Mathematical Modeling of the Main Processes of Chemical Production. High School, Moscow (1985)
9. Golovanchikov, A.B., Dulkina, N.A.: Modeling of the Flow Structures in Chemical Reactors. VolgGTU, Volgograd (2009). Thesis in publication
10. Gelperin, N.I., Pebalk, V.L., Kostanyan, A.E.: Flow Structure and Efficiency of Column Apparatus of the Chemical Industry. Chemistry, Moscow (1977)
11. Dobryakov, A.V., Golovanchikov, A.B.: Wastewater treatment from radioactive isotopes. Life Saf. **22**, 22–26 (2010)

Ion Exchange in Continuous Apparatus with Diffused Flow Structure in Liquid

Alexander Golovanchikov and Nikolay Merentsov(✉)

Volgograd State Technical University, Lenin Avenue, 28, Volgograd 400005, Russia
steeple@mail.ru

Abstract. The authors have studied the physical and mathematical models of continuous ion exchange apparatus having diffused flow structure model in liquid phase and counterflow of ionite granules in ideal displacement mode. The study included comparative calculations of continuous ion exchange apparatuses having ideal displacement in both phases, ideal and non-ideal mixing with different diffusion Peclet numbers in liquid phase.

Keywords: Continuous ion exchange apparatus · Diffusion · Peclet number · Ideal displacement model · Diffusion model · Flow structure · Ideal mixing · Ionite

1 Introduction

Calculations for continuous ion exchange apparatuses are based on ideal displacement models with counterflow of both phases: liquid phase and solid grainy phase with a dense layer of ionite, with ideal displacement modes in both phases, or liquid phase and solid phase with a pseudofluidized layer of both phases having ideal mixing modes in both phases [1–3]. However, experimental studies related to registration of response curves for liquid phase in continuous mass exchange apparatuses having counterflow of both phases indicate that the flow structure in liquid phase is far from both ideal displacement and ideal mixing [4].

In this case, diffusion model of flow structures is widely used for chemical reactors; this model allows determining longitudinal diffusion Peclet number from the value of response curve dispersions; substituting that number into the mathematical model of the chemical reactor with diffused flow structure one can calculate the rate of conversion or final concentration of the reacting component, the value of which is between the final concentrations of reactors with ideal mixing C_m ($Pe \to 0, D_e \to \infty$) and ideal displacement C_d ($Pe \to \infty, D_e \to 0$) [5–9]. Similar to chemical reactors with diffused flow structures, it is necessary to calculate mass exchange processes with the same flow structure, particularly, continuous ion exchange apparatuses with counterflow of liquid and solid grainy phase when the structure of the liquid flows is modelled by one-parameter diffusion model and of the solid grainy phase of ionite – by ideal displacement model [5–10].

V. Murgul and M. Pasetti (Eds.): EMMFT-2018, AISC 983, pp. 645–652, 2019.
https://doi.org/10.1007/978-3-030-19868-8_62

2 Calculation

We can consider physical model of such ion exchange process. Schematic view of its material and concentration flows at layer height z with elementary height dz is provided in Fig. 1.

Let us make an elementary material balance for the extracted component for column section having height dz (see Fig. 1)

$$\upsilon SC_a + \upsilon_D S = \upsilon S\left(C_a + \frac{dC_a}{dz}dz\right) + S\left(\upsilon_D + \frac{d\upsilon_D}{dz}dz\right) + K_\upsilon Sdz\left(C_a - C_a^*\right). \quad (1)$$

Here in the left part is the inflow of the extracted component into the elementary layer having height dz due to convection with a velocity υ and due to longitudinal diffusion with a velocity υ_D – in the right part is the outflow of the extracted component, correspondingly, due to convection, longitudinal diffusion and mass transfer during ion exchange of the extracted component from liquid phase into ionite granules.

For differential equation of longitudinal diffusion and for differential equation of molecular diffusion (Fick's first law) we can make it

$$\upsilon_D = -D_l \frac{dC_a}{dz}, \quad (2)$$

so the initial differential Eq. (1) with account of formula (2) is transformed into

$$D_l \frac{d^2C_a}{dz^2} = \upsilon \frac{dC_a}{dz} + K_\upsilon\left(C_a - C_a^*\right). \quad (3)$$

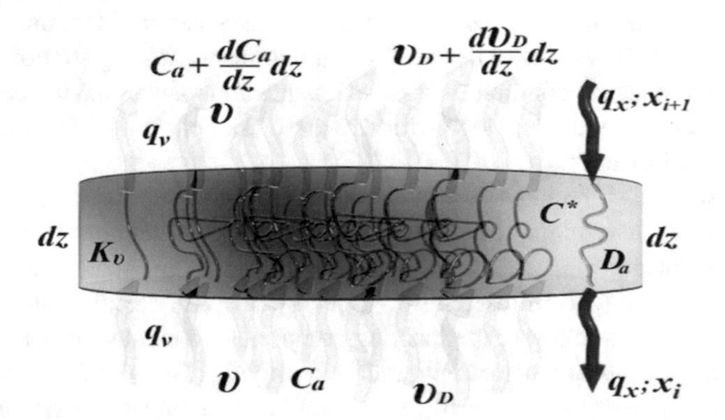

Fig. 1. Schematic view of changes in concentration C of the component extracted from liquid at height element dz in continuous ion exchange column with diffused flow structure model in liquid phase.

We obtained design differential equation of second order

$$\frac{d^2C}{dh^2} = Pe\frac{dC}{dh} + K_{\upsilon}\tau Pe(C - C^*) \tag{4}$$

with boundary conditions:

$$\left.\begin{aligned} h = 0, \quad 1 &= C_n - \left(\tfrac{dC}{dh}\right)_n \tfrac{1}{Pe} \\ h = 1, \quad\quad \mathrm{C} &= \mathrm{C}_f \end{aligned}\right\} \tag{5}$$

For continuous ion exchange apparatus with counterflow of liquid and solid grainy phases having diffused flow structure model for liquid concentration of the extracted component described with differential Eq. (4) can be as follows:

1. Process parameters of the ion exchange column are calculated as per the known mathematical model of continuous ion exchange process having ideal displacement in both phases are fictitious velocity of liquid υ, apparatus diameter D_a, number of transfer units *NTU*, height of the ionite layer *H*, ionite flow q_x and final concentration C_f in it of the component extracted from liquid [1, 3] (see Fig. 2);
2. We select spacing ΔC_n for possible change interval of initial concentration $C_n < 1$ within ($C_f \div 1$), for example $\Delta C_n = (1 - C_f)/200$;
3. Varying in sequence

   $$C_n = 1 - \Delta C_n \cdot i_n; \qquad 1 \leq i_n \leq 200,$$

   for given *Pe* number we can find the initial concentration gradient as per the first formula in system (5)

   $$g_n = -Pe(C_0 - C_i);$$

4. Using the equilibrium line equation of the ion exchange process

   $$x_n^* = \frac{aC_n}{1 + bC_n}$$

   we can determine the initial equilibrium concentration x_n^*, and using the material balance equation we can determine minimum flow of ionite

   $$q_{xm} = q_v C_{a0}(1 - C_f)/(x_n^* - x_n).$$

   Then we calculate the process ionite flow [1, 2]

   $$q_x = (1.1 \div 1.3)q_{xm};$$

   and determine the final concentration in ionite

   $$x_f = x_i + \frac{q_v(1 - C_f)C_{a0}}{q_x}$$

for managing the tie line (see Fig. 1, line 2). Using the formula

$$NTU = \int_{C_f}^{C_i} \frac{dC}{C - C^*}$$

we can determine the number of transfer units, height of the ionite layer moving downwards

$$H = q_v NTU / \left(\frac{\pi D_a^2}{4} K_v \right)$$

and average time of the presence of liquid in the layer

$$\tau = H/\upsilon;$$

5. For the given Pe number selected for varying as per item 3 of the algorithm C_n can be determined using formula (3), design value of concentration of the extracted component C_{df} and we can compare it with the given value of C_f. For that in numerical calculations we divide the height H into equal intervals Δh, for example $\Delta h = 0.01$.
6. For the first iteration:

$$i = 1, \quad C_1 = C_n, \quad g_1 = g_n, \; x_1 = x_f, \; C_1^* = \frac{1}{C_{a0}} [x_1/(a - bx_1)].$$

From Eq. (4) when shifting to finite differences Δh

$$g_2 = g_1 + g_1 Pe\Delta h + K_\upsilon \tau \cdot Pe \cdot \Delta h\left(C_1 - C_1^*\right);$$
$$C_2 = C_1 + g_1 \Delta h; \qquad x_2 = x_1 - \left(x_f - x_n\right)\left(C_1 - C_2\right)/\left(C_n - C_f\right).$$

Making substitutions we receive for some i iteration

$$C_i^* = [x_i/(a - bx_i)]/C_{a0}; \qquad g_{i+1} = g_i(1 + Pe\Delta h) + K_\upsilon \tau \cdot Pe\Delta h\left(C_i - C_i^*\right);$$
$$C_{i+1} = C_i + g_i \Delta h; \qquad x_{i+1} = x_i - \frac{\left(x_f - x_n\right)}{\left(C_n - C_f\right)}\left(C_i - C_{i+1}\right).$$

If the last value $C_i = 100 = C_{df}$ differs from the given C_f not more than by the given relative value, for example $\delta = (C_{df} - C_f)/C_f = 0.01$, the calculation is complete. If the last condition is not met, we continue taking smaller values of initial concentration C_n, repeating items 3 ÷ 6 of the algorithm.

The table provides the results of calculation of continuous ion exchange column following the suggested algorithm and comparison with counterflow ion exchange column with ideal displacement and ideal mixing flow structures [1, 2]. The last calculations have been made for condition $C_n = C_f$ for tie line that corresponds to the horizontal straight line (3) in the diagram of the equilibrium and tie lines (see Fig. 2) by the following formulas

$$x_n^* = \frac{aC_f}{1 + bC_f};\ q_{xm}^* = \frac{q_v(C_n - C_f)}{x_n^* - x_n};\ x_{xm} = x_n + \frac{q_v(C_n - C_f)}{q_{xm}};$$

$$NTU_x = \int_{x_n}^{x_{fm}} \frac{dx}{x_n^* - x};\qquad H_m = \frac{q_{xm} NTU_x}{\frac{\pi}{4} D_a^2 K_\upsilon \rho_n}.$$

Initial and reference data have been taken the same as in the sample calculations of continuous cation-ion exchange column with a suspended layer for treatment of water from Na+ cations [1, 3]. The Peclet number Pe = 3.42 has been received after registration of the response curve in the laboratory column with a grained ionite layer, calculation of its dispersion and diffused Peclet number as per the formula [5–9] (Table 1).

Table 1. Initial and reference data and calculation results for industrial ion exchange column with ideal displacement flow structure for both phases compared with ideal displacement flow structure for dispersed phase and diffused flow structure for liquid phase.

No	Parameter	Units	Symbol	Value
1	2	3	4	5
Initial and reference data				
1	Treated water capacity	m^3/h	q_υ	10
2	Initial concentration of sodium cation concentration	kg/m^3	C_n	0.1
3	Final concentration of sodium cation concentration	kg/m^3	C_f	0.005
4	Porosity of the layer	m^3/m^3	ε_0	0.4
5	Molecular mass of sodium cations	kg/kmol	M	23
6	Average diameter of ionite particles	m	d	$9 \cdot 10^{-4}$
7	Bulk density of ionite	kg/m^3	ρ_n	800
8	Equilibrium constant of cation exchanger KU-2	–	K_p	1.2
9	Water density	kg/m^3	ρ	1000
10	Water viscosity	Pa	μ	10^{-3}
11	Full exchange capacity of cation exchanger KU-2	kg/kg	x_0	0.109
12	Effective diffusion factor for ion diffusion in an ionite particle	m^2/s	D_x	$2.3 \cdot 10^{-10}$
I General calculation parameters				
1	Diameter of a standard apparatus	m	D_a	1.5
2	Fictitious velocity of water in the apparatus	m/s	υ	$1.57 \cdot 10^{-3}$
3	External mass output factor	m/s	β	$1.96 \cdot 10^{-5}$
4	Volumetric factor of mass output	$1/s$	K_v	$7.85 \cdot 10^{-2}$
5	Biot number	–	Bi	$5.61 \cdot 10^{-2}$

(continued)

Table 1. (*continued*)

No	Parameter	Units	Symbol	Value
1	2	3	4	5
6	Equilibrium line factor	m^3/kg m^3/kgA	a b	1.31 2
II Calculation parameters for ideal displacement for treated water				
7	Equilibrium concentration in ionite corresponding to initial concentration of extracted Na+ ions in gas	kgA/kg	x_n^*	0.1092
8	Final concentration of extracted Na+ ions in ionite	kgA/kg	x_{fd}	0.1038
9	Operation flow of ionite	kg/h	q_{xd}	9.16
10	Number of transfer units for concentration of Na + ions in water	–	NTU	11.94
11	Height of the moving layer of ionite	m	H_d	0.24
12	Velocity of the ionite layer moving downwards	m/sec	υ_{dx}	$1.8{\cdot}10^{-6}$
13	Time of ionite grains presence in the apparatus	s	τ_{dx}	132 968
14	Time of water presence in the ionite layer	s	τ_{dw}	152.2
III Calculation parameters for ideal mixing for treated water				
15	Equilibrium concentration in ionite corresponding to the concentration of extracted Na+ ions in the apparatus and at its output	kgA/kg	x_f^*	$6.49{\cdot}10^{-3}$
16	Final concentration of Na+ ions in ionite	kgA/kg	x_{fm}	$6.16{\cdot}10^{-3}$
17	Ionite flow	kg/h	q_{xm}	154.16
18	Number of transfer units calculated via concentrations of extracted ions in ionite	–	NTU_x	4.96
19	Height of the layer of ionite	m	H_m	$1.92{\cdot}10^{-3}$
20	Time of ionite grains presence in the apparatus	s	τ_{mx}	63.2
21	Velocity of ionite grains moving downwards	m/sec	υ_{mx}	$3.03{\cdot}10^{-5}$
22	Time of treated water presence in the ionite layer	s	τ_{mw}	1.22
23	Dispersion of response curve for treated water	–	σ^2	0.47
24	Peclet number, diffusion	–	Pe^*	3.42
25	Initial relative concentration of Na+ ions	–	C_n	0.792
26	Final concentration of Na+ ions in ionite	kg/kg	x_f	$8.52{\cdot}10^{-2}$
27	Ionite flow	kg/h	q_x	11.16
28	Height of the layer of ionite	m	H	0. 232
29	Time of ionite presence in the apparatus	s	τ_x	105 760
30	Velocity of ionite grains moving downwards	m/s	υ_x	$2.1{\cdot}10^{-6}$
31	Time of treated water presence in the ionite layer	s	τ_w	147.2

$$\sigma^2 = \frac{2}{Pe} - \frac{2}{Pe^2}\left(1 - e^{-Pe}\right).$$

Using the results of calculations performed as per the algorithm described above, the authors have created diagrams of tie and equilibrium lines and dependences between initial relative concentration, layer height, ionite flow and Peclet number (see Fig. 2-4).

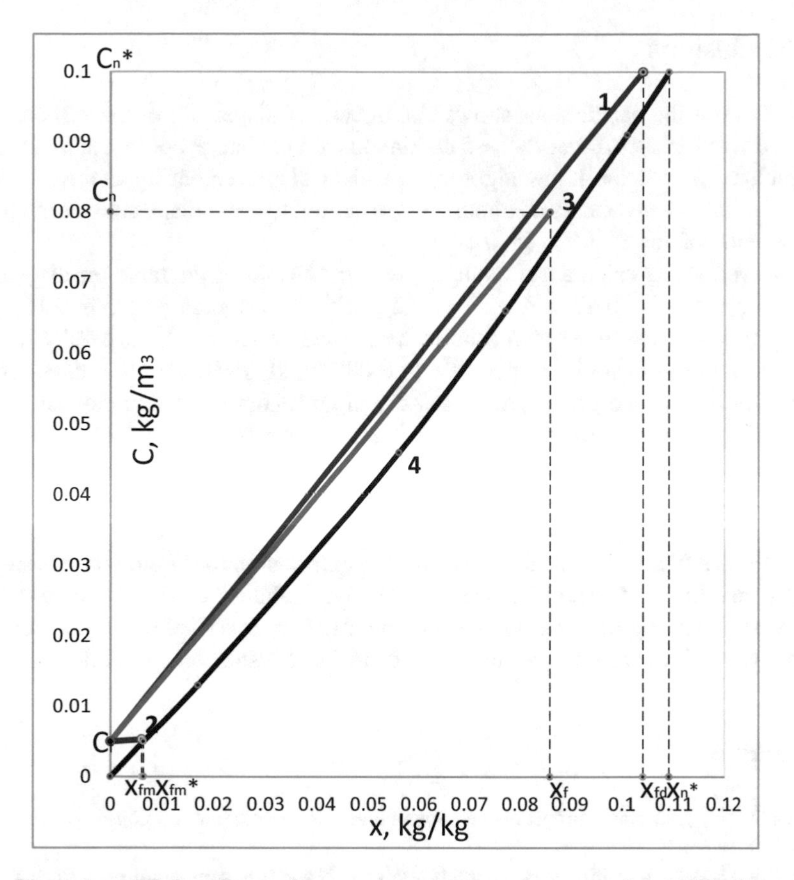

Fig. 2. Tie and equilibrium lines of ion exchange process in counterflow of liquid and solid grainy phase of ionite: 1 – tie line of ideal displacement for liquid; 2 – tie line of ideal mixing for liquid; 3 – tie line for liquid at Pe^{*} = 3.42 (diffused flow structure); 4 – equilibrium line [1, 3].

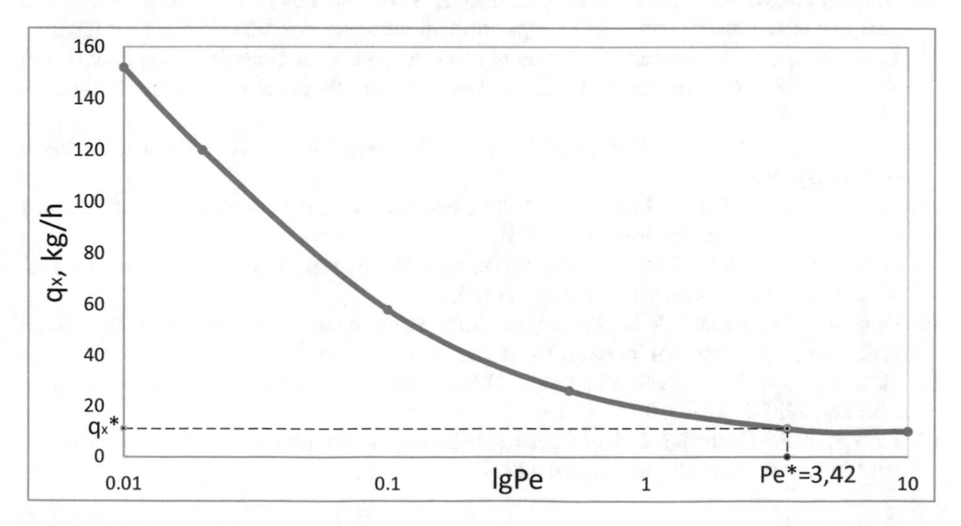

Fig. 3. Dependence between ionite flow and Peclet number in continuous counterflow ion exchange column.

3 Conclusions

The analysis of the calculations shows that in case of dispersion of the response curve of treated water $\sigma^2 < 0.1$ and $Pe^* > 5$ continuous ion exchange column apparatuses can be calculated as per the know algorithm of ideal displacement apparatuses [11, 12]. This is similar to the conclusion about approximation to ideal displacement of chemical reactors with values $\sigma^2 < 0.1$ [5–9].

However, in the calculation example given in the end of the table for dispersion of response curve $\sigma^2 = 0.47$ and $Pe^* = 3.42$ ionite flow increases from 9.16 kg/h to 11.6 kg/h, that is 1.2 times more; the increase is reasoned by the necessity to reduce final concentration in ionite from 0.104 to 0.085 kg/kg so that tie line does not cross equilibrium line 4 (see Fig. 2), at that at the input relative concentration of extracted ions abruptly decreases from 1 to 0.79 because of reverse longitudinal diffusion.

Thus, calculations for diffusion model of flow structures for liquid phase in continuous ion exchange counterflow column are required when Peclet number is less than 5 (response curve dispersion for liquid phase is $\sigma^2 > 0.12$). In other cases, it can be considered that like in chemical reactors the flow structures for liquid phase is similar to the flow structures of ideal displacements, and calculation of the parameters can be performed following the known calculation method for continuous ion exchange apparatuses with counterflow of both phases in ideal displacement mode (Fig. 3).

References

1. Dytnerskiy, Y.I.: Basic processes and apparatuses of chemical technology. Alians, Moscow (2008)
2. Rodionov, A.I., Klushin, V.N., Torocheshnikov, N.S.: Environmental engineering. Chemistry, Moscow (1989
3. Timonin, A.S.: Engineering and ecological reference book, vol. 1, Kaluga (2013)
4. Golovanchikov, A.B., Dulkina, N.A., Aristova, Y.V.: Modeling of hydromechanical and heat and mass transfer processes in apparatus and reactors. VolgGTU, Volgograd (2013)
5. Levenshpil, O.: Engineering description of chemical processes. Chemistry, Moscow (1966)
6. Kafarov, V.V.: Cybernetics Methods in chemistry and chemical technology. Chemistry, Moscow (1985)
7. Tyabin, N.V., Golovanchikov, A.B.: Cybernetics methods in rheology and chemical technology, Volgograd (1983)
8. Zakheim, A.Y., Glebov, M.B. Mathematical modeling of the main processes of chemical production. High school, Moscow (1985)
9. Golovanchikov, A.B., Dulkina, N.A.: Modeling of the flow structures in chemical reactors. Thesis Publication, VolgGTU, Volgograd (2009)
10. Gelperin, N.I., Pebalk, V.L., Kostanyan, A.E.: Flow structure and efficiency of column apparatus of the chemical industry. Chemistry, Moscow (1977)
11. Dobryakov, A.V., Golovanchikov, A.B.: Wastewater treatment from radioactive isotopes. Life Saf. **22**, 22–26 (2010)
12. Pavlov, K.F.: Examples and problems for the course on processes and apparatuses of chemical technology. Alians, Moscow (2013)

Numerical Simulation of Neutralization of Nitrogen Oxides in the Exhaust Gases of Electric Arc Installation

Sergey Aulchenko[1,2], Vitaly Belyaev[1(✉)], Sergey Vashenko[1], and Oleg Kovalev[1]

[1] Khristianovich Institute of Theoretical and Applied Mechanics SB RAS, Novosibirsk 630090, Russia
belyaev@itam.nsc.ru

[2] Novosibirsk State University of Architecture and Civil Engineering, Novosibirsk 630008, Russia

Abstract. In this paper, we simulated two ways to reduce the level of nitric oxide at the outlet of the gas-dynamic path of an arc installation. In the framework of the kinetic approach based on a coaxial air supply to optimize the average mass temperature, it was shown that with optimal selection of flow rates and temperatures of the central and ejected jets, the content of nitric oxide at the outlet can be reduced more than three times. As part of the chemical approach, when distributed methane gas is supplied at the inlet to the neutralizer, it has been shown that, with an optimal selection of the degree of cooling of the working gas and pressure, the content of nitric oxide can be reduced more than four times.

Keywords: Electric arc installations · Nitrogen oxides · Numerical simulation

1 Introduction

The main sources of nitrogen oxide emissions in the atmosphere are both natural phenomena, such as lightning and forest fires, and anthropogenic—vehicle exhaust gases, flue gases from thermal power plants and gas turbines, etc. [1–3]. Nitrogen oxides combining with water vapor in the air form nitric acid and along with sulfur oxides cause "acid rain". The maximum permissible concentration (MPC) of NOx according to [1] is 0.1 mg/m3 (or 0.053 ppm (parts per million)). Operating modes of electric arc installations (EI) are characterized by high temperatures of the working flow, as a rule, air (working temperatures of electric arc heaters vary from 3500 to 5000 K), therefore the maximum amount of NO can reach 10%. As a result, the content of toxic nitrogen oxides in the exhaust gases of installations can exceed the MPC in one hundred times, even taking into account the ejection into the exhaust air flow with a flow rate that exceeds many times the main flow rate. The main mechanism of formation of nitrogen oxides in EI is the thermal mechanism (Zeldovich mechanism) [4]. On this basis, two ways of reducing the content of nitrogen oxides in exhaust gases of electric arc installations were considered: kinetic and chemical.

V. Murgul and M. Pasetti (Eds.): EMMFT-2018, AISC 983, pp. 653–660, 2019.
https://doi.org/10.1007/978-3-030-19868-8_63

2 Formulation of the Problem

To evaluate the proposed methods for controlling the process of neutralization of nitrogen oxides in gas-dynamic installations with electric arc heaters, mathematical modeling of thermal and gas-dynamic processes in the installation path, namely in the nozzle, working part, diffuser and partially in the neutralizer was carried out.

Figure 1 shows a scheme of one of these installations. In addition to the gas entering the working path of the installation, the diagram shows the places of possible air supply (coaxial jet with flow rate G_{air}) or methane (distributed supply, G_m) for the implementation of kinetic or chemical methods of neutralizing nitric oxide, respectively. $\dot{X} = X/L, \dot{Y} = Y/L$, L—length of the installation.

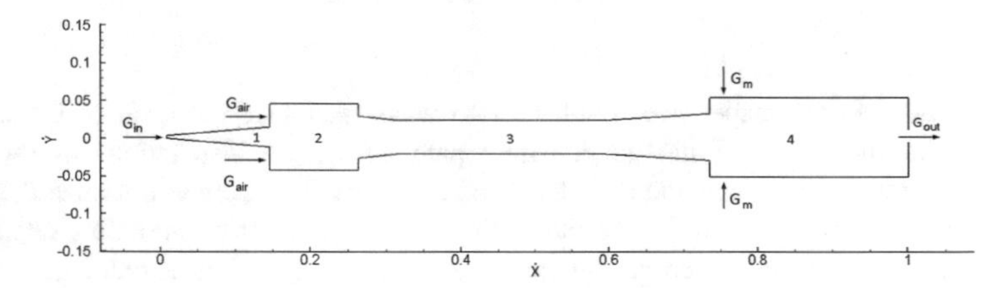

Fig. 1. A scheme of installation: 1 - nozzle; 2 - working chamber; 3 - diffuser; 4 - neutralizer; G_{in} - the inlet gas flow rate; G_{air} - air flow; G_m - methane consumption; G_{out} - gas flow rate at the outlet.

2.1 Kinetic Approach

As the estimates showed, the limiting factor determining the NO concentration at the exit is the gas-dynamic time scale for the flow through the working path of the installation. Taking into account the operating parameters of the electric arc heaters and the channel geometry, the time taken to pass through the working path according to calculations does not exceed 10–15 ms. Moreover, the main time the flow is in the neutralizer, where nitrogen oxide enters, the concentration of which is almost equal to the original ("frozen"), because at the temperature level in the working part and in the diffuser, the time to establish an equilibrium concentration is much longer than the residence time of gas in this area of the installation, see Table 1.

Table 1. The time to establish an equilibrium concentration τ_{NO} at normal air density.

Parameter	Calculation results, see below formulae (1, 2)						
T, K	1700	2000	2100	2300	2600	3000	4000
τ_{NO}, s	140	1.0	0.27	$3.1 \cdot 10^{-2}$	$2.2 \cdot 10^{-3}$	$1.4 \cdot 10^{-4}$	$1.5 \cdot 10^{-6}$

Therefore, to reduce the concentration of nitric oxide in the neutralizer, it is necessary that the average mass temperature of the air flow at the entrance should be 2400–

2600 K (see Fig. 2), since the characteristic time to establish an equilibrium concentration is 5–10 ms, taking into account the actual gas density in the neutralizer.

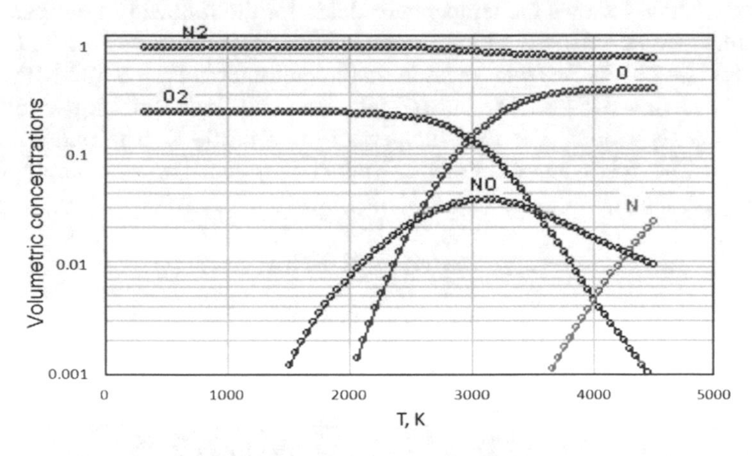

Fig. 2. Volumetric concentrations of the components of the gas mixture, obtained using the software complex thermodynamic calculation of the composition of the phases "Terra" [5].

In this regard, it seems appropriate to choose such a ratio of the flow rate of the working gas of an electric arc heater G_{in} and cold air G_{air} supplied coaxially at the exit section of the supersonic nozzle so that the average mass temperature does not exceed 2400–2600 K.

2.2 Chemical Approach

The chemical approach is based on the burning of free oxygen with methane in the mode of excess of the latter in the neutralizer, which may allow to reduce the concentration of nitrogen oxides. It should be borne in mind that when methane is burned at a high temperature, not only carbon dioxide is produced, but also carbon monoxide.

3 Math Modeling

To assess the effectiveness of the proposed methods for controlling the process of neutralization of oxides, mathematical modeling of thermal and gas-dynamic processes in the nozzle, the working part, the diffuser and partially in the neutralizer was carried out.

The medium is assumed to be compressible and viscous, obeying the two-parameter equation of state. The flow is turbulent. It is possible to take into account different amounts of environmental components depending on the nature of the chemical reactions. The simulation was performed using the ANSYS Fluent software package [6] based on the averaged Navier-Stokes equations, supplemented by the SST turbulence model. For one of the typical operating modes of the installation, the fields of full and static enthalpies, temperatures, and flow rates were obtained, both in the absence and in the presence of coaxial jets with different air flow rates (from 1 kg/s to 4 kg/s).

Figures 3 and 4 shows the temperature fields for the following flow parameters.

The main jet: rate G_{in} = 3.19 kg/s, composition—O_2 – 20%, NO – 5%, O – 2%, N – 0.001%, N_2 – 72.9%. The total pressure in the preheater is P_0 = 3.622 MPa, the total temperature is T_0 = 3254 K. The coaxial jet: rate G_{air} 0 kg/s and 1 kg/s, composition —O_2 – 20%, N_2 – 80%. The total temperature is T = 300 K. The outlet pressure is p = 34 kPa. The following reactions are taken into account, with appropriate rate constants:

$$O + N_2 \leftrightarrow N + NO,$$

$$N + O_2 \leftrightarrow O + NO,$$

$$k_1 = 1.8 \cdot 10^8 e^{-38370/T},\ k_{-1} = 3.8 \cdot 10^7 e^{-425/T},$$

$$k_2 = 1.8 \cdot 10^4 T e^{-4680/T},\ k_{-2} = 3.8 \cdot 10^3 T e^{-20820/T}.$$

The generalized thermal effect of these four reactions is −90 kJ/mol.

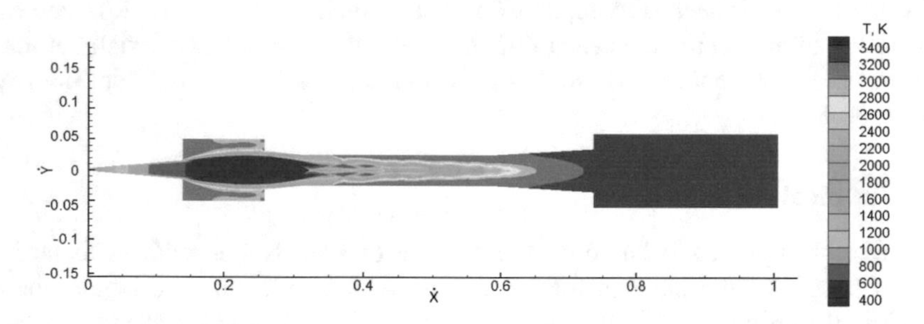

Fig. 3. Static temperature. Coaxial jet rate G_{air} = 0 kg/s.

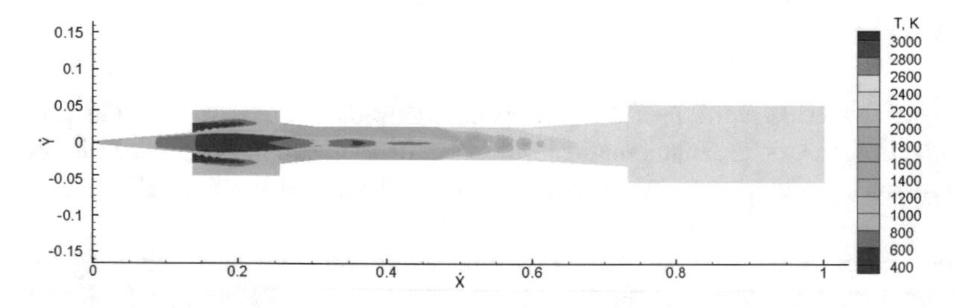

Fig. 4. Static temperature. Coaxial jet rate G_{air} = 1 kg/s.

The calculation results presented in Figs. 3 and 4 allows us to estimate the degree of mixing of the central and coaxial jets and the level of the gas temperature in the neutralizer. With a coaxial jet gas flow rate G_{in} = 0.5–1.0 kg/s, the gas temperature in the neutralizer can be obtained in the range of 2400–2600 K, at which the residence time of the gas in the neutralizer is comparable to the equilibrium concentration establishment time corresponding to this temperature. The obtained values of temperature and density allow us to estimate the time to establish equilibrium concentrations of nitric oxide (C_{NO}) and their level in the neutralizer. These estimates can be obtained from the following relations [7]:

$$\frac{d\eta c_{NO}}{dt} = k\eta^2 \cdot \left[(c_{NO})^2 - c_{NO}^2\right],$$

$$\eta = \rho/\rho_0, \tag{1}$$

$$(c_{NO}) \approx 4.619 \cdot [c_{N_2} \cdot c_{O_2}]^{0.5} \cdot \exp(-10750/T)$$

Based on (1), the characteristic time to establish an equilibrium concentration (in seconds) is:

$$\tau_{NO} = [\mathrm{k} \cdot (\mathrm{C_{NO}})\eta]^{-1} = 2.06 \cdot 10^{-12} \exp(53760/\mathrm{T})/(\eta \cdot \mathrm{C_{N2}})^{0.5}. \tag{2}$$

It should be noted that the time to establish equilibrium concentrations of nitric oxide significantly depends on the ratio of densities η. Since the density of the flow in the neutralizer is small, this leads to an increase in the settling time. For the considered variants of the flow in the path (η = 0.03–0.05), this time increases by approximately five times as compared with the case of normal air density. Table 2 shows data on changes in the concentration C_{NO} and rate G_{NO} for three values of the coaxial jets rates. The change in the concentration of nitric oxide in the neutralizer was calculated by formulas (1), which took into account the resulting actual density of the gas.

Table 2. Nitric oxide concentration and rate.

G_{air}, kg/s	C_{NO}, inlet	G_{NO}, inlet, kg/s	C_{NO}, neutralizer inlet	C_{NO}, outlet	G_{NO}, outlet, kg/s	ΔG_{NO}, %
0.5	0.05	0.16	0.043	0.040	0.148	−6.77
1.0	0.05	0.16	0.038	0.029	0.123	−22.50
2.0	0.05	0.16	0.031	0.030	0.158	−0.43

From the obtained results, we can conclude that it is possible to select such a ratio of the working gas of the electric arc heater and cold air supplied coaxially at the exit section of the supersonic nozzle so that the average mass temperature at the exit of the supersonic diffuser does not exceed 2500–2700 K and the residence time of the gas in the neutralizer does not exceed the time to establish an equilibrium concentration. In calculations, when the flow rate of the working gas is G_{in} = 3.19 kg/s and cold air is G_{air} = 1 kg/s, the temperature in the neutralizer is ~2500–2600 K, the residence time of the gas is ~0.01 s, and the time to establish the equilibrium concentration is ~0.011 s. In this variant (with an initial NO concentration of 0.05), the NO emission was reduced by ~1.3 times. The optimal selection of rates ratios can reduce the temperature in the neutralizer by 100–150°, while maintaining an acceptable ratio of gas transit time and the time to establish an equilibrium concentration. Its value may decrease, for example, from 0.029 to 0.02, which will give a decrease in NO ~2 times. If the initial concentration of nitric oxide is ~0.1, then according to the proposed scheme, the reduction in NO emissions can be more than 3 times.

As part of the chemical method of neutralizing nitric oxide, methane combustion in a neutralizer was simulated for different values of the initial concentration of nitric oxide at the inlet to the neutralizer, the pressures in the neutralizer and the degree of cooling of the inlet stream.

Simulation of the flow with methane combustion is carried out in the computational domain, which occupies a part of the neutralizer volume. As the boundary conditions in the inlet section, the flow parameters obtained when calculating the flow in the entire path are used. Components CH_4, H_2O, CO_2 are added to the existing components O_2, NO, O, N, N_2 of the mixture, and the concentration of methane in the inlet section is set with a slight excess over stoichiometry.

The following reactions were considered in the gas:

$$\begin{gathered} \mathrm{O} + \mathrm{N_2} \leftrightarrow \mathrm{N} + \mathrm{NO} \\ \mathrm{N} + \mathrm{O_2} \leftrightarrow \mathrm{O} + \mathrm{NO} \\ \mathrm{CH_4} + 2 \cdot \mathrm{O_2} = \mathrm{CO_2} + 2 \cdot \mathrm{H_2O} \end{gathered} \tag{3}$$

Table 3 presents the results of calculations of changes in the concentration and rate of nitric oxide during the flow of gas through the neutralizer. If there is a heat exchanger in the neutralizer, as well as other objects that impede the flow, most of the calculations were carried out taking into account a possible decrease in the gas temperature and an increase in pressure. The specified values of pressure and temperature are given in the fourth and fifth columns, respectively. In addition, part of the calculations was carried out for the initial concentration of NO which was almost doubled. It should be noted that the five reactions (3) do not exhaustively describe all the chemical transformations occurring in the stream. For example, does not take into account the possible dissociation of a number of components of the mixture. On the one hand, this can lead to a decrease in the outlet temperature, on the other hand, to a less significant drop in the concentration of nitric oxide.

Table 3. Nitric oxide concentration and rate, G_{in} = 3.2 kg/s, G_m = 0.19 kg/s.

No.	T inlet, K	P outlet, kPa	C_{NO}, inlet	G_{NO}, inlet, kg/s	C_{NO}, outlet	G_{NO}, outlet, kg/s	ΔG_{NO}, %	T outlet, K
1	2200	34	0.05	0.166	0.028	0.09	−42.7	3849
2	2200	50	0.05	0.166	0.014	0.05	−71.5	3938
3	1700	50	0.05	0.166	0.023	0.08	−53.2	3405
4	1200	50	0.05	0.166	0.034	0.11	−31.5	2878
5	2200	50	0.09	0.318	0.023	0.08	−75.2	4144
6	1700	50	0.09	0.318	0.040	0.13	−57.4	3582
7	1200	50	0.09	0.318	0.062	0.21	−34	2985

From Table 3 it follows that increasing the pressure leads to a greater decrease in the concentration of NO (cf. calculations No. 1 and No. 2). This happens both by increasing the density of the medium, and by reducing the flow rate and, consequently, increasing the residence time of the gas in the neutralizer. Lowering the temperature naturally reduces the rate of reactions (cf. calculations No. 2 and No. 3). Calculations also show that practically all molecular oxygen and most of methane have time to react in a narrow region near the inlet section. A further decrease in the concentration of methane occurs when it reacts with oxygen formed during the decomposition of nitric oxide in the second reaction (3). Atomic oxygen for this reaction is formed during the decomposition of NO in the first reaction (3). As noted above, depending on the gas temperature, this scheme can only approximately describe the processes occurring in it.

In the framework of this approximation and the options considered, the emission of nitric oxide can be reduced by about 4 times. Table 4 shows the calculation results for options No. 1, 2, 6 of Table 3 with double the length of the neutralizer, from which it follows that the emission of nitric oxide can be reduced even more. However, it should be borne in mind that along with a decrease in the content of nitric oxide when methane is burned at a temperature above 2000 K, the content of carbon monoxide increases. Therefore, when using a chemical approach with a temperature in the neutralizer above 2500 K, the gases leaving the neutralizer are afterburned.

Table 4. Nitric oxide concentration and rate (with double the length of the neutralizer), G_{in}= 3.2 kg/s, G_m= 0.19 kg/s.

No.	*T* inlet, K	*P* outlet, kPa	C_{NO}, inlet	G_{NO}, inlet, kg/s	C_{NO}, outlet	G_{NO}, outlet, kg/s	ΔG_{NO}, %	*T* outlet, K
9	2200	34	0.05	0.166	0.015	0.052	−68.5	3926
10	2200	50	0.05	0.166	0.004	0.015	−91	3996
11	1700	50	0.09	0.318	0.019	0.063	−80.2	4152

4 Conclusion

Using the methods of mathematical and numerical modeling in the framework of the kinetic approach, the dependences of the concentrations of nitrogen oxides at the outlet of the neutralizer on the flow rate of the coaxial jet are obtained. It is shown that with optimal selection of the ratio of flow rates and temperatures of the central and coaxial jets, the content of nitric oxide in the gas at the exit from the working path of the installation can be reduced more than three times. In the framework of the chemical approach, with the distributed supply of methane to the gas stream in the neutralizer, dependences of the concentrations of nitrogen oxide at the outlet of the neutralizer on the degree of gas cooling in the neutralizer and pressure in it are obtained. It is shown that with optimal selection of the degree of cooling of exhaust gases at the inlet to the neutralizer and its pressure, the content of nitrogen oxide at the outlet of the working path of the installation can be reduced more than four times, but this raises the problem of afterburning of carbon monoxide, which requires installation in gas tract additional equipment.

Acknowledgement. The research was partially carried out within the Program of Fundamental Scientific Research of the state academies of sciences in 2013–2020 (project No. AAAA-A17-117030610120-2).

References

1. U.S. Environmental Protection Agency, Nitrogen Oxides (NOx) Why and How They Are Controlled. EPA-456/F-99-006R, November 1999
2. Chameides, W.L., Stedman, D.H., et al.: NOx production in lightning. J. Atmos. Sci. **34**, 143–149 (1977)
3. Lefebvre, A.H., Ballal, D.R.: Gas Turbine Combustion: Alternative Fuels and Emissions, 3rd edn, pp. 359–440. CRC Press, Boca Raton (2010)
4. Zel'dovich, Ya.B., Sadovnikov, P.Ya., Frank-Kamenetskiy, D.A.: Okisleniye azota pri gorenii. Izd-vo AN SSSR (1947). (in Russian)
5. Trusov, B.G.: Program system TERRA for simulation phase and chemical equilibrium. In: Proceedings of the 14 International Symposium on Chemical Thermodynamics, St-Petersburg, pp. 483–484 (2002)
6. Computational fluid dynamics (CFD) code ANSYS Fluent, academic version 15.0 (2013)
7. Rayzer, U.P.: Obrazovaniye okislov azota v udarnoy volne pri sil'nom vzryve v vozdukhe. Zhurnal fizicheskoy khimii **33**(3), 700–709 (1959). (in Russian)

Background Technology of Finish-Strengthening Part Processing in Granulated Actuation Media

Mikhail Tamarkin(✉), Elina Tishchenko, Alexander Melnikov, and Evgeny Chernyshev

Don State Technical University, Gagarin sq., Rostov-on-Don 344000, Russia
tehn_rostov@mail.ru

Abstract. The research results of the finishing and hardening processes in granular media are given. Single interaction of the medium particles and the workpiece surface is considered. The parameters of a single trace, which is an ellipsoid, are estimated. The dependence for determining the maximum depth of the medium particle penetration into the part surface is given. The surface roughness is studied. The dependence for calculating the arithmetic average surface roughness under the finish-strengthening treatment in granular media, considering the effect of the process fluid on the treatment, is determined. The estimation of the depth of the hardened layer and the deformation ratio is studied. The dependences enable to describe a plastically deformed area extending to a certain depth around the residual dent (plastic indentation), and the relationship between different hardness numbers. The process time formula, which considers the deformation features in the treatment area and reflects the phenomena occurring under the medium particle – workpiece surface contact, is developed. All the dependences obtained are tested experimentally. Their adequacy is established. The resulting set of theoretical models enables to create the mathematical support for the preproduction planning of optimal design solutions. Engineering guidelines are given on the development of a CAD system for part processing in the granular actuation media.

Keywords: Finish-strengthening treatment in granular actuation media · Surface condition · Computer-assisted design

1 Introduction

Methods of finishing and strengthening part processing in granular media (FSP GM) are widely used in the modern engineering production at the final stages of the product manufacture, and they are up-to-date and advanced techniques for treatment through the surface plastic deformation (SPD). They are characterized by plastic deformation of the surface layer and the absence of chip formation. This enables to have an impact on the increase in durability of the machined parts, corrosion hardening when the part is working in an aggressive environment, and, eventually, to extend the product life cycle.

V. Murgul and M. Pasetti (Eds.): EMMFT-2018, AISC 983, pp. 661–669, 2019.
https://doi.org/10.1007/978-3-030-19868-8_64

2 Materials and Methods

According to GOST 18296-72 [1], all SPD methods are divided into static and dynamic by the applied load. The present study is devoted to the dynamic SPD methods through which the treatment is performed in an environment of free bodies (most often they are polished steel balls), which perform a repeated applied shock on the work surface while leaving a single trace. Both the entire surface of the part and its particular areas can be processed. It may be exemplified by the methods of vibration finishing-strengthening treatment, centrifugal-rotational processing in the medium of steel balls, treatment by shot, etc.

Herewith, the increase in fatigue and durability of the machined part, its wear resistance and contact stiffness caused by the formation of compressive residual stresses in the surface layer of the machined part, the growth of microhardness, as well as by the depth and deformation rate, raise competitive capacity of products obtained through SPD processing in granular media as compared with other finishing and hardening treatment techniques. Thus, it can be argued that the studies conducted are relevant and reasoned.

When analyzing the phenomena in the machining area, it is necessary to consider the process of a single interaction of the particle of the hardening medium and the work surface. We make some assumptions: the ball, moving at α angle to the work surface, penetrates into it and covers a small distance. An imprint is formed, which is an ellipsoid whose semi-axis (a is large and b is small) can be determined by the following dependences [2]:

$$b = \sqrt{R^2 - (R - h_{max})^2}, \tag{1}$$

$$a = \frac{\pi}{2} \cdot (ctg\alpha - f) \cdot h_{\max} + b, \tag{2}$$

where R is the balls' radius; $h_{\max}$ is maximum depth of penetration of the balls into the workpiece surface; α is the angle of ball - workpiece surface interaction; f is the friction coefficient of the ball sliding on the workpiece material. During the implementation of the SPD process, the superposition of single elliptical prints occurs, and the surface roughness profile is formed. To simplify the interaction scheme, it is assumed that the diameters of all the balls are the same. Only the ball gliding over the surface during the penetration excluding possible rolling is considered; only the average probabilistic values of various parameters of a single interaction are analyzed.

The maximum depth of penetration of the balls into the workpiece surface is determined as follows [2]:

$$h_{max} = 2 \cdot V_{эф} \cdot R \cdot sin\alpha \cdot \sqrt{\frac{\rho_{ш}}{3 \cdot k_s \cdot c \cdot \sigma_T}}, \tag{3}$$

where $V_{эф}$ is effective impact rate of the ball and the workpiece surface; $\rho_{ш}$ is ball material density; k_s is coefficient considering the effect of the part surface roughness on the actual contact area; c is bearing capacity factor of the contact surface; σ_T is yield strength of the workpiece material. The change in the surface roughness of the machined part occurs from the initial to some characteristic for this SPD method, under the specific operation modes. Such roughness is called steady, since under the repeated superpositions of single traces, it is reproduced on the surface until the occurrence of the over-peening. The operation is carried out using the process-fluid-wash. The coefficient characterizing its effect is introduced. The arithmetic average of the steady surface roughness under finishing and hardening treatment in granular media can be determined by the following formula:

$$Ra_{уст} = k \cdot k_{ТЖ} \sqrt{\frac{h_{\max} \cdot a \cdot b \cdot l_{ед}}{R^2}}, \tag{4}$$

where k is empirical coefficient; $k_{ТЖ}$ is coefficient considering the effect of the process fluid.

The analytical calculation of the depth of h_H hardened layer and ε deformation rate enables to evaluate the increase in performance properties of the machined part, such as fatigue strength and durability. The hardened layer thickness determines the area on the workpiece surface, in which there are residual deformations of the grains and dislocations of the crystal lattice, which are formed as a result of the external load application. The analytical determination of the hardened layer depth and the deformation rate depending on the mechanical-and-physical properties of the part material and the process parameters is a very difficult task, and it is considered in the papers of many researchers [2–5]. When conducting our own study, we took into account that with the transition from static loading to shock, the material resistance to penetration of the indenters, which are steel balls, increases. In this case, an increase in the level of material hardness occurs, as a result of which HD dynamic hardness appears to be greater than the plastic hardness under HD static loading. A quantitative measure of the increase in hardness under dynamic loading is η dynamic coefficient of hardness [6], which is $\eta = \frac{HD_{д}}{HD}$ ratio.

The dependence of the dynamic coefficient of hardness on the indenter penetration rate under steel processing, which allows determining the dynamic hardness of steels by their static hardness and initial impact velocity both for the case of a sphere - plane interaction, and for contacting bodies of arbitrary shape and curvature is given in Sidyakin's paper [6]:

$$\eta = 0.5\left(1 - \frac{137V_0}{HD} + \sqrt{1 + \frac{2250 \cdot V_0}{HD}}\right) \tag{5}$$

where HD is plastic hardness under static loading; Vo is initial impact velocity.

In practice, it is often necessary to convert some hardness numbers into others. This can be done strictly on the basis of empirical dependencies, which do not fully allow describing the internal relations between different numbers of hardness. In [5], the author proposes to apply one of the most frequently used dependences to translate Brinell hardness into dynamic hardness:

$$\mathrm{HB} = 0.2 \cdot \mathrm{HD}^{0.89}$$

With the presented dependence and hardness coefficient, as well as with Mises-Hencky plasticity condition [5, 6], when describing a plastically deformed area extending to a certain depth around a residual dent (plastic imprint), the following formulas are obtained for calculating the depth of the hardened layer and the deformation rate for SPD processing methods in granular media:

$$h_n = k \cdot k_{TЖ} \cdot R \cdot \left[1 - \frac{1}{2}\left(1 - \frac{b}{a}\right)^4\right] \cdot \sqrt{V_{эф} \cdot \sin\alpha} \cdot \sqrt[4]{\frac{\rho_ш}{k_s \cdot c \cdot \sigma_T}} \tag{6}$$

$$\varepsilon = k \cdot k_{TЖ} \cdot \sqrt{V_{эф}} \cdot \sqrt[4]{\frac{\rho_ш}{\eta \cdot HB^{1,12}}} \tag{7}$$

where HB is Brinell hardness.

In the development of workpart procedure, the analytical calculation of the time to reach the specified roughness, on which the production efficiency directly depends, assumes great importance. The formula for determining the machining time should consider the deformation features in the work area and reflect the phenomena that occur under the medium particle - workpiece surface contact. Such a dependency is obtained, and it is as follows [5]:

$$t = \frac{4 \cdot h_{max} \cdot F \cdot R^2}{V_S \cdot f_в} \tag{8}$$

where F is number of repeated shocks at the same point of the working surface; V_S is volume of the deformable metal with a single interaction of the ball and the working surface; $f_в$ is cycle rate of the working medium impact on the workpiece surface. According to the author's study [5], the selection of F value under the SPD processing methods in granular media depends on the material hardness. F number of repeated impacts varies from 10 to 20; large values are given to the lower hardness of the workpiece material. Considering that exceeding the reasonable processing time leads to over-peening of the surface, experimental studies were conducted to determine the time of over-peening formation on samples of various materials depending on the processing modes and working media characteristics. Further on, this allowed us to select such conditions for finishing and hardening methods in granular media, in which the required processing time is ensured not to reach the time of the over-peening formation.

3 Results

To test the adequacy of the above theoretical dependences, a significant scope of experimental research was carried out to determine the arithmetic average surface roughness, the hardened layer depth and the deformation ratio. Samples from various materials (steels, copper and aluminium alloys) using polished steel balls of different diameters as the working medium were subjected to the finishing-strengthening treatment in granular media. Process liquid was supplied to the work area. After treatment, the samples were removed, washed, and dried. Modern probing devices were used to conduct measurements.

The results of theoretical and experimental studies were compared; and the corresponding graphs, some of which are shown in Fig. 1, 2, 3, 4, 5 and 6, were plotted. The theoretical curve constructed from the above dependences is shown in solid line. Dots mark the experimentally obtained values. Confidential intervals are plotted on the graph (confidence coefficient is 95%).

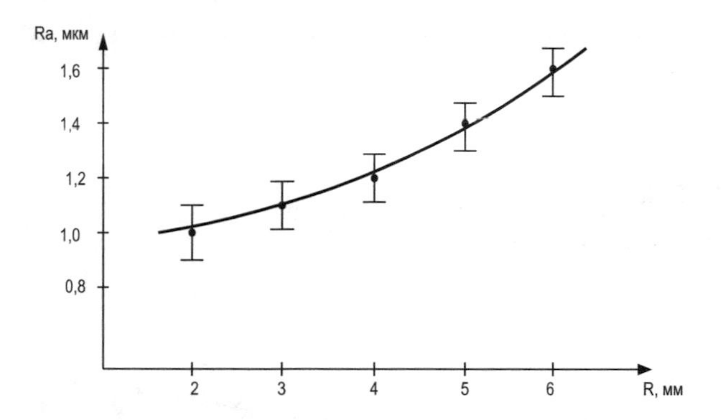

Fig. 1. Dependence of surface roughness on ball radius for vibration finishing and hardening treatment (sample from AVT aluminium alloy). Process conditions: chamber frequency is 26.7 Hz, oscillation amplitude is 0.002 m.

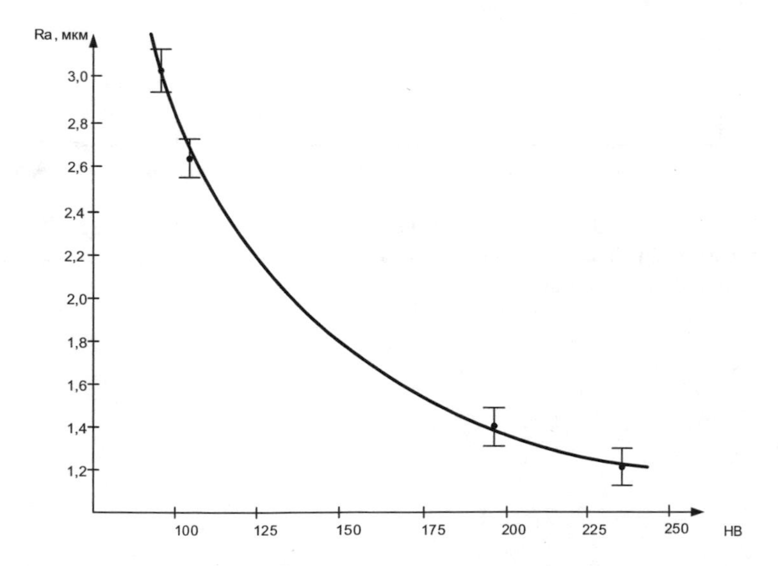

Fig. 2. Dependence of surface roughness on Brinell hardness. Process conditions: chamber frequency is 30 Hz, oscillation amplitude is 0.0025 m.

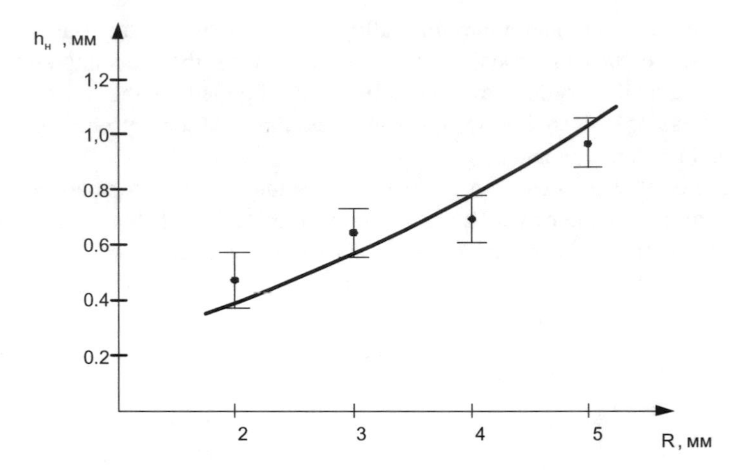

Fig. 3. Dependence of hardened layer depth on ball size for vibration finishing and hardening treatment (sample from HVG steel). Process conditions: chamber frequency is 30 Hz, oscillation amplitude is 0.002 m.

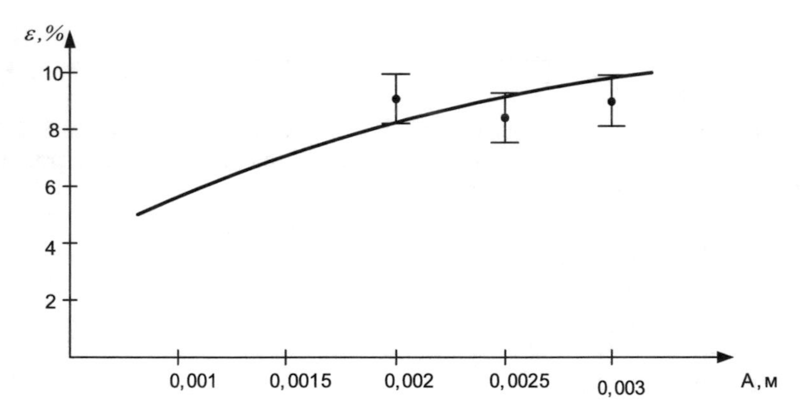

Fig. 4. Dependence of deformation ratio on chamber oscillation amplitude for vibration finishing and hardening treatment (sample from 45 steel). Process conditions: chamber frequency is 30 Hz, ball radius is 0.005 m.

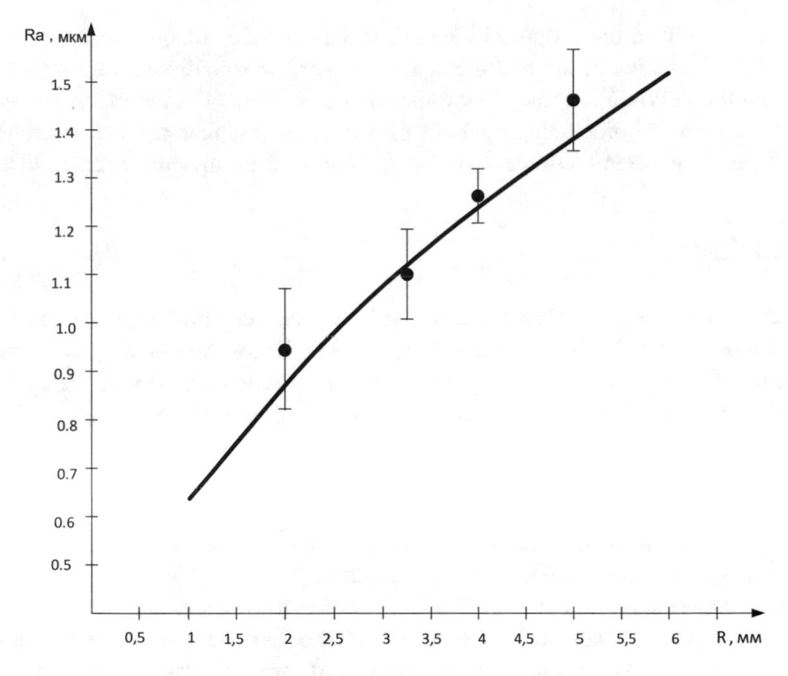

Fig. 5. Dependence of surface roughness on ball radius for centrifugal-rotational finishing and hardening treatment (sample from 12X18H10T stainless steel). Rotor speed is 10 Hz.

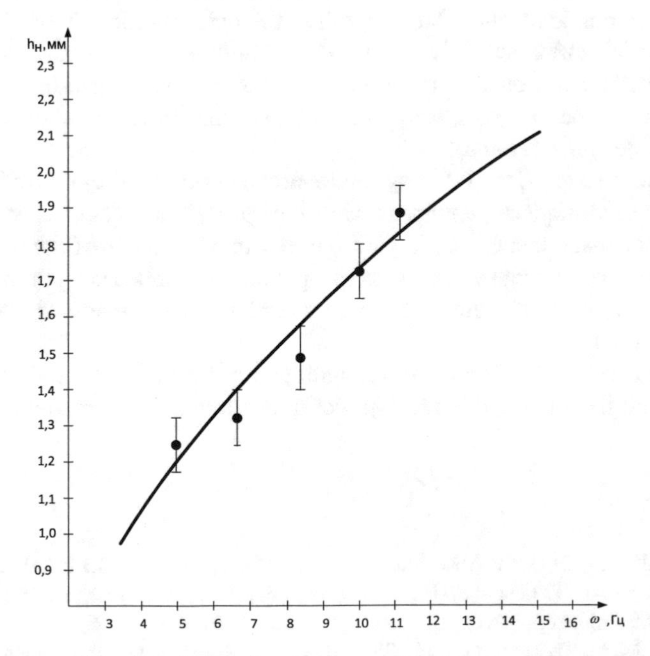

Fig. 6. Dependence of hardened layer depth on rotor speed for centrifugal-rotary finishing and hardening treatment (sample from D16 aluminium alloy). Ball radius is 0.065 m.

The adequacy of the proposed theoretical dependences of the process parameters effect on the formation of the arithmetic average surface roughness, the hardened layer depth, and the deformation rate is established. The empirical value of Fisher criterion, which for all studied processing methods proved to be less than critical, is determined. The difference between theoretical and experimental data does not exceed 20%.

4 Conclusion

The proposed set of theoretical models that have been experimentally tested for adequacy can be used as the basis for developing CAD TP for the finishing and hardening part treatment in granular working environments, which enables to design highly efficient versions of processes through optimizing operation modes and media characteristics.

Since the production methods of finishing and hardening treatment in granular media can have a large number of variants of parameter combinations leading to the desired processing results, an important task is to optimize technological solutions and to develop a design methodology using computers.

The obtained set of theoretical models enables to develop software for the production design of optimal solutions. It allows us to describe a great many technological situations, arising from various combinations of process conditions and working environment characteristics, and to select those that enable treatment with the lowest labor-output ratio or cost of price.

You can select the best processing quality, high performance, or low cost of price as the target economic function that determines the optimization criteria. The system of theoretical models presented above allows for the ranking of processing conditions and media characteristics. It enables to select those that can be defined by a unique value and those that can be set by an array to iterate over the combination of options and to form a set of design solutions.

At the initial stage of the CAD system operation, structural optimization is carried out (selection of workpiece parameters, choice of processing methods, selection of the number of processing steps, supply with the minimum number of the affected equipment). At the second stage of the system operation, parametric optimization is performed (selection of combinations of process conditions and media characteristics for partial operations).

The developed CAD TP enables to significantly increase the process design productivity of the finishing and hardening workpiece treatment in granular media.

References

1. GOST 18296-1972: USSR State Standards Committee, Moscow, 13 p (1972)
2. Tamarkin, M.A., Tishchenko, E.E., Fedorov, V. P.: IOP Conference Series: Materials Science and Engineering, vol. 124, no. 1, Article no. 012169 (2016)
3. Tamarkin, M.A., Tishchenko, E.E., Shvedova, A.S., Isaev, A.G.: Strengthening Technologies and Coatings, No. 11 (131) (2015)

4. Tamarkin, M.A., Tishchenko, E.E., Kazakov, D.V., Isaev, A.G.: Russ. Eng. Res. **37**(4), 326–329 (2017)
5. Shvedova, A.S.: Canadian Science (Engineering) Dissertation, p. 146 (2008)
6. Sidyakin, Yu.I.: Doctor of Science (Engineering) Dissertation author's abstract, p. 34 (2006)
7. Tamarkin, M.A., Tishchenko, E.E., Kazakov, D.V., Grebenkin, R.V.: VII International Science-Practical Conference, Kemerovo, pp. 186–190 (2015)
8. Tamarkin, M.A., Tishchenko, E.E., Sadovaya, I.V., Isaev, A.G.: XII International Science-Practical Conference, Rostov-on-Don, pp. 317–322 (2015)
9. Shvedova, A.S.: Vestnik of DSTU 15(1(80)), 114–120 (2015)
10. Tamarkin, M.A., Shvedova, A.S., Grebenkin, R.V., Novokreshchenov, S.A.: Vestnik of DSTU 163(3(86)), 46–52 (2016)
11. Tamarkin, M.A., Shvedova, A.S., Kazakov, D.V.: IV International Science-Technology Conference, Togliatti, pp. 150–155 (2015)
12. Shvedova, A.S., Stelmakh, A.V.: VII International Science-Technology Conference, Rostov-on-Don, pp. 7–17 (2015)
13. Shvedova, A.S., Isaev, A.G., Novokreshchenov, S.A.: XVI All-Russian Science-Technology Conference, Perm, pp. 240–244 (2015)
14. Shvedova, A.S., Golub, D.I., Razdorskiy, S.A., Rudenko, A.A.: XII International Science-Technology Conference, Rostov-on-Don, pp. 332–338 (2015)
15. Tamarkin, M.A., Tishchenko, E.E., Shvedova, A.S.: STIN (3), 26–28 (2018)
16. Tamarkin, M.A., Tishchenko, E.E., Shvedova, A.S.: Russ. Eng. Res. **38**(9), 726–727 (2018)

Characterization of Thermoelectric Generators for Cathodic Protection of Pipelines of the City Heating

Vladimir Yezhov(✉), Natalia Semicheva, Ekaterina Pakhomova, Aleksey Burtsev, Artem Brezhnev, and Nikita Perepelitsa

Southwest State University, 50 Let Oktyabrya Street, 94, 305040 Kursk, Russia
{vl-ezhov,nsemicheva,egpakhomova}@yandex.ru,
{ap_burtsev,nikperepelicza}@mail.ru,
brarvik@icloud.com

Abstract. In the scientific article, the design of an experimental thermoelectric power source for the cathodic protection station of the heat network pipeline from electrochemical corrosion is proposed, which allows utilizing low-potential heat with subsequent direct conversion to electricity in thermoelectric converters consisting of two metals different in their thermionic properties, using the effect of thermoelectricity. Thermoelectricity is the phenomenon of direct conversion of heat into electricity in conductors, as well as the reverse phenomenon of direct heating and cooling of junctions of two conductors by passing current. The transition of thermal energy into electrical energy occurs in thermoelectric converters. A thermoelectric transducer is a pair of conductors made of different materials connected at one end. Thermoelectric converters are used to produce electricity by directly converting heat into electricity. When the thermoelectric module connected to the electrical circuit is heated, electricity is generated. At the same time, the generated electric current can be used as an Autonomous power supply to the cathodic protection station. The technique of the experiment, as well as experimental studies with subsequent analysis of the main characteristics of the thermoelectric generator, was conducted. As a result, the optimal design of the thermoelectric generator for the cathodic protection station was chosen.

Keywords: Heat supply · Power generation · Energy efficiency · Pipeline · Thermoelectric material · Thermoelectricity

1 Introduction

Currently, thermoelectric modules are widely used in high-tech areas such as telecommunications, space, precision weapons, medicine, etc. Thermoelectric modules are also actively implemented in household appliances: portable refrigerators, freezers, coolers for drinking water and beverages, compact air conditioners, etc.

Thermoelectricity is the phenomenon of the direct conversion of heat into electricity in conductors as well as the reverse phenomenon of the direct heating and

V. Murgul and M. Pasetti (Eds.): EMMFT-2018, AISC 983, pp. 670–678, 2019.
https://doi.org/10.1007/978-3-030-19868-8_65

cooling of junctions of two conductors by the current passing through them. The transfer of thermal energy into electrical one occurs in thermoelectric converter. A thermoelectric converter is a pair of conductors made of different materials joint at one end. When one of the junctions of an element is heated more than the other, a thermoelectric effect occurs.

Thermoelectric converter is used to produce electricity by direct converting heat into electricity. Electricity is generated when heating a thermoelectric module connected to an electrical circuit.

According to the authors, heat supply systems, namely the recovery of low-grade heat from the heating network pipelines and its direct conversion into electrical energy to ensure the independence of the cathodic protection station in heating network pipelines for protection against electrochemical corrosion may be considered as one of the applications.

2 Research Methods

To solve this problem, an experimental setup was developed and experiments were carried out, on the basis of which a method for calculating the main characteristics of thermoelectric elements was developed [1].

The installation diagram is shown in Fig. 1.

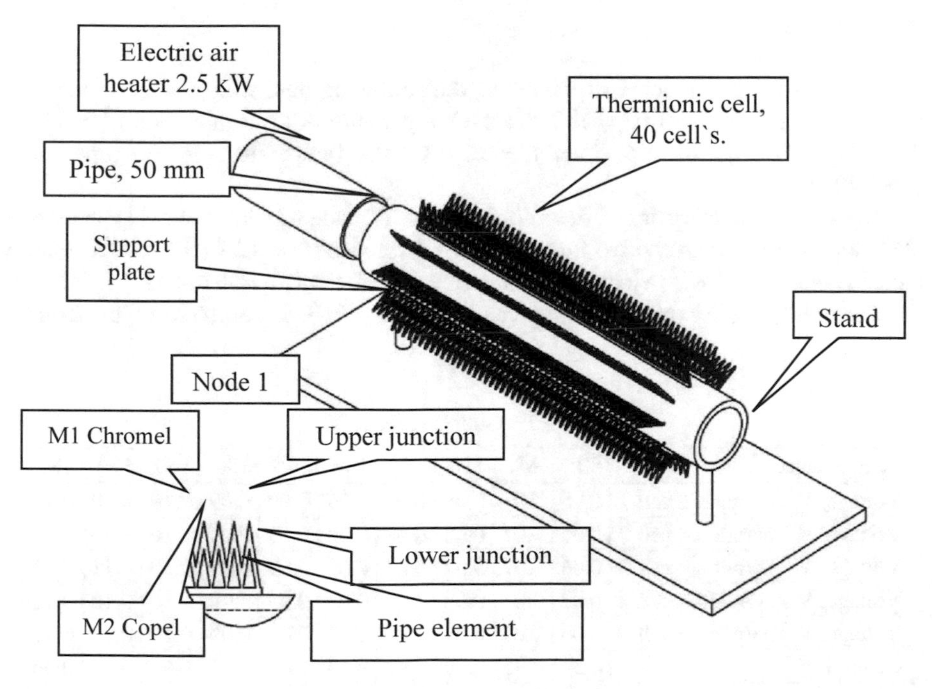

Fig. 1. Experimental setup diagram.

The heated air from the electric calorifier was used as a coolant (working medium) in the experimental installation. Experiments on the study of heat transfer between the thermoelectric section and the pipeline were carried out in the following sequence [2].

1. turning on the heater (construction Hairdryer) with a fixed flow rate of heated air, followed by heating until the steady state.
2. installation of calibrated mercury thermometers at the inlet and outlet of the thermoelectric Section.
3. record thermometer readings.
4. recording of anemometer readings relative to the velocities at the output of the thermoelectric sections.

The flow temperature was determined as the arithmetic mean value of mercury thermometer readings at the points of entry and exit from the channels of thermoelectric sections.

The thermoelectric converter was made of 400 couples of pieces made of different metals M1 and M2 (M1 – Chromel, M2 – Copel), [**Error! Reference source not found.**] joint in a zigzag manner, the ends of which were flattened and tightly pressed together and located in the area of heating and cooling, close to the outer edge and the outer surface of the pipe section; the unconnected ends of the thermoelectric sections of each zigzag row were connected to the collectors with the same charges.

3 Results

As a result of the conducted experiments, the following data were obtained: the temperature of hot junctions t_1^{hc} = 180 °C; the temperature of cold junctions t_2^{cl} = 80 °C; the temperature at the pipe inlet t_1 = 180 °C; the temperature at the pipe outlet t_2 = 140 °C.

The main characteristics of thermionic elements made of Chromel (M1) and Copel (M2) are: thermoelectromotive force (TEMF) coefficient $\alpha = 12.97 \cdot 10^{-3}$ V/K; quality factor $Z = 2.8 \cdot 10^{-3}$ K^{-1}; electrical conductivity coefficient $\sigma = 8 \cdot 104$ Ω^{-1} m^{-1}.

The results of the experiment are given in Tables 1, 2, 3, 4, 5 and 6 and Figs. 2 and 3.

Table 1. Output voltage, V

Temperature, °C	80	90	110	120	130	140	150	160	170	180
Voltage, V Thermionic cell 1	0.06	0.07	0.08	0.08	0.09	0.09	0.09	0.10	0.10	0.11
Voltage, V Thermionic cell 2	0.06	0.07	0.08	0.08	0.09	0.09	0.09	0.10	0.10	0.11
Voltage, V Thermionic cell 3	0.06	0.07	0.07	0.08	0.09	0.09	0.09	0.10	0.11	0.12
Voltage, V Thermionic cell 4	0.06	0.07	0.08	0.08	0.09	0.09	0.09	0.10	0.10	0.11
Voltage, V Thermionic cell 5	0.06	0.07	0.07	0.08	0.08	0.09	0.09	0.10	0.10	0.11
General, V	0.32	0.35	0.39	0.41	0.43	0.46	0.48	0.51	0.53	0.56

Table 2. Amperage, mA

Temperature, °C	80	90	110	120	130	140	150	160	170	180
Amperage, mA Thermionic cell 1	11.2	12.4	13.8	15.3	17.0	17.9	18.9	19.9	20.9	22
Amperage, mA Thermionic cell 2	11.2	12.4	13.8	15.3	17.0	17.9	18.9	19.9	20.9	22
Amperage, mA Thermionic cell 3	12.7	14.1	15.7	17.4	19.3	20.4	21.4	22.6	23.8	25
Amperage, mA Thermionic cell 4	12.2	13.5	15.0	16,.7	18.6	19.5	20.6	21.7	22.8	24
Amperage, mA Thermionic cell 5	12.2	13.5	15.0	16.7	18.6	19.5	20.6	21.7	22.8	24
General, mA	59.4	66.0	73.3	81.5	90.5	95.3	100.3	105.6	111.2	117

Table 3. Output voltage, V

Temperature, °C	80	90	110	120	130	140	150	160	170	180
Voltage, V Thermionic cell 1	0.07	0.08	0.08	0.09	0.09	0.10	0.10	0.11	0.11	0.12
Voltage, V Thermionic cell 2	0.07	0.08	0.08	0.09	0.09	0.10	0.10	0.11	0.11	0.12
Voltage, V Thermionic cell 3	0.06	0.06	0.07	0.07	0.08	0.08	0.09	0.09	0.10	0.1
Voltage, V Thermionic cell 4	0.06	0.06	0.07	0.07	0.08	0.08	0.09	0.09	0.10	0.1
Voltage, V Thermionic cell 5	0.05	0.05	0.06	0.06	0.06	0.07	0.07	0.07	0.08	0.08
General, V	0.29	0.33	0.36	0.38	0.40	0.42	0.45	0.47	0.49	0.52

Table 4. Amperage, mA

Temperature, °C	80	90	110	120	130	140	150	160	170	180
Amperage, mA Thermionic cell 1	14.2	15.8	17.5	19.5	21.7	22.8	24.0	25.3	26.6	28
Amperage, mA Thermionic cell 2	15.2	16.9	18.8	20.9	23.2	24.4	25.7	27.1	28.5	30
Amperage, mA Thermionic cell 3	10.2	11.3	12.5	13.9	15.5	16.3	17.1	18.1	19.0	20
Amperage, mA Thermionic cell 4	10.2	11.3	12.5	13.9	15.5	16.3	17.1	18.1	19.0	20
Amperage, mA Thermionic cell 5	12.2	13.5	15.0	16.7	18.6	19.5	20.6	21.7	22.8	24
General, mA	61.9	68.8	76.5	85.0	94.4	99.4	104.6	110.1	115.9	122

Table 5. The total thermoelectric voltage, V

Temperature, °C	80	90	110	120	130	140	150	160	170	180
The total voltage, V	0.43	0.48	0.53	0.56	0.59	0.62	0.65	0.69	0.72	0.76

Table 6. The total amperage, mA

Temperature, °C	80	90	110	120	130	140	150	160	170	180
The total amperage, mA	115.2	128.0	142.3	158.1	175.6	184.9	194.6	204.9	215.7	227

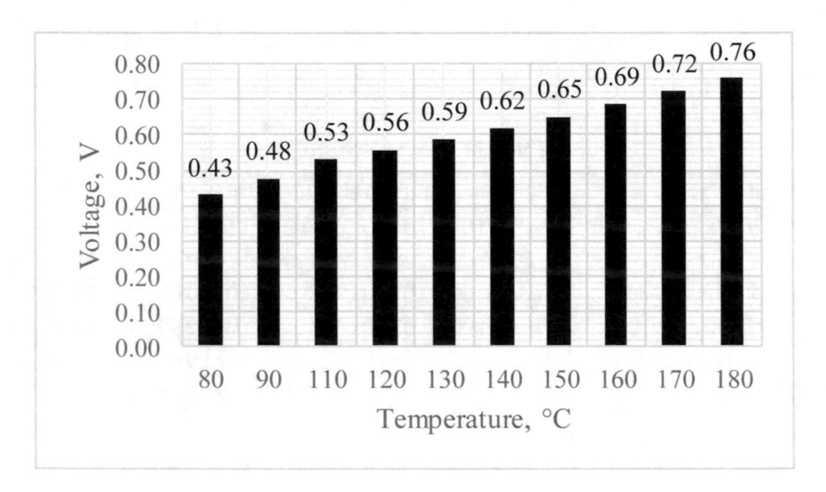

Fig. 2. Voltage versus temperature of heated pipe surface curve

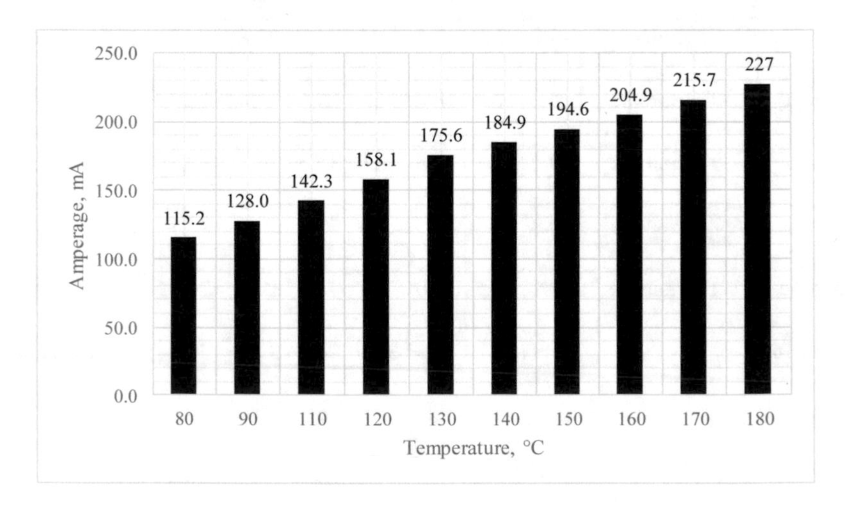

Fig. 3. Amperage versus temperature of heated pipe surface curve

4 Discussion of Results

The experimental results [3] show that with increasing hot junctions temperature, voltage and current parameters increase. [4] Determination of the main characteristics of the thermoelectric EMF source:

Average heat flow temperature, °C:

$$\Delta t = (t_1 + t_2)/2 \tag{1}$$

$$\Delta t = (180 + 273 + 140 + 273)/2 = 433\ ^\circ\mathrm{C}$$

Density of air, kg/m^3:

$$p_{hc} = P_{hc}/(287.4\ \Delta t) \tag{2}$$

$$p_{hc} = 101325/(287.4 \cdot 433) = 0.814\ \mathrm{kg/m^3}$$

Air heat capacity, J/(kg K):.

$$Cp_{hc} = \left[1.0005 + 1.1904 \cdot 10^{-4} \cdot (\Delta t - 273)\right] \cdot 10^3 \tag{3}$$

$$Cp_{hc} = \left[1.0005 + 1.1904 \cdot 10^{-4} \cdot (433 - 273)\right] \cdot 10^3 = 1019.5\ \mathrm{J/(kg\ K)}$$

Air thermal conductivity, W/(m K):

$$\lambda_{hc} = 2.44 \cdot 10^{-2} \cdot (\Delta t/273)^{0.82} \tag{4}$$

$$\lambda_{hc} = 2.44 \cdot 10^{-2} \cdot (433\,/\,273)^{0.82} = 0.0356\ \mathrm{W/(m\ K)}$$

Air dynamic viscosity coefficient, Pa s:

$$\mu_{hc} = 1.717 \cdot 10^{-5} \cdot (\Delta t/273)^{0.683} \tag{5}$$

$$\mu_{hc} = 1.717 \cdot 10^{-5} \cdot (433/273)^{0.683} = 2.35 \cdot 10^{-5}\mathrm{Pa\ s}$$

Air he kinematic viscosity coefficient, m^2/s:

$$\upsilon_{hc} = \mu_{hc}/p_{hc} \tag{6}$$

$$\upsilon_{hc} = 2.35 \cdot 10^{-5}/0.814 = 2.88 \cdot 10^{-6}\,\mathrm{m^2/s}$$

Thermal diffusivity coefficient, m^2/s:

$$\alpha_{hc} = \lambda_{hc}/(Cp_{hc} \cdot p_{hc}) \tag{7}$$

$$\alpha_{hc} = 0.03256/(1019.5 \cdot 0.0814) = 0.43 \cdot 10^{-6}\,\mathrm{m^2/s}$$

Prandtl number:

$$Pr_{hc} = \upsilon_{hc}/\alpha_{hc} \quad (8)$$

$$Pr_{hc} = 2.88 \cdot 10^{-6}/0.43 \cdot 10^{-6} = 0.674$$

Reynolds criterion:

$$Re_{hc} = (\omega_{hc} \cdot D)/\upsilon_{hc} \quad (9)$$

$$Re_{hc} = \left(8 \cdot 59 \cdot 10^{-3}\right)/2.88 \cdot 10^{-6} = 16334$$

$Re_{hc} > 2320$ is the turbulent flow.
Nusselt criterion:

$$Nu_{hc} = 0.037 \cdot Re_{hc}^{0.8} \cdot Pr_{hc}^{0.43} \quad (10)$$

$$Nu_{hc} = 0.037 \cdot 16334^{0.8} \cdot 0.674^{0.43} = 73.3$$

Heat transfer coefficient, W/(m^2 K):

$$k = (Nu_{hc} \cdot \lambda_{hc})/D \quad (11)$$

$$k = (73.3 \cdot 0.0356)/59 \cdot 10^{-3} = 44.2\,\mathrm{W/\left(m^2\,K\right)}$$

Amount of heat perceived by hot junctions, W:

$$Q_h = F_h \cdot k \cdot \left(t_1^{hc} - t_2^{hc}\right) \quad (12)$$

$$Q_h = 0.02 \cdot 44.2 \cdot (453 - 413) = 33.3\,\mathrm{W}$$

To calculate a thermoelectric generator [7], it is first necessary to determine the auxiliary thermoelectromotive force coefficient, [8] which depends on the quality factor of the metals used to make a thermoelectric generator and the total resistance of thermoelectric sections [9–11].

1. Auxiliary coefficient:

$$m = \left(1 + 0.5 \cdot Z \cdot \left(t_1^{hc} - t_2^{cl}\right)\right)^{0.5} \quad (13)$$

$$m = \left(1 + 0.5 \cdot 2.8\,10^{-3} \cdot (453 - 353)\right)^{0.5} = 1.068$$

2. Thermionic converter resistance, Ohm:

$$R = \left(\alpha \cdot \left(t_1^{hc} - t_2^{cl}\right) \cdot N\right) / \left(I \cdot \left[\left(1 + 0.5 \cdot Z \cdot \left(t_1^{hc} - t_2^{cl}\right) - 1\right)^{0.5}\right]\right. \qquad (14)$$

$$R = \left(12.96 \cdot 10^{-3} \cdot (453 - 353) \cdot 400\right) / \left(0.227 \cdot \left[\left(1 + 0.5 \cdot 2.8\,10^{-3} \cdot (453 - 353) - 1\right)^{0.5}\right]\right.$$

$$R = 6108.2\ \Omega$$

3. Thermocouple resistance, Ohm:

$$R_0 = R/N \qquad (15)$$

$$R_0 = 6108.2/400 = 15.27\ \Omega$$

4. Electrical power delivered to the external circuit, W:

$$Q = \left[\left((2 \cdot N \cdot \alpha)^2 \cdot \left(t_1^{hc} - t_2^{cl}\right)\right)/R\right] \cdot \left(m/(m+1)^2\right) \qquad (16)$$

$$Q = \left[\left((2 \cdot 400 \cdot 12.97\,10^{-3})^2 \cdot (453 - 353)\right)/6108.2\right] \cdot \left(1.068/(1.068 + 1)^2\right)$$

$$Q = 0.44\ \mathrm{W}$$

5. Efficiency, %:

$$\eta = Q/Q_h \qquad (17)$$

$$\eta = 0.44/33.3 = 1.32\%$$

Cathodic protection station NGK-IPPCZ-Euro-0.2 (24) (oil-gas play) was used as an example [10]:

- rated output voltage is 24 V;
- rated output current is 8 A;
- power consumption is 226 W.

Pipeline diameter is 300 mm. To ensure independence of the cathodic protection station, 26.545 pcs of thermoelectric elements of the studied thermoelectric EMF source (couple material is Chromel – Copel) are needed; the length of the covered pipeline is 4.2 m, the heat exchange area is 0.13 m^2.

As a result of studying the corresponding scientific literature, it was found that when creating a thermoelectric generator, it is possible to use other couples, for example, iron – constantan with the thermoelectric coefficient of 15–17 mV/°C, tungsten – rhenium with the thermoelectric coefficient of 34 mV/°C, platinum-rhodium – platinum with the thermoelectric coefficient of 11–14 mV/°C [5]; by using these couples, it is possible to significantly reduce the dimensions of the thermoelectric EMF source.

5 Conclusions

- The design of an independent thermoelectric EMF source for a cathodic protection station is proposed;
- Data obtained in the experiments conducted using the laboratory setup for thermoelectricity generation are given;
- Calculation equations to determine the main characteristics of a thermoelectric EMF source are proposed;
- The proposed design of the EMF source allows adjusting the parameters of the current depending on the flow rate and temperature of the pumped gas (liquid);
- The experiment and calculation results show that the use of the thermoelectric effect makes it possible to provide an independent power supply of the cathodic protection station, which increases the reliability and effectiveness of the protection of pipelines against corrosion.

References

1. Guo, J.Q., et al.: Development of skutterudite thermoelectric materials and modules. J. Electron. Mater. **41**(6), 1036–1042 (2012)
2. Ezhov, V., et al.: Investigation of technical characteristics of thermoelectric add-on for pressure jet burners. J. Appl. Eng. Sci. **14**(4), 461–464 (2016)
3. Ali Bashir, M.B., et al.: A review on thermoelectric renewable energy: principle parameters that affect their performance. Renew. Sustain. Energy Rev. **30**, 337–355 (2013)
4. Bartholomé, K., et al.: Thermoelectric modules based on half-heusler materials produced in large quantities. J. Electron. Mater. **43**(6), 1775–1781 (2014)
5. Ezhov, V.S., et al.: Independant power supply source for the station of cathodic protection of pipelines against corrosion. J. Appl. Eng. Sci. **15**(4), 501–504 (2017)
6. Salvador, J.R., et al.: Conversion efficiency of skutterudite-based thermoelectric modules. Phys. Chem. Chem. Phys. **16**(24), 12510–12520 (2014)
7. Schmechel, R., et al.: Concepts for medium-high to high temperature thermoelectric heat-to-electricity conversion: a review of selected materials and basic considerations of module design. Transl. Mater. Res. **2**(2), 025001 (2015)
8. Shi, X., et al.: Multiple-filled skutterudites: High thermoelectric figure of merit through separately optimizing electrical and thermal transports. J. Am. Chem. Soc. **133**(20), 7837–7846 (2011)
9. Yezhov, V., et al.: Direct heat energy conversion into electrical energy: An experimental study. J. Appl. Eng. Sci. **13**(4), 265–270 (2015)

Substantiation of Technological Solutions for the Repair of the Anodic Tank Grounding

Aleksandr Tarasenko[1](✉), Petr Chepur[1], and Alesya Gruchenkova[2]

[1] Industrial University of Tyumen, Volodarskogo Street 38, 625000 Tyumen, Russia
a.a.tarasenko@gmail.com
[2] Surgut Oil and Gas Institute, Entuziastov Street 38, 628405 Surgut, Russia

Abstract. The article reviews the features of applying the technology of repairing the bottom and anodic earthing of a vertical steel cylindrical tank without dismantling the floating roof during its reconstruction. The main non-trivial task of this project was to carry out work inside the tank in the cramped conditions of the sub-pontoon space. To replace the bottom of the tank, along with the extended anodic earthing devices, located at a depth of 0.9 m from the surface, a technological corridor free from floating roof racks was proposed, as well as a technological corridor for moving dismantled materials, soil, new bottom sheets towards the installation opening in the wall. As part of the proposed installation diagram, the floating roof was in the installation position provided for by the tank construction project with the provision of a 2.2 m high sub-pontoon space. To solve the problem of ensuring the strength of the structure, a finite element model of a floating roof was built in the ANSYS software package, which allows it to be analyzed at different types of mounting loads. The results obtained using the model formed the basis of the adopted design decisions. Using FEM [4–6, 8–10, 12–15], the design of a temporary reinforcement frame was theoretically justified and proposed, which ensured a fivefold safety margin sufficient for construction and installation works.

Keywords: Tank · Repair · Floating roof

1 Introduction

In this paper, the authors consider the technological aspects of the repair of the bottom and anodic earthing of a vertical steel tank RVSPK-20000 m^3 (Fig. 1) without dismantling the floating roof; the tank was removed out of service for repair in 2014 based on the results of a full technical diagnostics (Fig. 2).

The operating conditions were complicated by the fact that the facility is located 20 km away from the Caspian Sea (Republic of Kazakhstan), where air and soil have high salt content, and, consequently, the corrosion rate of metal structures increases significantly [1–6].

In the course of technical diagnostics of the tank, full-scale magnetic scanning and ultrasonic testing were carried out in accordance with the requirements of the current

V. Murgul and M. Pasetti (Eds.): EMMFT-2018, AISC 983, pp. 679–685, 2019.
https://doi.org/10.1007/978-3-030-19868-8_66

Fig. 1. Vertical steel tank with floating roof.

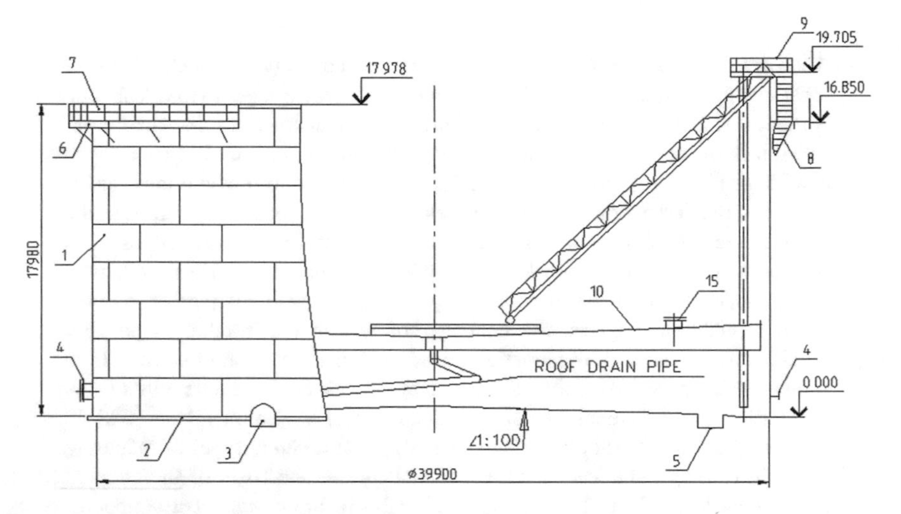

Fig. 2. General view of the tank RVSPK-20000 m^3: 1 – wall; 2 – annular plate and bottom; 3 – cleanout box; 4 – pipe branches and hatches on the wall; 5 – cleanout sump; 6 – wind girder; 7 – fencing and roofing; 8 – stairs; 9 – transition to a rolling ladder; 10 – floating roof; 11 – backing; 12 – rolling ladder.

regulatory documentation. The diagnostic results showed significant corrosion damage over the entire bottom area and the tank annular plate on the ground side, and faults (breakage) of 70% of graphite anodic earthing devices located under the tank bottom were also found. Upon selective opening of the central part of the bottom, corrosion damage to the metal was confirmed. The project documentation for the construction of this tank does not provide for anti-corrosion protection of metal in the central part of the bottom from the ground side. When opening the faulty anodic earthing devices in selective pits, it was found that the wear of the anodic earthing devices does not exceed 5%, and their failure is due to a break in the electrical contact of the cable in the field connection to the anodic earthing device with contact melting.

It could happen due to the excess of the protection current parameters specified for such a connection. Opening the pits also made it possible to establish the salinity of the sand bottom of the tank base. Damage to the remaining elements of the tank, such as a double-deck steel floating roof, wall, foundation, roof pillars, and process piping in general, were not significant and corresponded to the life of the tank. The base of the tank is a reinforced concrete bowl on piles. Inside the bowl is a leakage monitoring system, extended anodic earthing devices of the ECP and soil backfill. After analyzing the results of the diagnostic examination, the following conclusions were made:

- the ECP system of the tank bottom failed due to excessive protection current relative to design and factory parameters of the anodic earthing devices;
- an increase in the protection current of the anodic earthing devices is due to the lack of corrosion protection (hydrophobic layer) of the tank bottom from the ground side, which sharply reduced the resistivity of the protected structure and, as a consequence, increased the protection current.

During the operation of the tank, the services revealed the presence of water in the "bowl" of the tank base in the soil pores of the backfill that was entering the system of catching oil leaks when the tank is filled.

According to the results of the full technical diagnostics of the tank, it was necessary to replace the valve gate and water discharge of the floating roof.

The results of the survey of the tank foundation paving also showed the need for its replacement. Thus, for the tank repair project the Customer has formulated the following tasks:

- repair of base metal defects based on the results of full technical diagnostics; complete replacement of the central part of the bottom and annular plates; replacement of the branch pipes of the first wall belt due to the technological need when working on the replacement of the bottom annular plates; replacement of the corner weld joint strip of the first belt of the wall, 500 mm high, due to the technological necessity in the course of work on the replacement of the bottom annular plates; complete replacement of the anodic earthing devices; repair of the floating roof without dismantling; elimination of precipitation paths in the soil backfill of the RVSPK base.

Taking into account the characteristics of the tank base, it was possible to perform electrochemical protection of the RVSPK bottom against ground corrosion only with the use of the extended anodic earthing devices located under the tank bottom [4]. The main non-trivial task of this project was to carry out work inside the tank in the cramped conditions of the sub-pontoon space.

To replace the bottom of the tank, along with the extended anodic earthing devices, located at a depth of 0.9 m from the surface, a technological corridor free from floating roof racks was proposed, as well as a technological corridor for moving dismantled materials, soil, new bottom sheets towards the installation opening in the wall. At the same time, the floating roof was in the installation position provided for by the tank construction project with the provision of a 2.2 m high sub-pontoon space.

2 Experimental Part

To solve this problem, a finite element model of a floating roof was built in the ANSYS [7–11] software package, making it possible to assess its stress-strain state for various installation loads. The results obtained using the model formed the basis of the adopted design decisions described below [10, 12, 13].

Laying anodic earthing devices was carried out in a trench with a depth of 0.9 m in 2 m increments (Fig. 3). Anodic earthing devices are made with the factory anodic-to-cable connection to avoid burning the contact. To ensure the operation of the ECP in the normal mode, an insulating layer of the protected structure (bottom) with high resistivity to current spreading is required.

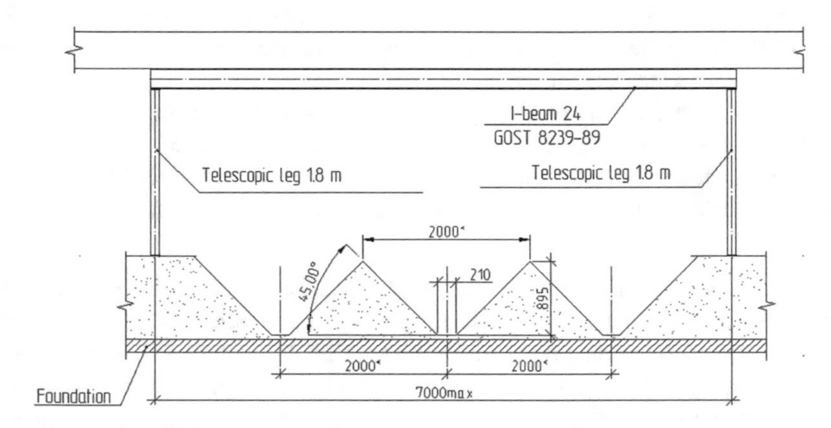

Fig. 3. Temporary structures ensuring the rigidity of the floating roof membrane.

The project provides for the installation of a hydrophobic (waterproofing) layer between the ground bed of the tank base (ground filler of the bowl) and the metal of the central part of the bottom.

When working in cramped conditions, the most appropriate as well as economically viable option for the hydrophobic layer of the central part of the bottom is a sand-bitumen mixture.

Under the central part of the RVSPK bottom, the hydrophobic layer is made of a sand-bitumen mixture of the following composition:

- sand, with a particle size from 0.1 to 2 mm - 70%;
- sandy dust and clay particles with a particle size of less than 0.1 mm - 20%;
- binding agent - 10%. Liquid petroleum bitumens and tar can be used as a binder.

The presence of acids and free sulfur in the binder is not allowed. The compaction coefficient of the hydrophobic layer should not be lower than 0.99. The thickness of the hydrophobic layer under the central part of the bottom is 50 mm. These waterproofing materials in addition to isolating ground moisture from the metal of the bottom provide the required value of resistivity to current spreading - 5·104 Ω m^2. Waterproofing of the bottom annular plates is carried out by pasting the material Gidroizol GI-G in accordance with GOST 7415-86, 2.5 mm thick.

To perform work inside the tank, the technology of the through movement of crews was specially developed, special technical devices and small-sized equipment were selected, and the project map was developed as part of the project [14]. In order to prevent deformation of the metal structures of the tank when performing work related to cutting out fragments of the central part of the bottom, it was necessary to install elements of additional rigidity - braces.

Braces were made of the corners L70x8 with a length of 1410 mm and were welded to the wall and annular plates through the backing plates.

They were installed at a pitch of 2090 mm (60 pieces in total) along the inner radius of the wall, with an offset in the area of installation openings. After installing new annular plates, the braces were removed; special attention was paid to preventing damage to the base metal of the wall [15–17].

After the installation of the braces, the part of the bottom to be cut out to make a trench for laying the ECP anodes was marked. To prevent deformation of the floating roof of the RVSPK, temporary racks and stiffness frames were used (Fig. 4).

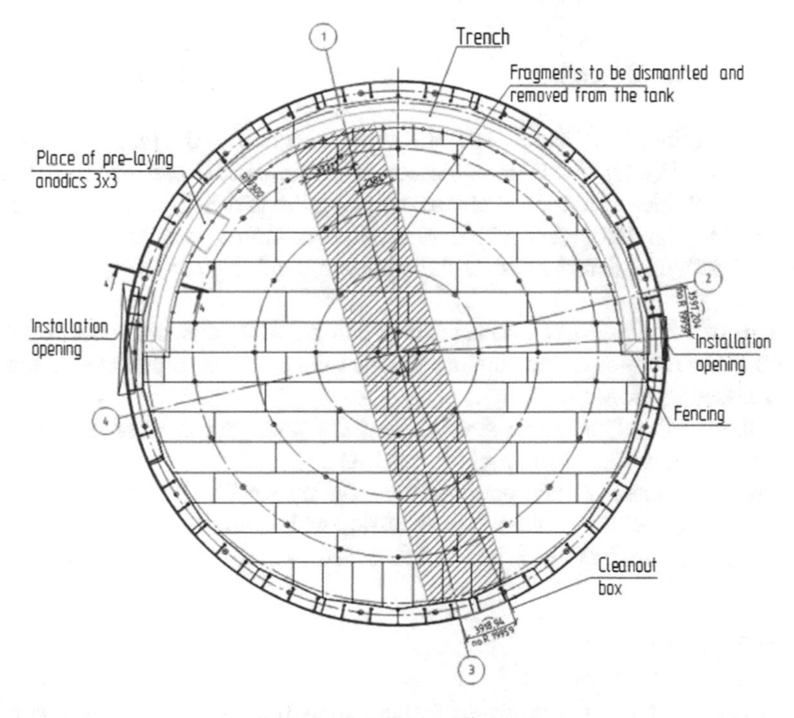

Fig. 4. Installation diagram.

A telescopic jack-up leg, allowing for a load of at least 20 kN, and an I-beam 24 was adopted as a supporting structure. The maximum span between the legs was 7 m, the construction step was not more than 3 m. The beam was mounted to the leg through a removable plug.

After the installation of the supporting elements, the support posts of the floating roof were dismantled [18, 19]. Further, cutting off the dismantled bottom fragments was made, and they were removed through the installation openings with the help of a mini-loader of the Bobcat type.

The next stage was making a trench for laying the ECP anodes. Excavation of the soil was carried out on a tarpaulin.

During the development of the trench, the section was cleared from the elements of the old ECP, and the possibility of damage to the pipelines of the leakage control system was also eliminated.

Next was laying the ECP anodes. The trench was filled with the existing soil in layers of 200 mm each with layer-by-layer compaction with manual vibrating plates. Above the compacted soil, a hydrophobic layer was made of a sand-bitumen mixture, 20 mm thick, with a slope 1:100 from the center of the tank to the wall, over the compacted soil. The supporting structures, previously installed roof support legs, were completely dismantled. After the installation of the hydrophobic layer at the work site, the layout of the rows of the central part of the bottom was made.

3 Results and Discussion

Replacement of the entire bottom was carried out similarly to the indicated technology by the grippers (Fig. 5). After the replacement of the central part of the bottom and annular plates, works were carried out on replacing the first belt of the tank wall using previously tested technologies with horizontal distribution stiffening ribs based on vertical legs to avoid deformation of the wall when cutting out fragments to replace sheets.

Sections of stiffeners were also calculated using a finite element model of the tank wall, taking into account various options for the joint action of both external loads and internal stresses in the wall.

After the repair of the main structures of the tank, the process equipment was replaced according to the Customer's instructions.

The specified repair method provided cost savings for the Customer (by 30%) and allowed reducing the repair time (up to 50%), which, under the conditions of the existing tank farm production, is a significant factor confirming its effectiveness.

4 Conclusion

The article substantiates the proposed technical solutions for the reconstruction of a steel tank with a capacity of 20000 m^3 without dismantling the floating roof during its reconstruction.

In the ANSYS software package, which implements the finite element method, a numerical elemental model of a floating roof is constructed, which makes it possible to analyze its stress-strain state under various installation loads. The results obtained became the basis for making design decisions.

Also, the design of a temporary reinforcing frame was theoretically justified and proposed ensuring a margin of safety sufficient for construction and installation works.

References

1. GOST 31385-2008. The vertical cylindrical steel tanks for oil and oil-products. General technical conditions. Standartinform, Moscow (2010)
2. RD 23.020.00-KTN-283-09. The rules of the repair and reconstruction of tanks for oil storage capacity of 1000–50000 cubic meters. OAO AK "Transneft", Moscow (2009)
3. RD-23.020.00-KTN-018-14, Main pipeline transport of oil and oil products. Vertical steel tanks for oil storage capacity of 1000–50000 cubic meters. The design standards. OAO AK "Transneft", Moscow (2014)
4. Russian Standard GOST 52910-2008, The vertical cylindrical steel tanks for oil and oil-products. General technical conditions. Standartinform, Moscow (2008)
5. Nekhaev, G.A.: Design and calculation of steel cylindrical tanks and tanks low pressure. The Association of Construction Universities, Moscow (2005)
6. Yasin, E.M., Rasshchepkin, K.E.: The upper zones stability of vertical cylindrical tanks for oil-products storage. J. Oil Ind. **3**, 57–59 (2008)
7. Korobkov, G.E., Zaripov, R.M., Shammazov, I.A.: Numerical modeling of the stress-strain state and stability of pipelines and tanks in complicated operating conditions. Nedra, St. Petersburg (2009)
8. Bruyaka, V.A., Fokin, V.G., Soldusova, E.A., Glazunova, N.A., Adeyanov, I.E.: Engineering analysis in ANSYS Workbench, Samara State Technical University, Samara (2010)
9. Beloborodov, A.V.: Evaluation of the finite element model construction quality in ANSYS. Ural State Technical University, Ekaterinburg (2005)
10. Tarasenko, A., Chepur, P., Gruchenkova, A.: Determining deformations of the central part of a vertical steel tank in the presence of the subsoil base inhomogeneity zones. In: AIP Conference Proceedings, vol. 1772, p. 060011 (2016)
11. Tarasenko, A., Gruchenkova, A., Chepur, P.: Joint deformation of metal structures in the tank and gas equalizing system while base settlement progressing. J. Procedia Eng. **165**, 1125–1131 (2016)
12. Chirkov, S.V.: Finite element model of a vertical steel tank with reinforcing elements during its lifting by hydraulic jacks. J. Fundam. Res. **9**(5), 1003–1007 (2014)
13. Chirkov, S.V.: Determination of the optimal number of cables to support the bottom when lifting the tank. J. News High. Educ. Inst. Oil Gas **5**, 72–78 (2014)
14. Bilenko, G.A.: The use of SYSWELD to study welding deformation. J. CAD Graph. **1**, 28–32 (2011)
15. Efemenkov, I.V.: Modelling and optimization of the welding metallurgical process technological parameters software SYSWELD. J. Success Modern Sci. **9**, 108–111 (2016)
16. Bilenko, G.A.: Analysis of structures residual stresses and deformations after welding working under pressure in the program SYSWELD. J. Metall. **8**, 32–34 (2012)
17. Konovalov, P.A., Mangushev, R.A., Sotnikov, S.N., Zemlyansky, A.A.: Foundations Steel Tanks and Deformation of It's Bases. The Association of Construction Universities, Moscow (2009)

Development of a Technology for Diagnosing Steel Tanks Without Removing the Protective Coating

Petr Chepur[1], Aleksandr Tarasenko[1](✉), Evgeniy Tikhanov[2], Vadim Krivorotov[2], and Alesya Gruchenkova[3](✉)

[1] Industrial University of Tyumen, Volodarskogo Street 38, 625000 Tyumen, Russia
a.a.tarasenko@gmail.com
[2] Ural Federal University Named After the First President of Russia B. N. Yeltsin, Mira Street 19, 620002 Ekaterinburg, Russia
[3] Surgut Oil and Gas Institute, Entuziastov Street 38, 628405 Surgut, Russia

Abstract. The technical and economic efficiency of the technology of conducting full technical diagnostics of vertical steel tanks (VST/RVS) without removing the anti-corrosion coating is substantiated. Requirements of the existing regulatory documentation for the diagnostic examination of large tanks are analyzed. A diagnostic complex is proposed that allows using the acoustic emission, magnetic and ultrasonic methods to carry out a comprehensive diagnostics of the VST metal structures without removing the anti-corrosion coating in the extent prescribed by the current regulatory documentation. A comparison of the technical and economic indicators of the diagnostics using traditional (with the removal of the protective coating) and innovative methods (without removing the protective coating) is presented. Since the main item of expenditure in the diagnostics of tanks is the removal, cleaning and application of the anti-corrosion coating on a large wall area, the proposed technology allows reducing the costs of operating organizations for the diagnostics of the VST up to 9 times. Based on the actual experience of implementing the technology when diagnosing the largest tank in the Russian Federation - RVSPK-100000, conclusions were made about the possibility of extending the proposed method to the tanks of the most common type: RVS-5000, RVS-10000, RVS-20000.

Keywords: Tank · Diagnostics · Non-destructive testing

1 Introduction

To ensure the conditions for reliable and safe operation of the oil pipeline transportation system, it is necessary to timely conduct and organize technical diagnostics of its main elements [1–4]. Being an integral part in the technological chain of oil pipeline transportation, large-sized vertical steel tanks must be in a trouble-free state for the entire period of operation. For this, the domestic industry regulations define the terms and extent of the regular diagnostic examination of the VST. The normative-technical

V. Murgul and M. Pasetti (Eds.): EMMFT-2018, AISC 983, pp. 686–692, 2019.
https://doi.org/10.1007/978-3-030-19868-8_67

documentation (NTD) [5, 6] defines the frequency (for a VST with a lifetime of less than and more than 20 years) and the extent (partial, full) of the mandatory technical diagnostics of tanks. Table 1 summarizes information about the frequency of the VST diagnostics.

Table 1. Frequency of technical diagnostics of vertical steel tanks.

Partial technical diagnostics	Full technical diagnostics	Lifetime
Once every five years after the construction, last diagnostics or repair	Once every ten years after the last repair or five years after the partial technical diagnostics	Up to 20 years
Once every four years after the last diagnostics or repair	Once every eight years after the last repair or four years after the partial technical diagnostics	More than 20 years

Analyzing the NTD requirements, it can be concluded that the full technical diagnostics of the tank is carried out at intervals of at least once every ten years, and partial - once every five years. The existing extent of diagnosing tanks make us think about the technical and economic aspect of conducting these tests [7].

Full technical diagnostics includes: visual inspection control (VIC), ultrasonic thickness gauging (UST), ultrasound scanning (USS), magnetic control (MC), radiographic control (RC), acoustic emission control (AEC), etc. According to [1], up to 80% of the cost of a full diagnostic examination of vertical steel tanks is associated with the removal and restoration of anti-corrosion coating.

However, the development of methods for non-destructive testing and modern equipment with the latest software [8] allow diagnostics of tanks without removing the protective coating with a given accuracy and quality according to the documentation. Figure 1 shows a tank with stripped areas of near-weld zones. Also, the photo of a real facility RVS-10000 demonstrates the process of restoring the anti-corrosion coating (ACC). The authors propose a technical complex consisting of specialized equipment and software that allows for full technical diagnostics of vertical steel tanks without compromising the integrity of the protective coating.

Within the framework of this complex are integrated:

- acoustic emission systems “Disp” and “Samos” with preamplifiers and transducers “PAC”,
- ultrasonic measuring device with phased aperture array technology (PAA) “OmniScan”,
- magnetic diagnostic complex Introkor M150.

The standard equipment for diagnosing VSTs with coating removal is also used:

- ultrasonic thickness gauge NDT MG2/D799 “Panametrics”,
- ultrasound scanner “Scanner”,
- magnetic thickness gauge MT2007,
- electrical spark detector Krona-2, X-ray unit Arina-5.

Fig. 1. RVS-10000 tank during the restoration of the ACC with stripped wall areas of near-weld zones.

2 Materials and Methods

The acoustic emission complex allows revealing the presence of defects in the metal of the tank wall due to the use of multichannel systems providing simultaneous recording and processing of the parameters of AE signals and their forms. The method of recording sound pulsed waves emitted by metal structures during loading allows determining the location of AE sources in the areas of the VST wall inaccessible to traditional control methods. This allows determining additional areas of anomalies that need to be analyzed at the next stage using other physical methods to refine the coordinates, estimate the size of defects, etc. The use of equipment based on the PAA technology (ultrasonic flaw detector “OmniScan MX2”) allows controlling the amplitude and phase of the excitation pulses of individual piezoelements in a multi-element converter. The excitation of the piezoelectric elements is carried out in such a way that makes it possible to control the parameters of the ultrasonic beam such as angle, focal length, and focal spot size by means of a computer program. The use of PAA technology due to the dense beam of ultrasonic radiation ensures high accuracy of the search for anomalies and defects in the metal in the presence of a protective coating. In Figs. 2 and 3, diagrams of ultrasonic flaw detection using the PAA technology are presented, as well as output signals when scanning the bottom of the VST.

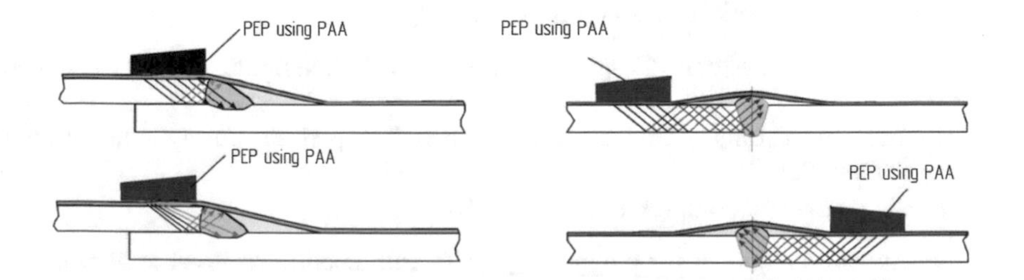

Fig. 2. Diagram of ultrasonic flaw detection using PAA.

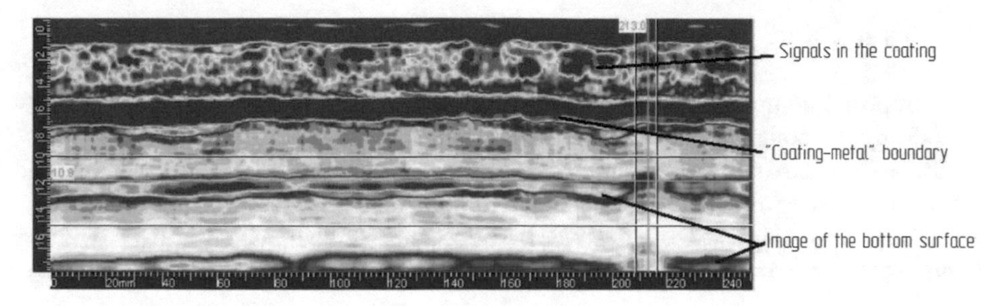

Fig. 3. Ultrasonic scanning of the VST bottom with a protective coating by the method of phased aperture arrays.

To detect defects of metal continuity of the bottom and annular plates, corrosion and fatigue cracks, cavities, pits, pitting defects from the side of the hydrophobic layer, it is proposed to use a complex consisting of the INTROKOR M150 magnetic flaw detector and the Wintrocor software product that allows interpretation of diagnostic data [9–12]. The flaw detector is based on the method of detecting magnetic fields of scattering from defects (MFL-method) during bottom magnetization [13, 14].

The magnetic relief is read by scanning the magnetic field with a multi-element transducer. Given the fact that the control is carried out without removing the ACC, the method allows 100% control of defects, residual thickness of metal sheets and a protective coating without gaps and "dead zones".

Also, with the use of this complex it is possible to determine the location and size of the defect and plot the bottom on the common "map" in automatic mode.

Table 2 shows the capabilities of the AE, UT, and magnetic control technology in the diagnosis of VSTs without removing the protective coatings. The integration of these methods within the framework of a single complex will allow technical diagnostics of VSTs in the extent prescribed by the NTD [1, 6], and with quality that is not inferior to the traditional method with tank stripping.

Table 2. Possibilities of various physical methods in diagnosing a VST [15, 16].

Operation	AE	UT	MC
Control of 100% wall/bottom area	+	–	–
Detection of internal and external defects	+	+	+
Determination of the location of defects	–	+	+
Determination of the coordinates of defects	–	+	+
Evaluation of defect sizes	–	+	+
Determination of the residual metal thickness	–	+	+
Evaluation of the ACC thickness	–	+	+

3 Results and Discussion

The proposed complex was tested during the full technical diagnostics of RVSPK-100000 by the company OOO NPP Simplex (Fig. 4).

Based on the data of testing the complex, the authors analyzed the economic efficiency of using the diagnostic method without removing the protective coating.

When calculating the cost-effectiveness of diagnostic examination of tanks without removing the protective coatings, the following parameters were taken:

- the investment phase is considered, consisting in the acquisition of the necessary technological equipment, and the subsequent eight-year period for the diagnostics and operation of the tank farm.
- annually, 20 tanks are diagnosed, including five RVS-5000 tanks, ten RVS-10000 tanks, five RVS-20000 tanks.

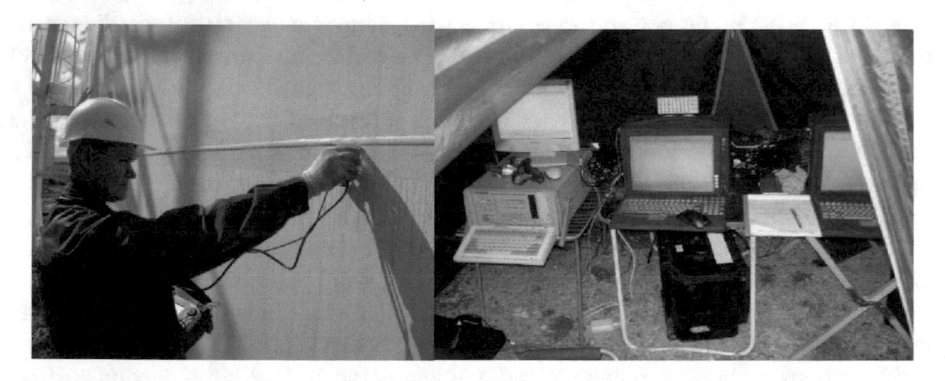

Fig. 4. Ultrasonic flaw detection of the RVSPK-100000 wall using the PAA method, Diagnostic complex.

The average annual revenue from the tank operation minus current costs is 50 million rubles. The calculation step is one year. The discount rate is taken at the level of 14% with a 100% share of own funds [1].

The total cash flow of the operating organization, taking into account the discounting for the nine years under consideration, with the amount of the initial investment in equipment equal to 7.7 million rubles, in case of applying the innovative method will be 29.4% or 1020.6 million rubles more than the corresponding figure when using the traditional method of tank diagnostics.

For the customer organization, the cost of performing technical diagnostics is reduced by 5 to 9 times, depending on the size of the VST. This is due to the fact that there is no need to carry out the most costly operations to remove and restore the ACC when using the proposed method [17, 18].

Figure 5 shows a chart comparing the cost of a full diagnostic examination by the traditional (with the removal of the ACC) and innovative (without removing the ACC) methods at current prices (2019) for the most common tanks in the Russian Federation: RVS-5000, 10000, 20000.

Summing up, we can say that by investing a relatively small amount of funds in the acquisition of new technological equipment that allows diagnostics of vertical steel tanks without removing the anti-corrosion coating, operating organizations get the economic effect of cost savings, tens and even hundreds of times more than the amount of investment made.

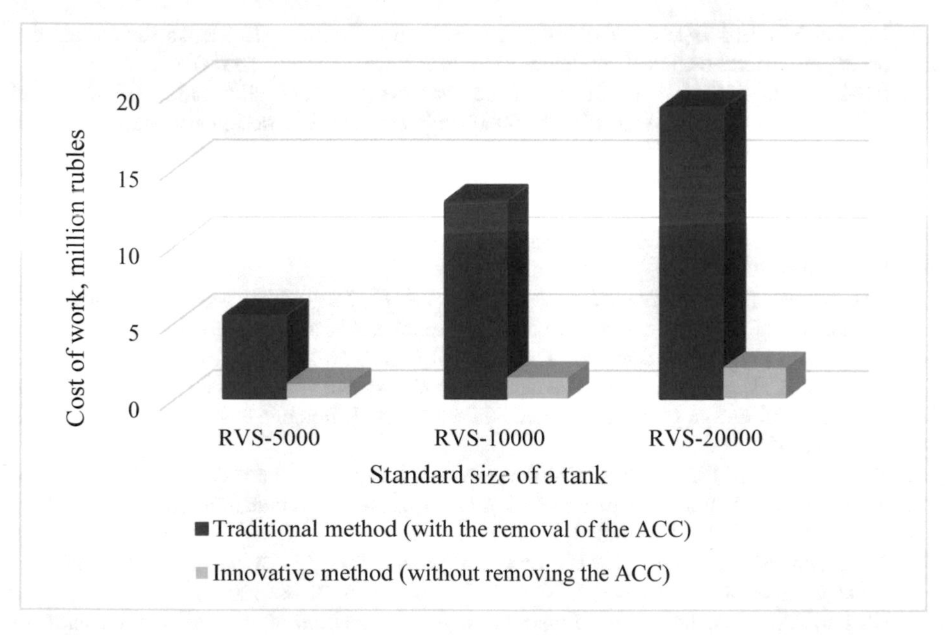

Fig. 5. Comparison of the cost of a comprehensive diagnostics of a VST for the customer by the traditional and innovative methods.

4 Conclusion

A diagnostic complex has been proposed that allows the use of acoustic emission, magnetic and ultrasonic methods to carry out comprehensive diagnostics of the VST metal structures of various sizes without removing the protective anti-corrosion coating in the extent and with the accuracy prescribed by the current regulatory documentation [1, 6]. Under the guidance and direct participation of the authors, full technical diagnostics of the RVSPK-100000 tank was performed. Pilot-production testing of the complex confirmed the operational suitability of the proposed complex and methods for diagnosing real industrial facilities. The technical and economic efficiency of the technology of conducting full technical diagnostics of vertical steel tanks without removing the anti-corrosion coating is substantiated. A comparison of technical and economic indicators of the diagnostics by traditional (with the removal of the protective coating) and innovative methods (without removing the protective coating) is presented.

References

1. Tikhanov, E.A.: Assessment of economic efficiency of capital repairs of the base of the vertical steel tank by the method of movement. J. Fundam. Res. **6**(2), 330–334 (2014)
2. Russian Standard GOST 31385-2008, The vertical cylindrical steel tanks for oil and oil-products. General technical conditions. Standartinform, Moscow (2010)
3. Russian Standard RD-23.020.00-KTN-283-09. The rules of the repair and reconstruction of tanks for oil storage capacity of 1000-50000 cubic meters. OAO AK "Transneft", Moscow (2009)
4. Russian Standard GOST 52910-2008, The vertical cylindrical steel tanks for oil and oil-products. General technical conditions. Standartinform, Moscow (2008)
5. RD-23.020.00-KTN-018-14, Main pipeline transport of oil and oil products. Vertical steel tanks for oil storage capacity of 1000–50000 cubic meters. The design standards, OAO AK "Transneft", Moscow (2014)
6. RD-23.020.00-KTN-271-10. Rules of technical diagnostics of tanks. OAO AK "Transneft", Moscow (2010)
7. Vasilev, G.G., Salnikov, A.P.: Analysis of causes of accidents with vertical steel tanks. J. Oil Ind. **2**, 106–108 (2015)
8. Sukhorukov, D.V., Slesarev, D.A., Abakumov, A.A., Polyakhov, M.Yu.: Technology of diagnostics of bottoms and walls of vertical steel tanks with the use of high-resolution scanning magnetic flaw detectors. J. Sphere Oil Gas **2**, 162–167 (2010)
9. Semin, E.E., Tarasenko, A.A.: The use of software systems in the assessment of technical condition and design of repairs of vertical steel tanks. J. Pipeline Transp.: Theory Pract. **4**, 84–87 (2006)
10. Tarasenko, A.A., Chepur, P.V., Chirkov, S.V.: Justification of the need to take into account the history of loading the structure during the repair of the foundation with the rise of the tank. J. Ind. Saf. **5**, 60–63 (2014)
11. Tarasenko, M.A., Silnickiy, P.F., Tarasenko, A.A.: Analysis of the results of corrosion damage defectoscopy of tanks. J. News Univ. Oil Gas **5**, 78–82 (2010)
12. Chepur, P.V., Astakhov, A.M., Tarasenko, D.A.: The method of calculation of distances of departure of the clearing device from the pipeline when the explosion of the gas mixture. J. Fundam. Res. **9**(2), 283–287 (2014)
13. Chepur, P.V., Tarasenko, A.A.: Methods of determining the need for repair of the tank in the sediments of the base. J. Fundam. Res. **8**(6), 1336–1340 (2014)
14. Chepur, P.V., Tarasenko, A.A., Tarasenko, D.A.: Study of the influence of the size of the ridge on the stress-strain state of the vertical steel cylindrical tank in the development of non-numbered precipitation of the outer contour of the bottom. J. Fundam. Res. **10**(15), 3441–3445 (2013)
15. Turin, D.V.: Modeling of oil steel cylindrical tanks. J. News Univ. Oil Gas **4**, 65–69 (2001)
16. Slepnev, I.V.: Stress-strain elastic-plastic state of steel vertical cylindrical tanks with inhomogeneous base settlement. Moscow Engineering and Building Institute, Moscow (1988)
17. Konovalov, P.A., Mangushev, R.A., Sotnikov, S.N., Zemlyansky, A.A.: Foundations steel tanks and deformation of its bases. The Association of construction universities, Moscow (2009)
18. Zemlyansky, A.A.: The design principles and experimental and theoretical studies of large tanks, Balakovo (2006)

Energy-Efficient Construction

The Increase of the Energy Efficiency of Protecting Constructions of Buildings

Andrei Ovsiannikov(✉), Vladimir Bolgov, Anna Vorotyntseva, and Alexey Efimiev

Voronezh State Technical University,
Moscow Avenue, 14, Voronezh 394026, Russia
ovsyannikovas@yandex.ru

Abstract. Modern society consumes increasing amount of energy. According to the climatic conditions, the cost of fuel both for providing the population with heat and for production in Russia is the highest. It can be said that Russia is the coldest country in the world, both in terms of the length of the heating season and the proportion of the population living in areas where negative average annual temperatures are observed. Reducing the energy consumption of buildings is one of the promising areas of the energy saving system. To reduce energy consumption for heating facilities, it is proposed to improve existing walling structures of ventilated facades by replacing the traditional facing layer (porcelain stoneware) with light-transmitting structures. This replacement will allow accumulation of solar energy during daytime and its subsequent use at night. The proposed protective constructions of building are an alternative to the traditional construction of the ventilation facade. When using the proposed protecting construction, there is a reduction in capital expenditures on the installation of a heating system for and object and a reduction in subsequent operating costs for heating and object in the autumn-winter period. The article compares the economic efficiency of the use of the light-transmitting structure, taking into account the capital costs for the construction of structures and the cost of operating costs compared with the traditional design of the ventilated facade.

Keywords: Increase · Energy efficiency · Protecting

1 Introduction

The location of the Russian Federation in northern latitudes implies cold long-lasting winters and a large amount of precipitation. This is the reason why the waste of warm energy per unit of living space is increased on 2–3 times comparing to Europe countries. Wide housing construction carried out in Russia in previous years with low energy prices has led to the fact that the heat-shielding characteristics of the building envelope are much lower than in countries close to Russia in climatic conditions. This led to significant costs for heating buildings [1–3].

One of the most effective ways to save energy is the reduction of heat loss through the building envelope (exterior walls) of buildings and structures.

V. Murgul and M. Pasetti (Eds.): EMMFT-2018, AISC 983, pp. 695–703, 2019.
https://doi.org/10.1007/978-3-030-19868-8_68

External additional insulation of enclosing structures provides a reduction in the cost of heating the building up to 40–50%.

Saving of energy resources is one of the most serious problems of this century. The place of our society among other developed countries depends on solving this problem. Russia has all the necessary natural resources and intellectual potential to successfully solve its energy problems, and it is objectively a resource base for European and Asian countries, exporting oil, petroleum products and natural gas in volumes that are strategically important for importing countries. A large amount of fuel and energy resources in our country should not envisage energy drainage, since the only energy efficient management is the most important factor of competitiveness of Russian goods and services in an open market economy.

Energy conservation should be attributed to the strategic objectives of the state, being both the main method of ensuring energy security and the only real way to maintain high revenues from the export of hydrocarbons.

The strategic goal of energy saving is one and it follows from its definition - an increase in energy efficiency in all sectors, in all settlements and in the whole country.

The goals of energy saving coincide with other goals, such as improving the environmental situation, improving the efficiency of energy supply systems, etc.

The solving of the increase of energy efficiency problems on current stage when there is a big reserve of low-cost events matches with the most of strategic aims of the government and economic entities.

One example of improving the energy efficiency of objects can be the replacement of existing external enclosing structures of ventilated facades with a facade device using light-transmitting structures to store solar energy. In this system, the cladding structure is capable of transmitting a significant amount of solar energy, which heats the supporting structures and must necessarily be taken into account when making thermophysical calculations at the facility [4–9].

The essence of this technology is to replace the already traditional design of the ventilation facade on the facade of the proposed technology, which can be made the necessary size and configurations [10].

2 Materials and Methods

Ventilated facade is a technology for the implementation of the facade, a system consisting of cladding materials that are attached by using galvanized steel, stainless steel or aluminum frame to the supporting layer of the wall. Air freely circulates on the gap between the facing and the wall, which removes condensate and moisture from the structures. All fastening elements of the ventilated facade system are universal, which allows solving complex architectural and design tasks from classical to cutting-edge [5]. For additional building walls, warming mineral wool insulation is fastened by using dish plugs. On the basement of the building, extrusion (polystyrene) insulation is used. It does not pass and does not absorb moisture. The size of the gap between the insulation and the facade of the building should not be less than 40 mm. This allows the ascending air flow to circulate between the cladding material and the wall, drying the layer of insulation in case of moisture on it. In order to prevent the blowing of air from the insulation, it is covered with a windproof, vapor-permeable membrane (film). The use of a ventilated facade with insulation makes the "dew point" outside the bearing walls of the building into the insulation zone. This ensures the durability of buildings, minimizes the likelihood of wetting of the walls, mold on the wall inside the building, increases the thermal insulation of the premises.

The systems of ventilated facades, providing insulation, can significantly reduce the amount of building material required for the construction or restoration of the wall, which contributes to significant savings in financial resources. In addition, a smaller amount of heavy building materials allows for a reduction in the overall weight of the wall and the entire structure, which means it will increase the number of floors of the building [11]. One of the directions of development of ventilated facade systems is the use of enclosing light-transmitting structures (double-glazed windows) instead of traditional cladding elements - ceramic granite tiles, facade cassettes as enclosing elements (Fig. 1, Table 1).

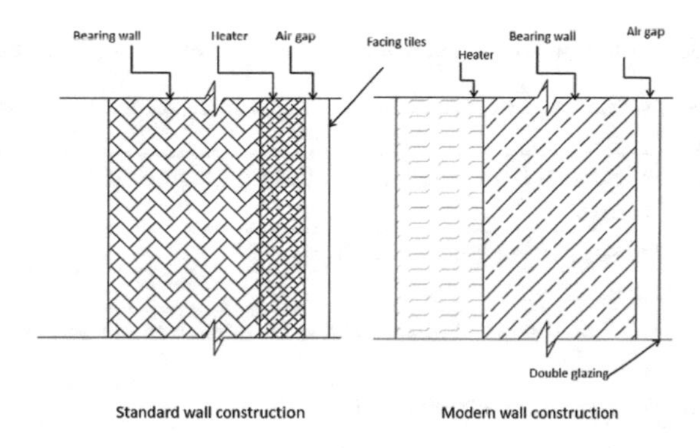

Fig. 1. Comparing the installation of enclosing structures.

Table 1. Exterior walling options.

Type of material used	Traditional option	Suggested option
Brick wall (external, supporting structure)		
Brick	Construction thickness	
	0.54 m	0.38 m
Insulation (mineral wool)	Construction thickness	
	0.1 m	0.1 m
Exterior walling	Ventilated facade	Double glazing
Monolithic reinforced concrete wall (external, supporting structure)		
Monolithic reinforced concrete	Construction thickness	
	0.5 m	0.3 m
Insulation (mineral wool)	Construction thickness	
	0.1 m	0.1 m
Exterior walling	Ventilated facade	Double glazing

As can be seen from the table, at the construction stage, the savings are due to a reduction in the thickness of the supporting structure without reducing the enclosing and heat-shielding capacity of the wall. Cost parameters of various types of enclosing structures are shown in Table 2. According to Table 2, it can be concluded that the cost of erecting a new building envelope is more expensive than the traditional one, on average, 1.50–1.75 times. Consequently, the design will be effective only if the operating costs for it will be significantly lower than during the operation of the traditional design. [2] When constructing light-transmitting enclosing structures, the heat transfer resistance of the multi-layer structure changes. The calculations are made according to the set of rules for the design of thermal protection of buildings. («SP 23-101-2004. Design of thermal protection of buildings»). This normative document consists of design methods, calculating of warm-technical characteristic of defending constructions methods recommendations and reference materials allowing implementation of the requirements of SNiP 23-02-2003 "Thermal protection of buildings".

Table 2. Comparison of options for the construction of supporting structures 100 m^2 of the facade area.

Name of works	Cost
1. Construction of a brick wall 0.54 m thick with the installation of a porcelain stoneware ventilated façade	734 800
1.1 Construction of the supporting structure cost	498 808
1.2 External walling cost	235 992
2. Construction of a brick wall with a thickness of 0.38 m with the installation of the facade of glass	1 276 133
2.1 Construction of the supporting structure cost	350 960
2.2 External walling cost	925 173
3. Construction of a monolithic reinforced concrete wall 0.5 m thick with the installation of ventilation facades	889 992
3.1 Construction of the supporting structure cost	654 001
3.2 External walling cost	235 992
4. Construction of a monolithic reinforced concrete wall with a thickness of 0.3 m and with the installation of the facade of the glass	1 348 979
4.1 Construction of the supporting structure cost	423 806
4.2 External walling cost	925 173

Heat transfer resistance R0, m2 °C/W, of homogeneous single-layer or multilayer enclosing structures with homogeneous layers or enclosing structures at a distance of at least two thicknesses of the enclosing structure should be determined by the formula:

$$R_0 = Rsi + Rk + Rse, \tag{1}$$

where Rsi if determined by the formula

$$Rsi = 1/\alpha int, \quad (2)$$

where αint – if a heat transfer coefficient inside the enclosing structures, W/(m2 °C), according to the SNiP Table 23-02-2003;

Rse if determined by the formula (3):

$$Rse = 1/\alpha bxt, \quad (3)$$

where αbxt - heat transfer coefficient of the outer surface of the building envelope for the cold period conditions, Vt/(m2 °C), according to the table SP 23-101-2004;

Rk – is a thermal resistance of the building envelope, defined as the sum of the thermal resistances of the individual layers by the formula (4)

$$Rk = R1 + R2 + \ldots + Rn, \quad (4)$$

where R1 + R2 +… + Rn – thermal resistance of individual layers of the building envelope, determined by the formula

$$R = \delta/\lambda, \quad (5)$$

where δ – if a layer thickness, m; λ – the calculated thermal conductivity of the material layer, Vt(m °C).

Heat transfer resistance of enclosing structures were determined depending on the number and materials of the layers. At the same time, thermal conductivity coefficients λA of the used materials were determined for operating conditions of used materials in Voronezh Region of Russian Federation:

- estimated temperature in the cold season for the conditions of Voronezh t_{ext} = −26 °C;
- duration of the heating period z_{ht} = 196 days;
- average outdoor temperature t_{ht} = −3,1 °C for warming period;
- degree-days of the heating period at an internal temperature of 20 °C - 4528 for the heating period.

After the calculations of the resistance to heat transfer for various types of enclosing structures, the following data on thermal conductivity was obtained:

1. Brick wall:
   - ventilated facade with granite cladding – 3.58 m^2 °C/W;
   - ventilated facade with light-transmitting structures – 4.33 m^2 °C/W.
2. Monolithic reinforced concrete wall:
   - ventilated facade with granite cladding – 3.42 m^2 °C/W;
   - ventilated facade with light-transmitting structures – 4.15 m^2 °C/W.

3 Results

The calculations have shown the possibility of reducing the required number of heating appliances during the construction of a facility with light-transmitting enclosing structures [4].

A comprehensive comparison of the cost of building facades and arrangement of the heating system is given in the Table 3.

Table 3. Comparison of costs for the construction of the supporting structure, installation of fencing and installation of heating devices on the 100 m^2 of the facade area.

Name of the construction	Traditional construction	Suggested option	Economy (waste)
1. For brick construction	782 791	1 314 526	−531 735
1.1 Construction of the supporting structure cost	498 808	350 960	147 848
1.2 External walling cost	23 5992	9251 73	−689 181
1.3. Installation of heating devices	47 991	38 393	9 598
2. For monolithic reinforced concrete housing construction	937 983	1 387 372	−449 389
2.1 Construction of the supporting structure cost	654 000	423 806	230 194
2.2 External walling cost	235 992	925 173	−689 181
2.3 Installation of heating devices	47 991	38 393	9 598

Despite the reduction in the cost of building a heating system, the option with translucent walling is much more expensive. Therefore, the rationale for the effectiveness of its implementation can be obtained only when determining the cost of its operation.

Figure 1 shows the construction schemes of a traditional enclosing structure and a variant with light-transmitting enclosing structures. From the presented figures it is clear that in the traditional version the insulating layer goes outside the building, therefore, it is exposed to climatic and weather loads, and in the studied version it is inside the building, therefore, its service life increases [6].

It should be noted that theoretical studies on the service life of various options for building envelopes showed:

1. On average, the service life of ventilated facades is 25–30 years, then it is necessary to replace the insulation layer and part of the facing elements;
2. The service life of the investigated variant with light-transmitting glass enclosing structures can reach 25–50 years.
3. The service life of the heating system in modern facilities is 20–25, after which it is desirable to overhaul these engineering communications.

Such high service life goes beyond the planning horizons and may not be considered further in our work.

4 Discussions

When operating various facilities, significant differences may occur when the costs of heating are compared. These operating costs in the annual operating cycle are unevenly distributed. According to the above data, the heating period in the Voronezh region is standardly defined by a length of 196 days. As a rule, its beginning falls on the last days of September - the beginning of October. It continues until the beginning - mid-April, depending on current weather conditions. The cost of heat for the 2018–2019 heating period is set at 1,700 rubles/Gcal. on average for the main large heat suppliers of the city of Voronezh. Reducing the cost of heating the facility during the heating period is shown in Table 4 (Figs. 2 and 3).

Table 4. Reducing the cost of heating the object per 100 m^2 of building envelope.

Month of a year	Heat loss, Gcal.		Heat loss, rub.		Cost reduction, rub.
	Traditional construction	Investigated design	Traditional construction	Investigated construction	
October	18.711	15.470	31 809	26 299	5 510
November	25.336	20.948	43 071	35 611	7 460
December	32.286	26.694	54 886	45 379	9 507
January	35.532	29.377	60 404	49 941	10 463
February	34.692	28.682	58 976	48 760	10 216
March	28.181	23.299	47 908	39 609	8 299
April	16.191	13.386	27 524	22 757	4 767
TOTAL for the heating period			324 578	268 356	56 222

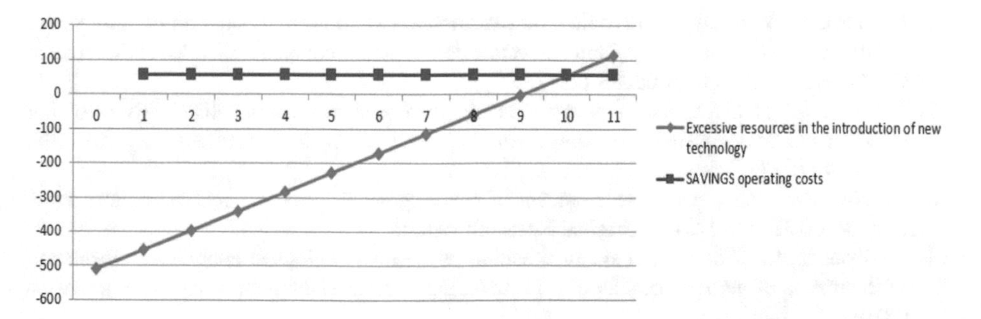

Fig. 2. Payback period for new technology in brick construction.

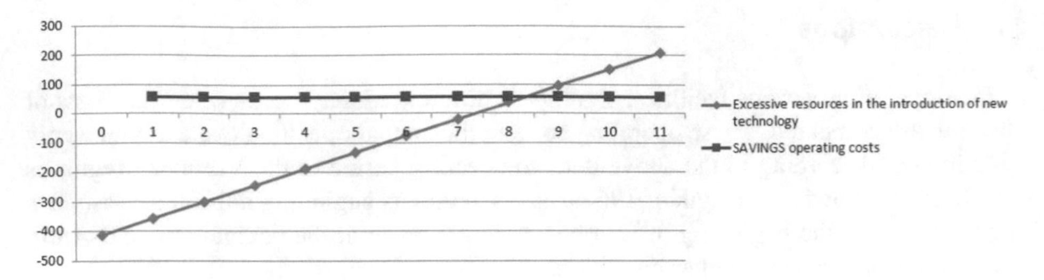

Fig. 3. Payback period for new technology in monolithic construction.

The payback period of the studied defensing structures with light-transmitting glass for various types of load-bearing walls will be respectively:

- Brick wall – 10.2 years;
- Monolithic reinforced concrete wall – 8.6 years.

5 Conclusion

The cost of enclosing structures according to the modern technology of ventilated facade, with ceramic granite cladding, is significantly lower than the implementation of the proposed option with light-transmitting enclosing structures. At the same time, the higher energy efficiency of the proposed option with light-transmitting walling leads to a reduction in the cost of heating the object. Therefore, when applying technology with light-transmitting enclosing structures, we will receive a reduction in total costs and an increase in the level of energy efficiency of a capital construction object.

References

1. Belyaeva, S., Voronov, D., Erypalov, S.: Methodical principles of evaluation of competitive ability of construction industry and real estate development companies. In: MATEC Web of Conferences, vol. 106, p. 08033 (2017)
2. Bukreev, A., et al.: Specifics of the strategic management of innovative activity of big development and construction companies. In: MATEC Web of Conferences, vol. 106, p. 08034 (2017)
3. Didkovskaya, O.V., et al.: Development of cost engineering system in construction. Proc. Eng. **153**, 131–135 (2016). Original Research Article
4. Lukmanova, I., Sizova, E., et al.: Justification of the methodological approach to formation of the parties of innovation activities. In: MATEC Web of Conferences, vol. 239, p. 04014 (2018)
5. Shchukina, T.V.: The Absorptive capacity of external building envelopes for passive use of solar radiation. Ind. Civ. Eng. **9**, 166–168 (2012)
6. Shchukina, T.V., et al.: Passive solar heating: how to control the heating regime. Int. J. Environ. Sci. Educ. **11**(18), 11361–11373 (2016)

7. Sheps, R.A., Shchukina, T.V., Akulova, I.I.: IOP Conference Series: Materials Science and Engineering, p. 012054 (2017)
8. Sheps, R.A., Yaremenko, S.A., Agafonov, M.V.: Accounting of solar energy in the design of thermal protection of buildings. Housing **1**, 29–33 (2017)
9. Sizova, E., Zhutaeva, E., Chugunov, A.: Methodical bases of selection and evaluation of the effectiveness of the projects. In: E3S Web of Conferences, vol. 33, p. 03016 (2018)
10. Ahmad, T., Aibinu, A., Thaheem, M.J.: The effects of high-rise residential construction on sustainability of housing systems. Proc. Eng. **180**, 1695–1704 (2017). Research article
11. Uvarova, S., Belyaeva, S., Myshovskaya, L.: In: IOP Conference Series: Earth and Environmental Science, vol. 90, no. 1, p. 012169 (2017)
12. Zahirovich-Herbert, V., Gibler, K.M.: The effect of new residential construction on housing prices. J. Hous. Econ. **26**, 1–18 (2014)

Mathematical Modeling of the Wind Vector Field When Flowing Around Artificial Structures

Olga Sotnikova[1(✉)], Lidiia Murashova[1], Mikhail Semenov[1], Alcksandr Malcnyov[2], and Vladimir Koida[2]

[1] Voronezh State Technical University, Moscow Avenue, 14, Voronezh 394026, Russia
hundred@vgasu.vrn.ru

[2] Military Educational and Scientific Center of the Air Force «N.E. Zhukovsky and Y.A. Gagarin Air Force Academy», Starykh Bol'shevikov St., 54 «A», Voronezh 394064, Russia

Abstract. Atmospheric thermo - and hydrodynamics is one of the defining factors affecting the work of engineering systems (in particular, ventilation systems) along with impact of wind and temperature fluctuations. Very noticeable is the dependence of the wind pressure and the density of its distribution on the surfaces of the external enclosing structures on the wind speed and the nature of its flow. The results of the analysis of the available modern representations of topical issue - assessment of influence of a wind flow, are presented in this article. Knowledge of the qualitative and quantitative characteristics will allow predicting features of operation of forced-air and exhaust ventilation systems by preventive model operation of parameters of a wind stream on the plane of windward or leeward enclosing structures of buildings. The results of a series of numerical experiments of the model offered by the authors of this article are given in stationary and non-stationary problem definition. It is shown that the information obtained during the operation of the numerical model is quite complete when making and justifying design decisions in the field of structural design, since this is typical for buildings, structures, and engineering ventilation systems.

Keywords: Wind vector field · Artificial structures · Mathematical modeling

1 Introduction

The wind stream when flowing around buildings and structures – one of conditions that must be considered in urban planning and design of buildings [1]. Construction of the new building changes a microclimate in its vicinity. Therefore, the wind comfort and safety become important requirements in the field of urban planning and design. The high-rise building has a great influence on a wind stream, therefore information on the speed field, which is formed when flowing around buildings, can be important for architects and town-planners.

V. Murgul and M. Pasetti (Eds.): EMMFT-2018, AISC 983, pp. 704–719, 2019.
https://doi.org/10.1007/978-3-030-19868-8_69

The wind stream can be simulated in the wind tunnel, nevertheless, with development of computer facilities and numerical methods, it is possible to model precisely the same conditions in the virtual environment by means of CFD-models [2–11], which can provide the considerable economic advantage at assessment of engineering design decisions. CFD-model operation can be used for assessment of the speed field of wind around buildings and also for assessment of thermal comfort, ventilation of air, influence of microclimatic conditions [9]. For the last two decades, along with CFD-model operation perfecting, many researches are concentrated on numerical model operation of an airflow around one building, their researches show the difficult phenomena of a stream [4, 7, 8]. CFD-model operation demands larger time expenditure, very shallow grid, and the universal models of turbulence [12] for the considered class of tasks. Also, there are classical analytical ways of calculation of interaction of wind streams with structural constructions [13, 14].

2 Materials and Methods

2.1 Mathematical Model on the Basis of a Method of Discrete Whirlwinds

Example of Application of Mathematical Model. In this work, we will consider the mathematical model based on application of hydrodynamic model of a true non-viscous liquid. This model is developed on the basis of a method of discrete whirlwinds that allows building a picture of streamlines, receiving the speed field of a wind stream around one building, several buildings or architectural constructions.

Let's consider detachable non-stationary flow of the building and small architectural construction (Fig. 1) with breaks forming, which have the characteristic geometrical sizes: length - l1, l2; height - h1, h2 and located apart from each other on Δl in the flat Oxy system. The wind stream moves with constant speed $\vec{U}_{\infty}(\tau)$ along a positive axis of Ox. Size l = h1, where h1 - height of the first construction is taken for the characteristic size; for the characteristic speed – value of speed of an incident flow $\vec{U}_{\infty}(\tau)$, where τ – the dimensionless time defined as $\vec{U}_{\infty}(\tau) \cdot t/l$, t - physical time; $\Delta\tau \approx 1/n$ – the dimensionless step on time; n – number of the bound vortexes that model the surface of the building.

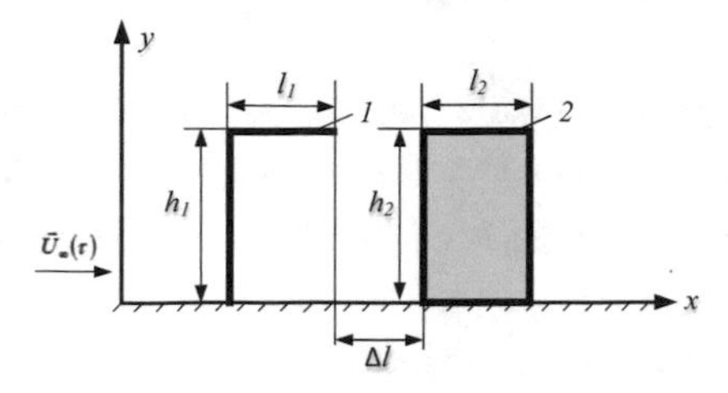

Fig. 1. Scheme of settlement area: 1 – a small architectural construction, 2 – the building.

For modeling the process of air flow around the buildings, we use a method of discrete whirlwinds [15, 16], which allows calculating structure of a vortex stream, a form of streamlines, and the field of speed under various conditions of placement of buildings [17, 18]. For calculation of the separation flow around the structures under consideration at the earth's surface, the main and mirrored vortex systems are used [15, 16]. Each of them consists of the bound discrete vortexes, which circulations Γμ are equal in size and are opposite according to the sign, and the free $\delta_1^\tau, \delta_2^\tau, \delta_3^\tau, \delta_4^\tau$ vortex wakes descending from breaks of surfaces in an instant τ (Fig. 2).

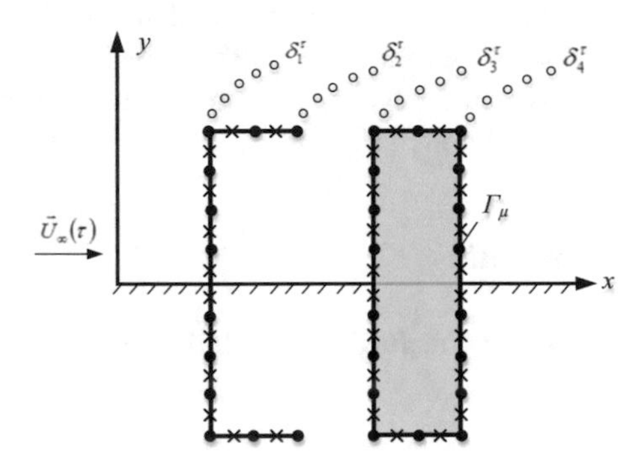

Fig. 2. Modeling of the surface of bodies by a system of discrete vortices (•) and reference points (×).

The motion of the latter in both vortex systems occurs symmetrically. With such a splitting into lines of symmetry Ox, the condition of no course is automatically satisfied at any time, which is equivalent to the presence of the earth's surface here. The scheme of splitting surfaces of constructions includes: n – reference points where the boundary condition of no course and n – discrete whirlwinds is satisfied. The first settle down on lines ν, and the second on lines μ between reference points, both in the main and mirrored vortex system (Fig. 2).

Circulation of the bound vortexes Γμ at the k-th time step are determined from the solution of a system of linear algebraic Eq. (1), which expresses the condition of non-flowing surfaces of structures and the condition of non-circulating flow [17–20].

$$\sum\nolimits_{v=0}^{n}\left[\sum\nolimits_{\mu=0}^{n}\Gamma_\mu\left(a_{v\mu}-a'_{v\mu}\right)=-cos\left(\vec{U}_\infty,\vec{n}\right)_v-\sum\nolimits_{kk=1}^{4}\sum\nolimits_{i=1}^{k}\delta_{kk,i}\left(a_{kk,i}-a'_{kk,i}\right)_v\right] \quad (1)$$

where Γμ – circulation μ- bound vortex; $a_{v\mu}, a'_{v\mu}$ – a normal component of speed in ν - reference point from μ - the bound vortex of the main and mirrored vortex systems; $\left(a_{kk,i}\right)_v, \left(a'_{kk,i}\right)_v$ – normal components of speed in ν – reference point from i of the free whirlwind of the main and mirrored vortex systems; kk – ordinal value of a vortex wake; k – quantity of the free whirlwinds in a stream in an instant τ in the corresponding vortex wake of kk; $cos\left(\vec{U}_\infty,\vec{n}\right)_v$ – a cosine of the angle between a flow rate vector $\vec{U}_\infty(\tau)$ and normal $\vec{n}$ to the surface of a body in each reference point.

The calculated values of circulation of the bound discrete vortexes Γμ in places of a break of surfaces define circulation of the free whirlwinds $\delta_1^i, \delta_2^i, \delta_3^i, \delta_4^i$ in settlement instants τ, $i = \overline{1,k}$.

The new provision of the free whirlwinds is determined by a method of Euler by [15, 17] ratios:

$$\begin{cases} x_{kk,i}^{\tau+1} = x_{kk,i}^{\tau} + \left[\sum_{\mu=0}^{n} \left(V_x - V_x'\right)_{i\mu} + \sum_{j=0}^{k} \left(V_x - V_x'\right)_{ij} + U(\tau)\right] \Delta t, \\ y_{kk,i}^{\tau+1} = y_{kk,i}^{\tau} + \left[\sum_{\mu=0}^{n} \left(V_y - V_y'\right)_{i\mu} + \sum_{j=0}^{k} \left(V_y - V_y'\right)_{ij}\right] \Delta t; \quad i = \overline{1,k}; \quad kk = \overline{1,4} \end{cases} \tag{2}$$

where $(V_x)_{i\mu}, (V_x')_{i\mu}, (V_y)_{i\mu}, \left(V_y'\right)_{i\mu}, (V_x)_{ij}, (V_x')_{ij}, (V_y)_{ij}, \left(V_y'\right)_{ij}$ – these components of speed can be found from a ratio (3)

$$\begin{cases} V_{xik}' = -\frac{1}{2\pi} \frac{y_k - y_i}{2\pi (x_k - x_i)^2 + (y_k - y_i)^2} \\ V_{yik} = \frac{1}{2\pi} \frac{y_k - y_i}{2\pi (x_k - x_i)^2 + (y_k - y_i)^2} \end{cases} \tag{3}$$

where k – a reference point in which speed is defined; i – a whirlwind which induces speed.

Having received the developed vortex stream, it is possible to construct streamlines [18] about the considered bodies (4).

$$\frac{dx}{V_x} = \frac{dx}{V_y} \tag{4}$$

where V_x, V_y – speed components in the considered point q from all vortex system: the bound vortexes of the main and mirrored vortex system of the first and second construction μ1, μ2; the free whirlwinds descending from sharp edges of both constructions $\delta_1, \delta_1', \delta_2, \delta_2', \delta_3, \delta_3', \delta_4, \delta_4'$

$$\begin{cases} (V_x)_q = \sum_{r=1}^{p} \sum_{i=0}^{n} \Gamma_{\mu_r} \left(V_{x\mu_r} - V_{x\mu_r}'\right)_i + \sum_{rr=1}^{pp} \sum_{j=1}^{k} \Gamma_{\delta_{rr}} \left(V_{x\delta_{rr}} - V_{x\delta_{rr}}'\right)_j + U_\infty(\tau) \\ \left(V_y\right)_q = \sum_{r=1}^{p} \sum_{i=0}^{n} \Gamma_{\mu_r} \left(V_{y\mu_r} - V_{y\mu_r}'\right)_i + \sum_{rr=1}^{pp} \sum_{j=1}^{k} \Gamma_{\delta_{rr}} \left(V_{y\delta_{rr}} - V_{y\delta_{rr}}'\right)_j \end{cases} \tag{5}$$

where Γ_{μ_r} – circulation μ – bound vortex of r-th construction, $\Gamma_{\delta_{rr}}$ – circulation of the free whirlwind of rr-th vortex wake; $U_\infty(\tau)$ – speed of an incident flow; $(V_x)_q, \left(V_y\right)_q$ – speed components in the considered plane point (x_q, y_q), which can be calculated as the sum of the corresponding components of speed from all vortex system (3): the bound vortexes μ_r (components $V_{x\mu_r}, V_{y\mu_r}$) which model surfaces of constructions, and the free whirlwinds δ_{rr} (components $V_{x\delta_{rr}}, V_{y\delta_{rr}}$) descended from sharp edges, on a settlement instant τ the main and mirrored vortex system.

Results of Model Operation. We use the described mathematical model for calculation of structure of a vortex stream, streamlines and the field of speed of a wind stream when flowing around the building and small architectural construction of identical length of l1 = l2 and height of h1 = h2.

Let's consider three options: Δl = 3l, Δl = 2l, Δl = l, where Δl – distance between the building and small architectural construction, l – the characteristic size. The results of calculation of structure of a vortex stream are presented in Figs. 3, 4 and 5 that corresponds to 50 steps on time or for an instant of 20 s.

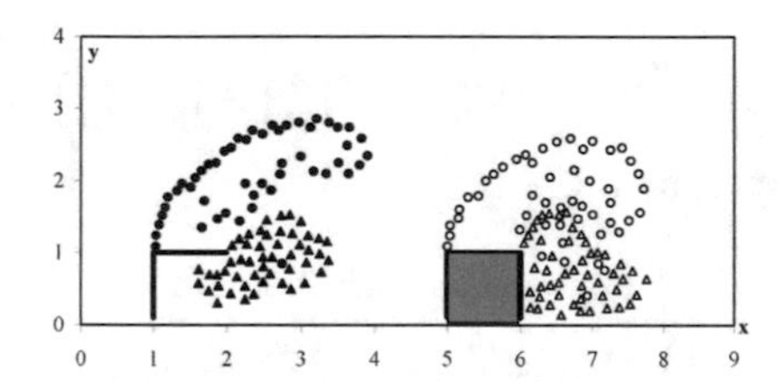

Fig. 3. Structure of a vortex stream Δl = 3l.

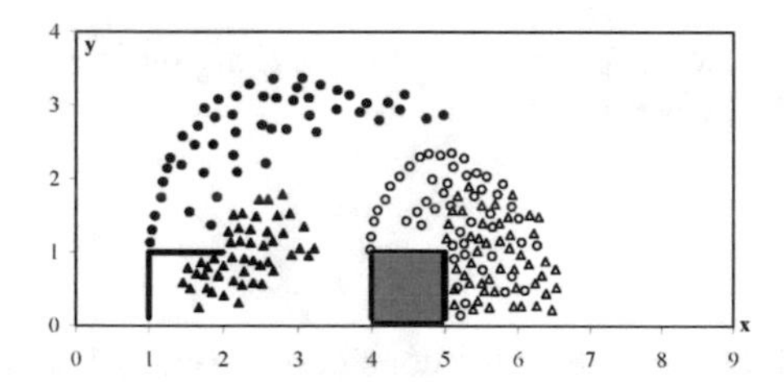

Fig. 4. Structure of a vortex stream Δl = 2l.

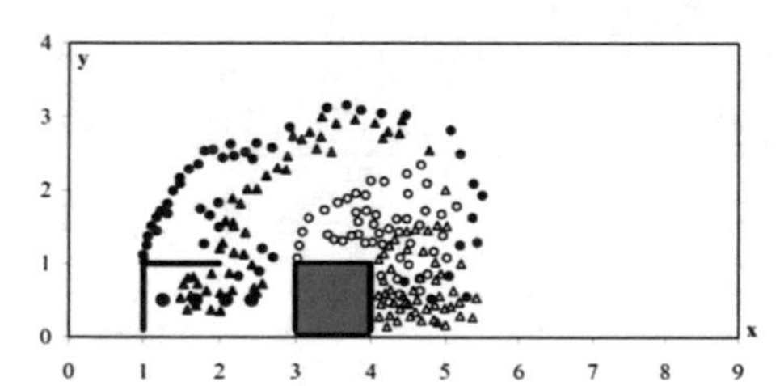

Fig. 5. Structure of a vortex stream Δl = l.

Analyzing the structure of a vortex stream presented in Figs. 3, 4 and 5, it is possible to see that if the considered objects are at sufficient distance from each other Δl = 3l (Fig. 3), that there is a powerful vortex stream in a small construction and on lee side of the main building, which promotes formation of hold-up spots. However, since the vortex shrouds coming down from the sharp edges of both objects do not interact with each other, over time, the vortex flow around the small structure is stretched and taken out of its limits, which contributes to good ventilation of the intercase space. At decrease of distance Δl = 2l, it is visible (Fig. 4) that gradually the vortex wakes descending from a small construction get into a vortex stream, which is formed near the main building, and at Δl = l (Fig. 5), powerful wind gusts (turbulence of big intensity) both in a gap between the considered objects and in a small construction will be formed.

In Fig. 6, the form of streamlines is presented, according to the third scenario of arrangement of constructions that confirms formation of zones of a returnable current.

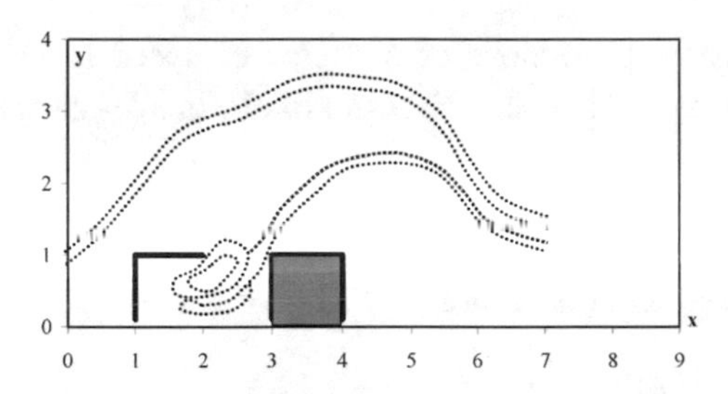

Fig. 6. Form of streamlines Δl = l.

2.2 An Example of Application of Mathematical Model for a Three-Dimensional Task in Stationary Statement

Microscale Meteorological Model. Let's consider three-dimensional microscale meteorological model on the example of a research of a detailed wind situation in the neighborhood of the airport, taking into account influence of structures of various storeys number, the available extended sites of vegetation (a forest belt and group of separate trees) and inhomogeneities of properties of the spreading surface (asphalt or a vegetable cover). In spite of the fact that the maximum height of buildings is limited and depends on distance to a runway, quite significant width and a grouping in one area is characteristic of their sizes. Thus, all this complex of buildings (the airport, hangars, warehouses) with rather small height represents a wide barrier on the way of ground wind. Air flows over the buildings from the top and from the side, which leads to a change in wind speed values along the runway. At the same time, influence of obstacles on the prevailing air flow depends on many factors, the most important of which is the speed of wind and its direction concerning an obstacle and also the scale of obstacles in relation to the runway sizes.

Let's consider three-dimensional stationary turbulent isothermal driving of the incompressible environment over the non-uniform spreading surface with elements of large-scale roughness. Elements of roughness represent rectangular obstacles which sizes are commensurable with sizes of area of a research. Let's consider that there are two types of the fixed obstacles: buildings, impenetrable for a stream, and permeable massifs of vegetation (forest belt) or separate trees. The spreading surface: asphalt or vegetable cover. The mathematical model includes equation of continuities and the equations of Navier-Stokes, average on Reynolds:

$$\frac{\partial \langle u_i \rangle}{\partial x_i} = 0 \tag{6}$$

$$\frac{\partial \langle u_i \rangle \langle u_j \rangle}{\partial x_j} = -\frac{1}{\rho}\frac{\partial \langle p \rangle}{\partial x_i} + \frac{\partial}{\partial x_j}\left(\nu \frac{\partial \langle u_i \rangle}{\partial x_j}\right) - \frac{\partial}{\partial x_j}\langle u_i^{'} u_j^{'} \rangle + FM_i, \quad i,j = 1,2,3 \tag{7}$$

Here $\langle u_i \rangle$ – average projections of a vector of speed to Ox coordinatei, $\langle p \rangle$ – pressure, ρ – density, ν – a kinematic viscosity of air, $\langle u_i^{'} u_j^{'} \rangle$ – stress tensor of Reynolds, FM_i – the function describing influence of vegetation on an aerodynamics. The axis of Ox3 is sent upright to wind.

Short circuit of a set of Eq. (2) is carried out with the use of two-parameter "k − ε" - model and a gradient and diffusion hypothesis of Bussinesk:

$$\frac{\partial \langle u_j \rangle k}{\partial x_j} = \frac{\partial}{\partial x_j}\left(\left(\nu + \frac{\nu_T}{\sigma_k}\right)\frac{\partial k}{\partial x_j}\right) + P - \varepsilon + FK \tag{8}$$

$$\frac{\partial \langle u_j \rangle \varepsilon}{\partial x_j} = \frac{\partial}{\partial x_j}\left(\left(\nu + \frac{\nu_T}{\sigma_k}\right)\frac{\partial \varepsilon}{\partial x_j}\right) + \frac{\varepsilon}{k}(C_{\varepsilon 1}P - C_{\varepsilon 2}\varepsilon) + FE \tag{9}$$

$$\nu_T = C_\mu \frac{k^2}{\varepsilon}, \quad -\left\langle u_i^{'} u_j^{'} \right\rangle = \nu_T\left(\tfrac{\partial \langle u_i \rangle}{\partial x_j} + \tfrac{\partial \langle u_j \rangle}{\partial x_i}\right) - \tfrac{2}{3}k\delta_{ij} \tag{10}$$

where νT – turbulent viscosity, k – a kinetic energy of turbulence, ε – a dissipation of a turbulent kinetic energy, $P = -\langle u_i^{'} u_j^{'} \rangle \frac{\partial \langle u_i \rangle}{\partial x_j}$ – oscillation of energy of turbulence, FK, FE – the functions describing influence of vegetation on a turbulent kinetic energy and a dissipation, σ = 1k, σ = 1.3ε, C = 1.44, C = 1.92, C = 0.09ε1ε2μ.

Terms in transport equations for model operation of influence of vegetation:

The equation of Reynolds – $FM_i = -\eta C_d a \langle u_i \rangle |V|$,

The equation of a kinetic energy of turbulence – $FK = \eta C_d a\left(\beta_P |V|^3 - \beta_d |V| k\right)$,

The equation of a dissipation of a kinetic energy of turbulence – $FE = C_{\varepsilon 4} \frac{\varepsilon}{k} FK$

Here η – a share of the spreading surface covered with trees, Cd – a resistance coefficient, a = a(x3)– vegetation density in the forest area (for example, for the array of pine trees η = 1, C d = 0.2, a = 0.3125 m^2/m^3), βP = 1 – a share of an average kinetic energy of a stream which was transformed to a turbulent kinetic energy because of vegetation resistance, and coefficient βd = 2.5 dissipation k share because of cascade process of transfer of turbulence in vegetation, Cε4 = 1.5 – an empirical constant, |V| – the speed vector module. Thus, influence of vegetation is considered by means of additional source terms in the average equations of Navier-Stokes and in transport equations of model of turbulence.

For a task of values of turbulent parameters near the spreading surface and a surface of elements of roughness, we will use method of wall functions [21]. The choice of such way of a task of boundary conditions for k, ε and turbulent tension is caused by the fact that turbulent characteristics near a surface (in the buffer layer and the viscous sublayer) have heavy gradients. The description of such behavior requires the significant amount of nodal points at a terminating and volume (difference) way of the decision. At the same time, it is known that in a zone of developed turbulence change of a tangent component of speed depending on distance from a surface is well described by the logarithmic law, and energies of turbulence – the linear low. Therefore, for determination of values of parameters near a wall, we will use a method of wall functions [21].

The boundary conditions at the outlet of the flow from the computational region and at the open side boundaries are the equality to zero of the derivatives along the normal. In some situations, when the boundary conditions at the entrance to the computational region are unknown, the following are used: power profile for wind speed $u_{ref}\left(x_3/z_{ref}\right)^{0.16}$, for a kinetic energy $k = 3/2\left(u_{ref}Tu\right)^2$, $\varepsilon = \left(c_\mu^{3/4} \cdot k^{3/4}\right)/l$, where u_{ref} – value of the module of a vector of speed at height of z_{ref}, l – the turbulent scale of length, Tu – intensity of turbulence. Roughness height over the territory with an asphalt covering we will accept equal 0.001 m, for other territory −0.05 m.

When calculating currents around buildings, we use the method of fictitious areas, the essence of which is that the values of vector and scalar values in the obstacle area are zero, and there are no diffusion flows at the boundaries of fictitious final volumes, and the friction of the streamlined surface is taken into account using the wall functions method.

The numerical solution of the above system of partial differential equations is carried out on the basis of the finite volume method using an exploded difference grid, when the values of the velocity components are determined on the edges of finite volumes, and the scalar characteristics are in the center. To reliably take into account the effect of obstacles on the direction and strength of the surface wind near the streamlined surfaces, additional grid thickening was carried out.

After splitting the computational region in the described way, each differential equation is integrated over each finite volume. In calculating the integrals, piecewise and polynomial interpolation is used for x1, x2, x3 -dependent quantities. The approximation of the convective terms of the transport equation is carried out using the counterflow scheme MLU of Van Lear.

The approximation of diffusion terms is carried out using a second-order central difference scheme. The result of discretization is an implicit difference scheme of the second order approximation in space.

Results of Modeling. Let's apply the considered microscale model to calculation of ground distribution of a field of vectors of wind and turbulent parameters over the territory of a runway (Figs. 7 and 8).

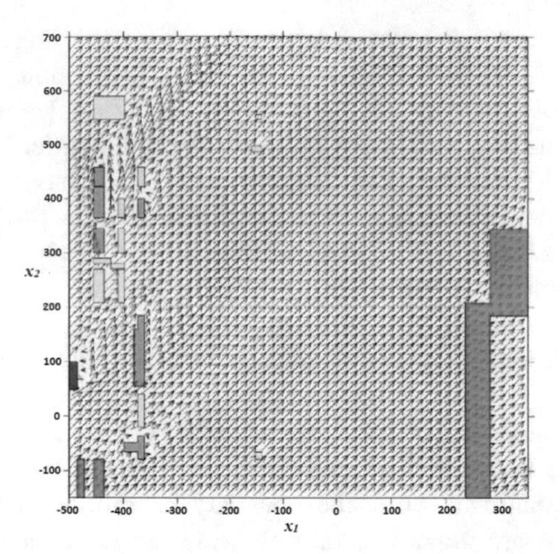

Fig. 7. The top view on the first area of a research (above). The asphalted territory is marked in gray color, rectangular figures with figures represent buildings with the indication of height (*m*), sites of a forest belt are marked in dark color. The field of vectors of a horizontal component of wind at the height of 5 m is presented below. The main direction of wind – western.

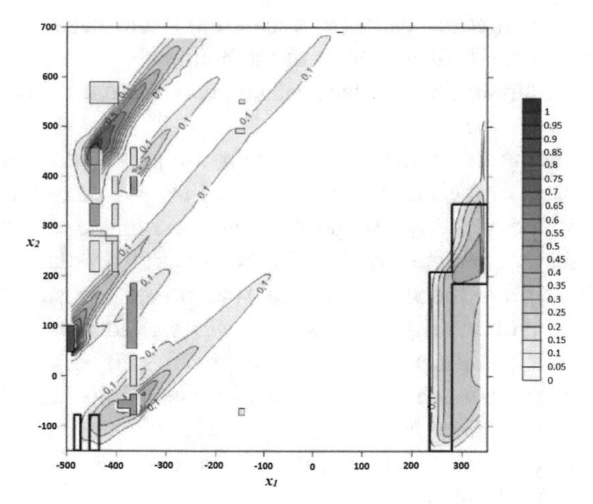

Fig. 8. Isolines of the module of a vector of the relative flow rate of W_{12} (m/s) in comparison with a stream in the absence of buildings and forests at the western direction of wind at the height of 10 m.

For the analysis of influence of the resolved elements of large-scale roughness for all considered directions of the main wind stream, the isolines of the module of the relative flow rate of W12 received on the following formula were constructed:

$$W_{12} = \left(\left(\langle u_1 \rangle - \langle u_1^0 \rangle \right)^2 + \left(\langle u_2 \rangle - \langle u_2^0 \rangle \right)^2 \right)^{0.5}$$

where speed components with a superscript "0" are calculated on condition of lack of buildings and a forest belt. Results of modeling are given in Figs. 1 and 2.

2.3 An Example of Application of Mathematical Model for a Three-Dimensional Task in Non-stationary Statement

Model of Three-Dimensional Non-stationary Eddy of the Incompressible Environment. At the present stage of development of the theory of turbulence, model operation of sinuous flows in a surrounding medium is carried out, generally with use of the equations of Navier-Stokes and transfer, average on Reynolds (Reynolds Averaged Navier – Stokes – RANS approach) for which it is required to solve a short circuit problem by engaging of semiempirical models of various level of complexity. However, despite significant progress in the development of these models, there are certain difficulties in the description of unsteady turbulent flows near poorly flowing bodies. First of all, it is caused by particular features of separated flows, namely, existence of the organized coherent structures determined by parameters of a current and geometry of area of the research. The method of large whirlwinds (Large Eddy Simulation is LES approach) is more preferable at model operation of sinuous separated flows as allows us to describe non-stationary structure of a sinuous flow, directly predicting behavior of large whirlwinds and cascade process of power transmission to more shallow whirlwinds with scales up to the size of a cell of a settlement grid. At the same time, smaller whirlwinds which can be considered isotropic are modelled by means of this or that tray model.

Model operation of a sinuous flow is carried out with use of a method of large whirlwinds which main idea consists in the formal mathematical division of large and shallow whirlwinds means of high-frequency filters. At the same time, large whirlwinds are resolved obviously, and the small-scale turbulence will be parametrized, i.e. is defined by characteristics of large-scale whirlwinds. Opportunities to it are opened by the theory of the universal equilibrium of Kolmogorov, according to which if there is no opportunity for permission of all scales of eddy at numerical model operation, then it is necessary to model small-scale whirlwinds as approximately isotropic structures in comparison with the counted obviously non-isotropic large-scale whirlwinds.

For division of large and minute structures, operation of filtration is used. It is possible to use the apparent filtration, according to which variables of cell-like scale are defined by the equation

$$\bar{f}(\vec{x},t) = \int_{R^3} G\left(\vec{x},\vec{x}'\right) f\left(\vec{x}',t\right) d\vec{x}', \quad \int_{R^3} G(\vec{x}) d\vec{x} = 1, \tag{11}$$

where f – the size which is subject to filtration, G – function of the filter with the characteristic scale of length Δ, or implicit filtration of the equations of Navier-Stokes, when the used difference grid acts as the filter.

The received mathematical model of three-dimensional non-stationary eddy of the incompressible environment includes the filtered equation of continuities and Navier-Stokes:

$$\frac{\partial \bar{u}_i}{\partial x_i} = 0; \tag{12}$$

$$\frac{\partial \bar{u}_i}{\partial t} + \frac{\partial \bar{u}_i \bar{u}_j}{\partial x_j} = -\frac{1}{\rho}\frac{\partial \bar{p}}{\partial x_i} + \frac{\partial}{\partial x_j}\left(\nu \frac{\partial \bar{u}_i}{\partial x_j}\right) - \frac{\partial \tau_{ij}}{\partial x_j} \tag{13}$$

$$\tau_{ij} = \overline{u_i u_j} - \bar{u}_i \bar{u}_j, \quad i,j = 1,2,3, \tag{14}$$

and also the filtered differential equation describing transfer of passive gaseous impurity:

$$\frac{\partial \bar{C}}{\partial t} + \frac{\partial \bar{u}_j \bar{C}}{\partial x_j} = \frac{\partial}{\partial x_j}\left(\frac{\nu}{Sc}\frac{\partial \bar{C}}{\partial x_j}\right) - \frac{\partial q_j}{\partial x_j} + F; \tag{15}$$

$$q_j = \overline{C u_j} - \bar{C}\bar{u}_j, \quad j = 1,2,3. \tag{16}$$

Here $\bar{u}_i$ – the filtered components of the instantaneous field of speed; $\bar{p}$ – the instantaneous value of pressure; ν – a kinematic viscosity coefficient; ρ – density; $\bar{C}$ – concentration of impurity; F – the function describing a source distribution; τij – a tensor of tray tension, qj – a tray mass flux; Sc – Schmidt's number. On the repeating index j toting is carried out.

The set of Eqs. (2)–(6) remains not selfcontained as expressions (4) and (6) contain the initial not filtered components of speed and concentration. τij and qj characterize influence of small-scale whirlwinds on evolution of large-scale whirlwinds, and they need to be modelled, as in approach of Bussinesk through establishment of their communication with speeds $\bar{u}_i$ and concentration $\bar{C}$. The last also makes an essence of a technique of tray model operation.

Nowadays, there is a large number of approaches to model operation of tray scales [26, 28], but the models based on use of turbulent viscosity (Eddy Viscosity Models, EVM) for not allowed obviously turbulent pulsations were most widely adopted. In such models, the stress tensor and a turbulent flow of weight are calculated by formulas

$$\tau_{ij} - \frac{1}{3}\delta_{ij}\tau_{ii} = -2K_T\overline{S}_{ij}\,, \quad \overline{S}_{ij} = \tfrac{1}{2}\left(\tfrac{\partial \overline{u}_i}{\partial x_j} + \tfrac{\partial \overline{u}_j}{\partial x_i}\right), \quad q_j = -\tfrac{K_T}{Sc_\tau}\tfrac{\partial \overline{C}}{\partial x_j}\,, \quad i,j = 1,2,3. \tag{17}$$

Here $\overline{S}_{ij}$ – the deformation speed tensor constructed on the filtered field of speed $\overline{u}_i$, $K_T = K_T(\overline{u}, \vec{x}, t) \geq 0$ – the coefficient of turbulent viscosity depending on the decision Sc_τ – turbulent number of Schmidt.

Choice of dependence $K_T = K_T(\overline{u}, \vec{x}, t)$ is extremely various. For example, for Smagorinsky's model and its modifications, the turbulent viscosity is defined on the basis of local flow parameters, and for dynamic and differential models (and their combinations) – from background of development of a current. More detailed information on models of turbulent viscosity can be found in works [26, 27, 28, 29].

In this work, Smagorinsky [30] model was used, in which influence of small-scale whirlwinds on evolution of large-scale is approximated by expression $\tau_{ij}^{Smog} = -2C_S^2\Delta_g^2\overline{S}\overline{S}_{ij}$, where Δ_g – model grid step, $|\overline{S}| = \sqrt{2\overline{S}_{ij}\overline{S}_{ij}}$ – norm of a tensor of speed of deformation, CS – Smagorinsky's constant.

Sticking and non-flowing conditions on solid boundaries were used as the boundary conditions for speed, the velocity profile was specified at the input, and the normalization of zero derivatives at the output. For the impurity concentration on the walls and the exit, the equality to zero of the derivatives along the normal is set, and the input values are zero.

Because spatial resolution for the considered case of a sinuous flow doesn't allow us to describe processes in a viscous boundary layer, total influence of a wall has to be considered by means of pristenochny model. It is known that in a zone of developed turbulence, the change of a longitudinal component of speed depending on distance from a surface is possible with a good accuracy to approximate logarithmic dependence. Therefore, for the correct description of behavior of average sizes, we use the prime pristenochny model, which isn't going out of the first estimated layer, which was successfully applied in other works and it is described in [23].

Neglecting a speed profile deviation from the logarithmic law, the speed in the first checkout over a rough surface can be determined using the following dependence:

$$\overline{u} = \frac{\overline{u}_\tau}{k} ln\left(\frac{z}{z_0}\right) \tag{18}$$

where $\overline{u}_\tau = \sqrt{\tau_w/\rho}$ – dynamic speed, τ_w – sliding friction tension on a wall, z_0 –a parameter of roughness, k ≈ 0.4 – a constant of the Pocket, z – distance to a surface. The assumption of logarithmic distribution of a profile of speed was used at assessment of the impact of tension of a sliding friction at a wall.

Approximation of a Differential Task. The numerical solution of the system of differential equations given above in partial derivatives is carried out on the basis of a method of final volume with use of the carried difference grid when values of components of speed are defined on sides of final volumes, and scalar characteristics – in the center. After splitting calculated area in the described way, each differential equation is integrated on each final volume. In calculating the integrals, piecewise-polynomial interpolation is used for x1, x2, x3 -dependent quantities. Approximation of convective members of the equation of transfer is carried out with use of the QUICK directional scheme. Approximation of diffusion terms is carried out using the central-difference scheme of the second order. The apparent time scheme (Adams – Bashford) is used to solve the transfer equations. The apparent difference scheme of the second order of approximation on time and space, which is conditionally steady, is a result of sampling. In the hydrodynamic part of the model, the predictor – corrector scheme was used to match the velocity and pressure fields, according to which the apparent scheme of Adams – Bashford for the equations of motion served as a predictor, and the correction of the velocity field satisfying the continuity equation on the new time layer equations for pressure. To solve a system of linear algebraic equations for finding pressure, we used the conjugate gradient method (CG) using the preconditioning method of upper relaxation with red-black ordering.

When calculating currents in areas of complex geometry, we used in this work the method of fictitious areas, the essence of which is that the values of vector and scalar values in the area of the barrier are zero and there are no diffusion streams at the boundaries of fictitious final volumes.

Results of Mathematical Model Operation. It necessary to check adequacy of constructed mathematical model to real sinuous flows on the example of a problem of flow of a cylinder of the square section (Fig. 1). The considered case has quite simple geometry, but is of great interest in terms of model operation of process of formation of large whirlwinds with their subsequent disintegration and dissipations in the arising unstable trace behind a cylinder (Figs. 9 and 10).

In Fig. 2, the speed field of vectors against the background of the field of pressure for three various instants is presented. Illustrations show ability of eddy-resolving model operation to describe non-stationary structure of a sinuous flow. It is shown that on the windward side of a cylinder, the area of positive pressure is formed, and in the field of turbulence pressure is, on the contrary, lower than in the neighborhood of the formed or forming whirlwind that will be completely coordinated with the mechanism of course of real physical processes (Figs. 11 and 12).

The presented simulation results are in good agreement with the experiment performed for fluid flow around a cylinder. On the basis of the constructed mathematical model of the turbulent flow of an incompressible medium, we will carry out the calculation for the three-dimensional model of a street canyon. The geometry of the study area is shown in Fig. 3.

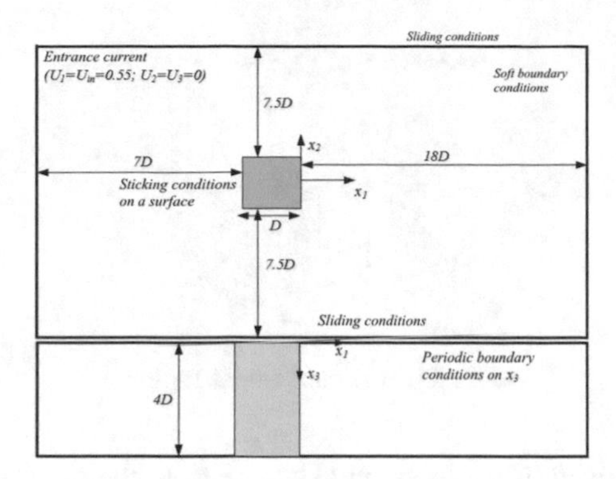

Fig. 9. Computational region.

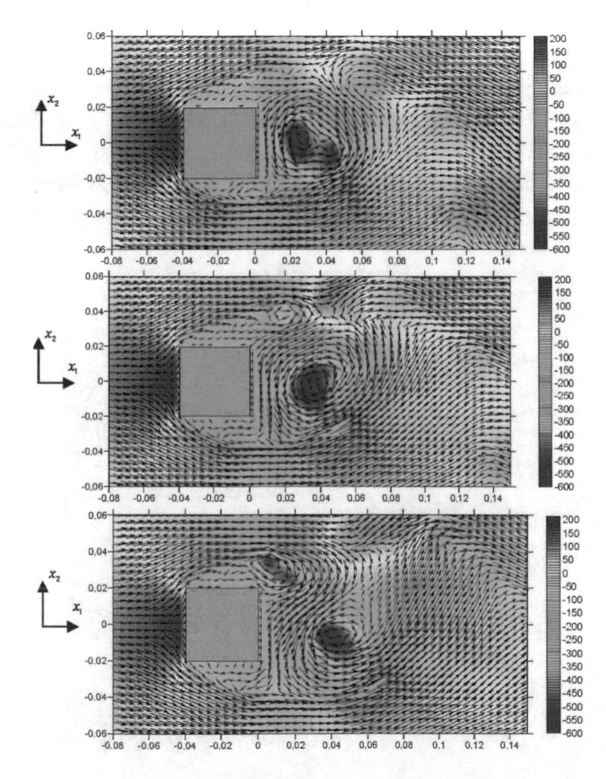

Fig. 10. A field of vectors of speed and the card of pressure in t instants $t_1 < t_2 < t_3$ where $t = t + 0.2T_{21}$, $t = t + 0.4T_{31}$, T is the whirlwind separation period.

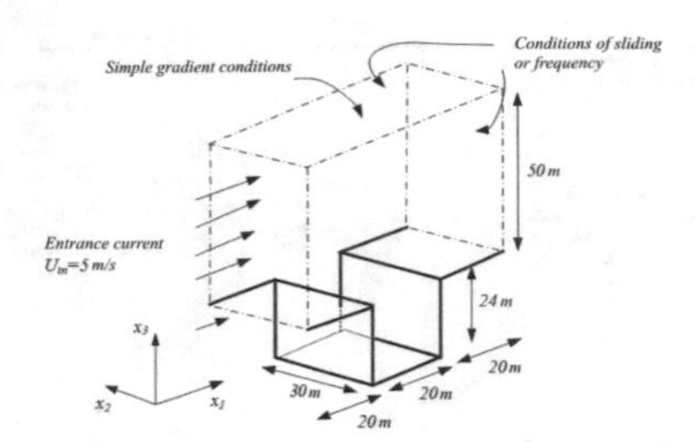

Fig. 11. Computational region.

The issue was solved in the following statement. In the transversal direction to the main stream, frequency conditions were used. In this case, it is necessary to position the lateral boundaries of the computational region so that the length of the region in the transverse direction is greater than the size of the largest vortex in the canyon. For the considered case, the largest whirlwind is limited to canyon width W = 20 m, then the cross length of a canyon can be taken equal 30 m (L/W = 1.5). Periodic boundary conditions were set in a lengthwise direction with the purpose to imitate the infinite series of canyons. Calculations were carried out on a grid 182 × 54 × 180. The source of impurity of constant intensity was located near a surface at height h z = 0.125 m over the center of foundation of the considered region. Results of model operation are given in Fig. 4.

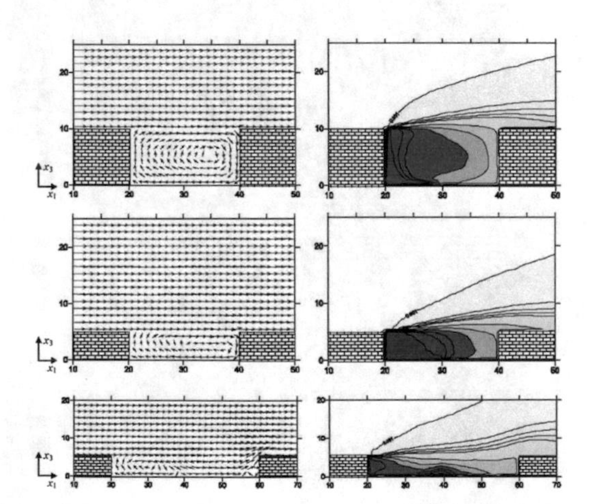

Fig. 12. Isolines of concentration of the impurity average along Ox_2 axis, field of vectors of speed, U_{in} = 5 m/s; impurity source (x_1 = 30 m, x_2 = 15 m, x_3 = 0.125 m).

3 Conclusion

The following were proposed: a mathematical model based on the discrete vortex method, which allows us to construct a picture of current lines, to obtain a wind stream velocity field near buildings or architectural structures; three-dimensional microscale meteorological model on the example of studying the wind situation in the vicinity of the airport, taking into account the influence of buildings of different heights and heterogeneity of the underlying surface properties.

References

1. Wu, H.: Designing for pedestrian comfort in response to local, 394–407 (2012)
2. Blocken, B.: Pedestrian wind comfort around a large football stadium in an urban environment **97**(5–6), 255–270 (2009)
3. Blocken, B.: Modification of pedestrian wind comfort in the silvertop tower passages by an automatic control **92**(10), 849–873 (2004)
4. Bosch, G.: Simulation of vortex shedding past a square cylinder with different turbulence models **28**(4), 601–616 (1998)
5. Franke, J.: Recommendations on the use of CFD in wind engineering **81**(1–3), 295–309 (2004)
6. Mohamed, S.F.: CFD simulation for wind comfort and safety in urban area: a case study of coventry university central campus **2**, 131–143 (2013)
7. Murakami, S.: Overview of turbulence models applied in CWE **1**, 74–76 (1998)
8. Paterson, D.A.: Computation of wind flows over three - dimensional buildings **24**(3), 193–213 (1986)
9. Stathopoulos, T.: Pedestrian level winds and outdoor human **94**(11), 769–780 (2006)
10. Wu, H.: Designing for pedestrian comfort in response to local **104–106**, 394–407 (2012)
11. Yoshie, R.: Cooperative project for CFD prediction of pedestrian wind environment in the architectural institute of Japan **95**(9–11), 1551–1578 (2007)
12. Shih, T.A.: New k-e eddy viscosity model for high Reynolds number turbulent **24**(3), 227–238 (1995)
13. Belov, I.A.: Interaction of unsteady flows with barriers. Leningrad (1983)
14. Retter, E.I.: Architectural and structural aerodynamics, Moscow (1984)
15. Belotserkovsky, S.M.: Mathematical model operation of plane-parallel detachable flow of bodies, Moscow (1988)
16. Belotserkovsky, S.M.: Model operation of detachable flow of a cylinder near the screen **2**, 78–84 (1986)
17. Rusakova, T.I.: A numerical research of structure of a vortex stream about high-rise constructions **1**, 154–160 (2006)
18. Rusakova, T.I.: Problem of numerical calculation of flow of buildings with an airflow **11**, 53–58 (2007)
19. Samarsky, A.A.: Mathematical model operation, Moscow (2001)
20. Goman, O.G.: Numerical model operation of rotationally separated flows of incompressible liquid, Moscow (1993)
21. Launder, B.E., Spalding, D.B.: The numerical computation of turbulent flows **3**(2), 269–289 (1974)

Principles and Implementation of Daylighting Systems in Classrooms

Milan Tanić[1(✉)], Danica Stanković[1], Slavisa Kondić[1], and Vadim Kankhva[2]

[1] Faculty of Civil Engineering and Architecture, University of Niš, Aleksandra Medvedeva 14, 18000 Niš, Serbia
milan.tanic@gaf.ni.ac.rs
[2] Moscow State University of Civil Engineering, Yaroslavskoye Shosse 26, 129337 Moscow, Russia

Abstract. Based on the architectural characteristics of a classroom, this paper examines the principles of design and the possibility of implementing daylighting systems in classrooms. The quality of daylight in classrooms depends on a whole range of factors, and it is necessary to analyze and evaluate daylighting systems in each specific case. This paper analyses architectural concepts and daylighting systems that appropriately admit natural daylight into the classroom and also eliminate possible negative effects. In addition to passive systems, special attention is paid to the advanced systems that can actively control light levels. The basic factors that directly affect the quality of daylight in classrooms are defined and are related to: classroom layout, natural lighting modes, applied daylighting systems for dispersion of natural lighting and the control of the luminous environment in the room.

Keywords: Daylighting systems · Illuminance · Classroom · School building

1 Introduction

The primary light source in the past was daylight, and it is to a considerable extent today. Our estimates of how things should look like are based primarily on their appearance in daylight, and artificial lighting often does not support the improvement of their visibility. This is also due to the fact that artificial lighting sources often produce too many visual effects. Without neglecting the significance and role of artificial lighting, natural light has remained a mainstay in architectural design, especially in those buildings which purpose indicates a certain connection with the outside environment. Daylighting strategies and architectural design strategies are inseparable. Daylight not only replaces artificial lighting, but also influences both heating and cooling loads [1]. Daylighting in buildings has direct and indirect impact on the quality of space, health of the occupants, and energy efficiency of the building. Compared with artificial lighting, daylighting provides an ideal color rendering environment, as well as psychological and physiological benefits for the occupants [2–6].

The nature of teaching and learning activities in school buildings, especially in classrooms, points out to the extraordinary importance of daylight. Orientation towards

V. Murgul and M. Pasetti (Eds.): EMMFT-2018, AISC 983, pp. 720–731, 2019.
https://doi.org/10.1007/978-3-030-19868-8_70

the sunny side is of great importance, especially when it comes to lower school classes, where the classroom is a place where students spend most of their time at school. The fact that most of the activities take place inside the classroom, during the day, classifies natural daylight into a group of the most significant psychological and physiological factors of working conditions in the school building [7]. By taking advantage of daylight in schools or the educational environment, classroom light amount and physical health level will increase, stress will decrease, success will improve and, as a result, students efficiency increases as observed and shown in the results of various research works [8, 9]. This paper examines the principles of architectural design and the possibility of implementing daylighting systems in classrooms. The question of the daylight ambient in the classrooms means to provide adequate conditions for maximizing the use of natural light in the building, at different times, in order to make relatively unified brightness throughout the space. Architectural concepts and daylighting systems are considered, which appropriately admit daylight into the classroom, but also eliminate the possible negative side effects which, besides positive ones, directly affect the quality of daylight in classrooms.

2 The Quality of Daylight in the Classroom

In order to achieve the optimum quality of daylight in classrooms, the illuminance is as important as the harmonious relationship between factors that directly affect the quality of the light ambience. These factors, unlike daylight levels, can be regulated by the design of the building, which includes the issue of facing the building towards the appropriate side of the world, the way of natural lighting, unifying light levels in each part of the classroom, controlling the light environment, etc. [10].

Systematizing these factors, the quality of daylight in the classroom can be determined by:

- disposition of classrooms/school buildings,
- the way of natural sunlight admission into classrooms,
- technical means applied in the dispersion of daylight,
- controlling the light ambient inside the room.

3 Disposition of Classroom

Depending on the season and the weather during the day the position of the sun is different, so it is important that the school building is located and oriented in a way that takes the most favorable position in relation to the course of the sun's movement.

In addition to natural conditions, the intensity of the insulation and the degree of heating of the school premises will depend on the position of the school facility at the selected location, the distance and height of adjacent buildings and the greenery around the school building, the projected dimensions and orientations of the spatial units of the school. Therefore, the orientation, the shape and the position of the school building stand out as very important factors for achieving the appropriate quality of daylight in classrooms. It is necessary to conduct such an analysis at an early stage in planning and designing a school building in order to make the most of available daylight.

3.1 The Position of the School Building

The optimum insolation of the school building is a basic precondition for achieving a quality lighting environment in the classrooms. In that sense, ensuring adequate daylight access into classrooms is one of the most significant factors in determining the most favorable position of the school building on the selected location.

In addition to urbanistic parameters that define the position of the school building, in the context of a proper insolation, a special attention is paid to the position of the school building in relation to adjacent buildings as well as to already existing and planned vegetation. The procedure for determining the position of the building in relation to adjacent objects must be in accordance with the condition that the incident angle of daylight, on the furthest desktop workplace from the window, is at least 23° [11]. In addition to this condition, in order to provide the scattered sunlight to be visible from each workplace, it is considered that, regardless of the existent shelter, every workplace, even the furthest one, should be provided with suitable visual contact with the celestial vault. In addition, the angle that includes the field of view in the furthest place in the classroom is 5° (see Fig. 1) [12].

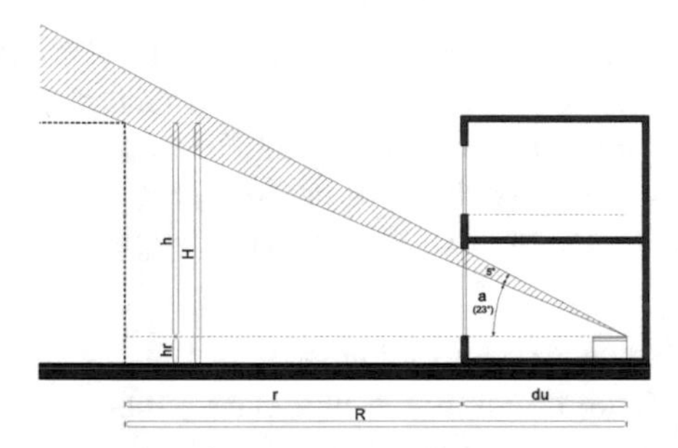

Fig. 1. Distance from the school building to the adjacent facility/obstacle [12].

The distance from the school building to the adjacent buildings is measured at a horizontal level 80 cm from the floor, at the height of the workplace. Having in mind the conditions set, the position of the school building can be defined as follows:

$$R = \frac{h}{tg\,\alpha} \tag{1}$$

and as

$$R = r + du \text{ and } H = hr + h \tag{2}$$

then follows that

$$r = \frac{H - hr}{tg\,\alpha} - du \tag{3}$$

r – the minimum distance from the school building to the neighboring building,
R – the distance from the school building to the furthest workplace in the classroom,
du – the depth of the classroom,
hr – the height of desktop level,
h – the height of the object from the horizontal plane at the height of the desktop level to the top of the obstacle,
H – the total height of the adjacent object, or the obstacle,
α – the angle of daylight at the farthest desktop from the window in the classroom [12].

For different heights of the opposing objects, using a pattern for determining the position of the school building, the rule is that the distance from the classroom to the opposite building should be more than 1.5 times of the height of the building, while the shortest distance shouldn't be less than 12 m.

3.2 The Orientation of the Building

The orientation of school spaces influences numerous factors of the architectural design of a school building, and it directly depends on the specific microclimatic conditions that prevail in the area of the site, the topographic characteristics of the school ground, the layout of the surrounding buildings, the type of natural sun-lighting and the ventilation of the school premises. The requirements in regard to the orientation of school premises, depending on their function, normatively prescribe optimal sunlight for each school space individually or a group of them. In doing so, their orientation must be most favorable for each specific location.

In European weather conditions, it is considered that the most favorable orientation of the general rooms for teaching (classroom) is south or south-east. This conviction starts from: first - students, while in school, spend most of their time in the classroom; second - the southern side has the highest intensity of bactericidal activity of the sun's air. The advantage of the southern orientation is reflected in the fact that it allows for a large interval of daily sunshine, which is very important during the winter, while during the summer, due to the high position of the sun, sunshine and, consequently, room heating is incomparably less compared to the eastern and western orientation. As these are factors that should not be deviated, in the design process they directly and most importantly influence the disposition of the classroom.

In rooms oriented towards east or west, due to the relatively low position of the sun and intense sunshine, there is often excessive heating as well as the appearance of glare, harmful shaded shadows and strong contrast of light. Such occurrences are the result of direct position towards the sun, as well as a diffuse skylightning. When it comes to these two orientations, it is necessary, at the time of intensive insolation of rooms, to take appropriate measures of protection in order to minimize the occurrence of such negative effects.

3.3 Form of Building

The method of grouping the classrooms, from the aspect of improving the quality of daylight, has caused the appearance of other school building systems derived from the two basic, corridors and halls'. The need for a better lighting of classrooms, on the one hand, with the best possible ventilation, on the other hand, has created an extremely large variety in choosing the potential form of a school building. The most favorable daylight admission can be achieved by the composition of a school building in which the long and narrow pavilions dominate, where the space between the school building pavilions should be of a sufficient depth to avoid mutual shading (see Fig. 2). This implies that such forms are not too dispersed but organized so they form a unique spatial structure, in accordance with the conditions that the site provides.

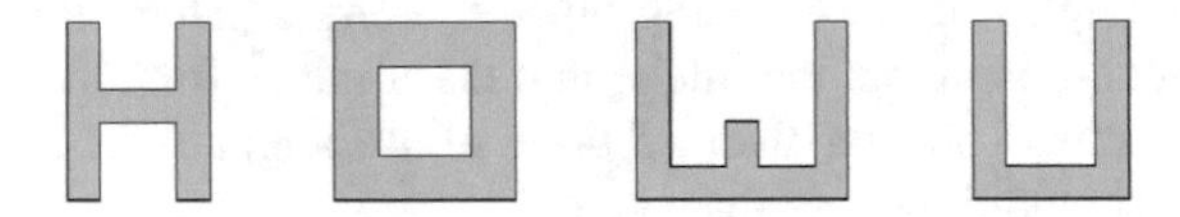

Fig. 2. Simplified examples of possible school building forms with a high daylight access [13].

The application of a solution with atrium units is a characteristic of the compact school building design, especially when the location conditions do not allow the dispersed form of a school building. In such cases, a group of classrooms, to which natural lighting is of special importance, is planned in the central zone of the building and organized around the atrium unit, taking into account the required orientation and optimal depth of each room separately.

4 The Modes of the Natural Lighting

Daylight levels in school facilities are significantly influenced by the architectural design, that is, the shape, height and depth of the classroom. Depending on these factors, an appropriate way of daylight admission into the classrooms is envisaged. There are several possible ways of natural lighting of the classroom:

- one-sided lighting,
- double-sided lighting,
- multi-sided lighting,
- zenithal lighting.

Dimensions of openings through which the daylight into the classroom enters are limited by the height of the room, so that the determination of these factors are interdependent. The modern hygienic-technical standards for the school space require the following relationship between the surface of the window and the classroom:

- one-sided luminosity: 1: 3 to 1: 4,
- double-sided luminosity: 1: 2. [11, 12] (Fig. 3).

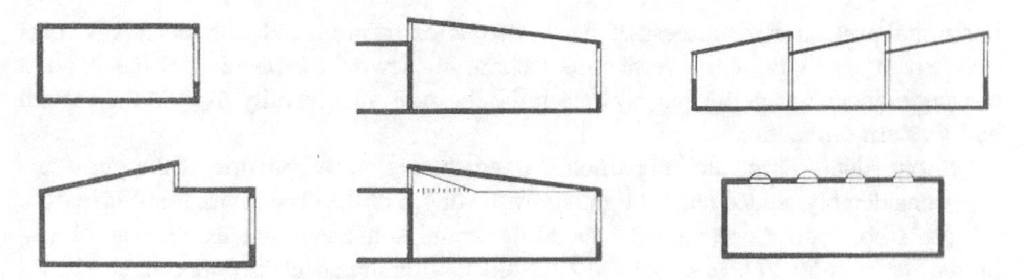

Fig. 3. Schematic cross-sections of one-sided, double-sided, multi-sided and zenith-lit classrooms [14].

5 The Application of Technical Means for Dispersion of Daylight

Preventing direct sunlight into the classrooms in order to eliminate the conditions for creating glare or overheating is done with fixed or movable sunscreen elements (shelters, blinds …) at times of high daylight intensity (see Fig. 4). In addition, the construction and position of the sun protection elements are such that they neither reduce the useful area of the window nor allow the harmful shadow. The optimum window design should integrate all function [5].

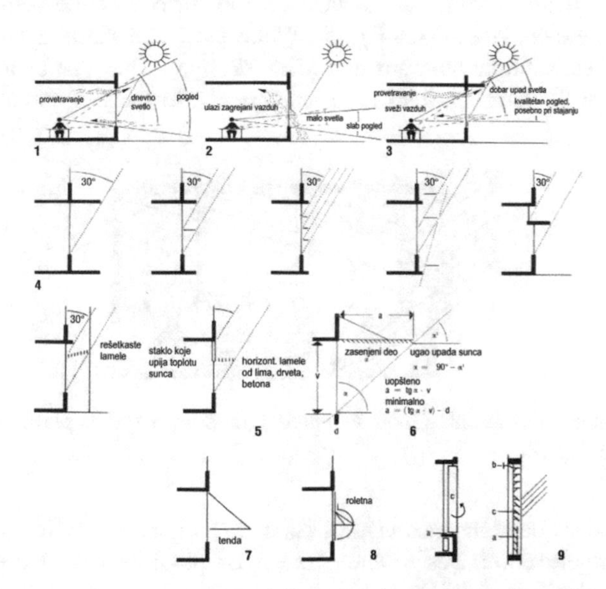

Fig. 4. The examples of sun protection elements [14].

The choice of technical sunscreen elements and their effectiveness depends on the orientation of the object. The sun protection is different when glass surfaces are oriented towards the south from those oriented towards the east or west. As a rule, horizontal protection, regardless of their construction, is projected with facade openings oriented to the south, while vertical protection, in view of the position of the sun and the angle under which the sun rays reach the room, is successfully used in the eastern and western orientations.

In one-sided illuminated classrooms, fixed shades on the outside of the glass surface considerably reduce natural lighting by the depth of the classroom. For this reason, the most commonly used are the mobile external sunscreen shades such as blinds, protective curtains etc. In case of double-sided illuminated classrooms, fixed shades increase the uniformity of lighting and reduce glare in the vicinity of the window.

The affordable lighting quality is achieved by unifying the intensity in each part of the classroom. The basic purpose of this procedure is to use a corresponding architectural solution of the exterior classroom wall to break a part of the daytime light and instead of directly transferring it into the classroom, sending it towards the ceiling of the classroom, and then from there it reflects on the desktop work surfaces [15]. Light redirecting strategies are used to refract light into the unlit areas to improve the daylight penetration. Systems that utilize this strategy include light shelves, anidolic mirrors, and prismatic panels [2, 16].

The uniform lighting of the classroom through the dispersion of sun rays can be accomplished in many ways, using different technical devices.

Some of the standard solutions indicate the application of the appropriate type of glass, glass prism or other glass-based products in the upper part of the facade openings so that one part of the daylight is reflected on the surface of the ceiling and on the opposite wall of the classroom (see Fig. 5). When used as a shading system, prismatic panels reflect direct sunlight but transmit diffuse skylight. They can be applied in many different ways, in fixed or sun-tracking arrangements, to facades and skylights [17].

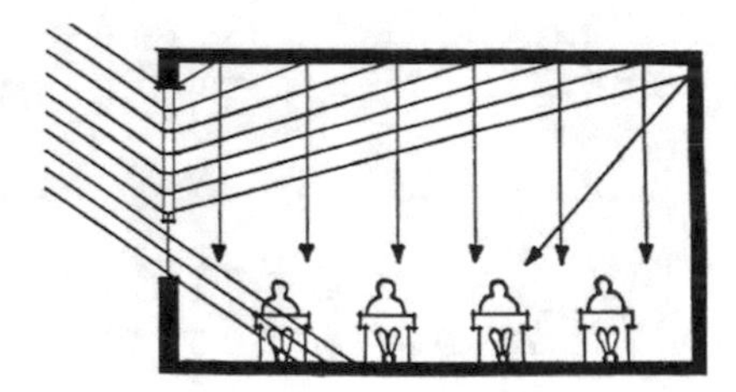

Fig. 5. A single-luminous teaching room in which a glass brick slab is placed in the upper part of the façade opening [12].

Another modern daylight treatment of classrooms is based on the application of a solution that largely eliminates a whole range of possible negative effects that, in addition to effective one, daylight brings.

Bearing in mind the justification of the use of certain technical sunscreen protection elements in view of the required south orientation, the quality of the lighting environment in the classrooms is especially contributed by the use of light shelves, that is, the horizontal shelf that the façade openings divides into the upper and lower areas [18]. Therefore, a plate as a kind of a shelf between the upper and lower pairs of window sills has got double purpose - to prevent the penetration of direct sunlight in depth, excessive glare and heating, and second, as a reflector of light upward in the direction of the ceiling and then indirectly and diffusely in the depth of the room (see Fig. 6).

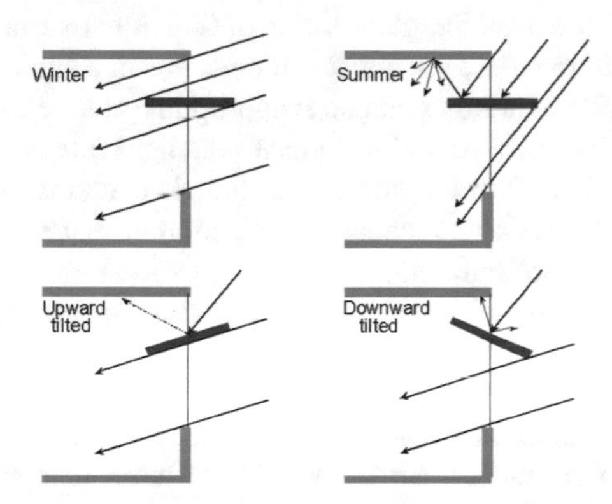

Fig. 6. The position of the light shelves [1].

Light shelves affect the architectural and structural design of a building and must be considered at the beginning of the designing phase because they require a relatively high ceiling to function effectively. Light shelves should be specifically designed for each window orientation, room configuration, and latitude. They are typically positioned to avoid glare and maintain a view outside [1].

The concept of light shelves results in the effect of a uniformly illuminated space over the upper zone of the window and an open view from the inside, protected by a shade of direct light, in the lower zone of the window.

6 Controlling the Light Ambient Inside the Room

In the context of the previously discussed daylighting systems for achieving optimum access to natural light and the corresponding light level, special attention is given to controlling the lighting environment inside the classroom. On this basis, in accordance with the nature of activity in the classrooms, two types of control can be distinguished:

- control of internal surface reflection, i
- integration of the daylight and artificial lighting system.

6.1 Internal Surface Reflection Control

In order to achieve the optimum quality of daylight in classrooms, given the nature of the "behavior" of light, it is necessary to harmonize the surface characteristics of all elements of the interior, especially in relation to light reflection and potential possibilities for creating glare, light contrast and shadow.

The reflection factor is expressed by the percentage of light that is reflected from one surface, and depends on the type of material, color and method of finishing surfaces in the classroom.

The glare may badly affect the quality of working conditions in the classrooms. Glare occurs due to uneven contrast at the view on desktop and surrounding surfaces, and it is directly dependent on the glare factor of individual surfaces in the classroom. The primary sources of the glare are direct sun rays, then the sun rays reflected from the facade walls of the school building and an inappropriate source of artificial lighting, as a supplement to daylight in a period of reduced daylight intensity.

We can significantly reduce the intensity of the glare, and it can be even reduced to the smallest possible extent by matching the selection of materials and color elements of the interior in the school building.

In order to avoid the possibility of creating a shadow, which would possibly impede teaching activities in the classroom, it is necessary to anticipate as small as possible cross-sections of the window frame and interstellar columns.

Hence, it is inadmissible to allow the glare, shadows and strong light-dark contrast in the visible area of a student. Stronger daylight admission into the depth of the classroom is facilitated by favorable window layout and with as much as possible shade diffusion on the desktop surface. Bright and vivid pastel colors much better reflect light than dark ones. Therefore, it is necessary to use the reflection table (Table 1) when choosing the colours of the ceilings, walls, floors and furniture. Based on the values shown in this table, in the choice of colors, it is preferable that ceilings are white and the walls green, yellow, light blue or light gray. Classrooms facing north can have ceilings in yellow tones.

Table 1. Values of the reflection percentage for certain elements of interior and for some colors [14].

PERCENTAGE OF REFLEXION				
Ceiling	Walls	Floor	Blackboard	Tables
70-80%	50-60%	15-30%	~ 20%	~ 40%
PERCENTAGE OF REFLEXION - COLOURS				
White				86%
Light yellow				76%
Golden yellow – yellow green				60%
Bright red, pink				61%
Light brown, light blue, silver gray				27 – 40%
Brown, Olive Green				20 – 30%
Dark brown, dark red				10 – 20%
Dark blue				6%
Black				4%

The school boards mustn't be shiny but it can be in different colors. Green is the best choice. For desks or tables the most appropriate colour is green or wood colour (reflection rate 35–50%).

6.2 The Integration of the Daylight and Artificial Lighting System

Improving the daylighting performance of buildings also contributes in reducing the need for artificial lighting with improved distribution of daylight throughout the interior space [2, 19]. In a period of reduced natural light intensity, daylight can not provide the most favorable luminance conditions in the classroom. In such cases, in order to maintain the quality of the light ambient within the limits of optimal values, it is necessary, as a form of supplementation, to include or integrate an artificial lighting system. Artificial lighting should replace daylight until the average light levels on desktops are reached. When choosing a type of artificial lighting, systems with the color of the light and the angle at which the light rays fall approximate to the characteristics of daylight are preferred. The position of the lighting fixtures in the classroom should ensure uniform illuminance on all desktop surface areas and prevents the creation of larger light contrasts.

Solutions that deal with the integration and automatisation of daylight and artificial lighting systems function with the help of appropriate sensors, based on which input data are collected, and then according to certain parameters, light levels are controlled throughout the classroom from one or several control points (see Fig. 7).

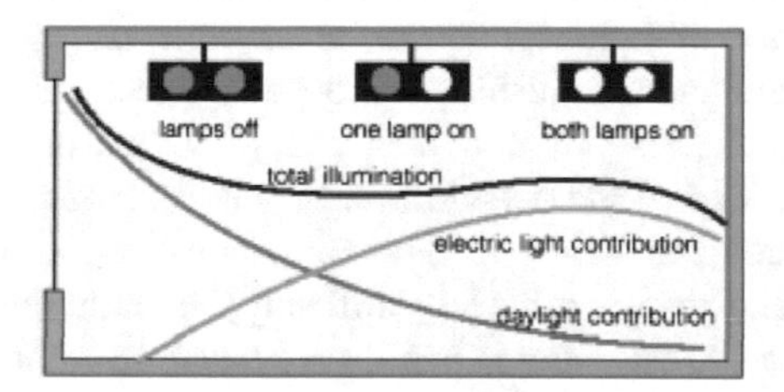

Fig. 7. The integration of daylighting and artificial lighting provides uniform illuminance levels in the room [13].

Combining daylighting and artificial lighting systems through, for example, a combined control strategy or the integration of lamps in an interior light shelf, is a design option in retrofits as well as new construction [1].

Automated lighting systems can improve classroom lighting solutions. In such systems, computers are used to control and manage a large number of parameters. This leads to the term "intelligent building", up to the moment when according to the predetermined rules, all energy parameters in one building are automatically regulated.

7 Conclusion

The consideration of the quality of the daylight ambience in the classrooms points to a set of principles and systems that should be applied in the process of architectural design of the school building. Applied solutions should, in addition to maximizing the daylighting in the building, also ensure uniform illuminance of space.

Requirements related to the illuminance of enclosed spaces regulate the minimum illuminance level in one workplace, depending on the activity that takes place in them. When it comes to school spaces, modern technical standards require that the average value of the daylight factor is from 2% to 5% of the minimum light intensity in an open space of is 5000 lx, measured at the workplace that is most distant from the window.

Given the optimal daylight levels, the basic factors that directly affect the quality of daylight in classrooms are defined, and they relate to: classroom layout, natural lighting modes, applied daylighting systems for dispersion of natural lighting and the control of the luminous environment inside the classroom. The classroom disposition is one of the passive methods for achieving adequate daylight. The orientation and position of the school building are determined by the sides of the world as well as by the distance of the potential shades. The form of the building is preconditioned by the way classrooms are grouped, with their consistent or dispersed form, that is, the central or peripheral position, to a large extent determine the measure of daylight admission. Depending on the shape, height and depth of the classroom, the appropriate way of admitting daylight into classrooms is determined. On this basis, several possible ways of natural lighting of classrooms are distinguished: one-sided, double-sided, multi-sided and zenithal lighting. The application of technical means for daylight dispersion is classified as the passive methods of luminance adjustment in rooms. The main purpose of this procedure is to unify the brightness level in each part of the classroom, that is, through a certain architectural design of the outside wall of the classroom, a part of the daylight, instead of being directly transferred to the room, is broken and directed towards the ceiling of the room, and then is reflected (diffusely) towards the desktop surfaces.

Lighting control inside the room is based on procedures that should harmonize the surface characteristics of all interior elements (internal surface reflection control) to prevent the possibility of creating excessively large glare, contrasts and harmful shadows due to direct or diffuse lightning. As there is natural light only during a day, in the amount and intensity that can not always be foreseen, it is necessary to integrate an artificial lighting system in a period of reduced sunlight intensity in order to maintain the quality of the light ambient within the limits of optimal values. This method can be classified into active daylighting systems. Therefore, the quality of daylight in classrooms is influenced by a number of factors, and it is necessary to analyze and evaluate the implementation of daylighting systems in each specific case.

References

1. Aschehoug, Ø., Edmonds, I., Christoffersen, J., Jakobiak, R.A.: Daylight in Buildings - A Source Book on Daylighting Systems and Components. Lawrence Berkeley National Laboratory, Washington (2000)

2. Park, D., Kim, P., Alvarenga, J., Jin, K., Aizenberg, J., Bechthold, M.: Dynamic daylight control system implementing thin cast arrays of polydimethylsiloxane-based millimeter-scale transparent louvers. Build. Environ. **82**, 87–96 (2014)
3. Galasiu, A.D., Veitch, J.A.: Occupant preferences and satisfaction with the luminous environment and control systems in daylit offices: a literature review. Energy Build. **38**(7), 728–742 (2006)
4. Hocheng, H., Huang, T.Y., Chou, T.H., Yang, W.H.: A brighter place: overview of microstructured sunlight guide. J. Achiev. Mater. Manuf. Eng. **43**(1), 409–417 (2010)
5. Huang, T.Y., Hocheng, H., Chou, T.H., Yang, W.H.: Bring free light to buildings: overview of daylighting system. In: Mendez-Vilas, A. (ed.) Materials and Processes for Energy: Communicating Current Research and Technological Developments, Formatex Research Center (2013)
6. Veitch, J.A., Galasiu, A.D.: The Physiological and Psychological Effects of Windows, Daylight, and View at Home: Review and Research Agenda. National Research Council of Canada, Ottawa (2012)
7. Fathutdinova, L.R., Chernyshova, E.P.: Psihologiya vospriyatiya ploshchadi. In: Aktual'nye problemy arhitektury, stroitel'stva i dizajna, pp. 120–128. Magnitogorskij gosudarstvennyj tekhnicheskij universitet im. G.I. Nosova, Magnitogorsk (2014)
8. Moazzeni, M.H., Ghiabaklou, Z.: Investigating the influence of light shelf geometry parameters on daylight performance and visual comfort, a case study of educational space in Tehran Iran. Buildings **6**, 1–16 (2016)
9. Boubekri, M.: Daylighting, Architecture and Health: Building Design Strategies. Routledge, London (2008)
10. Futrell, B.J., Ozelkan, E.C., Brentrup, D.: Optimizing complex building design for annual daylighting performance and evaluation of optimization algorithms. Energy Build. **92**, 234–245 (2015)
11. Anđelković, M.: Školske zgrade. Građevinski fakultet u Nišu, Niš (1995)
12. Bajbutović, Z.: Arhitektura školske zgrade. Svjetlost – Zavod za udžbenike i nastavna sredstva, Sarajevo (1983)
13. Pasini, I.: Daylighting guide for Canadian commercial buildings. Travaux Publics et Services Gouvernementaux, Ontario (2002)
14. Auf-Franić, H.: Osnovne škole. Golden marketing – Tehnička knjiga; Arhitektonski fakultet sveučilišta u Zagrebu, Zagreb (2004)
15. Boubekri, M.: Daylighting Design: Planning Strategies and Best Practice Solutions. Birkhäuser, Basel (2014)
16. Ullah, I., Shin, S.: Uniformly illuminated efficient daylighting system. Smart Grid Renew. Energy **4**(2), 161–166 (2013)
17. Rayaz, S., Rubab, S.: Review of advanced daylighting systems. Mater. Sci. Forum **760**, 79–84 (2013)
18. Meresi, A.: Evaluating daylight performance of light shelves combined with external blinds in south-facing classrooms in Athens. Greece. Energy Build. **116**, 190–205 (2016)
19. Krarti, M., Erickson, P.M., Hillman, T.C.: A simplified method to estimate energy savings of artificial lighting use from daylighting. Build. Environ. **40**(6), 747–754 (2005)

Flexible Design of Small Residential Spaces: Implementation Possibilities in Serbia

Slavisa Kondic(✉), Milica Živkovic, and Vojislav Nikolic

Faculty of Civil Engineering and Architecture, University of Nis, Niš, Serbia
skondic555@gmail.com

Abstract. Due to economic crisis and impoverishment of population, housing context in the Republic of Serbia at the end of 20th and the beginning of 21st century is often characterized by too small dwellings in comparison to the households using them. This can be concluded by the analysis of the data from The Census of Population, Households and Dwellings in the Republic of Serbia in 2011. These results show deficit of large apartments with 3 and more bedrooms in comparison to the number of households of 4 and more members, indicating lower quality of life and housing conditions of the tenants. Flexible design of the apartment could be a viable solution to this problem, improving the quality of life of the users of these overpopulated apartments. Flexibility allows changing the configuration of the apartment to adapt it to changing user needs. This paper proposes models for the improvement of living conditions by using flexible design.

Keywords: Flexibility · Architectural design · Small apartments

1 Introduction

General housing context in the Republic of Serbia at the end of 20th and the beginning of 21st century is characterized by severe economic crisis, reflecting in impoverishment of population. This created market conditions that became one of two key determinants of spatial processes [1]. Due to the low financial potency of future tenants, the market demands smaller apartments with reduced areas. This reflects on the requirements of the investors and defines architectural design principles.

The results of The Census of Population, Households and Dwellings in the Republic of Serbia in 2011 show significant deficit of large apartments with 3 and more bedrooms (22.59%) in comparison to the number of households of 4 and more members (34.13%). These results show that the number of rooms often does not correspond to the number of the users of the dwelling. The principle of one bedroom for parents and separate bedroom for each other member of the family, that is necessary for comfortable usage of the dwelling, is obviously not accomplished. The households of 4 and more members are often forced to use smaller apartments with two or often even one bedroom, due to the lack of finances. Previous research had shown that large number of households live in inadequate overcrowded living space. Often, some members do not have their own bedroom (usually parents) or adult children share a common room [2]. These are unfortunate consequences of negative social and

V. Murgul and M. Pasetti (Eds.): EMMFT-2018, AISC 983, pp. 732–739, 2019.
https://doi.org/10.1007/978-3-030-19868-8_71

economic processes in Serbia that must be accepted as reality. But it is necessary to find adequate response to these problems that would facilitate and improve the quality of life of households that are unable to provide adequate spatial comfort for healthy living. Some researchers formulated mathematical models of process optimization for the architectural layout of small apartments [3]. These models could be valuable as a theoretical research, but have very limited practical value and application, due to their complexity. Also some of the authors were considering minimal dwellings as a totally new way of living [4, 5].

Flexible design could be a viable solution to the problem of small dwellings that do not respond to the needs of the tenants. In 1927 Mics van der Rhoe wrote: "Today the factor of economy makes rationalization and standardization imperative in rental housing. On the other hand, the increased complexity of our requirements demands flexibility." [3]. This quote is limited to rental housing, but economy and rationalization are also important demands for contemporary multi-family housing in Serbia today. Different authors considered flexibility as very important aspect of architectural design, essential to the resilience of the buildings and sustainable city development [6], ability and potential of a building to change, adapt and reorganize itself in response to the changes [2], attempting to propel architecture into a more efficient and sustainable cultural product for a dynamic society [1]. Some researchers even go a step beyond and regard flexible architecture as buildings that are intended to respond to evolving situations in their form, operation, or location, crossing the boundaries between architecture, interior design, product design, and furniture design [8], architecture that adapts rather than stagnates; responds to change rather than rejects it; is motive rather than static [7].

Flexible approach to architectural design could be a viable model of improving housing quality in given market and economic conditions. Flexibility can also be considered as a necessary means in design of small dwellings. "If there was to be less space, then that space needed to be used in as efficient and flexible manner as possible" [12]. Changing capacity of the space by horizontal partitioning had been recognized as a solution to the problem of inadequate size of the dwelling by some researchers [5]. Also, flexibility is recognized as a means of achieving privacy in compact dwellings [13], as well as a solution to the lack of space enhancing life quality [11]. Applying the complex system of movable walls and foldable beds and other pieces of furniture enables the multifunctional and efficient use of space [16], which is very important for small residential spaces.

2 Methodology

In the beginning of the research it is necessary to identify the problem. This would be achieved by analysis of the data from The Census of Population, Households and Dwellings in the Republic of Serbia in 2011 [20]. These data would show ratio between dwellings of certain spatial structure (defined by number of bedrooms) and number of households that could comfortably use these dwellings. The ratio under 1 shows deficit of apartments of adequate size and spatial structure for given structure of household.

After the problem is identified potential improvement models should be formulated and classified through analysis of relevant examples of optimized small residential spaces as well as available literature on this subject. Flexibility, as an improvement model for the design of minimal dwelling, should be explored and evaluated.

3 Research Results

The results of previous research were summarized in two subsections. The first subsection shows the results of the analysis of the data from The Census of Population, Households and Dwellings in the Republic of Serbia in 2011. These results were summarized in Table 1. Second subsection contains the analysis of the relevant examples of flexible design from the literature.

3.1 Analysis of the Data from the Census

The results of the analysis of the data from The Census of Population, Households and Dwellings in the Republic of Serbia in 2011 are shown in Table 1. The number of dwellings with one bedroom, suitable for one or two users corresponds to the number of adequate households. But with larger apartments suitable for families there are obvious discrepancies.

Table 1. Number of dwellings according to the number of bedrooms and occupants in the Republic of Serbia

Dwellings according to the number of bedrooms	1	2	3	4	5 and more	Total
	1173883	701899	318333	117585	111508	2423208
	48.44%	28,97%	13.14%	4.85%	4.60%	
Dwellings according to the number of occupants	1 or 2	3	4	5	6 and more	Total
	1136498	459719	441108	202688	183195	2423208
	46.90%	18.97%	18.20%	8.36%	7.57%	
Ratio	1.03	1.53	0.72	0.58	0.61	

The number of smaller two bedroom family apartments is much larger than the number of corresponding three member households (most commonly parents with one child). The percentage of two bedroom apartments is 29.97% and the percentage of three member households is 18.97%. On the other hand, the percentage of dwellings with three or more bedrooms is 22.59% and the percentage of corresponding four and more member households is 34.13%. Ratio between the number of dwellings and number of corresponding households is 1.53 for two bedroom dwellings and ranges from 0.58 to 0.72 for larger dwellings. This obvious discrepancy implies that more than 10% of households in Serbia use dwellings with inadequate structure (smaller number of rooms than needed).

3.2 Relevant Examples of Flexible Design

Flexible design of the apartment, allowing transformations of the apartment according to households changing needs, could be a feasible model for solving the problem of inadequate apartment structure, taking into account the real market context. Milan Lojanica explores possibilities of changing spatial structure of the apartment over time, to accommodate changing family need, structure and size (Fig. 1). Possibility of adapting the apartment in order to accommodate the evolution of tenants needs and demands by changing its secondary structure – partitions and equipment, gives the apartment higher durability and use value.

Fig. 1. Milan Lojanica – the study of flexible apartment.

Besides these long term changes of the apartment structure, there is also a large number of flexible dwelling designs that are changeable during short period of time, allowing multifunctional use of space during the day (Fig. 2).

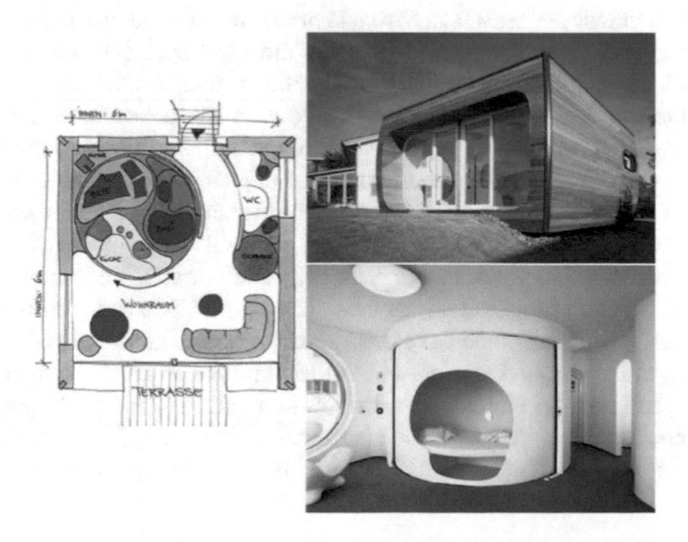

Fig. 2. Luigi Colani – Hanse Colani Rotor House.

Shröeder House by Gerrit Rietveld is one of the best known examples. Mobile partitions allow fast and simple changes in the spatial configuration following daily life rhythm of tenants. Following intense technological development flexible apartment concepts also evolve, allowing the integration of multiple functions in a small space. Moveable partitions, integrated with the elements of the equipment and home appliances, manually or even electro motor operated, allow fast and easy changes of apartment spatial configuration, minimizing spatial requirements.

One of the most interesting examples of such dwelling is Hans Colani Rotor House. Rotating cylinder containing three parts – kitchen, bathroom and a bed is integrated into a small 6 by 6 m plan. Each of this three segments can be integrated with the rest of the unit, changing space configuration and the way it is used.

4 Discussion

This research had shown significant discrepancy between the number of dwellings of certain spatial structure defined by number of bedrooms, and number of households that could comfortably use these dwellings. Large number of households in the Republic of Serbia live in inadequate overcrowded dwellings. Parents often use living rooms for sleeping, and children don't have separate rooms.

Flexible approach to the apartments design is often used to adapt living space to changing user needs. This is particularly important in small apartments, without sufficient space to satisfy basic needs of tenants and needed level of spatial comfort. The number of rooms in relation to the number of household members had been adopted as a primary indicator of the potential of the dwelling to provide satisfying quality of life. Since financial potency of tenants in Serbia is often very limited, usually the apartments that are designed for smaller number of users are used by lager households and are overcrowded. As families change, grow and reduce, their spatial needs also change. But their limited resources often do not allow them to purchase new, adequate dwellings. Apartments, initially purchased for a smaller two-member families are used for larger, three or four member households. In this situation flexibility is often very important, in order to provide all of the household members private space – separate rooms, and keep common space for socializing. Figure 3 shows model of the apartment A, with adequate flexibility potential to change its configuration and adapt to the size of the household ranging from 2 to 4 members. The area of the apartment is only 56 m^2.

Initially, the apartment could have been purchased for a household of two members – young couple. Its size and area are not too big and are suitable for such household. This initial configuration is given as configuration A1. As family changes and gains new members, flexible potential of the apartment allows changing its configuration to accommodate three (configuration A2) or four member family (configuration A3).

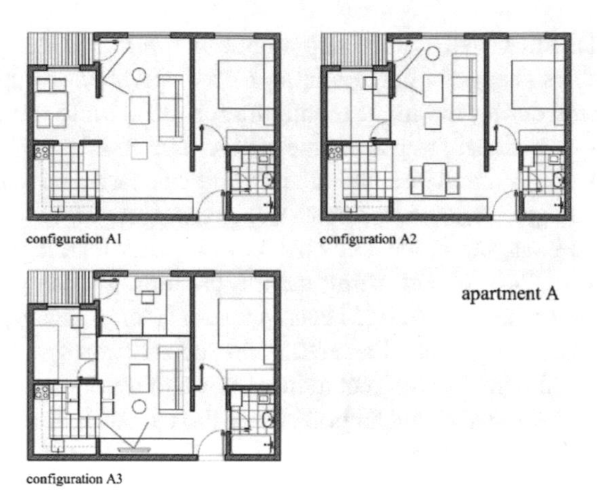

Fig. 3. Apartment A – possible spatial configurations.

This is a theoretical model of the apartment with flexible potential to change its configuration over time and adapt to changes in the size and structure of the household.

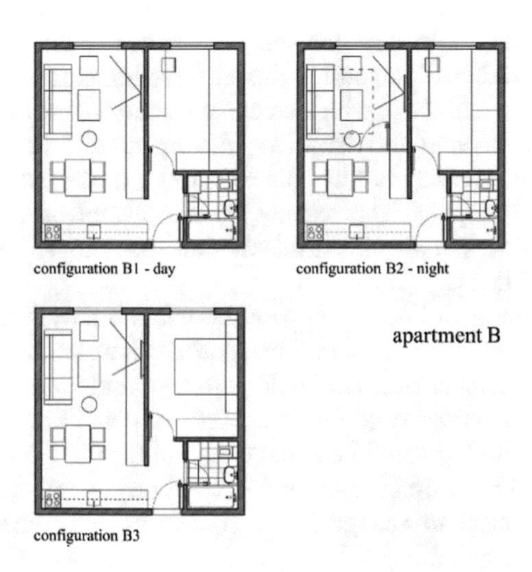

Fig. 4. Apartment B – possible spatial configurations.

Another theoretical model of the apartment is shown in Fig. 4. This is the model of the apartment B, of only 40 m^2, with its configuration changing over a period of one day. This is the apartment that can initially be purchased and used by a couple (configuration B3). But apartment of that size and configuration are often used by a larger households of three and even four members. In this case living area is used as a sleeping area for parents, while bedroom becomes a room for one or even two children (configuration B1). But installation of one mobile partition between living and dining in living area enables change in configuration of the apartment forming another small bedroom for parents during the night (configuration B2), while during the day living area can function as socializing area for the whole household.

Apartments A and B are theoretical models that can easily be used in practice and can change their configurations. Apartment A changes configuration over longer period of time adapting to changing household size. On the other hand apartment B changes its configuration over 24 h period, using mobile partition to change between day and night modes. Configurations A3 and B2 cannot be considered as high quality solutions, since available space is very limited and common socialization space is small and does not have direct natural light, but indirect lighting through transoms in the partitions. But in the context of very limited financial potency of the tenants that are not in position to provide dwelling of adequate size these models can be considered as justified and even improving life quality. They allow each member of the household to have private space and at the same time provide common space for socialization, although with limited quality.

5 Conclusions

Political and economic crisis in the Republic of Serbia at the end of 20th and the beginning of 21st century influenced housing real estate market. Low financial potency of future tenants conditioned demand for smaller apartments that future users could afford. These dwelling are often with inadequate structure and size. The number of household members is often too large for the apartment structure, so that every household member often does not have his own room or the users are often forced to use living room for sleeping. This cannot be considered as adequate solution for housing of good quality, but small apartments must be accepted as reality that comes from marketing conditions.

Data from The Census of Population, Households and Dwellings in the Republic of Serbia in 2011 show significant deficit of the apartments with three and more bedrooms in comparison to the number of households with four and more members.

In this context it is necessary to find a model that could improve quality of life of these households. Flexibility could be a viable model for that. Through flexible design the use of space could be optimized and living conditions could be improved. Flexible potential of the apartment to change its configuration over time in order to satisfy changing household size and needs is essential in this process. Changes in apartment configuration in order to provide separate rooms for household members could significantly improve quality of life of the tenants. These changes could be long-term changes following changes in the size and structure of the household, but also short-term changes that follow daily rhythm of life, allowing multifunctional and thus optimal use of limited space.

This research defines two theoretical models of flexible apartment, one for each of these two cases. Of course, there are many potential models that should be explored, and the aim of this research is to initiate further work on this subject. Flexibility, as a means of improving the quality of life through multifunctional use of space should be explored on many levels, from the level of the apartment, level of typical floor assembly to the level of the whole building. Progress in this area that would have practical application could significantly improve quality of life of large number of households that live in small dwelling of inadequate size.

References

1. Acharya, L.: Flexible architecture for the dynamic societies reflection on a journey from the 20th century into the future. Master's thesis in Art History Faculty of Humanities, Social Sciences and Education University of Tromsø (2013)
2. Estaji, H.: A review of flexibility and adaptability in housing design. Int. J. Contemp. Architect. "The New ARCH" **4**(2), 37–49 (2017)
3. Frampton, K.: Modern Architecture: A Critical History, 4th edn. Thames and Hudson, London (2007)
4. Habraken, N.J.: Design for flexibility. Build. Res. Inf. **36**(3), 290–296 (2008)
5. Kisharini, R.: Functionality and adaptability of low cost apartment space design: a case of Surabaya Indonesia. Technische Universiteit Eindhoven, Eindhoven (2015)
6. Kondrateva, L., Volkov, S.: Minimizing the residential indoor spaces for comfort living. World Appl. Sci. J. (Problems of Architecture and Construction) **23**, 207–211 (2013)
7. Kronenburg, R.: Flexible Architecture: the cultural impact of responsive building. In: 10th International Conference on Open Building, Paris (2004)
8. Kronenburg, R.: Flexible: Architecture that Responds to Change. Larence King Publishing Ltd., London (2007)
9. Lojanica, M.: The apartment is used and changed, Housing 1. Institute for Architecture and Urbanism of Serbia, Belgrade (1975)
10. Nedovic-Budic, Z., Djordjevic, D., Dabovic, T.: The mornings after… Serbian spatial planning legislation in context. Eur. Plan. Stud. **19**(3), 429–455 (2011)
11. Raviz, S.R.H., Eteghad, A.N., Guardiola, E.U., Aira, A.A.: Flexible housing: the role of spatial organization in achieving functional efficiency. Int. J. Architect. Res. **9**(2), 65–76 (2015)
12. Schneider, T., Till, J.: Flexible Housing. Architectural press, London (2007)
13. Shabani, M.M., Tahir, M.M., Arjmandi, H., Che-Ani, A.I., Abdullah, A.G., Usman, I.M.S.: Achieving privacy in the iranian contemporary compact apartment through flexible design. In: Fujita, H., Sasaki, J. (eds.) Selected Topics in Power Systems and Remote Sensing, 10th WSEAS/IASME International Conference on Electric Power Systems, High Voltages, Electric Machines (POWER 2010), pp. 285–296. Iwate Prefectural University, Japan (2010)
14. Szczegielniak, A., Fabianowski, D.: Optimisation of functional layout of small apartments in dense city centra housing areas. Procedia Eng. **161**, 1690–1696 (2016)
15. Teige, K.: The Minimum Dwelling. The MIT Press, Cambridge (2002)
16. Zivkovic, M., Jovanovic, G., Kondic, S.: Flexible planning strategies of sustainable city development. Facta Universitatis Series: Architecture and Civil Engineering **12**(3), 273–286 (2014)

Interior Acoustic Materials and Systems

Aleksey Zhukov[1], Ekaterina Shokodko[1](✉), Ekaterina Bobrova[2], Igor Bessonov[3], Gulzar Dosanova[4], and Nikita Ushakov[1]

[1] Moscow State University of Civil Engineering, Yaroslavskoye sh. 26, 129337 Moscow, Russia
bezuglova-e@inbox.ru

[2] Higher School of Economics, Myasnitskaya str. 20, 101000 Moscow, Russia

[3] Research Institute of Building Physics of the Russian Academy of Architecture and Building Sciences, Lokomotivniy tr. 21, 127238 Moscow, Russia

[4] Karakalpak State University Named After Berdakh, Abdirov str. 1, 230112 Nukus, Uzbekistan

Abstract. The aim of the research was to study the regularities of the influence of porosity parameters on the properties of decorative acoustic materials of various structures, including the combined structure using the granular or fibrous fillers and the cellular matrix. The results of study are presented in the article. The technological conditions for regulating the properties of decorative acoustical materials are formulated. The method for solving technological problems in the field of decorative acoustical materials is formulated. It was tested in solving the technological problems in obtaining materials with the cellular and mixed structure. Exemplified by the gypsum-containing materials, the influence of porosity on the sound absorption, as well as the regulation of the comfortable state of the internal environment in the room, is revealed. The article provides the expedience for the use of the foamed gypsum as a basis for the manufacture of multifunctional products. One of the varieties of such materials are the porous gypsum-zeolite products with a reinforcing component – the expanded vermiculite. It has high rates of sound absorption, fire resistance, environmental cleanliness and the ability to ion exchange, ensuring a reduction in the concentration of harmful substances in the environment.

1 Introduction

Decorative acoustic materials (DAM) are used in interior decoration. Systems with the use of DAM contribute to the formation of acoustic comfort in rooms, as well as contribute to the realization of different design solutions.

The functional prerogatives of decorative acoustic materials used in interior cladding can be divided into the following groups. First, it is the creation of decorative (architectural, design) cladding; secondly, it is a way to regulate the acoustic characteristics of the room; thirdly, it is the ability to control the comfortable state of the internal environment in the premises. Experience in the production and use of the DAM shows that the main direction of the DAM development should be the production of the single-layer porous materials of the factory readiness. The single-layer DAM confirmed

V. Murgul and M. Pasetti (Eds.): EMMFT-2018, AISC 983, pp. 740–747, 2019.
https://doi.org/10.1007/978-3-030-19868-8_72

their advantage in manufacturability, cost-effectiveness, the possibility of using non-deficient materials and a number of other indicators [1–4].

In the study and development of new types of DAM, the structure of the material is considered as a complex concept that characterizes the interposition of the components and their interrelation. The structure of the material is most often determined by the main matrix-forming material and is adjusted by the type and the amount of the binder component. As the main matrix-forming components for DAM, it is possible to use granular and fibrous materials, as well as powdered substances (mineral binders) that are porous during the processing. In this case, the most often cellular structure of the material is provided [5–8].

An important feature that determines the structural features of DAM may be the porosity - a quantitative and qualitative combination of air cavities in the material, i.e. the volume of the porosity, pore-size distribution, the prevailing character of pores (cellular, granular, interfiber, mixed).

The material structure has a great influence on the strength and performance characteristics of products. The materials of the fibrous, cellular and granular structure have a different character of adhesion of the constituent elements and, consequently, a different character of destruction. The choice of a rational, scientifically based combination of components required the study of the main types of structures, primarily in terms of their strength characteristics. Performance characteristics, such as hygroscopicity, moisture resistance, flammability, biostability, dustiness, durability and others, also depend on the type of structure and on the properties of the mineral components forming this structure.

The porosity of the finishing gypsum-containing materials, the presence of highly porous fillers and reinforcing fillers allow to optimize the sound-absorbing properties of the material as well as improve the ability of the coating to adsorb harmful emissions that accumulate in the rooms between vents. Gypsum-containing materials are also considered as an element of fire retardant barriers. The introduction of non-combustible reinforcing materials improves this indicator.

2 Materials and Methods

The sound absorption of the front elements of suspended ceilings is one of the most important functions of the acoustic plates. For the study of the sound absorption, the products with the density of 400 kg/m^3 and the zeolite content of 0, 10 and 20% by weight of the gypsum binder were used; without perforation and with perforation in the amount of 15% of the sample area; with the content of the reinforcing component (the expanded vermiculite) in the amount of 10% of the gypsum binder.

In accordance with GOST 16297-80, during tests, the air in the room had the temperature of 22 °C. The relative humidity of the air was 52%. Before testing, porous gypsum-containing materials were conditioned in the testing room for 3 h. The tests were carried out at the acoustic interferometer (Fig. 1).

Fig. 1. Acoustic interferometer.

A sample with a diameter of 100 mm and a height of 25 mm was inserted into the interferometer casing so that its non-face surface was on a rigid piston, and the front surface was at the cut-off clip level. The edges of the front side of the sample were smeared with clay and the holder was fixed in a pipe. Tests were carried out sequentially at frequencies of 63, 80, 100, 125, 160, 200, 250, 315, 400–1600 and 2000 Hz.

3 Results

The results of the sound absorption studies of the foam-gypsum materials are presented in Figs. 2, 3 and 4.

The Fig. 2 shows the dependence of the sound absorption characteristics of the material on the sound frequency, the content of highly porous component (zeolite) and the presence of the reinforcing component. The figure shows that the highest value of the sound absorption coefficient has a material with a zeolite content of 10%, with a reinforcing component. A material containing 10% zeolite, without a reinforcing component, has less sound absorption than a material without zeolite. The material containing highly porous component (zeolite) in an amount of 20%, has the sound absorption at low and medium frequencies lower. At high frequencies, the material containing zeolite in the amount of 20% has the sound absorption significantly higher than the material containing 10% zeolite. At the same time, both zeolite-containing materials are inferior in terms of sound absorption to material with the reinforcing component (expanded vermiculite). In terms of sound absorption at high frequencies, a material containing 20% zeolite is slightly inferior to a material containing the reinforcing component.

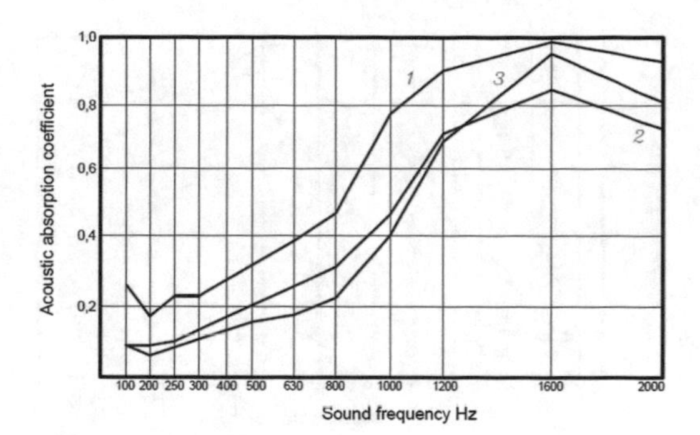

Fig. 2. The sound absorption characteristics of the products with a density of 400 kg/m^3. The content of the highly porous component is: 1–10% (with the reinforcing component); 2–10% (without the reinforcing component); 3–20% (without the reinforcing component).

The Fig. 3 shows the dependences of the sound absorption of the perforated material on the sound frequency, the amount of the zeolite and the presence of the reinforcing component. The highest sound absorption coefficient is observed for the material containing the highly porous component in the amount of 10% and the reinforcing component in the amount of 10%. At the same time, at the level of low and high frequencies, materials containing the zeolite in the amount of 20% approach the optimum value of sound absorption of the material.

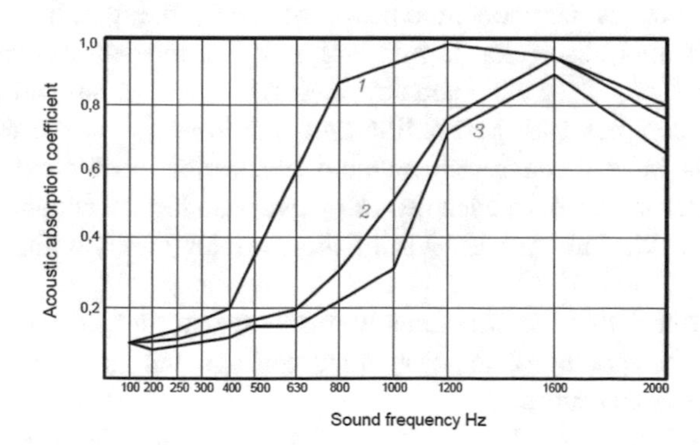

Fig. 3. The sound absorption of the perforated foam gypsum slabs with a density of 400 kg/m^3. The content of the highly porous component is: 1–10% (with the reinforcing component); 2–10% (without the reinforcing component); 3–20% (without the reinforcing component).

Thus, the presence of perforation in combination with the presence of a reinforcing component affects the sound absorption coefficient of materials containing a highly porous component.

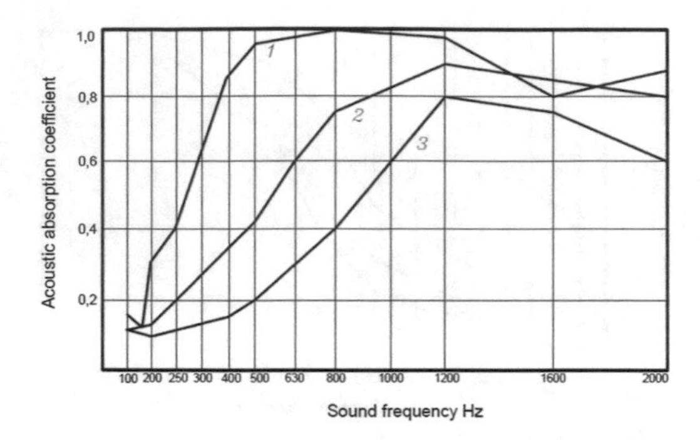

Fig. 4. The sound absorption of the perforated foam gypsum slabs with a density of 400 kg/m^3 with a space 50 mm from the ceiling surface. The content of the highly porous component is: 1–10% (with the reinforcing component); 2–10% (without the reinforcing component); 3–20% (without the reinforcing component).

The Fig. 4 shows the dependence of the sound absorption of the perforated material with providing a space 50 mm from the ceiling surface on the sound frequency, the amount of the zeolite and the presence of the reinforcing component. Spacing and perforation has a significant impact on the sound absorption coefficient of all materials at low and medium frequencies. The material with the zeolite content of 20% has the lowest values of sound absorption. For materials containing a reinforcing component, the spacing and perforation significantly improves the sound absorption. For materials with the highly porous component content of 10%, the graph clearly shows an improvement in sound absorption at mid and low frequencies. For material with the zeolite content of 20%, the spacing and perforation improves sound absorption slightly.

The study of samples with perforation and providing the space of 50 mm shows that the materials have a much better sound absorption coefficient. This is clearly visible at low and medium frequencies. However, at high frequencies, their sound absorption is somewhat lower than that of the samples tested without providing the space.

It should be noted that the structures with the spacing are of the greatest interest, since in practice the spacing of the suspended ceilings and slabs is carried out at the distance of 100 mm and more.

4 Discussion

The sound-absorbing properties of materials primarily depend on the type of pores on the surface of the material and the nature of the porosity inside. The technology of building materials distinguishes between open and closed pores, interconnected and closed porosity. Based on this classification, to assess the shape of pores, their separation according to acoustic activity into acoustically active and acoustically passive was made (Fig. 5). Acoustically active are the open pores, whose dimensions are commensurate with the wavelength [9–12]. The acoustically passive are the closed pores, which do not have the direct access to the surface of the material. Dead-end porosity and open non-communicating pores are classified as semi-passive.

Among open pores, there are hydraulically correct pores characterized by low values of input resistance. Such porosity is characteristic of the materials with the granular structure. Pores with constant resistance are inherent in materials with the fibrous structure. Hydraulically irregular porosity has a high input resistance. Such porosity is characteristic of materials with a cellular structure.

From the point of view of effective sound absorption, materials with a fibrous structure that have only the open porosity are most suitable. The porosity of cellular and granular materials should be adjusted in the direction of increasing the number of communicating pores - that is a mixed porosity. It is technologically expedient to form the mixed porosity by introducing into the composition of cellular mixtures porous fillers, such as zeolites, expanded perlite, vermiculite, etc.

It is also obvious that the reduction in pore size must be to a certain size, otherwise they will go into the category of acoustically passive pores, which will correspond to the presence of closed porosity. To determine the minimum dimensions of acoustically active pores, calculations were performed in which the attenuation of sound waves was considered as a loss of pressure during air movement along cylindrical pores.

The determination of the minimum pore sizes, participating in the damping of sound waves, was carried out from the condition that the excess sound pressure (Δ Ps) must be equal to the friction losses (Δ Pf). According to the Darcy-Weissbach formula:

$$\Delta P_f = \lambda \frac{1}{\mathrm{d_e}} \frac{\mathrm{V}^2}{2} \rho_t \tag{1}$$

ΔP_f – friction losses, λ – coefficient of friction on the pore walls, l - length of pores, de - equivalent diameter of pores, V – air velocity, ρ_t – air density.

When calculating, it is assumed that friction losses ΔP_f should compensate for the sound pressure corresponding to the upper threshold, i.e. Pmax = 107 N/m^2, the channel length, taking into account the average thickness of similar materials, is assumed to be 20 mm, V = 340 m/s, ρ_t = 1.28 kg/m^3. The coefficient of the air friction in small pore channels is in the range of 0.05–0.1. From formula (1) it follows that

$$d_e = \lambda \frac{1}{\Delta P_f} \frac{\mathrm{V}^2}{2} \rho_t \tag{2}$$

Calculations performed using this formula showed that, with the values adopted, the minimum diameter dmin is 60–120 μm. Taking into account that the maximum sound pressure was taken in the calculations, we can assume that the minimum pore size, which is actively involved in sound suppression, is 100 μm or more. As follows from the formula (2), the decrease in sound pressure provides effective damping of sound waves in large pores. The obtained data are corresponding the data obtained in the study of granular structures, where it was shown that the best pore size for sound-absorbing porous materials should be 100–400 μm.

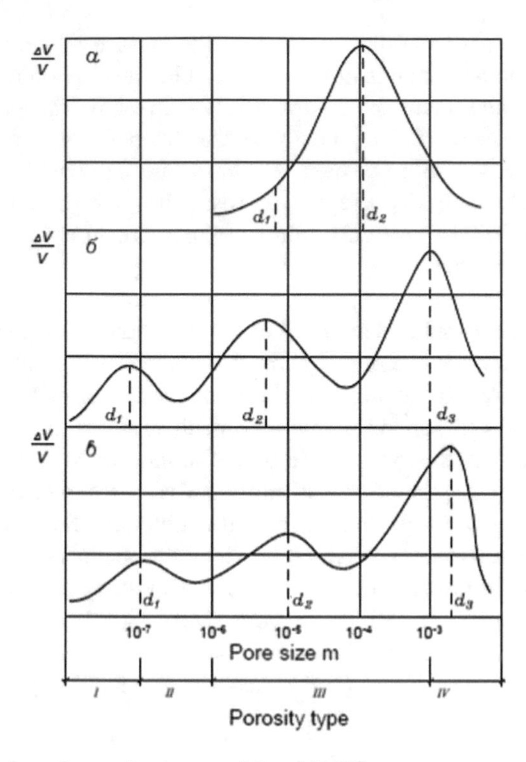

Fig. 5. The distribution of pore size in materials with different structure: a - fibrous; b - cellular; c - granular. Porosity type: I - gel; II - contraction; III - capillary; IV – macroporosity.

The pore size is closely related to the frequency of the sound. So an increase in sound absorption at low frequencies can be provided by relatively large pores, and the sound absorption at high frequencies can be provided by small pores. Therefore, a combination of large and small pores is necessary for sound absorption in a wide range of frequencies. There are significantly less data characterizing the lower limit of pore size [14–16].

Reducing the pore size of less than 50 microns leads to a sharp decrease in sound absorption. This is due to the deterioration of the conditions for the passage of sound waves into the material. Studies of the porosity of materials with high sound-absorbing properties have shown that they have fine-grained, polydisperse porosity with a pore size of 100–250 microns. At the same time, the through porosity is in the range of 70–90%. A larger percentage of interconnected porosity reduces sound absorption due to a decrease in viscous air friction in the material.

5 Conclusions

The results of studying the influence of porosity parameters on the sound absorption made it possible to formulate technological conditions for adjusting the properties: the creation of the multimodal porosity providing the sound suppression at medium and high frequencies; efficient transfer of the sound energy into the heat due to the air friction on the pore walls (increase in a clear component).

The study of the influence of highly porous granular components, as well as reinforcing components, on the sound-absorbing ability of porous gypsum-containing materials made it possible to establish that increasing the amount of highly porous filler (zeolite) reduces the sound-absorbing abilities of the materials. This may be due to the penetration of zeolite grains into the pores of the material, consequently the penetration of the sound wave through the thickness of the material through the communicating channels encounters resistance and reduces the sound-absorbing capacity of the material. So, when examining samples with zeolite content of 10 and 20% of the amount of gypsum binder, it was found that samples with 0% zeolite content have a better sound absorption coefficient at medium frequencies. At the same time, provision of the reinforcing component and perforation of the slabs, increases sound absorption. This is due to the improved conditions for the filtration regime in a highly porous gypsum matrix

The perforating up to 15% of the total area with a diameter of 3 and 5 mm increase the sound absorbing capacity of the materials under study. This is especially noticeable at mid frequencies for materials containing a reinforcing component.

References

1. Rumyantsev, B.M., Zhukov, A.D., Barybin, A.A., Bondar, D.D.: Sci. Rev. **7**, 32–35 (2017)
2. Rumyantsev, B.M., Zhukov, A.D., Bobrova, E.Yu.: Innovations Life **1**(20), 67–75 (2017)
3. Rumyantsev, B.M., Zhukov, A.D., Bobrova, E.Yu.: Innovations Life **1**(20), 17–24 (2017)
4. Korovyakov, V.F.: Improving the efficiency of production and use of gypsum materials and products. In: Proceedings All-Russia Sem, pp. 51–56. RAACS, Moscow (2002)
5. Telichenko, V.I., Oreshkin, D.V.: Ecol. Urb. Terr. **2**, 31–33 (2015)
6. Zhuk, P.M., Zhukov, A.D.: Eco. Ind. Russia **22**(4), 52–57 (2018)
7. Rumyantsev, B.M., Zhukov, A.D., Smirnova, T.V.: Internet-Bulletin VolgGASU **4**(35), 3 (2014)
8. Zhukov, A.D., Chugunkov, A.B.: Proc. MSUCE **1–2**, 273–278 (2011)
9. Vasilik, P.G., Buryanov, A.F.: Improving the efficiency of production and use of gypsum materials and products. In: Proceedings All-Russia Sem, pp. 30–36. "De-Nova" Publishing House, Moscow (2016)
10. Rumyantsev, B.M., Zhukov, A.D., Bobrova, E.Yu., Romanova, I.P.: MATEC Web Conf. **86**, 03022 (2016)
11. Rumyantsev, B.M.: Technology of decorative acoustic materials, MGSU, Moscow (2010)
12. Bessonov, I.V.: Improving the efficiency of production and use of gypsum materials and products. In: Proceedings All-Russia Sem, pp. 82–87. RAACS, Moscow (2002)
13. Kodzoev, M.-B., Isachenko, S., Kosarev, S., Basova, A., Skvortzov, A., Asamatdinov, M., Zhukov, A.: MATEC Web Conf. **170**, 03022 (2018)
14. Asamatdinov, M.O., Medvedev, A.A., Zhukov, A.D., Zarmanyan, E.V., Poserenin, A.I.: MATEC Web Conf. **193**, 03045 (2018)
15. Zhukov, A.D., Bobrova, E.Yu., Zelenshchikov, D.B., Mustafaev, R.M., Khimich, A.O.: Adv. Mater. Struct. Mech. Eng. **1025–1026**, 1031–1034 (2014)
16. Rumyantsev, B.M., Zhukov, A.D., Zelenshikov, D.B., Chkunin, A.S., Ivanov, K.K., Sazonova, Yu.V.: MATEC Web Conf. **86**, 04027 (2016)

Prediction of the Occurrence of Plastic Deformations in Structural Elements Made of Polyethylene

Pavel Bozhanov[1(✉)] and Aleksandr Treshchev[2]

[1] LLC “Engineering Center of Industrial Design”, Demonstracii st., 1-g, of. 408, 300041 Tula, Russia
bozhanovl30776@yandex.ru

[2] Tula State University, Lenin Avenue, 92, 300012 Tula, Russia

Abstract. The widespread introduction and distribution of pipes from polyethylene with high and low density necessitates a more detailed study of the properties of these materials. Since the strength characteristics of the materials under consideration depend on the type of stress state, as evidenced by numerous experimental data, studies in this direction are an actual engineering problem. In this paper, experimental data on the occurrence of a limit state in low density polyethylene and high density polyethylene under complex stress states are processed. The mathematical form of recording the onset of the limit state is formulated. On the basis of statistical processing of the obtained values, an approximation of the function of the stress state type was made, which allows determining the boundaries of the elastic deformation of the materials under study and their transition into the plastic range. Comparison of the obtained results with results that do not take into account the tendency of polyethylene to different resistance and dilatation was made. It was found that ignoring the specified specific properties of the materials under consideration leads to significant discrepancies between theoretical and experimental parameters of deformed states. Whereas the use of the approach outlined in this paper gives a good consistency of experimental and theoretical values, which makes it possible to use the considered method when solving applied engineering problems of calculating building structures.

Keywords: Strength criterion · Materials of different resistance · Dilatation · Polymers · Polyethylene

1 Introduction

High density polyethylene (HDPE) and low density polyethylene (LDPE) are widely used in the construction industry. It is impossible to imagine the laying of water engineering networks without the use of pipes manufactured on the basis of these materials. There is also no special need to explain how numerous and versatile is the stress state of pipelines operating in different soil conditions, under variable internal pressure and various types of external load. Certain technical standards and various kinds of guidelines for designing networks of the materials under consideration,

V. Murgul and M. Pasetti (Eds.): EMMFT-2018, AISC 983, pp. 748–757, 2019.
https://doi.org/10.1007/978-3-030-19868-8_73

developed mainly by manufacturers of one or another brand of products, were developed in a natural way, which basically generalize the experience of exploiting the structural elements themselves and don't base on a specific theoretical platform. On the other hand, any basis for making design decisions should be a theoretical basis that closely matches experimental research. Therefore, this paper proposes an approach that generalizes, on the one hand, the processing of experimental diagrams of the material when it reaches the limit states, and, on the other hand, the theoretical approximation of the identified relationships and the reduction of the dependencies to a rigorous mathematical formulation. Moreover, as will be shown later, the obtained equation for the dependence of the onset of a limit state on the type of stress state of a material extends to the full range of stress states characteristic of the material under consideration. In this case, an important aspect when choosing a theoretical approximation of the function of the type of stress state will be the parameter that determines this material state for a specific type of loading. Polyethylene is known to belong to the class of partially crystalline polymeric materials. These materials show a significant discrepancy between the yield stresses when stretching and compressing samples. Moreover, the onset of the limit state for these materials depends not only on the general form of the stress state, whether it is tension, compression or shear, but also on the specific ratio of stress tensor components that determine the transition of the material from the elastic to plastic phase. Such experimental dependences make it possible to assign polyethylene to the class of differently resisting dilating materials, for which the use of classical approaches, which do not take into account these properties, leads to significant deviations when performing engineering calculations. Experimental studies of the behavior of polyethylene under various stress states were carried out in [1–3]. The experimentally established diagrams of the dependence of the stresses on the longitudinal deformations obtained by stretching the samples from HDPE at different hydrostatic pressures are shown in Fig. 1. Diagrams 1–6 in Fig. 1 were obtained at hydrostatic pressures of 0.1 MPa, 30 MPa, 50 MPa, 100 MPa, 150 MPa, and 200 MPa, respectively. Similar experimental dependences obtained by stretching LDPE samples at different hydrostatic pressures are shown in Fig. 2. Diagrams 1–6 in Fig. 2 were obtained at hydrostatic pressures of 0.1 MPa, 30 MPa, 50 MPa, 100 MPa, 150 MPa, and 200 MPa, respectively. By analyzing the diagrams presented, it can be stated with absolute certainty that the plastic properties of the studied materials depend on the type of stress state. Moreover, it should be especially noted that these materials have a considerable plasticity resource, which allows using a model of an ideally elastic-plastic body. It should be noted that the peak of the divergence of the diagrams falls on the sector where the stress state of the samples is in front of the limiting area of the elastic state and the transition state beyond which plastic deformations are formed. On the basis of observational data, it can be concluded that taking into account the dependence of the deformation characteristics on the type of stress state generally becomes relevant at a sufficiently high level of stress beyond the elastic limits.

It is a well-known fact that the phenomenon of multi-resistance does not introduce significant effects to the operation of structures under uniaxial stress states, therefore the proposed definitions of criteria describing limit states should have practical meaning at any complexity level of the stress-strain state of the body. On the other hand, for many structures, the predominant factor, when describing their work, is the transition to a plastic state and the exhaustion of a plasticity resource.

The criteria for the limit state known to date are constructed using various hypotheses. First of all, the plasticity conditions for materials with different resistances are indirectly based on the modification of the Huber – Mises – Hencky or Tresca – Saint-Venant hypotheses using various forms of taking into account the influence of the ball tensor. In this regard, formulas have been proposed that suggest the dependence of the limit state on the hydrostatic pressure. Among the works based on this assumption are the criteria of Coulomb – Mohr, P.P. Balandin, Schleicher, I.N. Mirolyubov, Yu.I. Yagn, and some other scientists. However, the influence of one hydrostatic pressure on the onset of the limit state of differently resisting dilating materials cannot be considered satisfactory in the general case, and the considered diagrams clearly confirm this conclusion. In view of the foregoing, one cannot expect the criteria of the limit state, which imply a similar approach, to have universal use in practical calculations. Although, they can be recommended for use for individual materials in specific stress states. In the general case, the limiting dependences must take into account the influence of both the hydrostatic pressure and the type of stress state upon the transition of the material to the plastic deformation stage. It is obvious that an arbitrary stress state can be determined by several parameters depending on the type of stress state. However, an introduction to the defining relations of all the qualitative parameters is not always justified, since it significantly complicates the equations. Often, fairly accurate results can be obtained by characterizing the type of stress state on average, using only one parameter. Some researchers indicate the feasibility of taking into account the influence of the type of stress deviator by introducing a phase invariant into the plasticity condition. It should be noted that the phase invariant does not allow one to unambiguously distinguish particular types of the stress state of the material in their infinite spectrum, since it has the same values for absolutely different stress-strain states and is directly related to the Lode-Nadai parameter. It is not difficult to make sure that with the equality of the two main stresses, regardless of the value of the third main stress, the value of the Lode–Nadai parameter will remain equal to unity, i.e. the indicated value will be maintained in cases of triaxial tension, biaxial tension, compression, etc. In addition, the use of the Lode-Nadai parameter or phase invariant under plasticity conditions significantly complicates the calculations of the stress-strain state of structural elements. On the other hand, the amendment introduced by the inclusion of these parameters often has the same order as the variation of the experimental data. Although, when formulating the conditions of plasticity and strength, the use of the Lode-Nadai parameter or phase invariant may be appropriate in some cases, but often in combination with other qualitative parameters of the stress state.

The above analysis of the existing criteria of strength and plasticity allows concluding that at this stage of development of the mechanics of a deformable solid body, there is no single approach to determine the limit state of structural elements from dilating differently resistant materials. Thus, this paper is devoted to the actual problem of solid mechanics - the formulation of conditions for the limit state of dilating materials of different resistance and their practical use in applied engineering calculations of structures made of these materials, which, as was emphasized earlier, include high and low density polyethylene.

2 Materials and Methods

In order to formulate dependencies to determine the transition of a material from the elastic stage to the plastic range, it is necessary to first determine the numerical values of the corresponding components of the stress state at which the limit state occurs. Thus, from the graphs shown in Figs. 1 and 2, it is possible to determine σ_t – tensile stresses corresponding to the formation of plastic deformations (yield stress).

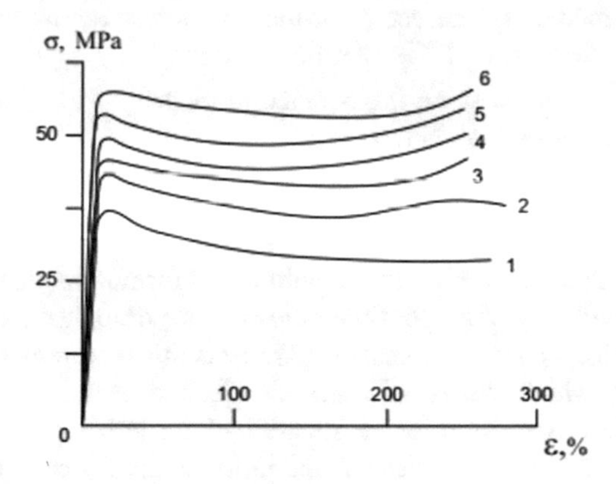

Fig. 1. Tensile stresses corresponding to the formation of plastic deformations

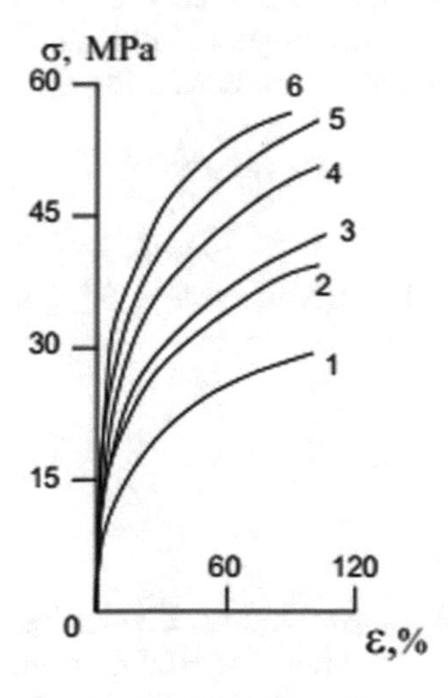

Fig. 2. Tensile stresses

To generalize the approach to the description of the stress-strain state of differently resisting dilating materials, adhering to the research carried out in [8, 9], the condition of plasticity can be formulated in a fairly general form:

$$F(\sigma_{ij}) = \tau \cdot f(\xi) = k_\tau \tag{1}$$

where $f(\xi)$– function of the stress state;

$\xi = \sigma / S_0$ octahedral normal normalized stress;
$S_0 = \sqrt{\sigma^2 + \tau^2}$ module of the vector of the total stress on the octahedral plane;
$\sigma = \sigma_{ij} \cdot \delta_{ij}/3$ – mean stress; δ_{ij} – Kronecker delta;
$\tau = \sqrt{S_{ij} \cdot S_{ij}/3}$ shear stress on the octahedral plane;
$S_{ij} = \sigma_{ij} - \delta_{ij} \cdot \sigma$ – stress deviator;
$k_\tau = \sqrt{2/3} \cdot \tau_S$; τ_S – net shear yield stress.

Earlier, it was emphasized that it is important to choose such a function as a parameter of the type of stress state that would allow formulating the onset of the limit state of a material with any of all possible ratios of the components of the stress tensor. The quality parameter ξ used in expression (1) meets this requirement, since it varies in the interval [–1; 1], which allows achieving the desired result.

The mathematical expression for the function characterizing the type of stress state - $f(\xi)$ included in condition (1) is determined individually for each material during the processing of experimental diagrams at the time of plastic deformation in the widest range of stress-strain state (SSS). Earlier, using this approach, specific mathematical formulations of the function of the stress state type were obtained for some polymeric materials. In particular, for polymethyl methacrylate, an exponential approximation of the function $f(\xi)$ in the form (2) was proposed in [3], and a piecewise linear function in the form (3) was obtained for the polycarbonate [3].

$$f(\xi) = e^{0{,}424 \cdot \xi} \tag{2}$$

$$\left.\begin{array}{ll} f(\xi) = 1 + 0{,}377\xi, & when \;\; -1 \leq \xi < 0 \\ f(\xi) = 1 - 0{,}339\xi, & when \;\; 0 \leq \xi \leq 1 \end{array}\right\} \tag{3}$$

3 Results

3.1 HDPE

Based on the experimental data shown in Fig. 1 and using the approach outlined in [8–10], we obtain the plasticity condition for HDPE in the form of criterion (1). The expression of the function of the stress state type entering into the plasticity condition (1) is determined on the basis of ensuring minimal differences when approximating the experimental values of the function of the stress state type in the entire range of the SSS change.

The initial values for determining the values of the parameter ξ and the corresponding values of the function of the stress state type will be the stress values experimentally determined from the diagrams, corresponding to the onset of plastic deformations σ_t. The value of the average octahedral stress will be determined by the expression (4):

$$\sigma = \sigma_t/3 - \mathrm{p} \tag{4}$$

The value of the tangential stress on the octahedral plane τ will be determined in the same way by the expression (5):

$$\tau = \sqrt{2}/3 \cdot |\sigma_t| \tag{5}$$

The value of the constant k_τ into condition (1) was assumed to be 18.5 MPa. The results of calculations for HDPE are summarized in Table 1.

For the values in Table 1, the approximation of the function of the type of stress state for HDPE $f(\xi)$ can be taken as follows:

$$f(\xi) = 1 + 0.1883\xi \tag{6}$$

Table 1. .

#	p, MPa	σ_t, MPa	τ, MPa	σ, MPa	S_0, MPa	ξ	$f_f(\xi)$
1.1	0.1	37.0	17.4	12.2	21.3	0.575	1.064
1.2	30	42.0	20.0	−16.0	25.4	−0.630	0.937
1.3	50	45.0	21.2	−35.0	40.9	−0.856	0.875
1.4	100	50.0	23.5	−83.3	86.6	−0.963	0.787
1.5	150	55.0	25.8	−131.7	134.2	−0.981	0.716
1.6	200	60.0	28.2	−180.0	182.2	−0.988	0.656

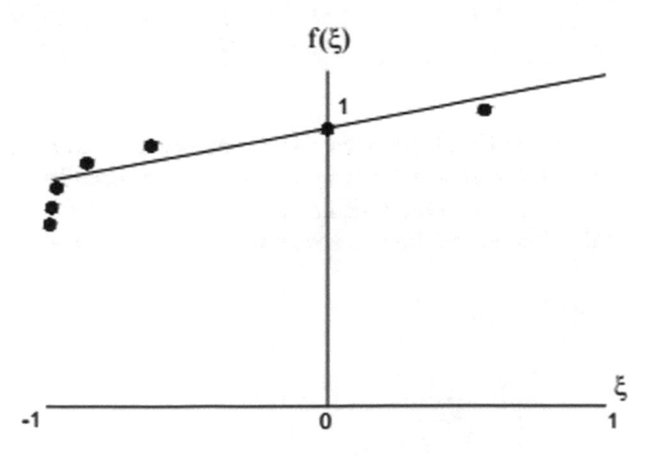

Fig. 3. Based on the experimental data

In Table 1, the first index of the point number corresponds to the number of the figure, which shows the experimental data for HDPE, and the second index corresponds to the number of the diagram in the specified figure.

The function graph of the stress state type $f(\xi)$ for HDPE is shown in Fig. 3.

3.2 LDPE

By analogy with Sect. 3.1, we obtain the expression of the function of the stress state type, which is included in the plasticity condition (1) on the basis of the condition of minimal discrepancies between the experimental values given in Fig. 2 and the values of the function of the stress state type under the chosen approximation. The initial values for determining the values of the parameter ξ and the corresponding values of the function of the stress state type, just as in Sect. 3.1, will be the stress values experimentally determined from the diagrams, corresponding to the beginning of the development of plastic deformations σ_t, and the mean σ and tangential τ stresses on the octahedral plane are determined by the expressions (4) and (5), respectively. The constant k_τ included into condition (1) reaches 6.6 MPa for LDPE. The results of calculations for LDPE are summarized in Table 2.

According to the values of the parameters from Table 2, the approximation of the function of the stress state type for LDPE is taken in the form of a power dependence:

$$f(\xi) = (1+\xi)^{0,12} \tag{7}$$

Table 2. To the values of the parameters.

#	p, MPa	σ_t, MPa	τ, MPa	σ, MPa	S_0, MPa	ξ	$f_f(\xi)$
2.1	0.1	12.0	5.60	3.90	6.86	0.569	1.170
2.2	30	18.0	8.50	−24.0	25.4	−0.943	0.780
2.3	50	20.0	9.40	−43.3	44.3	−0.977	0.702
2.4	100	24.0	11.3	−92.0	92.7	−0.993	0.585
2.5	150	28.0	13.2	−140.3	141.3	−0.996	0.502
2.6	200	32.0	15.0	−189.3	189.9	−0.997	0.439

In Table 2, the first index of the point number corresponds to the number of the figure, which shows the experimental data for LDPE, and the second index corresponds to the number of the diagram in the specified figure.

The graph of the function of the stress state type for LDPE is shown in Fig. 4.

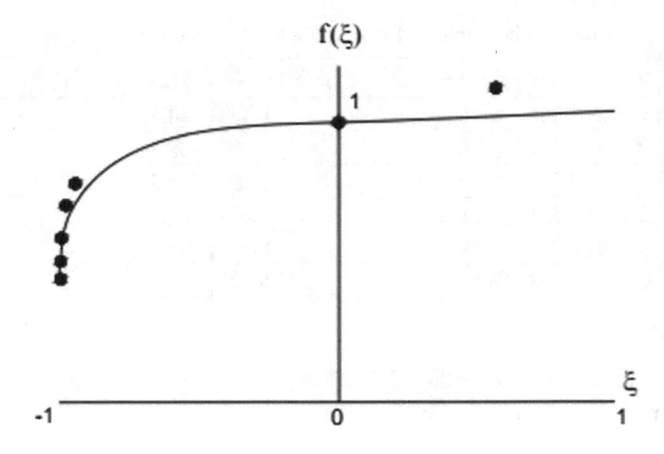

Fig. 4. The graph of the function of the stress state type for LDPE

4 Discussion

Table 3 shows the numerical discrepancies of the function values of the stress state type for the adopted approximation in the form of (6) from the actually obtained experimental data for HDPE, as well as the deviations of the experimental values of the function of the stress state type from the theoretical values in the case of disregarding the properties of different resistance of the considered material, i.e. the onset of the limit state defined by the classical Huber – Mises – Hencky criterion, which is obtained from condition (1) if we take the function $f_{cl}(\xi) = 1$ (const).

Table 3. The numerical discrepancies of the function values of the stress state type

#	$f_f(\xi)$	$f(\xi)$	$\lvert f_f(\xi) - f(\xi) \rvert / f_f(\xi)$, %	$f_{cl}(\xi)$	$\lvert f_f(\xi) - f_{cl}(\xi) \rvert / f_f(\xi)$ %
1.1	1.064	1.108	4.20	1.000	6.01
1.2	0.937	0.881	5.01	1.000	6.72
1.3	0.875	0.839	4.10	1.000	14.3
1.4	0.787	0.819	4.00	1.000	27.1
1.5	0.716	0.815	9.80	1.000	39.7
1.6	0.656	0.814	19.4	1.000	52.4

Table 4 shows the numerical discrepancies of the function values of the stress state type for the adopted approximation in the form of (7) from the actually obtained experimental data for LDPE, and also shows the deviations of the experimental values of the function of the stress state type from the theoretical values in the case of disregarding the properties of different resistance of the considered material, i.e. the onset of plastic states was determined by the classical Huber – Mises – Hencky criterion, which is obtained from condition (1) with $f_{cl}(\xi) = 1$ (const).

Table 4. The numerical discrepancies of the function values.

#	$f_f(\xi)$	$f(\xi)$	$\lvert f_f(\xi) - f(\xi)\rvert / f_f(\xi)$, %	$f_{cl}(\xi)$	$\lvert f_f(\xi) - f_{cl}(\xi)\rvert / f_f(\xi)$, %
2.1	1.170	1.056	9.70	1.000	14.5
2.2	0.780	0.709	9.10	1.000	28.2
2.3	0.702	0.636	9.40	1.000	42.5
2.4	0.585	0.551	5.81	1.000	70.9
2.5	0.502	0.515	2.59	1.000	99.2
2.6	0.439	0.498	13.4	1.000	127.8

Thus, the analysis of the obtained results allows us to conclude that the obtained expressions of the function of the stress state type for HDPE and LDPE in the form of (4) and (5), respectively, are sufficiently accurate approximations that allow more accurate determination of the limit states of the materials under study. The calculation error taking into account the specified approximations for the majority of the considered cases of the stress-strain state does not exceed 10%, which can be considered a sufficient result when using this condition in engineering and applied calculations. A discrepancy of more than 10% is observed only with a high level of hydrostatic pressure. Under such conditions, the glass transition of the amorphous part of the materials under study occurs, and a transition from one phase state to another is observed. The described changes occur when the value of the normalized octahedral stress $\xi \approx -0.990$. It would be possible to accept the interval approximation of the functions of the stress state type with separation into zones of change of the parameter ξ. At the same time, in the first section, when $-0.990 \leq \xi \leq 1$, the function of the stress state type would be defined by expression (6) for HDPE and expression (7) for LDPE, and for the second section, when $-1 \leq \xi < -0.990$, it would be necessary to select a separate approximation, and then the graph of the function would have two sections, and the errors obtained in experiments №№ 1.6 and 2.6 would also not exceed 10%. Obviously, these calculations do not pose any difficulty. However, given that such states occur only at very significant values of hydrostatic pressure exceeding 150 MPa, which is almost an unattainable limit in operation of structural elements made of these materials. Therefore, the authors consider it inexpedient to include additional sections in the obtained approximations of the functions of the stress state type. In support of this conclusion is the fact that the additionally allocated interval when $-1 \leq \xi < -0.990$ is 0.5% of the total interval of change of the parameter ξ. Although, from a theoretical point of view, the considered interval of change of the parameter $-1 \leq \xi < -0.990$ and the associated phase transitions of the materials under study may be of some practical interest. But, since the goal of this work is to predict the onset of a limit state in structural elements with naturally existing, rather than artificially simulated, external load values, the authors consider it sufficient to perform applied engineering calculations and choose the approximation of the functions of the stress state type in the form of (6) for HDPE and in the form of (7) for LDPE, even at extreme values of hydrostatic pressure, since the resulting errors for experiments 1.6 and 2.6 are 19.4% and 13.4% in practical calculations of real structural elements are compensated for by the use of appropriate safety factors. A much more interesting fact is that if the properties of the different resistances of the materials under study and their tendency to dilation are not taken into account, the calculation errors in some cases exceed 100%, which is unacceptable both from a theoretical point of view and from the point of view of the applicability of classical approaches for applied engineering calculations of structures made of polymeric materials.

5 Conclusions

In general, based on the results obtained, we can draw the following conclusions on the work:

1. The calculation of the limit states of HDPE and LDPE cannot be performed according to classical strength criteria, since errors with this kind of approach amount to more than 100% of the available experimental data.
2. The choice of the octahedral normal normalized stress ξ as a parameter of the stress state type is a practically reasonable decision, since the interval of its change is [–1; 1].
3. The proposed plasticity condition in the form (1) is a universal criterion, since the function of the stress state type is determined over the entire interval of change of the parameter of the stress state type.
4. The obtained specific equations of the function of the stress state type in the form of (6) for HDPE and in the form of (7) for LDPE are closely match experimental data and can be recommended for performing applied engineering calculations of structural elements.

References

1. Treschev, A.A.: Dependence of limit states of structural materials on the type of stress state. Construction **10**, 13–18 (1999). News of universities
2. Treschev, A.A., Bozhanov, P.V.: On the limitations of the use of classical approaches in the study of plastic bending of plates from dilating materials. Probl. Mech. Eng. Autom. **2**, 53–57 (2004)
3. Bozhanov, P.V., Treshchev, A.A.: Determination of strength criteria in the occurrence of plastic deformations in polycarbonate. Innov. Investments **12**, 323–326 (2018)

Thermal Protection of Multi-layer Exterior Walls with an Expanded Polystyrene Core

Ekaterina Ibe(✉), Galina Shibaeva, Denis Portnyagin, and Elena Afanasyeva

Khakass Technical Institute, Siberian Federal University, Komarova 15, 655017 Abakan, Khakasia, Russia
katerina.ibe@mail.ru

Abstract. In the article, the results of the study of temperature and humidity conditions of a multilayered structure in climatic conditions of Southern Siberia are stated. The complex analysis of a heat-shielding of a building is made. The humidity condition of multilayered external walls is studied. The basic construction units with the purpose of revealing of cold bridges are researched. Fields of temperature deformations of external walls are received. Unsuccessful experience of application of 3D-panels is shown, and ways of their elimination from the point of view of increase of a heat-shielding are offered. Methods of research represent modeling by means of ELCUT and SCAD Office software packages. Recommendations for the design of construction units of a building using 3D-panel technology are proposed in accordance with the requirements of not only normative documents, but also from the perspective of a comfortable thermal environment of the rooms.

Keywords: Energy efficiency · External wall · Temperature field · Thermal bridge · Multilayered structure

1 Introduction

Russia is a laggard in the global ranking on the energy efficiency of civil engineering. The problem of finding new inexpensive energy-effective building technologies is very actual today. The main attention is paid to structural and technological activities aimed at increasing the thermal resistance of protections [1–3]. Energy efficiency building envelopes (walls, floors, facades) should meet the requirements of regulatory documents. By the way, the majority of the insulation in envelopes doesn't conform to regulatory documents.

There are three conditions according to SP 50.13330.2012 which are needed to consider when a design project is making. In general, the thermal insulations of building envelopes hold true, but the thermal insulations of building sites don't hold true. The building envelopes have thermal diversities that thermal bridges pass through.

Usually the thermal performance calculations are made without taking into account important factors:

V. Murgul and M. Pasetti (Eds.): EMMFT-2018, AISC 983, pp. 758–767, 2019.
https://doi.org/10.1007/978-3-030-19868-8_74

1. Heat losses of buildings have a significant effect on the thermal protection properties and durability of structures. The heat losses have increased substantially with increasing of a filler structure's humidity and where there are places with a low thermal insulation.
2. It is not taken into account a real humidity of outdoor air. Usually the calculation is made with normative values of the humidity of outdoor air by region.
3. It is not taken into account evaporation inside the room when the calculation is made.
4. It is not taken into account that lightweight filler constructions have a small thermal inertia and process of moisture transfer is made quicker.

The main part of heat losses and the thermal bridges meets in the structural sites of exterior walls [4–12].

Nowadays, multilayered wall structures are very popular in the residential buildings. Multilayered protecting designs are widely applied to decrease in heat losses of a building and satisfaction of requirements of normative documents last years in construction, including with accommodation inside of thickness of a wall of a thermal insulation layer. The thermal protection of multilayer enclosing structures directly depends on the territory climatic region. In II–IV climatic regions of Russia, there are no sharp fluctuations in temperature, which increases the period of effective operation of multilayer enclosing structures. However, in I (cold) climatic region of Russia, durability of enclosing structures is reduced due to the sharply continental climate. Long period of time with negative temperatures leads to the formation of moisture accumulation in the external wall, which adversely affects the heat-shielding properties [13, 14]. Due to the fact that the multilayer structures consist of dissimilar materials, the joint operation at significant temperature differences may be disturbed. It is noted [15–17] that the main disadvantage of multilayer walls is the appearance of cracks at negative temperatures. The multilayer building technology with 3D-panel develops in Russia (Krasnoyarsk) [18–20]. The combination of concrete and polystyrene layers is very interesting from the standpoint of thermal protection, especially the junctions of the external walls and floors, which are very important for a cold climate. Heat-conducting inclusions in multilayer fences described above create a thermally heterogeneous structure, which creates negative conditions for the durability of the external walls. The object of this researching is the temperature-humidity regime of the external multilayered enclosing structures made of expanded polystyrene and concrete in the conditions of the cold climate.

2 Materials and Methods

Multilayered enclosing structures of residential buildings – "3D-panel" – are use as construction example in the Southern Siberia conditions, Krasnoyarsk territory.

The 3D-panel size $1,2 \times 3$ m is made in factory conditions. These wall structures (see Fig. 1) contain an expanded polystyrene layer (100–150 mm) which is fixed reinforcement steel mesh ∅3 mm on either side of expanded polystyrene layers. Steel rods pass through the expanded polystyrene layer and connect the reinforcement steel mesh. The crushed-stone concrete width 50 mm is applied to the reinforcement steel mesh by pressure concreting. So, at first, the buildings are constructed from polystyrene, and then concrete works are carried out. The 3D-panels are made by two methods: one-layer 3D-panel and two-layer 3D-panel (see Fig. 2 and Table 1).

Fig. 1. The 3D-panel.

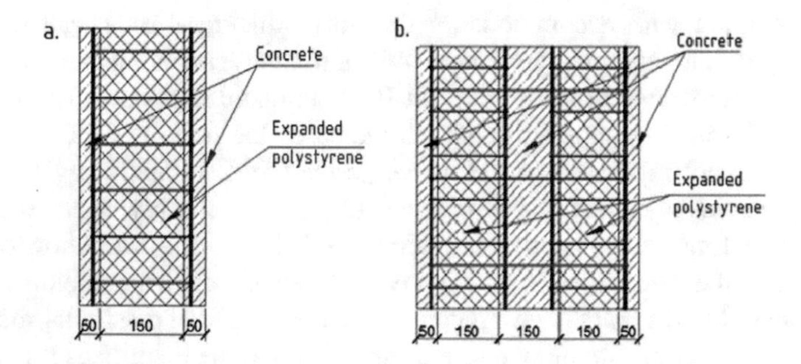

Fig. 2. The scheme of 3D-panel: a – one-layer 3D- panel; b – two-layer 3D- panel.

Table 1. Characteristic of materials.

Material	Material density ρ_0, kg/m^3	Thermal conductivity λ, W/(m K)	Vapor permeability coefficient μ, mg/(m h Pa)	Thickness of the layer δ, m
Crushed-stone concrete (Russian standard GOST 26633)	2500	1.74	0.03	0.05
Expanded polystyrene	15	0.037	0.05	0.15

In order to assess the influence of designs of multilayer exterior walls with concrete and expanded polystyrene, the complex analysis was carried out for the thermal protection of buildings, which includes:

- studying of the moisture regime of the external walls;
- studying of the main sites' thermal protection;
- studying of temperature deformations of external walls.

Mathematical modeling of the thermal regime of the structure in the cold season was carried out under the following boundary conditions:

- inside air temperature in the building t_{int} = 20 °C (Russian standard GOST 30494–2011);
- outside air temperature t_{ext} = –37 °C (Russian construction norm SP 131.13330.2012);
- heat transfer coefficient at the internal surface of the envelope α_{si} = 8.7 W/(m^2 K) (Russian construction norm SP 50.13330.2012);
- heat transfer coefficient at the external surface of the envelope α_{se} = 23 W/(m^2 K) (Russian construction norm SP 50.13330.2012).

The research methods constitute modeling with packages ElCUT and SCAD Office. The designing two-dimensional stationary temperature field and the thermal deformation of the sites of 3D-panel are presented. Calculations of water vapor in multi-layered wall structures are made according to SP 50.13330.2012.

3 Results

The most serious problem for building external envelopes is the risk of water vapor condensation in the thermal insulation layer. The vapor permeation calculation of the multilayer structure shows that the structure has a moistening zone. Distribution curves are shown on Fig. 3 intersect and there is a humidification zone in the wall.

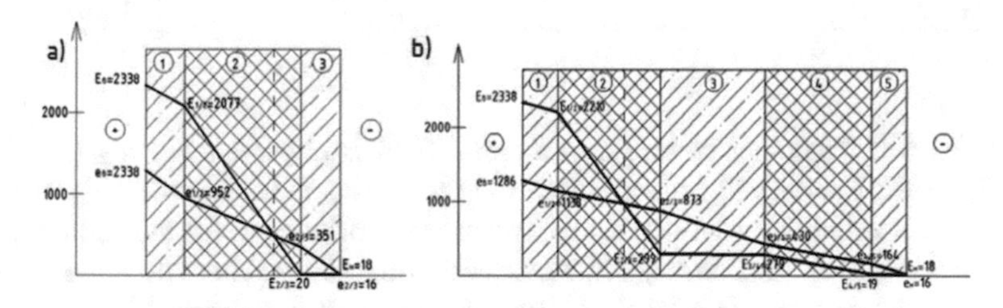

Fig. 3. Distribution of real (*e*) and maximum (*E*) water-vapor pressure in exterior wall: (a) – one-layer 3D- panel; (b) – two-layer 3D- panel.

Construction of external wall was calculated by the ElcutProfessional program for heat transfer by the stationary finite element method. The heat capacity, thermal conductivity and density of materials are taken from the reference book. Based on the input data, a temperature field is obtained.

The multilayered design of the 3D-panel is enough effective for local geographical area of construction (see Fig. 4, Table 2). Due to presence of heat-conducting inclusions in the form of a metal skeleton, a thermal field not uniform, but thus the temperature of the wall internal surface has practically stable values. Difference of temperatures does not exceed 0.5 °C.

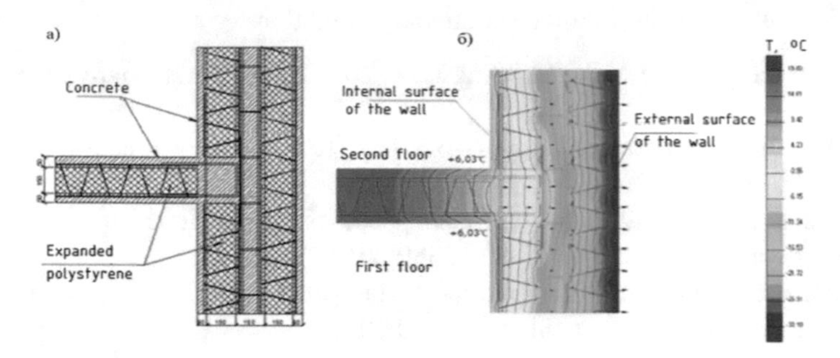

Fig. 4. Coupling joint between the wall (two-layer 3D- panel) and the floor slab: a – cross-section; b – temperature field.

Table 2. Calculation result of coupling joint between the wall (two-layer 3D-panel) and the floor slab

Performance features	Subscript, measurement unit	Value
Internal surface of the wall temperature	T_{is}, °C	+6.47
External surface of the wall temperature	T_{es}, °C	−31.20
Surface of the floor slab temperature	T_{sf}, °C	+16.47
Coupling joint between the wall and the floor slab temperature	T_{wf}, °C	+6.03
Heat flow	F, W	0.267

An absolutely another picture of a thermal field is observed in apex (see Fig. 5, Table 3). Apex is executed from concrete and on the fact is heat-conducting inclusion. The basic feature of this heat-conducting inclusion is occurrence of the additional heat flow which is taking place through it. These additional thermal heat flow to increase in heat losses and decrease in a level of a heat-shielding of a building.

By the results of temperature fields calculation on the internal wall surface in the field of this unit, significant decrease in temperature (up to −9.62 °C) in comparison with the temperature calculated on a surface of a wall - far from concrete apex is observed.

In Fig. 5, it is shown that within the limits of this area, there is a distortion of a temperature field due to influence of this heat-conducting inclusion. Around of concrete apex, the symmetric temperature field with the negative temperature, missing concentric lines on thickness of a wall is formed. It is obvious, what heat-conducting inclusion in the form of concrete apex promotes an intensification of a heat flow through a wall therefore in a place of an adjunction of 3D-panels rise in temperature is observed on 5 °C in comparison with temperature of an internal surface of a wall on the remote site of a wall.

In Fig. 6 distribution of temperature fields in a vertical section of basement part is resulted. As approaching a corner of an isotherm are displaced to an internal surface of a corner. It shows that the temperature of an internal surface of a wall in an angular zone becomes below temperature of a surface of a wall on a surface. The minimal value of temperature on a surface of angular interface it is equal 0.7 °C, that doesn't meet the requirements of normative documents, leads to occurrence of a dew-point, occurrence of a condensate and development of a mould at operation of a building.

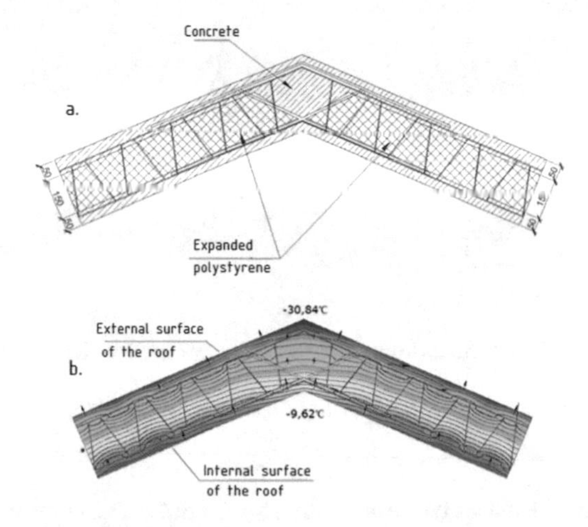

Fig. 5. The apex of the mansard roof: a – cross-section; b – temperature field.

Table 3. Calculation result of 3D- panel.

Performance features	Subscript, measurement unit	Value
Internal surface of the floor slab temperature	T_{if}, °C	−3.3
External surface of the floor slab temperature	T_{ef}, °C	−29.7
External surface of the apex of the mansard roof temperature	T_{wf}, °C	−9.62
Heat flow	F, W	0.357

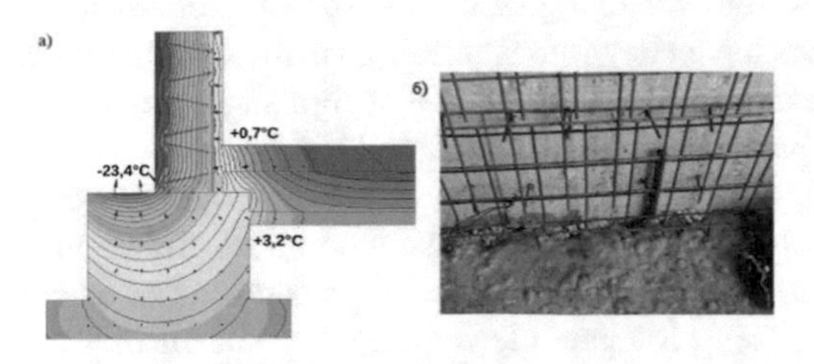

Fig. 6. The basement unit: a – temperature field; b – photo.

The study of temperature deformations of the multilayer construction showed that the joint work of concrete and polystyrene is not ensured. Since concrete is an absolutely inelastic material, it doesn't work on stretching. The results of calculating the angle-joint model showed that significant deformations occur in the internal concrete layer from the side of the room (see Fig. 7). The formation of cracks will promote the infiltration of air through the enclosing structure in conjunction with the capillary migration of moisture. Until now, there is no consensus on the mechanism of moisture migration to the freezing front. To explain this situation, scientists have put forward a capillary, osmotic, hydrodynamic, adsorption, crystallization-film theory.

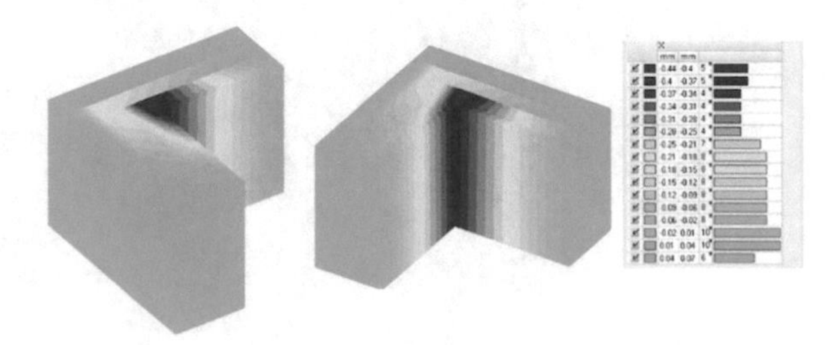

Fig. 7. Thermal deformations of the external walls angle.

4 Discussion

The joint work of all layers of external walls should provide the required parameters of thermal protection in frequent temperature changes conditions. However, the provision of a heat-saving condition raises doubts, because the described construction has a large number of heat-conducting inclusions. The main problems are thermal bridges make the heat transfer from rooms to outside through the steel rods and 3D-panel's sites.

The 3D-panel has a moisture-proof thermal insulation layer, that's why moisture conservation is impossible in this layer. The moisture accumulation will be carried out in a thin concrete layer and on the boundary layers polystyrene – concrete. So, there will be a migration of moisture through the external wall from both sides – inside and outside. The vapor condensation can occur near the steel mesh and the thermal insulation function will be reduced. In addition, the absence of water vapor removal from the interior through the envelope in the winter period, when the air ventilation in the interior is usually limited, can lead to an undesirable increase of relative humidity in the external wall and to the worsening of the internal microclimate.

Thus, researches have shown that freezing of the external walls built by 3D-panel technology, occurs as a result of errors on a design stage and installation. Also, there is a deterioration of parameters of a microclimate in rooms.

From the standpoint of energy saving, an important factor is heat losses through the external contour of the building. It is known when the building has low heat losses this reduces of energy for normal microclimate of the rooms. The overall heat losses for the residential building using 3D panel technology are shown in Table 4.

Table 4. Overall heat losses

Heat losses	Q_{walls}	$Q_{windows}$	Q_{roof}	Q_{floor}
Value, W	313.96	2.18	523.26	272.92

It is important to consider the real time of effective exploitation of the thermal insulation in the design for providing the raw heat-shielding properties of the thermal insulation materials. Because of fact that 3D-panel is multilayered construction there are the thermal insulation materials in the middle of concrete layers. That's why this construction is unrepairable. But the real time of effective exploitation of the expanded polystyrene is about 20–25 years. Siberia territory has a sharply continental climate with very cold winter. Therefore, often changing temperature differences occur in the joint area of concrete and expanded polystyrene. As a result, concrete cracks appear because of different linear expansion coefficients of concrete (0.000012 °C-1) and expanded polystyrene (0.00006 °C-1).

The essence of the capillary migration process of moisture in the 3D-panel is as follows. At a certain value of the air relative humidity in the room and a certain temperature of the internal surface of the room corner, a dew point appears. For this reason, the moisture of the internal concrete layer rises in the room corner. According to the heat engineering calculations, heat front through the concrete layer (see 4 in Fig. 8) forms a freezing front in this place (see 5 in Fig. 8), to which a capillary migration flow of moisture is organized. As a result, the average humidity of the enclosing structure materials increases. This leads to a loss in the design of heat-shielding properties, which will accordingly affect the reliability and durability of the structure.

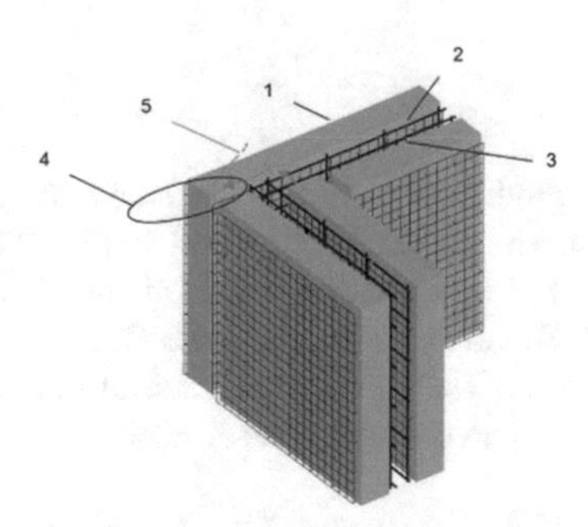

Fig. 8. The model of the external walls angle

5 Conclusions

Based on the results of the estimated evaluation of the thermal protection properties of the buildings filler constructions constructed by 3D-panel system, it is established:

1. It was found that the design meets the requirements of the norms for the climatic conditions of South Siberia. The value of the reduced resistance to the transfer

$R_0 = 3.92\,(m^2\ °C)/W$ turned out to be greater than the required $R0 = 3.42\,(m^2\ °C)/W$. However, when calculating the design of a 3D-panel for vapor permeability and condensation of water vapor in the period of variable temperature-humidity effects, it is revealed that the vaporous moisture passing through the enclosure accumulates in the layers of the filler structure. The layer of expanded polystyrene has acted as a barrier-store of moisture in the structure. It significantly reduces the heat-shielding properties of the external walls.

2. In conditions of constantly changing temperatures, the joint operation of concrete and expanded polystyrene will not be ensured due to different coefficients of linear expansion, as well as a high coefficient of thermal conductivity of concrete 1.74 W/m °C. As a result, cracks will appear in the concreting zone, contributing to the infiltration of cold air through the enclosing structure. This leads to a decrease in the durability of the exterior walls.
3. Estimated assessment of thermal protection of buildings made it possible to determine the most important defects in the thermal envelope of a building.
4. The obtained results make it possible to assess the damage caused to the customer as a result of an inefficient design decision and are the basis for bringing the building's technical solution in line with the requirements of the normative documentation.
5. At designing buildings, it is necessary to carry out thermal calculation not only rectilinear sites of walls, but also and thermal calculation of units of an adjunction and supporting designs. As a result of calculations, it is found out that value of a thermal stream through units of an adjunction and supporting designs more than in 2 times more than on rectilinear sites of a wall. Owing to what the corner of a room freezes through.

References

1. Pavlík, Z., Černý, R.: Hygrothermal performance study of an innovative interior thermal insulation system. J. Appl. Thermal Eng. **10**, 1941–1946 (2009)
2. Dylewski, R., Adamczyk, J.: Economic and environmental benefits of thermal insulation of building external walls. J. Build. Environ. **12**, 2615–2623 (2011)
3. Diao, R.D., Sun, L., Yang, F.: Thermal performance of building wall materials in villages and towns in hot summer and cold winter zone in China. J. Appl. Thermal Eng. **128**, 517–530 (2018)
4. Vasilyev, G.P., Lichman, V.A., Yurchenko, I.A., Kolesova, M.V.: Method of thermotechnical uniformity coefficient evaluation by analyzing thermograms. J. Mag. Civil Eng. **6**, 60–67 (2016)
5. Korniyenko, S.V., Vatin, N.I., Gorshkov, A.S.: Thermophysical field testing of residential buildings made of autoclaved aerated concrete blocks. J. Mag. Civil Eng. **4**, 10–25 (2016)
6. Kornienko, S.V.: Kompleksnaya ocenka ehnergoehffektivnosti i teplovoj zashchity zdanij. J Stroitel'stvo unikal'nyh zdanij i sooruzhenij **11**, 33–48 (2014)
7. Khalimov, O.Z., Shibaeva, G.N., Portnyagin, D.G., Ibe, E.E.: The improvement of aseismic horizontal frame's thermal insulations. In: IOP Conference Series: Materials Science and Engineering, vol. 365, p. 042071. IOP Publishing (2018)

8. Samarin, O.D.: Temperature in linear elements of enclosing structures. J. Mag. Civil Eng. **4**, 3–10 (2017)
9. Vedishcheva, I.S., Ananin, M.Y., Al, A.M., Vatin, N.I.: Influence of heat conducting inclusions on reliability of the system “sandwich panel –metal frame”. J. Mag. Civil Eng. **2**, 116–127 (2018)
10. Vasilyev, G.P., et al.: Study of heat engineering homogeneity fragments of enclosing structures in the climatic chamber. In: MATEC Web of Conferences, vol. 40, p. 05003. EDP Sciences (2016)
11. Anan’ev, A.I., Anan’ev, A.A.: Teplozashchitnye svojstva i dolgovechnost’ neprozrachnyh fasadnyh sistem zdanij. J. Vestnik MGSU **3**, 146–151 (2011)
12. Papadopoulos, A.M., Giama, E.: Environmental performance evaluation of thermal insulation materials and its impact on the building. J. Build. Environ. **5**(42), 2178–2187 (2007)
13. Yu, J., et al.: Experimental study on the thermal performance of a hollow block ventilation wall. J. Renew. Energy **122**, 619–631 (2018)
14. Kornienko, S.V., Vatin, N.I., Gorshkov, A.S.: Ocenka vlazhnostnogo rezhima sten s fasadnymi teploizolyacionnymi kompozicionnymi sistemami. J. Stroitel’stvo unikal’nyh zdanij i sooruzhenij **6**, 34–54 (2016)
15. Zimin, S.S., Romanov, N.P., Romanova, O.V.: Mechanisms for the formation of vertical cracks in the corner zones at the intersection of the exposed walls. J. Constr. Unique Build. Struct. **9**(36), 33–43 (2015)
16. Kuznecova, G.: Sloistye kladki v karkasno-monolitnom domostroenii. J. Tekhnologii stroitel’stva **1**, 10–13 (2009)
17. Vasilyev, G.P., et al.: Definition of the heat resistance of three-layer ferroconcrete panel. In: MATEC Web of Conferences, vol. 40, p. 5004. EDP Sciences (2016)
18. Korol’, E.A., Pugach, E.M., Nikolaev, A.E.: Tekhnologicheskaya ehffektivnost’ vozvedeniya ograzhdenij zdanij iz trekhslojnyh ehlementov. J. Modern Ind. Civil Constr. **3**, 157–163 (2007)
19. Kurlykova, E.S., Magaramova, N.S.: Vozvedenie zdanij iz mnogoslojnyh 3D-panelej. In: Nauka i molodezh’: problemy, poiski, resheniya: sbornik trudov konferencii. Izd-vo SibSIU, Novokuznetsk, pp. 234–237 (2015)
20. Chugaev, A.A.: Stroitel’stvo domov s primeneniem 3D paneli. In: Molodezhnaya nauka 2017: tekhnologi i innovacii: sbornik trudov konferencii. Izd-vo Prokrost, Perm’, pp. 244–246 (2017)

Building-Integrated Photovoltaics Technology for the Facades of High-Rise Buildings

Elena Generalova(✉) and Viktor Generalov

Samara State Technical University, Molodogvardeyskaya str. 244, 443100 Samara, Russia
generalova-a@yandex.ru

Abstract. The article deals with innovative and promising design of energy-efficient envelopes of high-rise buildings. The aim of the research is to study modern technologies and methods of integrating the energy producing photovoltaic modules into the envelope structures of skyscrapers. The research methodology is based on a system analysis of the world's best practices in the design and construction of energy-efficient high-rise buildings and structures. The article considers special features of varieties of silicon based photovoltaic modules. The paper analyses the efficiency of applying different types of solar panels along with the functional, structural and space-planning solutions of high-rise structures. The issues of creating the plastic of a facade taking into account the efficiency of photovoltaic panels are discussed. As a result, the study emphasizes the extremely important role of high-rise building envelope structures in using positive and reducing negative impacts of climatic factors on the energy balance of a building. In conclusions the article shows the prospective applications of BIPV technologies in high-rise construction.

Keywords: High-rise building · Eco Skyscrapers · Building Integrated Photovoltaics · Photovoltaic glass

1 Introduction

The inefficient use of energy resources is a common concern of all mankind and one of the greatest problems of the XXIst century. In this regard, the term "energy efficiency" is becoming increasingly popular. Its use in the context of energy-efficient architecture requires more attention to selecting the criteria by which the building can be considered to be energy-efficient. Today Eco Design has found much application in architecture and construction. It is an attitude to design, that results in organic and natural interaction of a building and nature in close connection with the unique characteristics of the specific place it is created for [1–5]. The term Eco Skyscrapers is used in the same context and characterizes the typology of innovative high-rise buildings that are built taking into account personal approach to the application of new structural systems, technologies and building materials. The integration of such buildings into the urban environment can be called «system integration» implying the rational use of water systems, energy, waste, sewage, etc. The paper proposes to focus on the envelope of

V. Murgul and M. Pasetti (Eds.): EMMFT-2018, AISC 983, pp. 768–777, 2019.
https://doi.org/10.1007/978-3-030-19868-8_75

Eco Skyscrapers, namely on the capability of envelope structures not only to protect a structure from negative environmental impact but on the ability to produce energy as well.

2 Materials and Methods

The specific feature of using solar panels in the envelope structures of high-rise buildings is of particular interest. The main function of solar photovoltaic modules is to convert sunlight into electric current. The output of the photovoltaic module generates constant electric current, which can be used both directly and accumulated in batteries for further use. At first glance, the simplest and most common solution is to install solar panels on the roofs of buildings. However, by using this method of placement, the efficiency of solar panels can be reduced under the impact of various environmental factors, such as dust, dirt and snow. In addition, the roof area of a high-rise building is not comparable to the area of its envelope. In this regard, solar panels integrated in the envelope facade structures seem to be the best solution [7–12].

Having analyzed the world experience, there can be classified two main directions of integrating the solar panels in buildings- BAPV (Building Applied Photovoltaics) are different ways of mounting photovoltaic modules on top of the building envelope; BIPV (Building Integrated Photovoltaics) are photovoltaic modules that become a part or completely form the building envelope. The experience of applying the technology of BIPV (Building Integrated Photovoltaics) is analyzed in the article. To understand the specifics and prospects of making energy-efficient envelopes for modern high-rise buildings it is necessary to have an idea about the existing variety of Photovoltaic glass (PV Glass). Various kinds of this glass differ by composition and production technology. So it directly affects the design and, ultimately, the architecture of the building. The vast majority of solar modules produced nowadays are based on silicon. There exist two major types of Photovoltaic glass made from crystalline and amorphous silicon.

1. Photovoltaic modules from Crystalline Silicon PV glass represent two transparent glass layers with photovoltaic silicon cells between. The transparency of the glass is determined by the size of the gaps between the cells. Under direct sunlight, crystalline Silicon generates twice + more power than amorphous Silicon. Accordingly, this PV Glass technology is ideal for those projects that seek for maximum energy generation, and are well oriented towards the Sun.
2. Thin-film photovoltaic modules from amorphous silicon. As a base for amorphous modules not only glass can be used but other flexible transparent materials as well. For example, there are modules on flexible basis that can be rolled into rolls for transportation and modules integrated into various household items such as clothes, bags, hats, etc. Modules on glass basis are integrated into envelope structures of buildings. Glass is used to protect the back side of the modules. Amorphous Silicon PV glass has homogeneous structure on the whole surface. It produces more power than crystalline Silicon glass when under diffuse light conditions (overcast) and high temperatures. This type of PV glass is of great interest from an

architectural point of view. It provides a variety of design solutions and differs in texture, colour and degree of transparency. This type of PV glass is perfect for facade applications.

Currently, production companies are able to meet the needs of the construction industry in a variety of non-standard products of PV glass. Photovoltaic modules are offered in the form of flat or flexible surfaces with cellular or multilayer structure. For example, being internationally recognized leader in the production of transparent photovoltaic glass, Spanish company Onyx Solar offers a wide range of Amorphous Silicon and Crystalline Silicon PV Glass that differ in colour, degree of transparency, standard and nonstandard sizes including the most available on the market PV Glass of 4 × 2 m.

Alongside with transparent products the opaque PV Glass (spandrel glass) is on demand for facade solutions. It is used when it is necessary to hide some structural elements on the facade. It should be noted that spandrel glass means higher solar cell density that eventually results in high energy efficiency.

3 Results

The efficiency and prospects of using BIPV technology in different climatic conditions are proposed to be considered at specific sites. For example, London that is located in temperate climate zone is not characterized by stable sunny weather. There are not more than 68 sunny days a year. At that the experiments with using PV Glass in high-rise structures is considered to be rather successful.

Fig. 1. «Salesforce Tower» (formerly «Heron Tower»), 230.0 m, completion 2011, London.

London skyscraper «Salesforce Tower» (or «Heron Tower», 230.0 m, completion 2011, design – Kohn Pedersen Fox Associates) has a heterogeneous energy-efficient shell, the solution of which is linked to the orientation of the facades to the cardinal points (see Fig. 1). Eastern and western facades represent a bioclimatic ventilated facade with automatic integral blinds. The core of the building's rigidity is oriented to the South. For protecting the building from excessive heat and at the same time for gaining additional energy the southern facade system is represented by a large solar shield from 48000 photovoltaic crystalline Silicon arrays of 3.374 square meters that hide double-decker lifts and fire exits [6].

Fig. 2. «FKI Tower», 245.5 m, completion 2011, Seoul.

Another energy-efficient skyscraper «FKI Tower» (245.5 m, completion 2011, design – Adrian Smith + Gordon Gill Architecture) is located in Seoul (see Fig. 2). This is also a temperate climate zone but the number of Sunny days a year is up to 200, which is a lot more than in London. The building of Federation of Korean Industries (FKI) has one of the most solar energy efficient electrical facades in the world. Innovative external wall was designed for this project and represents integrated architectural and engineering design solutions. On every floor the structure of the external wall from southeast and southwest sides is the combination of two panels of different type that are connected at an angle. By angling the top panel of Crystalline Silicon PV glass 30 degrees toward the Sun, the amount of accumulated energy is maximized. The lower panel (Low-E Insulated Glass) is angled 15 degrees toward the ground minimizing the amount of direct sun radiation and glare. From the aesthetic standpoint the whole building gains a unique folded exterior texture. The local electric utility company (KEPCO) provided a favorable 5-to-1 buy-back rate for onsite green-energy generation. The payback for the BIPV panels, which would have typically been 30–35 years, was reduced to about seven years, due to these incentives [13, 14].

In Dubai the climate conditions are different. It is in tropical zone with arid climate of hot deserts. Dubai is considered to be one of the hottest cities in the world as most days of the year are sunny and dry there. In 2018 in Zabeel Park a unique building «Dubai Frame» of 150 m high was completed (see Fig. 3). The bridge connecting two parts of this building serves as an observation deck of 105 m length with a sweeping city view. The façade of 1200 square meters is constructed from amorphous silicon photovoltaic glass of golden yellow colour. For envelope there were manufactured 2500 modules of 485 × 985 mm with a semi-transparency degree of 20% (L-vision). The façade system helps to regulate the microclimate inside the building. At that the total installed power capacity of the system allows generating a significant amount of energy necessary for its operation [15].

Fig. 3. «Dubai Frame», 150.0 m, completion 2018, Dubai.

Talking about the use of Photovoltaic (PV) systems in the construction of high-rise buildings Singapore experience should not be ignored. This city-state is located in the equatorial climate zone and is one of the sunniest cities in the world. A good example is the modern complex «Tanjong Pagar Centre», the tallest building in the country (290.0 m, completion 2016, design – Architects 61; Skidmore, Owings & Merrill LLP). Here not the façade system is used but a massive photovoltaic pergola located at the entrance of the building (see Fig. 4). The pergola covering 2600 square meters consists of 850 amorphous silicon photovoltaic glass modules of 2.456 × 1.245 mm with a semi-transparency degree of 10% (M vision) [16].

World experience shows that BIPV technologies can be used not only in new structures, but in the modernization of existing buildings. The example is the famous Chicago tower «Willis Tower» (formerly «Sears Tower», 424.1 m, completion 1974, Retrofit Start 2016, Retrofit End 2019, design – Skidmore, Owings & Merrill LLP). Being started in 2016 the modernization project of the building provides for substitution of old windows by new type of photovoltaic glass, developed by Israeli firm Pythagoras Solar. The 56-storey building was chosen for the pilot project. If the experiment is successful, all the windows on the southern facade will be replaced [17].

The new windows known as high-power photovoltaic glass units (PVGU) represent a smart hybrid technology that places monocrystalline silicon solar cells horizontally between two layers of glass. An internal plastic reflective prism directs angled sunlight onto the solar cells and allows diffuse daylight and horizontal light through (see Fig. 5). This will change the tallest building in North America into a huge urban vertical solar farm. The building has enough space to accommodate the equivalent of a 10-acre (4 ha) solar power plant.

Fig. 4. «Tanjong Pagar Centre», 290.0 m, completion 2016, Singapore [16].

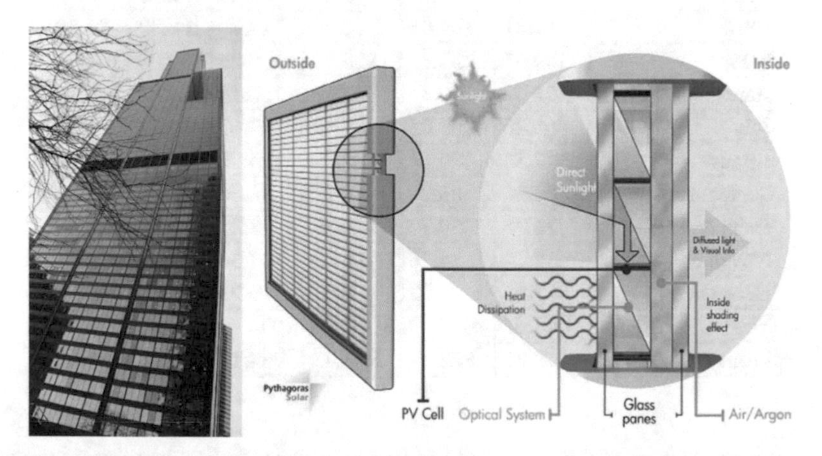

Fig. 5. «Willis Tower» (formerly «Sears Tower»), 424.1 m, completion 1974, Chicago; photovoltaic glass units (PVGU) [https://www.solarpowerworldonline.com/2012/06/out-of-the-incubator-pythagoras-solar-emerged-with-breakthrough-bipv-technology/].

4 Discussion

Analysis of the world's best practices in the design and construction of energy-efficient buildings showed that the technology of integrating the photovoltaic system in the envelope structures of skyscrapers (BIPV) is used in different climatic zones.

For the climatic conditions of the North-West region of Russia, seasonal use of solar energy is advisable. Estimates of the solar radiation resource arrival to variously oriented surfaces of the photovoltaic module in the weather conditions of St. Petersburg were performed. Calculations were performed for each hour of the year for surfaces oriented to the south, southeast (southwest), east (west), northeast (northwest), and located at the following angles to the horizon: 20°, 30°, 40°, 50°, 60°, and 90° [18, 19].

The integral values of the energy of the flux of solar radiation arriving at the south-oriented surface during the month and year are given in Table 1.

The analysis of the calculation results showed that the optimum angles of inclination of the surface to the horizon are: angle of 40° - ensuring the maximum arrival of solar radiation energy during the whole year, angle of 30° - ensuring the maximum arrival of solar radiation energy from April 1 to October 1. For the indicated angles of inclination, Table 2 presents the results of calculations for other orientations.

Table 1. Arrival of solar radiation energy, kWh/m^2

Month of the year	Angle of inclination to the horizon					
	20°	30°	40°	50°	60°	90°
January	13.48	15.24	16.69	17.77	18.46	17.99
February	32.65	35.82	38.25	39.86	40.62	37.61
March	82.17	87.77	91.5	93.24	92.95	80.29
April	111.81	114.44	114.74	112.7	108.37	84.29
May	159.95	159.7	156.55	150.23	140.93	101.24
June	171.97	169.3	164.36	156.14	144.88	100.37
July	165.26	163.63	159.43	152.07	141.77	99.93
August	127.58	129.1	128.04	124.42	118.34	89.08
September	82.3	86.2	88.31	88.56	86.96	71.92
October	36.7	39.7	41.89	43.2	43.58	39.25
November	13.73	15.26	16.49	17.36	17.87	17.04
December	7.02	7.93	8.67	9.23	9.59	9.38
Year	1004.62	1024.09	1024.92	1004.78	964.32	748.39
From 01.04–01.10	818.87	822.37	811.43	784.12	741.25	546.83

The integral values of the energy of the flux of solar radiation arriving at the south-oriented surface during the month are shown in Fig. 6.

Table 2 presents data on the annual solar radiation energy arrival and in the period from April 1 to October 1, and also data on electrical energy generation from 1 m^2 of the surfaces under consideration with an inclination angle of 30° and 40°. The efficiency was taken at the rate of 15%.

The maximum arrival of solar radiation is observed on the south-oriented surfaces inclined at 40°. If it is necessary to obtain the maximum arrival of the radiation energy in the period from April 1 to October 1, the surface should be inclined at an angle of 30° to the horizon.

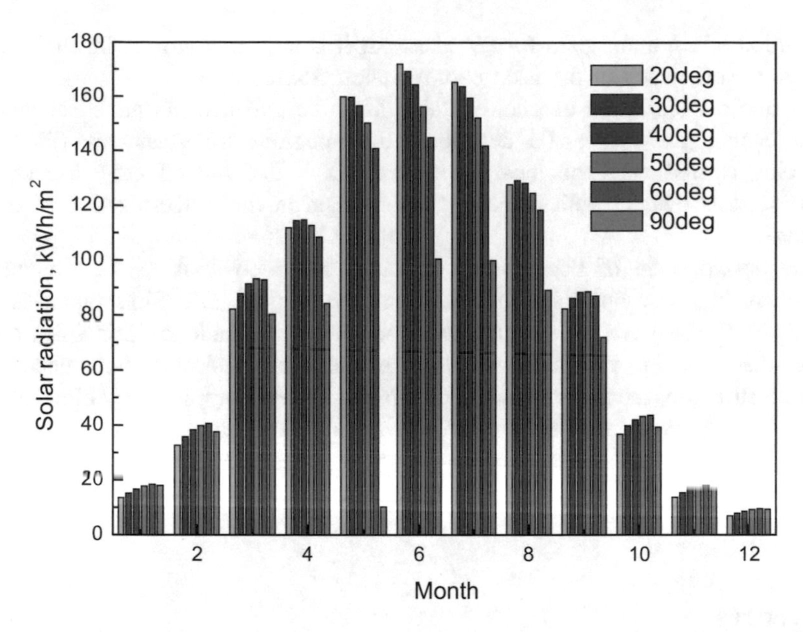

Fig. 6. The arrival of solar radiation energy on a south-oriented surface, kWh/m^2

Table 2. Arrival of solar radiation energy and electrical energy generation from 1 m^2, kWh/m^2

Surface orientation	Angle of inclination, °	Arrival of energy, kWh/m^2		Electrical energy generation, kWh/m^2	
		Per year	1.04–1.10	Per year	1.04–1.10
South	30	1024.09	822.37	153.6	123.4
	40	1024.92	811.43	153.7	121.7
Southwest (East)	30	987.65	804.51	148.1	120.7
	40	983.81	793.7	147.6	119.1
West East)	30	882.16	739.33	132.3	110.9
	40	861.44	720.22	129.2	108.0
Northwest (east)	30	746.33	639.74	111.9	96.0
	40	692.86	592.31	103.9	88.8

5 Conclusions

Modern photovoltaic systems designed for a building envelope structures (BIPV) provide energy savings, reduce heating and air conditioning costs and regulate indoor climate. This fact is the most important but not the only advantage. A wide range of colors and the ability to produce elements of individual size expands the boundaries of architectural imagination. Innovative envelope of high-rise buildings is not simply a substitution of common glass for PV glass but it is an active work with the facade to improve the efficiency of the used photovoltaic system.

Despite a wide range of photovoltaic glass, the experiments aimed at inventing hybrid technologies such as the described above photovoltaic glass units (PVGU) are still being carried out. The most important fact is that the effective use of BIPV technologies is evident both in new structures and in the modernization of existing buildings.

The development of Eco Design allows to take new look at the structure and function of high-rise buildings in the urban environment. Eco Skyscrapers are considered as a whole energy system based on energy dependence principles and the system that uses the potential of nature as the source of renewable energy. The envelope structure, the facade system of a high-rise building, is a key element in the concept of climate adaptation and energy saving. The study shows that high-rise buildings are a promising platform for the applications of BIPV, as they have a large area of envelope structures and are in dire need of reducing energy consumption for operation.

References

1. Wood, A.: Rethinking the skyscraper in the ecological age: design principles for a new high-rise vernacular. In: Proceedings of the CTBUH 2014 Shanghai Conference «Future Cities: Towards Sustainable Vertical Urbanism», Shanghai, China, pp. 26–38 (2014)
2. Generalov, V.P., Generalova, E.M.: Sustainable architecture, energy efficiency and sustainability of affordable housing on the example of Hong Kong. Vestnik of SSUACE. Town Planning Archit. **4**(21), 23–29 (2015)
3. Holdsworth, B.: Ecological high-rise: solar architects of the 21st century: Dr. Ken Yeang. Refocus, vol. 6, no. 1, pp. 58–60 (2005)
4. Generalov, V.P., Generalova, E.M.: Revealing the special features of the concepts «comfortable living» and «comfortable living environment». Urban Constr. Archit. **2**(23), 85–90 (2016)
5. Broduch, M.M.: Engineering equipment of high-rise buildings. AVOK-Press, Moscow (2007)
6. Salesforce Tower. http://www.skyscrapercenter.com/building/110-bishopgate/966. Accessed 12 Nov 2018
7. Generalov, V.P., Generalova, E.M.: High-rise residential buildings and complexes. Singapore. Experience in high-rise housing design and construction, Samara (2013)
8. Generalova, E.M., Generalov, V.P.: Designing high-rise housing: the Singapore experience. CTBUH J. (IV), 40–45 (2014)
9. Brodach, M.M., Shilkin, N.V.: Double glass facades. Sustain. Building Technol. 35–45 (2015)
10. Lotfabadi, P.: High-rise buildings and environmental factors. Renew. Sustain. Energy Rev. **38**(C), 285–295 (2014)
11. Frontini, F., Friesen, T.: Photovoltaic modules integrated into the building envelope. Sustainable Building Technol. 86–91 (2013)
12. Generalova, E.M., Generalov, V.P.: Innovative solutions for building envelopes of bioclimatic high-rise buildings. Environ. Technol. Resour. **1**, 103–108 (2017)

13. Betancur, J.: Multitasking façade: how to combine BIPV with passive solar mitigation strategies in a high-rise curtain wall system. Int. J. High-Rise Buildings **6**(4), 307–313 (2017)
14. FKI Tower. http://www.skyscrapercenter.com/building/fki-tower/8829. Accessed 12 Nov 2018
15. Dubai Frame Photovoltaic Façade. https://www.onyxsolar.com/dubai-frame. Accessed 12 Nov 2018
16. Tanjong Pagar Center. https://www.som.com/projects/tanjong_pagar_centre. Accessed 12 Nov 2018
17. Chicago's Willis Tower to Harness Sunlight. https://www.seeker.com/chicagos-willis-tower-to-harness-sunlight-1765195380.html. Accessed 12 Nov 2018
18. Aronova, E., Vatin, N., Murgul, V.: Design energy-plus-house for the climatic conditions of macedonia. Proc. Eng. **117**, 766–774 (2015). https://doi.org/10.1016/j.proeng.2015.08.231
19. Murgul, V., Vatin, N., Aronova, E.: Autonomous systems of solar energy supply under the weather conditions of Montenegro. Appl. Mech. Mater. **680**, 486–493 (2014). https://doi.org/10.4028/www.scientific.net/AMM.680.486

The Development of Energy Efficient Facing Composite Material Based on Technogenic Waste

Irina Vitkalova, Anastasiya Torlova, Evgeniy Pikalov(✉), and Oleg Selivanov

Vladimir State University named after A.G. and N.G. Stoletovs, 87, Gor'kogo str., 600000 Vladimir, Russia
Evgeniy-pikalov@mail.ru

Abstract. The research presents the results of the composition and method development for producing energy efficient facing composite material. The method includes waste grinding and drying, followed by dissolving of unplasticized polyvinyl chloride waste in methylene chloride and further cold mixing of the resulting solution with cullet, which serves here as a filler. The produced mixture is formed using one-stage cold pressing and the billets are heat-treated at the temperature exceeding the boiling point of methylene chloride. The described method permits complex utilization of two waste types at the reduced production energy intensity due to the decrease of the processing temperature. The processing temperature reduction is also important, as it eliminates thermal destruction of polyvinyl chloride, characterized by low thermal stability. Waste dissolving also allows to simplify its granulometric composition and to reduce the cost of grinding. An additional advantage of the developed compositions and method is the possibility to produce energy efficient material regarding thermal conductivity, which in terms of frost resistance and water absorption can be used both for indoor and outdoor cladding. The compressive and bending strength of the material is relatively low, but it is satisfactory during its performance without applying high mechanical stress. Thus, the developed material reduces multilayer walls thickness in the constructions, increasing the available space and reducing the load on the foundation. The paper presents the research results regarding compressive strength and water absorption of the developed material, depending on the components proportion in the raw mixture composition and pressing pressure.

Keywords: Polymer waste · Polyvinyl chloride · Cullet · Polymer composite material · Facing material · Polymers dissolving

1 Introduction

Until recently composite materials, including polymer binders, were not used in the construction industry in Russia, but since 2013 the state authorities started to support the application of composite materials in transport infrastructure, construction, housing and communal services in our country. It depends on the advantages of composite

V. Murgul and M. Pasetti (Eds.): EMMFT-2018, AISC 983, pp. 778–785, 2019.
https://doi.org/10.1007/978-3-030-19868-8_76

materials for construction purposes, including high strength, low weight, chemical resistance, low water absorption, as well as simple production, application and repair.

In the production of the construction polymer composite materials, the possibility of joint utilization of polymer and other waste used as fillers, binders and functional additives is particularly essential [1], as it allows to reduce the rate of waste accumulation and to expand the range of construction materials and products using cheap secondary resources [2, 3].

The research objective is to develop the method of producing energy efficient polymer composite material, used in manufacturing of facing and finishing construction products. The research has suggested using polyvinyl chloride waste solution in methylene chloride as binder and window cullet as filler.

The application of thermoplastic polymer can reduce the production cycle duration, as thermoplastic is cooled faster than thermoset is cured. Moreover thermoplastic polymer can be recycled, thus making it possible to use waste instead of the primary polymer, as a lot of polymer waste is accumulated due to the large scale production of polyvinyl chloride products. Besides it should be taken into account that polyvinyl chloride is characterized by strength, low water absorption, chemical resistance, good recyclability and the fact that this polymer does not support combustion; and this property is essentially important for construction materials [4].

Methylene chloride has been chosen for dissolving polyvinyl chloride waste due to the affinity of this solvent with the used waste and high penetration of methylene chloride thanks to its small molar volume, which accelerates the polymers dissolving. In addition, methylene chloride is one of the most commonly used solvents as it refers to the hardly combustible liquids of low toxicity (hazard class 4) and low cost. The high solvent volatility contributes to the rapid transition of polymer from the solution into the glassy state. The polymer transfer into a viscous state by dissolving has a number of advantages over melting. For the polymers dissolving it is not necessary to control the particle size strictly after preliminary grinding. The use of a polymer solution allows cold mixing and molding, and also allows to achieve better homogeneity of the raw mixture compared to the powder polymer.

Additional advantage is the simplification of the heat treatment mode. Firstly the exposure temperature decreases, since methylene chloride boiling point (45–50 °C) is significantly lower than the melting polyvinyl chloride point (150–220 °C [5]). Besides it should be noted that during temperature-time mode variation or violation thermal degradation of polymer is hardly possible. It is particularly important as polyvinyl chloride refers to non-thermostable polymers (degradation temperature starts at 135–140 °C [5]). The main disadvantage of polymer solutions is the loss of solvent during volatilization and therefore, the developed method suggests heat treatment of the resulting composite material and solvent vapors removal for their subsequent condensation and reuse. It helps to reduce solvent costs and to avoid environment pollution by methylene chloride vapor.

The use of the cullet is stipulated by the Vladimir region conditions. According to studies the region is characterized by the considerable accumulation of glass waste (up to 10% of the total waste) as a result of the local glass enterprises activities and household consumption of glass products [6–8]. The main advantages of cullet, used as filler, are strength, chemical resistance, water resistance, heat resistance and incombustibility.

2 Materials and Methods

To produce thermoplastic binder, the waste of unplasticized polyvinyl chloride (C2H3Cl)n was used. This waste includes the residues of constructional profiles (docking profiles and skirting boards) and finishing wall panels left from various sources. To transfer the thermoplastic binder to a liquid state, methylene chloride CH2C12 of the first grade according to GOST 9968-86 was used, where the basic substance content amounted 98.8 wt. %. Window sheet cullet of the following composition (by wt. %): SiO_2 = 73.5; CaO = 7.4; MgO = 1.9; Na_2O = 11.1; K_2O = 5.2; Al_2O_3 = 0.9 was used as a filer [6].

Before the application polyvinyl chloride waste and window cullet were pre-dried to a constant mass, and then crushed with subsequent fraction selection of less than 0,63 mm particle size. Afterwards the polyvinyl chloride powder was dissolved in methylene chloride, and the resulting solution was mixed with a cullet to obtain a homogeneous mass, used for making the samples of the studied composite material by single-stage pressing. After the pressing the samples were heated at the temperature of 45–50 °C with 45 min exposure to evaporate the solvent. Samples of each raw mixture composition were made in five samples batches.

Water absorption (W, %), frost resistance (FR, cycles), compressive strength (σcs, MPa) and bending (σbs, MPa), thermal conductivity (λ, W/m °C) were determined in the cladding composite material samples according to the standard methods for construction materials.

3 Results

The initial research stage was devoted to the preparing of polymer solutions of different proportions of the unplasticized polyvinyl chloride (PVC) waste and methylene chloride (MC) under the pressing pressure of 8 MPa.

The certain pressure was chosen due to the fact that under the selected pressure, cubic samples of 50 mm side can be obtained, which were used for determining the compressive strength of the resulting facing composite material.

The research stated that at the proportion of PVC:MC less than 1:1.2 there was almost instantaneous binder gelation, which made its use impossible for producing a raw mixture. At the proportion of PVC:MC less than 1:1.5 the solvent amount was not enough to achieve the sufficient binder fluidity for effective mixing. Moreover at the low MC, PVC quickly transfers into the glassy state due to the high solvent volatility. When PVC:MC proportion was over 1:2.5 the solvent amount led to the excessive moisture of the raw mixture after mixing the solution with the filler and to the adhesion of the mixture to the mold surfaces. Besides MC excess leads to the insufficient PVC amount after solvent evaporation thus reaching the required properties of the resulting cladding composite material.

When selecting the amount of filler, it has been stated that when mixing the binder solutions at the proportion of PVC:MC from 1:1.5 to 1:2.5 with less than 35 wt. % cullet, raw material mixture of high humidity is produced, which is excessively compacted under the pressure and can be squeezed out of the mold through the technological gaps for air discharge, thus causing sample deformation or even mold jamming. When using cullet more than 75 wt. % molded samples possess poor mechanical strength (less than 8 MPa) and crumble.

The results of determining compressive strength and water absorption of the samples produced on the basis of the studied compositions are shown in Figs. 1 and 2.

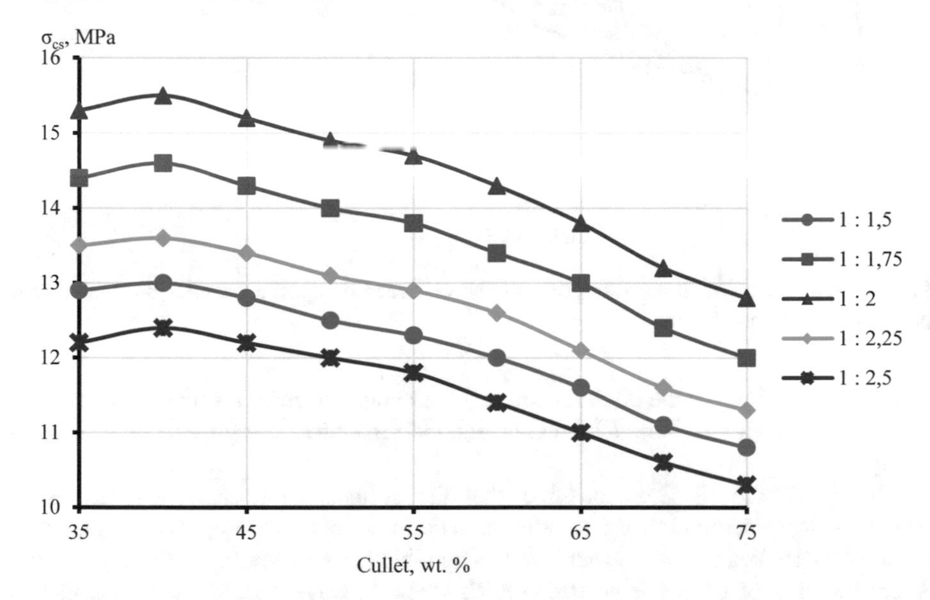

Fig. 1. Dependence of compression strength of the produced composite material on components proportion in the raw mixture

The obtained data shows that the compressive strength rises with the increase of solvent amount in the binder to a ratio of 1:2, further increase of MC leads to the weakening of the considered property (see **Error! Reference source not found.**). This effect can be explained by the lack of MC, binder solution of high viscosity causes rapid transition of PVC into a glassy state thus not allowing to achieve raw mixture homogeneity by stirring and a sufficient degree of compaction during compression. The solvent excess facilitates mixing and compacting, prolongs the binder dissolved state, but after the solvent removal, the PVC amount is not enough to form a strong structure of the composite material.

It is worth noting that the maximum compressive strength is achieved when the cullet amounts from 35 to 45 wt. %. Such dependence is explained by the fact that with less amount of filler, the material strength is primarily determined by PVC strength but cullet particles are dispersed in the binder and do not form rigid structures. When cullet

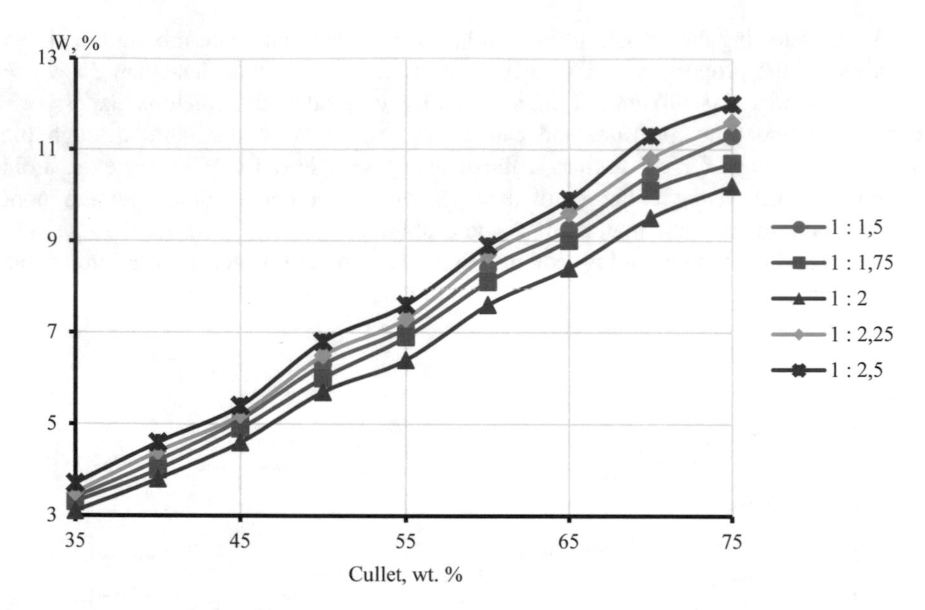

Fig. 2. Dependence of water absorption of the produced composite material on components proportion in the raw mixture

amounts over 45 wt. % the binder amount is not enough to form a structure from cullet particles, interconnected through PVC layers, their thickness can provide high material strength.

Besides data in Fig. 2 demonstrates that almost linear water absorption increase is observed alongside an increase of cullet amount in the raw mixture, and the minimum value of the studied property is achieved using a binder with a ratio of PVC:MC = 1:2. As in the case of compressive strength, the lack of solvent makes it impossible to achieve uniformity in mixing, which leads to the formation of an open-porous structure that facilitates the penetration of water into the sample volume. Similar to the case of compressive strength, the lack of solvent makes it impossible reach uniformity by mixing. It causes the formation of an open-porous structure that facilitates the penetration of water into the sample depth. If there is an excess of solvent and the filler amount increases, an open-porous structure is also formed, but only due to the lack of PVC after MC evaporation.

Judging by the obtained data, the subsequent experiments were conducted using the binder composition with the ratio of PVC:MC = 1:2. Beside the composition of the raw material mixture, the pressing pressure affects the properties of the resulting facing composite material, since the sample compaction degree during molding depends on this parameter. The temperature and time of the following samples processing practically do not influence the resulting material characteristics, as heat treatment does not affect the material composition and structure, but only solvent evaporation is observed. In this respect, from all technological parameters the research requires only to study the pressing pressure effect on the samples compressive strength and water absorption.

The experiment stated that under the pressing pressure of up to 5 MPa inclusive, the compaction degree is not enough for producing sufficiently dense and durable samples, the material crumbled on the edges and front surfaces. Under the pressing pressure of 11 MPa and over, the material was adhering to the mold surface, followed by the tearing of the samples fragments during demolding, which depends on the binder extrusion from the sample under the excessive pressure. The samples compressive strength and water absorption under pressing pressure from 6 to 10 MPa and with the different filler composition are shown in Figs. 3 and 4.

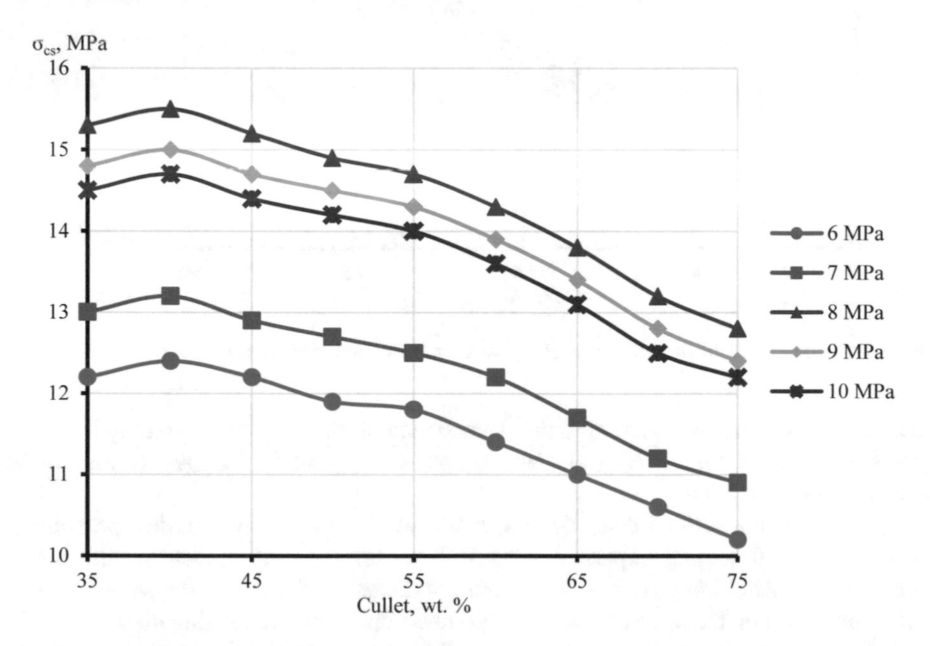

Fig. 3. Dependence of the produced composite material compression strength on pressing pressure

The presented data demonstrates that the compressive strength of the samples produced under the pressure of 6 and 7 MPa is inferior to the samples produced under higher pressure. Herewith, the samples produced under the pressure of 9 and 10 MPa are slightly inferior in strength to the samples produced under the pressure of 8 MPa. Such dependence can be explained by the fact that under the pressure below 8 MPa the compaction degree is insufficient to form a rigid structure ensure during the MC evaporation, but under the pressure above 8 MPa a part of the binder is squeezed onto the surface during the compression, thus reducing the structure homogeneity. Another factor reducing the strength includes defects appearing at fragments tearing off the sample during demolding.

Besides Fig. 4 shows that the pressing pressure only slightly affects water absorption, which, as at the previous research stage, evenly increases alongside the increase of the filler amount in the raw mixture. In this case, according to the similar

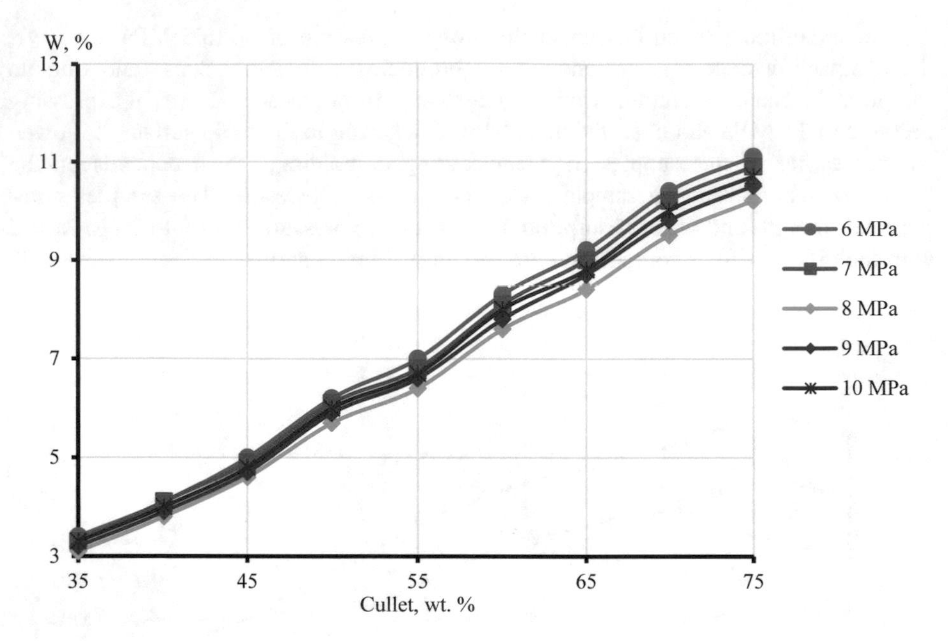

Fig. 4. Dependence of the produced composite material water absorption on pressing pressure

reasons as listed above concerning the dependence of the compression strength and the pressing pressure, the lowest water absorption was typical for samples formed under the pressure of 8 MPa.

Basing on the studied dependences, it was decided to apply pressing pressure of 8 MPa in the following experiments as well as for the development of the binder composition. Additional studies have been conducted to determine the samples properties produced on the basis of the raw mixture composition comprising 40 wt. % of the cullet, in order to assess the possibility of using the developed composite material for the production of cladding products (see Table 1).

Table 1. The properties of the studied composite materials.

Water absorption, %	Frost resistance, cycles	Compression strength, MPa	Bending strength, MPa	Thermal conductivity, Wt/m^2 °C
3,8	52	15.5	3.7	0.295

The received data proves that the developed composite material is characterized by poor water absorption and thermal conductivity, high frost resistance, though its strength is relatively low.

4 Conclusions

The research results testify that composite material, corresponding to the requirements of GOST 13996-93 concerning water absorption and frost resistance for ceramic facade tiles, can be produced basing on 40 wt. % of cullet, 20 wt. % of PVC and 40 wt. % of MC under the pressure of 8 MPa. Since frost resistance of the developed composite material exceeds 50 cycles, it can be used for cladding both walls and facade basements of buildings and structures. The material is characterised be relatively low strength, but it can be quite enough, since facing and finishing materials are not subjected to heavy mechanical stress during the performance. It should be noted that the developed material strength is approximately consistent with the ceramic brick brand M150 characterised by the compressive and bending strength of 15 and 2.8 MPa respectively. As for the thermal conductivity the developed material belongs to the group of effective materials ($0.24 < \lambda < 0.36$).

Thus, the developed method proves that polymer and glass waste can be recycled for producing energy-efficient cladding and finishing materials for outdoor and indoor application in buildings and structures.

References

1. Shakhova, V., Vorobyeva, A., Vitkalova, I., Torlova, A., Pikalov, E.: Modern technologies of polymer waste disposal and the problem of their application. Mod. High Technol. **11–2**, 320–325 (2016)
2. Ershova, O., Ivanovskiy, S., Chuprova, L., Bakhaeva, A.: Modern composite materials based on polymer matrix. Int. J. Appl. Fundam. Res. **4–1**, 14–18 (2015)
3. Snezhkov, V., Retsits, G.: Polymer waste - into ready made products. Solid Waste **1**, 16–19 (2011)
4. Sokolskaya, K., Kolosova, A., Vitkalova, I., Torlova, A., Pikalov, E.: Binders for producing modern polymer composite materials. Fundam. Res. **10–2**, 290–295 (2017)
5. Kuznetsov, E., Prokhorova, I., Faizullina, D.: Album of technological schemes of polymers and plastic masses production on their basis. Chemistry, Moscow (2007
6. Shakhova, V., Vitkalova, I., Torlova, A. Pikalov, E., Selivanov, O.: Development of composite ceramic material using cullet. In: MATEC Web of Conferences, vol. 193, p. 03032 (2018). https://doi.org/10.1051/matecconf/201819303032
7. Vitkalova, I., Torlova, A., Pikalov, E., Selivanov, O.: Industrial waste utilization in the panel production for high building facade and socle facing. In: E3S Web of Conferences, vol. 33, p. 02062 (2018). https://doi.org/10.1051/e3sconf/20183302062
8. Torlova, A., Vitkalova, I., Pikalov, E., Selivanov, O., Sysoev, É., Chukhlanov, V.: Glass Ceram. **74**(3–4), 107–109 (2017)

Energy Efficiency Improving of Construction Ceramics, Applying Polymer Waste

Irina Vitkalova, Anastasiya Torlova, Evgeniy Pikalov(✉), and Oleg Selivanov

Vladimir State University Named After A.G. and N.G. Stoletovs, 87, Gor'kogo Str., 600000 Vladimir, Russia
evgeniy-pikalov@mail.ru

Abstract. The paper presents the results of the charge composition development on the basis of low-plasticity clay with the addition of 15 wt. % of the unplasticized polyvinyl chloride waste as a combustion additive to increase ceramics porosity and to reduce its density, thus decreasing the load on the foundation during the construction, and reducing thermal conductivity and consequently increasing material energy efficiency. This waste was chosen as its processing, using other methods, is very complicated due to low thermal stability of polyvinyl chloride. Boric acid is also introduced into the charge composition and it serves as a fusing agent, reducing the liquid-phase sintering temperature and forming a vitreous phase, which improves product strength characteristics, reduces water absorption and increases its frost resistance. However, the vitreous phase amount is negligible and the material retains its porosity, sufficient to produce relatively effective products in terms of their thermal performance. The resulting ceramics meets all other basic performance requirements for ordinary ceramic wall bricks. Consequently the developed charge composition allows producing rather high quality construction ceramics at a low production cost on the basis of low-demand primary raw materials and technogenic waste and use it to construct multilayer walls corresponding to modern thermal engineering standards applying rational consumption of construction materials. Herewith the properties of the developed ceramic material allow to some extent combining the functions of the base, thermal insulation and partly facing layers, thus expanding the spheres of using such products.

Keywords: Polymer waste · Polyvinyl chloride · Low-plastic clay · Boric acid · Flux · Energy efficiency

1 Introduction

Recently energy saving in the construction sphere has been paid great attention. It depends on the rise of prices for heat transfer and on the increase of heat engineering standards. Therefore, the use of traditional ceramic products becomes inconvenient, as it is necessary to increase the masonry thickness to meet the new standards requirements, but it causes the increases the construction cost and reduces the free inner space. To comply with the heat regulations and to ensure the efficiency of construction materials, multi-layer walls are constructed using energy-efficient products. The most

V. Murgul and M. Pasetti (Eds.): EMMFT-2018, AISC 983, pp. 786–794, 2019.
https://doi.org/10.1007/978-3-030-19868-8_77

common option is the construction of three-layer walls, with the inner load-bearing layer made of solid ceramic brick, the middle layer made of thermal insulation material, providing heat requirements, and the outer layer made of facing brick or other cladding materials. The application of energy-efficient ceramic bricks for constructing load-bearing layers is important, as it allows further reduction of the masonry thickness in compliance with thermal engineering standards [1].

Energy efficiency of ceramic products implies their low thermal conductivity, which allows keeping heat inside the buildings. The easiest way to reduce thermal conductivity of the products is to increase their porosity, which in turn is easily provided by using of combustion additives [2].

This research considers the possibility of improving the wall ceramics energy efficiency produced using low-plastic clay due to the application polymer waste serving as a combustion additive. As low-plasticity clay application is limited because of the poor strength, frost resistance and crack resistance of the products, produced on its basis therefore a fusing agent is introduced into the charge together with the combustion additive for reducing the liquid-phase sintering temperature of the material in order to increase its strength and reduce water absorption.

The polymer waste is used because of its large-capacity, resistivity to decomposition and low density, and therefore it is accumulated in large amounts. Speaking about industrial waste, the simplest and most effective recycling method is their recovery into the production process, where they are generated. Consumption waste recovery becomes more complicated as secondary polymer raw materials are characterized by instability and poor physical and mechanical properties, compared to primary polymer raw materials due to partial destruction and the possible presence of non-polymer inclusions and impurities of other polymers [3, 4].

In this regard, the selective accumulation and sorting or the development of processing methods for polymer waste mixtures are necessary for the conversion of polymer waste into products [3, 5]. Further they are mixed with the primary polymer raw material to produce products which do not require high quality, to produce composite materials, mainly for construction purposes [3]. However, the problem of processing high destruction degree polymer waste has not been solved yet. Nowadays it being solved only by incineration is as it allows disposing polymer waste of any homogeneity and purity degree. In most cases, the incineration method is used in cases where polymer waste is used as a fuel [3, 4], but this method makes it possible to use polymer waste as burning additives in the production of ceramic products. In most cases, the incineration method is used when polymer waste is used as a fuel [3, 4], but this method makes it possible to use polymer waste as combustion additives in the ceramics production.

The research authors have previously developed charge composition based on low-plasticity clay to produce walling and facing materials [6–9]. Two groups of additives have been used in the mentioned compositions to increase the ceramics strength and reduce water absorption: additives serving as vitreous phase sources and additives reducing the liquid-phase sintering temperature and increase the vitreous phase amount. In this research, it has been decided to limit the introduction of the flux to reduce the liquid-phase sintering temperature, since the considerable formation of vitreous phase significantly reduces porosity leading to the decrease of product energy efficiency.

2 Materials and Methods

The principle charge component, serving as a possible basis for the wall ceramics developing, was low-plasticity clay from Suvorotskoye deposit of the Vladimir region of the following composition (wt. %): SiO_2 = 67.5; Al_2O_3 = 10.75; Fe_2O_3 = 5.85; CaO = 2.8; MgO = 1.7; K_2O = 2.4; Na_2O = 0.7. This clay is characterized by low plasticity because it contains sufficiently high quantities of aluminum, calcium and magnesium oxides. The clay plasticity index, determined by the standard method, was 5.2, and, consequently, in compliance with GOST 9169-75, it refers to the low-plasticity type [7].

Waste from unplasticized polyvinyl chloride (UPVC) - the waste from building profiles and finishing panels consumption in particular - was used as a combustion additive in the developed composition of the charge. On the one hand, this waste is associated with large production and consumption volumes of profiles and panels, resulting in large amount of waste. On the other hand, UPVC is a thermally unstable polymer so it additionally complicates its recycling, thus its application as a combustible additive can become a disposal alternative.

It should be noted that UPVC combustion emits toxic gases, including hydrochloric acid vapor, so it is necessary to process UPVC combustion products additionally by reburning them at the temperature of 1200–1400 °C, when toxic compounds decompose, or to clean the flue gases using sorption methods of cleaning or dry cleaning, i.e. to introduce quicklime, magnesium oxide or sodium hydroxide into the dust gases, which interact with toxic incineration products and form harmless compounds. For instance, the lime neutralizes the hydrogen chloride with the formation of calcium chloride [3]:

$$CaO + 2HCl = CaCl_2 + H_2O$$

In this research, boric acid B grade 2 (GOST 18704-78) with a basic material mass fraction of 98.6% was used as a fuse. Its efficiency has been tested in the previous experiments [7–9], their result stated that this additive can increase the strength and reduce the water absorption of ceramics based on the low-plasticity clay.

Semi-dry pressing technology was used to obtain samples of the developed ceramics [7]. Clay and polymer waste were pre-dried reaching the constant mass and crushed with further selection of max 0.63 mm fraction. Afterwards the charge components were weighed in the required ratio and mixed first in a dry state, and then with water to obtain a molding mass with a moisture content of 8 wt. %. The samples were made from the resulting molding mass at a specific pressing pressure of 15 MPa and a maximum firing temperature of 1050 °C. Due to low humidity of the molding mass, the samples were not dried. Samples were made in batches of three samples each.

For the samples performance and energy efficiency assessment of the developed wall ceramics the following characteristics have been determined according to the standard methods for ceramic materials: compressive strength (σ_{cs}, MPa) and bending (σ_{bs}, MPa), water absorption (W, %), frost resistance (FR, cycles), density (ρ, kg/m^3), thermal conductivity (λ, W/(m^2 °C)), open (P_{opn}, %) and total (P_{tot}, %) porosity.

3 Results

The first experimental stage was devoted to the study of separate impact of polymer waste and boric acid on the compression strength and water absorption of ceramics on the basis of the low plasticity clay. According to the data, obtained after the polymer waste introduction into the charge (see Fig. 1) alongside the increase the considered additive amount, the compressive strength is significantly reduced but water absorption is sufficiently increased, resulting in frost resistance decrease. It is connected with the considerable formation of pores and voids in the material depth at the polymer waste combustion during the firing.

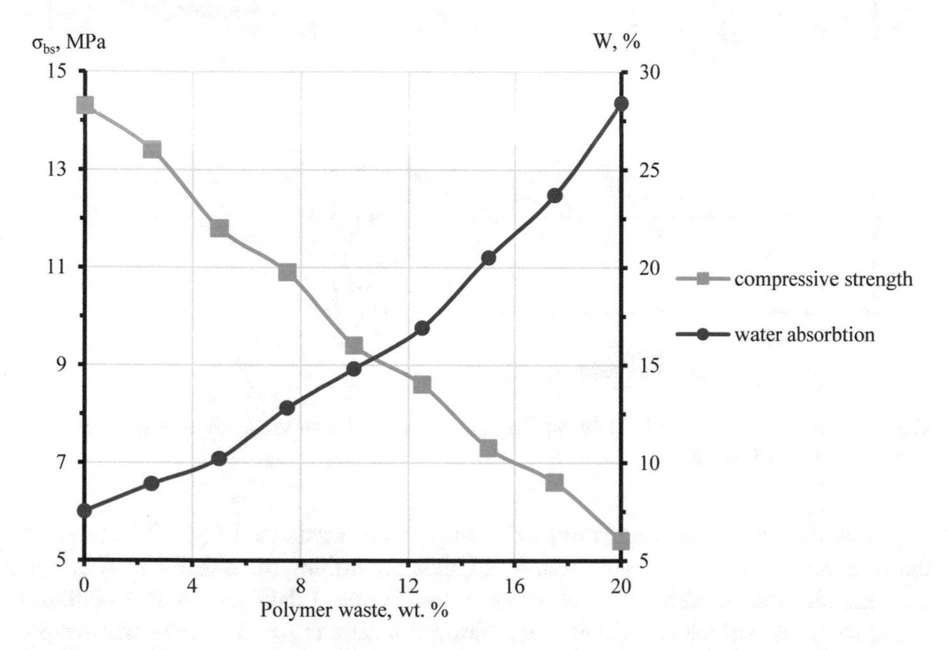

Fig. 1. Dependence of ceramics properties on polymer waste amount in the charge composition

Thus, a separate introduction of polymer waste does not allow producing ceramic material with sufficient properties for the walling production on the basis of the considered clay. Therefore, boric acid was introduced into the charge together with polymer waste at the second experimental stage to increase the strength and reduce water absorption of the developed ceramics. It should be highlighted that the previous experiments results revealed that without the introduction of additives forming the vitreous phase, the amount of boric acid in the charge should be limited to 2.5 wt. %, as higher concentrations of the additive cause the decrease of ceramics environmental safety, associated with the 3rd hazard class of boric acid [7]. In this connection, the influence of the polymer waste amount on the basic properties of wall ceramics was studied, when boric acid 2,5 wt. % was introduced into the charge.

As data from Fig. 2 shows similar dependencies resulting in the weakening of the compression and bending strength are observed during joint introduction of polymer waste and boric acid in order to increase these properties.

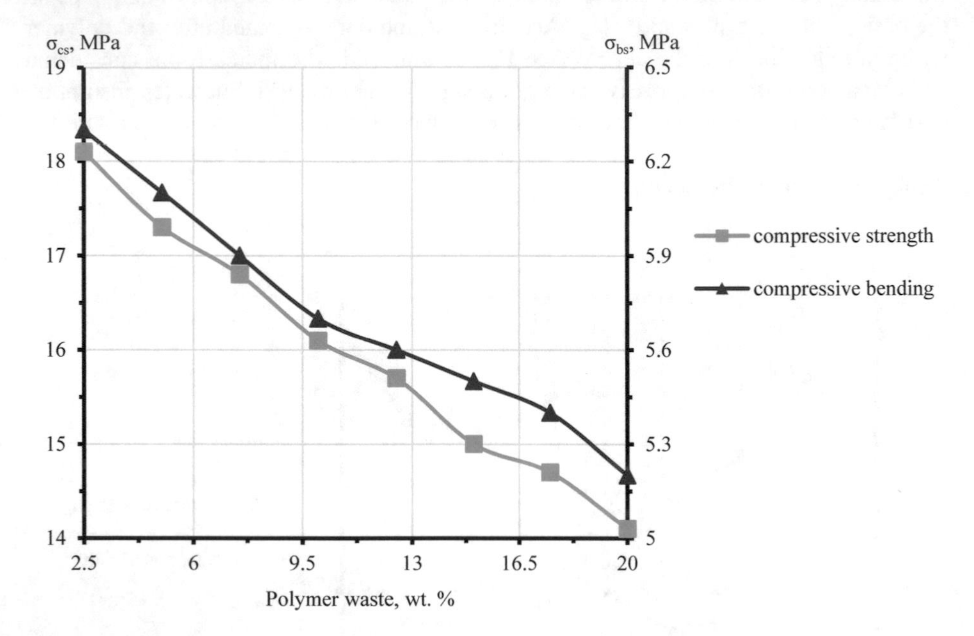

Fig. 2. Relation of compression and bending strength on polymer waste amount with boric acid introduction of 2.5 wt. %

However, unlike separate introduction of polymer waste (see Fig. 1), in this case, the strength weakening is not so significant, thus confirming the effectiveness of boric acid introduction in addition to polymer waste. Figure 3 data proves that additional introduction of boric acid reduces water absorption, if compared to the ceramics produced only with the polymer waste, and consequently allows to achieve sufficiently high frost resistance of the produced wall ceramics.

Figure 4 data proves that the combined introduction of polymer waste and boric acid with an increase of polymer waste amount into the charge results in the reducing of density and thermal conductivity of the developed wall ceramics, which in turn causes product energy efficiency improvement and decrease of the load on the construction foundation. The density and thermal conductivity decrease occurs due to the increase of the open and total porosity of the developed material as a result of the combustion of polymer waste during firing (see Fig. 5).

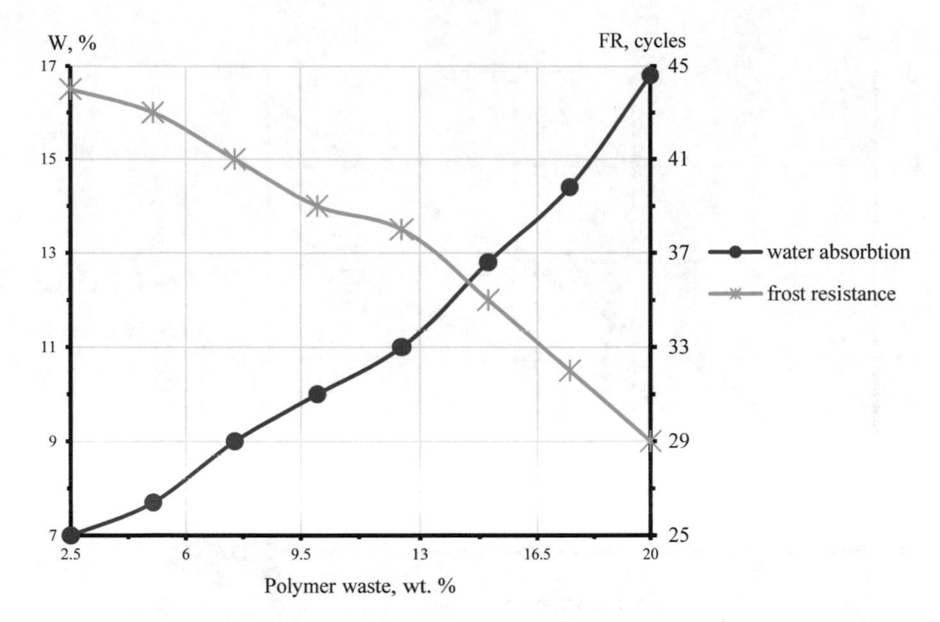

Fig. 3. Relation water absorption and frost resistance on polymer waste amount with boric acid introduction of 2.5 wt. %

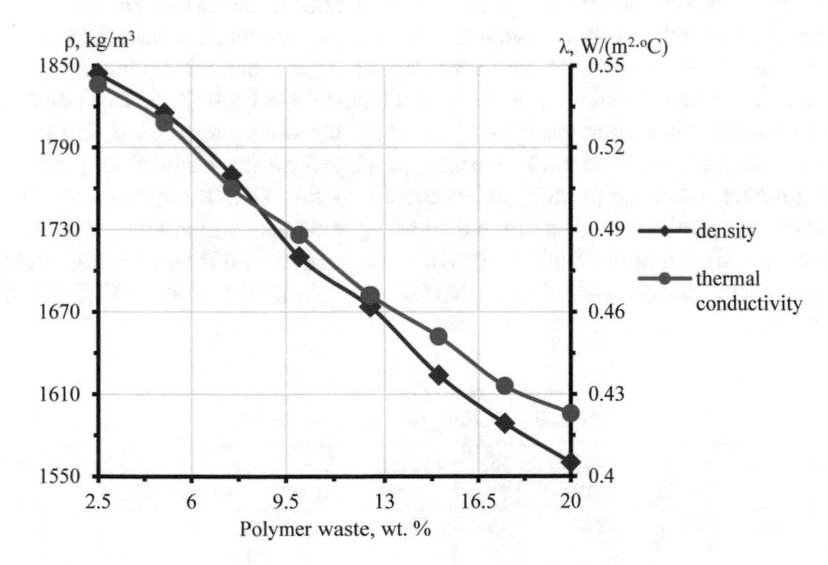

Fig. 4. The dependence of density and thermal conductivity on polymer waste amount and introduction of 2.5 wt. % boric acid

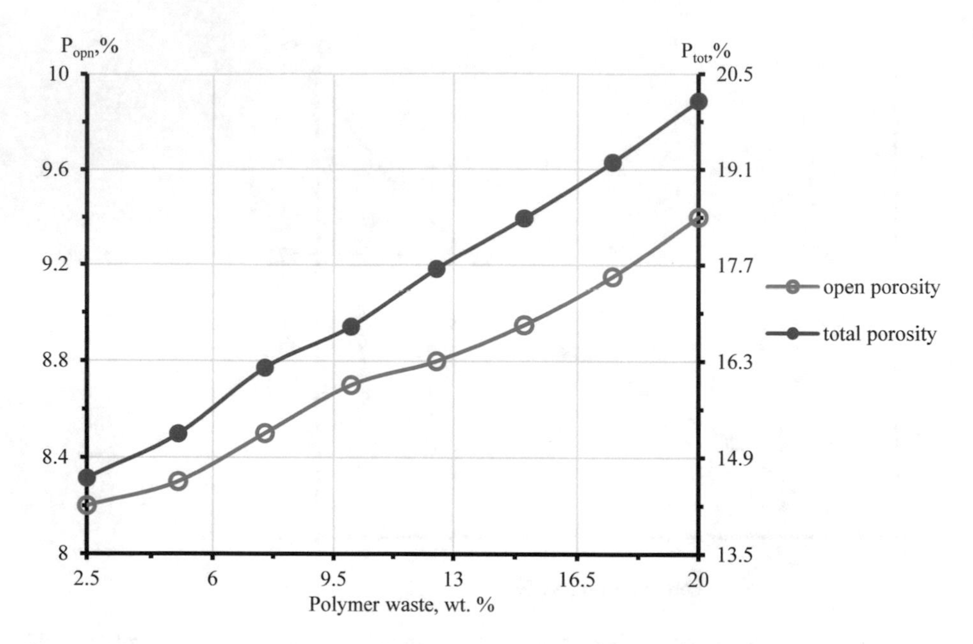

Fig. 5. The dependence of open and closed porosity on polymer waste amount and introduction of 2.5 wt. % boric acid

Having summarized the data in Figs. 2, 3, 4 and 5, and taking into account that compressive strength is very important for the walling products, it was decided to limit the amount of polymer waste to 15 wt. %. This polymer waste amount provides compressive strength equal to 15 MPa, which meets GOST 530-2012 requirements for ordinary solid ceramic brick brand M150, used for the walls construction. The principle properties of the developed wall ceramics, produced on the basis of the charge at a given polymer waste proportion, as compared to this GOST requirements and the properties of ceramics, based on the studied low-plastic clay without the introduction of additives, are shown in the Table 1. The data on strength and thermal conductivity for energy-efficient solid brick brand M150 are presented for GOST 530-2012 requirements.

Table 1. Walling ceramics properties.

Composition	σ_{cs}, MPa	σ_{bs}, MPa	W, %	FR, cycles	ρ, kg/m^3	λ, Wt/(m^2 °C)	P_{opn}, %	P_{tot}, %
GOST 530-2012	15.0	2.8	>5	>25	-	<0.46	-	-
Without additives	14.3	7.4	7.5	41	1936.4	0.562	8.0	13.8
Developed	15.0	5.5	12.8	35	1623.8	0.451	8.95	18.38

Thus, the developed charge composition allows improving the products quality based on low-plasticity clay and obtaining property meeting the requirements of GOST 530-2012 for energy-efficient solid ceramic bricks for the walls construction.

4 Conclusions

The research resulted in the development of the charge composition on the basis of low-plasticity clay, including 15 wt. % polymer waste as a combustion additive and 2.5 wt. % boric acid as a flux. The charge of such composition facilitates the application of low-plasticity clay, which has a limited consumption due to the poor products quality, produced on its basis, and also the disposal of polymer waste, which is hard to process because of its low physical and mechanical properties, different degree of purity and destruction, as well as low thermal stability, as in case with UPVC, used for the developed composition.

The principle properties of the developed wall ceramics meet GOST 530-2012 requirements, applied for the ceramic brick and stone for the walls construction in buildings and structures. The material frost resistance not only meets the requirements of the specified GOST, but also allows the use of the developed wall ceramics in the production of facial ceramic bricks, with the minimum frost resistance requirement of 35 cycles. As for the thermal conductivity, the developed wall ceramics corresponds to the conditionally effective products in terms of thermal characteristics, with thermal conductivity in the range from 0.36 to 0.46 $W/(m^2\,^{\circ}C)$. The rest properties of the developed wall ceramics are average for ceramic bricks.

Thus, the developed wall ceramics application allows producing energy-efficient products that can be used in the construction of multi-layer walls of buildings, providing modern thermal standards and rational consumption of building materials. Herewith, the developed material properties, to some extent, allow combining the functions of the bearing, thermal insulation and partly facing layers, thus expanding the possibilities of products application.

References

1. Badyin, G., S'ychev, S.: Modern technologies of buildings construction and reconstruction. BHV-Petersburg, St. Petersburg (2013)
2. Moroz, I.: Technology of construction ceramics. 3nd edn. rev. and ad.; Repr. izd. 1980. ECOLIT, Moskow (2011)
3. Shakhova, V., Vorobyeva, A., Vitkalova, I., Torlova, A., Pikalov, E.: Modern technologies of polymer waste disposal and the problem of their application. Mod. High Technol. **11–2**, 320–325 (2016)
4. Bazunova, V., Prochukhan, Y.: The disposal methods of polymer waste. Bashkir University Vestnic **13**(4), 875–885 (2008)
5. Snezhkov, V., Retsits, G.: Polymer waste - into ready made products. Solid Waste **1**, 16–19 (2011)

6. Vorob'eva, A., Shakhova, V., Pikalov, E., Selivanov, O., Sysoev, É., Chukhlanov, V.: Production of facing ceramic with a glazing effect based on low-plastic clay and technogenic waste from Vladimir Oblast. Glass Ceram. **75**(1–2), 51–54 (2018)
7. Vitkalova, I., Torlova, A., Pikalov, E., Selivanov, O.: Industrial waste utilization in the panels production for high buildings facade and socle facing. In: E3S Web of Conferences, vol. 33, p. 02062 (2018). https://doi.org/10.1051/e3sconf/20183302062
8. Vitkalova, I., Torlova, A., Pikalov, E., Selivanov, O.: Development of environmentally safe acid-resistant ceramics using heavy metals containing waste. In: MATEC Web of Conferences, vol. 193, p. 03035 (2018). https://doi.org/10.1051/matecconf/201819303035
9. Shakhova, V., Vitkalova, I., Torlova, A., Pikalov, E., Selivanov, O.: Development of composite ceramic material using cullet. In: MATEC Web of Conferences, vol. 193, p. 03032 (2018). https://doi.org/10.1051/matecconf/201819303032

Main Hydraulic Characteristics of Counter-Vortex Flows

Genrikh Orekhov(✉)

Moscow State University of Civil Engineering, Yaroslavskoe shosse, 26, Moscow 129337, Russia
orehov_genrih@mail.ru

Abstract. In fluid and gas mechanics, various types of flows are considered. The most common in practice are the longitudinal axial, circulating and circulating longitudinal (swirled) types of flows. This article is devoted to the analysis of the main hydraulic characteristics of the so-called counter-vortex flows. This kind of flows is an artificially created current and does not occur in nature. It is characterized by a complicate spatial form of distribution of velocity components and specific structural parameters. The description is given of an approximate pattern of the organization of counter-vortex flows in a circular cylindrical chamber. The characteristic profiles of structural parameters of the initial high-velocity circulation-longitudinal flows, which are the basis for forming the counter-vortex flows, are considered. Examples of distribution of axial and meridian velocities in the interaction chamber obtained by calculation for the laminar motion mode and on a physical model using up-to-date laser equipment are given. The vortex structures of various types of currents are compared: longitudinally axial, circulation-longitudinal and counter-vortex. Extremely high values of coefficients of the flow initial energy dissipation and hydraulic resistance of the entire system are noted. The results of the computational comparison of the hydraulic resistance coefficients of two ways of the flow excess energy dissipation are shown: sudden expansion and counter-vortex interaction. The main features of the counter-vortex flows are formulated and summarized. Some practical applications have been proposed this basis in hydraulic engineering and aeration systems for water masses.

Keywords: Hydraulics · Kinds of liquid and gas flows · Distribution of velocity components · Flow structure · Flow energy dissipation · Counter-vortex flows · Hydraulic spillways · Aeration

1 Introduction

To solve engineering hydraulic problems accurate and reliable information on the characteristics of a particular kind of fluid or gas flow is required making it possible to be properly taken into account in designing and operating the industrial installations. This article is devoted to counter-vortex flows and to identifying those features that can be used in practice when creating various structures, devices and apparatus in technological processes.

V. Murgul and M. Pasetti (Eds.): EMMFT-2018, AISC 983, pp. 795–806, 2019.
https://doi.org/10.1007/978-3-030-19868-8_78

The swirled flows of liquid and gas have found wide application in up-to-date technologies due to their many aerodynamic, thermodynamic and hydro mechanical qualities, which make it possible to repeatedly intensify the processes of energy-, mass-, and heat exchange [1–4]. The offer to use coaxially arranged, interacting, oppositely swirled flows gave an essential impetus in the direction of further research in this area [5]. This circumstance put forward to researchers the tasks of scientific analysis of a complicated hydrodynamic phenomenon, which we call counter-vortex flows [6]. The proposal to use interacting swirled flows was caused by the search for the ways to intensify the methods of mixing liquids and gases and dissipation of the excess kinetic energy of high-velocity water flows in the spillways of high-head hydraulic systems [7]. In this regard, mention should be made of the work of the researchers who noted that the water flow moving from the upper pool to the tail water can extinguish its energy either by doing useful work (installing a hydraulic turbine) or by overcoming the reactive forces that arise when the flow passes through the energy dissipators (when the flow kinetic energy is converted to heat) [8–11].

Counter-vortex flow is a spatial non-uniform flow with interacting, oppositely rotating, coaxially arranged in a cylindrical channel layers of liquid or gas. In contrast to the longitudinal axial and circulation longitudinal flows widespread in nature and technology, the counter-vortex flows are not observed in nature. They are characterized by a complicated configuration and specific hydraulic characteristics.

Counter-vortex flow is a term formed in the course of studying this phenomenon [12, 13]. A concise phrase – counter-vortex flow – contains a deep meaning of those hydrodynamic processes that occur in this flow. It reflects the fact that in a circular-cylindrical chamber two or more longitudinal-circulating layers of liquid or gas meet and begin interacting forming initial vortices in the flow part whose scale is comparable with the initial, oppositely swirled coaxial layers. In the cylindrical chamber, these layers superimpose each other and their spatial interact. A high gradient of circumferential velocities along the radius, almost tending to infinity in the transverse shear layer at the boundary of the macro-vortices, is created. As a result of such a superposition of interacting layers, a complicated spatial flow arises, whose characteristic features differ from all other known flows. Intensive mass and energy exchange occurs within the entire volume of the liquid due to artificially created turbulence. Initial macro-vortices equal to the diameters of oppositely rotating layers start generating secondary vortices of a smaller scale with the energy of the initial flow of coaxial oppositely swirled layers transferring through the vortex flows to the energy of artificial turbulence of a high degree of intensity [14, 15]. In this sense, the flow under consideration is a counter-vortex (with oppositely rotating vortices of different scale) throughout its existence up to its complete degeneration and transition to the ordinary longitudinal uniform current. The cascade structure of the vortices rather quickly decays due to energy dissipation, therefore the counter-vortex flow is uneven, decaying. Devices that use the effects of a counter-vortex flow are called counter-vortex devices or structures. For example, a hydraulic spillway that uses the peculiarities of a counter-vortex flow is called a counter-vortex spillway.

2 Objective

The purpose of this work is to formulate the main integral hydraulic characteristics of counter-vortex flows obtained on the basis of numerous studies on physical models. The studies were conducted on models using water as a working fluid. The counter-vortex flows are characterized by a complicated flow pattern in the zone of interaction of the layers with an opposite swirling, an extremely high degree of energy dissipation, velocity and pressure pulsations. Under these conditions, knowledge of the integral values of the coefficients of hydraulic resistance and energy dissipation.

The task of the research was to substantiate the main hydraulic characteristics of the counter-vortex flows based on the conducted physical modeling.

3 Method

Counter-vortex flows of fluid or gas are complicated, which include recirculation zones with high turbulence intensity in the interaction of oppositely rotating fluid layers, a cellular cascade structure of vortex fields. Therefore, a physical experiment practically is the only way to solve any problems of hydraulics of such flows. To determine the main integral hydraulic characteristics of the counter-vortex flows, numerous experiments were carried out on models of different scale and in the prototype. The models were of different design but the general concept of organization of the counter-vortex flow was the same.

Figure 1 shows one of possible schemes for the organization of a two-layer counter-vortex flow in a circular cylindrical chamber (pipe). The entire flow path can be divided into four zones: *A, B, C* and *D*.

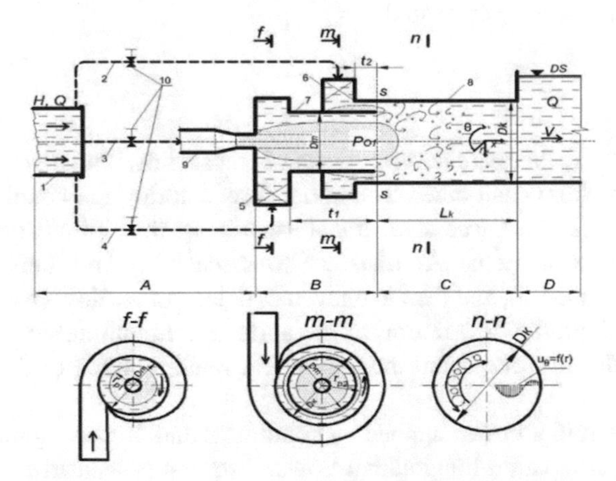

Fig. 1. An approximate scheme of organization of counter-vortex flow in the circular cylindrical chamber. Two-layer flow. Zone *A* – pressure longitudinal-axial flow; zone *B* – circulation and circulation longitudinal free flow; zone *C* – counter-vortex flow; zone *D* – longitudinal axial free uniform flow in the tail water; 1 – pressure pipe; 2, 3, 4 – supply pipelines; 5 – swirling device of the inner circulation-longitudinal layer; 6 – swirling device of the outer circulation-longitudinal layer; 7 – intermediate pipe; 8 – interaction chamber of oppositely swirled layers; 9 – central pipeline for forming the central jet; 10 – locking devices.

A device providing such a flow consists of a supply pressure pipe 1 (Fig. 1) with the following initial parameters: head - *H* and flow rate - *Q*. The total flow *Q* is divided into two parts with the help of pipelines 2 and 4, which supply fluid to the swirling devices 5 and 6. In zone *A*, to which we refer the flow in the supply pipelines to their outlet sections, located in the counter-vortex device, there is a uniform pressure fluid flow. Using the locking devices 10, the total flow *Q* can be regulated.

Swirling devices 5 and 6 can be of any type and design, depending on a certain task: tangential, scapular, slotted, auger, belt. Applicably to hydraulic installations tangential swirlers of tubeless and cylindrical types, as well as diagonal ones [16], are used, as a rule.

In this diagram (Fig. 1), tangential swirling devices with a single-lead helix (cross sections *f-f* and *m-m*) are shown as swirling devices. The swirling devices forms, longitudinal-circulating fluid layers at the exit from the spiral (zone *B*), which meet with each other in a circular cylindrical interaction chamber 8 with a diameter of D_k. The counter-vortex flow forms in process of interaction of oppositely swirled and coaxially located layers of fluid, which extends over the entire length of zone *C*.

Cross section *n-n* illustrates the circulation of the interacting layers at its beginning. Towards the end of the area of zone *C*, the counter-vortex flow degenerates and passes to ordinary longitudinal-axial turbulent flow. The diagram of Fig. 1 in zone *D* shows, as an example, the transition of the counter-vortex flow in chamber 8 to free flow in a channel with flow rate *Q* forming the downstream of a hydraulic structure. However, the exit of the stream from chamber 8 may be in the form of the outflow to the atmosphere.

The studies of the counter-vortex currents were carried out on the models of different scale and in the prototype structures of *H* head from a few meters to 70 m of water table. The flow rate *Q* on different models ranged from several dozen liters per second to 5 cubic meters per second.

4 Results

As seen from Fig. 1, there are several forms of transit flow movement that are fundamentally different from each other. The initial flow entering the counter-vortex device undergoes a series of successive transformations: the initial pressure uniform longitudinal-axial flow (zone *A*) transfers to circulation and uniform circulation-longitudinal flow (zone *B*), and then to uneven counter-vortex flow (zone *C*). As a result of damping the latter, the flow returns to the uniform longitudinal-axial flow (zone *D*).

Let us consider in more detail the nature and main features of the flows in zones *B* and *C*.

In zone *B*, the initial flows supplied by conduits 2 and 4, passing through swirlers 5 and 6, form two circulation-longitudinal flows. The first is located in a circular cylindrical pipe 7 of D_m diameter and t_1 length (Fig. 1) and creates an internal rotating layer of the counter-vortex flow. The second is located in a shorter pipe of t_2 length, the diameter of which coincides with the diameter D_k of the interaction chamber. It creates an outer rotating layer of the subsequent counter-vortex flow. The forms and nature of circulation-longitudinal flows in zone *B* may be different, but numerous experiments have established [6] that the circulation-longitudinal layers corresponding to the model of quasi-potential flow are the most suitable for implementation of counter-vortex flows.

Such high-velocity turbulent swirling flows acquire the properties of non-viscous non-eddying (potential) flows. The non-eddying flow is incompatible with viscosity, in other words, the non-eddying (potential) motion is impossible for a viscous (real) fluid. Despite this, in practice rather commonly realized is this simplest circulation-longitudinal movement due to the fact that at high flow rates or large dimensions of structures or main process equipment (for example, large spillway structures) the Reynolds numbers are very high and, therefore, viscosities are small compared to other forces. In fact, the swirling flow of real fluid is in reality everywhere vortex. In reality, the swirling flow of actual fluid is everywhere of vortex pattern. However, an increase in Reynolds numbers in turbulent flows due to an increase in motion velocity can be considered as a decrease in the influence of viscous forces. Here we can talk about almost potentiality − the quasi-potentiality of the flow. Due to high centrifugal accelerations along the axis of rotation, a cylindrical cavity forms in such high-velocity flows—a vortex core of a variable radial length r_{p1}, r_{p2} (cross section *f-f* and *m-m* in Fig. 1). The axial vortex core can be filled with gas (air) or vapor of the fluid itself. That is why it is sometimes called the cavitation cavity. Thus, the live section of the high-velocity swirled flow is annular between the channel walls and the vortex core, and the flow itself is free-flow with a free surface at the boundary of the vortex core. The vapor pressure of the fluid in the vortex core can be below atmospheric pressure (vacuum, up to absolute), or equal to the atmospheric one in case of connection of the vortex core with the atmosphere. Figure 2 shows the radial distributions of the main flow parameters corresponding to the quasi-potential flow model:

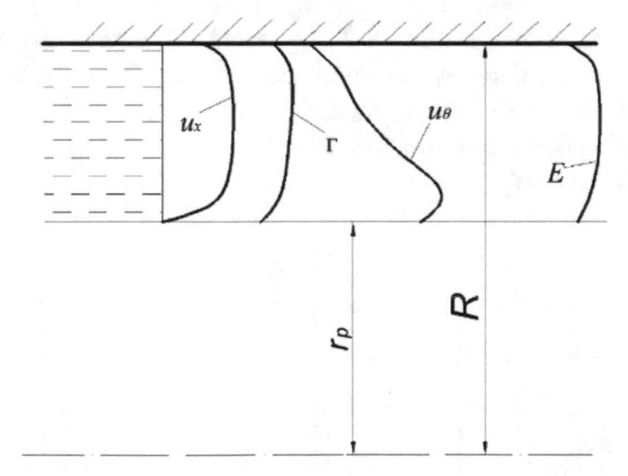

Fig. 2. Characteristic profiles of structural parameters of high-velocity swirled flow in a cylindrical channel [19].

– consistency along the reference radius of peripheral velocities circulation

$$G(r) = 2\pi ru; \tag{1}$$

– consistency along the reference radius of specific flow energy

$$E(r) = \frac{p}{\rho} + \frac{U^2}{2} - P = Const, \;\; where\; U = \sqrt{u_r^2 + u_\theta^2 + u_x^2}; \tag{2}$$

– consistency along the reference radius of axial flow velocities

$$u_x(r) = \frac{Q}{S} = \frac{Q}{\pi\left(R^2 - r_p^2\right)} = V = Const; \tag{3}$$

– u_θ vs circular channel radius

$$Uo = \frac{G}{2\eta R}\; u_\theta = \frac{\Gamma}{2\pi R} \tag{4}$$

Here u_θ – circumferential velocity component, u_r – radial component, u_x – axial component directed along the channel axis, P – pressure, U – total flow velocity, P – potential of outer mass forces, ρ – density, S – live cross-sectional area of the flow, R – circular channel radius, r_p – vortex core radius.

The counter-vortex flow forms in zone *C*. Interaction of oppositely swirling layer starts at the line passing through the section of the internal pipe 7 (line s-s in Fig. 1). There is a flow featuring complicated structure in this area. The distribution (normalized by mean flow rate velocity) of azimuthal (circumferential) u_θ and axial (longitudinal) u_x velocities is shown in Fig. 3. The presented velocity distributions along the radius and length of the core are obtained theoretically (by calculations) for laminar flow regime with Re = 500 (the Reynolds number is determined by D_k) [15]. The counter-vortex flow with approximately equal torques of oppositely swirled coaxial layers is shown $M_1 = -M_2$.

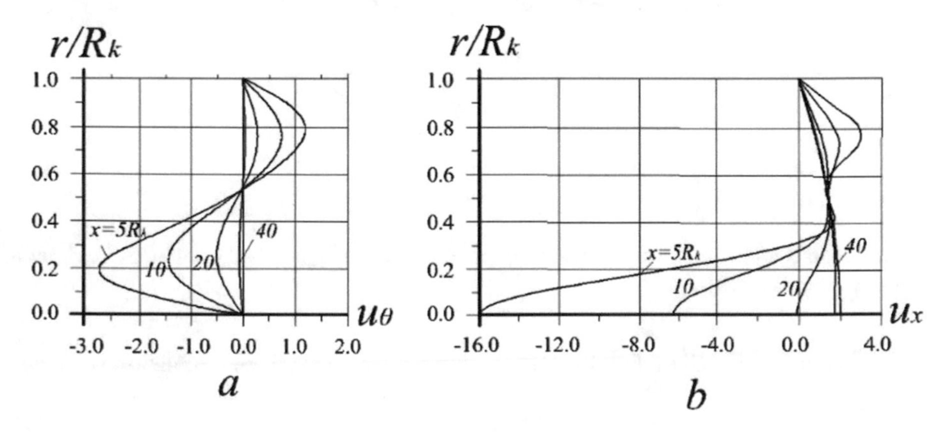

Fig. 3. Distribution of calculated normalized velocities in the laminar counter-vortex flow: *a* – azimuthal u_θ, *b* – axial u_x. The design parameters of the counter-vortex flow: the initial circulation at the entrance to the chamber is $G_0 = -1.1$; the initial angular velocity of rotation of the fluid $\Omega_0 = 3.8$; initial value of Rossby number $Ro = 0$

The counter-vortex flow (within the active zone *C)* transforms to a longitudinal one, that is, without swirl (Fig. 3a), where the length of intensive viscous diffusion of the circulation of the interacting layers is approximately equal to 40 pipe radii. The fact of intensive diffusion of circulation is obvious, because it is determined by the mutual suppression of coaxial oppositely swirled layers. This is facilitated by the cascade structure of vortex fields, which, as mentioned above, forms in the counter-vortex flows. Recirculation (return) flows with significant negative velocities (Fig. 3b) can be observed in the distributions of axial velocities u_x, in the sections close to the beginning of the active zone, in the axial layers. Recycling zones also exist in circulation-longitudinal flows of fluid and gas.

The presence of these zones is a characteristic feature of any swirling flow. Outside the recurrent paraxial flow in the thickness of the flow, axial velocities are significantly higher than the average (standardized value $V = 1$), that maintain the balance of the volume flow. There is a return paraxial flow with the counter-vortex interaction of two coaxial layers at a length of up to twenty radii of the interaction chamber. The change in the Reynolds number due to fluid viscosity, flow velocity, or pipe size affect directly proportional on the length of the backflow section.

When calculating turbulent circulation-longitudinal fluid flows, good convergence with the experimental data is obtained using a turbulent analog of the Reynolds number calculated from the vortex viscosity of a turbulent medium [6]

$$Re_t = \frac{VR}{\varepsilon_t} = \frac{1}{\chi}\sqrt{\frac{8}{\lambda}}, \tag{5}$$

where ε_t – eddy viscosity, λ – coefficient of hydraulic resistance by length, $\chi = 0.2$ – universal constant (for water).

If in actual conditions coefficient λ varies in the range of 0.011–0.03, then the turbulent Reynolds number will have values in the range $Re_t = 80-135$. Comparing with the obtained results, it is easy to see that with counter-vortex interaction of turbulent swirled flows, the length of the active zone will reduce 3.5–6 times, i.e. will be approximately 7–11 radii of the interaction chamber.

Measurements on physical models showed that the distribution of velocity components in the interaction area of oppositely swirled layers give a flow pattern similar to the theoretical one [24]. Figure 4 shows, as an example, the diagrams of axial u_x and azimuthal u_θ, velocities at different lengths of the interaction chamber. Measurements were made using the method of tracer visualization of the flow (PIV method). The set of equipment implementing this method allows for velocity fields of laminar and turbulent flows of gases and fluids to be obtained. During the studies the flow turbulent regime was established with $Re = 6.9\ 10^5$. The specific velocity distributions shown along the length of the active zone apply a number of features to the counter-vortex flow, and the velocity structure of the counter-vortex flow radically differs from all known types of flows.

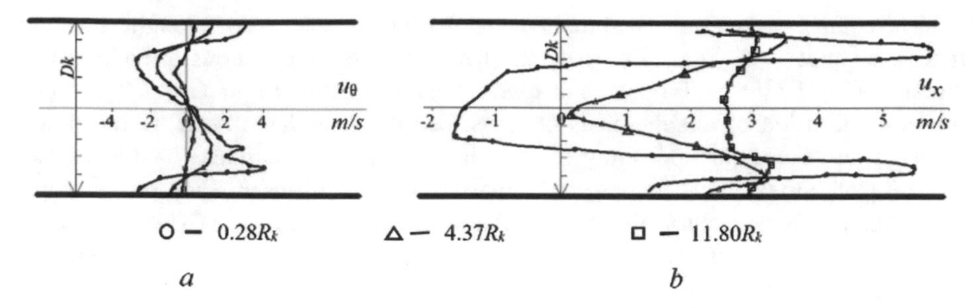

Fig. 4. Distribution of azimuthal and axial velocities on the physical model: *a* – azimuthal u_θ; *b* – axial u_x

The developed cellular structure of vortex fields cannot be observed either in longitudinal-axial flows, where the generation of vorticity of only one sign is only due to viscous drag of the fluid in the near-wall layer, or in circulation-longitudinal flow with unidirectional swirling of fluid layers in which there are two zones of generation of vortices of mutually opposite sign: a wall region of viscous drag, as in the longitudinal-axial flow, and an axial vortex zone in the flow core [14]. Thus, the structure of the counter-vortex flows is formed under the predominant influence of internal processes in the zones of generation of cascade vortex fields caused by viscous forces. In general, this is a fundamental difference between the counter-vortex flows from circulation-longitudinal and longitudinal-axial ones. The comparison shows that the counter-vortex flows have an exceptionally high value of the vortex intensity. Azimuthal ($rot_\theta U$) and axial ($rot_x U$) vortices generated at the entrance to the active zone reach values of more than 500 normalized normal units (laminar flow regime). This is several times greater than the maximum values of vortex fields generated at the same Reynolds numbers in circulation-longitudinal flows with any degree of swirling flow, and, moreover, in longitudinal flows, where the only longitudinal component is $rot_x U = 4$ (two orders of magnitude less than counter-vortex ones).

This flow pattern generates artificial intensive turbulence. At the beginning of zone C (Fig. 1), the maximum of the standard of pressure pulsations is 9–10%, which is 4–5 times higher than the value of the conventional longitudinal-axial flow, which is about 2%. The level of velocity pulsations in the initial section of the interaction chamber reaches extremely high values equal to $\sigma_\theta = 0.8$ in the azimuthal and $\sigma_x = 0.55$ in the axial direction. The processes of intensive turbulent mass and energy exchange with counter-vortex flow are the result of exceptionally high angular velocity gradients ($\Omega = u_\theta / r$).

It should be noted that the characteristics considered do not depend on external conditions, since they are determined by the field of mass centrifugal forces formed by circulation-longitudinal fluid layers constituting the counter-vortex flow.

One of the most important practical issues arising when considering the hydraulics of the counter-vortex devices is their energy-dissipation ability. The energy-dissipation ability can be determined by the energy dissipation factor reduced to the total head on the swirling devices.

$$\eta = 1 - \frac{V^2}{2gH}, \tag{6}$$

where, V is the average flow rate velocity, H is the total head, defined as the difference in pressures the flows ahead of the swirling devices (zone A in Fig. 1) and in the race channel (zone D in Fig. 1).

In addition to the energy dissipation coefficient, the effectiveness of counter-vortex systems can also be evaluated by a conventional method through the hydraulic resistance coefficient.

$$\varsigma = \frac{2g\Delta h_{w\Sigma}}{V^2}, \tag{7}$$

where $\Delta h_{w\Sigma}$ – total hydraulic losses in the water conductor system including zones B and C in Fig. 1.

The results of the studies show that the efficiency of flow energy dissipation in the interaction of coaxial oppositely swirled flows is rather high reaching maximum values equal to η = 90–98% of the head on the studied models and in the prototypes. Moreover, the length L_k, at which the total energy is dissipated, is relatively small and amounts to about 6–8 $D_k/2$ for numerous physical models studied. The flow energy dissipation coefficient can be adjusted in a fairly wide range depending on the set tasks. Figure 5 shows relation curves illustrating the dependence of the flow energy dissipation coefficient η and the hydraulic resistance coefficient ζ on the scale Λ of the physical model.

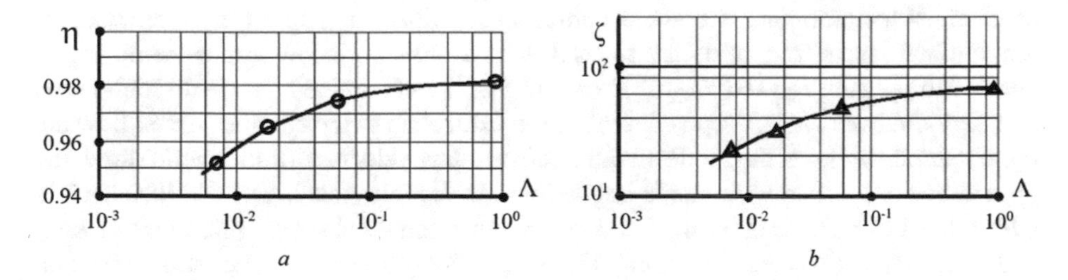

Fig. 5. Resistance coefficients of the counter-vortex flow: a – $\eta = f(\Lambda)$; $b - \varsigma = f(\Lambda)$

Figure 6 shows the experimental dependences of the coefficients of hydraulic resistance and counter-vortex flow energy dissipation on the relative vacuum in the axial zone of the counter-vortex flow.

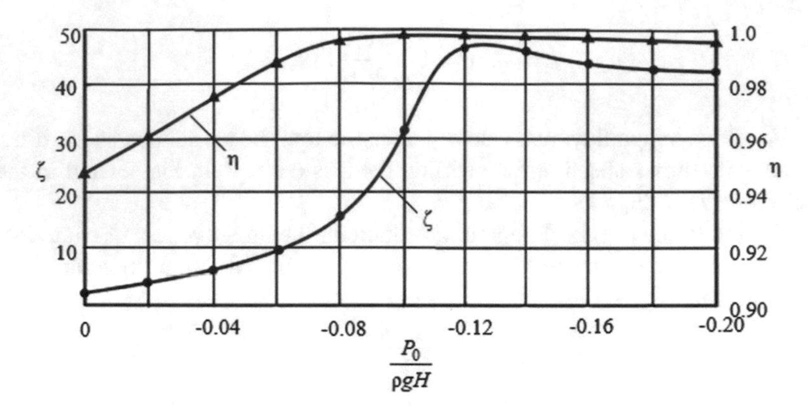

Fig. 6. Integral experimental distributions $\varsigma = f(P_0/\rho gH)$ and $\eta = f(P_0/\rho gH)$

It can be seen that the relative vacuum in the axial zone of the counter-vortex flow strongly influences the coefficient of hydraulic resistance of the entire system. The curves show the dependence $\varsigma = f(P_0/\rho gH)$ for two modes: 1 – two pipelines 2 and 4 are working; 2 – three pipelines 2, 4 and central one 3 are working (Fig. 1).

As already noted, the pressure in the circulating flow decreases to the axis of rotation, where a vacuum can be observed. The vacuum can be deep to the vapor pressure of saturation (physical limit), when a flow rupture occurs in the form of a hollow vortex core. These phenomena are associated with centrifugal forces that provide a positive radial pressure gradient and tend to break the circulation flow in the central (paraxial) zone, throwing it to the periphery - to the walls of the cylindrical channel. When forming a counter-vortex flow with the help of two circulation-longitudinal layers, two areas are formed with a pressure below atmospheric. These areas with pressure P_{01} and P_{02} are shown in Fig. 1 with dotted lines. With high heads and high fluid velocities, deep vacuums in the axial zone of the counter-vortex flow are quite natural, while on the models with relatively low velocities of the model flow, the appearance of limiting vacuums is impossible. At the full-scale high-head installations with a head of more than 50 m, the vacuum can reach an absolute value (−98.1 kPa) with the relative vacuum reaching the value $P_0/\rho gH = -0.2$. The structures and dynamics of the circulation flow without a core (that is, the flow at which the flow fills the entire cross section of the conduit) and with the core drastically differ. Therefore, when simulating a counter-vortex flow, it is necessary to provide relative vacuum in the near-axial flow zone the same as on the full-scale installations and also to simulate the cavity of the vortex core itself. Both are simulated by the air inlet to the axial zone of the counter-vortex flow. For example, if in the course of the experiments it was found that in the axial area of the model counter-vortex flow, the vacuum scaled to the prototype exceeds the physically possible value, then the vacuum in the model should be reduced to the required values for the scale recalculation. At that, as practice demonstrated, the linear dimensions of the vortex core on the model also come to the scale relationship with the size of the core in the prototype.

It is of interest to compare the efficiency of the flow energy dissipation in the counter-vortex dissipator and in other types of dissipators, for example, in the dissipator with a sudden expansion of the flow in the stilling chamber. Its efficiency can be assessed by the hydraulic resistance coefficient calculated according to the Borda formula

$$\varsigma = \left(\frac{S_2}{S_2} - 1 \right)^2. \tag{8}$$

Here, S_1 is the total cross-sectional area of the conduits to be closed by the gates (contracted discharge section) and S_2 the cross-sectional area of the pressure stilling chamber (wide section).

The results of comparative calculations show that the hydraulic resistance coefficient of the counter-vortex flow energy dissipator is in the range 4.61–45.68 (for modes without the central jet in operation), and for a sudden expansion 0.25–12.25. The counter-vortex method of transit flow energy dissipation has a significantly higher efficiency compared to the sudden expansion of the flow. Moreover, the difference between the efficiency of the counter-vortex dissipator and the dissipator by Borda is the more noticeable; the greater is the ratio S_2/S_1 in (8). Moreover, for efficient operation of the counter-vortex dissipator, no flooding of the outlet section of the mixing chamber is required, while the pressure stilling chamber of sudden expansion must necessarily be flooded from the downstream.

5 Conclusion

Based on the above, we can formulate the main features of the counter-vortex flows:

- extremely high degree of flow energy dissipation due to viscous forces, the highest of all known types of flows;
- vortex cascade structure in the entire volume of the flow with a high degree of vortex intensity, which is two orders of magnitude greater than the intensity in conventional longitudinal currents;
- high intensity of artificial turbulence, in which the pulsations of velocity and pressure 4–5 times exceed similar values in ordinary longitudinal currents;
- the presence of vacuum in the axial zone of the counter-vortex flow, which can be adjusted depending on the task;
- an extremely high degree of mass- and energy-exchange processes and the high mixing capacity associated with it.

The listed hydraulic characteristics of the counter-vortex flows and features of the structural parameters ensure the effectiveness of their application in practice in the most diverse branches of up-to-date technologies and technological processes.

References

1. Sazhin, B.S., Sazhina, M.B., Aparushkina, M.A., et al.: News of the universities. Technol. Text. Ind. **1**(343), 135–138 (2013)
2. Parra-Santos, T., Perez, R., Szasz, R.Z., Gutkowski, A.N.: Castro F EPJ Web of Confereces, p. 114 02087 (2016)
3. Ianiro, A., Lynch, K.P., Violato, D.: Cardone G J. Fluid Mech. **843**, 180–210 (2018)
4. Akhmetov, V.K., Volshanic, V.V., Zuikov, A.L., Orekhov, G.V.: Simulation and calculation of counter-rotating flows. Monograph. Moscow State University of Civil Engineering Publ., p. 252 (2012)
5. Huang, G.-B., Hu, H., Wang, C.-H., Du, L.: J. Hydrodyn. **29**(B), 504–509 (2017)
6. Xu, B., Chen, D., Tolo, S., Patelli, E., Jiang, Y.: Int. J. Electr. Power Energy Syst. **95**, 156–165 (2018)
7. Guo, W.: Energies **11**(12), 3340 (2018)
8. Chen, H.-Y., Chen, H.-Y., Deng, J., Nlu, Z.-Y., Wang, W.: J. Hydrodyn. Ser. B **22**, 582–589 (2010)
9. Gadge, P.P., Jothiprakash, V.: ISH J. Hydraul. Engin. **22**, 227–232 (2016)
10. Orekhov, G.V.: IOP Conf. Ser.: Mater. Sci. Eng. **365**, 042023 (2018)
11. Hashimoto, H.: Rep. Inst. Yigh Speed Mech. Tohoku Univ., vol. 19, pp. 241–257 (1967)
12. Zuikov, A.L.: Int. J. Comput. Civil Struct. Eng. **2**(8), 82–96 (2012)
13. Zhang, Z., Liao, R., Liuo, J.: Adv. Mech. Eng. **10**(9), 1–11 (2018)
14. Ruan, Z., Baars, W.J., Abbassi, M.R., Hutchins, N., Marusic, I.: 21st Australasian Fluid Mechanics Conference, 10–13 December, Adelaide, Australia (2018)
15. Mohammad, Z.A., Rashid Sarkar, M.A.: Int. J. Eng. Mater. Manuf. **3**(3), 122–133 (2018)
16. Churin, P., Kapustin, S., Orehov, G.: Bull. Eurasian Sci. **4**(17), 53TBH413 (2013)

Method of Modifying Portland Slag Cement with Ultrafine Component

Svetlana Samchenko[(✉)], Irina Kozlova, Olga Zemskova, Tatiana Nikiporova, and Sergey Kosarev

Moscow State University of Civil Engineering, Yaroslavskoe shosse, 26, Moscow 129337, Russia
vskanhva@mail.ru

Abstract. The paper presents the ways of modifying Portland slag cement with ultrafine substance for increasing its early strength. Blast-furnace granulated slag was used as ultrafine component having been preliminary ground in a laboratory airflow jet mill under restriction of upper limits of grinding to particle size of 1 μm. The obtained ultrafine slag was added into Portland slag cement in the dry state by dry blending and then in the suspension state instead of mixing water. Slag suspensions were prepared in water and water-polymer dispersion media. Plasticizer based on polycarboxylate resins was used as a stabilizer. Acoustic cavitation was used to stabilize water and water-polymer slag suspensions. The comparison of physical-and-mechanical and structural characteristics of modified samples detected that introducing slag suspension stabilized with plasticizer and acoustic cavitation instead of mixing water shows better efficiency. The effect of electrostatic factor is based on the formation of a double electric layer with the participation of functional groups of plasticizer at the surface of slag particles. The effect of structural-mechanical factor is conditioned by the formation of gelatinous films of main chain radicals on the surface of slag particles. Additionally, the stabilization of slag particles is provided by side branches of the plasticizer main chain, creating a spatial effect. Acoustic cavitation applied to water-polymer slag suspension results in the strengthening of electrostatic and structural-mechanical factors of aggregative stability and homogeneity of the suspension. The uniform distribution of stabilized slag particles in the hardening system leads to concentration of new formations on its surface and production of solid and firm structure of cement stone.

Keywords: Portland slag cement · Ultrafine slag · Slag suspension · Plasticizer · Acoustic cavitation · Strength · Porosity · Aggregative stability factors

1 Introduction

Today researches are pursued on developing new types of building materials and improving the existing production technologies in order to increase the scale of construction and to reduce its cost. Even at the beginning of the last century, concrete from Portland slag cement was used to reduce construction costs. Slag cements have a number of advantages over non-additive cements. They are weather-resistant,

V. Murgul and M. Pasetti (Eds.): EMMFT-2018, AISC 983, pp. 807–816, 2019.
https://doi.org/10.1007/978-3-030-19868-8_79

corrosion-resistant, heat-resistant and highly moisture-resistant [1–5]. With all the advantages of Portland slag cement, it also has a disadvantage associated with slow strength development within the early period of hardening.

Many scientists are engaged in the issue of increasing the early strength of Portland slag cement [6–13], [14]. One of the ways to increase the early strength is two-stage grinding of Portland slag cement, which allows increasing the strength of cement in 1–7 days of hardening by 1.5 times [8]. The use of thermal activation of slag leads to the increase in its activity and, consequently, in the strength of lime-slag binder. Thermal activation of slag consists in the combination of a sudden heating of the source material to a temperature of 500 °C, a relatively short isothermal holding and a rapid cooling. Thermally activated slag after its thermal activation was crushed to a specific surface of 310–320 m^2/g and was used for the production of slag-alkaline binder [9, 10]. The increase in strength of concrete from Portland slag cement was helped by the use of heat and moisture treatment [11]. Many studies are dedicated to the use of hardening accelerators and plasticizers in the production of slag cements. The use of plasticizers improves the early strength of cement by 40–50% [12]. In the work [13], to increase the early strength of Portland slag cement, a mixture of slag with sodium aluminate was used, which increases the compressive strength in 1 day by 49%, in 2 days - by 44%. Application of complex additive in the production of highly functional concrete based on Portland slag cement, which contains plasticizer, metakaolin or microsilica, promotes accelerated hardening of concrete and the formation of its microstructure mainly from stable hydration tumors that are not subjected to recrystallization with cyclic changes in the environment.

It was shown that the increase in the early strength of cement stone can be achieved if ultrafine component is used in the composition of cement. During the hardening of binder, the ultrafine additive will act as a substrate for the nucleation and growth of crystal hydrate neoplasms, which will subsequently form a dense and durable structure of the cement stone. However, ultrafine additives tend to aggregate, which leads to the opposite effect: uneven distribution in the volume of binder and the decrease in strength. To prevent aggregation, ultrafine additives are stabilized, for example, with surfactants or using the acoustic cavitation method.

Based on the above, in order to increase the early strength of Portland slag cement, it is proposed to introduce ultrafine addition of blast-furnace granulated slag by several ways:

- dry blending of the material with ultrafine component, followed by mixing of the composition with water;
- mixing Portland slag cement with aqueous suspension of ultrafine slag instead of water;
- mixing Portland slag cement with aqueous suspension of ultrafine slag prone to acoustic cavitation;
- mixing Portland slag cement with aqueous suspension of ultrafine slag stabilized with a plasticizer;
- mixing Portland slag cement with aqueous suspension of ultrafine slag stabilized with plasticizer and acoustic cavitation.

The effectiveness of the method of modifying the ultrafine component will be evaluated by the physical-mechanical and structural characteristics of the cement stone.

2 Materials and Methods

The subject of the present research is Portland slag cement modified by ultrafine slag.

Portland slag cement was obtained by joint grinding in a laboratory mill of 60% clinker, 40% slag and 5% natural gypsum to a residue R008 = 6% on a sieve. The chemical and mineralogical composition of clinker is presented in Tables 1 and 2 respectively.

Table 1. Chemical composition of clinker.

Chemical composition of clinker	CaO	SiO_2	Al_2O_3	Fe_2O_3	MgO	SO_3	R_2O
The percentage of component in clinker, %	63.89	20.63	5.62	5.15	3.68	0.59	1.36

Table 2. Mineralogical composition of clinker.

Mineralogical composition of clinker	Alite	Belite	The aluminate phase	The ferrite phase
The percentage of mineral in clinker, %	63	11	6	15

Blast-furnace granulated slag was used as the main component of Portland slag cement and as its reinforcing component, which was obtained in a laboratory airflow jet mill LHL-1 that has upper limit boundaries of grinding to a particle size of 1 μm. Characteristics of granulated blast furnace slag are presented in Table 3, the particle size distribution of ultrafine slag is given in Table 4.

Table 3. Specifics of blast-furnace granulated slag.

Chemical composition, %	CaO	SiO_2	Al_2O_3	MgO	Fe_2O_3+FeO	others
	45.40	38.20	8.10	3.20	0.80	4.30
Phase composition, %	Glass			Crystalline phase		
	93.20			6.80		

Table 4. Granulometric composition of ultrafine slag (reinforcing component).

Size of particles, μm	0–1	1–5	5–7	7–10	10–15	15–20	20–30
Fraction content, weight %	9	37	17	14	10	8	5

Portland slag cement compositions with reinforcing component were prepared in a laboratory ceramic mill by mixing ultrafine slag with cement. The additive was introduced in the amount of 1; 3; 5% of cement. Mixing compositions was carried out for 1 h.

For the introduction of reinforcing component in the composition of cement instead of mixing water, suspensions of ultrafine slag were prepared with concentration of 10; 30; 50 g/l.

A plasticizer based on esters of polycarboxylate resins was used as a stabilizer for slag suspension. The plasticizer was added in an amount of 5 g/l of suspension.

In order to stabilize water and water-polymer slag suspensions, acoustic dispersion was performed at a frequency of ultrasonic vibrations of 44 kHz; dispersion temperature - 25 ± 2 °C; time dispersion - 10–15 min. A constant temperature of dispersion was maintained by means of temperature control of suspensions.

Series of samples based on Portland slag cement (Table 5) were prepared for research and evaluation of modification methods. The strength of cement was determined by the national standard GOST 30744-2001. The porosity of cement stone was determined by saturating the samples with inert liquid.

Table 5. Compositions of samples under study on the basis of Portland slag cement.

№ of series	№ of a sample	Additive composition	Concentration of ultrafine slag, *%/**g/L	Modification method	Sample grouting fluid
1	1-0	–	–	–	Water
	1-1	Ultrafine slag	1*	Dry blending	Water
	1-2	Ultrafine slag	3*	Dry blending	Water
	1-3	Ultrafine slag	5*	Dry blending	Water
2	2-1	Slag suspension (water + ultrafine slag)	10**	Adding instead of hardening water	Slag suspension
	2-2	Slag suspension (water + ultrafine slag)	30**	Adding instead of hardening water	Slag suspension
	2-3	Slag suspension (water + ultrafine slag)	50**	Adding instead of hardening water	Slag suspension
3	3-1	Slag suspension (water + ultrafine slag)	10**	Acoustic cavitation + adding instead of hardening water	slag suspension
	3-2	Slag suspension (water + ultrafine slag)	30**	Acoustic cavitation + adding instead of hardening water	Slag suspension
	3-3	Slag suspension (water + ultrafine slag)	50**	Acoustic cavitation + adding instead of hardening water	Slag suspension

(*continued*)

Table 5. (*continued*)

№ of series	№ of a sample	Additive composition	Concentration of ultrafine slag, *%/**g/L	Modification method	Sample grouting fluid
4	4-0	–	–	–	Water + plasticizer
	4-1	Slag suspension (water + plasticizer + ultrafine slag)	10**	Adding instead of hardening water	Water-polymer slag suspension
	4-2	Slag suspension (water + plasticizer + ultrafine slag)	30**	Adding instead of hardening water	Water-polymer slag suspension
	4-3	Slag suspension (water + plasticizer + ultrafine slag)	50**	Adding instead of hardening water	Water-polymer slag suspension
5	5-0	–	–	Acoustic cavitation	Water + plasticizer
	5-1	Slag suspension (water + plasticizer + ultrafine slag)	10**	Acoustic cavitation + adding instead of hardening water	Water-polymer slag suspension
	5-2	Slag suspension (water + plasticizer + ultrafine slag)	30**	Acoustic cavitation + adding instead of hardening water	Water-polymer slag suspension
	5-3	Slag suspension (water + plasticizer + ultrafine slag)	50**	Acoustic cavitation + adding instead of hardening water	Water-polymer slag suspension

3 Results and Discussion

To assess the method of modifying Portland slag cement with ultrafine slag, studies on physical-mechanical and structural characteristics of the samples were carried out. The graphs of physical-mechanical (compressive strength) and structural (porosity) characteristics were plotted against the hydration time in accordance with Table 5 (Fig. 1). Figure 1a, b shows the values of strength and porosity of samples that contain 1% ultrafine slag or 10 g/l of suspension; Fig. 2c, d: 3% ultrafine slag or 30 g/l of suspension; Fig. 1d, e: 5% ultrafine slag or 50 g/l of suspension.

The charts show that the introduction of ultrafine component into Portland slag cement contributes to the increase in strength and decrease in porosity of cement stone, regardless of the method of introducing the additive into the material. With the introduction of ultrafine component in Portland slag cement by dry mixing the material with the additive, the strength of the samples increased in the early periods of hardening by an average of 50%, at the grade age – by 59%. At the same time, the porosity decreased by 19% during 28 days of hardening. For the analysis, the authors used the curves in Fig. 1 corresponding to samples 1-0; 1-1; 1-2; 1-3. With the introduction of the ultrafine component in the composition of Portland slag cement in the state of suspension, the strength of the samples increased on average by 60%; at grade age - by

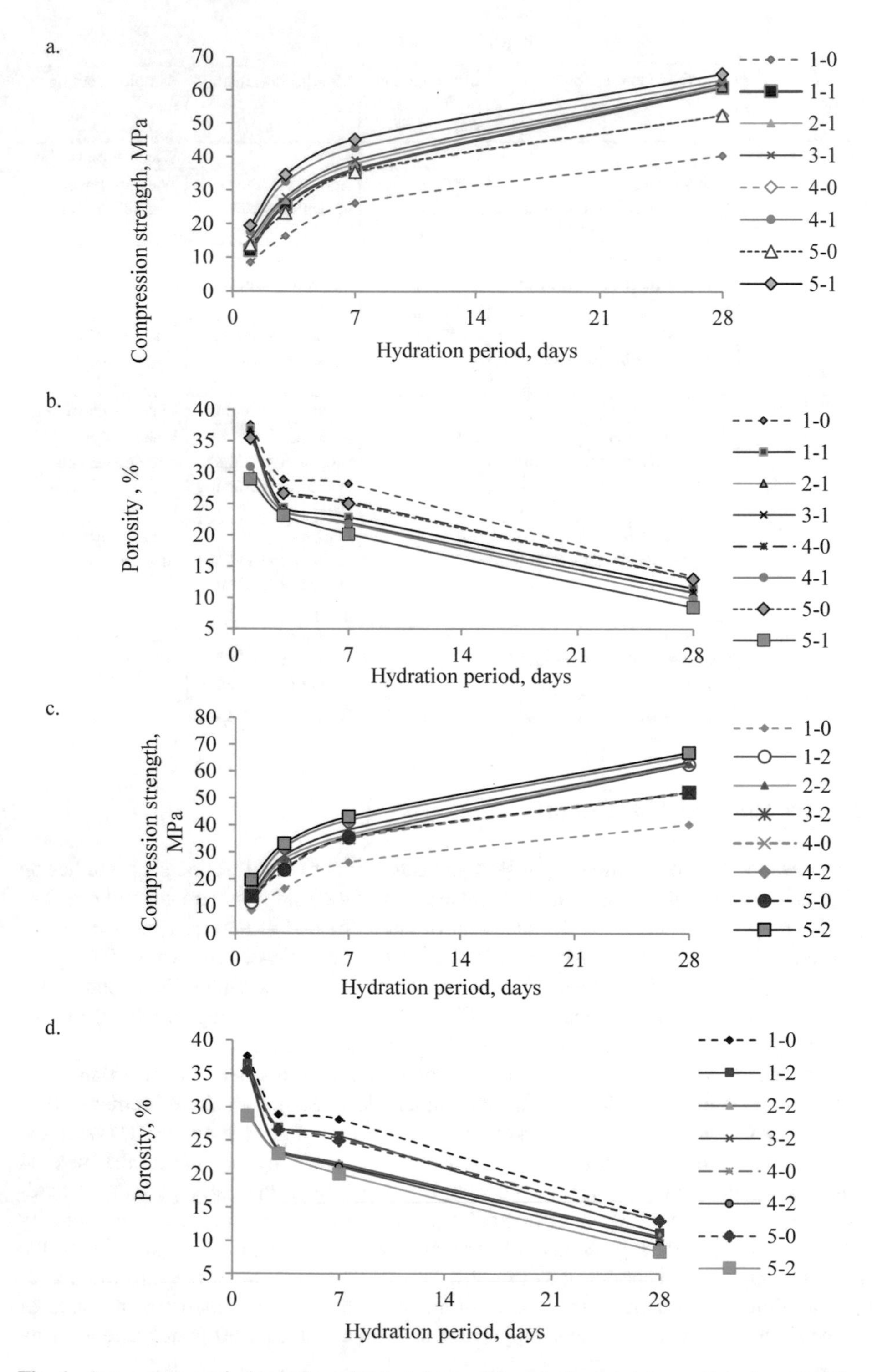

Fig. 1. Dependence of physical-mechanical (strength) and structural (porosity) characteristics on the period of hydration of samples in accordance with Table 5.

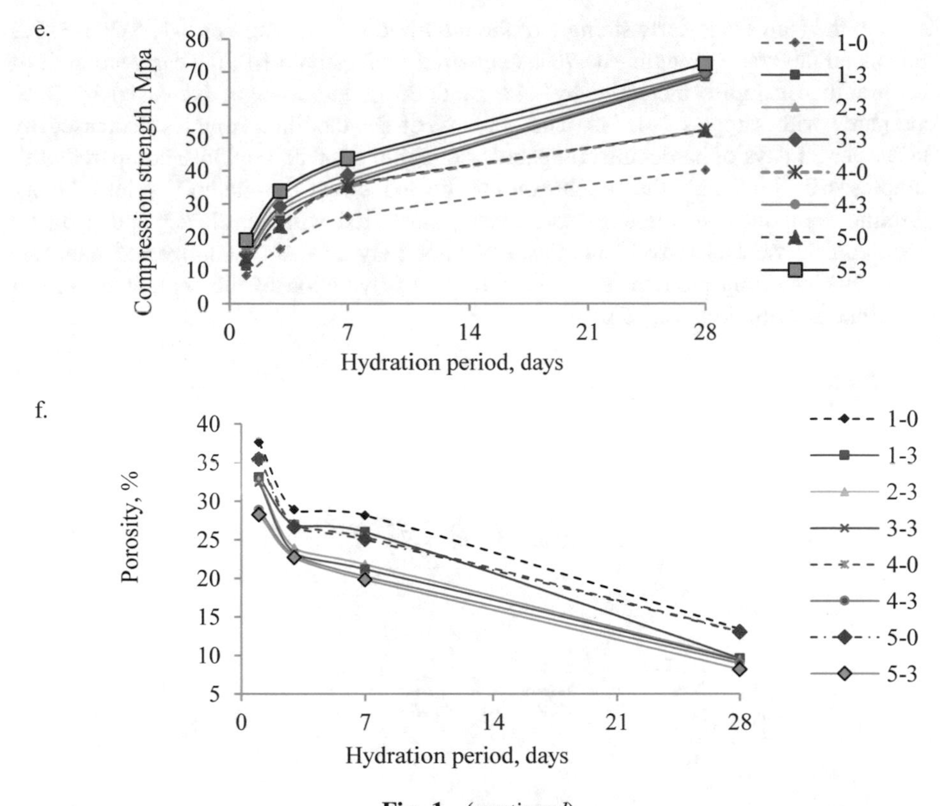

Fig. 1. (*continued*)

60%, which is comparable to dry mixing, the porosity decreased by an average of 23%. For the analysis, we used the curves in Fig. 1 corresponding to samples 1-0; 2-1; 2-2; 2-3. When treating slag suspensions by acoustic cavitation, an increase in the early strength of the samples was noted on average by 74%; at grade age - by 61%, which is also comparable to dry blending. The porosity of cement stone with the addition of ultrafine slag in the 28 days of hardening decreased by 24%. For the analysis, the authors used the curves in Fig. 1 corresponding to samples 1-0; 3-1; 3-2; 3-3. As noted above, the use of plasticizers leads to the increase in early and grade strength, which is confirmed by curves 4-0, 5-0 in Fig. 1.

Meanwhile, curves 4-0 and 5-0 show that acoustic cavitation does not affect the plasticizer and, accordingly, the characteristics of cement stone. When comparing the characteristics of cement stone modified by water-polymer slag suspension (curves 4-1; 4-2; 4-3) with the characteristics of additive-free samples (1-0; 4-0), there is an increase in strength compared with sample 1-0 in early terms of hardening by 2 times, at grade age by 64%. If comparing to sample 4-0, the early strength increased by 30%, and the grade strength increased by 27%. The porosity of modified samples decreased by 30% during 28 days of hardening compared with sample 1-0 and by 28% compared with sample 4-0. Treatment of water-polymer slag suspensions with acoustic cavitation leads to the increase in early strength of the modified samples (curves 5-1; 5-2; 5-3) 2.2 times, and the grade strength - by 70% compared with sample 1-0; the early strength of the modified samples increased by 44%, and the grade strength increased by 30% compared with samples 4-0, 5-0. The porosity of the modified samples decreased by 38% after 28 days of hardening compared with sample 1-0 and by 36% compared with samples 4-0; 5-0. Thus, the conclusion can be made that the method of introducing ultrafine component in the suspension state, stabilized with plasticizer and acoustic cavitation, is more effective. The plasticizer not only acts as a stabilizer of slag particles, but also fully participates in the process of hydration of the cement system in combination with slag (Fig. 2).

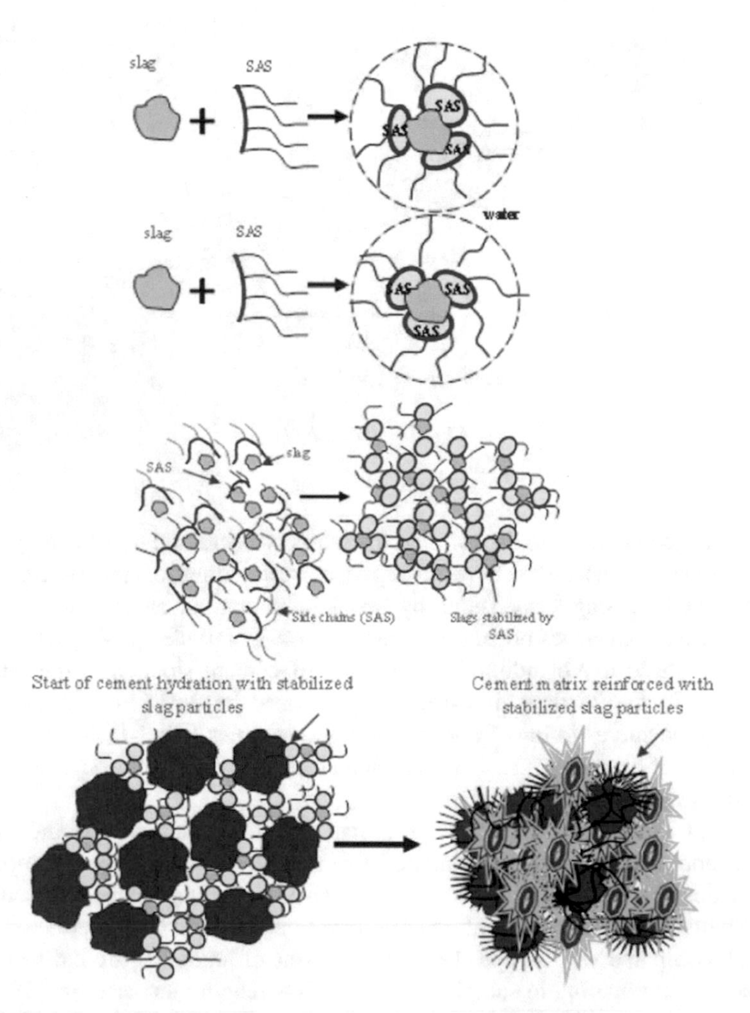

Fig. 2. Schematic representation of slag stabilization process with polycarboxylates (a) and the part of polydioxy-stabilized ultrafine slag additive in formation of the cement stone frame (b).

The plasticizer has polar and non-polar parts. The polar component of the plasticizer molecule interacts with the surface layer of slag, causing the electrostatic aggregate stability factor. The non-polar part forms a gelatinous barrier, which prevents slag particles from coming together and prohibits the coagulation process, conditioning the structural-mechanical factor of aggregative stability. As a stabilizer of slag particles, a plasticizer on a polycarboxylate basis was used, in the non-polar chain of which there are side branches, creating a spatial effect and providing additional stabilization of slag. The application of acoustic cavitation to water-polymer slag suspension enhances the effect of electrostatic and structural-mechanical factors of aggregative stability due to the increase in pressure in the medium caused by ultrasonic waves. This leads to destruction of slag aggregates and the formation of double electrical layer on new interfacial surfaces of slag grains and to uniform distribution of slag particles within the volume of the material. Thus, slag particles stabilized with a plasticizer and acoustic cavitation in a hardening system not only act as centers of directional crystallization, concentrating new formations on their surface, but also interact with calcium hydroxide formed as a result of hydrolysis and hydration of clinker minerals, which ensures intensive generation and growth of crystalline hydrates. Primary hydrated compounds involving stabilized slag suspensions form a stable colloidal system, where processes of self-reinforcement pass, associated with recrystallization of particles of neoplasms of colloidal sizes into larger ones. Such crystalline hydrates are involved in the construction of a spatial framework, which is compacted and strengthened due to prolonged hydration, thereby enhancing the strength characteristics of cement stone.

4 Conclusions

The studies showed that the early strength of cement stone and of concrete that involves Portland slag cement can be increased, if slag suspension stabilized with plasticizer and acoustic cavitation is used instead of mixing water. The research revealed, that the early strength of modified samples increased by 2.2 times, their grade strength – by 70% compared with non-additive Portland slag cement and by 44% and 30% respectively, compared with the sample containing plasticizer. Moreover, the porosity of cement stone at the grade age decreased by 36–38% compared with otherwise samples. The increased physical-mechanical and structural characteristics of cement stone of the modified samples are associated with the uniform distribution of ultrafine substance within the volume of cement matrix. Slag suspension is stabilized due to the effect of electrostatic and structural-mechanical factors of aggregative stability. The first factor is associated with the formation of a double electric layer at the surface of slag particles with the participation of polar component of the plasticizer. The effect of structural-mechanical factor is based on the formation of non-polar component of the plasticizer of gelatinous films that prevent coagulation processes in suspension. Applying acoustic cavitation to the water-polymer slag suspension leads to strengthening of the electrostatic and structural-mechanical factors of aggregative stability and homogeneity of the suspension. The uniform distribution of stabilized slag particles in a hardening system leads to concentration of new formations on its surface and production of solid and firm structure of cement stone.

References

1. Lesovik, V.S., Ageeva, M.S., Ivanov, A.V.: Granulirovannye shlaki v proizvodstve kompozicionnyh vyazhushchih. Vestnik Belgorodskogo gosudarstvennogo tekhnologicheskogo universiteta im. V.G. Shuhova **3**, 29–32 (2011)
2. Pan, Z.H., Zhou, J.L., Jiang, X.: Investigating the effects of steel slag powder on the properties of self-compacting concrete with recycled aggregates. Constr. Build. Mater. **200**, 570–577 (2019). https://doi.org/10.1016/j.conbuildmat.2018.12.150
3. Nedeljkovic, M., Ghiassi, B., Laan, S., Li, Z.M., Ye, G.: Effect of curing conditions on the pore solution and carbonation resistance of alkali-activated fly ash and slag pastes. Cem. Concr. Res. **116**, 146–158 (2019). https://doi.org/10.1016/j.cemconres.2018.11.011
4. Yan, X.C., Jiang, L.H., Guo, M.Z., Chen, Y.J., Song, Z.J., Bian, R.: Evaluation of sulfate resistance of slag contained concrete under steam curing. Constr. Build. Mater. **195**, 231–237 (2019). https://doi.org/10.1016/j.conbuildmat.2018.11.073
5. Pu, L., Unluer, C.: Durability of carbonated MgO concrete containing fly ash and ground granulated blast-furnace slag. Constr. Build. Mater. **192**, 403–415 (2018). https://doi.org/10.1016/j.conbuildmat.2018.10.121
6. Farhan, N.A., Sheikh, M.N., Hadi, M.N.S.: Investigation of engineering properties of normal and high strength fly ash based geopolymer and alkali-activated slag concrete compared to ordinary Portlandcement concrete. Constr. Build. Mater. **196**, 26–42 (2019). https://doi.org/10.1016/j.conbuildmat.2018.11.083
7. Norrarat, P., Tangchirapat, W., Songpiriyakij, S., Jaturapitakkul, C.: Evaluation of strengths from cement hydration and slag reaction of mortars containing high volume of ground river sand and GGBF slag. Adv. Civ. Eng. Article No. 4892015 (2019). https://doi.org/10.1155/2019/4892015
8. Chousidis, N., Ioannou, I., Batis, G.: Utilization of Electrolytic Manganese Dioxide (EMD) waste in concrete exposed to salt crystallization. Constr. Build. Mater. **158**, 708–718 (2018). https://doi.org/10.1016/j.conbuildmat.2017.10.036
9. Norhasri, M.S.M., Hamidah, M.S., Fadzil, A.M., Megawati, O.: Inclusion of nano metakaolin as additive in ultrahigh performance concrete (UHPC). Constr. Build. Mater. **127**, 167–175 (2016). https://doi.org/10.1016/j.conbuildmat.2016.09.127
10. Roychand, R., De Silva, S., Law, D., Setunge, S.: Micro and nano engineered high volume ultrafine fly ash cement composite with and without additives. Int. J. Concr. Struct. Mater. **10** (1), 113–124 (2016). https://doi.org/10.1007/s40069-015-0122-7
11. Radulova, G.M., Slavova, T.G., Kralchevsky, P.A., Basheva, E.S., Marinova, K.G., Danov, K.D.: Encapsulation of oils and fragrances by core-in-shell structures from silica particles, polymers and surfactants: the brick-and-mortar concept. Colloids Surf. A-physicochem. Eng. Asp. **559**, 351–364 (2018). https://doi.org/10.1016/j.colsurfa.2018.09.079
12. Rad, M.D., Telmadarreie, A., Xu, L., Dong, M.Z., Bryant, S.L.: Insight on methane foam stability and texture via adsorption of surfactants on oppositely charged nanoparticles. Langmuir **34**(47), 14274–14285 (2018). https://doi.org/10.1021/acs.langmuir.8b01966
13. Samchenko, S., Kozlova, I., Zemskova, O.: Model and mechanism of carbon nanotube stabilization with plasticizer. In: MATEC Web of Conferences, vol. 193, p. 03050 (2018). https://doi.org/10.1051/matecconf/201819303050

Complex Method of Stabilizing Slag Suspension

Svetlana Samchenko, Irina Kozlova(✉), Olga Zemskova, Dmitry Zamelin, and Angelina Pepelyaeva

Moscow State University of Civil Engineering, Yaroslavskoe shosse, 26, Moscow 129337, Russia
iv.kozlova@mail.ru

Abstract. The paper presents several methods of obtaining ultrafine slag, its stabilization and introduction into the composition of cement. Ultrafine slag with polifraction composition, which has particle size of 0.5–1.0 μm, was obtained through grinding blast-furnace granulated slag in a laboratory airflow jet mill. The authors present the method for introducing ultrafine slag in the state of suspension into the cement composition instead of water. The aggregates of ultrafine slag are destroyed under ultrasound, while a lot of new active centers of slag surface are formed. In the slag surface layer, the double electric layer with the participation of plasticizer is formed faster, the stabilization of slag particles in suspension is enhanced. The effectiveness of the stabilization of slag suspensions by complex methods is confirmed by the results of physical-mechanical tests of modified samples from Portland slag cement. In early periods of hardening, the strength of experimental samples increased by 2 times, at grade age – by 70% compared to Portland slag cement without additives. It was also established that the strength of modified daily samples from Portland slag cement is comparable to the strength of a daily sample from additive-free Portland cement, which can expand the scope of use of Portland slag cement in the construction industry.

Keywords: Ultrafine slags · Suspension · Plasticizer · Acoustic cavitation · Sulfonaphalene formaldehyde · Sedimentation and aggregative stability · Double electric layer · Factors of aggregative stability · Strength

1 Introduction

Today scientific community pays special attention to the issues of improving the environment, waste treatment, application of energy-saving and resource-saving technologies. The construction industry gains the lead in applying waste as secondary raw material when producing building materials. Replacing natural raw materials with production wastes solves several environmental and economic problems:

– wastes of production are utilized [1–5];
– land plots are cleared;
– material consumption of large-scale production decreases;

V. Murgul and M. Pasetti (Eds.): EMMFT-2018, AISC 983, pp. 817–827, 2019.
https://doi.org/10.1007/978-3-030-19868-8_80

- demands for raw materials are met in areas that do not have or have limited natural resources;
- increase in the number of operational characteristics of building materials [6–9].

Blast-furnace slags (byproducts of cast-iron production) are widely spread in the construction industry. They became valuable raw material especially in the production of Portland cement. With the introduction of blast-furnace slag into composition of the raw mix, the productivity of the furnaces increases and the fuel consumption reduces. The use of slag as an active mineral additive can improve a number of construction and technical properties of cement, such as corrosion resistance, weather resistance, heat resistance, etc. However, slag cements have a disadvantage associated with slow strength gaining in early hardening periods. Increase in strength after the first day of hardening can be achieved by introducing nano or ultrafine additives into the composition of cements [10–15], which in the hardening system will act as centers of crystallization of new formations and ensure the formation of the strong and dense structure of cement stone. When using nano or ultrafine materials there is not always a positive effect. Nanoscale and submicron particles tend to aggregate. They stick together and form aggregates that do not allow to distribute uniformly in the volume of the cement system and obtain stable characteristics of cement stone. To eliminate this disadvantage, nano and ultrafine additives should be stabilized before the introduction into the cement. Stabilization of nano and ultrafine additives can be provided by the introduction of surfactants, acoustic or hydrodynamic cavitation. Thus, the purpose and objectives of the study are defined as it follows:

- to obtain ultrafine additives, for example, slag;
- to develop a method for stabilizing ultrafine slag;
- to study strength of cement samples made from Portland slag cement, modified with ultrafine slag.

2 Materials and Methods

The subject of the research is granulated blast-furnace slag, ground to ultrafine state; suspension of ultrafine slag; Portland slag cement (PSC), modified ultrafine slag.

Characteristics of granulated blast-furnace slag are presented in Table 1, which says that CaO and SiO_2 predominate in the slag composition. Therefore, the presence of basic calcium orthosilicates, predominantly Ca_2SiO_4, is assumed.

Obtaining ultrafine slag was carried out in a laboratory vortex jet mill LHL-1 with the restriction of upper boundaries of grinding to a particle size of 1 μm.

The particle size of the ultrafine slag was determined on laser diffraction particle size analyzer Mastersizer 3000. The distribution curve of slag particles by fractions and the particle size distribution of slag are shown in Table 2.

Slag suspensions were prepared at a concentration of 10; 30; 50 g/l.

For the preparation of slag suspensions, tap (drinking) water with pH = 6.7 was used. The total salt content in water does not exceed 5000 mg/dm^3, including the content of sulfates not exceeding 2,700 mg/dm^3 (in terms of SO_4^{2-}).

Table 1. Specifics of blast-furnace granulated slag.

Chemical composition, %	CaO	SiO_2	Al_2O_3	MgO	Fe_2O_3 + FeO	Others	Slag quality ratio Qr	Lime factor Fl
	45.40	38.20	8.10	3.20	0.80	4.30		
Phase composition, %	Glass			Crystalline phase			1.45	1.1
	93.20			6.80				

Table 2. Particle size composition of ultrafine slag.

Particle size, μm	0–0.5	0.5–1.0	1.0–1.5	1.5–2.0	2.0–3.0	3.0–4.0	4.0–5.0
Fraction content, weight.%	8.21	30.44	23.35	17.20	11.77	5.89	3.14

Plasticizer based on sulfurized naphthalene formaldehyde resins, which is a long-chain anionic surfactant, was used as a stabilizer for the slag suspension. Its molecules have non-polar and polar parts. Polar groups of sulfonaphthalene formaldehyde (R-SO_3^-) are formed by the anions SO_3^-. The plasticizer was introduced in the amount of 1 to 10 g/l of suspension.

Acoustic dispersion of suspensions was carried out with sonicator UZDN-I under the frequency of ultrasonic vibrations - 44 kHz; dispersion temperature - 25 ± 2 °C; time of dispersion – 10–15 min. The constant temperature of dispersion was maintained by temperature control of suspensions.

Portland slag cement was obtained by combined grinding of 60% clinker, 40% slag and 5% natural gypsum in a laboratory mill to the residue R008 = 6% on a sieve. The chemical and mineralogical composition of clinker is presented in Table 3.

Table 3. Chemical and mineralogical composition of clinker.

Chemical composition of clinker, %		Mineralogical composition of clinker, %	
CaO	63.89	Alite	63
SiO_2	20.63		
Al_2O_3	5.62	Belite	11
Fe_2O_3	5.15		
MgO	3.68	The aluminate phase	6
SO_3	0.59		
R_2O	1.36	The ferrite phase	15

Cement samples for research were prepared by mixing Portland slag cement with stabilized suspensions of ultrafine slag in water-polymer dispersion medium. The strength of cement was determined by the national standard GOST 30744-2001.

3 Results and Discussion

3.1 Substantiation of the Use of Integrated Method of Stabilizing Slag Suspensions

To conduct research, the authors used blast-furnace granulated slag, ground in the laboratory jet mill LHL-1 with the restriction of upper boundaries of grinding to a particle size of 1 μm. Slag of polyfraction composition with a predominant particle size of 0.5–1.0 μm was obtained (Table 2).

Earlier, the authors carried out studies on the introduction of ultrafine slag into the composition of cement by dry mixing. Samples with enhanced strength characteristics were obtained. However, there was a large dispersion of strength values in early periods of hardening. Therefore, particles of ultrafine slag in the cement system are unevenly distributed. Thus, slag particles are so small that they tend to aggregate. The enlargement of slag particles occurs due to coagulation and is accompanied by a loss of aggregative stability.

To increase the aggregative stability of ultrafine slag particles and ensure their uniform distribution in the cement composition, it is proposed to add an additive to the cement composition in the state of suspension instead of mixing water, similarly to carbon nanotube suspensions.

Studies were conducted to establish the aggregative and sedimentation stability of suspensions of ultrafine slag in an aqueous dispersion medium [24]. It is determined that the process of sedimentation of slag particles is divided into 3 periods. The fastest sedimentation of slag particles occurs within the first period (6–8 min). To increase the stability of slag suspensions, especially in the first sedimentation period, the following ought to be done:

1. introduce stabilizer into slag suspension;
2. affect the slag suspension, containing stabilizer, with acoustic cavitation.

The use of plasticizer based on sulfurized naphthalene formaldehyde resins for stabilization of carbon nanotube suspensions can ensure uniform distribution of nanoparticles in the volume of the cement matrix and the formation of the structure of cement stone with enhanced physical, mechanical and structural characteristics. The choice of the method of dispersion is conditioned by the fact that acoustic cavitation, based on the passage of high-intensity sound waves through a layer of suspension, is capable of destroying the aggregates of ultrafine slag and homogenizing the suspension. Sound waves cause the formation and collapse of vapor bubbles in suspension, thus creating short pulses (up to 108 Pa and more), capable of destroying even very firm materials. Thus, the uniform distribution of ultrafine slag in the volume of the cement matrix and increasing the strength of cement stone from slag cement, especially in early periods of hardening, can be achieved by using slag suspensions, stabilized using a complex method in a water-polymer dispersion medium under conditions of acoustic cavitation, instead of water. Stabilization of slag suspensions with plasticizer.

It was previously determined that in water dispersion medium ultrafine slag particles form two types of micelles with negative and positive proton number. Initially, hydration products from silicic acid groups of various basicities are formed on the surface of the slag grains, which charge the slag surface negatively. The further process of hydration on the surface of the slag particles leads to the accumulation of Ca^{2+} ions, which charge the slag surface positively. The presence of a charge on the surface of slag particles determines the aggregative and sedimentation stability achievable due to the electrostatic factor of aggregative stability. To enhance the stabilization of positively charged particles of slag, plasticizer beased on sulfonaphthalene formaldehyde was introduced into water suspension.

To determine the plasticizer concentration that ensures the stability of the slag suspension the protective number of the suspension was determined experimentally. For the slag suspension stabilized by sulfonaphthalene formaldehyde, the protective number was 0.021 g/l, the optimum concentration of sulfonaphthalene formaldehyde was 5 g/l at the concentration of slag in the suspension 50 g/l (Table 4).

Table 4. Results of studies on aggregative stability of slag suspensions stabilized by sulfonaphalene formaldehyde.

Name of indicator	№ of test									
	1	2	3	4	5	6	7	8	9	10
Slag concentration in suspension, g/l	50	50	50	50	50	50	50	50	50	50
Stabilizer concentration, g/l	1	2	3	4	5	6	7	8	9	10
Protective suspension number, $*10^{-3}$ g/l	0.8	3.4	7.5	13.2	21.0	30.0	41.0	54.0	67.5	83.0
End time of complete slag settling, h-min	3-30	4-50	6-00	7-20	9-00	9-00	9-10	9-20	9-20	9-30

The effectiveness of the stabilization of slag suspensions by sulfonaphthalene formaldehyde is confirmed by the results presented in Table 5.

Table 5. Time and speed of slag particle sedimentation in water and water-polymer dispersion media.

Experimental Conditions *	Slag concentration, g/l	Particle sedimentation period					
		I		II		III	
		Particle sedimentation duration, h : min : s	Speed of particle sedimentation, 10^{-6} m/s	Particle sedimentation duration, h : min : s	Speed of particle sedimentation, 10^{-6} m/s	Particle sedimentation duration, h : min : s	Speed of particle sedimentation, 10^{-6} m/s
1	10	00:08:05	302.00	00:30:00	82.20	02:30:00	16.40
	30	00:06:20	344.00	00:20:00	110.00	02:25:00	16.80
	50	00:06:10	395.00	00:16:00	143.00	02:20:00	17.20
2	10	03:40:00	11.80	05:00:00	9.69	08:20:00	5.57
	30	03:00:00	13.40	04:30:00	9.93	08:10:00	5.67
	50	02:30:00	17.80	04:00:00	13.10	07:30:00	6.71
V_1/V_2	10	25.6		8.5		2.9	
	30	25.6		11.1		3.0	
	50	22.2		10.9		2.6	

Where: *1 – without sulfonaphthalene formaldehyde; 2 – with sulfonaphthalene formaldehyde.

It follows from the above data that with the introduction of sulfonaphthalene formaldehyde into water dispersion medium, the aggregative and sedimentation stability of slag suspensions increases. In the first period, on average, sustainability increases 24 times, in the second period - 10 times; in the third period - 3 times.

The interaction of ultrafine slag with sulfonaphthalene formaldehyde occurs according to the equation:

$$Ca_2SiO_4 + R\text{-}SO_3\text{-}Na + HOH = CaHSiO_4\text{-}OSO_2\text{-}R + NaOH \quad (1)$$

In water dispersion medium, sulfonaphthalene formaldehyde dissociates to form surface-active anions — functional sulfonated groups of R-SO3. In the surface layer of positively charged slag grains, the process of adsorption begins, accompanied by the fixation of the functional groups of sulfonaphthalene formaldehyde. The slag surface is negatively charged, and the H + ions are in the diffuse layer. The electric double layer is formed with the participation of sulfonaphthalene formaldehyde (an electrostatic aggregate stability factor). The slag stabilization mechanism by sulfonaphthalene formaldehyde is shown in Fig. 1; the structure of slag micelles stabilized by sulfonaphthalene formaldehyde is shown in Fig. 2

Formation of the adsorption layer at the surface of slag particles surrounded by plasticizer radicals is affected by its hydrocarbon chain that is outbound to them. Moreover, sulfonaphthalene formaldehyde radicals form strong and elastic gelatinous films on the surface of slag particles (Fig. 3 (6)), enhancing the stabilization of slag grains due to the structural-mechanical factor of aggregative stability.

Thus, studies have shown that the introduction of sulfonaphthalene formaldehyde in slag suspension enhances the stabilization of slag particles in suspension. Stabilization is enhanced due to the effect of electrostatic and structural-mechanical factors of aggregative stability on the system at the same time.

3.2 Stabilization of Slag Suspensions by the Complex Method

Acoustic cavitation was applied to enhance the stabilization of ultrafine slag particles in water-polymer suspension. The time and speed of sedimentation of slag particles in a water-polymer suspension after acoustic treatment (Table 6) were determined.

A comparative analysis of the stabilization of slag suspensions in the sedimentation rate of the slag particles in the suspension after carrying out acoustic cavitation and without its use has been carried out. The results of the analysis are presented in Table 7. From the above data it follows that the slag suspension, stabilized with a plasticizer and acoustic cavitation, becomes more stable. The stability of the slag suspensions stabilized by the complex method has increased:

- on average in the first period - 93 times; in the II period - 47 times, in the III period 15 times in comparison with the stability of the aqueous slag suspensions;
- on average, during all periods of sedimentation, 3–4 times compared with the stability of water-polymer slag suspensions.

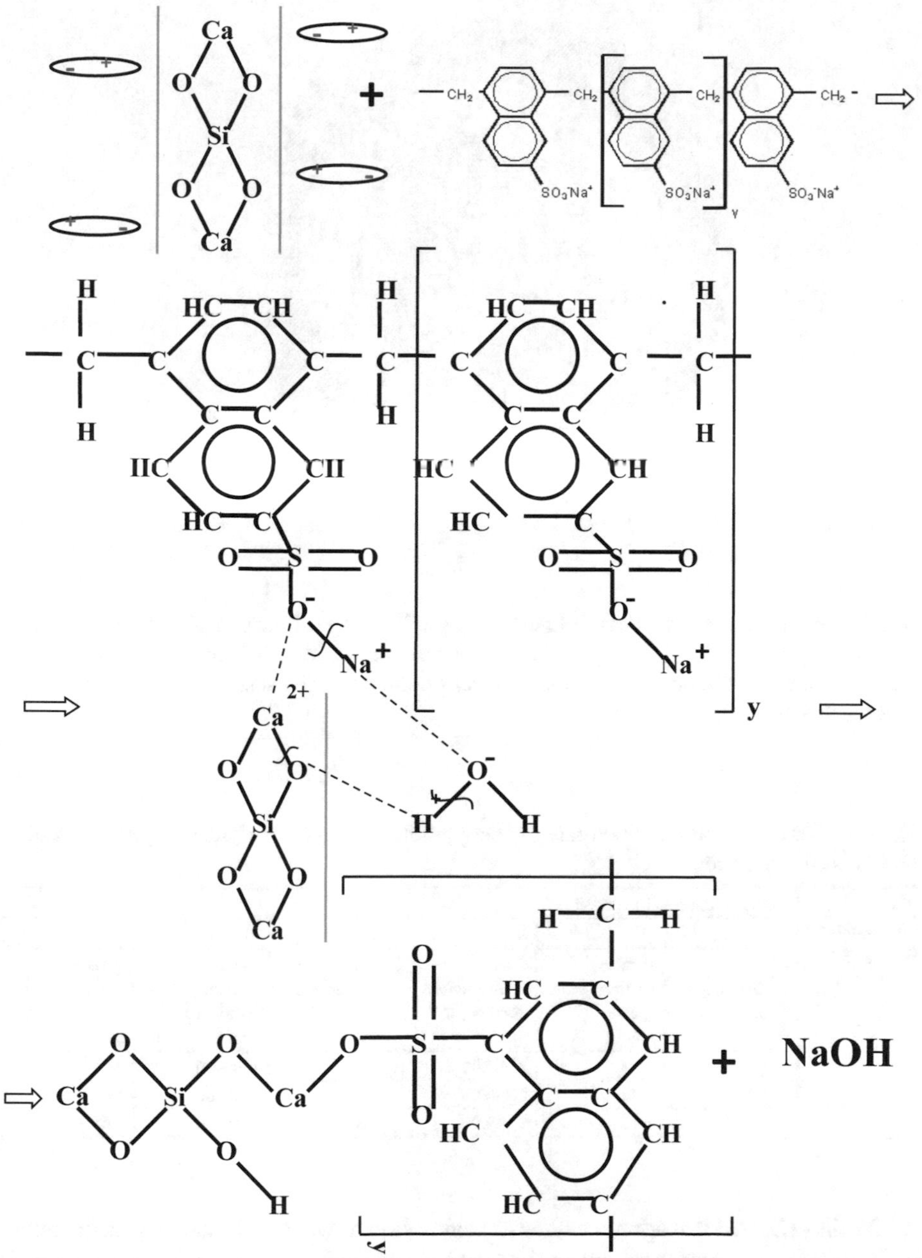

Fig. 1. Mechanism of stabilization of ultrafine slag by sulfonaphthalene formaldehyde.

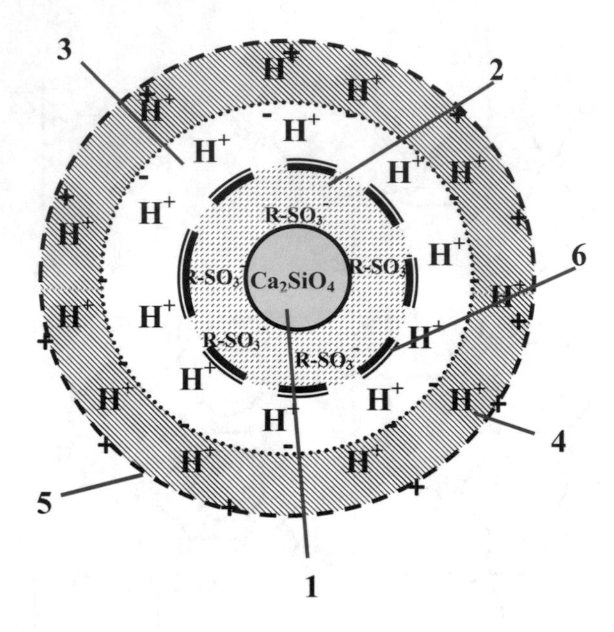

Fig. 2. The structure of micelles of ultrafine slag stabilized by sulfonaphthalene formaldehyde: 1 - slag aggregate (Ca_2SiO_4); 2 - layer of potential-determining ions; 3 - counterions of the dense part of double electric layer; 4 - diffusion layer counterions; 5 - slag micelle; 6 - gel film of radicals R-SO_3^-.

Table 6. Time and sedimentation rate of slag particles in a water-polymer dispersion medium after acoustic cavitation.

The concentration of slag, g/l	Particle sedimentation period					
	I		II		III	
	Particle sedimentation duration, h : min: s	Particle sedimentation speed, 10^{-6} m/s	Particle sedimentation duration, h : min: s	Particle sedimentation speed, 10^{-6} m/s	Particle sedimentation duration, h : min: s	Particle sedimentation speed, 10^{-6} m/s
10	14:20:00	3.02	28:00:00	1.56	46:20:00	0.96
30	12:00:00	3.36	17:30:00	2.48	41:00:00	1.09
50	08:40:00	5.19	13:30:00	3.23	30:00:00	1.45

Studies showed that applying acoustic cavitation to water-polymer slag suspensions leads to greater aggregative and sedimentation stability. This is due to the increase in pressure in the medium caused by ultrasonic vibrations. Ultrafine slag aggregates are destroyed by ultrasound. There is an intensive formation of new surfaces, including positively charged ones. Its interaction with the sulfo-groups of the plasticizer is carried out faster due to increasing the active centers of the slag surface. The subsequent formation of an double electrical layer on new interfacial surfaces not only leads to a uniform distribution of slag particles in the dispersion medium, but also to a uniform

Table 7. Comparative analysis of the sedimentation rate of slag particles in suspensions.

The ratio of the sedimentation rate of slag particles in suspensions *	The concentration of slag in suspension, g/l	Particle sedimentation period		
		I	II	III
V_1/V_3	10	100	53	17
	30	102	44	15
	50	76	44	12
V_2/V_3	10	3.9	4.0	4.2
	30	4.0	2.6	3.7
	50	3.4	2.5	3.3

*1 – without sulphonaphthalene formaldehyde; 2 – with sulphonaphthalene formaldehyde; 3 – the use of the complex method (stabilizer + acoustic cavitation).

distribution of the stabilizer on the surface of slag particles, thereby causing the increase in electrostatic and structural-mechanical factors of the aggregative stability of slag suspensions.

3.3 Study of the Influence of Stabilized Slag Suspensions on Properties of Portland Slag Cement

To assess the uniformity of distribution of ultrafine slag in the composition of Portland slag cement, as well as assessing the stabilization of slag suspensions using a complex method, physical and mechanical tests of the samples were carried out. Samples were prepared by mixing Portland slag cement with suspension of ultrafine slag stabilized by the complex method. The results are presented in Table 8.

Table 8. Strength of modified samples.

№ of sample	Additive composition	Acoustic cavitation	Strength, MPa after			
			1 days	3 days	7 days	28 days
1	-		8.5	16.4	26.2	40.2
2	Plasticizer (5 g/l)		7.9	18.2	31.4	49.9
3	Slag suspension: plasticizer (5 g/l); slag (10 g/l)	+	17.7	32.5	43.1	63.9
4	Slag suspension: plasticizer (5 g/l); slag (30 g/l)	+	17.8	31.8	42.3	65.9
5	Slag suspension: plasticizer (5 g/l); slag (50 g/l)	+	17.1	32.7	42.9	71.5

Table 8 implies that the strength of samples mixed with slag suspension stabilized by applying the complex method increases approximately 2 times in early period of hardening; at a grade age – by 70% compared with non-additive Portland slag cement. It means that introducing slag suspensions stabilized by the complex method into the composition of Portland slag cement contributes to the increase in the strength of cement stone in periods of early hardening. The average value of the strength of the modified samples during the first days of hardening (17.6 MPa) is close to the strength of additive-free Portland cement, the daily strength of which is 18.2 MPa. The results obtained also allow assuming the uniform distribution of slag particles in the cement matrix. Slag particles stabilized by plasticizer ast as centers of directional solidification in the hardening system, concentrating new formations on their surface that are involved in building a dense and durable framework, thereby contributing to the improvement of the strength characteristics of cement stone. This confirms the feasibility of using stabilized ultrafine slags in the composition of cement and allows, firstly, solving the problem of slow hardening of cement stone from Portland slag cement in early periods of hardening and, secondly, expanding the scope of its application.

4 Conclusions

The studies showed the effectiveness of the complex method applied for stabilization of slag suspensions. The use of plasticizer based on sulfurized naphthalene formaldehyde resins as a stabilizer contributes to the increase in aggregative and sedimentation stability of suspensions due to the effect of electrostatic and structural-mechanical factors of aggregative stability.

The effect of the first factor is based on the formation of double electrical layer with the participation of functional groups of the plasticizer ($SO3^-$) at the surface of slag particles. The effect of structural-mechanical factor of aggregative stability is entailed by the formation of firm elastic gelatinous films of the main chain radicals of plasticizer on the surface of slag particles.

The use of acoustic cavitation to water-polymer slag suspensions provides 93 times increase in stabilization of slag particles in suspension within period I, 47 times within period II, 15 times within period III, if comparing with the stability of water slag suspensions. Moreover, there is 3–4 times increase within all periods of sedimentation if compared with the stability of water-polymer slag suspensions.

Effect enhancement for electrostatic and structural-mechanical factors of aggregative stability is entailed by the increase of the pressure in suspension, caused by ultrasonic waves. For that reason, the aggregates of ultrafine slag are destroyed and new active centers involved in the formation of double electrical layer are formed at the surface of slag particles.

The introduction of slag suspension stabilized by the complex method into the composition of cement instead of mixing water allows achieving the uniform distribution of additive within the material volume and increasing the physical-mechanical characteristics of cement stone.

References

1. Franco de Carvalho, J.M., Melo, T.V.D., Fontes, W.C., Batista, J.O.D.S., Brigolini, G.J., Peixoto, R.A.F.: More eco-efficient concrete: an approach on optimization in the production and use of waste-based supplementary cementing materials. Constr. Build. Mater. **206**, 397–409 (2019). https://doi.org/10.1016/j.conbuildmat.2019.02.054
2. Shapovalov, N.A., Zagorodnyuk, L.H., Tikunova, I.V., ShcHekina, A.Y., Shkarin, A.V.: SHlaki metallurgicheskogo proizvodstva – ehffektivnoe syr'yo dlya polucheniya suhih stroitel'nyh smesej. Fundamental'nye issledovaniya **1–1**, 167–172 (2013)
3. Cao, L., Shen, W., Huang, J., Yang, Y., Zhang, D., Huang, X., Lv, Z., Ji, X.: Process to utilize crushed steel slag in cement industry directly: multi-phased clinker sintering technology. J. Clean. Prod. **217**, 520–529 (2019). https://doi.org/10.1016/j.jclepro.2019.01.260
4. Wang, Y., Suraneni, P.: Experimental methods to determine the feasibility of steel slags as supplementary cementitious materials. Constr. Build. Mater. **204**, 458–467 (2019). https://doi.org/10.1016/j.conbuildmat.2019.01.196
5. Luhar, S., Luhar, I.: Potential application of E-wastes in construction industry: A review. Constr. Build. Mater. **203**, 222–240 (2019). https://doi.org/10.1016/j.conbuildmat.2019.01.080
6. Manh Do, T., Kang, G.-O., Kim, Y.-S.: Development of a new cementless binder for controlled low strength material (CLSM) using entirely by-products. Constr. Build. Mater. **206**, 576–589 (2019). https://doi.org/10.1016/j.conbuildmat.2019.02.088
7. Yakovlev, G.I., Pervushin, G.N., Korzhenko, A., Bur'yanov, A.F., Kerene, Y.A., MaevaI, S., Hazeev, D.R., Pudov, I.A., Sen'kov, S.A.: Primenenie dispersij mnogoslojnyh uglerodnyh nanotrubok pri proizvodstve silikatnogo gazobetona avtoklavnogo tverdeniya. Stroitel'nye materialy **2**, 25–29 (2013)
8. Karpova, E.A., Ali, E.M., Skripkyunas, G., Kerene, Y.A., Kichajte, A., Yakovlev, G.I., Maciyauskas, M., Pudov, I.A., Aliev, E.V., Sen'kov, S.A.: Modifikaciya cementnogo betona kompleksnymi dobavkami na osnove ehfirov polikarboksilata, uglerodnyh nanotrubok i mikrokremnezema. Stroitel'nye materialy **2**, 40–48 (2015)
9. Petrunin, S., Vaganov, V., Sobolev, K.: Cement composites reinforced with functionalized carbon nanotubes. J. Soc. Am. Music. **1611**, 2 (2014)
10. Yakovlev, G., Pervushin, G., Maeva, I., Pudov, I., Shaybadullina, A., Keriene, J., Buryanov, A., Korzhenko, A., Senkov, S.: Modification of construction materials with multi-walled carbon nanotubes. Procedia Engineering "Modern Building Materials, Structures and Techniques", pp. 407–413 (2013)
11. Konsta-Gdoutos, M.: Highly dispersed carbon nanotubes reinforced cement based materials. Cem. Concr. Res. **40**, 1052–1059 (2010)
12. Samchenko, S.V., Zemskova, O.V., Kozlova, I.V.: The efficiency of application of physical and chemical methods on the homogeneous dispersion of carbon nanotubes in water suspension. Cement-Wapno-Beton **XX/LXXXII**(5), 322–327 (2015). WOS:000365377000006
13. Samchenko, S.V., Zemskova, O.V., Kozlova, I.V.: Stabilization of carbon nanotubes with superplasticizers basedon polycarboxylate resin ethers. Russ. J. Appl. Chem. **87**(12), 1872–1876 (2014). https://doi.org/10.1134/S1070427214120131
14. Samchenko, S., Kozlova, I., Zemskova, O.: Model and mechanism of carbon nanotube stabilization with plasticizer. In: MATEC Web of Conferences, vol. 193, p. 03050 (2018). https://doi.org/10.1051/matecconf/201819303050
15. Samchenko, S., Zemskova, O., Kozlova, I.: Ultradisperse slag suspensions aggregative and sedimentative stability. In: MATEC Web of Conferences, vol. 106, p. 03017 (2017). SPbWOSCE-2016 3017. https://doi.org/10.1051/matecconf/20171060

Modelling of Technology of Mineral Wool Products

Alexey Zhukov[1], Ekaterina Bobrova[2], Andrey Medvedev[1](✉), Nikita Ushakov[1], Diana Beniya[1], and Alexey Poserenin[3]

[1] Moscow State University of Civil Engineering, Yaroslavskoye shosse, 26, Moscow 129337, Russia
medvedev747@yandex.ru
[2] Higher School of Economics, Myasnitskaya Str. 20, Moscow 101000, Russia
[3] Russian State Geological Prospecting University, Miklukcho-Maklaya, 23, Moscow 117997, Russia

Abstract. The article describes the methodology for solving technological problems using experimental statistical modeling. This method of modeling is characterized by the universality of the experimental data collection methodology. Such models describe with known accuracy (degree of adequacy) the relationship between the inputs and outputs of the system without analyzing the internal structure of this system. Processing of the results by mathematical methods and identification of the regression equations as algebraic polynomials allowed to use the method of local optimization for the analysis of the obtained dependences. The method of local optimization is the development of existing methods of optimization of technological processes and methods for determining the unqualified absolute extremum of the function in the study of models of multifactor processes with unstable extrema and non-matching areas of the optimum. The technology of mineral wool products became the object of both research and application of mathematical techniques. Carrying out the experiment on the space of 14 variable factors allowed to estimate the influence of each factor on the result; to construct the level of regression for the most significant factors and to realize their optimization. The obtained optimized dependences and their graphical interpretation became the basis for the nomogram, which allows to predict the properties of the differences. The obtained dependences became the basis of the algorithm of the computer program, which allows determining and predicting the properties of products depending on the values of variable factors.

Keywords: Mineral wool products · Mathematical methods · Modelling

1 Introduction

Analytical methods and modelling are widely used in the material properties studying and in the study of technological processes, that is, in technological analysis. Technological modeling is one of the applied aspects of system-dynamic modeling, which is widely used for the analysis and prediction of complex processes of different nature and the creation of specialized information and analytical decision-making systems.

V. Murgul and M. Pasetti (Eds.): EMMFT-2018, AISC 983, pp. 828–838, 2019.
https://doi.org/10.1007/978-3-030-19868-8_81

The application of simulation methods in the industrial systems studying allows to solve the problems of business planning of logistics and cargo handling, improving operational reliability and optimization of energy and water supply systems [1–6].

Implementation of system modeling providing allows us to investigate the technological risks of industrial systems; to analyze the accidents risk; to assess the conditions for ensuring the performance of technological installations while maintaining a predetermined level of performance. Based on the use of methods of survivability theory, it became possible to assess the safety of industrial technologies; ensuring the integration of spatial information and expertise knowledge in the modeling of natural and technical complexes [7–11].

The basis of the system analysis of technological processes is mathematical modeling, in particular with the methods of mathematical statistics using. Therefore, the article presents the classical schemes of statistical models based on D-optimal plans and the least squares method. The acquisition of skills in the analysis of a priori information, the choice of the desired plan of experiments, the construction of a mathematical description of the process in the field of experimentation, statistical analysis, the choice of the shortest path to the optimum and the implementation of the movement along this path is the sum of knowledge necessary for each experimenter.

2 Experimental

The research of the influence of various factors on the production process of modified mineral wool plates was based on the General method of technological analysis. Based on the analysis of the technology, a group of factors affecting to the result was established. The choice of individual factors for subsequent experiments took into account the requirements of the statistical analysis methodology, in particular, their mutual independence. As a result, the scheme of the process is formed, which combines two technological redistribution (Fig. 1): formation of mineral wool carpet and heat treatment of the carpet with subsequent cooling.

There are identified 14 factors: the average density of the carpet (X_1); the thickness of the carpet (X_2); the content of the binder (X_3); the uniformity of carpet (X_4); the compressibility of the carpet under the load (X_5); fiber diameter (X_6); fiber length (X_7); relative thickness of carpet cutting (X_8); degree of compaction of the coating layer (X_9); the degree of compaction of two-layer carpet (X_{10}); coolant temperature (X_{11}); the speed of the sucking action of the coolant (X_{12}); duration of heat treatment (X_{13}); the duration of cooling (X_{14}).

The effect of material particles in this research was not considered since modern multi – roll centrifuges allow to obtain mineral wool with a low content of non-fibrous inclusions-no more than 5%. This content does not have much influence on the thermal conductivity and other performance characteristics.

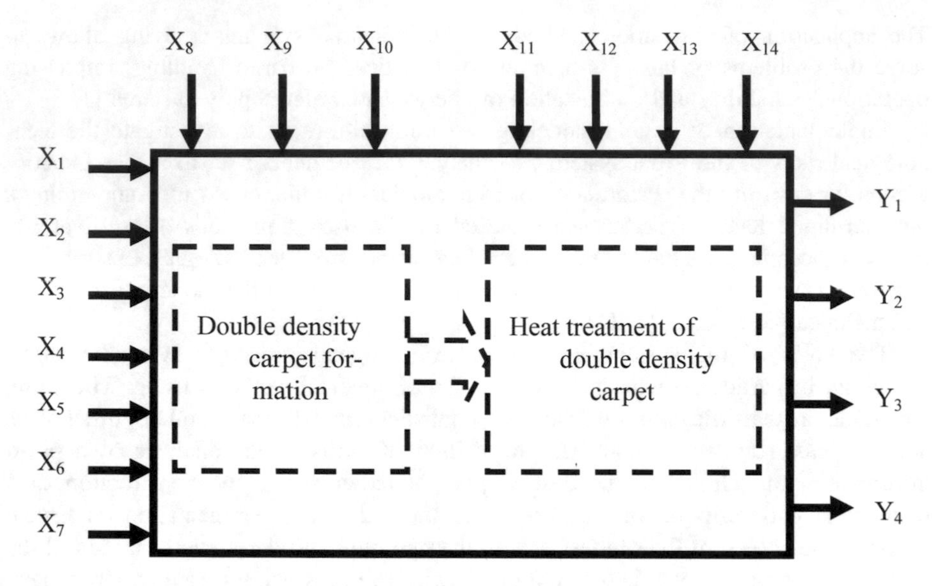

Fig. 1. Structural scheme of the technology.

Factors X_8–X_{10} are the parameters of control the process of forming a double-density carpet. The last group of factors characterizes the process of heat treatment. As response functions were adopted: compressive strength at 10% deformation (Y1), kPa; pull strength (Y_2), kPa; average plate density (Y_3) kg/m^3; thermal conductivity (Y_4), W/(m·K). For convenience of calculations and interpretation the auxiliary function Y_4 equal to $\lambda \cdot 10^3$.

3 Results

The experiment was conducted in two stages. At the first stage, the influence of each of the variable factors on the responses was evaluated. Conducting the experiment of the first stage and processing its results allowed to determine the degree of influence of each factor on the result. The thickness of the carpet (X_2), the relative thickness of its cutting (X_8) and the parameters of heat treatment (X_{11}, X_{12}, X_{13}, X_{14}) were in the category of insignificant factors. All coefficients with these factors turned out to be less than the confidence intervals (Δb).

This result does not mean that these factors should not be taken into account when optimizing the technology. The matter is in the technological regulations, which significantly limits the areas of variation of these factors, that is, the established framework for the change of factors is in itself close to the optimum and is characterized by low gradients. In order to assess the contribution of this group of factors, a different modeling system and other models studied in the third Chapter were used.

The higher the coefficient of absolute value, the higher the influence of this factor on this result. According to the results of the first stage, the following regression equations are obtained:

$$Y_1 = 40 + 20X_1 + 7X_3 + 5X_5 + 6X_6 + 8X_{10} + 5X_{11}$$

$$Y_2 = 7 + 2,9X_1 + 1,1X_3 + 1,8X_7 + 0,7X_9 + 0,9X_{10} + 0,7X_{11}$$

$$Y_3 = 100 + 40X_1 + 10X_2 - 12X_4 + 8X_7 + 9X_{10}$$

$$Y_4 = 38 + 1,0X_1 - 0,4X_4 + 0,3X_6 + 0,2X_7 + 0,4X_{10}$$

It has been established that the following factors have the greatest influence on the results of the experiment (response functions): the average density of the carpet (X_1); binder content (X_3); fiber diameter (X_6); fiber length (X_7); the degree of compaction of the two-layer carpet (X_{10}), coolant temperature, °C (X_{11}). The plan of the second stage of the experiment was based on the areas of variation of these factors and in accordance with the conditions defined above.

Subject matter of study at the second stage were the factors whose influence on the result was manifested to the greatest extent. The planning matrix of experiments was built based on the plans. The intervals of variation of the factors remained the same as in the experiment of the first stage. As a result, after evaluating the significance of the coefficients and checking the adequacy of the models, the following regression equations are obtained:

$$Y_1 = 38 + 18X_1 + 6X_3 + 5X_6 + 5X_7 + 5X_{10} + 4X_{11} + 4X_1X_{10} - 2X_1X_7 + 3X_1X_{11} - 2X_6X_{10}$$

At the confidence interval Δb = 1,6.

$$Y_2 = 6,5 + 2,2X_1 + 0,9X_3 + 1,2X_7 + 0,9X_{10} + 0,8X_{11} - 0,3X_1X_7 + 0,3X_1X_{11} + 0,4X_3X_7 - 0,5X_{10}^2$$

At the confidence interval Δb = 0,2

$$Y_3 = 96 + 36X_1 + 9X_7 + 8X_{10} + 4X_1X_{10} - 5X_7^2$$

At the confidence interval Δb = 3,2

$$Y_4 = 37,5 + 0,8X_1 + 0,3X_6 + 0,4X_{10} + 0,2X_1X_{10} - 0,2X_6^2$$

At the confidence interval Δb = 0,12

The principal nature of the influence of factors on the results of the second stage has not changed compared with the results of the first stage. At the same time, some pair and quadratic interactions, which are significant in predicting the properties of products (solving interpolation problems), the parameters of their manufacture and the choice of optimization strategy, appeared.

4 Discussion

The proposed method of local optimization is the development of existing methods of optimization of technological processes and is the development of methods for determining the unqualified absolute extremum of the function (equating to zero all partial derivatives of the gradient vector of the basis function) in the study of models of multifactor processes with unstable extrema and non-matching areas of the optimum [12–15].

Determination of the local extremum of the basic functions for each of the factors is possible by taking the partial derivative of the function by this factor and equating it to zero:

$$dX_k/dX_i = dF(X_1, X_2, \ldots X_i)_/dX_i = 0; dX_l/dX_j = dF\left(X_1, X_2, \ldots X_j\right)/dX_j = 0$$

The optimization dependence obtained in this way can have numerical fixed expressions $[X_i]$ = *const* и $[X_j]$ = *const* or be a function of one, two or more variable factors:

$$[X_i] = f(X_1, X_2, \ldots X_{i-1}); [X_j] = f(X_1, X_2, \ldots X_{j-1})$$

The mutual relations between the factors determined by the optimization dependence make it possible to estimate only approximately the result or interdependence of the factors and are not optimization equations [16, 17].

The solution of the basis function with the substitution of the optimization dependence allows to obtain an optimization equation that allows to adequately predict the results of the experiment with the optimized values of the variables. Possible solutions can be presented in the form of equations

$$X_k([X_i]) = F(X_1, X_2, \ldots [X_i]); X_l([X_j]) = F(X_1, X_2, \ldots [X_j])$$

$$X_k([X_i], [X_j]) = F(X_1, X_2, \ldots [X_i], [X_j]); X_k([X_i], [X_i]) = F(X_1, X_2, \ldots [X_i]\,[X_j])$$

Optimization of localization of the result is carried out by setting the limit of the value of the level of response change required by the experimenter and determining the functional dependence between the variable factors satisfying the introduced condition: X_k = *const*, или X_k $([X_i])$ = *const*.

Optimization of the result localization can be carried out both on the basis function and on any other response function.

According to the results of the experiment of the second sage three functions have acquired extreme character. Namely, the function of the strength of the separation of layers (Y_2) allows us to expect an extreme factor of the degree of compaction of a two – layer carpet in the heat treatment chamber (X_{10}), the function of the average density of plates (Y_3) – along the length of the fiber (X_7), and the function of thermal conductivity (Y_4) – along the fiber diameter (X_6). Carrying out a number of necessary calculations gave the definition of the value of the factors corresponding to the optimum functions. Further, these values are translated from coded into natural form. The results of optimization processing are given below.

1. We use the equation for the strength of the separation of layers and determine the extremum of the function $Y_2(X_1, X_3, X_7, X_{10})$ by the factor X_{10}:

$$\frac{\partial Y_2}{\partial X_{10}} = 0,9 - 1,0X_{10} = 0, \text{therefore } X_{10} = 0,9$$

   Accordingly, the natural value of the factor $\widetilde{X_{10}} = 13 + 3.0,9 = 13,7 \mp 0,8\%$

2. We use the equation for the average density and determine the extremum of the function $Y_3(X_1, X_7, X_{10})$ by the factor X_7:

$$\frac{\partial Y_3}{\partial X_7} = 9 - 10X_7 = 0, \text{ therefore } X_7 = 0,9$$

   Accordingly, the natural value of the factor $\widetilde{X_7} = 40 + 10 \cdot 0,9 = 49 \mp 5$ mm

3. We use the equation for the strength of the separation of layers and determine the extremum of the function $Y_4(X_1, X_6, X_{10})$ by the factor X_{10}:

$$\frac{\partial Y_4}{\partial X_6} = 0,3 - 0,4X_6 = 0, \text{ therefore } X_6 = 0,75$$

   Accordingly, the natural value of the factor $\widetilde{X_{10}} = 4 + 1 \cdot 0,75 = 4,75 \mp 0,2$ μ.

Thus, it is established that in the given intervals of variation three factors can be optimized analytically. The best values for strength, density and thermal conductivity correspond to the following values of these factors: the diameter of the fiber should be in the range from 4.55 to 4.95 μ; fiber length-from 44 to 54 mm the degree of compaction of the two – layer carpet in the heat treatment chamber should be 12.9–14.5%.

We determine the final form of the regression equations taking into account the local optima obtained. To do this, substitute the optimal values of the factors determined analytically $X_{10} = 0,9; X_7 = 0,9; X_6 = 0,75$ in the regression equation:

1. Optimization of the compressive strength equation at 10% strain

$$Y_1 = 38 + 18X_1 + 6X_3 + 5 \cdot 0,75 + 5 \cdot 0,9 + 5 \cdot 0,9 + 4X_{11} + 4X_1 \cdot 0,9 - 2X_1 \cdot 0,9 + 3X_1X_{11} - 2 \cdot 0,75 \cdot 0,9$$

   The optimized equation of compressive strength at 10% deformation:

$$Y_1 = 49 + 20X_1 + 6X_3 + 4X_{11} + 3X_1X_{11}$$

2. Optimization of equations of the pull-off strength of layers

$$Y_2 = 6,5 + 2,2X_1 + 0,9X_3 + 1,2 \cdot 0,9 + 0,9 \cdot 0,9 + 0,8X_{11} - 0,3X_1 \cdot 0,9 + 0,3X_1X_{11} + 0,4X_3 \cdot 0,9 - 0,5 \cdot 0,81$$

The optimized equation for pull-off strength of layers:

$$Y_2 = 7,0 + 2,4X_1 + 1,3X_3 + 0,8X_{11} + 0,3X_1X_{11}$$

3. Optimization of the mean density equation

$$Y_3 = 96 + 36X_1 + 9 \cdot 0,9 + 8 \cdot 0,9 + 4X_1 \cdot 0,9 - 5 \cdot 0,81$$

The optimized equation for the average density:

$$Y_3 = 106 + 40X_{1-1}2X_4$$

4. Optimization of the heat conduction equation $(Y_4 = \lambda \cdot 10^3)$

$$Y_4 = 37,5 + 0,8X_1 + 0,3 \cdot 0,75 + 0,4 \cdot 0,9 + 0,2X_1 \cdot 0,9 - 0,2 \cdot (0,75)^2$$

The optimized equation for heat conduction:

$$Y_4 = 37,0 + 1,2X_1 - 0,4X_4$$

The solution of interpolation problems and the use of optimization results can be carried out analytically or graphically. Analytical interpretation is carried out according to the algorithm with the creation of computational mini-programs. The obtained dependences became the basis of the algorithm of the computer program, allowing by calculation to determine and predict the properties of products depending on the values of variable factors.

Graphical interpretation of optimization solutions is shown in Figs. 2 and 3. The dependence of the strength of plates at 10% compression (R_{10}) on the technological parameters density of the carpet after the corrugators (X_1), the flow rate of the binder (X_3) and the temperature of the coolant (X_{11}) is shown. In the second graphic scheme, the dependence of the strength of the plates on the separation of layers (R_{oc}) from the same technological parameters is estimated.

Graphic interpretation of optimization decisions based on the results of the first and second stages of the experiment, as well as analytical optimization, is presented in Fig. 4. In the nomogram in sector I, the dependence of the average density of the plates on the average density of the carpet after the corrugator and the degree of compaction of the carpet in the heat treatment chamber is established.

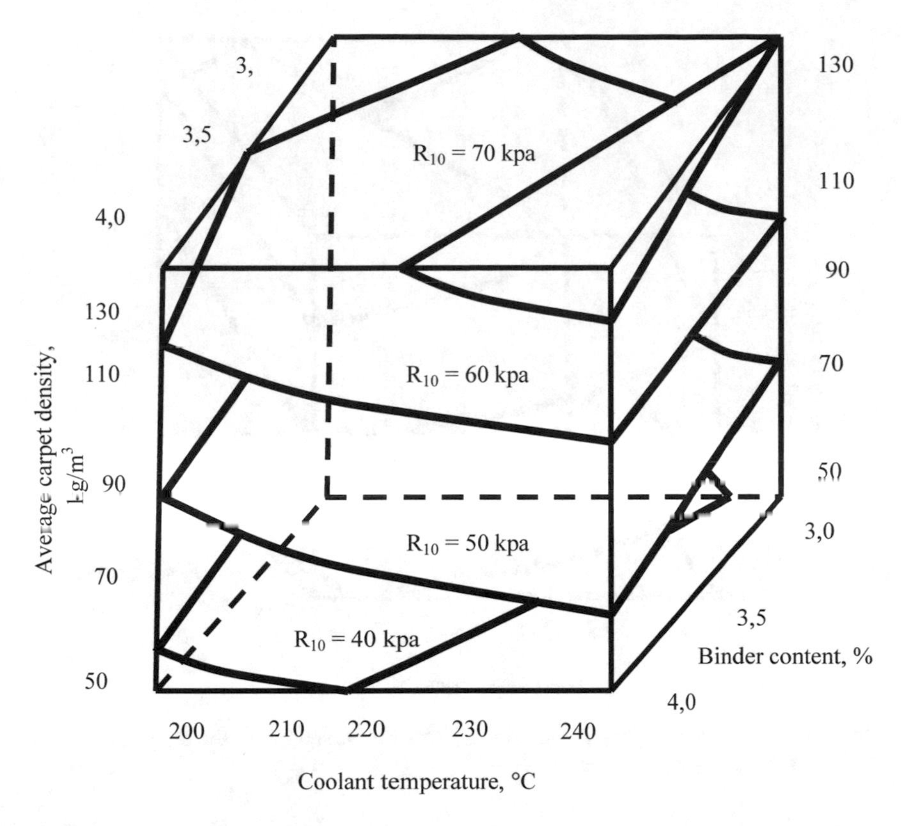

Fig. 2. Dependence of strength of plates at 10% compression ($R_{10}(X_1, X_3, X_{11})$) on technological parameters.

In sector III of the nomogram, it is possible to determine the dependence of the strength of the plates under compression at 10% strain on the average density of the carpet and the consumption of binder. In the IV sector - the dependence of the strength of the layers on the average density of thc carpet and the length of the fiber.

As a result, the four sectors of the nomogram can establish a relationship between the properties of the fiber, the main characteristics of the carpet, binder consumption, molding parameters and the main quality indicators of finished products. In this case, the nomogram can be solved as a direct problem: to predict the properties of finished products depending on the set values of different factors, and the inverse problem: to choose the values of the process parameters, in order to obtain the requirements for strength, density, thermal conductivity.

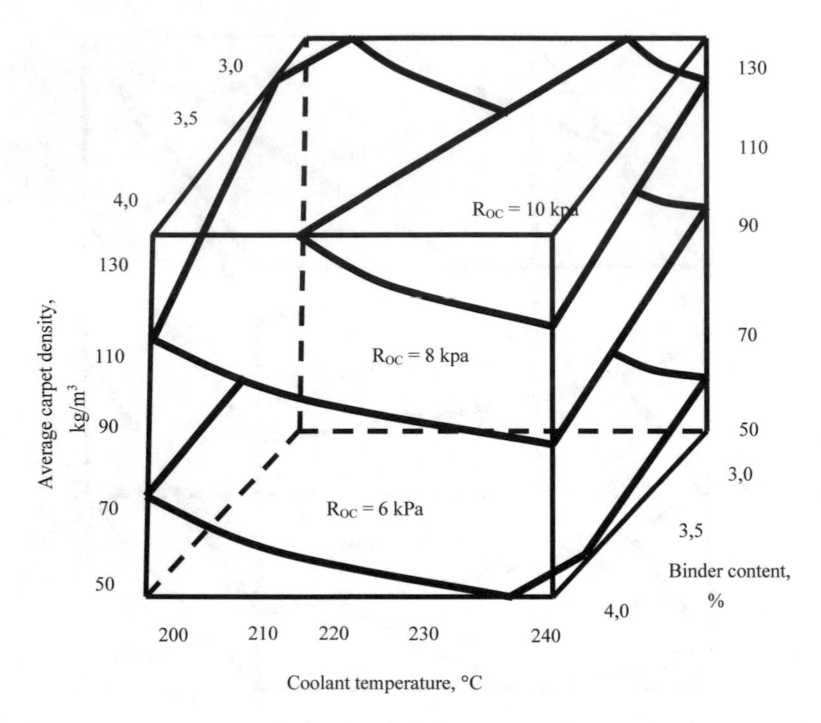

Fig. 3. The dependence of the strength of the plates the separation of the layers (Roc (X1, X3, X11)) from technological parameters.

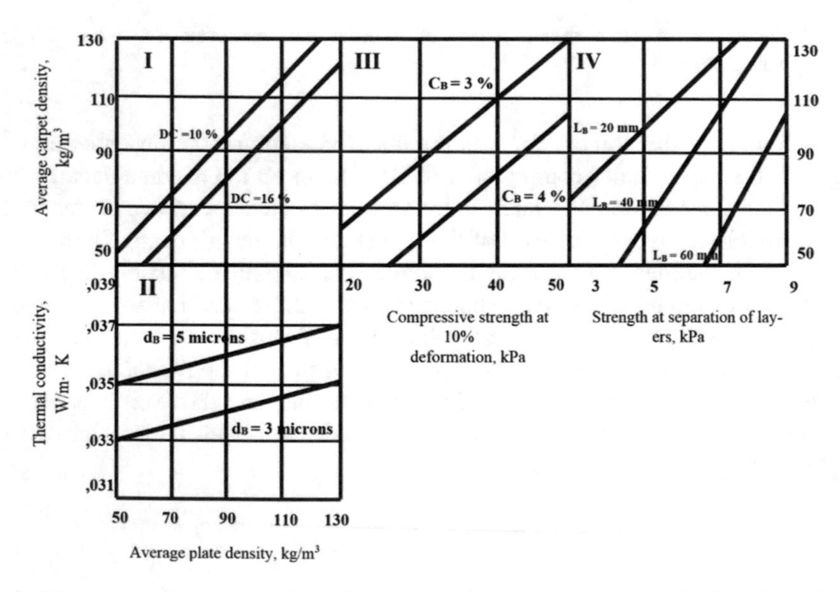

Fig. 4. Nomogram for the selection of parameters for the manufacture of mineral wool plates. Dependence of the main quality indicators of products on the average density of the carpet, the degree of compaction of the carpet (DC), the consumption of binder (C), diameter (d) and length (l) fiber.

5 Conclusions

Empiric-mathematical simulation of technology of mineral wool products and the analytical optimization of process parameters allowed us to study the extent and completeness of the influence of technological factors on the properties of the resulting products, their strength characteristics, density, and conductivity.

As a result of optimization, a clear relationship between the average density of products and their thermal conductivity was established. Both of these functions depend on the average density of the carpet after the corrugator, and these dependences are linear, and the tangents of the angle of inclination of these direct dependences to the abscissa axis have a ratio of 1:0,25. This pattern was taken into account when compiling the II sector of the nomogram, where the dependence of the thermal conductivity of products on their average density and diameter of mineral fibers was determined.

The obtained optimized dependences and their graphical interpretation became the basis for drawing up a nomogram that allows how to predict the properties of products. The obtained dependences became the basis of the algorithm of the computer program, which allows to determine and predict the properties of products depending on the values of variable factors.

References

1. Bessonov, I.V., Starostin, A.V., Oskina, V.M.: About dimensionally stable fibrous insulation. J. Herald MSUCE **3**, 134–139 (2011)
2. Bobrov, Yu.L.: Express method and laboratory equipment for quality evaluation of porous materials. Modern methods of quality control problems and their solutions, pp. 43–44 (1985)
3. Zhuk, H., Zhukov, A.: Normative legal base of environmental assessment of building materials: prospects for improvement. Ecol. Ind. Russia **4**, 52–57 (2018)
4. Arquis, E., Cicasu, C.: Convection phenomen in mineral wool in-stalled on vertical walls. In: Proceedings of the International Scientific-Practical Conference "Effective Heat and Sound Insulating Materials in Modern Construction and Housing and Communal Services", pp. 18–21. Publishing house MSUCE (2006)
5. Asamatdinov, M., Medvedev, A., Zhukov, A., Zarmanyan, E., Poserenin, A.: MATEC Web of Conferences, vol. 193 (2018)
6. Pyataev, E., Medvedev, A., Poserenin, A., Burtseva, M., Mednikova, E., Mukhametzyanov, V.: MATEC Web of Conferences, vol. 251 (2018)
7. Shmelev, S.E.: Ways to choose the optimal set of energy-saving measures. Build. Mater. **3**, 7–9 (2013)
8. Gnip, I., Vaitkus, S., Kersulis, V., Vejelis, S.: Long-term prediction of creep strains of mineral wool slabs under constant compressive stress. Mech. Time Depend Mater. **16**, 31–46 (2012)
9. Medvedev, A., Bobrova, E., Poserenin, A., Zarmanyan, E.: MATEC Web of Conferences, vol. 170 (2018)
10. Gnip, I.Ya., Vaitkus, S.I.: Analytical description of mineral wool creeping deformation during prolonged compression. Build. Mater. **11**, 57–62 (2013)
11. Zhukov, A., Smirnova, T., Zelenshchikov, D., Khimich, A.: Thermal treatment of the mineral wool mat. Adv. Mat. Res. **838–841**, 196–200 (2014)

12. Rumiantcev, B., Zhukov, A., Zelenshikov, D., Chkunin, A., Ivanov, K., Sazonova, Y.: MATEC Web of Conference, vol. 86 (2016)
13. Bessonov, I.V., Zhukov, A.D., Bobrova, E.Yu.: Building systems and features of the use of thermal insulation materials. Mag. Hous. Constr. **7**, 49–52 (2015)
14. Zhukov, A., Bobrova, E., Zelenshchikov, D., Mustafaev, R., Khimich, A.: Insulation systems and green sustainable construction. Adv. Mat. Str. Mech. Eng. **1025–1026**, 1031–1034 (2014)
15. Hlevchuk, V.R., Bessonov, I.V.: On current thermal performance of mineral wool. Problems of construction of thermal physics, climate systems, and energy efficiency in buildings, pp. 127–135 (1998)
16. Rumiantcev, B., Zhukov, A., Bobrova, E., Romanova, I., Zelenshikov, D., Smirnova, T.: The systems of insulation and a methodology for assessing the durability. In: MATEC Web of Conference, vol. 86 (2016)
17. Zhukov, A.D., Konoval'tseva, T.V., Bobrova, E.Yu., Zinovieva, E.A., Ivanov K.K.: MATEC Web of Conference, vol. 251 (2018)

Numeric Verification of the Weak Solutions to the Typical Crocco Limit Problem Using Implicit Difference Scheme of the Second Order

Mikhail Petrichenko, Vitaly Sergeev, Evgeny Kotov, Darya Nemova(✉), and Darya Tarasova

Peter the Great St. Petersburg Polytechnic University, 29, Politekhnicheskaya Street, 195251 St. Petersburg, Russian Federation
darya0690@mail.ru

Abstract. The goal of the current article is to verify the weak analytic solution of the Crocco equation on the [0, 1] interval by comparing it with the numeric solution. The digital experiment has been conducted using the implicit difference scheme of the second order. We demonstrate convergence of the obtained results.

Keywords: Energy efficiency · Sustainable building · Ventilated facade · Porous environment · Filtration · Crocco limit problem · Fluid · Filtration current

1 Introduction

To increase the sustainability of construction it is necessary to consider reducing both the energy consumption by improving the building assemblies that make up the building envelope. This envelope has to guarantee the quality of the environment inside the building, since the exchange between the inner and outer environment takes place through it [1]. Filtration theory that describes locomotion of liquids, gases and mixtures of the above – fluids – in porous and fractured solid (under deformation) environments from the continuum mechanics point of view lays a theoretical basis for the underground hydromechanics and hydrogeology. One of the main methods of investigation of filtration processes in various porous, fractured and fractured porous environments is mathematical modelling that allows describing experimental data, verifying hypotheses, determining consistent patterns, performing numeric modelling and providing information about the objects that are impossible or just too expensive to research experimentally. Topicality of the research of history of development of mathematical models in the filtration theory is based on the notion of importance of underground hydromechanics in such areas of human activity as utilization of underground waters, oil and gas development, design and operation of hydraulic structures, melioration and multiple others. Many scientists have conducted research in the area [1–35]. The goal of this article is to verify the weak analytical solution of the Crocco equation on the [0, 1] interval by comparing it with the numeric solution.

V. Murgul and M. Pasetti (Eds.): EMMFT-2018, AISC 983, pp. 839–848, 2019.
https://doi.org/10.1007/978-3-030-19868-8_82

2 Materials and Methods

2.1 Mathematical Solution to the Typical Crocco Limit Problem

A typical Crocco limit problem is formally stated as follows:

$$\begin{aligned} & yy'' + \gamma x = 0,\ D(y) = (x : x_0 < x < 1), \\ & y'(x_0) = y(1) = 0, \mathrm{Im}(y) = (y : y_0 > y > 0), \end{aligned} \tag{1}$$

where $y_0 := y(x_0) > 0$.

In the classic state of the typical limit problem γ = 1/2, x0 = 0, y0 := y(0). We will further consider only this case. In hydrodynamic applications y(x) (dimensionless quantity) is friction, x (dimensionless quantity) is a longitudinal component of speed in the border layer of the plane that is subject to flat flow in longitudinal direction. Then y0 represents shear friction stress of the wall (Blasius constant) [2]. In the hydraulic filtration theory x (dimensionless quantity) is depth of filtration current through scalar (uniform and isotropic) porous environment, y is Crocco potential that is defined as $y(x) = \int_x^1 s dx', y(1) = y'(0) = 0$, where s is a longitudinal coordinate that is measured along the filtration current. In the filtration problems, the y0 constant is proportional to filtration flow in the current cross-section at the point of exit to the environment [3].

The following statements apply:

1. The Crocco equation has two solution branches, a positive one y + (x) and a negative one y – (x). The negative branch is determined as a solution to the limit problem:

$$\begin{aligned} & 2y_- y_-'' + x = 0,\ D(y_-) = (x : 0 < x < 1), \\ & y_-'(0) = y(1) = 0, \mathrm{Im}(y_-) = (y_- : -y_0 < y_- < 0). \end{aligned}$$

   Meanwhile $y_+(x) + y_-(x) = 0$, $0 < x < 1$. The proof is evident.
   Further we will only consider the positive branch of the solution of the Crocco equation, the $y := y_+$.
2. The solution to the typical Crocco limit problem (1) has the following properties:

$$\begin{aligned} & y'(x) < 0, y''(x) < 0; \\ & y'(x) \underset{x \to 1-0}{\to} -\infty. \end{aligned}$$

Thus y0 > y(x), 0 < x < 1. In order to prove it, we will formally decrease the order of the Crocco equation and reduce it to the following integral equation:

$$2y' = -\int_0^x \frac{tdt}{y(t)} \to y' \le 0, 0 \le x < 1,$$

The integral on the right can be calculated using Bonnet average theorem. We will obtain the following:

$$2yy' = -1/2\left(1-\theta^2\right)x^2, \tag{2}$$

where θ is a proper fraction. We need to transition to the limit where x → 1 − 0.

The solution of the Eq. (2) so that y(1) = 0 is:

$$y^2(x;\theta) = 1/6\left(1-\theta^2\right)\left(1-x^3\right). \tag{3}$$

This solution continuously depends on the magnitude of the θ fraction. Average in relation to θ solution of the Eq. (3) is a weak solution to the typical Crocco limit task that is interpreted as a distribution by θ with distribution density y(x; θ).

Because of (3), the weak solution to the typical Crocco limit problem is:

$$y(x) = 1/3\sqrt{1-x^3}, \tag{4}$$

and $y_0 = y(0) = 1/3$, that is a close rational approximation for the Blasius constant. We can see from the formula (4) that the weak solution can be continued over the negative values of x keeping continuity and smoothness of the solution at x = 0.

The solution to the typical Crocco limit problem is tied to solving the nonlinear integral equation:

$$y(x) = 1/2\left\{\int\limits_0^1 \frac{(1-s)sds}{y(s)} - \int\limits_0^x \frac{(x-s)sds}{y(s)}\right\}. \tag{5}$$

From the Eq. (5), we derive the following expression for the Blasius constant:

$$y_0 := y(0) = 1/2\int\limits_0^1 \frac{(1-s)sds}{y(s)}.$$

The solution of the Eq. (5) can be expressed as the Lagrange series [3]. It is already proven that convergence radius of the Lagrange series is less than 1, and that the series diverges at x → 1 − 0. The alternative method to using Lagrange series can be the following iterative process:

$$y_k(x) = 1/2\left\{\int\limits_0^1 \frac{(1-s)sds}{y_{k-1}(s)} - \int\limits_0^x \frac{(x-s)sds}{y_{k-1}(s)}\right\}, k = 1(1)\infty,$$

where the bottom index denotes the iteration number. Iterated values of the Blasius constant can be determined from the sequence:

$$y_k(0) = 1/2\int\limits_0^1 \frac{(1-s)sds}{y_{k-1}(s)}.$$

We consequently obtain:

$$k = 1 : y_0(x) = y_0 = \sqrt{1/12} = 0,2887;$$
$$k = 2 : y_1(x) \cdot y_0 = 1/12(1 - x^3), y_1(x) = (1 - x^3)/\sqrt{12}, y_1(0) = 1/\sqrt{12};$$
$$k = 3 : y_2(x) = \sqrt{3}\left(\int_0^1 \frac{(1-s)sds}{1-s^3} - \int_0^x \frac{(x-s)sds}{1-s^3}\right) = \sqrt{3}\left\{\begin{array}{l} \ln\sqrt{3} - \frac{x+2}{3}\ln\sqrt{1+x+x^2} \\ + 1/3(1-x)\ln\left(\frac{1}{1-x}\right) - \frac{\pi}{6\sqrt{3}} \\ + \frac{1}{\sqrt{3}}\left(arctg\frac{2x+1}{\sqrt{3}} - \frac{\pi}{6}\right) \end{array}\right\},$$
$$y_2(0) = \sqrt{3}\left(\ln\sqrt{3} - \frac{\pi}{6\sqrt{3}}\right),$$

and so forth. Thus, the first three iterative values of the Blasius constant form the sequence of values:

$$y_0(0) = 1/\sqrt{12} = 0,2887..., y_1(0) = 0,2887..., y_3(0) = 0,4278...,$$

and, on average, within the first three iterations 0.3299 < y(0) < 0.3344. The iteration process calls for trivial and lengthy calculations that can already be seen on the third iteration. It is obvious that any iterative solution has all the main properties of the solution to the limit problem (1):

$$\forall x \in (0,1), \forall k = 1(1)\infty, y_k'(x) < 0, y_k''(x) < 0, y'(x) \underset{x \to 1-0}{\to} -\infty.$$

The inconvenience of the iterative process lies in the cumbersomeness of the iterative solution expressions and in the lack of proof of convergence of the iterative process. Both of this obstacles can be overcome utilizing the difference approximation of the limit problem (1).

2.2 Numeric Solution to the Crocco Limit Problem

During the numeric solution to the problem (1) on the $x \in (0,1)$ interval, the calculated area consists of N intervals with the constant step $h = 1/N\ (x_j = jh,\ j = 0.1, \ldots, N)$. During the sampling process of the Eq. (1) we use the difference scheme of the second order:

$$\frac{y_{j-1} - 2y_j + y_{j+1}}{h^2} + \gamma\frac{x_j}{y_j} = 0 \tag{6}$$

The Eq. (6) is linear relative to y_{j+1} component, thus, if we already know the components y_{j-1}, y_j, j = 1(1)N of the $\boldsymbol{y}$ vector, in order to calculate y_{j+1} we produce a system of linear algebraic equations.

The boundary conditions in the limit problem (1) during the sampling process become:

$$\frac{3y_0 - 4y_1 + y_2}{2h} = 0,\ y_N = 0 \tag{7}$$

If you define differences in the Eqs. (6) and (7) as:

$$\begin{cases} f_0 = 3y_0 - 4y_1 + y_2, \\ f_j = y_{j-1} - 2y_j + y_{j+1} + \gamma h^2 \frac{x_j}{y_j}, \\ f_N = y_N, \end{cases} \tag{8}$$

then the problem (6)–(8) can be written in the equivalent form of the system of linear algebraic equations $\mathbf{F}(\mathbf{y}) = \mathbf{0}$, where $\mathbf{F}$ and $\mathbf{y}$ are vectors:

$$\mathbf{F} = [f_0 \; f_1 \ldots f_N]^T,$$

$$\mathbf{y} = [y_0 \; y_1 \ldots y_N]^T,$$

In order to solve the resulting nonlinear system we use the iterative Newton's method:

$$\mathbf{y}^{(k+1)} = \mathbf{y}^{(k)} + \Delta\mathbf{y}^{(k)}.$$

where $\Delta\mathbf{y}^{(k)}$, $\Delta\mathbf{y}^{(k)} = \left[\Delta y_0^{(k)} \; \Delta y_1^{(k)} \ldots y_N^{(k)}\right]^T$ is a residual vector that is obtained by solving the linearized matrix equation with a Jacobian matrix $\boldsymbol{J}_F(\mathbf{y})$ of the $N + 1$ order:

$$\mathbf{J}_F\left(\mathbf{y}^{(k)}\right)\Delta\mathbf{y}^{(k)} = -\mathbf{F}\left(\mathbf{y}^{(k)}\right) \tag{9}$$

$$J_F\left(\mathbf{y}^{(k)}\right) = \frac{\partial(f_0, \ldots, f_N)}{\partial(y_0, \ldots, y_N)}. \tag{10}$$

We assume that the $\boldsymbol{J}_F(\mathbf{y})$ matrix is well conditioned. Then the system (10) is correct and has the sole solution: $\Delta\mathbf{y}^{(k)} = -\mathbf{J}_F^{-1}(\mathbf{y}^{(k)})\mathbf{F}(\mathbf{y}^{(k)})$.

Substituting (8) to the Eq. (9), considering (10) we get the following:

$$3\Delta y_0^{(k)} - 4\Delta y_1^{(k)} + \Delta y_2^{(k)} = -f_0^{(k)} \tag{11}$$

$$f_0^{(k)} = 3y_0^{(k)} - 4y_1^{(k)} + y_2^{(k)} \tag{12}$$

$$a_j \Delta y_{j-1}^{(k)} + b_j \Delta y_j^{(k)} + c_j \Delta y_{j+1}^{(k)} = -f_j^{(k)} \tag{13}$$

$$a_j = 1, b_j = -2 - \gamma h^2 \frac{x_j}{\left(y_j^{(k)}\right)^2}, c_j = 1, \tag{14}$$

$$f_j^{(k)} = y_{j-1}^{(k)} - 2y_j^{(k)} + y_{j+1}^{(k)} + \gamma h^2 \frac{x_j}{y_j^{(k)}},$$

$$\Delta y_N^{(k)} = -y_N^{(k)}. \tag{15}$$

It is obvious that the system of equations (11)–(15) contains three unknowns in each of the equations and is similar to a tridiagonal system. Usually in such systems of equations, the first and the last equations contain just two unknowns. But in this system the first equation contains three unknowns $\Delta y_0^{(k)}$, $\Delta y_1^{(k)}$, $\Delta y_2^{(k)}$.

In order to exclude $\Delta y_0^{(k)}$, we can present (11) as:

$$\Delta y_0^{(k)} = \frac{1}{3}\left[4\Delta y_1^{(k)} - \Delta y_2^{(k)} - f_0^{(k)}\right].$$

Then, substituting (16) and (14) in (15), when $j = 1$, we get the expression

$$\widehat{b_1}\Delta y_1^{(k)} + \widehat{c_1}\Delta y_2^{(k)} = -\widehat{f_1^{(k)}} \tag{16}$$

where

$$\widehat{b_1} = b_1 + \frac{4}{3}a_1,\ \widehat{c_1} = c_1 - \frac{1}{3}a_1,\ \widehat{f_1^{(k)}} = f_1^{(k)} - \frac{1}{3}f_0^{(k)} \tag{17}$$

The matrix of the system of equations (11), (15) and (16) is tridiagonal. The system can be solved varying j:

$$\Delta y_j^{(k)} = p_j - q_j\Delta y_{j+1}^{(k)} \tag{18}$$

From the Eq. (16) we get:

$$\Delta y_1^{(k)} = \left(-\frac{\widehat{f_1^{(k)}}}{\widehat{b_1}}\right) - \left(\frac{\widehat{c_1}}{\widehat{b_1}}\right)$$

from which we can conclude that

$$p_1 = -\frac{\widehat{f_1^{(k)}}}{\widehat{b_1}}, q_1 = \frac{\widehat{c_1}}{\widehat{b_1}} \tag{19}$$

From the Eqs. (15) and (19), we get:

$$a_j\left(p_{j-1} - q_{j-1}\Delta y_j^{(k)}\right) + b_j\Delta y_j^{(k)} + c_j\Delta y_{j+1}^{(k)} = -f_j^{(k)}$$

that can be re-written as:

$$(b_j - a_jq_{j-1})\Delta y_j^{(k)} + c_j\Delta y_{j+1}^{(k)} = -f_j^{(k)} - a_jp_{j-1},$$

from which we get the following:

$$p_j = \frac{-f_j^{(k)} - a_jp_{j-1}}{b_j - a_jq_{j-1}},\ q_j = \frac{c_j}{b_j - a_jq_{j-1}},\ j = 2, 3, \ldots, N-1 \tag{20}$$

Considering the boundary condition $y_N = 0$ for every k, we get $y_N^{(k)} \equiv 0$ and $\Delta y_N^{(k)} \equiv 0$. Calculating p_j and q_j for $j = 1, 2, \ldots, N-1$ using expressions (18) and (19), we can calculate $\Delta y_j^{(k)}$ for $j = N-1, N-2, \ldots, 0$ using the expression (18).

We keep calculating until the predetermined accuracy ε is achieved:

$$\|\mathbf{\Delta y}^{(k)}\| < \varepsilon,$$

where ||*|| is sup – norm of the residual vector.

3 Results and Discussion

We present the numeric solution to the problem (4)–(5) on the interval $x \in [0, 1]$ with $\gamma = 1$ with the varying amount of steps N with $\varepsilon = 10^{-6}$ in the Fig. 1. As an initial approximation we use the expression $y_0(x) = \frac{\sqrt{1-x^2}}{2}$. The weak solution (4) with Blasius constant equaling 0.4714 (the exact number is 0,47) is shown with the bold solid line.

The Table 1 contains solutions for y(0) at $\gamma = 1$ and varying amount of steps N, and the solutions obtained by other authors [1–3].

Table 1. Calculated values of $y(0)$

N	$\gamma = 0.5$	$\gamma = 1$
$N = 100$	0.339566	0.472865
$N = 1000$	0.335198	0.471984
$N = 10000$	0.332051	0.470430
$N = 100000$	0.332053	0.469855
$N = 1000000$	0.332053	0.469676
Performance [1]	0.332057	0.469600
Performance [2]	0.3320573362	0.4695999889
Performance [3]	0.332057	0.469599

In order to continue the solution to the problem (1) into the area $x < 0$ we use the differential scheme of the second order (6) with the following boundary conditions:

$$y(0) = \tilde{y}_0, y'(0) = 0 \tag{21}$$

where $\tilde{y}_0$ is the value $y(0)$ from the solution obtained on the interval $x \in [0, 1]$.

After the sampling process, the boundary conditions become (21):

$$y_0 = \tilde{y}_0, \ \frac{y_0 - y_{-1}}{h} = 0$$

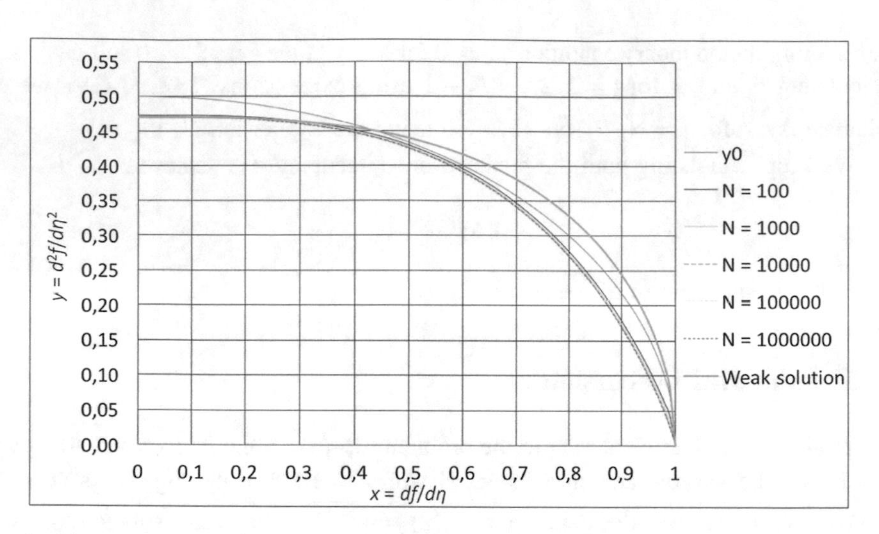

Fig. 1. The numeric solution to the Crocco problem on the interval $x \in [0, 1]$, $\gamma = 1$, y (0) = 0,47.

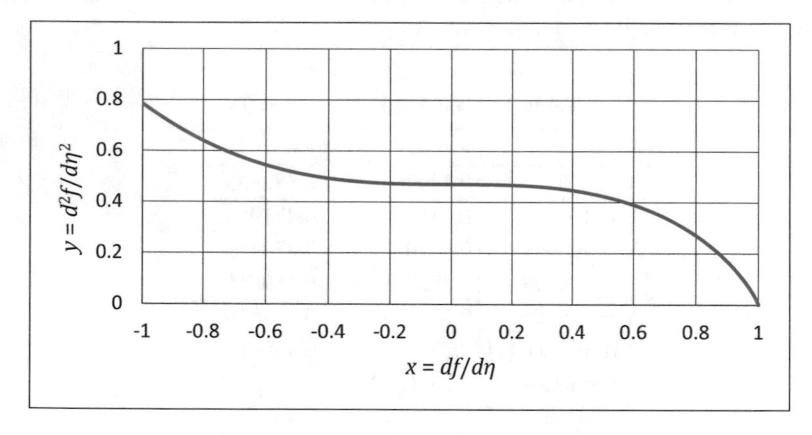

Fig. 2. The solution to the Crocco problem on the interval $x \in [-1, 1]$ with $\gamma = 1$

Where you can notice that $y_{-1} = y_0 = \tilde{y}_0$.

Considering (11):

$$y_j = 2y_{j+1} - y_{j+2} - \gamma h^2 \frac{x_{j+1}}{y_{j+1}},\ j = -2, -3, \ldots, -M \tag{22}$$

where M is the amount of calculation steps in the area $x < 0$.

In the Fig. 2 we demonstrate the numeric solution to the limit problem (1) extended on the interval $x \in [-1, 1]$ with $\gamma = 1$.

4 Conclusions

In this article we have conducted the verification of a weak analytic solution of the Crocco equation on the [0, 1] interval by comparing it with the numeric solution.

Using the results obtained in this article we can make the following conclusions:

- on the interval [0, 1) the numeric solution approximates the weak one on the norm C1 [0, 1);
- at the right endpoint, $x \to 1 - 0$, the numeric solution does not feature a derivative gap at any magnitude of the sampling interval of the [0, 1] range interval. Possibly, the reason for that lies in the low order of the difference scheme and its symmetry.

The results of the work can be applied in various technical fields [26, 27, 33] . One direct application is the development of new ventilated façade systems [35].

References

1. Loktionova, E.A., Miftakhova, D.R.: Fluid filtration in the clogged pressure pipelines. Mag. Civ. Eng. **76**(8), 214–224 (2017)
2. Pu, T.M., Zhou, C.H.: An immersed boundary/wall modeling method for RANS simulation of compressible turbulent flows. Int. J. Numer. Methods Fluids **87**(5), 217–238 (2018)
3. Varin, V.P.: Asymptotic expansion of Crocco solution and the Blasius constant. Comput. Math. Math. Phys. **58**(4), 517–528 (2018)
4. Asaithambi, A.: A series solution of the Falkner-Skan equation using the Crocco-Wang transformation. Int. J. Mod. Phys. C **28**(11), 1750139 (2017)
5. Yacouba, S., Tambue, A.: Null controllability and numerical method for Crocco equation with incomplete data based on an exponential integrator and finite difference-finite element method. Comput. Math Appl. **74**(5), 1043–1058 (2017)
6. Polyanin, A.D., Zaitsev, V.F.: Transformations, properties, and exact solutions of unsteady axisymmetric boundary layer equations for non-Newtonian fluids. Theor. Found. Chem. Eng. **51**(4), 437–447 (2017)
7. Gao, Z.-X., Jiang, C.-W., Lee, C.-H.: A new wall function boundary condition including heat release effect for supersonic combustion flows. Appl. Therm. Eng. **92**, 62–70 (2016)
8. Asaithambi, A.: Numerical solution of the Blasius equation with Crocco-Wang transformation. J. Appl. Fluid Mech. **9**(5), 2595–2603 (2016)
9. Gong, S., Guo, Y., Wang, Y.-G.: Boundary layer problems for the two-dimensional compressible Navier-Stokes equations. Anal. Appl. **14**(1), 1–37 (2016)
10. Duan, L., Martín, M.P.: Direct numerical simulation of hypersonic turbulent boundary layers. Part 4. Effect of high enthalpy. J. Fluid Mech. **684**, 25–59 (2011)
11. Zuppardi, G., Morsa, L.: Evaluation of non-equilibrium by the Crocco theorem. Proc. Inst. Mech. Eng. Part G J. Aerosp. Eng. **225**(9), 985–994 (2011)
12. Ivanov, N., Zasimova, M., Smirnov, E., Markov, D.: Evaluation of mean velocity and mean speed for test ventilated room from RANS and LES CFD modeling. In: E3S Web of Conferences, vol. 85, p. 0200 (2019)
13. Evaluation of mean velocity and mean speed for test ventilated room from RANS and LES CFD modeling. E3S Web Conf. **85** (2019). Article no. 0200
14. Baranov, M.A., Klimchitskaya, G.L., Mostepanenko, V.M., Velichko, E.N.: Fluctuation-induced free energy of thin peptide films. Phys. Rev. E **99**(2), 022410 (2019)

15. Shishkina, E.V., Gavrilov, S.N., Mochalova, Y.A.: Non-stationary localized oscillations of an infinite Bernoulli-Euler beam lying on the Winkler foundation with a point elastic inhomogeneity of time-varying stiffness. J. Sound Vib. **440**, 174–185 (2019)
16. Klimchitskaya, G.L., Mostepanenko, V.M., Nepomnyashchaya, E.K., Velichko, E.N.: Impact of magnetic nanoparticles on the Casimir pressure in three-layer systems. Phys. Rev. B **99**(4), 045433 (2019)
17. Smirnov, S.I., Bulovich, S.V., Smirnov, E.M.: Calculation of unsteady two-phase quasi-onedimensional channel flow based on the two-fluid model and the artificialviscosity numerical scheme. J. Phys. Conf. Ser. **1135**(1), 012103 (2018)
18. Matyushenko, A.A., Garbaruk, A.V.: Validation of the SST-HL turbulence model for separated flows and flows around airfoils. J. Phys. Conf. Ser. **1135**(1), 012097 (2018)
19. Ivanov, N.G., Zasimova, M.A.: Mean air velocity correction for thermal comfort calculation: assessment of velocity-to-speed conversion procedures using large Eddy simulation data. J. Phys. Conf. Ser. **1135**(1), 012106 (2018)
20. Stabnikov, A.S., Garbaruk, A.V.: Analysis of the abilities of algebraic laminar-turbulent transition models. J. Phys. Conf. Ser. **1135**(1), 012104 (2018)
21. Guseva, E.K., Strelets, M.K., Travin, A.K., Burnazzi, M., Knopp, T.: Zonal RANS-IDDES and RANS computations of turbulent wake exposed to adverse pressure gradient. J. Phys. Conf. Ser. **1135**(1), 012092 (2018)
22. Iben, U., Makhnov, A., Schmidt, A.: Numerical study of a vapor bubble collapse near a solid wall. J. Phys. Conf. Ser. **1135**(1), 012096 (2018)
23. Grebenikova, N.M., Smirnov, K.J., Davydov, V.V., Rud, V.Y.U., Artemiev, V.V.: Features of monitoring the state of the liquid medium by refractometer. J. Phys. Conf. Ser. **1135**(1), 012055 (2018)
24. Smirnov, M., Kirillov, A., Lapshin, K., Porshnev, G., Laskin, A.: Design criteria for novel supersonic nozzles with high pitch-chord ratio. MATEC Web Conf. **245**, 09012 (2018)
25. Chernyshov, M., Tyapko, A.: Optimal regular reflection of shock and blast waves. MATEC Web Conf. **245**, 12005 (2018)
26. Egorov, M., Suslov, V.: Fluid mechanics tests of separator-reheater turbines. MATEC Web Conf. **245**, 09009 (2018)
27. Porubov, A.V., Bondarenkov, R.S., Bouche, D., Fradkov, A.L.: Two-step shock waves propagation for isothermal Euler equations. Appl. Math. Comput. **332**, 160–166 (2018)
28. Smirnov, E.M., Smirnovsky, A.A., Schur, N.A., Zaitsev, D.K., Smirnov, P.E.: Comparison of RANS and IDDES solutions for turbulent flow and heat transfer past a backward-facing step. Heat and Mass Transfer/Waerme- und Stoffuebertragung **54**(8), 2231–2241 (2018)
29. Boldyrev, Y.Y.: Variational Rayleigh problem of gas lubrication theory. Low compressibility numbers. Fluid Dyn. **53**(4), 471–478 (2018)
30. Guseva, E.K., Gritskevich, M.S., Garbaruk, A.V.: Assessment of two approaches to accelerate RANS to les transition in shear layers in the framework of ANSYS-FLUENT. J. Phys. Conf. Ser. **1038**(1), 012134 (2018)
31. Ivanov, N.G., Zasimova, M.A.: Large Eddy simulation of airflow in a test ventilated room. J. Phys. Conf.Ser. **1038**(1), 012136 (2018)
32. Iben, U., Makhnov, A., Schmidt, A.: Numerical investigation of cavitating flows with liquid degassing. J. Phys. Conf. Ser. **1038**(1), 012128 (2018)
33. Logunov, S.E., Davydov, V.V., Vysoczky, M.G., Mazing, M.S.: New method of researches of the magnetic fields force lines structure. J. Phys. Conf. Ser. **1038**(1), 012093 (2018)
34. Bulovich, S.V., Smirnov, E.M.: Experience in using a numerical scheme with artificial viscosity at solving the Riemann problem for a multi-fluid model of multiphase flow. AIP Conf. Proc. **1959**, 050007 (2018)
35. Iben, U., Makhnov, A., Schmidt, A.: Three-dimensional numerical simulations of turbulent cavitating flow in a rectangular channel. AIP Conf. Proc. **1959**, 050013 (2018)

Decision Support System for Wood Fuel Production and Logistics

Anton Sokolov(✉) and Vladimir Syunev

Petrozavodsk State University, Lenin str. 33, 185910 Petrozavodsk, Russia
a_sokolov@psu.karelia.ru

Abstract. The topic of this article is the use of novel informational technologies and mathematical methods for the improvement of wood fuel production and logistics. Overall set of algorithms, optimization models, and interface is proposed aiming to support of bioenergy development in Russia. The developed decision support system for wood fuel production and logistics supports three different methods: roadside chipping uses a machinery set, including a mobile chipper and chip trucks; terminal chipping uses the transporting of not chips, but loose logging residues; bundling - the logging residues are compressed to composite residue logs using a mobile bundler in the terrain. Since the decision support system takes into account economic aspects and, if necessary, warns of machines wantage and gives recommendations for production organizing, companies increase productivity and improve the economy of the entire supply chain of wood fuel. The decision support system is suitable for various levels of production planning, including solving of strategic infrastructure tasks related to the development of bioenergy in Russia.

Keywords: GIS · Simulation · Optimization · Wood harvesting · Woody biomass · Wood fuels · Bioenergy

1 Introduction

Russia has a big potential in the field of forest bioenergy. Only in the North-West of the Russian Federation it is possible to harvest significant volumes of woody biomass in the form of firewood, logging residues (waste), stump and root wood, small diameter stems from thinning, which could allow the produce the wood fuel (WF) in the amount of 208 TWh, sufficient to cover 20% of the total energy needs of the region [1, 2].

The collection of woody biomass, production and transportation of wood fuel are closely related to roundwood harvesting operations. Method of the roundwood harvesting determines which kind of biomass can be used as a raw material for producing fuel chips, at what stage and in what amounts the wood biomass is formed, as well as the biomass distribution by area [3, 4]. Depending on the machinery set and the source of woody biomass, about 12 alternative organization schemes of wood fuel production process are most common [2] and 7 of them are based on cut-to-length harvesting, 3 - on harvesting of the whole trees, and two - on harvesting of stumps and small diameter stems from thinning.

V. Murgul and M. Pasetti (Eds.): EMMFT-2018, AISC 983, pp. 849–857, 2019.
https://doi.org/10.1007/978-3-030-19868-8_83

To support the decision-making on the selection of suitable wood biomass harvesting and processing methods, a decision support system for the production and logistics of wood fuel was developed [5]. This system has become a new part of an already existing system for optimization of forest roads network and timber transportation [6–9]. The core of the developed system is the simulation model of the wood fuel initial processing and transportation. This model will be described in the article.

2 Methods and Models

The implementation of the simulation model depends on the method used for initial processing and transportation of the wood fuel. The developed system supports three different methods:

1. **Roadside** - chipping uses a machinery set, including a mobile chipper and chip trucks.
2. **Terminal chipping** - uses the transporting of not chips, but loose logging residues for further processing by consumers. In this case, all consumers must have a chipper. This method uses residues trucks or chip trucks.
3. **Bundling** - the logging residues are compressed to composite residue logs (CRLs) using a mobile bundler in the terrain. These CRLs have similar dimensions as round wood, and are forwarded to roadside, and transported by timber trucks to the end users, where chipping takes place. In this case, all consumers also must have a chipper.

At the same time, the work of several machinery sets using one or several methods can be simulated. Therefore, to ensure the modeling, at first, information about the number of such sets, their type and characteristics should be inputted into the system. This is carried out with the help of a special dialogue (Fig. 1), where all the machinery sets are arranged in order of priority. Setting the priority is necessary, since modeling of the machinery set work in the system is performed sequentially, first for the first in the list, then for the second, etc. Thus, with insufficient raw materials on the harvesting sites, the probability of being idle is higher for sets located lower in the priority list.

Input or adjustment of the parameters of the machinery sets is carried out in the dialog shown in Fig. 2. Here you need to input the name of the set, select the method and input the parameters, as well as the working schedule: the number of shifts, the duration of the shift, the average utilization rate. Additionally, the current position of the machinery set can be inputted. Any machinery set may be temporarily excluded from the simulation.

Figure 2 shows the state of the dialogue for input the characteristics of a machinery set with a mobile chipper at the roadside. For this kind of machinery set, a model of the chipper and its average hourly productivity in bulk m^3 of chips per hour, the model of chip trucks, their number, the amount of transported chips, the garage in which the trucks are based, as well as the time required for unloading should be inputted.

When defining the characteristics of the terminal chipping machinery set, only the parameters of the chip trucks are inputted. Unlike the technology using the chipper, the

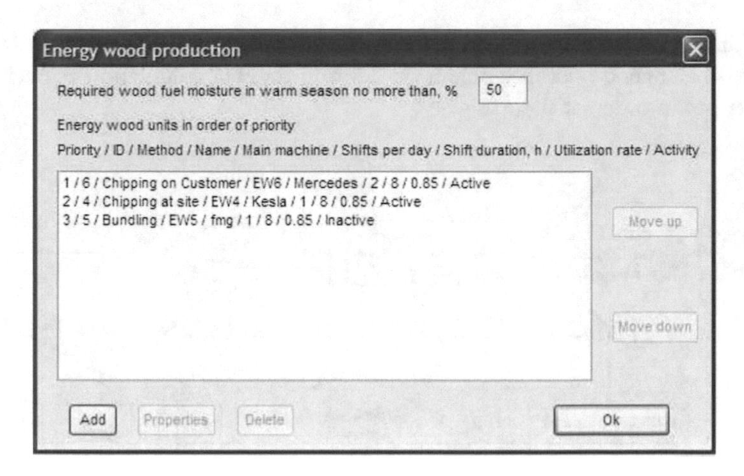

Fig. 1. Energy wood production dialogue.

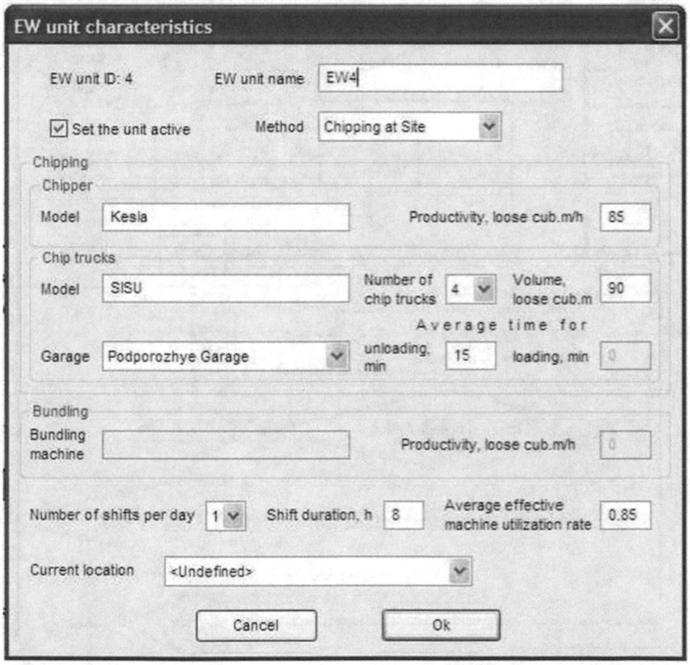

Fig. 2. Wood fuel production machinery sets characteristics.

average truck loading time is additionally required here, and the current position is not required.

When the bundler used, its model, productivity and, if necessary, its current position are inputted.

The diagram of the developed simulation algorithm is presented in Fig. 3. The algorithm is launched once per model day during a given calculation period for each machinery set in order of their priority.

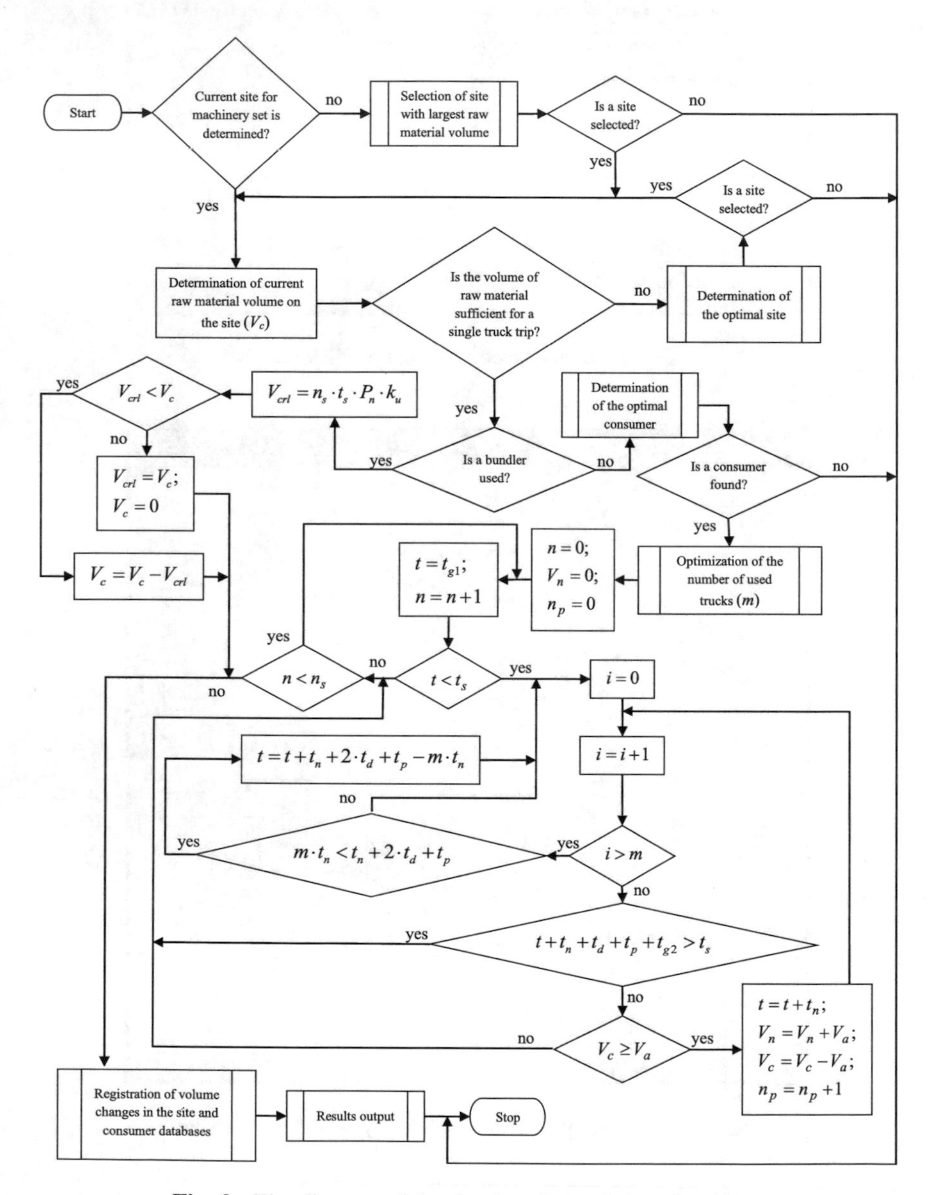

Fig. 3. The diagram of the developed simulation algorithm.

In the diagram, the following notation is used:

- V_c – the current volume of raw materials (logging residues and/or firewood) at the site;
- V_{crl} – the volume of composite residue logs (CRLs) produced by the bundler for the model day;
- n_s – shifts number;
- t_s – shifts duration;
- P_n – bundler average hourly productivity;
- k_u – bundler effective utilization rate;
- m – optimal number of trucks for chips or logging residues transportation;
- n – current shift identifier;
- Vn – current volume of wood chips or logging residues produced by the machinery set;
- n_r – current number of runs made by trucks;
- t – the current time from the beginning of the shift;
- t_{g1} – driving time from the garage to the current site;
- t_{g2} – driving time from the optimal consumer to the garage;
- t_n – loading time of one truck;
- t_d – driving time from the current site to the optimal consumer;
- t_p – unloading time of one truck;
- V_a – volume carried by one truck.

For the correct implementation of this algorithm, the user must previously define the harvesting sites on which the woody biomass will be harvested and processed. If the use of logging residues is required, it is necessary to set the corresponding switch in the appropriate position in the harvesting site characteristics dialog (marked in Fig. 4). If a chipper is used and it is required to process wood into chips at the harvesting site, in this dialogue should not input the consumer for firewood, then their processing into wood chips along with or separately from logging residues will be simulated. If it is necessary to process firewood into chips at harvesting site separately from logging residues, you should change the position of the switch and not input consumers for firewood. If at least one consumer of firewood is inputted, the transportation of firewood to a specified consumer by timber trucks along with roundwood will be simulated, and its processing into chips will not be simulated.

The implementation of the algorithm begins with the definition of a harvesting site, on which the current machinery set will work during this model day. If in the previous model days the machinery set was idle, the site for harvesting of woody biomass is determined. It is the site, on which the harvesting of roundwood is completed and the woody biomass is dried, and the volume of raw materials suitable for this machinery set is the largest. In this case, for the machinery set with the chipper, the optimization problem (1) is solved. For machinery set with the bundler, as well as for transporting of logging residues, problem (2) is solved. If in the previous days the machinery set worked on any of the sites, it keeps on this site.

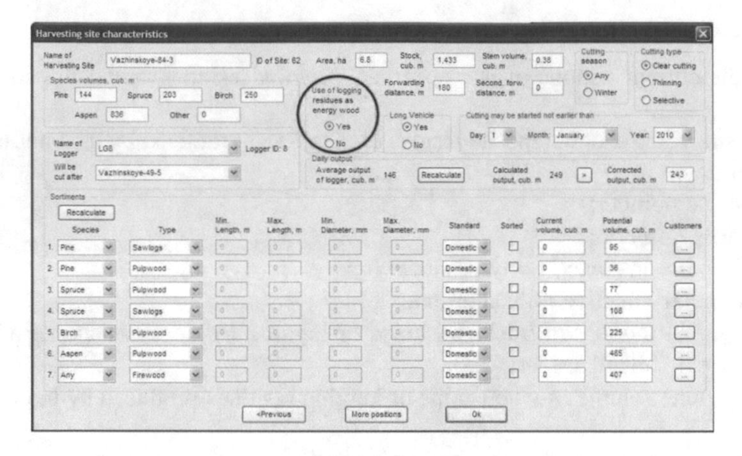

Fig. 4. Harvesting sites characteristics dialogue.

$$\begin{cases} \sum_{i=1}^{N} x_i \cdot (\mu_i \cdot V_{ri} + V_{di}) \to max; \\ \sum_{i=1}^{N} x_i \cdot M_i \le M_{max} \\ \sum_{i=1}^{N} x_i \cdot V_{1i} = 0; \\ \sum_{i=1}^{N} x_i \cdot \phi_i = 0; \\ \sum_{i=1}^{N} x_i \le 1; \\ x_i \in \{0,1\}; i = 1,2,\ldots,N; \\ \mu_i \in \{0,1\}; i = 1,2,\ldots,N; \\ \phi_i \in \{0,1\}; i = 1,2,\ldots,N, \{ \end{cases} \quad (1)$$

where

- N – total number of harvesting sites;
- x_i – controlled variables taking the value 1, if the i-th site is included in the harvesting plan at this stage, and 0 otherwise;
- μ_i – variables that take the value 1 if the user has assigned the i-th site to harvesting of logging residues and 0 otherwise;
- V_{ri} – current volume of logging residues in the i-th site;
- V_{di} – current volume of firewood in the i-th site;
- M_i – current humidity of biomass on the i-th site;
- $M_{\max}$ – maximum allowable humidity of woody biomass (in the warm season);
- V_{1i} – current volume of roundwood in the i-th site;
- φ_i – variables that take the value 1, if the machinery set currently works on the i-th site at the harvesting and initial processing of the fuel wood and 0 otherwise;
- t_i – driving time from the current site to the i-th one;
- V_{ci} – total amount of raw materials suitable for processing by the current machinery set on the i-th site;
- V_a – volume transported by one truck of the current machinery set in one run.

$$\begin{cases} \sum_{i=1}^{N} x_i \cdot \mu_i \cdot V_{ri} \rightarrow max; \\ \sum_{i=1}^{N} x_i \cdot M_i \leq M_{max} \\ \sum_{i=1}^{N} x_i \cdot V_{1i} = 0; \\ \sum_{i=1}^{N} x_i \cdot \phi_i = 0; \\ \sum_{i=1}^{N} x_i \leq 1; \\ x_i \in \{0,1\}; i = 1,2,\ldots,N; \\ \mu_i \in \{0,1\}; i = 1,2,\ldots,N; \\ \phi_i \in \{0,1\}; i = 1,2,\ldots,N. \end{cases} \tag{2}$$

At the next step of the algorithm, the volume of raw materials at the current site V_c is determined. When using a chipper, it is the sum of the volumes of logging residues and firewood located currently at the site, and in cases of using a bundler or transporting of logging residues, – only the amount of logging residues.

Further, the available volume V_c is compared with the volume transported in one run by trucks belonging to this machinery set. If V_c is less than the volume transported in one run, the machinery set is transferred to another harvesting site, and V_c is compared to the load capacity of the truck with the lowest capacity among all the machinery sets. If V_c is less than this value, - the harvesting site is closed. When the harvesting site is changing, the machinery set is transferred to a site that is closest in time, where the harvesting of roundwood is completed, and the volume of raw materials of allowable humidity is sufficient to carry out at least one trucks run. To do this, the optimization problem (3) is solved.

$$\begin{cases} \sum_{i=1}^{N} x_i \cdot \mu_i \cdot t_i \rightarrow min; \\ \sum_{i=1}^{N} x_i \cdot V_{ci} \geq V_a; \\ \sum_{i=1}^{N} x_i \cdot M_i \leq M_{max} \\ \sum_{i=1}^{N} x_i \cdot V_{1i} = 0; \\ \sum_{i=1}^{N} x_i \cdot \phi_i = 0; \\ \sum_{i=1}^{N} x_i \leq 1; \\ x_i, \mu_i, \phi_i \in \{0,1\}; i = 1,2,\ldots,N. \end{cases} \tag{3}$$

The continuation of the algorithm depends on the type of machinery set being modeled. In the case of a bundler, the volume of composite residue logs (CRLs) produced by it per day is determined in accordance with a given productivity. This volume is recorded at the database of the harvesting site and becomes available for further transportation. After this, the algorithm ends.

In cases with a chipper or logging residues transportation, the optimum consumer is determined:

$$\begin{cases} \sum_{i=1}^{K} y_i \cdot t_i \rightarrow min; \\ \sum_{i=1}^{K} y_i \cdot (V_{mi} - V_{mni}) \geq V_a; \\ \sum_{i=1}^{K} y_i \leq 1; \\ y_i \in \{0,1\}; \; i = 1,2,\ldots,K, \end{cases} \tag{4}$$

where

- K – total number of consumers;
- y_i – controlled variables taking the value 1 if the i-th consumer becomes an unloading point for the current harvesting site, and 0 otherwise;
- t_i – driving time from the current site to the i-th consumer;
- V_{mi} – the volume of production (chips or logging residues), scheduled to receiving by the *i*-th consumer in the current month;
- V_{mni} – the volume of production (chips or logging residues), already received by the i-th consumer in the current month.

At the next step of the algorithm, the optimal number of trucks used for the transportation of chips or logging residues is determined.

In the case of using the chipper, the task of maximizing its load is solved with synchronous minimizing the waiting time of trucks in the loading queue:

$$\begin{cases} k_{lf} = \left| \begin{array}{l} \frac{m \cdot t_{п}}{t_{п}+2 \cdot t_d+t_{р}}, \text{ при } \frac{m \cdot t_{п}}{t_{п}+2 \cdot t_d+t_{р}} \leq 1 \\ 1, \text{ при } \frac{m \cdot t_{п}}{t_{п}+2 \cdot t_d+t_{р}} > 1 \end{array} \right. \rightarrow max; \\ t_q = t_{п} + 2 \cdot t_d + t_{р} - m \cdot t_{п} \rightarrow min; \\ k_{lf} \leq 1; \\ m \leq n_a, \{ \end{cases} \tag{5}$$

where

- k_{lf} – chipping machine load factor in cycle;
- t_q – waiting time of trucks in the loading queue;
- m – controlled variable - the optimal number of trucks involved;
- n_a – total number of trucks in the machinery set.

When using the logging residues transportation, the waiting time of trucks in the loading queue are minimized:

$$\begin{cases} t_q = t_{п} + 2 \cdot t_d + t_{р} - m \cdot t_{п} \rightarrow min; \\ m \leq n_a. \end{cases} \tag{6}$$

The loading time t_n is determined in accordance with the capacity of the chipper (in case of its use) or is set by the user (in the case of logging residues transportation).

After this, the run-by-run simulation of the machinery sets operation is performed as shown in the diagram (see Fig. 3). Harvested and transported volumes, load time of machines and other indicators are determined. The implementation of the algorithm ends when a point in time is reached which corresponds to the end of the last change of model days.

3 Conclusion

The logistic approach to the wood fuel procurement has not yet been well developed in Russia. Decision support systems developed in countries with long-term experience in using wood fuel in bioenergy, such as Finland and Sweden, are not always applicable in the Russian context. The reason for this is the specific organizational structure of logging companies, poor condition and maintenance of roads, etc.

The decision support system developed using GIS technologies is a tool that helps the wood harvesting companies to make the most profitable decisions on the organization of the harvesting, transportation and processing of woody biomass to wood fuel. The use of this system allows to increase the efficiency of introducing the wood fuel production, to reduce the cost of harvesting and transportation and to improve the use of the machinery fleet. Wood harvesting companies receives enough information to make reliable operational, tactical and strategic decisions.

Since the decision support system takes into account economic aspects and, if necessary, warns of machines wantage and gives recommendations for production organizing, companies increase productivity and improve the economy of the entire supply chain of wood fuel. The decision support system is suitable for various levels of production planning, including solving of strategic infrastructure tasks related to the development of bioenergy in Russia.

References

1. Gerasimov, Y., Karjalainen, T.: Energy wood resources in Northwest Russia. Biomass Bioenerg. **35**, 1655–1662 (2011)
2. Goltsev, V., Ilavský, J., Karjalainen, T., Gerasimov, Y.: Potential of energy wood resources and technologies for their supply in Tihvin and Boksitogorsk districts of the Leningrad region. Biomass Bioenerg. **34**, 1440–1448 (2010)
3. Kanzian, C., Holzleitner, F., Stampfer, K., Ashton, S.: Regional energy wood logistics – optimizing local fuel supply. Silva Fennica **43**(1), 113–128 (2009)
4. Yamamoto, H., Junichi, F., Kenji, Y.: Evaluation of bioenergy potential with a multi-regional global-land-use-and-energy model. Biomass Bioenerg. **21**, 185–203 (2001)
5. Gerasimov, Y., Sokolov, A.: Support system of production and logistics of wood fuel: methodology and models. Conifers Boreal Area **31**(1–2), 200–207 (2012)
6. Gerasimov, Y.Y., Sokolov, A.P., Karjalainen, T.: GIS-based decision-support program for planning and analyzing short-wood transport in Russia. Croatian J. Forest Eng. **29**, 163–175 (2008)
7. Gerasimov, Y., Sokolov, A., Syunev, V.: Optimization of industrial and fuel wood supply chain associated with cut-to-length harvesting. Syst. Methods Technol. **3**(11), 118–124 (2011)
8. Sokolov, A.P., Syunev, V.S.: Comparative study of wood fuel supply chains in Russian Karelia. In: Proceedings of 18th International Multidisciplinary Scientific Geoconference SGEM 2018, vol. 18, no. 4.1, pp. 227–234 (2018)
9. Gerasimov, Yu.Yu., Sokolov, A.P.: Decision making toolset for woody biomass supply chain in Karelia. Appl. Mech. Mater. **459**(4), 319–324 (2014)

Cold-Bonded Fly Ash Lightweight Aggregate Concretes with Low Thermal Transmittance: Review

Yurii Barabanshchikov[1], Iyliia Fedorenko[2], Sergey Kostyrya[2], and Kseniia Usanova[1(✉)]

[1] Peter the Great St. Petersburg Polytechnic University, 195251 St. Petersburg, Russia
plml@mail.ru

[2] Vedeneev VNIIG, JSC, 195220 St. Petersburg, Russia

Abstract. The paper presents a brief study of low thermal transmittance concretes with fly ash aggregate obtained from fly ash of coal-fired thermal power plants. The research analyzes the cold-bonded fly ash aggregate from the Novosibirskaya GRES Thermal Power Plant (Novosibirsk, Russia). The paper experimentally investigates bulk density, water absorption, bulk crushing resistance, resistance to freezing and thawing. The experimental study shows that the water absorption of the coarse aggregate reaches 6.1%, while a cylinder compressive strength of aggregate accounts for 6.2 MPa. The obtained characteristics show that the analyzed aggregates can be effectively used in wide classes of structural concrete.

Keywords: Fly ash aggregate · Cold-bonded fly ash aggregate · Sintered fly ash aggregate · Pelletized fly ash · Granulated fly ash · Concrete · Cement · Bulk density · Water absorption · Bulk crushing resistance · Resistance to freezing · Thermal transmittance

1 Introduction

1.1 Cold-Bonded and Sintered Fly Ash Aggregate

Fly ash aggregates are used in high performance concrete, self-compacting concrete and lightweight concrete with low thermal transmittance. Fly ash aggregates are synthesize in two ways. The first one is the pelletization of fly ash, followed by a sintering fresh aggregate pellets at high temperatures in furnaces (sintered fly ash aggregates). The second one is the cold bonding pelletization of fly ash through moistening in a revolving tilted pan (cold-bonded fly ash aggregates).

The combined use of sintered fly ash aggregates and cold-bonded fly ash aggregates in concrete mixtures was studied in the paper [1]. Studies [2–5] investigated the differences between properties of the lightweight concretes including either cold bonded or sintered fly ash aggregates.

V. Murgul and M. Pasetti (Eds.): EMMFT-2018, AISC 983, pp. 858–866, 2019.
https://doi.org/10.1007/978-3-030-19868-8_84

1.2 Cold-Bonded Fly Ash Lightweight Aggregate Concretes

Cold-bonded fly ash lightweight aggregate concretes were previously investigated in [6–24]. In these articles, the following characteristics of concrete were most commonly studied: compressive strength, split tensile strength, flexural tensile strength, stiffness and drying shrinkage (Table 1). In addition, the workability of the concrete mixture was studied in researches [8, 13]. However, such properties of the concrete mixture as stratification, water segregation and concrete segregation were not studied in the works presented in the review.

The concrete properties that are considered in the reviewed articles are listed in Table 1.

Table 1. Properties of concrete, which are considered in the reviewed articles.

Concrete properties	Reference to article
Compressive strength	[6–9, 11–16, 18–22, 24]
Split tensile strength	[7, 11, 14–16, 21, 22]
Flexural tensile strength	[16, 22]
Water penetration	[18]
Stiffness	[14–16, 21, 22]
Water absorption	[9, 20]
Gas permeability	[6, 18]
Chloride ion permeability	[6]
Porosity	[9]
Bond between concrete and steel	[10, 11]
Drying shrinkage	[7, 12, 14, 15]
Creep	[14–16]

Based on the abovementioned articles, some important findings can be obtained:

- lightweight concrete with a 100% fly ash aggregate as coarse aggregate and a specific gravity of 1773 kg/m^3 reaches the strength of 21.3 MPa within 24 h [6];
- the use of steel fiber significantly increases split tensile strength, flexural tensile strength, fracture energy and characteristic length of the concrete. However, the addition of steel fiber does not significantly change the compressive strength of concrete [22];
- increasing the content of fly ash aggregate reduces the amount of superplasticizers for workability [23];
- shrinkage cracking of lightweight concrete is considerably lower than that of normal weight concrete [15];
- a mathematical model was proposed in research article [10] for predicting the bond strength between a reinforcing steel bar and the cold-bonded fly ash lightweight aggregate concretes. The results revealed that the utilization of steel fiber enhanced the bond strength and the ductility of pull-out failure;

- the investigation [14] deals with the reduction shrinkage cracking of the cold-bonded fly ash lightweight aggregate concretes. Results indicated that the improvement of the lightweight aggregate properties extended the cracking time of the concretes resulting in finer cracks associated with the lower free shrinkage. Moreover, there was a noticeable increase in the compressive and splitting tensile strengths, and the modulus of elasticity;
- the paper [13] develops empirical models for the workability and compressive strength of cold-bonded fly ash aggregate concrete in terms of mixture proportioning variables, such as cement content, water content and volume fraction of cold-bonded aggregate. The models are developed employing statistically designed experiments based on the Response Surface Methodology;
- authors [7, 17] developed approximation dependencies for the obtained experimental results of fly ash lightweight aggregate concrete.

1.3 Sintered Fly Ash Lightweight Aggregate Concretes

Sintered fly ash lightweight aggregate concretes were previously investigated [25–43]. In these articles, the following characteristics of concrete were most commonly studied: compressive strength, split tensile strength, flexural tensile strength, stiffness, chloride ion permeability and drying shrinkage (Table 2). In addition, the workability of the concrete mix was studied in [27, 33, 35]. However, such properties of the concrete mixture as stratification, water segregation and concrete segregation were not studied in the presented studies.

Concrete properties, which are considered in the reviewed articles are shown in Table 2.

Table 2. Properties of concrete, which are considered in the reviewed articles.

Concrete properties	Reference to article
Compressive strength	[26–30, 32, 33, 35–43]
Split tensile strength	[27, 28, 30, 32, 35, 37, 40]
Flexural tensile strength	[28, 35, 37, 38, 40]
Freeze-thaw resistance	[30]
Water penetration	[36, 40]
Stiffness	[27, 28, 30, 32, 37]
Long time performance	[41]
Water absorption	[35]
Chloride ion permeability	[27, 30, 31, 40]
Drying shrinkage	[28, 33, 37, 41]

Analyzing the presented articles, some substantial conclusions can be made:

- sintered fly ash lightweight aggregate concretes are around 22% lighter and, at the same time, 20% stronger than normal weight aggregate concrete. Drying shrinkage is around 33% less than that of normal weight concrete. Thus it is possible to reduce the amount of cement by as much as 20% without affecting the required strength [41];

- the aim of the research article [42] is to reduce the water-cement ratio in structural lightweight concrete as a result of mixing water absorption by the lightweight aggregate. Although the porous aggregate is the weakest element in structural lightweight concrete, in this case its higher content may be excessively compensated by a stronger cement matrix resulting from the reduction of the water-cement ratio;
- the ratios of splitting tensile strength to compressive strength of fly ash lightweight aggregate concrete were found to be similar to that of normal weight concrete [30]. All the 28- and 56-day concrete specimens had a durability factor greater than 85 and 90, respectively, which met the requirement for freezing and thawing durability;
- the investigation [26] deals with the Bolomey equation for relating the cement-water ratio to compressive strengths of concrete containing normal weight aggregates. Authors suggested a modified Bolomey equation, considering indirectly the presence of fly ash aggregates. This equation was verified using the experimental data published by other researchers.

1.4 Fly Ash Lightweight Aggregate Concretes with Fiber Reinforcement

Some of the above reviewed articles use steel fiber [10, 11, 18, 22, 28, 32, 37] polypropylene fiber [29], or combinations of these two [29, 32] in the concrete mixes.

1.5 Concrete Mixtures with Plasticizers and Modifiers and Fly Ash Aggregate

Microsilica, nanosilica and superplasticizer are recent additions to concrete admixtures [44–48]. Some studies use microsilica [4, 11, 15, 16, 33, 36], nanosilica [18], or superplasticizer [23, 33, 36] in the concrete mixes.

1.6 Review Findings

During the extensive studies in recent years, mathematical models for mechanical properties of concrete have been obtained. These models are based on empirical data, such as the modified Bolomey equation and regression dependencies. However, the reviewed papers contain no strong theory for the formation of the basic properties of such concretes depending on the design formula, which allows predicting properties in a wide range of variable parameters of the matrix and concrete aggregates. This provides a strong reason for a further research.

This paper investigates the cold-bonded fly ash lightweight aggregate concretes from the fly ash of the Novosibirskaya GRES Thermal Power Plant (Novosibirsk, Russia). This thermal power plant is chosen because of the large amount of the produced fly ash and the lack of research on it.

2 Materials and Methods

The Fly ash aggregate was tested in Vedeneev VNIIG, JSC, Saint Petersburg, Russia. In accordance with the Russian standard GOST 9758-2012 "Non-organic porous aggregates for construction work. Test methods", the following tests were carried out:

- determination of loose bulk density;
- determination of water absorption for the coarse aggregate;
- determination of the aggregate size;
- determination of bulk crushing resistance;
- a freeze-thaw durability test for the coarse aggregate in a sodium sulfate solution.

3 Test Results and Discussion

The test results of the cold-bonded fly ash lightweight aggregate are presented in Table 3.

Table 3. The test results of cold-bonded fly ash lightweight aggregate.

Characteristics		Units	Values
Size fraction		mm	5–15
Bulk density		kg/m^3	970
Water absorption		%	6.1
Bulk crushing resistance		MPa	6.2
Grading, aggregate size	20 mm	%	0
	15 mm		4.8
	12.5 mm		26.6
	10 mm		29.7
	5 mm		37.2
	Less than 5 mm		1.7
Resistance to freezing and thawing on Russian standard GOST 9758-2012		-	Not less than F25

The size fraction of 5–15 mm is a characteristic of cold-bonded fly ash aggregate and it is close to the size fraction in the investigations [16, 36].

According to EN 13055:2016 "Lightweight aggregates", the granular material of a mineral origin has a particle density not exceeding 2000 kg/m^3 (2.00 mg/m^3) or a loose bulk density not exceeding 1200 kg/m^3 (1.20 mg/m^3). Thus, the fly ash aggregate with loose bulk density of 970 kg/m^3 meets the requirements of this standard.

The method for determining the cylinder compressive strength according to the Russian state standard GOST 9758-2012 "Non-organic porous aggregates for construction work. Test methods" correlates to the method of Annex C of EN 13055:2016. The Compressive strength is 6.2 MPa.

The water absorption of the fly ash aggregate is 6.1%, which is significantly greater than that of other types of traditional aggregates, except for expanded clay gravel with a value of 8–20%.

The fly ash aggregate in accordance with its size fraction and bulk crushing resistance is similar to the aggregate in the investigations [6–24].

4 Conclusions

A brief review of publications on this topic was made. The characteristics of the fly ash aggregate from the Novosibirskaya GRES Thermal Power Plant are investigated. Based on results obtained, the following conclusions can be underlined:

- the number of publications on this topic has increased since 2002, reflecting the growing interest of researchers in this topic. The largest number of publications was made by researchers from India and Turkey;
- the accumulated experimental data are not sufficient to develop a strong theory or dependencies to predict the mechanical properties in a wide class of structural concretes. The existing attempts to derive the calculated dependencies are similar to the developed approximations (regression analysis) [21, 30] or the refinement of the coefficients for the Bolomey equation [26].
- the experimental studies of cold-bonded fly ash aggregate performed by the authors showed a water absorption of the coarse aggregate of 6.1% and a bulk crushing resistance of 6.2 MPa;
- the obtained characteristics of the aggregates demonstrate their applicability in low thermal transmittance concretes.

References

1. Kumar, P.P., Rama Mohan Rao, P.: Packing density of self compacting concrete using normal and lightweight aggregates. Int. J. Civ. Eng. Technol. **8**, 1156–1166 (2017)
2. Kockal, N.U., Ozturan, T.: Durability of lightweight concretes with lightweight fly ash aggregates. Constr. Build. Mater. **25**, 1430–1438 (2011). https://doi.org/10.1016/j.conbuildmat.2010.09.022
3. Gomathi, P., Sivakumar, A.: Accelerated curing effects on the mechanical performance of cold bonded and sintered fly ash aggregate concrete. Constr. Build. Mater. **77**, 276–287 (2015). https://doi.org/10.1016/j.conbuildmat.2014.12.108
4. Güneyisi, E., Gesoğlu, M., Pürsünlü, Ö., Mermerdaş, K.: Durability aspect of concretes composed of cold bonded and sintered fly ash lightweight aggregates. Compos. B Eng. **53**, 258–266 (2013). https://doi.org/10.1016/j.compositesb.2013.04.070
5. Kirubakaran, D., Joseravindraraj, B.: Utilization of pelletized fly ash aggregate to replace the natural aggregate: a review. Int. J. Civ. Eng. Technol. **9**, 147–154 (2018)
6. Gesoğlu, M., Güneyisi, E., Ali, B., Mermerdaş, K.: Strength and transport properties of steam cured and water cured lightweight aggregate concretes. Constr. Build. Mater. **49**, 417–424 (2013). https://doi.org/10.1016/j.conbuildmat.2013.08.042

7. Kockal, N.U., Ozturan, T.: Strength and elastic properties of structural lightweight concretes. Mater. Des. **32**, 2396–2403 (2011). https://doi.org/10.1016/j.matdes.2010.12.053
8. Gomathi, P., Sivakumar, A.: Synthesis of geopolymer based class-F fly ash aggregates and its composite properties in concrete. Arch. Civ. Eng. **60**, 55–75 (2014). https://doi.org/10.2478/ace-2014-0003
9. Al Bakri, A.M.M., Kamarudin, H., Binhussain, M., Nizar, I.K., Rafiza, A.R., Zarina, Y.: Comparison of geopolymer fly ash and ordinary Portland cement to the strength of concrete. Adv. Sci. Lett. **19**, 3592–3595 (2013). https://doi.org/10.1166/asl.2013.5187
10. Güneyisi, E., Gesoğlu, M., Ipek, S.: Effect of steel fiber addition and aspect ratio on bond strength of cold-bonded fly ash lightweight aggregate concretes. Constr. Build. Mater. **47**, 358–365 (2013). https://doi.org/10.1016/j.conbuildmat.2013.05.059
11. Gesoğlu, M., Güneyisi, E., Alzeebaree, R., Mermerdaş, K.: Effect of silica fume and steel fiber on the mechanical properties of the concretes produced with cold bonded fly ash aggregates. Constr. Build. Mater. **40**, 982–990 (2013). https://doi.org/10.1016/j.conbuildmat.2012.11.074
12. Priyadharshini, P., Mohan Ganesh, G., Santhi, A.S.: Effect of cold bonded fly ash aggregates on strength & restrained shrinkage properties of concrete. In: IEEE-International Conference on Advances in Engineering, Science and Management, ICAESM-2012, pp. 160–164 (2012)
13. Joseph, G., Ramamurthy, K.: Workability and strength behaviour of concrete with cold-bonded fly ash aggregate. Mater. Struct./Materiaux et Constructions **42**, 151–160 (2009). https://doi.org/10.1617/s11527-008-9374-x
14. Gesoğlu, M., Özturan, T., Güneyisi, E.: Effects of cold-bonded fly ash aggregate properties on the shrinkage cracking of lightweight concretes. Cem. Concr. Compos. **28**, 598–605 (2006). https://doi.org/10.1016/j.cemconcomp.2006.04.002
15. Gesoglu, M., Özturan, T., Güneyisi, E.: Shrinkage cracking of lightweight concrete made with cold-bonded fly ash aggregates. Cem. Concr. Res. **34**, 1121–1130 (2004). https://doi.org/10.1016/j.cemconres.2003.11.024
16. Gesoğlu, M., Özturan, T., Güneyisi, E.: Effect of coarse aggregate properties on the ductility of lightweight concretes. In: Role of Cement Science in Sustainable Development - Proceedings of the International Symposium - Celebrating Concrete: People and Practice, pp. 537–546 (2003)
17. Kockal, N.U., Ozturan, T.: Properties of lightweight concretes made from lightweight fly ash aggregates. In: Excellence in Concrete Construction Through Innovation - Proceedings of the International Conference on Concrete Construction, pp. 251–261 (2009)
18. Their, J.M., Özakça, M.: Developing geopolymer concrete by using cold-bonded fly ash aggregate, nano-silica, and steel fiber. Constr. Build. Mater. **180**, 12–22 (2018). https://doi.org/10.1016/j.conbuildmat.2018.05.274
19. Narattha, C., Chaipanich, A.: Phase characterizations, physical properties and strength of environment-friendly cold-bonded fly ash lightweight aggregates. J. Clean. Prod. **171**, 1094–1100 (2018). https://doi.org/10.1016/j.jclepro.2017.09.259
20. Venkata Suresh, G., Pavan Kumar Reddy, P., Karthikeyan, J.: Effect of GGBS and fly ash aggregates on properties of geopolymer concrete. J. Struct. Eng. (India) **43**, 436–444 (2016)
21. Thomas, J., Harilal, B.: Mechanical properties of cold bonded quarry dust aggregate concrete subjected to elevated temperature. Constr. Build. Mater. **125**, 724–730 (2016). https://doi.org/10.1016/j.conbuildmat.2016.08.093
22. Güneyisi, E., Gesoglu, M., Özturan, T., Ipek, S.: Fracture behavior and mechanical properties of concrete with artificial lightweight aggregate and steel fiber. Constr. Build. Mater. **84**, 156–168 (2015). https://doi.org/10.1016/j.conbuildmat.2015.03.054

23. Gesoglu, M., Güneyisi, E., Ozturan, T., Oz, H.O., Asaad, D.S.: Shear thickening intensity of self-compacting concretes containing rounded lightweight aggregates. Constr. Build. Mater. **79**, 40–47 (2015). https://doi.org/10.1016/j.conbuildmat.2015.01.012
24. Gopi, R., Revathi, V., Kanagaraj, D.: Light expanded clay aggregate and fly ash aggregate as self curing agents in self compacting concrete. Asian J. Civ. Eng. **16**, 1025–1035 (2015)
25. Shivaprasad, K.N., Das, B.B.: Effect of duration of heat curing on the artificially produced fly ash aggregates. IOP Conf. Ser. Mater. Sci. Eng. **431** (2018). https://doi.org/10.1088/1757-899x/431/9/092013
26. Rajamane, N.P., Ambily, P.S.: Modified Bolomey equation for strengths of lightweight concretes containing fly ash aggregates. Mag. Concr. Res. **64**, 285–293 (2012). https://doi.org/10.1680/macr.11.00157
27. Dinakar, P.: Properties of fly-ash lightweight aggregate concretes. Proc. Inst. Civ. Eng. Constr. Mater. **166**, 133–140 (2013). https://doi.org/10.1680/coma.11.00046
28. Domagala, L.: Modification of properties of structural lightweight concrete with steel fibres. J. Civ. Eng. Manage. **17**, 36–44 (2011). https://doi.org/10.3846/13923730.2011.553923
29. Harish, K.V., Dattatreya, J.K., Neelamegam, M.: Properties of sintered fly ash aggregate concrete with and without fibre and latex. Indian Concr. J. **85**, 35–42 (2011)
30. Kockal, N.U., Ozturan, T.: Effects of lightweight fly ash aggregate properties on the behavior of lightweight concretes. J. Hazard. Mater. **179**, 954–965 (2010). https://doi.org/10.1016/j.jhazmat.2010.03.098
31. Kayali, O., Zhu, B.: Chloride induced reinforcement corrosion in lightweight aggregate high-strength fly ash concrete. Constr. Build. Mater. **19**, 327–336 (2005). https://doi.org/10.1016/j.conbuildmat.2004.07.003
32. Kayali, O., Haque, M.N., Zhu, B.: Some characteristics of high strength fiber reinforced lightweight aggregate concrete. Cem. Concr. Compos. **25**, 207–213 (2003). https://doi.org/10.1016/S0958-9465(02)00016-1
33. Nair, H.K., Ramamurthy, K.: Behaviour of concrete with sintered fly ash aggregate. Indian Concr. J. **84**, 33–38 (2010)
34. Kikuchi, M., Mukai, T.: Properties of structural light-weight concrete containing sintered fly ash aggregate and clinker ash. Trans. Jpn. Concr. Inst. **8**, 45–50 (1986)
35. Satpathy, H.P., Patel, S.K., Nayak, A.N.: Development of sustainable lightweight concrete using fly ash cenosphere and sintered fly ash aggregate. Constr. Build. Mater. **202**, 636–655 (2019). https://doi.org/10.1016/j.conbuildmat.2019.01.034
36. Dash, S., Kar, B., Mukherjee, P.S.: Pervious concrete using fly ash aggregate as coarse aggregate-an experimental study. AIP Conf. Proc. **1953** (2018). https://doi.org/10.1063/1.5032808
37. Babu, B.R., Thenmozhi, R.: An investigation of the mechanical properties of sintered fly ash lightweight aggregate concrete (SFLWAC) with steel fibers. Arch. Civ. Eng. **64**, 73–85 (2018). https://doi.org/10.2478/ace-2018-0005
38. Bursa, C., Tanriverdi, M., Çiçek, T.: Use of fly ash aggregates in production of light-weight concrete. In: IMCET 2017: New Trends in Mining - Proceedings of 25th International Mining Congress of Turkey, pp. 469–476 (2017)
39. Wasserman, R., Bentur, A.: Effect of lightweight fly ash aggregate microstructure on the strength of concretes. Cem. Concr. Res. **27**, 525–537 (1997). https://doi.org/10.1016/S0008-8846(97)00019-7
40. Cerny, V., Kocianova, M., Drochytka, R.: Possibilities of lightweight high strength concrete production from sintered fly ash aggregate. Procedia Eng. **195**, 9–16 (2017). https://doi.org/10.1016/j.proeng.2017.04.517
41. Kayali, O.: Fly ash lightweight aggregates in high performance concrete. Constr. Build. Mater. **22**, 2393–2399 (2008). https://doi.org/10.1016/j.conbuildmat.2007.09.001

42. Domagała, L.: The effect of lightweight aggregate water absorption on the reduction of water-cement ratio in fresh concrete. Procedia Eng. **108**, 206–213 (2015). https://doi.org/10.1016/j.proeng.2015.06.139
43. Černý, V., Sokol, P., Drochytka, R.: Production possibilities of concrete based on artificial fly ash aggregates. Adv. Mater. Res. **923**, 130–133 (2014). https://doi.org/10.4028/www.scientific.net/AMR.923.130
44. Fediuk, R.S., Lesovik, V.S., Svintsov, A.P., Mochalov, A.V., Kulichkov, S.V., Stoyushko, N.Y., Gladkova, N.A., Timokhin, R.A.: Self-compacting concrete using pretreatmented rice husk ash. Mag. Civ. Eng. **79**, 66–76 (2018). https://doi.org/10.18720/MCE.79.7
45. Denisov, A.V.: The impact of superplasticizers on the radiation changes in Portland cement stone and concretes. Mag. Civ. Eng. **73**, 70–87 (2017). https://doi.org/10.18720/MCE.73.7
46. Barabanshchikov, Y.G., Belyaeva, S.V., Arkhipov, I.E., Antonova, M.V., Shkolnikova, A. A., Lebedeva, K.S.: Influence of superplasticizers on the concrete mix properties. Mag. Civ. Eng. **74**, 140–146 (2017). https://doi.org/10.18720/MCE.74.11
47. Akimov, L., Ilenko, N., Mizharev, R., Cherkashin, A., Vatin, N., Chumadova, L.: Composite concrete modifier CM 02-10 and its impact on the strength characteristics of concrete. MATEC Web Conf. **53** (2016). https://doi.org/10.1051/matecconf/20165301022
48. Smirnova, O.M.: Compatibility of Portland cement and polycarboxylate-based superplasticizers in high-strength concrete for precast constructions. Mag. Civ. Eng. **66**, 12–22 (2016). https://doi.org/10.5862/MCE.66.2

Calculation of Shear Stability of Conjugation of the Main Pillars with the Foundation in Wooden Frame Buildings

Vladimir Rimshin[1(✉)], Boris Labudin[2], Vladimir Morozov[2], Alexandr Orlov[2], Aram Kazarian[3], and Vagan Kazaryan[3]

[1] Moscow State University of Civil Engineering, Yaroslavskoe shosse 26, 129337 Moscow, Russia
v.rimshin@niisf.ru
[2] Northern (Arctic) Federal University named after M. V. Lomonosov, Severnaya Dvina Emb. 17, 163002 Arkhangelsk, Russia
[3] JSC Research Center of Construction, 2nd Institutskaya 6, 109428 Moscow, Russia

Abstract. The authors give a list of wood advantages as a building material and main glued-wooden constructions. Glued-wooden frame constructions, their elements and manufacturing methods are considered. The advantages and disadvantages of glued-plywood and glued-plank wooden uprights and the area of their application are listed. The issues related to the design and interface node manufacture (support node) of these structures with foundation are considered. The authors note the ways of increasing the stiffness, strength and reliability of these nodes and their advantages. As a node connector of glued-wooden upright with foundation, a steel claw washer C1-type is investigating. The variants of connecting such nodes for glued-plywood and glued-plank wooden uprights are offered, the example of calculation is given. As a material for the study, two variants for fastening wooden patch plate to glued-wooden upright (rectangular and box section) are adopted: the 1st - by bolts; the 2nd - by claw connectors. For each fasten variant, the required number of connectors is calculated. The calculation indicated that claw connectors C1-type have a greater load-bearing capacity than a bolted connection 16 mm in diameter. As a result of the research, it was found that the use of claw connectors reduces steel intensity of the interface node in 2.1–2.3 times, as well as provides the strength, stiffness and operational reliability of the connection.

1 Introduction

Wood is an environmentally friendly material of natural origin. Wood serves as a raw material for various wooden composite materials, for example, glued wood, LVL-timber, CLT-panels etc. which are main elements of load-bearing enclosing building structures.

Glued wood is widely used in building structures - beams, frames, uprights, trusses, arches, etc. They have high strength and resistance to chemical attack of many corrosive medium, allow reducing the structures weight and shorting construction time.

V. Murgul and M. Pasetti (Eds.): EMMFT-2018, AISC 983, pp. 867–876, 2019.
https://doi.org/10.1007/978-3-030-19868-8_85

The acoustic properties, aesthetic qualities and architectural advantages of wooden structures are important.

Glued-wooden frame structures are generally used in construction of single-span buildings. They consist of an upright and a girder of solid or box section. The manufacture of these elements can be carried out simultaneously by bending the multilayer bag during the pressing process, or separately, when the frame members (girder and upright) are connected by toothed connectors with curved inserts [1]. Another variant of the frame structure is a beam-and-upright solution. It consists of a girder, hinged or elastically supported on an upright (column). The article considers the solution for such building constructions.

The glued-plywood and glued-plank wooden uprights of rectangular (box) section are manufacturability, have a greater fire resistance grading in comparison with the uprights of other designs, for example, lattice columns. At high height ($h > 6$ m), these uprights are distinguished by a considerable material intensity. A cross-section change for an I-beam reduces their material intensity and, at the same time, significantly reduces the manufacturability, which ultimately leads to an increase in cost.

The glued-plywood and lattice uprights have the least material intensity. The glued-plywood uprights have smooth surfaces (it is especially important for buildings with a chemically aggressive environment), good exterior and have sufficiently high stiffness and strength at low weight. However, their use is limited to fire-fighting requirements, unless adequate fire-resisting measures are taken. Such uprights are suitable for use in prefabricated buildings, and as end-type frame uprights of buildings for various purposes at high height.

One of the most important issues related to the design of glued-wooden uprights is providing the stiffness, strength and load-bearing capacity of upright-to-foundation interface node (support node). As a rule, the support node is made by simply resting of upright in a steel shoe, fixed to the foundation with anchor bolts. Uprights are attached to the shoe by bolts. Their diameter and number are determined by the calculation and design requirements [2].

Increasing the strength and stiffness, consequently, the operational reliability of glued constructions main uprights can be achieved by increasing the shear rigidity in joints of the wooden structure elements. Various types of special connectors can be used in the form of claw connectors, shearing rings, dowel pins groups etc. for achieving this result [3]. This eliminates the use of a metal shoe to fixing the column in the foundation.

The advantages of joints with claw washers include simplicity of pressing the joint, increased strength, stiffness and bearing capacity of the joint, due to the force is fractionally distributed over a large wood crushing surface and provides high stiffness of the joint [4], sufficient safety of bending dowel pin, wood crushing. Such joints can be very effectively used in the manufacture of trusses, composite beams and uprights, can be used for strengthening of wooden structures. Therefore, they are widely used in world practice, and are applied in our country from recent times [5].

Otreshko, Filippov, Konstantinov, Svetozarova, Dushechkin, Serov, Sannikov, Serov, Karateev, Labudin [6], Naychuk, Leshchuk [7], Johansen [8], Blass [9], Endzhievsky, Hillson [10], Wilkinson [11], Smith [12], Roche, Robeller, Humbert, Weinand [13], Šmak, Straka [14] and some other researchers have addressed to the issues of design and construction of the above mentioned glued structures. Lennov, Naumov, Mironov, Kotlov, Ariskin, Galakhov, Karelskiy [15], Ishmaeva [16], Smirnov, Xu Yun, Chernykh [17], Turkovski, Pogoreltsev, Arkaev [18], Malinowski [19], Yao Wei, V.M. Vdovin, E. Meghlat, M. Oudjene, J. Karadelis, P. Brown and some other researchers have investigated various types of connectors and joints of wooden structures elements.

Modern regulations still use research results of the 30–50s of the last century. Thus, the use of modern connector types is relevant, and there is a need to continue research in this direction.

2 Methods

The constructive concept of upright-to-foundation interface node is developed for two types of wooden structures - glued-plywood and glued-plank wooden structures.

It is advisable to use claw connectors in the area of fixing glued-plywood and glued-plank wooden frame upright with foundation, to increase the strength and stiffness of such joints. Figures 1 and 2 show variants of such fixing using claw connectors.

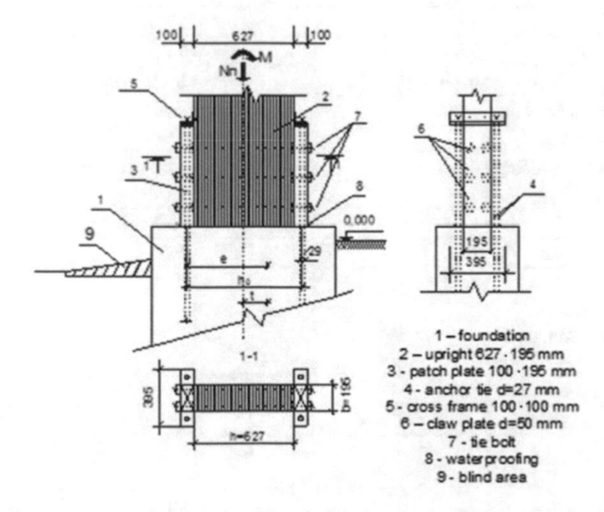

Fig. 1. The variant of using claw washers in the design of fixing rectangular glued-plank wooden upright with a foundation

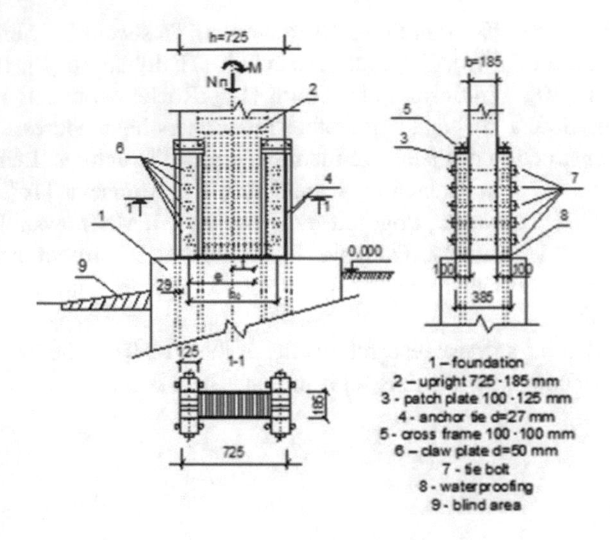

Fig. 2. The variant of using claw washers in the design of fixing box glued-plywood upright with a foundation

Table 1. The initial data

Characteristic	Rectangular glued-plank wooden upright (Fig. 1)	Box glued-plywood upright (Fig. 2)
Step between transverse frames (B), m	6	6
Frame span (L), m	15	15
Upright length (H), m	7.5	7.5
Upright cross-sections, mm	b · h = 195.627	b · h = 185.725
Ultimate bending moment at the base of upright (M), kN m	103.7	80
Constant vertical load, including proper weight (N_n), kN	41.3	52.8
Coefficient ξ	0.93	0.81
Diameter of bolts (d_b), mm	16	16
Coefficient, that takes into account the quantity of Nn attributable to the tension area of upright (e/t)	2.78	2.77
Arm of resisting couple (e), mm	568	644
Maximum load in the anchors, kN	$N_p = \frac{M}{e \cdot \xi} - \frac{N_n}{m} = \frac{103.7}{0.568 \cdot 0.93} - \frac{41.3}{2.78} = 181.5\,\text{kN}$	$N_p = \frac{M}{h_0 \cdot \xi} - \frac{N_n}{2} = \frac{80}{0.60 \cdot 0.81} - \frac{52.8}{2} = 138.2\,\text{kN}$
Bearing capacity of a bolt per shear, kN	$T_H = 2.5d_b^2 = 2.5 \cdot 16^2 = 6.4\,\text{kN}$	$T_H = 2.5d_b^2 = 2.5 \cdot 16^2 = 6.4\,\text{kN}$
Number of bolts	$n = \frac{N_p}{T_H} = \frac{181.5}{6.4} = 28$	$n = \frac{N_p}{T_H} = \frac{138.2}{6.4} = 22$
Number of bolts and connectors	12	10

The offered variants for two cases upright-to-foundation interface node are considered: the 1st - the patch plate is fastened to the upright by bolts; the 2nd - the patch plate is fastened to the upright by bolts with claw washers.

The initial data for calculation are given in Table 1.

For the second case of fastening a patch plate to upright, calculation is carried out according to [24]. The diameter of bolts, their bearing capacity ($F_{v,Rk,bolt}$), the upright thickness (t1), the patch plate thickness (t2) and the constant load intensity (Np) are adopted similarly to the first case. A claw washer C1-type [24] was selected as a connector (Fig. 3). Previously, two variants of connectors were considered for lattice uprights: round (C1) and square (C8).

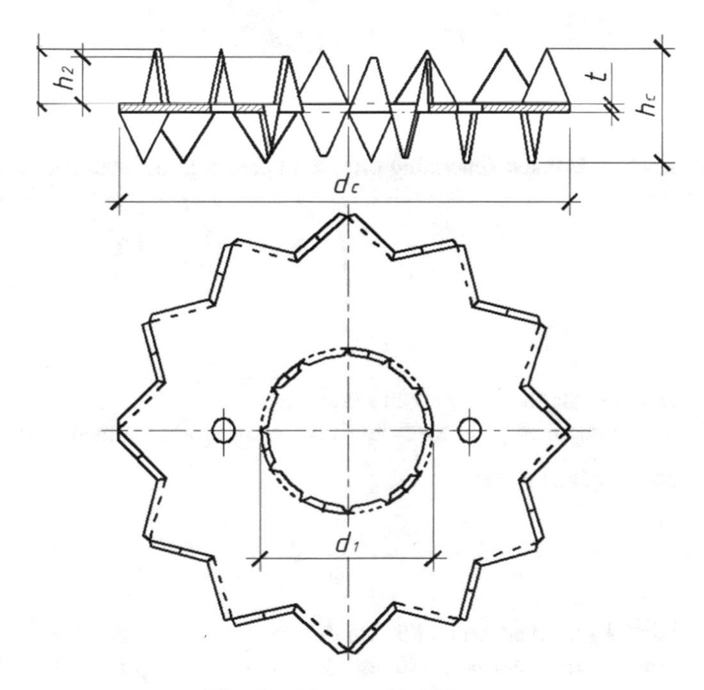

Fig. 3. Claw washer C1-type

The characteristics of claw washer C1-type are shown in Table 2.

Table 2. The characteristics of claw washer C1-type [24]

Claw washer type	Diameter d_c, mm	Height h_c, mm	Thickness plate t, mm	Diameter of central hole d_1, mm
C1	50	13	1.0	17

The characteristic (normative) value of joint bearing capacity with a toothed plate is determined from the expression:

$$F_{v,Rk,con} = F_{v,Rk} + F_{v,Rk,bolt} \tag{1}$$

$F_{v,Rk,con}$ - the characteristic value of joint bearing capacity with a toothed plate, N;
$F_{v,Rk}$ - the characteristic value of toothed plate bearing capacity, N;
$F_{v,Rk,bolt}$ - the characteristic value of tie bolt bearing capacity, N.

The characteristic value of bearing capacity $F_{v,Rk}$ of a single two-sided toothed plate in the joint is determined from expression:

$$F_{v,Rk} = 25k_1k_2k_3d_c^{1.5} \tag{2}$$

d_c – diameter of a toothed plate.

The coefficient k_1 is taken depending on joined element thickness and is determined from expression:

$$k_1 = \begin{cases} 1 \\ \frac{t_1}{3h_e} \\ \frac{t_2}{5h_e} \end{cases} \tag{3}$$

t_1 – upright thickness; t_2 – patch plate thickness;
h_e – impression depth of plate teeth into considered joint element, mm.

For two-sided toothed plate:

$$h_e = \frac{(h_c - t)}{2}, \tag{4}$$

The coefficient k_2 is taken depending on distance a_3, t to a loaded end grain. If the distance to a loaded end grain is less than $2d_c$, the value of k_2 should be determined from expression:

$$k_2 = \min\begin{cases} 1 \\ \frac{a_{3,t}}{1.5d_c} \end{cases}, \tag{5}$$

$$a_{3,t} = \max\begin{cases} 1.1d_c \\ 7d \\ 80\,\text{mm}, \end{cases} \tag{6}$$

d – bolt diameter, mm.

The coefficient k_3 is taken depending on wood density and is determined from expression:

$$k_3 = min\begin{cases} 1 \\ \frac{\rho_k}{350}, \end{cases} \quad (7)$$

ρ_k – the characteristic (normative) value of wood density in joint, kg/m^3. Spruce is taken as a upright material, the second-class quality, ρk = 500 kg/m^3.

3 Results

Substituting numerical values into formulas (1–7), the following results are obtained.

The number of C1-type connectors with tie bolts 16 mm in diameter for the rectangular glued-plank wooden upright is calculated [28].

1. $k_1 = \begin{cases} 1 \\ \frac{100}{3 \cdot 6} = 5.6 \\ \frac{627}{5 \cdot 6} = 20.9. \end{cases}$

2. $a_{3,t} = \max\begin{cases} 1.1 \cdot 50 = 55\,\text{mm} \\ 7 \cdot 16 = 112\,\text{mm} \\ 80\,\text{mm}. \end{cases}$

3. $k_2 = \min\begin{cases} 1 \\ \frac{112}{1.5 \cdot 50} = 1.49. \end{cases}$

4. $k_3 = \min\begin{cases} 1 \\ \frac{500}{350} = 1.4. \end{cases}$

5. $F_{v,Rk} = 25 \cdot 1 \cdot 1 \cdot 1 \cdot 50^{1.5} = 8838\,\text{N} = 8.8\,\text{kN}.$
6. $F_{v,Rk,con} = 8838 + 6400 = 15238\,\text{N} = 15.2\,\text{kN}.$
7. The number of claw washer with bolts for fastening a patch plate will be:

$$n = \frac{N_p}{F_{v,Rk,con}} = \frac{181.5}{15.2} = 11.9.$$

The number of connectors n = 12 (set in two ranges) is accepted.

The number of C1-type connectors with a bolt 16 mm in diameter for the box glued-plywood upright is calculated:

8. $k_1 = \begin{cases} 1 \\ \frac{100}{3 \cdot 6.5} = 5.1 \\ \frac{185}{5 \cdot 6.5} = 5.7. \end{cases}$

9. $a_{3,t} = \max\begin{cases} 1.1 \cdot 50 = 55\,\text{mm} \\ 7 \cdot 16 = 112\,\text{mm} \\ 80\,\text{mm}. \end{cases}$

10. $k_2 = \min\begin{cases} 1 \\ \frac{98}{1.5 \cdot 50} = 1.3. \end{cases}$

11. $k_3 = \min\begin{cases} 1 \\ \frac{500}{350} = 1.4. \end{cases}$

12. $F_{v,Rk} = 25 \cdot 1 \cdot 1 \cdot 1 \cdot 50^{1.5} = 8838\,\text{N} = 8.8\,\text{kN}.$
13. $F_{v,Rk,con} = 8838 + 6400 = 15238\,\text{N} = 15.2\,\text{kN}.$

The number of claw washer with bolts for fastening a patch plate will be:

$$n = \frac{N_p}{F_{v,Rk,con}} = \frac{138.2}{15.2} = 9.1.$$

The number of connectors n = 10 is accepted.

4 Conclusions

As a result of the fixing calculation of the main glued-plywood and glued-plank wooden uprights with a reinforced concrete foundation, the following are concluded:

1. 28 bolts 16 mm in diameter for the glued-plank wooden uprights and 22 bolts - for the glued-plywood wooden uprights are required to ensure a reliable fixing in the first case (bolt connection only).
2. 12 bolts 16 mm in diameter and 12 double-sided C1-type connectors 50 mm in diameter are required for the glued-plan wooden uprights, 10 bolts 16 mm in diameter and 10 double-sided C1-type connectors 50 mm in diameter for the glued-plywood wooden uprights are required to ensure reliable coupling in the second case (bolts with double-sided toothed connectors).
3. Joints consisting of bolts and claw connectors C1-type have the load-bearing capacity in 2.4 times large than the load-bearing capacity of the joint by only bolts 16 mm in diameter.
4. The use of claw connectors C1-type in the fixing of the glued-plywood and glued-plank wooden uprights reduces the steel intensity in the structural elements node in 2.1–2.3 times.

References

1. Vidy derevyannykh kleyenykh konstruktsiy v stroitelstve [Types of wooden glued constructions] [Elektronnyy resurs]. http://engstroy.spbstu.ru/autors/literature_samples.pdf. Accessed 25 Dec 2017
2. Uzly stoyek [Nodes of racks] [Elektronnyy resurs]. https://studfiles.net/preview/5847511/page:20/. Accessed 25 Dec 2017
3. Labudin, B.V., Morozov, V.S., Karel'skiy, A.V, Petrova, A.L., Orlov, A.O.: Sovershenstvovaniye konstruktivnykh resheniy osnovnykh stoyek karkasnykh zdaniy [Perfection of constructive solutions of the main racks of frame buildings]. In: Sbornik materialov konferentsii "Stroitelnaya nauka XXI-vek: teoriya, obrazovaniye, praktika, innovatsii Severo-Arkticheskomu regionu". Arkhangelsk, pp. 118–128 (2017). (rus)
4. Popov, Ye.V.: Sovershenstvovaniye konstruktsii i tekhnologii izgotovleniya derevokompozitnykh plito-rebristykh izdeliy dlya domostroyeniya [Perfection of a design and manufacturing technology of wood-composite slab-ribbed products for house-building]. Dissertatsiya. Arkhangelsk, 175 p. (2017). (rus)
5. Rimshin, V.I., Labudin, B.V., Melekhov, V.I., Popov Ye.V., Roshchina S.I.: Soyedineniya elementov derevyannykh konstruktsiy na shponkakh i shaybakh [Connections of elements of wooden structures on dowels and washers]. Vestnik MGSU, no. 9, pp. 35–50 (2016). (rus)
6. Labudin, B.V.: Sovershenstvovaniye kleyenykh derevyannykh konstruktsiy s prostranstvenno-regulyarnoy strukturoy: monografiya [Perfection of wood structures with a spatially regular structure: monograph]. Arkhangelsk. Arkhang. Gos. Tekhn. Un-t, 267 p. (2007). (rus)
7. Naychuk, A.Ya., Leshchuk, Ye.V.: Chislennoye issledovaniye napryazhennogo sostoyaniya drevesiny v zone vintov, rabotayushchikh na vydergivaniye poperek volokon [Numerical study of stress state of wood in zone of screws working on pulling across the fibers]. In: Sb. trudov mezhdunarodnoy nauchno-tekhnicheskoy konf. "Stroitelnaya nauka – XXI vek: teoriya, obrazovaniye, praktika, innovatsii severo - arkticheskomu regionu". Arkhangelsk, pp. 276–283 (2014). (rus)
8. Johanson, K.W.: Theory of timber connections. International Association for Bridge and Structural Engineering, no. 9, pp. 249–262 (1949)
9. Blass, H.J., Schädle, P.: Ductility aspects of reinforced and non-reinforced timberjoints. Eng. Struct. **33**, 3018–3026 (2011)
10. Hilson, B.O.: Joints with Dowel-type Fasteners—Theory. Paper C3: Timber Engineering Step 1: Basis of Design, Material Properties, Structural Components and Joints. Centrum Hout, Almere, The Netherlands
11. Wilkinson, T.L.: Dowel bearing strength, 12 p. Res. Pap. FPL-Rp-505. US DA Forest Serv. Forest Prod. Lab, Madison, WI (1991)
12. Smith, I.: Short-term load-deformation relationship for joints with dowel type connectors. Ph.D. thesis, CNAA (1983)
13. Roche, S., Robeller, C., Humbert, L., Weinand, Y.: On the semi-rigidity of dovetail joint for the joinery of LVL panels. Eur. J. Wood Wood Prod. **73**(5), 667–675 (2015)
14. Šmak, M., Straka, B.: Development of new types of timber structures based on theoretical analysis and their real behaviour. Wood Res. **59**(3), 459–470 (2014)
15. Karelskiy, A.V., Zhuravleva, T.P., Labudin, B.V.: Ispytaniye na izgib derevyannykh sostavnykh balok, soyedinennykh metallicheskimi zubchatymi plastinami, razrushayushchey nagruzkoy [The test for bending of wooden composite beams connected by metal toothed plates by destroying load]. Inzhenerno-stroitelnyy zhurnal **2**(54), 77–85 (2015). (rus)

16. Ishmayeva, D.D.: Zhestkiye uzlovyye soyedineniya na kleyenykh stal'nykh shaybakh v balochnykh strukturakh iz kleyenykh derevyannykh elementov [Rigid junctions on glued steel washers in beam structures made of glued wooden elements]. Dissertatsiya. Penza, 171 p. (2014). (rus)
17. Chernykh, A.G., Danilov, Ye.V.: Metody issledovaniya soyedineniy derevyannykh konstruktsiy na kogtevykh shponkakh [Methods for studying of wooden structures on nails]. Sovremennyye problemy nauki i obrazovaniya, no. 2, pp. 150–157 (2013). (rus)
18. Arkayev, M.A.: Usileniye derevyannykh konstruktsiy s ispolzovaniyem stalnykh vitykh krestoobraznykh sterzhney [Strengthening of wooden structures using steel twisted cross-shape rods]. Dissertatsiya. Orenburg, 190 p. (2017). (rus)
19. Malinowski, C.: Zur Geschichteder Verbindungstechnik-Verbinderaus Stahlblech. Bauenmit Holz. Bd. 11, pp. 776–779; Bd. 12, pp. 872–877 (1989)
20. Yao, V.: Razrabotka i raschet uzlovykh soyedineniy nesushchikh, prostranstvennykh sterzhnevykh konstruktsiy iz bambuka [Development and calculation of nodal connections of load-bearing, three dimensional core structures from bamboo]. Dissertatsiya. SPb., 163 p. (2015). (rus)
21. Vdovin, V.M., Mukhayev, A.I., Ariskin, M.V.K.: Otsenke napryazhenno-deformirovannogo sostoyaniya derevyannykh elementov, soyedinennykh tsentrovymi vkleyennymi shponkami [Evaluation of the stressed-deformed state of the wooden elements connected by central glued dowels]. Regionalnaya arkhitektura i stroitelstvo, no. 2, pp. 81–90 (2013). (rus)
22. Meghlat, E., Oudjene, M., Ait-Aider, H., Batoz, J.: A new approach to model nailed and screwed timber joints using the finite element method. Constr. Build. Mater. **41**, 263–269 (2013)

Comparison of Thermal Insulation Characteristics of PIR, Mineral Wool, Carbon Fiber, and Aerogel

Nikolai Vatin[1], Shukhrat Sultanov[2](✉), and Anastasia Krupina[2]

[1] South Ural State University, Chelyabinsk 454080, Russia
[2] Peter the Great St. Petersburg Polytechnic University, St. Petersburg 195251, Russia
sultanov_sht@spbstu.ru

Abstract. The performance of a thermally insulating material is mainly determined by its thermal conductivity, which depends on the density of the material, porosity, moisture content, and average temperature difference. This paper presents the results of experimental studies of the coefficient of thermal conductivity of mineral wool, polyisocyanurate foam, aerogel, and carbon in a stationary thermal regime. Thermal insulation indicators of insulation are determined on a flat sample using the device PIT 2.1. The results should be of great importance for manufacturers of materials, building owners and designers when choosing suitable insulation materials and correctly predicting the thermal and energy performance of buildings and their energy efficiency.

Keywords: Thermal insulation material · Polyisocyanurate foam · Mineral wool · Thermal conductivity · Aerogel · Carbon · Construction

1 Introduction

Global warming is the most immediate threat to global sustainable development, primarily due to CO_2 emissions. Achieving optimal thermal design of buildings can cause significant reductions in CO_2, as well as other numerous benefits. The construction industry is developing along the path of reducing material consumption and labor intensity, reducing the time and financial costs for the construction of buildings [1–4]. The most effective way to solve the problems described above is the use of highly efficient thermal insulation materials in construction. Today, energy-efficient thermal insulation materials include insulators that have a thermal conductivity not higher than 0.06 W/(m °C). At the same time, these materials should be characterized by the availability of raw materials, low energy consumption and low production costs [5], have water and frost resistance [6], mechanical strength, environmental [7] and fire safety. The durability of insulation materials also has a significant impact on the life cycle cost of a building. About 70% of this cost relates to the stage of operation of the building, and the lion's share of this number falls on heating and cooling. If, over time,

V. Murgul and M. Pasetti (Eds.): EMMFT-2018, AISC 983, pp. 877–883, 2019.
https://doi.org/10.1007/978-3-030-19868-8_86

insulation products do not retain their characteristics and the heat transfer through the building envelope increases, energy costs can also increase significantly. Repair or replacement of insulation products before the estimated service life will require significant additional costs, since access to the insulation layer is not always easy. At present [8–10], in various branches of industrial production, including the construction industry, a search is being made for high-tech insulating materials that exceed thermal insulation materials in the most common insulation materials - mineral wool and extruded polystyrene foam.

Mineral wool (Fig. 1) is a thermal insulation material that has existed on the market since 1937. Due to the high thermal insulation properties, it had no competitors for a long time. The exceptional thermal, fire and acoustic properties of mineral wool derive from the mat of fibres that prevents the movement of air, and from mineral wool's inert chemical composition. The thermal characteristics of mineral wool are mainly due to the prevention of convection by the release of air in a rough matrix of open-pore material. Static air has a low thermal conductivity. Heat transfer is also reduced because the material acts as a physical barrier to radiation processes. The fact is that mineral wool well absorbs moisture, so that to reduce the amount of absorbed moisture and to bind the components together, the fibers are soaked with special organic additives. These additives, like all organics, are flammable. The more densely used mineral wool - the more organic matter in it. The most combustible mineral wool is used for insulating flat roofs and as the core of sandwich panels [11, 12].

Fig. 1. Types of mineral wool

PIR (Fig. 2) plate based on polyisocyanurate as a heat insulating material with one of the lowest thermal conductivity is used worldwide. Polyurethane foam (PUR) and polyisocyanurate (PIR) are two classes of related polymers produced by the reaction of several components. PIR has higher flame retardant qualities than traditional PUR insulation. The operating temperature of the PIR reaches 140 °C, whereas PUR can only be used at temperatures below 100 °C. Both PUR and PIR have high moisture-resistant qualities and are practically vapor-proof. An important factor is also the fact that the density of polyurethane foam is ten times less than that of wood, i.e. in a unit of volume less than the product to be burned. The construction experience over the past ten years shows that the use of sandwich panels and PPU heat insulation plates (PUR and PIR) is by far the most effective and promising both in terms of ease of installation and minimization of costs, and in terms of energy saving [13–16].

Fig. 2. Types of PIR

Carbon and aerogel (Fig. 3) are materials that have not been well studied and, presumably, will be able to push out PIR and mineral wool from leading places. Materials are typically characterized by low density solid, low optical index of refraction, low thermal conductivity, low speed of sound through materials, high surface area, and low dielectric constant. They have great potential in a wide range of applications as energy efficient insulation, windows, acoustics, and so forth. In works [17–20] a comparative analysis of physical parameters was carried out and some advantages and disadvantages of each of the considered types of heaters were revealed.

Fig. 3. Aerogel and Carbon.

There are a number of factors that must be considered when considering the possibility of using certain types of thermal insulation [21]. One of these is the fire hazard of a building material. Fire hazard is assessed according to various fire-fighting characteristics, such as: flammability, flame spread over the surface, smoke-generating ability, toxicity and flammability. The combination of these indicators allows you to assign a certain fire hazard class to any building material. Papers [22, 23] evaluate the PIR behavior in case of fire.

2 Methods

The essence of the method is to create a stationary heat flux passing through a flat sample of a certain thickness and directed perpendicular to the front (largest) edges of the sample, measuring the density of this heat flux, the temperature of the opposite front faces and the sample thickness. The relative error in determining the effective thermal conductivity and thermal resistance by this method does not exceed ±3%, if the test was carried out in full compliance with the requirements of this standard. A sample is made in the form of a rectangular parallelepiped, the largest (front) whose faces have the shape of a square with a side equal to the side of the working surfaces of the instrument plates. The sample is dried to constant weight at a temperature specified in the regulatory document on the material or product. The sample is considered dried to constant weight, if the loss of its mass after the next drying for 0.5 h does not exceed 0.1%. The sample to be tested is placed in the instrument. The heat flux through the test sample is considered steady (stationary) if the thermal resistance values of the sample, calculated from the results of five consecutive measurements of signals from temperature sensors and heat flux density, differ from each other by less than 1%, while these values do not increase and do not decrease monotone.

Materials studied:

- Mineral wool with dimensions of 250 × 250 mm and a thickness of 50 mm;
- PIR plates with dimensions of 250 × 250 mm and 50 mm thick with double-sided cladding with aluminum foil 50 microns thick;
- PIR plates with dimensions of 250 × 250 mm and 50 mm thick without cladding;
- PIR plates with dimensions of 250 × 250 mm and 30 mm thick with double-sided cladding with aluminum foil 50 microns thick;
- PIR plates with dimensions of 250 × 250 mm and 30 mm thick without cladding;
- Aerogel plates with dimensions of 250 × 250 mm and a thickness of 50 mm
- Carboniferous plates with dimensions of 250 × 250 mm and a thickness of 50 mm

Thermal insulation materials are presented in Fig. 4.

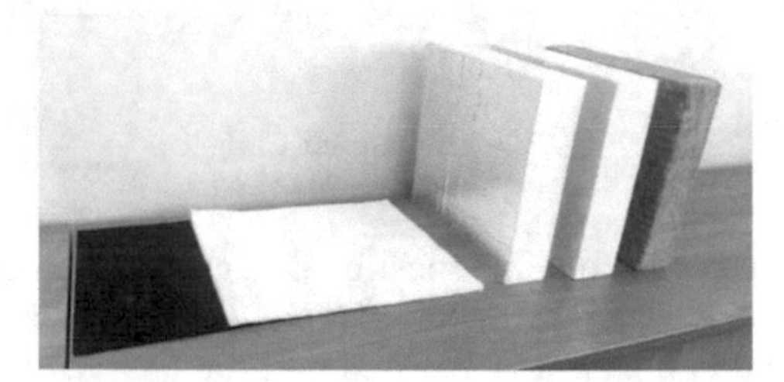

Fig. 4. Thermal insulation materials

Regulations:

- GOST 7076-99 Building materials and products. Method of determination of steady-state thermal conductivity and thermal resistance.

3 Results and Discussion

The thermal conductivity of thermal insulation materials was determined using the PIT-2.1 device (IzTekh LLC) (Fig. 5). In accordance with GOST 7076 at an average sample temperature of (25 ± 1) C ((298 ± 1) K).

Fig. 5. The device for measuring thermal conductivity PIT-2.1

The results are shown in Table 1.

Table 1. Determination of thermal conductivity of samples of mineral wool, PIR, carbon and aerogel.

Sample	W (dried)
	t_{cf} = 25 °C
Mineral wool ρ = 130 kg/m^3, δ = 50 mm	0.0351
Mineral wool ρ = 130 kg/m^3, δ = 50 mm	0.0331
PIR with facing of foil, δ = 50 mm	0.0220
PIR with facing of foil, δ = 50 mm	0.0210
PIR without facing, δ = 50 mm	0.0242
PIR without facing, δ = 50 mm	0.0239
PIR with facing of foil, δ = 30 mm	0.0200
PIR with facing of foil, δ = 30 mm	0.0227
PIR without facing, δ = 30 mm	0.0244
PIR without facing, δ = 30 mm	0.0242
Carbon, δ = 2 mm	0.0140
Carbon, δ = 2 mm	0.0100
Carbon, δ = 2 mm	0.0110
Aerogel, δ = 5 mm	0.0163
Aerogel, δ = 5 mm	0.0152
Aerogel, δ = 5 mm	0.0167

4 Conclusions

As a result of the tests, materials with the lowest thermal conductivity were identified: carbon fiber and aerogel plates. A careful analysis of the literature shows that, together with the loss of the thermal insulation ability of mineral wool during wetting and flammability of PIR panels, these two materials are quite capable of replacing mineral wool and foam polyisocyanurate panels on the market.

References

1. Sulakatko, V., Lill, I., Liisma, E.: Analysis of on-site construction processes for Effective External Thermal Insulation Composite System (ETICS) installation. Procedia Econ. Financ. **21**, 297–305 (2015). https://doi.org/10.1016/S2212-5671(15)00180-X
2. Csoknyai, T., Hrabovszky-horváth, S., Georgiev, Z.: Building stock characteristics and energy performance of residential buildings in Eastern-European countries. **132**, 39–52 (2016). https://doi.org/10.1016/j.enbuild.2016.06.062
3. Vereecken, E., Roels, S.: Capillary active interior insulation systems for wall retrofitting: a more nuanced story. Int. J. Archit. Herit. **10**, 558–569 (2016). https://doi.org/10.1080/15583058.2015.1009575
4. Gorshkov, A.S., Vatin, N.I., Rymkevich, P.P., Kydrevich, O.O.: Payback period of investments in energy saving. Mag. Civ. Eng. **78**, 65–75 (2018). https://doi.org/10.18720/MCE.78.5
5. Petrichenko, M., Nemova, D., Kotov, E., Tarasova, D., Sergeev, V.: Ventilated facade integrated with the HVAC system for cold climate, pp. 47–58 (2018). https://doi.org/10.18720/mce.77.5
6. Kim, J.-H., Kim, S.M., Kim, J.T.: Simulation performance of building wall with vacuum insulation panel. Procedia Eng. **180**, 1247–1255 (2017). https://doi.org/10.1016/j.proeng.2017.04.286
7. Papadopoulos, A.M., Giama, E.: Environmental performance evaluation of thermal insulation materials and its impact on the building. Build. Environ. **42**, 10
8. Vilches, A., Garcia-Martinez, A., Sanchez-Montañes, B.: Life cycle assessment (LCA) of building refurbishment: a literature review. Energy Build. **135** (2017). https://doi.org/10.1016/j.enbuild.2016.11.042
9. Saber, H.H., Swinton, M.C., Kalinger, P., Paroli, R.M.: Long-term hygrothermal performance of white and black roofs in North American climates. Build. Environ. **50**, 141–154 (2012). https://doi.org/10.1016/j.buildenv.2011.10.022
10. Asdrubali, F., Ferracuti, B., Lombardi, L., Guattari, C., Evangelisti, L., Grazieschi, G.: A review of structural, thermo-physical, acoustical, and environmental properties of wooden materials for building applications. Build. Environ. **114**, 307–332 (2017). https://doi.org/10.1016/j.buildenv.2016.12.033
11. Berge, A., Johansson, P.: Literature review of high performance thermal insulation. Civ. Environ. Eng. (2012). https://doi.org/10.1097/bsd.0000000000000074
12. Yang, S.J., Zhang, L.W.: Research on properties of rock-mineral wool as thermal insulation material for construction. Adv. Mater. Res. **450–451**, 618–622 (2012). https://doi.org/10.4028/www.scientific.net/AMR.450-451.618
13. Kirpluks, M., Cabulis, U., Avots, A.: Flammability of bio-based rigid polyurethane foam as sustainable thermal insulation material. In: Insulation Materials in Context of Sustainability (2016)

14. Gravit, M., Kuleshin, A., Khametgalieva, E., Karakozova, I.: Technical characteristics of rigid sprayed PUR and PIR foams used in construction industry. In: IOP Conference Series: Earth and Environmental Science (2017)
15. Wu, L., Gemert, J. Van, Camargo, R.E.: Rheology Study in Polyurethane Rigid Foams
16. Pescari, S., Tudor, D., Tölgyi, S., M.C.: Study concerning the thermal insulation panels with double-side anti-condensation foil on the exterior and polyurethane foam or polyisocyanurate on the interior. Key Eng. Mater. **660**, 244–248 (2015)
17. Jelle, B.P., Baetens, R., Gustavsen, A.: Aerogel Insulation for Building Applications. In: The Sol-Gel Handbook (2015)
18. Liu, Z.H., Wang, F., Deng, Z.P.: Thermal insulation material based on SiO_2 aerogel. Constr. Build. Mater. (2016). https://doi.org/10.1016/j.conbuildmat.2016.06.096
19. Schiavoni, S., D'Alessandro, F., Bianchi, F., Asdrubal, I.F.: Insulation materials for the building sector: a review and comparative analysis (2016)
20. Asdrubali, F., D'Alessandro, F., Schiavoni, S.: A review of unconventional sustainable building insulation materials. Sustain. Mater. Technol. (2015). https://doi.org/10.1016/j.susmat.2015.05.002
21. Gorshkov, A.S., Rymkevich, P.P.: A diagram method of describing the process of non-stationary heat transfer. Mag. Civ. Eng. **60**, 68–82 (2015). https://doi.org/10.5862/MCE.60.8
22. Zhang, Z., Ashida, K.: Amide-modified polyisocyanurate foams having high thermal stability. J. Cell. Plast. **33**, 487–501 (1997). https://doi.org/10.1177/0021955X9703300505
23. Wang, Y.C., Foster, A.: Experimental and numerical study of temperature developments in PIR core sandwich panels with joint. Fire Saf. J. (2017) https://doi.org/10.1016/j.firesaf.2017.03.003

The Design of a Timber-Metal Arch with an Additional Lattice Taking Ductility of Nodal Joints into Account

Ivan Inzhutov[1], Victor Zhadanov[2], Ilya Polyakov[1](✉), Maria Plyasunova[1], and Nikolai Lyakh[1]

[1] Siberian Federal University, 82 Svobodny Prospect, Krasnoyarsk 660041, Russia
Polyakov_ilya@bk.ru
[2] Orenburg State University, 13 Pobedy Street, Orenburg 460018, Russia

Abstract. The article is dedicated to the rational efficient design of a three-hinged arched coating. It is noted that as a result of a significant decrease in the bending moment under asymmetric snow load, the structure possesses a good rate of basic materials consumption. The numerical analysis of the arch stress-strain state has been done. The results of the arch stress-strain state under asymmetric and symmetric loads, including comparison with the stress-strain state of a classical three-hinged arch, have been given.

Keywords: Arched structures · Ductility · Timber construction

1 Introduction

In the Russian and international construction practical experience, arched coverings are widely used [1]. This is caused by such advantages of supporting arches as a lack or insignificant values of bending moments (depending on the structure geometry) under uniformly distributed load along the length [2]. However, in case of one-side or asymmetric loads of the same strength, significant bending moments occur.

This peculiarity makes designers to search for a solution that will allow reducing the effect of asymmetric loads significantly.

The analysis of patents on inventions has revealed the following techniques [3–5]:

- installing a pre-stressed metal tightening into the structure of a three-hinged arch;
- installing additional inclined flexible rods and rigid bars of the lattice;
- installing a diaphragm with the stiffness of 10 to 20 times bigger than the upper belt linear stiffness in the middle of the arch span;
- increasing the height of the cross section of the upper belt by using lattice section made of still elements.

We should note that along with the arch technological loads (such as crane and equipment loads), asymmetric loads occur as a result of uneven snow distribution along the arch length. In the modern standards regulating snow load on arched surfaces there are the following unloading schemes: Option No. 1 – parabolic load distribution with

V. Murgul and M. Pasetti (Eds.): EMMFT-2018, AISC 983, pp. 884–891, 2019.
https://doi.org/10.1007/978-3-030-19868-8_87

the maximal intensity in the middle of the arch span and zero intensity at the point of 60° slope of the arch; Option No. 2 – linear load distribution with zero intensity in the middle of the arch span and maximum intensity on one side of the arch at the point of 30° slope; on the other hand, the intensity at the 30° slope point is half of the maximum (Fig. 1) [6].

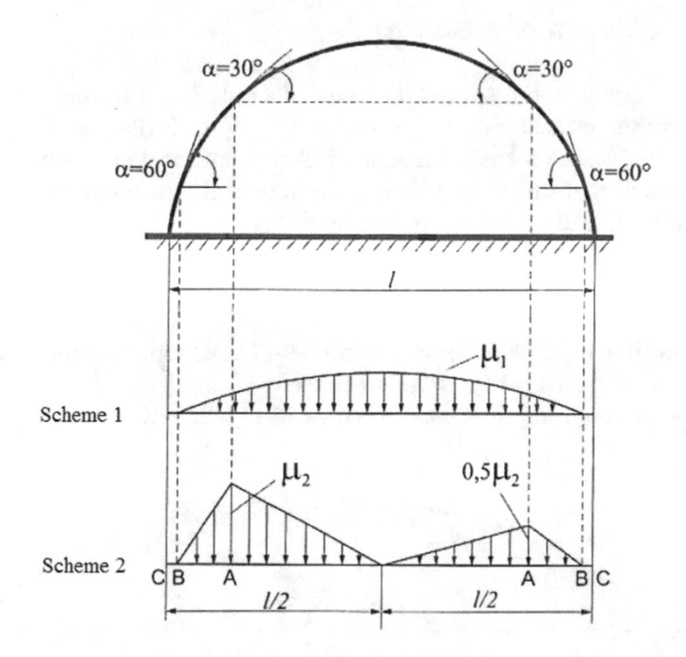

Fig. 1. The scheme of snow load distribution on the arch structure.

Widely spread one-sided and asymmetric loads for various types of buildings with arched coverings causes the necessity of designing new technological solutions that will allow reducing arch bending moments and thus decreasing material consumption and increasing economic attractiveness of applying arched structures.

To decrease bending moments Prof. V.I. Zhadanov and Prof. P.A. Dmitriev have proposed a new rational scheme of an arched structure (Fig. 2, b) [7].

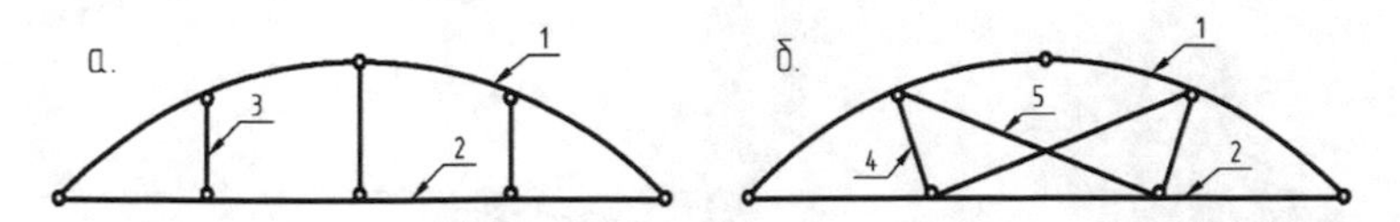

Fig. 2. The schemes of classical (a) and rational (b) arches: 1 – upper circular belt, 2 – tightening, 3 – suspensions, 4 – struts, 5 – flexible rods.

By using struts and flexible rods, the reduction of bending moments under one-sided loads has been assured. A timber-metal circular arch with an additional lattice is a coating structure that includes the upper circular belt, tightening that connects the arch bearing joints, struts and flexible rods. The struts are installed symmetrically in the 1/4 of the arch span and directed along the radius perpendicular to the belt. The connection of the struts with the upper belt and tightening is hinged.

2 Design Solution of a New Arch

The arched covering with a span of 10 m has been designed for the region with the following climate characteristics: a snow area - V (2.5 kPa), a wind area – III (0.38 kPa). The distance between the arches is 3 m. Laminated plywood wooden plates have been used as roof structures: they have been insulated with mineral wool; the roof covering has been made of metal composite panels.

The design of the new three-hinged arch (Fig. 3, a) includes the upper belt, assembled of two curved wooden beams, wooden struts attached to the upper belt in the quarters of the span and installed in the arch plane in the direction perpendicular to its axis, metal flexible rods that connect the neighbor struts' tops and bottoms, the lower belt made of steel products of circular cross-section.

The main nodes of the structure are shown in Fig. 3, b–f.

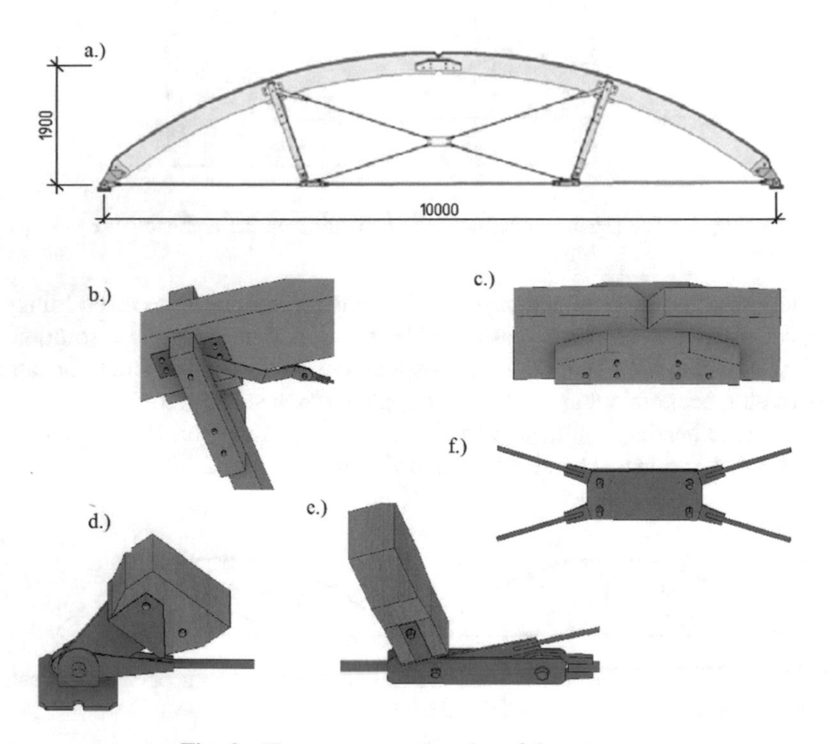

Fig. 3. The structure and nodes of the new arc.

The proposed arch design with the span of 10 m has been developed before the album stage of working drawings and it is characterized by the following values of main materials consumption: the mass of metal elements per one arch is 89.64 kg; the amount of wood needed to produce one arch is 0.545 m^3.

3 Methodology of Numerical Research

To study stress-strain state of the arch, the SCAD software has been used. The elements of a three-hinged circular arch with a span of 10 m, a lifting boom of 1.9 m and a curvature radius of 13 m has been modelled on the design scheme by the finite rod elements. The struts positions have varied, taking distance a from the reference node to adjunction to the upper belt where: a = 2.5 m (scheme 1), a = 3 m (scheme 2) and a = 2 m (scheme 3).

On the design scheme, the upper belt has been divided into 8 equal sections, for scheme 2 and 3, one of the sections has been additionally divided at the point of the intersection with the strut. The upper belt on the design scheme has been divided into 8 equal sections, for scheme 2 and 3 one of the sections has been additionally divided at the point of the intersection with the strut.

During the arch stress-strain state four different versions of the load have been considered:

1. Dead load of the load-bearing structures, including the roof. The load is evenly distributed along the length of the arch; the values of the loads from the dead load of enclosing structures have been measured at the width of the load area – the distance between the arches of 3 m. The dead load of the arch bearing elements has been assigned by the SCAD internal tools based on the pre-assigned sections of the elements;
2. Snow loads. The intensity of snow loads has been adopted for the V snow area that is 2.5 kPa:
   - Option 1, the load value has been calculated for each node of the arch. The evenly distributed load is trapezoidal. The load intensity has been calculated for the extreme points of each arch element; the load value between the points has been determined by linear interpolation.
   - Option 2, the load values have been calculated at the points with a 30° slope and at the extreme points of the arch; the intermediate values have been determined by interpolation. The distributed load for the arch rods has been set similarly to Option 1.
3. Wind loads. The wind load intensity for snow area III is 0.38 кPa:

All the loads, except for the snow load, have been applied perpendicular to the support line. The load value has been calculated for the distance between the arches in the longitudinal direction of 3 m.

The stress in the considered arch has been compared to the stress in a classical three-hinged circular arch for the schemes of snow and wind loads regulated by Construction Rules and Regulations 20.13330.2016 «Loads and affects».

Figure 4 shows the schemes of load cases.

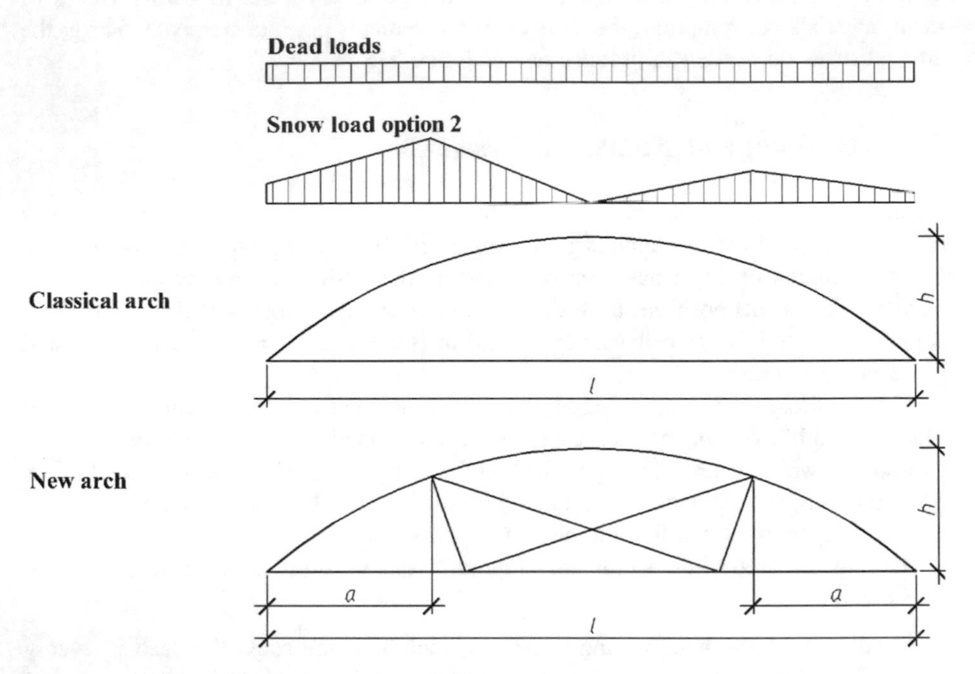

Fig. 4. The scheme of load cases. *h* – a lift boom, *l* – span, *a* – distance from the supporting node to the connection point of the strut and the upper belt.

a has been measured in the following values: $a_1 = \frac{1}{4}$ L (2,5 m), $a_2 = \frac{1}{3}$ L (3 m), $a_3 = \frac{1}{5}$ L (2 m).

$$E_y = \frac{E}{1 + \delta_o \cdot E \cdot A/(N \cdot l)} \tag{1}$$

where E is the initial modulus of timber elements' elasticity (E = 10000 MPa);

δ_o is the calculated limit value of the ductility deformations that is taken in accordance with the limit deformation of the nodal joint (which is 1.5 mm at the frontal felling and end to end; 2.0 mm at the dowels of all types; 3.0 mm in the junctions across the fibers) and the degree of the bearing capacity use, respectively; *A* is a cross-sectional area of the rod, m^2; *N* is a force effecting the rod, кN; *l* is a length of the rod, m.

The calculation of the limit ductility deformation value δ_o:

$$\delta_o = \delta_1 + \delta_2 \tag{2}$$

where δ_1 is a ductility deformation of the bolted connection, mm; δ_2 is a shear deformation of timber elements, mm.

4 Analyzing the Obtained Results

The results of the design scheme statistical analysis show that the most unfavorable load combination is the combined effect of the structure own weight and asymmetric snow load (Option 2). In this case the flexible rod adjusted to the arch upper belt at the point of the maximum load (in our case it is on the left side) turns out to be compressed. Since the flexible rod was not initially intended to perceive compressive forces it has been excluded at the next stage of the calculations in the design scheme.

The maximum values of the moments, transverse and longitudinal forces in the different arch elements are shown in Table 1.

Table 1. Maximum values of the moments (M, $kN \cdot m$), transverse (Q, kN) and longitudinal forces (N, kN)

Maximum values of forces Design scheme	M, $kN \cdot m$	Q, kN	N, kN
A classical three-hinged arch consisting of an upper belt and a tightening	25.89	25.52	112.43
Shifting the strut from the support joint by 2,5 m	15.55	19.67	126.98
Shifting the strut from the support joint by 3 m	15.92	19.88	127.26
Shifting the strut from the support joint by 2 m	16.70	20.29	126.44

It has been determined that the maximum values of the moments and transverse forces are observed in a classical arch; when the timber struts and flexible rods are installed the value of the calculated moment reduces by 40…35.5% (depending on the strut location), at the same time the transverse forces decrease by 23…20.5% (Fig. 5).

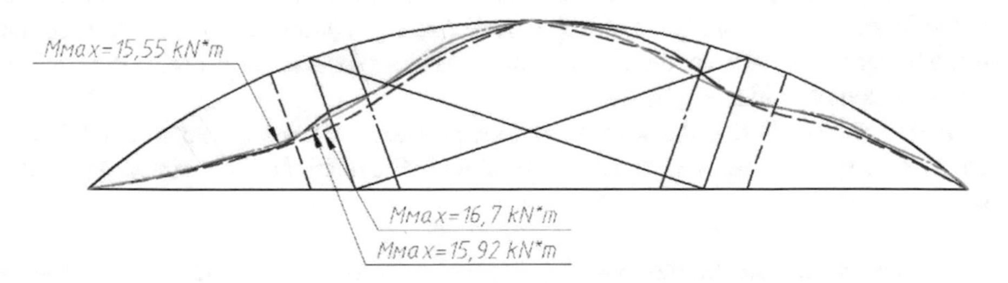

Fig. 5. Diagrams of moments under various arch design schemes.

The maximal values of forces in the new design arch occur in the variant when the strut is shifted from the support joint by 2 m. Besides, it has been noted that when the strut is shifted away from the support joint the longitudinal forces in the rods reduce while the stress in the struts increases.

According to the analysis of the diagrams, it has been concluded that the most efficient option is shifting the strut away from the support joint by 2.5 m, as it has the most favorable stress values and it is the most relevant in terms of design and esthetics.

When the strut is shifted closer to the center of the span, the maximum moment is increased by 2.5%, while shifting it towards to the support joint gives the increase by 7.5%.

To take the ductility of the nodal joints into account, the elements with nominal elasticity modulus have been introduced to the design scheme. Figure 6 shows the diagrams of moments My, kN · m, where the dotted line shows the diagram gained taking the ductility calculations into account, and the solid line shows the diagram without taking ductility into account. The calculation has been made under the most unfavorable load combination.

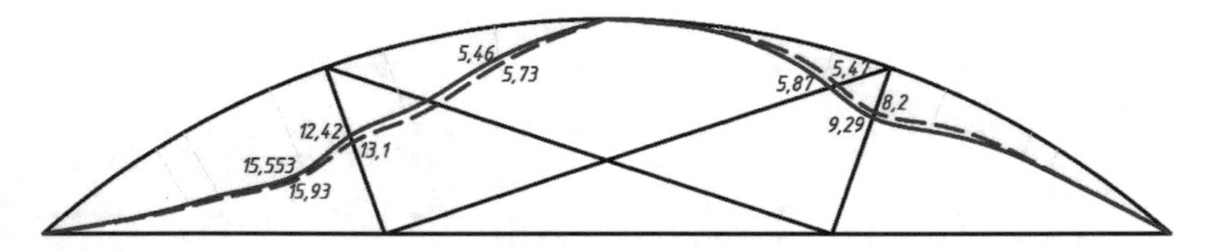

Fig. 6. Diagrams of moments M_y, $\kappa N \cdot m$.

The arch calculation taking the nodal joints' ductility into account shows that the value of the maximum moment has increased from 15.553 up to 15.93 kN m. The value of the maximum longitudinal force has turned out to be less than the value calculated without taking the ductility into account: 124.333 kN versus 124.548 kN.

5 Conclusion

As a result of the conducted research, it has been determined that the use of a new three-hinged arch design allows reducing the calculated bending moment at the section of the upper belt by 40% under asymmetrical load and decreasing timber consumption for the structure production by 18%.

As a result of shifting the struts to various positions, it has been determined that in terms of the stress-strain state, the struts installed in the quarters of the arch span are the most efficient.

Acknowledgements. In the research, the students Astafyev, A.S., Tuchin, D.V., Siberian Federal University, participated.

References

1. Pischl, R., Schickhofer, G.: The Mur River Wooden Bridge, Austria. Struct. Eng. Int. **3**(4), 217–219 (1993)
2. Stoyanov, V.V., Boyadzhi, A.A.: Experimental research of strength and deformability of a combined metal-timber arch structure. Timber J. (3), 93–103 (2015)

3. Shengeliya, A.K., Kasabyan, L.V., Danilevskiy, P.M., Levin, Yu.I.: Patent of the Russian Federation № 2687439/29-33, 22 November 78. A three-hinged timber arch, patent of Russia № 767317. Statement № 36 (1980)
4. Shapiro, A.V., Sarafanova, M.N., Glezerov, E.I., Grigoryev, V.A.: Patent of the Russian Federation № 4834734/33, 04 April 90. A construction arch, Patent of Russia № 1756485A1. Statement № 31 (1992)
5. Kuznetsov, I.L., Safin, R.K.: Patent of the Russian Federation № 2742128/29-33, 28 March 79. A latticed arch, Patent of Russia № 844701. Statement № 25 (1981)
6. СП 20.13330.2016. Loads and impacts. Actual edition воздействия. Актуализированная редакция of the Construction Rules and Regulations 2 January 07-85* — Introduction 01 January 2017. Standards Publishing House, Moscow (2017). 80
7. Zhadanov, V.I., Dmitriev, P.A., Mikhailenko, O.A., Arkaev, M.A.: Patent of the Russian Federation № 2498026/20, 23 April 2012. An arched structure for building roofing, Patent of Russia № 2498026 Statement № 31 (2013)
8. Inzhutov, I.S.: Block-trusses based on wood for building coatings. Ph.D. thesis in Engineering Novosibirsk State Academy of Civil Engineering (Novosibirsk) 282 (1995)
9. Deriglazov, O.Y.: Developing, designing and researching of a timber ribbed ring dome with stiffness units and demountable nodes: Ph.D. thesis in engineering/Tomsk State University of Civil Construction and Architecture, Tomsk, 183 (2007)
10. Rozhkov, A.F.: Managed block sections with prestressed wooden elements, Ph.D. thesis in Engineering Krasnoyarsk State Academy of Architecture and Building, Krasnoyarsk, 191 (2006)
11. Tur, V.I., Tur, A.V.: Influence of the compliance of nodal connections on the stress-strain state of a metal net dome. Fundam. Res. **6**(Part 6), 1165–1168 (2014)
12. Cannizzaro, F., Greco, A., Caddemi, S., Caliò, I.: Closed form solutions of a multi-cracked circular arch under static loads. Int. J. Solids Struct. **121**, 191–200 (2017)

Fuzzy Controller of Rotation Angle of Blades of Horizontal-Axial Wind Power Generation Plant

Yura Bulatov[1(✉)], Andrey Kryukov[2,3], Van Huan Nguyen[3], and Duy Hung Tran[4]

[1] Bratsk State University, Bratsk, Russia
bulatovyura@yandex.ru
[2] Irkutsk State Transport University, Irkutsk, Russia
and_kryukov@mail.ru
[3] Irkutsk National Research Technical University, Irkutsk, Russia
huanco.k7a@gmail.com
[4] Military Industrial College, Viet Tri, Phu Tho Province, Socialist Republic of Vietnam
tranduyhung67@yahoo.com

Abstract. Currently, power generation industry transition to a new technological platform is underway. The platform is based on the smart grids concept. It provides for a broader use of the distributed generation plants that use such renewable energy sources as wind power generation plants. The distributed generation plants can be operated in the existing grids or be combined into microgrids to enhance consumers power supply reliability. Efficiency of wind power generators in microgrids can be raised using automatic control systems. However, when solving issues of wind power generators modes control, problems arise that cannot be resolved by traditional methods. A wind power generation plant is a non-linear and non-stationary facility for which fuzzy controllers can be used. Currently, there is no unified technique for tuning such controllers. This article deals with simulation and tuning of the fuzzy control system for horizontal-axial slow-speed wind power generation plant that can be operated both independently for a selected load and in the microgrid. Simulation results in MATLAB system show that fuzzy power control of the wind power generation plant allows its operation stability at variation of consumer loads. A rule database has been generated based on the suggested tuning technique that ensures wind power generator effective work both in independent mode and in the microgrids designed to enhance power supply reliability of railroad transport consumers.

Keywords: Wind power generation plant · Fuzzy controller · Simulation · Railroad power supply system · Microgrid

V. Murgul and M. Pasetti (Eds.): EMMFT-2018, AISC 983, pp. 892–901, 2019.
https://doi.org/10.1007/978-3-030-19868-8_88

1 Introduction

Recently, interest to the use of renewable, green wind energy has grown considerably. The installed power of wind generation plants (WGP) connected to power grids from 2013 to 2017 has grown in the world from 297 GW to 539 GW. In addition, many countries have developed their long-term ambitious plans for wind power generation development for 2020…2030 [1]. The main impediment limiting the use of wind as a power source is its velocity variability. However, WGP designs perfection is constantly underway. Currently, energy generation industry transition to a new technological platform based on smart grids concept is underway [2–5]. This concept provides for a broader use of distributed generation plants (DGP) that can operate either in existing grids or be combined into microgrids [6–14]. The use of WGP in the microgrid would allow voltage stabilization during peak load hours, and reduction in man-induced loads on the environment. Efficiency of WGP in microgrids can be raised using automatic control systems. However, when solving issues of WGP modes control, problems arise that can be resolved by traditional methods with difficulties. A wind power generation plant is a non-linear and non-stationary facility for which fuzzy controllers (FC) can be used [15–19]. Nevertheless, currently, there is no unified technique for tuning such controllers. This article deals with simulation and tuning of the fuzzy control system for horizontal-axial slow-speed WGP that can be operated both independently for a selected load, and in the microgrid.

2 A Method for Horizontal-Axial WGP Power Control

When controlling wind generation plant, an issue of source data fuzziness is encountered, since WGP operates on a casual schedule. The necessity for accounting of air flow uncertain velocities is WGP automatic control system's distinct feature.

Mechanical power developed by horizontal-axial wind turbine is determined using the formula:

$$P_m = \frac{1}{8}\rho\pi D^2 V^3 C_p(\lambda,\beta), \tag{1}$$

where ρ – air density, kg/m^3; D – diameter of the area swept by the wind wheel, m; V – wind velocity, m/s; C_p – wind power efficiency (or power coefficient).

Wind power efficiency depends on wind wheel design features, in particular, the blades rotation angle β and specific speed λ, determined as [20]

$$\lambda = \frac{\omega \cdot r}{V}, \tag{2}$$

where ω – wind turbine rotor rotational frequency; r – wind wheel radius.

One can find dependences of power ratio on specific speed for different rotational angles of the blades from which it can be seen that with a large deviation of the blades from the rotation plane of the wind wheel, the power generated by the wind turbine decreases [20]. The wind turbine power can be controlled by changing the blades length and their rotation angle. This work analyses the method of output power control, and consequently, rotational frequency of horizontal-axial WGP via the mechanism of wind wheel blades rotation by the electrical drive. In this case, it is assumed that the wind turbine is designed for wind and possibility of balloon rotation possibility is not considered. MATLAB integrated environment is used for WGP and the blades rotational angle control system simulation. It is proposed that servo drive with reduction gear should be used as the blade rotational mechanism, the reduction gear model being created in Simulink package of MATLAB system (Fig. 1). In this case, speed loop transfer function can be represented as aperiodic link with amplification gain K_s and time constant T_s. The reduction gear is represented in a simplified form as the amplifier with Kr gain. To determine the angle of deviation, it is necessary to integrate the signal over frequency. The servo drive feature is the presence of a negative feedback that would allow for maintaining the deviation angle set value on the input (*Angle*).

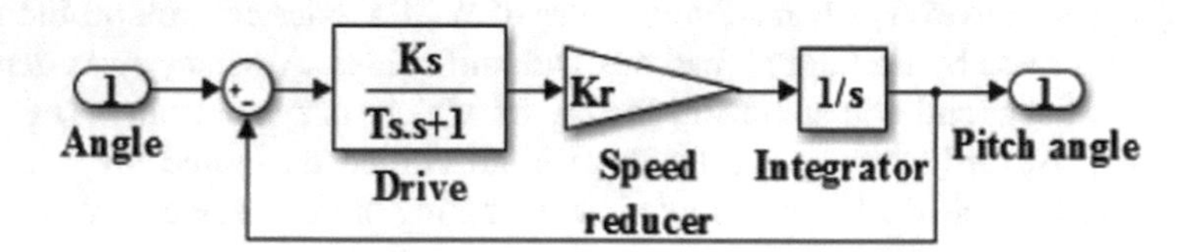

Fig. 1. Structural diagram of the blades' rotational mechanism model.

In order to determine the blades rotation angle set value in different modes of WGP operation, the use of a fuzzy controller model is proposed which is created with Simulink packages and Fuzzy Logic Toolbox of MATLAB system.

3 Technique for Synthesis of WGP Frequency and Power Fuzzy Controller

The system of a fuzzy logical inference is a counterpart of any fuzzy controller, the system including: fuzzification and defuzzification units; knowledge base containing the base of rules and fuzzy variables; the output unit (Fig. 2).

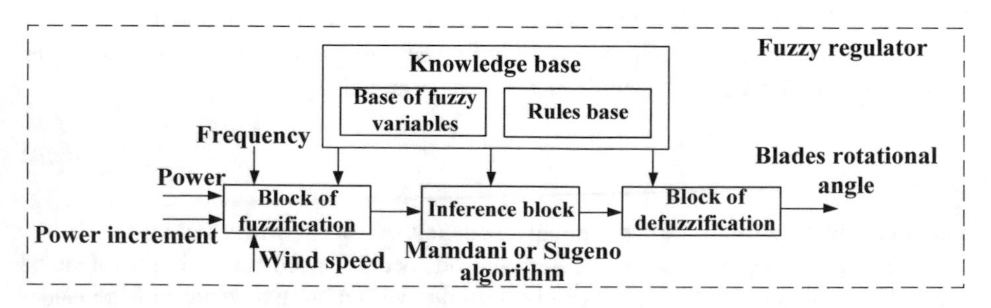

Fig. 2. The system of fuzzy logical output for WGP controller.

Information coming to the system of fuzzy inference input corresponds to the control process variables. The control, signals are generated at the output of the fuzzy inference system. In order to ensure WGP frequency and power control it is suggested that Mamdani and Sugeno algorithms of fuzzy logical inference should be used that imply the following stages performance:

1. Formation of the fuzzy inference system's rules database. The form of IF 'Condition', THEN 'Conclusion' (F) is the most frequently case for the rules database representation. Here F determines weight coefficient of the respective rule and can accept values from interval [0, 1]. When forming the rules database of the fuzzy productions, it is necessary to determine the following sets: rules of the fuzzy productions, input and output linguistic variables.
2. Input variables fuzzification. Fuzzification stage purpose is establishing correspondence between a specific value of an individual input variable of the fuzzy inference system and membership function value of its corresponding term of the input linguistic variable.
3. Aggregation of subconditions in fuzzy productions rules. Aggregation is a procedure of conditions truth degree determining for each of the fuzzy inference system rule.
4. Activation or composition of sub-conclusions in fuzzy productions rules. Activation in the fuzzy inference systems is a procedure or a process for determining the truth degree for each of the fuzzy productions rules sub-conclusion. Min-activation method is used in the proposed fuzzy controller:

$$\mu'(y) = \min\{c_i, \mu(y)\}, \tag{3}$$

   where c_i – sub-conclusions truth degree values for each of the rules included into the considered rules database of the fuzzy inference system; $\mu(y)$ – membership function value of the term which is a value of some output variable set on the universal Y range.
5. Accumulation of the production fuzzy rules conclusions. The aim of accumulation is to integrate all truth degrees of conclusions (sub-conclusions) to obtain membership function for each of the output variables. The reason for this stage performance is that sub-conclusions associated with one and the same output linguistic variable belong to different rules of the fuzzy inference system. Accumulation is performed via the method of the fuzzy sets max-integrating. It should be noted that this stage is absent for Sugeno algorithm, since calculations are performed with vulgar real numbers.
6. Output variables defuzzification. The aim of defuzzification is to obtain a common quantitative value for each of the output variables, while using accumulation results of all linguistic variables, this quantitative value having the ability of being used by special devices that are external with respect to the fuzzy inference system. The gravity center method is used in the proposed fuzzy inference system [21].

No mathematical description of the controlled object is required for FC tuning. One has to have an idea of an object behavior only. Rules data base shall be generated based on this knowledge as per the 'if … then' form. This circumstance is the main advantage of FC before classical controllers. The fuzzy controller synthesis, in compliance with fuzzy logic inference stages, implies determining ranges of input and output values changes, choice of fuzzy variables membership function and their linguistic evaluation. The following signals are proposed to be given in order to control WGP power and frequency: deviation of rotor rotational frequency from the rated value $d\omega$, WGP set power value P_z (current value of generator electrical power can also be used P_e), wind velocity, and wind turbine mechanical power gain dPm, determined as the difference between current value of mechanical power and power at the preceding moment of time: $dP_m = P_m(n) - P_m(n-1)$. FC signal is a signal of setting the blades rotation angle.

The following fuzzy term-sets are used in the proposed WGP fuzzy control system:

1. for linguistic variable 'deviation of rotor rotational frequency $d\omega$': *NB* – negative big; *NS* – negative small; *Z* – zero; *PS* – positive small; *PB* – positive big.
2. for linguistic variable 'the set WGP power value *Pz*': *VS* – very small; *S* – small; *A* – average; *B* – big; *M* – maximal.
3. for linguistic variable 'wind velocity *V*': *W* – week; *B* – basic; *S* – strong.
4. for linguistic variable 'mechanical power gain dP_m': *N* – negative; *Z* – zero; *P* – positive.
5. for linguistic variable 'blades rotational angle *Angle*': *Z* – zero; *VS* – very small; *S* – small; *A* – average; *B* – big; *VB* – very big; *L* – limiting.

To generate fuzzy rules expert knowledge base, it is suggested that dependences of mechanical power from rotor rotational frequency for blades different rotation angles and wind velocity, should be obtained experimentally. The obtained dependences are divided into intervals corresponding to term-sets of linguistic variables. The relevant rules database is created based on these dependences that takes into account the main criterion – approximation to the WGP maximal power point in these conditions.

In FIS Editor program of MATLAB system, the system of fuzzy logical inference for WGP control was developed and tuned in accordance with technique described, while using Sugeno algorithm. The relevant pieces of the rules databases and surfaces of the fuzzy inference are represented in Fig. 3.

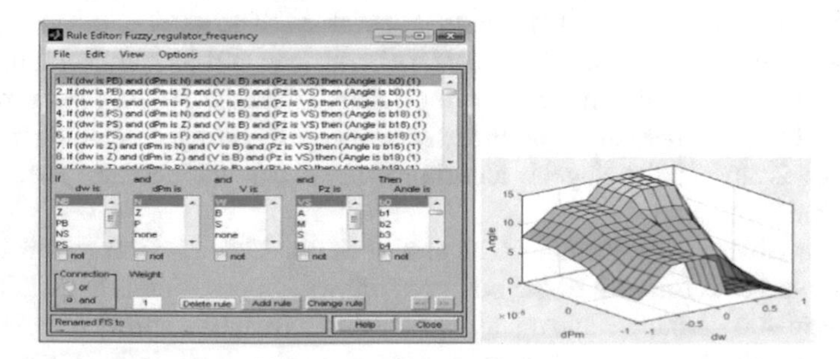

Fig. 3. Pieces without rules and the fuzzy inference surfaces of WGP control system for basic wind velocity: Sugeno fuzzy logical inference system.

4 Description of the Model and the Simulation Results

Standard MATLAB system model was used for the wind wheel in which the following equation of turbine characteristics was used [22]:

$$C_p(\lambda, \beta) = C_1\left(\frac{C_2}{\lambda_i} - C_3\beta - C_4\right)e^{-\frac{C_5}{\lambda_i}} + C_6\lambda, \tag{4}$$

where $\frac{1}{\lambda_i} = \frac{1}{\lambda + 0.08\beta} - \frac{0.035}{\beta^3 + 1}$; constant coefficients used in the model: $C_1 = 0.5176$, $C_2 = 116$, $C_3 = 0.4$, $C_4 = 5$, $C_5 = 21$, $C_6 = 0.0068$.

The wind wheel model allows construction of power coefficient characteristics dependence on specific speed for different values of the blades rotational angle β and dependence of turbine power on rotor rotational frequency for the blades' different rotational angles and wind velocities. These characteristics were used for tuning the fuzzy controller. A synchronous machine model with damper winding was used for WGP generator which is a component of the SimPowerSystems package. Generator excitation system was modeled by first-order device with k_f coefficient, time constant T_f and voltage limiting unit. In addition, amplifier with ka coefficient with Ta time constant were taken into account. The following numerical parameter values were accepted: $k_a = 1$; $T_a = 0.001$; $k_f = 1$; $T_f = 0.025$. In order to achieve voltage stabilization, a model of generator automatic excitation regulator was used on terminals of WGP generator which is a proportional-integral differential (PID) regulator whose detailed description is given in work [23]. The studies conducted on WGP model that operates in standalone mode for a selected load (Fig. 4) have proved working efficiency of the fuzzy controller tuned as per the technique suggested which is nothing but maintaining a consumer's set power value, frequency and voltage when connecting an additional load of 200 KW. The rated power of the generator used equaled to 1 MW·A, WGP initial loading equaled to 75% (Fig. 5).

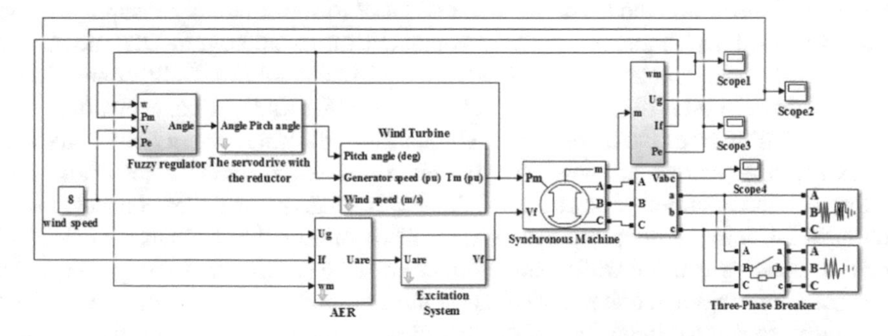

Fig. 4. The diagram of horizontal-axial slow-speed WGP controlled model in MATLAB.

As consequence, WGP operates virtually similarly when controlled both by Mamdani fuzzy logical inference system and Sugeno. However, when Sugeno algorithm is used, machine time consumption is significantly reduced when calculating control actions. Relevant oscillograms for Sugeno system are provided in Fig. 5a, which allow seeing that stabilization of generated power and voltage is observed when additional load is connected at the time moment of 4 s. In contrast, if the blades rotational angle does not change, i.e. WGP operates without power control, the system becomes unstable (Fig. 5b).

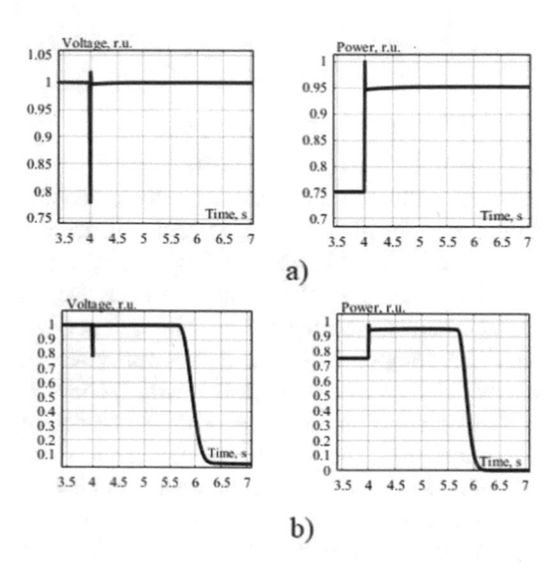

Fig. 5. Oscillograms of WGP voltage and power change when additional load is connected: (a) WGP operates with the fuzzy controller of blades rotational angle; (b) without the blades rotational angle control (deviation angle $\beta = 20^{\circ}$); r.u. – relative units.

The studies were also conducted with regard to structural scheme of railroad power supply system provided in Fig. 6. An individual power supply region (PSR) of non-tractive consumers was modeled that included DGP (turbogenerator), a supplying load group with total active power of 5 MW combined into a microgrid. DGP power was equal to 2.5 MW. Contemporary converter equipment allows DGP connection to electrical power systems (EPS) using DC links (DCL). Such a concept limits short circuit in DGP sources buses, ensures high quality of electrical energy, thus providing consumers with guaranteed power supply. Wind power generating plant whose maximal power was assumed to be equal to 1 MW at modeling stage, was connected to the microgrid DC buses. Energy storage unit was also connected to DC buses of WGP the model of which was made while using Battery unit of SimPowerSystems package. The power of energy storage unit made based on lithium-ion battery was equal to 1.5 MW. To control rotational frequency of DGP turbogenerator rotor, automatic excitation regulator (AER) and automatic regulator of rotor speed (ARRS) models were used whose description is given in [23]. The search of turbogenerator regulators optimal setting was performed using the method of harmonized setting using genetic algorithm [24]. Shut down of PSR main supply was used as a disturbance for a period of 0.5 s, due to which DGP and WGP generator with total power of 3.5 MW operate for a load of 5 MW which causes the consumer's supply voltage dip. In this case, DGP and WGP with the fuzzy controller allowed voltage maintaining on non-tractive consumers buses at a level close to the rated one and smoothen the resulted fluctuations. In case WGP is disconnected, voltage cannot be maintained at a rated level, and the supplied power quality is reduced significantly. The use of constantly connected electrical energy storage together with WGP allows maintaining the consumers power supply in case of short-time shut-down of the PSR main supply and reducing voltage dip virtually to zero. Relevant voltage oscillograms for non-tractive consumer buses confirming these conclusions are represented in Fig. 7.

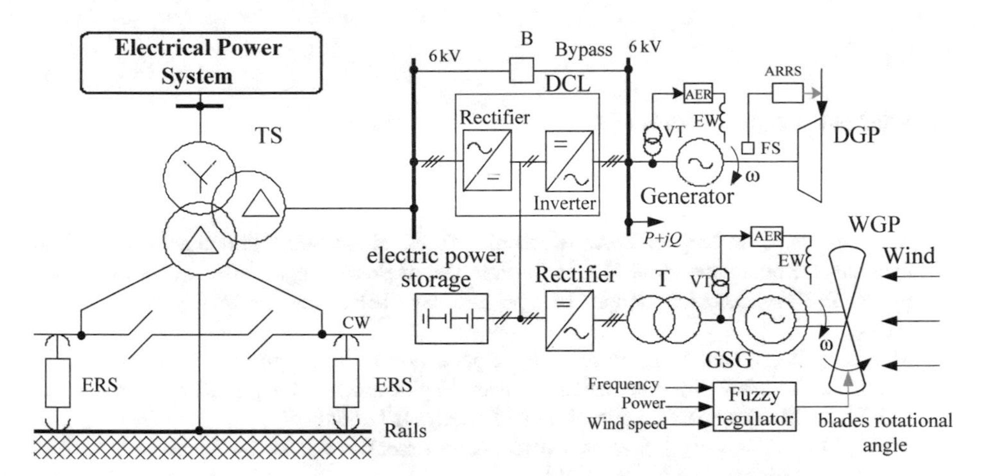

Fig. 6. Fragment of the power supply system of the railway: TS – traction substation; ERS – electro rolling stock; CW – contact wire; DCL – DC link; B - breaker; GSG – gearless synchronous generator; T – transformer; EW – excitation winding; VT – voltage transformer; FS – frequency sensor; AER – automatic excitation regulator; ARRS – automatic regulator of rotor speed.

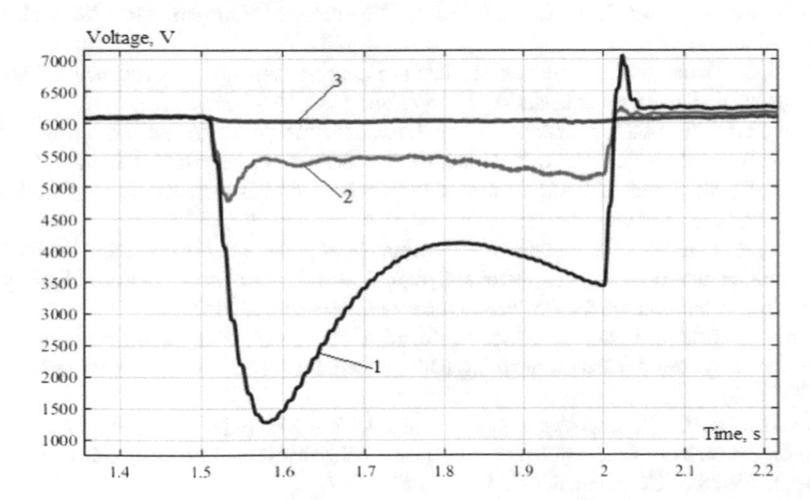

Fig. 7. Oscillograms of the existing voltage value on non-tractive consumers buses during short-time power supply shut-down: 1 – WGP is shut down (maximal voltage dip makes $\delta U = 79\%$); 2 – DGP and WGP operate in the microgrid ($\delta U = 21\%$); 3 – DGP, WGP and electrical power storage unit operate in the microgrid ($\delta U = 1.3\%$).

5 Conclusion

Computer modeling results indicate that WGP power control allows maintaining the stability of its operation not only in case of wind velocity variation, but in case of consumers loads changing. The suggested technique for the WGP fuzzy controller tuning allows formation of a universal rules database to ensure WGP efficient operation both in a standalone mode and in the microgrids designed for enhancing power supply of railroad transport stationary consumers.

References

1. Wind power capacity reaches 539 GW, 52.6 GW added in 2017. https://wwindea.org/blog/2018/02/12/2017-statistics. Accessed 24 Jan 2019
2. Wang, J., Huang, A.Q., Sung, W., Liu, Y., Baliga, B.J.: Smart grid technologies. IEEE Ind. Electron. Mag. **2**(3), 16–23 (2009)
3. Morzhin, Yu.I., Shakaryan, Yu.G., Kucherov, Yu.N., et al.: Smart grid concept for unified national electrical network of Russia. In: Preprints of Proceedings of IEEE PES Innovative Smart Grid Technologies Europe 2011, pp. 1–5. GB: IEEE, The University of Manchester (2011)
4. Mohsen, F.N., Amin, M.S., Hashim, H.: Application of smart power grid in developing countries. In: IEEE 7th International Power Engineering and Optimization Conference (PEOCO) (2013). https://doi.org/10.1109/PEOCO.2013.6564586
5. Bernd, M.B., Zbigniew, A.S.: Smart Grids – Fundamentals and Technologies in Electricity Networks. Springer, Heidelberg (2014)
6. Barker, Ph.P., De Mello, R.W.: Determining the impact of distributed generation on power systems: part 1 - radial distribution systems. In: 2000 IEEE PES Summer Meeting, Seattle, WA, USA, 11–15 July, pp. 222–233 (2000)
7. Voropai, N.I., Stychinsky, Z.A.: Renewable Energy Sources: Theoretical Foundations, Technologies, Technical Characteristics, Economics. Otto-von-Guericke-Universität, Magdeburg (2010)
8. Torriti, J.: Demand side management for the European supergrid: occupancy variances of European single-person households. Energ. Policy **44**, 199–206 (2012)
9. Ellabban, O., Abu-Rub, H., Blaabjerg, F.: Renewable energy resources: current status, future prospects and their enabling technology. Renew. Sustain. Energy Rev. **39**, 748–764 (2014)
10. Magdi, S., AL-Sunni, M.F.M.: Control and Optimization of Distributed Generation Systems. Springer, Cham (2015)
11. Suslov, K., Solonina, N., Stepanov, V.: A principle of power quality control in the intelligent distribution networks. In: International Symposium on Smart Electric Distribution Systems and Technologies. EDST 2015. Proceedings, pp. 260–264 (2015)
12. Martínez Ceseña, E.A., Capuder, T., Mancarella, P.: Flexible distributed multienergy generation system expansion planning under uncertainty. IEEE Trans. Smart Grid **7**, 348–357 (2016)
13. Kryukov, A.V., Kargapol'cev, S.K., Bulatov, Y., Skrypnik, O.N., Kuznetsov, B.F.: Intelligent control of the regulators adjustment of the distributed generation installation. Far East J. Electron. Commun. **5**(17), 1127–1140 (2017)

14. Bulatov, Y.N., Kryukov, A.V.: Neuro fuzzy control system for distributed generation plants. In: Proceedings of the Vth International Workshop Critical Infrastructures: Contingency Management, Intelligent, Agent-based, Cloud Computing and Cyber Security, vol. 158, pp. 13–19. Atlantis Press. Advances in Intelligent Systems Research (2018)
15. Haiguo, P., Zhixin, W.: Simulation Research of Fuzzy-PID Synthesis Yaw Vector Control System of Wind Turbine, pp. 469–476 (2007)
16. Jahmeerbacus, I., Bhurtun, C.: Fuzzy control of a variable-speed wind power generating system. Energize 41–45 (2008)
17. Jianzhong, Z., Cheng, M., Chen, Z., Fu, X.: Pitch angle control for variable speed wind turbines. In: DRPT 2008. Nanjing China, 6–9 April (2008)
18. Adzic, E., Ivanovic, Z., Adzic, M., Katic, V.: Maximum power search in wind turbine based on fuzzy logic control. Acta Polytech. Hung. **1**(6), 131–149 (2009)
19. Wu, K.C., Joseph, R.K., Thupili, N.K.: Evaluation of classical and fuzzy logic controllers for wind turbine yaw control. In: Mechanical and Industrial Engineering Department The University of Texas at El Paso, pp. 254–258. IEEE (2009)
20. Roy, A., Bandyopadhyay, S.: Wind Power Based Isolated Energy Systems. Springer, Cham (2019). https://doi.org/10.1007/978-3-030-00542-9
21. Pusdekar Roshana, M., Bawaskar Anil, B.: VLSI architecture of centre of gravity based defuzzifier unit. Int. J. Eng. Innovative Technol. (IJEIT) **10**(4), 50–52 (2015)
22. Heier, S.: Grid Integration of Wind Energy Conversion Systems. Wiley, New York (1998)
23. Bulatov, Y.N., Kryukov, A.V., Van Huan, N.: Automatic prognostic regulators of distributed generators. In: 2018 International Multi-Conference on Industrial Engineering and Modern Technologies (FarEastCon), pp. 1–4. IEEE Conference Publications (2018)
24. Bulatov, Y.N., Kryukov, A.V.: Optimization of automatic regulator settings of the distributed generation plants on the basis of genetic algorithm. In: 2nd International Conference on Industrial Engineering, Applications and Manufacturing (ICIEAM), pp. 1–6. IEEE Conference Publications (2016)

Modeling the Effect of Sorption Moisturizing of Silicate Ceramics on the Thermal Conductivity of Products and Structures

Vitaly Beregovoi(✉)

Penza State University of Architecture and Construction,
28, Titova str., 440028 Penza, Russia
techbeton@pguas.ru

Abstract. The influence of the relative humidity of the ambient air on the kinetics of sorption moisturizing of cellular ceramics with a density of 450…600 kg/m^3 has been studied. A model was obtained for assessing the effect of sorption moisturizing on the heat-conducting properties of the material. The dependence of the calculation of thermal conductivity with regard to changes in the properties of bound water is given. The average value of thermal conductivity increment for 1% sorption moisturizing has been established. The list of phase composition modifiers to reduce the sorption capacity of ceramics has been determined. A comparison was made of the heat-conducting properties of porous silicate ceramics and existing products based on traditional clay.

Keywords: Model of thermal conductivity · Sorption · Ceramics · Natural silicates · Modifiers

1 Introduction

One of the main criteria for the quality of cellular ceramics is the heat-conducting properties of the material and the magnitude of their increment under operating conditions. According to current requirements, the thermal conductivity of building materials is normalized in a dry state, as well as at a relative ambient humidity of 75 and 97%. The thermal conductivity of a building material is a structurally sensitive indicator. In addition to the parameters of the structure, it is significantly influenced by many interdependent and complementary factors, accompanying the process of operating products in enclosing structure of building [1, 2]

$$\lambda(t) = \lambda_0 - \Delta\lambda(t), \tag{1}$$

where λ_0 – the thermal conductivity of ceramics after fabrication; $\Delta\lambda$ – the change in thermal conductivity, that occurs during operation.

To assess the sorption capacity of cellular ceramics obtained by firing a pre-expanded raw mass based on natural silicate [3] an experiment was carried out. The determination of the maximum sorption moisturizing of the samples was carried out according to the standard method. To do this, fragments, weighing 100…120 g, were

V. Murgul and M. Pasetti (Eds.): EMMFT-2018, AISC 983, pp. 902–910, 2019.
https://doi.org/10.1007/978-3-030-19868-8_89

collected from the middle part of the sample of light ceramics. They were placed in a drying cabinet with a temperature of 90 °C and dried to constant weight. Dried material was transferred to a glass desiccator, at the bottom of which was a solution of sulfuric acid with a concentration, that created relative air humidity (φ) of 75 and 97%.

2 Results and Discussion

The moisture content of the material (by mass) was determined by the difference in mass of the material in a dry state and in the state of equilibrium moisture content, the results of the experiments are shown in Figs. 1 and 2.

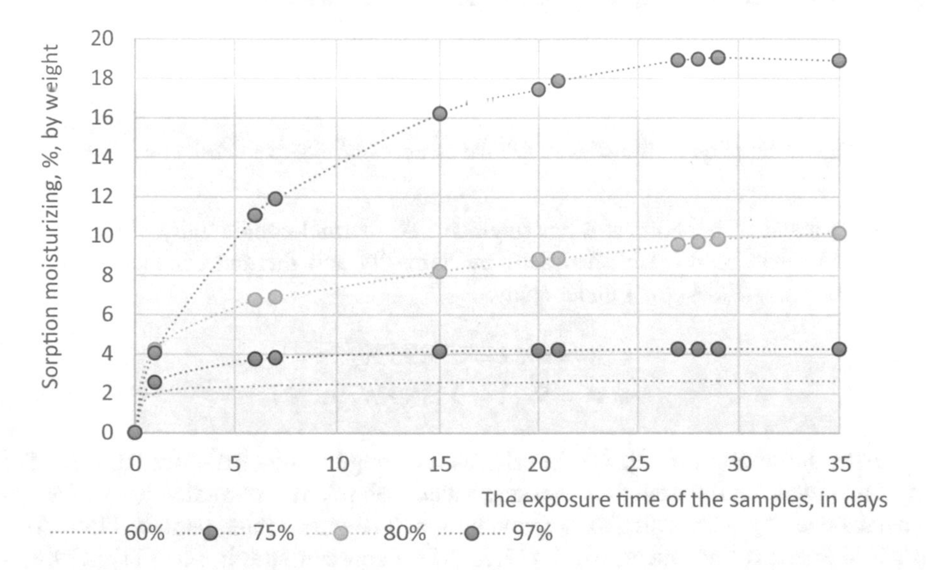

Fig. 1. The influence of the relative humidity of the ambient air on the sorption moisturizing of ceramics

Data analysis shows, that the maximum sorption moisturizing of cellular ceramics after exposure for 35 days at a relative humidity of 97% is 18…20%.

According to the obtained values, the average value of the thermal conductivity increment per 1% moisture content of the material was calculated:

$$\Delta\lambda = \frac{\lambda_w - \lambda_0}{w} \tag{2}$$

where λ_0 and λ_w – the thermal conductivity of ceramics in a dry and moist state, W/(m °C) respectively; w – the moisture content of the material by mass, %.

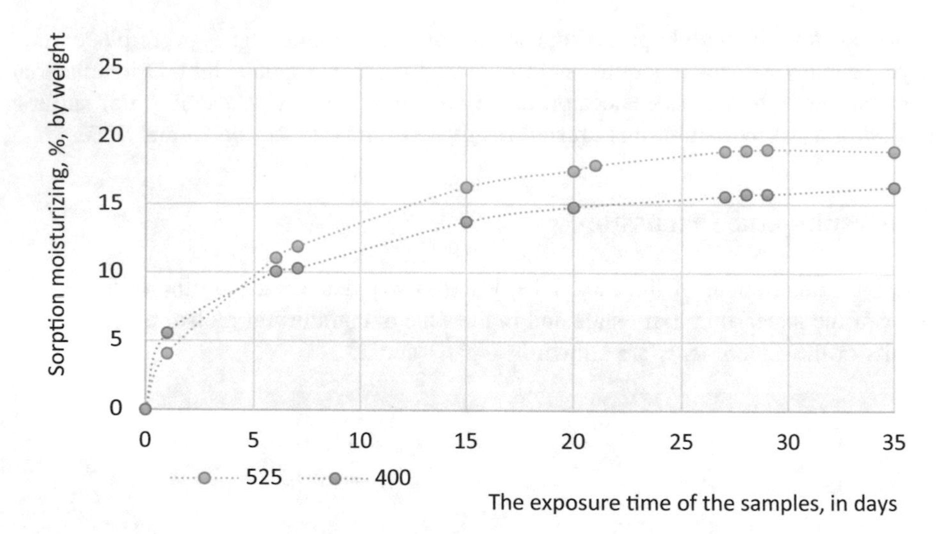

Fig. 2. Sorption moisturizing of cellular ceramics of different density (φ = 97%)

As a result of experimental measurements of thermal conductivity, the following relationship has been established between humidity and thermal conductivity of the ceramics (ρ_m = 400 kg/m^3) under study:

$$-\text{at } w = 0\%\ \lambda = 0.0925\,\text{W}/(\text{m}\,^\circ\text{C})$$
$$-\text{at } w = 5\%\ \lambda = 0.1189\,\text{W}/(\text{m}\,^\circ\text{C})$$

After substitution in Eq. (2) the calculated average value of $\Delta\lambda$ was 0.00528. The material, which is operated at elevated values of relative humidity (φ = 97%), is characterized by more complex sorption curves. They have three periods: No. 1 (0… 12%) – intensive increment; No. 2 (12…17%) – constant speed; No. 3 (17…19%) – damping speed.

The tests have shown that the dependence of the thermal conductivity coefficient of silicate ceramics on its moisture also has a non-linear form and the nature of the process is largely similar to the sorption moisturizing: at the first stage, the increment of the value is significantly higher than in subsequent stages [1, 3].

In this connection, it becomes necessary to correct the linear relationship given below, taking into account the more complex effect of the volume moisture on the thermal conductivity of cellular ceramics:

$$\lambda_w = \lambda_o \cdot (1 + \Delta\lambda^o \cdot 0.01 w_0), \tag{3}$$

where $\Delta\lambda^o$ – the increment of the thermal conductivity coefficient by 1% of the volume humidity, %; w_0 – moisture content of the material, %, by volume.

Table 1 shows the values of the correction factor for some porous materials.

Table 1. Value w_o for lightweight building materials

Materials	Average density of material, kg/m^3			
	300	400	500	600
Fiberboard and chipboard	9.8	7,5	6	5
Arbolit	8.2	6.4	5.2	4.3
Cellular concrete	17.2	15	13.4	12.2
Cellular concrete	1.5…2.0	—	2.9…6.2	5.2…12.0

For the purpose of more correct description of the patterns of change in the heat-conducting properties, it is required to consider in more details features of formation of thin layers of water on a surface of cells of silicate ceramics.

Based on the idea of the surface mosaicity of the crystals of the ceramic matrix, on the border with the water film are the K^+ cations, which are the poles of the packet dipoles $K^+{-}OH^-$ with the electric dipole moment $P = 18D$ (D – debye). They create an electric field in a film with an intensity in the first monolayer of water ($\varepsilon = 1$, $r = 0.35$ nm) [4].

$$E = \frac{2P}{4\pi \cdot \varepsilon \cdot \varepsilon_0 \cdot r^3} \tag{4}$$

Under the action of an electric field, water molecules orient themselves normally to the surface of crystals, forming a strong adsorption-bonded layer, different from the structure of bulk water, formed by hydrogen bonds between molecules. Due to the strong electric field, the bound water molecules are immobile, and the hydrogen bonds are deformed and substantially weakened. Earlier in [5], it was found that the thermal conductivity of water films in crystals is 1.5 orders of magnitude greater, than that of bulk water.

Clarification of the dependence given in Eq. (3) was obtained as a result of analyzing the features of the passage of heat flux through the cellular structure of ceramics. It was assumed, that the thin-film adsorption moisture is concentrated on the inner surface of the cells, forming an additional highly heat-conducting element. In this regard, the calculation is divided into two stages:

1. The thermal conductivity of the first component of the cellular material system «air-adsorbed moisture» is calculated

$$\lambda_{comp.1} = \frac{k \cdot \lambda_w . \lambda_{air}}{\lambda_{air} \cdot (1 - \sqrt[3]{V_{air}}) + k \cdot \lambda_w \cdot \sqrt[3]{V_{air}}} \cdot \sqrt[3]{V_{air}^2} + k \cdot \lambda_{air} \cdot \left(1 - \sqrt[3]{V_{air}^2}\right) \tag{5}$$

where k – the coefficient, taking into account the increase in thermal conductivity of bound water ($k \approx 6$); V_{air} and V_w – the relative content of air and water in the cell of the material ($V_{air} + V_w = 1$).

For materials within the range of sorption moisturizing 0…12% (by volume) the value of the coefficient k was determined by the least squares method, using experimental data [1].

2. Presenting the moistened cellular material in the form of the «first component-ceramic matrix» system, the final thermal conductivity is calculated:

$$\lambda_{cel.mat.} = \frac{\lambda_{matr} \cdot \lambda_{comp.1}}{\lambda_{comp.1} \cdot (1 - \sqrt[3]{P_{sum}}) + \lambda_{matr.} \cdot \sqrt[3]{P_{sum}}} \cdot \sqrt[3]{P^2_{sum}} + \lambda_{matr.} \cdot \left(1 - \sqrt[3]{P^2_{sum}}\right) \quad (6)$$

where P_{sum} – total porosity of cellular ceramics, rel. units; λ_{matr} – thermal conductivity of the matrix of cellular material; ρ_m and ρ – average and true density of ceramics.

In Fig. 3 are shown the experimental (1) and calculated (2) dependences of the thermal conductivity of the cellular material, obtained by Eq. (6).

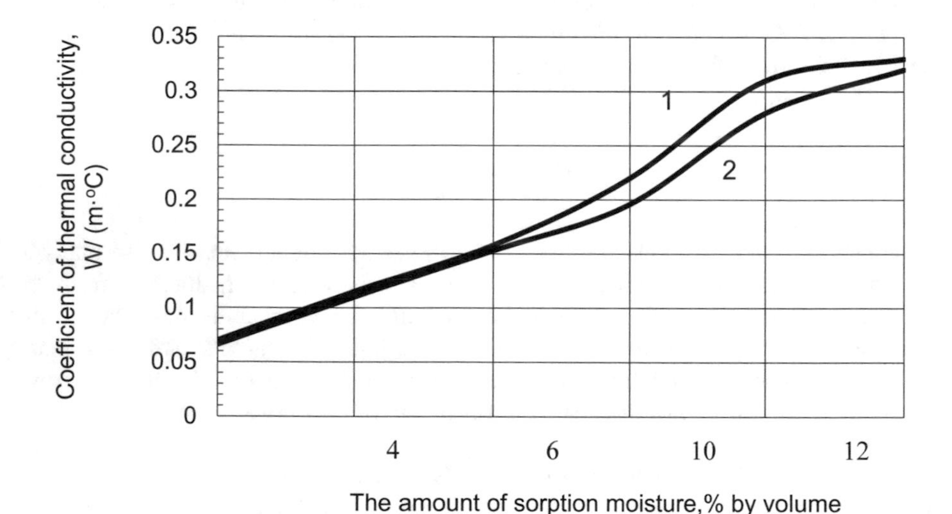

Fig. 3. The dependence of the thermal conductivity of a cellular material (ρ_m = 350…400 kg/m^3) on sorption moisturizing: 1 – experimental; 2 – calculated

Analysis of graphs shows sufficient convergence of experimental and calculated data obtained from the model, taking into account the nonlinearity of the increment of the thermal conductivity of water.

The preservation of the heat insulating ability of products made of silicate ceramics during operation is possible by reducing the magnitude of the sorption moisturizing of the material. This was achieved by introducing into the basic composition the chemicals Na_2CO_3, $Na_2B_4O_7$, $Na_2P_2O_7$, Na_2SiO_3, saturating the raw mass with low-melting oxides Na_2O, K_2O, B_2O_3. The formation of low-temperature eutectic leads to the formation of a developed vitreous phase with minimal sorption of water vapor.

The structure-forming processes occurring during the synthesis of glass in the structure of the silicon-matrix are described by state diagrams of silicate melts of Na_2O (R_2O)-CaO-SiO_2 [6]. The calculation of the temperature of the formation of the eutectic is based on the sequential determination of the melting temperature [7]:

– simple (two-component) subsystems

$$T_i = \frac{T}{[1 - (\ln x_i / N_i)]} \tag{7}$$

where T_i – the melting temperature of the mixture at a given concentration of thc *i*-th component; T – the melting temperature of the *i*-th component; x_i – the molar fraction of the *i*-th component; N_i – the number of atoms in the molecule of the *i*-th component;

– three-component systems:

$$T_n(n) = T_m \cdot \left(\frac{T_m}{T_{m-1}}\right)^{(m-1)\left(1-\frac{m}{n}\right)}, \tag{8}$$

where $T_n(n)$ – the minimum eutectic temperature in the *n*-component system; n – the number of components in the system under study; m – minimum eutectic temperatures in systems with a smaller number of components, than in the system under study, $2 \leq m \leq (n-1)$.

Equations (7) and (8) make it possible to select fluxing additives, that effectively reduce the temperature of the beginning of the formation of the glass melt. The results of the calculations are given in Table 2.

Table 2. The temperature and the amount of silicate melt

Composition				Melt		
SiO_2	CaO	Na_2O	K_2O	Designation	Amount, %	Temperature, °C
of the raw mix						
83.5	9.5	2.5	1.5			
of the eutectic						
73.5	5.2	21.3		N_1	11.7	725
73.4	12.9	13.7		N_2	18.2	1047
73.0	1.9		25.1	K_1	6.0	720
52.6	10.2		37.2	K_2	4.0	869
				$N_1 + K_1$	17.7	723.3
				$N_2 + K_2$	22.3	1014.8

The calculation results show that the introduction of 3.0… 5.5% of modifiers leads to the formation of a silicate melt in the amount of 8…27% in the firing process at temperatures of 720…1200 °C. This amount is sufficient to significantly improve the sorption properties of ceramics (Fig. 4).

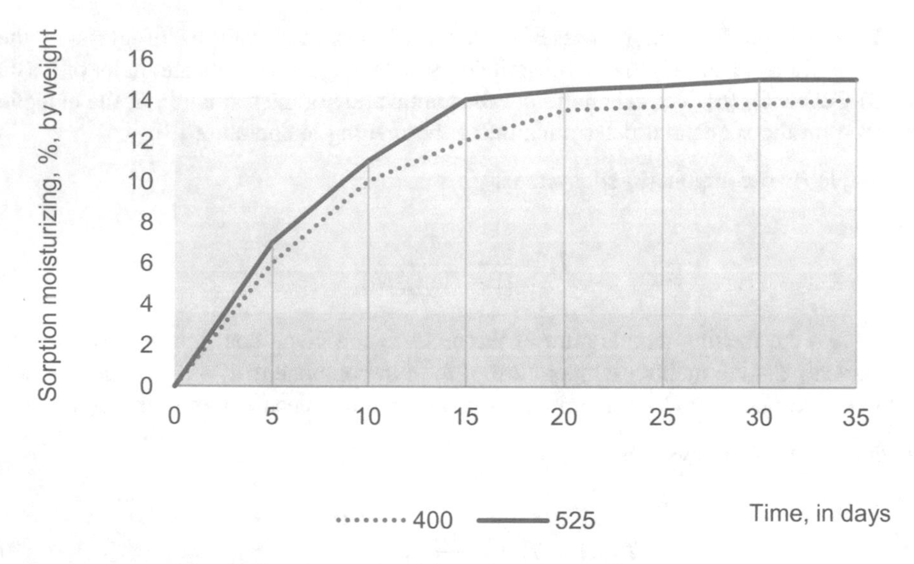

Fig. 4. Sorption moisturizing of modified silicate ceramics

As additives to accelerate the process of melt formation Na_2CO_3, $Na_2B_4O_7$, $Na_2P_2O_7$, Na_2SiO_3 was used in an amount of 1…10%. The duration of the process of forming the places of sintering of the vitreous and crystalline phases during the roasting of siliceous ceramics is expressed by the equation [8]:

$$\tau_2 = \left(\frac{r_2}{r_1}\right)^{\gamma} \cdot \tau_1, \tag{9}$$

where τ_2 – the sintering time of particles of size r_2; τ_1 – the sintering time of particles of size r_1; γ – a coefficient depending on the type of sintering mechanism.

To assess the effect of diluents on the change in the fractional composition of silicate particles in the aqueous medium, the method of sedimentation analysis was used. Gaizes suspensions are kinetically unstable to microheterogeneous systems with a medium or coarse dispersion. In studies of the suspension of mineral, particles with a radius of up to 5 microns was considered as a transitional system, which is applicable to the mechanism of aggregate and sedimentation stability, developed for colloids. Sodium pyrophosphate ($Na_4P_2O_7$) and soda (Na_2CO_3) were used as liquefying additives. The results of experiments for sodium pyrophosphate are shown in the Fig. 5.

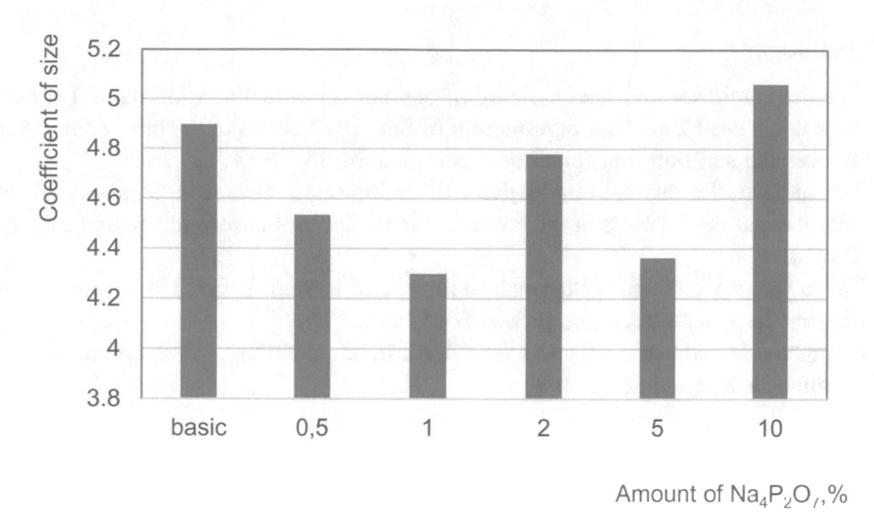

Fig. 5. The coefficient of size of suspensions containing additive $Na_4P_2O_7$

When analyzing the obtained dependences, it is necessary to take into account the appearance of a double electric layer on SiO_2 particles. The surface of SiO_2 crystals is partially hydrated with the formation of silicic acid capable of ionizing according to the scheme:

$$H_2SiO_3 \leftrightarrow H + HSiO^{3-}$$

$HSiO^{3-}$ ions, «related» to SiO_2, are selectively adsorbed on the surface of SiO_2, giving it a negative charge, and hydrogen ions pass into the solution

As shown in Fig. 5, the introduction of the $Na_4P_2O_7$ in an amount of up to 1,0% leads to an increase in the relative content of particles of small fractions due to the peptization of large aggregates.

In addition, the thin films separating the solid surfaces of the similarly charged particles, wedging pressure due to the adsorption-solvate mechanism determined by the presence on the surface of the Na+ diffuse layer of hydrated shells of oriented dipoles of water with high viscosity and elasticity.

3 Conclusions

For a comparative assessment of the obtained indicators porous ceramics, based on traditional clay, was used as an industrial standard. It is the most competitive product among modern heat-efficient firing wall materials. Such materials are produced in the form of large-format blocks (Poroton, Porotherm, etc.), the best of which are characterized by a density of 650…660 kg/m^3, thermal conductivity 0.15…0.16 W/(m °C) and moisture 18 ± 2%.

Summary:

1. The maximum sorption moisturizing of cellular ceramics based on natural silicates does not exceed 20%. The introduction of fluxing additives in the base composition reduces the sorption moisturizing of ceramics by 15…20%.
2. To calculate the thermal conductivity of cellular ceramics, it is necessary to take into account the increase in the thermal conductivity of adsorption-bound moisture ($k = 6$).
3. The average value of the thermal conductivity increment per 1% sorption moisturizing for silicate ceramics is $5.28 \cdot 10^{-3}$ W/(m °C).
4. Compared to industrial products from traditional clay, the thermal conductivity of porous silicate ceramics is 40% less.

References

1. Beregovoi, V.A., Korolev, E.V., Bazhenov, Y.M.: Efficient Heat-Insulating Foam Ceramic Concretes. MGSU, Moscow (2011)
2. Beregovoi, V.A., Snadin, E.V.: Nat. Sil. Cell. Cer. J. Izvestiya vuzov. Construction **2**, 13 (2018)
3. Beregovoi, V.A., Beregovoi, A.M.: L. Cer. Mater. J. Solid State Phenomena **284**, 90 (2018)
4. Karnakov, V.A., Ezhova, Ya.V., Marchuk, S.D.: An. prop. abs. films sil. J. Solid State Physics, 11, p. 1946 (2006)
5. Metsik, M.S., Perevertaev, V.D., Liopo, V.A., Timoshtchenco, G.T., Kiselev, A.B.: J. Colloid Interface Sci. **43**, 663 (2006)
6. Pavlushkin, N.M.: Glass. Moscow, Stroyizdat (2008)
7. Maslennikova, G.N., Haritonov, F.Y., Dubov, I.V.: The Calculations in Technologies of the Ceramics. Stroyizdat, Moscow (2008)
8. Strelov, K.K.: Fundamental Theory of Technology of Refractories. Metall, Moscow (2005)

Strengthening of Concrete Beams with the Use of Carbon Fiber

Ekaterina Kuzina[1(✉)] and Vladimir Rimshin[1,2]

[1] Scientific-Research Institute of Building Physics of the Russian Academy Architecture and Construction Sciences, Lokomotivny proezd 21, 127238 Moscow, Russia
kkuzzina@mail.ru

[2] Moscow State University of Civil Engineering, Yaroslavskoe shosse, 26, 129337 Moscow, Russia

Abstract. The article considers the engineering calculation for proving the need to reinforce concrete bent beams, and it is also proposed a method for reinforcement concrete bent beams by external reinforcement with carbon fiber materials, presented destruction models of reinforced concrete beams that was obtained experimentally. The effectiveness of this reinforcement type with fiber composite materials on the beams bearing capacity is evaluated. Experimental data on the strength and deformability of bending reinforced concrete beams reinforced with carbon-based material were obtained from the results of the studies, the effect of various reinforcement schemes of bending reinforced concrete beams was experimentally proved using carbon fiber bandages, the destruction capacity of concrete beams reinforced with carbon fiber was obtained and analyzed. Considering the experience of using this reinforcement technology abroad, it can be said confidently that composite materials reinforced with fiber composite materials will be in leading places in Russia in the near future.

Keywords: Experiments · Carbon fiber reinforcement · Concrete beams · Increasing the bearing capacity

1 Introduction

In recent years, fiber-reinforced composite materials have been used more often for increasing the bearing capacity of reinforced concrete structures. Fiber-reinforced composite materials usually consist of high-strength fibers (such as aramid, carbon and fiberglass) bonded with epoxy or other adhesives, often called dies. Fiber-reinforced composite materials have a linear relationship between stresses and strains up to fracture when stretched. Their properties are mainly determined by the type, orientation and number of reinforcing fibers [1–3]. Experimental studies were carried out using external reinforcement with carbon fiber materials for concrete beams. But there still remains a lack of experimental studies of such structure's behavior [4, 5].

V. Murgul and M. Pasetti (Eds.): EMMFT-2018, AISC 983, pp. 911–919, 2019.
https://doi.org/10.1007/978-3-030-19868-8_90

2 Methods

The results are generated by the ARBAT program. The reliability coefficient γ_n is equal to 1. The reliability coefficient (the second limiting state) is equal to 1. The reinforced concrete beam constructive solution is presented in Fig. 1.

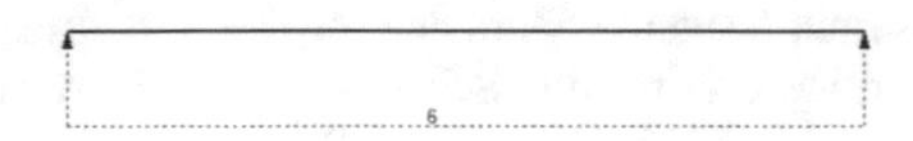

Fig. 1. The reinforced concrete beams constructive solution.

The cross section of the reinforced concrete beam with the reinforcing bars location is shown in Fig. 2.

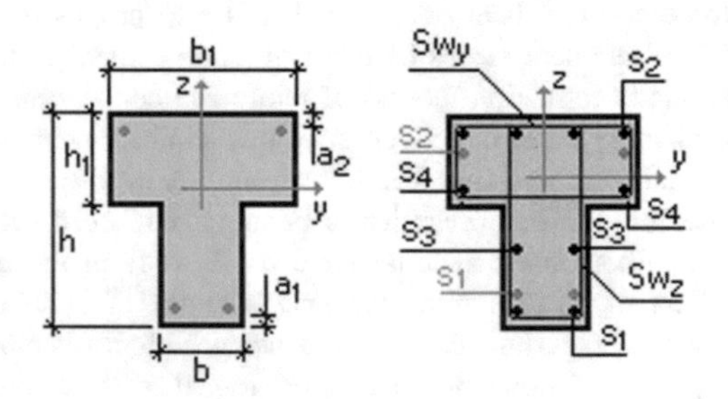

Fig. 2. Cross section of reinforced concrete beams.

Cross section characteristics:
$b = 341$ mm. $h = 220$ mm. $b_1 = 1200$ mm. $h_1 = 38$ mm. $a_1 = 30$ mm. $a_2 = 32$ mm. Table 1 presents the reinforcement characteristics.

Table 1. Reinforcement characteristics.

Reinforcement	Class	Coefficient of working conditions
Longitudinal	A240	1
Transverse	A240	1

There are the characteristics of a given reinforcement in Table 2.

Table 2. Reinforcement characteristics.

Area	Length (m)	Reinforcement	Section
1	6	$S_1 - 6\varnothing12$	

Type of concrete is heavy. Concrete class is B30. Concrete density is 2.5 t/m^3. Coefficients of concrete working conditions: long-acting loads consideration is $\gamma b1 = 0.9$, the resulting coefficient without $\gamma b1$ is 1. Air humidity is 40–75%. Requirements for the width of crack opening are selected from the condition of reinforcement safety. The admissible crack opening width: with a short opening is 0.4 mm, with a long opening - 0.3 mm. Table 3 presents the value and type of constant load.

Table 3. Loading 1 - permanent.

Load type	Size	
Evenly distributed	0.859	T/m

Load 1 is permanent. The reliability coefficient is 1.1. The coefficient of the long-term part is 1. Figures 3, 4, 5, 6, 7, 8, 9 and 10 shows maximum and minimum values of the shear force and bending moment with constant loading

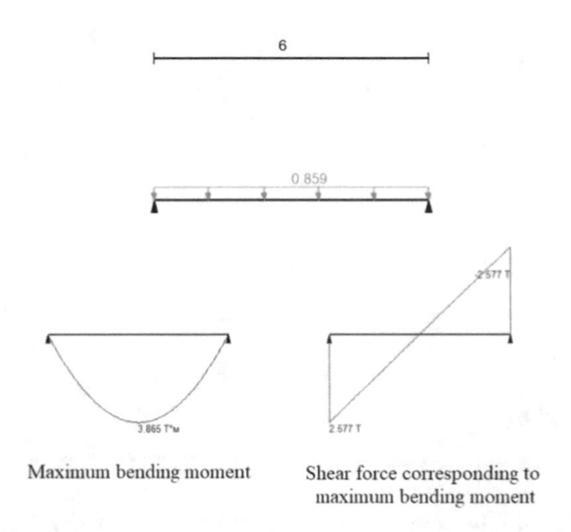

Fig. 3. The envelope value M_{max} according to design loads.

There are the support reactions in a reinforced concrete beam in Table 4.

There is the plot of materials on the bending moment in Fig. 11.

According to the results of the calculation, it was found that the beam has insufficient bearing capacity, therefore, a method of reinforcing concrete beam with carbon fiber was proposed. Although the technology of reinforcement concrete structures with external carbon fiber reinforcement is still not fully understood, it has been proved after some experiments that composite materials are more progressive than traditional ones and are actively used in projects around the world. Scientists have established their competitive advantages observing the behavior of materials at sites [6, 7]. For example, the bending strength of reinforced concrete slabs and beams can be increased by gluing

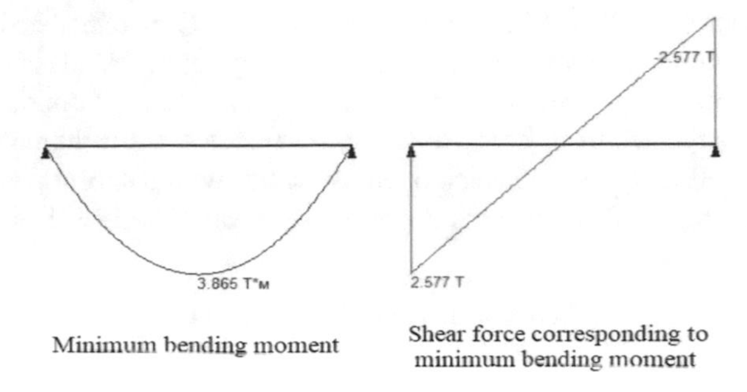

Fig. 4. The envelope value M_{min} according to design loads.

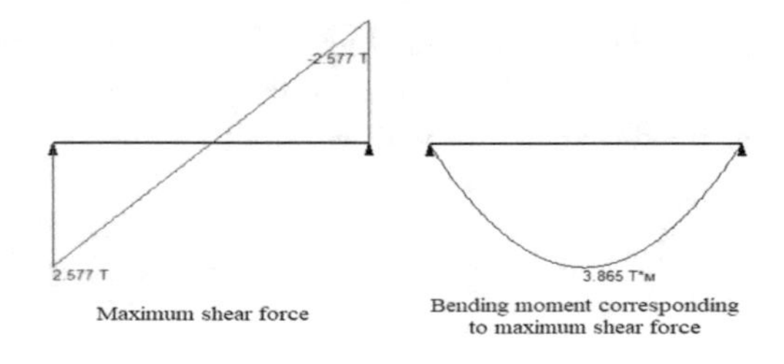

Fig. 5. The envelope value Q_{max} according to design loads.

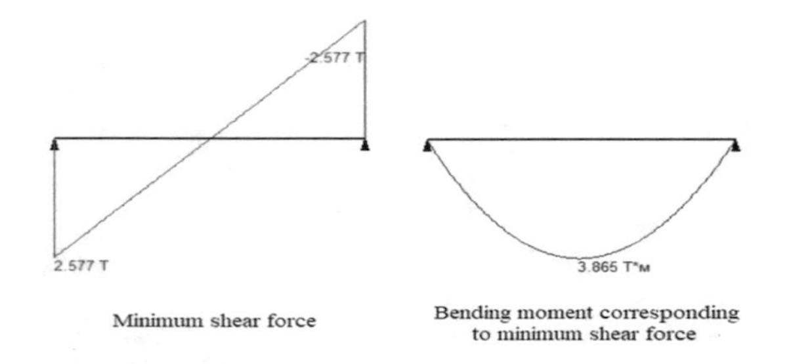

Fig. 6. The envelope value Q_{min} according to design loads.

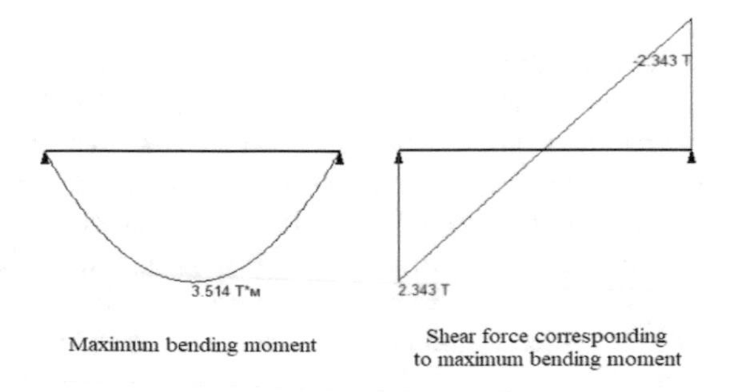

Fig. 7. The envelope value M_{max} according to normative loads.

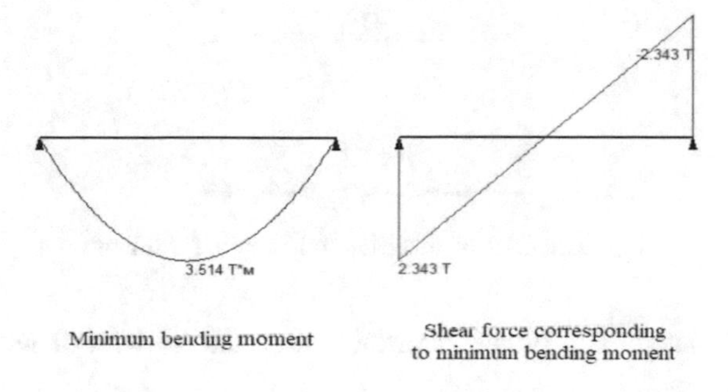

Fig. 8. The envelope value M_{min} according to normative loads.

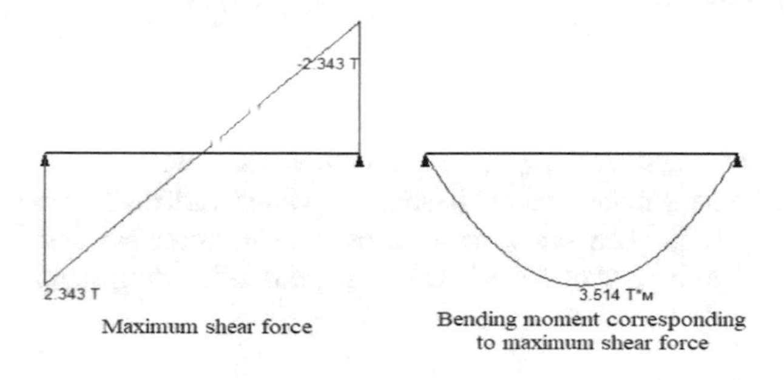

Fig. 9. The envelope value Q_{max} according to normative loads.

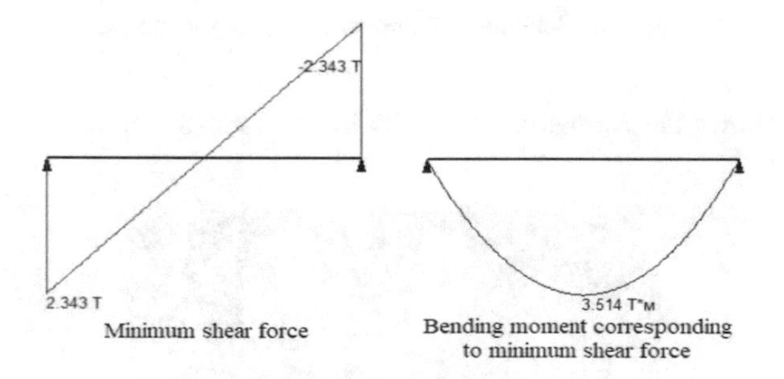

Fig. 10. The envelope value Q_{min} according to normative loads.

Table 4. Loading 1 - permanent.

	Strength in support 1	Strength in support 2
	T	T
By criterion M_{max}	2.577	2.577
By criterion M_{min}	2.577	2.577
By criterion Q_{max}	2.577	2.577
By criterion Q_{min}	2.577	2.577

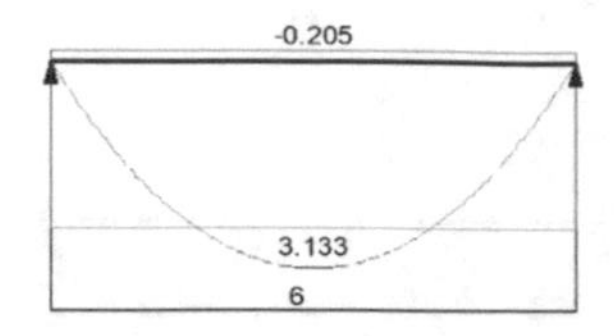

Fig. 11. Plot of materials on bending moment.

fiber-reinforced matcrials to the solid surface of structures with or without additional fasteners [8, 9].

3 Experiment

Reinforced concrete beams with a length of 1290 mm and cross-sectional dimensions of 120 × 190 mm, 4 series - 10 beams in each were used as prototypes. Every beam was loaded until the first cracks appeared. Then the beams were reinforced with carbon canvases. The layers number and the bandages location varied. 2–3 beams were tested as control samples in each series, that is, brought to complete destruction without reinforcement. Figure 12 shows a schematic diagram of loading samples.

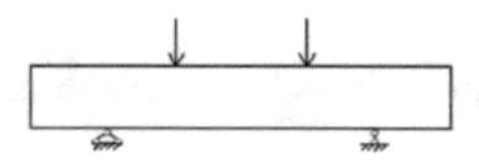

Fig. 12. Schematic diagram of loading samples.

Figure 13 shows the formation of cracks in the control sample.

Fig. 13. Cracking in the control sample.

At each stage of loading, mid-span deflections and deformations were measured in some beam's sections. The effect of variables on bending, the deformations distribution in concrete, the cracks formation, the fracture type, the load type that lead to appearance of first crack, and the ultimate strain value was studied. Figure 14 shows the destruction schemes for beams of 1–4 series.

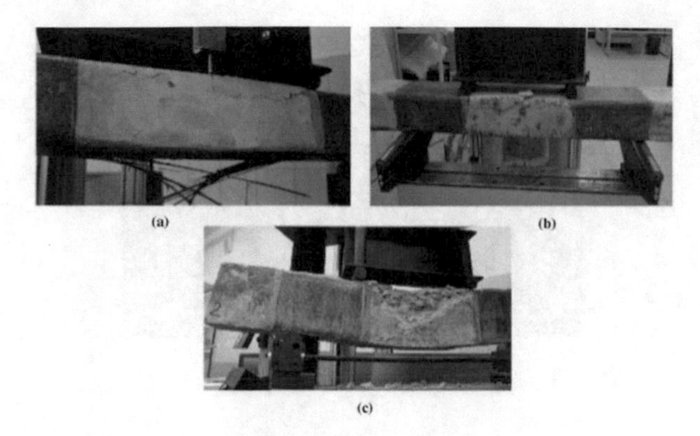

Fig. 14. Destruction scheme: (a) fiber delamination; (b) chipping concrete and fiber delamination; (c) the destruction of the concrete in compressed zone.

4 Results

The obtained destruction types prove the destruction models proposed by foreign scientists. They are presented in Fig. 15.

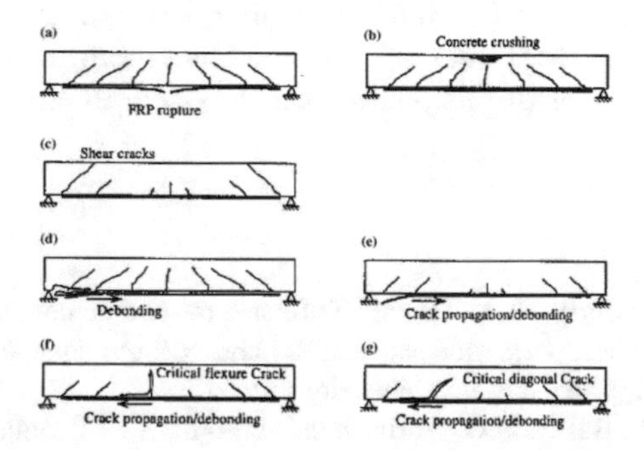

Fig. 15. The destruction models reinforced with fiber composite materials [9].

A mid-span fiber gap was also obtained when the number of layers was more than three in this experimental work (Fig. 16).

An increase in the bearing capacity of beams by more than 3 times has been experimentally proven. The beams destruction without reinforcement occurred at a load of 4–7 tons, while reinforcement was 8–21 tons. Strengthening by external reinforcement with carbon fiber is one of the few low-cost, but at the same time it is effective method.

Fig. 16. Fiber break.

5 Conclusions

Experimental data on the strength and deformability of bending reinforced concrete beams reinforced with carbon-based material were obtained from the results of the studies, the effect of various reinforcement schemes of bending reinforced concrete beams was experimentally proved using carbon fiber bandages, the destruction capacity of concrete beams reinforced with carbon fiber was obtained and analyzed.

Considering the experience of using this technology abroad, it can be said confidently that composite materials reinforced with fibers will be in leading places in Russia in the near future. Every year, scientists conduct more and more tests confirming the indisputable advantages of this reinforcement system in construction [10–18].

References

1. Kachlakev, D.I., McCurry, D.D.: Testing of full-size reinforced concrete beams strengthened with FRP composites: experimental results and design methods verification. Final Report SPR 387.11, Oregon State University (2016)
2. Sayed Ahmed, E.Y., Bakay, N.G.: Shrive bond strength of FRP laminates to concrete: state-of-the-art review. Electron. J. Struct. Eng. **9**, 45–61 (2017)
3. Kuzina, E., Cherkas, A., Rimshin, V.: Technical aspects of using composite materials for strengthening constructions. In: IOP Conference Series: Materials Science and Engineering 21, Construction - The Formation of Living Environment (2018)
4. Kuzina, E., Rimshin, V.: Deformation monitoring of road transport structures and facilities using engineering and geodetic techniques. In: Advances in Intelligent Systems and Computing, vol. 692 (2018)
5. Telichenko, V., Rimshin, V., Eremeev, V., Kurbatov, V.: Mathematical modeling of groundwaters pressure distribution in the underground structures by cylindrical form zone. In: MATEC Web of Conferences 27. Cep. “27th R-S-P Seminar, Theoretical Foundation of Civil Engineering (27RSP), TFoCE 2018” (2018)
6. Varlamov, A.A., Rimshin, V.I., Tverskoi, S.Y.: Planning and management of urban environment using the models of degradation theory. In: IOP Conference Series: Earth and Environmental Science 3. Cep. “International Conference on Sustainable Cities” (2018)
7. Cherkas, A., Rimshin, V.: Application of composite reinforcement for modernization of buildings and structures. In: MATEC Web of Conferences, vol. 117, p. 00027 (2017)

8. Telichenko, V.I., Rimshin, V.I., Karelskii, A.V., Labudin, B.V., Kurbatov, V.L.: Strengthening technology of timber trusses by patch plates with toothed-plate connectors. J. Ind. Pollut. Control **33**, 1034–1041 (2017)
9. Shubin, I.L., Zaitsev, Y.V., Rimshin, V.I., Kurbatov, V.L., Sultygova, P.S.: Fracture of high performance materials under multiaxial compression and thermal effect. Eng. Solid Mech. **5**(2), 139–144 (2017)
10. Krishan, A.L., Rimshin, V.I., Telichenko, V.I., Rakhmanov, V.A., Narkevich, M.Yu.: Practical implementation of the calculation of the bearing capacity trumpet-concrete column. Izvestiya Vysshikh Uchebnykh Zavedenii (2017)
11. Korotaev, S.A., Kalashnikov, V.I., Rimshin, V.I., Erofeeva, I.V., Kurbatov, V.L.: The impact of mineral Aggregates on the thermal conductivity of cement composites. Ecol. Environ. Conserv. **22**(3), 1159–1164 (2016)
12. Bazhenov, Y.M., Erofeev, V.T., Rimshin, V.I., Markov, S.V., Kurbatov, V.L.: Changes in the topology of a concrete porous space in interactions with the external medium. Eng. Solid Mech. **4**(4), 219–225 (2016)
13. Krishan, A.L., Troshkina, E.A., Rimshin, V.I., Rahmanov, V.A., Kurbatov, V.L.: Load-bearing capacity of short concrete-filled steel tube columns of circular cross section. Res. J. Pharm. Biol. Chem. Sci. **7**(3), 2518–2529 (2016)
14. Erofeev, V.T., Zavalishin, E.V., Rimshin, V.I., Kurbatov, V.L., Stepanovich, M.B.: Frame composites based on soluble glass. Res. J. Pharm. Biol. Chem. Sci. **7**(3), 2506–2517 (2016)
15. Erofeev, V., Kalashnikov, V., Karpushin, S., Tretiakov, I., Matvievskiy, A.: Physical and mechanical properties of the cement stone based on biocidal Portland cement with active mineral additive. Solid State Phenom. **871**, 28–32 (2016)
16. Antoshkin, V.D., Travush, V.I., Erofeev, V.T., Rimshin, V.I., Kurbatov, V.L.: The problem optimization triangular geometric line field. Mod. Appl. Sci. **9**(3), 46 (2015)
17. Krishan, A., Rimshin, V., Erofeev, V., Kurbatov, V., Markov, S.: The energy integrity resistance to the destruction of the long-term strength concrete. Procedia Eng. **117**(1), 211–217 (2015)
18. Erofeev, V.T., Bogatov, A.D., Smirnov, V.F., Rimshin, V.I., Kurbatov, V.L.: Bioresistant building composites on the basis of glass wastes. Biosci. Biotechnol. Res. Asia **12**(1), 661–669 (2015)

Ice Fog Protection for Ventilation Systems to Be Used in Extreme North Regions

Andrew Arbatskiy[1](✉), Andrew Garyaev[1], Vasiliy Glasov[1], and Sergey Polyakov[2]

[1] Moscow Power Engineering Institute, Krasnokazarmennaya 14, 111250 Moscow, Russia
arbatsky1985@mail.ru
[2] SiberCool Laboratory, Rusklimat, Russia

Abstract. Currently, a fairly large number of solutions exist to complete ventilation plants for use in north regions. In addition to essential requirements for the housing of a ventilation plant, such as heat insulation minimum of 45 mm, body elements withstanding low temperatures, heating up and heat insulation of valves, there are special requirements for a composition of filtration and ventilation plant heating sections, namely: not permitted installation of a hot-water calorifer as the first heating stage at estimated temperatures below −35 °C (because calorifer frost protection is activated too late) in regions, where ice fog is highly likely to occur, such as Yakutia, Khanty-Mansiysk, Chukotka etc., it is required to either install a finless heat exchanger upstream of a filter, or that one of higher interfin spacers. This paper discusses in detail the problem related to ice fog protection of filters. A fully automated filter protection system with no additional consumption of energy and materials is proposed, as in the case of heat exchangers with high interfin spacers. The system has been described in the paper (1), which also contains summarized results of research. This paper contains details on the design of an ice fog protection system that was not previously considered.

Keywords: Ventilation · Extreme North · Ventilation plants · Ice fog · Filtration

1 System Description

The ice fog protection system represents a package of air return pipes laid from a fan section to a filtration section (valve-filter unit – Fig. 1) with further air distribution over the entire filter surface. A layout of a valve-filter unit with an air distribution pipe is presented in Figs. 2 and 3. The system operates on the basis of an algorithm proposed as follows:

– Valves “C” and “D” open, by turn, with a time interval required for through-drying of the filter;
– The filter downstream of the closed valve is blown with warm air to be distributed via a special pipe (Fig. 3). Due to this, ice fog on the filter melts down and evaporates.

V. Murgul and M. Pasetti (Eds.): EMMFT-2018, AISC 983, pp. 920–928, 2019.
https://doi.org/10.1007/978-3-030-19868-8_91

In such a way, we can highlight two research purposes:

1. A time shall be determined, required for through-drying of the filter;
2. The best geometrics and arrangement of holes of the distribution pipe shall be determined.

This stage of research works covers through-drying of the filters only without defrosting. This is because there are some parameters that remain unknown: average consumption of ice fog settled down onto the filter, specific weight of ice solids at various temperatures. These parameters can be determined by means of full-scale tests only, in situ, and they will be implemented with further research works.

Air consumption via air return pipes depends only upon diameters of various pipes and pressure differences between pressure and suction sides with due consideration of pressure losses in intermediate elements of ventilation plants. And this can be calculated according to an equation [2]:

$$G = F\sqrt{\frac{2\Delta P}{\rho}} \tag{1}$$

where: F - cross-section area of an air return pipe, m^2; ΔP – pressure differences on pipe ends, Pa; ρ – air density, kg/m^3.

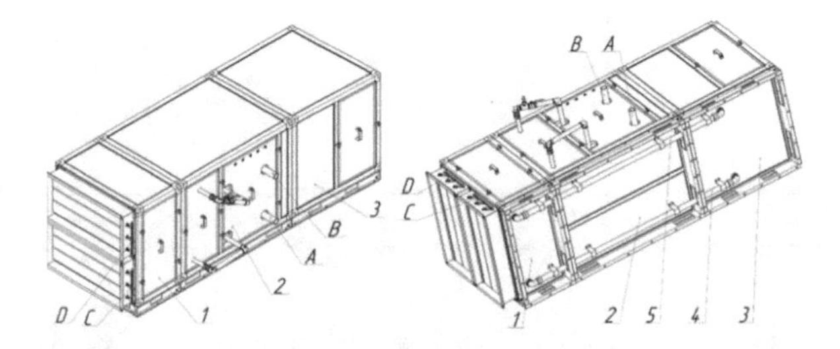

Fig. 1. Order of connection of sections and air return pipes: 1 – valve-filter unit; 2 – heating unit; 3 – fan section; 4, 5 – air return pipes; A, B – hot heat carrier supply; C, D – air valves.

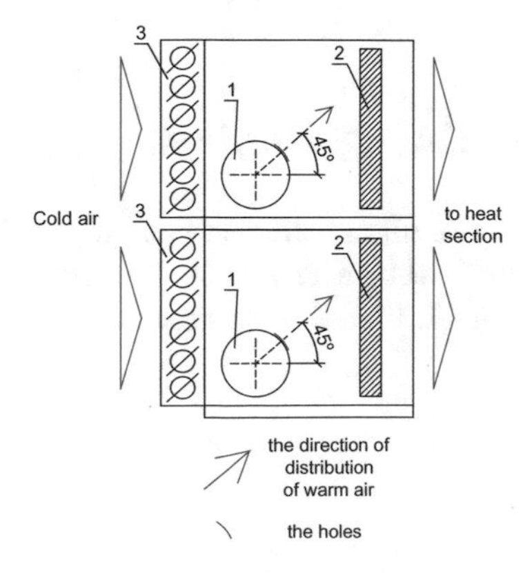

Fig. 2. Valve-filter unit layout diagram: 1 – air distribution pipe (Fig. 3b), 2 – filter, 3 – air valve.

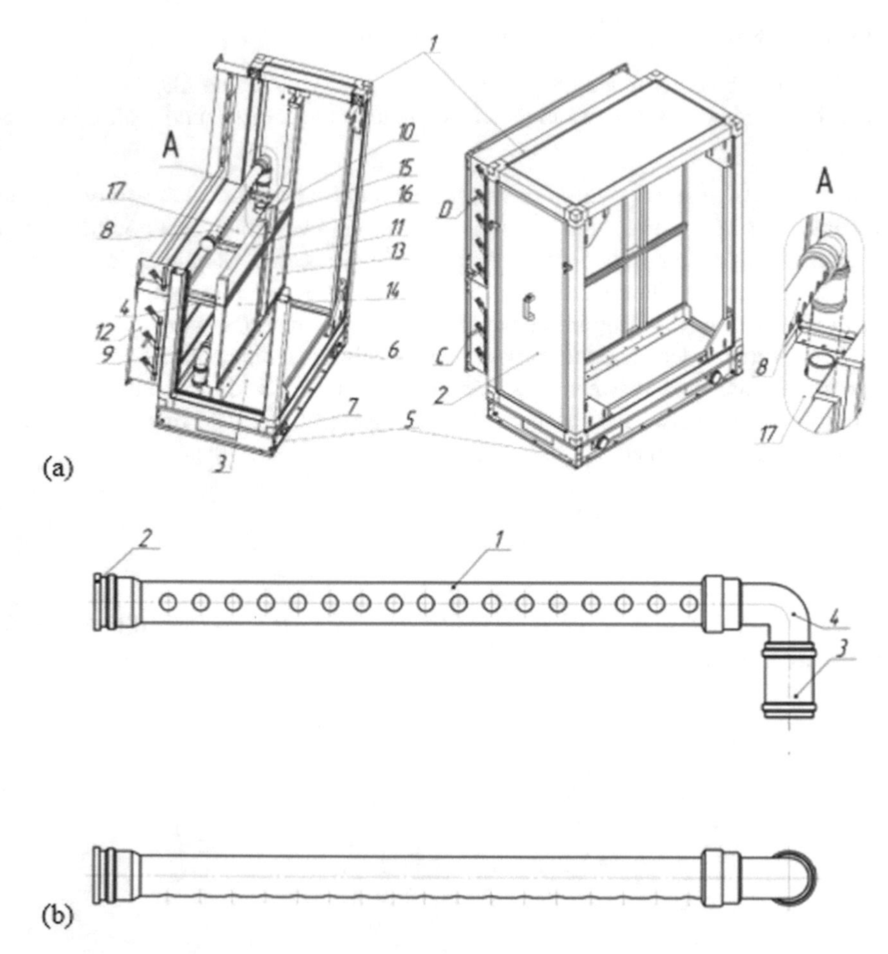

Fig. 3. (a) Valve-filter unit structural drawing: 1 – Unit casing, 2 – Service door, 3 – Unit bottom, 4 – Filter housing, 5 – Section base frame, 6, 7 – Air return pipes, 8, 9 – Distribution pipe, 10, 11 – Filter finishing element, 12 – Air valve, 13, 14, 15, 16 – Filters, 17 – Unit partition wall, C, D – Electric drives stems; (b) Air distribution pipe: 1 – PVC-pipe with holes, 2 – Plug, 3 – Base, 4 – Elbow

2 Research of Filters Through-Drying

To investigate through-drying of filters, one has built a ventilation plant with a valve-filter unit, a design of which is shown in Fig. 3. The composition of sections corresponds to that presented in Fig. 1. Figure 4 represents a general view of the ventilation plant:

Fig. 4. General view of a valve-filter unit testing facility

A dimensional drawing is in Fig. 5:

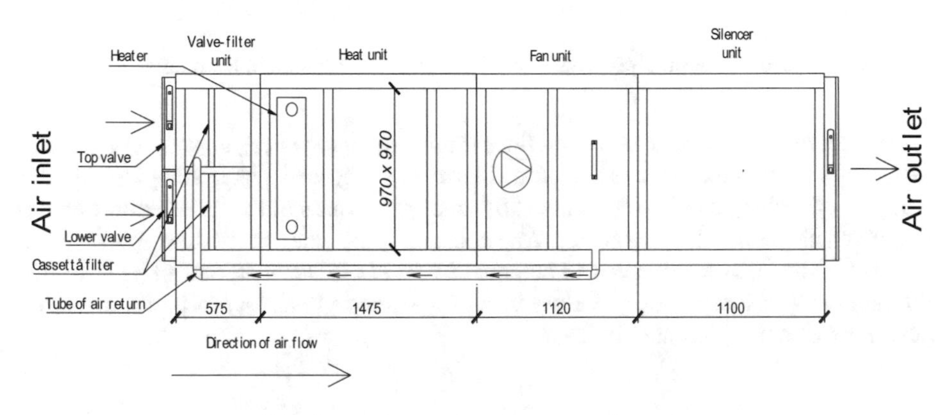

Fig. 5. Testing facility dimensional drawing

Subjects of research were: geometrics and positioning of an air return pipe, as presented in Fig. 6.

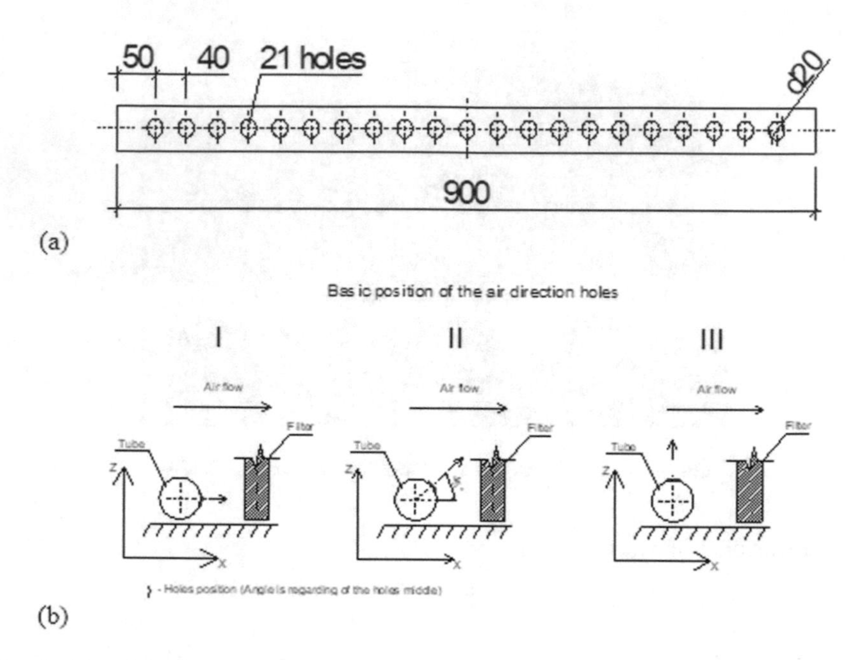

Fig. 6. (a) Air distribution pipe geometrics, (b) Options to arrange holes relative to a filter.

Subject to research were: a soaked filter (immersed into water), as well as a partially wetted one (with water sprayed upon a filter using a special water spraying device). A degree of soaking shall be determined by an original mass of the filter (source data is presented in Tables 2, 3 and 4). Air consumed by the ventilation plant made L = 30 000 m^3/h, at a fan speed to be equal to 50 Hz. Filter material: glass fibre, G4 filtration class in line with [3]. Values of air consumed via air return pipes at various valve positions are presented in Table 1:

Table 1. Air consumed via air return pipes at various valve positions, m^3/h

Valves positions	Lower valve blowing pipe	Upper valve blowing pipe
Upper valve closed	91.8	88.3
Lower valve closed	116.6	45.2

Measurements of details, as well as comparison of measured and estimated values will be presented in further papers. The measurements taken have resulted in the data as follows:

Table 2. Source data and results of measurements of filter drying rates with holes arrangement I according to Fig. 6(b)

Source data			
Upper valve closed **(upper valve subject to measurements)**		Lower valve closed **(lower valve subject to measurements)**	
Dry filter weight, kg	**2.465**	Dry filter weight, kg	**2.437**
High humidity filter weight (soaked filter), kg	3.045	High humidity filter weight (soaked filter), kg	3.065
Low humidity filter weight (filter sprayed with water), kg	2.829	Low humidity filter weight (filter sprayed with water), kg	2.717
Air temperature by the time of drying, C	23.3	Air temperature by the time of drying, C	21.9
Air humidity by the time of drying, %	42.1	Air humidity by the time of drying, %	44.5
Measurements results for water immersed (soaked) filter (high humidity)			
Drying time, s	Filter weight, kg	Drying time, s	Filter weight, kg
300	2.642	300	2.585
900	2.514	900	2.483
1500	2.484	1500	2.451
2100	2.474	2100	2.440
2700	**2.465**	2700	**2.437**
Measurements results for filter sprayed with water (low humidity)			
Drying time, s	Filter weight, kg	Drying time, s	Filter weight, kg
300	2.558	300	2.525
900	2.496	900	2.465
1500	2.472	1500	2.448
2100	**2.465**	2100	2.438
2400	–	2400	**2.437**

Table 3. Source data and results of measurements of filter drying rates with holes arrangement II according to Fig. 6(b)

Source data			
Upper valve closed **(upper valve subject to measurements)**		Lower valve closed **(lower valve subject to measurements)**	
Dry filter weight, kg	**2.462**	Dry filter weight, kg	**2.432**
High humidity filter weight (soaked filter), kg	3.511	High humidity filter weight (soaked filter), kg	3.656

(*continued*)

Table 3. (*continued*)

Source data			
Low humidity filter weight (filter sprayed with water), kg	2.882	Low humidity filter weight (filter sprayed with water), kg	2.782
Air temperature by the time of drying, C	20.2	Air temperature by the time of drying, C	20.6
Air humidity by the time of drying, %	40.1	Air humidity by the time of drying, %	41.7
Measurements results for water immersed (soaked) filter (high humidity)			
Drying time, s	Filter weight, kg	Drying time, s	Filter weight, kg
300	2.732	300	2.757
900	2.577	900	2.606
1500	2.531	1500	2.54
2100	2.507	2100	2.497
2700	2.469	2700	2.458
3000	**2.462**	3000	2.442
3300	–	3300	**2.432**
Measurements results for filter sprayed with water (low humidity)			
Drying time, s	Filter weight, kg	Drying time, s	Filter weight, kg
300	2.569	300	2.535
900	2.497	900	2.463
1500	2.471	1500	2.443
2100	**2.462**	2100	**2.432**

Table 4. Source data and results of measurements of filter drying rates with holes arrangement III according to Fig. 6(b)

Source data			
Upper valve closed **(upper valve subject to measurements)**		Lower valve closed **(lower valve subject to measurements)**	
Dry filter weight, kg	**2.462**	Dry filter weight, kg	**2.432**
High humidity filter weight (soaked filter), kg	3.352	High humidity filter weight (soaked filter), kg	3.376
Low humidity filter weight (filter sprayed with water), kg	2.771	Low humidity filter weight (filter sprayed with water), kg	2.798
Air temperature by the time of drying, C	24	Air temperature by the time of drying, C	23.7

(*continued*)

Table 4. (*continued*)

Source data			
Air humidity by the time of drying, %	44.8	Air humidity by the time of drying, %	49.4
Measurements results for water immersed (soaked) filter (high humidity)			
Drying time, s	Filter weight, kg	Drying time, s	Filter weight, kg
300	2.765	300	2.672
900	2.603	900	2.52
1500	2.535	1500	2.47
2100	2.490	2100	2.440
3000	**2.462**	3000	**2.432**
Measurements results for filter sprayed with water (low humidity)			
Drying time, s	Filter weight, kg	Drying time, s	Filter weight, kg
300	2.544	300	2.550
900	2.478	900	2.474
1500	**2.462**	1500	2.451
2100	–	2100	2.441
2400	–	2400	**2.432**

Completion of drying was determined by a point of time, when a filter weight became equal to a previously measured one. Slight deviations from the original filter weights can be explained by different conditions of ambient air humidity during measurements, as well as wear of the filtering material in the course of measurement procedures. Slight differences between weights of soaked filters refer to the same. But slight deviations of weights are negligible, and the measurement results provide true and accurate information on dynamics of filter drying processes at various positions of the air distribution pipe.

3 Conclusions

Tables 2, 3 and 4 show that an approximate drying time of a water sprayed filter does not exceed 40 min. The best results of water sprayed filter drying time are with arrangement of holes to distribute air at an angle of 45 °C (Position II according to Fig. 6b).

Soaked filters drying time: the best measurement results are obtained with blowing of the lower part (position I in line with Fig. 6b). An explanation of this would be as follows: in a soaked filter (immersed into water), water flows down into its lower part, and with a sprayed filter, it stays in filter fibres.

In such a way, we have obtained the data on dynamics of filters drying by means of blowing via an air distribution pipe of a fixed geometry. Essential operating efficiency of the proposed system to prevent ventilation plants filters from clogging with ice fog in the environment of Extreme North regions is considered to be properly proven.

The data can be further used for survey investigations in this field.

References

1. Arbatskiy, A., Polyakov, S.: Solutions to problems relating to completing of ventilation plants to be used in Extreme North regions. Association of HVAC Engineers: Ventilation, Heating, Air Conditioning, Heat Supply Systems and Construction Thermophysics, no. 6, pp. 62–65 (2018)
2. Pavlov, N.N., Shiller, Y.N., Stroyizdat, M. (eds.): Sanitary-engineering installation for indoor use: Project designer's reference book (1992)
3. Air Filters. Classification. Marking. GOST R 51251-99

Validation of the Temperature Gradient Simulation in Steel Structures in SOFiSTiK

Marina Gravit, Ivan Dmitriev(✉), and Yurij Lazarev

Peter the Great St. Petersburg Polytechnic University,
Politechnicheskaya st. 29, 195251 St. Petersburg, Russia
i.i.dmitriev@yandex.ru

Abstract. There is a temperature gradient investigation of the steel hollow circular column. This dependence was carried out from simulated section in the PC SOFiSTiK and experimental data from the manual. The comparison of the modelling and manual results gives excellent convergence for both unprotected section and with fire protection of cement-sand plaster. The average relative deviation for most of the values does not exceed 5%. Modeling of the temperature gradient in steel structures can be considered validated. Software complex SOFiSTiK can be used to pre-assess the heating of sections of building structures.

Keywords: Fire protection · Fire resistance · Steel structures · Temperature gradient · Thermal gradients · Computer simulation · Modeling · Validation · Finite element method

1 Introduction

The calculating system for building structures has been currently improving at a fundamentally different level. The construction process, especially design, must comply with the BIM standards [1, 2]. The possibility of using modern software systems based on finite element algorithms makes feasible to refine experimental dependencies and computational methods. The finite element analysis allows us to estimate the temperature gradient distribution over the considered cross-section of the building structure, including complex composite. This approach provides a reliable preliminary assessment of the structure fire resistance and saves the fire test's cost.

The finite element analysis is not limited by the assumptions, which are made to simplify the calculated theoretical dependencies. In theoretical calculations, according to [3–9], a balance temperature distribution over the metal structure cross-section is assumed due to the relatively high thermal metal diffusivity. It is possible to consider the dynamics of the temperature gradient development by using electronic computers. The presence of this assumption may cause significant deviations of the actual fire resistance from the calculated, especially critical for thick-walled structures, including metal.

Nowadays, modeling of thermophysical processes in building structures is commonly carried out in software packages Ansys (USA) and Abaqus (USA), which are considered as the most appropriate programs for solving the problems of fluid and gas

V. Murgul and M. Pasetti (Eds.): EMMFT-2018, AISC 983, pp. 929–938, 2019.
https://doi.org/10.1007/978-3-030-19868-8_92

mechanics, heat transfer and heat exchange, including fire resistance [10–16]. We consider the possibility of using the SOFiSTiK software package (Germany) for analyzing the distribution of temperature fields for various building structures [17, 18].

The article's aim is to substantiate the convergence of the obtained temperature gradient values for the simulated and real object – a steel hollow circular column. In this paper, the validation of the results obtained in the SOFiSTiK software package with the results of fire tests, which are presented in the manual [7–9].

2 Methods

The Hydra module of the SOFiSTiK software package (ver. 2018) was used to analyze the temperature gradient of the section. Data entry was carried out through a text editor Teddy.

The standard fire mode was considered according to ISO 834 [19] and described by dependency (1) in the simulation.

$$T_t - T_0 = 345\log(8t + 1) \tag{1}$$

- T – time, min;
- T_t – temperature in the furnace at time t, °C;
- T_0 – temperature in the furnace at t = 0, °C.

The initial temperature inside the room adopted T_0 = +20 °C.

In the work, hollow circular sections were considered according to Russian national standard GOST 58064-2018 with different reduced thickness. The sections 1 and 8 are taken additionally to match the given thickness (Table 1).

Table 1. Simulated cross sections and their characteristics.

No	External diameter D, mm	Inner diameter d, mm	Wall thickness δ_s, mm	Cross-sectional area A_s, cm^2	Heating perimeter P_s, cm	Reduced thickness δ_{red}, mm
1	30	25.6	2.2	192.0	94.2	2.0
2	73	62	5.5	1165.7	398.8	5.1
3	140	118	11	4455.7	439.6	10.1
4	245	213	16	11505.0	769.3	15.0
5	426	384	21	26705.7	1337.6	20.0
6	530	468	31	48572.7	1664.2	29.2
7	1020	938	41	126036.5	3202.8	39.4
8	1420	1294	63	268441.7	4458.8	60.2

Adopted heating section from all sides (around the perimeter of the column). The reduced thickness δ_{red} for a hollow circular section is calculated by the formula (2):

$$\delta_{red} = \frac{A_s}{P_s} = \frac{\pi(D^2 - d^2)}{4\pi D} = \frac{\left(D^2 - (D - 2\delta_s)^2\right)}{4D} = \frac{\delta_s(D - \delta_s)}{D} \tag{2}$$

Thermophysical characteristics of steel according to [20]:

$$\lambda_t = 78 - 0.041t \tag{3}$$

$$c_t = 310 + 0.48t \tag{4}$$

Where:

- λ_t – coefficient of thermal conductivity $[\frac{W}{\mathrm{m \cdot K}}]$;
- c_t – coefficient of heat capacity $[\frac{\mathrm{J}}{\mathrm{kg \cdot K}}]$;
- t – material heating temperature, K.

3 Results and Discussion

3.1 Simulation of a Steel Structure Without Fire Protection

The simulated temperature distributions for the considered time points are shown in Figs. 1a–l.

The temperature of the heating section is taken as the arithmetic average between the outside and inside temperature. The values obtained in the simulation are summarized in a table, part of which is shown in intermediate Tables 2 and 3.

Table 2. The temperature of the outer and inner surface of the simulated tube (partly).

No	Time, min									
	5		10		15		20		25	
1	462.3	461.9	651.3	651.1	725.6	725.5	772.9	772.8	808.5	808.4
2	391.0	389.6	546.1	544.8	683.1	682.2	753.2	752.6	796.7	796.2
3	183.0	180.0	389.8	385.8	559.4	555.3	677.0	673.5	752.9	750.2
4	139.5	135.1	304.9	298.6	461.0	453.7	589.5	582.1	686.6	680.0
5	114.7	108.8	250.6	242.2	388.5	378.2	512.5	501.3	616.4	605.1
6	90.6	81.9	193.2	180.7	304.0	288.4	411.9	394.0	511.1	491.7
7	77.0	65.5	158.7	142.0	249.6	228.8	341.8	317.5	430.7	403.8
8	64.4	47.4	123.6	98.8	190.9	160.0	261.6	225.4	333.0	291.9

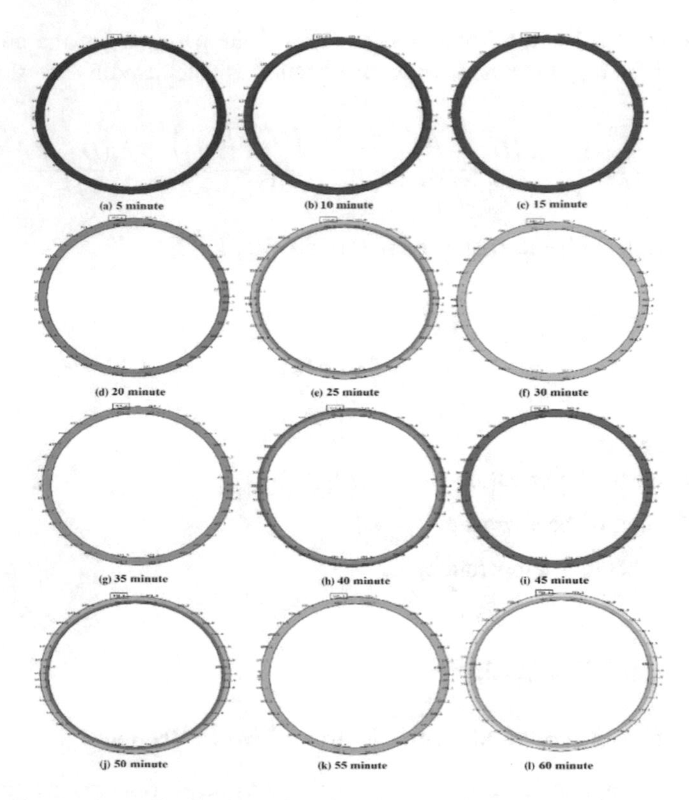

Fig. 1. a–l. Temperature distributions for time 5, 10, 15 … 60 min (program SOFiSTiK), Section 8 (1420 × 63).

Table 3. The temperature of the heating sections depending on the time for the standard fire mode (program SOFiSTiK).

SOFiSTiK section	Time, min											
	5	10	15	20	25	30	35	40	45	50	55	60
ISO	576.4	678.4	738.6	781.4	814.6	841.8	864.8	884.7	902.3	918.1	932.3	945.3
30 × 2.2	462.1	651.2	725.6	772.9	808.5	837.0	860.9	881.4	899.5	915.7	930.2	943.4
73 × 5.5	390.3	545.5	682.7	752.9	796.5	828.4	854.3	875.5	894.9	911.6	926.7	940.3
140 × 11	181.5	387.8	557.4	675.3	751.6	801.6	836.7	863.4	885.2	903.7	919.9	934.4
245 × 16	137.3	301.8	457.4	585.8	683.3	753.7	804.3	841.3	869.6	892.1	910.9	927.0
426 × 21	111.8	246.4	383.4	506.9	610.8	693.5	757.3	806.0	843.4	872.7	896.2	915.7
530 × 31	86.3	187.0	296.2	403.0	501.4	588.5	663.1	725.2	776.5	818.2	852.4	880.6
1020 × 41	71.3	150.4	239.2	329.7	417.3	499.1	573.7	640.1	698.1	748.2	790.9	809.3
1420 × 63	55.9	111.2	175.5	243.5	312.5	380.2	445.5	507.5	565.3	618.6	667.4	711.5

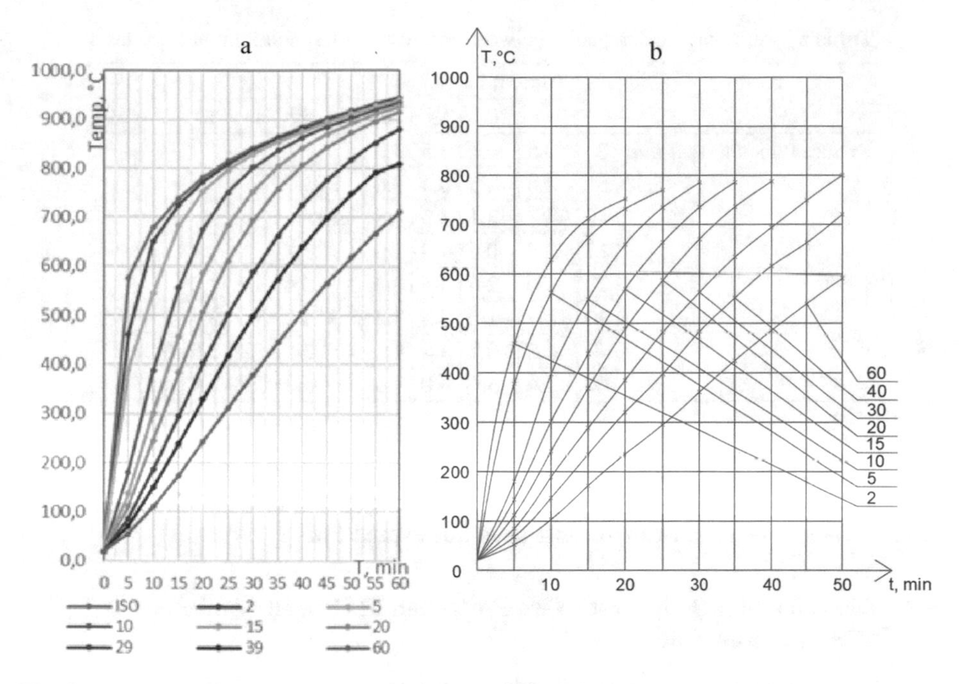

Fig. 2. Changes in the heating temperature of the structure depending on the time of exposure to the standard fire mode for various reduced metal thicknesses, mm — SOFiSTiK program.

The values from the graphs shown in Fig. 2b are approximated in Tables 4 and 5.

Table 4. The heating temperature of the cross sections depending on the time for the standard fire mode.

Data according to [8]		Time, min										
		0	5	10	15	20	25	30	35	40	45	50
Reduced thicknesses, mm	2	20.0	460.0	620.0	700.0	750.0						
	5	20.0	380.0	560.0	680.0	730.0	770.0					
	10	20.0	190.0	400.0	550.0	670.0	730.0					
	15	20.0	140.0	300.0	435.0	580.0	670.0	730.0				
	20	20.0	110.0	240.0	370.0	490.0	590.0	670.0	730.0	790.0		
	30	20.0	85.0	190.0	295.0	390.0	485.0	565.0	630.0	700.0	750.0	800.0
	40	20.0	70.0	150.0	230.0	315.0	400.0	480.0	540.0	605.0	660.0	710.0
	60	20.0	55.0	105.0	165.0	225.0	300.0	360.0	415.0	490.0	540.0	590.0

Table 5. Percentage discrepancy between the results of simulation and fire tests.

		Time, min									
		5	10	15	20	25	30	35	40	45	50
Reduced thicknesses, mm	2	0.5	4.8	3.5	3.0						
	5	2.6	−2.7	0.4	3.0	3.3					
	10	−4.7	−3.1	1.3	0.8	2.9					
	15	−2.0	0.6	4.9	1.0	1.9	3.1				
	20	1.6	2.6	3.5	3.3	3.4	3.4	3.6	2.0		
	30	1.4	−1.6	0.4	3.2	3.3	4.0	5.0	3.5	3.4	2.2
	40	1.8	0.2	3.8	4.4	4.1	3.8	5.9	5.5	5.5	5.1
	60	1.6	5.6	6.0	7.6	4.0	5.3	6.8	3.4	4.5	4.6

Modeling the temperature distribution over the cross section in the PC SOFiSTiK gives excellent convergence of the results with the graph (Fig. 2b). The average relative deviation for most of the results does not exceed 5%.

3.2 Modeling of a Design with Fire Protection of Cement-Sand Plaster (CSP), 20 mm Thick

The simulated sections of the construction with fire protection of cement-sand plaster 20 mm for the considered points of time are shown in Figs. 3a–j (Tables 6, 7 and 8).

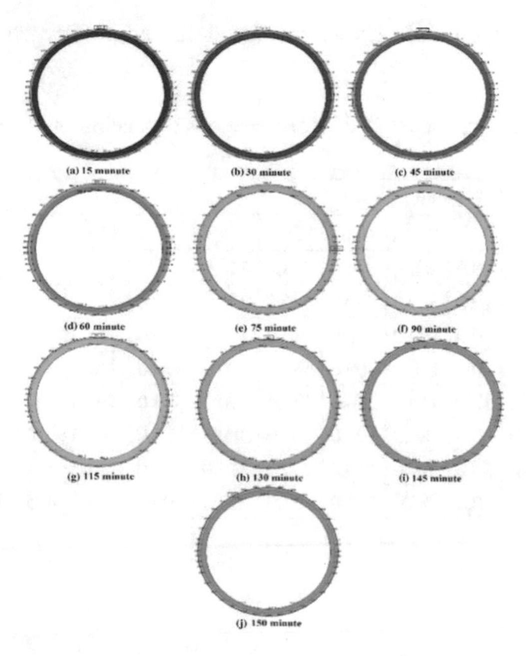

Fig. 3. a–j. Temperature distributions of fireproof construction for a time of 15, 30, 45 … 150 min (program SOFiSTiK), Section 8 (1420 × 63) + 20 mm CSP.

Table 6. The temperature of the steel column heating with fire protection of CSP 20 mm depending on the time for the standard fire mode. (SOFiSTiK program).

SOFiSTiK section	Time, min											
	5	10	15	20	25	30	35	40	45	50	55	60
ISO	738.6	841.8	902.3	945.3	978.7	1006.0	1029.1	1049.0	1066.7	1082.4	738.6	841.8
30 × 2.2	383.3	637.3	764.9	846.5	903.9	945.7	978.4	1005.0	1029.0	1049.0	383.3	637.3
73 × 5.5	264.1	499.2	644.0	740.4	812.1	868.6	913.3	949.5	979.8	1006.0	264.1	499.2
140 × 11	190.5	390.0	529.2	630.0	708.0	769.9	821.8	865.7	903.2	935.5	190.5	390.0
245 × 16	148.5	317.4	445.6	544.1	622.5	686.6	740.3	786.7	827.5	863.6	148.5	317.4
426 × 21	121.9	267.0	383.7	477.1	553.5	617.5	672.0	719.2	760.7	798.0	121.9	267.0
530 × 31	95.0	211.8	312.1	396.1	467.4	528.6	582.0	629.1	670.9	708.6	95.0	211.8
1020 × 41	76.0	171.9	257.8	332.5	397.5	454.4	505.3	550.8	591.7	629.1	76.0	171.9
1420 × 63	57.6	126.8	193.3	253.6	308.1	357.6	402.7	443.9	481.7	516.6	57.6	126.8

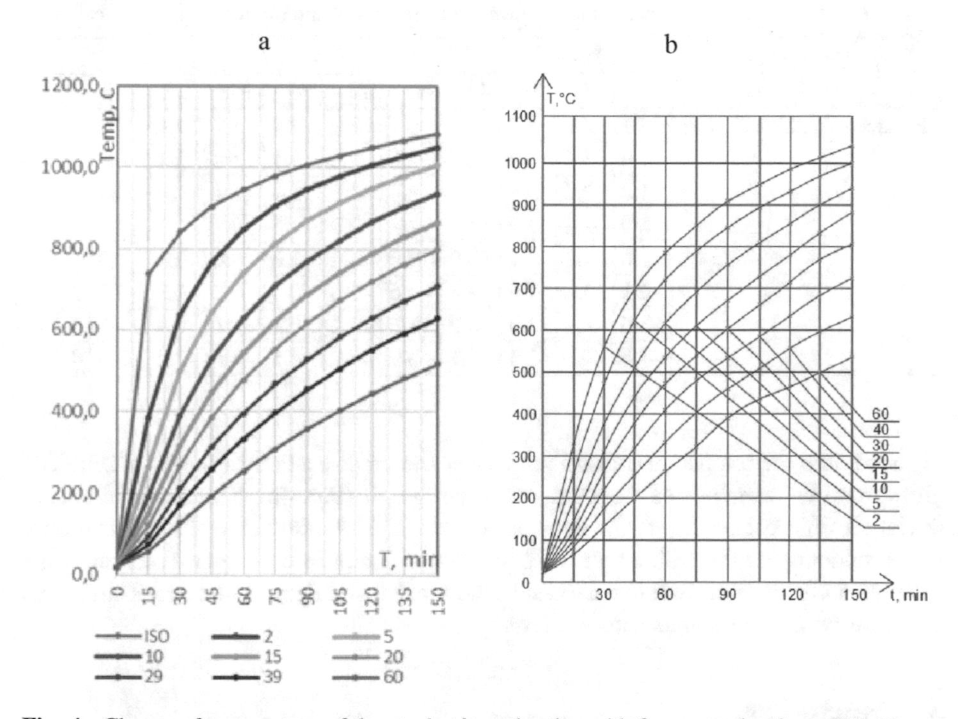

Fig. 4. Change of temperature of the steel column heating with fire protection from CSP 20 mm depending on the exposure time of the standard fire mode for various metal thickness, mm (a) SOFiSTiK program (b) According to [8, Fig. 9.3.4].

Table 7. The heating temperature of the cross sections depending on the time for the standard fire mode.

Data according to [8]		Time, min										
		0	15	30	45	60	75	90	105	120	135	150
Reduced thicknesses, mm	2	738.6	841.8	902.3	945.3	978.7	1006.0	1029.1	1049.0	1066.7	1082.4	–
	5	329.0	561.0	697.0	785.0	850.0	909.0	950.0	986.0	1013.0	1041.0	–
	10	259.0	472.0	623.0	721.0	789.0	846.0	894.0	932.0	969.0	1000.0	–
	15	194.0	376.0	510.0	616.0	691.0	756.0	810.0	856.0	900.0	939.0	–
	20	147.0	310.0	440.0	535.0	611.0	671.0	729.0	781.0	833.0	883.0	–
	30	120.0	264.0	377.0	473.0	539.0	604.0	661.0	715.0	769.0	806.0	800.0
	40	97.0	219.0	317.0	410.0	470.0	534.0	585.0	634.0	683.0	725.0	710.0
	60	78.0	174.0	254.0	333.0	404.0	460.0	509.0	557.0	596.0	632.0	590.0

Table 8. Percentage discrepancy between the results of simulation and fire tests

		Time, min									
		15	30	45	60	75	90	105	120	135	150
Reduced thicknesses, mm	14.2	12.0	8.9	7.3	6.0	3.9	2.9	1.9	1.6	0.8	–
	1.9	5.4	3.3	2.6	2.8	2.6	2.1	1.8	1.1	0.6	–
	–1.8	3.6	3.6	2.2	2.4	1.8	1.4	1.1	0.4	–0.4	–
	1.0	2.3	1.3	1.7	1.8	2.3	1.5	0.7	–0.7	–2.2	–
	1.6	1.1	1.7	0.9	2.6	2.2	1.6	0.6	–1.1	–1.0	–
	–2.1	–3.4	–1.6	–3.5	–0.6	–1.0	–0.5	–0.8	–1.8	–2.3	2.2
	–2.6	–1.2	1.5	–0.2	–1.6	–1.2	–0.7	–1.1	–0.7	–0.5	5.1
	–2.4	–4.1	–2.4	–3.7	–2.9	–4.9	–4.5	–4.5	–3.4	–3.6	4.6

Modeling the temperature distribution over the cross section in the PC SOFiSTiK gives good convergence of the results with the graph (Fig. 4b). The average relative deviation for most of the results does not exceed 5%. A noticeable discrepancy only when modeling a tube with a reduced thickness of 2 mm in the larger direction of the calculated parameters. To refine the results, additional simulations of various structures are required and compared with fire tests.

4 Conclusions

A good convergence of the temperature gradient simulation in the PC SOFiSTiK with fire test data for an unprotected and with fire protection of cement-sand plaster steel circular hollow column was obtained. The relative deviation of the results does not exceed 5% for most of the modeling options. Modeling of the temperature gradient in steel structures can be considered validated. Software program SOFiSTiK can be used to pre-assess the heating of sections of building structures.

References

1. Saknite, T., Serdjuks, D., Goremikins, V., Pakrastins, L., Vatin, N.I.: Fire design of arch-type timber roof. Mag. Civ. Eng. **64**(4), 26–39 (2016). https://doi.org/10.5862/MCE.64.3
2. Lazarevska, M., Cvetkovska, M., Knezevic, M., Trombeva Gavriloska, A., Milanovic, M., Murgul, V., Vatin, N.: Neural network prognostic model for predicting the fire resistance of eccentrically loaded RC columns. Appl. Mech. Mater. **627**, 276–282 (2014). https://doi.org/10.4028/www.scientific.net/AMM.627.276
3. EN 1991-1-2:2002 Eurocode 1: Actions on structures. Part 1-2: General actions – Actions on structures exposed to fire
4. EN 1992-1-2:2004 Eurocode 2: Design of concrete structures. Part 1-2: General rules. Structural fire design
5. EN 1993-1-2:2009 Eurocode 3: Design of steel structures. Part 1-2. General rules. Structural fire design
6. EN 1994-1-2:2005 Eurocode 4: Design of composite steel and concrete structures. Part 1-2: General rules - Structural fire design
7. Lennon, T., Moore, D.B., Wang, Y.C., Bailey, C.G.: Designers' guides to the Eurocodes. Designers' guide to EN 1991-1-2, 1992-1-2, 1993-1-2 and 1994-1-2 handbook for the fire design of steel, composite and concrete structures to the Eurocodes. Thomas Telford Ltd, 136 (2007)
8. Roitman, V.M.: Engineering solutions for assessing the fire resistance of buildings being designed and reconstructed. Association "Fire Safety and Science", Moscow, Russia, p. 382 (2001)
9. Twilt, L., et al.: Design guide for structural hollow section columns exposed to fire. Koln Verl. TUV Rheinland (1994). ISBN 3-8249-0171-4
10. Heinisuo, M., Jokinen, T.: Tubular composite columns in a non-symmetrical fire. Mag. Civ. Eng. **5**(49), 107–120 (2014). https://doi.org/10.5862/MCE.49.11
11. Schaumann, P., Tabeling, F., Weisheim, W.: Numerical simulation of the heating behaviour of steel profiles with intumescent coating adjacent to trapezoidal steel sheets in fire. J. Struct. Fire Eng. **7**(2), 158–167 (2016)
12. Schaumann, P., Bahr, O., Kodur, V.: Numerical studies on HSC-filled steel columns exposed to fire. Tubular Struct. **XI**, 411–416 (2017). https://doi.org/10.1201/9780203734964-50
13. Hamins, A., Mcgrattan, K., Prasad, K., Maranghides, A., Mcallister, T.: Experiments and modeling of unprotected structural steel elements exposed to a fire. Fire Saf. Sci. **8**, 189–200 (2005). https://doi.org/10.3801/IAFSS.FSS.8-189
14. Tondini, N., Hoang, V.L., Demonceau, J.-F., Franssen, J.-M.: Experimental and numerical investigation of high-strength steel circular columns subjected to fire. J. Constr. Steel Res. **80**, 57–81 (2013)
15. Tondini, N., Demonceau, J.-F.: Numerical analysis of the fire resistance of high-strength steel circular columns. In: Proceedings of Eurosteel, vol. 1 (2-3), pp. 2563–2571 (2017). https://doi.org/10.1002/cepa.305
16. Koh, S.K., Mensinger, M., Meyer, P., Schaumann, P.: Fire in hollow spaces: short circuit as ignition sources and the role of ventilation. In: 2nd International Fire Safety Symposium, p. 781. Doppiavoce, Napoli, Italy (2017)
17. Gravit, M., Lyulikov, V., Fatkullina, A.: Possibilities of modern software complexes in simulation fire protection of constructions structures with Sofistik. In: MATEC Web of Conferences, vol. 193, p. 03026 (2018). https://doi.org/10.1051/matecconf/201819303026

18. Gravit, M., Gumerova, E., Lulikov, V.: Computer modelling of fire resistant solutions for structures in high-rise buildings with using of new fire-retardant materials. In: SHS Web of Conferences, vol. 44 (2018). https://doi.org/10.1051/shsconf/20184400035
19. ISO 834-1: Fire resistance tests – elements of building construction. Part 1: general requirements, international organization for standardization ISO 834. Geneva, Switzerland (1999)
20. Organization standard ADSC 11251254.001-018-03 Design of fire protection of load-bearing steel structures using various types of linings. Association for the Development of Steel Construction. Moscow: Axiom Graphics Union, 72 (2018)

Polymer Additives for Cement Systems Based on Polycarboxylate Ethers

Linar Talipov(✉) and Evgeny Velichko

Moscow State University of Civil Engineering,
Yaroslavskoe shosse, 26, Moscow 129337, Russia
nakifulu@mail.ru

Abstract. The brief historical background on superplasticizers - polymer additives based on polycarboxylate ethers (PCE) are exposed in the paper. Prerequisites for changing the classification of plasticizers are represented. A range of main chemical components included in modern polymeric plasticizing agents is shown. The generalized structure of PCE molecules are considered, as well as special PCE with various modifications designed to solve specific problems. The architecture of PCE polymer molecules developed in recent years is presented. The main synthesis methods of PCE superplasticizers are summarized and represented. Commercial production process of PCE polymers is schematically shown. It is noted that the conservation of concrete mix is one of the main problems in the production and delivery of ready-mixed concrete at the moment. This problem is especially urgent in countries with hot climatic conditions. Various concepts developed in order to solve this problem are presented; also, some shortcomings of the mentioned methods are pointed out.

1 Introduction

Plasticizing polymer additives for concrete and concrete mixtures, mortars, and also for dry construction mixtures are hydrophilic surfactants. According to Russian regulatory documents, PCE additives are classified as superplasticizers according to the plasticization degree, but the effect of their plasticizing ability significantly exceeds the plasticization of other types of superplasticizers (naphthalene and melamine-containing). Therefore, a number of PCE scientists described it as a "hyperplasticizer." In this connection, it would be rational to revise the classification of plasticizing/water reducing chemical additives in the Russian regulatory and technical documentation. The polycarboxylates were firstly mentioned in 1981. The additive was developed in Japan by Nippon Shokubai Co. Ltd. as a copolymer with a certain percentage of hydrophilic functional groups [1]. Then the same company applied for 50 patents for the synthesis of polycarboxylate additives in the solvent system. PCE gradually began to crowd out other plasticizing/water reducing additives. At the moment, polymer plasticizers have almost completely replaced traditional ones (second generation plasticizers). The main advantage of PCE over traditional plasticizers is the programmable architecture of their molecules.

The first PCE products were copolymers of sodium methacrylate and methoxy polyethylene glycol methacrylate macromonomers (the so-called MPEG type).

V. Murgul and M. Pasetti (Eds.): EMMFT-2018, AISC 983, pp. 939–946, 2019.
https://doi.org/10.1007/978-3-030-19868-8_93

However, many other PCE products are currently on the market, including allyl ether (APEG), metellil ether (HPEG) and isoprenyl ether (IPEG or TPEG) [2]. PCE plasticizers can be divided into two groups – ordinary and special ones. Ordinary PCEs usually consist of a comb-like copolymer structure. The main anionic chain of such structure is mainly polymethacrylate, which holds adsorbing carboxyl groups and non-absorbent ester side chains. The chains mainly consist of polyethylene oxide (PEO) or polyethylene glycol (PEG). A particularity of PCE is that the comb core can be designed and modified in accordance with various requirements for concrete structures and for the concrete mix, as well as in accordance with the compatibility of materials used for the concrete mix. PCEs are characterized by various molecular weight and molecular mass distribution, the length of main chain, the length of graft side chain, type and density of functional groups. A change in any of the abovementioned characteristics may change the final molecular conformation and subsequently have a specific, targeted effect on the properties of the concrete mix. The correlation between the density, the length of side chains and the density, the length of the main chain in PCE molecules and, accordingly, the effect of plasticization and the plasticization persistence of cement pastes over time was studied in the research [3]. In addition, the regression dependence equations were obtained (1 and 2), where y_1 is the density of lateral chain, y_2 is the length of the main chain as the number of C-C bonds, n is the polymerization degree of the macromer.

$$y_1 = 244.7n^{-0.594} \cdot 100\% \tag{1}$$

$$y_2 = 350.1e^{-0.013n} \tag{2}$$

The structure of ordinary PCE by Gay and Raphael [4] is represented in Fig. 1, additions to which are reflected in the research [5] for this type of comb-shaped homopolymer. In this structure, n is the number of repeating structural units, each of which contains N monomers in the main chain and P monomers in the side chain. Figure 2 represents the chemical formula of ordinary PCE – a copolymer of polymethacrylic acid and polyethylene oxide, where x is the number of monomers in the side chain, w is the number of side chains in the copolymer, z is the number of carboxyl groups in the copolymer main chain.

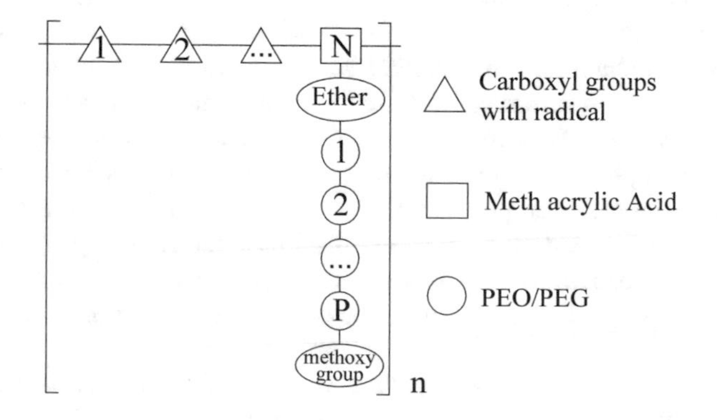

Fig. 1. The structure of ordinary PCE

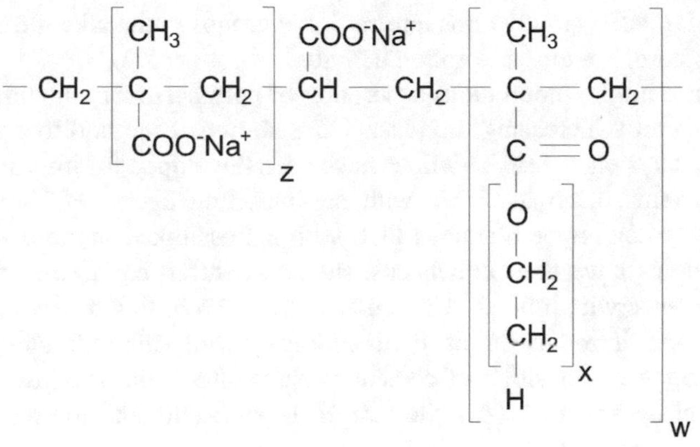

Fig. 2. The chemical formula of ordinary PCE which is a copolymer of polymethacrylic acid and polyethylene oxide

Special PCEs are obtained by modifying the molecular architecture of molecules. Relatively recently, a star-shaped PCE was synthesized [6]. It is reported that PCE with a star-shaped structure has a higher adsorption efficiency due to a higher content of polar groups per unit volume of the molecule. Moreover, a star-shaped PCE compared with the comb one gives a more effective plasticization of cement pastes and suspensions at equal concentrations, allowing significant reduction of the additive amount. All these effects are achieved due to the special star-like structure of PCE molecules.

One of the problems in the production and delivery of concrete mixture to the final consumer is its persistence. Persistence is the time, during which the mixture loses its castability when curing after the end of mixing. In order to increase the persistence, specialists from China synthesized the so-called "Slow-Release" PCE [7, 8]. The structure of such molecules is also comb-shaped, but there are additional chains of hydroxyl ester groups (HEG) (Fig. 3). The mechanism of such molecules is based on the fact that more adsorption-active molecules are used simultaneously with them. While active PCEs are being adsorbed on clinker minerals, displacing PCE molecules

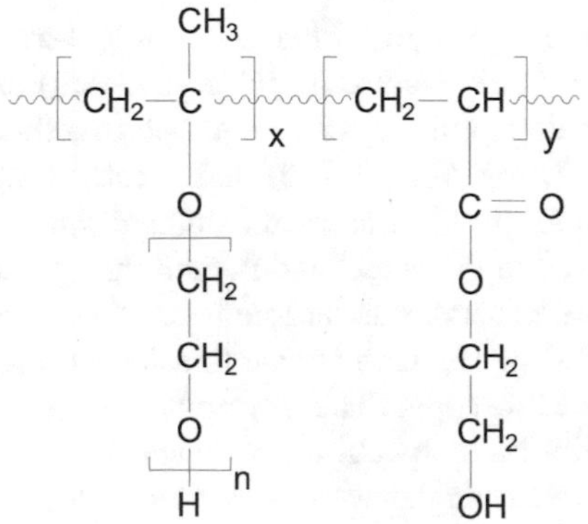

Fig. 3. Chemical structure of slowly releasing PCE.

containing HEG, HEG slowly turns into carboxyl groups in the alkaline environment of the cement system, gaining adsorption activity.

Slow release implies continuous adsorption of polymers from the liquid phase for a longer time, thereby increasing the plasticizing ability of the additive and the persistence of cement systems. Also, solutions have been developed for improved persistence of cement systems, such as PCEs with an inhibiting agent, PCEs with a release component, or a new, hyperbranched PCE with a crosslinked monomer. However, all these technologies have their drawbacks, such as reduced early strength of concrete, reduced persistence with limited PCE concentration. According to the research [9], it is reported that the introduction of PCE additives with different atomic masses of molecules into the composition of cement systems affects the increase in persistence. Also, a parallel is drawn between the size of the molecule and the persistence of the cement system, namely it is reported that the higher is the atomic mass of the molecule, the longer is the persistence. However, the research conducted is the evidence to the contrary.

2 Methods

Three commercial PCE polymers for concrete mixtures produced in Russia were investigated in order to determine the correlation between the structural characteristics of molecules and the effect and properties of the concrete mixture declared by the manufacturer. All samples had a predominant content of one fraction. The results of the studying the structural characteristics are shown in Table 1.

Table 1. Results of structural studies

Name	C/E	M_n, Da	M_w, Da	PDI	Persistance, min
PCE 1	2.1	8225	11376	1.38	up to 90
PCE 2	3.3	15958	>46616	>2.92	up to 70
PCE 3	2.5	>22570	>280156	–	20–40

The average molecular weight (Mn) and weight-average molecular weight (Mw) were determined by size-exclusion (gel permeation) chromatography at 30 °C using a Waters chromatographic system equipped with a cascade of columns (Ultrahydrogel 120 and Ultrahydrogel 250) and a refractometric detector. A 0.1 M aqueous solution of sodium nitrate was used as eluent; the flow rate was 1.0 ml/min. All samples were dissolved in the eluent and filtered through a PVDF membrane filter. The columns were calibrated using standard samples of polyethylene glycol (molecular weight range 20600-960 Da). The carboxylate to ester ratio (C/E) was studied by 13C nuclear using NMR spectroscopy. The researches were performed on a Bruker Avance III NanoBay 300 MHz spectrometer using standard Bruker techniques, in thermostatic mode at 25 °C. Samples were dissolved in deuterated water in a 1:1 ratio. The values were determined from the integral intensities in the 13C spectrum. The signals of the ester groups are taken as peaks in the range of 175–179 ppm, the signals of the carboxylate are taken as peaks in the area of 180–183 ppm.

3 Results

According to the data in Table 1, the larger the size of the molecules, the less the persistence. Even heavier molecules over 1 million Da were considered in the study [9]. Therefore, it can be assumed that the correlation between the persistence and atomic mass of molecules is nonlinear with a certain peak value. However, the persistence dependence of the cement systems should be considered on a number of structural characteristics of PCE molecules, such as atomic mass, the carboxylate to ester ratio, charge density, length of side and main chains, and the nature of side chain distribution of along the molecule skeleton, etc.

4 Discussion

There are mainly two following polymer synthesis methods of PCE production: a method of radical copolymerization and a method of grafting or esterification. Polymers synthesized by the grafting method are more commonly used as additives for cement systems, since this method leads to the formation of PCEs with a statistical distribution of carboxylate groups. Therefore, PCE molecule covers a large area of clinker mineral, reducing the concentration of the additive [10].

There are two main stages in the production of PCE plasticizers: the esterification (Fig. 4a) and polymerization (Fig. 4b).

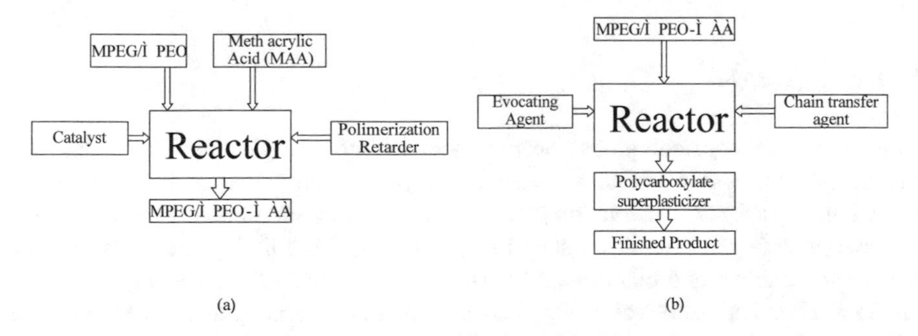

Fig. 4. Main stages in the production of PCE plasticizers

PCEs are mainly used as surface-active substances, such as superplasticizers for various cement systems including mainly concrete mixes, of which high-strength concrete mixes and self-compacting concrete mixes (SCC) are particularly distinguished [11]. It was found that the dispersing ability of PCE is due to adsorption (generally physical) on the areas of cement particles with a positive charge and mainly on aluminate and ferrite phases (C3A and C4AF) and their initial hydration products like ettringite and monosulfonic aluminum [12]. Negatively charged carboxyl COO groups bind to Ca^{2+} ions, which are subsequently incorporated into the crystal lattice of the hydrate phases [13]. The dispersion efficiency of PCE is primarily due to the charge neutralization of positively charged clusters, and then it is due to the steric repulsion created by the side chains (polyglycol) [14, 15]. Without superplasticizer, a significant part of the water becomes immobilized between cement agglomerates (flocculi), which in turn makes the cement system consistency more rigid. PCE disperses such flocculi, thereby releasing the immobilized water and increasing the flowability of the cement system.

Studies are also being conducted on the use of PCE as additives to reduce energy consumption when grinding cement clinker [16]. This is especially important given that up to a third of the energy costs during cement production accounts for grinding.

Studies on the use of PCEs as part of a complex anti-corrosion additive for steel reinforcement in concrete are actively carried out. The operation principle of such additive is complex, the mechanism of PCE in the additive composition is based on selective and competitive adsorption, and is also associated with persistence of concrete mix [17].

One of the common PCE problems is the incompatibility of materials. Therefore, experimental test studies are necessary prior to the use of PCE. As a rule, this incompatibility is due to the fact that the dispersed particles in the cement systems are of different composition types, mosaic of slots, texture and other parameters of surface quality. As a result, the selective adsorption of PCE may occur. It is known that PCEs are incorporating into the layered structure of clay by side chains. Therefore, their dispersing ability is limited by the content of clay impurities [18]. PCE adsorption in cement systems containing dispersed silicon dioxide occurs predominantly on silica fume [19]. According to the research [20], air entrainment up to 6.7% was recorded, which can significantly affect the strength properties and durability of cement systems.

5 Conclusions

Along with the constant global development of technologies, the technologies of cement systems and concrete are also developing. Requirements to materials and technologies used for construction change. Special attention is paid to environmental and energy-saving factors at all stages of producing the building materials and their usage, and concrete and other cement systems occupy one of the leading positions in this area, chemical additives being their concomitants. And plasticizing and water-reducing additives occupy the biggest volume part on application of all additives produced. In their evolution, PCE additives were upgraded from ordinary comb polymers to new ones with different molecular structures dynamically adjusted to different requirements for cement systems. In addition, PCEs began to find the pre-requisites for using them to grind clinker and for the composition of complex anti-corrosion systems for steel reinforcement in concrete, which is used in various corrosive environments. However, compatibility issues with various types of binders for specific cases are solved for a relatively long time, since this process requires a lot of research. In order to solve such problems in the future, it is possible to create special software with various scientific fields and computational methods integrated in it. This, in turn, will ensure the rapid determination of the structural characteristics and dosages of PCE for specific materials.

References

1. Hirata, T.: Cement dispersant 84 2022 JP Patent (S59–018338) (1981)
2. Plank, J.: PCE superplasticizers – chemistry, application and perspectives. Ibausil **1**, 91–102 (2012)
3. Wang, Z.-M., Lu, Z.-C., Liu, X.: Optimization of the structural parameters and properties of PCE based on the length of grafted side chain. In: Proceedings of the 11th International Conference on Superplasticizers and Other Chemical Admixtures in Concrete, vol. 302, Issue 20, pp. 265–278 (2015)
4. Gay, C., Raphael, E.: Comb-like polymers inside nanoscale pores. Adv. Colloid Interface Sci. **94**, 229–236 (2001)
5. Yaphary, Y.L., Raymond, H.W., Lau, D.: Chemical technologies for modern concrete production. Procedia Eng. **172**, 1270–1277 (2017)
6. Liu, X., Wang, Z., Zhu, J., Zhao, M., Liu, W., Yin, D.: Preparation and characterization of star-shaped polycarboxylate superplasticizer. In: Proceedings of the 11th International Conference on Superplasticizers and Other Chemical Admixtures in Concrete, vol. 302, Issue 14, pp. 183–198 (2015)
7. Liu, J., Liu, J., Yang, Y., Zhou, D., Ran Q.: Preparation and mechanism study of slow-release polycarboxylate superplasticizers. In: Proceedings of the 11th International Conference on Superplasticizers and Other Chemical Admixtures in Concrete, vol. 302, Issue 18, pp. 243–252 (2015)
8. Li, M., Wang, Y., Jiang, H., Zheng, C., Guo, Z.: Synthesis, characterization and mechanism of polycarboxylate superplasticizer with slump retention capability. In: IOP Conference Series: Materials Science and Engineering, vol. 182(012036) (2017)
9. Platel, D., Suau, J., Chosson, C., Matter, Y.: New additive to enhance the slump retention. In: Proceedings of the 11th International Conference on Superplasticizers and Other Chemical Admixtures in Concrete, vol. 302, Issue 04, pp. 53–62 (2015)
10. Pickelmann, J., Li, H., Baumann, R., Plank, J.: A 13C NMR spectroscopic study on the reparation of acid and ester groups in MPEG type PCEs prepared via radical copolymerization and grafting techniques. In: Proceedings of the 11th International Conference on Superplasticizers and Other Chemical Admixtures in Concrete, vol. 302, Issue 02, pp. 25–38 (2015)
11. Okamura, H., Ozawa, K.: Mix-design for self-compacting concrete. Concr. Libr. JSCE **25**, 107–120 (1995)
12. Bonen, D., Shondeep, S.: The superplasticizer adsorption capacity of cement pastes, pore solution composition, and parameters affecting flow loss. Cem. Concr. Res. **7**(25), 1423–1434 (1995)
13. Sakai, E., Yamada, K., Ohta, A.: Molecular structure and dispersion-adsorption mechanisms of comb-type superplasticizers used in Japan. J. Adv. Concr. Technol. **1**(1), 16–25 (2003)
14. Marchona, D., Sulserb, U., Eberhardtb, A., Flatt, R.J.: Molecular design of comb-shaped polycarboxylate dispersants for environmentally friendly concrete. Soft Matter **9**(45), 10719–10728 (2013)
15. Yoshioka, K., Sakai, E., Daimon, M., Kitahara, A.: Role of steric hindrance in the performance of superplasticizers for concrete. J. Am. Ceram. Soc. **80**(10), 2667–2671 (2005)

16. Mishra, R., Heinz, H., Zimmermann, J., Müller, T., Flatt, R.J.: Understanding the effectiveness of polycarboxylates as grinding aids. In: Conference: American Concrete Institute Symposium Series, vol. 288, Issue 16 (2012)
17. Talipov, L., Velichko, E.: Effect of polycarboxylate and polyarylate surfactants on corrosion of the steel reinforcement embedded in the concrete. In: VI International Scientific Conference (IPICSE-2018), vol. 251, p. 01026 (2018)
18. Lei, L., Plank, J.: A concept for a polycarboxylate superplasticizer possessing enhanced clay tolerance. Cem. Concr. Res. **42**(10), 1299–1306 (2012)
19. Schröfl, C., Gruber, M., Plank, J.: Preferential adsorption of polycarboxylate superplasticizers on cement and silica fume in ultra-high performance concrete (UHPC). Cem. Concr. Res. **42**(11), 1401–1408 (2012)
20. Winnefeld, F., Becker, S., Pakusch, J., Gotz, T.: Effects of the molecular architecture of comb-shaped superplasticizers on their performance in cementitious systems. Cement Concr. Compos. **29**(4), 251–262 (2007)

Assessing the Effect of Polymer's Type and Concentration on the Properties of the Polymer-Bitumen Binders (PBB) on Road Construction

Marina Vysotskaya(✉), Dmitriy Kuznetsov, Aleksandr Obukhov, Artem Shiryaev, and Dmitry Litovchenko

Belgorod State Technological University named after V.G. Shukhov (BSTU), Kostyukova str. 46, 308012 Belgorod, Russia
roruri@rambler.ru

Abstract. The use of polymers for modifying bitumen allows obtaining high-tech binders with a given set of operational and technical properties. The object is to study the influence of polymers, which appeared on the construction market, on the properties of bitumen, such as softening temperature, brittleness and plasticity. Also, the structure-forming ability of polymers in a bitumen environment and their contribution to the aging process of the modified binder were assessed. Seven various modifiers were studied, such as Dorso 46-02, polystyrene (PS) – 525, LG EVA 28400, SBS L 30-01 A, Calprene 501M, Globalprene 3501. The system was filled with polymer in concentration limits 1.0–5.0%, while the content of the plasticizing component remained unchanged at 2.0%. In order to assess the structure-forming ability of polymers, the surface morphology of each of them was studied, and the properties of the obtained polymer-bitumen binder (PBB) samples were analyzed before and after aging. The surface morphology of polymers was studied using the scanning electron microscopy (SEM) by the means of TESCAN MIRA3 LMU autoemission electron microscope. The aging intensity of bitumens modified by the studied polymers was estimated by the change in the penetration rates at 25 °C and 0 °C, the softening temperature and brittleness after the RTFOT furnace. The surface morphology of the considered polymers and their structure-forming ability with different bitumen filling were analyzed. The analyses showed that samples with a loose structure interact with the bitumen components more effectively and quickly, and form a stable homogeneous structure of the modified binder.

Keywords: Bitumen · Polymer · PBB · Structure-forming ability · Thermal destruction

V. Murgul and M. Pasetti (Eds.): EMMFT-2018, AISC 983, pp. 947–959, 2019.
https://doi.org/10.1007/978-3-030-19868-8_94

1 Introduction

The methods used to modify bitumen are different, but the common goal is to create high-tech binders allowing developing structural materials and products with a given set of operational and technical properties. Recently, there has been a significant amount of interest in bitumens modified by various polymers. Polymer-bitumen binders (PBB) are organic binders consisting of bitumen, a plasticizer, a polymer, and, if necessary, a complex of chemical additives (for example, a cross-linking agent).

When it comes to the quality of PBB and their properties, it is necessary to take into account the key role of polymer bitumen modifiers with specific properties caused by chemical and topological structure, as well as physical structure. Organic polymers can be long linear, branched, mesh, or with other topologies. They are huge chain macromolecules with a molecular weight from tens of thousands to millions. The bonds between the atoms of macromolecules are covalent in nature, very strong and stable. Due to the thermodynamic flexibility of polymer macromolecules, they can be in solid or liquid states depending on temperature. Similar transitions are also typical for bitumen, which provokes thermal destruction processes in the system, and then the general aging of the PBB. Such difficulties have caused a keen research interest of scientists in this area [1–10]. The object of the research is to study the influence of polymers, which appeared on the construction market, on the properties of bitumen, as well as to study the structure-forming capacity of these polymers and to evaluate their contribution to the aging processes of PBB. Before considering polymer modifiers, it is worth noting that the original bitumen makes a significant contribution to the final result of the modification. In order to eliminate this factor, bitumen of one batch was used.

2 Objects and Methods

The BND 90/130 grade bitumen of the Moscow Refinery was used as the research object, as well as plasticizer component in the amount of 2.0%. The content of plasticizer does not make significant changes in the indicators of the properties of bitumen, but is necessary for the preparation of PBB. The list of research objects also includes polymers of the following brands: Dorso 46-02, polystyrene (PS) – 525, LG EVA 28400, SBS L 30-01 A, Calprene 501M, Globalprene 3501. The system was filled with polymer in concentration limits of 1.0–5.0%. After that, the structure-forming ability of the polymer and the thermal destruction tendency of the system were evaluated. The maximum polymer content was established by the technical possibility of its introduction and dissolution in the maltene part of the binder.

The polymer surface morphology is an important aspect in the way of obtaining high-quality PBB. In order to study the structure of polymers, the method of scanning electron microscopy (SEM) was used. The study was carried out using a TESCAN MIRA3 LMU autoemission electron microscope, which is technically able to examine non-conductive samples in their natural state, without conductive layer sputtering. It allows obtaining surfaces of the considered object with high resolution.

The preparation of PBB was carried out in the following technological sequence: a plasticizing component was introduced into the pre-dried bitumen, mixed and heated to a temperature of 170 °C, then the weight of the polymer was introduced. Immediately after introduction the polymer has been grinded for 20 min. Resulting composition was stirred for 3 h for better dilution of the polymer in bitumen and ending of the reactions between the modifier and the bitumen-based hydrocarbons. At the end of mixing, the containers with the finished PBB were placed in a drying cabinet and kept for 2 h until the complete maturation, at the end of which standard properties of the PBB samples were determined and the tendency to thermal destruction was assessed.

3 Discussion

The morphology of the surfaces was studied in order to preliminary predict the behavior of polymers when combined with bitumen and their dissolution (Fig. 1). According to their physical properties, the studied polymers belong to different groups, such as plastomers (Polystyrene (PS) - 525, LG EVA 28400) and elastomers (Dorso 46-02, SBS L 30-01 A, Calprene 501M, Globalprene 3501)

Fig. 1. Photomicrograph of the polymer surface of the following brands: (a) Polystyrene (PS) - 525, (b) LG EVA 28400, (c) Calprene 501M, (d) Globalprene 3501, (e) SBS L 30-01 A.

As can be seen from Fig. 1a–b, the presented plastomers have more dense packing of molecules, which is natural for their crystalline structure. Such polymers are characterized by low mobility of structural units. However, polymer LG EVA 28400 has a more prominent surface microstructure, unlike polystyrene. The presence of micropores on the surface of the granules (Fig. 1c–d) will lead to a greater selective diffusion of the bitumen components. Due to this loose structure, the rubbers are softer and more flexible. The branched form of radial polymers contributes to a greater number of contacts on the surface, which can be seen on the surface of the polymer in Fig. 1.

Depending on the orderliness degree of macromolecules, two following phase state types of polymers are distinguished [11]: amorphous and crystalline. The amorphous phase is characterized by the chaotic arrangement of the macromolecule with a certain structural orderliness, maintained at relatively small distances, commensurable with the size of the macromolecule. The crystalline phase is characterized by the orderly arrangement of macromolecules in the polymer, and the orderliness is maintained at distances exceeding the size of the macromolecule hundreds and thousands of times. The dissolution rate of the polymer in bitumen depends on the specific surface area of their contact. Thus, only functional groups located on the surface of polymer granules immediately react in contact with bitumen. For the functional groups that are not located on the surface, aromatic hydrocarbons of bitumen must be diffused beforehand. The duration of such diffusion is determined by many factors, and one of them is the packing density of polymer macromolecules. Various availability of functional groups of macromolecules or their links located in the amorphous and crystalline regions can affect the homogeneity of the reaction products and, consequently, degrade the quality of the modified binder. Thus, analyzing the surface morphology of the polymers under consideration, it can be assumed that Calprene 501M, Globalprene 3501 and SBS L 30-01 A samples have an amorphous structure. Due to its looseness, they will more efficiently and quickly interact with the bitumen components, structure it and form a stable and uniform PBB. The BND 90/130 grade bitumen was used for making the PBB, indicators of its properties and the properties of the modified bitumen with various polymer filling are presented in Table 1.

Under the given variation conditions of the experiment, the compositions of PBV 60 meeting the requirements (GOST R 52056-2003) were not developed with all polymers, since determining the optimal formulation requires varying component composition and, first of all, a plasticizer, which allows increasing the proportion of maltene part in the binder composition. However, the purpose of the study was developed at the very beginning and the conditions of the experiment were indicated. Therefore, the component composition of the modified binders remained unchanged.

Table 1. The results of testing the PBB compositions.

Polymer content, %	Test name					
	Softening temperature according to ball-and-ring method, °C	Needle penetration depth, 0.1 mm		Stretchability, cm	Elasticity, %	Fraas brittle point, °C
		at 25 °C	at 0 °C	at 25 °C	at 25 °C	
1	2	3	4	5	6	7
BND 90/130						
0.0	44.0	106.0	28.0	100.0	0.0	−22
Requirements for PBB 60 according to GOST R 52056-2003						
-	54	60	32	25	80	−20
PBB with "Polystyrene (PS) - 525" polymer						
1.0	69.4	35.0	24.0	20.0	35.0	−18.0
2.0	67.0	31.0	22.0	15.0	28.0	−19.0
3.0	70.0	29.0	20.0	10.0	20.0	−19.0
PBB with "LG EVA 28400" polymer						
1.0	50.0	60.0	30.0	57.0	50.0	−22.0
2.0	53.0	56.0	26.0	55.0	50.0	−21.0
3.0	54.0	54.0	27.0	39.0	40.0	−22.0
4.0	60.0	54.0	28.0	18.0	43.0	−22.0
PBB with "Calprene 501M" polymer						
1.0	49.0	77.0	31.0	65.0	60.0	−21.0
2.0	52.0	68.0	31.0	40.0	73.0	−21.0
3.0	56.0	62.0	32.0	38.0	76.0	−21.0
4.0	61.0	47.0	31.0	38.0	84.0	−21.0
5.0	65.0	49.0	30.0	35.0	89.0	−22.0
PBB with "Globalprene 350" polymer						
1.0	47.0	92.0	32.0	41.0	62.0	−22.0
2.0	52.0	86.0	32.0	39.0	72.0	−22.0
3.0	56.0	78.0	32.0	39.0	78.0	−22.0
4.0	64.0	65.0	31.0	39.0	85.0	−22.0
5.0	67.0	59.0	31.0	38.0	86.0	−23.0
PBB with "SBS L 30-01 A" polymer						
1.0	47	95	33.0	49.0	62.0	−23.0
2.0	51	87	32.0	41.0	72.0	−23.0
3.0	54	74	32.0	39.0	78.0	−22.0
4.0	59	66	31.0	36.0	85.0	−22.0
5.0	63	58	31.0	30.0	86.0	−21.0

The test results show that the bitumen penetration decreases after the introduction of the polymer, and the bitumen corresponds to a more viscous brand. The bitumen structuring processes were most pronounced with the introduction of “Polystyrene (PS) - 525” and “LG EVA 28400” polymers, with dense packaging of polymer macromolecules, Fig. 1a–b. As can be seen from the Table 1, it was not possible to obtain PBB 60 using these polymers in the variation range of 1.0–4.0%. Therefore, one should be apprehensive about a sharp deterioration in the technological properties of the binders with a low plasticizer content and an increased content of the polymer modifier, as well as a decrease in the productivity of plants due to high viscosity of binders. Obviously, using these polymers for making the PBB requires a greater content of plasticizer or a complex introduction with another polymer.

The polymers Calprene 501M, Globalprene 3501, SBS L 30-01 A, characterized by a loose structure have a «softer» structure-forming effect. At the same time, it is possible to obtain PBB 60 with a 3.0% content based on them, according to a set of indicators of properties in the established conditions.

However, the effectiveness assessment of the obtained PBBs can be made only under the condition that the consumer’s requirements for the modified binder are known, with a possible adjustment of the composition by introducing the additional polymer content and increasing the plasticizer content.

In general, all samples show an increase in the softening temperature, but the change in this indicator is much less than penetration. Low-temperature characteristics change insignificantly.

Stretchability at 25 °C, as a rule, decreases with an increase in the content of the polymer, compared with the initial bitumen. The elastic spatial polymer mesh forms, which increases the granularity of the system. Elasticity becomes more important with equal content of the polymer in the case of using the elastomers. There is also an inverse correlation, the greater the elasticity of the PBB, the less its cohesion. This confirms the rule, according to which, the elasticity of polymers decreases with the increase of their hardness [12]. It is also shown by the obtained data (Table 1).

A distinctive feature of organic binders, comparing to other types of binders, is the tendency to aging under the influence of external factors and, above all, temperatures. The aging mechanism of polymer-bitumen binders is more complex than one of bitumen. Both the system as a whole and each component of it are subjected to aging. Thus, the polymer is subjected to spontaneous and irreversible changes in properties due to the destruction of bonds in the chains of macromolecules. And although the aging of PBB occurs last in its system, the importance of the processes occurring during this period is great and determines the properties of the binder, such as homogeneity, delaminatability, elasticity, etc. PBB is a complex material, and a change in the quality of one component inevitably entails a change in the quality of the entire system. Aging of the polymer inevitably leads to a decrease in the proportion of effective polymer in the binder, which leads to undesirable changes in the properties of PBB, and affects the quality of the asphalt concrete pavement. The crack resistance decreases, as well as deformability, ductility and elasticity at low temperatures, etc.

In order to assess the thermal destruction tendency of the polymer in the composition of the PBB, the prepared compositions of binders (Table 1) have undergone aging processes. The obtained data is presented in Figs. 2, 3, 4, 5, 6 and 7.

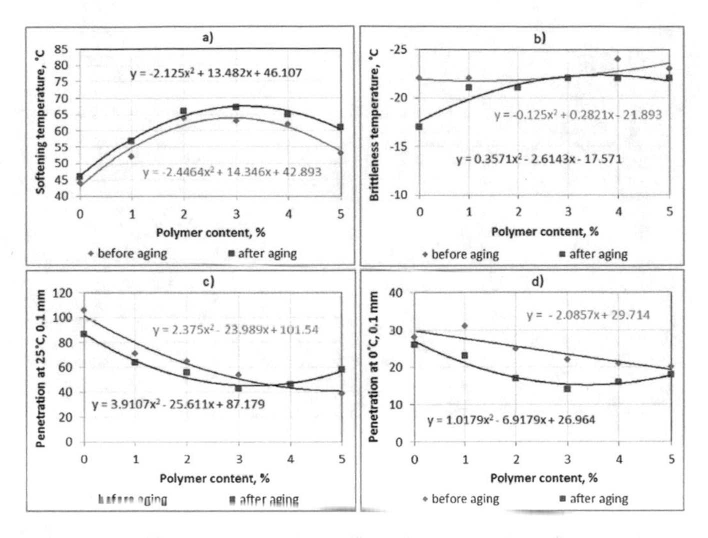

Fig. 2. Changes in the properties of PBB with polymer additive "Dorso 46-02" (a, b) softening temperature and brittleness; (c, d) depth of needle penetration at 25 °C and 0 °C.

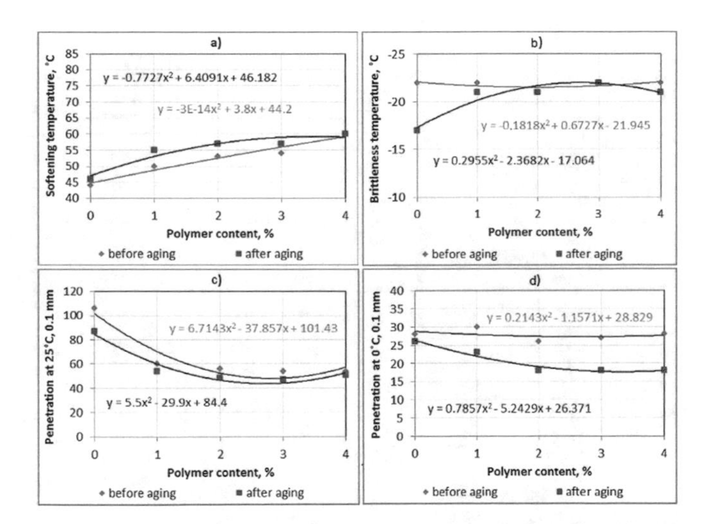

Fig. 3. Changes in the properties of PBB with polymer additive "LG EVA 28400" (a, b) softening temperature and brittleness; (c, d) depth of needle penetration at 25 °C and 0 °C.

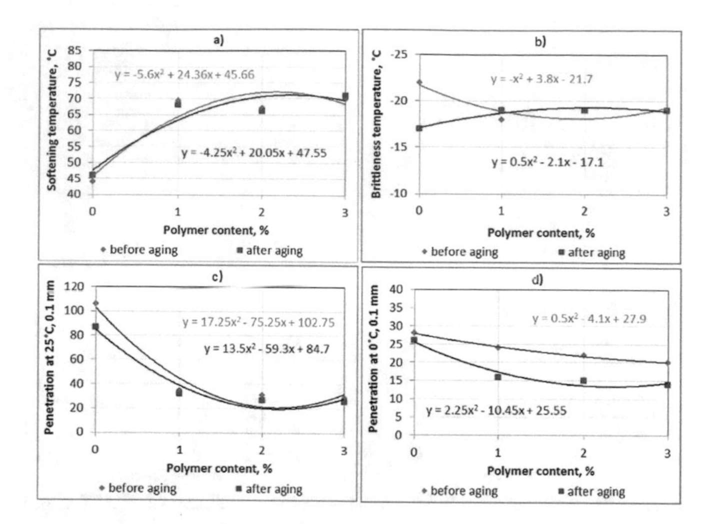

Fig. 4. Changes in the properties of PBB with polymer additive "PS - 525" (a, b) softening temperature and brittleness; (c, d) depth of needle penetration at 25 °C and 0 °C.

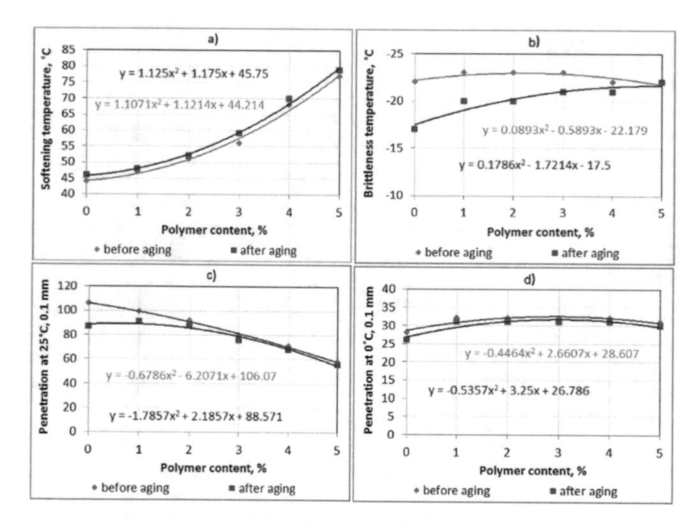

Fig. 5. Changes in the properties of PBB with polymer additive "SBS L 30-01 A" (a, b) softening temperature and brittleness; (c, d) depth of needle penetration at 25 °C and 0 °C.

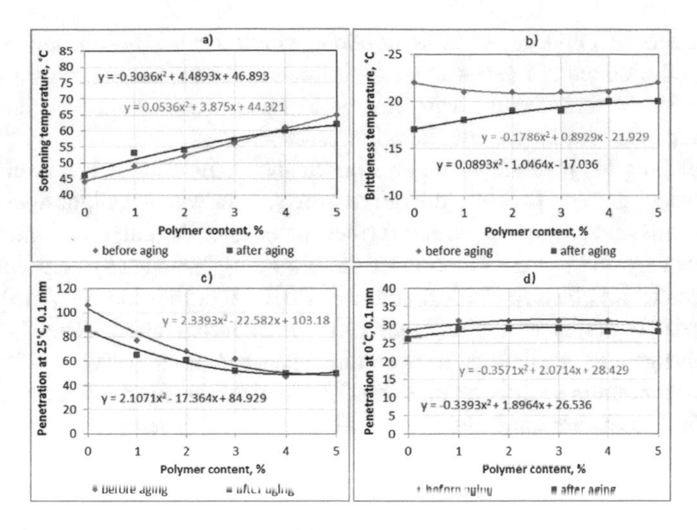

Fig. 6. Change the properties of PBB with polymer additive "Calprene 501" (a, b) softening temperature and brittleness; (c, d) depth of needle penetration at 25 °C and 0 °C.

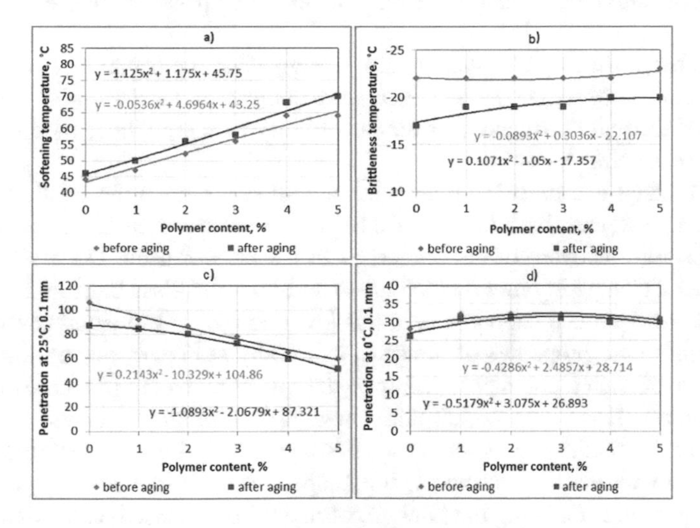

Fig. 7. Changes in the properties of PBB with polymer additive "Globalprene 350" (a, b) softening temperature and brittleness; (c, d) depth of needle penetration at 25 °C and 0 °C.

At the molecular level, polymer aging is caused by a change in molecular weight and its distribution (MWD), which are the most important characteristics of polymers determining their unique properties [13]. Based on the physical properties of polymers, they can be divided into two following large groups [14]:

- plastomers, which are characterized by increased strength, high elasticity modulus and weak extensibility;
- elastomers, with a small modulus of elasticity and high resilience.

Copolymers of ethylene, such as ethylene vinyl acetate (EVA), are of particular relevance and production interest among plastomers for the modification of bitumens. Let us consider the formulations using this type of polymer (Figs. 2, 3) and highlight its inherent properties during the modification process.

The softening temperature curve on Fig. 2a shows anomalous behavior. When the polymer content goes up to 3.0%, the curve grows, after which a slight decrease of this indicator begins and progresses with a further increase of the polymer content to 5.0%. This is obviously due to the saturation of the bitumen-plasticizer system with polymer macromolecules involved in the formation of PBB. A further increase in the polymer content leads to the system filling without the expected effect. This is shown by minimal changes in brittleness temperature, penetration at 0 °C and 25 °C after polymer concentration is more than 3.0%. Linear relationship is detected when using ethylene vinyl acetate manufactured by LG, the softening temperature of the binder increases with increasing polymer content.

According to the Fig. 2c, the penetration is noticeably reduced as the polymer is added. During the aging process, a break point is formed with a polymer content of 3.0%. However, this phenomenon cannot be detected according to Fig. 3c. A sharp decrease in penetration is observed with an increase in the polymer content to 2.0%, after which no significant changes occur. The curve of penetration after aging retains this trend.

Such behavior of the PBB system allows suggesting the occurrence of plasticizing effects. Their causes might be the increased content of Dorso 46-02 polymer and processes of thermal destruction, or incompatibility of the polymer modifier with the bitumen or plasticizer used.

Ethylene vinyl acetate (EVA), does not significantly affect the elastic recovery of bitumen as a plastomer. Introduction of a higher polymer content is necessary in order to achieve elastic recovery of the binder, as in the case of using Dorso 46-02, which will provoke a change in other indicators. In addition, the glass transition temperature of ethylene vinyl acetate copolymers is not low enough [14] to significantly improve the low-temperature properties of bitumen, especially at high concentrations of EVA. Thus, the use of such a polymer to improve the low-temperature properties of bitumen is rather limited. It is worth noting that low-temperature properties are a problem of domestic organic binders, to which close attention is paid. Low-temperature properties are traditionally estimated according to the temperature of brittleness and penetration at 0 °C, the latter indicator being the linchpin of many PBBs. Based on the data obtained, it can be concluded that the maximum loss of the penetration index at 0 °C after aging is common for the LG EVA 28400 and PS-525 polymers.

Plastomers also include so-called plastics, for example, polystyrene. However, it is believed that these thermoplastic polymers occupy an intermediate place, and they are called elastoplastics [14]. Let us consider the composition where this type of polymer was used (Fig. 4) and highlight its inherent properties.

Trend line of penetration for a sample modified with polystyrene (Fig. 4c), is similar to the sample with an EVA 28400, but is in the region of lower values. Due to the high structuring ability of the polymer, a sharp decrease in penetration to values less than 40 mm-1 is observed with the introduction of only 1.0%, which probably can lead to unsatisfactory cohesion of such PBB.

Thermoplastic elastomers are usually more effective than plastomers for bitumen modification. The most popular type of bitumen modifier is thermoplastic elastomers SBS. Let us consider the formulations prepared using it. According to the graphs in Figs. 5, 6 and 7c, it can be noted that the increase in the polymer content causes a decrease in the value of penetration at 25 °C, this correlation is common for all samples of PBB. In the case of the application of "Calprene 501", the curve goes on a plateau with the additive content 4.0–5.0%. No anomalies are observed, the curves after aging repeat the behavior of the curves before aging. The softening temperature increases in a linear relationship for all compositions with increasing polymer content in the system.

Minimal destructive processes in the PBB composition after aging, regardless of the volumetric system filling with polymer, are common for such polymers as SBS L 30-01 A, Calprene 501M, as well as Globalprene 3501. Moreover, the nature of the changing properties curves of the latter 2 after aging is identical. This is primarily due to the fact that all three polymers are block copolymers consisting of blocks of styrene and polybutadiene (S-B-S). The difference lies in the content of bound styrene. Thus, it reaches up to 30% in SBS L 30-01 A, and in Calprene 501M and Globalprene 3501 it is 31% and 32%, respectively.

4 Conclusions

1. The strongly structured binder can be obtained by adding polymers belonging to the class of plastomers. The obtained binder is characterized by increased rigidity and, as a result, lower temperature sensitivity. However, the manufacturability of the product is lost, which limits its use. It was not possible to obtain PBB 60 in the 1–4% variation range by using polymers "Polystyrene (PS) - 525" and "LG EVA 28400". Perhaps these polymers will be effective for developing PBB40 formulations, as well as for mastic and waterproofing compounds due to their high structure-forming ability.
2. Elastomer polymers structure the system to a lesser extent, thus maintaining the physical and mechanical properties at a level sufficient to obtain a product that meets the requirements of GOST. The resulting binder is characterized by high elasticity. The polymers Calprene 501M, Globalprene 3501, SBS L 30-01 A, characterized by a loose structure have a "soft" structure-forming effect on bitumen. A set of indicators corresponding to the PBB 60 was obtained with a polymer content of 3%.
3. During aging of the sample with the polymer "Dorso 46-02", the viscosity change is the opposite. With a high destruction degree and low molecular weight, the polymer is transformed into a viscous-flow plastic state from a highly elastic one, providing a plasticizing effect. This phenomenon was noted in numerous researches. The correlations of the remaining aged samples, although they are in the region of lower (higher) values, follow the curve trend of the original binders. However, the intensity of the occurring processes is reduced. This can be estimated by the slope of the curves.

4. Penetration at 25 °C for all compositions decreases with increasing polymer content due to the formation of a spatial polymer grid. The use of polystyrene 525 has the strongest effect on the system. The cases when a plateau is formed and the penetration values do not change can be explained by the achievement of the optimal maltene/asphalt proportion parts due to adsorption by the first polymer.
5. A change in the depth of a steel needle immersion at 0 °C indicates that a polymer concentration increase does not have a significant effect on this indicator. A significant change in penetration at 0 °C will most likely be caused by the appearance and content of thc plasticizing component, which acts as a constant in the framework of research.
6. Analyzing the graphs revealed a certain relationship between the softening temperature and penetration. This is clearly seen in many formulations. The polymer concentration, at which the minimum penetration is observed on the graph, is the maximum point on the softening temperature graph. On the basis of the road surface working conditions, the compositions of PBB that have an equal penetration rate at 25 °C and a higher softening temperature are more valuable.
7. In general terms, the addition of a polymer contributes to the hardening of the system, and lowering the brittleness temperature usually occurs due to the introduction of a plasticizer. However, it is worth paying attention to the fact that the use of SBS-type polymers has a greater influence on the change of this indicator in the process of aging, which is due to the structure of these modifiers.
8. Based on the studies performed, it can be concluded that the most informative indicators with maximum deviations from the properties of the original compounds are low-temperature characteristics of the binders. The tendency of PBBs to destructive processes was assessed using the RTFOT method.

References

1. Singh, S.K., Kumar, Y., Ravindranath, S.S.: Thermal degradation of SBS in bitumen during storage: Influence of temperature, SBS concentration, polymer type and base bitumen. Mech. Time Depend. Mater. **147**, 64–75 (2018)
2. Singh, B., Saboo, N., Kumar, P.: Use of Fourier transform infrared spectro copy to study ageing characteristics of asphalt binders. Pet. Scence Technol. **35**(16), 1648–1654 (2017)
3. Hofko, B., Alavi, M.Z., Grothe, H., Jones, D., Harvey, J.: Repeatability and sensitivity of FTIR ATR spectral analysis methods for bituminous binders. Mater. Struct./Materiaux et Constructions **50**(3), 187 (2017)
4. Zhu, C., Zhang, H., Shi, C., Li, S.: Effect of nano-zinc oxide and organic expanded vermiculite on rheological properties of different bitumens before and after aging. Constr. Build. Mater. **146**, 30–37 (2017)
5. Iwanski, M., Cholewinska, M., Mazurek, G.: Impact of the ageing on viscoelastic properties of bitumen with the liquid surface active agent at operating temperatures. In: IOP Conference Series: Materials Science and Engineering, vol. 245, no. 3 (2017)
6. Vysotskaya, M.A., Shekhovtsova, S.Y., Kindeev, O.N.: The destruction of polymer modified binder and prescription factors her determining. Int. J. Pharm. Technol. **8**, 18200–18211 (2016)

7. Tarsi, G, Varveri, A.: Effects of different aging methods on chemical and rheological properties of bitumen. J. Mater. Civ. Eng. **30**(3) (2018)
8. Olabemiwo, O.M., Esan, A.O., Bakare, H.O., Agunbiade, F.O.: Polymer modified-natural bitumen thermal aging resistance studies. Int. J. Pavement Eng. 1–9 (2017)
9. Baglieri, O., Santagata, E., Sapora, A., Cornetti, P., Carpinteri, A.: Fractional viscoelastic modeling of antirutting response of bituminous binders. J. Eng. Mech. **143**(5), D4016002 (2017)
10. Yuliestyan, A., Cuadri, A.A., García-Morales, M., Partal, P.: Influence of polymer melting point and Melt Flow Index on the performance of ethylene-vinyl-acetate modified bitumen for reduced-temperature application. Mater. Des. **96**, 180–188 (2016)
11. Trofimov, V.A.: Ximiya vy'sokomolekulyarny'x soedinenij, Nizhnij Tagil (2008). (in Russian)
12. Reologiya bitumov. http://www.dorlab-ltd.ru/informatciya-2/svoystvaneftebitumov/reologiya-bitumov. (in Russian)
13. Tager, A.A.: Fiziko-ximiya polimerov, Moscow (2007). (in Russian)
14. Spravochnik ximika 21. Ximiya i ximicheskaya texnologiya. http://chem21.info. (in Russian)

Dynamics of Changes in the Properties of Municipal Solid Waste in the Environment

Oksana Paramonova(✉), Vadim Bespalov, Oksana Gurova, Lyudmila Alekseenko, and Irina Tsarevskaya

Don State Technical University,
Sotsialisticheskaya str. 162, 344022 Rostov-on-Don, Russia
paramonova_oh@mail.ru

Abstract. The problem of waste management has been the most actual as a result of intensive growth in production and consumption in most countries of the world for decades. Having realized the senselessness of combating raising volumes of waste, increasing the scale of their disposal or incineration, mankind has come to the conclusion that the most promising way to solve this problem is to use several types of waste for several times (for example, as packaging) or recycling with return of materials in the turnover. However, in cases where it is impossible to implement these directions, it is advisable to incinerate the waste generated in an environmentally safe manner and then deposit the slag residues. From the standpoint of ensuring the ecological safety of urban economy, the main aspect in solving the problem of waste management is the environmental aspect, although in the above-mentioned concepts the definition of criteria for the environmental friendliness of waste management systems is not fully disclosed. In addition, an indirect assessment of environmental friendliness in these concepts can only be revealed during the implementation of the waste management stages, i.e. in the process of reducing pollution. The authors propose a fundamentally new approach based on the physical-energy concept of describing pollution processes and reducing environmental pollution in urban areas with municipal solid waste (MSW). The basic principle of the physical-energy concept is to consider MSW as a dispersed system that continuously changes its properties throughout the entire life cycle of waste.

Keywords: Municipal solid wastes · Urban territory · Municipal solid waste management system in urban areas · Environmental pollution · Environmental waste reduction · Environmental safety · Properties of municipal solid waste · Changes in waste properties · Waste life cycle · Waste management stages

1 Introduction

The problem of handling MSW is very relevant, since it affects the main spheres of human activity (economic, social and environmental). Numerous scientific papers [1–6] are devoted to this topic, in which the listed aspects are considered separately to some extent. However, in our opinion, the environmental aspect should be in the first

V. Murgul and M. Pasetti (Eds.): EMMFT-2018, AISC 983, pp. 960–970, 2019.
https://doi.org/10.1007/978-3-030-19868-8_95

place, although not enough attention has been paid to it so far. That is why the optimal solution to most aspects of this problem, mostly ecological, has not yet been found.

Our analysis of scientific principles [1–5] showed that: first, in the development process, MSW management systems acquired all new functions, which is associated with different opinions regarding the starting point of the beginning of the life cycle of the waste and the change in this connection of the quantity and sequence of implementation of the waste management stages; secondly, diverse climatic, physiographic features and resource and economic significance of the territories of populated areas led to the formation of various kinds of concepts that offer solutions to the problem of MSW management through a more detailed consideration of one or another aspect (either technological or economic, social or environmental), and not a comprehensive study of them; thirdly, the dynamics of the properties of MSW at each stage of their life cycle is not fully represented in any of the existing concepts, which, in our opinion, is the most important from the point of view of ensuring the environmental safety of urban areas.

Our findings emphasize the relevance of the research problem. The main objective of the MSW management system functioning in urban areas is to ensure environmental safety, therefore, when developing a physical-energy concept [12, 13] based on the basic principles of the theory of the dispersed systems stability, we focus on studying the dynamics of the MSW properties and their behavior in their life cycle environment. At the same time, the main distinctive features of the physical-energy concept of the above-mentioned concepts, in our opinion, are:

- the starting point of the beginning of the MSW life cycle is considered to be their formation at the source of occurrence, as a rule, in the internal volumes of the premises;
- consideration of the waste life cycle as two closely interrelated processes of environmental pollution of MSW [12, 13] and pollution reduction [12–16] through the implementation of the stages/system of waste management;
- the study of the MSW properties in dynamics and in the relationship of waste parameters with environmental parameters.

Thus, the results of our analysis allowed us to reveal the many-sided nature of the waste management problem and the diversity of approaches to its solution, as well as to bring to the forefront the environmental aspect as the most important from the point of view of ensuring the urban areas environmental safety.

2 Theory or Experimental Methods

The physical and energy concept we propose consists in considering two interrelated processes: environmental pollution with municipal solid waste (MSW) and reduction of environmental pollution with municipal solid waste (MSW).

We believe that the process of environmental pollution should be presented in the form of internal and external processes, since formation, accumulation and distribution of MSW can be carried out both indoors (residential, public, etc.), and outside (that is, located directly close to the area where the MSW formation occurs).

At the same time, the internal process implemented in the premises, in our opinion, should be considered as a set of stages of the formation of MSW at the source of their occurrence, internal accumulation of individual elements of MSW on the working surface and their internal distribution in the volume of the room. The external process, implemented already outside the premises, is advisable to be presented as the stages of external separation of individual elements of MSW from the room into the environment and external distribution, i.e. spatial movement of individual elements of the MSW in the area adjacent to the object.

For example, the formation of food (organic) waste is carried out as a result of cooking. If it is not removed immediately after cooking or is not cleaned up (which depends mainly on the cleanliness and mentality of the citizens), then food debris will be accumulated on the working surface and/or may fall on the underlying surface of the room. Due to the leakiness of the room when opening doors and/or windows, the air masses movement and the entry (discharge) of individual elements of food debris outside the room into the external environment and spread there on the territory adjacent to the object is possible.

The scheme of spatial displacement of MSW is presented in Fig. 1.

In the first case, an important feature of the interior spaces in addition to the geometric characteristics (length, width, height) and microclimate parameters (humidity, mobility, air pressure) is the need to provide them with comfortable sanitary and hygienic conditions for the citizens' livelihood, which depends exclusively on the cleanliness of the citizens and their mentality, and predetermines, thereby, the peculiarities of solving the "internal task" of reducing indoor environmental pollution.

In the second case, the individual elements (losses) of MSW inevitably fall into the environment outside the premises and are distributed over the adjacent territory, which has no local boundaries and can be considered as unlimited space.

Moreover, the flow of MSW into the environment can occur either in the process of distribution of individual elements of MSW from the internal room volume to the adjacent territory, when the sources are located inside the room (a), but these volumes are connected with each other with various kinds of openings, or in the process of formation and allocation of MSW when sources are located outside the premises (b). In real conditions, both options are equally probable.

In addition, the number of parameters determining the state of the environment of the adjacent territory far exceeds the number of parameters determining the state of a closed or partially closed room volume. The combination of complex characteristics, state parameters and boundary conditions for the environment adjacent to the source of the MSW formation territory determines the solution of issues of reducing environmental pollution outside the premises and constitute a single "external task".

When solving the "external problem", it is necessary to use peculiar methodological approaches to the study of the formation processes, isolation and distribution of MSW in the environment.

Therefore, if in solving the "internal problem" the main parameters are the immutable boundary conditions, then in the case of the "external problem" these parameters become functionally dependent.

Thus, analyzing the process of environmental pollution [14–16], we can conclude that this process is a set of successive stages of interaction of MSW with various

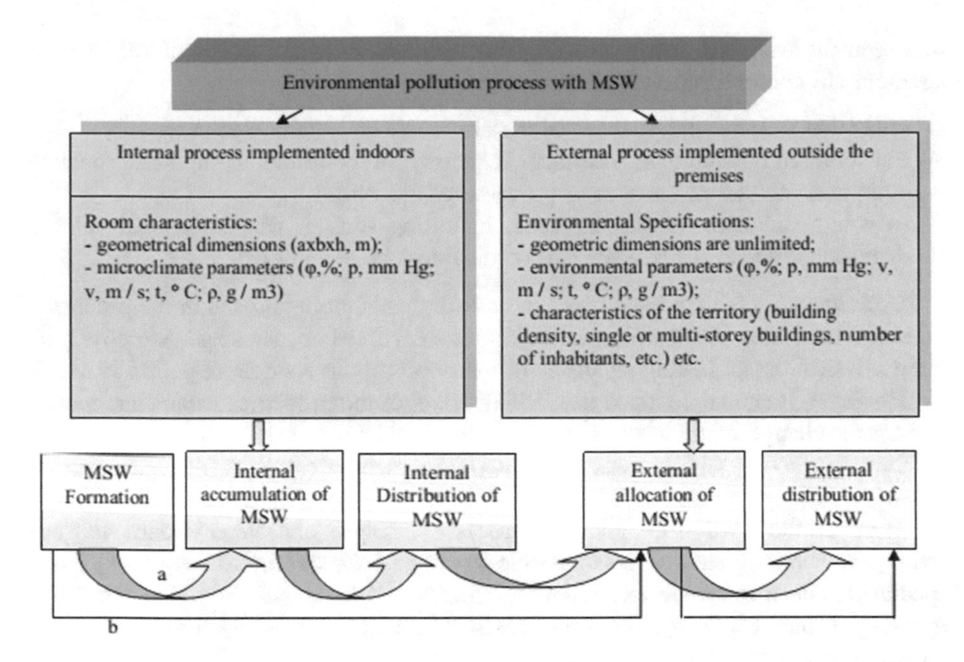

Fig. 1. Scheme of MSW spatial movement.

objects (equipment, underlying surface), each of which enters this interaction at a certain stage and leads to a change in the MSW properties.

Consequently, the degree of environmental pollution is determined by the properties and behavior of individual elements of MSW entering it.

Further, exploring the process of reducing pollution of the environment of MSW in a similar way, we have obtained evidence that this process should be presented as a set of sequential and targeted stages, and the stages of the process of reducing environmental pollution should correspond to each stage of the pollution process, described above.

Based on the foregoing, it is logical to assume that the process of reducing environmental pollution of MSW consists of:

- the collection process-1, which should be organized at the stage of formation, accumulation and distribution of MSW inside the premises due to the cleaning of working and underlying surfaces;
- the collection process-2, which should be organized at the stage of isolation and accumulation of individual items of MSW in the form of losses during transportation throughout the city also by means of cleaning. The more MSW losses are generated at each stage, the more difficult it is to ensure the ecological safety of the territory;
- processes of transportation and utilization - at the stage of their accumulation and distribution in the environment.

Since the cleaning of internal volumes of premises depends solely on the citizens themselves and therefore cannot be applied to the organization of the waste management system in urban areas, in our opinion, the main stages of MSW management are appropriate to consider in enlarged form:

- collection (implies the presence of certain waste collection sites in the city)
- transportation (export at a certain frequency of collected waste with specially equipped vehicles to the sites of processing and landfill);
- recycling (involves various methods, including sorting, where the final stage is disposal, which is realized for non-waste fractions at landfills).

Consequently, both in the process of environmental pollution and in the process of pollution reduction, various physical objects are involved at each stage. Moreover, the main physical object linking all stages of the processes in a single sequence is MSW.

However, it should be noted that, MSW changes the properties parameters values, undergo a change of these parameters and consistently transition from one state to another, interacting with various technological and physical objects at each process stage.

Thus, on the basis of the pollution processes analysis and the environmental pollution reduction we studied, it is possible to suggest for the future the structure of a special engineering-ecological system for the MSW treatment, in which the corresponding external influences will be consistently and purposefully provided to individual elements of MSW.

Since the starting point of the MSW life cycle is its formation, which occurs under the influence of external conditions and/or the effects of various physical objects, the next step in our research is to study the MSW properties parameters at the initial stage.

3 Experimental Section

Methods of examination are based on the main statements of the dispersion systems theory, the system analysis, analytical generalization of the known scientific and practical results.

4 Results and Discussion Section

It is known that in the operation of the waste management system, as well as in determining the amount of possible formation of secondary material resources, it is necessary to take into account the quantitative and qualitative composition of MSW.

The qualitative characteristics of MSW include the following parameters [17]:

- morphological, fractional and chemical composition;
- physical properties (humidity, φ, %, density, ρ, mg/m^3);
- special properties (mechanical connectivity, abrasivity, caking);
- compression characteristic;
- hazardous properties (toxicity, fire hazard, high reactivity).

It is also known that the composition of MSW is influenced by such factors as the climatic zone, the degree of improvement of the housing stock (the presence of garbage chutes, gas, water supply, sewage, heating systems), the number of the buildings floors, the development of catering, trade culture, etc. equally important is the lifestyle and the population well-being.

There is no doubt that each of the listed characteristics plays an important role, since affects both the choice of methods and design solutions for the implementation of the MSW collection, transportation, and the technology of their subsequent disposal. However, most often the properties of waste are considered in the aggregate, because in the Russian Federation, MSW are collected in a mixed way, without any reference to the waste management stages. For the purpose of considering MSW from the point of view of their application as secondary resources (raw materials), from the standpoint of their behavior in the environment, the impact on the environment at each stage of their life cycle, it is necessary, in our opinion, to study each component separately.

In addition, as we have shown earlier, the existing approaches to the study of the properties of MSW do not fully reveal the relationship between the characteristics of raw materials, waste produced, and environmental characteristics [1–6].

This relationship can be traced, if we consider any pollutant as a dispersed system. Then MSW, in our opinion, in most cases can be represented as a polydisperse system consisting of several solid dispersed phases (separate MSW fractions) and a gaseous dispersion medium (air gaps between the MSW fractions).

Consideration of MSW as a dispersed system allows us to conclude that the description of its behavior in the environment can be formed on the basis of a comprehensive sequential study of its state and an orderly analysis of the parameters characterizing the properties of the dispersed phase and the dispersion medium.

The processes of formation, isolation and distribution of MSW, as shown above, cause environmental pollution and in real conditions can occur either sequentially or simultaneously in space and in time [12–16]. Thus, they should be considered together, not separately from each other. The unity of the nature of such processes emphasizes the narrow relationship between them. In this case, we can only talk about the prevalence of one of the processes' influence. This effect depends on a number of parameters such as properties of dispersed phases (d.p.) and dispersion medium (d.m.) of waste, environmental characteristics, etc.

For this purpose, the parameters determining the properties of the dispersed phases (d.p.) and the dispersion medium (d.m.) must be grouped taking into account the physical essence of the processes and phenomena to which the MSW is subjected.

At the stage of the MSW formation, the changes in its properties occur as a result of partial or complete loss of its quantitative and/or qualitative characteristics. Depending on the nature of the changes, all the processes/phenomena can be divided into the following groups: physical, chemical, physicochemical and biological.

Physical are the processes occurring when exposed to external factors (temperature, humidity, light, gas composition, mechanical effects, etc.). These include moisture evaporation (shrinkage), moistening, temperature changes (heating, freezing, cooling), deformation (crushing, combating, acquiring an unusual form, etc.), sorption of volatile substances.

Chemical are the processes that cause changes in chemical substances and their properties under the influence of external factors (air, water or light oxygen) and internal reactions. For example, in food waste most often the rancid fats, the chemical reaction of product acids with packaging metals (metal cans) are found.

Biochemical are the processes that cause changes in chemicals with the participation of enzymes. These processes, in turn, are divided into hydrolytic, redox and synthetic processes.

Microbiological are the processes occurring with the participation of microorganisms (fermentation, rotting, molding).

The analysis of these processes allows us to identify the main parameters of the properties of the MSW components and their groups (Table 1).

Such a generalization of the parameters defining the properties of the solid chemical treatment (Table 1) allows, firstly, to supplement each group of parameters of both the dispersed phase and the dispersion medium with new characteristics in the process of developing the theoretical foundations; secondly, to conduct a focused and consistent assessment of all aspects of the dynamics of the formation, separation, distribution and disposal of the MSW individual components.

Table 1. Properties of the main components of MSW at the stage of formation in the process of environmental pollution.

	Properties of MSW components at the stage of their formation					
MSW component name	The main phenomena/processes to which the components of the MSW are exposed	The main groups of parameters of properties that change under the influence of the process	Parameter name	Parameter symbol	Parameter units	Value
Paper, cardboard	Mass change (density)	Geometric parameters (GP)	dispersion	$d_{(p)}$	mm	50–350
		Physical and chemical parameters (PCP)	density	ρ	kg/m^3	20–165
	Hydrokinetic and capillary phenomena	Hydrodynamic parameters (HDP)	relative humidity	$\varphi_{(p)}$	%	avg.58
	Temperature change (thermal processes)	Thermophysical parameters (TPP)	lower calorific value per working mass	Ql_{ow}	kJ/kg	9490
			calorific value		kcal/kg	1800–2750
			heat capacity (specific heat)	Q	kJ/kg °C	2000–2500
			thermal conductivity	K	W/m °C	
Food waste	Change in mass (density)	Geometric parameters (GP)	dispersion	$d_{(p)}$	mm	50–250
		Physical and chemical parameters (PCP)	density	ρ	kg/m^3	90–110
	Adhesion to the surface (adhesion)	Physical and chemical parameters (PCP)	adhesion coefficient	k_{adh}		
	Hydrokinetic phenomena (diffusion, sorption, desorption)	Hydrodynamic parameters (HDP)	diffusion rate	v_{dif}		
			relative humidity	$\varphi_{(p)}$	%	70–80

(continued)

Table 1. (*continued*)

	Properties of MSW components at the stage of their formation					
MSW component name	The main phenomena/processes to which the components of the MSW are exposed	The main groups of parameters of properties that change under the influence of the process	Parameter name	Parameter symbol	Parameter units	Value
	Biological processes (decay, fermentation, molding, aging)	Physical and chemical parameters (PCP)	*ash content on the working mass*		*% dry weight*	*maj.40*
		Agrochemical parameters	*total nitrogen, phosphorus, potassium, calcium*		*% dry weight*	*Is given in general on MSW*
			high bacteriological hazard			
	Temperature change (thermal processes)	Thermophysical parameters (TPP)	lower calorific value per working mass	Q_{low}	kJ/kg	3430
			calorific value		kcal/kg	750–900
			thermal conductivity	K	W/m °C	
		Physical and chemical parameters (PCP)	strength		MPa	
			tensile strength		MPa	
			density	ρ	kg/m^3	15–2200
	Temperature change (thermal processes, melting)	Thermophysical parameters (TPP)	heat capacity	Q	kJ/kg °C	23370
			calorific value		kJ/kg	38580
Metal	Change in mass (density) (deformation, compaction, abrasion)	Geometric parameters (GP)	dispersion	$d_{(p)}$	mm	50–250
		Physical and chemical parameters (PCP)	abrasiveness			
			density	ρ	kg/m^3	18–95
	Temperature change (thermal processes, melting)	Thermophysical parameters (TPP)	heat capacity		°C	400–860
	Magnetization	Electromagnetic Parameters (EMP)	magnetics susceptibility		kg^{-1}	
Glass	Change in mass, density	Geometric parameters (GP)	dispersion	$d_{(p)}$	mm	50–250
		Physical and chemical parameters (PCP)	abrasiveness			
			strength		MPa	500–2000
			tensile strength-stretching- for bending		MPa	35–100 0,2
			density	ρ	kg/m^3	2,24·10^3–2,9·10^3

(*continued*)

Table 1. (*continued*)

	Properties of MSW components at the stage of their formation					
MSW component name	The main phenomena/processes to which the components of the MSW are exposed	The main groups of parameters of properties that change under the influence of the process	Parameter name	Parameter symbol	Parameter units	Value
	Temperature change (thermal processes, melting)	Thermophysical parameters (TPP)	softening temperature	t_{soft}	°C	550–1500
			thermal expansion		kJ/kg	$9 \cdot 10^6$–$15 \cdot 10^6$
			heat capacity (specific heat)	*Q*	kJ/kg °К	0,63–1,05
			thermal conductivity	К	W/m °C	0,5–1
	Change in mass, density; biological processes (decay)	Geometric parameters (GP)	dispersion	$d_{(p)}$	mm	50–350
		Physical and chemical parameters (PCP)	density	ρ	kg/m^3	47–318
Wood/wood waste	Hydrokinetic phenomena	Hydrodynamic parameters (HDP)	relative humidity	$\varphi_{(p)}$	%	maj.28
	Temperature change (thermal processes)	Thermophysical parameters (TPP)	heat capacity	*Q*	kJ/kg °К	2000–2500
			thermal conductivity	К	W/m °C	
			calorific value		kcal/kg	3200–3400
	Change in mass, density; biological processes (decay)	Geometric parameters (GP)	dispersion	$d_{(p)}$	mm	50–350
		Physical and chemical parameters (PCP)	density	ρ	kg/m^3	80–116
Textile	Hydrokinetic phenomena	Hydrodynamic parameters (GDP)	relative humidity	$\varphi_{(p)}$	%	15–27
	Temperature change (thermal processes)	Thermophysical parameters (TPP)	lower calorific value per working mass	Ql_{ow}	kJ/kg	15720
			calorific value		kcal/kg	2900–3400
	Change of mass, density (for example, deformation, violation of integrity)	Geometric parameters (GP)	dispersion	$d_{(p)}$	mm	50–250
		Physical and chemical parameters (PCP)	density	ρ	kg/m^3	65–237
Leather, rubber	Hydrokinetic phenomena	Hydrodynamic parameters (GDP)	relative humidity	$\varphi_{(p)}$	%	0,75–5
	Temperature change (thermal processes)	Thermophysical parameters (TPP)	heat capacity	*Q*	kJ/kg °К	
			lower calorific value per working mass	Ql_{ow}	kJ/kg	25790
			calorific value		kcal/kg	5000–6000

5 Conclusion

The study of the scientific regulatory and reference literature [1–17] allowed us to obtain the characteristics of the properties of the MSW main components (represented by the largest fractions of the total mass).

Each of the MSW component properties undergoes a change of parameters within their life cycle during a sequential transition from one state to another, which needs to be studied step by step in order to find the informed design solutions for the implementation of each stage of the MSW treatment system in an urban area.

Since a certain stage of the process of environmental pollution corresponds to a certain stage of the reducing pollution process, the description of the above properties of the MSW components and the phenomena to which they are exposed depends on the initial stage - the MSW formation (Table 1), which allows to find the best technical solutions for each subsequent stage implementation.

Thus, the studies performed allowed us to streamline the sequence of technological operations for handling solid waste, to identify the relationship of the pollution processes and to reduce environmental pollution of solid waste, to study changes in the properties of components of solid waste within their life cycle, to give the opportunity to choose engineering and environmental measures to reduce the negative impact of MSW.

References

1. Europeans practice of waste management: problems, solutions and prospects (2005)
2. Borovsky, E.E.: Industrial and domestic waste: environmental problems (2007)
3. Weisman, Ya., Korotaev, V.N., Slyusarev N.N.: Waste management. Collection, transport, compaction, sorting municipal solid waste (2012)
4. Murray, R.: Target - zero waste (2004)
5. Bendere, R.: Waste management (2003)
6. McDougall, F.R., White, P.R., Franke, M., Hindle, P.: Integrated solid waste management: a life cycle inventory (2008)
7. Babanine, I.: Recycle revolution. How to solve the waste problem at minimal cost (2008)
8. Beldeeva, L.N., Lazutkina, S., Komarova, L.F.: Environmentally sound management of waste (2006)
9. Shmandy, V.M., Klimenko, N.A., Golik, S.: Environmental security (2013)
10. Starostin, V.Y., Ulanova, O.V.: Modern problems of science and education **5** (2013)
11. Ulanova, O.V., Starostina, V.Yu.: Ecology, production **4**, 10–16 (2012)
12. Bespalov, V.I., Paramonova, O.N.: Eur. Appl. Sci. Int. German/English/Russian Lang. **1** (1/1), 202–204 (2013). Peer-Reviewed Journal and is Published Monthly
13. Paramonova, O., Bespalov, V., Gurova, O., Krivtsova N.: MATEC Web Conference, vol. 170 (2018). International Science Conference SPbWOSCE-2017 "Business Technologies for Sustainable Urban Development". https://www.matec-conferences.org/articles/matecconf/abs/2018/29/matecconf_spbwosce2018_04014/matecconf_spbwosce2018_04014.html
14. Bespalov, V.I., Paramonova, O.N.: Collection of Scientific Papers SWorld. Scientific research and its practical application. Current status and development trends 2012, Issue 3, vol. 9, 89 (2012)

15. Bespalov, V.I., Paramonova, O.N.: Modern problems and solutions in science, transport, manufacturing and education in 2012, pp. 76–80 (2012)
16. Bespalov, V.I., Paramonova, O.N.: Modern scientific achievements in 2013, pp. 92–96 (2013)
17. Peace, A.N.: Communal Ecology: Encyclopedic Handbook (2007)

Methodical Aspects of Inventory Management in a Construction Company

Ruslan Minnullin(✉) and Inna Nekrasova

Industrial University of Tyumen, Volodarskogo St., 38, Tyumen 625000, Russia
minnullinaay@yandex.ru

Abstract. The article proposes combined use of ABC and XYZ classifications. The analyzed methods allow estimating the main parameters of inventory management: demand, order, supply, warehousing. To confirm effectiveness of its use, it was tested in a road construction company. The authors have analyzed the results of proposed recommendations; they have showed their applicability in practice.

Keywords: Inventory management · ABC classification · XYZ classification

1 Introduction

Almost any organization must have the necessary level of inventories in order to conduct successful economic activity. The presence of at least one of the following factors leads to the need to have reserves:

- change in demand for manufactured products;
- fluctuation in timing of production;
- uneven timing of delivery of raw materials;
- fluctuation of production rates;
- certain conditions of demand and equipment performance requiring batch production;
- presence of costs associated with creation, storage and shortage of resources.

In most cases, there is a joint manifestation of these factors.

Large stocks in warehouses of enterprises reduce the amount of work on operational regulation of logistics of production and, therefore, create prerequisites for reducing administrative and management costs. However, increase in stocks requires an increase in storage space, which inevitably causes increase in general expenses due to the costs of maintaining buildings and structures, as well as depreciation. The more stocks, the longer the time the materials are in stock, the greater the loss due to damage and expiration. As a result, there is increase in costs for materials and semi-finished products, increase in general expenses in terms of wages with warehouse personnel charges, maintenance of stocks, and increase in non-production costs due to losses from material damage during storage.

The complexity of the task of inventory management is due to complex interrelationship of various elements of production system, which includes inventory

V. Murgul and M. Pasetti (Eds.): EMMFT-2018, AISC 983, pp. 971–980, 2019.
https://doi.org/10.1007/978-3-030-19868-8_96

management subsystem. According to the system approach, to increase efficiency of functioning of inventory management system, it is necessary to form it on the basis of production system operating at the enterprise.

Inventories are among the volumes that require significant cash and therefore represent one of the factors affecting the size and structure of cost of distribution.

Recently there was published a large number of scientific researches on inventory management. The results of such studies are presented in works of Russian and foreign scientists, including N.V. Belkina, Yu.A. Belyaev, G.L. Brodetsky, N.N. Goldobinu, V.M. Gromenko, A.M. Zevakova, P. Zermati, K.V. Inyutinu, Li, M. Linders, A.A. Pervozvansky, B.K. Plotkina, N. Prabhu, S.S. Prasnikova, Yu.I. Ryzhikova, V.A. Sakovich, G. White, G. Firon, J. Hadley, F. Hansmenna et al. [1–4].

From the point of view of logistics, each of the functional subsystems of an organization requires a certain accumulation of material resources in order to weaken direct dependence between all the market actors, including purchasing subsystem. As noted earlier, the first subsystem of supply, which is most often represented by input material flow, is a process of moving material resources from the procurement market to the organization's warehouses, where raw materials are formed to ensure safety of material and technical supply, its flexible operation and uninterrupted operation of production as a whole.

The intensity of material flow also determines the need for formation of stocks.

There is also probability of changing the intensity of input material flow in case of violation of the established schedule of deliveries. In this case, the stock is necessary so that production process does not stop. This is especially important for organizations with an unbroken production cycle.

In the absence of stocks, intensity of material flows in the distribution system (output material flows) fluctuates in accordance with changes in intensity of production (internal material flow) [5, 6].

Thus, implementation of effective production activities involves a streamlined system for managing the movement of material flows, as well as material resources with use of modern production and information technologies and competent methods for developing management decisions.

2 Methods

To continuously monitor the current level of material resources, it is necessary to form a system of inventory management in the organization [7–10].

The general scheme of inventory management model is shown in Fig. 1.

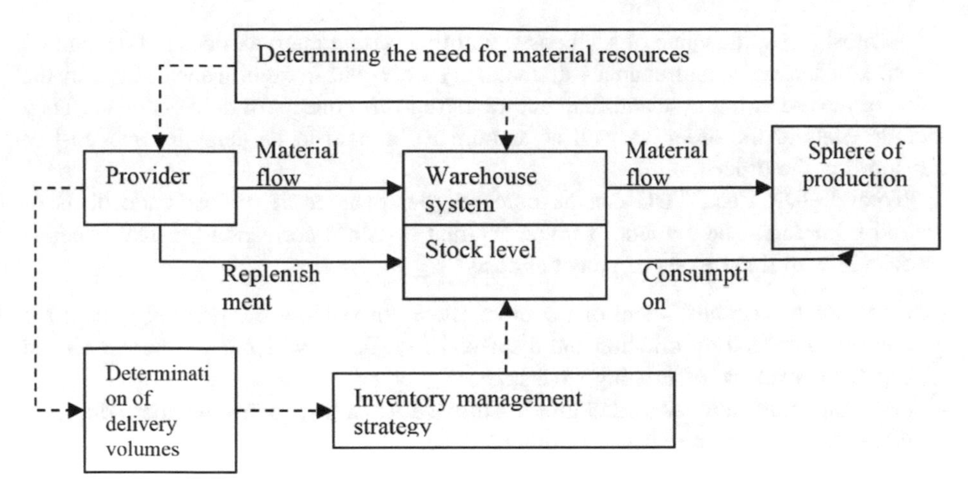

Fig. 1. General scheme of model of inventory management

The main parameters of inventory management, taken into account in general scheme:

– Demand (consumption) parameters: demand intensity, demand function, time characteristics of discrete demand;
– Order parameters: order size, order time, time interval between two adjacent orders;
– Delivery parameters: size of the delivery schedule, time of delivery, time interval between two adjacent deliveries, delay time of delivery (order fulfillment);
– Stock level in warehouse: current, average, maximum, insurance [11–13].

When building inventory management system, it is necessary to take into account a number of costs.

Total Carrying Cost (TTC) is a much more complex measure than it is usually represented. Storage of stocks involves appearance of two types of costs:

(a) Costs associated with physical presence of goods;
(b) Costs due to immobilization of financial resources in stocks.

Total Order Cost (TOC) means costs associated with placing an order to a supplier. It includes following costs:

– cost of finding a potential supplier (or consumer);
– administrative costs associated with registration of contractual terms for the purchase of raw materials and delivery of finished products, negotiating with suppliers (consumers), business trips, long-distance calls, etc.;
– cost of processing orders;
– freight forwarding costs and handling costs;
– costs of carrying out quantitative and qualitative acceptance of products, registration of a certificate of quality and sale of finished products, etc.

In most cases, the value of such costs, within a certain interval, does not depend on the size of the order and changes dramatically only with a significant change in the value of the order. In the scientific literature there is the concept of delivery costs. They include costs of the order, as well as amount to be paid to the supplier and cost of transporting the order.

Total Deficit Cost, TDC can be much higher than profit of lost trade deals or unrealized orders. The amount of inventory that does not correspond to real needs is associated with three types of potential costs:

- costs due to non-fulfillment of the order (delay in sending the ordered goods) are additional costs for promotion and dispatch of the order, which could not be carried out at the expense of existing stocks;
- costs in connection with sales losses - in case when a regular customer addresses competing with manufacturer of this product;
- costs in connection with loss of customer - in cases when the lack of stocks turns out to be not only the loss of a transaction, but also the fact that customer starts looking for other sources of supply.

The need to control the state of stocks is due to increased costs in case if the actual size of stock goes beyond the limits stipulated by the norms of stock. Stock control is carried out on the basis of inventory accounting data and can be carried out continuously or after certain periods.

From a variety of inventory management systems, in practice, a limited number of models are used, which, as a rule, do not require significant amount of initial information and provide relatively simple methods for regulating and monitoring inventory parameters, ordering (Table 1).

Analyzed inventory management systems work adequately under certainty. However, in conditions of constant changes in demand in the periods between orders, inventory management system is subject to a deficit state or an increase in costs in case of a significant excess of the standard value of current inventory level. In this case, it is recommended to conduct continuous or periodic monitoring and control of inventory level based on the inventory data.

A widely used method for the control and management of multi-product stocks, proposed both in scientific literature and practice of leading companies, is the ABC method based on the 80:20 rule. The established empiric V. Pareto asserts in 1897 that the limited number of elements in the amount of 20% constituting the phenomenon, in most cases, 80% determine its appearance [14].

Inventories can vary significantly in terms of unit value, and total value of a particular position. The rule of "80:20" is confirmed by the practice of various companies: reserves can be divided into three groups, while the following trend is observed: the nomenclature positions of the most expensive (80% of financial means) material resources in the reserves are much less (20%) than the average, and Medium is much less than cheap:

Table 1. Comparative characteristics of the main models of inventory management systems

Current stock level	Inventory management system models	Terms of use	Advantages	Disadvantages
Periodic check	Operational inventory management systems; uniform delivery systems; replenishment systems to the maximum level; systems with two levels	Low maintenance costs; good demand predictability; the need for lots of different volumes	No need for constant monitoring of stock availability	High level of stocks
Continuous check	Systems with a fixed order size; systems with two levels	Large losses from the lack of stocks; high costs of maintaining stocks; high degree of uncertainty in demand	Allow to work at a low level of a stock, prevent deficiency	The need for constant monitoring of inventory levels

- A – the volume of the nomenclature positions of material resources is about 20% and in the total cost is about 80%;
- B – the volume of the nomenclature positions of material resources with an average price of about 30% and in the total cost is about 15%;
- C – the volume of the nomenclature positions of material resources with a low price is about 50% and in total cost is about 5%.

In addition to the classification of the stock list by the ABC method, the XYZ method can be used to quantify a parameter such as the rate of consumption of stocks.

The consumption rate is estimated through the coefficient of variation of the statistical series, which value allows distributing the nomenclature positions into groups:

- X is a group characterized by a stable value of consumption with the value of the coefficient of variation of statistical series of shipments up to 25%;
- Y – the need for these resources is characterized by known trends and the value of the coefficient of variation of statistical series of shipments from 25 to 50%;
- Z – irregular use of resources, difficulty of predicting the magnitude of consumption, the value of the coefficient of variation of statistical series of shipments of more than 50%.

As a result of this classification, the concept of minimizing inventory levels can be applied to group X by achieving corresponding characteristics of supply and stockpiling at a level close to insurance. For group Y, the concept of optimizing inventory levels can be used. Group Z stock levels cannot be optimized. In this case, the strategy of minimizing or maximizing inventory levels is applicable.

This classification, together with the ABC groups, allows us to split the reserves into nine blocks, each of them has two characteristics: the cost of reserves and the accuracy of forecasting the demand (Fig. 2).

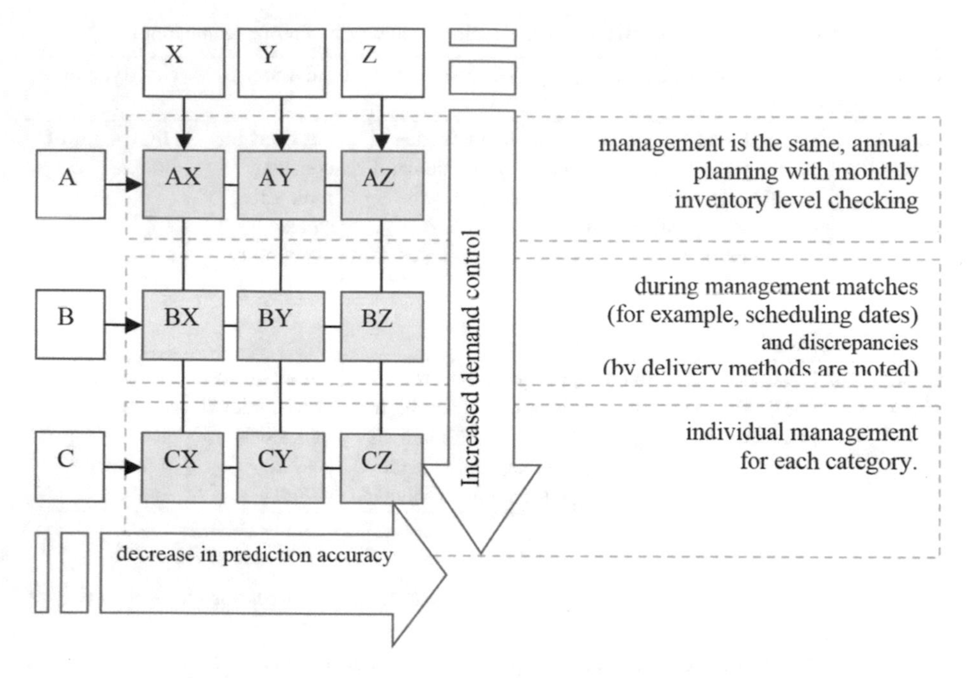

Fig. 2. Combining the results of ABC-and XYZ-classifications

Features of inventory management of material resources by category are as follows:

CX, CY, CZ – management is the same, annual planning with monthly checking of inventory level;
BX, BY, BZ – during management, coincidences (for example, in terms of planning periods) and discrepancies (in delivery methods) are noted;
AX, AY, AZ – individual management for each category. For example, for the category AX, you can calculate the optimal amount of purchases and apply a model with a fixed order size, and for AZ this cannot be done, so you have to create a reserve stock and use the inventory management model with established replenishment periodicity to a constant level.

3 Results

In order to test the proposed recommendations, an analysis of inventory management was conducted at the road construction organization of the city of Tyumen, OJSC TODEP.

The Tyumen Region Road Operations Company (OJSC TODEP) was established in 1994 through the reorganization of Tyumenavtodor, and currently is part of the network of road automobile facilities of the Ministry of Transport of the Russian Federation. The main activities of the organization are: maintenance, repair, construction and reconstruction of public roads and road structures. The organization's balance sheet annually hosts 16 thousand kilometers of federal, regional and municipal roads.

The main tasks of OJSC "TODEP" are:

- timely and high-quality execution of tasks for construction and commissioning of facilities and construction objects;
- steady increase in efficiency of construction production on the basis of its intensification, economy of all types of resources used, introduction of achievements of science and technology;
- improving organization of construction industry, reducing the cost of construction and installation works;
- use and increase effectiveness of capital investments in construction, reconstruction of its own facilities.

The organization has more than 200 units of a dumping park: 50 units of excavators, 150 units of self-propelled road rollers, 28 complexes for laying asphalt-concrete mixture, due to these complexes it carries out all types of road-building works only on its own.

This organization works on transit conditions with all suppliers of material resources. At the same time, storage terminals are not involved, since most deliveries are carried out using railway transport and/or from a closely located manufacturer. The complexity of distribution of material resources lies in accurately determining the needs of each department, which has access railway tracks in conjunction with the needs of departments, located close to this branch.

The ranking of material resources used by OJSC "TODEP" for construction, repair and maintenance of roads is based on the proposed ABC and XYZ analysis and is presented in Tables 2, 3, 4 and 5.

Thus, Table 2 presents the primary list of material resources used by OJSC TODEP for production of road construction works for construction, repair and maintenance of roads for ranking by the ABC method.

With the help of Microsoft Excel spreadsheet processor, primary list has been streamlined and material resources used by the ABC method have been ranked by OJSC TODEP. The results are presented in Table 3.

Table 2. The primary list of material resources used by JSC "TODEP" for production of road construction works for construction, repair and maintenance of roads according to the ABC method

No.	Supplier name	Amount of purchases, thousand rubles			Position size, thousand rubles	Share of positions in the stock, %
		1 period	2 period	3 period		
1	Bitumen	169 404	390 504	423720	983 628	27.476%
2	Rubble	460 320	682 893	769510	1912 723	53.429%
3	River sand	140 853	174 865	185743	501 461	14.008%
4	Technical salt	4 347	5 184	5008,5	14 540	0.406%
5	Paints	26 010	24 017	25861	75 888	2.120%
6	Thermoplastic	8 903,55	18 775	9021	36 700	1.025%
7	Glass beads	2 160,50	8 992	4150	15 303	0.427%
8	Brick	680,22	592	601,11	1 873	0.052%
9	Pipe, metal. constructions	1007	1 411,05	1038,91	3 457	0.097%
10	Road signs	3204	4015	3568	10 787	0.301%
11	Barrier fences	587	790,68	622,72	2 000	0.056%
12	Border	1 048,57	1423	1129,275	3 601	0.101%
13	Reinforced concrete products	326,11	399,89	392,2	1 118	0.031%
14	Cement	175	246,70	193,70	615	0.017%
15	Lumber	128	163,56	252,68	544	0.015%
16	Filler	158,90	102,00	190,85	452	0.013%
17	Tow	12,00	56,08	41,07	109	0.003%
18	Oil paint/ linseed oil	192,10	148,00	196,10	536	0.015%
19	Grass seed for lawn	121,40	316,68	405,30	843	0.024%
20	Brizol	72,10	68,90	85,10	226	0.006%
21	Mineral fertilizers	18,90	38,87	45,60	103	0.003%
Total		823 420	1 319 804	1 436 696	3 579 920	100,0%

In accordance with the ABC method, all material resources used for construction, repair and maintenance of roads are divided into three unequal groups according to the volume of purchases of each position in the total volume. This analysis of consumption of material resources revealed a group of key material resources - crushed stone, river sand, bitumen.

To determine the frequency and regularity of consumption of material resources, an XYZ analysis was performed.

Table 3. An ordered list of material resources used by JSC "TODEP" for production of road construction works on construction, repair and maintenance of roads according to the ABC method

No.	Supplier name	Item number	Share of positions in the stock, %	The share of positions in the stock with a cumulative total, %	Group
1	Bitumen	2	53.4%	53.4%	**A**
2	Rubble	1	27.476%	80.9%	**B**
3	River sand	3	14.008%	94.9%	**B**
4	Technical salt	5	2.120%	97.0%	**C**
5	Paints	6	1.025%	98.1%	**C**
6	Thermoplastic	7	0.427%	98.5%	**C**
7	Glass beads	4	0.406%	98.9%	**C**
8	Brick	15	0.375%	99.3%	**C**
9	Pipe, metal. constructions	10	0.301%	99.6%	**C**
10	Road signs	12	0.101%	99.7%	**C**
11	Barrier fences	9	0.097%	99.8%	**C**
12	Border	11	0.056%	99.8%	**C**
13	Reinforced concrete products	8	0.052%	99.9%	**C**
14	Cement	13	0.031%	99.9%	**C**
15	Lumber	14	0.017%	99.9%	**C**
16	Filler	16	0.0152%	100.0%	**C**
17	Tow	19	0.0150%	100.0%	**C**
18	Oil paint/linseed oil	17	0.013%	100.0%	**C**
19	Grass seed for lawn	21	0.006%	100.0%	**C**
20	Brizol	18	0.0030%	100.0%	**C**
21	Mineral fertilizers	22	0.0029%	100.0%	**C**
Total		1 319 804		100.0%	

4 Conclusions

1. Technology of arrangement of reinforced concrete slabs have recently been developed in the direction of increasing the specific weight of prefabricated monolithic constructive and technological systems. In Russia, the national innovative prefabricated monolithic system MARCO has emerged and is developing which we have taken as the basis for further improvement.
2. Considering the unfavorable conditions for the delivery of prefabricated elements of the MARCO system to our region, as well as the possibilities of using local materials to fill the inter-block space, an improved construction of prefabricated monolithic slab has been proposed and presented in the work.

3. The design and experimental production of the proposed structural and technological prefabricated monolithic system in the conditions of low-rise suburban development of the central part of Crimea has been made.
4. As a result of the processing of the data of the camera recording, timing and physiological condition of the construction workers, the expediency of using the proposed technology based on an assessment of the indicators of the severity of their labor in comparison with the arrangement of monolithic reinforced concrete slabs has been shown.

References

1. Moder, J., Phillips, S.: The method of network planning in the organization of work, p. 314 (1966)
2. Meskon, M., Albert, M., Hedouri, F.: Fundamentals of Management, p. 799 (2004)
3. Sternik, G.M.: Property Policy Issues - Practical Experience, vol. 5 (104), pp. 67–83 (2010)
4. Abbasov, M.E.: Issues of property policy - practical experience, vol. 6 (105), pp. 14–28 (2010)
5. Gribovsky, C.B., Barinov, N.P.: Property Relations in the Russian Federation, vol. 5, pp. 96–106 (2006)
6. Minnullina, A.: Efficiency assessment of procurement by the organization and planning of civil works. In: MATEC Web of Conference. International Science Conference SPbWOSCE-2016 “Smart City”, p. 08067 (2017). https://doi.org/10.1051/matecconf/201710608067
7. Truntsevsky, Y.V., Lukiny, I.I., Sumachev, A.V., Kopytova, A.V.: MATEC Web of Conferences, vol. 170, p. 01067 (2018). https://doi.org/10.1051/matecconf/201817001067
8. Vikhansky, O.S., Naumov, A.I.: Management, p. 576 (2014)
9. Razu., M.L.: Basics of Project Management, p. 768 (2006)
10. Frolova, O., Kopytova, A., Matys, E.: MATEC Web of Conferences, vol. 170, p. 01064 (2018). https://doi.org/10.1051/matecconf/201817001064
11. Izvin, D., Lez’Er, V., Kopytova, A.: MATEC Web of Conferences, vol. 170, p. 01065 (2018). https://doi.org/10.1051/matecconf/201817001065
12. Minnullina, A.Yu.: Economics, vol. 8 (57), pp. 307–311 (2009)
13. Minnullina, A.Yu.: Modern problems of management: a collection of materials of the All-Russian scientific and practical, vol. 5, pp. 97–102 (2015)
14. Minnullina, A., Vasiliev, V.: Determining the supply of material resources for high-rise construction: scenario approach. In: E3S Web of Conferences,vol. 33, p. 03060 (2018). https://doi.org/10.1051/e3sconf/20183303060

Author Index

V. Murgul and M. Pasetti (Eds.): EMMFT-2018, AISC 983, pp. 981–984, 2019.
https://doi.org/10.1007/978-3-030-19868-8